Quadratic Formula

If $ax^2 + bx + c = 0$, where $a \neq 0$, then

$$x = \frac{-b \pm \sqrt{b^2 - 4ac}}{2a}$$

Straight Lines

$$m = \frac{y_2 - y_1}{x_2 - x_1} \quad \text{(slope formula)}$$
$$y - y_1 = m(x - x_1) \quad \text{(point-slope form)}$$
$$y = mx + b \quad \text{(slope-intercept form)}$$
$$x = \text{constant} \quad \text{(vertical line)}$$
$$y = \text{constant} \quad \text{(horizontal line)}$$

Inequalities

If $a < b$, then $a + c < b + c$.
If $a < b$ and $c > 0$, then $ac < bc$.
If $a < b$ and $c > 0$, then $a(-c) > b(-c)$.

Logarithms

$\log_b x = y$ if and only if $x = b^y$

$\log_b(mn) = \log_b m + \log_b n$

$$\log_b \frac{m}{n} = \log_b m - \log_b n$$

$\log_b m^r = r \log_b m$

$\log_b 1 = 0$

$\log_b b = 1$

$\log_b b^r = r$

$b^{\log_b m} = m$

$$\log_b m = \frac{\log_a m}{\log_a b}$$

Counting

$$_nP_r = \frac{n!}{(n - r)!}$$

$$_nC_r = \frac{n!}{r!(n - r)!}$$

Greek Alphabet

alpha	A	α	nu	N	ν
beta	B	β	xi	Ξ	ξ
gamma	Γ	γ	omicron	O	o
delta	Δ	δ	pi	Π	π
epsilon	E	ϵ	rho	P	ρ
zeta	Z	ζ	sigma	Σ	σ
eta	H	η	tau	T	τ
theta	Θ	θ	upsilon	Υ	υ
iota	I	ι	phi	Φ	ϕ, φ
kappa	K	κ	chi	X	χ
lambda	Λ	λ	psi	Ψ	ψ
mu	M	μ	omega	Ω	ω

Business Relations

$$\text{Interest} = (\text{principal})(\text{rate})(\text{time})$$
$$\text{Total cost} = \text{variable cost} + \text{fixed cost}$$
$$\text{Average cost per unit} = \frac{\text{total cost}}{\text{quantity}}$$
$$\text{Total revenue} = (\text{price per unit})(\text{number of units sold})$$
$$\text{Profit} = \text{total revenue} - \text{total cost}$$

Ordinary Annuity Formulas

$$A = R\frac{1 - (1 + r)^{-n}}{r} = Ra_{\overline{n}|}r \qquad \text{(present value)}$$

$$S = R\frac{(1 + r)^{n} - 1}{r} = Rs_{\overline{n}|}r \qquad \text{(future value)}$$

Graphs of Elementary Functions

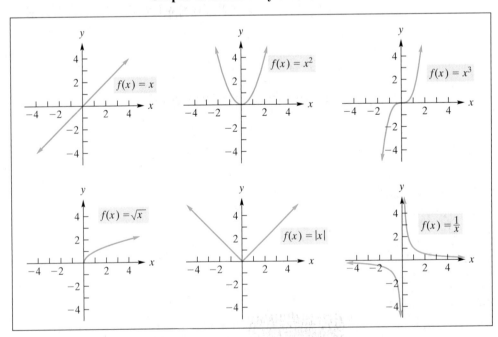

INTRODUCTORY MATHEMATICAL ANALYSIS

For Business, Economics, and the Life and Social Sciences

TWELFTH EDITION

Ernest F. Haeussler, Jr.
The Pennsylvania State University

Richard S. Paul
The Pennsylvania State University

Richard J. Wood
Dalhousie University

PEARSON
Prentice
Hall

PEARSON EDUCATION INTERNATIONAL

Acquisitions Editor: *Chuck Synovec*
Vice President and Editorial Director, Mathematics: *Christine Hoag*
Project Manager: *Michael Bell*
Production Editor: *Debbie Ryan*
Senior Managing Editor: *Linda Mihatov Behrens*
Executive Managing Editor: *Kathleen Schiaparelli*
Manufacturing Buyer: *Maura Zaldivar*
Manufacturing Manager: *Alexis Heydt-Long*
Marketing Manager: *Wayne Parkins*
Marketing Assistant: *Jennifer de Leeuwerk*
Editorial Assistant/Print Supplements Editor: *Joanne Wendelken*
Art Director: *Maureen Eide*
Interior Designer: *Dina Curro*
Cover Designer: *Kris Carney*
Art Editor: *Thomas Benfatti*
Creative Director: *Juan R. López*
Director of Creative Services: *Paul Belfanti*
Cover Photo: *Ian Cumming/Axiom Photographic Agency/Getty Images*
Manager, Cover Visual Research & Permissions: *Karen Sanatar*
Director, Image Resource Center: *Melinda Patelli*
Manager, Rights and Permissions: *Zina Arabia*
Manager, Visual Research: *Beth Brenzel*
Image Permission Coordinator: *Nancy Seise*
Photo Researcher: *Rachel Lucas*
Art Studio: *Laserwords*

Printed in the United States of America

10 9 8 7 6 5 4 3 2 1

ISBN-13: 978-0-13-242435-6
ISBN-10:   0-13-242435-5

Pearson Education Ltd. *London*
Pearson Education Australia Pty. Ltd. *Sydney*
Pearson Education Singapore, Pte. Ltd.
Pearson Education North Asia Ltd. *Hong Kong*
Pearson Education Canada, Ltd. *Toronto*
Pearson Educación de Mexico, S.A. de C.V.
Pearson Education-Japan, *Tokyo*
Pearson Education Malaysia, Pte. Ltd.
Pearson Education, Upper Saddle River, New Jersey

CONTENTS

PREFACE

The twelfth edition of *Introductory Mathematical Analysis* continues to provide a mathematical foundation for students in business, economics, and the life and social sciences. It begins with noncalculus topics such as functions, equations, mathematics of finance, matrix algebra, linear programming, and probability. Then it progresses through both single-variable and multivariable calculus, including continuous random variables. Technical proofs, conditions, and the like are sufficiently described but are not overdone. Our guiding philosophy led us to include those proofs and general calculations that shed light on how the corresponding calculations are done in applied problems. Informal intuitive arguments are often given too.

Organization Changes to the Twelfth Edition

The few organization changes for this edition reflect comments of adoptees and reviewers. The material in the former Appendix A (as it appeared in Editions 9 through 11) has been folded into the body of the text. In particular, Summation Notation now appears as Section 5 in Chapter 1. The former Summation section from Chapter 14 has also been included in the new Section 5 of Chapter 1. Many instructors found the introduction of summation notation at the same time as a major concept, namely the integral, to be distracting. Our intention is that locating summation in Chapter 1 should accord the topic a more appropriate status. Summation notation is simple but to the extent that it will be new for many students it should brighten a chapter that is otherwise review work for most students. Having summation notation at hand so early in the text allows us to practice with it several times, notably in the work on combinatorics and probability (Chapter 8), before it becomes indispensible in conjunction with the integral (Chapter 14).

The topic of Interest Compounded Continuously has been moved from Chapter 10 to become Section 3 in Chapter 5 on the Mathematics of Finance. Since exponential functions, and the number *e*, are introduced in Chapter 4 this is quite a natural move and allows a more unified treatment of interest rates. A number of instructors thought it important to be able to compare continuously compounded interest with ordinary compound interest while the latter is fresh in the minds of students. However, continuous annuities are still in Chapter 15 as an application of integration.

Finally, Differentiability and Continuity, formerly a separate section, is now included as part of Section 1 of Chapter 11 and we have removed A Comment on Homogeneous Functions from Chapter 17.

Applications

An abundance and variety of applications for the intended audience appear throughout the book; students continually see how the mathematics they are learning can be used. These applications cover such diverse areas as business, economics, biology, medicine, sociology, psychology, ecology, statistics, earth science, and archaeology. Many of these real-world situations are drawn from literature and are documented by references, sometimes from the Web. In some, the background and context are given in order to stimulate interest. However, the text is self-contained, in the sense that it assumes no prior exposure to the concepts on which the applications are based. The **Principles in Practice** element provides students with further applications. Located in the margins of Chapters 1 through 17, these additional exercises give students real-world applications and more opportunities to see the chapter material put into practice. An icon indicates Principles in Practice problems that can be solved using a graphing calculator. Answers to Principles in Practice problems appear at the end of the text and complete solutions to these problems are found in the Solutions Manuals.

Simplified Language and Terminology

In this edition a special effort has been made to choose good terminology without introducing simultaneously an alternative word or phrase connected by the word *or*. For example, when introducing terminology for a point (a, b) in the plane, "We call a the *abscissa or x-coordinate* ..." has been replaced by "We call a the *x-coordinate* ...". In a similar vein, we now speak of a *Simplex Table* rather than of a *Simplex Tableau*. In general, we have tried to be more colloquial when doing so can be done without sacrificing mathematical precision.

Improved Pedagogy

We noticed in reviewing Section 9.3, on Markov Chains, that considerable simplification to the problem of finding steady state vectors is obtained by writing state vectors as columns rather than rows. This does necessitate that a transition matrix $\mathbf{T} = [t_{ij}]$ have

$$t_{ij} = \text{probability that next state is } i \text{ given that current state is } j$$

but avoids artificial transpositions later.

In Chapter 13 on Curve Sketching we have greatly expanded the use of *sign charts*. In particular, a sign chart for a first derivative is always accompanied by a further line interpreting the results for the function which is to be graphed. Thus, on an interval where we record '+' for f' we also record '/' for f and on an interval where we record '−' for f' we also record '\' for f. The resulting strings of such elements, say /\/, with further embellishments that we describe in the text, provide a very preliminary sketch of the curve in question. We freely acknowledge that this is a blackboard technique used by many instructors, but it appears too rarely in textbooks.

Throughout the text we have retained the popular "Now work Problem n" feature of other Prentice Hall books. The idea is that after a worked example, students are directed to an end of section problem that reinforces the ideas of the worked example. For the most part, problems led to this way are odd-numbered so that students can check their answers in the back of the book or find complete solutions in the Student Solutions Manual.

In the same vein, we have expanded the use of cautionary warnings to the student. These are represented by *CAUTION* icons that point out commonly made errors. As before, **Definitions** are clearly stated and displayed. Key concepts, as well as important rules and formulas, are boxed to emphasize their importance.

Each chapter (except Chapter 0) has a review section that contains a list of important terms and symbols, a chapter summary, and numerous review problems. In the twelfth edition we have referenced key examples with each group of important terms and symbols.

Answers to odd-numbered problems appear at the end of the book. For many of the differentiation problems, the answers appear in both 'unsimplified' and 'simplified' forms. (Of course 'simplified' is in any event a subjective term when applied to mathematical expressions, that tends to presuppose the nature of subsequent calculations with such expressions.) This allows students to readily check their work.

Examples and Exercises

More than 850 examples are worked out in detail. Some include a *strategy* that is specifically designed to guide the student through the logistics of the solution before the solution is obtained. An abundant number of diagrams (almost 500) and exercises (more than 5000) are included. Of the exercises, more than 900 are new for the twelfth edition. In each exercise set, grouped problems are given in increasing order of difficulty. In many exercise sets the problems progress from the basic mechanical drill-type to more interesting thought-provoking problems. Many real-world type problems with real data are included. Considerable effort has been made to produce a proper balance between the drill-type exercises and the problems requiring the integration of the concepts learned.

Technology

In order that students appreciate the value of current *technology,* optional graphing calculator material appears throughout the text both in the exposition and exercises. It appears for a variety of reasons: as a mathematical tool, to visualize a concept, as a computing aid, and to reinforce concepts. Although calculator displays for a TI-83 Plus accompany the corresponding technology discussion, our approach is general enough so that it can be applied to other graphing calculators.

In the exercise sets, graphing calculator problems are indicated by an icon. To provide flexibility for an instructor in planning assignments, these problems are usually placed at the end of an exercise set.

Course Planning

Because instructors plan a course outline to serve the individual needs of a particular class and curriculum, we will not attempt to provide detailed sample outlines. However, depending on the background of the students, some instructors will choose to omit Chapter 0 (Algebra Refresher). A considerable number of courses can be served by the book.

A program that allows three quarters of Mathematics for well-prepared Business students can start a first course with Chapter 1 and choose such topics as are of interest, up to and including Chapter 9. For example, if the students are concurrently taking a Finance course then it may be desirable to exclude Chapter 5 on the Mathematics of Finance (to avoid duplication of material for distinct credits). Others will find that Chapter 7 on Linear Programming includes more material than their students need. In this case, specific sections such as 7.3, 7.5, and 7.8 can be excluded without loss of continuity. On the other hand, Section 1.1 introduces some business terms, such as total revenue, fixed cost, variable cost and profit that recur throughout the book. Similarly, Section 3.2 introduces the notion of supply and demand equations, and Section 3.6 discusses the equilibrium point and the break-even point, all of fundamental importance for business applications.

A second, single quarter, course on Differential Calculus will use Chapter 10 on Limits and Continuity, followed by the three 'differentiation chapters', 11 through 13 inclusive. Here, Section 12.6 on Newton's Method can be omitted without loss of continuity while some instructors may prefer to review Chapter 4 on Exponential and Logarithmic Functions prior to their study as differentiable functions.

Finally, Chapters 14 through 17 inclusive could define a third, single quarter, course on Integral Calculus with an introduction to Multivariable Calculus. In an applied course it is well to stress the use of tables to find integrals and thus the techniques of 'parts' and 'partial fractions', in 15.1 and 15.2 respectively, should be considered optional. Chapter 16 is certainly not needed for Chapter 17 and Section 15.7 on Improper Integrals can be safely omitted if Chapter 16 is not covered.

Schools which have two academic terms per year tend to give business students a term devoted to Finite Mathematics and a term devoted to Calculus. For these schools we recommend Chapters 1 through 9 for the first course, starting wherever the preparation of the students allows, and Chapters 10 through 17 for the second—deleting most optional material.

Supplements

The *Instructor's Solution Manual* has worked solutions to all problems, including those in the Principles in Practice elements and in the Mathematical Snapshots.

The *Test Item File,* used by some instructors, provides over 1700 test questions, keyed to chapter and section. It includes an editing feature that allows questions to be added or changed. Also for instructors is *TestGen,* an algorithmic test generator that allows for multiple tests to be created. It is fully editable.

The *Student Solutions Manual* includes worked solutions for all odd-numbered problems and all Principles in Practice problems.

Acknowledgments

We express our appreciation to the following colleagues who contributed comments and suggestions that were valuable to us in the evolution of this text:

E. Adibi (*Chapman University*); R. M. Alliston (*Pennsylvania State University*); R. A. Alo (*University of Houston*); K. T. Andrews (*Oakland University*); M. N. de Arce (*University of Puerto Rico*); E. Barbut (*University of Idaho*); G. R. Bates (*Western Illinois University*); D. E. Bennett (*Murray State University*); C. Bernett (*Harper College*); A. Bishop (*Western Illinois University*); P. Blau (*Shawnee State University*); R. Blute (*University of Ottawa*); S. A. Book (*California State University*); A. Brink (*St. Cloud State University*); R. Brown (*York University*); R. W. Brown (*University of Alaska*); S. D. Bulman-Fleming (*Wilfrid Laurier University*); D. Calvetti (*National College*); D. Cameron (*University of Akron*); K. S. Chung (*Kapiolani Community College*); D. N. Clark (*University of Georgia*); E. L. Cohen (*University of Ottawa*); J. Dawson (*Pennsylvania State University*); A. Dollins (*Pennsylvania State University*); G. A. Earles (*St. Cloud State University*); B. H. Edwards (*University of Florida*); J. R. Elliott (*Wilfrid Laurier University*); J. Fitzpatrick (*University of Texas at El Paso*); M. J. Flynn (*Rhode Island Junior College*); G. J. Fuentes (*University of Maine*); L. Gerber (*St. John's University*); T. G. Goedde (*The University of Findlay*); S. K. Goel (*Valdosta State University*); G. Goff (*Oklahoma State University*); J. Goldman (*DePaul University*); J. T. Gresser (*Bowling Green State University*); L. Griff (*Pennsylvania State University*); F. H. Hall (*Pennsylvania State University*); V. E. Hanks (*Western Kentucky University*); R. C. Heitmann (*The University of Texas at Austin*); J. N. Henry (*California State University*); W. U. Hodgson (*West Chester State College*); B. C. Horne, Jr. (*Virginia Polytechnic Institute and State University*); J. Hradnansky (*Pennsylvania State University*); P. Huneke (*The Ohio State University*); C. Hurd (*Pennsylvania State University*); J. A. Jiminez (*Pennsylvania State University*); W. C. Jones (*Western Kentucky University*); R. M. King (*Gettysburg College*); M. M. Kostreva (*University of Maine*); G. A. Kraus (*Gannon University*); J. Kucera (*Washington State University*); M. R. Latina (*Rhode Island Junior College*); P. Lockwood-Cooke (*West Texas A&M University*); J. F. Longman (*Villanova University*); I. Marshak (*Loyola University of Chicago*); D. Mason (*Elmhurst College*); F. B. Mayer (*Mt. San Antonio College*); P. McDougle (*University of Miami*); F. Miles (*California State University*); E. Mohnike (*Mt. San Antonio College*); C. Monk (*University of Richmond*); R. A. Moreland (*Texas Tech University*); J. G. Morris (*University of Wisconsin-Madison*); J. C. Moss (*Paducah Community College*); D. Mullin (*Pennsylvania State University*); E. Nelson (*Pennsylvania State University*); S. A. Nett (*Western Illinois University*); R. H. Oehmke (*University of Iowa*); Y. Y. Oh (*Pennsylvania State University*); J. U. Overall (*University of La Verne*); A. Panayides (*William Patterson University*); D. Parker (*University of Pacific*); N. B. Patterson (*Pennsylvania State University*); V. Pedwaydon (*Lawrence Technical University*); E. Pemberton (*Wilfrid Laurier University*); M. Perkel (*Wright State University*); D. B. Priest (*Harding College*); J. R. Provencio (*University of Texas*); L. R. Pulsinelli (*Western Kentucky University*); M. Racine (*University of Ottawa*); N. M. Rice (*Queen's University*); A. Santiago (*University of Puerto Rico*);

J. R. Schaefer (*University of Wisconsin–Milwaukee*); S. Sehgal (*The Ohio State University*); W. H. Seybold, Jr. (*West Chester State College*); G. Shilling (*The University of Texas at Arlington*); S. Singh (*Pennsylvania State University*); L. Small (*Los Angeles Pierce College*); E. Smet (*Huron College*); J. Stein (*California State University, Long Beach*); M. Stoll (*University of South Carolina*); T. S. Sullivan (*Southern Illinois University Edwardsville*); E. A. Terry (*St. Joseph's University*); A. Tierman (*Saginaw Valley State University*); B. Toole (*University of Maine*); J. W. Toole (*University of Maine*); D. H. Trahan (*Naval Postgraduate School*); J. P. Tull (*The Ohio State University*); L. O. Vaughan, Jr. (*University of Alabama in Birmingham*); L. A. Vercoe (*Pennsylvania State University*); M. Vuilleumier (*The Ohio State University*); B. K. Waits (*The Ohio State University*); A. Walton (*Virginia Polytechnic Institute and State University*); H. Walum (*The Ohio State University*); E. T. H. Wang (*Wilfrid Laurier University*); A. J. Weidner (*Pennsylvania State University*); L. Weiss (*Pennsylvania State University*); N. A. Weigmann (*California State University*); S. K. Wong (*Ohio State University*); G. Woods (*The Ohio State University*); C. R. B. Wright (*University of Oregon*); C. Wu (*University of Wisconsin–Milwaukee*); B. F. Wyman (*Ohio State University*).

Some exercises are taken from problem supplements used by students at Wilfrid Laurier University. We wish to extend special thanks to the Department of Mathematics of Wilfrid Laurier University for granting Prentice Hall permission to use and publish this material, and also to Prentice Hall, who in turn allowed us to make use of this material.

We again express our sincere gratitude to the faculty and course coordinators of The Ohio State University and Columbus State University who took a keen interest in this and other editions, offering a number of invaluable suggestions.

Special thanks are due to Cindy Trimble of C Trimble & Associates for her careful checking of the manuscript solutions manuals, and answer pages. Her work was extraordinarily detailed and helpful to the authors.

Ernest F. Haeussler, Jr.
Richard S. Paul
Richard J. Wood

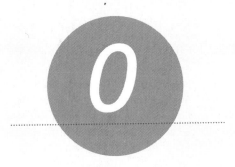

REVIEW OF ALGEBRA

 Mathematical Snapshot   Modeling Load Cell Behavior

esley Griffith works for a yacht supply company in Antibes, France. Often, she needs to examine receipts in which only the total paid is reported and then determine the amount of the total which was French value-added tax, known as TVA for "Taxe à la Value Ajouté". The French TVA rate is 19.6%. A lot of Lesley's business comes from Italian suppliers and purchasers, so she also must deal with the similar problem of receipts containing Italian sales tax at 18%.

A problem of this kind cries out for a formula, but many people are able to work through a particular instance of the problem, using specified numbers, without knowing the formula. Thus if Lesley has a 200 Euro French receipt, she might reason as follows: If the item cost 100 Euros before tax then the receipt total would be for 119.6 Euros with tax of 19.6—and then with a leap of faith—*tax in a receipt total of 200 is to 200 as 19.6 is to 119.6*. Stated mathematically,

$$\frac{\text{tax in } 200}{200} = \frac{19.6}{119.6} \approx 16.4\%$$

At this point it is fairly clear that the amount of TVA in a 200 Euro receipt is about 16.4% of 200 Euros, which is about 32.8 Euros. In fact, many people will now guess that

$$\text{tax in } R = R\left(\frac{p}{100 + p}\%\right)$$

gives the tax in a receipt R, when the tax rate is p%. Thus, if Lesley feels comfortable with her deduction, she can multiply her Italian receipts by $\frac{18}{118}$% to determine the tax they contain.

Of course, most people do not remember formulas for very long and are uncomfortable basing a monetary calculation on a leap of faith. The purpose of this chapter is to review the algebra necessary for you to construct your own formulas, *with confidence*, as needed. In particular, we will derive Lesley's formula, without a mysterious invocation of proportion, from principles with which everybody is familiar. This usage of algebra will appear throughout the book, in the course of making *general calculations with variable quantities*.

In this chapter we will review real numbers and algebraic expressions and the basic operations on them. The chapter is designed to give you a brief review of some terms and methods of symbolic calculation. No doubt you have been exposed to much of this material before. However, because these topics are important in handling the mathematics that comes later, an immediate second exposure to them may be beneficial. Devote whatever time is necessary to the sections in which you need review.

To become familiar with sets, the classification of real numbers, and the real-number line.

0.1 Sets of Real Numbers

A **set** is a collection of objects. For example, we can speak of the set of even numbers between 5 and 11, namely, 6, 8, and 10. An object in a set is called an **element** of that set. If this sounds a little circular, do not worry. The words *set* and *element* are like *line* and *point* in plane geometry. We cannot hope to define them in more primitive terms. It is only with practice in using them that we come to understand their meaning. The situation is also rather like the way in which a child learns a first language. Without knowing any words, a child infers the meaning of a few very simple words and ultimately uses these to build a working vocabulary. None of us needs to understand the mechanics of this process in order to learn how to speak. In the same way, it is possible to learn practical mathematics without becoming embroiled in the issue of undefined primitive terms.

One way to specify a set is by listing its elements, in any order, inside braces. For example, the previous set is {6, 8, 10}, which we could denote by a letter such as *A*, allowing us to write $A = \{6, 8, 10\}$. Note that {8, 10, 6} also denotes the same set, as does {10, 10, 6, 8}. *A set is determined by its elements* and neither repetitions nor rearrangements in a listing affect the set. A set *A* is said to be a subset of a set *B* if and only if every element of *A* is also an element of *B*. For example, if $A = \{6, 8, 10\}$ and $B = \{6, 8, 10, 12\}$, then *A* is a subset of *B*.

Certain sets of numbers have special names. The numbers 1, 2, 3, and so on form the set of **positive integers:**

$$set\ of\ positive\ integers = \{1, 2, 3, \ldots\}$$

The three dots mean that the listing of elements is unending, although we do know what the elements are.

The positive integers, together with 0 and the **negative integers** $-1, -2, -3, \ldots$, form the set of **integers:**

$$set\ of\ integers = \{\ldots, -3, -2, -1, 0, 1, 2, 3, \ldots\}$$

![caution icon] **CAUTION**

The reason for $q \neq 0$ is that we cannot divide by zero.

Every integer is a rational number.

The real numbers consist of all decimal numbers.

The set of **rational numbers** consists of numbers, such as $\frac{1}{2}$ and $\frac{5}{3}$, that can be written as a quotient of two integers. That is, a rational number is a number that can be written as $\frac{p}{q}$, where *p* and *q* are integers and $q \neq 0$. (The symbol "$\neq$" is read "is not equal to.") For example, the numbers $\frac{19}{20}$, $\frac{-2}{7}$, and $\frac{-6}{-2}$ are rational. We remark that $\frac{2}{4}, \frac{1}{2}, \frac{3}{6}, \frac{-4}{-8}$, 0.5, and 50% all represent the same rational number. The integer 2 is rational, since $2 = \frac{2}{1}$. In fact, every integer is rational.

All rational numbers can be represented by decimal numbers that *terminate*, such as $\frac{3}{4} = 0.75$ and $\frac{3}{2} = 1.5$, or by *nonterminating repeating decimal numbers* (composed of a group of digits that repeats without end), such as $\frac{2}{3} = 0.666\ldots$, $\frac{-4}{11} = -0.3636\ldots$, and $\frac{2}{15} = 0.1333\ldots$. Numbers represented by *nonterminating nonrepeating* decimals are called **irrational numbers.** An irrational number cannot be written as an integer divided by an integer. The numbers π (pi) and $\sqrt{2}$ are examples of irrational numbers. Together, the rational numbers and the irrational numbers form the set of **real numbers.**

Real numbers can be represented by points on a line. First we choose a point on the line to represent zero. This point is called the *origin*. (See Figure 0.1.) Then a standard measure of distance, called a *unit distance*, is chosen and is successively marked off both to the right and to the left of the origin. With each point on the line we associate a directed distance, which depends on the position of the point with respect

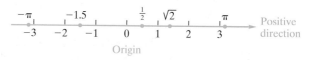

FIGURE 0.1   The real-number line.

to the origin. Positions to the right of the origin are considered positive (+) and positions to the left are negative (−). For example, with the point $\frac{1}{2}$ unit to the right of the origin there corresponds the number $\frac{1}{2}$, which is called the **coordinate** of that point. Similarly, the coordinate of the point 1.5 units to the left of the origin is −1.5. In Figure 0.1, the coordinates of some points are marked. The arrowhead indicates that the direction to the right along the line is considered the positive direction.

To each point on the line there corresponds a unique real number, and to each real number there corresponds a unique point on the line. For this reason, we say that there is a *one-to-one correspondence* between points on the line and real numbers. We call such a line with coordinates marked off a **real-number line.** We feel free to treat real numbers as points on a real-number line and vice versa.

Problems 0.1

In Problems 1–12, determine the truth of each statement. If the statement is false, give a reason why that is so.

1. −13 is an integer.

2. $\frac{-2}{7}$ is rational.

3. −3 is a natural number.

4. 0 is not rational.

5. 5 is rational.

6. $\frac{7}{0}$ is a rational number.

7. $\sqrt{25}$ is not a positive integer.

8. $\sqrt{2}$ is a real number.

9. $\frac{0}{0}$ is rational.

10. $\sqrt{3}$ is a natural number.

11. −3 is to the right of −4 on the real-number line.

12. Every integer is positive or negative.

OBJECTIVE

To name, illustrate, and relate properties of the real numbers in terms of their operations.

0.2 Some Properties of Real Numbers

We now state a few important properties of the real numbers. Let a, b, and c be real numbers.

1. The Transitive Property of Equality

$$\text{If } a = b \text{ and } b = c, \text{ then } a = c.$$

Thus, two numbers that are both equal to a third number are equal to each other. For example, if $x = y$ and $y = 7$, then $x = 7$.

2. The Closure Properties of Addition and Multiplication

For all real numbers a and b, there are unique real numbers $a + b$ and ab.

This means that any two numbers can be added and multiplied, and the result in each case is a real number.

3. The Commutative Properties of Addition and Multiplication

$$a + b = b + a \quad \text{and} \quad ab = ba$$

This means that two numbers can be added or multiplied in any order. For example, $3 + 4 = 4 + 3$ and $7(-4) = (-4)(7)$.

4. The Associative Properties of Addition and Multiplication

$$a + (b + c) = (a + b) + c \quad \text{and} \quad a(bc) = (ab)c$$

This means that in addition or multiplication, numbers can be grouped in any order. For example, $2 + (3 + 4) = (2 + 3) + 4$; in both cases, the sum is 9. Similarly, $2x + (x + y) = (2x + x) + y$ and $6(\frac{1}{3} \cdot 5) = (6 \cdot \frac{1}{3}) \cdot 5$.

5. The Identity Properties

There are unique real numbers denoted 0 and 1 such that, for each real number a,

$$0 + a = a \quad \text{and} \quad 1a = a$$

6. The Inverse Properties

For each real number a, there is a unique real number denoted $-a$ such that

$$a + (-a) = 0$$

The number $-a$ is called the **negative** of a.

For example, since $6 + (-6) = 0$, the negative of 6 is -6. The negative of a number is not necessarily a negative number. For example, the negative of -6 is 6, since $(-6) + (6) = 0$. That is, the negative of -6 is 6, so we can write $-(-6) = 6$.

For each real number a, *except 0*, there is a unique real number denoted a^{-1} such that

$$a \cdot a^{-1} = 1$$

The number a^{-1} is called the **reciprocal** of a.

CAUTION

Zero does not have a reciprocal because there is no number that, when multiplied by 0, gives 1.

Thus, all numbers *except 0* have a reciprocal. You may recall that a^{-1} can be written $\frac{1}{a}$. For example, the reciprocal of 3 is $\frac{1}{3}$, since $3(\frac{1}{3}) = 1$. Hence, $\frac{1}{3}$ is the reciprocal of 3. The reciprocal of $\frac{1}{3}$ is 3, since $(\frac{1}{3})(3) = 1$. *The reciprocal of 0 is not defined.*

7. The Distributive Properties

$$a(b + c) = ab + ac \quad \text{and} \quad (b + c)a = ba + ca$$

For example, although $2(3 + 4) = 2(7) = 14$, we can also write

$$2(3 + 4) = 2(3) + 2(4) = 6 + 8 = 14$$

Similarly,

$$(2 + 3)(4) = 2(4) + 3(4) = 8 + 12 = 20$$

and

$$x(z + 4) = x(z) + x(4) = xz + 4x$$

The distributive property can be extended to the form

$$a(b + c + d) = ab + ac + ad$$

In fact, it can be extended to sums involving any number of terms.

Subtraction is defined in terms of addition:

$$a - b \quad \text{means} \quad a + (-b)$$

where $-b$ is the negative of b. Thus, $6 - 8$ means $6 + (-8)$.

In a similar way, we define **division** in terms of multiplication. If $b \neq 0$, then $a \div b$, or $\dfrac{a}{b}$, or a/b is defined by

$$\frac{a}{b} = a(b^{-1})$$

Since $b^{-1} = \dfrac{1}{b}$

$\dfrac{a}{b}$ means a times the reciprocal of b.

$$\frac{a}{b} = a(b^{-1}) = a\left(\frac{1}{b}\right)$$

Thus, $\frac{3}{5}$ means 3 times $\frac{1}{5}$, where $\frac{1}{5}$ is the reciprocal of 5. Sometimes we refer to $a \div b$ or $\dfrac{a}{b}$ as the *ratio* of a to b. We remark that since 0 does not have a reciprocal, **division by 0 is not defined.**

The following examples show some manipulations involving the preceding properties.

● EXAMPLE 1   Applying Properties of Real Numbers

a. $x(y - 3z + 2w) = (y - 3z + 2w)x$, by the commutative property of multiplication.

b. By the associative property of multiplication, $3(4 \cdot 5) = (3 \cdot 4)5$. Thus, the result of multiplying 3 by the product of 4 and 5 is the same as the result of multiplying the product of 3 and 4 by 5. In either case, the result is 60.

NOW WORK PROBLEM 9

● EXAMPLE 2   Applying Properties of Real Numbers

a. *Show that* $2 - \sqrt{2} = -\sqrt{2} + 2$.

Solution: By the definition of subtraction, $2 - \sqrt{2} = 2 + (-\sqrt{2})$. However, by the commutative property of addition, $2 + (-\sqrt{2}) = -\sqrt{2} + 2$. Hence, by the transitive property of equality, $2 - \sqrt{2} = -\sqrt{2} + 2$. More concisely, we omit intermediate steps and directly write

$$2 - \sqrt{2} = -\sqrt{2} + 2$$

b. *Show that* $(8 + x) - y = 8 + (x - y)$.

Solution: Beginning with the left side, we have

$$(8 + x) - y = (8 + x) + (-y) \qquad \text{(definition of subtraction)}$$
$$= 8 + [x + (-y)] \qquad \text{(associative property)}$$
$$= 8 + (x - y) \qquad \text{(definition of subtraction)}$$

Hence, by the transitive property of equality,

$$(8 + x) - y = 8 + (x - y)$$

c. *Show that* $3(4x + 2y + 8) = 12x + 6y + 24$.

Solution: By the distributive property,

$$3(4x + 2y + 8) = 3(4x) + 3(2y) + 3(8)$$

But by the associative property of multiplication,

$$3(4x) = (3 \cdot 4)x = 12x \quad \text{and similarly} \quad 3(2y) = 6y$$

Thus, $3(4x + 2y + 8) = 12x + 6y + 24$

NOW WORK PROBLEM 21

● EXAMPLE 3   Applying Properties of Real Numbers

a. *Show that* $\dfrac{ab}{c} = a\left(\dfrac{b}{c}\right)$ *for* $c \neq 0$.

Solution: By the definition of division,

$$\frac{ab}{c} = (ab) \cdot \frac{1}{c} \quad \text{for } c \neq 0$$

But by the associative property,

$$(ab) \cdot \frac{1}{c} = a\left(b \cdot \frac{1}{c}\right)$$

However, by the definition of division, $b \cdot \dfrac{1}{c} = \dfrac{b}{c}$. Thus,

$$\frac{ab}{c} = a\left(\frac{b}{c}\right)$$

We can also show that $\dfrac{ab}{c} = \left(\dfrac{a}{c}\right)b$.

b. *Show that* $\dfrac{a+b}{c} = \dfrac{a}{c} + \dfrac{b}{c}$ *for* $c \neq 0$.

Solution: By the definition of division and the distributive property,

$$\frac{a+b}{c} = (a+b)\frac{1}{c} = a \cdot \frac{1}{c} + b \cdot \frac{1}{c}$$

However,

$$a \cdot \frac{1}{c} + b \cdot \frac{1}{c} = \frac{a}{c} + \frac{b}{c}$$

Hence,

$$\frac{a+b}{c} = \frac{a}{c} + \frac{b}{c}$$

NOW WORK PROBLEM 27 ◖●●

Finding the product of several numbers can be done by considering products of numbers taken just two at a time. For example, to find the product of x, y, and z, we could first multiply x by y and then multiply that product by z; that is, we find $(xy)z$. Or, alternatively, we could multiply x by the product of y and z; that is, we find $x(yz)$. The associative property of multiplication guarantees that both results are identical, regardless of how the numbers are grouped. Thus, it is not ambiguous to write xyz. This concept can be extended to more than three numbers and applies equally well to addition.

Not only should you be able to manipulate real numbers; you should also be aware of, and familiar with, the terminology involved.

The following list states important properties of real numbers that you should study thoroughly. Being able to manipulate real numbers is essential to your success in mathematics. A numerical example follows each property. All denominators are different from zero.

Property	*Example(s)*
1. $a - b = a + (-b)$	$2 - 7 = 2 + (-7) = -5$
2. $a - (-b) = a + b$	$2 - (-7) = 2 + 7 = 9$
3. $-a = (-1)(a)$	$-7 = (-1)(7)$
4. $a(b + c) = ab + ac$	$6(7 + 2) = 6 \cdot 7 + 6 \cdot 2 = 54$
5. $a(b - c) = ab - ac$	$6(7 - 2) = 6 \cdot 7 - 6 \cdot 2 = 30$
6. $-(a + b) = -a - b$	$-(7 + 2) = -7 - 2 = -9$
7. $-(a - b) = -a + b$	$-(2 - 7) = -2 + 7 = 5$
8. $-(-a) = a$	$-(-2) = 2$
9. $a(0) = 0$	$2(0) = 0$
10. $(-a)(b) = -(ab) = a(-b)$	$(-2)(7) = -(2 \cdot 7) = 2(-7) = -14$
11. $(-a)(-b) = ab$	$(-2)(-7) = 2 \cdot 7 = 14$
12. $\dfrac{a}{1} = a$	$\dfrac{7}{1} = 7, \dfrac{-2}{1} = -2$

Property	*Example(s)*
13. $\dfrac{a}{b} = a\left(\dfrac{1}{b}\right)$	$\dfrac{2}{7} = 2\left(\dfrac{1}{7}\right)$
14. $\dfrac{a}{-b} = -\dfrac{a}{b} = \dfrac{-a}{b}$	$\dfrac{2}{-7} = -\dfrac{2}{7} = \dfrac{-2}{7}$
15. $\dfrac{-a}{-b} = \dfrac{a}{b}$	$\dfrac{-2}{-7} = \dfrac{2}{7}$
16. $\dfrac{0}{a} = 0$ when $a \neq 0$	$\dfrac{0}{7} = 0$
17. $\dfrac{a}{a} = 1$ when $a \neq 0$	$\dfrac{2}{2} = 1, \ \dfrac{-5}{-5} = 1$
18. $a\left(\dfrac{b}{a}\right) = b$	$2\left(\dfrac{7}{2}\right) = 7$
19. $a \cdot \dfrac{1}{a} = 1$ when $a \neq 0$	$2 \cdot \dfrac{1}{2} = 1$
20. $\dfrac{a}{b} \cdot \dfrac{c}{d} = \dfrac{ac}{bd}$	$\dfrac{2}{3} \cdot \dfrac{4}{5} = \dfrac{2 \cdot 4}{3 \cdot 5} = \dfrac{8}{15}$
21. $\dfrac{ab}{c} = \left(\dfrac{a}{c}\right)b = a\left(\dfrac{b}{c}\right)$	$\dfrac{2 \cdot 7}{3} = \dfrac{2}{3} \cdot 7 = 2 \cdot \dfrac{7}{3}$
22. $\dfrac{a}{bc} = \left(\dfrac{a}{b}\right)\left(\dfrac{1}{c}\right) = \left(\dfrac{1}{b}\right)\left(\dfrac{a}{c}\right)$	$\dfrac{2}{3 \cdot 7} = \dfrac{2}{3} \cdot \dfrac{1}{7} = \dfrac{1}{3} \cdot \dfrac{2}{7}$
23. $\dfrac{a}{b} = \left(\dfrac{a}{b}\right)\left(\dfrac{c}{c}\right) = \dfrac{ac}{bc}$ when $c \neq 0$	$\dfrac{2}{7} = \left(\dfrac{2}{7}\right)\left(\dfrac{5}{5}\right) = \dfrac{2 \cdot 5}{7 \cdot 5}$
24. $\dfrac{a}{b(-c)} = \dfrac{a}{(-b)(c)} = \dfrac{-a}{bc} =$ $\dfrac{-a}{(-b)(-c)} = -\dfrac{a}{bc}$	$\dfrac{2}{3(-5)} = \dfrac{2}{(-3)(5)} = \dfrac{-2}{3(5)} =$ $\dfrac{-2}{(-3)(-5)} = -\dfrac{2}{3(5)} = -\dfrac{2}{15}$
25. $\dfrac{a(-b)}{c} = \dfrac{(-a)b}{c} = \dfrac{ab}{-c} =$ $\dfrac{(-a)(-b)}{-c} = -\dfrac{ab}{c}$	$\dfrac{2(-3)}{5} = \dfrac{(-2)(3)}{5} = \dfrac{2(3)}{-5} =$ $\dfrac{(-2)(-3)}{-5} = -\dfrac{2(3)}{5} = -\dfrac{6}{5}$
26. $\dfrac{a}{c} + \dfrac{b}{c} = \dfrac{a+b}{c}$	$\dfrac{2}{9} + \dfrac{3}{9} = \dfrac{2+3}{9} = \dfrac{5}{9}$
27. $\dfrac{a}{c} - \dfrac{b}{c} = \dfrac{a-b}{c}$	$\dfrac{2}{9} - \dfrac{3}{9} = \dfrac{2-3}{9} = \dfrac{-1}{9}$
28. $\dfrac{a}{b} + \dfrac{c}{d} = \dfrac{ad+bc}{bd}$	$\dfrac{4}{5} + \dfrac{2}{3} = \dfrac{4 \cdot 3 + 5 \cdot 2}{5 \cdot 3} = \dfrac{22}{15}$
29. $\dfrac{a}{b} - \dfrac{c}{d} = \dfrac{ad-bc}{bd}$	$\dfrac{4}{5} - \dfrac{2}{3} = \dfrac{4 \cdot 3 - 5 \cdot 2}{5 \cdot 3} = \dfrac{2}{15}$
30. $\dfrac{\frac{a}{b}}{\frac{c}{d}} = \dfrac{a}{b} \div \dfrac{c}{d} = \dfrac{a}{b} \cdot \dfrac{d}{c} = \dfrac{ad}{bc}$	$\dfrac{\frac{2}{3}}{\frac{7}{5}} = \dfrac{2}{3} \div \dfrac{7}{5} = \dfrac{2}{3} \cdot \dfrac{5}{7} = \dfrac{2 \cdot 5}{3 \cdot 7} = \dfrac{10}{21}$
31. $\dfrac{a}{\frac{b}{c}} = a \div \dfrac{b}{c} = a \cdot \dfrac{c}{b} = \dfrac{ac}{b}$	$\dfrac{2}{\frac{3}{5}} = 2 \div \dfrac{3}{5} = 2 \cdot \dfrac{5}{3} = \dfrac{2 \cdot 5}{3} = \dfrac{10}{3}$
32. $\dfrac{\frac{a}{b}}{c} = \dfrac{a}{b} \div c = \dfrac{a}{b} \cdot \dfrac{1}{c} = \dfrac{a}{bc}$	$\dfrac{\frac{2}{3}}{5} = \dfrac{2}{3} \div 5 = \dfrac{2}{3} \cdot \dfrac{1}{5} = \dfrac{2}{3 \cdot 5} = \dfrac{2}{15}$

Property 23 could be called the **fundamental principle of fractions,** which states that *multiplying or dividing both the numerator and denominator of a fraction by the same nonzero number results in a fraction that is equal to the original fraction.* Thus,

$$\frac{7}{\frac{1}{8}} = \frac{7 \cdot 8}{\frac{1}{8} \cdot 8} = \frac{56}{1} = 56$$

By Properties 28 and 23, we have

$$\frac{2}{5} + \frac{4}{15} = \frac{2 \cdot 15 + 5 \cdot 4}{5 \cdot 15} = \frac{50}{75} = \frac{2 \cdot 25}{3 \cdot 25} = \frac{2}{3}$$

We can also do this problem by converting $\frac{2}{5}$ and $\frac{4}{15}$ into fractions that have the same denominators and then using Property 26. The fractions $\frac{2}{5}$ and $\frac{4}{15}$ can be written with a common denominator of $5 \cdot 15$:

$$\frac{2}{5} = \frac{2 \cdot 15}{5 \cdot 15} \quad \text{and} \quad \frac{4}{15} = \frac{4 \cdot 5}{15 \cdot 5}$$

However, 15 is the *least* such common denominator and is called the *least common denominator* (LCD) of $\frac{2}{5}$ and $\frac{4}{15}$. Thus,

$$\frac{2}{5} + \frac{4}{15} = \frac{2 \cdot 3}{5 \cdot 3} + \frac{4}{15} = \frac{6}{15} + \frac{4}{15} = \frac{6+4}{15} = \frac{10}{15} = \frac{2}{3}$$

Similarly,

$$\frac{3}{8} - \frac{5}{12} = \frac{3 \cdot 3}{8 \cdot 3} - \frac{5 \cdot 2}{12 \cdot 2} \qquad (\text{LCD} = 24)$$

$$= \frac{9}{24} - \frac{10}{24} = \frac{9 - 10}{24}$$

$$= -\frac{1}{24}$$

Problems 0.2

In Problems 1–10, determine the truth of each statement.

1. Every real number has a reciprocal.

2. The reciprocal of $\dfrac{7}{3}$ is $\dfrac{3}{7}$.

3. The negative of 7 is $\dfrac{-1}{7}$.

4. $2(3 \cdot 4) = (2 \cdot 3)(2 \cdot 4)$

5. $-x + y = -y + x$

6. $(x + 2)(4) = 4x + 8$

7. $\dfrac{x + 2}{2} = \dfrac{x}{2} + 1$

8. $3\left(\dfrac{x}{4}\right) = \dfrac{3x}{4}$

*9. $x(5 \cdot y) = (x5) \cdot (xy)$

10. $x(4y) = 4xy$

In Problems 11–20, state which properties of the real numbers are being used.

11. $2(x + y) = 2x + 2y$

12. $(x + 5) + y = y + (x + 5)$

13. $2(3y) = (2 \cdot 3)y$

14. $\dfrac{5}{11} = \dfrac{1}{11} \cdot 5$

15. $5(b - a) = (a - b)(-5)$

16. $y + (x + y) = (y + x) + y$

17. $8 - y = 8 + (-y)$

18. $5(4 + 7) = 5(7 + 4)$

19. $(8 + a)b = 8b + ab$

20. $(-1)[-3 + 4] = (-1)(-3) + (-1)(4)$

In Problems 20–26, show that the statements are true by using properties of the real numbers.

*21. $2x(y - 7) = 2xy - 14x$

22. $(a - b) + c = a + (c - b)$

23. $(x + y)(2) = 2x + 2y$

24. $2[27 + (x + y)] = 2[(y + 27) + x]$

25. $x[(2y + 1) + 3] = 2xy + 4x$

26. $(1 + a)(b + c) = b + c + ab + ac$

*27. Show that $x(y - z + w) = xy - xz + xw$.
[*Hint:* $b + c + d = (b + c) + d$.]

Simplify each of the following if possible.

28. $-2 + (-4)$ 29. $-6 + 2$ 30. $6 + (-4)$

31. $7 - 2$ 32. $7 - (-4)$ 33. $-5 - (-13)$

34. $-a - (-b)$ 35. $(-2)(9)$ 36. $7(-9)$

37. $(-2)(-12)$ 38. $19(-1)$ 39. $\dfrac{-1}{\dfrac{-1}{9}}$

40. $-(-6 + x)$ 41. $-7(x)$ 42. $-12(x - y)$

43. $-[-6 + (-y)]$ 44. $-3 \div 15$ 45. $-9 \div (-27)$

46. $(-a) \div (-b)$ 47. $2(-6 + 2)$ 48. $3[-2(3) + 6(2)]$

49. $(-2)(-4)(-1)$ **50.** $(-12)(-12)$

52. $3(x-4)$ **53.** $4(5+x)$

55. $0(-x)$ **56.** $8\left(\dfrac{1}{11}\right)$

58. $\dfrac{14x}{21y}$ **59.** $\dfrac{3}{-2x}$

61. $\dfrac{a}{c}(3b)$ **62.** $(5a)\left(\dfrac{7}{5a}\right)$

64. $\dfrac{7}{y}\cdot\dfrac{1}{x}$ **65.** $\dfrac{2}{x}\cdot\dfrac{5}{y}$

51. $X(1)$

54. $-(x-2)$

57. $\dfrac{5}{1}$

60. $\dfrac{2}{3}\cdot\dfrac{1}{x}$

63. $\dfrac{-aby}{-ax}$

66. $\dfrac{1}{2}+\dfrac{1}{3}$

67. $\dfrac{5}{12}+\dfrac{3}{4}$ **68.** $\dfrac{3}{10}-\dfrac{7}{15}$ **69.** $\dfrac{4}{5}+\dfrac{6}{5}$

70. $\dfrac{X}{\sqrt{5}}-\dfrac{Y}{\sqrt{5}}$ **71.** $\dfrac{3}{2}-\dfrac{1}{4}+\dfrac{1}{6}$ **72.** $\dfrac{2}{5}-\dfrac{3}{8}$

73. $\dfrac{\frac{6}{x}}{y}$ **74.** $\dfrac{\frac{l}{3}}{m}$ **75.** $\dfrac{\frac{-x}{y^2}}{\frac{z}{xy}}$

76. $\dfrac{7}{0}$ **77.** $\dfrac{0}{7}$ **78.** $\dfrac{0}{0}$

79. $0\cdot 0$

OBJECTIVE

To review positive integral exponents, the zero exponent, negative integral exponents, rational exponents, principal roots, radicals, and the procedure of rationalizing the denominator.

 CAUTION

Some authors say that 0^0 is not defined. However, $0^0 = 1$ is a consistent and often useful definition.

0.3 Exponents and Radicals

The product $x\cdot x\cdot x$ of 3 x's is abbreviated x^3. In general, for n a positive integer, x^n is the abbreviation for the product of n x's. The letter n in x^n is called the *exponent*, and x is called the *base*. More specifically, if n is a positive integer, we have

1. $x^n = \underbrace{x\cdot x\cdot x\cdot\cdots\cdot x}_{n\text{ factors}}$

2. $x^{-n} = \dfrac{1}{x^n} = \dfrac{1}{\underbrace{x\cdot x\cdot x\cdot\cdots\cdot x}_{n\text{ factors}}}$ for $x\neq 0$

3. $\dfrac{1}{x^{-n}} = x^n$

4. $x^0 = 1$

● EXAMPLE 1   Exponents

a. $\left(\dfrac{1}{2}\right)^4 = \left(\dfrac{1}{2}\right)\left(\dfrac{1}{2}\right)\left(\dfrac{1}{2}\right)\left(\dfrac{1}{2}\right) = \dfrac{1}{16}$

b. $3^{-5} = \dfrac{1}{3^5} = \dfrac{1}{3\cdot 3\cdot 3\cdot 3\cdot 3} = \dfrac{1}{243}$

c. $\dfrac{1}{3^{-5}} = 3^5 = 243$

d. $2^0 = 1,\ \pi^0 = 1,\ (-5)^0 = 1$

e. $x^1 = x$

NOW WORK PROBLEM 5

If $r^n = x$, where n is a positive integer, then r is an *nth root* of x. Second roots, the case $n = 2$, are called *square roots*; and third roots, the case $n = 3$, are called *cube roots*. For example, $3^2 = 9$, so 3 is a square root of 9. Since $(-3)^2 = 9$, -3 is also a square root of 9. Similarly, -2 is a cube root of -8, since $(-2)^3 = -8$, while 5 is a fourth root of 625 since $5^4 = 625$.

Some numbers do not have an nth root that is a real number. For example, since the square of any real number is nonnegative, there is no real number that is a square root of -4.

The **principal nth root**[1] of x is the nth root of x that is positive if x is positive and is negative if x is negative and n is odd. We denote the principal nth root of x by $\sqrt[n]{x}$. Thus,

$$\sqrt[n]{x}\text{ is }\begin{cases}\text{positive if }x\text{ is positive}\\ \text{negative if }x\text{ is negative and }n\text{ is odd}\end{cases}$$

For example, $\sqrt[2]{9} = 3$, $\sqrt[3]{-8} = -2$, and $\sqrt[3]{\dfrac{1}{27}} = \tfrac{1}{3}$. We define $\sqrt[n]{0} = 0$.

---

[1]Our use of "principal nth root" does not agree with that in advanced books.

The symbol $\sqrt[n]{x}$ is called a **radical.** Here n is the *index*, x is the *radicand*, and $\sqrt{}$ is the *radical sign*. With principal square roots, we usually omit the index and write $\sqrt{x}$ instead of $\sqrt[2]{x}$. Thus, $\sqrt{9} = 3$.

If x is positive, the expression $x^{p/q}$, where p and q are integers, with no common factors, and q is positive is defined to be $\sqrt[q]{x^p}$. Hence,

$$x^{3/4} = \sqrt[4]{x^3}; \quad 8^{2/3} = \sqrt[3]{8^2} = \sqrt[3]{64} = 4$$

$$4^{-1/2} = \sqrt[2]{4^{-1}} = \sqrt{\frac{1}{4}} = \frac{1}{2}$$

Here are the basic laws of exponents and radicals:[2]

Law	*Example(s)*
1. $x^m \cdot x^n = x^{m+n}$	$2^3 \cdot 2^5 = 2^8 = 256;\ x^2 \cdot x^3 = x^5$
2. $x^0 = 1$	$2^0 = 1$
3. $x^{-n} = \dfrac{1}{x^n}$	$2^{-3} = \dfrac{1}{2^3} = \dfrac{1}{8}$
4. $\dfrac{1}{x^{-n}} = x^n$	$\dfrac{1}{2^{-3}} = 2^3 = 8;\ \dfrac{1}{x^{-5}} = x^5$
5. $\dfrac{x^m}{x^n} = x^{m-n} = \dfrac{1}{x^{n-m}}$	$\dfrac{2^{12}}{2^8} = 2^4 = 16;\ \dfrac{x^8}{x^{12}} = \dfrac{1}{x^4}$
6. $\dfrac{x^m}{x^m} = 1$	$\dfrac{2^4}{2^4} = 1$
7. $(x^m)^n = x^{mn}$	$(2^3)^5 = 2^{15};\ (x^2)^3 = x^6$
8. $(xy)^n = x^n y^n$	$(2 \cdot 4)^3 = 2^3 \cdot 4^3 = 8 \cdot 64 = 512$
9. $\left(\dfrac{x}{y}\right)^n = \dfrac{x^n}{y^n}$	$\left(\dfrac{2}{3}\right)^3 = \dfrac{2^3}{3^3} = \dfrac{8}{27}$
10. $\left(\dfrac{x}{y}\right)^{-n} = \left(\dfrac{y}{x}\right)^n$	$\left(\dfrac{3}{4}\right)^{-2} = \left(\dfrac{4}{3}\right)^2 = \dfrac{16}{9}$
11. $x^{1/n} = \sqrt[n]{x}$	$3^{1/5} = \sqrt[5]{3}$
12. $x^{-1/n} = \dfrac{1}{x^{1/n}} = \dfrac{1}{\sqrt[n]{x}}$	$4^{-1/2} = \dfrac{1}{4^{1/2}} = \dfrac{1}{\sqrt{4}} = \dfrac{1}{2}$
13. $\sqrt[n]{x}\,\sqrt[n]{y} = \sqrt[n]{xy}$	$\sqrt[3]{9}\sqrt[3]{2} = \sqrt[3]{18}$
14. $\dfrac{\sqrt[n]{x}}{\sqrt[n]{y}} = \sqrt[n]{\dfrac{x}{y}}$	$\dfrac{\sqrt[3]{90}}{\sqrt[3]{10}} = \sqrt[3]{\dfrac{90}{10}} = \sqrt[3]{9}$
15. $\sqrt[m]{\sqrt[n]{x}} = \sqrt[mn]{x}$	$\sqrt[3]{\sqrt[4]{2}} = \sqrt[12]{2}$
16. $x^{m/n} = \sqrt[n]{x^m} = (\sqrt[n]{x})^m$	$8^{2/3} = \sqrt[3]{8^2} = (\sqrt[3]{8})^2 = 2^2 = 4$
17. $(\sqrt[m]{x})^m = x$	$(\sqrt[8]{7})^8 = 7$

● **EXAMPLE 2** **Exponents and Radicals**

a. By Law 1,

$$x^6 x^8 = x^{6+8} = x^{14}$$

$$a^3 b^2 a^5 b = a^3 a^5 b^2 b^1 = a^8 b^3$$

$$x^{11} x^{-5} = x^{11-5} = x^6$$

$$z^{2/5} z^{3/5} = z^1 = z$$

$$x x^{1/2} = x^1 x^{1/2} = x^{3/2}$$

---

[2]Although some laws involve restrictions, they are not vital to our discussion.

b. By Law 16,

$$\left(\frac{1}{4}\right)^{3/2} = \left(\sqrt{\frac{1}{4}}\right)^3 = \left(\frac{1}{2}\right)^3 = \frac{1}{8}$$

c. $\left(-\frac{8}{27}\right)^{4/3} = \left(\sqrt[3]{\frac{-8}{27}}\right)^4 = \left(\frac{\sqrt[3]{-8}}{\sqrt[3]{27}}\right)^4$ (Laws 16 and 14)

$$= \left(\frac{-2}{3}\right)^4$$

$$= \frac{(-2)^4}{3^4} = \frac{16}{81}$$ (Law 9)

d. $(64a^3)^{2/3} = 64^{2/3}(a^3)^{2/3}$ (Law 8)

$$= (\sqrt[3]{64})^2 a^2$$ (Laws 16 and 7)

$$= (4)^2 a^2 = 16a^2$$

NOW WORK PROBLEM 39

Rationalizing the denominator of a fraction is a procedure in which a fraction having a radical in its denominator is expressed as an equivalent fraction without a radical in its denominator. We use the fundamental principle of fractions, as Example 3 shows.

⬤ EXAMPLE 3 Rationalizing Denominators

a. $\dfrac{2}{\sqrt{5}} = \dfrac{2}{5^{1/2}} = \dfrac{2 \cdot 5^{1/2}}{5^{1/2} \cdot 5^{1/2}} = \dfrac{2 \cdot 5^{1/2}}{5^1} = \dfrac{2\sqrt{5}}{5}$

b. $\dfrac{2}{\sqrt[6]{3x^5}} = \dfrac{2}{\sqrt[6]{3} \cdot \sqrt[6]{x^5}} = \dfrac{2}{3^{1/6}x^{5/6}} = \dfrac{2 \cdot 3^{5/6}x^{1/6}}{3^{1/6}x^{5/6} \cdot 3^{5/6}x^{1/6}}$

$$= \dfrac{2(3^5 x)^{1/6}}{3x} = \dfrac{2\sqrt[6]{3^5 x}}{3x}$$

NOW WORK PROBLEM 63

The following examples illustrate various applications of the laws of exponents and radicals. All denominators are understood to be nonzero.

⬤ EXAMPLE 4 Exponents

a. *Eliminate negative exponents in* $\dfrac{x^{-2}y^3}{z^{-2}}$.

Solution:

$$\frac{x^{-2}y^3}{z^{-2}} = x^{-2} \cdot y^3 \cdot \frac{1}{z^{-2}} = \frac{1}{x^2} \cdot y^3 \cdot z^2 = \frac{y^3 z^2}{x^2}$$

By comparing our answer with the original expression, we conclude that we can bring a factor of the numerator down to the denominator, and vice versa, by changing the sign of the exponent.

b. *Simplify* $\dfrac{x^2 y^7}{x^3 y^5}$.

Solution:

$$\frac{x^2 y^7}{x^3 y^5} = \frac{y^{7-5}}{x^{3-2}} = \frac{y^2}{x}$$

c. *Simplify* $(x^5 y^8)^5$.

Solution:

$$(x^5 y^8)^5 = (x^5)^5 (y^8)^5 = x^{25} y^{40}$$

d. *Simplify* $(x^{5/9} y^{4/3})^{18}$.

Solution:

$$(x^{5/9} y^{4/3})^{18} = (x^{5/9})^{18} (y^{4/3})^{18} = x^{10} y^{24}$$

e. *Simplify* $\left(\dfrac{x^{1/5} y^{6/5}}{z^{2/5}} \right)^5$.

Solution:

$$\left(\frac{x^{1/5} y^{6/5}}{z^{2/5}} \right)^5 = \frac{(x^{1/5} y^{6/5})^5}{(z^{2/5})^5} = \frac{x y^6}{z^2}$$

f. *Simplify* $\dfrac{x^3}{y^2} \div \dfrac{x^6}{y^5}$.

Solution:

$$\frac{x^3}{y^2} \div \frac{x^6}{y^5} = \frac{x^3}{y^2} \cdot \frac{y^5}{x^6} = \frac{y^3}{x^3}$$

NOW WORK PROBLEM 51 ●●●

●●EXAMPLE 5   **Exponents**

a. *Eliminate negative exponents in* $x^{-1} + y^{-1}$ *and simplify.*

Solution:

$$x^{-1} + y^{-1} = \frac{1}{x} + \frac{1}{y} = \frac{y + x}{xy}$$

b. *Simplify* $x^{3/2} - x^{1/2}$ *by using the distributive law.*

Solution:

$$x^{3/2} - x^{1/2} = x^{1/2}(x - 1)$$

c. *Eliminate negative exponents in* $7x^{-2} + (7x)^{-2}$.

Solution:

$$7x^{-2} + (7x)^{-2} = \frac{7}{x^2} + \frac{1}{(7x)^2} = \frac{7}{x^2} + \frac{1}{49x^2}$$

d. *Eliminate negative exponents in* $(x^{-1} - y^{-1})^{-2}$.

Solution:

$$(x^{-1} - y^{-1})^{-2} = \left(\frac{1}{x} - \frac{1}{y} \right)^{-2} = \left(\frac{y - x}{xy} \right)^{-2}$$

$$= \left(\frac{xy}{y - x} \right)^2 = \frac{x^2 y^2}{(y - x)^2}$$

e. *Apply the distributive law to* $x^{2/5}(y^{1/2} + 2x^{6/5})$.

Solution:

$$x^{2/5}(y^{1/2} + 2x^{6/5}) = x^{2/5} y^{1/2} + 2x^{8/5}$$

NOW WORK PROBLEM 41 ●●

● EXAMPLE 6   **Radicals**

a. *Simplify* $\sqrt[4]{48}$.

Solution:

$$\sqrt[4]{48} = \sqrt[4]{16 \cdot 3} = \sqrt[4]{16}\,\sqrt[4]{3} = 2\sqrt[4]{3}$$

b. *Rewrite* $\sqrt{2+5x}$ *without using a radical sign.*

Solution:

$$\sqrt{2+5x} = (2+5x)^{1/2}$$

c. *Rationalize the denominator of* $\dfrac{\sqrt[5]{2}}{\sqrt[3]{6}}$ *and simplify.*

Solution:

$$\frac{\sqrt[5]{2}}{\sqrt[3]{6}} = \frac{2^{1/5} \cdot 6^{2/3}}{6^{1/3} \cdot 6^{2/3}} = \frac{2^{3/15}6^{10/15}}{6} = \frac{(2^3 6^{10})^{1/15}}{6} = \frac{\sqrt[15]{2^3 6^{10}}}{6}$$

d. *Simplify* $\dfrac{\sqrt{20}}{\sqrt{5}}$.

Solution:

$$\frac{\sqrt{20}}{\sqrt{5}} = \sqrt{\frac{20}{5}} = \sqrt{4} = 2$$

NOW WORK PROBLEM 71

● EXAMPLE 7   **Radicals**

a. *Simplify* $\sqrt[3]{x^6 y^4}$.

Solution:

$$\sqrt[3]{x^6 y^4} = \sqrt[3]{(x^2)^3 y^3 y} = \sqrt[3]{(x^2)^3} \cdot \sqrt[3]{y^3} \cdot \sqrt[3]{y}$$
$$= x^2 y \sqrt[3]{y}$$

b. *Simplify* $\sqrt{\dfrac{2}{7}}$.

Solution:

$$\sqrt{\frac{2}{7}} = \sqrt{\frac{2 \cdot 7}{7 \cdot 7}} = \sqrt{\frac{14}{7^2}} = \frac{\sqrt{14}}{\sqrt{7^2}} = \frac{\sqrt{14}}{7}$$

c. *Simplify* $\sqrt{250} - \sqrt{50} + 15\sqrt{2}$.

Solution:

$$\sqrt{250} - \sqrt{50} + 15\sqrt{2} = \sqrt{25 \cdot 10} - \sqrt{25 \cdot 2} + 15\sqrt{2}$$
$$= 5\sqrt{10} - 5\sqrt{2} + 15\sqrt{2}$$
$$= 5\sqrt{10} + 10\sqrt{2}$$

d. *If x is any real number, simplify* $\sqrt{x^2}$.

Solution:

$$\sqrt{x^2} = \begin{cases} x & \text{if } x \geq 0 \\ -x & \text{if } x < 0 \end{cases}$$

Thus, $\sqrt{2^2} = 2$ and $\sqrt{(-3)^2} = -(-3) = 3$.

NOW WORK PROBLEM 75   ●●●

Problems 0.3

In Problems 1–14, simplify and express all answers in terms of positive exponents.

1. $(2^3)(2^2)$

2. $x^6 x^9$

3. $w^4 w^8$

4. $z^3 z z^2$

*5. $\dfrac{x^3 x^5}{y^9 y^5}$

6. $(x^{12})^4$

7. $\dfrac{(a^3)^7}{(b^4)^5}$

8. $\left(\dfrac{x^2}{y^3}\right)^5$

9. $(2x^2 y^3)^3$

10. $\left(\dfrac{w^2 s^3}{y^2}\right)^2$

11. $\dfrac{x^9}{x^5}$

12. $\left(\dfrac{2a^4}{7b^5}\right)^6$

13. $\dfrac{(x^3)^6}{x(x^3)}$

14. $\dfrac{(x^2)^3 (x^3)^2}{(x^3)^4}$

In Problems 15–28, evaluate the expressions.

15. $\sqrt{25}$

16. $\sqrt[4]{81}$

17. $\sqrt[7]{-128}$

18. $\sqrt{0.04}$

19. $\sqrt[4]{\dfrac{1}{16}}$

20. $\sqrt[3]{-\dfrac{8}{27}}$

21. $(49)^{1/2}$

22. $(64)^{1/3}$

23. $9^{3/2}$

24. $(9)^{-5/2}$

25. $(32)^{-2/5}$

26. $(0.09)^{-1/2}$

27. $\left(\dfrac{1}{32}\right)^{4/5}$

28. $\left(-\dfrac{64}{27}\right)^{2/3}$

In Problems 29–40, simplify the expressions.

29. $\sqrt{50}$

30. $\sqrt[3]{54}$

31. $\sqrt[3]{2x^3}$

32. $\sqrt{4x}$

33. $\sqrt{16x^4}$

34. $\sqrt[4]{\dfrac{x}{16}}$

35. $2\sqrt{8} - 5\sqrt{27} + \sqrt[3]{128}$

36. $\sqrt{\dfrac{3}{13}}$

37. $(9z^4)^{1/2}$

38. $(16y^8)^{3/4}$

*39. $\left(\dfrac{27t^3}{8}\right)^{2/3}$

40. $\left(\dfrac{256}{x^{12}}\right)^{-3/4}$

In Problems 41–52, write the expressions in terms of positive exponents only. Avoid all radicals in the final form. For example,

$$y^{-1}\sqrt{x} = \dfrac{x^{1/2}}{y}$$

*41. $\dfrac{a^5 b^{-3}}{c^2}$

42. $\sqrt[5]{x^2 y^3 z^{-10}}$

43. $5m^{-2} m^{-7}$

44. $x + y^{-1}$

45. $(3t)^{-2}$

46. $(3-z)^{-4}$

47. $\sqrt[5]{5x^2}$

48. $(X^3 Y^{-3})^{-3}$

49. $\sqrt{x} - \sqrt{y}$

50. $\dfrac{u^{-2} v^{-6} w^3}{v w^{-5}}$

*51. $x^2 \sqrt[4]{xy^{-2} z^3}$

52. $\sqrt[4]{a^{-3} b^{-2}} a^5 b^{-4}$

In Problems 53–58, rewrite the exponential forms using radicals.

53. $(2a - b + c)^{2/3}$

54. $(ab^2 c^3)^{3/4}$

55. $x^{-4/5}$

56. $2x^{1/2} - (2y)^{1/2}$

57. $3w^{-3/5} - (3w)^{-3/5}$

58. $[(x^{-4})^{1/5}]^{1/6}$

In Problems 59–68, rationalize the denominators.

59. $\dfrac{6}{\sqrt{5}}$

60. $\dfrac{3}{\sqrt[4]{8}}$

61. $\dfrac{4}{\sqrt{2x}}$

62. $\dfrac{y}{\sqrt{2y}}$

*63. $\dfrac{1}{\sqrt[3]{3x}}$

64. $\dfrac{2}{3\sqrt[3]{y^2}}$

65. $\dfrac{\sqrt{12}}{\sqrt{3}}$

66. $\dfrac{\sqrt{18}}{\sqrt{2}}$

67. $\dfrac{\sqrt[5]{2}}{\sqrt[4]{a^2 b}}$

68. $\dfrac{\sqrt{2}}{\sqrt[3]{3}}$

In Problems 69–90, simplify. Express all answers in terms of positive exponents. Rationalize the denominator where necessary to avoid fractional exponents in the denominator.

69. $2x^2 y^{-3} x^4$

70. $\dfrac{3}{u^{5/2} v^{1/2}}$

*71. $\dfrac{\sqrt{243}}{\sqrt{3}}$

72. $\{[(3a^3)^2]^{-5}\}^{-2}$

73. $\dfrac{2^0}{(2^{-2} x^{1/2} y^{-2})^3}$

74. $\dfrac{\sqrt{s^5}}{\sqrt[3]{s^2}}$

*75. $\sqrt[3]{x^2 y z^3} \sqrt[3]{xy^2}$

76. $(\sqrt[4]{3})^8$

77. $3^2 (32)^{-2/5}$

78. $(\sqrt[5]{x^2 y})^{2/5}$

79. $(2x^{-1} y^2)^2$

80. $\dfrac{3}{\sqrt[3]{y}\sqrt[4]{x}}$

81. $\sqrt{x}\sqrt{x^2 y^3}\sqrt{xy^2}$

82. $\sqrt{75k^4}$

83. $\dfrac{(ab^{-3} c)^8}{(a^{-1} c^2)^{-3}}$

84. $\sqrt[3]{7(49)}$

85. $\dfrac{(x^2)^3}{x^4} \div \left[\dfrac{x^3}{(x^3)^2}\right]^2$

86. $\sqrt{(-6)(-6)}$

87. $-\dfrac{8s^{-2}}{2s^3}$

88. $(a^5 b^{-3}\sqrt{c})^3$

89. $(3x^3 y^2 \div 2y^2 z^{-3})^4$

90. $\dfrac{1}{\left(\dfrac{\sqrt{2}x^{-2}}{\sqrt{16}x^3}\right)^2}$

To add, subtract, multiply, and divide algebraic expressions. To define a polynomial, to use special products, and to use long division to divide polynomials.

0.4 Operations with Algebraic Expressions

If numbers, represented by symbols, are combined by any or all of the operations of addition, subtraction, multiplication, division, exponentiation, and extraction of roots, then the resulting expression is called an *algebraic expression*.

⬤ **EXAMPLE 1   Algebraic Expressions**

a. $\sqrt[3]{\dfrac{3x^3 - 5x - 2}{10 - x}}$ is an algebraic expression in the variable x.

b. $10 - 3\sqrt{y} + \dfrac{5}{7 + y^2}$ is an algebraic expression in the variable y.

c. $\dfrac{(x + y)^3 - xy}{y} + 2$ is an algebraic expression in the variables x and y.

The algebraic expression $5ax^3 - 2bx + 3$ consists of three *terms:* $+ 5ax^3$, $-2bx$, and $+3$. Some of the *factors* of the first term, $5ax^3$, are 5, a, x, x^2, x^3, $5ax$, and ax^2. Also, $5a$ is the coefficient of x^3, and 5 is the *numerical coefficient* of ax^3. If a and b represent fixed numbers throughout a discussion, then a and b are called *constants.*

Algebraic expressions with exactly one term are called *monomials.* Those having exactly two terms are *binomials,* and those with exactly three terms are *trinomials.* Algebraic expressions with more than one term are called *multinomials.* Thus, the multinomial $2x - 5$ is a binomial; the multinomial $3\sqrt{y} + 2y - 4y^2$ is a trinomial.

A *polynomial in x* is an algebraic expression of the form[3]

$$c_n x^n + c_{n-1} x^{n-1} + \cdots + c_1 x + c_0$$

where n is a nonnegative integer and the coefficients $c_0, c_1, \ldots, c_n$ are constants with $c_n \neq 0$. We call n the *degree* of the polynomial. Hence, $4x^3 - 5x^2 + x - 2$ is a polynomial in x of degree 3, and $y^5 - 2$ is a polynomial in y of degree 5. A nonzero constant is a polynomial of degree zero; thus, 5 is a polynomial of degree zero. The constant 0 is considered to be a polynomial; however, no degree is assigned to it.

In the following examples, we illustrate operations with algebraic expressions.

CAUTION

The words *polynomial* and *multinomial* should not be used interchangeably. For example, $\sqrt{x} + 2$ is a multinomial, but not a polynomial. On the other hand, $x + 2$ is a multinomial and a polynomial.

EXAMPLE 2 Adding Algebraic Expressions

Simplify $(3x^2 y - 2x + 1) + (4x^2 y + 6x - 3)$.

Solution: We first remove the parentheses. Next, using the commutative property of addition, we gather all similar terms together. *Similar terms* are terms that differ only by their numerical coefficients. In this example, $3x^2 y$ and $4x^2 y$ are similar, as are the pairs $-2x$ and $6x$, and 1 and -3. Thus,

$$(3x^2 y - 2x + 1) + (4x^2 y + 6x - 3) = 3x^2 y - 2x + 1 + 4x^2 y + 6x - 3$$
$$= 3x^2 y + 4x^2 y - 2x + 6x + 1 - 3$$

By the distributive property,

$$3x^2 y + 4x^2 y = (3 + 4)x^2 y = 7x^2 y$$

and

$$-2x + 6x = (-2 + 6)x = 4x$$

Hence, $(3x^2 y - 2x + 1) + (4x^2 y + 6x - 3) = 7x^2 y + 4x - 2$

NOW WORK PROBLEM 7

EXAMPLE 3 Subtracting Algebraic Expressions

Simplify $(3x^2 y - 2x + 1) - (4x^2 y + 6x - 3)$.

Solution: Here we apply the definition of subtraction and the distributive property:

$$(3x^2 y - 2x + 1) - (4x^2 y + 6x - 3)$$
$$= (3x^2 y - 2x + 1) + (-1)(4x^2 y + 6x - 3)$$
$$= (3x^2 y - 2x + 1) + (-4x^2 y - 6x + 3)$$

---

[3]The three dots indicate all other terms that are understood to be included in the sum.

$$= 3x^2y - 2x + 1 - 4x^2y - 6x + 3$$
$$= 3x^2y - 4x^2y - 2x - 6x + 1 + 3$$
$$= (3 - 4)x^2y + (-2 - 6)x + 1 + 3$$
$$= -x^2y - 8x + 4$$

NOW WORK PROBLEM 13

 EXAMPLE 4 Removing Grouping Symbols

Simplify $3\{2x[2x + 3] + 5[4x^2 - (3 - 4x)]\}$.

Solution: We first eliminate the innermost grouping symbols (the parentheses). Then we repeat the process until all grouping symbols are removed—combining similar terms whenever possible. We have

$$3\{2x[2x + 3] + 5[4x^2 - (3 - 4x)]\} = 3\{2x[2x + 3] + 5[4x^2 - 3 + 4x]\}$$
$$= 3\{4x^2 + 6x + 20x^2 - 15 + 20x\}$$
$$= 3\{24x^2 + 26x - 15\}$$
$$= 72x^2 + 78x - 45$$

NOW WORK PROBLEM 15

The distributive property is the key tool in multiplying expressions. For example, to multiply $ax + c$ by $bx + d$ we can consider $ax + c$ to be a single number and then use the distributive property:

$$(ax + c)(bx + d) = (ax + c)bx + (ax + c)d$$

Using the distributive property again, we have

$$(ax + c)bx + (ax + c)d = abx^2 + cbx + adx + cd$$
$$= abx^2 + (ad + cb)x + cd$$

Thus, $(ax + c)(bx + d) = abx^2 + (ad + cb)x + cd$. In particular, if $a = 2, b = 1$, $c = 3$, and $d = -2$, then

$$(2x + 3)(x - 2) = 2(1)x^2 + [2(-2) + 3(1)]x + 3(-2)$$
$$= 2x^2 - x - 6$$

We now give a list of special products that may be obtained from the distributive property and are useful in multiplying algebraic expressions.

Special Products

1. $x(y + z) = xy + xz$ (distributive property)

2. $(x + a)(x + b) = x^2 + (a + b)x + ab$

3. $(ax + c)(bx + d) = abx^2 + (ad + cb)x + cd$

4. $(x + a)^2 = x^2 + 2ax + a^2$ (square of a binomial)

5. $(x - a)^2 = x^2 - 2ax + a^2$ (square of a binomial)

6. $(x + a)(x - a) = x^2 - a^2$ (product of sum and difference)

7. $(x + a)^3 = x^3 + 3ax^2 + 3a^2x + a^3$ (cube of a binomial)

8. $(x - a)^3 = x^3 - 3ax^2 + 3a^2x - a^3$ (cube of a binomial)

● EXAMPLE 5   **Special Products**

a. By Rule 2,

$$(x+2)(x-5) = [x+2][x+(-5)]$$
$$= x^2 + (2-5)x + 2(-5)$$
$$= x^2 - 3x - 10$$

b. By Rule 3,

$$(3z+5)(7z+4) = 3 \cdot 7z^2 + (3 \cdot 4 + 5 \cdot 7)z + 5 \cdot 4$$
$$= 21z^2 + 47z + 20$$

c. By Rule 5,

$$(x-4)^2 = x^2 - 2(4)x + 4^2$$
$$= x^2 - 8x + 16$$

d. By Rule 6,

$$(\sqrt{y^2+1}+3)(\sqrt{y^2+1}-3) = (\sqrt{y^2+1})^2 - 3^2$$
$$= (y^2+1) - 9$$
$$= y^2 - 8$$

e. By Rule 7,

$$(3x+2)^3 = (3x)^3 + 3(2)(3x)^2 + 3(2)^2(3x) + (2)^3$$
$$= 27x^3 + 54x^2 + 36x + 8$$

NOW WORK PROBLEM 19 ●●

● EXAMPLE 6   **Multiplying Multinomials**

Find the product $(2t-3)(5t^2+3t-1)$.

Solution: We treat $2t-3$ as a single number and apply the distributive property twice:

$$(2t-3)(5t^2+3t-1) = (2t-3)5t^2 + (2t-3)3t - (2t-3)1$$
$$= 10t^3 - 15t^2 + 6t^2 - 9t - 2t + 3$$
$$= 10t^3 - 9t^2 - 11t + 3$$

NOW WORK PROBLEM 35 ●●

In Example 3(b) of Section 0.2, we showed that $\dfrac{a+b}{c} = \dfrac{a}{c} + \dfrac{b}{c}$. Similarly, $\dfrac{a-b}{c} = \dfrac{a}{c} - \dfrac{b}{c}$. Using these results, we can divide a multinomial by a monomial by dividing each term in the multinomial by the monomial.

● EXAMPLE 7   **Dividing a Multinomial by a Monomial**

a. $\dfrac{x^3+3x}{x} = \dfrac{x^3}{x} + \dfrac{3x}{x} = x^2 + 3$

b. $\dfrac{4z^3 - 8z^2 + 3z - 6}{2z} = \dfrac{4z^3}{2z} - \dfrac{8z^2}{2z} + \dfrac{3z}{2z} - \dfrac{6}{2z}$

$$= 2z^2 - 4z + \frac{3}{2} - \frac{3}{z}$$

NOW WORK PROBLEM 47 ●●

Long Division

To divide a polynomial by a polynomial, we use so-called long division when the degree of the divisor is less than or equal to the degree of the dividend, as the next example shows.

● EXAMPLE 8   Long Division

Divide $2x^3 - 14x - 5$ by $x - 3$.

Solution: Here $2x^3 - 14x - 5$ is the *dividend* and $x - 3$ is the *divisor*. To avoid errors, it is best to write the dividend as $2x^3 + 0x^2 - 14x - 5$. Note that the powers of x are in decreasing order. We have

$$
\begin{array}{r}
2x^2 + 6x + 4 \leftarrow \text{quotient} \\
\text{divisor} \rightarrow x - 3 \overline{)\,2x^3 + 0x^2 - 14x - 5\,} \leftarrow \text{dividend} \\
\underline{2x^3 - 6x^2} \\
6x^2 - 14x \\
\underline{6x^2 - 18x} \\
4x - 5 \\
\underline{4x - 12} \\
7 \leftarrow \text{remainder}
\end{array}
$$

Note that we divided x (the first term of the divisor) into $2x^3$ and got $2x^2$. Then we multiplied $2x^2$ by $x - 3$, getting $2x^3 - 6x^2$. After subtracting $2x^3 - 6x^2$ from $2x^3 + 0x^2$, we obtained $6x^2$ and then "brought down" the term $-14x$. This process is continued until we arrive at 7, the *remainder*. We always stop when the remainder is 0 or is a polynomial whose degree is less than the degree of the divisor. Our answer may be written as

$$2x^2 + 6x + 4 + \frac{7}{x - 3}$$

That is, the answer to the question

$$\frac{\text{dividend}}{\text{divisor}} = ?$$

has the form

$$\text{quotient} + \frac{\text{remainder}}{\text{divisor}}$$

A way of checking a division is to verify that

$$(\text{quotient})(\text{divisor}) + \text{remainder} = \text{dividend}$$

By using this equation, you should be able to verify the result of the example.

NOW WORK PROBLEM 51

Problems 0.4

Perform the indicated operations and simplify.

1. $(8x - 4y + 2) + (3x + 2y - 5)$

2. $(6x^2 - 10xy + 2) + (2z - xy + 4)$

3. $(8t^2 - 6s^2) + (4s^2 - 2t^2 + 6)$

4. $(\sqrt{x} + 2\sqrt{x}) + (\sqrt{x} + 3\sqrt{x})$

5. $(\sqrt{a} + 2\sqrt{3b}) - (\sqrt{c} - 3\sqrt{3b})$

6. $(3a + 7b - 9) - (5a + 9b + 21)$

*7. $(6x^2 - 10xy + \sqrt{2}) - (2z - xy + 4)$

8. $(\sqrt{x} + 2\sqrt{x}) - (\sqrt{x} + 3\sqrt{x})$

9. $(\sqrt{x} + \sqrt{2y}) - (\sqrt{x} + \sqrt{3z})$

10. $4(2z - w) - 3(w - 2z)$

11. $3(3x + 3y - 7) - 3(8x - 2y + 2)$

12. $(u - 3v) + (-5u - 4v) + (u - 3)$

*13. $5(x^2 - y^2) + x(y - 3x) - 4y(2x + 7y)$

14. $2 - [3 + 4(s - 3)]$

*15. $2\{3[3(x^2 + 2) - 2(x^2 - 5)]\}$

16. $4\{3(t + 5) - t[1 - (t + 1)]\}$

17. $-5(4x^2(2x + 2) - 2(x^2 - (5 - 2x)))$

18. $-\{-3[2a + 2b - 2] + 5[2a + 3b] - a[2(b + 5)]\}$

*19. $(x + 4)(x + 5)$ **20.** $(u + 2)(u + 5)$

21. $(w + 2)(w - 5)$ **22.** $(z - 7)(z - 3)$

23. $(2x + 3)(5x + 2)$ **24.** $(t - 5)(2t + 7)$

25. $(X + 2Y)^2$ **26.** $(2x - 1)^2$

27. $(x - 5)^2$ **28.** $(\sqrt{x} - 1)(2\sqrt{x} + 5)$

29. $(\sqrt{3x} + 5)^2$ **30.** $(\sqrt{y} - 3)(\sqrt{y} + 3)$

31. $(2s - 1)(2s + 1)$ **32.** $(z^2 - 3w)(z^2 + 3w)$

33. $(x^2 - 3)(x + 4)$ **34.** $(x + 1)(x^2 + x + 3)$

*35. $(x^2 - 4)(3x^2 + 2x - 1)$ **36.** $(3y - 2)(4y^3 + 2y^2 - 3y)$

37. $x\{2(x + 5)(x - 7) + 4[2x(x - 6)]\}$

38. $[(2z + 1)(2z - 1)](4z^2 + 1)$

39. $(x + y + 2)(3x + 2y - 4)$

40. $(x^2 + x + 1)^2$ **41.** $(2a + 3)^3$

42. $(3y - 2)^3$ **43.** $(2x - 3)^3$

44. $(x + 2y)^3$ **45.** $\dfrac{z^2 - 18z}{z}$

46. $\dfrac{2x^3 - 7x + 4}{x}$ ***47.** $\dfrac{6x^5 + 4x^3 - 1}{2x^2}$

48. $\dfrac{(3y - 4) - (9y + 5)}{3y}$

49. $(x^2 + 5x - 3) \div (x + 5)$

50. $(x^2 - 5x + 4) \div (x - 4)$

*51. $(3x^3 - 2x^2 + x - 3) \div (x + 2)$

52. $(x^4 + 2x^2 + 1) \div (x - 1)$

53. $x^3 \div (x + 2)$

54. $(6x^2 + 8x + 1) \div (2x + 3)$

55. $(3x^2 - 4x + 3) \div (3x + 2)$

56. $(z^3 + z^2 + z) \div (z^2 - z + 1)$

OBJECTIVE

To state the basic rules for factoring and apply them to factor expressions.

0.5 Factoring

If two or more expressions are multiplied together, the expressions are called *factors* of the product. Thus, if $c = ab$, then a and b are both factors of the product c. The process by which an expression is written as a product of its factors is called *factoring*.

Listed next are rules for factoring expressions, most of which arise from the special products discussed in Section 0.4. The right side of each identity is the factored form of the left side.

Rules for Factoring

1. $xy + xz = x(y + z)$ (common factor)

2. $x^2 + (a + b)x + ab = (x + a)(x + b)$

3. $abx^2 + (ad + cb)x + cd = (ax + c)(bx + d)$

4. $x^2 + 2ax + a^2 = (x + a)^2$ (perfect-square trinomial)

5. $x^2 - 2ax + a^2 = (x - a)^2$ (perfect-square trinomial)

6. $x^2 - a^2 = (x + a)(x - a)$ (difference of two squares)

7. $x^3 + a^3 = (x + a)(x^2 - ax + a^2)$ (sum of two cubes)

8. $x^3 - a^3 = (x - a)(x^2 + ax + a^2)$ (difference of two cubes)

When factoring a polynomial, we usually choose factors that themselves are polynomials. For example, $x^2 - 4 = (x + 2)(x - 2)$. We will not write $x - 4$ as $(\sqrt{x} + 2)(\sqrt{x} - 2)$ unless it allows us to simplify other calculations.

Always factor as completely as you can. For example,

$$2x^2 - 8 = 2(x^2 - 4) = 2(x + 2)(x - 2)$$

● EXAMPLE 1   Common Factors

a. *Factor* $3k^2x^2 + 9k^3x$ *completely.*

Solution: Since $3k^2x^2 = (3k^2x)(x)$ and $9k^3x = (3k^2x)(3k)$, each term of the original expression contains the common factor $3k^2x$. Thus, by Rule 1,

$$3k^2x^2 + 9k^3x = 3k^2x(x + 3k)$$

Note that although $3k^2x^2 + 9k^3x = 3(k^2x^2 + 3k^3x)$, we do not say that the expression is completely factored, since $k^2x^2 + 3k^3x$ can still be factored.

b. *Factor $8a^5x^2y^3 - 6a^2b^3yz - 2a^4b^4xy^2z^2$ completely.*

Solution:

$$8a^5x^2y^3 - 6a^2b^3yz - 2a^4b^4xy^2z^2 = 2a^2y(4a^3x^2y^2 - 3b^3z - a^2b^4xyz^2)$$

NOW WORK PROBLEM 5 ◖●●

●EXAMPLE 2 **Factoring Trinomials**

a. *Factor $3x^2 + 6x + 3$ completely.*

Solution: First we remove a common factor. Then we factor the resulting expression completely. Thus, we have

$$3x^2 + 6x + 3 = 3(x^2 + 2x + 1)$$
$$= 3(x + 1)^2 \qquad \text{(Rule 4)}$$

b. *Factor $x^2 - x - 6$ completely.*

Solution: If this trinomial factors into the form $(x + a)(x + b)$, which is a product of two binomials, then we must determine the values of a and b. Since $(x + a)(x + b) = x^2 + (a + b)x + ab$, it follows that

$$x^2 + (-1)x + (-6) = x^2 + (a + b)x + ab$$

By equating corresponding coefficients, we want

$$a + b = -1 \quad \text{and} \quad ab = -6$$

If $a = -3$ and $b = 2$ then both conditions are met and hence:

$$x^2 - x - 6 = (x - 3)(x + 2)$$

As a check, it is wise to multiply the right side to see if it agrees with the left side.

c. *Factor $x^2 - 7x + 12$ completely.*

Solution:

$$x^2 - 7x + 12 = (x - 3)(x - 4)$$

NOW WORK PROBLEM 9 ◖●●

●EXAMPLE 3 **Factoring**

The following is an assortment of expressions that are completely factored. The numbers in parentheses refer to the rules used.

a. $x^2 + 8x + 16 = (x + 4)^2$ — (4)

b. $9x^2 + 9x + 2 = (3x + 1)(3x + 2)$ — (3)

c. $6y^3 + 3y^2 - 18y = 3y(2y^2 + y - 6)$ — (1)

$\qquad\qquad\qquad = 3y(2y - 3)(y + 2)$ — (3)

d. $x^2 - 6x + 9 = (x - 3)^2$ — (5)

e. $z^{1/4} + z^{5/4} = z^{1/4}(1 + z)$ — (1)

f. $x^4 - 1 = (x^2 + 1)(x^2 - 1)$ — (6)

$\qquad\qquad = (x^2 + 1)(x + 1)(x - 1)$ — (6)

g. $x^{2/3} - 5x^{1/3} + 4 = (x^{1/3} - 1)(x^{1/3} - 4)$ — (2)

h. $ax^2 - ay^2 + bx^2 - by^2 = a(x^2 - y^2) + b(x^2 - y^2)$ — (1), (1)

$\qquad\qquad\qquad\qquad = (x^2 - y^2)(a + b)$ — (1)

$\qquad\qquad\qquad\qquad = (x + y)(x - y)(a + b)$ — (6)

i. $8 - x^3 = (2)^3 - (x)^3 = (2 - x)(4 + 2x + x^2)$ (8)

j. $x^6 - y^6 = (x^3)^2 - (y^3)^2 = (x^3 + y^3)(x^3 - y^3)$ (6)

$\qquad = (x + y)(x^2 - xy + y^2)(x - y)(x^2 + xy + y^2)$ (7), (8)

Note in Example 3(f) that $x^2 - 1$ is factorable, but $x^2 + 1$ is not. In Example 3(h), note that the common factor of $x^2 - y^2$ was not immediately evident.

Problems 0.5

Factor the following expressions completely.

1. $2ax + 2b$

2. $6y^2 - 4y$

3. $10xy + 5xz$

4. $3x^2 y - 9x^3 y^3$

*5. $8a^3 bc - 12ab^3 cd + 4b^4 c^2 d^2$

6. $6u^3 v^3 + 18u^2 vw^4 - 12u^2 v^3$

7. $z^2 - 49$

8. $x^2 - x - 6$

*9. $p^2 + 4p + 3$

10. $s^2 - 6s + 8$

11. $16x^2 - 9$

12. $x^2 + 2x - 24$

13. $a^2 + 12a + 35$

14. $4t^2 - 9s^2$

15. $x^2 + 6x + 9$

16. $y^2 - 15y + 50$

17. $5x^2 + 25x + 30$

18. $3t^2 + 12t - 15$

19. $3x^2 - 3$

20. $9y^2 - 18y + 8$

21. $6y^2 + 13y + 2$

22. $4x^2 - x - 3$

23. $12s^3 + 10s^2 - 8s$

24. $9z^2 + 30z + 25$

25. $u^{13/5} v - 4u^{3/5} v^3$

26. $9x^{4/7} - 1$

27. $2x^3 + 2x^2 - 12x$

28. $x^2 y^2 - 4xy + 4$

29. $(4x + 2)^2$

30. $2x^2 (2x - 4x^2)^2$

31. $x^3 y^2 - 14x^2 y + 49x$

32. $(5x^2 + 2x) + (10x + 4)$

33. $(x^3 - 4x) + (8 - 2x^2)$

34. $(x^2 - 1) + (x^2 - x - 2)$

35. $(y^4 + 8y^3 + 16y^2) - (y^2 + 8y + 16)$

36. $x^3 y - 4xy + z^2 x^2 - 4z^2$

37. $b^3 + 64$

38. $x^3 - 1$

39. $x^6 - 1$

40. $27 + 8x^3$

41. $(x + 3)^3 (x - 1) + (x + 3)^2 (x - 1)^2$

42. $(a + 5)^3 (a + 1)^2 + (a + 5)^2 (a + 1)^3$

43. $P(1 + r) + P(1 + r)r$

44. $(X - 3I)(3X + 5I) - (3X + 5I)(X + 2I)$

45. $x^4 - 16$

46. $81x^4 - y^4$

47. $y^8 - 1$

48. $t^4 - 4$

49. $X^4 + 4X^2 - 5$

50. $x^4 - 10x^2 + 9$

51. $x^4 y - 2x^2 y + y$

52. $4x^3 - 6x^2 - 4x$

OBJECTIVE

To simplify, add, subtract, multiply, and divide algebraic fractions. To rationalize the denominator of a fraction.

0.6 Fractions

Students should take particular care in studying *fractions*. In everyday life, numerical fractions often disappear from view with the help of calculators. However, understanding how to manipulate fractions of algebraic expressions is an essential prerequisite for calculus. Most calculators are of no help!

Simplifying Fractions

By using the fundamental principle of fractions (Section 0.2), we may be able to simplify algebraic expressions that are fractions. That principle allows us to multiply or divide both the numerator and the denominator of a fraction by the same nonzero quantity. The resulting fraction will be equal to the original one. The fractions that we consider are assumed to have nonzero denominators. Thus, all the factors of the denominators in our examples are assumed to be nonzero. This will often mean that certain values are excluded for the variables that occur in the denominators.

● EXAMPLE 1 **Simplifying Fractions**

a. *Simplify* $\dfrac{x^2 - x - 6}{x^2 - 7x + 12}$.

Solution: First, we completely factor the numerator and denominator:

$$\frac{x^2 - x - 6}{x^2 - 7x + 12} = \frac{(x - 3)(x + 2)}{(x - 3)(x - 4)}$$

Dividing both numerator and denominator by the common factor $x - 3$, we have

$$\frac{(x-3)(x+2)}{(x-3)(x-4)} = \frac{1(x+2)}{1(x-4)} = \frac{x+2}{x-4}$$

Usually, we just write

$$\frac{x^2 - x - 6}{x^2 - 7x + 12} = \frac{\overset{1}{\cancel{(x-3)}}(x+2)}{\underset{1}{\cancel{(x-3)}}(x-4)} = \frac{x+2}{x-4}$$

or

$$\frac{x^2 - x - 6}{x^2 - 7x + 12} = \frac{(x-3)(x+2)}{(x-3)(x-4)} = \frac{x+2}{x-4}$$

The process of eliminating the common factor $x - 3$ is commonly referred to as "cancellation."

b. *Simplify* $\dfrac{2x^2 + 6x - 8}{8 - 4x - 4x^2}$.

Solution:

$$\frac{2x^2 + 6x - 8}{8 - 4x - 4x^2} = \frac{2(x^2 + 3x - 4)}{4(2 - x - x^2)} = \frac{2(x-1)(x+4)}{4(1-x)(2+x)}$$

$$= \frac{2(x-1)(x+4)}{2(2)[(-1)(x-1)](2+x)}$$

$$= \frac{x+4}{-2(2+x)} = -\frac{x+4}{2(x+2)}$$

Note that $1 - x$ was written as $(-1)(x - 1)$ to facilitate cancellation.

NOW WORK PROBLEM 3

Multiplication and Division of Fractions

The rule for multiplying $\dfrac{a}{b}$ by $\dfrac{c}{d}$ is

$$\frac{a}{b} \cdot \frac{c}{d} = \frac{ac}{bd}$$

● EXAMPLE 2   Multiplying Fractions

a. $\dfrac{x}{x+2} \cdot \dfrac{x+3}{x-5} = \dfrac{x(x+3)}{(x+2)(x-5)}$

b. $\dfrac{x^2 - 4x + 4}{x^2 + 2x - 3} \cdot \dfrac{6x^2 - 6}{x^2 + 2x - 8} = \dfrac{[(x-2)^2][6(x+1)(x-1)]}{[(x+3)(x-1)][(x+4)(x-2)]}$

$$= \frac{6(x-2)(x+1)}{(x+3)(x+4)}$$

NOW WORK PROBLEM 9

To divide $\dfrac{a}{b}$ by $\dfrac{c}{d}$, where $c \neq 0$, we have

In short, to divide by a fraction we invert the divisor and multiply.

$$\frac{a}{b} \div \frac{c}{d} = \frac{\dfrac{a}{b}}{\dfrac{c}{d}} = \frac{a}{b} \cdot \frac{d}{c}$$

● EXAMPLE 3   Dividing Fractions

a. $\dfrac{x}{x+2} \div \dfrac{x+3}{x-5} = \dfrac{x}{x+2} \cdot \dfrac{x-5}{x+3} = \dfrac{x(x-5)}{(x+2)(x+3)}$

b. $\dfrac{\dfrac{x-5}{x-3}}{2x} = \dfrac{\dfrac{x-5}{x-3}}{\dfrac{2x}{1}} = \dfrac{x-5}{x-3} \cdot \dfrac{1}{2x} = \dfrac{x-5}{2x(x-3)}$

c. $\dfrac{\dfrac{4x}{x^2-1}}{\dfrac{2x^2+8x}{x-1}} = \dfrac{4x}{x^2-1} \cdot \dfrac{x-1}{2x^2+8x} = \dfrac{4x(x-1)}{[(x+1)(x-1)][2x(x+4)]}$

$$= \dfrac{2}{(x+1)(x+4)}$$

NOW WORK PROBLEM 11

Rationalizing the Denominator

Sometimes the denominator of a fraction has two terms and involves square roots, such as $2 - \sqrt{3}$ or $\sqrt{5} + \sqrt{2}$. The denominator may then be rationalized by multiplying by an expression that makes the denominator a difference of two squares. For example,

$$\dfrac{4}{\sqrt{5} + \sqrt{2}} = \dfrac{4}{\sqrt{5} + \sqrt{2}} \cdot \dfrac{\sqrt{5} - \sqrt{2}}{\sqrt{5} - \sqrt{2}}$$

$$= \dfrac{4(\sqrt{5} - \sqrt{2})}{(\sqrt{5})^2 - (\sqrt{2})^2} = \dfrac{4(\sqrt{5} - \sqrt{2})}{5 - 2}$$

$$= \dfrac{4\sqrt{5} - \sqrt{2})}{3}$$

Rationalizing the *numerator* is a similar procedure.

● EXAMPLE 4 Rationalizing Denominators

a. $\dfrac{x}{\sqrt{2} - 6} = \dfrac{x}{\sqrt{2} - 6} \cdot \dfrac{\sqrt{2} + 6}{\sqrt{2} + 6} = \dfrac{x(\sqrt{2} + 6)}{(\sqrt{2})^2 - 6^2}$

$$= \dfrac{x(\sqrt{2} + 6)}{2 - 36} = -\dfrac{x(\sqrt{2} + 6)}{34}$$

b. $\dfrac{\sqrt{5} - \sqrt{2}}{\sqrt{5} + \sqrt{2}} = \dfrac{\sqrt{5} - \sqrt{2}}{\sqrt{5} + \sqrt{2}} \cdot \dfrac{\sqrt{5} - \sqrt{2}}{\sqrt{5} - \sqrt{2}}$

$$= \dfrac{(\sqrt{5} - \sqrt{2})^2}{5 - 2} = \dfrac{5 - 2\sqrt{5}\sqrt{2} + 2}{3} = \dfrac{7 - 2\sqrt{10}}{3}$$

NOW WORK PROBLEM 53

Addition and Subtraction of Fractions

In Example 3(b) of Section 0.2, it was shown that $\dfrac{a}{c} + \dfrac{b}{c} = \dfrac{a+b}{c}$. That is, if we add two fractions having a common denominator, then the result is a fraction whose denominator is the common denominator. The numerator is the sum of the numerators of the original fractions. Similarly, $\dfrac{a}{c} - \dfrac{b}{c} = \dfrac{a-b}{c}$.

● EXAMPLE 5 Adding and Subtracting Fractions

a. $\dfrac{p^2 - 5}{p - 2} + \dfrac{3p + 2}{p - 2} = \dfrac{(p^2 - 5) + (3p + 2)}{p - 2}$

$$= \dfrac{p^2 + 3p - 3}{p - 2}$$

b. $\dfrac{x^2 - 5x + 4}{x^2 + 2x - 3} - \dfrac{x^2 + 2x}{x^2 + 5x + 6} = \dfrac{(x-1)(x-4)}{(x-1)(x+3)} - \dfrac{x(x+2)}{(x+2)(x+3)}$

$$= \dfrac{x-4}{x+3} - \dfrac{x}{x+3} = \dfrac{(x-4) - x}{x+3} = -\dfrac{4}{x+3}$$

c. $\dfrac{x^2 + x - 5}{x - 7} - \dfrac{x^2 - 2}{x - 7} + \dfrac{-4x + 8}{x^2 - 9x + 14} = \dfrac{x^2 + x - 5}{x - 7} - \dfrac{x^2 - 2}{x - 7} + \dfrac{-4}{x - 7}$

$$= \dfrac{(x^2 + x - 5) - (x^2 - 2) + (-4)}{x - 7}$$

$$= \dfrac{x - 7}{x - 7} = 1$$

NOW WORK PROBLEM 29 ●●●

To add (or subtract) two fractions with *different* denominators, use the fundamental principle of fractions to rewrite the fractions as fractions that have the same denominator. Then proceed with the addition (or subtraction) by the method just described.

For example, to find

$$\frac{2}{x^3(x - 3)} + \frac{3}{x(x - 3)^2}$$

we can convert the first fraction to an equal fraction by multiplying the numerator and denominator by $x - 3$:

$$\frac{2(x - 3)}{x^3(x - 3)^2}$$

and we can convert the second fraction by multiplying the numerator and denominator by x^2:

$$\frac{3x^2}{x^3(x - 3)^2}$$

These fractions have the same denominator. Hence,

$$\frac{2}{x^3(x - 3)} + \frac{3}{x(x - 3)^2} = \frac{2(x - 3)}{x^3(x - 3)^2} + \frac{3x^2}{x^3(x - 3)^2}$$

$$= \frac{3x^2 + 2x - 6}{x^3(x - 3)^2}$$

We could have converted the original fractions into equal fractions with *any* common denominator. However, we chose to convert them into fractions with the denominator $x^3(x - 3)^2$. This is the **least common denominator (LCD)** of the fractions $2/[x^3(x - 3)]$ and $3/[x(x - 3)^2]$.

In general, to find the LCD of two or more fractions, first factor each denominator completely. *The LCD is the product of each of the distinct factors appearing in the denominators, each raised to the highest power to which it occurs in any single denominator.*

● EXAMPLE 6   **Adding and Subtracting Fractions**

a. *Subtract:* $\dfrac{t}{3t + 2} - \dfrac{4}{t - 1}$.

Solution: The LCD is $(3t + 2)(t - 1)$. Thus, we have

$$\frac{t}{(3t + 2)} - \frac{4}{t - 1} = \frac{t(t - 1)}{(3t + 2)(t - 1)} - \frac{4(3t + 2)}{(3t + 2)(t - 1)}$$

$$= \frac{t(t - 1) - 4(3t + 2)}{(3t + 2)(t - 1)}$$

$$= \frac{t^2 - t - 12t - 8}{(3t + 2)(t - 1)} = \frac{t^2 - 13t - 8}{(3t + 2)(t - 1)}$$

b. *Add:* $\dfrac{4}{q-1} + 3.$

Solution: The LCD is $q-1$.

$$\frac{4}{q-1} + 3 = \frac{4}{q-1} + \frac{3(q-1)}{q-1}$$

$$= \frac{4 + 3(q-1)}{q-1} = \frac{3q+1}{q-1}$$

NOW WORK PROBLEM 33 ●●●

● EXAMPLE 7 **Subtracting Fractions**

$$\frac{x-2}{x^2+6x+9} - \frac{x+2}{2(x^2-9)}$$

$$= \frac{x-2}{(x+3)^2} - \frac{x+2}{2(x+3)(x-3)} \qquad [\text{LCD} = 2(x+3)^2(x-3)]$$

$$= \frac{(x-2)(2)(x-3)}{(x+3)^2(2)(x-3)} - \frac{(x+2)(x+3)}{2(x+3)(x-3)(x+3)}$$

$$= \frac{(x-2)(2)(x-3) - (x+2)(x+3)}{2(x+3)^2(x-3)}$$

$$= \frac{2(x^2-5x+6) - (x^2+5x+6)}{2(x+3)^2(x-3)}$$

$$= \frac{2x^2-10x+12-x^2-5x-6}{2(x+3)^2(x-3)}$$

$$= \frac{x^2-15x+6}{2(x+3)^2(x-3)}$$

NOW WORK PROBLEM 39 ●●●

Example 8 shows two methods of simplifying a "complicated" fraction.

● EXAMPLE 8 **Combined Operations with Fractions**

Simplify $\dfrac{\dfrac{1}{x+h} - \dfrac{1}{x}}{h}$, where $h \neq 0$.

Solution: First we combine the fractions in the numerator and obtain

$$\frac{\dfrac{1}{x+h} - \dfrac{1}{x}}{h} = \frac{\dfrac{x}{x(x+h)} - \dfrac{x+h}{x(x+h)}}{h} = \frac{\dfrac{x-(x+h)}{x(x+h)}}{h}$$

$$= \frac{\dfrac{-h}{x(x+h)}}{\dfrac{h}{1}} = \frac{-h}{x(x+h)h} = -\frac{1}{x(x+h)}$$

The original fraction can also be simplified by multiplying the numerator and denominator by the LCD of the fractions involved in the numerator (and denominator), namely, $x(x+h)$:

$$\frac{\dfrac{1}{x+h} - \dfrac{1}{x}}{h} = \frac{\left[\dfrac{1}{x+h} - \dfrac{1}{x}\right] x(x+h)}{h[x(x+h)]}$$

$$= \frac{x-(x+h)}{x(x+h)h} = \frac{-h}{x(x+h)h} = -\frac{1}{x(x+h)}$$

NOW WORK PROBLEM 47 ●●●

Problems 0.6

In Problems 1–6, simplify.

1. $\dfrac{a^2 - 9}{a^2 - 3a}$

2. $\dfrac{x^2 - 3x - 10}{x^2 - 4}$

*3. $\dfrac{x^2 - 9x + 20}{x^2 + x - 20}$

4. $\dfrac{3x^2 - 27x + 24}{2x^3 - 16x^2 + 14x}$

5. $\dfrac{6x^2 + x - 2}{2x^2 + 3x - 2}$

6. $\dfrac{12x^2 - 19x + 4}{6x^2 - 17x + 12}$

In Problems 7–48, perform the operations and simplify as much as possible.

7. $\dfrac{y^2}{y - 3} \cdot \dfrac{-1}{y + 2}$

8. $\dfrac{t^2 - 9}{t^2 + 3t} \cdot \dfrac{t^2}{t^2 - 6t + 9}$

*9. $\dfrac{ax - b}{x - c} \cdot \dfrac{c - x}{ax + b}$

10. $\dfrac{x^2 - y^2}{x + y} \cdot \dfrac{x^2 + 2xy + y^2}{y - x}$

*11. $\dfrac{2x - 2}{x^2 - 2x - 8} \div \dfrac{x^2 - 1}{x^2 + 5x + 4}$

12. $\dfrac{x^2 + 2x}{3x^2 - 18x + 24} \div \dfrac{x^2 - x - 6}{x^2 - 4x + 4}$

13. $\dfrac{\frac{X^2}{8}}{\frac{X}{4}}$

14. $\dfrac{\frac{3x^2}{7x}}{\frac{x}{14}}$

15. $\dfrac{\frac{2m}{n^2}}{\frac{6m}{n^3}}$

16. $\dfrac{\frac{c + d}{c}}{\frac{c - d}{2c}}$

17. $\dfrac{\frac{4x}{3}}{\frac{2x}{}}$

18. $\dfrac{4x}{\frac{3}{2x}}$

19. $\dfrac{-9x^3}{\frac{x}{3}}$

20. $\dfrac{\frac{-12Y^4}{Y}}{4}$

21. $\dfrac{\frac{x - 3}{x^2 - 7x + 12}}{x - 4}$

22. $\dfrac{\frac{x^2 + 6x + 9}{x}}{x + 3}$

23. $\dfrac{\frac{10x^3}{x^2 - 1}}{\frac{5x}{x + 1}}$

24. $\dfrac{\frac{x^2 - x - 6}{x^2 - 9}}{\frac{x^2 - 4}{x^2 + 2x - 3}}$

25. $\dfrac{\frac{x^2 + 7x + 10}{x^2 + 6x + 5}}{\frac{x^2 - 2x - 8}{x^2 - 3x - 4}}$

26. $\dfrac{\frac{(x + 3)^2}{4x - 3}}{\frac{7x + 21}{9 - 16x^2}}$

27. $\dfrac{\frac{4x^2 - 9}{x^2 + 3x - 4}}{\frac{2x - 3}{1 - x^2}}$

28. $\dfrac{\frac{6x^2 y + 7xy - 3y}{xy - x + 5y - 5}}{\frac{x^3 y + 4x^2 y}{xy - x + 4y - 4}}$

*29. $\dfrac{x^2}{x + 3} + \dfrac{5x + 6}{x + 3}$

30. $\dfrac{2}{x + 2} + \dfrac{x}{x + 2}$

31. $\dfrac{2}{t} + \dfrac{1}{3t}$

32. $\dfrac{9}{X^3} - \dfrac{1}{X^2}$

*33. $1 - \dfrac{x^3}{x^3 - 1}$

34. $\dfrac{4}{s + 4} + s$

35. $\dfrac{4}{2x - 1} + \dfrac{x}{x + 3}$

36. $\dfrac{x + 1}{x - 1} - \dfrac{x - 1}{x + 1}$

37. $\dfrac{1}{x^2 - 2x - 3} + \dfrac{1}{x^2 - 9}$

38. $\dfrac{4}{2x^2 - 7x - 4} - \dfrac{x}{2x^2 - 9x + 4}$

*39. $\dfrac{4}{x - 1} - 3 + \dfrac{-3x^2}{5 - 4x - x^2}$

40. $\dfrac{2x - 3}{2x^2 + 11x - 6} - \dfrac{3x + 1}{3x^2 + 16x - 12} + \dfrac{1}{3x - 2}$

41. $(1 + x^{-1})^2$

42. $(x^{-1} + y^{-1})^2$

43. $(x^{-1} - y)^{-1}$

44. $(a + b^{-1})^2$

45. $\dfrac{7 + \frac{1}{x}}{5}$

46. $\dfrac{\frac{x + 3}{x}}{x - \frac{9}{x}}$

*47. $\dfrac{3 - \frac{1}{2x}}{x + \frac{x}{x + 2}}$

48. $\dfrac{\frac{x - 1}{x^2 + 5x + 6} - \frac{1}{x + 2}}{3 + \frac{x - 7}{3}}$

In Problems 49 and 50, perform the indicated operations, but do not rationalize the denominators.

49. $\dfrac{3}{\sqrt[3]{x + h}} - \dfrac{3}{\sqrt[3]{x}}$

50. $\dfrac{a\sqrt{a}}{\sqrt{5 + a}} + \dfrac{1}{\sqrt{a}}$

In Problems 51–60, simplify, and express your answer in a form that is free of radicals in the denominator.

51. $\dfrac{1}{2 + \sqrt{3}}$

52. $\dfrac{1}{1 - \sqrt{2}}$

*53. $\dfrac{\sqrt{2}}{\sqrt{3} - \sqrt{6}}$

54. $\dfrac{5}{\sqrt{6} + \sqrt{7}}$

55. $\dfrac{2\sqrt{2}}{\sqrt{2} - \sqrt{3}}$

56. $\dfrac{2\sqrt{5}}{\sqrt{3} - \sqrt{7}}$

57. $\dfrac{3}{t + \sqrt{7}}$

58. $\dfrac{x - 3}{\sqrt{x} - 1} + \dfrac{4}{\sqrt{x} - 1}$

59. $\dfrac{5}{2 + \sqrt{3}} - \dfrac{4}{1 - \sqrt{2}}$

60. $\dfrac{4}{\sqrt{x} + 2} \cdot \dfrac{x^2}{3}$

OBJECTIVE

To discuss equivalent equations and to develop techniques for solving linear equations, including literal equations as well as fractional and radical equations that lead to linear equations.

0.7 Equations, in Particular Linear Equations

Equations

An **equation** is a statement that two expressions are equal. The two expressions that make up an equation are called its **sides**. They are separated by the **equality sign**, $=$.

● EXAMPLE 1   **Examples of Equations**

a. $x + 2 = 3$

b. $x^2 + 3x + 2 = 0$

c. $\dfrac{y}{y - 4} = 6$

d. $w = 7 - z$

In Example 1, each equation contains at least one variable. A **variable** is a symbol that can be replaced by any one of a set of different numbers. The most popular symbols for variables are letters from the latter part of the alphabet, such as x, y, z, w, and t. Hence, equations (a) and (c) are said to be in the variables x and y, respectively. Equation (d) is in the variables w and z. In the equation $x + 2 = 3$, the numbers 2 and 3 are called *constants*. They are fixed numbers.

Here we discuss restrictions on variables.

We *never* allow a variable in an equation to have a value for which any expression in that equation is undefined. For example, in

$$\frac{y}{y - 4} = 6$$

y cannot be 4, because this would make the denominator zero; while in

$$\sqrt{x - 3} = 9$$

we must have $x \geq 3$, so that the expression under the square root symbol is non-negative. (We cannot divide by zero and we cannot take square roots of negative numbers.) In some equations, the allowable values of a variable are restricted for physical reasons. For example, if the variable t represents time, negative values of t may not make sense. In that case we should assume that $t \geq 0$.

To *solve* an equation means to find all values of its variables for which the equation is true. These values are called *solutions* of the equation and are said to *satisfy* the equation. When only one variable is involved, a solution is also called a *root*. The set of all solutions is called the *solution set* of the equation. Sometimes a letter representing an unknown quantity in an equation is simply called an *unknown*. Let us illustrate these terms.

● EXAMPLE 2   **Terminology for Equations**

a. In the equation $x + 2 = 3$, the variable x is the unknown. The only value of x that satisfies the equation is obviously 1. Hence, 1 is a root and the solution set is $\{1\}$.

b. -2 is a root of $x^2 + 3x + 2 = 0$ because substituting -2 for x makes the equation true: $(-2)^2 + 3(-2) + 2 = 0$. Hence -2 is an element of the solution set but in this case it is not the only one. There is one more. Can you find it?

c. $w = 7 - z$ is an equation in two unknowns. One solution is the pair of values $w = 4$ and $z = 3$. However, there are infinitely many solutions. Can you think of another?

NOW WORK PROBLEM 1 ●●●

Equivalent Equations

Two equations are said to be *equivalent* if they have exactly the same solutions, which means, precisely, that the solution set of one is equal to the solution set of the other. Solving an equation may involve performing operations on it. We prefer that any such operation result in an equivalent equation. Here are three operations that guarantee equivalence:

1. Adding (subtracting) the same polynomial to (from) both sides of an equation, where the polynomial is in the same variable as that occurring in the equation.

    For example, if $-5x = 5 - 6x$, then adding $6x$ to both sides gives the equivalent equation $-5x + 6x = 5 - 6x + 6x$, which in turn is equivalent to $x = 5$.

C A U T I O N

Equivalence is not guaranteed if both sides are multiplied or divided by an expression involving a variable.

2. Multiplying (dividing) both sides of an equation by the same nonzero constant.

    For example, if $10x = 5$, then dividing both sides by 10 gives the equivalent equation $\dfrac{10x}{10} = \dfrac{5}{10}$, equivalently, $x = \dfrac{1}{2}$.

3. Replacing either side of an equation by an equal expression.

    For example, if the equation is $x(x + 2) = 3$, then replacing the left side by the equal expression $x^2 + 2x$ gives the equivalent equation $x^2 + 2x = 3$.

    We repeat: Applying Operations 1–3 guarantees that the resulting equation is equivalent to the given one. However, sometimes in solving an equation we have to apply operations other than 1–3. These operations may *not* necessarily result in equivalent equations. They include the following:

Operations That May Not Produce Equivalent Equations

4. Multiplying both sides of an equation by an expression involving the variable.
5. Dividing both sides of an equation by an expression involving the variable.

Operation 6 includes taking roots of both sides.

6. Raising both sides of an equation to equal powers.

    Let us illustrate the last three operations. For example, by inspection, the only root of $x - 1 = 0$ is 1. Multiplying each side by x (Operation 4) gives $x^2 - x = 0$, which is satisfied if x is 0 or 1. (Check this by substitution.) But 0 *does not* satisfy the *original* equation. Thus, the equations are not equivalent.

    Continuing, you may check that the equation $(x - 4)(x - 3) = 0$ is satisfied when x is 4 or when x is 3. Dividing both sides by $x - 4$ (Operation 5) gives $x - 3 = 0$, whose only root is 3. Again, we do not have equivalence, since in this case a root has been "lost." Note that when x is 4, division by $x - 4$ implies division by 0, an invalid operation.

    Finally, squaring each side of the equation $x = 2$ (Operation 6) gives $x^2 = 4$, which is true if $x = 2$ or if $x = -2$. But -2 is not a root of the given equation.

    From our discussion, it is clear that when Operations 4–6 are performed, we must be careful about drawing conclusions concerning the roots of a given equation. Operations 4 and 6 *can* produce an equation with more roots. Thus, you should check whether or not each "solution" obtained by the these operations satisfies the *original* equation. Operation 5 *can* produce an equation with fewer roots. In this case, any "lost" root may never be determined. Thus, avoid Operation 5 whenever possible.

    In summary, an equation may be thought of as a set of restrictions on any variable in the equation. Operations 4–6 may increase or decrease the number of restrictions, giving solutions different from those of the original equation. However, Operations 1–3 never affect the restrictions.

A graphing calculator can be used to test for a root. For example, suppose we want to determine whether 3/2 is a root of the equation

$$2x^3 + 7x^2 = 19x + 60$$

First, we rewrite the equation so that one side is 0. Subtracting $19x + 60$ from both sides gives the equivalent equation

$$2x^3 + 7x^2 - 19x - 60 = 0$$

For a TI-83 Plus graphing calculator, we enter the expression $2x^3 + 7x^2 - 19x - 60$ as Y_1 and then evaluate Y_1 at $x = 3/2$. Figure 0.2 shows that the result is -66, which is not 0. Thus, 3/2 is not a root. However, Y_1 evaluated at $x = -5/2$ *is* 0. So $-5/2$ is a root of the original equation.

It is worth noting that if the original equation had been in terms of the variable t, that is,

$$2t^3 + 7t^2 = 19t + 60$$

then we must replace t by x because the calculator evaluates Y_1 at a specified value of x, not t.

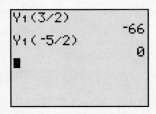

FIGURE 0.2   For $2x^3 + 7x^2 - 19x - 60 = 0$, $\frac{3}{2}$ is not a root, but $\frac{-5}{2}$ is a root.

Linear Equations

The principles presented so far will now be demonstrated in the solution of a *linear equation*.

> **DEFINITION**
>
> A *linear equation* in the variable x is an equation that can be written in the form
>
> $$ax + b = 0 \qquad\qquad (1)$$
>
> where a and b are constants and $a \neq 0$.

A linear equation is also called a first-degree equation or an equation of degree one, since the highest power of the variable that occurs in Equation (1) is the first.

To solve a linear equation, we perform operations on it until we have an equivalent equation whose solutions are obvious. This means an equation in which the variable is isolated on one side, as the following examples show.

● EXAMPLE 3   Solving a Linear Equation

Solve $5x - 6 = 3x$.

Solution: We begin by getting the terms involving x on one side and the constant on the other. Then we solve for x by the appropriate mathematical operation. We have

$$5x - 6 = 3x$$

$$5x - 6 + (-3x) = 3x + (-3x) \qquad \text{(adding } -3x \text{ to both sides)}$$

$$2x - 6 = 0 \qquad \text{(simplifying, that is, Operation 3)}$$

$$2x - 6 + 6 = 0 + 6 \qquad \text{(adding 6 to both sides)}$$

$$2x = 6 \qquad \text{(simplifying)}$$

$$\frac{2x}{2} = \frac{6}{2} \qquad \text{(dividing both sides by 2)}$$

$$x = 3$$

Clearly, 3 is the only root of the last equation. Since each equation is equivalent to the one before it, we conclude that 3 must be the only root of $5x - 6 = 3x$. That is, the solution set is {3}. We can describe the first step in the solution as moving a term from one side of an equation to the other while changing its sign; this is commonly called *transposing*. Note that since the original equation can be put in the form $2x + (-6) = 0$, it is a linear equation.

NOW WORK PROBLEM 23

EXAMPLE 4   Solving a Linear Equation

Solve $2(p + 4) = 7p + 2$.

Solution: First, we remove parentheses. Then we collect similar terms and solve. We have

$$2(p + 4) = 7p + 2$$
$$2p + 8 = 7p + 2 \qquad \text{(distributive property)}$$
$$2p = 7p - 6 \qquad \text{(subtracting 8 from both sides)}$$
$$-5p = -6 \qquad \text{(subtracting } 7p \text{ from both sides)}$$
$$p = \frac{-6}{-5} \qquad \text{(dividing both sides by } -5)$$
$$p = \frac{6}{5}$$

NOW WORK PROBLEM 27

EXAMPLE 5   Solving a Linear Equation

Solve $\dfrac{7x + 3}{2} - \dfrac{9x - 8}{4} = 6$.

Solution: We first clear the equation of fractions by multiplying *both* sides by the least common denominator (LCD), which is 4. Then we use various algebraic operations to obtain a solution. Thus,

$$4 \left(\frac{7x + 3}{2} - \frac{9x - 8}{4} \right) = 4(6)$$
$$4 \cdot \frac{7x + 3}{2} - 4 \cdot \frac{9x - 8}{4} = 24 \qquad \text{(distributive property)}$$
$$2(7x + 3) - (9x - 8) = 24 \qquad \text{(simplifying)}$$
$$14x + 6 - 9x + 8 = 24 \qquad \text{(distributive property)}$$
$$5x + 14 = 24 \qquad \text{(simplifying)}$$
$$5x = 10 \qquad \text{(subtracting 14 from both sides)}$$
$$x = 2 \qquad \text{(dividing both sides by 5)}$$

NOW WORK PROBLEM 31

 CAUTION

The distributive property requires that *both* terms within the parentheses be multiplied by 4.

Every linear equation has exactly one root.

Each equation in Examples 3–5 has one and only one root. This is true of every linear equation in one variable.

Literal Equations

Equations in which some of the constants are not specified, but are represented by letters, such as a, b, c, or d, are called **literal equations,** and the letters are called **literal constants**. For example, in the literal equation $x + a = 4b$, we may consider a and b to be literal constants. Formulas, such as $I = Prt$, that express a relationship between certain quantities may be regarded as literal equations. If we want to express a particular letter in a formula in terms of the others, this letter is considered the unknown.

● EXAMPLE 6 Solving Literal Equations

a. *The equation* $I = Prt$ *is the formula for the simple interest* I *on a principal of* P *dollars at the annual interest rate of* r *for a period of* t *years. Express* r *in terms of* I, P, *and* t.

Solution: Here we consider r to be the unknown. To isolate r, we divide both sides by Pt. We have

$$I = Prt$$

$$\frac{I}{Pt} = \frac{Prt}{Pt}$$

$$\frac{I}{Pt} = r \text{ so } r = \frac{I}{Pt}$$

When we divided both sides by Pt, we assumed that $Pt \neq 0$, since we cannot divide by 0. Similar assumptions will be made in solving other literal equations.

b. *The equation* $S = P + Prt$ *is the formula for the value* S *of an investment of a principal of* P *dollars at a simple annual interest rate of* r *for a period of* t *years. Solve for* P.

Solution:

$$S = P + Prt$$

$$S = P(1 + rt) \qquad \text{(factoring)}$$

$$\frac{S}{1 + rt} = P \qquad \text{(dividing both sides by } 1 + rt)$$

NOW WORK PROBLEM 87 ●●

● EXAMPLE 7 Solving a Literal Equation

Solve $(a + c)x + x^2 = (x + a)^2$ *for* x.

Solution: We first simplify the equation and then get all terms involving x on one side:

$$(a + c)x + x^2 = (x + a)^2$$

$$ax + cx + x^2 = x^2 + 2ax + a^2$$

$$ax + cx = 2ax + a^2$$

$$cx - ax = a^2$$

$$x(c - a) = a^2$$

$$x = \frac{a^2}{c - a}$$

NOW WORK PROBLEM 89 ●●

● EXAMPLE 8 Solving the "Tax in a Receipt" Problem

We will recall Lesley Griffith's problem from the opening paragraphs of this chapter. We will generalize the problem so as to illustrate further the use of literal equations. Lesley has a receipt for an amount R. She knows that the percentage sales tax rate is p. She wants to know the amount that was paid in sales tax. Certainly,

$$\text{price} + \text{tax} = \text{receipt}$$

Writing P for the price (which she does not yet know), the tax is $(p/100)P$ so that she has

$$P + \frac{p}{100}P = R$$

$$P\left(1 + \frac{p}{100}\right) = R$$

$$P = \frac{R}{\left(1 + \dfrac{p}{100}\right)}$$

$$= \frac{R}{\dfrac{100 + p}{100}}$$

$$= \frac{100\,R}{100 + p}$$

It follows that the tax paid is

$$R - P = R - \frac{100\,R}{100 + p} = R\left(1 - \frac{100}{100 + p}\right) = R\left(\frac{p}{100 + p}\right)$$

where you should check the manipulations with fractions, supplying more details if necessary. Recall that the French tax rate is 19.6% and the Italian tax rate is 18%. We conclude that Lesley has only to multiply a French receipt by $\frac{19.6}{119.6} \approx 0.16388$ to determine the tax it contains, while for an Italian receipt she should multiply the amount by $\frac{18}{118}$.

NOW WORK PROBLEM 107

Fractional Equations

A **fractional equation** is an equation in which an unknown is in a denominator. We illustrate that solving such a nonlinear equation may lead to a linear equation.

● EXAMPLE 9   Solving a Fractional Equation

Solve $\dfrac{5}{x - 4} = \dfrac{6}{x - 3}$.

Solution:

Strategy   We first write the equation in a form that is free of fractions. Then we use standard algebraic techniques to solve the resulting equation.

An alternative solution that avoids multiplying both sides by the LCD is as follows:

$$\frac{5}{x-4} - \frac{6}{x-3} = 0$$

Assuming that x is neither 3 nor 4 and combining fractions gives

$$\frac{9-x}{(x-4)(x-3)} = 0$$

A fraction can be 0 only when its numerator is 0 and its denominator is not. Hence, $x = 9$.

Multiplying both sides by the LCD, $(x - 4)(x - 3)$, we have

$$(x - 4)(x - 3)\left(\frac{5}{x - 4}\right) = (x - 4)(x - 3)\left(\frac{6}{x - 3}\right)$$

$$5(x - 3) = 6(x - 4) \qquad \text{(linear equation)}$$

$$5x - 15 = 6x - 24$$

$$9 = x$$

In the first step, we multiplied each side by an expression involving the *variable x*. As we mentioned in this section, this means that we are not guaranteed that the last equation is equivalent to the *original* equation. Thus, we must check whether or not 9 satisfies the *original* equation. Since

$$\frac{5}{9 - 4} = \frac{5}{5} = 1 \quad \text{and} \quad \frac{6}{9 - 3} = \frac{6}{6} = 1$$

we see that 9 indeed satisfies the original equation.

NOW WORK PROBLEM 55

Some equations that are not linear do not have any solutions. In that case, we say that the solution set is the **empty set**, which we denote by $\emptyset$. Example 10 will illustrate.

● EXAMPLE 10   Solving Fractional Equations

a. *Solve* $\dfrac{3x+4}{x+2} - \dfrac{3x-5}{x-4} = \dfrac{12}{x^2-2x-8}$.

Solution: Observing the denominators and noting that

$$x^2 - 2x - 8 = (x+2)(x-4)$$

we conclude that the LCD is $(x+2)(x-4)$. Multiplying both sides by the LCD, we have

$$(x+2)(x-4)\left(\frac{3x+4}{x+2} - \frac{3x-5}{x-4}\right) = (x+2)(x-4) \cdot \frac{12}{(x+2)(x-4)}$$

$$(x-4)(3x+4) - (x+2)(3x-5) = 12$$

$$3x^2 - 8x - 16 - (3x^2 + x - 10) = 12$$

$$3x^2 - 8x - 16 - 3x^2 - x + 10 = 12$$

$$-9x - 6 = 12$$

$$-9x = 18$$

$$x = -2 \tag{2}$$

However, the *original* equation is not defined for $x = -2$ (we cannot divide by zero), so there are no roots. Thus, the solution set is $\emptyset$. Although -2 is a solution of Equation (2), it is not a solution of the *original* equation.

b. *Solve* $\dfrac{4}{x-5} = 0$.

Solution: The only way a fraction can equal zero is for the numerator to equal zero (and the denominator to not equal zero). Since the numerator, 4, is never 0, the solution set is $\emptyset$.

NOW WORK PROBLEM 49   ●●

● EXAMPLE 11   Literal Equation

If $s = \dfrac{u}{au+v}$, *express u in terms of the remaining letters; that is, solve for u.*

Solution:

Strategy   Since the unknown, u, occurs in the denominator, we first clear fractions and then solve for u.

$$s = \frac{u}{au+v}$$

$$s(au+v) = u \qquad \text{(multiplying both sides by } au+v)$$

$$sau + sv = u$$

$$sau - u = -sv$$

$$u(sa - 1) = -sv$$

$$u = \frac{-sv}{sa-1} = \frac{sv}{1-sa}$$

NOW WORK PROBLEM 91   ●●

Radical Equations

A **radical equation** is one in which an unknown occurs in a radicand. The next two examples illustrate the techniques employed to solve such equations.

●EXAMPLE 12   Solving a Radical Equation

Solve $\sqrt{x^2 + 33} - x = 3$.

Solution: To solve this radical equation, we raise both sides to the same power to eliminate the radical. This operation does *not* guarantee equivalence, so we must check any resulting "solutions." We begin by isolating the radical on one side. Then we square both sides and solve using standard techniques. Thus,

$$\sqrt{x^2 + 33} = x + 3$$
$$x^2 + 33 = (x + 3)^2 \qquad \text{(squaring both sides)}$$
$$x^2 + 33 = x^2 + 6x + 9$$
$$24 = 6x$$
$$4 = x$$

You should show by substitution that 4 is indeed a root.

NOW WORK PROBLEM 79

With some radical equations, you may have to raise both sides to the same power more than once, as Example 13 shows.

●EXAMPLE 13   Solving a Radical Equation

Solve $\sqrt{y - 3} - \sqrt{y} = -3$.

The reason we want one radical on each side is to eliminate squaring a binomial with two different radicals.

Solution: When an equation has two terms involving radicals, first write the equation so that one radical is on each side, if possible. Then square and solve. We have

$$\sqrt{y - 3} = \sqrt{y} - 3$$
$$y - 3 = y - 6\sqrt{y} + 9 \qquad \text{(squaring both sides)}$$
$$6\sqrt{y} = 12$$
$$\sqrt{y} = 2$$
$$y = 4 \qquad \text{(squaring both sides)}$$

Substituting 4 into the left side of the *original* equation gives $\sqrt{1} - \sqrt{4}$, which is -1. Since this does not equal the right side, -3, there is no solution. That is, the solution set is $\emptyset$.

NOW WORK PROBLEM 77

Problems 0.7

In Problems 1–6, determine by substitution which of the given numbers, if any, satisfy the given equation.

*1. $9x - x^2 = 0; 1, \ 0$

2. $12 - 7x = -x^2; 4, 3$

3. $z + 3(z - 4) = 5; \frac{17}{4}, \ 4$

4. $2x + x^2 - 8 = 0; 2, -4$

5. $x(6 + x) - 2(x + 1) - 5x = 4; -2, \ 0$

6. $x(x + 1)^2(x + 2) = 0; 0, -1, \ 2$

In Problems 7–16, determine what operations were applied to the first equation to obtain the second. State whether or not the operations guarantee that the equations are equivalent. Do not solve the equations.

7. $x - 5 = 4x + 10; x = 4x + 15$

8. $8x - 4 = 16; x - \frac{1}{2} = 2$

9. $x = 4; x^3 = 64$

10. $2x^2 + 4 = 5x - 7; x^2 + 2 = \frac{5}{2}x - \frac{7}{2}$

11. $x^2 - 2x = 0; x - 2 = 0$

12. $\dfrac{2}{x-2} + x = x^2; 2 + x(x-2) = x^2(x-2)$

13. $\dfrac{x^2-1}{x-1} = 3; x^2 - 1 = 3(x-1)$

14. $(x+3)(x+11)(x+7) = (x+3)(x+2);$
$(x+11)(x+7) = x+2$

15. $\dfrac{2x(3x+1)}{2x-3} = 2x(x+4); 3x + 1 = (x+4)(2x-3)$

16. $2x^2 - 9 = x; x^2 - \dfrac{1}{2}x = \dfrac{9}{2}$

In Problems 17–46, solve the equations.

17. $4x = 10$ **18.** $0.2x = 7$

19. $3y = 0$ **20.** $2x - 4x = -5$

21. $-8x = 12 - 20$ **22.** $4 - 7x = 3$

*23.** $5x - 3 = 9$ **24.** $\sqrt{2}x + 3 = 8$

25. $7x + 7 = 2(x+1)$ **26.** $4s + 3s - 1 = 41$

*27.** $5(p-7) - 2(3p-4) = 3p$

28. $t = 2 - 2[2t - 3(1-t)]$

29. $\dfrac{x}{5} = 2x - 6$ **30.** $\dfrac{5y}{7} - \dfrac{6}{7} = 2 - 4y$

*31.** $7 + \dfrac{4x}{9} = \dfrac{x}{2}$ **32.** $\dfrac{x}{3} - 4 = \dfrac{x}{5}$

33. $r = \dfrac{4}{3}r - 5$ **34.** $\dfrac{3x}{5} + \dfrac{5x}{3} = 9$

35. $3x + \dfrac{x}{5} - 5 = \dfrac{1}{5} + 5x$ **36.** $y - \dfrac{y}{2} + \dfrac{y}{3} - \dfrac{y}{4} = \dfrac{y}{5}$

37. $\dfrac{2y-3}{4} = \dfrac{6y+7}{3}$ **38.** $\dfrac{t}{4} + \dfrac{5}{3}t = \dfrac{7}{2}(t-1)$

39. $w - \dfrac{w}{2} + \dfrac{w}{6} - \dfrac{w}{24} = 120$ **40.** $\dfrac{7 + 2(x+1)}{3} = \dfrac{6x}{5}$

41. $\dfrac{x+2}{3} - \dfrac{2-x}{6} = x - 2$ **42.** $\dfrac{x}{5} + \dfrac{2(x-4)}{10} = 7$

43. $\dfrac{9}{5}(3-x) = \dfrac{3}{4}(x-3)$

44. $\dfrac{2y-7}{3} + \dfrac{8y-9}{14} = \dfrac{3y-5}{21}$

45. $\dfrac{4}{3}(5x-2) = 7[x - (5x-2)]$

46. $(2x-5)^2 + (3x-3)^2 = 13x^2 - 5x + 7$

47. $\dfrac{5}{x} = 25$ **48.** $\dfrac{4}{x-1} = 2$

*49.** $\dfrac{7}{3-x} = 0$ **50.** $\dfrac{3x-5}{x-3} = 0$

51. $\dfrac{3}{5-2x} = \dfrac{7}{2}$ **52.** $\dfrac{x+3}{x} = \dfrac{2}{5}$

53. $\dfrac{q}{5q-4} = \dfrac{1}{3}$ **54.** $\dfrac{4p}{7-p} = 1$

*55.** $\dfrac{1}{p-1} = \dfrac{2}{p-2}$ **56.** $\dfrac{2x-3}{4x-5} = 6$

57. $\dfrac{1}{x} + \dfrac{1}{7} = \dfrac{3}{7}$ **58.** $\dfrac{2}{x-1} = \dfrac{3}{x-2}$

59. $\dfrac{3x-2}{2x+3} = \dfrac{3x-1}{2x+1}$ **60.** $\dfrac{x+2}{x-1} + \dfrac{x+1}{3-x} = 0$

61. $\dfrac{y-6}{y} - \dfrac{6}{y} = \dfrac{y+6}{y-6}$ **62.** $\dfrac{y-2}{y+2} = \dfrac{y-2}{y+3}$

63. $\dfrac{-5}{2x-3} = \dfrac{7}{3-2x} + \dfrac{11}{3x+5}$

64. $\dfrac{1}{x-3} - \dfrac{3}{x-2} = \dfrac{4}{1-2x}$ **65.** $\dfrac{9}{x-3} = \dfrac{3x}{x-3}$

66. $\dfrac{x}{x+3} - \dfrac{x}{x-3} = \dfrac{3x-4}{x^2-9}$

67. $\sqrt{x+5} = 4$ **68.** $\sqrt{z-2} = 3$

69. $\sqrt{3x-4} - 8 = 0$ **70.** $4 - \sqrt{3x+1} = 0$

71. $\sqrt{\dfrac{x}{2} + 1} = \dfrac{2}{3}$ **72.** $(x+6)^{1/2} = 7$

73. $\sqrt{4x-6} = \sqrt{x}$ **74.** $\sqrt{4+3x} = \sqrt{2x+5}$

75. $(x-5)^{3/4} = 27$ **76.** $\sqrt{y^2-9} = 9 - y$

*77.** $\sqrt{y} + \sqrt{y+2} = 3$ **78.** $\sqrt{x} - \sqrt{x+1} = 1$

*79.** $\sqrt{z^2+2z} = 3 + z$ **80.** $\sqrt{\dfrac{1}{w}} - \sqrt{\dfrac{2}{5w-2}} = 0$

In Problems 81–92, express the indicated symbol in terms of the remaining symbols.

81. $I = Prt; r$

82. $P\left(1 + \dfrac{p}{100}\right) - R = 0; \; P$

83. $p = 8q - 1; \; q$

84. $p = -3q + 6; \; q$

85. $S = P(1 + rt); \; r$

86. $r = \dfrac{2mI}{B(n+1)}; \; I$

*87.** $A = \dfrac{R[1 - (1+i)^{-n}]}{i}; \; R$

88. $S = \dfrac{R[(1+i)^n - 1]}{i}; \; R$

*89.** $r = \dfrac{d}{1-dt}; \; t$

90. $\dfrac{x-a}{b-x} = \dfrac{x-b}{a-x}; \; x$

*91.** $r = \dfrac{2mI}{B(n+1)}; \; n$

92. $\dfrac{1}{p} + \dfrac{1}{q} = \dfrac{1}{f}; \; q$

93. Geometry   Use the formula $P = 2l + 2w$ to find the length l of a rectangle whose perimeter P is 660 m and whose width w is 160 m.

94. Geometry   Use the formula $V = \pi r^2 h$ to find the height h of a cola can whose volume V is 355 ml and whose radius r is 2 cm.

95. Sales Tax   A salesperson needs to calculate the cost of an item with a sales tax of 8.25%. Write an equation that represents the total cost c of an item costing x dollars.

96. Revenue   A day care center's total monthly revenue from the care of x toddlers is given by $r = 450x$, and its total monthly costs are given by $c = 380x + 3500$. How many toddlers need to be enrolled each month to break even? In other words, when will revenue equal costs?

97. Straight-Line Depreciation If you purchase an item for business use, in preparing your income tax you may be able to spread out its expense over the life of the item. This is called *depreciation*. One method of depreciation is *straight-line depreciation*, in which the annual depreciation is computed by dividing the cost of the item, less its estimated salvage value, by its useful life. Suppose the cost is C dollars, the useful life is N years, and there is no salvage value. Then the value V (in dollars) of the item at the end of n years is given by

$$V = C\left(1 - \frac{n}{N}\right)$$

If new office furniture is purchased for $3200, has a useful life of 8 years, and has no salvage value, after how many years will it have a value of $2000?

98. Radar Beam When radar is used on a highway to determine the speed of a car, a radar beam is sent out and reflected from the moving car. The difference F (in cycles per second) in frequency between the original and reflected beams is given by

$$F = \frac{vf}{334.8}$$

where v is the speed of the car in miles per hour and f is the frequency of the original beam (in megacycles per second).

Suppose you are driving along a highway with a speed limit of 65 mi/h. A police officer aims a radar beam with a frequency of 2500 megacycles per second at your car, and the officer observes the difference in frequency to be 495 cycles per second. Can the officer claim that you were speeding?

99. Savings Bronwyn and Steve want to buy a house, so they have decided to save one-fifth of each of their salaries. Bronwyn earns $27.00 per hour and receives an extra $18.00 a week because she declined company benefits, and Steve earns $35.00 per hour plus benefits. They want to save at least $550.00 each week. How many hours must they each work each week to achieve their goal?

100. Predator–Prey Relation To study a predator–prey relationship, an experiment[4] was conducted in which a blindfolded subject, the "predator," stood in front of a 3-ft-square table on which uniform sandpaper discs, the "prey," were placed. For 1 minute the "predator" searched for the discs by tapping with a finger. Whenever a disc was

found, it was removed and searching resumed. The experiment was repeated for various disc densities (number of discs per 9 ft²). It was estimated that if y is the number of discs picked up in 1 minute when x discs are on the table, then

$$y = a(1 - by)x$$

where a and b are constants. Solve this equation for y.

101. Prey Density In a certain area, the number y of moth larvae consumed by a single predatory beetle over a given period of time is given by

$$y = \frac{1.4x}{1 + 0.09x}$$

where x is the *prey density* (the number of larvae per unit of area). What prey density would allow a beetle to survive if it needs to consume 10 larvae over the given period?

102. Store Hours Suppose the ratio of the number of hours a video store is open to the number of daily customers is constant. When the store is open 8 hours, the number of customers is 92 less than the maximum number of customers. When the store is open 10 hours, the number of customers is 46 less than the maximum number of customers. Write an equation describing this situation, and find the maximum number of daily customers.

103. Travel Time The time it takes a boat to travel a given distance upstream (against the current) can be calculated by dividing the distance by the difference of the speed of the boat and the speed of the current. Write an equation that calculates the time t it takes a boat moving at a speed r against a current c to travel a distance d. Solve your equation for c.

104. Wireless Tower A wireless tower is 100 meters tall. An engineer determines electronically that the distance from the top of the tower to a nearby house is 1 meter greater than the horizontal distance from the base of the tower to the house. Write an equation for the difference in terms of the horizontal distance from the base of the tower to the house. Solve the equation and subsequently determine the distance from the top of the tower to the house.

105. Automobile Skidding Police have used the formula $s = \sqrt{30fd}$ to estimate the speed s (in miles per hour) of a car if it skidded d feet when stopping. The literal number f is the coefficient of friction, determined by the kind of road (such as concrete, asphalt, gravel, or tar) and whether the road is wet or dry. Some values of f are given in Table 0.1. At 45 mi/h, about how many feet will a car skid on a dry concrete road? Give your answer to the nearest foot.

TABLE 0.1

	Concrete	Tar
Wet	0.4	0.5
Dry	0.8	1.0

[4]C. S. Holling, "Some Characteristics of Simple Types of Predation and Parasitism," *The Canadian Entomologist*, XCI, no. 7 (1959), 385–98.

106. Interest Earned Allison Bennett discovers that she has $1257 in an off-shore account that she has not used for a year. The interest rate was 7.3% compounded annually. How much interest did she earn from that account over the last year?

*107. **Tax in a Receipt** In Nova Scotia consumers pay HST, *harmonized sales tax*, of 15%. Tom Wood travels from Alberta, which has only federal GST, *goods and services tax*, of 7% to Nova Scotia for a chemistry conference. When he later submits his expense claims in Alberta, the comptroller is puzzled to find that her usual multiplier of $\frac{7}{107}$ to determine tax in a receipt is not producing correct results. What percentage of Tom's Nova Scotia receipts are HST?

In Problems 108–111, use a graphing calculator to determine which of the given numbers, if any, are roots of the given equation.

108. $112x^2 = 6x + 1$; $\dfrac{1}{8}, -\dfrac{2}{5}, -\dfrac{1}{14}$

109. $8x^3 + 11x + 21 = 58x^2$; $5, -\dfrac{1}{2}, \dfrac{2}{3}$

110. $\dfrac{3.1t - 7}{4.8t - 2} = 7$; $\sqrt{6}, -\dfrac{47}{52}, \dfrac{14}{61}$

111. $\left(\dfrac{v}{v+3}\right)^2 = v$; $0, \dfrac{27}{4}, \dfrac{13}{3}$

OBJECTIVE

To solve quadratic equations by factoring or by using the quadratic formula.

0.8 Quadratic Equations

To learn how to solve certain classical problems, we turn to methods of solving *quadratic equations*.

> **DEFINITION**
>
> A **quadratic equation** in the variable x is an equation that can be written in the form
>
> $$ax^2 + bx + c = 0 \tag{1}$$
>
> where a, b, and c are constants and $a \neq 0$.

A quadratic equation is also called a *second-degree equation* or an *equation of degree two*, since the highest power of the variable that occurs is the second. Whereas a linear equation has only one root, a quadratic equation may have two different roots.

Solution by Factoring

A useful method of solving quadratic equations is based on factoring, as the following example shows.

● EXAMPLE 1 **Solving a Quadratic Equation by Factoring**

a. *Solve $x^2 + x - 12 = 0$.*

Solution: The left side factors easily:

$$(x - 3)(x + 4) = 0$$

Think of this as two quantities, $x - 3$ and $x + 4$, whose product is zero. **Whenever the product of two or more quantities is *zero*, at least one of the quantities *must* be zero.** This means that either

$$x - 3 = 0 \quad \text{or} \quad x + 4 = 0$$

Solving these gives $x = 3$ and $x = -4$, respctively. Thus, the roots of the original equation are 3 and -4, and the solution set is $\{-4, 3\}$.

b. *Solve $6w^2 = 5w$.*

Solution: We write the equation as

$$6w^2 - 5w = 0$$

so that one side is 0. Factoring gives

$$w(6w - 5) = 0$$

CAUTION

We do not divide both sides by w (a variable) since equivalence is not guaranteed and we may "lose" a root.

Setting each factor equal to 0, we have

$$w = 0 \quad \text{or} \quad 6w - 5 = 0$$
$$w = 0 \quad \text{or} \quad 6w = 5$$

Thus, the roots are $w = 0$ and $w = \frac{5}{6}$. Note that if we had divided both sides of $6w^2 = 5w$ by w and obtained $6w = 5$, our only solution would be $w = \frac{5}{6}$. That is, we would lose the root $w = 0$. This confirms our discussion of Operation 5 in Section 0.7.

NOW WORK PROBLEM 3

CAUTION

You should approach a problem like this with caution. If the product of two quantities is equal to -2, it is not true that at least one of the quantities must be -2. Why?

● EXAMPLE 2   Solving a Quadratic Equation by Factoring

Solve $(3x - 4)(x + 1) = -2$.

Solution: We first multiply the factors on the left side:

$$3x^2 - x - 4 = -2$$

Rewriting this equation so that 0 appears on one side, we have

$$3x^2 - x - 2 = 0$$
$$(3x + 2)(x - 1) = 0$$
$$x = -\frac{2}{3}, \ 1$$

NOW WORK PROBLEM 7

Some equations that are not quadratic may be solved by factoring, as Example 3 shows.

● EXAMPLE 3   Solving a Higher-Degree Equation by Factoring

a. *Solve* $4x - 4x^3 = 0$.

Solution: This is called a *third-degree equation*. We proceed to solve it as follows:

$$4x - 4x^3 = 0$$
$$4x(1 - x^2) = 0 \qquad \text{(factoring)}$$
$$4x(1 - x)(1 + x) = 0 \qquad \text{(factoring)}$$

Setting each factor equal to 0 gives $4 = 0$ (impossible), $x = 0$, $1 - x = 0$, or $1 + x = 0$. Thus,

$$x = 0 \text{ or } x = 1 \text{ or } x = -1$$

so that the solution set is $\{-1, 0, 1\}$.

CAUTION

Do not neglect the fact that the factor x gives rise to a root.

b. *Solve* $x(x + 2)^2(x + 5) + x(x + 2)^3 = 0$.

Solution: Factoring $x(x + 2)^2$ from both terms on the left side, we have

$$x(x + 2)^2[(x + 5) + (x + 2)] = 0$$
$$x(x + 2)^2(2x + 7) = 0$$

Hence, $x = 0$, $x + 2 = 0$, or $2x + 7 = 0$, from which it follows that the solution set is $\{-\frac{7}{2}, -2, 0\}$.

NOW WORK PROBLEM 23

● EXAMPLE 4   **A Fractional Equation Leading to a Quadratic Equation**

Solve

$$\frac{y+1}{y+3} + \frac{y+5}{y-2} = \frac{7(2y+1)}{y^2+y-6} \tag{2}$$

Solution: Multiplying both sides by the LCD, $(y+3)(y-2)$, we get

$$(y-2)(y+1) + (y+3)(y+5) = 7(2y+1) \tag{3}$$

Since Equation (2) was multiplied by an expression involving the variable y, remember (from Section 0.7) that Equation (3) is not necessarily equivalent to Equation (2). After simplifying Equation (3), we have

$$2y^2 - 7y + 6 = 0 \qquad \text{(quadratic equation)}$$

$$(2y-3)(y-2) = 0 \qquad \text{(factoring)}$$

We have shown that *if* y satisfies the original equation *then* $y = \frac{3}{2}$ or $y = 2$. Thus, $\frac{3}{2}$ and 2 are the only *possible* roots of the given equation. But 2 cannot be a root of Equation (2), since substitution leads to a denominator of 0. However, you should check that $\frac{3}{2}$ does indeed satisfy the *original* equation. Hence, its only root is $\frac{3}{2}$.

NOW WORK PROBLEM 53 ●●●

● EXAMPLE 5   **Solution by Factoring**

Solve $x^2 = 3$.

Solution:

$$x^2 = 3$$

$$x^2 - 3 = 0$$

Factoring, we obtain

$$(x - \sqrt{3})(x + \sqrt{3}) = 0$$

Thus $x - \sqrt{3} = 0$ or $x + \sqrt{3} = 0$, so $x = \pm\sqrt{3}$.

NOW WORK PROBLEM 9 ●●●

CAUTION

Do not hastily conclude that the solution of $x^2 = 3$ consists of $x = \sqrt{3}$ only.

A more general form of the equation $x^2 = 3$ is $u^2 = k$. In the same manner as the preceding, we can show that

$$\text{If } u^2 = k \qquad \text{then} \qquad u = \pm\sqrt{k}. \tag{4}$$

Quadratic Formula

Solving quadratic equations by factoring can be quite difficult, as is evident by trying that method on $0.7x^2 - \sqrt{2}x - 8\sqrt{5} = 0$. However, there is a formula called the *quadratic formula* that gives the roots of any quadratic equation.

Quadratic Formula

The roots of the quadratic equation $ax^2 + bx + c = 0$, where a, b, and c are constants and $a \neq 0$, are given by

$$x = \frac{-b \pm \sqrt{b^2 - 4ac}}{2a}$$

Actually, the quadratic formula is not hard to derive if one first writes the quadratic equation in the form

$$x^2 + \frac{b}{a}x + \frac{c}{a} = 0$$

and then as

$$\left(x + \frac{b}{2a}\right)^2 - K^2 = 0$$

for a number K, as yet to be determined. This leads to

$$\left(x + \frac{b}{2a} - K\right)\left(x + \frac{b}{2a} + K\right) = 0$$

which in turn leads to $x = -\frac{b}{2a} + K$ or $x = -\frac{b}{2a} - K$ by the methods already under consideration. It is not hard to see what K must be, but it requires further thought to understand how one could discover the value of K without knowing the answer in advance.

● EXAMPLE 6   **A Quadratic Equation with Two Real Roots**

Solve $4x^2 - 17x + 15 = 0$ by the quadratic formula.

Solution: Here $a = 4$, $b = -17$, and $c = 15$. Thus,

$$x = \frac{-b \pm \sqrt{b^2 - 4ac}}{2a} = \frac{-(-17) \pm \sqrt{(-17)^2 - 4(4)(15)}}{2(4)}$$

$$= \frac{17 \pm \sqrt{49}}{8} = \frac{17 \pm 7}{8}$$

The roots are $\dfrac{17 + 7}{8} = \dfrac{24}{8} = 3$ and $\dfrac{17 - 7}{8} = \dfrac{10}{8} = \dfrac{5}{4}$.

NOW WORK PROBLEM 31   ●●

● EXAMPLE 7   **A Quadratic Equation with One Real Root**

Solve $2 + 6\sqrt{2}y + 9y^2 = 0$ by the quadratic formula.

Solution: Look at the arrangement of the terms. Here $a = 9$, $b = 6\sqrt{2}$, and $c = 2$. Hence,

$$y = \frac{-b \pm \sqrt{b^2 - 4ac}}{2a} = \frac{-6\sqrt{2} \pm \sqrt{0}}{2(9)}$$

Thus,

$$y = \frac{-6\sqrt{2} + 0}{18} = -\frac{\sqrt{2}}{3} \quad \text{or} \quad y = \frac{-6\sqrt{2} - 0}{18} = -\frac{\sqrt{2}}{3}$$

Therefore, the only root is $-\dfrac{\sqrt{2}}{3}$.

NOW WORK PROBLEM 33   ●●

● EXAMPLE 8   **A Quadratic Equation with No Real Solution**

Solve $z^2 + z + 1 = 0$ by the quadratic formula.

Solution: Here $a = 1$, $b = 1$, and $c = 1$. The roots are

$$z = \frac{-b \pm \sqrt{b^2 - 4ac}}{2a} = \frac{-1 \pm \sqrt{-3}}{2}$$

Now, $\sqrt{-3}$ denotes a number whose square is -3. However, no such real number exists, since the square of any real number is nonnegative. Thus, the equation has no real roots.[5]

NOW WORK PROBLEM 37 ●●●

This describes the nature of the roots of a quadratic equation.

From Examples 6–8, you can see that a quadratic equation has either two different real roots, exactly one real root, or no real roots, depending on whether $b^2 - 4ac > 0, = 0$, or < 0, respectively.

TECHNOLOGY

Using the program feature of a graphing calculator, one can create a program that gives the real roots of the quadratic equation $Ax^2 + Bx + C = 0$. Figure 0.3 shows such a program for the TI-83 graphing calculator. To execute it for

$$20x^2 - 33x + 10 = 0$$

you are prompted to enter the values of A, B, and C. (See Figure 0.4.) The resulting roots are $x = 1.25$ and $x = 0.4$.

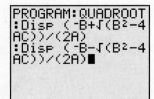

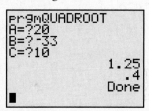

FIGURE 0.3 Program to find real roots of $Ax^2 + Bx + C = 0$.

FIGURE 0.4 Roots of $20x^2 - 33x + 10 = 0$.

Quadratic-Form Equation

Sometimes an equation that is not quadratic can be transformed into a quadratic equation by an appropriate substitution. In this case, the given equation is said to have **quadratic-form**. The next example will illustrate.

● EXAMPLE 9 **Solving a Quadratic-Form Equation**

Solve $\dfrac{1}{x^6} + \dfrac{9}{x^3} + 8 = 0$.

Solution: This equation can be written as

$$\left(\frac{1}{x^3}\right)^2 + 9\left(\frac{1}{x^3}\right) + 8 = 0$$

so it is quadratic in $1/x^3$ and hence has quadratic-form. Substituting the variable w for $1/x^3$ gives a quadratic equation in the variable w, which we can then solve:

$$w^2 + 9w + 8 = 0$$

$$(w + 8)(w + 1) = 0$$

$$w = -8 \quad \text{or} \quad w = -1$$

CAUTION

Do not assume that -8 and -1 are solutions of the *original* equation.

[5] $\frac{-1\pm\sqrt{-3}}{2}$ can be expressed as $\frac{-1\pm i\sqrt{3}}{2}$, where $i = \sqrt{-1}$ is called the *imaginary unit*. We emphasize that $i = \sqrt{-1}$ is not a real number. Complex numbers are numbers of the form $a + ib$, with a and b real, but they are not discussed in this book.

Returning to the variable x, we have

$$\frac{1}{x^3} = -8 \quad \text{or} \quad \frac{1}{x^3} = -1$$

Thus,

$$x^3 = -\frac{1}{8} \text{ or } x^3 = -1$$

from which it follows that

$$x = -\frac{1}{2} \text{ or } x = -1$$

Checking, we find that these values of x satisfy the original equation.

NOW WORK PROBLEM 49

Problems 0.8

In Problems 1–30, solve by factoring.

1. $x^2 - 4x + 4 = 0$

2. $t^2 + 3t + 2 = 0$

*3. $t^2 - 8t + 15 = 0$

4. $x^2 + 3x - 10 = 0$

5. $x^2 - 2x - 3 = 0$

6. $x^2 - 16 = 0$

*7. $u^2 - 13u = -36$

8. $3w^2 - 12w + 12 = 0$

*9. $x^2 - 4 = 0$

10. $3u^2 - 6u = 0$

11. $t^2 - 5t = 0$

12. $x^2 + 9x = -14$

13. $4x^2 + 1 = 4x$

14. $2z^2 + 9z = 5$

15. $v(3v - 5) = -2$

16. $2 + x - 6x^2 = 0$

17. $-x^2 + 3x + 10 = 0$

18. $\frac{1}{7}y^2 = \frac{3}{7}y$

19. $2p^2 = 3p$

20. $-r^2 - r + 12 = 0$

21. $x(x + 4)(x - 1) = 0$

22. $(w - 3)^2(w + 1)^2 = 0$

*23. $t^3 - 49t = 0$

24. $x^3 - 4x^2 - 5x = 0$

25. $6x^3 + 5x^2 - 4x = 0$

26. $(x + 1)^2 - 5x + 1 = 0$

27. $(x - 3)(x^2 - 4) = 0$

28. $5(x^2 + x - 12)(x - 8) = 0$

29. $p(p - 3)^2 - 4(p - 3)^3 = 0$ **30.** $x^4 - 3x^2 + 2 = 0$

In Problems 31–44, find all real roots by using the quadratic formula.

*31. $x^2 + 2x - 24 = 0$

32. $x^2 - 2x - 15 = 0$

*33. $4x^2 - 12x + 9 = 0$

34. $q^2 - 5q = 0$

35. $p^2 - 2p - 7 = 0$

36. $2 - 2x + x^2 = 0$

*37. $4 - 2n + n^2 = 0$

38. $2x^2 + x = 5$

39. $4x^2 + 5x - 2 = 0$

40. $w^2 - 2w + 1 = 0$

41. $0.02w^2 - 0.3w = 20$

42. $0.01x^2 + 0.2x - 0.6 = 0$

43. $2x^2 + 4x = 5$

44. $-2x^2 - 6x + 5 = 0$

In Problems 45–54, solve the given quadratic-form equation.

45. $x^4 - 5x^2 + 6 = 0$

46. $X^4 - 3X^2 - 10 = 0$

47. $\frac{3}{x^2} - \frac{7}{x} + 2 = 0$

48. $x^{-2} + x^{-1} - 12 = 0$

*49. $x^{-4} - 9x^{-2} + 20 = 0$

50. $\frac{1}{x^4} - \frac{9}{x^2} + 8 = 0$

51. $(X - 5)^2 + 7(X - 5) + 10 = 0$

52. $(3x + 2)^2 - 5(3x + 2) = 0$

*53. $\frac{1}{(x - 2)^2} - \frac{12}{x - 2} + 35 = 0$

54. $\frac{2}{(x + 4)^2} + \frac{7}{x + 4} + 3 = 0$

In Problems 55–76, solve by any method.

55. $x^2 = \frac{x + 3}{2}$

56. $\frac{x}{2} = \frac{7}{x} - \frac{5}{2}$

57. $\frac{3}{x - 4} + \frac{x - 3}{x} = 2$

58. $\frac{2}{2x + 1} - \frac{6}{x - 1} = 5$

59. $\frac{3x + 2}{x + 1} - \frac{2x + 1}{2x} = 1$

60. $\frac{6(w + 1)}{2 - w} + \frac{w}{w - 1} = 3$

61. $\frac{2}{r - 2} - \frac{r + 1}{r + 4} = 0$

62. $\frac{2x - 3}{2x + 5} + \frac{2x}{3x + 1} = 1$

63. $\frac{t + 1}{t + 2} + \frac{t + 3}{t + 4} = \frac{t + 5}{t^2 + 6t + 8}$

64. $\frac{2}{x + 1} + \frac{3}{x} = \frac{4}{x + 2}$

65. $\frac{2}{x^2 - 1} - \frac{1}{x(x - 1)} = \frac{2}{x^2}$

66. $5 - \frac{3(x + 3)}{x^2 + 3x} = \frac{1 - x}{x}$

67. $\sqrt{2x - 3} = x - 3$

68. $3\sqrt{x + 4} = x - 6$

69. $q + 2 = 2\sqrt{4q - 7}$

70. $x + \sqrt{4x - 5} = 0$

71. $\sqrt{z + 3} - \sqrt{3z} - 1 = 0$

72. $\sqrt{x} - \sqrt{2x - 8} - 2 = 0$

73. $\sqrt{x} - \sqrt{2x + 1} + 1 = 0$

74. $\sqrt{y-2}+2=\sqrt{2y+3}$

75. $\sqrt{x+3}+1=3\sqrt{x}$

76. $\sqrt{\sqrt{t}+2}=\sqrt{3t-1}$

In Problems 77 and 78, find the roots, rounded to two decimal places.

77. $0.04x^2-2.7x+8.6=0$

78. $0.01x^2+0.2x-0.6=0$

79. Geometry The area of a rectangular picture with a width 2 inches less than its length is 48 square inches. What are the dimensions of the picture?

80. Temperature The temperature has been rising X degrees per day for X days. X days ago it was 15 degrees. Today it is 51 degrees. How much has the temperature been rising each day? How many days has it been rising?

81. Economics One root of the economics equation

$$\overline{M}=\frac{Q(Q+10)}{44}$$

is $-5+\sqrt{25+44\overline{M}}$. Verify this by using the quadratic formula to solve for Q in terms of $\overline{M}$. Here Q is real income and $\overline{M}$ is the level of money supply.

82. Diet for Rats A group of biologists studied the nutritional effects on rats that were fed a diet containing 10% protein.[6] The protein was made up of yeast and corn flour. By changing the percentage P (expressed as a decimal) of yeast in the protein mix, the group estimated that the average weight gain g (in grams) of a rat over a period of time was given by

$$g=-200P^2+200P+20$$

What percentage of yeast gave an average weight gain of 60 grams?

83. Drug Dosage There are several rules for determining doses of medicine for children when the adult dose has been specified. Such rules may be based on weight, height, and so on. If A is the age of the child, d is the adult dose, and c is the child's dose, then here are two rules:

$$\text{Young's rule:}\qquad c=\frac{A}{A+12}d$$

$$\text{Cowling's rule:}\qquad c=\frac{A+1}{24}d$$

At what age(s) are the children's doses the same under both rules? Round your answer to the nearest year. Presumably, the child has become an adult when $c=d$. At what age does the child become an adult according to Cowling's rule? According to Young's rule? If you know how to graph

functions, graph both $Y(A)=\dfrac{A}{A+12}$ and $C(A)=\dfrac{A+1}{24}$ as functions of A, for $A\geq 0$, in the same plane. Using the graphs, make a more informed comparison of Young's rule and Cowling's rule than is obtained by merely finding the age(s) at which they agree.

84. Delivered Price of a Good In a discussion of the delivered price of a good from a mill to a customer, DeCanio[7] arrives at and solves the two quadratic equations

$$(2n-1)v^2-2nv+1=0$$

and

$$nv^2-(2n+1)v+1=0$$

where $n\geq 1$.

(a) Solve the first equation for v.

(b) Solve the second equation for v if $v<1$.

85. Motion Suppose the height h of an object thrown straight upward from the ground is given by

$$h=39.2t-4.9t^2$$

where h is in meters and t is the elapsed time in seconds.

(a) After how many seconds does the object strike the ground?

(b) When is the object at a height of 68.2 m?

In Problems 86–91, use a program to determine any real roots of the equation. Round answers to three decimal places. For Problems 86 and 87, confirm your result algebraically.

86. $2x^2-3x-27=0$

87. $8x^2-18x+9=0$

88. $10x^2+5x-2=0$

89. $27x^2-\dfrac{11}{8}x+5=0$

90. $\dfrac{9}{2}z^2-6.3=\dfrac{z}{3}(1.1-7z)$

91. $(\pi t-4)^2=4.1t-3$

---

[6]Adapted from R. Bressani, "The Use of Yeast in Human Foods," in R. I. Mateles and S. R. Tannenbaum (eds.), *Single-Cell Protein* (Cambridge, MA: MIT Press, 1968).

[7]S. J. DeCanio, "Delivered Pricing and Multiple Basing Point Equilibria: A Revolution," *Quarterly Journal of Economics*, XCIX, no. 2 (1984), 329–49.

Mathematical Snapshot

Modeling Load Cell Behavior[8]

A *load cell* is a device that measures a force, such as weight, by translating it into an electrical signal. Load cells are found in many applications, for example, in bathroom scales. When you stand on the scale, the force that your body exerts on the platform is translated by the load cell into an electrical signal of variable voltage depending upon your weight. The electrical signal is further converted to a digital output on the screen of the scale.

As measuring devices, load cells have to behave predictably and consistently. While an unreliable bathroom scale would not enjoy good sales for long, it would probably not be regarded as a dangerous product. Reliability for other applications of load cells is often considerably more serious. Lifting equipment, such as cranes, may contain load cells that provide warning when the equipment is reaching its rated safe-operating limit. In such an application an error might be disastrous.

A common requirement for load cells is that the voltage output, V, be related to the force input, F, by a linear equation as discussed in Section 0.7:

$$V = aF + b$$

A linear response permits a simple translation of the voltage into a digital output.

Suppose that a company which manufactures load cells for cranes puts a sample cell through a calibration test and obtains the following data (with force measured in thousands of pounds and voltage measured in volts):

Force	Voltage	Force	Voltage
150.000	0.11019	1650.000	1.20001
300.000	0.21956	1800.000	1.30822
450.000	0.32949	1950.000	1.41599
600.000	0.43899	2100.000	1.52399
750.000	0.54803	2250.000	1.63194
900.000	0.65694	2400.000	1.73947
1050.000	0.76562	2550.000	1.84646
1200.000	0.87487	2700.000	1.95392
1350.000	0.98292	2850.000	2.06128
1500.000	1.09146	3000.000	2.16844

If the load cell is behaving properly, a linear equation will be a good *model* of the data. In other words, when the data values are plotted as points on a graph, it should be possible to draw a line that passes through all the points, within an acceptable tolerance of error.

[8]Based on Section 4.6.1 of *Engineering Statistics Handbook*, National Institute of Standards and Technology/SEMATECH, www.nist.gov/itl/div898/ handbook/pmd/section6/pmd61.htm.

The mathematics for finding the equation of the line that best fits a given collection of data is quite involved. Fortunately, a graphing calculator can do it automatically. The result for our data is

$$V = 0.0007221\,F + 0.006081368 \qquad (5)$$

Plotting the data and the equation together produces the result shown in Figure 0.5.

It looks as if the linear model is a very good fit indeed. But is it as good as it can be? Let us look at the differences between the measured voltages and the respective values predicted by the linear model. For each force value in the data table, we subtract from the measured voltage in the table the corresponding voltage predicted by Equation (5).

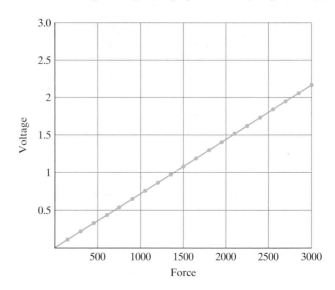

FIGURE 0.5 The linear model.

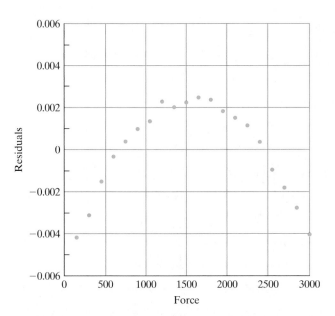

FIGURE 0.6   Plot of residuals.

For example, for force value, 450.000, we calculate

$$0.32949 - (0.0007221(450.000) + 0.006081368) = -0.00154$$

The difference we calculated is called a *residual*.

If, for each force value, we calculate the corresponding residual then we can plot points with horizontal coordinates given by force and vertical coordinates given by residual as in (Figure 0.6).

For example, one of the plotted points is (450.000, −0.00154).

Apparently, the data points in the middle of Figure 0.5 actually lie slightly above the line (positive residuals), while those toward the ends of the line lie slightly below it (negative residuals). The pattern of data points, in other words, has a slight curvature to it, which becomes evident only when we plot the residuals and "zoom in" on the vertical scale.

The plot of the residuals looks like a parabola (see Chapter 3). Because a parabola has an equation containing a squared term, we can expect that a quadratic equation will be a better model of the data than a linear one. Using the quadratic regression function on a graphing calculator produces the equation

$$V = (-3.22693 \times 10^{-9})F^2 + 0.000732265F + 0.000490711$$

The small coefficient on the F-squared term indicates a slight nonlinearity in the data.

For the manufacturer of the load cell, the slight nonlinearity will call for a decision. On the one hand, a nonlinear load cell response might lead to dangerously inaccurate measurements in some applications, especially if the cell is having to measure forces well outside the range of the test data. (Cranes mounted on cargo ships sometimes carry loads of up to 5000 tons, or 10 million pounds.) On the other hand, all manufacturing processes involve compromises between what is ideal and what is practically feasible.

Problems

1. Enter the force and voltage values as two separate lists on a graphing calculator, and then use the Linear Regression function on the Statistics menu to generate a regression equation. Compare your result to the linear equation in the preceding discussion.

2. On most graphing calculators, if you multiply the list of forces by 0.0007221 and add 0.006081368, and then subtract the result from the list of voltages, you will have the list of residuals. Why does this work? Store the residuals as a new list; then plot them and compare your result with Figure 0.6.

3. Use the Quadratic Regression function on a graphing calculator to generate a new regression equation. Compare your result with the quadratic equation in the preceding discussion.

4. The quadratic model also has *residuals*. Explain what this means. When these are plotted they look like:

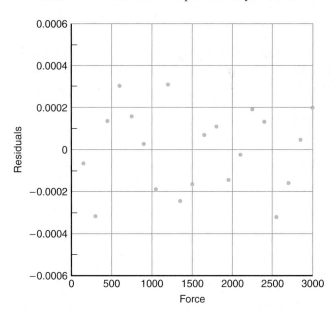

Compare the scale on the vertical axis with the one in Figure 0.6 for the linear model. What does the comparison suggest? What does the pattern of data points for the quadratic residuals suggest?

1

APPLICATIONS AND MORE ALGEBRA

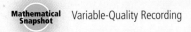 **Mathematical Snapshot** Variable-Quality Recording

In this chapter, we will apply equations to various real-life situations. We will then do the same with inequalities, which are statements that one quantity is less than ($<$), greater than ($>$), less than or equal to ($\leq$), or greater than or equal to ($\geq$) some other quantity.

One application of inequalities is the regulation of sporting equipment. Dozens of baseballs are used in a typical major league game and it would be unrealistic to expect that every ball weigh exactly $5\frac{1}{8}$ ounces. But it is reasonable to require that each one weigh no less than 5 ounces and no more than $5\frac{1}{4}$ ounces, which is how the official rules read (www.majorleaguebaseball.com). Note that *no less than* is a synonym for *greater than or equal to* while *no more than* is a synonym for *less than or equal to*. In translating English statements into mathematics, we advocate avoiding the negative wordings as a first step. In any event, we have

$$\text{ball weight} \geq 5 \text{ ounces} \quad \text{and} \quad \text{ball weight} \leq 5\frac{1}{4} \text{ ounces}$$

which can be combined to give

$$5 \text{ ounces} \leq \text{ball weight} \leq 5\frac{1}{4} \text{ ounces}$$

which reads nicely as saying that the ball must weigh between 5 and $5\frac{1}{4}$ ounces (where *between* here includes the extreme values).

Another inequality applies to the sailboats used in the America's Cup race, which takes place every three to four years. The International America's Cup Class (IACC) for yachts is defined by the following rule:

$$\frac{L + 1.25\sqrt{S} - 9.8\sqrt[3]{DSP}}{0.679} \leq 24.000 \text{ m}$$

The "$\leq$" signifies that the expression on the left must come out as less than or equal to the 24 m on the right. The L, S, and DSP are themselves specified by complicated formulas, but roughly, L stands for length, S for sail area, and DSP for displacement (the hull volume below the waterline).

The IACC formula gives yacht designers some latitude. Suppose a yacht has $L = 20.2$ m, $S = 282$ m^2, and $DSP = 16.4$ m^3. Since the formula is an inequality, the designer could reduce the sail area while leaving the length and displacement unchanged. Typically, however, values of L, S, and DSP are used that make the expression on the left as close to 24 m as possible.

In addition to looking at applications of linear equations and inequalities, this chapter will review the concept of absolute value and introduce summation notation.

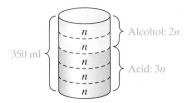

FIGURE 1.1   Chemical solution (Example 1).

   C A U T I O N

Note that the solution to an equation is not necessarily the solution to the problem posed.

1.1 Applications of Equations

In most cases, to solve practical problems you must translate the relationships stated in the problems into mathematical symbols. This is called *modeling*. The following examples illustrate basic techniques and concepts. Examine each of them carefully before going to the problems.

EXAMPLE 1   Mixture

A chemist must prepare 350 ml *of a chemical solution made up of two parts alcohol and three parts acid. How much of each should be used?*

Solution: Let n be the number of milliliters in each part. Figure 1.1 shows the situation. From the diagram, we have

$$2n + 3n = 350$$
$$5n = 350$$
$$n = \frac{350}{5} = 70$$

But $n = 70$ is *not* the answer to the original problem. Each *part* has 70 ml. The amount of alcohol is $2n = 2(70) = 140$, and the amount of acid is $3n = 3(70) = 210$. Thus, the chemist should use 140 ml of alcohol and 210 ml of acid. This example shows how helpful a diagram can be in setting up a word problem.

NOW WORK PROBLEM 5

EXAMPLE 2   Observation Deck

A rectangular observation deck overlooking a scenic valley is to be built. [See Figure 1.2(a).] The deck is to have dimensions 6 m *by* 12 m. *A rectangular shelter with area* 40 m² *is to be centered over the deck. The uncovered part of the deck is to be a walkway of uniform width. How wide should this walkway be?*

Solution: A diagram of the deck is shown in Figure 1.2(b). Let w be the width (in meters) of the walkway. Then the part of the deck under the shelter has dimensions $12 - 2w$ by $6 - 2w$. Since its area must be 40 m², where area = (length)(width), we have

$$(12 - 2w)(6 - 2w) = 40$$
$$72 - 36w + 4w^2 = 40 \quad \text{(multiplying)}$$
$$4w^2 - 36w + 32 = 0$$
$$w^2 - 9w + 8 = 0 \quad \text{(dividing both sides by 4)}$$
$$(w - 8)(w - 1) = 0$$
$$w = 8,\ 1$$

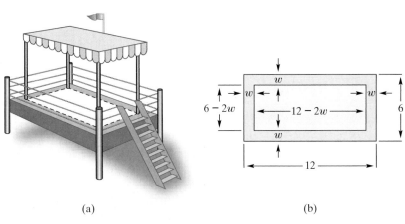

(a)   (b)

FIGURE 1.2   Deck walkway (Example 2).

Although 8 is a solution to the equation, it is *not* a solution to our problem, because one of the dimensions of the deck itself is only 6 m. Thus, the only possible solution is that the walkway be 1 m wide.

NOW WORK PROBLEM 7

The key words introduced here are *fixed cost, variable cost, total cost, total revenue*, and *profit*. This is the time for you to gain familiarity with these terms because they recur throughout the book.

In the next example, we refer to some business terms relative to a manufacturing firm. **Fixed cost** is the sum of all costs that are independent of the level of production, such as rent, insurance, and so on. This cost must be paid whether or not output is produced. **Variable cost** is the sum of all costs that are dependent on the level of output, such as labor and material. **Total cost** is the sum of variable cost and fixed cost:

$$\text{total cost} = \text{variable cost} + \text{fixed cost}$$

Total revenue is the money that the manufacturer receives for selling the output:

$$\text{total revenue} = (\text{price per unit})\,(\text{number of units sold})$$

Profit is total revenue minus total cost:

$$\text{profit} = \text{total revenue} - \text{total cost}$$

● EXAMPLE 3 Profit

The Anderson Company produces a product for which the variable cost per unit is $6 and fixed cost is $80,000. Each unit has a selling price of $10. Determine the number of units that must be sold for the company to earn a profit of $60,000.

Solution: Let q be the number of units that must be sold. (In many business problems, q represents quantity.) Then the variable cost (in dollars) is $6q$. The *total* cost for the business is therefore $6q + 80,000$. The total revenue from the sale of q units is $10q$. Since

$$\text{profit} = \text{total revenue} - \text{total cost}$$

our model for this problem is

$$60,000 = 10q - (6q + 80,000)$$

Solving gives

$$60,000 = 10q - 6q - 80,000$$
$$140,000 = 4q$$
$$35,000 = q$$

Thus, 35,000 units must be sold to earn a profit of $60,000.

NOW WORK PROBLEM 9

● EXAMPLE 4 Pricing

Sportcraft manufactures women's sportswear and is planning to sell its new line of slacks to retail outlets. The cost to the retailer will be $33 per pair of slacks. As a convenience to the retailer, Sportcraft will attach a price tag to each pair. What amount should be marked on the price tag so that the retailer may reduce this price by 20% during a sale and still make a profit of 15% on the cost?

Solution: Here we use the fact that

Note that price = cost + profit.

$$\text{selling price} = \text{cost per pair} + \text{profit per pair}$$

Let p be the tag price per pair, in dollars. During the sale, the retailer actually receives $p - 0.2p$. This must equal the cost, 33, plus the profit, $(0.15)(33)$. Hence,

$$\text{selling price} = \text{cost} + \text{profit}$$

$$p - 0.2p = 33 + (0.15)(33)$$

$$0.8p = 37.95$$

$$p = 47.4375$$

From a practical point of view, Sportcraft should mark the price tag at $47.44.

NOW WORK PROBLEM 13 ◖●●

●EXAMPLE 5 Investment

A total of $10,000 was invested in two business ventures, A and B. At the end of the first year, A and B yielded returns of 6% and $5\frac{3}{4}$%, respectively, on the original investments. How was the original amount allocated if the total amount earned was $588.75?

Solution: Let x be the amount (in dollars) invested at 6%. Then $10,000 - x$ was invested at $5\frac{3}{4}$%. The interest earned from A was $(0.06)(x)$, and from B it was $(0.0575)(10,000 - x)$, which total 588.75. Hence,

$$(0.06)x + (0.0575)(10,000 - x) = 588.75$$

$$0.06x + 575 - 0.0575x = 588.75$$

$$0.0025x = 13.75$$

$$x = 5500$$

Thus, $5500 was invested at 6%, and $10,000 - \$5500 = \$4500$ was invested at $5\frac{3}{4}$%.

NOW WORK PROBLEM 11 ◖●●

●EXAMPLE 6 Bond Redemption

The board of directors of Maven Corporation agrees to redeem some of its bonds in two years. At that time, $1,102,500 will be required. Suppose the firm presently sets aside $1,000,000. At what annual rate of interest, compounded annually, will this money have to be invested in order that its future value be sufficient to redeem the bonds?

Solution: Let r be the required annual rate of interest. At the end of the first year, the accumulated amount will be $1,000,000 plus the interest, $1,000,000r$, for a total of

$$1,000,000 + 1,000,000r = 1,000,000(1 + r)$$

Under compound interest, at the end of the second year the accumulated amount will be $1,000,000(1 + r)$ plus the interest on this, which is $1,000,000(1 + r)r$. Thus, the total value at the end of the second year will be

$$1,0000,000(1 + r) + 1,000,000(1 + r)r$$

This must equal $1,102,500:

$$1,000,000(1 + r) + 1,000,000(1 + r)r = 1,102,500 \qquad (1)$$

Since $1,000,000(1 + r)$ is a common factor of both terms on the left side, we have

$$1,000,000(1 + r)(1 + r) = 1,102,500$$

$$1,000,000(1 + r)^2 = 1,102,500$$

$$(1 + r)^2 = \frac{1,102,500}{1,000,000} = \frac{11,025}{10,000} = \frac{441}{400}$$

$$1 + r = \pm\sqrt{\frac{441}{400}} = \pm\frac{21}{20}$$

$$r = -1 \pm \frac{21}{20}$$

Thus, $r = -1 + (21/20) = 0.05$, or $r = -1 - (21/20) = -2.05$. Although 0.05 and -2.05 are roots of Equation (1), we reject -2.05, since we want r to be positive. Hence, $r = 0.05$ so the desired rate is 5%.

NOW WORK PROBLEM 15

At times there may be more than one way to model a word problem, as Example 7 shows.

EXAMPLE 7   Apartment Rent

A real-estate firm owns the Parklane Garden Apartments, which consist of 96 apartments. At $550 per month, every apartment can be rented. However, for each $25 per month increase, there will be three vacancies with no possibility of filling them. The firm wants to receive $54,600 per month from rents. What rent should be charged for each apartment?

Solution:

Method I. Suppose r is the rent (in dollars) to be charged per apartment. Then the increase over the $550 level is $r - 550$. Thus, the number of $25 increases is $\dfrac{r - 550}{25}$. Because each $25 increase results in three vacancies, the total number of vacancies will be $3\left(\dfrac{r - 550}{25}\right)$. Hence, the total number of apartments rented will be $96 - 3\left(\dfrac{r - 550}{25}\right)$. Since

$$\text{total rent} = (\text{rent per apartment})(\text{number of apartments rented})$$

we have

$$54{,}600 = r\left[96 - \frac{3(r - 550)}{25}\right]$$

$$54{,}600 = r\left[\frac{2400 - 3r + 1650}{25}\right]$$

$$54{,}600 = r\left[\frac{4050 - 3r}{25}\right]$$

$$1{,}365{,}000 = r(4050 - 3r)$$

Thus,

$$3r^2 - 4050r + 1{,}365{,}000 = 0$$

By the quadratic formula,

$$r = \frac{4050 \pm \sqrt{(-4050)^2 - 4(3)(1{,}365{,}000)}}{2(3)}$$

$$= \frac{4050 \pm \sqrt{22{,}500}}{6} = \frac{4050 \pm 150}{6} = 675 \pm 25$$

Hence, the rent for each apartment should be either $650 or $700.

Method II. Suppose n is the number of $25 increases. Then the increase in rent per apartment will be $25n$ and there will be $3n$ vacancies. Since

$$\text{total rent} = (\text{rent per apartment})(\text{number of apartments rented})$$

we have

$$54{,}600 = (550 + 25n)(96 - 3n)$$

$$54{,}600 = 52{,}800 + 750n - 75n^2$$

$$75n^2 - 750n + 1800 = 0$$

$$n^2 - 10n + 24 = 0$$

$$(n - 6)(n - 4) = 0$$

Thus, $n = 6$ or $n = 4$. The rent charged should be either $550 + 25(6) = \$700$ or $550 + 25(4) = \$650$.

NOW WORK PROBLEM 29

Problems 1.1

1. Fencing A fence is to be placed around a rectangular plot so that the enclosed area is 800 ft² and the length of the plot is twice the width. How many feet of fencing must be used?

2. Geometry The perimeter of a rectangle is 300 ft, and the length of the rectangle is twice the width. Find the dimensions of the rectangle.

3. Tent Caterpillars One of the most damaging defoliating insects is the tent caterpillar, which feeds on foliage of shade, forest, and fruit trees. A homeowner lives in an area in which the tent caterpillar has become a problem. She wishes to spray the trees on her property before more defoliation occurs. She needs 145 oz of a solution made up of 4 parts of insecticide *A* and 5 parts of insecticide *B*. The solution is then mixed with water. How many ounces of each insecticide should be used?

4. Concrete Mix A builder makes a certain type of concrete by mixing together 1 part portland cement (made from lime and clay), 3 parts sand, and 5 parts crushed stone (by volume). If 765 ft³ of concrete are needed, how many cubic feet of each ingredient does he need?

*5. Furniture Finish According to *The Consumer's Handbook* [Paul Fargis, ed. (New York: Hawthorn, 1974)], a good oiled furniture finish contains two parts boiled linseed oil and one part turpentine. If you need a pint (16 fluid oz) of this furniture finish, how many fluid ounces of turpentine are needed?

6. Forest Management A lumber company owns a forest that is of rectangular shape, 1 mi by 2 mi. If the company cuts a uniform strip of trees along the outer edges of this forest, how wide should the strip be if $\frac{3}{4}$ sq mi of forest is to remain?

*7. Garden Pavement A rectangular plot, 4 m by 8 m, is to be used for a garden. It is decided to put a pavement inside the entire border so that 12 m² of the plot is left for flowers. How wide should the pavement be?

8. Ventilating Duct The diameter of a circular ventilating duct is 140 mm. This duct is joined to a square duct system as shown in Figure 1.3. To ensure smooth airflow, the areas of the circle and square sections must be equal. To the nearest millimeter, what should the length *x* of a side of the square section be?

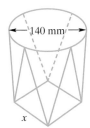

FIGURE 1.3   Ventilating duct (Problem 8).

*9. Profit A corn refining company produces corn gluten cattle feed at a variable cost of $82 per ton. If fixed costs are $120,000 per month and the feed sells for $134 per ton, how many tons must be sold each month for the company to have a monthly profit of $560,000?

10. Sales The Smith Company management would like to know the total sales units that are required for the company to earn a profit of $150,000. The following data are available: unit selling price of $50; variable cost per unit of $25; total fixed cost of $500,000. From these data, determine the required sales units.

*11. Investment A person wishes to invest $20,000 in two enterprises so that the total income per year will be $1440. One enterprise pays 6% annually; the other has more risk and pays $7\frac{1}{2}$% annually. How much must be invested in each?

12. Investment A person invested $20,000, part at an interest rate of 6% annually and the remainder at 7% annually. The total interest at the end of 1 year was equivalent to an annual $6\frac{3}{4}$% rate on the entire $20,000. How much was invested at each rate?

*13. Pricing The cost of a product to a retailer is $3.40. If the retailer wishes to make a profit of 20% on the selling price, at what price should the product be sold?

14. Bond Retirement In three years, a company will require $1,125,800 in order to retire some bonds. If the company now invests $1,000,000 for this purpose, what annual rate of interest, compounded annually, must it receive on that amount in order to retire the bonds?

*15. Expansion Program In two years, a company will begin an expansion program. It has decided to invest $3,000,000 now so that in two years the total value of the investment will be $3,245,000, the amount required for the expansion. What is the annual rate of interest, compounded annually, that the company must receive to achieve its purpose?

16. Business A company finds that if it produces and sells *q* units of a product, its total sales revenue in dollars is $100\sqrt{q}$. If the variable cost per unit is $2 and the fixed cost is $1200, find the values of *q* for which

$$\text{total sales revenue} = \text{variable cost} + \text{fixed cost}$$

(That is, profit is zero.)

17. Dormitory Housing A college dormitory houses 210 students. This fall, rooms are available for 76 freshmen. On the average, 95% of those freshmen who request room applications actually reserve a room. How many room applications should the college send out if it wants to receive 76 reservations?

18. Poll A group of people were polled, and 20%, or 700, of them favored a new product over the best-selling brand. How many people were polled?

19. Prison Guard Salary It was reported that in a certain women's jail, female prison guards, called matrons, received 30% (or $200) a month less than their male counterparts, deputy sheriffs. Find the yearly salary of a deputy sheriff. Give your answer to the nearest dollar.

20. Striking Drivers A few years ago, cement drivers were on strike for 46 days. Before the strike, these drivers earned $7.50 per hour and worked 260 eight-hour days a year. What percentage increase is needed in yearly income to make up for the lost time within 1 year?

21. Break Even A manufacturer of video-games sells each copy for $21.95. The manufacturing cost of each copy is $14.92. Monthly fixed costs are $8500. During the first month of sales of a new game, how many copies must be sold in order for the manufacturer to break even (that is, in order that total revenue equal total cost)?

22. Investment Club An investment club bought a bond of an oil corporation for $4000. The bond yields 7% per year. The club now wants to buy shares of stock in a hospital supply company. The stock sells at $15 per share and earns a dividend of $0.60 per share per year. How many shares should the club buy so that its total investment in stocks and bonds yields 6% per year?

23. Vision Care As a fringe benefit for its employees, a company established a vision-care plan. Under this plan, each year the company will pay the first $35 of an employee's vision-care expenses and 80% of all additional vision-care expenses, up to a maximum *total* benefit payment of $100. For an employee, find the total annual vision-care expenses covered by this program.

24. Quality Control Over a period of time, the manufacturer of a caramel-center candy bar found that 3.1% of the bars were rejected for imperfections.

(a) If c candy bars are made in a year, how many would the manufacturer expect to be rejected?

(b) This year, annual consumption of the candy is projected to be six hundred million bars. Approximately how many bars will have to be made if rejections are taken into consideration?

25. Business Suppose that consumers will purchase q units of a product when the price is $(80 - q)/4$ dollars *each*. How many units must be sold in order that sales revenue be $400?

26. Investment How long would it take to triple an investment at simple interest with a rate of 4.5% per year? [*Hint*: See Example 6(a) of Section 0.7, and express 4.5% as 0.045.]

27. Business Alternatives The inventor of a new toy offers the Kiddy Toy Company exclusive rights to manufacture and sell the toy for a lump-sum payment of $25,000. After estimating that future sales possibilities beyond one year are nonexistent, the company management is reviewing an alternative proposal to give a lump-sum payment of $2000 plus a royalty of $0.50 for each unit sold. How many units must be sold the first year to make this alternative as economically attractive to the inventor as the original request? [*Hint*: Determine when the incomes under both proposals are the same.]

28. Parking Lot A company parking lot is 120 ft long and 80 ft wide. Due to an increase in personnel, it is decided to double the area of the lot by adding strips of equal width to one end and one side. Find the width of one such strip.

∗29. Rentals You are the chief financial advisor to a corporation that owns an office complex consisting of 50 units. At $400 per month, every unit can be rented. However, for each $20 per month increase, there will be two vacancies with no possibility of filling them. The corporation wants to receive a total of $20,240 per month from rents in the complex. You are asked to determine the rent that should be charged for each unit. What is your reply?

30. Investment Six months ago, an investment company had a $3,100,000 portfolio consisting of blue-chip and glamour stocks. Since then, the value of the blue-chip investment increased by $\frac{1}{10}$, whereas the value of the glamour stocks decreased by $\frac{1}{10}$. The current value of the portfolio is $3,240,000. What is the *current* value of the blue-chip investment?

31. Revenue The monthly revenue of a certain company is given by $R = 800p - 7p^2$, where p is the price in dollars of the product the company manufactures. At what price will the revenue be $10,000 if the price must be greater than $50?

32. Price–Earnings Ratio The *price–earnings ratio (P/E)* of a company is the ratio of the market value of one share of the company's outstanding common stock to the earnings per share. If *P/E* increases by 10% and the earnings per share increase by 20%, determine the percentage increase in the market value per share of the common stock.

33. Market Equilibrium When the price of a product is p dollars each, suppose that a manufacturer will supply $2p - 10$ units of the product to the market and that consumers will demand to buy $200 - 3p$ units. At the value of p for which supply equals demand, the market is said to be in equilibrium. Find this value of p.

34. Market Equilibrium Repeat Problem 33 for the following conditions: At a price of p dollars each, the supply is $2p^2 - 3p$ and the demand is $20 - p^2$.

35. Security Fence For security reasons, a company will enclose a rectangular area of 11, 200 ft^2 in the rear of its plant. One side will be bounded by the building and the other three sides by fencing. (See Figure 1.4.) If 300 ft of fencing will be used, what will be the dimensions of the rectangular area?

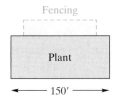

36. **Package Design**   A company is designing a package for its product. One part of the package is to be an open box made from a square piece of aluminum by cutting out a 2-in. square from each corner and folding up the sides. (See Figure 1.5.) The box is to contain 50 in³. What are the dimensions of the square piece of aluminum that must be used?

37. **Product Design**   A candy company makes the popular Dandy Bar. The rectangular-shaped bar is 10 centimeters (cm) long, 5 cm wide, and 2 cm thick. (See Figure 1.6.) Because of increasing costs, the company has decided to cut the volume of the bar by a drastic 28%. The thickness will be the same, but the length and width will be reduced by equal amounts. What will be the length and width of the new bar?

38. **Product Design**   A candy company makes a washer-shaped candy (a candy with a hole in it); see Figure 1.7. Because of increasing costs, the company will cut the volume of candy in each piece by 22%. To do this, the firm will keep the same

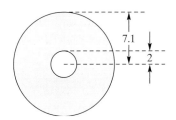

thickness and outer radius but will make the inner radius larger. At present the thickness is 2.1 millimeters (mm), the inner radius is 2 mm, and the outer radius is 7.1 mm. Find the inner radius of the new-style candy. [*Hint*: The volume V of a solid disc is $\pi r^2 h$, where r is the radius and h is the thickness of the disc.]

39. **Compensating Balance**   *Compensating balance* refers to that practice wherein a bank requires a borrower to maintain on deposit a certain portion of a loan during the term of the loan. For example, if a firm takes out a $100,000 loan that requires a compensating balance of 20%, it would have to leave $20,000 on deposit and would have the use of $80,000. To meet the expenses of retooling, the Barber Die Company needs $195,000. The Third National Bank, with whom the firm has had no prior association, requires a compensating balance of 16%. To the nearest thousand dollars, what amount of loan is required to obtain the needed funds? Now solve the general problem of determining the amount L of a loan that is needed to handle expenses E if the bank requires a compensating balance of $p\%$.

40. **Incentive Plan**   A machine company has an incentive plan for its salespeople. For each machine that a salesperson sells, the commission is $40. The commission for *every* machine sold will increase by $0.04 for each machine sold over 600. For example, the commission on each of 602 machines sold is $40.08. How many machines must a salesperson sell in order to earn $30,800?

41. **Real Estate**   A land investment company purchased a parcel of land for $7200. After having sold all but 20 acres at a profit of $30 per acre over the original cost per acre, the company regained the entire cost of the parcel. How many acres were sold?

42. **Margin of Profit**   The *margin of profit* of a company is the net income divided by the total sales. A company's margin of profit increased by 0.02 from last year. Last year the company sold its product at $3.00 each and had a net income of $4500. This year it increased the price of its product by $0.50 each, sold 2000 more, and had a net income of $7140. The company never has had a margin of profit greater than 0.15. How many of its product were sold last year and how many were sold this year?

43. **Business**   A company manufactures products A and B. The cost of producing each unit of A is $2 more than that of B. The costs of production of A and B are $1500 and $1000, respectively, and 25 more units of A are produced than of B. How many of each are produced?

To solve linear inequalities in one variable and to introduce interval notation.

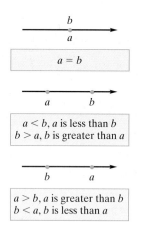

FIGURE 1.8 Relative positions of two points.

Keep in mind that the rules also apply to $\leq$, $>$, and $\geq$.

1.2 Linear Inequalities

Suppose a and b are two points on the real-number line. Then either a and b coincide, or a lies to the left of b, or a lies to the right of b. (See Figure 1.8.)

If a and b coincide, then $a = b$. If a lies to the left of b, we say that a is less than b and write $a < b$, where the *inequality symbol* "$<$" is read "is less than." On the other hand, if a lies to the right of b, we say that a is greater than b, written $a > b$. The statements $a > b$ and $b < a$ are equivalent. (If you have trouble keeping these symbols straight, it may help to notice that $<$ looks somewhat like the letter L for *left* and that we have $a < b$ precisely when a lies to the left of b.)

Another inequality symbol "$\leq$" is read "is less than or equal to" and is defined as follows: $a \leq b$ if and only if $a < b$ or $a = b$. Similarly, the symbol "$\geq$" is defined as follows: $a \geq b$ if and only if $a > b$ or $a = b$. In this case, we say that a is greater than or equal to b.

We often use the words *real numbers* and *points* interchangeably, since there is a one-to-one correspondence between real numbers and points on a line. Thus, we can speak of the points -5, -2, 0, 7, and 9 and can write $7 < 9$, $-2 > -5$, $7 \leq 7$, and $7 \geq 0$. (See Figure 1.9.) Clearly, if $a > 0$, then a is positive; if $a < 0$, then a is negative.

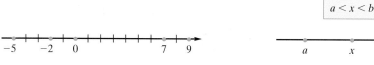

FIGURE 1.9 Points on a number line.　　　**FIGURE 1.10** $a < x$ and $x < b$.

Suppose that $a < b$ and x is between a and b. (See Figure 1.10.) Then not only is $a < x$, but also, $x < b$. We indicate this by writing $a < x < b$. For example, $0 < 7 < 9$. (Refer back to Figure 1.9.)

The next definition is stated in terms of the less-than relation ($<$), but it applies also to the other relations ($>$, $\leq$, $\geq$).

> **DEFINITION**
>
> An *inequality* is a statement that one number is less than another number.

Of course, we represent inequalities by means of inequality symbols. If two inequalities have their inequality symbols pointing in the same direction, then the inequalities are said to have the *same sense*. If not, they are said to be *opposite in sense*. Hence, $a < b$ and $c < d$ have the same sense, but $a < b$ has the opposite sense of $c > d$.

Solving an inequality, such as $2(x-3) < 4$, means finding all values of the variable for which the inequality is true. This involves the application of certain rules, which we now state.

Rules for Inequalities

1. If the same number is added to or subtracted from both sides of an inequality then another inequality results, having the same sense as the original inequality. Symbolically,

$$\text{If } a < b, \text{ then } a + c < b + c \text{ and } a - c < b - c.$$

For example, $7 < 10$ so $7 + 3 < 10 + 3$.

2. If both sides of an inequality are multiplied or divided by the same **positive** number then another inequality results, having the same sense as the original

inequality. Symbolically,

$$\text{If } a < b \text{ and } c > 0, \text{ then } ac < bc \text{ and } \frac{a}{c} < \frac{b}{c}.$$

For example, $3 < 7$ and $2 > 0$ so $3(2) < 7(2)$ and $\frac{3}{2} < \frac{7}{2}$.

3. If both sides of an inequality are multiplied or divided by the same **negative** number then another inequality results, having the opposite sense of the original inequality. Symbolically,

$$\text{If } a < b \text{ and } c < 0, \text{ then } a(c) > b(c) \text{ and } \frac{a}{c} > \frac{b}{c}.$$

For example, $4 < 7$ and $-2 < 0$ so $4(-2) > 7(-2)$ and $\frac{4}{-2} > \frac{7}{-2}$.

4. Any side of an inequality can be replaced by an expression equal to it. Symbolically,

$$\text{If } a < b \text{ and } a = c, \text{ then } c < b.$$

For example, if $x < 2$ and $x = y + 4$, then $y + 4 < 2$.

5. If the sides of an inequality are either both positive or both negative and reciprocals are taken on both sides, then another inequality results, having the opposite sense of the original inequality. Symbolically,

$$\text{If } 0 < a < b \text{ or } a < b < 0, \text{ then } \frac{1}{a} > \frac{1}{b}.$$

For example, $2 < 4$ so $\frac{1}{2} > \frac{1}{4}$ and $-4 < -2$ so $\frac{1}{-4} > \frac{1}{-2}$.

6. If both sides of an inequality are positive and each side is raised to the same positive power, then another inequality results, having the same sense as the original inequality. Symbolically,

$$\text{If } 0 < a < b \text{ and } n > 0, \text{ then } a^n < b^n.$$

For n a positive integer, this rule further provides

$$\text{If } 0 < a < b, \text{ then } \sqrt[n]{a} < \sqrt[n]{b}$$

For example, $4 < 9$ so $4^2 < 9^2$ and $\sqrt{4} < \sqrt{9}$.

A pair of inequalities will be said to be *equivalent inequalities* if when either is true then the other is true. When any of Rules 1–6 are applied to an inequality, it is easy to show that the result is an equivalent inequality. We will now apply Rules 1–4 to a *linear inequality*.

DEFINITION

A *linear inequality* in the variable x is an inequality that can be written in the form

$$ax + b < 0$$

where a and b are constants and $a \neq 0$.

We should expect that the inequality will be true for some values of x and false for others. To **solve** an inequality involving a variable is to find all values of the variable for which the inequality is true.

CAUTION

Multiplying or dividing an inequality by a negative number gives an inequality of the opposite sense.

The definition also applies to $\leq$, $>$, and $\geq$.

SOLVING A LINEAR INEQUALITY

A salesman has a monthly income given by $I = 200 + 0.8S$, where S is the number of products sold in a month. How many products must he sell to make at least $4500 a month?

$x < 5$

FIGURE 1.11   All real numbers less than 5.

SOLVING A LINEAR INEQUALITY

A zoo veterinarian can purchase four different animal foods with various nutrient values for the zoo's grazing animals. Let x_1 represent the number of bags of food 1, x_2 represent the number of bags of food 2, and so on. The number of bags of each food needed can be described by the following equations:

$$x_1 = 150 - x_4$$
$$x_2 = 3x_4 - 210$$
$$x_3 = x_4 + 60$$

Assuming that each variable must be nonnegative, write four inequalities involving x_4 that follow from these equations.

 CAUTION

Dividing both sides by -2 reverses the sense of the inequality.

● EXAMPLE 1   **Solving a Linear Inequality**

Solve $2(x - 3) < 4$.

Solution:

Strategy   We shall replace the given inequality by equivalent inequalities until the solution is evident.

$$2(x - 3) < 4$$
$$2x - 6 < 4 \qquad \text{(Rule 4)}$$
$$2x - 6 + 6 < 4 + 6 \qquad \text{(Rule 1)}$$
$$2x < 10 \qquad \text{(Rule 4)}$$
$$\frac{2x}{2} < \frac{10}{2} \qquad \text{(Rule 2)}$$
$$x < 5 \qquad \text{(Rule 4)}$$

All of the foregoing inequalities are equivalent. Thus, the original inequality is true for *all* real numbers x such that $x < 5$. For example, the inequality is true for $x = -10, -0.1, 0, \frac{1}{2}$, and 4.9. We may write our solution simply as $x < 5$ and can geometrically represent it by the colored half-line in Figure 1.11. The parenthesis indicates that 5 is *not included* in the solution.

NOW WORK PROBLEM 9 ●○○

In Example 1 the solution consisted of a set of numbers, namely, all numbers less than 5. It is common to use the term **interval** to describe such a set. In the case of Example 1, the set of all x such that $x < 5$ can be denoted by the *interval notation* $(-\infty, 5)$. The symbol $-\infty$ is not a number, but is merely a convenience for indicating that the interval includes all numbers less than 5.

There are other types of intervals. For example, the set of all numbers x for which $a \leq x \leq b$ is called a **closed interval** and includes the numbers a and b, which are called *endpoints* of the interval. This interval is denoted by $[a, b]$ and is shown in Figure 1.12(a). The square brackets indicated that a and b *are included* in the interval. On the other hand, the set of all x for which $a < x < b$ is called an **open interval** and is denoted by (a, b). The endpoints are *not* part of this set. [See Figure 1.12(b).] Extending these concepts, we have the intervals shown in Figure 1.13.

● EXAMPLE 2   **Solving a Linear Inequality**

Solve $3 - 2x \leq 6$.

Solution:

$$3 - 2x \leq 6$$
$$-2x \leq 3 \qquad \text{(Rule 1)}$$
$$x \geq -\frac{3}{2} \qquad \text{(Rule 3)}$$

The solution is $x \geq -\frac{3}{2}$, or, in interval notation, $[-\frac{3}{2}, \infty)$. This is represented geometrically in Figure 1.14.

NOW WORK PROBLEM 7 ●○○

Closed interval $[a, b]$ Open interval (a, b)

(a)   (b)

FIGURE 1.12   Closed and open intervals.

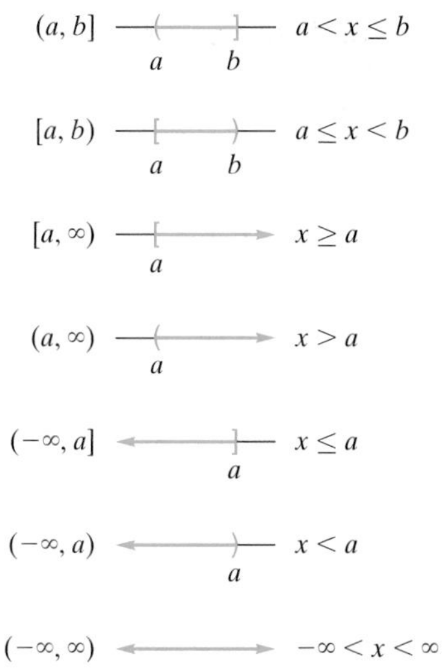

$(a, b]$ ⊶———⊷ $a < x \le b$
             a b

$[a, b)$ ⊷———⊶ $a \le x < b$
             a b

$[a, \infty)$ ⊷———→ $x \ge a$
             a

(a, ∞) ⊶———→ $x > a$
             a

$(-\infty, a]$ ←———⊷ $x \le a$
                a

$(-\infty, a)$ ←———⊶ $x < a$
                a

$(-\infty, \infty)$ ←———→ $-\infty < x < \infty$

FIGURE 1.13   Intervals.

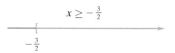

FIGURE 1.14   The interval $[-\frac{3}{2}, \infty)$.

● **EXAMPLE 3**   **Solving a Linear Inequality**

Solve $\frac{3}{2}(s - 2) + 1 > -2(s - 4)$.

Solution:

$$\frac{3}{2}(s - 2) + 1 > -2(s - 4)$$

$$2\left[\frac{3}{2}(s - 2) + 1\right] > 2[-2(s - 4)] \qquad \text{(Rule 2)}$$

$$3(s - 2) + 2 > -4(s - 4)$$

$$3s - 4 > -4s + 16$$

$$7s > 20 \qquad \text{(Rule 1)}$$

$$s > \frac{20}{7} \qquad \text{(Rule 2)}$$

FIGURE 1.15   The interval $(\frac{20}{7}, \infty)$.

The solution is $(\frac{20}{7}, \infty)$; see Figure 1.15.

NOW WORK PROBLEM 19 ●●●

● **EXAMPLE 4**   **Solving Linear Inequalities**

a. *Solve* $2(x - 4) - 3 > 2x - 1$.

   Solution:

$$2(x - 4) - 3 > 2x - 1$$

$$2x - 8 - 3 > 2x - 1$$

$$-11 > -1$$

   Since it is never true that $-11 > -1$, there is no solution, and the solution set is ∅.

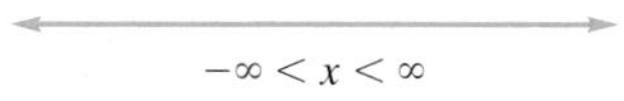

b. *Solve* $2(x - 4) - 3 < 2x - 1$.

   Solution: Proceeding as in part (a), we obtain $-11 < -1$. This is true for all real numbers x, so the solution is $(-\infty, \infty)$; see Figure 1.16.

      $-\infty < x < \infty$

FIGURE 1.16   The interval $(-\infty, \infty)$.

NOW WORK PROBLEM 15 ●●●

Problems 1.2

In Problems 1–34, solve the inequalities. Give your answer in interval notation, and indicate the answer geometrically on the real-number line.

1. $3x > 12$

2. $4x < -2$

3. $5x - 11 \leq 9$

4. $5x \leq 0$

5. $-4x \geq 2$

6. $2y + 1 > 0$

*7.** $5 - 7s > 3$

8. $4s - 1 < -5$

*9.** $3 < 2y + 3$

10. $4 \leq 3 - 2y$

11. $x + 5 \leq 3 + 2x$

12. $-3 \geq 8(2 - x)$

13. $3(2 - 3x) > 4(1 - 4x)$

14. $8(x + 1) + 1 < 3(2x) + 1$

*15.** $2(4x - 2) > 4(2x + 1)$

16. $4 - (x + 3) \leq 3(3 - x)$

17. $x + 2 < \sqrt{3} - x$

18. $\sqrt{2}(x + 2) > \sqrt{8}(3 - x)$

*19.** $\dfrac{5}{6}x < 40$

20. $-\dfrac{2}{3}x > 6$

21. $\dfrac{9y + 1}{4} \leq 2y - 1$

22. $\dfrac{3y - 2}{3} \geq \dfrac{1}{4}$

23. $-3x + 1 \leq -3(x - 2) + 1$

24. $0x \leq 0$

25. $\dfrac{1 - t}{2} < \dfrac{3t - 7}{3}$

26. $\dfrac{3(2t - 2)}{2} > \dfrac{6t - 3}{5} + \dfrac{t}{10}$

27. $2x + 13 \geq \dfrac{1}{3}x - 7$

28. $3x - \dfrac{1}{3} \leq \dfrac{5}{2}x$

29. $\dfrac{2}{3}r < \dfrac{5}{6}r$

30. $\dfrac{7}{4}t > -\dfrac{8}{3}t$

31. $\dfrac{y}{2} + \dfrac{y}{3} > y + \dfrac{y}{5}$

32. $9 - 0.1x \leq \dfrac{2 - 0.01x}{0.2}$

33. $0.1(0.03x + 4) \geq 0.02x + 0.434$

34. $\dfrac{3y - 1}{-3} < \dfrac{5(y + 1)}{-3}$

35. Savings Each month last year, Brittany saved more than $50, but less than $150. If S represents her total savings for the year, describe S by using inequalities.

36. Labor Using inequalities, symbolize the following statement: The number of labor hours x to produce a product is not less than $2\frac{1}{2}$ nor more than 4.

37. Geometry In a right triangle, one of the acute angles x is less than 3 times the other acute angle plus 10 degrees. Solve for x.

38. Spending A student has $360 to spend on a stereo system and some compact discs. If she buys a stereo that costs $219 and the discs cost $18.95 each, find the greatest number of discs she can buy.

OBJECTIVE

To model real-life situations in terms of inequalities.

1.3 Applications of Inequalities

Solving word problems may sometimes involve inequalities, as the following examples illustrate.

● EXAMPLE 1 Profit

For a company that manufactures aquarium heaters, the combined cost for labor and material is $21 per heater. Fixed costs (costs incurred in a given period, regardless of output) are $70,000. If the selling price of a heater is $35, how many must be sold for the company to earn a profit?

Solution:

Strategy Recall that

$$\text{profit} = \text{total revenue} - \text{total cost}$$

We will find total revenue and total cost and then determine when their difference is positive.

Let q be the number of heaters that must be sold. Then their cost is $21q$. The total cost for the company is therefore $21q + 70,000$. The total revenue from the sale of q heaters will be $35q$. Now,

$$\text{profit} = \text{total revenue} - \text{total cost}$$

and we want profit > 0. Thus,

$$\text{total revenue} - \text{total cost} > 0$$

$$35q - (21q + 70,000) > 0$$

$$14q > 70,000$$

$$q > 5000$$

Since the number of heaters must be a non-negative integer, we see that at least 5001 heaters must be sold for the company to earn a profit.

NOW WORK PROBLEM 1

EXAMPLE 2   Renting versus Purchasing

A builder must decide whether to rent or buy an excavating machine. If he were to rent the machine, the rental fee would be $3000 per month (on a yearly basis), and the daily cost (gas, oil, and driver) would be $180 for each day the machine is used. If he were to buy it, his fixed annual cost would be $20,000, and daily operating and maintenance costs would be $230 for each day the machine is used. What is the least number of days each year that the builder would have to use the machine to justify renting it rather than buying it?

Solution:

Strategy   We will determine expressions for the annual cost of renting and the annual cost of purchasing. We then find when the cost of renting is less than that of purchasing.

Let d be the number of days each year that the machine is used. If the machine is rented, the total yearly cost consists of rental fees, which are $(12)(3000)$, and daily charges of $180d$. If the machine is purchased, the cost per year is $20,000 + 230d$. We want

$$\text{cost}_{\text{rent}} < \text{cost}_{\text{purchase}}$$

$$12(3000) + 180d < 20,000 + 230d$$

$$36,000 + 180d < 20,000 + 230d$$

$$16,000 < 50d$$

$$320 < d$$

Thus, the builder must use the machine at least 321 days to justify renting it.

NOW WORK PROBLEM 3

EXAMPLE 3   Current Ratio

The *current ratio* of a business is the ratio of its current assets (such as cash, merchandise inventory, and accounts receivable) to its current liabilities (such as short-term loans and taxes payable).

After consulting with the comptroller, the president of the Ace Sports Equipment Company decides to take out a short-term loan to build up inventory. The company has current assets of $350,000 and current liabilities of $80,000. How much can the company borrow if the current ratio is to be no less than 2.5? (Note: The funds received are considered as current assets and the loan as a current liability.)

Solution: Let x denote the amount the company can borrow. Then current assets will be $350,000 + x$, and current liabilities will be $80,000 + x$. Thus,

$$\text{current ratio} = \frac{\text{current assets}}{\text{current liabilities}} = \frac{350,000 + x}{80,000 + x}$$

We want

$$\frac{350,000 + x}{80,000 + x} \geq 2.5$$

Although the inequality that must be solved is not linear, it leads to a linear inequality.

Since x is positive, so is $80,000 + x$. Hence, we can multiply both sides of the inequality by $80,000 + x$ and the sense of the inequality will remain the same. We have

$$350,000 + x \geq 2.5(80,000 + x)$$

$$150,000 \geq 1.5x$$

$$100,000 \geq x$$

Consequently, the company may borrow as much as $100,000 and still maintain a current ratio greater than or equal to 2.5.

● EXAMPLE 4   Publishing

A publishing company finds that the cost of publishing each copy of a certain magazine is $1.50. The revenue from dealers is $1.40 per copy. The advertising revenue is 10% of the revenue received from dealers for all copies sold beyond 10,000. What is the least number of copies that must be sold so as to have a profit for the company?

Solution:

Strategy   We have

$$\text{profit} = \text{total revenue} - \text{total cost}$$

so we find an expression for profit and then set it greater than 0.

Let q be the number of copies that are sold. The revenue from dealers is $1.40q$, and the revenue from advertising is $(0.10)[(1.40)(q - 10,000)]$. The total cost of publication is $1.50q$. Thus,

$$\text{total revenue} - \text{total cost} > 0$$

$$1.40q + (0.10)[(1.40)(q - 10,000)] - 1.50q > 0$$

$$1.4q + 0.14q - 1400 - 1.5q > 0$$

$$0.04q - 1400 > 0$$

$$0.04q > 1400$$

$$q > 35,000$$

Therefore, the total number of copies sold must be greater than 35,000. That is, at least 35,001 copies must be sold to guarantee a profit.

NOW WORK PROBLEM 5

Problems 1.3

*1. **Profit**   The Davis Company manufactures a product that has a unit selling price of $20 and a unit cost of $15. If fixed costs are $600,000, determine the least number of units that must be sold for the company to have a profit.

2. **Profit**   To produce 1 unit of a new product, a company determines that the cost for material is $2.50 and the cost of labor is $4. The fixed cost, regardless of sales volume, is $5000. If the cost to a wholesaler is $7.40 per unit, determine the least number of units that must be sold by the company to realize a profit.

*3. **Leasing versus Purchasing**   A businesswoman wants to determine the difference between the costs of owning and leasing an automobile. She can lease a car for $420 per month (on an annual basis). Under this plan, the cost per mile (gas and oil) is $0.06. If she were to purchase the car, the fixed annual expense would be $4700, and other costs would amount to $0.08 per mile. What is the least number of miles she would have to drive per year to make leasing no more expensive than purchasing?

4. **Shirt Manufacturer**   A T-shirt manufacturer produces N shirts at a total labor cost (in dollars) of $1.3N$ and a total

material cost of 0.4*N*. The fixed cost for the plant is $6500. If each shirt sells for $3.50, how many must be sold by the company to realize a profit?

*5. **Publishing**   The cost of publication of each copy of a magazine is $0.55. It is sold to dealers for $0.60 per copy. The amount received for advertising is 10% of the amount received for all magazines sold beyond 30,000. Find the least number of magazines that can be published without loss—that is, such that profit ≥ 0—assuming that 90% of the issues published will be sold.

6. **Production Allocation**   A company produces alarm clocks. During the regular workweek, the labor cost for producing one clock is $2.00. However, if a clock is produced on overtime, the labor cost is $3.00. Management has decided to spend no more than a total of $25,000 per week for labor. The company must produce 11,000 clocks this week. What is the minimum number of clocks that must be produced during the regular workweek?

7. **Investment**   A company invests a total of $30,000 of surplus funds at two annual rates of interest: 5% and $6\frac{3}{4}$%. It wishes an annual yield of no less than $6\frac{1}{2}$%. What is the least amount of money that the company must invest at the $6\frac{3}{4}$% rate?

8. **Current Ratio**   The current ratio of Precision Machine Products is 3.8. If the firm's current assets are $570,000, what are its current liabilities? To raise additional funds, what is the maximum amount the company can borrow on a short-term basis if the current ratio is to be no less than 2.6? (See Example 3 for an explanation of current ratio.)

9. **Sales Allocation**   At present, a manufacturer has 2500 units of product in stock. The product is now selling at $4 per unit. Next month the unit price will increase by $0.50. The manufacturer wants the total revenue received from the sale of the 2500 units to be no less than $10,750. What is the maximum number of units that can be sold this month?

10. **Revenue**   Suppose consumers will purchase *q* units of a product at a price of $\frac{100}{q} + 1$ dollars per unit. What is the minimum number of units that must be sold in order that sales revenue be greater than $5000?

11. **Hourly Rate**   Painters are often paid either by the hour or on a per-job basis. The rate they receive can affect their working speed. For example, suppose they can work either for $9.00 per hour or for $320 plus $3 for each hour less than 40 if they complete the job in less than 40 hours. Suppose the job will take *t* hours. If *t* ≥ 40, clearly the hourly rate is better. If *t* < 40, for what values of *t* is the hourly rate the better pay scale?

12. **Compensation**   Suppose a company offers you a sales position with your choice of two methods of determining your yearly salary. One method pays $35,000 plus a bonus of 3% of your yearly sales. The other method pays a straight 5% commission on your sales. For what yearly sales amount is it better to choose the first method?

13. **Acid Test Ratio**   The *acid test ratio* (or *quick ratio*) of a business is the ratio of its liquid assets—cash and securities plus accounts receivable—to its current liabilities. The minimum acid test ratio for a financially healthy company is around 1.0, but the standard varies somewhat from industry to industry. If a company has $450,000 in cash and securities and has $398,000 in current liabilities, how much does it need to be carrying as accounts receivable in order to keep its acid test ratio at or above 1.3?

1.4 Absolute Value

To solve equations and inequalities involving absolute values.

The absolute value of a real number is the number obtained when its sign is ignored.

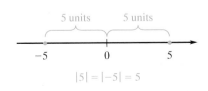

FIGURE 1.17   Absolute value.

Absolute-Value Equations

On the real-number line, the distance of a number *x* from 0 is called the **absolute value** of *x* and is denoted by $|x|$. For example, $|5| = 5$ and $|-5| = 5$ because both 5 and −5 are 5 units from 0. (See Figure 1.17.) Similarly, $|0| = 0$. Notice that $|x|$ can never be negative; that is, $|x| \geq 0$.

If *x* is positive or zero, then $|x|$ is simply *x* itself, so we can omit the vertical bars and write $|x| = x$. On the other hand, consider the absolute value of a negative number, like $x = -5$.

$$|x| = |-5| = 5 = -(-5) = -x$$

Thus, if *x* is negative, then $|x|$ is the positive number −*x*. The minus sign indicates that we have changed the sign of *x*. The geometric definition of absolute value as a distance is equivalent to the following:

DEFINITION

The *absolute value* of a real number x, written $|x|$, is defined by

$$|x| = \begin{cases} x, & \text{if } x \geq 0 \\ -x, & \text{if } x < 0 \end{cases}$$

Observe that $|-x| = |x|$ follows from the definition.

Applying the definition, we have $|3| = 3, |-8| = -(-8) = 8$, and $|\frac{1}{2}| = \frac{1}{2}$. Also, $-|2| = -2$ and $-|-2| = -2$.

Also, $|-x|$ is not necessarily x and, thus, $|-x - 1|$ is not necessarily $x + 1$. For example, if we let $x = -3$, then $|-(-3)| \neq -3$, and

$$|-(-3) - 1| \neq -3 + 1$$

● **EXAMPLE 1 Solving Absolute-Value Equations**

a. *Solve* $|x - 3| = 2$.

Solution: This equation states that $x - 3$ is a number 2 units from 0. Thus, either

$$x - 3 = 2 \quad \text{or} \quad x - 3 = -2$$

Solving these equations gives $x = 5$ or $x = 1$.

b. *Solve* $|7 - 3x| = 5$.

Solution: The equation is true if $7 - 3x = 5$ or if $7 - 3x = -5$. Solving these equations gives $x = \frac{2}{3}$ or $x = 4$.

c. *Solve* $|x - 4| = -3$.

Solution: The absolute value of a number is never negative, so the solution set is $\emptyset$.

NOW WORK PROBLEM 19

We can interpret $|a - b| = |-(b - a)| = |b - a|$ as the distance between a and b. For example, the distance between 5 and 9 can be calculated via

$$\begin{aligned} \text{either} & \quad |9 - 5| = |4| = 4 \\ \text{or} & \quad |5 - 9| = |-4| = 4 \end{aligned}$$

Similarly, the equation $|x - 3| = 2$ states that the distance between x and 3 is 2 units. Thus, x can be 1 or 5, as shown in Example 1(a) and Figure 1.18.

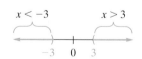

FIGURE 1.18 The solution of $|x - 3| = 2$ is 1 or 5.

Absolute-Value Inequalities

Let us turn now to inequalities involving absolute values. If $|x| < 3$, then x is less than 3 units from 0. Hence, x must lie between -3 and 3, that is, on the interval $-3 < x < 3$. [See Figure 1.19(a).] On the other hand, if $|x| > 3$, then x must be greater than 3 units from 0. Hence, there are two intervals in the solution: Either $x < -3$ or $x > 3$. [See Figure 1.19(b).] We can extend these ideas as follows: If $|x| \leq 3$, then $-3 \leq x \leq 3$; if $|x| \geq 3$, then $x \leq -3$ or $x \geq 3$. Table 1.1 gives a summary of the solutions to absolute-value inequalities.

(a) Solution of $|x| < 3$

$-3 < x < 3$

(b) Solution of $|x| > 3$

$x < -3 \qquad x > 3$

FIGURE 1.19 Solutions of $|x| < 3$ and $|x| > 3$.

TABLE 1.1

Inequality ($d > 0$)	Solution
$|x| < d$	$-d < x < d$
$|x| \leq d$	$-d \leq x \leq d$
$|x| > d$	$x < -d$ or $x > d$
$|x| \geq d$	$x \leq -d$ or $x \geq d$

● EXAMPLE 2   Solving Absolute-Value Inequalities

a. *Solve* $|x - 2| < 4$.

Solution: The number $x - 2$ must be less than 4 units from 0. From the preceding discussion, this means that $-4 < x - 2 < 4$. We may set up the procedure for solving this inequality as follows:

$$-4 < x - 2 < 4$$
$$-4 + 2 < x < 4 + 2 \qquad \text{(adding 2 to each member)}$$
$$-2 < x < 6$$

Thus, the solution is the open interval $(-2, 6)$. This means that all numbers between -2 and 6 satisfy the original inequality. (See Figure 1.20.)

b. *Solve* $|3 - 2x| \leq 5$.

Solution:

$$-5 \leq 3 - 2x \leq 5$$
$$-5 - 3 \leq -2x \leq 5 - 3 \qquad \text{(subtracting 3 throughout)}$$
$$-8 \leq -2x \leq 2$$
$$4 \geq x \geq -1 \qquad \text{(dividing throughout by } -2\text{)}$$
$$-1 \leq x \leq 4 \qquad \text{(rewriting)}$$

Note that the sense of the original inequality was *reversed* when we divided by a negative number. The solution is the closed interval $[-1, 4]$.

NOW WORK PROBLEM 29   ◖◗●

● EXAMPLE 3   Solving Absolute-Value Inequalities

a. *Solve* $|x + 5| \geq 7$.

Solution: Here $x + 5$ must be *at least* 7 units from 0. Thus, either $x + 5 \leq -7$ *or* $x + 5 \geq 7$. This means that either $x \leq -12$ *or* $x \geq 2$. Thus, the solution consists of two intervals: $(-\infty, -12]$ and $[2, \infty)$. We can abbreviate this collection of numbers by writing

$$(-\infty, -12] \cup [2, \infty)$$

where the connecting symbol $\cup$ is called the *union* symbol. (See Figure 1.21.) More formally, the **union** of sets A and B is the set consisting of all elements that are in either A or B (or in both A and B).

b. *Solve* $|3x - 4| > 1$.

Solution: Either $3x - 4 < -1$ *or* $3x - 4 > 1$. Thus, either $3x < 3$ *or* $3x > 5$. Therefore, $x < 1$ *or* $x > \frac{5}{3}$, so the solution consists of all numbers in the set $(-\infty, 1) \cup (\frac{5}{3}, \infty)$.

NOW WORK PROBLEM 31   ◖◗●

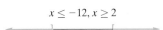

$-2 < x < 6$

FIGURE 1.20 The solution of $|x - 2| < 4$ is the interval $(-2, 6)$.

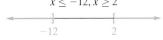

$x \leq -12, x \geq 2$

FIGURE 1.21 The union $(-\infty, -12] \cup [2, \infty)$.

⚠ CAUTION

The inequalities $x \leq -12$ or $x \geq 2$ in (a) and $x < 1$ or $x > \frac{5}{3}$ in (b) do not give rise to a single interval as in Examples 1 and 2.

PRINCIPLES IN PRACTICE 1

ABSOLUTE-VALUE NOTATION

Express the following statement using absolute-value notation: The actual weight w of a box of cereal must be within 0.3 oz of the weight stated on the box, which is 22 oz.

● EXAMPLE 4   Absolute-Value Notation

Using absolute-value notation, express the following statements:

a. *x is less than 3 units from 5.*

Solution:

$$|x - 5| < 3$$

b. *x differs from 6 by at least 7.*

Solution:

$$|x - 6| \geq 7$$

c. $x < 3$ and $x > -3$ simultaneously.

Solution:

$$|x| < 3$$

d. x is more than 1 unit from -2.

Solution:

$$|x - (-2)| > 1$$
$$|x + 2| > 1$$

e. x is less than σ (a Greek letter read "sigma") units from μ (a Greek letter read "mu").

Solution:

$$|x - \mu| < \sigma$$

NOW WORK PROBLEM 11

Properties of the Absolute Value

Five basic properties of the absolute value are as follows:

1. $|ab| = |a| \cdot |b|$
2. $\left|\dfrac{a}{b}\right| = \dfrac{|a|}{|b|}$
3. $|a - b| = |b - a|$
4. $-|a| \le a \le |a|$
5. $|a + b| \le |a| + |b|$

For example, Property 1 states that the absolute value of the product of two numbers is equal to the product of the absolute values of the numbers. Property 5 is known as *the triangle inequality*.

● EXAMPLE 5   Properties of Absolute Value

a. $|(-7) \cdot 3| = |-7| \cdot |3| = 21$

b. $|4 - 2| = |2 - 4| = 2$

c. $|7 - x| = |x - 7|$

d. $\left|\dfrac{-7}{3}\right| = \dfrac{|-7|}{|3|} = \dfrac{7}{3}; \left|\dfrac{-7}{-3}\right| = \dfrac{|-7|}{|-3|} = \dfrac{7}{3}$

e. $\left|\dfrac{x-3}{-5}\right| = \dfrac{|x-3|}{|-5|} = \dfrac{|x-3|}{5}$

f. $-|2| \le 2 \le |2|$

g. $|(-2) + 3| = |1| = 1 \le 5 = 2 + 3 = |-2| + |3|$

NOW WORK PROBLEM 5

Problems 1.4

In Problems 1–10, evaluate the absolute value expression.

1. $|-13|$

2. $|2^{-1}|$

3. $|8 - 2|$

4. $|(-4 - 6)/2|$

*5. $|2(-\frac{7}{2})|$

6. $|3 - 5| - |5 - 3|$

7. $|x| < 4$

8. $|x| < 10$

9. $|2 - \sqrt{5}|$

10. $|\sqrt{5} - 2|$

*11. Using the absolute-value symbol, express each fact.

(a) x is less than 3 units from 7.

(b) x differs from 2 by less than 3.

(c) x is no more than 5 units from 7.

(d) The distance between 7 and x is 4.

(e) $x + 4$ is less than 2 units from 0.

(f) x is between -3 and 3, but is not equal to 3 or -3.

(g) $x < -6$ or $x > 6$.

(h) The number x of hours that a machine will operate efficiently differs from 105 by less than 3.

(i) The average monthly income x (in dollars) of a family differs from 850 by less than 100.

12. Use absolute-value notation to indicate that $f(x)$ and L differ by less than ϵ.

13. Use absolute-value notation to indicate that the prices p_1 and p_2 of two products may differ by no more than 9 (dollars).

14. Find all values of x such that $|x - \mu| \le 2\sigma$.

In Problems 15–36, solve the given equation or inequality.

15. $|x| = 7$

16. $|-x| = 2$

17. $\left|\dfrac{x}{5}\right| = 7$

18. $\left|\dfrac{5}{x}\right| = 12$

*19. $|x - 5| = 8$

20. $|4 + 3x| = 6$

21. $|5x - 2| = 0$

22. $|7x + 3| = x$

23. $|7 - 4x| = 5$

24. $|5 - 3x| = 2$

25. $|x| < M$, for $M > 0$

26. $|-x| < 3$

27. $\left|\dfrac{x}{4}\right| > 2$

28. $\left|\dfrac{x}{3}\right| > \dfrac{1}{2}$

*29. $|x + 9| < 5$

30. $|2x - 17| < -4$

*31. $\left|x - \dfrac{1}{2}\right| > \dfrac{1}{2}$

32. $|1 - 3x| > 2$

33. $|5 - 8x| \le 1$

34. $|4x - 1| \ge 0$

35. $\left|\dfrac{3x - 8}{2}\right| \ge 4$

36. $\left|\dfrac{x - 7}{3}\right| \le 5$

In Problems 37–38, express the statement using absolute-value notation.

37. In a science experiment, the measurement of a distance d is 35.2 m, and is accurate to ± 20 cm.

38. The difference in temperature between two chemicals that are to be mixed must be no less than 5 degrees and no more than 10 degrees.

39. **Statistics** In statistical analysis, the Chebyshev inequality asserts that if x is a random variable, μ is its mean, and σ is its standard deviation, then

$$(\text{probability that } |x - \mu| > h\sigma) \ge \frac{1}{h^2}$$

Find those values of x such that $|x - \mu| > h\sigma$.

40. **Manufacturing Tolerance** In the manufacture of widgets, the average dimension of a part is 0.01 cm. Using the absolute-value symbol, express the fact that an individual measurement x of a part does not differ from the average by more than 0.005 cm.

OBJECTIVE

To write sums in summation notation and evaluate such sums.

1.5 Summation Notation

There was a time when school teachers made their students add up all the positive integers from 1 to 105 (say), perhaps as punishment for unruly behaviour while the teacher was out of the classroom. In other words, the students were to find

$$1 + 2 + 3 + 4 + 5 + 6 + 7 + \cdots + 104 + 105 \tag{1}$$

A related exercise was to find

$$1 + 4 + 9 + 16 + \cdots + 81 + 100 + 121 \tag{2}$$

The three dots notation is supposed to convey the idea of continuing the task, using the same pattern, until the last explicitly given terms have been added too. With this notation there are no hard and fast rules about how many terms at the beginning and end are to be given explicitly. The custom is to provide as many as are needed to ensure that the intended reader will find the expression unambiguous. This is too imprecise for many mathematical applications.

Suppose that for any positive integer i we define $a_i = i^2$. Then, for example, $a_6 = 36$ and $a_8 = 64$. The instruction, "Add together the numbers a_i, for i taking on the integer values 1 through 11 inclusive" is a precise statement of Equation (2). It would be precise regardless of the formula defining the values a_i, and this leads to the following:

> **DEFINITION**
>
> If, for each positive integer i, there is given a unique number a_i, and m and n are positive integers with $m \le n$, then **the sum of the numbers a_i, with i successively taking on the values m through n is denoted**
>
> $$\sum_{i=m}^{n} a_i$$

Thus

$$\sum_{i=m}^{n} a_i = a_m + a_{m+1} + a_{m+2} + \cdots + a_n \tag{3}$$

The $\sum$ is the Greek capital letter sigma, from which we get the letter S. It stands for "sum," and the expression $\sum_{i=m}^{n} a_i$, can be read as the the sum of all numbers a_i, where i ranges from m to n (through positive integers being understood). The description of a_i may be very simple. For example, in Equation (1) we have $a_i = i$ and

$$\sum_{i=1}^{105} i = 1 + 2 + 3 + \cdots + 105 \tag{4}$$

while Equation (2) is

$$\sum_{i=1}^{11} i^2 = 1 + 4 + 9 + \cdots + 121 \tag{5}$$

We have merely defined a notation, which is called **summation notation.** In Equation (3), i is the *index of summation* and m and n are called the *bounds of summation*. It is important to understand from the outset that the name of the index of summation can be replaced by any other so that we have

$$\sum_{i=m}^{n} a_i = \sum_{j=m}^{n} a_j = \sum_{\alpha=m}^{n} a_\alpha = \sum_{N=m}^{n} a_N$$

for example. In each case, replacing the index of summation by the positive integers m through n successively and adding results in

$$a_m + a_{m+1} + a_{m+2} + \cdots + a_n$$

We now illustrate with some concrete examples.

● EXAMPLE 1   **Evaluating Sums**

Evaluate the given sums.

a. $\displaystyle\sum_{n=3}^{7}(5n - 2)$

Solution:

$$\sum_{n=3}^{7}(5n - 2) = [5(3) - 2] + [5(4) - 2] + [5(5) - 2] + [5(6) - 2] + [5(7) - 2]$$

$$= 13 + 18 + 23 + 28 + 33$$

$$= 115$$

b. $\displaystyle\sum_{j=1}^{6}(j^2 + 1)$

Solution:

$$\sum_{j=1}^{6}(j^2 + 1) = (1^2 + 1) + (2^2 + 1) + (3^2 + 1) + (4^2 + 1) + (5^2 + 1) + (6^2 + 1)$$

$$= 2 + 5 + 10 + 17 + 26 + 37$$

$$= 97$$

NOW WORK PROBLEM 5 ●●●

● EXAMPLE 2 Writing a Sum Using Summation Notation

Write the sum $14 + 16 + 18 + 20 + 22 + \cdots + 100$ *in summation notation.*

Solution: There are many ways to express this sum in sigma notation. One method is to notice that the values being added are $2n$, for $n = 7$ to 50. The sum can thus be written as

$$\sum_{n=7}^{50} 2n$$

Another method is to notice that the values being added are $2k + 12$, for $k = 1$ to 44. The sum can thus be represented as

$$\sum_{k=1}^{44} (2k + 12)$$

NOW WORK PROBLEM 9 ●●

Since summation notation is used to express the addition of terms, we can use the properties of addition when performing operations on sums written in summation notation. By applying these properties, we can create a list of properties, and formulas for summation notation.

By the distributive property of addition,

$$ca_1 + ca_2 + \cdots + ca_n = c(a_1 + a_2 + \cdots + a_n)$$

So, in summation notation,

$$\sum_{i=m}^{n} ca_i = c \sum_{i=m}^{n} a_i \tag{6}$$

Note that c must be constant with respect to i for Equation (6) to be used.

By the commutative property of addition,

$$a_1 + b_1 + a_2 + b_2 + \cdots + a_n + b_n = a_1 + a_2 + \cdots + a_n + b_1 + b_2 + \cdots + b_n$$

So we have

$$\sum_{i=m}^{n} (a_i + b_i) = \sum_{i=m}^{n} a_i + \sum_{i=m}^{n} b_i \tag{7}$$

Sometimes we want to change the bounds of summation.

$$\sum_{i=m}^{n} a_i = \sum_{i=p}^{p+n-m} a_{i+m-p}. \tag{8}$$

A sum of 37 terms can be regarded as the sum of the first 17 terms plus the sum of the next 20 terms. The next rule generalzes this observation.

$$\sum_{i=m}^{p-1} a_i + \sum_{i=p}^{n} a_i = \sum_{i=m}^{n} a_i \tag{9}$$

In addition to these four basic rules, there are some other rules worth noting. The first two follow, respectively, from Equations (6) and (7):

$$\sum_{i=1}^{n} c = cn \tag{10}$$

$$\sum_{i=m}^{n} (a_i - b_i) = \sum_{i=m}^{n} a_i - \sum_{i=m}^{n} b_i \tag{11}$$

Establishing the next three formulas is best done by a proof method called mathematical induction, which we will not demonstrate here.

$$\sum_{i=1}^{n} i = \frac{n(n+1)}{2} \tag{12}$$

$$\sum_{i=1}^{n} i^2 = \frac{n(n+1)(2n+1)}{6} \tag{13}$$

$$\sum_{i=1}^{n} i^3 = \frac{n^2(n+1)^2}{4} \tag{14}$$

However, we can deduce Equation (12). If we add the following equations, "vertically," term by term,

$$\sum_{i=1}^{n} i = 1 + 2 + 3 + \cdots + n$$

$$\sum_{i=1}^{n} i = n + (n-1) + (n-2) + \cdots + 1$$

we get

$$2\sum_{i=1}^{n} i = (n+1) + (n+1) + (n+1) + \cdots + (n+1)$$

and since there are n terms on the right, we conclude

$$\sum_{i=1}^{n} i = \frac{n(n+1)}{2}$$

Observe that if a teacher assigns the task of finding

$$1 + 2 + 3 + 4 + 5 + 6 + 7 + \cdots + 104 + 105$$

as a *punishment* and if he or she knows the formula given by Equation (12), then a student's work can be checked quickly by

$$\sum_{i=1}^{105} i = \frac{105(106)}{2} = 105 \cdot 53 = 5300 + 265 = 5565$$

● EXAMPLE 3 **Applying the Properties of Summation Notation**

Evaluate the given sums.

a. $\displaystyle\sum_{j=30}^{100} 4$ **b.** $\displaystyle\sum_{k=1}^{100}(5k+3)$ **c.** $\displaystyle\sum_{k=1}^{200} 9k^2$

Solutions:

a.

$$\sum_{j=30}^{100} 4 = \sum_{j=1}^{71} 4 \qquad \text{[by Equation (8)]}$$

$$= 4 \cdot 71 \qquad \text{[by Equation (10)]}$$

$$= 284$$

b.

$$\sum_{k=1}^{100}(5k+3) = \sum_{k=1}^{100} 5k + \sum_{k=1}^{100} 3 \qquad \text{[by Equation (7)]}$$

$$= 5\left(\sum_{k=1}^{100} k\right) + 3\left(\sum_{k=1}^{100} 1\right) \qquad \text{[by Equation (6)]}$$

$$= 5 \left(\frac{100 \cdot 101}{2} \right) + 3(100) \qquad \text{[by Equations (12) and (10)]}$$

$$= 25,250 + 300$$

$$= 25,550$$

c.

$$\sum_{k=1}^{200} 9k^2 = 9 \sum_{k=1}^{200} k^2 \qquad \text{[by Equation (6)]}$$

$$= 9 \left(\frac{200 \cdot 201 \cdot 401}{6} \right) \qquad \text{[by Equation (13)]}$$

$$= 24,180,300$$

NOW WORK PROBLEM 19

Problems 1.5

In Problems 1 and 2, give the bounds of summation and the index of summation for each expression.

1. $\displaystyle\sum_{t=12}^{17} (8t^2 - 5t + 3)$

2. $\displaystyle\sum_{m=3}^{450} (8m - 4)$

In Problems 3–6, evaluate the given sums.

3. $\displaystyle\sum_{i=1}^{7} 6i$

4. $\displaystyle\sum_{p=0}^{4} 10p$

*5. $\displaystyle\sum_{k=3}^{9} (10k + 16)$

6. $\displaystyle\sum_{n=7}^{11} (2n - 3)$

In Problems 7–12, express the given sums in summation notation.

7. $36 + 37 + 38 + 39 + \cdots + 60$

8. $1 + 4 + 9 + 16 + 25$

*9. $5^3 + 5^4 + 5^5 + 5^6 + 5^7 + 5^8$

10. $11 + 15 + 19 + 23 + \cdots + 71$

11. $2 + 4 + 8 + 16 + 32 + 64 + 128 + 256$

12. $10 + 100 + 1000 + \cdots + 100,000,000$

In Problems 13–26, evaluate the given sums.

13. $\displaystyle\sum_{k=1}^{43} 100$

14. $\displaystyle\sum_{k=35}^{135} 2$

15. $\displaystyle\sum_{k=1}^{n} \left(5 \cdot \frac{1}{n} \right)$

16. $\displaystyle\sum_{k=1}^{200} (k - 100)$

17. $\displaystyle\sum_{k=51}^{100} 10k$

18. $\displaystyle\sum_{k=1}^{n} \frac{n}{n+1} k^2$

*19. $\displaystyle\sum_{k=1}^{20} (5k^2 + 3k)$

20. $\displaystyle\sum_{k=1}^{100} \frac{3k^2 - 200k}{101}$

21. $\displaystyle\sum_{k=51}^{100} k^2$

22. $\displaystyle\sum_{k=1}^{50} (k + 50)^2$

23. $\displaystyle\sum_{k=1}^{10} \left\{ \left[4 - \left(\frac{2k}{10} \right)^2 \right] \left(\frac{2}{10} \right) \right\}$

24. $\displaystyle\sum_{k=1}^{100} \left\{ \left[4 - \left(\frac{2}{100} k \right)^2 \right] \left(\frac{2}{100} \right) \right\}$

25. $\displaystyle\sum_{k=1}^{n} \left\{ \left[5 - \left(\frac{3}{n} \cdot k \right)^2 \right] \frac{3}{n} \right\}$

26. $\displaystyle\sum_{k=1}^{n} \frac{k^2}{(n+1)(2n+1)}$

1.6 Review

Important Terms and Symbols   Examples

Section 1.1	**Applications of Equations**	
	fixed cost variable cost total cost total revenue profit	Ex. 3, p. 48
Section 1.2	**Linear Inequalities**	
	$a < b$ $a \leq b$ $a > b$ $a \geq b$ $a < x < b$	Ex. 1, p. 56
	inequality sense of an inequality	Ex. 2, p. 56
	equivalent inequalities linear inequality	Ex. 1, p. 56
	interval open interval closed interval endpoint	
	(a, b) $[a, b]$ $(-\infty, b)$ $(-\infty, b]$ (a, ∞) $[a, \infty)$ $(-\infty, \infty)$	Ex. 3, p. 57

Summary

With a word problem, an equation is not given to you. Instead, you must set it up by translating verbal statements into an equation (or inequality). This is *mathematical modelling*. It is important that you first read the problem more than once so that you clearly understand what facts are given and what you are asked to find. Then choose a letter to represent the unknown quantity that you want to find. Use the relationships and facts given in the problem and translate them into an equation involving the letter. Finally, solve the equation, and see if your solution answers what was asked. Sometimes the solution to the *equation* will not be the answer to the *problem,* but it may be useful in obtaining that answer.

Some basic relationships that are used in solving business problems are as follows:

$$\text{total cost} = \text{variable cost} + \text{fixed cost}$$

$$\text{total revenue} = (\text{price per unit})(\text{number of units sold})$$

$$\text{profit} = \text{total revenue} - \text{total cost}$$

The inequality symbols $<, \le, >,$ and $\ge$ are used to represent an inequality, which is a statement that one number is, for example, less than another number. Three basic operations that, when applied to an inequality, guarantee an equivalent inequality are as follows:

1. Adding (or subtracting) the same number to (or from) both sides.

2. Multiplying (or dividing) both sides by the same positive number.

3. Multiplying (or dividing) both sides by the same negative number and reversing the sense of the inequality.

An algebraic definition of absolute value is

$$|x| = x \text{ if } x \ge 0 \quad \text{and} \quad |x| = -x \text{ if } x < 0$$

We interpret $|a - b|$ or $|b - a|$ as the distance between a and b. If $d > 0$, then the solution to the inequality $|x| < d$ is the interval $(-d, d)$. The solution to $|x| > d$ consists of two intervals and is given by $(-\infty, -d) \cup (d, \infty)$. Some basic properties of the absolute value are as follows:

1. $|ab| = |a| \cdot |b|$
2. $\left|\dfrac{a}{b}\right| = \dfrac{|a|}{|b|}$
3. $|a - b| = |b - a|$
4. $-|a| \le a \le |a|$
5. $|a + b| \le |a| + |b|$

Review Problems

Problem numbers shown in color indicate problems suggested for use as a practice chapter test.

In Problems 1–15, solve the equation or inequality.

1. $5x - 2 \ge 2(x - 7)$

2. $2x - (7 + x) \le x$

3. $-(5x + 2) < -(2x + 4)$

4. $-2(x + 6) > x + 4$

5. $3p(1 - p) > 3(2 + p) - 3p^2$

6. $3\left(5 - \dfrac{7}{3}q\right) < 9$

7. $\dfrac{x + 5}{3} - \dfrac{1}{2} \le 2$

8. $\dfrac{x}{3} - \dfrac{x}{4} > \dfrac{x}{5}$

9. $\dfrac{1}{4}s - 3 \le \dfrac{1}{8}(3 + 2s)$

10. $\dfrac{1}{3}(t + 2) \ge \dfrac{1}{4}t + 4$

11. $|3 - 2x| = 7$

12. $\left|\dfrac{5x - 6}{13}\right| = 0$

13. $|2z - 3| < 5$

14. $4 < \left|\dfrac{2}{3}x + 5\right|$

15. $|3 - 2x| \ge 4$

16. Evaluate $\displaystyle\sum_{i=1}^{5}(i + 2)^3$ by first cubing the binomial and then using Equations (10), (12), (13), and (14) of Section 1.5.

17. Evaluate $\displaystyle\sum_{i=3}^{7} i^3$ by using $\displaystyle\sum_{i=1}^{7} i^3 - \sum_{i=1}^{2} i^3$. Explain why this works quoting any equations from Section 1.5 that are used. Explain why the answer is necessarily the same as that in Problem 16.

18. **Profit**   A profit of 40% on the selling price of a product is equivalent to what percent profit on the cost?

19. **Stock Exchange**   On a certain day, there were 1132 different issues traded on the New York Stock Exchange. There were 48 more issues showing an increase than showing a decline, and no issues remained the same. How many issues suffered a decline?

20. **Sales Tax**   The sales tax in a certain state is 6.5%. If a total of $3039.29 in purchases, including tax, is made in the course of a year, how much of it is tax?

21. **Production Allocation**   A company will manufacture a total of 10,000 units of its product at plants A and B. Available data are as follows:

	Plant A	Plant B
Unit cost for labor and material	$5	$5.50
Fixed cost	$30,000	$35,000

Between the two plants the company has decided to allot no more than $117,000 for total costs. What is the minimum number of units that must be produced at plant A?

22. **Propane Tanks**   A company is replacing two propane tanks with one new tank. The old tanks are cylindrical, each 25 ft high. One has a radius of 10 ft and the other a radius of 20 ft.

The new tank is essentially spherical and it will have the same volume as the old tanks combined. Find the radius of the new tank. [*Hint*: The volume V of a cylindrical tank is $V = \pi r^2 h$, where r is the radius of the circular base and h is the height of the tank. The volume of a spherical tank is $W = \frac{4}{3}\pi R^3$, where R is the radius of the tank.]

23. **Operating Ratio**   The *operating ratio* of a retail business is the ratio, expressed as a percentage, of operating costs (everything from advertising expenses to equipment depreciation) to net sales (i.e., gross sales minus returns and allowances). An operating ratio less than 100% indicates a profitable operation, while an operating ratio in the 80%–90% range is extremely good. If a company has net sales of $236,460 in one period, write an inequality describing the operating costs that would keep the operating ratio below 90%.

Mathematical Snapshot

Variable-Quality Recording[1]

There is a bewildering array of technological equipment for a consumer to use to record movies, television programs, computer programs, games, and songs. Whether the equipment be iPod, DVD, CD, or even VHS, it is nearly always possible to record at variable compression ratios with variable quality.

(If you are using an older tape device with different *speeds,* then the fastest speed is ratio 1 to 1 while a lower speed that allows you to store r times as much has a compression ratio of r to 1. For example, the VHS standard, SP, is 1 to 1, while LP is 2 to 1 and EP is 3 to 1.)

The storage medium might be a disc or a tape (or something not yet marketed), but there is always a quantity versus quality trade-off that is inherent in any recording device imaginable. For a given medium, the more information that is stored on it by increased compression, the poorer the quality becomes.

Suppose, for argument's sake, that you want to record a movie that is 210 minutes long on a DVD. To get it all to fit on a single disc at a fixed compression ratio, you will need the ratio that allows between 3 and 4 hours of recording time. The ratio that is judged to be movie quality will only allow about 2 hours of recording time and so will not suffice alone. However, you might want to record as much as you can at the better quality by switching from one ratio to another at a precalculated time.

We will solve the problem of finding the time *to switch* in a general way that will allow for all possible applications of this kind. We want to store M minutes of entertainment on a device that with a compression of 1 to 1 will store m minutes. We have available recording compression ratios of r to 1 and R to 1, say with $1 < r < R$, so that R corresponds to packing in more at lower quality. The number $\dfrac{M}{r}$ gives the number of 1 to 1 minutes that will be needed to store M

minutes at ratio r to 1. If the number $\dfrac{M}{r}$ is greater than m, then we cannot store all M minutes on our device at ratio r. Assuming that $\dfrac{M}{R}$ *is* less than m, we want to find the time t when we need to switch from r to R in order to record all M minutes of the entertainment.

If t minutes are recorded at ratio r, then they will consume $\dfrac{t}{r}$ of the available m 1 to 1 minutes of recording time. The remaining $M - t$ minutes of entertainment will consume a further $\dfrac{M - t}{R}$ of the the available m 1 to 1 minutes at ratio R. Thus in order to use up *all* the available recording space, we should find t so that

$$\frac{t}{r} + \frac{M - t}{R} = m$$

While this equation, being completetly literal, might appear complicated, it is very simple with respect to t, the variable for which we want to solve. Indeed it is *linear* in t, and there are only a few steps needed to a get a general solution.

$$\frac{t}{r} + \frac{M}{R} - \frac{t}{R} = m$$

$$\left(\frac{1}{r} - \frac{1}{R}\right) t = m - \frac{M}{R}$$

$$\left(\frac{R - r}{rR}\right) t = \frac{mR - M}{R}$$

$$t = \frac{mR - M}{R} \cdot \frac{rR}{R - r}$$

$$t = \frac{r(mR - M)}{R - r}$$

Notice that the formula is not symmetric with respect to r and R. It tells us how many minutes after recording at high quality we should switch to the poorer quality in order to complete the recording in the available space. If you want to save the higher-quality component for the *end* of the recording, you will have to adjust the formula. See Problems 1, 2, 3, 4, and 7. It should be emphasized that the formula need not be memorized (unless you plan to use it a lot). The method is far more important. The existence of the general solution ensures that the method will always work. In the Problems, try to set up and solve specific problems using the method rather than substituting into the formula.

To learn more about compression schemes, visit wikipedia.org and look up "data compression" and related terms.

[1] Adapted from Gregory N. Fiore, "An Application of Linear Equations to the VCR," *Mathematics Teacher*, 81 (October 1988), 570–72. By permission of the National Council of Teachers of Mathematics.

Problems

A VCR using standard T-120 tape will record for 2 hours in SP mode. Thus $m = 120$ for standard VCR equipment. Use this value in Problems 1–4.

1. If LP and SP modes on a VCR are used, in that order, to record a $2\frac{1}{2}$-hour movie, how long after the start of the movie should the switch *from* LP *to* SP be made?

2. If EP and SP modes are used, in that order, to record a $2\frac{1}{2}$-hour show, how many minutes after the start of the show should the switch from EP to SP be made?

3. If LP and SP modes are used, in that order, to record a movie of length M minutes, how long after the start of the movie should the switch from LP to SP be made?

4. EP and SP modes are used, in that order, to record a movie of length M minutes, how long after the start of the movie should the switch from EP to SP be made?

5. For a standard CD, the value of m is about 74. Use the Solver function on a graphing calculator to solve the equation

$$\frac{x}{12} + \frac{1080 - x}{20} = 74$$

Then, in a similar manner, solve the equation

$$\frac{x}{15} + \frac{1590 - x}{24} = 74$$

6. What, in the context of recording compressed audio on CDs, do each of the equations in Problem 5 represent?

7. Derive the general formula for finding the time to switch recording ratios if the higher quality (ratio r) is to be saved for the end of the recording.

2

FUNCTIONS AND GRAPHS

 A Taxing Experience!

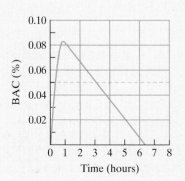

Suppose a 180-pound man drinks four beers in quick succession. We know that his blood alcohol concentration, or BAC, will first rise, then gradually fall back to zero. But what is the best way to describe how quickly the BAC rises, where it peaks, and how fast it falls again?

If we obtain measured BAC values for this particular drinker, we can display them in a table, as follows:

Time (h)	1	2	3	4	5	6
BAC(%)	0.0820	0.0668	0.0516	0.0364	0.0212	0.0060

However, a table can show only a limited number of values and so does not really give the overall picture.

We might instead relate the BAC to time t using a combination of linear and quadratic equations (recall Chapter 0):

$$BAC = -0.1025t^2 + 0.1844t \qquad \text{if } t \le 0.97$$

$$BAC = -0.0152t + 0.0972 \qquad \text{if } t > 0.97$$

As with the table, however, it is hard to look at the equations and quickly understand what is happening with BAC over time.

Probably the best description of changes in the BAC over time is a graph, like the one on the left. Here we easily see what happens. The blood alcohol concentration climbs rapidly, peaks at 0.083% after about an hour, and then gradually tapers off over the next five-and-a-half hours. Note that for most of three hours, this male drinker's BAC is above 0.05%, the point at which one's driving skills typically begin to decline. The curve will vary from one drinker to the next, but women are generally affected more severely than men, not only because of weight differences but also because of the different water contents in men's and women's bodies.

The relationship between time and blood alcohol content is an example of a function. This chapter deals in depth with functions and their graphs.

To understand what a function is and to determine domains and function values.

2.1 Functions

In the 17th century, Gottfried Wilhelm Leibniz, one of the inventors of calculus, introduced the term *function* into the mathematical vocabulary. The concept of a function is one of the most basic in all of mathematics. In particular, it is essential to the study of calculus.

In everyday speech we often hear educated people say things like "Interest rates are a function of oil prices" or "Pension income is a function of years worked" or "Blood alcohol concentration after drinking beer is a function of time." Sometimes such usage agrees with the mathematical usage—but not always. We have to be more careful with our use of the word *function* in order to make it mathematically useful, but there are features of everyday usage that are nevertheless worth highlighting.

For example, the spirit of the preceding examples seems to be covered by "Quantities of type *Y* are a function of quantities of type *X*." There are two types of quantities—although it is possible for *Y* to be the same as *X*—and the *X*-value seems somehow to *determine* the *Y*-value. In general, the usage is not symmetric in *X* and *Y*. To illustrate: the phrase "Oil prices are a function of interest rates" just does not ring true. Most people would not believe that even the Federal Reserve's manipulation of interest rates can determine oil prices. Most economists would be able to recall a time when the Fed's interest rate was 6% and the price of oil was $30 a barrel and another time when the interest rate was also 6% and yet oil was $40 a barrel. A given interest rate does not ensure a unique oil price. Hence the *input* quantity, interest rate, does not *determine* the *output* quantity, oil price. On the other hand, suppose we bring into a testing facility a person who has just drunk five beers and test her blood alcohol concentration then and each hour thereafter for six hours. For each of the time values {0, 1, 2, 3, 4, 5, 6}, the measurement of blood alcohol concentration will produce *exactly one value*.

For our purposes this last example provides the key to making precise our usage of the word *function: For each input value, x, (a time), there is exactly one output value, y (a blood alcohol concentration).*

In fact, with this criterion it is *not* correct to say that "Interest rates are a function of oil prices." Much as we might like to think that (high) oil prices are the *cause* of economic woes, it is not the case that a given particular oil price will give rise to a unique interest rate. To see this you might want to look at

```
http://www.wtrg.com/oil_graphs/oilprice1947.gif
```

and

```
http://www.goldeagle.com/editorials_00/leopold011400.html
```

From the first you should be able to determine two different (fairly recent) times when the price of oil was the same. If the two times give rise to different interest rates in the second, then you have evidence that a particular oil price does not give rise to a unique interest rate. It is also not true that "Pension income is a function of years worked." If the value of "years worked" is 25, then the value of "pension income" is not yet determined. In most organizations, a CEO and a systems manager will retire with quite different pensions after 25 years of service. However, in this example we might be able to say that, *for each job description,* pension income is a function of years worked.

If $100 is invested at, say, 6% simple interest, then the interest earned *I* is a function of the length of time *t* that the money is invested. These quantities are related by

$$I = 100(0.06)t \tag{1}$$

Here, for each value of t, there is exactly one value of I given by Equation (1). In a situation like this we will often write $I(t) = 100(0.06)t$ to reinforce the idea that the I-value is determined by the t-value. Sometimes we write $I = I(t)$ to make the claim that I is a function of t even if we do not know a formula rule that makes this so. Formula (1) assigns the output 3 to the input $\frac{1}{2}$ and the output 12 to the input 2. We can think of Formula (1) as defining a *rule*: Multiply t by 100(0.06). The rule assigns to each input number t exactly one output number I, which we symbolize by the following arrow notation:

$$t \;\mapsto\; I \quad \text{or} \quad t \;\mapsto\; 100(0.06)t$$

A formula provides a way of describing a rule to cover potentially infinitely many cases, but if there are only finitely many values of the input variable, as in the chapter opening paragraph, then the *rule* as provided by the observations recorded in the table there may not be part of any recognizable *formula*. We use the word *rule* rather than *formula* below to allow us to capture this useful generality.

DEFINITION
A ***function*** is a rule that assigns to each input number exactly one output number. The set of all input numbers to which the rule applies is called the ***domain*** of the function. The set of all output numbers is called the ***range***.

For the interest function defined by Equation (1), the input number t cannot be negative, because negative time makes no sense in this example. Thus, the domain consists of all nonnegative numbers—that is, all $t \geq 0$, where the variable gives the time elapsed from when the investment was made.

We have been using the term *function* in a restricted sense because, in general, the inputs or outputs do not have to be numbers. For example, a list of states and their capitals assigns to each state its capital (exactly one output). Hence, a function is implied. However, for the time being, we will consider only functions whose domains and ranges consist of real numbers.

A variable that represents input numbers for a function is called an **independent variable.** A variable that represents output numbers is called a **dependent variable** because its value *depends* on the value of *the independent variable.* We say that the dependent variable is a *function of the independent variable.* That is, output is a function of input. Thus, for the interest formula $I = 100(0.06)t$, the independent variable is t, the dependent variable is I, and I is a function of t.

As another example, the equation

$$y = x + 2 \tag{2}$$

defines y as a function of x. The equation gives the rule, "Add 2 to x." This rule assigns to each input x exactly one output $x + 2$, which is y. If $x = 1$, then $y = 3$; if $x = -4$, then $y = -2$. The independent variable is x and the dependent variable is y.

Not all equations in x and y define y as a function of x. For example, let $y^2 = x$. If x is 9, then $y^2 = 9$, so $y = \pm 3$. Hence, to the input 9, there are assigned not one, but *two*, output numbers: 3 and -3. This violates the definition of a function, so y is **not** a function of x.

On the other hand, some equations in two variables define either variable as a function of the other variable. For example, if $y = 2x$, then for each input x, there is exactly one output, $2x$. Thus, y is a function of x. However, solving the equation for x gives $x = y/2$. For each input y, there is exactly one output, $y/2$. Consequently, x is a function of y.

Usually, the letters f, g, h, F, G, and so on are used to represent function rules. For example, Equation (2), $y = x + 2$, defines y as a function of x, where the rule is "Add 2 to the input." Suppose we let f represent this rule. Then we say that f is the function. To indicate that f assigns the output 3 to the input 1, we write $f(1) = 3$,

CAUTION

In $y^2 = x$, x and y are related, but the relationship does not give y as a function of x.

which is read "f of 1 equals 3." Similarly, $f(-4) = -2$. More generally, if x is any input, we have the following notation:

input
↓
$f(x)$
↑
output

$f(x)$, which is read "f of x," means the output number in the range of f that corresponds to the input number x in the domain.

$f(x)$ is an output number.

Thus, the output $f(x)$ is the same as y. But since $y = x + 2$, we can write $y = f(x) = x + 2$ more simply,

$$f(x) = x + 2$$

For example, to find $f(3)$, which is the output corresponding to the input 3, we replace each x in $f(x) = x + 2$ by 3:

$$f(3) = 3 + 2 = 5$$

Likewise,

$$f(8) = 8 + 2 = 10$$

$$f(-4) = -4 + 2 = -2$$

  CAUTION

*$f(x)$ does **not** mean f times x. $f(x)$ is the output that corresponds to the input x.*

Function notation is used extensively in calculus.

Output numbers such as $f(-4)$ are called **function values.** Keep in mind that they are in the range of f.

Quite often, functions are defined by "function notation." For example, the equation $g(x) = x^3 + x^2$ defines the function g that assigns the output number $x^3 + x^2$ to an input number x:

$$g: x \mapsto x^3 + x^2$$

In other words, g adds the cube and the square of an input number. Some function values are

$$g(2) = 2^3 + 2^2 = 12$$

$$g(-1) = (-1)^3 + (-1)^2 = -1 + 1 = 0$$

$$g(t) = t^3 + t^2$$

$$g(x + 1) = (x + 1)^3 + (x + 1)^2$$

The idea of replacement is very important in determining function values.

Note that $g(x + 1)$ was found by replacing each x in $x^3 + x^2$ by the input $x + 1$.

When we refer to the function g defined by $g(x) = x^3 + x^2$, we shall feel free to call the equation itself a function. Thus, we speak of "the function $g(x) = x^3 + x^2$," and, similarly, "the function $y = x + 2$."

Let's be specific about the domain of a function. Unless otherwise stated, the domain consists of all real numbers for which the rule of the function makes sense; that is, the set of all real numbers for which the rule gives function values that are real numbers.

For example, suppose

$$h(x) = \frac{1}{x - 6}$$

Here any real number can be used for x except 6, because the denominator is 0 when x is 6. So the domain of h is understood to be all real numbers except 6.

Equality of Functions

To say that two functions f and g are equal, denoted $f = g$, is to say that

1. The domain of f is equal to the domain of g;
2. For every x in the domain of f and g, $f(x) = g(x)$.

Requirement 1 says that a number x is in the domain of f if and only if it is in the domain of g. Thus, if we have $f(x) = x^2$, with no explicit mention of domain, and $g(x) = x^2$ for $x \geq 0$, then $f \neq g$. For here the domain of f is the whole real line $(-\infty, \infty)$ and the domain of g is $[0, \infty)$. On the other hand, if we have $f(x) = (x+1)^2$ and $g(x) = x^2 + 2x + 1$, then for both f and g the domain is understood to be $(-\infty, \infty)$ and the issue for deciding if $f = g$ is whether, for each real number x, we have $(x + 1)^2 = x^2 + 2x + 1$. But this is true; it is a special case of item 4 in the Special Products of Section 0.4. In fact, older textbooks refer to statements like $(x+1)^2 = x^2 + 2x + 1$ as "identities," to indicate that they are true for any admissible value of the variable and to distinguish them from statements like $(x+1)^2 = 0$, which are true for some values of x.

Given functions f and g, it follows that we have $f \neq g$ if *either* the domain of f is different from the domain of g *or* there is some x for which $f(x) \neq g(x)$.

● EXAMPLE 1 Determining Equality of Functions

Determine which of the following functions are equal.

a. $f(x) = \dfrac{(x + 2)(x - 1)}{(x - 1)}$

b. $g(x) = x + 2$

c. $h(x) = \begin{cases} x + 2 & \text{if } x \neq 1 \\ 0 & \text{if } x = 1 \end{cases}$

d. $k(x) = \begin{cases} x + 2 & \text{if } x \neq 1 \\ 3 & \text{if } x = 1 \end{cases}$

Solution: The domain of f is the set of all real numbers other than 1, while that of g is the set of all real numbers. (For these we are following the convention that the domain is the set of all real numbers for which the rule makes sense.) We will have more to say about functions like h and k that are defined by *cases* in Example 4 of Section 2.2. Here we observe that the domain of h and the domain of k are both $(-\infty, \infty)$, since for both we have a rule that makes sense for each real number. The domains of g, h, and k are equal to each other but that of f is different. So by requirement 1 for equality of functions, $f \neq g$, $f \neq h$ and $f \neq k$. By definition, $g(x) = h(x) = k(x)$ for all $x \neq 1$, so the matter of equality of g, h and k depends on their values at 1. Since $g(1) = 3$, $h(1) = 0$ and $k(1) = 3$, we conclude that $g = k$ and $g \neq h$ (and $h \neq k$). While this example might appear to be contrived, it is typical of an issue that arises frequently in calculus.

NOW WORK PROBLEM 3 ●●●

PRINCIPLES IN PRACTICE 1

FINDING DOMAINS

The area of a circle depends on the length of the radius of the circle.

a. Write a function $a(r)$ for the area of a circle when the length of the radius is r.

b. What is the domain of this function out of context?

c. What is the domain of this function in the given context?

● EXAMPLE 2 Finding Domains

Find the domain of each function.

a. $f(x) = \dfrac{x}{x^2 - x - 2}$

Solution: We cannot divide by zero, so we must find any values of x that make the denominator 0. These *cannot* be input numbers. Thus, we set the denominator equal to 0 and solve for x:

$$x^2 - x - 2 = 0 \qquad \text{(quadratic equation)}$$

$$(x - 2)(x + 1) = 0 \qquad \text{(factoring)}$$

$$x = 2, -1$$

Therefore, the domain of f is all real numbers *except* 2 and -1.

b. $g(t) = \sqrt{2t - 1}$

Solution: $\sqrt{2t - 1}$ is a real number if $2t - 1$ is greater than or equal to 0. If $2t - 1$ is negative, then $\sqrt{2t - 1}$ is not a real number. (It is an *imaginary number*.)

Since our function values, at least for now, must be real numbers, we must assume that

$$2t - 1 \geq 0$$

$$2t \geq 1 \qquad \text{(adding 1 to both sides)}$$

$$t \geq \frac{1}{2} \qquad \text{(dividing both sides by 2)}$$

Thus, the domain is the interval $[\frac{1}{2}, \infty)$.

NOW WORK PROBLEM 7

● EXAMPLE 3 Finding Domain and Function Values

Let $g(x) = 3x^2 - x + 5$. Any real number can be used for x, so the domain of g is all real numbers.

a. *Find $g(z)$.*

Solution: Replacing each x in $g(x) = 3x^2 - x + 5$ by z gives

$$g(z) = 3(z)^2 - z + 5 = 3z^2 - z + 5$$

b. *Find $g(r^2)$.*

Solution: Replacing each x in $g(x) = 3x^2 - x + 5$ by r^2 gives

$$g(r^2) = 3(r^2)^2 - r^2 + 5 = 3r^4 - r^2 + 5$$

c. *Find $g(x + h)$.*

Solution:

$$g(x + h) = 3(x + h)^2 - (x + h) + 5$$
$$= 3(x^2 + 2hx + h^2) - x - h + 5$$
$$= 3x^2 + 6hx + 3h^2 - x - h + 5$$

NOW WORK PROBLEM 31(a)

 C A U T I O N

Don't be confused by notation. In Example 3(c), we find $g(x + h)$ by replacing each x in $g(x) = 3x^2 - x + 5$ by the input $x + h$. $g(x + h), g(x) + h$ and $g(x) + g(h)$ are all quite different quantities.

The difference quotient of a function is an important mathematical concept.

● EXAMPLE 4 Finding a Difference Quotient

If $f(x) = x^2$, find $\dfrac{f(x + h) - f(x)}{h}$.

Solution: The expression $\dfrac{f(x + h) - f(x)}{h}$ is referred to as a **difference quotient.** Here the numerator is a difference of function values. We have

$$\frac{f(x + h) - f(x)}{h} = \frac{(x + h)^2 - x^2}{h}$$
$$= \frac{x^2 + 2hx + h^2 - x^2}{h} = \frac{2hx + h^2}{h}$$
$$= \frac{h(2x + h)}{h} = 2x + h$$

NOW WORK PROBLEM 35

In some cases, the domain of a function is restricted for physical or economic reasons. For example, the previous interest function $I = 100(0.06)t$ has $t \geq 0$ because t represents time. Example 5 will give another illustration.

DEMAND FUNCTION

Suppose the weekly demand function for large pizzas at a local pizza parlor is $p = 26 - \dfrac{q}{40}$.

a. If the current price is $18.50 per pizza, how many pizzas are sold each week?

b. If 200 pizzas are sold each week, what is the current price?

c. If the owner wants to double the number of large pizzas sold each week (to 400), what should the price be?

EXAMPLE 5   Demand Function

Suppose that the equation $p = 100/q$ describes the relationship between the price per unit p of a certain product and the number of units q of the product that consumers will buy (that is, demand) per week at the stated price. This equation is called a *demand equation* for the product. If q is an input number, then to each value of q there is assigned exactly one output number p:

$$q \;\mapsto\; \frac{100}{q} = p$$

For example,

$$20 \;\mapsto\; \frac{100}{20} = 5$$

that is, when q is 20, p is 5. Thus, price p is a function of quantity demanded, q. This function is called a **demand function**. The independent variable is q, and p is the dependent variable. Since q cannot be 0 (division by 0 is not defined) and cannot be negative (q represents quantity), the domain is all values of q such that $q > 0$.

NOW WORK PROBLEM 43

We have seen that a function is essentially a *correspondence* whereby to each input number in the domain there is assigned exactly one output number in the range. For the correspondence given by $f(x) = x^2$, some sample assignments are shown by the arrows in Figure 2.1. The next example discusses a functional correspondence that is not given by an algebraic formula.

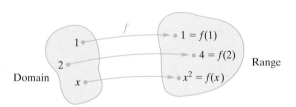

FIGURE 2.1   Functional correspondence for $f(x) = x^2$.

SUPPLY SCHEDULE

p Price per Unit in Dollars	q Quanity Supplied per Week
500	11
600	14
700	17
800	20

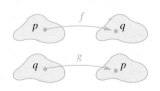

FIGURE 2.2   Supply schedule and supply functions.

EXAMPLE 6   Supply Schedule

The table in Figure 2.2 is a *supply schedule*. Such a table gives a correspondence between the price p of a certain product and the quantity q the producers will supply per week at that price. For each price, there corresponds exactly one quantity, and vice versa.

If p is the independent variable, then q is a function of p, say, $q = f(p)$, and

$$f(500) = 11 \quad f(600) = 14 \quad f(700) = 17 \quad \text{and} \quad f(800) = 20$$

Notice that as price per unit increases, the producers are willing to supply more units per week.

On the other hand, if q is the independent variable, then p is a function of q, say $p = g(q)$, and

$$g(11) = 500 \quad g(14) = 600 \quad g(17) = 700 \quad \text{and} \quad g(20) = 800$$

We speak of f and g as **supply functions.**

NOW WORK PROBLEM 53

Function values are easily computed with a graphing calculator. For example, suppose

$$f(x) = 17x^4 - 13x^3 + 7$$

and we wish to find $f(0.7)$, $f(-2.31)$, and $f(10)$. With a TI-83 Plus, we first enter the function as Y_1:

$$Y_1 = 17X^\wedge 4 - 13X^\wedge 3 + 7$$

After pressing the TABLE key, we successively enter the x-values .7, -2.31, and 10. The results are shown in

Figure 2.3. We remark that there are other methods of determining function values with the TI-83 Plus.

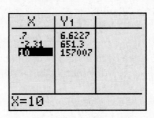

FIGURE 2.3 Table of function values for $f(x) = 17x^4 - 13x^3 + 7$.

Problems 2.1

In Problems 1–4, determine whether the given functions are equal.

1. $f(x) = \sqrt{x^2}; g(x) = x$

2. $G(x) = (\sqrt{x+1})^2; H(x) = x + 1$

***3.** $h(x) = \dfrac{|x|}{x}; k(x) = \begin{cases} 1 & \text{if } x \geq 0 \\ -1 & \text{if } x < 0 \end{cases}$

4. $f(x) = \begin{cases} \dfrac{x^2 - 4x + 3}{x - 3} & \text{if } x \neq 3 \\ 2 & \text{if } x = 3 \end{cases}$;

$g(x) = x - 1$

In Problems 5–16, give the domain of each function.

5. $f(x) = \dfrac{8}{x}$

6. $g(x) = \dfrac{x}{5}$

***7.** $h(x) = \sqrt{x - 3}$

8. $K(z) = \dfrac{1}{\sqrt{z - 1}}$

9. $f(z) = 3z^2 + 2z - 4$

10. $H(x) = \dfrac{x}{x + 8}$

11. $f(x) = \dfrac{9x - 9}{2x + 7}$

12. $g(x) = \sqrt{4x + 3}$

13. $g(y) = \dfrac{4}{y^2 - 4y + 4}$

14. $\phi(x) = \dfrac{x + 5}{x^2 + x - 6}$

15. $h(s) = \dfrac{4 - s^2}{2s^2 - 7s - 4}$

16. $G(r) = \dfrac{2}{r^2 + 1}$

In Problems 17–28, find the function values for each function.

17. $f(x) = 2x + 1; f(0), f(3), f(-4)$

18. $H(s) = 5s^2 - 3; H(4), H(\sqrt{2}), H\left(\dfrac{2}{3}\right)$

19. $G(x) = 2 - x^2; G(-8), G(u), G(u^2)$

20. $F(x) = -5x; F(s), F(t + 1), F(x + 3)$

21. $\gamma(u) = 2u^2 - u; \gamma(-2), \gamma(2v), \gamma(x + a)$

22. $h(v) = \dfrac{1}{\sqrt{v}}; h(16), h\left(\dfrac{1}{4}\right), h(1 - x)$

23. $f(x) = x^2 + 2x + 1; f(1), f(-1), f(x + h)$

24. $H(x) = (x + 4)^2; H(0), H(2), H(t - 4)$

25. $k(x) = \dfrac{x - 7}{x^2 + 2}; k(5), k(3x), k(x + h)$

26. $k(x) = \sqrt{x - 3}; k(4), k(3), k(x + 1) - k(x)$

27. $f(x) = x^{4/3}; f(0), f(64), f\left(\dfrac{1}{8}\right)$

28. $g(x) = x^{2/5}; g(32), g(-64), g(t^{10})$

In Problems 29–36, find (a) $f(x + h)$ and (b) $\dfrac{f(x + h) - f(x)}{h}$; simplify your answers.

29. $f(x) = 4x - 5$

30. $f(x) = \dfrac{x}{2}$

***31.** $f(x) = x^2 + 2x$

32. $f(x) = 3x^2 - 2x - 1$

33. $f(x) = 3 - 2x + 4x^2$

34. $f(x) = x^3$

***35.** $f(x) = \dfrac{1}{x}$

36. $f(x) = \dfrac{x + 8}{x}$

37. If $f(x) = 5x + 3$, find $\dfrac{f(3 + h) - f(3)}{h}$.

38. If $f(x) = 2x^2 - x + 1$, find $\dfrac{f(x) - f(2)}{x - 2}$.

In Problems 39–42, is y a function of x? Is x a function of y?

39. $9y - 3x - 4 = 0$

40. $x^2 + y = 0$

41. $y = 7x^2$

42. $x^2 + y^2 = 1$

***43.** The formula for the area of a circle of radius r is $A = \pi r^2$. Is the area a function of the radius?

44. Suppose $f(b) = a^2 b^3 + a^3 b^2$. (a) Find $f(a)$. (b) Find $f(ab)$.

45. Value of Business A business with an original capital of \$25,000 has income and expenses each week of \$6500 and \$4800, respectively. If all profits are retained in the business, express the value V of the business at the end of t weeks as a function of t.

46. Depreciation If a \$30,000 machine depreciates 2% of its original value each year, find a function f that expresses the machine's value V after t years have elapsed.

47. Profit Function If q units of a certain product are sold (q is nonnegative), the profit P is given by the equation $P = 1.25q$. Is P a function of q? What is the dependent variable? the independent variable?

48. Demand Function Suppose the yearly demand function for a particular actor to star in a film is $p = \dfrac{1,200,000}{q}$, where q is the number of films he stars in during the year. If the actor currently charges \$600,000 per film, how many films does he star in each year? If he wants to star in four films per year, what should his price be?

49. Supply Function   Suppose the weekly supply function for a pound of house-blend coffee at a local coffee shop is $p = \dfrac{q}{48}$, where q is the number of pounds of coffee supplied per week. How many pounds of coffee per week will be supplied if the price is \$8.39 a pound? How many pounds of coffee per week will be supplied if the price is \$19.49 a pound? How does the amount supplied change as the price increases?

50. Hospital Discharges   An insurance company examined the records of a group of individuals hospitalized for a particular illness. It was found that the total proportion discharged at the end of t days of hospitalization is given by

$$f(t) = 1 - \left(\frac{300}{300 + t}\right)^3$$

Evaluate (a) $f(0)$, (b) $f(100)$, and (c) $f(900)$. (d) At the end of how many days was half $(1/2 = 0.500)$ of the group discharged?

51. Psychology   A psychophysical experiment was conducted to analyze human response to electrical shocks.[1] The subjects received a shock of a certain intensity. They were told to assign a magnitude of 10 to this particular shock, called the standard stimulus. Then other shocks (stimuli) of various intensities were given. For each one, the response R was to be a number that indicated the perceived magnitude of the shock relative to that of the standard stimulus. It was found that R was a function of the intensity I of the shock (I in microamperes) and was estimated by

$$R = f(I) = \frac{I^{4/3}}{2500} \qquad 500 \le I \le 3500$$

Evaluate (a) $f(1000)$ and (b) $f(2000)$. (c) Suppose that I_0 and $2I_0$ are in the domain of f. Express $f(2I_0)$ in terms of $f(I_0)$. What effect does the doubling of intensity have on response?

52. Psychology   In a paired-associate learning experiment,[2] the probability of a correct response as a function of the number n of trials has the form

$$P(n) = 1 - \frac{1}{2}(1 - c)^{n-1} \qquad n \ge 1$$

where the estimated value of c is 0.344. Find $P(1)$ and $P(2)$ by using this value of c.

*53. Demand Schedule   The following table is called a *demand schedule*. It gives a correspondence between the price p of a product and the quantity q that consumers will demand (that is, purchase) at that price. (a) If $p = f(q)$, list the numbers in the domain of f. Find $f(2900)$ and $f(3000)$. (b) If $q = g(p)$, list the numbers in the domain of g. Find $g(10)$ and $g(17)$.

Price per Unit, p	Quantity Demanded per Week, q
\$10	3000
12	2900
17	2300
20	2000

In Problems 54–57, use your calculator to find the indicated values for the given function. Round answers to two decimal places.

54. $f(x) = 2.03x^3 - 5.27x^2 - 13.71$; (a) $f(1.73)$, (b) $f(-5.78)$, (c) $f(\sqrt{2})$

55. $f(x) = \dfrac{14.7x^2 - 3.95x - 15.76}{24.3 - x^3}$; (a) $f(4)$, (b) $f(-17/4)$, (c) $f(\pi)$

56. $f(x) = (20.3 - 3.2x)(2.25x^2 - 7.1x - 16)^4$; (a) $f(0.3)$, (b) $f(-0.02)$, (c) $f(1.9)$

57. $f(x) = \sqrt{\dfrac{\sqrt{2}x^2 + 7.31(x + 1)}{5.03}}$; (a) $f(12.35)$, (b) $f(-123)$, (c) $f(0)$

To introduce constant functions, polynomial functions, rational functions, case-defined functions, the absolute-value function, and factorial notation.

2.2 Special Functions

In this section, we look at functions having special forms and representations. We begin with perhaps the simplest type of function there is: a *constant function*.

PRINCIPLES IN PRACTICE 1

CONSTANT FUNCTIONS

Suppose the monthly health insurance premiums for an individual are \$125.00.

a. Write the monthly health insurance premiums as a function of the number of visits the individual makes to the doctor.

b. How do the health insurance premiums change as the number of visits to the doctor increases?

c. What kind of function is this?

● EXAMPLE 1   Constant Functions

Let $h(x) = 2$. The domain of h is all real numbers. All function values are 2. For example,

$$h(10) = 2 \qquad h(-387) = 2 \qquad h(x + 3) = 2$$

We call h a *constant function* because all the function values are the same. More generally, we have this definition:

A function of the form $h(x) = c$, where c is a *constant*, is called a **constant function.**

NOW WORK PROBLEM 19 ●●●

---

[1] Adapted from H. Babkoff, "Magnitude Estimation of Short Electrocutaneous Pulses," *Psychological Research*, 39, no. 1 (1976), 39–49.

[2] D. Laming, *Mathematical Psychology* (New York: Academic Press, 1983).

A constant function belongs to a broader class of functions, called *polynomial functions*. In general, a function of the form

$$f(x) = c_n x^n + c_{n-1} x^{n-1} + \cdots + c_1 x + c_0$$

where n is a nonnegative integer and $c_n, c_{n-1}, \ldots, c_0$ are constants with $c_n \neq 0$, is called a **polynomial function** (in x). The number n is called the **degree** of the polynomial, and c_n is the **leading coefficient.** Thus,

$$f(x) = 3x^2 - 8x + 9$$

is a polynomial function of degree 2 with leading coefficient 3. Likewise, $g(x) = 4 - 2x$ has degree 1 and leading coefficient -2. Polynomial functions of degree 1 or 2 are called **linear** or **quadratic functions,** respectively. For example, $g(x) = 4 - 2x$ is linear and $f(x) = 3x^2 - 8x + 9$ is quadratic. Note that a nonzero constant function, such as $f(x) = 5$ [which can be written as $f(x) = 5x^0$], is a polynomial function of degree 0. The constant function $f(x) = 0$ is also considered a polynomial function but has no degree assigned to it. The domain of any polynomial function is the set of all real numbers.

> Each term in a polynomial function is either a constant or a constant times a positive integral power of x.

PRINCIPLES IN PRACTICE 2

POLYNOMIAL FUNCTIONS

The function $d(t) = 3t^2$, for $t \geq 0$, represents the distance in meters a car will go in t seconds when it has a constant acceleration of 6 m per second.

a. What kind of function is this?

b. What is its degree?

c. What is its leading coefficient?

● EXAMPLE 2 **Polynomial Functions**

a. $f(x) = x^3 - 6x^2 + 7$ is a polynomial (function) of degree 3 with leading coefficient 1.

b. $g(x) = \dfrac{2x}{3}$ is a linear function with leading coefficient $\dfrac{2}{3}$.

c. $f(x) = \dfrac{2}{x^3}$ is *not* a polynomial function. Because $f(x) = 2x^{-3}$ and the exponent for x is not a nonnegative integer, this function does not have the proper form for a polynomial. Similarly, $g(x) = \sqrt{x}$ is not a polynomial, because $g(x) = x^{1/2}$.

NOW WORK PROBLEM 3 ●●●

A function that is a quotient of polynomial functions is called a **rational function.**

● EXAMPLE 3 **Rational Functions**

a. $f(x) = \dfrac{x^2 - 6x}{x + 5}$ is a rational function, since the numerator and denominator are each polynomials. Note that this rational function is not defined for $x = -5$.

b. $g(x) = 2x + 3$ is a rational function, since $2x + 3 = \dfrac{2x + 3}{1}$. In fact, every polynomial function is also a rational function.

NOW WORK PROBLEM 5 ●●●

> Every polynomial function is a rational function.

Sometimes more than one expression is needed to define a function, as Example 4 shows.

PRINCIPLES IN PRACTICE 3

CASE-DEFINED FUNCTION

To reduce inventory, a department store charges three rates. If you buy 0–5 pairs of socks, the price is \$3.50 per pair. If you buy 6–10 pairs of socks, the price is \$3.00 per pair. If you buy more than 10 pairs, the price is \$2.75 per pair. Write a case-defined function to represent the cost of buying n pairs of socks.

● EXAMPLE 4 **Case-Defined Function**

Let

$$F(s) = \begin{cases} 1 & \text{if } -1 \leq s < 1 \\ 0 & \text{if } 1 \leq s \leq 2 \\ s - 3 & \text{if } 2 < s \leq 8 \end{cases}$$

This is called a **case-defined function** because the rule for specifying it is given by rules for each of several disjoint cases. Here s is the independent variable, and the domain of F is all s such that $-1 \leq s \leq 8$. The value of s determines which expression to use.

Find $F(0)$: Since $-1 \le 0 < 1$, we have $F(0) = 1$.

Find $F(2)$: Since $1 \le 2 \le 2$, we have $F(2) = 0$.

Find $F(7)$: Since $2 < 7 \le 8$, we substitute 7 for s in $s - 3$.

$$F(7) = 7 - 3 = 4$$

NOW WORK PROBLEM 19

T E C H N O L O G Y

To illustrate how to enter a case-defined function with a TI-83 Plus, Figure 2.4 shows a key sequence for the function

$$f(x) = \begin{cases} 2x & \text{if } x < 0 \\ x^2 & \text{if } 0 \le x < 10 \\ -x & \text{if } x \ge 10 \end{cases}$$

Since $|x|$ provides a unique real number for each real number x, absolute value, $|-|$, is a function.

FIGURE 2.4   Entering a case-defined function.

● **EXAMPLE 5   Absolute-Value Function**

The function $|-|(x) = |x|$ is called the *absolute-value function*. Recall that the **absolute value,** of a real number x is denoted $|x|$ and is defined by

The absolute-value function can be considered a case-defined function.

$$|x| = \begin{cases} x & \text{if } x \ge 0 \\ -x & \text{if } x < 0 \end{cases}$$

Thus, the domain of $|-|$ is all real numbers. Some function values are

$$|16| = 16$$

$$\left|-\tfrac{4}{3}\right| = -\left(-\tfrac{4}{3}\right) = \tfrac{4}{3}$$

$$|0| = 0$$

NOW WORK PROBLEM 21

In our next examples, we make use of *factorial notation*.

The symbol $r!$, with r a positive integer, is read "r **factorial.**" It represents the product of the first r positive integers:

$$r! = 1 \cdot 2 \cdot 3 \cdots r$$

We also define

$$0! = 1$$

For each nonnegative integer n, $(-)!(n) = n!$ determines a unique number so it follows that $(-)!$ is a function whose domain is the set of nonnegative integers.

PRINCIPLES IN PRACTICE 4

FACTORIALS

Seven different books are to be placed on a shelf. How many ways can they be arranged? Represent the question as a factorial problem and give the solution.

● **EXAMPLE 6   Factorials**

a. $5! = 1 \cdot 2 \cdot 3 \cdot 4 \cdot 5 = 120$

b. $3!(6 - 5)! = 3! \cdot 1! = (3 \cdot 2 \cdot 1)(1) = (6)(1) = 6$

c. $\dfrac{4!}{0!} = \dfrac{1 \cdot 2 \cdot 3 \cdot 4}{1} = \dfrac{24}{1} = 24$

NOW WORK PROBLEM 27

● EXAMPLE 7   **Genetics**

Factorials occur frequently in probability theory.

Suppose two black guinea pigs are bred and produce exactly five offspring. Under certain conditions, it can be shown that the probability P that exactly r of the offspring will be brown and the others black is a function of r, P = P(r), where

$$P(r) = \frac{5! \left(\frac{1}{4}\right)^r \left(\frac{3}{4}\right)^{5-r}}{r!(5-r)!} \qquad r = 0, 1, 2, \ldots, 5$$

The letter P in P = P(r) is used in two ways. On the right side, P represents the function rule. On the left side, P represents the dependent variable. The domain of P is all integers from 0 to 5, inclusive. Find the probability that exactly three guinea pigs will be brown.

Solution: We want to find $P(3)$. We have

$$P(3) = \frac{5! \left(\frac{1}{4}\right)^3 \left(\frac{3}{4}\right)^2}{3!2!} = \frac{120 \left(\frac{1}{64}\right)\left(\frac{9}{16}\right)}{6(2)} = \frac{45}{512}$$

NOW WORK PROBLEM 35 ●●

Problems 2.2

In Problems 1–4, determine whether the given function is a polynomial function.

1. $f(x) = x^2 - x^4 + 4$

2. $f(x) = \dfrac{x^3 + 7x - 3}{3}$

*3. $g(x) = \dfrac{1}{x^2 + 2x + 1}$

4. $g(x) = 3^{-2}x^2$

In Problems 5–8, determine whether the given function is a rational function.

*5. $f(x) = \dfrac{x^2 + x}{x^3 + 4}$

6. $f(x) = \dfrac{3}{2x + 1}$

7. $g(x) = \begin{cases} 1 & \text{if } x < 5 \\ 4 & \text{if } x \geq 5 \end{cases}$

8. $g(x) = 4x^{-4}$

In Problems 9–12, find the domain of each function.

9. $h(z) = 19$

10. $f(x) = \sqrt{\pi}$

11. $f(x) = \begin{cases} 5x & \text{if } x > 1 \\ 4 & \text{if } x \leq 1 \end{cases}$

12. $f(x) = \begin{cases} 4 & \text{if } x = 3 \\ x^2 & \text{if } 1 \leq x < 3 \end{cases}$

In Problems 13–16, state (a) the degree and (b) the leading coefficient of the given polynomial function.

13. $F(x) = 7x^3 - 2x^2 + 6$

14. $g(x) = 7x$

15. $f(x) = \dfrac{1}{\pi} - 3x^5 + 2x^6 + x^7$

16. $f(x) = 9$

In Problems 17–22, find the function values for each function.

17. $f(x) = 8$; $f(2)$, $f(t + 8)$, $f(-\sqrt{17})$

18. $g(x) = |x - 3|$; $g(10)$, $g(3)$, $g(-3)$

*19. $F(t) = \begin{cases} 1 & \text{if } t > 0 \\ 0 & \text{if } t = 0 \\ -1 & \text{if } t < 0 \end{cases}$

$F(10)$, $F(-\sqrt{3})$, $F(0)$, $F\left(-\dfrac{18}{5}\right)$

20. $f(x) = \begin{cases} 4 & \text{if } x \geq 0 \\ 3 & \text{if } x < 0 \end{cases}$;

$f(3)$, $f(-4)$, $f(0)$

*21. $G(x) = \begin{cases} x - 1 & \text{if } x \geq 3 \\ 3 - x^2 & \text{if } x < 3 \end{cases}$;

$G(8)$, $G(3)$, $G(-1)$, $G(1)$

22. $F(\theta) = \begin{cases} 2\theta - 5 & \text{if } \theta < 2 \\ \theta^2 - 3\theta + 1 & \text{if } \theta > 2 \end{cases}$;

$F(3)$, $F(-3)$, $F(2)$

In Problems 23–28, determine the value of each expression.

23. $6!$

24. $0!$

25. $(4 - 2)!$

26. $6! \cdot 2!$

*27. $\dfrac{n!}{(n-1)!}$

28. $\dfrac{8!}{5!(8-5)!}$

29. Train Trip   A daily round-trip train ticket to the city costs $4.50. Write the cost of a daily round-trip ticket as a function of a passenger's income. What kind of function is this?

30. Geometry   A rectangular prism has length three more than its width and height one less than twice the width. Write the volume of the rectangular prism as a function of the width. What kind of function is this?

31. Cost Function   In manufacturing a component for a machine, the initial cost of a die is $850 and all other additional costs are $3 per unit produced. (a) Express the total cost C (in dollars) as a linear function of the number q of units produced. (b) How many units are produced if the total cost is $1600?

32. Investment   If a principal of P dollars is invested at a simple annual interest rate of r for t years, express the total accumulated amount of the principal and interest as a function of t. Is your result a linear function of t?

33. Sales   To encourage large group sales, a theater charges two rates. If your group is less than 12, each ticket costs $9.50. If your group is 12 or more, each ticket costs $8.75. Write a case-defined function to represent the cost of buying n tickets.

34. Factorials   The business mathematics class has elected a grievance committee of four to complain to the faculty about the introduction of factorial notation into the course. They decide that they will be more effective if they label themselves as members A, G, M, and S, where member A will lobby faculty with surnames A through F, member G will lobby faculty with surnames G through L, and so on. In how many ways can the committee so label its members? In how many ways can a committee of five label itself with five different labels?

*35. **Genetics**   Under certain conditions, if two brown-eyed parents have exactly three children, the probability that there will be exactly r blue-eyed children is given by the function $P = P(r)$, where

$$P(r) = \frac{3!\left(\frac{1}{4}\right)^r\left(\frac{3}{4}\right)^{3-r}}{r!(3-r)!}, \qquad r = 0, 1, 2, 3$$

Find the probability that exactly two of the children will be blue-eyed.

36. **Genetics**   In Example 7, find the probability that all five offspring will be brown.

37. **Bacteria Growth**   Bacteria are growing in a culture. The time t (in hours) for the bacteria to double in number (the generation time) is a function of the temperature T (in °C) of the culture. If this function is given by[3]

$$t = f(T) = \begin{cases} \dfrac{1}{24}T + \dfrac{11}{4} & \text{if } 30 \le T \le 36 \\[2mm] \dfrac{4}{3}T - \dfrac{175}{4} & \text{if } 36 < T \le 39 \end{cases}$$

(a) determine the domain of f and (b) find $f(30)$, $f(36)$, and $f(39)$.

In Problems 38–41, use your calculator to find the indicated function values for the given function. Round answers to two decimal places

38. $f(x) = \begin{cases} 0.19x^4 - 27.99 & \text{if } x \ge 5.99 \\ 0.63x^5 - 57.42 & \text{if } x < 5.99 \end{cases}$

   (a) $f(7.98)$ (b) $f(2.26)$ (c) $f(9)$

39. $f(x) = \begin{cases} 29.5x^4 + 30.4 & \text{if } x < 3 \\ 7.9x^3 - 2.1x & \text{if } x \ge 3 \end{cases}$

   (a) $f(2.5)$ (b) $f(-3.6)$ (c) $f(3.2)$

40. $f(x) = \begin{cases} 4.07x - 2.3 & \text{if } x < -8 \\ 19.12 & \text{if } -8 \le x < -2 \\ x^2 - 4x^{-2} & \text{if } x \ge -2 \end{cases}$

   (a) $f(-5.8)$ (b) $f(-14.9)$ (c) $f(7.6)$

41. $f(x) = \begin{cases} x/(x+3) & \text{if } x < -5 \\ x(x-4)^2 & \text{if } -5 \le x < 0 \\ \sqrt{2.1x+3} & \text{if } x \ge 0 \end{cases}$

   (a) $f(-\sqrt{30})$ (b) $f(46)$ (c) $f(-2/3)$

2.3 Combinations of Functions

OBJECTIVE

To combine functions by means of addition, subtraction, multiplication, division, multiplication by a constant, and composition.

There are several ways of combining two functions to create a new function. Suppose f and g are the functions given by

$$f(x) = x^2 \quad \text{and} \quad g(x) = 3x$$

Adding $f(x)$ and $g(x)$ gives

$$f(x) + g(x) = x^2 + 3x$$

This operation defines a new function called the *sum* of f and g, denoted $f + g$. Its function value at x is $f(x) + g(x)$. That is,

$$(f + g)(x) = f(x) + g(x) = x^2 + 3x$$

For example,

$$(f + g)(2) = 2^2 + 3(2) = 10$$

In general, for any functions f and g, we define the **sum** $f + g$, the **difference** $f - g$, the **product** fg, and the **quotient** $\dfrac{f}{g}$ as follows:[4]

$$(f + g)(x) = f(x) + g(x)$$
$$(f - g)(x) = f(x) - g(x)$$
$$(fg)(x) = f(x) \cdot g(x)$$
$$\frac{f}{g}(x) = \frac{f(x)}{g(x)} \quad \text{for } g(x) \ne 0$$

---

[3] Adapted from F. K. E. Imrie and A. J. Vlitos, "Production of Fungal Protein from Carob," in *Single-Cell Protein II*, ed. S. R. Tannenbaum and D. I. C. Wang (Cambridge, MA: MIT Press, 1975).

[4] In each of the four combinations, we assume that x is in the domains of both f and g. In the quotient, we also do not allow any value of x for which $g(x)$ is 0.

A special case of fg deserves separate mention. For any real number c and any function f, we define cf by

$$(cf)(x) = c \cdot f(x)$$

This restricted case of product is called **scalar product.** The scalar product tends to share properties with sums (and differences) that are not enjoyed by general products (and quotients).

For $f(x) = x^2$ and $g(x) = 3x$, we have

$$(f + g)(x) = f(x) + g(x) = x^2 + 3x$$

$$(f - g)(x) = f(x) - g(x) = x^2 - 3x$$

$$(fg)(x) = f(x) \cdot g(x) = x^2(3x) = 3x^3$$

$$\frac{f}{g}(x) = \frac{f(x)}{g(x)} = \frac{x^2}{3x} = \frac{x}{3} \quad \text{for } x \neq 0$$

$$(\sqrt{2} f)(x) = \sqrt{2} f(x) = \sqrt{2} x^2$$

● **EXAMPLE 1 Combining Functions**

If $f(x) = 3x - 1$ and $g(x) = x^2 + 3x$, find

a. $(f + g)(x)$

b. $(f - g)(x)$

c. $(fg)(x)$

d. $\dfrac{f}{g}(x)$

e. $(\frac{1}{2} f)(x)$

Solution:

a. $(f + g)(x) = f(x) + g(x) = (3x - 1) + (x^2 + 3x) = x^2 + 6x - 1$

b. $(f - g)(x) = f(x) - g(x) = (3x - 1) - (x^2 + 3x) = -1 - x^2$

c. $(fg)(x) = f(x)g(x) = (3x - 1)(x^2 + 3x) = 3x^3 + 8x^2 - 3x$

d. $\dfrac{f}{g}(x) = \dfrac{f(x)}{g(x)} = \dfrac{3x - 1}{x^2 + 3x}$

e. $\left(\dfrac{1}{2} f\right)(x) = \dfrac{1}{2}(f(x)) = \dfrac{1}{2}(3x - 1) = \dfrac{3x - 1}{2}$

NOW WORK PROBLEM 3(a)–(f) ●●●

Composition

We can also combine two functions by first applying one function to a number and then applying the other function to the result. For example, suppose $g(x) = 3x$, $f(x) = x^2$, and $x = 2$. Then $g(2) = 3 \cdot 2 = 6$. Thus, g sends the input 2 to the output 6:

$$2 \overset{g}{\mapsto} 6$$

Next, we let the output 6 become the input for f:

$$f(6) = 6^2 = 36$$

So f sends 6 to 36:

$$6 \overset{f}{\mapsto} 36$$

By first applying g and then f, we send 2 to 36:

$$2 \overset{g}{\mapsto} 6 \overset{f}{\mapsto} 36$$

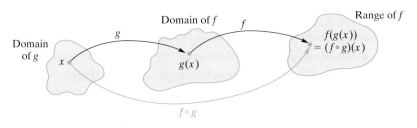

FIGURE 2.5   Composite of f with g.

To be more general, let's replace the 2 by x, where x is in the domain of g. (See Figure 2.5). Applying g to x, we get the number $g(x)$, which we shall assume is in the domain of f. By applying f to $g(x)$, we get $f(g(x))$, read "f of g of x," which is in the range of f. The operation of applying g and then applying f to the result is called *composition* and the resulting function, denoted $f \circ g$, is called the *composite* of f with g. This function assigns the output number $f(g(x))$ to the input number x. (See the bottom arrow in Figure 2.5.) Thus, $(f \circ g)(x) = f(g(x))$.

> **DEFINITION**
>
> If f and g are functions, the **composite of f with g is the function $f \circ g$ defined by**
>
> $$(f \circ g)(x) = f(g(x))$$
>
> where the domain of $f \circ g$ is the set of all those x in the domain of g such that $g(x)$ is in the domain of f.

For $f(x) = x^2$ and $g(x) = 3x$, we can get a simple form for $f \circ g$:

$$(f \circ g)(x) = f(g(x)) = f(3x) = (3x)^2 = 9x^2$$

For example, $(f \circ g)(2) = 9(2)^2 = 36$, as we saw before.

When dealing with real numbers and the operation of addition, 0 is special in that, for any real number a, we have

$$a + 0 = a = 0 + a$$

The number 1 enjoys a similar property with respect to multiplication. For any real number a, we have

$$a1 = a = 1a$$

For reference in Section 2.4, we note that the function I defined by $I(x) = x$ satisfies for any function f,

$$f \circ I = f = I \circ f$$

where here we mean equality of functions as defined in Section 2.1. Indeed, for any x,

$$(f \circ I)(x) = f(I(x)) = f(x) = I(f(x)) = (I \circ f)(x)$$

The function I is called the *identity* function.

PRINCIPLES IN PRACTICE 1

COMPOSITION

A CD costs x dollars wholesale. The price the store pays is given by the function $s(x) = x + 3$. The price the customer pays is $c(x) = 2x$, where x is the price the store pays. Write a composite function to find the customer's price as a function of the wholesale price.

● **EXAMPLE 2   Composition**

Let $f(x) = \sqrt{x}$ and $g(x) = x + 1$. Find

a. $(f \circ g)(x)$

b. $(g \circ f)(x)$

CAUTION

Generally, $f \circ g$ and $g \circ f$ are quite different. In Example 2,

$$(f \circ g)(x) = \sqrt{x+1}$$

but we have

$$(g \circ f)(x) = \sqrt{x} + 1$$

Observe that $(f \circ g)(1) = \sqrt{2}$, while $(g \circ f)(1) = 2$. Also, do not confuse $f(g(x))$ with $(fg)(x)$, which is the product $f(x)g(x)$. Here

$$f(g(x)) = \sqrt{x+1}$$

but

$$f(x)g(x) = \sqrt{x}(x+1)$$

Solution:

a. $(f \circ g)(x)$ is $f(g(x))$. Now g adds 1 to x, and f takes the square root of the result. Thus,

$$(f \circ g)(x) = f(g(x)) = f(x+1) = \sqrt{x+1}$$

The domain of g is all real numbers x, and the domain of f is all nonnegative reals. Hence, the domain of the composite is all x for which $g(x) = x+1$ is nonnegative. That is, the domain is all $x \geq -1$, or equivalently, the interval $[-1, \infty)$.

b. $(g \circ f)(x)$ is $g(f(x))$. Now f takes the square root of x, and g adds 1 to the result. Thus, g adds 1 to $\sqrt{x}$, and we have

$$(g \circ f)(x) = g(f(x)) = g(\sqrt{x}) = \sqrt{x} + 1$$

The domain of f is all $x \geq 0$, and the domain of g is all reals. Hence, the domain of the composite is all $x \geq 0$ for which $f(x) = \sqrt{x}$ is real, namely, all $x \geq 0$.

NOW WORK PROBLEM 7

Composition is *associative*, meaning that for any three functions f, g, and h,

$$(f \circ g) \circ h = f \circ (g \circ h)$$

EXAMPLE 3   Composition

If $F(p) = p^2 + 4p - 3$, $G(p) = 2p + 1$, *and* $H(p) = |p|$, *find*

a. $F(G(p))$

b. $F(G(H(p)))$

c. $G(F(1))$

Solution:

a. $F(G(p)) = F(2p+1) = (2p+1)^2 + 4(2p+1) - 3 = 4p^2 + 12p + 2 = (F \circ G)(p)$

b. $F(G(H(p))) = (F \circ (G \circ H))(p) = ((F \circ G) \circ H)(p) = (F \circ G)(H(p)) = (F \circ G)(|p|) = 4|p|^2 + 12|p| + 2 = 4p^2 + 12|p| + 2$

c. $G(F(1)) = G(1^2 + 4 \cdot 1 - 3) = G(2) = 2 \cdot 2 + 1 = 5$

NOW WORK PROBLEM 9

In calculus, it is necessary at times to think of a particular function as a composite of two simpler functions, as the next example shows.

PRINCIPLES IN PRACTICE 2

EXPRESSING A FUNCTION AS A COMPOSITE

Suppose the area of a square garden is $g(x) = (x+3)^2$. Express g as a composite of two functions, and explain what each function represents.

EXAMPLE 4   Expressing a Function as a Composite

Express $h(x) = (2x - 1)^3$ *as a composite.*

Solution:

We note that $h(x)$ is obtained by finding $2x - 1$ and cubing the result. Suppose we let $g(x) = 2x - 1$ and $f(x) = x^3$. Then

$$h(x) = (2x - 1)^3 = [g(x)]^3 = f(g(x)) = (f \circ g)(x)$$

which gives h as a composite of two functions.

NOW WORK PROBLEM 13

Two functions can be combined by using a graphing calculator. Consider the functions

$$f(x) = 2x + 1 \quad \text{and} \quad g(x) = x^2$$

which we enter as Y_1 and Y_2, as shown in Figure 2.6. The sum of f and g is given by $Y_3 = Y_1 + Y_2$ and the composite $f \circ g$ by $Y_4 = Y_1(Y_2)$. For example, $f(g(3))$ is obtained by evaluating Y_4 at 3.

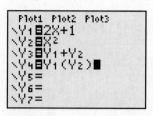

FIGURE 2.6 Y_3 and Y_4 are combinations of Y_1 and Y_2.

Problems 2.3

1. If $f(x) = x + 3$ and $g(x) = x + 5$, find the following.

(a) $(f + g)(x)$ **(b)** $(f + g)(0)$ **(c)** $(f - g)(x)$

(d) $(fg)(x)$ **(e)** $(fg)(-2)$ **(f)** $\dfrac{f}{g}(x)$

(g) $(f \circ g)(x)$ **(h)** $(f \circ g)(3)$ **(i)** $(g \circ f)(x)$

(j) $(g \circ f)(3)$

2. If $f(x) = 2x$ and $g(x) = 6 + x$, find the following.

(a) $(f + g)(x)$ **(b)** $(f - g)(x)$ **(c)** $(f - g)(4)$

(d) $(fg)(x)$ **(e)** $\dfrac{f}{g}(x)$ **(f)** $\dfrac{f}{g}(2)$

(g) $(f \circ g)(x)$ **(h)** $(g \circ f)(x)$ **(i)** $(g \circ f)(2)$

*3. If $f(x) = x^2 + 1$ and $g(x) = x^2 - x$, find the following.

(a) $(f + g)(x)$ **(b)** $(f - g)(x)$ **(c)** $(f - g)\left(-\frac{1}{2}\right)$

(d) $(fg)(x)$ **(e)** $\dfrac{f}{g}(x)$ **(f)** $\dfrac{f}{g}\left(-\frac{1}{2}\right)$

(g) $(f \circ g)(x)$ **(h)** $(g \circ f)(x)$ **(i)** $(g \circ f)(-3)$

4. If $f(x) = x^2 + 1$ and $g(x) = 5$, find the following.

(a) $(f + g)(x)$ **(b)** $(f + g)\left(\frac{2}{3}\right)$ **(c)** $(f - g)(x)$

(d) $(fg)(x)$ **(e)** $(fg)(7)$ **(f)** $\dfrac{f}{g}(x)$

(g) $(f \circ g)(x)$ **(h)** $(f \circ g)(12{,}003)$ **(i)** $(g \circ f)(x)$

5. If $f(x) = 3x^2 + 6$ and $g(x) = 4 - 2x$, find $f(g(2))$ and $g(f(2))$.

6. If $f(p) = \dfrac{4}{p}$ and $g(p) = \dfrac{p - 2}{3}$, find both $(f \circ g)(p)$ and $(g \circ f)(p)$.

*7. If $F(t) = t^2 + 7t + 1$ and $G(t) = \dfrac{2}{t - 1}$, find $(F \circ G)(t)$ and $(G \circ F)(t)$.

8. If $F(t) = \sqrt{t}$ and $G(t) = 3t^2 + 4t + 2$, find $(F \circ G)(t)$ and $(G \circ F)(t)$.

*9. If $f(v) = \dfrac{1}{v^2 + 1}$ and $g(v) = \sqrt{v + 2}$, find $(f \circ g)(v)$ and $(g \circ f)(v)$.

10. If $f(x) = x^2 + 2x - 1$, find $(f \circ f)(x)$.

In Problems 11–16, find functions f and g such that $h(x) = f(g(x))$.

11. $h(x) = 11x - 7$

12. $h(x) = \sqrt{x^2 - 2}$

*13. $h(x) = \dfrac{1}{x^2 - 2}$

14. $h(x) = (9x^3 - 5x)^3 - (9x^3 - 5x)^2 + 11$

15. $h(x) = \sqrt[4]{\dfrac{x^2 - 1}{x + 3}}$

16. $h(x) = \dfrac{2 - (3x - 5)}{(3x - 5)^2 + 2}$

17. Profit A coffeehouse sells a pound of coffee for $9.75. Expenses are $4500 each month, plus $4.25 for each pound of coffee sold.

(a) Write a function $r(x)$ for the total monthly revenue as a function of the number of pounds of coffee sold.

(b) Write a function $e(x)$ for the total monthly expenses as a function of the number of pounds of coffee sold.

(c) Write a function $(r - e)(x)$ for the total monthly profit as a function of the number of pounds of coffee sold.

18. Geometry Suppose the volume of a cube is $v(x) = (4x - 2)^3$. Express v as a composite of two functions, and explain what each function represents.

19. Business A manufacturer determines that the total number of units of output per day, q, is a function of the number of employees, m, where

$$q = f(m) = \dfrac{(40m - m^2)}{4}$$

The total revenue r that is received for selling q units is given by the function g, where $r = g(q) = 40q$. Find $(g \circ f)(m)$. What does this composite function describe?

20. Sociology Studies have been conducted concerning the statistical relations among a person's status, education, and income.[5] Let S denote a numerical value of status based on annual income I. For a certain population, suppose

$$S = f(I) = 0.45(I - 1000)^{0.53}$$

---

[5] R. K. Leik and B. F. Meeker, *Mathematical Sociology* (Englewood Cliffs, NJ: Prentice Hall, 1975).

Furthermore, suppose a person's income I is a function of the number of years of education E, where

$$I = g(E) = 7202 + 0.29 E^{3.68}$$

Find $(f \circ g)(E)$. What does this function describe?

In Problems 21–24, for the given functions f and g, find the indicated function values. Round answers to two decimal places.

21. $f(x) = (4x - 13)^2, g(x) = 0.2x^2 - 4x + 3$

(a) $(f + g)(4.5)$, (b) $(f \circ g)(-2)$

22. $f(x) = \sqrt{\dfrac{x - 3}{x + 1}}, g(x) = 11.2x + 5.39$

(a) $\dfrac{f}{g}(-2)$, (b) $(g \circ f)(-10)$

23. $f(x) = x^{4/5}, g(x) = x^2 - 8$

(a) $(fg)(7)$, (b) $(g \circ f)(3.75)$

24. $f(x) = \dfrac{5}{x + 3}, g(x) = \dfrac{2}{x^2}$

(a) $(f - g)(7.3)$, (b) $(f \circ g)(-4.17)$

2.4 Inverse Functions

OBJECTIVE

To introduce inverse functions, their properties, and their uses.

Just as $-a$ is the number for which

$$a + (-a) = 0 = (-a) + a$$

and, for $a \neq 0$, a^{-1} is the number for which

$$aa^{-1} = 1 = a^{-1}a$$

so, given a function f, we can inquire about the existence of a function g satisfying

$$f \circ g = I = g \circ f \qquad (1)$$

where I is the identity function, introduced in the subsection titled "Composition" of Section 2.3 and given by $I(x) = x$. Suppose that we have g as above and a function h that also satisfies the equations of (1) so that

$$f \circ h = I = h \circ f$$

Then

$$h = h \circ I = h \circ (f \circ g) = (h \circ f) \circ g = I \circ g = g$$

shows that there is at most one function satisfying the requirements of g in (1). In mathematical jargon, g is uniquely determined by f and is therefore given a name, $g = f^{-1}$, that reflects its dependence on f. The function f^{-1} is read as f **inverse** and called the **inverse** of f.

The additive inverse $-a$ exists for any number a; the multiplicative inverse a^{-1} exists precisely if $a \neq 0$. The existence of f^{-1} places a strong requirement on a function f. It can be shown that f^{-1} exists if and only if, for all a and b, whenever $f(a) = f(b)$, then $a = b$. It may be helpful to think that such an f can be *cancelled (on the left)*.

 **CAUTION**

Do not confuse f^{-1}, the inverse of f, and $\dfrac{1}{f}$, the multiplicative reciprocal of f. Unfortunately, the nomenclature for inverse functions clashes with the numerical use of $(-)^{-1}$. Usually, $f^{-1}(x)$ is different from $\dfrac{1}{f}(x) = \dfrac{1}{f(x)}$. For example, $I^{-1} = I$ (since $I \circ I = I$) so $I^{-1}(x) = x$, but $\dfrac{1}{I}(x) = \dfrac{1}{I(x)} = \dfrac{1}{x}$.

A function f that satisfies

for all a and b, if $f(a) = f(b)$ then $a = b$

is called a **one-to-one** function.

Thus, we can say that a function has an inverse precisely if it is one-to-one. An equivalent way to express the one-to-one condition is

for all a and b, if $a \neq b$ then $f(a) \neq f(b)$

so that distinct inputs give rise to distinct outputs. Observe that this condition is not met for many simple functions. For example, if $f(x) = x^2$, then $f(-1) = (-1)^2 = 1 = (1)^2 = f(1)$ and $-1 \neq 1$ shows that the squaring function is not one-to-one. Similarly, $f(x) = |x|$ is not one-to-one.

In general, the domain of f^{-1} is the range of f and the range of f^{-1} is the domain of f.

Let us note here that the the equations of (1) are equivalent to

$$f(f^{-1}(x)) = x = f^{-1}(f(x)) \qquad (2)$$

The first equation holds for all x in the domain of f^{-1} and the second equation holds for all x in the domain of f. In general, the domain of f^{-1}, which is equal to the range of f, can be quite different from the domain of f.

● EXAMPLE 1   Inverses of Linear Functions

According to Section 2.2, a function of the form $f(x) = ax + b$, where $a \neq 0$, is a linear function. *Show that a linear function is one-to-one. Find the inverse of $f(x) = ax + b$ and show that it is also linear.*

Solution: Assume that $f(u) = f(v)$, that is,

$$au + b = av + b \qquad (3)$$

To show that f is one-to-one, we must show that $u = v$ follows from this assumption. Subtracting b from both sides of (3) gives $au = av$, from which $u = v$ follows by dividing both sides by a. (We assumed that $a \neq 0$.) Since f is given by first multiplying by a and then adding b, we might expect that the effect of f can be undone by first subtracting b and then dividing by a. So consider $g(x) = \dfrac{x - b}{a}$. We have

$$(f \circ g)(x) = f(g(x)) = a\frac{x - b}{a} + b = (x - b) + b = x$$

and

$$(g \circ f)(x) = g(f(x)) = \frac{(ax + b) - b}{a} = \frac{ax}{a} = x$$

Since g satisfies the two requirements of (1), it follows that g is the inverse of f. That is, $f^{-1}(x) = \dfrac{x - b}{a} = \dfrac{1}{a}x + \dfrac{-b}{a}$ and the last equality shows that f^{-1} is also a linear function.

NOW WORK PROBLEM 1 ●●

● EXAMPLE 2   Identities for Inverses

Show that

a. *If f and g are one-to-one functions, the composite $f \circ g$ is also one-to-one and $(f \circ g)^{-1} = g^{-1} \circ f^{-1}$.*

b. *If f is one-to-one, then $(f^{-1})^{-1} = f$.*

Solution:

a. Assume $(f \circ g)(a) = (f \circ g)(b)$, that is, $f(g(a)) = f(g(b))$. Since f is one-to-one, $g(a) = g(b)$. Since g is one-to-one, $a = b$ and this shows that $f \circ g$ is one-to-one. The equations

$$(f \circ g) \circ (g^{-1} \circ f^{-1}) = f \circ (g \circ g^{-1}) \circ f^{-1} = f \circ I \circ f^{-1} = f \circ f^{-1} = I$$

and

$$(g^{-1} \circ f^{-1}) \circ (f \circ g) = g^{-1} \circ (f^{-1} \circ f) \circ g = g^{-1} \circ I \circ g = g^{-1} \circ g = I$$

show that $g^{-1} \circ f^{-1}$ is the inverse of $f \circ g$, which, in symbols, is the statement $g^{-1} \circ f^{-1} = (f \circ g)^{-1}$.

b. In Equations (2) replace f by f^{-1}. Taking g to be f shows that Equations (1) are satisfied, and this gives $(f^{-1})^{-1} = f$.

● EXAMPLE 3 Inverses Used to Solve Equations

Many equations take the form $f(x) = 0$, where f is a function. If f is a one-to-one function then the equation has $x = f^{-1}(0)$ as its unique solution.

Solution: Applying f^{-1} to both sides of $f(x) = 0$ gives $f^{-1}(f(x)) = f^{-1}(0)$, and $f^{-1}(f(x)) = x$ shows that $x = f^{-1}(0)$ is the only possible solution. Since $f(f^{-1}(0)) = 0$, $f^{-1}(0)$ is indeed a solution.

● EXAMPLE 4 Restricting the Domain of a Function

It may happen that a function f whose domain is the natural one, consisting of all numbers for which the defining rule makes sense, is not one-to-one, and yet a one-to-one function g can be obtained by restricting the domain of f.

Solution: For example, we have shown that the function $f(x) = x^2$ is not one-to-one but the function $g(x) = x^2$ *with domain explicitly given as* $[0, \infty)$ is one-to-one. Since $(\sqrt{x})^2 = x$ and $\sqrt{x^2} = x$, for $x \geq 0$, it follows that $\sqrt{\ }$ is the inverse of the restricted squaring function g. Here is a more contrived example. Let $f(x) = |x|$ (with its natural domain). Let $g(x) = |x|$ *with domain explicitly given as* $(-\infty, -1) \cup [0, 1]$. The function g is one-to-one and hence has an inverse.

● EXAMPLE 5 Finding the Inverse of a Function

To find the inverse of a one-to-one function f, solve the equation $y = f(x)$ for x in terms of y obtaining $x = g(y)$. Then $f^{-1}(x) = g(x)$. To illustrate, *find $f^{-1}(x)$ if $f(x) = (x - 1)^2$, for $x \geq 1$.*

Solution: Let $y = (x - 1)^2$, for $x \geq 1$. Then $x - 1 = \sqrt{y}$ and hence $x = \sqrt{y} + 1$. It follows that $f^{-1}(x) = \sqrt{x} + 1$.

NOW WORK PROBLEM 5 ●

Problems 2.4

In Problems 1–6, find the inverse of the given function.

*1. $f(x) = 3x + 7$

2. $g(x) = 2x + 1$

3. $F(x) = \frac{1}{2}x - 7$

4. $f(x) = (4x - 5)^2$, for $x \geq \frac{5}{4}$

*5. $A(r) = \pi r^2$, for $r \geq 0$

6. $V(r) = \frac{4}{3}\pi r^3$

In Problems 7–10, determine whether or not the function is one-to-one.

7. $f(x) = 5x + 12$

8. $g(x) = (5x + 12)^2$

9. $h(x) = (5x + 12)^2$, for $x \geq -\frac{12}{5}$

10. $F(x) = |x - 9|$

In Problems 11 and 12, solve each equation by finding an inverse function.

11. $(4x - 5)^2 = 23$, for $x \geq \frac{5}{4}$

12. $\frac{4}{3}\pi r^3 = 100$

13. Demand Function   The function

$$p = p(q) = \frac{1{,}200{,}000}{q} \qquad q > 0$$

expresses an actor's charge per film p as a function of the number of films q that she stars in. Express the number of films in which she stars in terms of her charge per film. Show that the expression is a function of p. Show that the resulting function is inverse to the function giving p in terms of q.

14. Supply Function   The weekly supply function for a pound of house-blend coffee at a coffee shop is

$$p = p(q) = \frac{q}{48} \qquad q > 0$$

where q is the number of pounds of coffee supplied per week and p is the price per pound. Express q as a function of p and demonstrate the relationship between the two functions.

To graph equations and functions in rectangular coordinates, to determine intercepts, to apply the vertical-line test and the horizontal-line test, and to determine the domain and range of a function from a graph.

2.5 Graphs in Rectangular Coordinates

A **rectangular coordinate system** allows us to specify and locate points in a plane. It also provides a geometric way to graph equations in two variables, in particular those arising from functions.

In a plane, two real-number lines, called *coordinate axes*, are constructed perpendicular to each other so that their origins coincide, as in Figure 2.7. Their point of intersection is called the *origin* of the coordinate system. For now, we will call the horizontal line the *x-axis* and the vertical line the *y-axis*. The unit distance on the *x*-axis need not be the same as on the *y*-axis.

The plane on which the coordinate axes are placed is called a *rectangular coordinate plane* or, more simply, an *x,y-plane*. Every point in the *x,y*-plane can be labeled to indicate its position. To label point *P* in Figure 2.8(a), we draw perpendiculars from *P* to the *x*-axis and *y*-axis. They meet these axes at 4 and 2, respectively. Thus, *P* determines two numbers, 4 and 2. We say that the **rectangular coordinates** of *P* are given by the **ordered pair** $(4, 2)$. The word *ordered* is important. In Figure 2.8(b), the point corresponding to $(4, 2)$ is not the same as that for $(2, 4)$:

$$(4, 2) \neq (2, 4)$$

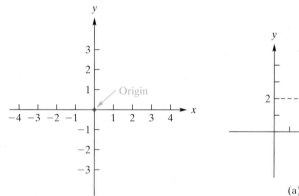

FIGURE 2.7 Coordinate axes.

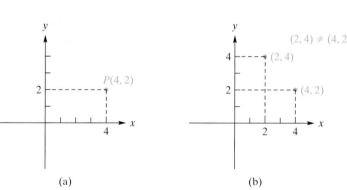

(a)   (b)

FIGURE 2.8 Rectangular coordinates.

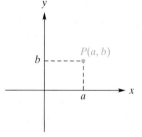

FIGURE 2.9 Coordinates of *P*.

In general, if *P* is any point, then its rectangular coordinates will be given by an ordered pair of the form (a, b). (See Figure 2.9.) We call *a* the *x-coordinate* of *P*, and *b* the *y-coordinate* of *P*.

Accordingly, with each point in a given coordinate plane, we can associate exactly one ordered pair (a, b) of real numbers. Also, it should be clear that with each ordered pair (a, b) of real numbers, we can associate exactly one point in that plane. Since there is a *one-to-one correspondence* between the points in the plane and all ordered pairs of real numbers, we refer to a point *P* with *x*-coordinate *a* and *y*-coordinate *b* simply as the point (a, b), or as $P(a, b)$. Moreover, we use the words *point* and *ordered pair* interchangeably.

In Figure 2.10, the coordinates of various points are indicated. For example, the point $(1, -4)$ is located one unit to the right of the *y*-axis and four units below the *x*-axis. The origin is $(0, 0)$. The *x*-coordinate of every point on the *y*-axis is 0, and the *y*-coordinate of every point on the *x*-axis is 0.

The coordinate axes divide the plane into four regions called *quadrants* (Figure 2.11). For example, quadrant I consists of all points (x_1, y_1) with $x_1 > 0$ and $y_1 > 0$. The points on the axes do not lie in any quadrant.

Using a rectangular coordinate system, we can geometrically represent equations in two variables. For example, let us consider

$$y = x^2 + 2x - 3 \tag{1}$$

A solution of this equation is a value of *x* and a value of *y* that make the equation true. For example, if $x = 1$, substituting into Equation (1) gives

$$y = 1^2 + 2(1) - 3 = 0$$

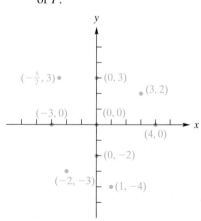

FIGURE 2.10 Coordinates of points.

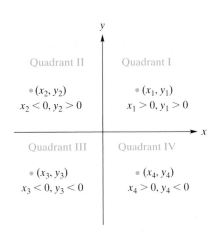

FIGURE 2.11   Quadrants.

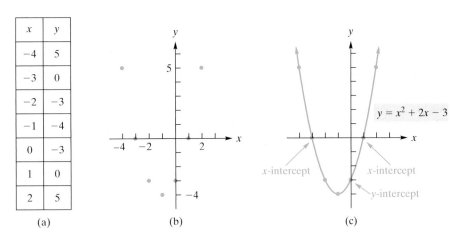

FIGURE 2.12   Graphing $y = x^2 + 2x - 3$.

Thus, $x = 1$, $y = 0$ is a solution of Equation (1). Similarly,

$$\text{if } x = -2 \quad \text{then} \quad y = (-2)^2 + 2(-2) - 3 = -3$$

and so $x = -2$, $y = -3$ is also a solution. By choosing other values for x, we can get more solutions. [See Figure 2.12(a).] It should be clear that there are infinitely many solutions of Equation (1).

Each solution gives rise to a point (x, y). For example, to $x = 1$ and $y = 0$ corresponds $(1, 0)$. The **graph** of $y = x^2 + 2x - 3$ is the geometric representation of all its solutions. In Figure 2.12(b), we have plotted the points corresponding to the solutions in the table.

Since the equation has infinitely many solutions, it seems impossible to determine its graph precisely. However, we are concerned only with the graph's general shape. For this reason, we plot enough points so that we may intelligently guess its proper shape. (The calculus techniques that we will study in Chapter 13 will make such "guesses" much more intelligent.) Then we join these points by a smooth curve wherever conditions permit. This gives the curve in Figure 2.12(c). Of course, the more points we plot, the better our graph is. Here we assume that the graph extends indefinitely upward, which is indicated by arrows.

The point $(0, -3)$ where the curve intersects the y-axis is called the y-*intercept*. The points $(-3, 0)$ and $(1, 0)$ where the curve intersects the x-axis are called the x-*intercepts*. In general, we have the following definition.

Often, we simply say that the y-intercept is -3 and the x-intercepts are -3 and 1.

DEFINITION
An x-***intercept*** of the graph of an equation in x and y is a point where the graph intersects the x-axis. A y-***intercept*** is a point where the graph intersects the y-axis.

To find the x-intercepts of the graph of an equation in x and y, we first set $y = 0$ and solve the resulting equation for x. To find the y-intercepts, we first set $x = 0$ and solve for y. For example, let us find the x-intercepts for the graph of $y = x^2 + 2x - 3$. Setting $y = 0$ and solving for x gives

$$0 = x^2 + 2x - 3$$
$$0 = (x + 3)(x - 1)$$
$$x = -3, \ 1$$

Thus, the x-intercepts are $(-3, 0)$ and $(1, 0)$, as we saw before. If $x = 0$, then

$$y = 0^2 + 2(0) - 3 = -3$$

So $(0, -3)$ is the y-intercept. Keep in mind that an x-intercept has its y-coordinate 0, and a y-intercept has its x-coordinate 0. Intercepts are useful because they indicate precisely where the graph intersects the axes.

EXAMPLE 1   Intercepts and Graph

Find the x- and y-intercepts of the graph of $y = 2x + 3$, and sketch the graph.

Solution: If $y = 0$, then

$$0 = 2x + 3 \quad \text{so that} \quad x = -\frac{3}{2}$$

Thus, the x-intercept is $(-\frac{3}{2}, 0)$. If $x = 0$, then

$$y = 2(0) + 3 = 3$$

So the y-intercept is $(0, 3)$. Figure 2.13 shows a table of some points on the graph and a sketch of the graph.

NOW WORK PROBLEM 9

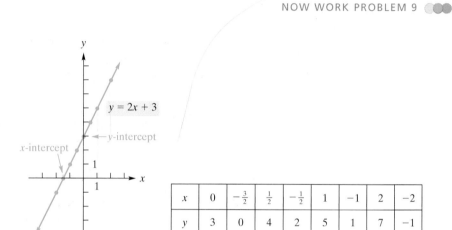

x	0	$-\frac{3}{2}$	$\frac{1}{2}$	$-\frac{1}{2}$	1	-1	2	-2
y	3	0	4	2	5	1	7	-1

FIGURE 2.13   Graph of $y = 2x + 3$.

EXAMPLE 2   Intercepts and Graph

Determine the intercepts, if any, of the graph of $s = \dfrac{100}{t}$, and sketch the graph.

Solution: For the graph, we will label the horizontal axis t and the vertical axis s (Figure 2.14). Because t cannot equal 0 (division by 0 is not defined), there is no s-intercept. Thus, the graph has no point corresponding to $t = 0$. Moreover, there is no t-intercept, because if $s = 0$, then the equation

$$0 = \frac{100}{t}$$

has no solution. Remember, the only way that a fraction can be 0 is by having its numerator 0. Figure 2.14 shows the graph. In general, the graph of $s = k/t$, where k is a nonzero constant, is called a *rectangular hyperbola*.

NOW WORK PROBLEM 11

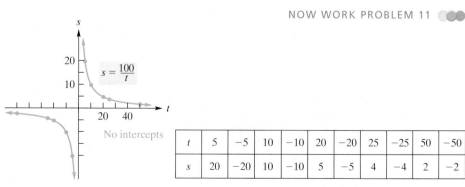

t	5	-5	10	-10	20	-20	25	-25	50	-50
s	20	-20	10	-10	5	-5	4	-4	2	-2

FIGURE 2.14   Graph of $s = \dfrac{100}{t}$.

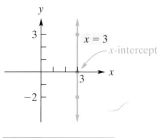

x	3	3	3
y	0	3	-2

FIGURE 2.15 Graph of $x = 3$.

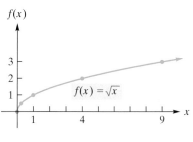

x	0	$\frac{1}{4}$	1	4	9
f(x)	0	$\frac{1}{2}$	1	2	3

FIGURE 2.16 Graph of $f(x) = \sqrt{x}$.

GRAPH OF THE
ABSOLUTE-VALUE FUNCTION

Brett rented a bike from a rental shop, rode at a constant rate of 12 mi/h for 2.5 hours along a bike path, and then returned along the same path. Graph the absolute-value-like function that represents Brett's distance from the rental shop as a function of time over the appropriate domain.

EXAMPLE 3 Intercepts and Graph

Determine the intercepts of the graph of $x = 3$, and sketch the graph.

Solution: We can think of $x = 3$ as an equation in the variables x and y if we write it as $x = 3 + 0y$. Here y can be any value, but x must be 3. Because $x = 3$ when $y = 0$, the x-intercept is $(3, 0)$. There is no y-intercept, because x cannot be 0. (See Figure 2.15.) The graph is a vertical line.

NOW WORK PROBLEM 13

Each function f gives rise to an equation, namely $y = f(x)$, which is a special case of the equations we have been graphing. Its graph consists of all points $(x, f(x))$, where x is in the domain of f. The vertical axis can be labeled either y or $f(x)$, where f is the name of the function, and is referred to as the **function-value axis**. *We always label the horizontal axis with the independent variable but note that economists label the vertical axis with the independent variable.* Observe that in graphing a function the "solutions" (x, y) that make the equation $y = f(x)$ true are handed to us. For each x in the domain of f, we have exactly one y obtained by evaluating $f(x)$. The resulting pair $(x, f(x))$ is a point on the graph and these are the only points on the graph of the equation $y = f(x)$.

A useful geometric observation is that the graph of a function has at most one point of intersection with any vertical line in the plane. Recall that the equation of a vertical line is necessarily of the form $x = a$, where a is a constant. If a is not in the domain of the function f, then $x = a$ will not intersect the graph of $y = f(x)$. If a is in the domain of the function f then $x = a$ will intersect the graph of $y = f(x)$ at the point $(a, f(a))$ and only there. Conversely, if a set of points in the plane has the property that any vertical line intersects the set at most once then the set of points is actually the graph of a function. (The domain of the function is the set of all real numbers a with the property that the line $x = a$ does intersect the given set of points and for such an a the corresponding function value is the y-coordinate of the unique point of intersection of the line $x = a$ and the given set of points.) This is the basis of the **vertical-line test** that we will discuss after Example 7.

EXAMPLE 4 Graph of the Square-Root Function

Graph $f(x) = \sqrt{x}$.

Solution: The graph is shown in Figure 2.16. We label the vertical axis as $f(x)$. Recall that $\sqrt{x}$ denotes the *principal* square root of x. Thus, $f(9) = \sqrt{9} = 3$, not ± 3. Also, we cannot choose negative values for x, because we don't want imaginary numbers for $\sqrt{x}$. That is, we must have $x \geq 0$. Let us now consider intercepts. If $f(x) = 0$, then $\sqrt{x} = 0$, or $x = 0$. Also, if $x = 0$, then $f(x) = 0$. Thus, the x-intercept and the vertical-axis intercept are the same, namely, $(0, 0)$.

NOW WORK PROBLEM 29

EXAMPLE 5 Graph of the Absolute-Value Function

Graph $p = G(q) = |q|$.

Solution: We use the independent variable q to label the horizontal axis. The function-value axis can be labeled either $G(q)$ or p. (See Figure 2.17.) Notice that the q- and p-intercepts are the same point, $(0, 0)$.

NOW WORK PROBLEM 31

DEFINITION

A *zero* of a function f is any value of x for which $f(x) = 0$.

q	0	1	-1	3	-3	5	-5
p	0	1	1	3	3	5	5

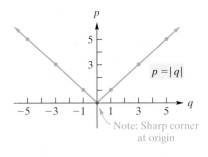

Note: Sharp corner at origin

FIGURE 2.17   Graph of $p = |q|$.

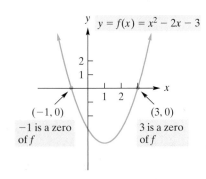

$(-1, 0)$
-1 is a zero of f

$(3, 0)$
3 is a zero of f

FIGURE 2.18   Zeros of function.

For example, a zero of the function $f(x) = 2x - 6$ is 3 because $f(3) = 2(3) - 6 = 0$. Here we call 3 a *real* zero because it is a real number. We note that zeros of f can be found by setting $f(x) = 0$ and solving for x. Thus, the zeros of a function are precisely the *x*-intercepts of its graph, because it is at these points that $f(x) = 0$.

To further illustrate, Figure 2.18 shows the graph of the function $y = f(x) = x^2 - 2x - 3$. The *x*-intercepts of the graph are -1 and 3. Hence, -1 and 3 are zeros of f, or equivalently, -1 and 3 are solutions to the equation $x^2 - 2x - 3 = 0$.

TECHNOLOGY

To solve the equation $x^3 = 3x - 1$ with a graphing calculator, we first express the equation in the form $f(x) = 0$:

$$f(x) = x^3 - 3x + 1 = 0$$

Next we graph f and then estimate the *x*-intercepts, either by using zoom and trace or by using the root operation. (See Figure 2.19.) Note that we defined our window for $-4 \le x \le 4$ and $-5 \le y \le 5$.

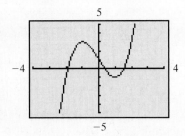

FIGURE 2.19   The roots of $x^3 - 3x + 1 = 0$ are approximately -1.88, 0.35, and 1.53.

Figure 2.20 shows the graph of a function $y = f(x)$. The point $(x, f(x))$ tells us that corresponding to the input number x on the horizontal axis is the output number $f(x)$ on the vertical axis, as indicated by the arrow. For example, corresponding to the input 4 is the output 3, so $f(4) = 3$.

From the shape of the graph, it seems reasonable to assume that, for any value of x, there is an output number, so the domain of f is all real numbers. Notice that the set of all *y*-coordinates of points on the graph is the set of all nonnegative numbers.

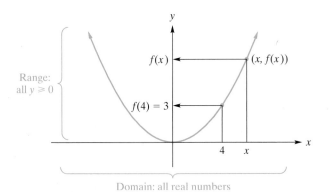

Range: all $y \ge 0$

$f(x)$

$(x, f(x))$

$f(4) = 3$

Domain: all real numbers

FIGURE 2.20   Domain, range, function values.

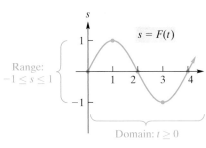

FIGURE 2.21 Domain, range, function values.

Thus, the range of f is all $y \geq 0$. This shows that we can make an "educated" guess about the domain and range of a function by looking at its graph. *In general, the domain consists of all x-values that are included in the graph, and the range is all y-values that are included.* For example, Figure 2.16 tells us that both the domain and range of $f(x) = \sqrt{x}$ are all nonnegative numbers. From Figure 2.17, it is clear that the domain of $p = G(q) = |q|$ is all real numbers and the range is all $p \geq 0$.

● **EXAMPLE 6**   **Domain, Range, and Function Values**

Figure 2.21 shows the graph of a function F. To the right of 4, assume that the graph repeats itself indefinitely. Then the domain of F is all $t \geq 0$. The range is $-1 \leq s \leq 1$. Some function values are

$$F(0) = 0 \quad F(1) = 1 \quad F(2) = 0 \quad F(3) = -1$$

NOW WORK PROBLEM 5 ●○

T E C H N O L O G Y

Using a graphing calculator, we can estimate the range of a function. The graph of

$$f(x) = 6x^4 - 8.1x^3 + 1$$

is shown in Figure 2.22. The lowest point on the graph corresponds to the minimum value of $f(x)$, and the range is all reals greater than or equal to this minimum. We can estimate this minimum y-value either by using trace and zoom or by selecting the "minimum" operation.

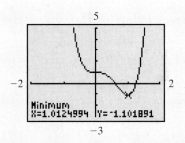

FIGURE 2.22 The range of $f(x) = 6x^4 - 8.1x^3 + 1$ is approximately $[-1.10, \infty)$.

PRINCIPLES IN PRACTICE 4

GRAPH OF A CASE-DEFINED FUNCTION

To encourage conservation, a gas company charges two rates. You pay $0.53 per therm for 0–70 therms and $0.74 for each therm over 70. Graph the case-defined function that represents the monthly cost of t therms of gas.

● **EXAMPLE 7**   **Graph of a Case-Defined Function**

Graph the case-defined function

$$f(x) = \begin{cases} x & \text{if } 0 \leq x < 3 \\ x - 1 & \text{if } 3 \leq x \leq 5 \\ 4 & \text{if } 5 < x \leq 7 \end{cases}$$

Solution: The domain of f is $0 \leq x \leq 7$. The graph is given in Figure 2.23, where the *hollow dot* means that the point is *not* included in the graph. Notice that the range of f is all real numbers y such that $0 \leq y \leq 4$.

NOW WORK PROBLEM 35 ●○

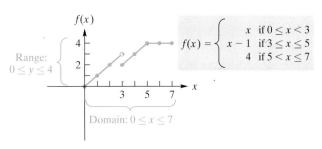

x	0	1	2	3	4	5	6	7
$f(x)$	0	1	2	2	3	4	4	4

FIGURE 2.23 Graph of case-defined function.

There is an easy way to tell whether a curve is the graph of a function. In Figure 2.24(a), notice that with the given x there are associated *two* values of y: y_1 and y_2. Thus, the curve is *not* the graph of a function of x. Looking at it another way, we have the following general rule, called the **vertical-line test.** If a *vertical* line L can be drawn that intersects a curve in at least two points, then the curve is *not* the graph of a function of x. When no such vertical line can be drawn, the curve *is* the graph of a function of x. Consequently, the curves in Figure 2.24 do not represent functions of x, but those in Figure 2.25 do.

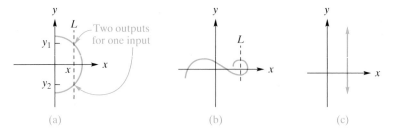

FIGURE 2.24 y is not a function of x.

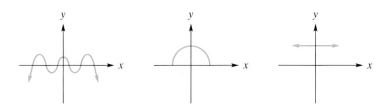

FIGURE 2.25   Functions of x.

● EXAMPLE 8   A Graph That Does Not Represent a Function of x

Graph $x = 2y^2$.

Solution: Here it is easier to choose values of y and then find the corresponding values of x. Figure 2.26 shows the graph. By the vertical-line test, the equation $x = 2y^2$ does not define a function of x.

NOW WORK PROBLEM 39

After we have determined whether a curve is the graph of a function, perhaps using the vertical-line test, there is an easy way to tell whether the function in question is one-to-one. In Figure 2.20 we see that $f(4) = 3$ and, apparently, also $f(-4) = 3$. Since the distinct input values -4 and 4 produce the same output, the function is not one-to-one. Looking at it another way, we have the following general rule, called the **horizontal-line test**. If a *horizontal* line L can be drawn that intersects the graph of a function in at least two points, then the function is *not* one-to-one. When no such horizontal line can be drawn, the function is one-to-one.

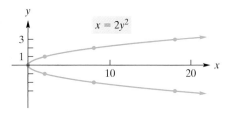

x	0	2	2	8	8	18	18
y	0	1	-1	2	-2	3	-3

FIGURE 2.26   Graph of $x = 2y^2$.

Problems 2.5

In Problems 1 and 2, locate and label each of the points, and give the quadrant, if possible, in which each point lies.

1. $(2, 7), (8, -3), (-\frac{1}{2}, -2), (0, 0)$

2. $(-4, 5), (3, 0), (1, 1), (0, -6)$

3. Figure 2.27(a) shows the graph of $y = f(x)$.

    (a) Estimate $f(0)$, $f(2)$, $f(4)$, and $f(-2)$.
    (b) What is the domain of f?
    (c) What is the range of f?
    (d) What is a real zero of f?

4. Figure 2.27(b) shows the graph of $y = f(x)$.

    (a) Estimate $f(0)$ and $f(2)$.
    (b) What is the domain of f?
    (c) What is the range of f?
    (d) What is a real zero of f?

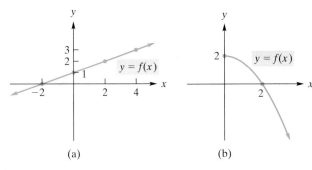

      (a)   (b)

FIGURE 2.27 Diagram for Problems 3 and 4.

*5. Figure 2.28(a) shows the graph of $y = f(x)$.

    (a) Estimate $f(0)$, $f(1)$, and $f(-1)$.
    (b) What is the domain of f?
    (c) What is the range of f?
    (d) What is a real zero of f?

6. Figure 2.28(b) shows the graph of $y = f(x)$.

    (a) Estimate $f(0)$, $f(2)$, $f(3)$, and $f(4)$.
    (b) What is the domain of f?
    (c) What is the range of f?
    (d) What is a real zero of f?

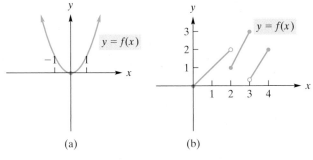

      (a)   (b)

FIGURE 2.28 Diagram for Problems 5 and 6.

In Problems 7–20, determine the intercepts of the graph of each equation, and sketch the graph. Based on your graph, is y a function of x, and, if so, is it one-to-one and what are the domain and range?

7. $y = 2x$

8. $y = x + 1$

*9. $y = 3x - 5$

10. $y = 3 - 2x$

*11. $y = x^4$

*13. $x = 0$

15. $y = x^3$

17. $x = -|y|$

19. $2x + y - 2 = 0$

12. $y = \frac{2}{x^3}$

14. $y = 4x^2 - 16$

16. $x = -9$

18. $x^2 = y^2$

20. $x + y = 1$

In Problems 21–34, graph each function and give the domain and range. Also, determine the intercepts.

21. $s = f(t) = 4 - t^2$

22. $f(x) = 5 - 2x^2$

23. $y = h(x) = 3$

24. $g(s) = -17$

25. $y = h(x) = x^2 - 4x + 1$

26. $y = f(x) = x^2 + 2x - 8$

27. $f(t) = -t^3$

28. $p = h(q) = 1 + 2q + q^2$

*29. $s = f(t) = \sqrt{t^2 - 9}$

30. $F(r) = -\frac{1}{r}$

*31. $f(x) = |2x - 1|$

32. $v = H(u) = |u - 3|$

33. $F(t) = \frac{16}{t^2}$

34. $y = f(x) = \frac{2}{x - 4}$

In Problems 35–38, graph each case-defined function and give the domain and range.

*35. $c = g(p) = \begin{cases} p + 1 & \text{if } 0 \le p < 7 \\ 5 & \text{if } p \ge 7 \end{cases}$

36. $\phi(x) = \begin{cases} 3x + 2 & \text{if } -1 \le x < 3 \\ 20 - x^2 & \text{if } x \ge 3 \end{cases}$

37. $g(x) = \begin{cases} x + 6 & \text{if } x \ge 3 \\ x^2 & \text{if } x < 3 \end{cases}$

38. $f(x) = \begin{cases} x + 1 & \text{if } 0 < x \le 3 \\ 4 & \text{if } 3 < x \le 5 \\ x - 1 & \text{if } x > 5 \end{cases}$

*39. Which of the graphs in Figure 2.29 represent functions of x?

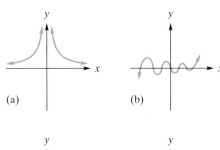

    (a)   (b)

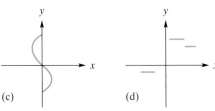

    (c)   (d)

FIGURE 2.29 Diagram for Problem 39.

40. Which of the graphs in Figure 2.30 represent one-to-one functions of x?

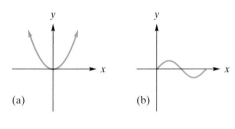

(a)   (b)

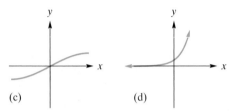

(c)   (d)

FIGURE 2.30 Diagram for Problem 40.

41. Debt Payments Tara has charged $2400 on her credit cards. She plans to pay them off at the rate of $275 per month. Write an equation to represent the amount she owes, excluding any finance charges, after she has made x payments and identify the intercepts.

42. Pricing To encourage an even flow of customers, a restaurant varies the price of an item throughout the day. From 6:00 P.M. to 8:00 P.M., customers pay full price. At lunch, from 10:30 A.M. until 2:30 P.M., customers pay half price. From 2:30 P.M. until 4:30 P.M., customers get a dollar off the lunch price. From 4:30 P.M. until 6:00 P.M., customers get $5.00 off the dinner price. From 8:00 P.M. until closing time at 10:00 P.M., customers get $5.00 off the dinner price. Graph the case-defined function that represents the cost of an item throughout the day for a dinner price of $18.

43. Supply Schedule Given the following supply schedule (see Example 6 of Section 2.1), plot each quantity–price pair by choosing the horizontal axis for the possible quantities. Approximate the points in between the data by connecting the data points with a smooth curve. The result is a *supply curve*. From the graph, determine the relationship between price and supply. (That is, as price increases, what happens to the quantity supplied?) Is price per unit a function of quantity supplied?

Quantity Supplied per Week, q	Price per Unit, p
30	$10
100	20
150	30
190	40
210	50

44. Demand Schedule The following table is called a *demand schedule*. It indicates the quantities of brand X that consumers will demand (that is, purchase) each week at certain prices per unit (in dollars). Plot each quantity–price pair by choosing the vertical axis for the possible prices. Connect the points with a smooth curve. In this way, we

approximate points in between the given data. The result is called a *demand curve*. From the graph, determine the relationship between the price of brand X and the amount that will be demanded. (That is, as price decreases, what happens to the quantity demanded?) Is price per unit a function of quantity demanded?

Quantity Demanded, q	Price per Unit, p
5	$20
10	10
20	5
25	4

45. Inventory Sketch the graph of

$$y = f(x) = \begin{cases} -100x + 1000 & \text{if } 0 \le x < 7 \\ -100x + 1700 & \text{if } 7 \le x < 14 \\ -100x + 2400 & \text{if } 14 \le x < 21 \end{cases}$$

A function such as this might describe the inventory y of a company at time x.

46. Psychology In a psychological experiment on visual information, a subject briefly viewed an array of letters and was then asked to recall as many letters as possible from the array. The procedure was repeated several times. Suppose that y is the average number of letters recalled from arrays with x letters. The graph of the results approximately fits the graph of

$$y = f(x) = \begin{cases} x & \text{if } 0 \le x \le 4 \\ \frac{1}{2}x + 2 & \text{if } 4 < x \le 5 \\ 4.5 & \text{if } 5 < x \le 12 \end{cases}$$

Plot this function.[6]

In Problems 47–50, use a graphing calculator to find all real roots, if any, of the given equation. Round answers to two decimal places.

47. $5x^3 + 7x = 3$

48. $x^2(x - 3) = 2x^4 - 1$

49. $(9x + 3.1)^2 = 7.4 - 4x^2$

50. $(x - 2)^3 = x^2 - 3$

In Problems 51–54, use a graphing calculator to find all real zeros of the given function. Round answers to two decimal places.

51. $f(x) = x^3 + 5x + 7$

52. $f(x) = 2x^4 - 1.5x^3 + 2$

53. $g(x) = x^4 - 1.7x^2 + 2x$

54. $g(x) = \sqrt{3}x^5 - 4x^2 + 1$

In Problems 55–57, use a graphing calculator to find (a) the maximum value of $f(x)$ and (b) the minimum value of $f(x)$ for the indicated values of x. Round answers to two decimal places.

55. $f(x) = x^4 - 4.1x^3 + x^2 + 10$ $1 \le x \le 4$

6 Adapted from G. R. Loftus and E. F. Loftus, *Human Memory: The Processing of Information* (New York: Lawrence Erlbaum Associates, Inc., distributed by the Halsted Press, Division of John Wiley & Sons, Inc., 1976).

56. $f(x) = x(2.1x^2 - 3)^2 - x^3 + 1 \quad -1 \le x \le 1$

57. $f(x) = \dfrac{x^2 - 4}{2x - 5} \quad 3 \le x \le 5$

58. From the graph of $f(x) = \sqrt{2}x^3 + 1.1x^2 + 4$, find (a) the range and (b) the intercepts. Round values to two decimal places.

59. From the graph of $f(x) = 1 - 4x^3 - x^4$, find (a) the maximum value of $f(x)$, (b) the range of f, and (c) the real zeros of f. Round values to two decimal places.

60. From the graph of $f(x) = \dfrac{x^3 + 1.1}{3.8 + x^{2/3}}$, find (a) the range of f and (b) the intercepts. (c) Does f have any real zeros? Round values to two decimal places.

61. Graph $f(x) = \dfrac{4.1x^3 + \sqrt{2}}{x^2 - 3}$ for $2 \le x \le 5$. Determine (a) the maximum value of $f(x)$, (b) the minimum value of $f(x)$, (c) the range of f, and (d) all intercepts. Round values to two decimal places.

2.6 Symmetry

OBJECTIVE

To study symmetry about the x-axis, the y-axis, and the origin, and to apply symmetry to curve sketching.

Examining the graphical behavior of equations is a basic part of mathematics. In this section, we examine equations to determine whether their graphs have *symmetry*. In a later chapter, you will see that calculus is a *great* aid in graphing because it helps determine the shape of a graph. It provides powerful techniques for determining whether or not a curve "wiggles" between points.

Consider the graph of $y = x^2$ in Figure 2.31. The portion to the left of the y-axis is the reflection (or mirror image) through the y-axis of that portion to the right of the y-axis, and vice versa. More precisely, if (a, b) is any point on this graph, then the point $(-a, b)$ must also lie on the graph. We say that this graph is *symmetric about the y-axis*.

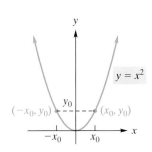

FIGURE 2.31 Symmetry about the y-axis.

DEFINITION

A graph is ***symmetric about the y-axis*** if and only if $(-a, b)$ lies on the graph when (a, b) does.

● EXAMPLE 1 *y*-Axis Symmetry

Use the preceding definition to show that the graph of $y = x^2$ is symmetric about the y-axis.

Solution: Suppose (a, b) is *any* point on the graph of $y = x^2$. Then

$$b = a^2$$

We must show that the coordinates of $(-a, b)$ satisfy $y = x^2$. But

$$(-a)^2 = a^2 = b$$

shows this to be true. Thus we have *proved* with simple algebra what the picture of the graph led us to believe: The graph of $y = x^2$ is symmetric about the y-axis.

NOW WORK PROBLEM 7 ●●

When one is testing for symmetry in Example 1, (a, b) can be any point on the graph. In the future, for convenience, we write (x, y) for a typical point on the graph. This means that a graph is symmetric about the y-axis if replacing x by $-x$ in its equation results in an equivalent equation.

Another type of symmetry is shown by the graph of $x = y^2$ in Figure 2.32. Here the portion below the x-axis is the reflection through the x-axis of that portion above the x-axis, and vice versa. If the point (x, y) lies on the graph, then $(x, -y)$ also lies on it. This graph is said to be *symmetric about the x-axis*.

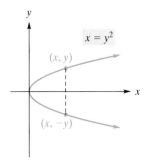

FIGURE 2.32 Symmetry about the x-axis.

DEFINITION

A graph is ***symmetric about the x-axis*** if and only if $(x, -y)$ lies on the graph when (x, y) does.

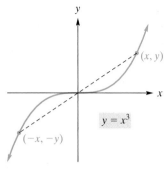

FIGURE 2.33 Symmetry about the origin.

Thus, the graph of an equation in x and y has x-axis symmetry if replacing y by $-y$ results in an equivalent equation. For example, applying this test to the graph of $x = y^2$, we see that $(-y)^2 = x$ if and only if $y^2 = x$, simply because $(-y)^2 = y^2$. Hence the graph of $x = y^2$ is symmetric about the x-axis.

A third type of symmetry, *symmetry about the origin*, is illustrated by the graph of $y = x^3$ (Figure 2.33). Whenever the point (x, y) lies on the graph, $(-x, -y)$ also lies on it.

DEFINITION

A graph is **symmetric about the origin** if and only if $(-x, -y)$ lies on the graph when (x, y) does.

Thus, the graph of an equation in x and y has symmetry about the origin if replacing x by $-x$ and y by $-y$ results in an equivalent equation. For example, applying this test to the graph of $y = x^3$ shown in Figure 2.33 gives

$$-y = (-x)^3$$
$$-y = -x^3$$
$$y = x^3$$

where all three equations are equivalent, in particular the first and last. Accordingly, the graph is symmetric about the origin.

Table 2.1 summarizes the tests for symmetry. When we know that a graph has symmetry, we can sketch it by plotting fewer points than would otherwise be needed.

TABLE 2.1 Tests for Symmetry

Symmetry about x-axis	Replace y by $-y$ in given equation. Symmetric if equivalent equation is obtained.
Symmetry about y-axis	Replace x by $-x$ in given equation. Symmetric if equivalent equation is obtained.
Symmetry about origin	Replace x by $-x$ and y by $-y$ in given equation. Symmetric if equivalent equation is obtained.

● EXAMPLE 2 Graphing with Intercepts and Symmetry

Test $y = \dfrac{1}{x}$ for symmetry about the x-axis, the y-axis, and the origin. Then find the intercepts and sketch the graph.

Solution:

Symmetry *x-axis:* Replacing y by $-y$ in $y = 1/x$ gives

$$-y = \frac{1}{x} \quad \text{equivalently} \quad y = -\frac{1}{x}$$

which is not equivalent to the given equation. Thus, the graph is *not* symmetric about the x-axis.

y-axis: Replacing x by $-x$ in $y = 1/x$ gives

$$y = \frac{1}{-x} \quad \text{equivalently} \quad y = -\frac{1}{x}$$

which is not equivalent to the given equation. Hence, the graph is *not* symmetric about the y-axis.

Origin: Replacing x by $-x$ and y by $-y$ in $y = 1/x$ gives

$$-y = \frac{1}{-x} \quad \text{equivalently} \quad y = \frac{1}{x}$$

which is the given equation. Consequently, the graph *is* symmetric about the origin.

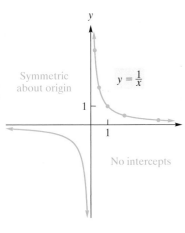

Symmetric
about origin

$y = \frac{1}{x}$

No intercepts

x	$\frac{1}{4}$	$\frac{1}{2}$	1	2	4
y	4	2	1	$\frac{1}{2}$	$\frac{1}{4}$

FIGURE 2.34 Graph of $y = \frac{1}{x}$.

Intercepts   Since x cannot be 0, the graph has no y-intercept. If y is 0, then $0 = 1/x$, and this equation has no solution. Thus, no x-intercept exists.

Discussion   Because no intercepts exist, the graph cannot intersect either axis. If $x > 0$, we obtain points only in quadrant I. Figure 2.34 shows the portion of the graph in quadrant I. By symmetry, we reflect that portion through the origin to obtain the entire graph.

NOW WORK PROBLEM 9

● **EXAMPLE 3**   **Graphing with Intercepts and Symmetry**

Test $y = f(x) = 1 - x^4$ for symmetry about the x-axis, the y-axis, and the origin. Then find the intercepts and sketch the graph.

Solution:

Symmetry   *x-axis:* Replacing y by $-y$ in $y = 1 - x^4$ gives

$$-y = 1 - x^4 \quad \text{equivalently} \quad y = -1 + x^4$$

which is not equivalent to the given equation. Thus, the graph is *not* symmetric about the x-axis.

y-axis: Replacing x by $-x$ in $y = 1 - x^4$ gives

$$y = 1 - (-x)^4 \quad \text{equivalently} \quad y = 1 - x^4$$

which is the given equation. Hence, the graph *is* symmetric about the y-axis.

Origin: Replacing x by $-x$ and y by $-y$ in $y = 1 - x^4$ gives

$$-y = 1 - (-x)^4 \quad \text{equivalently} \quad -y = 1 - x^4 \quad \text{equivalently} \quad y = -1 + x^4$$

which is not equivalent to the given equation. Thus, the graph is *not* symmetric about the origin.

Intercepts   Testing for x-intercepts, we set $y = 0$ in $y = 1 - x^4$. Then

$$1 - x^4 = 0$$
$$(1 - x^2)(1 + x^2) = 0$$
$$(1 - x)(1 + x)(1 + x^2) = 0$$
$$x = 1 \quad \text{or} \quad x = -1$$

The x-intercepts are therefore $(1, 0)$ and $(-1, 0)$. Testing for y-intercepts, we set $x = 0$. Then $y = 1$, so $(0, 1)$ is the only y-intercept.

Discussion   If the intercepts and some points (x, y) to the right of the y-axis are plotted, we can sketch the *entire* graph by using symmetry about the y-axis (Figure 2.35).

NOW WORK PROBLEM 19

x	y
0	1
$\frac{1}{2}$	$\frac{15}{16}$
$\frac{3}{4}$	$\frac{175}{256}$
1	0
$\frac{3}{2}$	$-\frac{65}{16}$

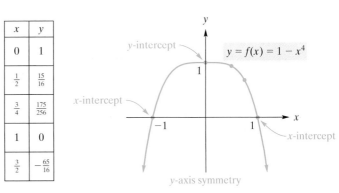

$y = f(x) = 1 - x^4$

y-intercept

x-intercept

x-intercept

y-axis symmetry

FIGURE 2.35   Graph of $y = 1 - x^4$.

The only *function* whose graph is
symmetric about the *x*-axis is the
function constantly 0.

The constant function $f(x) = 0$, for all x, is easily seen to be symmetric about the *x*-axis. In Example 3, we showed that the graph of $y = f(x) = 1 - x^4$ does not have *x*-axis symmetry. For any *function* f, suppose that the graph of $y = f(x)$ has *x*-axis symmetry. According to the definition, this means that we also have $-y = f(x)$. This tells us that for an arbitrary x in the domain of f we have $f(x) = y$ and $f(x) = -y$. Since for a function each *x*-value determines a unique *y*-value, we must have $y = -y$, and this implies $y = 0$. Since x was arbitrary, it follows that if the graph of a *function* is symmetric about the *x*-axis, then the function must be the constant 0.

● EXAMPLE 4   Graphing with Intercepts and Symmetry

Test the graph of $4x^2 + 9y^2 = 36$ for intercepts and symmetry. Sketch the graph.

Solution:

Intercepts   If $y = 0$, then $4x^2 = 36$, so $x = \pm 3$. Thus, the *x*-intercepts are $(3, 0)$ and $(-3, 0)$. If $x = 0$, then $9y^2 = 36$, so $y = \pm 2$. Hence, the *y*-intercepts are $(0, 2)$ and $(0, -2)$.

Symmetry   *x-axis:* Replacing y by $-y$ in $4x^2 + 9y^2 = 36$ gives

$$4x^2 + 9(-y)^2 = 36 \quad \text{equivalently} \quad 4x^2 + 9y^2 = 36$$

which is the original equation, so there is symmetry about the *x*-axis.

y-axis: Replacing x by $-x$ in $4x^2 + 9y^2 = 36$ gives

$$4(-x)^2 + 9y^2 = 36 \quad \text{equivalently} \quad 4x^2 + 9y^2 = 36$$

which is the original equation, so there is also symmetry about the *y*-axis.

Origin: Replacing x by $-x$ and y by $-y$ in $4x^2 + 9y^2 = 36$ gives

$$4(-x)^2 + 9(-y)^2 = 36 \quad \text{equivalently} \quad 4x^2 + 9y^2 = 36$$

which is the original equation, so the graph is also symmetric about the origin.

Discussion   In Figure 2.36, the intercepts and some points in the first quadrant are plotted. The points in that quadrant are then connected by a smooth curve. By symmetry about the *x*-axis, the points in the fourth quadrant are obtained. Then, by symmetry about the *y*-axis, the complete graph is found. There are other ways of graphing the equation by using symmetry. For example, after plotting the intercepts and some points in the first quadrant, we can obtain the points in the third quadrant by symmetry about the origin. By symmetry about the *x*-axis (or *y*-axis), we can then obtain the entire graph.

NOW WORK PROBLEM 23   ●●●

This fact can be a time-saving device in
checking for symmetry.

In Example 4, the graph is symmetric about the *x*-axis, the *y*-axis, and the origin. It can be shown that **for any graph, if any two of the three types of symmetry exist, then the remaining type must also exist.**

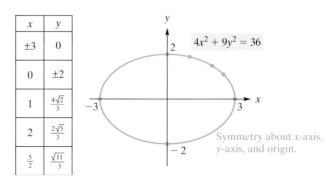

x	y
± 3	0
0	± 2
1	$\frac{4\sqrt{2}}{3}$
2	$\frac{2\sqrt{5}}{3}$
$\frac{5}{2}$	$\frac{\sqrt{11}}{3}$

Symmetry about *x*-axis,
y-axis, and origin.

FIGURE 2.36   Graph of $4x^2 + 9y^2 = 36$.

● EXAMPLE 5   **Symmetry about the Line $y = x$**

DEFINITION

A graph is ***symmetric about the the line*** $y = x$ if and only if (b, a) lies on the graph when (a, b) does.

Another way of stating the definition is to say that interchanging the roles of x and y in the given equation results in an equivalent equation.

Use the preceding definition to show that $x^2 + y^2 = 1$ is symmetric about the line $y = x$.

Solution: Interchanging the roles of x and y produces $y^2 + x^2 = 1$, which is equivalent to $x^2 + y^2 = 1$. Thus $x^2 + y^2 = 1$ is symmetric about $y = x$.

The point with coordinates (b, a) is the mirror image in the line $y = x$ of the point (a, b). If f is a one-to-one function, $b = f(a)$ if and only if $a = f^{-1}(b)$. Thus the graph of f^{-1} is the mirror image in the line $y = x$ of the graph of f. It is interesting to note that for *any* function f we can form the mirror image of the graph of f. However, the resulting graph need not be the graph of a function. For this mirror image to be itself the graph of a function, it must pass the vertical-line test. However vertical lines and horizontal lines are mirror images in the line $y = x$, and we see that for the mirror image of the graph of f to pass the vertical-line test is for the graph of f to pass the horizontal-line test. This last happens precisely if f is one-to-one, which is the case if and only if f has an inverse.

● EXAMPLE 6   **Symmetry and Inverse Functions**

Sketch the graph of $g(x) = 2x + 1$ and its inverse in the same plane.

Solution: As we shall study in greater detail in Chapter 3, the graph of g is the straight line with slope (see Section 3.1) 2 and y-intercept 1. This line, the line $y = x$, and the reflection of $y = 2x + 1$ in $y = x$ are shown in Figure 2.37.

NOW WORK PROBLEM 27   ●●

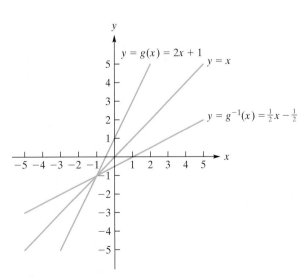

FIGURE 2.37   Graph of $y = g(x)$ and $y = g^{-1}(x)$.

Problems 2.6

In Problems 1–16, find the x- and y-intercepts of the graphs of the equations. Also, test for symmetry about the x-axis, the y-axis, the origin, and the line $y = x$. Do not sketch the graphs.

1. $y = 5x$

2. $y = f(x) = x^2 - 4$

3. $2x^2 + y^2 x^4 = 8 - y$

4. $x = y^3$

5. $16x^2 - 9y^2 = 25$

6. $y = 57$

*7. $x = -2$

8. $y = |2x| - 2$

*9. $x = -y^{-4}$

10. $y = \sqrt{x^2 - 25}$

11. $x - 4y - y^2 + 21 = 0$

12. $x^2 + xy + y^3 = 0$

13. $y = f(x) = \dfrac{x^3 - 2x^2 + x}{x^2 + 1}$

14. $x^2 + xy + y^2 = 0$

15. $y = \dfrac{3}{x^3 + 8}$

16. $y = \dfrac{x^4}{x + y}$

In Problems 17–24, find the x- and y-intercepts of the graphs of the equations. Also, test for symmetry about the x-axis, the y-axis, the origin, and the line $y = x$. Then sketch the graphs.

17. $3x + y^2 = 9$

18. $x - 1 = y^4 + y^2$

*19. $y = f(x) = x^3 - 4x$

20. $3y = 5x - x^3$

21. $|x| - |y| = 0$

22. $x^2 + y^2 = 16$

*23. $9x^2 + 4y^2 = 25$

24. $x^2 - y^2 = 4$

25. Prove that the graph of $y = f(x) = 5 - 1.96x^2 - \pi x^4$ is symmetric about the y-axis, and then graph the function. (a) Make use of symmetry, where possible, to find all intercepts. Determine (b) the maximum value of $f(x)$ and (c) the range of f. Round all values to two decimal places.

26. Prove that the graph of $y = f(x) = 2x^4 - 7x^2 + 5$ is symmetric about the y-axis, and then graph the function. Find all real zeros of f. Round your answers to two decimal places.

*27. Sketch the graph of $f(x) = -3x + 2$ and its inverse in the same plane.

OBJECTIVE

To become familiar with the shapes of the graphs of six basic functions and to consider translation, reflection, and vertical stretching or shrinking of the graph of a function.

2.7 Translations and Reflections

Up to now, our approach to graphing has been based on plotting points and making use of any symmetry that exists. But this technique is not necessarily the preferred way. Later in this book, we will analyze graphs by using other techniques. However, some functions and their associated graphs occur so frequently that we find it worthwhile to memorize them. Figure 2.38 shows six such functions.

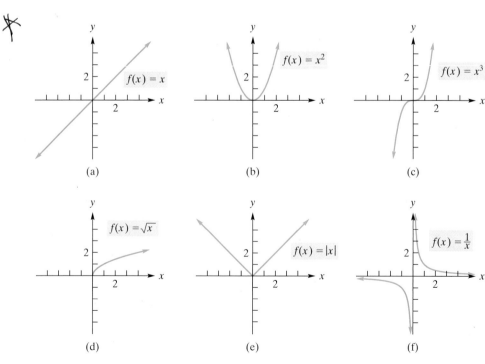

FIGURE 2.38   Functions frequently used.

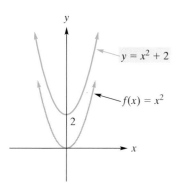

FIGURE 2.39   Graph of
$y = x^2 + 2$.

At times, by altering a function through an *algebraic* manipulation, the graph of the new function can be obtained from the graph of the original function by performing a *geometric* manipulation. For example, we can use the graph of $f(x) = x^2$ to graph $y = x^2 + 2$. Note that $y = f(x) + 2$. Thus, for each x, the corresponding ordinate for the graph of $y = x^2 + 2$ is 2 more than the ordinate for the graph of $f(x) = x^2$. This means that the graph of $y = x^2 + 2$ is simply the graph of $f(x) = x^2$, shifted, or *translated*, 2 units upward. (See Figure 2.39.) We say that the graph of $y = x^2 + 2$ is a *transformation* of the graph of $f(x) = x^2$. Table 2.2 gives a list of basic types of transformations.

TABLE 2.2 Transformations, $c > 0$

Equation	How to Transform Graph of $y = f(x)$ to Obtain Graph of Equation
$y = f(x) + c$	shift c units upward
$y = f(x) - c$	shift c units downward
$y = f(x - c)$	shift c units to right
$y = f(x + c)$	shift c units to left
$y = -f(x)$	reflect about x-axis
$y = f(-x)$	reflect about y-axis
$y = cf(x)$ $c > 1$	vertically stretch away from x-axis by a factor of c
$y = cf(x)$ $c < 1$	vertically shrink toward x-axis by a factor of c

● **EXAMPLE 1   Horizontal Translation**

Sketch the graph of $y = (x - 1)^3$.

Solution: We observe that $(x - 1)^3$ is x^3 with x replaced by $x - 1$. Thus, if $f(x) = x^3$, then $y = (x - 1)^3 = f(x - 1)$, which has the form $f(x - c)$, where $c = 1$. From Table 2.2, the graph of $y = (x - 1)^3$ is the graph of $f(x) = x^3$, shifted 1 unit to the right. (See Figure 2.40.)

NOW WORK PROBLEM 3 ●●

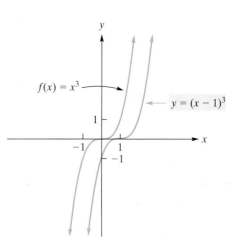

FIGURE 2.40   Graph of $y = (x - 1)^3$.

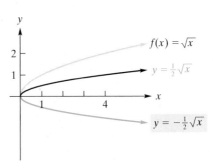

FIGURE 2.41   To graph $y = -\frac{1}{2}\sqrt{x}$, shrink $y = \sqrt{x}$ and reflect result about x-axis.

● EXAMPLE 2   Shrinking and Reflection

Sketch the graph of $y = -\frac{1}{2}\sqrt{x}$.

Solution: We can do this problem in two steps. First, observe that $\frac{1}{2}\sqrt{x}$ is $\sqrt{x}$ multiplied by $\frac{1}{2}$. Thus, if $f(x) = \sqrt{x}$, then $\frac{1}{2}\sqrt{x} = \frac{1}{2}f(x)$, which has the form $cf(x)$, where $c = \frac{1}{2}$. So the graph of $y = \frac{1}{2}\sqrt{x}$ is the graph of f shrunk vertically toward the x-axis by a factor of $\frac{1}{2}$ (transformation 8, Table 2.2; see Figure 2.41). Second, the minus sign in $y = -\frac{1}{2}\sqrt{x}$ causes a reflection in the graph of $y = \frac{1}{2}\sqrt{x}$ about the x-axis (transformation 5, Table 2.2; see Figure 2.41).

NOW WORK PROBLEM 5   ●●●

Problems 2.7

In Problems 1–12, use the graphs of the functions in Figure 2.38 and transformation techniques to plot the given functions.

1. $y = x^3 - 1$

2. $y = -x^2$

*3. $y = \dfrac{1}{x - 2}$

4. $y = \sqrt{x + 2}$

*5. $y = \dfrac{2}{3x}$

6. $y = |x| - 2$

7. $y = |x + 1| - 2$

8. $y = -\dfrac{1}{3}\sqrt{x}$

9. $y = 1 - (x - 1)^2$

10. $y = (x - 1)^2 + 1$

11. $y = \sqrt{-x}$

12. $y = \dfrac{5}{2 - x}$

In Problems 13–16, describe what must be done to the graph of $y = f(x)$ to obtain the graph of the given equation.

13. $y = -2f(x + 3) + 2$

14. $y = f(x + 3) - 4$

15. $y = f(-x) - 5$

16. $y = f(3x)$

17. Graph the function $y = \sqrt[3]{x} + k$ for $k = 0, 1, 2, 3, -1, -2,$ and -3. Observe the vertical translations compared to the first graph.

18. Graph the function $y = \sqrt[5]{x + k}$ for $k = 0, 1, 2, 3, -1, -2,$ and -3. Observe the horizontal translations compared to the first graph.

19. Graph the function $y = k\sqrt[3]{x}$ for $k = 1, 2, \frac{1}{2},$ and 3. Observe the vertical stretching and shrinking compared to the first graph. Graph the function for $k = -2$. Observe that the graph is the same as that obtained by stretching the reflection of $y = \sqrt[3]{x}$ about the x-axis by a factor of 2.

2.8 Review

Important Terms and Symbols   Examples

Section 2.1	**Functions**	Examples
	function domain range independent variable	Ex. 2, p. 78
	dependent variable function value, $f(x)$	Ex. 3, p. 79
	difference quotient, $\dfrac{f(x + h) - f(x)}{h}$	Ex. 4, p. 79
	demand function supply function	Ex. 5, Ex. 6 p. 80
Section 2.2	**Special Functions**	
	constant function polynomial function, (linear and quadratic)	Ex. 1, Ex. 2, p. 82, 83
	rational function case-defined function	Ex. 3, Ex. 4, p. 83
	absolute value, $|x|$ factorial, $r!$	Ex. 5, Ex. 6, p. 84

Summary

A function f is a rule of correspondence that assigns exactly one output number $f(x)$ to each input number x. Usually, a function is specified by an equation that indicates what must be done to an input x to obtain $f(x)$. To obtain a particular function value $f(a)$, we replace each x in the equation by a.

The domain of a function consists of all input numbers, and the range consists of all output numbers. Unless otherwise specified, the domain of f consists of all real numbers x for which $f(x)$ is also a real number.

Some special types of functions are constant functions, polynomial functions, and rational functions. A function that is defined by more than one expression is called a case-defined function.

A function has an inverse if and only if it is one-to-one.

In economics, supply (demand) functions give a correspondence between the price p of a product and the number of units q of the product that producers (consumers) will supply (buy) at that price.

Two functions f and g can be combined to form a sum, difference, product, quotient, or composite as follows:

$$(f + g)(x) = f(x) + g(x)$$
$$(f - g)(x) = f(x) - g(x)$$
$$(fg)(x) = f(x)g(x)$$
$$\left(\frac{f}{g}\right)(x) = \frac{f(x)}{g(x)}$$
$$(f \circ g)(x) = f(g(x))$$

A rectangular coordinate system allows us to represent equations in two variables, in particular these arising from functions, geometrically. The graph of an equation in x and y consists of all points (x, y) that correspond to the solutions of the equation. We plot a sufficient number of points and connect them (where appropriate) so that the basic shape of the graph is apparent. Points where the graph intersects the x- and y-axes are called x-intercepts and y-intercepts, respectively. An x-intercept is found by letting y be 0 and solving for x; a y-intercept is found by letting x be 0 and solving for y.

The graph of a function f is the graph of the equation $y = f(x)$ and consists of all points $(x, f(x))$ such that x is in the domain of f. The zeros of f are the values of x for which $f(x) = 0$. From the graph of a function, it is easy to determine the domain and range.

The fact that a graph represents a function can be determined by using the vertical-line test. A vertical line cannot cut the graph of a function at more than one point.

The fact that a function is one-to-one can be determined by using the horizontal-line test on its graph. A horizontal line cannot cut the graph of a one-to-one function at more than one point. When a function passes the horizontal-line test, the graph of the inverse can be obtained by reflecting the original graph in the line $y = x$.

When the graph of an equation has symmetry, the mirror-image effect allows us to sketch the graph by plotting fewer points than would otherwise be needed. The tests for symmetry are as follows:

Symmetry about x-axis	Replace y by $-y$ in given equation. Symmetric if equivalent equation is obtained.
Symmetry about y-axis	Replace x by $-x$ in given equation. Symmetric if equivalent equation is obtained.
Symmetry about origin	Replace x by $-x$ and y by $-y$ in given equation. Symmetric if equivalent equation is obtained.
Symmetry about $y = x$	Interchange x and y in given equation. Symmetric if equivalent equation is obtained.

Sometimes the graph of a function can be obtained from that of a familiar function by means of a vertical shift upward or downward, a horizontal shift to the right or left, a reflection about the x-axis or y-axis, or a vertical stretching or shrinking away from or toward the x-axis. Such transformations are indicated in Table 2.2 in Section 2.7.

Review Problems

Problem numbers shown in color indicate problems suggested for use as a practice chapter test.

In Problems 1–6, give the domain of each function.

1. $f(x) = \dfrac{x}{x^2 - 6x + 5}$

2. $g(x) = x^4 + 5|x - 1|$

3. $F(t) = 7t + 4t^2$

4. $G(x) = 18$

5. $h(x) = \dfrac{\sqrt{x}}{x - 1}$

6. $H(s) = \dfrac{\sqrt{s - 5}}{4}$

In Problems 7–14, find the function values for the given function.

7. $f(x) = 3x^2 - 4x + 7$; $f(0), f(-3), f(5), f(t)$

8. $h(x) = 7$; $h(4), h\left(\dfrac{1}{100}\right), h(-156), h(x + 4)$

9. $G(x) = \sqrt[4]{x - 3}$; $G(3), G(19), G(t + 1), G(x^3)$

10. $F(x) = \dfrac{x - 3}{x + 4}$; $F(-1), F(0), F(5), F(x + 3)$

11. $h(u) = \dfrac{\sqrt{u + 4}}{u}$; $h(5), h(-4), h(x), h(u - 4)$

12. $H(s) = \dfrac{(s - 4)^2}{3}$; $H(-2), H(7), H\left(\dfrac{1}{2}\right), H(x^2)$

13. $f(x) = \begin{cases} -3 & \text{if } x < 1 \\ 4 + x^2 & \text{if } x > 1 \end{cases}$;

$f(4), f(-2), f(0), f(1)$

14. $f(q) = \begin{cases} -q + 1 & \text{if } -1 \le q < 0 \\ q^2 + 1 & \text{if } 0 \le q < 5 \\ q^3 - 99 & \text{if } 5 \le q \le 7 \end{cases}$;

$f\left(-\dfrac{1}{2}\right), f(0), f\left(\dfrac{1}{2}\right), f(5), f(6)$

In Problems 15–18, find (a) $f(x + h)$ and (b) $\dfrac{f(x + h) - f(x)}{h}$, and simplify your answers.

15. $f(x) = 3 - 7x$

16. $f(x) = 11x^2 + 4$

17. $f(x) = 4x^2 + 2x - 5$

18. $f(x) = \dfrac{7}{x + 1}$

19. If $f(x) = 3x - 1$ and $g(x) = 2x + 3$, find the following.

(a) $(f + g)(x)$

(b) $(f + g)(4)$

(c) $(f - g)(x)$

(d) $(fg)(x)$

(e) $(fg)(1)$

(f) $\dfrac{f}{g}(x)$

(g) $(f \circ g)(x)$

(h) $(f \circ g)(5)$

(i) $(g \circ f)(x)$

20. If $f(x) = -x^2$ and $g(x) = 3x - 2$, find the following.

(a) $(f + g)(x)$

(b) $(f - g)(x)$

(c) $(f - g)(-3)$

(d) $(fg)(x)$

(e) $\dfrac{f}{g}(x)$

(f) $\dfrac{f}{g}(2)$

(g) $(f \circ g)(x)$

(h) $(g \circ f)(x)$

(i) $(g \circ f)(-4)$

In Problems 21–24, find $(f \circ g)(x)$ and $(g \circ f)(x)$.

21. $f(x) = \dfrac{1}{x^2}$, $g(x) = x + 1$

22. $f(x) = \dfrac{x + 1}{4}$, $g(x) = \sqrt{x}$

23. $f(x) = \sqrt{x + 2}$, $g(x) = x^3$

24. $f(x) = 2$, $g(x) = 3$

In Problems 25 and 26, find the intercepts of the graph of each equation, and test for symmetry about the x-axis, the y-axis, the origin, and $y = x$. Do not sketch the graph.

25. $y = 3x - x^3$

26. $\dfrac{x^2 y^2}{x^2 + y^2 + 1} = 4$

In Problems 27 and 28, find the x- and y-intercepts of the graphs of the equations. Also, test for symmetry about the x-axis, the y-axis, and the origin. Then sketch the graphs.

27. $y = 9 - x^2$

28. $y = 3x - 7$

In Problems 29–32, graph each function and give its domain and range. Also, determine the intercepts.

29. $G(u) = \sqrt{u + 4}$

30. $f(x) = |x| + 1$

31. $y = g(t) = \dfrac{2}{|t - 4|}$

32. $h(u) = \sqrt{-5u}$

33. Graph the following case-defined function, and give its domain and range:

$$y = f(x) = \begin{cases} 2 & \text{if } x \le 0 \\ 2 - x & \text{if } x > 0 \end{cases}$$

34. Use the graph of $f(x) = \sqrt{x}$ to sketch the graph of $y = \sqrt{x - 2} - 1$.

35. Use the graph of $f(x) = x^2$ to sketch the graph of $y = -\dfrac{1}{2}x^2 + 2$.

36. Trend Equation   The projected annual sales (in dollars) of a new product are given by the equation $S = 150{,}000 + 3000t$, where t is the time in years from 2001. Such an equation is called a *trend equation*. Find the projected annual sales for 2006. Is S a function of t?

37. In Figure 2.42, which graphs represent functions of x?

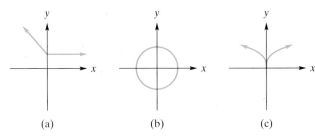

| (a) | (b) | (c) |

FIGURE 2.42   Diagram for Problem 37.

38. If $f(x) = (x^2 - x + 7)^3$, find (a) $f(2)$ and (b) $f(1.1)$. Round your answers to two decimal places.

39. Find all real roots of the equation

$$5x^3 - 7x^2 = 4x - 2$$

Round your answers to two decimal places.

40. Find all real roots of the equation

$$x^4 - 4x^3 = (2x - 1)^2$$

Round your answers to two decimal places.

41. Find all real zeros of

$$f(x) = x(2.1x^2 - 3)^2 - x^3 + 1$$

Round your answers to two decimal places.

42. Determine the range of

$$f(x) = \begin{cases} -2.5x - 4 & \text{if } x < 0 \\ 6 + 4.1x - x^2 & \text{if } x \geq 0 \end{cases}$$

43. From the graph of $f(x) = -x^3 + 0.04x + 7$, find (a) the range and (b) the intercepts. Round values to two decimal places.

44. From the graph of $f(x) = \sqrt{x+5}(x^2 - 4)$, find (a) the minimum value of $f(x)$ (b) the range of f, and (c) all real zeros of f. Round values to two decimal places.

45. Graph $y = f(x) = x^2 + x^k$ for $k = 0, 1, 2, 3,$ and 4. For which values of k does the graph have (a) symmetry about the y-axis? (b) symmetry about the origin?

A Taxing Experience!

Occasionally, you will hear someone complain that an unexpected source of income is going to *bump* him or her *into the next tax bracket* with the further speculation that this will result in a *reduction in take-home earnings.* It is true that U.S. federal income tax is prescribed by case-defined functions (the cases defining what are often called *brackets*), but we will see that there are no *jumps* in tax paid as a function of income. It is an urban myth that an increase in pretax income can result in a decrease in take-home income.

We will examine the 2006 federal tax rates for a married couple filing a joint return. The relevant document is Schedule Y-1, available at http://www.irs.gov/formspubs/article/0,,id=150856,00.html and reproduced in Figure 2.43.

FIGURE 2.43   Internal Revenue Service 2006 Schedule Y-1.

We claim that Schedule Y-1 defines a function, call it f, of income x, for $x \geq 0$. Indeed, for any $x \geq 0$, x belongs to exactly one of the intervals

$$[0, 15,100]$$
$$(15,100, 61,300]$$
$$(61,300, 123,700]$$
$$(123,700, 188,450]$$
$$(188,450, 336,550]$$
$$(336,550, \infty)$$

and as soon as the interval is determined, there is a single rule that applies to compute a unique value $f(x)$.

For example, to compute $f(79,500)$, tax on an income of \$79,500, observe first that 79,500 belongs to the interval $(61,300, 123,700]$ and for such an x the tax formula is $f(x) = 8,440 + 0.25(x - 61,300)$, since $x - 61,300$ is the amount over \$61,300 and it is taxed at the rate 25\% = 0.25.

Therefore,

$$f(79,500) = 8,440 + 0.25(79,500 - 61,300)$$
$$= 8,440 + 0.25(18,200)$$
$$= 8,440 + 4,550$$
$$= 12,990$$

To illustrate further we write out the entire Schedule Y-1 in our generic notation for a case-defined function.

$$f(x) = \begin{cases} 0.10x & \text{if } 0 \leq x \leq 15,100 \\ 1,510 + 0.15(x - 15,100) & \text{if } 15,100 < x \leq 61,300 \\ 8,440 + 0.25(x - 61,300) & \text{if } 61,300 < x \leq 123,700 \\ 24,040 + 0.28(x - 123,700) & \text{if } 123,700 < x \leq 188,450 \\ 42,170 + 0.33(x - 188,450) & \text{if } 188,450 < x \leq 336,550 \\ 91,043 + 0.35(x - 336,550) & \text{if } x > 336,550 \end{cases}$$

With these formulas, we can geometrically depict the income tax function, as in Figure 2.44.

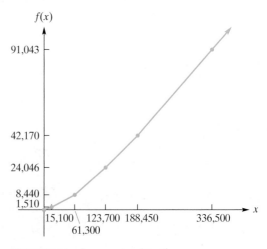

FIGURE 2.44   Income tax function.

Problems

Use the preceding income tax function f to determine the tax on the given taxable income in the year 2006.

1. $23,000

2. $85,000

3. $290,000

4. $462,000

5. Look up the most recent Schedule X at `http://www.irs.gov/formspubs/article/0,,id=150856,00.html` and repeat Problems 1–4 for a single person.

6. Why is it significant that $f(15,100) = \$1,510$, $f(61,300) = \$8,440$ and so on?

7. Define the function g by $g(x) = x - f(x)$. Thus $g = I - f$, where I is the identity function defined in Section 2.3. The function g gives, for each pretax income x, the amount that the taxpayer gets as take-home income and is, like f, a case-defined function. Write a complete description for g, in terms of cases, as we have done for f.

8. Graph the function g defined in Problem 7. Observe that if $a < b$, then $g(a) < g(b)$. This shows that if pretax income increases, then take-home income increases, irrespective of any jumping to a higher bracket (thereby debunking an urban myth).

LINES, PARABOLAS, AND SYSTEMS

 Mobile Phone Billing Plans

For the problem of industrial pollution, some people advocate a market-based solution: Let manufacturers pollute, but make them pay for the privilege. The more pollution, the greater the fee, or levy. The idea is to give manufacturers an incentive not to pollute more than necessary.

Does this approach work? In the figure below, curve 1 represents the cost per ton of cutting pollution. A company polluting indiscriminately can normally do some pollution reduction at a small cost. As the amount of pollution is reduced, however, the cost per ton of further reduction rises and ultimately increases without bound. This is illustrated by curve 1 rising indefinitely as the total tons of pollution produced approaches 0. (You should try to understand why this *model* is reasonably accurate.)

Line 2 is a levy scheme that goes easy on clean-running operations but charges an increasing per-ton fee as the total pollution amount goes up. Line 3, by contrast, is a scheme in which low-pollution manufacturers pay a high per-ton levy while gross polluters pay less per ton (but more overall). Questions of fairness aside, how well will each scheme work as a pollution control measure?

Faced with a pollution levy, a company tends to cut pollution *so long as it saves more in levy costs than it incurs in reduction costs*. The reduction efforts continue until the reduction costs exceed the levy savings.

The latter half of this chapter deals with systems of equations. Here, curve 1 and line 2 represent one system of equations, and curve 1 and line 3 represent another. Once you have learned how to solve systems of equations, you can return to this page and verify that the line 2 scheme leads to a pollution reduction from amount *A* to amount *B*, while the line 3 scheme fails as a pollution control measure, leaving the pollution level at *A*.

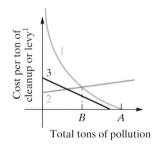

Total tons of pollution

---

[1]Technically, this is the *marginal* cost per ton (see Section 11.3).

OBJECTIVE

To develop the notion of slope and different forms of equations of lines.

3.1 Lines

Slope of a Line

Many relationships between quantities can be represented conveniently by straight lines. One feature of a straight line is its "steepness." For example, in Figure 3.1, line L_1 rises faster as it goes from left to right than does line L_2. In this sense, L_1 is steeper.

To measure the steepness of a line, we use the notion of *slope*. In Figure 3.2, as we move along line L from $(1, 3)$ to $(3, 7)$, the x-coordinate increases from 1 to 3, and the y-coordinate increases from 3 to 7. The average rate of change of y with respect to x is the ratio

$$\frac{\text{change in } y}{\text{change in } x} = \frac{\text{vertical change}}{\text{horizontal change}} = \frac{7 - 3}{3 - 1} = \frac{4}{2} = 2$$

The ratio of 2 means that for each 1-unit increase in x, there is a 2-unit *increase* in y. Due to the increase, the line *rises* from left to right. It can be shown that, regardless of which two points on L are chosen to compute the ratio of the change in y to the change in x, the result is always 2, which we call the *slope* of the line.

DEFINITION
Let (x_1, y_1) and (x_2, y_2) be two different points on a nonvertical line. The slope of the line is

$$m = \frac{y_2 - y_1}{x_2 - x_1} \left(= \frac{\text{vertical change}}{\text{horizontal change}} \right) \tag{1}$$

CAUTION

Having no slope does not mean having a slope of zero.

A vertical line does not have a slope, because any two points on it must have $x_1 = x_2$ [see Figure 3.3(a)], which gives a denominator of zero in Equation (1). For a horizontal line, any two points must have $y_1 = y_2$. [See Figure 3.3(b).] This gives a numerator of zero in Equation (1), and hence the slope of the line is zero.

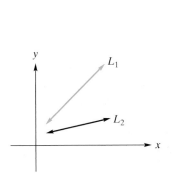

FIGURE 3.1 Line L_1 is "steeper" than L_2.

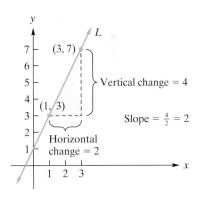

FIGURE 3.2 Slope of a line.

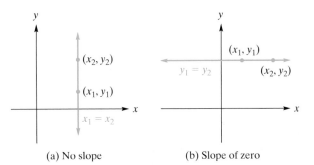

(a) No slope   (b) Slope of zero

FIGURE 3.3 Vertical and horizontal lines.

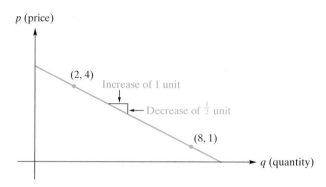

FIGURE 3.4   Price-quantity line.

This example shows how the slope can be interpreted

PRINCIPLES IN PRACTICE 1

PRICE-QUANTITY RELATIONSHIP

A doctor purchased a new car in 2001 for $32,000. In 2004, he sold it to a friend for $26,000. Draw a line showing the relationship between the selling price of the car and the year in which it was sold. Find and interpret the slope.

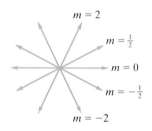

FIGURE 3.5   Slopes of lines.

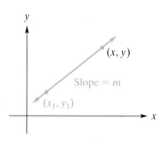

FIGURE 3.6   Line through (x_1, y_1) with slope m.

● EXAMPLE 1   Price-Quantity Relationship

The line in Figure 3.4 shows the relationship between the price p of a widget (in dollars) and the quantity q of widgets (in thousands) that consumers will buy at that price. Find and interpret the slope.

Solution: In the slope formula (1), we replace the x's by q's and the y's by p's. Either point in Figure 3.4 may be chosen as (q_1, p_1). Letting $(2, 4) = (q_1, p_1)$ and $(8, 1) = (q_2, p_2)$, we have

$$m = \frac{p_2 - p_1}{q_2 - q_1} = \frac{1 - 4}{8 - 2} = \frac{-3}{6} = -\frac{1}{2}$$

The slope is negative, $-\frac{1}{2}$. This means that, for each 1-unit increase in quantity (one thousand widgets), there corresponds a **decrease** in price of $\frac{1}{2}$ (dollar per widget). Due to this decrease, the line **falls** from left to right.

NOW WORK PROBLEM 3 ●●●

In summary, we can characterize the orientation of a line by its slope:

Zero slope:	horizontal line
Undefined slope:	vertical line
Positive slope:	line rises from left to right
Negative slope:	line falls from left to right

Lines with different slopes are shown in Figure 3.5. Notice that *the closer the slope is to 0, the more nearly horizontal is the line. The greater the absolute value of the slope, the more nearly vertical is the line.* We remark that two lines are parallel if and only if they have the same slope or are both vertical.

Equations of Lines

If we know a point on a line and the slope of the line, we can find an equation whose graph is that line. Suppose that line L has slope m and passes through the point (x_1, y_1). If (x, y) is *any* other point on L (see Figure 3.6), we can find an algebraic relationship between x and y. Using the slope formula on the points (x_1, y_1) and (x, y) gives

$$\frac{y - y_1}{x - x_1} = m$$
$$y - y_1 = m(x - x_1) \qquad (2)$$

Every point on L satisfies Equation (2). It is also true that *every* point satisfying Equation (2) must lie on L. Thus, Equation (2) is an equation for L and is given a special name:

$$y - y_1 = m(x - x_1)$$

*is a **point-slope form** of an equation of the line through (x_1, y_1) with slope m.*

POINT-SLOPE FORM

A new applied mathematics program at a university has grown in enrollment by 14 students per year for the last five years. If the program enrolled 50 students in its third year, what is an equation for the number of students S in the program as a function of the number of years T since its inception?

● EXAMPLE 2   Point-Slope Form

Find an equation of the line that has slope 2 and passes through $(1, -3)$.

Solution: Using a point-slope form with $m = 2$ and $(x_1, y_1) = (1, -3)$ gives

$$y - y_1 = m(x - x_1)$$
$$y - (-3) = 2(x - 1)$$
$$y + 3 = 2x - 2$$

which can be rewritten as

$$2x - y - 5 = 0$$

NOW WORK PROBLEM 9 ●●

An equation of the line passing through two given points can be found easily, as Example 3 shows.

DETERMINING A LINE FROM TWO POINTS

Find an equation of the line passing through the given points. A temperature of 41°F is equivalent to 5°C, and a temperature of 77°F is equivalent to 25°C.

Choosing $(4, -2)$ as (x_1, y_1) would give the same result.

● EXAMPLE 3   Determining a Line from Two Points

Find an equation of the line passing through $(-3, 8)$ *and* $(4, -2)$.

Solution:

Strategy   First we find the slope of the line from the given points. Then we substitute the slope and one of the points into a point-slope form.

The line has slope

$$m = \frac{-2 - 8}{4 - (-3)} = -\frac{10}{7}$$

Using a point-slope form with $(-3, 8)$ as (x_1, y_1) gives

$$y - 8 = -\frac{10}{7}[x - (-3)]$$
$$y - 8 = -\frac{10}{7}(x + 3)$$
$$7y - 56 = -10x - 30$$
$$10x + 7y - 26 = 0$$

NOW WORK PROBLEM 13 ●●

Recall that a point $(0, b)$ where a graph intersects the y-axis is called a y-intercept (Figure 3.7). If the slope m and y-intercept b of a line are known, an equation for the line is [by using a point-slope form with $(x_1, y_1) = (0, b)$]

$$y - b = m(x - 0)$$

Solving for y gives $y = mx + b$, called the *slope-intercept form* of an equation of the line:

$$y = mx + b$$

is the **slope-intercept form** *of an equation of the line with slope m and y-intercept b.*

FIGURE 3.7   Line with slope m and y-intercept b.

● EXAMPLE 4   Slope-Intercept Form

Find an equation of the line with slope 3 and y-intercept -4.

Solution: Using the slope-intercept form $y = mx + b$ with $m = 3$ and $b = -4$ gives

$$y = 3x + (-4)$$
$$y = 3x - 4$$

NOW WORK PROBLEM 17

FIND THE SLOPE AND
Y-INTERCEPT OF A LINE

One formula for the recommended dosage (in milligrams) of medication for a child t years old is

$$y = \frac{1}{24}(t+1)a$$

where a is the adult dosage. For an over-the-counter pain reliever, $a = 1000$. Find the slope and y-intercept of this equation.

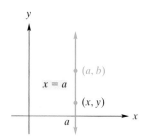

FIGURE 3.8 Vertical line through (a, b).

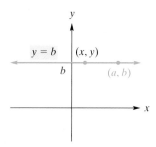

FIGURE 3.9 Horizontal line through (a, b).

 CAUTION

Do not confuse the forms of equations of horizontal and vertical lines. Remember which one has the form $x = $ constant and which one has the form $y = $ constant.

 EXAMPLE 5 Find the Slope and y-Intercept of a Line

Find the slope and y-intercept of the line with equation $y = 5(3 - 2x)$.

Solution:

Strategy We will rewrite the equation so it has the slope-intercept form $y = mx + b$. Then the slope is the coefficient of x and the y-intercept is the constant term.

We have

$$y = 5(3 - 2x)$$
$$y = 15 - 10x$$
$$y = -10x + 15$$

Thus, $m = -10$ and $b = 15$, so the slope is -10 and the y-intercept is 15.

NOW WORK PROBLEM 25

If a *vertical* line passes through (a, b) (see Figure 3.8), then any other point (x, y) lies on the line if and only if $x = a$. The y-coordinate can have any value. Hence, an equation of the line is $x = a$. Similarly, an equation of the *horizontal* line passing through (a, b) is $y = b$. (See Figure 3.9.) Here the x-coordinate can have any value.

EXAMPLE 6 Equations of Horizontal and Vertical Lines

a. An equation of the vertical line through $(-2, 3)$ is $x = -2$. An equation of the horizontal line through $(-2, 3)$ is $y = 3$.

b. The x-axis and y-axis are horizontal and vertical lines, respectively. Because $(0, 0)$ lies on both axes, an equation of the x-axis is $y = 0$, and an equation of the y-axis is $x = 0$.

NOW WORK PROBLEMS 21 AND 23

From our discussions, we can show that every straight line is the graph of an equation of the form $Ax + By + C = 0$, where A, B, and C are constants and A and B are not both zero. We call this a **general linear equation** (or an *equation of the first degree*) **in the variables x and y,** and x and y are said to be **linearly related.** For example, a general linear equation for $y = 7x - 2$ is $(-7)x + (1)y + (2) = 0$. Conversely, the graph of a general linear equation is a straight line. Table 3.1 gives the various forms of equations of straight lines.

TABLE 3.1 Forms of Equations of Straight Lines

Point-slope form	$y - y_1 = m(x - x_1)$
Slope-intercept form	$y = mx + b$
General linear form	$Ax + By + C = 0$
Vertical line	$x = a$
Horizontal line	$y = b$

Example 3 suggests that we could add another entry to the table. For if we know that points (x_1, y_1) and (x_2, y_2) are points on a line, then the slope of that line is $m = \dfrac{y_2 - y_1}{x_2 - x_1}$ and we could say that $y - y_1 = \dfrac{y_2 - y_1}{x_2 - x_1}(x - x_1)$ is **two-point form** for an equation of a line passing through points (x_1, y_1) and (x_2, y_2). Whether one chooses to remember many formulas or a few problem-solving principles is very much a matter of individual taste.

CONVERTING FORMS OF
EQUATIONS OF LINES

Find a general linear form of the
Fahrenheit–Celsius conversion
equation whose slope–intercept
form is $F = \dfrac{9}{5}C + 32$.

CAUTION

This illustrates that a general linear
form of a line is not unique

GRAPHING A GENERAL LINEAR
EQUATION

Sketch the graph of the
Fahrenheit–Celsius conversion
equation that you found in
Principles in Practice 5. How
could you use this graph to
convert a Celsius temperature to
Fahrenheit?

● EXAMPLE 7   **Converting Forms of Equations of Lines**

a. *Find a general linear form of the line whose slope-intercept form is*

$$y = -\frac{2}{3}x + 4$$

Solution: Getting one side to be 0, we obtain

$$\frac{2}{3}x + y - 4 = 0$$

which is a general linear form with $A = \frac{2}{3}$, $B = 1$, and $C = -4$. An alternative
general form can be obtained by clearing fractions:

$$2x + 3y - 12 = 0$$

b. *Find the slope-intercept form of the line having a general linear form*
$3x + 4y - 2 = 0$.

Solution: We want the form $y = mx + b$, so we solve the given equation for y.
We have

$$3x + 4y - 2 = 0$$
$$4y = -3x + 2$$
$$y = -\frac{3}{4}x + \frac{1}{2}$$

which is the slope-intercept form. Note that the line has slope $-\frac{3}{4}$ and
y-intercept $\frac{1}{2}$.

NOW WORK PROBLEM 37

● EXAMPLE 8   **Graphing a General Linear Equation**

Sketch the graph of $2x - 3y + 6 = 0$.

Solution:

Strategy   Since this is a general linear equation, its graph is a straight line. Thus,
we need only determine two different points on the graph in order to sketch it.
We will find the intercepts.

If $x = 0$, then $-3y + 6 = 0$, so the y-intercept is 2. If $y = 0$, then $2x + 6 = 0$, so
the x-intercept is -3. We now draw the line passing through $(0, 2)$ and $(-3, 0)$. (See
Figure 3.10.)

NOW WORK PROBLEM 27

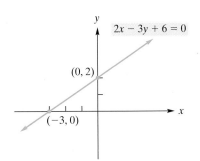

FIGURE 3.10   Graph of
$2x - 3y + 6 = 0$.

T E C H N O L O G Y

To graph the equation of Example 8 with a graphing calculator, we first express y in terms of x:

$$2x - 3y + 6 = 0$$
$$3y = 2x + 6$$
$$y = \frac{1}{3}(2x + 6)$$

Essentially, y is expressed as a function of x; the graph is
shown in Figure 3.11.

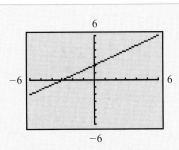

FIGURE 3.11   Calculator graph of $2x - 3y + 6 = 0$.

Parallel and Perpendicular Lines

As stated previously, there is a rule for parallel lines:

> **Parallel Lines** *Two lines are parallel if and only if they have the same slope or are both vertical.*

It follows that any line is parallel to itself.

There is also a rule for perpendicular lines. Look back to Figure 3.5 and observe that the line with slope $-\frac{1}{2}$ is perpendicular to the line with slope 2. The fact that the slope of either of these lines is the negative reciprocal of the slope of the other line is not a coincidence, as the following rule states.

> **Perpendicular Lines** *Two lines with slopes m_1 and m_2 are perpendicular to each other if and only if*
>
> $$m_1 = -\frac{1}{m_2}$$
>
> *Moreover, any horizontal line and any vertical line are perpendicular to each other.*

Rather than simply remembering this equation for the perpendicularity condition, you should observe why it makes sense. For if two lines are perpendicular, with neither vertical, then one will necessarily rise from left to right while the other will fall from left to right. Thus the slopes must have different signs. Also, if one is steep, then the other is relatively flat, which suggests a relationship such as is provided by reciprocals.

PRINCIPLES IN PRACTICE 7

PARALLEL AND PERPENDICULAR LINES

Show that a triangle with vertices at $A(0, 0)$, $B(6, 0)$, and $C(7, 7)$ is not a right triangle.

● EXAMPLE 9 Parallel and Perpendicular Lines

Figure 3.12 shows two lines passing through $(3, -2)$. One is parallel to the line $y = 3x + 1$, and the other is perpendicular to it. Find equations of these lines.

Solution: The slope of $y = 3x + 1$ is 3. Thus, the line through $(3, -2)$ that is *parallel* to $y = 3x + 1$ also has slope 3. Using a point-slope form, we get

$$y - (-2) = 3(x - 3)$$

$$y + 2 = 3x - 9$$

$$y = 3x - 11$$

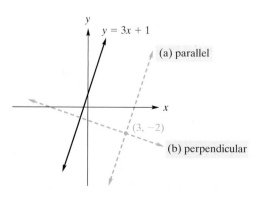

FIGURE 3.12 Lines parallel and perpendicular to $y = 3x + 1$ (Example 9).

The slope of a line *perpendicular* to $y = 3x + 1$ must be $-\frac{1}{3}$ (the negative reciprocal of 3). Using a point-slope form, we get

$$y - (-2) = -\frac{1}{3}(x - 3)$$

$$y + 2 = -\frac{1}{3}x + 1$$

$$y = -\frac{1}{3}x - 1$$

NOW WORK PROBLEM 55 ◖●◗

Problems 3.1

In Problems 1–8, find the slope of the straight line that passes through the given points.

1. $(4, 1), (7, 10)$

2. $(-2, 10), (5, 3)$

*3.** $(6, -2), (8, -3)$

4. $(2, -4), (3, -4)$

5. $(5, 3), (5, -8)$

6. $(0, -6), (3, 0)$

7. $(5, -2), (4, -2)$

8. $(1, -7), (9, 0)$

In Problems 9–24, find a general linear equation $(Ax + By + C = 0)$ of the straight line that has the indicated properties, and sketch each line.

*9.** Passes through $(-1, 7)$ and has slope -5

10. Passes through the origin and has slope 75

11. Passes through $(-2, 5)$ and has slope $-\frac{1}{4}$

12. Passes through $(-\frac{5}{2}, 5)$ and has slope $\frac{1}{3}$

*13.** Passes through $(-6, 1)$ and $(1, 4)$

14. Passes through $(5, 2)$ and $(6, -4)$

15. Passes through $(-3, -4)$ and $(-2, -8)$

16. Passes through $(0, 0)$ and $(2, 3)$

*17.** Has slope 2 and y-intercept 4

18. Has slope 5 and y-intercept -7

19. Has slope $-\frac{1}{2}$ and y-intercept -3

20. Has slope 0 and y-intercept $-\frac{1}{2}$

*21.** Is horizontal and passes through $(-5, -3)$

22. Is vertical and passes through $(-1, -1)$

*23.** Passes through $(2, -3)$ and is vertical

24. Passes through the origin and is horizontal

In Problems 25–34, find, if possible, the slope and y-intercept of the straight line determined by the equation, and sketch the graph.

*25.** $y = 4x - 6$

26. $x - 2 = 6$

*27.** $3x + 5y - 9 = 0$

28. $y + 4 = 7$

29. $x = -5$

30. $x - 9 = 5y + 3$

31. $y = 3x$

32. $y - 7 = 3(x - 4)$

33. $y = 3$

34. $6y - 24 = 0$

In Problems 35–40, find a general linear form and the slope–intercept form of the given equation.

35. $2x = 5 - 3y$

36. $3x + 2y = 6$

*37.** $4x + 9y - 5 = 0$

38. $3(x - 4) - 7(y + 1) = 2$

39. $-\frac{x}{2} + \frac{2y}{3} = -4\frac{3}{4}$

40. $y = \frac{1}{300}x + 8$

In Problems 41–50, determine whether the lines are parallel, perpendicular, or neither.

41. $y = 7x + 2, \ y = 7x - 3$

42. $y = 4x + 3, \ y = 5 + 4x$

43. $y = 5x + 2, \ -5x + y - 3 = 0$

44. $y = x, \ y = -x$

45. $x + 3y + 5 = 0, \ y = -3x$

46. $x + 3y = 0, \ x + 6y - 4 = 0$

47. $y = 3, \ x = -\frac{1}{3}$

48. $x = 3, \ x = -3$

49. $3x + y = 4, \ x - 3y + 1 = 0$

50. $x - 2 = 3, \ y = 2$

In Problems 51–60, find an equation of the line satisfying the given conditions. Give the answer in slope-intercept form if possible.

51. Passing through $(1, 1)$ and parallel to $y = -\frac{x}{4} - 2$

52. Passing through $(2, -8)$ and parallel to $x = -4$

53. Passing through $(2, 1)$ and parallel to $y = 2$

54. Passing through $(3, -4)$ and parallel to $y = 3 + 2x$

*55.** Perpendicular to $y = 3x - 5$ and passing through $(3, 4)$

56. Perpendicular to $y = -4$ and passing through $(1, 1)$

57. Passing through $(5, 2)$ and perpendicular to $y = -3$

58. Passing through $(4, -5)$ and perpendicular to the line $3y = -\frac{2x}{5} + 3$

59. Passing through $(-7, -5)$ and parallel to the line $2x + 3y + 6 = 0$

60. Passing through $(-4, 10)$ and parallel to the y-axis

61. A straight line passes through $(1, 2)$ and $(-3, 8)$. Find the point on it that has an x-coordinate of 5.

62. A straight line has slope 3 and y-intercept $(0, 1)$. Does the point $(-1, -2)$ lie on the line?

63. **Stock** In 1996, the stock in a computer hardware company traded for \$37 per share. However, the the company was in trouble and the stock price dropped steadily, to \$8 per share in 2006. Draw a line showing the relationship between the price per share and the year in which it traded for the time interval [1996, 2006], with years on the x-axis and price on the y-axis. Find and interpret the slope.

In Problems 64–65, find an equation of the line describing the following information.

64. Home Runs   In one season, a major league baseball player has hit 14 home runs by the end of the third month and 20 home runs by the end of the fifth month.

65. Business   A delicatessen owner starts her business with debts of $100,000. After operating for five years, she has accumulated a profit of $40,000.

66. Due Date   The length, L, of a human fetus more than 12 weeks old can be estimated by the formula $L = 1.53t - 6.7$, where L is in centimeters and t is in weeks from conception. An obstetrician uses the length of a fetus, measured by ultrasound, to determine the approximate age of the fetus and establish a due date for the mother. The formula must be rewritten to result in an age, t, given a fetal length, L. Find the slope and L-intercept of the equation.

67. Discus Throw   A mathematical model can approximate the winning distance for the Olympic discus throw by the formula $d = 184 + t$, where d is in feet and $t = 0$ corresponds to the year 1948. Find a general linear form of this equation.

68. Campus Map   A coordinate map of a college campus gives the coordinates (x, y) of three major buildings as follows: computer center, $(3.5, -1.5)$; engineering lab, $(0.5, 0.5)$; and library $(-1, -2.5)$. Find the equations (in slope-intercept form) of the straight-line paths connecting (a) the engineering lab with the computer center and (b) the engineering lab with the library. Are these two paths perpendicular to each other?

69. Geometry   Show that the points $A(0, 0)$, $B(0, 4)$, $C(2, 3)$, and $D(2, 7)$ are the vertices of a parallelogram. (Opposite sides of a parallelogram are parallel.)

70. Approach Angle   A small plane is landing at an airport with an approach angle of 45 degrees, or slope of -1. The plane begins its descent when it has an elevation of 3600 feet. Find the equation that describes the relationship between the craft's altitude and distance traveled, assuming that at distance 0 it starts the approach angle. Graph your equation on a graphing calculator. What does the graph tell you about the approach if the airport is 3800 feet from where the plane starts its landing?

71. Cost Equation   The average daily cost, C, for a room at a city hospital has risen by $59.82 per year for the years 1990 through 2000. If the average cost in 1996 was $1128.50, what is an equation which describes the average cost during this decade, as a function of the number of years, T, since 1990?

72. Revenue Equation   A small business predicts its revenue growth by a straight-line method with a slope of $50,000 per year. In its fifth year, it had revenues of $330,000. Find an equation that describes the relationship between revenues, R, and the number of years, T, since it opened for business.

73. Graph $y = -0.9x + 7.3$ and verify that the y-intercept is 7.3.

74. Graph the lines whose equations are

$$y = 1.5x + 1$$
$$y = 1.5x - 1$$

and

$$y = 1.5x + 2.5$$

What do you observe about the orientation of these lines? Why would you expect this result from the equations of the lines themselves?

75. Graph the line $y = 7.1x + 5.4$. Find the coordinates of any two points on the line, and use them to estimate the slope. What is the actual slope of the line?

76. Using the standard window, graph the lines with equations

$$0.1875x - 0.3y + 0.94 = 0$$

and

$$0.32x + 0.2y + 1.01 = 0$$

on the same viewing rectangle. Now, change the window to a square window (for example, on a TI-83, use ZOOM, ZSquare). Notice that the lines appear to be perpendicular to each other. Prove that this is indeed the case.

To develop the notion of demand and supply curves and to introduce linear functions.

3.2 Applications and Linear Functions

Many situations in economics can be described by using straight lines, as evidenced by Example 1.

PRINCIPLES IN PRACTICE 1

PRODUCTION LEVELS

A sporting-goods manufacturer allocates 1000 units of time per day to make skis and ski boots. If it takes 8 units of time to make a ski and 14 units of time to make a boot, find an equation to describe all possible production levels of the two products.

● **EXAMPLE 1**   **Production Levels**

Suppose that a manufacturer uses 100 lb of material to produce products A and B, which require 4 lb and 2 lb of material per unit, respectively. If x and y denote the number of units produced of A and B, respectively, then all levels of production are given by the combinations of x and y that satisfy the equation

$$4x + 2y = 100 \qquad \text{where } x, \ y \geq 0$$

Thus, the levels of production of A and B are linearly related. Solving for y gives

$$y = -2x + 50 \qquad \text{(slope-intercept form)}$$

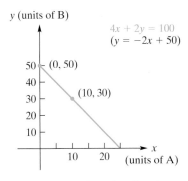

$4x + 2y = 100$
$(y = -2x + 50)$

y (units of B)

50 (0, 50)
40
30 (10, 30)
20
10

10   20
x (units of A)

FIGURE 3.13 Linearly related production levels.

 CAUTION

Typically, a demand curve falls from left to right and a supply curve rises from left to right. However, there are exceptions. For example, the demand for insulin could be represented by a vertical line, since this demand can remain constant regardless of price.

so the slope is -2. The slope reflects the rate of change of the level of production of B with respect to the level of production of A. For example, if 1 more unit of A is to be produced, it will require 4 more pounds of material, resulting in $\frac{4}{2} = 2$ *fewer* units of B. Accordingly, as x increases by 1 unit, the corresponding value of y decreases by 2 units. To sketch the graph of $y = -2x + 50$, we can use the y-intercept $(0, 50)$ and the fact that when $x = 10$, $y = 30$. (See Figure 3.13.)

NOW WORK PROBLEM 21 ●◖◗

Demand and Supply Curves

For each price level of a product, there is a corresponding quantity of that product that consumers will demand (that is, purchase) during some time period. Usually, the higher the price, the smaller is the quantity demanded; as the price falls, the quantity demanded increases. If the price per unit of the product is given by p and the corresponding quantity (in units) is given by q, then an equation relating p and q is called a **demand equation.** Its graph is called a **demand curve.** Figure 3.14(a) shows a demand curve. In keeping with the practice of most economists, the horizontal axis is the q-axis and the vertical axis is the p-axis. We will assume that the price per unit is given in dollars and the period is one week. Thus, the point (a, b) in Figure 3.14(a) indicates that, at a price of b dollars per unit, consumers will demand a units per week. Since negative prices or quantities are not meaningful, both a and b must be nonnegative. For most products, an increase in the quantity demanded corresponds to a decrease in price. Thus, a demand curve typically falls from left to right, as in Figure 3.14(a).

In response to various prices, there is a corresponding quantity of product that *producers* are willing to supply to the market during some time period. Usually, the higher the price per unit, the larger is the quantity that producers are willing to supply; as the price falls, so will the quantity supplied. If p denotes the price per unit and q denotes the corresponding quantity, then an equation relating p and q is called a **supply equation,** and its graph is called a **supply curve.** Figure 3.14(b) shows a supply curve. If p is in dollars and the period is one week, then the point (c, d) indicates that, at a price of d dollars each, producers will supply c units per week. As before, c and d are nonnegative. A supply curve usually rises from left to right, as in Figure 3.14(b). This indicates that a producer will supply more of a product at higher prices.

Observe that a function whose graph either falls from left to right or rises from left to right *throughout its entire domain* will pass the horizontal line test of Section 2.5. Certainly, the demand curve and the supply curve in Figure 3.15 are each cut at most once by any horizontal line. Thus if the demand curve is the graph of a function $p = D(q)$, then D will have an inverse and we can solve for q uniquely to get $q = D^{-1}(p)$. Similarly, if the supply curve is the graph of a function $p = S(q)$, then S is also one-to-one, has an inverse S^{-1}, and we can write $q = S^{-1}(p)$.

We will now focus on demand and supply curves that are straight lines (Figure 3.15). They are called *linear* demand and *linear* supply curves. Such curves have

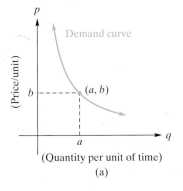

p

Demand curve

(Price/unit)

b ------ (a, b)

a
q

(Quantity per unit of time)

(a)

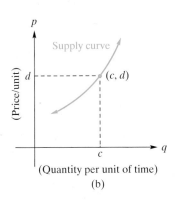

p

Supply curve

(Price/unit)

d ------------ (c, d)

c
q

(Quantity per unit of time)

(b)

FIGURE 3.14 Demand and supply curves.

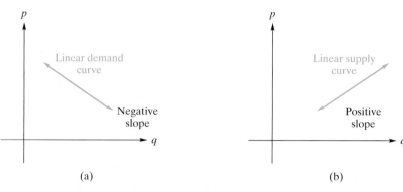

FIGURE 3.15 Linear demand and supply curves.

equations in which p and q are linearly related. Because a demand curve typically falls from left to right, a linear demand curve has a negative slope. [See Figure 3.15(a).] However, the slope of a linear supply curve is positive, because the curve rises from left to right. [See Figure 3.15(b).]

● **EXAMPLE 2** **Finding a Demand Equation**

Suppose the demand per week for a product is 100 *units when the price is* $58 *per unit and* 200 *units at* $51 *each. Determine the demand equation, assuming that it is linear.*

Solution:

Strategy Since the demand equation is linear, the demand curve must be a straight line. We are given that quantity q and price p are linearly related such that $p = 58$ when $q = 100$ and $p = 51$ when $q = 200$. Thus, the given data can be represented in a q,p-coordinate plane [see Figure 3.15(a)] by points (100, 58) and (200, 51). With these points, we can find an equation of the line—that is, the demand equation.

The slope of the line passing through (100, 58) and (200, 51) is

$$m = \frac{51 - 58}{200 - 100} = -\frac{7}{100}$$

An equation of the line (point-slope form) is

$$p - p_1 = m(q - q_1)$$
$$p - 58 = -\frac{7}{100}(q - 100)$$

Simplifying gives the demand equation

$$p = -\frac{7}{100}q + 65 \qquad (1)$$

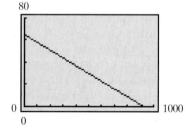

FIGURE 3.16 Graph of demand function $p = -\frac{7}{100}q + 65$.

Customarily, a demand equation (as well as a supply equation) expresses p, in terms of q and actually defines a function of q. For example, Equation (1) defines p as a function of q and is called the *demand function* for the product. (See Figure 3.16).

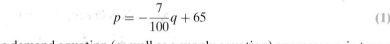

NOW WORK PROBLEM 15 ●●●

Linear Functions

A *linear function* was defined in Section 2.2 to be a polynomial function of degree 1. Somewhat more explicitly,

DEFINITION

A function f is a *linear function* if and only if $f(x)$ can be written in the form $f(x) = ax + b$, where a and b are constants and $a \neq 0$.

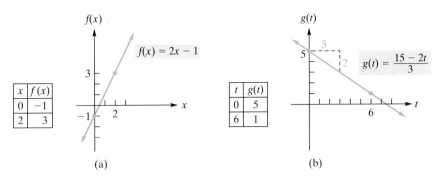

FIGURE 3.17   Graphs of linear functions.

Suppose that $f(x) = ax + b$ is a linear function, and let $y = f(x)$. Then $y = ax + b$, which is an equation of a straight line with slope a and y-intercept b. Thus, **the graph of a linear function is a straight line that is neither vertical nor horizontal.** We say that the function $f(x) = ax + b$ has slope a.

PRINCIPLES IN PRACTICE 3

GRAPHING LINEAR FUNCTIONS

A computer repair company charges a fixed amount plus an hourly rate for a service call. If x is the number of hours needed for a service call, the total cost of a call is described by the function $f(x) = 40x + 60$. Graph the function by finding and plotting two points.

● **EXAMPLE 3   Graphing Linear Functions**

a. *Graph* $f(x) = 2x - 1$.

Solution: Here f is a linear function (with slope 2), so its graph is a straight line. Since two points determine a straight line, we need only plot two points and then draw a line through them. [See Figure 3.17(a).] Note that one of the points plotted is the vertical-axis intercept, -1, which occurs when $x = 0$.

b. *Graph* $g(t) = \dfrac{15 - 2t}{3}$.

Solution: Notice that g is a linear function, because we can express it in the form $g(t) = at + b$.

$$g(t) = \frac{15 - 2t}{3} = \frac{15}{3} - \frac{2t}{3} = -\frac{2}{3}t + 5$$

The graph of g is shown in Figure 3.17(b). Since the slope is $-\frac{2}{3}$, observe that as t increases by 3 units, $g(t)$ *decreases* by 2.

NOW WORK PROBLEM 3 ●●

PRINCIPLES IN PRACTICE 4

DETERMINING A LINEAR FUNCTION

The height of children between the ages of 6 years and 10 years can be modeled by a linear function of age t in years. The height of one child changes by 2.3 inches per year, and she is 50.6 inches tall at age 8. Find a function that describes the height of this child at age t.

● **EXAMPLE 4   Determining a Linear Function**

Suppose f is a linear function with slope 2 and $f(4) = 8$. *Find* $f(x)$.

Solution: Since f is linear, it has the form $f(x) = ax + b$. The slope is 2, so $a = 2$, and we have

$$f(x) = 2x + b \tag{2}$$

Now we determine b. Since $f(4) = 8$, we replace x by 4 in Equation (2) and solve for b:

$$f(4) = 2(4) + b$$
$$8 = 8 + b$$
$$0 = b$$

Hence, $f(x) = 2x$.

NOW WORK PROBLEM 7 ●●

DETERMINING A LINEAR FUNCTION

An antique necklace is expected to be worth $360 after 3 years and $640 after 7 years. Find a function that describes the value of the necklace after x years.

● **EXAMPLE 5   Determining a Linear Function**

If $y = f(x)$ is a linear function such that $f(-2) = 6$ and $f(1) = -3$, find $f(x)$.

Solution:

Strategy   The function values correspond to points on the graph of f. With these points we can determine an equation of the line and hence the linear function.

The condition that $f(-2) = 6$ means that when $x = -2$, then $y = 6$. Thus, $(-2, 6)$ lies on the graph of f, which is a straight line. Similarly, $f(1) = -3$ implies that $(1, -3)$ also lies on the line. If we set $(x_1, y_1) = (-2, 6)$ and $(x_2, y_2) = (1, -3)$, the slope of the line is given by

$$m = \frac{y_2 - y_1}{x_2 - x_1} = \frac{-3 - 6}{1 - (-2)} = \frac{-9}{3} = -3$$

We can find an equation of the line by using a point-slope form:

$$y - y_1 = m(x - x_1)$$
$$y - 6 = -3[x - (-2)]$$
$$y - 6 = -3x - 6$$
$$y = -3x$$

Because $y = f(x)$, $f(x) = -3x$. Of course, the same result is obtained if we set $(x_1, y_1) = (1, -3)$.

NOW WORK PROBLEM 9 ●●●

In many studies, data are collected and plotted on a coordinate system. An analysis of the results may indicate a functional relationship between the variables involved. For example, the data points may be approximated by points on a straight line. This would indicate a linear functional relationship, such as the one in the next example.

● **EXAMPLE 6   Diet for Hens**

In testing an experimental diet for hens, it was determined that the average live weight w (in grams) of a hen was statistically a linear function of the number of days d after the diet began, where $0 \leq d \leq 50$. Suppose the average weight of a hen beginning the diet was 40 grams and 25 days later it was 675 grams.

a. *Determine w as a linear function of d.*

Solution: Since w is a linear function of d, its graph is a straight line. When $d = 0$ (the beginning of the diet), $w = 40$. Thus, $(0, 40)$ lies on the graph. (See Figure 3.18.) Similarly, $(25, 675)$ lies on the graph. If we set $(d_1, w_1) = (0, 40)$ and $(d_2, w_2) = (25, 675)$, the slope of the line is

$$m = \frac{w_2 - w_1}{d_2 - d_1} = \frac{675 - 40}{25 - 0} = \frac{635}{25} = \frac{127}{5}$$

Using a point-slope form, we have

$$w - w_1 = m(d - d_1)$$
$$w - 40 = \frac{127}{5}(d - 0)$$
$$w - 40 = \frac{127}{5}d$$
$$w = \frac{127}{5}d + 40$$

which expresses w as a linear function of d.

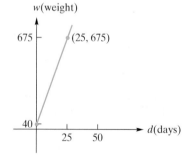

w(weight)

675 ⊢ •(25, 675)

40 ⊢

25   50 d(days)

FIGURE 3.18   Linear function describing diet for hens.

b. *Find the average weight of a hen when $d = 10$.*

> Solution: When $d = 10$, $w = \frac{127}{5}(10) + 40 = 254 + 40 = 294$. Thus, the average weight of a hen 10 days after the beginning of the diet is 294 grams.

NOW WORK PROBLEM 19

Problems 3.2

In Problems 1–6, find the slope and vertical-axis intercept of the linear function, and sketch the graph.

1. $y = f(x) = -4x$ **2.** $y = f(x) = x + 1$

*3. $h(t) = 5t - 7$ **4.** $f(s) = 3(5 - s)$

5. $h(q) = \dfrac{2 - q}{7}$ **6.** $h(q) = 0.5q + 0.25$

In Problems 7–14, find f(x) if f is a linear function that has the given properties.

*7. slope $= 4$, $f(2) = 8$ **8.** $f(0) = 3$, $f(4) = -5$

*9. $f(1) = 2$, $f(-2) = 8$ **10.** slope $= -2$, $f(\frac{2}{5}) = -7$

11. slope $= -\frac{2}{3}$, $f(-\frac{2}{3}) = -\frac{2}{3}$ **12.** $f(1) = 1$, $f(2) = 2$

13. $f(-2) = -1$, $f(-4) = -3$ **14.** slope $= 0.01$, $f(0.1) = 0.01$

*15. **Demand Equation** Suppose consumers will demand 40 units of a product when the price is $12.75 per unit and 25 units when the price is $18.75 each. Find the demand equation, assuming that it is linear. Find the price per unit when 37 units are demanded.

16. Demand Equation The demand per week for a CD is 26,000 copies when the price is $12 each, and 10,000 copies when the price is $18 each. Find the demand equation for the CD, assuming that it is linear.

17. Supply Equation A refrigerator manufacturer will produce 3000 units when the price is $940, and 2200 units when the price is $740. Assume that price, p, and quantity, q, produced are linearly related and find the supply equation.

18. Supply Equation Suppose a manufacturer of shoes will place on the market 50 (thousand pairs) when the price is 35 (dollars per pair) and 35 when the price is 30. Find the supply equation, assuming that price p and quantity q are linearly related.

*19. **Cost Equation** Suppose the cost to produce 10 units of a product is $40 and the cost of 20 units is $70. If cost, c, is linearly related to output, q, find a linear equation relating c and q. Find the cost to produce 35 units.

20. Cost Equation An advertiser goes to a printer and is charged $79 for 100 copies of one flyer and $88 for 400 copies of another flyer. This printer charges a fixed setup cost plus a charge for every copy of single-page flyers. Find a function that describes the cost of a printing job, if x is the number of copies made.

*21. **Electric Rates** An electric utility company charges residential customers 12.5 cents per kilowatt-hour plus a base charge each month. One customer's monthly bill comes to $51.65 for 380 kilowatt-hours. Find a linear function that describes the total monthly charges for electricity if x is the number of kilowatt-hours used in a month.

22. Radiation Therapy A cancer patient is to receive drug and radiation therapies. Each cubic centimeter of the drug to be used contains 210 curative units, and each minute of radiation exposure gives 305 curative units. The patient requires 2410 curative units. If d cubic centimeters of the drug and r minutes of radiation are administered, determine an equation relating d and r. Graph the equation for $d \geq 0$, and $r \geq 0$; label the horizontal axis as d.

23. Depreciation Suppose the value of a mountain bike decreases each year by 10% of its original value. If the original value is $1800, find an equation that expresses the value v of the bike t years after purchase, where $0 \leq t \leq 10$. Sketch the equation, choosing t as the horizontal axis and v as the vertical axis. What is the slope of the resulting line? This method of considering the value of equipment is called *straight-line depreciation*.

24. Depreciation A new television depreciates $120 per year, and it is worth $340 after four years. Find a function that describes the value of this television, if x is the age of the television in years.

25. Appreciation A new apartment building was sold for $960,000 five years after it was purchased. The original owners calculated that the building appreciated $45,000 per year while they owned it. Find a linear function that describes the appreciation of the building, if x is the number of years since the original purchase.

26. Appreciation A house purchased for $245,000 is expected to double in value in 15 years. Find a linear equation that describes the house's value after t years.

27. Repair Charges A business-copier repair company charges a fixed amount plus an hourly rate for a service call. If a customer is billed $159 for a one-hour service call and $287 for a three-hour service call, find a linear function that describes the price of a service call, where x is the number of hours of service.

28. Sheep Wool Length For sheep maintained at high environmental temperatures, respiratory rate, r (per minute), increases as wool length, l (in centimeters), decreases.[2] Suppose sheep with a wool length of 2 cm have an (average) respiratory rate of 160, and those with a wool length of 4 cm have a respiratory rate of 125. Assume that r

---

[2]Adapted from G. E. Folk, Jr., *Textbook of Environmental Physiology,* 2nd ed. (Philadelphia: Lea & Febiger, 1974).

and l are linearly related. (a) Find an equation that gives r in terms of l. (b) Find the respiratory rate of sheep with a wool length of 1 cm.

29. **Isocost Line** In production analysis, an *isocost line* is a line whose points represent all combinations of two factors of production that can be purchased for the same amount. Suppose a farmer has allocated $20,000 for the purchase of x tons of fertilizer (costing $200 per ton) and y acres of land (costing $2000 per acre). Find an equation of the isocost line that describes the various combinations that can be purchased for $20,000. Observe that neither x nor y can be negative.

30. **Isoprofit Line** A manufacturer produces products X and Y for which the profits per unit are $4 and $6, respectively. If x units of X and y units of Y are sold, then the total profit P is given by $P = 4x + 6y$, where $x, y \geq 0$. (a) Sketch the graph of this equation for $P = 240$. The result is called an *isoprofit line,* and its points represent all combinations of sales that produce a profit of $240. (b) Determine the slope for $P = 240$. (c) If $P = 600$, determine the slope. (d) Are isoprofit lines for products X and Y parallel?

31. **Grade Scaling** For reasons of comparison, a professor wants to rescale the scores on a set of test papers so that the maximum score is still 100, but the average is 65 instead of 56. (a) Find a linear equation that will do this. [*Hint:* You want 56 to become 65 and 100 to remain 100. Consider the points $(56, 65)$ and $(100, 100)$ and, more generally, (x, y), where x is the old score and y is the new score. Find the slope and use a point-slope form. Express y in terms of x.] (b) If 62 on the new scale is the lowest passing score, what was the lowest passing score on the original scale?

32. **Psychology** The result of Sternberg's psychological experiment[3] on information retrieval is that a person's reaction time R, in milliseconds, is statistically a linear function of memory set size N as follows:

$$R = 38N + 397$$

Sketch the graph for $1 \leq N \leq 5$. What is the slope?

33. **Psychology** In a certain learning experiment involving repetition and memory,[4] the proportion, p, of items recalled was estimated to be linearly related to the effective study time, t (in seconds), where t is between 5 and 9. For an effective study time of 5 seconds, the proportion of items recalled was 0.32. For each 1-second increase in study time, the proportion recalled increased by 0.059. (a) Find an equation that gives p in terms of t. (b) What proportion of items was recalled with 9 seconds of effective study time?

34. **Diet for Pigs** In testing an experimental diet for pigs, it was determined that the (average) live weight, w (in kilograms), of a pig was statistically a linear function of the number of days, d, after the diet was initiated, where $0 \leq d \leq 100$. If the weight of a pig beginning the diet was 21 kg, and thereafter the pig gained 6.3 kg every 10 days, determine w as a function of d, and find the weight of a pig 55 days after the beginning of the diet.

35. **Cricket Chirps** Biologists have found that the number of chirps made per minute by crickets of a certain species is related to the temperature. The relationship is very close to being linear. At 68°F, the crickets chirp about 124 times a minute. At 80°F, they chirp about 172 times a minute. (a) Find an equation that gives Fahrenheit temperature, t, in terms of the number of chirps, c, per minute. (b) If you count chirps for only 15 seconds, how can you quickly estimate the temperature?

To sketch parabolas arising from quadratic functions.

3.3 Quadratic Functions

In Section 2.2, a *quadratic function* was defined as a polynomial function of degree 2. In other words,

> **DEFINITION**
>
> A function f is a **quadratic function** if and only if $f(x)$ can be written in the form $f(x) = ax^2 + bx + c$, where a, b, and c are constants and $a \neq 0$.

For example, the functions $f(x) = x^2 - 3x + 2$ and $F(t) = -3t^2$ are quadratic. However, $g(x) = \dfrac{1}{x^2}$ is *not* quadratic, because it cannot be written in the form $g(x) = ax^2 + bx + c$.

---

[3]G. R. Loftus and E. F. Loftus, *Human Memory: The Processing of Information* (New York: Lawrence Erlbaum Associates, Inc., distributed by the Halsted Press, Division of John Wiley & Sons, Inc., 1976).

[4]D. L. Hintzman, "Repetition and Learning," in *The Psychology of Learning,* Vol. 10, ed. G. H. Bower (New York: Academic Press, Inc., 1976), p. 77.

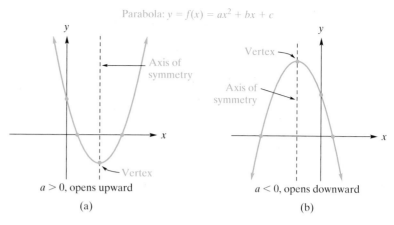

FIGURE 3.19   Parabolas.

The graph of the quadratic function $y = f(x) = ax^2 + bx + c$ is called a **parabola** and has a shape like the curves in Figure 3.19. If $a > 0$, the graph extends upward indefinitely, and we say that the parabola *opens upward* [Figure 3.19(a)]. If $a < 0$, the parabola *opens downward* [Figure 3.19(b)].

Each parabola in Figure 3.19 is *symmetric* about a vertical line, called the **axis of symmetry** of the parabola. That is, if the page were folded on one of these lines, then the two halves of the corresponding parabola would coincide. The axis (of symmetry) is *not* part of the parabola, but is a useful aid in sketching the parabola.

Figure 3.19 also shows points labeled **vertex,** where the axis cuts the parabola. If $a > 0$, the vertex is the "lowest" point on the parabola. This means that $f(x)$ has a minimum value at this point. By performing algebraic manipulations on $ax^2 + bx + c$ (referred to as *completing the square*), we can determine not only this minimum value, but also where it occurs. We have

$$f(x) = ax^2 + bx + c = (ax^2 + bx) + c$$

Adding and subtracting $\dfrac{b^2}{4a}$ gives

$$f(x) = \left(ax^2 + bx + \frac{b^2}{4a} \right) + c - \frac{b^2}{4a}$$

$$= a \left(x^2 + \frac{b}{a}x + \frac{b^2}{4a^2} \right) + c - \frac{b^2}{4a}$$

so that

$$f(x) = a \left(x + \frac{b}{2a} \right)^2 + c - \frac{b^2}{4a}$$

Since $\left(x + \dfrac{b}{2a} \right)^2 \geq 0$ and $a > 0$, it follows that $f(x)$ has a minimum value when $x + \dfrac{b}{2a} = 0$, that is, when $x = -\dfrac{b}{2a}$. The y-coordinate corresponding to this value of x is $f\left(-\dfrac{b}{2a} \right)$. Thus, the vertex is given by

$$\text{vertex} = \left(-\frac{b}{2a}, f\left(-\frac{b}{2a} \right) \right)$$

This is also the vertex of a parabola that opens downward ($a < 0$), but in this case $f\left(-\dfrac{b}{2a} \right)$ is the maximum value of $f(x)$. [See Figure 3.19(b).]

Observe that a function whose graph is a parabola is not one-to-one, in either the opening upward or opening downward case, since many horizontal lines will cut the graph twice. However, if we restrict the domain of a quadratic function to either

$\left[-\dfrac{b}{2a}, \infty \right)$ or $\left(-\infty, -\dfrac{b}{2a} \right]$, then the restricted function will pass the horizontal line test and therefore be one-to-one. (There are many other restrictions of a quadratic function that are one-to-one; however, their domains consist of more than one interval.) It follows that such restricted quadratic functions have inverse functions.

The point where the parabola $y = ax^2 + bx + c$ intersects the y-axis (that is, the y-intercept) occurs when $x = 0$. The y-coordinate of this point is c, so the y-intercept is c. In summary, we have the following.

Graph of Quadratic Function

The graph of the quadratic function $y = f(x) = ax^2 + bx + c$ is a parabola.

1. If $a > 0$, the parabola opens upward. If $a < 0$, it opens downward.
2. The vertex is $\left(-\dfrac{b}{2a}, f\left(-\dfrac{b}{2a} \right) \right)$.
3. The y-intercept is c.

We can quickly sketch the graph of a quadratic function by first locating the vertex, the y-intercept, and a few other points, such as those where the parabola intersects the x-axis. These *x-intercepts* are found by setting $y = 0$ and solving for x. Once the intercepts and vertex are found, it is then relatively easy to pass the appropriate parabola through these points. In the event that the x-intercepts are very close to the vertex or that no x-intercepts exist, we find a point on each side of the vertex, so that we can give a reasonable sketch of the parabola. Keep in mind that passing a (dashed) vertical line through the vertex gives the axis of symmetry. By plotting points to one side of the axis, we can use symmetry and obtain corresponding points on the other side.

● EXAMPLE 1   Graphing a Quadratic Function

Graph the quadratic function $y = f(x) = -x^2 - 4x + 12$.

Solution: Here $a = -1$, $b = -4$, and $c = 12$. Since $a < 0$, the parabola opens downward and thus has a highest point. The x-coordinate of the vertex is

$$-\frac{b}{2a} = -\frac{-4}{2(-1)} = -2$$

The y-coordinate is $f(-2) = -(-2)^2 - 4(-2) + 12 = 16$. Thus, the vertex is $(-2, 16)$, so the maximum value of $f(x)$ is 16. Since $c = 12$, the y-intercept is 12. To find the x-intercepts, we let y be 0 in $y = -x^2 - 4x + 12$ and solve for x:

$$0 = -x^2 - 4x + 12$$
$$0 = -(x^2 + 4x - 12)$$
$$0 = -(x + 6)(x - 2)$$

Hence, $x = -6$ or $x = 2$, so the x-intercepts are -6 and 2. Now we plot the vertex, axis of symmetry, and intercepts. [See Figure 3.20(a).] Since $(0, 12)$ is *two* units to the *right* of the axis of symmetry, there is a corresponding point *two* units to the *left* of the axis with the same y-coordinate. Thus, we get the point $(-4, 12)$. Through all points, we draw a parabola opening downward. [See Figure 3.20(b).]

NOW WORK PROBLEM 15

● EXAMPLE 2   Graphing a Quadratic Function

Graph $p = 2q^2$.

Solution: Here p is a quadratic function of q, where $a = 2$, $b = 0$, and $c = 0$. Since $a > 0$, the parabola opens upward and thus has a lowest point. The q-coordinate of

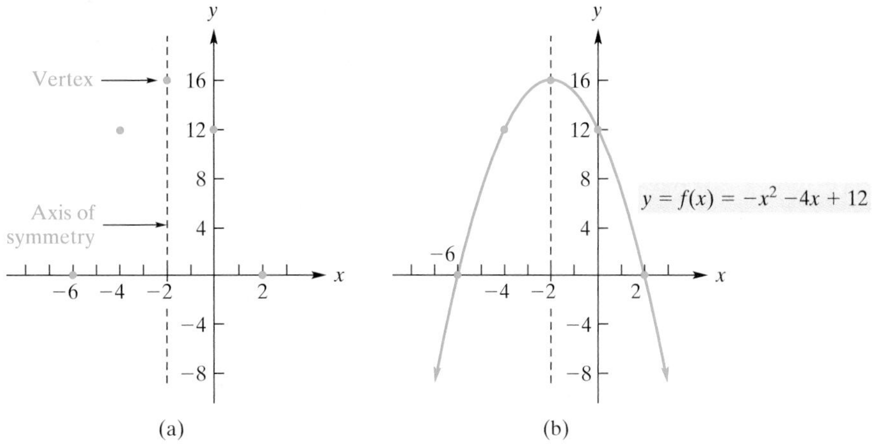

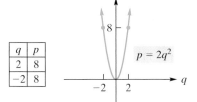

FIGURE 3.20   Graph of parabola $y = f(x) = -x^2 - 4x + 12$.

the vertex is

$$-\frac{b}{2a} = -\frac{0}{2(2)} = 0$$

and the p-coordinate is $2(0)^2 = 0$. Consequently, the *minimum* value of p is 0 and the vertex is $(0, 0)$. In this case, the p-axis is the axis of symmetry. A parabola opening upward with vertex at $(0, 0)$ cannot have any other intercepts. Hence, to draw a reasonable graph, we plot a point on each side of the vertex. If $q = 2$, then $p = 8$. This gives the point $(2, 8)$ and, by symmetry, the point $(-2, 8)$. (See Figure 3.21.)

NOW WORK PROBLEM 13

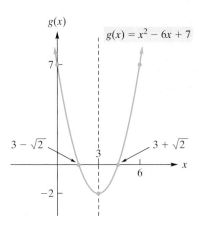

FIGURE 3.21   Graph of parabola $p = 2q^2$.

Example 3 illustrates that finding intercepts may require use of the quadratic formula.

GRAPHING A QUADRATIC FUNCTION

A man standing on a pitcher's mound throws a ball straight up with an initial velocity of 32 feet per second. The height h of the ball in feet t seconds after it was thrown is described by the function $h(t) = -16t^2 + 32t + 8$, for $t \geq 0$. Find the function's vertex and intercepts, and graph the function.

FIGURE 3.22   Graph of parabola $g(x) = x^2 - 6x + 7$.

● EXAMPLE 3   Graphing a Quadratic Function

Graph $g(x) = x^2 - 6x + 7$.

Solution: Here g is a quadratic function, where $a = 1, b = -6$, and $c = 7$. The parabola opens upward, because $a > 0$. The x-coordinate of the vertex (lowest point) is

$$-\frac{b}{2a} = -\frac{-6}{2(1)} = 3$$

and $g(3) = 3^2 - 6(3) + 7 = -2$, which is the minimum value of $g(x)$. Thus, the vertex is $(3, -2)$. Since $c = 7$, the vertical-axis intercept is 7. To find x-intercepts, we set $g(x) = 0$.

$$0 = x^2 - 6x + 7$$

The right side does not factor easily, so we will use the quadratic formula to solve for x:

$$x = \frac{-b \pm \sqrt{b^2 - 4ac}}{2a} = \frac{-(-6) \pm \sqrt{(-6)^2 - 4(1)(7)}}{2(1)}$$

$$= \frac{6 \pm \sqrt{8}}{2} = \frac{6 \pm \sqrt{4 \cdot 2}}{2} = \frac{6 \pm 2\sqrt{2}}{2}$$

$$= \frac{6}{2} \pm \frac{2\sqrt{2}}{2} = 3 \pm \sqrt{2}$$

Therefore, the x-intercepts are $3 + \sqrt{2}$ and $3 - \sqrt{2}$. After plotting the vertex, intercepts, and (by symmetry) the point $(6, 7)$, we draw a parabola opening upward in Figure 3.22.

NOW WORK PROBLEM 17

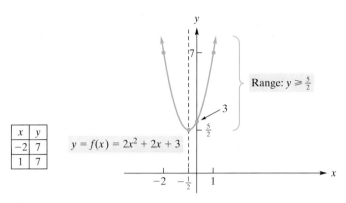

FIGURE 3.23   Graph of $y = f(x) = 2x^2 + 2x + 3$.

● EXAMPLE 4   Graphing a Quadratic Function

Graph $y = f(x) = 2x^2 + 2x + 3$ and find the range of f.

Solution: This function is quadratic with $a = 2$, $b = 2$, and $c = 3$. Since $a > 0$, the graph is a parabola opening upward. The x-coordinate of the vertex is

$$-\frac{b}{2a} = -\frac{2}{2(2)} = -\frac{1}{2}$$

and the y-coordinate is $2(-\frac{1}{2})^2 + 2(-\frac{1}{2}) + 3 = \frac{5}{2}$. Thus, the vertex is $(-\frac{1}{2}, \frac{5}{2})$. Since $c = 3$, the y-intercept is 3. A parabola opening upward with its vertex above the x-axis has no x-intercepts. In Figure 3.23 we plotted the y-intercept, the vertex, and an additional point $(-2, 7)$ to the left of the vertex. By symmetry, we also get the point $(1, 7)$. Passing a parabola through these points gives the desired graph. From the figure, we see that the range of f is all $y \geq \frac{5}{2}$, that is, the interval $[\frac{5}{2}, \infty)$.

NOW WORK PROBLEM 21 ●●

● EXAMPLE 5   Finding and Graphing an Inverse

For the parabola given by the function

$$y = f(x) = ax^2 + bx + c$$

determine the inverse of the restricted function given by $g(x) = ax^2 + bx + c$, for $x \geq -\dfrac{b}{2a}$. Graph g and g^{-1} in the same plane, in the case where $a = 2$, $b = 2$, and $c = 3$.

Solution: Following the procedure described in Example 5 of Section 2.4, we begin by solving $y = ax^2 + bx + c$, where $x \geq -\dfrac{b}{2a}$, for x in terms of y. We do this by applying the quadratic formula to $ax^2 + bx + c - y = 0$, which gives $x = \dfrac{-b \pm \sqrt{b^2 - 4a(c-y)}}{2a} = \dfrac{-b}{2a} \pm \dfrac{\sqrt{b^2 - 4a(c-y)}}{2a}$. Whenever $\sqrt{b^2 - 4a(c-y)}$ is defined (as a real number) it is nonnegative. Therefore, the sign of $\dfrac{\sqrt{b^2 - 4a(c-y)}}{2a}$ depends on a. It is nonnegative when a is positive, that is, when the parabola opens upward, and nonpositive when a is negative, that is, when the parabola opens downward. Thus, in order to satisfy $x \geq -\dfrac{b}{2a}$ we must take the $+$ in $\pm$ when $a > 0$ and the parabola opens upward and the $-$ in $\pm$ when $a < 0$ and the parabola opens downward. For definiteness now, let us deal with the case of $a > 0$. It follows, returning to the procedure of Example 5 of 2.4, that $g^{-1}(x) = \dfrac{-b + \sqrt{b^2 - 4a(c-x)}}{2a}$. The vertex of any paraola has y-coordinate given by $f\left(-\dfrac{b}{2a}\right) = a\left(-\dfrac{b}{2a}\right)^2 + b\left(-\dfrac{b}{2a}\right) + c = -\dfrac{b^2 - 4ac}{4a}$. The domain of g is by definition $\left[-\dfrac{b}{2a}, \infty\right)$. It is now apparent that in

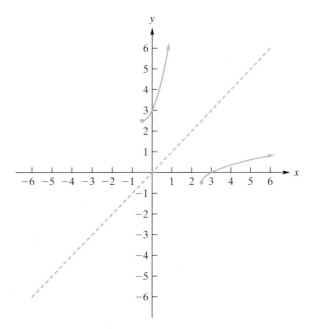

FIGURE 3.24   Graph of g and g^{-1}.

the upward-opening case, the range of g is $\left[-\dfrac{b^2 - 4ac}{4a}, \infty\right)$. As was stated in Section 2.4, it is a general fact that the domain of g^{-1} is the range of g. Let us verify that claim in this situation by considering the domain of $\dfrac{-b + \sqrt{b^2 - 4a(c - x)}}{2a}$ directly. The domain is the set of all x for which $b^2 - 4a(c - x) \geq 0$. Evidently, this inequality is equivalent to $b^2 - 4ac + 4ax \geq 0$, which in turn is equivalent to $4ax \geq -(b^2 - 4ac)$. In other words, $x \geq -\dfrac{b^2 - 4ac}{4a}$ as required.

To complete the exercise, observe that in Figure 3.23 we have provided the graph of $y = 2x^2 + 2x + 3$. For the task at hand, we redraw that part of the curve which lies to the right of the axis of symmetry. This provides the graph of g. Next we provide a dotted copy of the line $y = x$. Finally, we draw the mirror image of g in the line $y = x$ to obtain the graph of g^{-1} as in Figure 3.24.

NOW WORK PROBLEM 27

EXAMPLE 6   Maximum Revenue

The demand function for a manufacturer's product is $p = 1000 - 2q$, where p is the price (in dollars) per unit when q units are demanded (per week) by consumers. Find the level of production that will maximize the manufacturer's total revenue, and determine this revenue.

Solution:

Strategy   To maximize revenue, we must determine the revenue function, $r = f(q)$. Using the relation

$$\textbf{total revenue} = (\textbf{price})(\textbf{quantity})$$

we have

$$r = pq$$

Using the demand equation, we can express p in terms of q, so r will be a function of q.

We have

$$r = pq$$
$$= (1000 - 2q)q$$
$$r = 1000q - 2q^2$$

Note that r is a quadratic function of q, with $a = -2, b = 1000$, and $c = 0$. Since $a < 0$ (the parabola opens downward), r is maximum at the vertex (q, r), where

$$q = -\frac{b}{2a} = -\frac{1000}{2(-2)} = 250$$

The maximum value of r is given by

$$r = 1000(250) - 2(250)^2$$
$$= 250,000 - 125,000 = 125,000$$

Thus, the maximum revenue that the manufacturer can receive is $125,000, which occurs at a production level of 250 units. Figure 3.25(a) shows the graph of the revenue function. Only that portion for which $q \geq 0$ and $r \geq 0$ is drawn, since quantity and revenue cannot be negative.

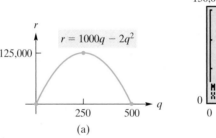

FIGURE 3.25 Graph of revenue function.

NOW WORK PROBLEM 29

TECHNOLOGY

The maximum (or minimum) value of a function can be conveniently found with a graphing calculator either by using trace and zoom or by using the "maximum" (or "minimum") feature. Figure 3.25(b) shows the display for the revenue function of Example 6, namely, the graph of $y = 1000x - 2x^2$. Note that we replaced r by y and q by x.

Problems 3.3

In Problems 1–8, state whether the function is quadratic.

1. $f(x) = 5x^2$

2. $g(x) = \dfrac{1}{2x^2 - 4}$

3. $g(x) = 7 - 6x$

4. $k(v) = 3v^2(v^2 + 2)$

5. $h(q) = (3 - q)^2$

6. $f(t) = 2t(3 - t) + 4t$

7. $f(s) = \dfrac{s^2 - 9}{2}$

8. $g(t) = (t^2 - 1)^2$

In Problems 9–12, do not include a graph.

9. (a) For the parabola $y = f(x) = -4x^2 + 8x + 7$, find the vertex. (b) Does the vertex correspond to the highest point or the lowest point on the graph?

10. Repeat Problem 9 if $y = f(x) = 8x^2 + 4x - 1$.

11. For the parabola $y = f(x) = x^2 + x - 6$, find (a) the y-intercept, (b) the x-intercepts, and (c) the vertex.

12. Repeat Problem 11 if $y = f(x) = 5 - x - 3x^2$.

In Problems 13–22, graph each function. Give the vertex and intercepts, and state the range.

*13. $y = f(x) = x^2 - 6x + 5$

14. $y = f(x) = -4x^2$

*15. $y = g(x) = -2x^2 - 6x$

16. $y = f(x) = x^2 - 4$

*17. $s = h(t) = t^2 + 6t + 9$

18. $s = h(t) = 2t^2 + 3t - 2$

19. $y = f(x) = -9 + 8x - 2x^2$

20. $y = H(x) = 1 - x - x^2$

*21. $t = f(s) = s^2 - 8s + 14$

22. $t = f(s) = s^2 + 6s + 11$

In Problems 23–26, state whether $f(x)$ has a maximum value or a minimum value, and find that value.

23. $f(x) = 49x^2 - 10x + 17$ **24.** $f(x) = -3x^2 - 18x + 7$

25. $f(x) = 4x - 50 - 0.1x^2$ **26.** $f(x) = x(x + 3) - 12$

In Problems 27 and 28, restrict the quadratic function to those x satisfying $x \geq v$, where v is the x-coordinate of the vertex of the parabola. Determine the inverse of the restricted function. Graph the restricted function and its inverse in the same plane.

*27. $f(x) = x^2 - 2x + 4$ **28.** $f(x) = -x^2 + 4x - 3$

*29. Revenue   The demand function for a manufacturer's product is $p = f(q) = 200 - 5q$, where p is the price (in dollars) per unit when q units are demanded (per day). Find the level of production that maximizes the manufacturer's total revenue and determine this revenue.

30. Revenue   The demand function for an office supply company's line of plastic rulers is $p = 0.85 - 0.00045q$, where p is the price (in dollars) per unit when q units are demanded (per day) by consumers. Find the level of production that will maximize the manufacturer's total revenue, and determine this revenue.

31. Revenue   The demand function for an electronics company's laptop computer line is $p = 2400 - 6q$, where p is the price (in dollars) per unit when q units are demanded (per week) by consumers. Find the level of production that will maximize the manufacturer's total revenue, and determine this revenue.

32. Marketing   A marketing firm estimates that n months after the introduction of a client's new product, $f(n)$ thousand households will use it, where

$$f(n) = \frac{10}{9}n(12 - n), \quad 0 \leq n \leq 12$$

Estimate the maximum number of households that will use the product.

33. Profit   The daily profit for the garden department of a store from the sale of trees is given by $P(x) = -x^2 + 18x + 144$, where x is the number of trees sold. Find the function's vertex and intercepts, and graph the function.

34. Psychology   A prediction made by early psychology relating the magnitude of a stimulus, x, to the magnitude of a response, y, is expressed by the equation $y = kx^2$, where k is a constant of the experiment. In an experiment on pattern recognition, $k = 2$. Find the function's vertex and graph the equation. (Assume no restriction on x.)

35. Biology   Biologists studied the nutritional effects on rats that were fed a diet containing 10% protein.[5] The protein consisted of yeast and corn flour. By varying the percentage, P, of yeast in the protein mix, the group estimated that the average weight gain (in grams) of a rat over a period of time was

$$f(P) = -\frac{1}{50}P^2 + 2P + 20, \quad 0 \leq P \leq 100$$

Find the maximum weight gain.

---

[5]Adapted from R. Bressani, "The Use of Yeast in Human Foods," in *Single-Cell Protein*, ed. R. I. Mateles and S. R. Tannenbaum (Cambridge, MA: MIT Press, 1968).

36. Height of Ball   Suppose that the height, s, of a ball thrown vertically upward is given by

$$s = -4.9t^2 + 62.3t + 1.8$$

where s is in meters and t is elapsed time in seconds. (See Figure 3.26.) After how many seconds will the ball reach its maximum height? What is the maximum height?

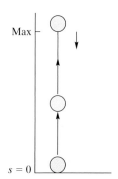

FIGURE 3.26   Ball thrown upward (Problem 36).

37. Archery   A boy standing on a hill shoots an arrow straight up with an initial velocity of 85 feet per second. The height, h, of the arrow in feet, t seconds after it was released, is described by the function $h(t) = -16t^2 + 85t + 22$. What is the maximum height reached by the arrow? How many seconds after release does it take to reach this height?

38. Toy Toss   A 6-year-old girl standing on a toy chest throws a doll straight up with an initial velocity of 16 feet per second. The height h of the doll in feet t seconds after it was released is described by the function $h(t) = -16t^2 + 16t + 4$. How long does it take the doll to reach its maximum height? What is the maximum height?

39. Rocket Launch   A toy rocket is launched straight up from the roof of a garage with an initial velocity of 80 feet per second. The height, h, of the rocket in feet, t seconds after it was released, is described by the function $h(t) = -16t^2 + 80t + 16$. Find the function's vertex and intercepts, and graph the function.

40. Area   Express the area of the rectangle shown in Figure 3.27 as a quadratic function of x. For what value of x will the area be a maximum?

FIGURE 3.27   Diagram for Problem 40.

41. Enclosing Plot   A building contractor wants to fence in a rectangular plot adjacent to a straight highway using the highway for one side, which will be left unfenced. (See Figure 3.28.) If the contractor has 500 feet of fence, find the dimensions of the maximum enclosed area.

FIGURE 3.28   Diagram for Problem 41.

42. Find two numbers whose sum is 78 and whose product is a maximum.

43. From the graph of $y = 1.4x^2 - 3.1x + 4.6$, determine the coordinates of the vertex. Round values to two decimal places. Verify your answer by using the vertex formula.

44. Find the zeros of $f(x) = -\sqrt{2}x^2 + 3x + 8.5$ by examining the graph of f. Round values to two decimal places.

45. Determine the number of real zeros for each of the following quadratic functions:
   (a) $f(x) = 4.2x^2 - 8.1x + 10.4$
   (b) $f(x) = 5x^2 - 2\sqrt{35}x + 7$
   (c) $f(x) = \dfrac{5.1 - 7.2x - x^2}{4.8}$

46. Find the maximum value (rounded to two decimal places) of the function $f(x) = 5.4 + 12x - 4.1x^2$ from its graph.

47. Find the minimum value (rounded to two decimal places) of the function $f(x) = 20x^2 - 13x + 7$ from its graph.

3.4 Systems of Linear Equations

OBJECTIVE

To solve systems of linear equations in both two and three variables by using the technique of elimination by addition or by substitution. (In Chapter 6, other methods are shown.)

Two-Variable Systems

When a situation must be described mathematically, it is not unusual for a *set* of equations to arise. For example, suppose that the manager of a factory is setting up a production schedule for two models of a new product. Model A requires 4 resistors and 9 transistors. Model B requires 5 resistors and 14 transistors. From its suppliers, the factory gets 335 resistors and 850 transistors each day. How many of each model should the manager plan to make each day so that all the resistors and transistors are used?

It's a good idea to construct a table that summarizes the important information. Table 3.2 shows the number of resistors and transistors required for each model, as well as the total number available.

TABLE 3.2

	Model A	Model B	Total Available
Resistors	4	5	335
Transistors	9	14	850

Suppose we let x be the number of model A made each day and y be the number of model B. Then these require a total of $4x + 5y$ resistors and $9x + 14y$ transistors. Since 335 resistors and 850 transistors are available, we have

$$\begin{cases} 4x + 5y = 335 & \text{(1)} \\ 9x + 14y = 850 & \text{(2)} \end{cases}$$

We call this set of equations a **system** of two linear equations in the variables x and y. The problem is to find values of x and y for which *both* equations are true *simultaneously*. A pair (x, y) of such values is called a *solution* of the system.

Since Equations (1) and (2) are linear, their graphs are straight lines; call these lines L_1 and L_2. Now, the coordinates of any point on a line satisfy the equation of that line; that is, they make the equation true. Thus, the coordinates of any point of intersection of L_1 and L_2 will satisfy both equations. This means that a point of intersection gives a solution of the system.

If L_1 and L_2 are drawn on the same plane, there are three situations that could occur:

CAUTION

Note that a *single* solution is given by a *pair* of values.

1. L_1 and L_2 may intersect at exactly one point, say, (a, b). (See Figure 3.29.) Thus, the system has the solution $x = a$ and $y = b$.

2. L_1 and L_2 may be parallel and have no points in common. (See Figure 3.30.) In this case, there is no solution.

3. L_1 and L_2 may be the same line. (See Figure 3.31.) Here the coordinates of any point on the line are a solution of the system. Consequently, there are infinitely many solutions.

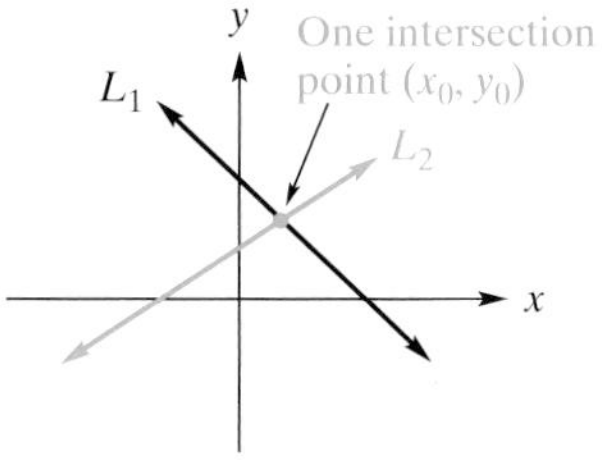

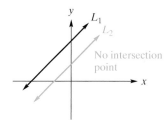

FIGURE 3.29 Linear system (one solution).

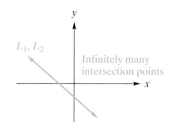

FIGURE 3.30 Linear system (no solution).

FIGURE 3.31 Linear system (infinitely many solutions).

Our main concern in this section is algebraic methods of solving a system of linear equations. We will successively replace the system by other systems that have the same solutions. Generalizing the terminology of Section 0.7, in the subsection titled "Equivalent Equations," we say that two systems are *equivalent* if their sets of solutions are equal. The replacement systems have progressively more desirable forms for determining the solution. More precisely, we seek an equivalent system containing an equation in which one of the variables does not appear. (In this case we say that the variable has been *eliminated*.) In dealing with systems of *linear* equations, our passage from a system to an equivalent system will always be accomplished by one of the following procedures:

1. Interchanging two equations
2. Multiplying one equation by a nonzero constant
3. Replacing an equation by itself plus a multiple of another equation

We will return to these procedures in more detail in Chapter 6. For the moment, since we will also consider nonlinear systems in this chapter, it is convenient to express our solutions in terms of the very general principles of Section 0.7 that guarantee equivalence of equations.

We will illustrate the elimination procedure for the system in the problem originally posed:

$$\begin{cases} 4x + 5y = 335 & \text{(3)} \\ 9x + 14y = 850 & \text{(4)} \end{cases}$$

To begin, we will obtain an equivalent system in which x does not appear in one equation. First we find an equivalent system in which the coefficients of the x-terms in each equation are the same except for their sign. Multiplying Equation (3) by 9 [that is, multiplying both sides of Equation (3) by 9] and multiplying Equation (4) by -4 gives

$$\begin{cases} 36x + 45y = 3015 & \text{(5)} \\ -36x - 56y = -3400 & \text{(6)} \end{cases}$$

The left and right sides of Equation (5) are equal, so each side can be *added* to the corresponding side of Equation (6). This results in

$$-11y = -385$$

which has only one variable, as planned. Solving gives

$$y = 35$$

so we obtain the equivalent system

$$\begin{cases} -36x - 56y = -3400 & \text{(7)} \\ y = 35 & \text{(8)} \end{cases}$$

Replacing y in Equation (7) by 35, we get

$$-36x - 56(35) = -3400$$
$$-36x - 1960 = -3400$$
$$-36x = -1440$$
$$x = 40$$

Thus, the original system is equivalent to

$$\begin{cases} x = 40 \\ y = 35 \end{cases}$$

We can check our answer by substituting $x = 40$ and $y = 35$ into *both* of the original equations. In Equation (3), we get $4(40) + 5(35) = 335$, or $335 = 335$. In Equation (4), we get $9(40) + 14(35) = 850$, or $850 = 850$. Hence, the solution is

$$x = 40 \quad \text{and} \quad y = 35$$

Each day the manager should plan to make 40 of model A and 35 of model B. Our procedure is referred to as **elimination by addition.** Although we chose to eliminate x first, we could have done the same for y by a similar procedure.

PRINCIPLES IN PRACTICE 1

ELIMINATION-BY-ADDITION METHOD

A computer consultant has $200,000 invested for retirement, part at 9% and part at 8%. If the total yearly income from the investments is $17,200, how much is invested at each rate?

● EXAMPLE 1   **Elimination-by-Addition Method**

Use elimination by addition to solve the system.

$$\begin{cases} 3x - 4y = 13 \\ 3y + 2x = 3 \end{cases}$$

Solution: Aligning the x- and y-terms for convenience gives

$$\begin{cases} 3x - 4y = 13 & \text{(9)} \\ 2x + 3y = 3 & \text{(10)} \end{cases}$$

To eliminate y, we multiply Equation (9) by 3 and Equation (10) by 4:

$$\begin{cases} 9x - 12y = 39 & \text{(11)} \\ 8x + 12y = 12 & \text{(12)} \end{cases}$$

Adding Equation (11) to Equation (12) gives $17x = 51$, from which $x = 3$. We have the equivalent system

$$\begin{cases} 9x - 12y = 39 & \text{(13)} \\ x = 3 & \text{(14)} \end{cases}$$

Replacing x by 3 in Equation (13) results in

$$9(3) - 12y = 39$$
$$-12y = 12$$
$$y = -1$$

so the original system is equivalent to

$$\begin{cases} y = -1 \\ x = 3 \end{cases}$$

The solution is $x = 3$ and $y = -1$. Figure 3.32 shows a graph of the system.

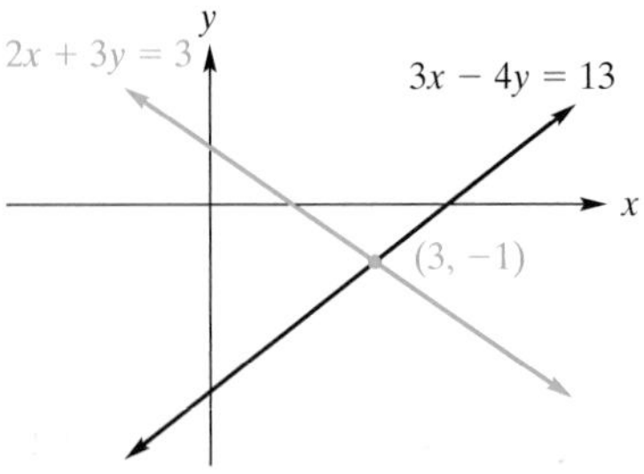

FIGURE 3.32   Linear system of Example 1: one solution.

NOW WORK PROBLEM 1   ●●●

The system in Example 1,

$$\begin{cases} 3x - 4y = 13 & \text{(15)} \\ 2x + 3y = 3 & \text{(16)} \end{cases}$$

can be solved another way. We first choose one of the equations—for example, Equation (15)—and solve it for one variable in terms of the other, say x in terms of y. Hence Equation (15) is equivalent to $3x = 4y + 13$, which is equivalent to

$$x = \frac{4}{3}y + \frac{13}{3}$$

and we obtain

$$\begin{cases} x = \dfrac{4}{3}y + \dfrac{13}{3} & (17) \\ 2x + 3y = 3 & (18) \end{cases}$$

Substituting the right side of Equation (17) for x in Equation (18) gives

$$2\left(\frac{4}{3}y + \frac{13}{3}\right) + 3y = 3 \qquad (19)$$

Thus, x has been eliminated. Solving Equation (19), we have

$$\frac{8}{3}y + \frac{26}{3} + 3y = 3$$

$$8y + 26 + 9y = 9 \qquad \text{(clearing fractions)}$$

$$17y = -17$$

$$y = -1$$

Replacing y in Equation (17) by -1 gives $x = 3$, and the original system is equivalent to

$$\begin{cases} x = 3 \\ y = -1 \end{cases}$$

as before. This method is called **elimination by substitution.**

PRINCIPLES IN PRACTICE 2

METHOD OF ELIMINATION BY SUBSTITUTION

Two species of deer, A and B, living in a wildlife refuge are given extra food in the winter. Each week, they receive 2 tons of food pellets and 4.75 tons of hay. Each deer of species A requires 4 pounds of the pellets and 5 pounds of hay. Each deer of species B requires 2 pounds of the pellets and 7 pounds of hay. How many of each species of deer will the food support so that all of the food is consumed each week?

● **EXAMPLE 2   Method of Elimination by Substitution**

Use elimination by substitution to solve the system

$$\begin{cases} x + 2y - 8 = 0 \\ 2x + 4y + 4 = 0 \end{cases}$$

Solution: It is easy to solve the first equation for x. Doing so gives the equivalent system

$$\begin{cases} x = -2y + 8 & (20) \\ 2x + 4y + 4 = 0 & (21) \end{cases}$$

Substituting $-2y + 8$ for x in Equation (21) yields

$$2(-2y + 8) + 4y + 4 = 0$$

$$-4y + 16 + 4y + 4 = 0$$

The latter equation simplifies to $20 = 0$. Thus, we have the system

$$\begin{cases} x = -2y + 8 & (22) \\ 20 = 0 & (23) \end{cases}$$

Since Equation (23) is *never* true, there is **no solution** of the original system. The reason is clear if we observe that the original equations can be written in slope-intercept form as

$$y = -\frac{1}{2}x + 4$$

and

$$y = -\frac{1}{2}x - 1$$

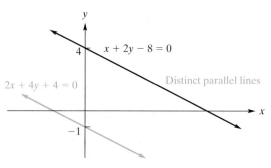

FIGURE 3.33 Linear system of Example 2: no solution.

These equations represent straight lines having slopes of $-\frac{1}{2}$, but different y-intercepts, namely, 4 and -1. That is, they determine different parallel lines. (See Figure 3.33.)

NOW WORK PROBLEM 9

PRINCIPLES IN PRACTICE 3

A LINEAR SYSTEM WITH INFINITELY MANY SOLUTIONS

Two species of fish, A and B, are raised in one pond at a fish farm where they are fed two vitamin supplements. Each day, they receive 100 grams of the first supplement and 200 grams of the second supplement. Each fish of species A requires 15 mg of the first supplement and 30 mg of the second supplement. Each fish of species B requires 20 mg of the first supplement and 40 mg of the second supplement. How many of each species of fish will the pond support so that all of the supplements are consumed each day?

EXAMPLE 3 A Linear System with Infinitely Many Solutions

Solve

$$\begin{cases} x + 5y = 2 & (24) \\ \dfrac{1}{2}x + \dfrac{5}{2}y = 1 & (25) \end{cases}$$

Solution: We begin by eliminating x from the second equation. Multiplying Equation (25) by -2, we have

$$\begin{cases} x + 5y = 2 & (26) \\ -x - 5y = -2 & (27) \end{cases}$$

Adding Equation (26) to Equation (27) gives

$$\begin{cases} x + 5y = 2 & (28) \\ 0 = 0 & (29) \end{cases}$$

Because Equation (29) is *always* true, any solution of Equation (28) is a solution of the system. Now let us see how we can express our answer. From Equation (28), we have $x = 2 - 5y$, where y can be any real number, say, r. Thus, we can write $x = 2 - 5r$. The complete solution is

$$x = 2 - 5r$$
$$y = r$$

where r is any real number. In this situation r is called a **parameter,** and we say that we have a one-parameter family of solutions. Each value of r determines a particular solution. For example, if $r = 0$, then $x = 2$ and $y = 0$ is a solution; if $r = 5$, then $x = -23$ and $y = 5$ is another solution. Clearly, the given system has infinitely many solutions.

It is worthwhile to note that by writing Equations (24) and (25) in their slope-intercept forms, we get the equivalent system

$$\begin{cases} y = -\dfrac{1}{5}x + \dfrac{2}{5} \\ y = -\dfrac{1}{5}x + \dfrac{2}{5} \end{cases}$$

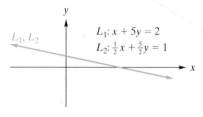

FIGURE 3.34 Linear system of Example 3: infinitely many solutions.

in which both equations represent the same line. Hence, the lines coincide (Figure 3.34), and Equations (24) and (25) are equivalent. The solution of the system consists of the coordinate pairs of all points on the line $x + 5y = 2$, and these points are given by our parametric solution.

NOW WORK PROBLEM 19

Graphically solve the system

$$\begin{cases} 9x + 4.1y = 7 \\ 2.6x - 3y = 18 \end{cases}$$

Solution: First we solve each equation for y, so that each equation has the form $y = f(x)$:

$$y = \frac{1}{4.1}(7 - 9x)$$

$$y = -\frac{1}{3}(18 - 2.6x)$$

Next we enter these functions as Y_1 and Y_2 and display them on the same viewing rectangle. (See Figure 3.35.) Finally, either using trace and zoom or using the

intersection feature, we estimate the solution to be $x = 2.52$, $y = -3.82$.

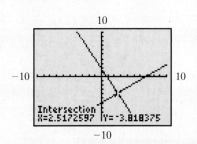

FIGURE 3.35 Graphical solution of system.

● EXAMPLE 4   **Mixture**

A chemical manufacturer wishes to fill an order for 500 liters of a 25% acid solution. (Twenty-five percent by volume is acid.) If solutions of 30% and 18% are available in stock, how many liters of each must be mixed to fill the order?

Solution: Let x and y be the number of liters of the 30% and 18% solutions respectively that should be mixed. Then

$$x + y = 500$$

To help visualize the situation, we draw the diagram in Figure 3.36. In 500 liters of a 25% solution, there will be $0.25(500) = 125$ liters of acid. This acid comes from two sources: $0.30x$ liters of it come from the 30% solution, and $0.18y$ liters of it come from the 18% solution. Hence,

$$0.30x + 0.18y = 125$$

These two equations form a system of two linear equations in two unknowns. Solving the first for x gives $x = 500 - y$. Substituting in the second gives

$$0.30(500 - y) + 0.18y = 125$$

Solving this equation for y, we find that $y = 208\frac{1}{3}$ liters. Thus, $x = 500 - 208\frac{1}{3} = 291\frac{2}{3}$ liters. (See Figure 3.37.)

NOW WORK PROBLEM 25 ●◖◗

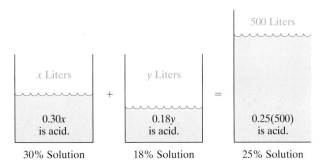

FIGURE 3.36 Mixture problem.

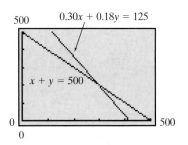

FIGURE 3.37 Graph for Example 4.

Three-Variable Systems

The methods used in solving a two-variable system of linear equations can be used to solve a three-variable system of linear equations. A **general linear equation in the three variables** x, y, **and** z is an equation having the form

$$Ax + By + Cz = D$$

where A, B, C, and D are constants and A, B, and C are not all zero. For example, $2x - 4y + z = 2$ is such an equation. Geometrically, a general linear equation in three variables represents a *plane* in space, and a solution to a system of such equations is the intersection of planes. Example 5 shows how to solve a system of three linear equations in three variables.

●EXAMPLE 5   Solving a Three-Variable Linear System

Solve

$$\begin{cases} 2x + \ y + \ z = 3 & (30) \\ -x + 2y + 2z = 1 & (31) \\ x - \ y - 3z = -6 & (32) \end{cases}$$

Solution: This system consists of three linear equations in three variables. From Equation (32), $x = y + 3z - 6$. By substituting for x in Equations (30) and (31), we obtain

$$\begin{cases} 2(y + 3z - 6) + y + z = 3 \\ -(y + 3z - 6) + 2y + 2z = 1 \\ \qquad\qquad\qquad x = y + 3z - 6 \end{cases}$$

Simplifying gives

$$\begin{cases} 3y + 7z = 15 & (33) \\ \ y - z = -5 & (34) \\ \quad x = y + 3z - 6 & (35) \end{cases}$$

Note that x does not appear in Equations (33) and (34). Since any solution of the original system must satisfy Equations (33) and (34), we will consider their solution first:

$$\begin{cases} 3y + 7z = 15 & (33) \\ \ y - \ z = -5 & (34) \end{cases}$$

From Equation (34), $y = z - 5$. This means that we can replace Equation (33) by

$$3(z - 5) + 7z = 15 \quad \text{that is,} \quad z = 3$$

Since z is 3, we can replace Equation (34) with $y = -2$. Hence, the previous system is equivalent to

$$\begin{cases} z = 3 \\ y = -2 \end{cases}$$

The original system becomes

$$\begin{cases} z = 3 \\ y = -2 \\ x = y + 3z - 6 \end{cases}$$

from which $x = 1$. The solution is $x = 1$, $y = -2$, and $z = 3$, which you should verify.

NOW WORK PROBLEM 15 ●●●

Just as a two-variable system may have a one-parameter family of solutions, a three-variable system may have a one-parameter or a two-parameter family of solutions. The next two examples illustrate.

● EXAMPLE 6   **One-Parameter Family of Solutions**

Solve

$$\begin{cases} x - 2y = 4 & (35) \\ 2x - 3y + 2z = -2 & (36) \\ 4x - 7y + 2z = 6 & (37) \end{cases}$$

Solution: Note that since Equation (35) can be written $x - 2y + 0z = 4$, we can view Equations (35) to (37) as a system of three linear equations in the variables x, y, and z. From Equation (35), we have $x = 2y + 4$. Using this equation and substitution, we can eliminate x from Equations (36) and (37):

$$\begin{cases} x = 2y + 4 \\ 2(2y + 4) - 3y + 2z = -2 \\ 4(2y + 4) - 7y + 2z = 6 \end{cases}$$

which simplifies to give

$$\begin{cases} x = 2y + 4 & (38) \\ y + 2z = -10 & (39) \\ y + 2z = -10 & (40) \end{cases}$$

Multiplying Equation (40) by -1 gives

$$\begin{cases} x = 2y + 4 \\ y + 2z = -10 \\ -y - 2z = 10 \end{cases}$$

Adding the second equation to the third yields

$$\begin{cases} x = 2y + 4 \\ y + 2z = -10 \\ 0 = 0 \end{cases}$$

Since the equation $0 = 0$ is always true, the system is equivalent to

$$\begin{cases} x = 2y + 4 & (41) \\ y + 2z = -10 & (42) \end{cases}$$

Solving Equation (42) for y, we have

$$y = -10 - 2z$$

which expresses y in terms of z. We can also express x in terms of z. From Equation (41),

$$x = 2y + 4$$
$$= 2(-10 - 2z) + 4$$
$$= -16 - 4z$$

Thus, we have

$$\begin{cases} x = -16 - 4z \\ y = -10 - 2z \end{cases}$$

Since no restriction is placed on z, this suggests a parametric family of solutions. Setting $z = r$, we have the following family of solutions of the given system:

$$x = -16 - 4r$$
$$y = -10 - 2r$$
$$z = r$$

Other parametric representations of the solution are possible.

where r can be any real number. We see, then, that the given system has infinitely many solutions. For example, setting $r = 1$ gives the particular solution $x = -20$,

$y = -12$, and $z = 1$. There is nothing special about the name of the parameter. In fact, since $z = r$, we could consider z to be the parameter.

NOW WORK PROBLEM 19

● EXAMPLE 7   **Two-Parameter Family of Solutions**

Solve the system

$$\begin{cases} x + 2y + \;\; z = 4 \\ 2x + 4y + 2z = 8 \end{cases}$$

Solution: This is a system of two linear equations in three variables. We will eliminate x from the second equation by first multiplying that equation by $-\frac{1}{2}$:

$$\begin{cases} x + 2y + z = 4 \\ -x - 2y - z = -4 \end{cases}$$

Adding the first equation to the second gives

$$\begin{cases} x + 2y + z = 4 \\ \qquad\qquad 0 = 0 \end{cases}$$

From the first equation, we obtain

$$x = 4 - 2y - z$$

Since no restriction is placed on either y or z, they can be arbitrary real numbers, giving us a two-parameter family of solutions. Setting $y = r$ and $z = s$, we find that the solution of the given system is

$$x = 4 - 2r - s$$

$$y = r$$

$$z = s$$

where r and s can be any real numbers. Each assignment of values to r and s results in a solution of the given system, so there are infinitely many solutions. For example, letting $r = 1$ and $s = 2$ gives the particular solution $x = 0$, $y = 1$, and $z = 2$. As in the last example, there is nothing special about the names of the parameters. In particular, since $y = r$ and $z = s$, we could consider y and z to be the two parameters.

NOW WORK PROBLEM 23

Problems 3.4

In Problems 1–24, solve the systems algebraically.

*1. $\begin{cases} x + 4y = 3 \\ 3x - 2y = -5 \end{cases}$

2. $\begin{cases} 4x + 2y = 9 \\ 5y - 4x = 5 \end{cases}$

3. $\begin{cases} 3x - 4y = 13 \\ 2x + 3y = 3 \end{cases}$

4. $\begin{cases} 2x - \;\; y = 1 \\ -x + 2y = 7 \end{cases}$

5. $\begin{cases} u + v = 5 \\ u - v = 7 \end{cases}$

6. $\begin{cases} 2p + \;\; q = 16 \\ 3p + 3q = 33 \end{cases}$

7. $\begin{cases} x - 2y = -7 \\ 5x + 3y = -9 \end{cases}$

8. $\begin{cases} 3x + 5y = 7 \\ 5x + 9y = 7 \end{cases}$

*9. $\begin{cases} 4x - 3y - 2 = 3x - 7y \\ x + 5y - 2 = y + 4 \end{cases}$

10. $\begin{cases} 5x + 7y + 2 = 9y - 4x + 6 \\ \frac{21}{2}x - \frac{4}{3}y - \frac{11}{4} = \frac{3}{2}x + \frac{2}{3}y + \frac{5}{4} \end{cases}$

11. $\begin{cases} \frac{2}{3}x + \frac{1}{2}y = 2 \\ \frac{3}{8}x + \frac{5}{6}y = -\frac{11}{2} \end{cases}$

12. $\begin{cases} \frac{1}{2}z - \frac{1}{4}w = \frac{1}{6} \\ \frac{1}{2}z + \frac{1}{4}w = \frac{1}{6} \end{cases}$

13. $\begin{cases} 5p + 11q = 7 \\ 10p + 22q = 33 \end{cases}$

14. $\begin{cases} 5x - 3y = 2 \\ -10x + 6y = 4 \end{cases}$

*15. $\begin{cases} 2x + y + 6z = 3 \\ x - y + 4z = 1 \\ 3x + 2y - 2z = 2 \end{cases}$

16. $\begin{cases} x + y + \;\; z = -1 \\ 3x + y + \;\; z = 1 \\ 4x - 2y + 2z = 0 \end{cases}$

17. $\begin{cases} x + 4y + 3z = 10 \\ 4x + 2y - 2z = -2 \\ 3x - \;\; y + \;\; z = 11 \end{cases}$

18. $\begin{cases} x + y + \;\; z = 18 \\ x - y - \;\; z = 12 \\ 3x + y + 4z = 4 \end{cases}$

*19. $\begin{cases} x - 2z = 1 \\ y + \;\; z = 3 \end{cases}$

20. $\begin{cases} 2y + 3z = 1 \\ 3x - 4z = 0 \end{cases}$

21. $\begin{cases} x - \;\; y + 2z = 0 \\ 2x + \;\; y - \;\; z = 0 \\ x + 2y - 3z = 0 \end{cases}$

22. $\begin{cases} x - 2y - \;\; z = 0 \\ 2x - 4y - 2z = 0 \\ -x + 2y + \;\; z = 0 \end{cases}$

*23. $\begin{cases} 2x + 2y - \;\; z = 3 \\ 4x + 4y - 2z = 6 \end{cases}$

24. $\begin{cases} 5x + y + z = 17 \\ 4x + y + z = 14 \end{cases}$

*25. **Mixture**   A chemical manufacturer wishes to fill an order for 800 gallons of a 25% acid solution. Solutions of 20% and

35% are in stock. How many gallons of each solution must be mixed to fill the order?

26. Mixture A gardener has two fertilizers that contain different concentrations of nitrogen. One is 3% nitrogen and the other is 11% nitrogen. How many pounds of each should she mix to obtain 20 pounds of a 9% concentration?

27. Fabric A textile mill produces fabric made from different fibers. From cotton, polyester, and nylon, the owners want to produce a fabric blend that will cost $3.25 per pound to make. The cost per pound of these fibers is $4.00, $3.00, and $2.00, respectively. The amount of nylon is to be the same as the amount of polyester. How much of each fiber will be in the final fabric?

28. Taxes A company has taxable income of $312,000. The federal tax is 25% of that portion left after the state tax has been paid. The state tax is 10% of that portion left after the federal tax has been paid. Find the federal and state taxes.

29. Airplane Speed An airplane travels 900 mi in 2 h, 55 min, with the aid of a tail wind. It takes 3 h, 26 min, for the return trip, flying against the same wind. Find the speed of the airplane in still air and the speed of the wind.

30. Speed of Raft On a trip on a raft, it took $\frac{1}{2}$ hour to travel 10 miles downstream. The return trip took $\frac{3}{4}$ hour. Find the speed of the raft in still water and the speed of the current.

31. Furniture Sales A manufacturer of dining-room sets produces two styles: early American and contemporary. From past experience, management has determined that 20% more of the early American styles can be sold than the contemporary styles. A profit of $250 is made on each early American set sold, whereas a profit of $350 is made on each contemporary set. If, in the forthcoming year, management desires a total profit of $130,000, how many units of each style must be sold?

32. Survey National Surveys was awarded a contract to perform a product-rating survey for Crispy Crackers. A total of 250 people were interviewed. National Surveys reported that 62.5% more people like Crispy Crackers than disliked them. However, the report did not indicate that 16% of those interviewed had no comment. How many of those surveyed liked Crispy Crackers? How many disliked them? How many had no comment?

33. Equalizing Cost United Products Co. manufactures calculators and has plants in the cities of Exton and Whyton.

At the Exton plant, fixed costs are $7000 per month, and the cost of producing each calculator is $7.50. At the Whyton plant, fixed costs are $8800 per month, and each calculator costs $6.00 to produce. Next month, United Products must produce 1500 calculators. How many must be made at each plant if the total cost at each plant is to be the same?

34. Coffee Blending A coffee wholesaler blends together three types of coffee that sell for $2.20, $2.30, and $2.60 per pound, so as to obtain 100 lb of coffee worth $2.40 per pound. If the wholesaler uses the same amount of the two higher priced coffees, how much of each type must be used in the blend?

35. Commissions A company pays its salespeople on a basis of a certain percentage of the first $100,000 in sales, plus a certain percentage of any amount over $100,000 in sales. If one salesperson earned $8500 on sales of $175,000 and another salesperson earned $14,800 on sales of $280,000, find the two percentages.

36. Yearly Profits In news reports, profits of a company this year (T) are often compared with those of last year (L), but actual values of T and L are not always given. This year, a company had profits of $25 million more than last year. The profits were up 30%. Determine T and L from these data.

37. Fruit Packaging The Ilovetiny.com Organic Produce Company has 3600 lb of Donut Peaches that it is going to package in boxes. Half of the boxes will be loose-filled, each containing 20 lb of peaches, and the others will be packed with eight lb clam-shells (flip-top plastic containers), each containing 2.2 lb of peaches. Determine the number of boxes and the number of clamshells that are required.

38. Investments A person made two investments, and the percentage return per year on each was the same. Of the total amount invested, $\frac{3}{10}$ of it plus $600 was invested in one venture, and at the end of 1 year the person received a return of $384 from that venture. If the total return after 1 year was $1120, find the total amount invested.

39. Production Run A company makes three types of patio furniture: chairs, rockers, and chaise lounges. Each requires wood, plastic, and aluminum, in the amounts shown in the following table. The company has in stock 400 units of wood, 600 units of plastic, and 1500 units of aluminum. For its end-of-the-season production run, the company wants to use up all the stock. To do this, how many chairs, rockers, and chaise lounges should it make?

	Wood	Plastic	Aluminum
Chair	1 unit	1 unit	2 units
Rocker	1 unit	1 unit	3 units
Chaise lounge	1 unit	2 units	5 units

40. Investments   A total of $35,000 was invested at three interest rates: 7, 8, and 9%. The interest for the first year was $2830, which was not reinvested. The second year the amount originally invested at 9% earned 10% instead, and the other rates remained the same. The total interest the second year was $2960. How much was invested at each rate?

41. Hiring Workers   A company pays skilled workers in its assembly department $16 per hour. Semiskilled workers in that department are paid $9.50 per hour. Shipping clerks are paid $10 per hour. Because of an increase in orders, the company needs to hire a total of 70 workers in the assembly and shipping departments. It will pay a total of $725 per hour to these employees. Because of a union contract, twice as many semiskilled workers as skilled workers must be employed. How many semiskilled workers, skilled workers, and shipping clerks should the company hire?

42. Solvent Storage   A 10,000-gallon railroad tank car is to be filled with solvent from two storage tanks, A and B. Solvent

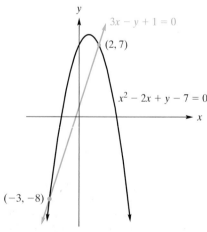

FIGURE 3.38   Nonlinear system of equations.

from A is pumped at the rate of 25 gal/min. Solvent from B is pumped at 35 gal/min. Usually, both pumps operate at the same time. However, because of a blown fuse, the pump on A is delayed 5 minutes. Both pumps finish operating at the same time. How many gallons from each storage tank will be used to fill the car?

43. Verify your answer to Problem 1 by using your graphing calculator.

44. Verify your answer to Problem 11 by using your graphing calculator.

45. Graphically solve the system
$$\begin{cases} 0.24x - 0.34y = 0.04 \\ 0.11x + 0.21y = 0.75 \end{cases}$$

46. Graphically solve the system
$$\begin{cases} x + y = 2 \\ \frac{1}{4}x + \frac{2}{5}y = \frac{3}{5} \end{cases}$$
Round the x- and y-values to two decimal places.

47. Graphically solve the system
$$\begin{cases} 0.5736x - 0.3420y = 0 \\ 0.8192x + 0.9397y = 20 \end{cases}$$
Round the x- and y-values to one decimal place.

OBJECTIVE

To use substitution to solve nonlinear systems.

3.5 Nonlinear Systems

A system of equations in which at least one equation is not linear is called a **nonlinear system.** We can often solve a nonlinear system by substitution, as was done with linear systems. The following examples illustrate.

EXAMPLE 1   Solving a Nonlinear System

Solve
$$\begin{cases} x^2 - 2x + y - 7 = 0 & \text{(1)} \\ 3x - y + 1 = 0 & \text{(2)} \end{cases}$$

Solution:

Strategy   If a nonlinear system contains a linear equation, we usually solve the linear equation for one variable and substitute for that variable in the other equation.

Solving Equation (2) for y gives
$$y = 3x + 1 \qquad \text{(3)}$$
Substituting into Equation (1) and simplifying, we have
$$x^2 - 2x + (3x + 1) - 7 = 0$$
$$x^2 + x - 6 = 0$$
$$(x + 3)(x - 2) = 0$$
$$x = -3 \text{ or } x = 2$$

If $x = -3$, then Equation (3) implies that $y = -8$; if $x = 2$, then $y = 7$. You should verify that each pair of values satisfies the given system. Hence, the solutions are $x = -3$, $y = -8$ and $x = 2$, $y = 7$. These solutions can be seen geometrically in the graph of the system in Figure 3.38. Notice that the graph of Equation (1) is a parabola and the graph of Equation (2) is a line. The solutions correspond to the intersection points $(-3, -8)$ and $(2, 7)$.

NOW WORK PROBLEM 1

● EXAMPLE 2   Solving a Nonlinear System

Solve

$$\begin{cases} y = \sqrt{x + 2} \\ x + y = 4 \end{cases}$$

Solution: Solving the second equation, which is linear, for y gives

$$y = 4 - x \qquad\qquad (4)$$

Substituting into the first equation yields

$$4 - x = \sqrt{x + 2}$$

$$16 - 8x + x^2 = x + 2 \qquad \text{(squaring both sides)}$$

$$x^2 - 9x + 14 = 0$$

$$(x - 2)(x - 7) = 0$$

Thus, $x = 2$ or $x = 7$. From Equation (4), if $x = 2$, then $y = 2$; if $x = 7$, then $y = -3$. Since we performed the operation of squaring both sides, we must check our results. Although the pair $x = 2$, $y = 2$ satisfies both of the original equations, this is not the case for $x = 7$ and $y = -3$. Thus, the solution is $x = 2$, $y = 2$. (See Figure 3.39.)

NOW WORK PROBLEM 13

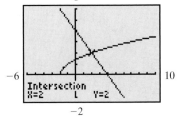

FIGURE 3.39   Nonlinear system of Example 2.

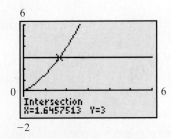

FIGURE 3.40   Solution of $0.5x^2 + x = 3$.

> ## T E C H N O L O G Y
>
> *Graphically solve the equation $0.5x^2 + x = 3$, where $x \geq 0$.*
>
> **Solution:** To solve the equation, we could find zeros of the function $f(x) = 0.5x^2 + x - 3$. Alternatively, we can think of this problem as solving the nonlinear system
>
> $$y = 0.5x^2 + x$$
>
> $$y = 3$$
>
> In Figure 3.40, the intersection point is estimated to be $x = 1.65$, $y = 3$. Note that the graph of $y = 3$ is a horizontal line. The solution of the given equation is $x = 1.65$.

CAUTION

This example illustrates the need for checking all "solutions."

Problems 3.5

In Problems 1–14, solve the given nonlinear system.

*1. $\begin{cases} y = x^2 - 9 \\ 2x + y = 3 \end{cases}$

2. $\begin{cases} y = x^3 \\ x - y = 0 \end{cases}$

3. $\begin{cases} p^2 = 5 - q \\ p = q + 1 \end{cases}$

4. $\begin{cases} y^2 - x^2 = 28 \\ x - y = 14 \end{cases}$

5. $\begin{cases} x = y^2 \\ y = x^2 \end{cases}$

6. $\begin{cases} p^2 - q + 1 = 0 \\ 5q - 3p - 2 = 0 \end{cases}$

7. $\begin{cases} y = 4x - x^2 + 8 \\ y = x^2 - 2x \end{cases}$

8. $\begin{cases} x^2 + 4x - y = -4 \\ y - x^2 - 4x + 3 = 0 \end{cases}$

9. $\begin{cases} p = \sqrt{q} \\ p = q^2 \end{cases}$

10. $\begin{cases} z = 4/w \\ 3z = 2w + 2 \end{cases}$

11. $\begin{cases} x^2 = y^2 + 13 \\ y = x^2 - 15 \end{cases}$

12. $\begin{cases} x^2 + y^2 - 2xy = 1 \\ 3x - y = 5 \end{cases}$

*13. $\begin{cases} x = y + 1 \\ y = 2\sqrt{x + 2} \end{cases}$

14. $\begin{cases} y = \dfrac{x^2}{x - 1} + 1 \\ y = \dfrac{1}{x - 1} \end{cases}$

15. Decorations   The shape of a paper streamer suspended above a dance floor can be described by the function $y = 0.01x^2 + 0.01x + 7$, where y is the height of the streamer (in feet) above the floor and x is the horizontal distance (in feet) from the center of the room. A rope holding up other decorations touches the streamer and is described by the function $y = 0.01x + 8.0$. Where does the rope touch the paper streamer?

16. **Awning** The shape of a decorative awning over a storefront can be described by the function $y = 0.06x^2 + 0.012x + 8$, where y is the height of the edge of the awning (in feet) above the sidewalk and x is the distance (in feet) from the center of the store's doorway. A vandal pokes a stick through the awning, piercing it in two places. The position of the stick can be described by the function $y = 0.912x + 5$. Where are the holes in the awning caused by the vandal?

17. Graphically determine how many solutions there are to the system
$$\begin{cases} y = \dfrac{1}{x} \\ y = x^2 - 4 \end{cases}$$

18. Graphically solve the system
$$\begin{cases} 2y = x^3 \\ y = 8 - x^2 \end{cases}$$
to one-decimal-place accuracy.

19. Graphically solve the system
$$\begin{cases} y = x^2 - 2x + 1 \\ y = x^3 + x^2 - 2x + 3 \end{cases}$$
to one-decimal-place accuracy.

20. Graphically solve the system
$$\begin{cases} y = x^3 - x + 1 \\ y = 3x + 2 \end{cases}$$
to one-decimal-place accuracy.

In Problems 21–23, graphically solve the equation by treating it as a system. Round answers to two decimal places.

21. $0.8x^2 + 2x = 6$ where $x \geq 0$

22. $\sqrt{x + 2} = 5 - x$

23. $x^3 - 3x^2 = x - 8$

OBJECTIVE

To solve systems describing equilibrium and break-even points.

3.6 Applications of Systems of Equations

Equilibrium

Recall from Section 3.2 that an equation that relates price per unit and quantity demanded (supplied) is called a *demand equation (supply equation)*. Suppose that, for product Z, the demand equation is

$$p = -\frac{1}{180}q + 12 \tag{1}$$

and the supply equation is

$$p = \frac{1}{300}q + 8 \tag{2}$$

where $q, p \geq 0$. The corresponding demand and supply curves are the lines in Figures 3.41 and 3.42, respectively. In analyzing Figure 3.41, we see that consumers will purchase 540 units per week when the price is \$9 per unit, 1080 units when the price is \$6, and so on. Figure 3.42 shows that when the price is \$9 per unit producers will place 300 units per week on the market, at \$10 they will supply 600 units, and so on.

When the demand and supply curves of a product are represented on the same coordinate plane, the point (m, n) where the curves intersect is called the **point of equilibrium.** (See Figure 3.43.) The price n, called the **equilibrium price,** is the price at which consumers will purchase the same quantity of a product that producers wish to sell at that price. In short, n is the price at which stability in the producer–consumer relationship occurs. The quantity m is called the **equilibrium quantity.**

To determine precisely the equilibrium point, we solve the system formed by the supply and demand equations. Let us do this for our previous data, namely, the system

$$\begin{cases} p = -\dfrac{1}{180}q + 12 & \text{(demand equation)} \\ p = \dfrac{1}{300}q + 8 & \text{(supply equation)} \end{cases}$$

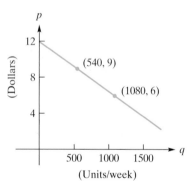

Demand equation: $p = -\frac{1}{180}q + 12$

FIGURE 3.41 Demand curve.

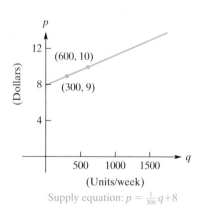

Supply equation: $p = \frac{1}{300}q + 8$

FIGURE 3.42 Supply curve.

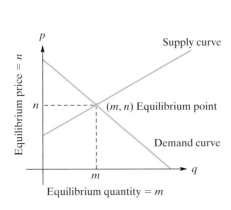

FIGURE 3.43   Equilibrium.

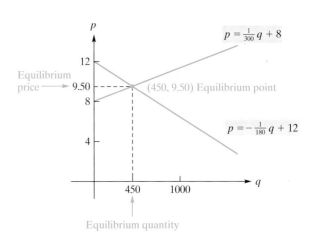

FIGURE 3.44   Equilibrium.

By substituting $\dfrac{1}{300}q + 8$ for p in the demand equation, we get

$$\frac{1}{300}q + 8 = -\frac{1}{180}q + 12$$

$$\left(\frac{1}{300} + \frac{1}{180}\right)q = 4$$

$$q = 450 \qquad \text{(equilibrium quantity)}$$

Thus,

$$p = \frac{1}{300}(450) + 8$$

$$= 9.50 \qquad \text{(equilibrium price)}$$

and the equilibrium point is (450, 9.50). Therefore, at the price of \$9.50 per unit, manufacturers will produce exactly the quantity (450) of units per week that consumers will purchase at that price. (See Figure 3.44.)

● EXAMPLE 1   Tax Effect on Equilibrium

Let $p = \dfrac{8}{100}q + 50$ be the supply equation for a manufacturer's product, and suppose the demand equation is $p = -\dfrac{7}{100}q + 65$.

a. *If a tax of \$1.50 per unit is to be imposed on the manufacturer, how will the original equilibrium price be affected if the demand remains the same?*

Solution: Before the tax, the equilibrium price is obtained by solving the system

$$\begin{cases} p = \dfrac{8}{100}q + 50 \\[2mm] p = -\dfrac{7}{100}q + 65 \end{cases}$$

By substitution,

$$-\frac{7}{100}q + 65 = \frac{8}{100}q + 50$$

$$15 = \frac{15}{100}q$$

$$100 = q$$

and

$$p = \frac{8}{100}(100) + 50 = 58$$

Thus, $58 is the original equilibrium price. Before the tax, the manufacturer supplies q units at a price of $p = \frac{8}{100}q + 50$ per unit. After the tax, he will sell the same q units for an additional $1.50 per unit. The price per unit will be $\left(\frac{8}{100}q + 50\right) + 1.50$, so the new supply equation is

$$p = \frac{8}{100}q + 51.50$$

Solving the system

$$\begin{cases} p = \dfrac{8}{100}q + 51.50 \\[2mm] p = -\dfrac{7}{100}q + 65 \end{cases}$$

will give the new equilibrium price:

$$\frac{8}{100}q + 51.50 = -\frac{7}{100}q + 65$$

$$\frac{15}{100}q = 13.50$$

$$q = 90$$

$$p = \frac{8}{100}(90) + 51.50 = 58.70$$

The tax of $1.50 per unit increases the equilibrium price by $0.70. (See Figure 3.45.) Note that there is also a decrease in the equilibrium quantity from $q = 100$ to $q = 90$, because of the change in the equilibrium price. (In the problems, you are asked to find the effect of a subsidy given to the manufacturer, which will reduce the price of the product.)

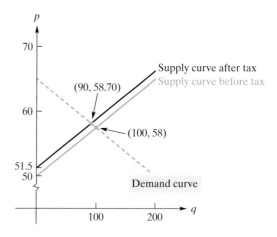

FIGURE 3.45 Equilibrium before and after tax.

NOW WORK PROBLEM 15

b. *Determine the total revenue obtained by the manufacturer at the equilibrium point both before and after the tax.*

Solution: If q units of a product are sold at a price of p dollars each, then the total revenue is given by

$$y_{TR} = pq$$

Before the tax, the revenue at $(100, 58)$ is (in dollars)

$$y_{TR} = (58)(100) = 5800$$

After the tax, it is

$$y_{TR} = (58.70)(90) = 5283$$

which is a decrease.

EXAMPLE 2 Equilibrium with Nonlinear Demand

Find the equilibrium point if the supply and demand equations of a product are $p = \dfrac{q}{40} + 10$ *and* $p = \dfrac{8000}{q}$, *respectively.*

Solution: Here the demand equation is not linear. Solving the system

$$\begin{cases} p = \dfrac{q}{40} + 10 \\ p = \dfrac{8000}{q} \end{cases}$$

by substitution gives

$$\frac{8000}{q} = \frac{q}{40} + 10$$

$$320{,}000 = q^2 + 400q \qquad \text{(multiplying both sides by } 40q\text{)}$$

$$q^2 + 400q - 320{,}000 = 0$$

$$(q + 800)(q - 400) = 0$$

$$q = -800 \quad \text{or} \quad q = 400$$

We disregard $q = -800$, since q represents quantity. Choosing $q = 400$, we have $p = (8000/400) = 20$, so the equilibrium point is $(400, 20)$. (See Figure 3.46.)

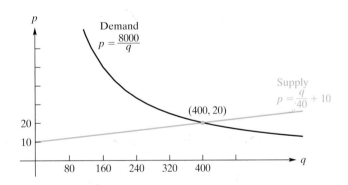

FIGURE 3.46 Equilibrium with nonlinear demand.

Break-Even Points

Suppose a manufacturer produces product A and sells it at $8 per unit. Then the total revenue y_{TR} received (in dollars) from selling q units is

$$y_{TR} = 8q \qquad \text{(total revenue)}$$

The difference between the total revenue received for q units and the total cost of q units is the manufacturer's profit (or loss if the difference is negative):

profit (or loss) = total revenue − total cost

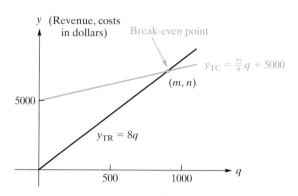

FIGURE 3.47 Break-even chart.

Total cost, y_{TC}, is the sum of total variable costs y_{VC} and total fixed costs y_{FC}:

$$y_{TC} = y_{VC} + y_{FC}$$

Fixed costs are those costs that, under normal conditions, do not depend on the level of production; that is, over some period of time they remain constant at all levels of output. (Examples are rent, officers' salaries, and normal maintenance.) **Variable costs** are those costs that vary with the level of production (such as the cost of materials, labor, maintenance due to wear and tear, etc.). For q units of product A, suppose that

$$y_{FC} = 5000 \qquad \text{(fixed cost)}$$

$$\text{and } y_{VC} = \frac{22}{9}q \qquad \text{(variable cost)}$$

Then

$$y_{TC} = \frac{22}{9}q + 5000 \qquad \text{(total cost)}$$

The graphs of total cost and total revenue appear in Figure 3.47. The horizontal axis represents the level of production, q, and the vertical axis represents the total dollar value, be it revenue or cost. The **break-even point** is the point at which total revenue equals total cost (TR = TC). It occurs when the levels of production and sales result in neither a profit nor a loss to the manufacturer. In the diagram, called a *break-even chart*, the break-even point is the point (m, n) at which the graphs of $y_{TR} = 8q$ and $y_{TC} = \frac{22}{9}q + 5000$ intersect. We call m the **break-even quantity** and n the **break-even revenue.** When total cost and revenue are linearly related to output, as in this case, for any production level greater than m, total revenue is greater than total cost, resulting in a profit. However, at any level less than m units, total revenue is less than total cost, resulting in a loss. At an output of m units, the profit is zero. In the following example, we will examine our data in more detail.

● **EXAMPLE 3** **Break-Even Point, Profit, and Loss**

A manufacturer sells a product at $8 per unit, selling all that is produced. Fixed cost is $5000 and variable cost per unit is $\frac{22}{9}$ (dollars).

a. *Find the total output and revenue at the break-even point.*

Solution: At an output level of q units, the variable cost is $y_{VC} = \frac{22}{9}q$ and the total revenue is $y_{TR} = 8q$. Hence,

$$y_{TR} = 8q$$

$$y_{TC} = y_{VC} + y_{FC} = \frac{22}{9}q + 5000$$

At the break-even point, total revenue equals total cost. Thus, we solve the system formed by the foregoing equations. Since

$$y_{TR} = y_{TC}$$

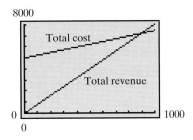

FIGURE 3.48 Equilibrium point (900, 7200).

we have

$$8q = \frac{22}{9}q + 5000$$

$$\frac{50}{9}q = 5000$$

$$q = 900$$

Hence, the desired output is 900 units, resulting in a total revenue (in dollars) of

$$y_{TR} = 8(900) = 7200$$

(See Figure 3.48.)

b. *Find the profit when* 1800 *units are produced.*

Solution: Since profit = total revenue − total cost, when $q = 1800$ we have

$$y_{TR} - y_{TC} = 8(1800) - \left[\frac{22}{9}(1800) + 5000\right]$$

$$= 5000$$

The profit when 1800 units are produced and sold is $5000.

c. *Find the loss when* 450 *units are produced.*

Solution: When $q = 450$,

$$y_{TR} - y_{TC} = 8(450) - \left[\frac{22}{9}(450) + 5000\right] = -2500$$

A loss of $2500 occurs when the level of production is 450 units.

d. *Find the output required to obtain a profit of* $10,000.

Solution: In order to obtain a profit of $10,000, we have

$$\text{profit} = \text{total revenue} - \text{total cost}$$

$$10,000 = 8q - \left(\frac{22}{9}q + 5000\right)$$

$$15,000 = \frac{50}{9}q$$

$$q = 2700$$

Thus, 2700 units must be produced.

NOW WORK PROBLEM 9

● **EXAMPLE 4** **Break-Even Quantity**

Determine the break-even quantity of XYZ Manufacturing Co., given the following data: total fixed cost, $1200; *variable cost per unit,* $2; *total revenue for selling q units,* $y_{TR} = 100\sqrt{q}$.

Solution: For q units of output,

$$y_{TR} = 100\sqrt{q}$$

$$y_{TC} = 2q + 1200$$

Equating total revenue to total cost gives

$$100\sqrt{q} = 2q + 1200$$

$$50\sqrt{q} = q + 600 \qquad \text{(dividing both sides by 2)}$$

Squaring both sides, we have

$$2500q = q^2 + 1200q + (600)^2$$

$$0 = q^2 - 1300q + 360,000$$

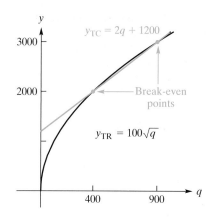

FIGURE 3.49 Two break-even points.

By the quadratic formula,

$$q = \frac{1300 \pm \sqrt{250,000}}{2}$$

$$q = \frac{1300 \pm 500}{2}$$

$$q = 400 \quad \text{or} \quad q = 900$$

Although both $q = 400$ and $q = 900$ are break-even quantities, observe in Figure 3.49 that when $q > 900$, total cost is greater than total revenue, so there will always be a loss. This occurs because here total revenue is not linearly related to output. Thus, producing more than the break-even quantity does not necessarily guarantee a profit.

NOW WORK PROBLEM 13

Problems 3.6

In Problems 1–8, you are given a supply equation and a demand equation for a product. If p represents price per unit in dollars and q represents the number of units per unit of time, find the equilibrium point. In Problems 1 and 2, sketch the system.

1. Supply: $p = \frac{4}{100}q + 3$, Demand: $p = -\frac{6}{100}q + 13$

2. Supply: $p = \frac{1}{1500}q + 4$, Demand: $p = -\frac{1}{2000}q + 9$

3. Supply: $35q - 2p + 250 = 0$, Demand: $65q + p - 537.5 = 0$

4. Supply: $246p - 3.25q - 2460 = 0$, Demand: $410p + 3q - 14{,}452.5 = 0$

5. Supply: $p = 2q + 20$, Demand: $p = 200 - 2q^2$

6. Supply: $p = (q + 10)^2$, Demand: $p = 388 - 16q - q^2$

7. Supply: $p = \sqrt{q + 10}$, Demand: $p = 20 - q$

8. Supply: $p = \frac{1}{4}q + 6$, Demand: $p = \frac{2240}{q + 12}$

In Problems 9–14, y_{TR} represents total revenue in dollars and y_{TC} represents total cost in dollars for a manufacturer. If q represents both the number of units produced and the number of units sold, find the break-even quantity. Sketch a break-even chart in Problems 9 and 10.

*9. $y_{TR} = 4q$
   $y_{TC} = 2q + 5000$

10. $y_{TR} = 14q$
    $y_{TC} = \frac{40}{3}q + 1200$

11. $y_{TR} = 0.05q$
    $y_{TC} = 0.85q + 600$

12. $y_{TR} = 0.25q$
    $y_{TC} = 0.16q + 360$

*13. $y_{TR} = 90 - \dfrac{900}{q + 3}$
    $y_{TC} = 1.1q + 37.3$

14. $y_{TR} = 0.1q^2 + 9q$
    $y_{TC} = 3q + 400$

*15. **Business** Supply and demand equations for a certain product are

$$3q - 200p + 1800 = 0$$

and

$$3q + 100p - 1800 = 0$$

respectively, where p represents the price per unit in dollars and q represents the number of units sold per time period.

(a) Find the equilibrium price algebraically, and derive it graphically.

(b) Find the equilibrium price when a tax of 27 cents per unit is imposed on the supplier.

16. **Business** A manufacturer of a product sells all that is produced. The total revenue is given by $y_{TR} = 7q$, and the total cost is given by $y_{TC} = 6q + 800$, where q represents the number of units produced and sold.

(a) Find the level of production at the break-even point, and draw the break-even chart.

(b) Find the level of production at the break-even point if the total cost increases by 5%.

17. **Business** A manufacturer sells a product at $8.35 per unit, selling all produced. The fixed cost is $2116 and the variable cost is $7.20 per unit. At what level of production will there be a profit of $4600? At what level of production will there be a loss of $1150? At what level of production will the break-even point occur?

18. **Business** The market equilibrium point for a product occurs when 13,500 units are produced at a price of $4.50 per unit. The producer will supply no units at $1, and the consumers will demand no units at $20. Find the supply and demand equations if they are both linear.

19. **Business** A manufacturer of a children's toy will break even at a sales volume of $200,000. Fixed costs are $40,000, and each unit of output sells for $5. Determine the variable cost per unit.

20. **Business** The Bigfoot Sandal Co. manufactures sandals for which the material cost is $0.85 per pair and the labor cost is $0.96 per pair. Additional variable costs amount to $0.32 per pair. Fixed costs are $70,500. If each pair sells for $2.63, how many pairs must be sold for the company to break even?

21. **Business** Find the break-even point for company X, which sells all it produces, if the variable cost per unit is $3, fixed costs are $1250, and $y_{TR} = 60\sqrt{q}$, where q is the number of units of output produced.

22. Business   A company has determined that the demand equation for its product is $p = 1000/q$, where p is the price per unit for q units produced and sold in some period. Determine the quantity demanded when the price per unit is (a) \$4, (b) \$2, and (c) \$0.50. For each of these prices, determine the total revenue that the company will receive. What will be the revenue regardless of the price? (*Hint:* Find the revenue when the price is p dollars.)

23. Business   Using the data in Example 1, determine how the original equilibrium price will be affected if the company is given a government subsidy of \$1.50 per unit.

24. Business   The Monroe Forging Company sells a corrugated steel product to the Standard Manufacturing Company and is in competition on such sales with other suppliers of the Standard Manufacturing Co. The vice president of sales of Monroe Forging Co. believes that by reducing the price of the product, a 40% increase in the volume of units sold to the Standard Manufacturing Co. could be secured. As the manager of the cost and analysis department, you have been asked to analyze the proposal of the vice president and submit your recommendations as to whether it is financially beneficial to the Monroe Forging Co. You are specifically requested to determine the following:

(a) Net profit or loss based on the pricing proposal
(b) Unit sales volume under the proposed price that is required to make the same \$40,000 profit that is now earned at the current price and unit sales volume

Use the following data in your analysis:

	Current Operations	Proposal of Vice President of Sales
Unit price	\$2.50	\$2.00
Unit sales volume	200,000 units	280,000 units
Variable cost		
Total	\$350,000	\$490,000
Per unit	\$1.75	\$1.75
Fixed cost	\$110,000	\$110,000
Profit	\$40,000	?

25. Business   Suppose products A and B have demand and supply equations that are related to each other. If q_A and q_B are the quantities produced and sold of A and B, respectively, and p_A and p_B are their respective prices, the demand equations are

$$q_A = 7 - p_A + p_B$$

and

$$q_B = 24 + p_A - p_B$$

and the supply equations are

$$q_A = -3 + 4p_A - 2p_B$$

and

$$q_B = -5 - 2p_A + 4p_B$$

Eliminate q_A and q_B to get the equilibrium prices.

26. Business   The supply equation for a product is

$$p = 0.4q^2 + 15.2$$

and the demand equation is

$$p = \frac{36.1}{1 + 0.4q}$$

Here p represents price per unit in dollars and q represents number of units (in thousands) per unit time. Graph both equations, and, from your graph, determine the equilibrium price and equilibrium quantity to one decimal place.

27. Business   For a manufacturer, the total-revenue equation is

$$y_{TR} = 20.5\sqrt{q + 4} - 41$$

and the total-cost equation is

$$y_{TC} = 0.02q^3 + 10.4,$$

where q represents (in thousands) both the number of units produced and the number of units sold. Graph a break-even chart and find the break-even quantity.

3.7 Review

Important Terms and Symbols   **Examples**

Section 3.6   **Applications of Systems of Equations**
point of equilibrium   equilibrium price   equilibrium quantity   Ex. 1, p. 151
break-even point   break-even quantity   break-even revenue   Ex. 3, p. 154

Summary

The orientation of a nonvertical line is characterized by the slope of the line given by

$$m = \frac{y_2 - y_1}{x_2 - x_1}$$

where (x_1, y_1) and (x_2, y_2) are two different points on the line. The slope of a vertical line is not defined, and the slope of a horizontal line is zero. Rising lines have positive slopes; falling lines have negative slopes. Two lines are parallel if and only if they have the same slope or are vertical. Two lines with slopes m_1 and m_2 are perpendicular to each other if and only if $m_1 = -\dfrac{1}{m_2}$. A horizontal line and a vertical line are perpendicular to each other.

Basic forms of equations of lines are as follows:

$$y - y_1 = m(x - x_1) \qquad \text{(point-slope form)}$$

$$y = mx + b \qquad \text{(slope-intercept form)}$$

$$x = a \qquad \text{(vertical line)}$$

$$y = b \qquad \text{(horizontal line)}$$

$$Ax + By + C = 0 \qquad \text{(general)}$$

The linear function

$$f(x) = ax + b \quad (a \neq 0)$$

has a straight line for its graph.

In economics, supply functions and demand functions have the form $p = f(q)$ and play an important role. Each gives a correspondence between the price p of a product and the number of units q of the product that manufacturers (or consumers) will supply (or purchase) at that price during some time period.

A quadratic function has the form

$$f(x) = ax^2 + bx + c \quad (a \neq 0)$$

The graph of f is a parabola that opens upward if $a > 0$ and downward if $a < 0$. The vertex is

$$\left(-\frac{b}{2a}, f\left(-\frac{b}{2a} \right) \right)$$

and the y-intercept is c. The axis of symmetry and the x- and y-intercepts, are useful in sketching the graph.

A system of linear equations may be solved with the method of elimination by addition or elimination by substitution. A solution may involve one or more parameters. Substitution is also useful in solving nonlinear systems.

Solving a system formed by the supply and demand equations for a product gives the equilibrium point, which indicates the price at which consumers will purchase the same quantity of a product that producers wish to sell at that price.

Profit is total revenue minus total cost, where total cost is the sum of fixed costs and variable costs. The break-even point is the point where total revenue equals total cost.

Review Problems

Problem numbers shown in color indicate problems suggested for use as a practice chapter test.

1. The slope of the line through $(2, 5)$ and $(3, k)$ is 4. Find k.

2. The slope of the line through $(5, 4)$ and $(k, 4)$ is 0. Find k.

In Problems 3–9, determine the slope-intercept form and a general linear form of an equation of the straight line that has the indicated properties.

3. Passes through $(-2, 3)$ and has y-intercept -1

4. Passes through $(-1, -1)$ and is parallel to the line $y = 3x - 4$

5. Passes through $(10, 4)$ and has slope $\frac{1}{2}$

6. Passes through $(3, 5)$ and is vertical

7. Passes through $(-2, 4)$ and is horizontal

8. Passes through $(1, 2)$ and is perpendicular to the line $-3y + 5x = 7$

9. Has y-intercept -3 and is perpendicular to $2y + 5x = 2$

10. Determine whether the point $(3, 13)$ lies on the line through $(1, 8)$ and $(-1, 2)$.

In Problems 11–16, determine whether the lines are parallel, perpendicular, or neither.

11. $x + 4y + 2 = 0, \quad 8x - 2y - 2 = 0$

12. $y - 2 = 2(x - 1), \quad 2x + 4y - 3 = 0$

13. $x - 3 = 2(y + 4), \quad y = 4x + 2$

14. $2x + 7y - 4 = 0, \quad 6x + 21y = 90$

15. $y = 3x + 5, \quad 6x - 2y = 7$

16. $y = 7x, \quad y = 7$

In Problems 17–20, write each line in slope-intercept form, and sketch. What is the slope of the line?

17. $3x - 2y = 4$ **18.** $x = -3y + 4$

19. $4 - 3y = 0$ **20.** $y = 2x$

In Problems 21–30, graph each function. For those that are linear, give the slope and the vertical-axis intercept. For those that are quadratic, give all intercepts and the vertex.

21. $y = f(x) = 17 - 5x$ **22.** $s = g(t) = 5 - 3t + t^2$

23. $y = f(x) = 9 - x^2$ **24.** $y = f(x) = 3x - 7$

25. $y = h(t) = t^2 - 4t - 5$ **26.** $y = k(t) = -3 - 3t$

27. $p = g(t) = -7t$ **28.** $y = F(x) = (2x - 1)^2$

29. $y = F(x) = -(x^2 + 2x + 3)$ **30.** $y = f(x) = \dfrac{x}{3} - 2$

In Problems 31–44, solve the given system.

31. $\begin{cases} 2x - y = 6 \\ 3x + 2y = 5 \end{cases}$ **32.** $\begin{cases} 8x - 4y = 7 \\ y = 2x - 4 \end{cases}$

33. $\begin{cases} 7x + 5y = 5 \\ 6x + 5y = 3 \end{cases}$ **34.** $\begin{cases} 2x + 4y = 8 \\ 3x + 6y = 12 \end{cases}$

35. $\begin{cases} \dfrac{1}{4}x - \dfrac{3}{2}y = -4 \\ \dfrac{3}{4}x + \dfrac{1}{2}y = 8 \end{cases}$

36. $\begin{cases} \dfrac{1}{3}x - \dfrac{1}{4}y = \dfrac{1}{12} \\ \dfrac{4}{3}x + 3y = \dfrac{5}{3} \end{cases}$

37. $\begin{cases} 3x - 2y + z = -2 \\ 2x + y + z = 1 \\ x + 3y - z = 3 \end{cases}$

38. $\begin{cases} 2x + \dfrac{3y + x}{3} = 9 \\ y + \dfrac{5x + 2y}{4} = 7 \end{cases}$

39. $\begin{cases} x^2 - y + 5x = 2 \\ x^2 + y = 3 \end{cases}$

40. $\begin{cases} y = \dfrac{18}{x + 4} \\ x - y + 7 = 0 \end{cases}$

41. $\begin{cases} x + 2z = -2 \\ x + y + z = 5 \end{cases}$

42. $\begin{cases} x + y + z = 0 \\ x - y + z = 0 \\ x + z = 0 \end{cases}$

43. $\begin{cases} x - y - z = 0 \\ 2x - 2y + 3z = 0 \end{cases}$

44. $\begin{cases} 2x - 5y + 6z = 1 \\ 4x - 10y + 12z = 2 \end{cases}$

45. Suppose a and b are linearly related so that $a = 1$ when $b = 2$ and $a = 5$ when $b = 3$. Find a general linear form of an equation that relates a and b. Also, find a when $b = 5$.

46. Temperature and Heart Rate When the temperature, T (in degrees Celsius), of a cat is reduced, the cat's heart rate, r (in beats per minute), decreases. Under laboratory conditions, a cat at a temperature of $36°C$ had a heart rate of 206, and at a temperature of $30°C$ its heart rate was 122. If r is linearly related to T, where T is between 26 and 38, (a) determine an equation for r in terms of T, and (b) determine the cat's heart rate at a temperature of $27°C$

47. Suppose f is a linear function such that $f(1) = 5$ and $f(x)$ decreases by four units for every three-unit increase in x. Find $f(x)$.

48. If f is a linear function such that $f(-1) = 8$ and $f(2) = 5$, find $f(x)$.

49. Maximum Revenue The demand function for a manufacturer's product is $p = f(q) = 200 - 2q$, where p is the price (in dollars) per unit when q units are demanded. Find the level of production that maximizes the manufacturer's total revenue, and determine this revenue.

50. Sales Tax The difference in price of two items before a 5% sales tax is imposed is $3.50. The difference in price after the sales tax is imposed is allegedly $4.10. Show that this scenario is not possible.

51. Equilibrium Price If the supply and demand equations of a certain product are $120p - q - 240 = 0$ and $100p + q - 1200 = 0$, respectively, find the equilibrium price.

52. Psychology In psychology, the term *semantic memory* refers to our knowledge of the meaning and relationships of words, as well as the means by which we store and retrieve such information.[6] In a network model of semantic memory, there is a hierarchy of levels at which information is stored.

In an experiment by Collins and Quillian based on a network model, data were obtained on the reaction time to respond to simple questions about nouns. The graph of the results shows that, on the average, the reaction time, R (in milliseconds), is a linear function of the level, L, at which a characterizing property of the noun is stored. At level 0, the reaction time is 1310; at level 2, the reaction time is 1460. (a) Find the linear function. (b) Find the reaction time at level 1. (c) Find the slope and determine its significance.

53. Break-Even Point A manufacturer of a certain product sells all that is produced. Determine the break-even point if the product is sold at $16 per unit, fixed cost is $10,000, and variable cost is given by $y_{VC} = 8q$, where q is the number of units produced (y_{VC} expressed in dollars).

54. Temperature Conversion Celsius temperature, C, is a linear function of Fahrenheit temperature, F. Use the facts that $32°F$ is the same as $0°C$ and $212°F$ is the same as $100°C$ to find this function. Also, find C when $F = 50$.

55. Pollution In one province of a developing nation, water pollution is analyzed using a supply-and-demand model. The *environmental supply equation* $L = 0.0183 - \dfrac{0.0042}{p}$ describes the levy-per-ton, L (in dollars), as a function of total pollution, p (in tons per square kilometer), for $p \geq 0.2295$. The *environmental demand equation*, $L = 0.0005 + \dfrac{0.0378}{p}$, describes the per-ton abatement cost as a function of total pollution for $p > 0$. Find the expected equilibrium level of total pollution to two decimal places.[7]

56. Graphically solve the linear system
$$\begin{cases} 3x + 4y = 20 \\ 7x + 5y = 64 \end{cases}$$

57. Graphically solve the linear system
$$\begin{cases} 0.3x - 0.4y = 2.5 \\ 0.5x + 0.7y = 3.1 \end{cases}$$
Round x and y to two decimal places.

58. Graphically solve the nonlinear system
$$\begin{cases} y = \dfrac{3}{7x} \quad \text{where } x > 0 \\ y = x^2 - 9 \end{cases}$$
Round x and y to two decimal places.

59. Graphically solve the nonlinear system
$$\begin{cases} y = x^3 + 1 \\ y = 2 - x^2 \end{cases}$$
Round x and y to two decimal places.

60. Graphically solve the equation
$$x^2 + 4 = x^3 - 3x$$
by treating it as a system. Round x to two decimal places.

[6]G. R. Loftus and E. F. Loftus, *Human Memory: The Processing of Information* (New York: Lawrence Erlbaum Associates, Inc., distributed by the Halsted Press, Division of John Wiley & Sons, Inc., 1976).

[7]See Hua Wang and David Wheeler, "Pricing Industrial Pollution in China: An Economic Analysis of the Levy System," World Bank Policy Research Working Paper #1644, September 1996.

Mathematical Snapshot

Mobile Phone Billing Plans

Selecting a mobile phone plan can be quite difficult. In most urban areas there are many service providers each offering a number of plans. The plans can include monthly access fees, free minutes, charges for additional airtime, regional roaming charges, national roaming charges, peak and off-peak rates, and long-distance charges (not to mention activation fees, cancellation fees, and the like). Even if a consumer has a fairly good knowledge of her typical mobile phone usage, she may have to do dozens of calculations to be absolutely sure of getting the best deal in town.

Mathematical modelling often involves making informed decisions about which factors in a problem are less important. These are then ignored to get a reasonably good approximate solution—in a reasonable amount of time. You may have heard the expression "simplifying assumptions." There are a lot of old jokes about this process. For example, a mathematically minded bookie who is trying to calculate the attributes of the horses in a given race should probably not assume that all the horses are perfectly spherical. We will simplify our comparison of mobile phone plans by considering just the number of "monthly home airtime minutes" available for the "monthly access fee" and the price per minute of "additional minutes." Many providers offer plans in terms of these basic parameters.

Examining Verizon's offerings for the Saddle River, New Jersey, area, in the spring of 2006, we found these America's Choice monthly plans:

P_1: 450 minutes for \$39.99 plus \$0.45 per additional minute

P_2: 900 minutes for \$59.99 plus \$0.40 per additional minute

P_3: 1350 minutes for \$79.99 plus \$0.35 per additional minute

P_4: 2000 minutes for \$99.99 plus \$0.25 per additional minute

P_5: 4000 minutes for \$149.99 plus \$0.25 per additional minute

P_6: 6000 minutes for \$199.99 plus \$0.20 per additional minute

where we have added the labels P_i, for $i = 1, 2, \ldots, 6$, for our further convenience. Thus, each entry above takes the form:

P_i: M_i minutes for \$$C_i$ plus \$$c_i$ per additional minute

where, for plan P_i, M_i is the number of airtime minutes available for the monthly access fee of C_i, with each additional minute costing c_i.

To represent these plans mathematically, we will write total monthly cost as a function of time, for each plan. In fact, we will write $P_i(t)$ for the monthly cost of t minutes using plan P_i. For each plan, the resulting function is a simple case-defined function with just two cases to consider. For each plan, we must consider $t \leq M_i$ and $t > M_i$. If $t \leq M_i$, then the cost is simply C_i but if $t > M_i$, then the number of additional minutes is $t - M_i$ and, since each of these costs c_i, the additional minutes cost $c_i(t - M_i)$, yielding in this case a total cost of $C_i + c_i(t - M_i)$.

Putting in the numerical values, we have the following six functions:

$$P_1(t) = \begin{cases} 39.99 & \text{if } t \leq 450 \\ 39.99 + 0.45(t - 450) & \text{if } t > 450 \end{cases}$$

$$P_2(t) = \begin{cases} 59.99 & \text{if } t \leq 900 \\ 59.99 + 0.40(t - 900) & \text{if } t > 900 \end{cases}$$

$$P_3(t) = \begin{cases} 79.99 & \text{if } t \leq 1350 \\ 79.99 + 0.35(t - 1350) & \text{if } t > 1350 \end{cases}$$

$$P_4(t) = \begin{cases} 99.99 & \text{if } t \leq 2000 \\ 99.99 + 0.25(t - 2000) & \text{if } t > 2000 \end{cases}$$

$$P_5(t) = \begin{cases} 149.99 & \text{if } t \leq 4000 \\ 149.99 + 0.25(t - 4000) & \text{if } t > 4000 \end{cases}$$

$$P_6(t) = \begin{cases} 199.99 & \text{if } t \leq 6000 \\ 199.99 + 0.20(t - 6000) & \text{if } t > 6000 \end{cases}$$

The graph of each function is easy to describe. In fact for the generic $P_i(t)$ we have, in the first quadrant, a horizontal line segment starting at $(0, C_i)$ and ending at (M_i, C_i). The graph continues, to the right of (M_i, C_i) as an infinite line segment, starting at (M_i, C_i) with slope c_i. However, to see how the functions P_i actually compare, we need to graph them all, in the same plane. We could do this by hand, but this is a good opportunity to use a handy capability of a graphing calculator. We enter the function $P_1(t)$ as

$$Y1 = 39.99 + 0.45(X - 450)(X > 450)$$

The > comes from the TEST menu, and the expression $(X > 450)$ equals either 1 or 0, depending on whether x is, or is not, greater than 450. Entering the other five functions in similar fashion and graphing them all together, we get the display shown in Figure 3.50.

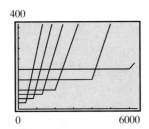

400

0   6000

FIGURE 3.50   Costs under different plans.

Which plan is best depends on the amount of calling time: For any given monthly airtime, the best plan is the one whose graph is lowest at that amount.

For very low calling times, the P_1 plan is best, but even at 495 minutes monthly usage it is more expensive than plan P_2 and remains so for any greater monthly usage. To find exactly the usage at which plans P_1 and P_2 cost the same, we of course solve

$$P_1(t) = P_2(t)$$

but because each is a case-defined function we really need the graphs to tell us *where to look for a solution*. From these it is clear that the intersection of the P_1 and P_2 curves occurs when P_1 is defined by its second branch and P_2 is defined by its first branch. Thus we must solve

$$39.99 + 0.45(t - 450) = 59.99$$

for t. To two decimal places this gives $t = 494.44$.

In fact, the graph suggests that it will be instructive to calculate $P_1(900)$ since $P_2(900)$ is still \$59.99, although of course the cost of P_2 increases for all $t > 900$. We find

$$P_1(900) = 39.99 + 0.45(900 - 450) = 39.99 + 0.45(450)$$

$$= 39.99 + 202.50 = 242.49$$

To research wireless phone service plans in your area, consult www.point.com.

Problems

1. If a person who actually uses a lot of airtime minutes a month, say 6000, is lured by low monthly access fees, calculate how much he will lose by using plan P_1 rather than plan P_6.

2. We have seen that for monthly usage less than 494.44 minutes, plan P_1 is best. Determine the interval of usage for which P_2 is best by finding the value of t for which $P_2(t) = P_3(t)$.

3. Repeat Problem 2 for Plan P_3.

4. Repeat Problem 2 for Plan P_4.

5. Repeat Problem 2 for Plan P_5.

6. Repeat Problem 2 for Plan P_6.

7. How can you be sure that for *any* value of t greater than that found in Problem 6, plan P_6 remains best? To put it another way, do P_5 and P_6 have any points of intersection on the *second* branches of *both* curves?

EXPONENTIAL AND LOGARITHMIC FUNCTIONS

Mathematical Snapshot Drug Dosages

Just as biological viruses spread through contact between organisms, so computer viruses spread when computers interact via the Internet. Computer scientists study how to fight computer viruses, which cause a lot of damage in the form of deleted and corrupted files. One thing computer scientists do is devise mathematical models of how quickly viruses spread. For instance, on Friday, March 26, 1999, the first cases of the virus known as Melissa were reported; by Monday, March 29, Melissa had reached over 100,000 computers.

Exponential functions, which this chapter discusses in detail, provide one plausible model. Consider a computer virus that hides in an e-mail attachment and, once the attachment is downloaded, automatically sends a message with a similar attachment to every address in the host computer's e-mail address book. If the typical address book contains 20 addresses, and if the typical computer user retrieves his or her e-mail once a day, then a virus on a single machine will have infected 20 machines after one day, $20^2 = 400$ machines after two days, $20^3 = 8000$ after three days, and, in general, after t days, the number N of infected computers will be given by the exponential function $N(t) = 20^t$.

This model assumes that all the computers involved are linked, via their address book lists, into a single, well-connected group. Exponential models are most accurate for small values of t; this model, in particular, ignores the slowdown that occurs when most e-mails start going to computers already infected, which happens as several days pass. For example, our model tells us that after eight days, the virus will infect $20^8 = 25.6$ billion computers—more computers than actually exist! But despite its limitations, the exponential model does explain why new viruses often infect many thousands of machines before antivirus experts have had time to react.

4.1 Exponential Functions

• • •

OBJECTIVE

To study exponential functions and their applications to such areas as compound interest, population growth, and radioactive decay.

CAUTION

Do not confuse the exponential function $y = 2^x$ with the *power function* $y = x^2$, which has a variable base and a constant exponent.

POINTER ➤

If you wish to review exponents, refer to Section 0.3.

There is a function that has an important role not only in mathematics, but also in business, economics, and other areas of study. It involves a constant raised to a variable power, such as $f(x) = 2^x$. We call such functions *exponential functions*.

DEFINITION

The function f defined by

$$f(x) = b^x$$

where $b > 0$, $b \neq 1$, and the exponent x is any real number, is called an **exponential function** with base b.[1]

Since the exponent in b^x can be any real number, you may wonder how we assign a value to something like $2^{\sqrt{2}}$, where the exponent is an irrational number. Stated simply, we use approximations. Because $\sqrt{2} = 1.41421\ldots$, $2^{\sqrt{2}}$ is approximately $2^{1.4} = 2^{7/5} = \sqrt[5]{2^7}$, which *is* defined. Better approximations are $2^{1.41} = 2^{141/100} = \sqrt[100]{2^{141}}$, and so on. In this way, the meaning of $2^{\sqrt{2}}$ becomes clear. A calculator value of $2^{\sqrt{2}}$ is (approximately) 2.66514.

When you work with exponential functions, it may be necessary to apply rules for exponents. These rules are as follows, where m and n are real numbers and a and b are positive.

✓ **Rules for Exponents**

1. $a^m a^n = a^{m+n}$

2. $\dfrac{a^m}{a^n} = a^{m-n}$

3. $(a^m)^n = a^{mn}$

4. $(ab)^n = a^n b^n$

5. $\left(\dfrac{a}{b}\right)^n = \dfrac{a^n}{b^n}$

6. $a^1 = a$

7. $a^0 = 1$

8. $a^{-n} = \dfrac{1}{a^n}$

Some functions that do not appear to have the exponential form b^x can be put in that form by applying the preceding rules. For example, $2^{-x} = 1/(2^x) = \left(\frac{1}{2}\right)^x$ and $3^{2x} = (3^2)^x = 9^x$.

PRINCIPLES IN PRACTICE 1

BACTERIA GROWTH

The number of bacteria in a culture that doubles every hour is given by $N(t) = A \cdot 2^t$, where A is the number originally present and t is the number of hours the bacteria have been doubling. Use a graphing calculator to plot this function for various values of $A > 1$. How are the graphs similar? How does the value of A alter the graph?

● **EXAMPLE 1   Bacteria Growth**

The number of bacteria present in a culture after t minutes is given by

$$N(t) = 300 \left(\frac{4}{3}\right)^t$$

Note that N(t) is a constant multiple of the exponential function $\left(\dfrac{4}{3}\right)^t$.

---

[1]If $b = 1$, then $f(x) = 1^x = 1$. This function is already known to us as a constant function.

a. *How many bacteria are present initially?*

Solution: Here we want to find $N(t)$ when $t = 0$. We have

$$N(0) = 300 \left(\frac{4}{3}\right)^0 = 300(1) = 300$$

Thus, 300 bacteria are initially present.

b. *Approximately how many bacteria are present after 3 minutes?*

Solution:

$$N(3) = 300 \left(\frac{4}{3}\right)^3 = 300 \left(\frac{64}{27}\right) = \frac{6400}{9} \approx 711$$

Hence, approximately 711 bacteria are present after 3 minutes.

NOW WORK PROBLEM 31

PRINCIPLES IN PRACTICE 2

GRAPHING EXPONENTIAL
FUNCTIONS WITH $b > 1$

Suppose an investment increases by 10% every year. Make a table of the factor by which the investment increases from the original amount for 0 to 4 years. For each year, write an expression for the increase as a power of some base. What base did you use? How does that base relate to the problem? Use your table to graph the multiplicative increase as a function of the number of years. Use your graph to determine when the investment will double.

Graphs of Exponential Functions

● **EXAMPLE 2** **Graphing Exponential Functions with $b > 1$**

Graph the exponential functions $f(x) = 2^x$ and $f(x) = 5^x$.

Solution: By plotting points and connecting them, we obtain the graphs in Figure 4.1. For the graph of $f(x) = 5^x$, because of the unit distance chosen on the y-axis, the points $(-2, \frac{1}{25})$, $(2, 25)$, and $(3, 125)$ are not shown.

We can make some observations about these graphs. The domain of each function consists of all real numbers, and the range consists of all positive real numbers. Each graph has y-intercept $(0, 1)$. Moreover, the graphs have the same general shape. Each *rises* from left to right. As x increases, $f(x)$ also increases. In fact, $f(x)$ increases without bound. However, in quadrant I, the graph of $f(x) = 5^x$ rises more quickly than that of $f(x) = 2^x$ because the base in 5^x is *greater* than the base in 2^x (that is, $5 > 2$). Looking at quadrant II, we see that as x becomes very negative, the graphs of both functions approach the x-axis.[2] This implies that the function values get very close to 0.

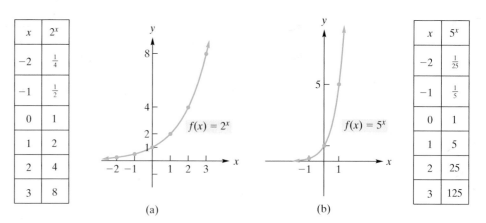

FIGURE 4.1 Graphs of $f(x) = 2^x$ and $f(x) = 5^x$.

NOW WORK PROBLEM 1

---

[2]We say that the x-axis is an *asymptote* for each graph.

The observations made in Example 2 are true for all exponential functions whose base b is greater than 1. Example 3 will examine the case for a base between 0 and 1 $(0 < b < 1)$.

PRINCIPLES IN PRACTICE 3

GRAPHING EXPONENTIAL FUNCTIONS WITH $0 < b < 1$

Suppose the value of a car depreciates by 15% every year. Make a table of the factor by which the value decreases from the original amount for 0 to 3 years. For each year, write an expression for the decrease as a power of some base. What base did you use? How does that base relate to the problem? Use your table to graph the multiplicative decrease as a function of the number of years. Use your graph to determine when the car will be worth half as much as its original price.

● **EXAMPLE 3** **Graphing Exponential Functions with $0 < b < 1$**

Graph the exponential function $f(x) = \left(\frac{1}{2}\right)^x$.

Solution: By plotting points and connecting them, we obtain the graph in Figure 4.2. Notice that the domain consists of all real numbers, and the range consists of all positive real numbers. The graph has y-intercept $(0, 1)$. Compared to the graphs in Example 2, the graph here *falls* from left to right. That is, as x increases, $f(x)$ decreases. Notice that as x becomes very positive, $f(x)$ takes on values close to 0 and the graph approaches the x-axis. However, as x becomes very negative, the function values are unbounded.

NOW WORK PROBLEM 3

There are two basic shapes for the graphs of exponential functions, and they depend on the base involved.

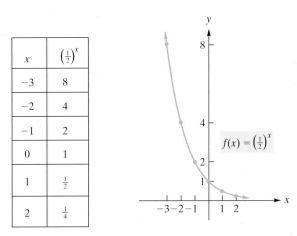

FIGURE 4.2 Graph of $f(x) = \left(\frac{1}{2}\right)^x$.

In general, the graph of an exponential function has one of two shapes, depending on the value of the base b. This is illustrated in Figure 4.3. It is important to observe that in either case the graph passes the horizontal line test. Thus all exponential functions are one-to-one. The basic properties of an exponential function and its graph are summarized in Table 4.1.

Recall from Section 2.7 that the graph of one function may be related to that of another by means of a certain transformation. Our next example pertains to this concept.

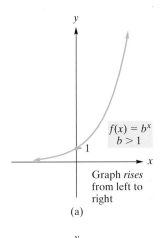

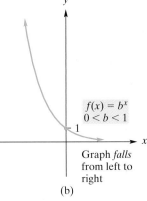

FIGURE 4.3 General shapes of $f(x) = b^x$.

TABLE 4.1 Properties of the Exponential Function $f(x) = b^x$

1. The domain of an exponential function consists of all real numbers.
   The range consists of all positive numbers.

2. The graph of $f(x) = b^x$ has y-intercept $(0, 1)$.
   There is no x-intercept.

3. If $b > 1$, the graph *rises* from left to right.
   If $0 < b < 1$, the graph *falls* from left to right.

4. If $b > 1$, the graph approaches the x-axis as x becomes more and more negative.
   If $0 < b < 1$, the graph approaches the x-axis as x becomes more and more positive.

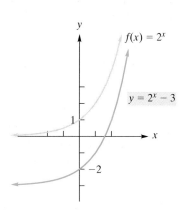

FIGURE 4.4 Graph of $y = 2^x - 3$.

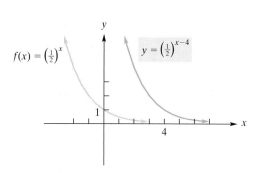

FIGURE 4.5 Graph of $y = \left(\frac{1}{2}\right)^{x-4}$.

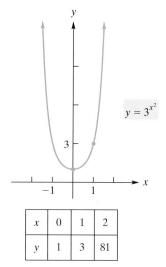

x	0	1	2
y	1	3	81

FIGURE 4.6 Graph of $y = 3^{x^2}$.

Example 4 makes use of transformations from Table 2.2 of Section 2.7.

● **EXAMPLE 4** **Transformations of Exponential Functions**

a. *Use the graph of $y = 2^x$ to plot $y = 2^x - 3$.*

Solution: The function has the form $f(x) - c$, where $f(x) = 2^x$ and $c = 3$. Thus its graph is obtained by shifting the graph of $f(x) = 2^x$ three units downward. (See Figure 4.4.)

b. *Use the graph of $y = \left(\frac{1}{2}\right)^x$ to graph $y = \left(\frac{1}{2}\right)^{x-4}$.*

Solution: The function has the form $f(x - c)$, where $f(x) = \left(\frac{1}{2}\right)^x$ and $c = 4$. Hence, its graph is obtained by shifting the graph of $f(x) = \left(\frac{1}{2}\right)^x$ four units to the right. (See Figure 4.5.)

NOW WORK PROBLEM 7 ●●

PRINCIPLES IN PRACTICE 4

TRANSFORMATIONS OF EXPONENTIAL FUNCTIONS

After watching his sister's money grow for three years in a plan with an 8% yearly return, George started a savings account with the same plan. If $y = 1.08^t$ represents the multiplicative increase in his sister's account, write an equation that will represent the multiplicative increase in George's account, using the same time reference. If George has a graph of the multiplicative increase in his sister's money at time t years since she started saving, how could he use the graph to project the increase in his money?

● **EXAMPLE 5** **Graph of a Function with a Constant Base**

Graph $y = 3^{x^2}$.

Solution: Although this is not an exponential function, it does have a constant base. We see that replacing x by $-x$ results in the same equation. Thus, the graph is symmetric about the y-axis. Plotting some points and using symmetry gives the graph in Figure 4.6.

NOW WORK PROBLEM 5 ●●

●T E C H N O L O G Y

If $y = 4^x$, consider the problem of finding x when $y = 6$. One way to solve this equation is to find the intersection of the graphs of $y = 6$ and $y = 4^x$. Figure 4.7 shows that x is approximately 1.29.

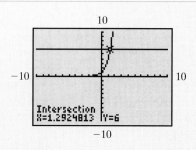

FIGURE 4.7 Solving the equation $6 = 4^x$.

⌐ Compound Interest

Exponential functions are involved in **compound interest,** whereby the interest earned by an invested amount of money (or **principal**) is reinvested so that it, too, earns interest. That is, the interest is converted (or *compounded*) into principal, and hence, there is "interest on interest."

For example, suppose that $100 is invested at the rate of 5% compounded annually. At the end of the first year, the value of the investment is the original principal ($100), plus the interest on the principal [100(0.05)]:

$$100 + 100(0.05) = \$105$$

This is the amount on which interest is earned for the second year. At the end of the second year, the value of the investment is the principal at the end of the first year ($105), plus the interest on that sum [105(0.05)]:

$$105 + 105(0.05) = \$110.25$$

Thus, each year the principal increases by 5%. The $110.25 represents the original principal, plus all accrued interest; it is called the **accumulated amount** or **compound amount.** The difference between the compound amount and the original principal is called the **compound interest.** Here the compound interest is $110.25 - 100 = 10.25$.

More generally, if a principal of P dollars is invested at a rate of $100r$ percent compounded annually (for example, at 5%, r is 0.05), the compound amount after 1 year is $P + Pr$, or, by factoring, $P(1+r)$. At the end of the second year, the compound amount is

$$P(1+r) + [P(1+r)]r = P(1+r)[1+r] \qquad \text{(factoring)}$$
$$= P(1+r)^2$$

Actually, the calculation above using factoring is not necessary to show that the componded amount after two years is $P(1+r)^2$. Since *any* amount P is worth $P(1+r)$ a year later, it follows that the amount of $P(1+r)$ is worth $P(1+r)(1+r) = P(1+r)^2$ a year later and one year later still the amount of $P(1+r)^2$ is worth $P(1+r)^2(1+r) = P(1+r)^3$).

This pattern continues. After four years, the compound amount is $P(1+r)^4$. In general, **the compound amount S of the principal P at the end of n years at the rate of r compounded annually** is given by

$$S = P(1+r)^n \qquad (1)$$

Notice from Equation (1) that, for a given principal and rate, S is a function of n. In fact, S involves an exponential function with base $1 + r$.

PRINCIPLES IN PRACTICE 5

COMPOUND AMOUNT AND COMPOUND INTEREST

Suppose $2000 is invested at 13% compounded annually. Find the value of the investment after five years. Find the interest earned over the first five years.

● EXAMPLE 6—Compound Amount and Compound Interest

Suppose $1000 *is invested for* 10 *years at* 6% *compounded annually.*

a. *Find the compound amount.*

Solution: We use Equation (1) with $P = 1000$, $r = 0.06$, and $n = 10$:

$$S = 1000(1 + 0.06)^{10} = 1000(1.06)^{10} \approx \$1790.85$$

Figure 4.8 shows the graph of $S = 1000(1.06)^n$. Notice that as time goes on, the compound amount grows dramatically.

b. *Find the compound interest.*

Solution: Using the results from part (a), we have

$$\text{compound interest} = S - P$$
$$= 1790.85 - 1000 = \$790.85$$

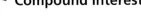

$S = 1000(1.06)^n$

FIGURE 4.8 Graph of $S = 1000(1.06)^n$.

NOW WORK PROBLEM 19 ●●●

Suppose the principal of $1000 in Example 6 is invested for 10 years as before, but this time the compounding takes place every three months (that is, *quarterly*) at the rate of $1\frac{1}{2}\%$ *per quarter*. Then there are four **interest periods** per year, and in 10 years there are $10(4) = 40$ interest periods. Thus, the compound amount with $r = 0.015$ is now

$$1000(1.015)^{40} \approx \$1814.02$$

and the compound interest is $814.02. Usually, the interest rate per interest period is stated as an annual rate. Here we would speak of an annual rate of 6% compounded quarterly, so that the rate per interest period, or the **periodic rate,** is $6\%/4 = 1.5\%$. This *quoted* annual rate of 6% is called the **nominal rate** or the **annual percentage rate (A.P.R.).** Unless otherwise stated, all interest rates will be assumed to be annual (nominal) rates. Thus a rate of 15% compounded monthly corresponds to a periodic rate of $15\%/12 = 1.25\%$.

On the basis of our discussion, we can generalize Equation (1). The formula

$$S = P(1 + r)^n \qquad (2)$$

gives **the compound amount S of a principal P at the end of n interest periods at the periodic rate of r.**

We have seen that for a principal of $1000 at a nominal rate of 6% over a period of 10 years, annual compounding results in a compound interest of $790.85, and with quarterly compounding the compound interest is $814.02. It is typical that for a given nominal rate, the more frequent the compounding, the greater is the compound interest. However, while increasing the compounding frequency always increases the amount of interest earned, the effect is not unbounded. For example, with weekly compounding the compound interest is

$$1000\left(1 + \frac{0.06}{52}\right)^{10(52)} - 1000 \approx \$821.49$$

and with daily compounding it is

$$1000\left(1 + \frac{0.06}{365}\right)^{10(365)} - 1000 \approx \$822.03$$

Sometimes the phrase "money is worth" is used to express an annual interest rate. Thus, saying that money is worth 6% compounded quarterly refers to an annual (nominal) rate of 6% compounded quarterly.

Population Growth

Equation (2) can be applied not only to the growth of money, but also to other types of growth, such as that of population. For example, suppose the population P of a town of 10,000 is increasing at the rate of 2% per year. Then P is a function of time t, in years. It is common to indicate this functional dependence by writing

$$P = P(t)$$

Here the letter P is used in two ways: On the right side, P represents the function; on the left side, P represents the dependent variable. From Equation (2), we have

$$P(t) = 10,000(1 + 0.02)^t = 10,000(1.02)^t$$

The abbreviation A.P.R. is a common one and is found on credit card statements and in advertising.

 CAUTION

A nominal rate of 6% does not necessarily mean that an investment increases in value by 6% in a year's time. The increase depends on the frequency of compounding.

PRINCIPLES IN PRACTICE 6

POPULATION GROWTH

A new company with five employees expects the number of employees to grow at the rate of 120% per year. Find the number of employees in four years.

●EXAMPLE 7 **Population Growth**

The population of a town of 10,000 grows at the rate of 2% per year. Find the population three years from now.

Solution: From the preceding discussion,

$$P(t) = 10,000(1.02)^t$$

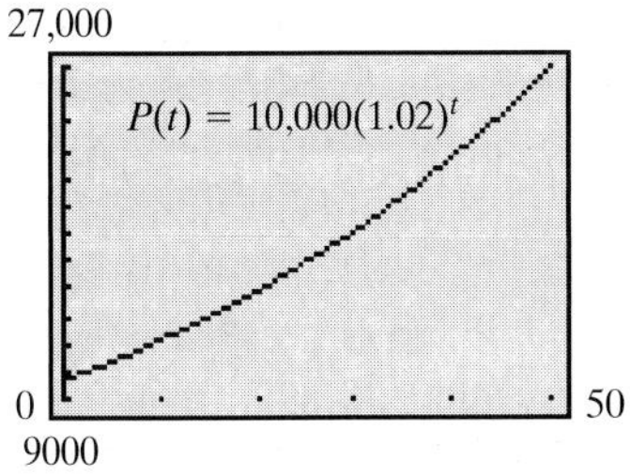

FIGURE 4.9 Graph of population function $P(t) = 10,000(1.02)^t$.

For $t = 3$, we have

$$P(3) = 10,000(1.02)^3 \approx 10,612$$

Thus, the population three years from now will be 10,612. (See Figure 4.9.)

NOW WORK PROBLEM 15

The Number e

It is useful to conduct a "thought experiment," based on the discussion following Example 6, to introduce an important number. Suppose that a single dollar is invested for one year with an A.P.R. of 100% (remember, this is a thought experiment!) compounded annually. Then the compound amount S at the end of the year is given by

$$S = 1(1 + 1)^1 = 2^1 = 2$$

Without changing any of the other data we now consider the effect of increasing the number of interest periods per year. If there are n interest periods per year, then the compound amount is given by

$$S = 1\left(1 + \frac{1}{n}\right)^n = \left(\frac{n+1}{n}\right)^n$$

In the following table we give approximate values for $\left(\dfrac{n+1}{n}\right)^n$ for some values of n.

TABLE 4.2 Approximations of e

n	$\left(\dfrac{n+1}{n}\right)^n$
1	$\left(\frac{2}{1}\right)^1 = 2.00000$
2	$\left(\frac{3}{2}\right)^2 = 2.25000$
3	$\left(\frac{4}{3}\right)^3 \approx 2.37037$
4	$\left(\frac{5}{4}\right)^4 \approx 2.44141$
5	$\left(\frac{6}{5}\right)^5 = 2.48832$
10	$\left(\frac{11}{10}\right)^{10} \approx 2.59374$
100	$\left(\frac{101}{100}\right)^{100} \approx 2.70481$
1000	$\left(\frac{1001}{1000}\right)^{1000} \approx 2.71692$
10,000	$\left(\frac{10,001}{10,000}\right)^{10,000} \approx 2.71815$
100,000	$\left(\frac{100,001}{100,000}\right)^{100,000} \approx 2.71827$
1,000,000	$\left(\frac{1,000,001}{1,000,000}\right)^{1,000,000} \approx 2.71828$

Apparently, the numbers $\left(\dfrac{n+1}{n}\right)^n$ increase as n does. However, they do not increase without bound. For example, it is possible to show that for any positive integer n, $\left(\dfrac{n+1}{n}\right)^n < 3$. In terms of our thought experiment this means that if you start with \$1.00 invested at 100% then, no matter how many interest periods there are per year, you will always have less than \$3.00 at the end of a year. There is a smallest real number that is greater than all of the numbers $\left(\dfrac{n+1}{n}\right)^n$. It is denoted by the letter e, in honour of the Swiss mathematician Leonhard Euler (1707–1783).

The number e is irrational so that its decimal expansion is nonrepeating, like those of π and $\sqrt{2}$ that we mentioned in Section 0.1. However, each of the numerical values for $\left(\dfrac{n+1}{n}\right)^n$ can be considered to be a decimal approximation of e. The approximate value $(\frac{1,000,001}{1,000,000})^{1,000,000} \approx 2.71828$ gives an approximation of e that is correct to 5 decimal places.

Exponential Function with Base e

The number

$$e \approx 2.718281828459$$

where the approximation is given correct to 12 decimal places, is used as the base for an exponential function. The exponential function with base e is called the **natural exponential function.**

Although e may seem to be a strange base, the natural exponential function has a remarkable property in calculus (which you will see in a later chapter) that justifies the name. It also occurs in economic analysis and problems involving growth or decay, such as population studies, compound interest, and radioactive decay. Approximate values of e^x can be found with a single key on most calculators. The graph of $y = e^x$ is shown in Figure 4.10. The accompanying table indicates y-values to two decimal places. Of course, the graph has the general shape of an exponential function with base greater than 1.

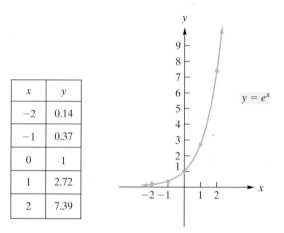

x	y
-2	0.14
-1	0.37
0	1
1	2.72
2	7.39

FIGURE 4.10 Graph of natural exponential function.

You should gain familiarity with the graph of the natural exponential function in Figure 4.10.

GRAPHS OF FUNCTIONS INVOLVING e

The multiplicative decrease in purchasing power P after t years of inflation at 6% can be modeled by $P = e^{-0.06t}$. Graph the decrease in purchasing power as a function of t years.

● **EXAMPLE 8** **Graphs of Functions Involving e**

a. *Graph $y = e^{-x}$.*

Solution: Since $e^{-x} = \left(\dfrac{1}{e}\right)^x$ and $0 < \dfrac{1}{e} < 1$, the graph is that of an exponential function falling from left to right. (See Figure 4.11.) Alternatively, we can consider the graph of $y = e^{-x}$ as a transformation of the graph of $f(x) = e^x$. Because $e^{-x} = f(-x)$, the graph of $y = e^{-x}$ is simply the reflection of the graph of f about the y-axis. (Compare the graphs in Figures 4.10 and 4.11.)

b. *Graph $y = e^{x+2}$.*

Solution: The graph of $y = e^{x+2}$ is related to that of $f(x) = e^x$. Since e^{x+2} is $f(x+2)$, we can obtain the graph of $y = e^{x+2}$ by horizontally shifting the graph of $f(x) = e^x$ two units to the left. (See Figure 4.12.)

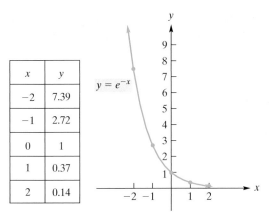

x	y
−2	7.39
−1	2.72
0	1
1	0.37
2	0.14

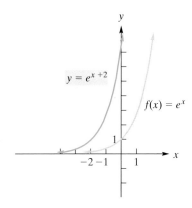

FIGURE 4.11 Graph of $y = e^{-x}$.

FIGURE 4.12 Graph of $y = e^{x+2}$.

● EXAMPLE 9   **Population Growth**

The projected population P of a city is given by

$$P = 100{,}000e^{0.05t}$$

where t is the number of years after 1990. Predict the population for the year 2010.

Solution: The number of years from 1990 to 2010 is 20, so let $t = 20$. Then

$$P = 100{,}000e^{0.05(20)} = 100{,}000e^1 = 100{,}000e \approx 271{,}828$$

NOW WORK PROBLEM 35 ●●●

In statistics, an important function used to model certain events occurring in nature is the **Poisson distribution function:**

$$f(n) = \frac{e^{-\mu}\mu^n}{n!} \quad n = 0, 1, 2, \ldots$$

The symbol μ (read "mu") is a Greek letter. In certain situations, $f(n)$ gives the probability that exactly n events will occur in an interval of time or space. The constant μ is the average, also called *mean*, number of occurrences in the interval. The next example illustrates the Poisson distribution.

● EXAMPLE 10   **Hemocytometer and Cells**

A hemocytometer is a counting chamber divided into squares and is used in studying the number of microscopic structures in a liquid. In a well-known experiment,[3] yeast cells were diluted and thoroughly mixed in a liquid, and the mixture was placed in a hemocytometer. With a microscope, the number of yeast cells on each square were counted. The probability that there were exactly x yeast cells on a hemocytometer square was found to fit a Poisson distribution with $\mu = 1.8$. Find the probability that there were exactly four cells on a particular square.

Solution: We use the Poisson distribution function with $\mu = 1.8$ and $n = 4$:

$$f(n) = \frac{e^{-\mu}\mu^n}{n!}$$

$$f(4) = \frac{e^{-1.8}(1.8)^4}{4!} \approx 0.072$$

---

[3]R. R. Sokal and F. J. Rohlf, *Introduction to Biostatistics* (San Francisco: W. H. Freeman and Company, 1973).

For example, this means that in 400 squares we would *expect* $400(0.072) \approx 29$ squares to contain exactly 4 cells. (In the experiment, in 400 squares the actual number observed was 30.)

Radioactive Decay

Radioactive elements are such that the amount of the element decreases with respect to time. We say that the element *decays*. It can be shown that, if N is the amount at time t, then

$$N = N_0 e^{-\lambda t} \tag{3}$$

where N_0 and λ (a Greek letter read "lambda") are positive constants. Notice that N involves an exponential function of t. We say that N follows an **exponential law of decay.** If $t = 0$, then $N = N_0 e^0 = N_0 \cdot 1 = N_0$. Thus, the constant N_0 represents the amount of the element present at time $t = 0$ and is called the **initial amount.** The constant λ depends on the particular element involved and is called the **decay constant.**

Because N decreases as time progresses, suppose we let T be the length of time it takes for the element to decrease to half of the initial amount. Then at time $t = T$, we have $N = N_0/2$. Equation (3) implies that

$$\frac{N_0}{2} = N_0 e^{-\lambda T}$$

We will now use this fact to show that over *any* time interval of length T, half of the amount of the element decays. Consider the interval from time t to $t + T$, which has length T. At time t, the amount of the element is $N_0 e^{-\lambda t}$, and at time $t + T$ it is

$$N_0 e^{-\lambda(t+T)} = N_0 e^{-\lambda t} e^{-\lambda T} = (N_0 e^{-\lambda T}) e^{-\lambda t}$$

$$= \frac{N_0}{2} e^{-\lambda t} = \frac{1}{2}(N_0 e^{-\lambda t})$$

which is half of the amount at time t. This means that if the initial amount present, N_0, were 1 gram, then at time T, $\frac{1}{2}$ gram would remain; at time $2T$, $\frac{1}{4}$ gram would remain; and so on. The value of T is called the **half-life** of the radioactive element. Figure 4.13 shows a graph of radioactive decay.

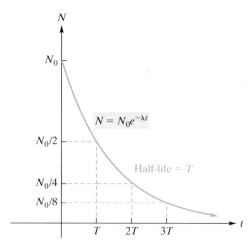

FIGURE 4.13 Radioactive decay.

● EXAMPLE 11   Radioactive Decay

A radioactive element decays such that after t days the number of milligrams present is given by

$$N = 100 e^{-0.062t}$$

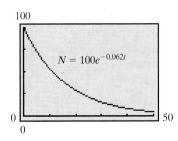

FIGURE 4.14 Graph of radioactive decay function $N = 100e^{-0.062t}$.

a. *How many milligrams are initially present?*

Solution: This equation has the form of Equation (3), $N = N_0 e^{-\lambda t}$, where $N_0 = 100$ and $\lambda = 0.062$. N_0 is the initial amount and corresponds to $t = 0$. Thus, 100 milligrams are initially present. (See Figure 4.14.)

b. *How many milligrams are present after 10 days?*

Solution: When $t = 10$,

$$N = 100e^{-0.062(10)} = 100e^{-0.62} \approx 53.8$$

Therefore, approximately 53.8 milligrams are present after 10 days.

NOW WORK PROBLEM 47

Problems 4.1

In Problems 1–12, graph each function.

*1. $y = f(x) = 4^x$

2. $y = f(x) = 3^x$

*3. $y = f(x) = \left(\frac{1}{3}\right)^x$

4. $y = f(x) = \left(\frac{1}{8}\right)^x$

*5. $y = f(x) = 2^{(x-1)^2}$

6. $y = f(x) = 3(2)^x$

*7. $y = f(x) = 3^{x+2}$

8. $y = f(x) = 2^{x-1}$

9. $y = f(x) = 2^x - 1$

10. $y = f(x) = 3^{x-1} - 1$

11. $y = f(x) = 3^{-x}$

12. $y = f(x) = \frac{1}{2}(2^{x/2})$

Problems 13 and 14 refer to Figure 4.15, which shows the graphs of $y = 0.4^x$, $y = 2^x$, and $y = 5^x$.

13. Of the curves A, B, and C, which is the graph of $y = 5^x$?

14. Of the curves A, B, and C, which is the graph of $y = 0.4^x$?

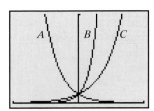

FIGURE 4.15 Diagram for Problems 13 and 14.

*15. **Population** The projected population of a city is given by $P = 125,000(1.11)^{t/20}$, where t is the number of years after 1995. What is the projected population in 2015?

16. Population For a certain city, the population P grows at the rate of 1.5% per year. The formula $P = 1,527,000(1.015)^t$ gives the population t years after 1998. Find the population in (a) 1999 and (b) 2000.

17. Paired-Associate Learning In a psychological experiment involving learning,[4] subjects were asked to give particular responses after being shown certain stimuli. Each stimulus was a pair of letters, and each response was either the digit 1 or 2. After each response, the subject was told the correct answer. In this so-called *paired-associate* learning experiment, the theoretical probability P that a subject makes a correct response on the nth trial is given by

$$P = 1 - \frac{1}{2}(1 - c)^{n-1}, \quad n \geq 1, \ 0 < c < 1$$

where c is a constant. Take $c = \frac{1}{2}$ and find P when $n = 1$, $n = 2$, and $n = 3$.

---
[4]D. Laming, *Mathematical Psychology* (New York: Academic Press, Inc., 1973).

18. Express $y = 2^{3x}$ as an exponential function in base 8.

In Problems 19–27, find (a) the compound amount and (b) the compound interest for the given investment and annual rate.

*19. $4000 for 7 years at 6% compounded annually

20. $5000 for 20 years at 5% compounded annually

21. $700 for 15 years at 7% compounded semiannually

22. $4000 for 12 years at $7\frac{1}{2}$% compounded semiannually

23. $3000 for 16 years at $8\frac{3}{4}$% compounded quarterly

24. $2000 for 12 years at 7% compounded quarterly

25. $5000 for $2\frac{1}{2}$ years at 9% compounded monthly

26. $500 for 5 years at 11% compounded semiannually

27. $8000 for 3 years at $6\frac{1}{4}$% compounded daily. (Assume that there are 365 days in a year.)

28. Investment Suppose $900 is placed in a savings account that earns interest at the rate of 4.5% compounded semiannually. (a) What is the value of the account at the end of five years? (b) If the account had earned interest at the rate of 4.5% compounded annually, what would be the value after five years?

29. Investment A certificate of deposit is purchased for $6500 and is held for six years. If the certificate earns 4% compounded quarterly, what is it worth at the end of six years?

30. Population Growth The population of a town of 5000 grows at the rate of 3% per year. (a) Determine an equation that gives the population t years from now. (b) Find the population three years from now. Give your answer to (b) to the nearest integer.

*31. **Bacteria Growth** Bacteria are growing in a culture, and their number is increasing at the rate of 5% an hour. Initially, 400 bacteria are present. (a) Determine an equation that gives the number, N, of bacteria present after t hours. (b) How many bacteria are present after one hour? (c) After four hours? Give your answers to (b) and (c) to the nearest integer.

32. Bacteria Reduction   A certain medicine reduces the bacteria present in a person by 10% each hour. Currently, 100,000 bacteria are present. Make a table of values for the number of bacteria present each hour for 0 to 4 hours. For each hour, write an expression for the number of bacteria as a product of 100,000 and a power of $\frac{9}{10}$. Use the expressions to make an entry in your table for the number of bacteria after t hours. Write a function N for the number of bacteria after t hours.

33. Recycling   Suppose the amount of plastic being recycled increases by 30% every year. Make a table of the factor by which recycling increases over the original amount for 0 to 3 years. For each year, write an expression for the increase as a power of some base. What base did you use? How does that base relate to the problem? Use your table to graph the multiplicative increase as a function of years. Use your graph to determine when the recycling will triple.

34. Population Growth   Cities A and B presently have populations of 70,000 and 60,000, respectively. City A grows at the rate of 4% per year, and B grows at the rate of 5% per year. Determine the difference in the populations of the cities at the end of five years. Give your answer to the nearest integer.

Problems 35 and 36 involve a declining population. If a population declines at the rate of r per time period, then the population after t time periods is given by

$$P = P_0(1 - r)^t$$

where P_0 is the initial population (the population when $t = 0$).

*35. **Population**   Because of an economic downturn, the population of a certain urban area declines at the rate of 1.5% per year. Initially, the population is 350,000. To the nearest person, what is the population after three years?

36. Enrollment   After a careful demographic analysis, a university forecasts that student enrollments will drop by 3% per year for the the next 12 years. If the university currently has 14,000 students, how many students will it have 12 years from now?

In Problems 37–40, use a calculator to find the value (rounded to four decimal places) of each expression.

37. $e^{1.5}$ **38.** $e^{3.4}$

39. $e^{-0.7}$ **40.** $e^{-2/3}$

In Problems 41 and 42, graph the functions.

41. $y = -e^{-(x+1)}$ **42.** $y = 2e^x$

43. Telephone Calls   The probability that a telephone operator will receive exactly x calls during a certain period is given by

$$P = \frac{e^{-3}3^x}{x!}$$

Find the probability that the operator will receive exactly three calls. Round your answer to four decimal places.

44. Normal Distribution   An important function used in economic and business decisions is the *normal distribution density function*, which, in standard form, is

$$f(x) = \frac{1}{\sqrt{2\pi}}e^{-\left(\frac{1}{2}\right)x^2}$$

Evaluate $f(0)$, $f(-1)$, and $f(1)$. Round your answers to three decimal places.

45. Express e^{kt} in the form b^t.   **46.** Express $\frac{1}{e^x}$ in the form b^x.

*47. **Radioactive Decay**   A radioactive element is such that N grams remain after t hours, where

$$N = 12e^{-0.031t}$$

(a) How many grams are initially present? To the nearest tenth of a gram, how many grams remain after (b) 10 hours? (c) 44 hours? (d) Based on your answer to part (c), what is your estimate of the half-life of this element?

48. Radioactive Decay   At a certain time, there are 75 milligrams of a radioactive substance. The substance decays so that after t years the number of milligrams present, N, is given by

$$N = 75e^{-0.045t}$$

How many milligrams are present after 10 years? Give your answer to the nearest milligram.

49. Radioactive Decay   If a radioactive substance has a half-life of 8 years, how long does it take for 1 gram of the substance to decay to $\frac{1}{16}$ gram?

50. Marketing   A mail-order company advertises in a national magazine. The company finds that, of all small towns, the percentage (given as a decimal) in which exactly x people respond to an ad fits a Poisson distribution with $\mu = 0.5$. From what percentage of small towns can the company expect exactly two people to respond? Round your answer to four decimal places.

51. Emergency-Room Admissions   Suppose the number of patients admitted into a hospital emergency room during a certain hour of the day has a Poisson distribution with mean 4. Find the probability that during that hour there will be exactly two emergency patients. Round your answer to four decimal places.

52. Graph $y = 17^x$ and $y = \left(\frac{1}{17}\right)^x$ on the same screen. Determine the intersection point.

53. Let $a > 0$ be a constant. Graph $y = 2^x$ and $y = 2^a \cdot 2^x$ on the same screen, for constant values $a = 2$ and $a = 3$. It appears that the graph of $y = 2^a \cdot 2^x$ is the graph of $y = 2^x$ shifted a units to the left. Prove algebraically that this is indeed true.

54. For $y = 7^x$, find x if $y = 4$. Round your answer to two decimal places.

55. For $y = 2^x$, find x if $y = 9$. Round your answer to two decimal places.

56. Cell Growth   Cells are growing in a culture, and their number is increasing at the rate of 7% per hour. Initially, 1000 cells are present. After how many full hours will there be at least 3000 cells?

57. Bacteria Growth   Refer to Example 1. How long will it take for 1000 bacteria to be present? Round your answer to the nearest tenth of a minute.

58. Demand Equation   The demand equation for a new toy is

$$q = 10{,}000(0.95123)^p$$

(a) Evaluate q to the nearest integer when $p = 10$.
(b) Convert the demand equation to the form

$$q = 10{,}000e^{-xp}$$

(*Hint:* Find a number x such that $0.95123 \approx e^{-x}$.)

(c) Use the equation in part (b) to evaluate q to the nearest integer when $p = 10$. Your answers in parts (a) and (c) should be the same.

59. Investment   If $2500 is invested in a savings account that earns interest at the rate of 4.3% compounded annually, after how many full years will the amount at least double?

OBJECTIVE

To introduce logarithmic functions and their graphs. Properties of logarithms will be discussed in Section 4.3.

POINTER ▶

To review inverse functions, refer to Section 2.4.

4.2 Logarithmic Functions

Since all exponential functions pass the horizontal line test, they are all one-to-one functions. It follows that each exponential function has an inverse. These functions, inverse to the exponential functions, are called the *logarithmic functions*.

More precisely, if $f(x) = b^x$, the exponential function base b (where $0 < b < 1$ or $1 < b$), then the inverse function $f^{-1}(x)$ is called the *logarithm function base b* and is denoted $\log_b x$. It follows from our general remarks about inverse functions in Section 2.4 that

$$y = \log_b x \quad \text{if and only if} \quad b^y = x$$

and we have the following fundamental equations:

$$\log_b b^x = x \tag{1}$$

and

$$b^{\log_b x} = x \tag{2}$$

where Equation (1) holds for all x in $(-\infty, \infty)$—which is the domain of the exponential function base b—and Equation (2) holds for all x in the range of the exponential function base b—which is $(0, \infty)$ and is necesarily the domain of the logarithm function base b. Said otherwise, given positive x, $\log_b x$ is the unique number with the property that $b^{\log_b x} = x$. The generalities about inverse functions also enable us to see immediately what the graph of a logarithmic function looks like.

In Figure 4.16 we have shown the graph of the particular exponential function $y = f(x) = 2^x$, whose general shape is typical of exponential functions $y = b^x$ for which the base b satisfies $1 < b$. We have added a (dashed) copy of the line $y = x$. The graph of $y = f^{-1}(x) = \log_2 x$ is obtained as the mirror image of $y = f(x) = 2^x$ in the line $y = x$.

In Table 4.3 we have tabulated the function values that appear as y-coordinates of the dots in Figure 4.16.

TABLE 4.3 Selected function values

x	2^x	x	$\log_2 x$
-2	$\frac{1}{4}$	$\frac{1}{4}$	-2
-1	$\frac{1}{2}$	$\frac{1}{2}$	-1
0	1	1	0
1	2	2	1
2	4	4	2
3	8	8	3

FIGURE 4.16   Graphs of $y = 2^x$ and $y = \log_2 x$.

Logarithmic and
exponential forms

Logarithm   Exponent
↓   ↓
$\log_2 8 = 3$ $2^3 = 8$
↑   ↑
base   base

FIGURE 4.17 A logarithm can
be considered an exponent.

It is clear that the exponential function base 2 and the logarithm function base 2 'undo' the effects of each other. Thus, for all x in the domain of 2^x, [which is $(-\infty, \infty)$], we have

$$\log_2 2^x = x$$

and, for all x in the domain of $\log_2 x$ [which is the range of 2^x, which is $(0, \infty)$], we have

$$2^{\log_2 x} = x$$

It cannot be said too often that

$$y = \log_b x \quad \text{means} \quad b^y = x$$

and conversely

$$b^y = x \quad \text{means} \quad y = \log_b x$$

In this sense, *a logarithm of a number is an exponent:* $\log_b x$ is the power to which we must raise b to get x. For example,

$$\log_2 8 = 3 \quad \text{because} \quad 2^3 = 8$$

We say that $\log_2 8 = 3$ is the **logarithmic form** of the **exponential form** $2^3 = 8$. (See Figure 4.17.)

PRINCIPLES IN PRACTICE 1

CONVERTING FROM
EXPONENTIAL TO LOGARITHMIC
FORM

If bacteria have been doubling every hour and the current amount is 16 times the amount first measured, then the situation can be represented by $16 = 2^t$. Represent this equation in logarithmic form. What does t represent?

EXAMPLE 1 Converting from Exponential to Logarithmic Form

		Exponential Form		*Logarithmic Form*
a.	Since	$5^2 = 25$	it follows that	$\log_5 25 = 2$
b.	Since	$3^4 = 81$	it follows that	$\log_3 81 = 4$
c.	Since	$10^0 = 1$	it follows that	$\log_{10} 1 = 0$

NOW WORK PROBLEM 1

PRINCIPLES IN PRACTICE 2

CONVERTING FROM
LOGARITHMIC TO EXPONENTIAL
FORM

An earthquake measuring 8.3 on the Richter scale can be represented by $8.3 = \log_{10}\left(\frac{I}{I_0}\right)$, where I is the intensity of the earthquake and I_0 is the intensity of a zero-level earthquake. Represent this equation in exponential form.

EXAMPLE 2 Converting from Logarithmic to Exponential Form

Logarithmic Form		*Exponential Form*
a. $\log_{10} 1000 = 3$	means	$10^3 = 1000$
b. $\log_{64} 8 = \dfrac{1}{2}$	means	$64^{1/2} = 8$
c. $\log_2 \dfrac{1}{16} = -4$	means	$2^{-4} = \dfrac{1}{16}$

NOW WORK PROBLEM 3

PRINCIPLES IN PRACTICE 3

GRAPH OF A LOGARITHMIC
FUNCTION WITH $b > 1$

Suppose a recycling plant has found that the amount of material being recycled has increased by 50% every year since the plant's first year of operation. Graph each year as a function of the multiplicative increase in recycling since the first year. Label the graph with the name of the function.

EXAMPLE 3 Graph of a Logarithmic Function with $b > 1$

Examine again the graph of $y = \log_2 x$ in Figure 4.16. This graph is typical for a logarithmic function with $b > 1$.

NOW WORK PROBLEM 9

GRAPH OF A LOGARITHMIC
FUNCTION WITH $0 < b < 1$

Suppose a boat depreciates 20% every year. Graph the number of years the boat is owned as a function of the multiplicative decrease in its original value. Label the graph with the name of the function.

● EXAMPLE 4　**Graph of a Logarithmic Function with** $0 < b < 1$

Graph $y = \log_{1/2} x$.

Solution: To plot points, we plot the inverse function $y = \left(\frac{1}{2}\right)^x$ and reflect the graph in the line $y = x$.

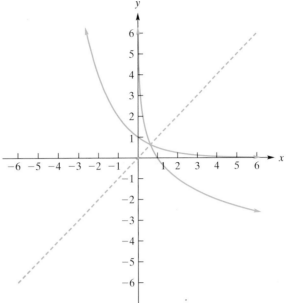

FIGURE 4.18　Graphs of $y = \left(\frac{1}{2}\right)^x$ and $y = \log_{1/2} x$.

From the graph, we can see that the domain of $y = \log_{1/2} x$ is the set of all positive reals, for that is the range of $y = \left(\frac{1}{2}\right)^x$ and the range of $y = \log_{1/2} x$ is the set of all real numbers, which is the domain of $y = \left(\frac{1}{2}\right)^x$. The graph falls from left to right. Numbers between 0 and 1 have positive base $\frac{1}{2}$ logarithms, and the closer a number is to 0, the larger is its base $\frac{1}{2}$ logarithm. Numbers greater than 1 have negative base $\frac{1}{2}$ logarithms. The logarithm of 1 is 0, *regardless of the base b*, and corresponds to the x-intercept $(1, 0)$. This graph is typical for a logarithmic function with $0 < b < 1$.

NOW WORK PROBLEM 11 ●●

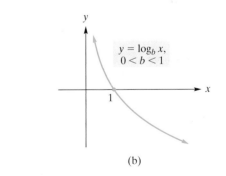

(a)　　　　　　　　　　　(b)

FIGURE 4.19　General shapes of $y = \log_b x$.

Summarizing the results of Examples 3 and 4, we can say that the graph of a logarithmic function has one of two general shapes, depending on whether $b > 1$ or $0 < b < 1$. (See Figure 4.19.) For $b > 1$, the graph rises from left to right; as x gets closer and closer to 0, the function values decrease without bound, and the graph gets closer and closer to the y-axis. For $0 < b < 1$, the graph falls from left to right; as x gets closer and closer to 0, the function values increase without bound, and the graph gets closer and closer to the y-axis. In each case, note that

1. The domain of a logarithmic function is the interval $(0, \infty)$. Thus, the logarithm of either a negative number or 0 does not exist.

2. The range is the interval $(-\infty, \infty)$.

3. The logarithm of 1 is 0, which corresponds to the x-intercept $(1, 0)$.

Logarithms to the base 10 are called **common logarithms.** They were frequently used for computational purposes before the calculator age. The subscript 10 is usually omitted from the notation:

$$\log x \quad \text{means} \quad \log_{10} x$$

Important in calculus are logarithms to the base e, called **natural logarithms.** We use the notation "ln" for such logarithms:

$$\ln x \quad \text{means} \quad \log_e x$$

You should gain familiarity with the graph of the natural logarithmic function in Figure 4.20.

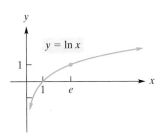

FIGURE 4.20   Graph of natural logarithmic function.

The symbol $\ln x$ may be read "log, natural, of x." Your calculator gives approximate values for natural and common logarithms. For example, verify that $\ln 2 \approx 0.69315$. This means that $e^{0.69315} \approx 2$. Figure 4.20 shows the graph of $y = \ln x$. Because $e > 1$, the graph has the general shape of that of a logarithmic function with $b > 1$ [see Figure 4.19(a)] and rises from left to right. While the conventions about log, with no subscript, and ln are well-established in elementary books, you should be careful when reading advanced books later. In advanced texts, $\log x$ usually means $\log_e x$, ln is not used at all, and logarithms base 10 are written explicitly as $\log_{10} x$.

FINDING LOGARITHMS

The number of years it takes for an amount invested at an annual rate of r and compounded continuously to quadruple is a function of the annual rate r given by $t(r) = \dfrac{\ln 4}{r}$. Use a calculator to find the rate needed to quadruple an investment in 10 years.

Remember the way in which a logarithm is an exponent.

● EXAMPLE 5   Finding Logarithms

a. *Find* $\log 100$.

Solution: Here the base is 10. Thus, $\log 100$ is the power to which we must raise 10 to get 100. Since $10^2 = 100$, $\log 100 = 2$.

b. *Find* $\ln 1$.

Solution: Here the base is e. Because $e^0 = 1$, $\ln 1 = 0$.

c. *Find* $\log 0.1$.

Solution: Since $0.1 = \frac{1}{10} = 10^{-1}$, $\log 0.1 = -1$.

d. *Find* $\ln e^{-1}$.

Solution: Since $\ln e^{-1}$ is the power to which e must be raised to obtain e^{-1}, clearly $\ln e^{-1} = -1$.

e. *Find* $\log_{36} 6$.

Solution: Because $36^{1/2}\ (= \sqrt{36})$ is 6, $\log_{36} 6 = \frac{1}{2}$.

NOW WORK PROBLEM 3   ●●●

Many equations involving logarithmic or exponential forms can be solved for an unknown quantity by first transforming from logarithmic form to exponential form or vice versa. Example 6 will illustrate.

SOLVING LOGARITHMIC AND EXPONENTIAL EQUATIONS

The multiplicative increase m of an amount invested at an annual rate of r compounded continuously for a time t is given by $m = e^{rt}$. What annual percentage rate is needed to triple the investment in 12 years?

● EXAMPLE 6   Solving Logarithmic and Exponential Equations

a. *Solve* $\log_2 x = 4$.

Solution: We can get an explicit expression for x by writing the equation in exponential form. This gives

$$2^4 = x$$

so $x = 16$.

b. *Solve* $\ln(x + 1) = 7$.

Solution: The exponential form yields $e^7 = x + 1$. Thus, $x = e^7 - 1$.

c. *Solve* $\log_x 49 = 2$.

> **Solution:** In exponential form, $x^2 = 49$, so $x = 7$. We reject $x = -7$ because a negative number cannot be a base of a logarithmic function.

d. *Solve* $e^{5x} = 4$.

> **Solution:** We can get an explicit expression for x by writing the equation in logarithmic form. We have

$$\ln 4 = 5x$$
$$x = \frac{\ln 4}{5}$$

NOW WORK PROBLEM 49

Radioactive Decay and Half-Life

From our discussion of the decay of a radioactive element in Section 4.1, we know that the amount of the element present at time t is given by

$$N = N_0 e^{-\lambda t} \qquad (3)$$

where N_0 is the initial amount (the amount at time $t = 0$) and λ is the decay constant. Let us now determine the half-life T of the element. At time T, half of the initial amount is present. That is, when $t = T$, $N = N_0/2$. Thus, from Equation (3), we have

$$\frac{N_0}{2} = N_0 e^{-\lambda T}$$

Solving for T gives

$$\frac{1}{2} = e^{-\lambda T}$$
$$2 = e^{\lambda T} \qquad \text{(taking reciprocals of both sides)}$$

To get an explicit expression for T, we convert to logarithmic form. This results in

$$\lambda T = \ln 2$$
$$T = \frac{\ln 2}{\lambda}$$

Summarizing, we have the following:

> *If a radioactive element has decay constant λ, then the half-life of the element is given by*

$$T = \frac{\ln 2}{\lambda} \qquad (4)$$

● EXAMPLE 7   Finding Half-Life

A 10-milligram sample of radioactive polonium 210 (which is denoted ^{210}Po) decays according to the equation

$$N = 10e^{-0.00501t}$$

where N is the number of milligrams present after t days. (See Figure 4.21.) Determine the half-life of ^{210}Po.

Solution: Here the decay constant λ is 0.00501. By Equation (4), the half-life is given by

$$T = \frac{\ln 2}{\lambda} = \frac{\ln 2}{0.00501} \approx 138.4 \text{ days}$$

NOW WORK PROBLEM 63

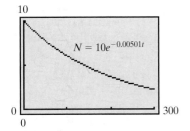

FIGURE 4.21   Radioactive decay function $N = 10e^{-0.00501t}$.

Problems 4.2

In Problems 1–8, express each logarithmic form exponentially and each exponential form logarithmically.

*1. $10^4 = 10,000$ 2. $2 = \log_{12} 144$

*3. $\log_2 64 = 6$ 4. $8^{2/3} = 4$

5. $e^3 = 20.0855$ 6. $e^{0.33647} = 1.4$

7. $\ln 3 = 1.09861$ 8. $\log 5 = 0.6990$

In Problems 9–16, graph the functions.

*9. $y = f(x) = \log_3 x$ 10. $y = f(x) = \log_4 2x$

*11. $y = f(x) = \log_{1/4} x$ 12. $y = f(x) = \log_{1/5} x$

13. $y = f(x) = \log_2 (x - 4)$ 14. $y = f(x) = \log_2 (-x)$

15. $y = f(x) = -2\ln x$ 16. $y = f(x) = \ln(x + 2)$

In Problems 17–28, evaluate the expression.

17. $\log_6 36$ 18. $\log_2 64$ 19. $\log_3 27$

20. $\log_{16} 4$ 21. $\log_7 7$ 22. $\log 10,000$

23. $\log 0.01$ 24. $\log_2 \sqrt[3]{2}$ 25. $\log_5 1$

26. $\log_5 \frac{1}{25}$ 27. $\log_2 \frac{1}{8}$ 28. $\log_4 \sqrt[5]{4}$

In Problems 29–48, find x.

29. $\log_3 x = 4$ 30. $\log_2 x = 8$

31. $\log_5 x = 3$ 32. $\log_4 x = 0$

33. $\log x = -1$ 34. $\ln x = 1$

35. $\ln x = -3$ 36. $\log_x 25 = 2$

37. $\log_x 8 = 3$ 38. $\log_x 3 = \frac{1}{2}$

39. $\log_x \frac{1}{6} = -1$ 40. $\log_x y = 1$

41. $\log_3 x = -3$ 42. $\log_x (2x - 3) = 1$

43. $\log_x (6 - x) = 2$ 44. $\log_8 64 = x - 1$

45. $2 + \log_2 4 = 3x - 1$ 46. $\log_3(x + 2) = -2$

47. $\log_x (2x + 8) = 2$ 48. $\log_x (6 + 4x - x^2) = 2$

In Problems 49–52, find x and express your answer in terms of natural logarithms.

*49. $e^{3x} = 2$ 50. $0.1e^{0.1x} = 0.5$

51. $e^{2x-5} + 1 = 4$ 52. $6e^{2x} - 1 = \frac{1}{2}$

In Problems 53–56, use your calculator to find the approximate value of each expression. Round your answer to five decimal places.

53. $\ln 5$ 54. $\ln 4.27$

55. $\ln 7.39$ 56. $\ln 9.98$

57. **Appreciation** Suppose an antique gains 10% in value every year. Graph the number of years it is owned as a function of the multiplicative increase in its original value. Label the graph with the name of the function.

58. **Cost Equation** The cost for a firm producing q units of a product is given by the cost equation

$$c = (3q \ln q) + 12$$

Evaluate the cost when $q = 6$. (Round your answer to two decimal places.)

59. **Supply Equation** A manufacturer's supply equation is

$$p = \log\left(10 + \frac{q}{2}\right)$$

where q is the number of units supplied at a price p per unit. At what price will the manufacturer supply 1980 units?

60. **Earthquake** The magnitude, M, of an earthquake and its energy, E, are related by the equation[5]

$$1.5M = \log\left(\frac{E}{2.5 \times 10^{11}}\right)$$

where M is given in terms of Richter's preferred scale of 1958 and E is in ergs. Solve the equation for E.

61. **Biology** For a certain population of cells, the number of cells at time t is given by $N = N_0(2^{t/k})$, where N_0 is the number of cells at $t = 0$ and k is a positive constant. (a) Find N when $t = k$. (b) What is the significance of k? (c) Show that the time it takes to have population N_1 can be written

$$t = k \log_2 \frac{N_1}{N_0}$$

62. **Inferior Good** In a discussion of an inferior good, Persky[6] solves an equation of the form

$$u_0 = A \ln(x_1) + \frac{x_2^2}{2}$$

for x_1, where x_1 and x_2 are quantities of two products, u_0 is a measure of utility, and A is a positive constant. Determine x_1.

*63. **Radioactive Decay** A 1-gram sample of radioactive lead 211 (^{211}Pb) decays according to the equation $N = e^{-0.01920t}$, where N is the number of grams present after t minutes. Find the half-life of ^{211}Pb to the nearest tenth of a minute.

64. **Radioactive Decay** A 100-milligram sample of radioactive actinium 227 (^{227}Ac) decays according to the equation

$$N = 100e^{-0.03194t}$$

where N is the number of milligrams present after t years. Find the half-life of ^{227}Ac to the nearest tenth of a year.

65. If $\log_y x = 3$ and $\log_z x = 2$, find a formula for z as an explicit function of y only.

66. Solve for y as an explicit function of x if

$$x + 3e^{2y} - 8 = 0$$

67. Suppose $y = f(x) = x \ln x$. (a) For what values of x is $y < 0$? (*Hint:* Determine when the graph is below the x-axis.) (b) Determine the range of f.

68. Find the x-intercept of $y = x^2 \ln x$.

69. Use the graph of $y = e^x$ to estimate $\ln 3$. Round your answer to two decimal places.

70. Use the graph of $y = \ln x$ to estimate e^2. Round your answer to two decimal places.

71. Determine the x-values of points of intersection of the graphs of $y = (x - 2)^2$ and $y = \ln x$. Round your answers to two decimal places.

---

[5]K. E. Bullen, *An Introduction to the Theory of Seismology* (Cambridge, U.K.: Cambridge at the University Press, 1963).

[6]A. L. Persky, "An Inferior Good and a Novel Indifference Map," *The American Economist*, XXIX, no. 1 (Spring 1985).

4.3 Properties of Logarithms

The logarithmic function has many important properties. For example,

1. $\log_b(mn) = \log_b m + \log_b n$

which says that the logarithm of a product of two numbers is the sum of the logarithms of the numbers. We can prove this property by deriving the exponential form of the equation:

$$b^{\log_b m + \log_b n} = mn$$

 CAUTION

Make sure that you clearly understand properties 1–3. They do *not* apply to the log of a sum $[\log_b(m+n)]$, to the log of a difference $[\log_b(m-n)]$, to the product of two logs $[(\log_b m)(\log_b n)]$, or to the quotient of two logs $\left[\dfrac{\log_b m}{\log_b n}\right]$.

Using first a familiar rule for exponents, we have

$$b^{\log_b m + \log_b n} = b^{\log_b m} b^{\log_b n}$$

$$= mn$$

where the second equality uses two instances of the fundamental equation (2) of Section 4.2. We will not prove the next two properties, since their proofs are similar to that of Property 1.

2. $\log_b \dfrac{m}{n} = \log_b m - \log_b n$

That is, the logarithm of a quotient is the difference of the logarithm of the numerator and the logarithm of the denominator.

3. $\log_b m^r = r \log_b m$

That is, the logarithm of a power of a number is the exponent times the logarithm of the number.

TABLE 4.4 Common Logarithms

x	$\log x$	x	$\log x$
2	0.3010	7	0.8451
3	0.4771	8	0.9031
4	0.6021	9	0.9542
5	0.6990	10	1.0000
6	0.7782	e	0.4343

Table 4.4 gives the values of a few common logarithms. Most entries are approximate. For example, $\log 4 \approx 0.6021$, which means $10^{0.6021} \approx 4$. To illustrate the use of properties of logarithms, we shall use this table in some of the examples that follow.

● **EXAMPLE 1   Finding Logarithms by Using Table 4.4**

a. *Find* $\log 56$.

Solution: Log 56 is not in the table. But we can write 56 as the product $8 \cdot 7$. Thus, by Property 1,

$$\log 56 = \log(8 \cdot 7) = \log 8 + \log 7 \approx 0.9031 + 0.8451 = 1.7482$$

Although the logarithms in Example 1 can be found with a calculator, we will make use of properties of logarithms.

b. *Find* $\log \frac{9}{2}$.

Solution: By Property 2,

$$\log \frac{9}{2} = \log 9 - \log 2 \approx 0.9542 - 0.3010 = 0.6532$$

c. *Find* $\log 64$.

Solution: Since $64 = 8^2$, by Property 3,

$$\log 64 = \log 8^2 = 2 \log 8 \approx 2(0.9031) = 1.8062$$

d. *Find* $\log \sqrt{5}$.

Solution: By Property 3, we have

$$\log \sqrt{5} = \log 5^{1/2} = \frac{1}{2} \log 5 \approx \frac{1}{2}(0.6990) = 0.3495$$

e. *Find* $\log \dfrac{16}{21}$.

Solution:

$$\log \dfrac{16}{21} = \log 16 - \log 21 = \log(4^2) - \log(3 \cdot 7)$$

$$= 2 \log 4 - [\log 3 + \log 7]$$

$$\approx 2(0.6021) - [0.4771 + 0.8451] = -0.1180$$

NOW WORK PROBLEM 3

 EXAMPLE 2   Rewriting Logarithmic Expressions

a. *Express* $\log \dfrac{1}{x^2}$ *in terms of* $\log x$.

Solution:

$$\log \dfrac{1}{x^2} = \log x^{-2} = -2 \log x \qquad \text{(Property 3)}$$

Here we have assumed that $x > 0$. Although $\log(1/x^2)$ is defined for $x \neq 0$, the expression $-2 \log x$ is defined only if $x > 0$. Note that we do have

$$\log \dfrac{1}{x^2} = \log x^{-2} = -2 \log |x|$$

for all $x \neq 0$.

b. *Express* $\log \dfrac{1}{x}$ *in terms of* $\log x$, *for* $x > 0$.

Solution: By Property 3,

$$\log \dfrac{1}{x} = \log x^{-1} = -1 \log x = -\log x$$

NOW WORK PROBLEM 19

From Example 2(b), we see that $\log(1/x) = -\log x$. Generalizing gives the following property:

4. $\log_b \dfrac{1}{m} = -\log_b m$

That is, the logarithm of the reciprocal of a number is the negative of the logarithm of the number.

For example, $\log \dfrac{2}{3} = -\log \dfrac{3}{2}$.

 EXAMPLE 3   Writing Logarithms in Terms of Simpler Logarithms

Manipulations such as those in Example 3 are frequently used in calculus.

a. *Write* $\ln \dfrac{x}{zw}$ *in terms of* $\ln x$, $\ln z$, *and* $\ln w$.

Solution:

$$\ln \dfrac{x}{zw} = \ln x - \ln(zw) \qquad \text{(Property 2)}$$

$$= \ln x - (\ln z + \ln w) \qquad \text{(Property 1)}$$

$$= \ln x - \ln z - \ln w$$

b. *Write* $\ln \sqrt[3]{\dfrac{x^5(x-2)^8}{x-3}}$ *in terms of* $\ln x$, $\ln(x-2)$, *and* $\ln(x-3)$.

Solution:

$$\ln \sqrt[3]{\frac{x^5(x-2)^8}{x-3}} = \ln \left[\frac{x^5(x-2)^8}{x-3}\right]^{1/3} = \frac{1}{3} \ln \frac{x^5(x-2)^8}{x-3}$$

$$= \frac{1}{3}\{\ln[x^5(x-2)^8] - \ln(x-3)\}$$

$$= \frac{1}{3}[\ln x^5 + \ln(x-2)^8 - \ln(x-3)]$$

$$= \frac{1}{3}[5\ln x + 8\ln(x-2) - \ln(x-3)]$$

NOW WORK PROBLEM 29

PRINCIPLES IN PRACTICE 1

COMBINING LOGARITHMS

The Richter scale measure of an earthquake is given by $R = \log\left(\dfrac{I}{I_0}\right)$, where I is the intensity of the earthquake and I_0 is the intensity of a zero-level earthquake. How much more on the Richter scale is an earthquake with intensity 900,000 times the intensity of a zero-level earthquake than an earthquake with intensity 9000 times the intensity of a zero-level earthquake? Write the answer as an expression involving logarithms. Simplify the expression by combining logarithms, and then evaluate the resulting expression.

● EXAMPLE 4   **Combining Logarithms**

a. *Write* $\ln x - \ln(x+3)$ *as a single logarithm.*

Solution:

$$\ln x - \ln(x+3) = \ln \frac{x}{x+3} \qquad \text{(Property 2)}$$

 b. *Write* $\ln 3 + \ln 7 - \ln 2 - 2\ln 4$ *as a single logarithm.*

Solution:

$$\ln 3 + \ln 7 - \ln 2 - 2\ln 4$$

$$= \ln 3 + \ln 7 - \ln 2 - \ln(4^2) \qquad \text{(Property 3)}$$

$$= \ln 3 + \ln 7 - [\ln 2 + \ln(4^2)]$$

$$= \ln(3 \cdot 7) - \ln(2 \cdot 4^2) \qquad \text{(Property 1)}$$

$$= \ln 21 - \ln 32$$

$$= \ln \frac{21}{32} \qquad \text{(Property 2)}$$

NOW WORK PROBLEM 37

Since $b^0 = 1$ and $b^1 = b$, by converting to logarithmic forms we have the following properties:

5. $\log_b 1 = 0$

6. $\log_b b = 1$

PRINCIPLES IN PRACTICE 2

SIMPLIFYING LOGARITHMIC EXPRESSIONS

If an earthquake is 10,000 times as intense as a zero-level earthquake, what is its measurement on the Richter scale? Write the answer as a logarithmic expression and simplify it. (See Principles in Practice 1 for formula.)

● EXAMPLE 5   **Simplifying Logarithmic Expressions**

a. *Find* $\ln e^{3x}$.

Solution: By the fundamental equation (1) of Section 4.2 with $b = e$, we have $\ln e^{3x} = 3x$.

b. *Find* $\log 1 + \log 1000$.

Solution: By Property 5, $\log 1 = 0$. Thus,

$$\log 1 + \log 1000 = 0 + \log 10^3$$

$$= 0 + 3 \qquad \text{(Fundamental equation (1) of}$$

$$= 3 \qquad \text{Section 4.2 with } b = 10)$$

 **c.** *Find* $\log_7 \sqrt[9]{7^8}$.

Solution:

$$\log_7 \sqrt[9]{7^8} = \log_7 7^{8/9} = \frac{8}{9}$$

d. *Find* $\log_3 \left(\dfrac{27}{81} \right)$.

Solution:

$$\log_3 \left(\frac{27}{81} \right) = \log_3 \left(\frac{3^3}{3^4} \right) = \log_3(3^{-1}) = -1$$

e. *Find* $\ln e + \log \dfrac{1}{10}$.

Solution:

$$\ln e + \log \frac{1}{10} = \ln e + \log 10^{-1}$$
$$= 1 + (-1) = 0$$

NOW WORK PROBLEM 41

Do not confuse $\ln x^2$ with $(\ln x)^2$. We have

$$\ln x^2 = \ln(x \cdot x)$$

but

$$(\ln x)^2 = (\ln x)(\ln x)$$

Sometimes $(\ln x)^2$ is written as $\ln^2 x$. This is not a new formula but merely a notation. More generally, some people write $f^2(x)$ for $(f(x))^2$. We recommend avoiding the notation $f^2(x)$.

● EXAMPLE 6 Use of Equation (2) of Section 4.2

a. *Find* $e^{\ln x^2}$.

Solution: By (2) with $b = e$, $e^{\ln x^2} = x^2$.

b. *Solve* $10^{\log x^2} = 25$ *for* x.

Solution:

$$10^{\log x^2} = 25$$
$$x^2 = 25 \qquad \text{[By Equation (2) of Section 4.2]}$$
$$x = \pm 5$$

NOW WORK PROBLEM 45

● EXAMPLE 7 Evaluating a Logarithm Base 5

Use a calculator to find $\log_5 2$.

Solution: Calculators typically have keys for logarithms in base 10 and base e, but not for base 5. However, we can convert logarithms in one base to logarithms in another base. Let us convert from base 5 to base 10. First, let $x = \log_5 2$. Then $5^x = 2$. Taking the common logarithms of both sides of $5^x = 2$ gives

$$\log 5^x = \log 2$$
$$x \log 5 = \log 2$$
$$x = \frac{\log 2}{\log 5} \approx 0.4307$$

If we had taken natural logarithms of both sides, the result would be $x = (\ln 2)/(\ln 5) \approx 0.4307$, the same as before.

Generalizing the method used in Example 7, we obtain the so-called *change-of-base* formula:

Change-of-Base Formula

7. $\log_b m = \dfrac{\log_a m}{\log_a b}$

Some students find the change-of-base formula more memorable when it is expressed in the form

$$(\log_a b)(\log_b m) = \log_a m$$

in which the two instances of b apparently cancel. Let us see how to prove this identity, for ability to see the truth of such statements greatly enhances one's ability to use them in practical applications. Since $\log_a m = y$ precisely if $a^y = m$, our task is equivalently to show that

$$a^{(\log_a b)(\log_b m)} = m$$

and we have

$$a^{(\log_a b)(\log_b m)} = \left(a^{\log_a b}\right)^{\log_b m}$$
$$= b^{\log_b m}$$
$$= m$$

using a rule for exponents and Fundamental Equation (2) twice.

The change-of-base formula allows logarithms to be converted from base a to base b.

● EXAMPLE 8   Change-of-Base Formula

Express $\log x$ in terms of natural logarithms.

Solution: We must transform from base 10 to base e. Thus, we use the change-of-base formula (Property 7) with $b = 10$, $m = x$, and $a = e$:

$$\log x = \log_{10} x = \frac{\log_e x}{\log_e 10} = \frac{\ln x}{\ln 10}$$

NOW WORK PROBLEM 49 ●●●

T E C H N O L O G Y

Problem: Display the graph of $y = \log_2 x$.

Solution: To enter the function, we must first convert it to base e or base 10. We choose base e. By Property 7,

$$y = \log_2 x = \frac{\log_e x}{\log_e 2} = \frac{\ln x}{\ln 2}$$

Now we graph $y = (\ln x)/(\ln 2)$, which is shown in Figure 4.22.

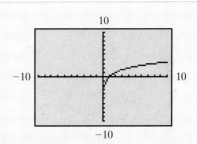

FIGURE 4.22   Graph of $y = \log_2 x$.

Problems 4.3

In Problems 1–10, let $\log 2 = a$, $\log 3 = b$, and $\log 5 = c$. Express the indicated logarithm in terms of a, b, and c.

1. $\log 30$

2. $\log 16$

*3. $\log \dfrac{2}{3}$

4. $\log \dfrac{5}{2}$

5. $\log \dfrac{8}{3}$

6. $\log \dfrac{6}{25}$

7. $\log 36$

8. $\log 0.00003$

9. $\log_2 3$

10. $\log_3 5$

In Problems 11–20, determine the value of the expression without the use of a calculator.

11. $\log_7 7^{48}$

12. $\log_5 (5\sqrt{5})^5$

13. $\log 0.0000001$

14. $10^{\log 3.4}$

15. $\ln e^{5.01}$

16. $\ln e$

17. $\ln \dfrac{1}{e^2}$

18. $\log_3 81$

*19. $\log \frac{1}{10} + \ln e^3$

20. $e^{\ln \pi}$

In Problems 21–32, write the expression in terms of $\ln x$, $\ln(x + 1)$, and $\ln(x + 2)$.

21. $\ln(x(x + 1)^2)$

22. $\ln \dfrac{\sqrt{x}}{x + 1}$

23. $\ln \dfrac{x^2}{(x + 1)^3}$

24. $\ln(x(x + 1))^3$

25. $\ln\left(\dfrac{x+1}{x+2}\right)^4$

26. $\ln\sqrt{x(x+1)(x+2)}$

27. $\ln\dfrac{x}{(x+1)(x+2)}$

28. $\ln\dfrac{x^2(x+1)}{x+2}$

*__29.__ $\ln\dfrac{\sqrt{x}}{(x+1)^2(x+2)^3}$

30. $\ln\dfrac{x}{(x+1)(x+2)}$

31. $\ln\left(\dfrac{1}{x+2}\sqrt[5]{\dfrac{x^2}{x+1}}\right)$

32. $\ln\sqrt[3]{\dfrac{x^3(x+2)^2}{(x+1)^3}}$

In Problems 33–40, express each of the given forms as a single logarithm.

33. $\log 6 + \log 4$

34. $\log_3 10 - \log_3 5$

35. $\log_2(2x) - \log_2(x+1)$

36. $2\log x - \dfrac{1}{2}\log(x-2)$

*__37.__ $5\log_2 10 + 2\log_2 13$

38. $5(2\log x + 3\log y - 2\log z)$

39. $2 + 10\log 1.05$

40. $\dfrac{1}{2}(\log 215 + 8\log 6 - 3\log 169)$

In Problems 41–44, determine the values of the expressions without using a calculator.

*__41.__ $e^{4\ln 3 - 3\ln 4}$

42. $\log_2[\ln(\sqrt{5+e^2}+\sqrt{5}) + \ln(\sqrt{5+e^2}-\sqrt{5})]$

43. $\log_6 54 - \log_6 9$

44. $\log_3\sqrt{3} - \log_2\sqrt[3]{2} - \log_5\sqrt[4]{5}$

In Problems 45–48, find x.

*__45.__ $e^{\ln(2x)} = 5$

46. $4^{\log_4 x + \log_4 2} = 3$

47. $10^{\log x^2} = 4$

48. $e^{3\ln x} = 8$

In Problems 49–53, write each expression in terms of natural logarithms.

*__49.__ $\log_2(2x+1)$

50. $\log_3(x^2+2x+2)$

51. $\log_3(x^2+1)$

52. $\log_5(9-x^2)$

53. If $e^{\ln z} = 7e^y$, solve for y in terms of z.

54. Statistics   In statistics, the sample regression equation $y = ab^x$ is reduced to a linear form by taking logarithms of both sides. Express $\log y$ in terms of x, $\log a$, and $\log b$ and explain what is meant by saying that the resulting expression is linear.

55. Military Compensation   In a study of military enlistments, Brown[7] considers total military compensation C as the sum of basic military compensation B (which includes the value of allowances, tax advantages, and base pay) and educational benefits E. Thus, $C = B + E$. Brown states that

$$\ln C = \ln B + \ln\left(1 + \dfrac{E}{B}\right)$$

Verify this.

56. Earthquake   According to Richter,[8] the magnitude M of an earthquake occurring 100 km from a certain type of seismometer is given by $M = \log(A) + 3$, where A is the recorded trace amplitude (in millimeters) of the quake. (a) Find the magnitude of an earthquake that records a trace amplitude of 10 mm. (b) If a particular earthquake has amplitude A_1 and magnitude M_1, determine the magnitude of a quake with amplitude $10A_1$ in terms of M_1.

57. Display the graph of $y = \log_6 x$.

58. Display the graph of $y = \log_4(x+2)$.

59. Display the graphs of $y = \log x$ and $y = \dfrac{\ln x}{\ln 10}$ on the same screen. The graphs appear to be identical. Why?

60. On the same screen, display the graphs of $y = \ln x$ and $y = \ln(4x)$. It appears that the graph of $y = \ln(4x)$ is the graph of $y = \ln x$ shifted upward. Determine algebraically the value of this shift.

61. On the same screen, display the graphs of $y = \ln(2x)$ and $y = \ln(6x)$. It appears that the graph of $y = \ln(6x)$ is the graph of $y = \ln(2x)$ shifted upward. Determine algebraically the value of this shift.

OBJECTIVE

To develop techniques for solving logarithmic and exponential equations.

4.4 Logarithmic and Exponential Equations

Here we solve *logarithmic* and *exponential equations*. A **logarithmic equation** is an equation that involves the logarithm of an expression containing an unknown. For example, $2\ln(x+4) = 5$ is a logarithmic equation. On the other hand, an **exponential equation** has the unknown appearing in an exponent, as in $2^{3x} = 7$.

To solve some logarithmic equations, it is convenient to use the fact that, for any base b, the function $y = \log_b x$ is one-to-one. This means of course that

$$\text{if}\quad \log_b m = \log_b n \quad\text{then}\quad m = n$$

This is visually apparent by inspecting the two possible shapes of $y = \log_b x$ given in Figure 4.19. In either event, it is clear that the function passes the horizontal line test of Section 2.5. On the other hand, we had already observed that the exponential

---

[7]C. Brown, "Military Enlistments: What Can We Learn from Geographic Variation?" *The American Economic Review,* 75, no. 1 (1985), 228–34.

[8]C. F. Richter, *Elementary Seismology* (San Francisco: W. H. Freeman and Company, 1958).

functions $y = b^x$ are one-to-one, meaning that

$$\text{if} \quad b^m = b^n \quad \text{then} \quad m = n$$

so each has an inverse, namely $y = \log_b x$, and since $(f^{-1})^{-1} = f$, each function $y = \log_b x$ has an inverse and is thus one-to-one. Also useful for solving logarithmic and exponential equations are the fundamental equations (1) and (2) in Section 4.2. (We have been deliberately repetitious here to review some basics.)

● EXAMPLE 1 Oxygen Composition

An experiment was conducted with a particular type of small animal.[9] The logarithm of the amount of oxygen consumed per hour was determined for a number of the animals and was plotted against the logarithms of the weights of the animals. It was found that

$$\log y = \log 5.934 + 0.885 \log x$$

where y is the number of microliters of oxygen consumed per hour and x is the weight of the animal (in grams). Solve for y.

Solution: We first combine the terms on the right side into a single logarithm:

$$\log y = \log 5.934 + 0.885 \log x$$
$$= \log 5.934 + \log x^{0.885} \qquad \text{(Property 3 of Section 4.3)}$$
$$\log y = \log(5.934 x^{0.885}) \qquad \text{(Property 1 of Section. 4.3)}$$

Since log is one-to-one, we have

$$y = 5.934 x^{0.885}$$

NOW WORK PROBLEM 1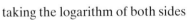

SOLVING AN EXPONENTIAL EQUATION

Greg took a number and multiplied it by a power of 32. Jean started with the same number and got the same result when she multiplied it by 4 raised to a number that was nine less than three times the exponent that Greg used. What power of 32 did Greg use?

● EXAMPLE 2 Solving an Exponential Equation

Find x if $(25)^{x+2} = 5^{3x-4}$.

Solution: Since $25 = 5^2$, we can express both sides of the equation as powers of 5:

$$(25)^{x+2} = 5^{3x-4}$$
$$(5^2)^{x+2} = 5^{3x-4}$$
$$5^{2x+4} = 5^{3x-4}$$

Since 5^x is a one-to-one function,

$$2x + 4 = 3x - 4$$
$$x = 8$$

NOW WORK PROBLEM 7

USING LOGARITHMS TO SOLVE AN EXPONENTIAL EQUATION

The sales manager at a fast-food chain finds that breakfast sales begin to fall after the end of a promotional campaign. The sales in dollars as a function of the number of days d after the campaign's end is given by $S = 800 \left(\dfrac{4}{3} \right)^{-0.1d}$. If the manager does not want sales to drop below 450 per day before starting a new campaign, when should he start such a campaign?

Some exponential equations can be solved by taking the logarithm of both sides after the equation is put in a desirable form. The following example illustrates.

● EXAMPLE 3 Using Logarithms to Solve an Exponential Equation

Solve $5 + (3)4^{x-1} = 12$.

Solution: We first isolate the exponential expression 4^{x-1} on one side of the equation:

$$5 + (3)4^{x-1} = 12$$
$$(3)4^{x-1} = 7$$
$$4^{x-1} = \frac{7}{3}$$

---

[9]R. W. Poole, *An Introduction to Quantitative Ecology* (New York: McGraw-Hill Book Company, 1974).

Now we take the natural logarithm of both sides:

$$\ln 4^{x-1} = \ln \frac{7}{3}$$

Simplifying gives

$$(x-1)\ln 4 = \ln \frac{7}{3}$$

$$x - 1 = \frac{\ln \frac{7}{3}}{\ln 4}$$

$$x = \frac{\ln \frac{7}{3}}{\ln 4} + 1 \approx 1.61120$$

NOW WORK PROBLEM 13 ●●●

In Example 3, we used natural logarithms to solve the given equation. However, logarithms in any base can be employed. Generally, natural or common logarithms are used if a decimal form of the solution is desired. If we used common logarithms, we would obtain

$$x = \frac{\log \frac{7}{3}}{\log 4} + 1 \approx 1.61120$$

T E C H N O L O G Y

Figure 4.23 shows a graphical solution of the equation $5 + (3)4^{x-1} = 12$ of Example 3. This solution occurs at the intersection of the graphs of $y = 5 + (3)4^{x-1}$ and $y = 12$.

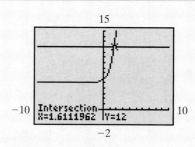

FIGURE 4.23   The solution of $5 + (3)4^{x-1} = 12$ is approximately 1.61120.

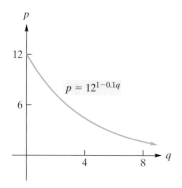

FIGURE 4.24   Graph of the demand equation $p = 12^{1-0.1q}$.

● **EXAMPLE 4   Demand Equation**

The demand equation for a product is $p = 12^{1-0.1q}$. Use common logarithms to express q in terms of p.

Solution: Figure 4.24 shows the graph of this demand equation for $q \geq 0$. As is typical of a demand equation, the graph falls from left to right. We want to solve the equation for q. Taking the common logarithms of both sides of $p = 12^{1-0.1q}$ gives

$$\log p = \log(12^{1-0.1q})$$

$$\log p = (1 - 0.1q)\log 12$$

$$\frac{\log p}{\log 12} = 1 - 0.1q$$

$$0.1q = 1 - \frac{\log p}{\log 12}$$

$$q = 10\left(1 - \frac{\log p}{\log 12}\right)$$

NOW WORK PROBLEM 43 ●●●

To solve some exponential equations involving base e or base 10, such as $10^{2x} = 3$, the process of taking logarithms of both sides may be combined with the identity $\log_b b^r = r$ (Fundamental Equation (1) from Section 4.2) to transform the equation into an equivalent logarithmic form. In this case, we have

$$10^{2x} = 3$$

$$2x = \log 3 \qquad \text{(logarithmic form)}$$

$$x = \frac{\log 3}{2} \approx 0.2386$$

● **EXAMPLE 5   Predator–Prey Relation**

In an article concerning predators and prey, Holling[10] refers to an equation of the form

$$y = K(1 - e^{-ax})$$

where x is the prey density, y is the number of prey attacked, and K and a are constants. Verify his claim that

$$\ln \frac{K}{K - y} = ax$$

Solution: To find ax, we first solve the given equation for e^{-ax}:

$$y = K(1 - e^{-ax})$$

$$\frac{y}{K} = 1 - e^{-ax}$$

$$e^{-ax} = 1 - \frac{y}{K}$$

$$e^{-ax} = \frac{K - y}{K}$$

Now we convert to logarithmic form:

$$\ln \frac{K - y}{K} = -ax$$

$$-\ln \frac{K - y}{K} = ax$$

$$\ln \frac{K}{K - y} = ax \qquad \text{(Property 4 of Section 4.3)}$$

as was to be shown.

NOW WORK PROBLEM 7 ●●

The Richter scale measure of an earthquake is given by $R = \log\left(\dfrac{I}{I_0}\right)$, where I is the intensity of the earthquake, and I_0 is the intensity of a zero-level earthquake. An earthquake that is 675,000 times as intense as a zero-level earthquake has a magnitude on the Richter scale that is 4 more than another earthquake. What is the intensity of the other earthquake?

Some logarithmic equations can be solved by rewriting them in exponential forms.

● **EXAMPLE 6   Solving a Logarithmic Equation**

Solve $\log_2 x = 5 - \log_2(x + 4)$.

Solution: Here we must assume that both x and $x + 4$ are positive, so that their logarithms are defined. Both conditions are satisfied if $x > 0$. To solve the equation, we first place all logarithms on one side so that we can combine them:

$$\log_2 x + \log_2(x + 4) = 5$$

$$\log_2[x(x + 4)] = 5$$

[10]C. S. Holling, "Some Characteristics of Simple Types of Predation and Parasitism," *The Canadian Entomologist,* 91, no. 7 (1959), 385–98.

In exponential form, we have

$$x(x+4) = 2^5$$
$$x^2 + 4x = 32$$
$$x^2 + 4x - 32 = 0 \quad \text{(quadratic equation)}$$
$$(x-4)(x+8) = 0$$
$$x = 4 \quad \text{or} \quad x = -8$$

Because we must have $x > 0$, the only solution is 4, as can be verified by substituting into the original equation. Indeed, replacing x by 4 in $\log_2 x$ gives $\log_2 4 = \log_2 2^2 = 2$ while replacing x by 4 in $5 - \log_2(x+4)$ gives $5 - \log_2(4+4) = 5 - \log_2(8) = 5 - \log_2 2^3 = 5 - 3 = 2$. Since the results are the same, 4 is a solution of the equation.

In solving a logarithmic equation, it is a good idea to check for extraneous solutions.

NOW WORK PROBLEM 5 ●●●

Problems 4.4

In Problems 1–36, find x. Round your answers to three decimal places.

*1. $\log(3x+2) = \log(2x+5)$ 2. $\log x - \log 5 = \log 7$

3. $\log 7 - \log(x-1) = \log 4$ 4. $\log_2 x + 3\log_2 2 = \log_2 \frac{2}{x}$

*5. $\ln(-x) = \ln(x^2 - 6)$ 6. $\ln(4-x) + \ln 2 = 2\ln x$

*7. $e^{2x} \cdot e^{5x} = e^{14}$ 8. $(e^{3x-2})^3 = e^3$ 9. $(81)^{4x} = 9$

10. $(27)^{2x+1} = \dfrac{1}{3}$ 11. $e^{2x} = 9$ 12. $e^{4x} = \frac{3}{4}$

*13. $2e^{5x+2} = 17$ 14. $5e^{2x-1} - 2 = 23$ 15. $10^{4/x} = 6$

16. $\dfrac{4(10)^{0.2x}}{5} = 3$ 17. $\dfrac{5}{10^{2x}} = 7$

18. $2(10)^x + (10)^{x+1} = 4$ 19. $2^x = 5$

20. $7^{2x+3} = 9$ 21. $7^{3x-2} = 5$

22. $4^{x/2} = 20$ 23. $2^{-2x/3} = \dfrac{4}{5}$

24. $5(3^x - 6) = 10$ 25. $(4)5^{3-x} - 7 = 2$

26. $\dfrac{7}{3^x} = 13$ 27. $\log(x-3) = 3$

28. $\log_2(x+1) = 4$ 29. $\log_4(9x-4) = 2$

30. $\log_4(2x+4) - 3 = \log_4 3$ 31. $\log(3x-1) - \log(x-3) = 2$

32. $\log(x-3) + \log(x-5) = 1$

33. $\log_2(5x+1) = 4 - \log_2(3x-2)$

34. $\log(x+2)^2 = 2$, where $x > 0$

35. $\log_2\left(\dfrac{2}{x}\right) = 3 + \log_2 x$ 36. $\ln(x-2) = \ln(2x-1) + 3$

37. **Rooted Plants**   In a study of rooted plants in a certain geographic region,[11] it was determined that on plots of size A (in square meters), the average number of species that occurred was S. When $\log S$ was graphed as a function of $\log A$, the result was a straight line given by

$$\log S = \log 12.4 + 0.26 \log A$$

Solve for S.

38. **Gross National Product**   In an article, Taagepera and Hayes refer to an equation of the form

$$\log T = 1.7 + 0.2068 \log P - 0.1334(\log P)^2$$

Here T is the percentage of a country's gross national product (GNP) that corresponds to foreign trade (exports plus imports), and P is the country's population (in units of 100,000).[12] Verify the claim that

$$T = 50 P^{(0.2068 - 0.1334 \log P)}$$

You may assume that $\log 50 = 1.7$. Also verify that, for any base b, $(\log_b x)^2 = \log_b(x^{\log_b x})$.

39. **Radioactivity**   The number of milligrams of a radioactive substance present after t years is given by

$$Q = 100e^{-0.035t}$$

(a) How many milligrams are present after 0 years?
(b) After how many years will there be 20 milligrams present? Give your answer to the nearest year.

40. **Blood Sample**   On the surface of a glass slide is a grid that divides the surface into 225 equal squares. Suppose a blood sample containing N red cells is spread on the slide and the cells are randomly distributed. Then the number of squares containing no cells is (approximately) given by $225e^{-N/225}$. If 100 of the squares contain no cells, estimate the number of cells the blood sample contained.

41. **Population**   In one city, the population P grows at the rate of 2% per year. The equation $P = 1,000,000(1.02)^t$ gives the population t years after 1998. Find the value of t for which the population is 1,500,000. Give your answer to the nearest tenth.

[11]R. W. Poole, *An Introduction to Quantitative Ecology* (New York: McGraw-Hill Book Company, 1974).

[12]R. Taagepera and J. P. Hayes, "How Trade/GNP Ratio Decreases with Country Size," *Social Science Research*, 6 (1977), 108–32.

42. Market Penetration In a discussion of market penetration by new products, Hurter and Rubenstein[13] refer to the function

$$F(t) = \frac{q - pe^{-(t+C)(p+q)}}{q[1 + e^{(t+C)(p+q)}]}$$

where p, q, and C are constants. They claim that if $F(0) = 0$, then

$$C = -\frac{1}{p+q}\ln\frac{q}{p}$$

Show that their claim is true.

*43. Demand Equation The demand equation for a consumer product is $q = 80 - 2^p$. Solve for p and express your answer in terms of common logarithms, as in Example 4. Evaluate p to two decimal places when $q = 60$.

44. Investment The equation $A = P(1.105)^t$ gives the value A at the end of t years of an investment of P dollars compounded annually at an annual interest rate of 10.5%. How many years will it take for an investment to double? Give your answer to the nearest year.

45. Sales After t years the number of units of a product sold per year is given by $q = 1000\left(\frac{1}{2}\right)^{0.8^t}$. Such an equation is called a *Gompertz equation* and describes natural growth in many areas of study. Solve this equation for t in the same manner as in Example 4, and show that

$$t = \frac{\log\left(\frac{3 - \log q}{\log 2}\right)}{\log 0.8}$$

Also, for any A and suitable b and a, solve $y = Ab^{a^x}$ for x and explain how the previous solution is a special case.

46. Learning Equation Suppose that the daily output of units of a new product on the tth day of a production run is given by

$$q = 500(1 - e^{-0.2t})$$

Such an equation is called a *learning equation* and indicates that as time progresses, output per day will increase. This may be due to a gain in a worker's proficiency at his or her job. Determine, to the nearest complete unit, the output on (a) the first day and (b) the tenth day after the start of a production run. (c) After how many days will a daily production run of 400 units be reached? Give your answer to the nearest day.

47. Verify that 4 is the only solution to the logarithmic equation in Example 6 by graphing the function

$$y = 5 - \log_2(x + 4) - \log_2 x$$

and observing when $y = 0$.

48. Solve $2^{3x+0.5} = 17$. Round your answer to two decimal places.

49. Solve $\ln(x + 2) = 5 - x$. Round your answer to two decimal places.

50. Graph the equation $(3)2^y - 4x = 5$. (*Hint:* Solve for y as a function of x.)

4.5 Review

Important Terms and Symbols — Examples

Section 4.1	**Exponential Functions**	
	exponential function, b^x, for $b > 1$ and for $0 < b < 1$	Ex. 2,3, p. 164,165
	compound interest principal compound amount	Ex. 6, p. 167
	interest period periodic rate nominal rate	
	e natural exponential function, e^x	Ex. 8, p. 170
	exponential law of decay initial amount decay constant half-life	Ex. 11, p. 172
Section 4.2	**Logarithmic Functions**	
	logarithmic function, $\log_b x$ common logarithm, $\log x$	Ex. 5, p. 178
	natural logarithm, $\ln x$	Ex. 5, p. 178
Section 4.3	**Properties of Logarithms**	
	change-of-base formula	Ex. 8, p. 185
Section 4.4	**Logarithmic and Exponential Equations**	
	logarithmic equation exponential equation	Ex. 1, p. 187

Summary

An exponential function has the form $f(x) = b^x$. The graph of $f(x) = b^x$ has one of two general shapes, depending on the value of the base b. (See Figure 4.3.) An exponential function is involved in the compound interest formula

$$S = P(1 + r)^n$$

where S is the compound amount of a principal of P at the end of n interest periods at the periodic rate r.

A frequently used base in an exponential function is the irrational number $e \approx 2.71828$. This base occurs in economic analysis and many situations involving growth or decay, such as population studies and radioactive decay. Radioactive elements follow the exponential law of decay,

$$N = N_0 e^{-\lambda t}$$

[13]A. P. Hurter, Jr., A. H. Rubenstein, et al., "Market Penetration by New Innovations: The Technological Literature," *Technological Forecasting and Social Change*, 11 (1978), 197–221.

where N is the amount of an element present at time t, N_0 is the initial amount, and λ is the decay constant. The time required for half of the amount of the element to decay is called the half-life.

The logarithmic function is the inverse function of the exponential function, and vice versa. The logarithmic function with base b is denoted $\log_b$, and $y = \log_b x$ if and only if $b^y = x$. The graph of $y = \log_b x$ has one of two general shapes, depending on the value of the base b. (See Figure 4.19.) Logarithms with base e are called natural logarithms and are denoted ln; those with base 10 are called common logarithms and are denoted log. The half-life T of a radioactive element can be given in terms of a natural logarithm and the decay constant: $T = (\ln 2)/\lambda$.

Some important properties of logarithms are the following:

$$\log_b(mn) = \log_b m + \log_b n$$

$$\log_b \frac{m}{n} = \log_b m - \log_b n$$

$$\log_b m^r = r \log_b m$$

$$\log_b \frac{1}{m} = -\log_b m$$

$$\log_b 1 = 0$$

$$\log_b b = 1$$

$$\log_b b^r = r$$

$$b^{\log_b m} = m$$

$$\log_b m = \frac{\log_a m}{\log_a b}$$

Moreover, if $\log_b m = \log_b n$, then $m = n$. Similarly, if $b^m = b^n$, then $m = n$. Many of these properties are used in solving logarithmic and exponential equations.

Review Problems

Problem numbers shown in color indicate problems suggested for use as a practice chapter test.

In Problems 1–6, write each exponential form logarithmically and each logarithmic form exponentially.

1. $3^5 = 243$ **2.** $\log_5 625 = 4$ **3.** $\log_{81} 3 = \frac{1}{4}$

4. $10^5 = 100,000$ **5.** $e^4 = 54.598$ **6.** $\log_9 9 = 1$

In Problems 7–12, find the value of the expression without using a calculator.

7. $\log_5 125$ **8.** $\log_4 16$ **9.** $\log_3 \frac{1}{81}$

10. $\log_{1/4} \frac{1}{64}$ **11.** $\log_{1/3} 9$ **12.** $\log_4 2$

In Problems 13–18, find x without using a calculator.

13. $\log_5 625 = x$ **14.** $\log_x \frac{1}{81} = -4$ **15.** $\log_2 x = -5$

16. $\ln \frac{1}{e} = x$ **17.** $\ln(2x + 3) = 0$ **18.** $e^{\ln(x+4)} = 7$

In Problems 19 and 20, let $\log 2 = a$ and $\log 3 = b$. Express the given logarithm in terms of a and b.

19. $\log 8000$ **20.** $\log \frac{9}{\sqrt{2}}$

In Problems 21–26, write each expression as a single logarithm.

21. $3 \log 7 - 2 \log 5$ **22.** $5 \ln x + 2 \ln y + \ln z$

23. $2 \ln x + \ln y - 3 \ln z$ **24.** $\log_6 2 - \log_6 4 - 9 \log_6 3$

25. $\frac{1}{2} \log_2 x + 2 \log_2(x^2) - 3 \log_2(x + 1) - 4 \log_2(x + 2)$

26. $4 \log x + 2 \log y - 3(\log z + \log w)$

In Problems 27–32, write the expression in terms of $\ln x$, $\ln y$, and $\ln z$.

27. $\ln \frac{x^3 y^2}{z^{-5}}$ **28.** $\ln \frac{\sqrt{x}}{(yz)^2}$ **29.** $\ln \sqrt[3]{xyz}$

30. $\ln \left[\frac{xy^3}{z^2} \right]^4$ **31.** $\ln \left[\frac{1}{x} \sqrt{\frac{y}{z}} \right]$ **32.** $\ln \left[\left(\frac{x}{y} \right)^2 \left(\frac{x}{z} \right)^3 \right]$

33. Write $\log_3(x + 5)$ in terms of natural logarithms.

34. Write $\log_2(7x^3 + 5)$ in terms of common logarithms.

35. Suppose that $\log_2 19 = 4.2479$ and $\log_2 5 = 2.3219$. Find $\log_5 19$.

36. Use natural logarithms to determine the value of $\log_4 5$.

37. If $\ln 3 = x$ and $\ln 4 = y$, express $\ln(16\sqrt{3})$ in terms of x and y.

38. Express $\log \dfrac{x^2 \sqrt[3]{x + 1}}{\sqrt[5]{x^2 + 2}}$ in terms of $\log x$, $\log(x + 1)$, and $\log(x^2 + 2)$.

39. Simplify $10^{\log x} + \log 10^x + \log 10$.

40. Simplify $\log 10^2 + \log 1000 - 5$.

41. If $\ln y = x^2 + 2$, find y.

42. Sketch the graphs of $y = 3^x$ and $y = \log_3 x$.

43. Sketch the graph of $y = 2^{x+3}$.

44. Sketch the graph of $y = -2 \log_2 x$.

In Problems 45–52, find x.

45. $\log(5x + 1) = \log(4x + 6)$ **46.** $\log 3x + \log 3 = 2$

47. $3^{4x} = 9^{x+1}$ **48.** $4^{3-x} = \frac{1}{16}$

49. $\log x + \log(10x) = 3$ **50.** $\log_2(x+4) = \log_2(x-2) + 3$

51. $\ln(\log_x 3) = 2$ **52.** $\log_2 x + \log_4 x = 3$

In Problems 53–58, find x. Round your answers to three decimal places.

53. $e^{3x} = 14$ **54.** $10^{3x/2} = 5$ **55.** $3(10^{x+4} - 3) = 9$

56. $7e^{3x-1} - 2 = 1$ **57.** $4^{x+3} = 7$ **58.** $3^{5/x} = 2$

59. Investment If $2600 is invested for $6\frac{1}{2}$ years at 6% compounded quarterly, find (a) the compound amount and (b) the compound interest.

60. Investment Find the compound amount of an investment of $4000 for five years at the rate of 11% compounded monthly.

61. Find the nominal rate that corresponds to a periodic rate of $1\frac{1}{6}\%$ per month.

62. Bacteria Growth Bacteria are growing in a culture, and their number is increasing at the rate of 5% an hour. Initially, 600 bacteria are present. (a) Determine an equation that gives the number, N, of bacteria present after t hours. (b) How many bacteria are present after one hour? (c) After five hours? Give your answer to (c) to the nearest integer.

63. Population Growth The population of a small town *grows* at the rate of -0.5% per year because the outflow of people to nearby cities in search of jobs exceeds the birth rate. In

2006 the population was 6000. (a) Determine an equation that gives the population, P, t years from 2006. (b) Find what the population will be in 2016 (be careful to express your answer as an integer).

64. **Revenue**   Due to ineffective advertising, the Kleer-Kut Razor Company finds that its annual revenues have been cut sharply. Moreover, the annual revenue, R, at the end of t years of business satisfies the equation $R = 200,000e^{-0.2t}$. Find the annual revenue at the end of two years and at the end of three years.

65. **Radioactivity**   A radioactive substance decays according to the formula

$$N = 10e^{-0.41t}$$

where N is the number of milligrams present after t hours. (a) Determine the initial amount of the substance present. (b) To the nearest tenth of a milligram, determine the amount present after 2 hours. (c) After 10 hours. (d) To the nearest tenth of an hour, determine the half-life of the substance, and (e) determine the number of hours for 1 milligram to remain.

66. **Radioactivity**   If a radioactive substance has a half-life of 10 days, in how many days will $\frac{1}{8}$ of the initial amount be present?

67. **Marketing**   A marketing-research company needs to determine how people adapt to the taste of a new cough drop. In one experiment, a person was given a cough drop and was asked periodically to assign a number, on a scale from 0 to 10, to the perceived taste. This number was called the *response magnitude*. The number 10 was assigned to the initial taste. After conducting the experiment several times, the company estimated that the response magnitude is given by

$$R = 10e^{-t/40}$$

where t is the number of seconds after the person is given the cough drop. (a) Find the response magnitude after 20 seconds. Give your answer to the nearest integer. (b) After how many seconds does a person have a response magnitude of 5? Give your answer to the nearest second.

68. **Sediment in Water**   The water in a midwestern lake contains sediment, and the presence of the sediment reduces the transmission of light through the water. Experiments indicate that the intensity of light is reduced by 10% by passage through 20 cm of water. Suppose that the lake is uniform with respect to the amount of sediment contained by the water. A measuring instrument can detect light at the intensity of 0.17% of full sunlight. This measuring instrument is lowered into the lake. At what depth will it first cease to record the presence of light? Give your answer to the nearest 10 cm.

69. **Body Cooling**   In a discussion of the rate of cooling of isolated portions of the body when they are exposed to low temperatures, there occurs the equation[14]

$$T_t - T_e = (T_t - T_e)_o e^{-at}$$

where T_t is the temperature of the portion at time t, T_e is the environmental temperature, the subscript o refers to the initial temperature difference, and a is a constant. Show that

$$a = \frac{1}{t} \ln \frac{(T_t - T_e)_o}{T_t - T_e}$$

70. **Depreciation**   An alternative to straight-line depreciation is *declining-balance* depreciation. This method assumes that an item loses value more steeply at the beginning of its life than later on. A fixed percentage of the value is subtracted each month. Suppose an item's initial cost is C and its useful life is N months. Then the value, V (in dollars), of the item at the end of n months is given by

$$V = C\left(1 - \frac{1}{N}\right)^n$$

so that each month brings a depreciation of $\frac{100}{N}$ percent. (This is called *single declining-balance depreciation;* if the annual depreciation were $\frac{200}{N}$ percent, then we would speak of *double-declining-balance.*) A notebook computer is purchased for $1800 and has a useful life of 48 months. It undergoes double declining-balance depeciation. After how many months, to the nearest integer, does its value drop below $700?

71. If $y = f(x) = \dfrac{\ln x}{x}$, determine the range of f. Round values to two decimal places.

72. Determine the points of intersection of the graphs of $y = \ln(x + 2)$ and $y = x^2 - 7$. Round your answers to two decimal places.

73. Solve $\ln x = 6 - 2x$. Round your answer to two decimal places.

74. Solve $6^{3-4x} = 15$. Round your answer to two decimal places.

75. Display the graph of $y = \log_2(x^2 + 1)$, observing that symmetry with respect to the y-axis and the range of this function allows you to restrict your display to the first quadrant.

76. Display the graph of the equation $(6)5^y + x = 2$. (*Hint:* Solve for y as an explicit function of x.)

77. Graph $y = 3^x$ and $y = \dfrac{3^x}{9}$ on the same screen. It appears that the graph of $y = \dfrac{3^x}{9}$ is the graph of $y = 3^x$ shifted two units to the right. Prove algebraically that this is indeed true.

---

[14]R. W. Stacy et al., *Essentials of Biological and Medical Physics* (New York: McGraw-Hill Book Company, 1955).

Mathematical Snapshot

Drug Dosages[15]

Determining and prescribing drug dosages are extremely important aspects of the medical profession. Quite often, caution must be taken because of possible adverse side or toxic effects of drugs.

Many drugs are used up by the human body in such a way that the amount present follows an *exponential law of decay* as studied in Section 4.1. That means, if $N(t)$ is the amount of the drug present in the body at time t, then

$$N = N_0 e^{-kt} \qquad (1)$$

where k is a positive constant and N_0 is the amount present at time $t = 0$. If H is the *half-life* of such a drug, meaning the time H for which $N(H) = N_0/2$, then again from Section 4.1,

$$H = (\ln 2)/k \qquad (2)$$

Note that H completely determines the constant k since we can rewrite Equation (2) as $k = (\ln 2)/H$.

Suppose that you want to analyze the situation whereby equal doses of such a drug are introduced into a patient's system every I units of time until a therapeutic level is attained, and then the dosage is reduced to maintain the therapeutic level. The reason for *reduced* maintenance doses is frequently related to the toxic effects of drugs.

In particular, assume that

(i) There are d doses of P units each;

(ii) A dose is given at times $t = 0, I, 2I, \ldots,$ and $(d-1)I$; and

(iii) The therapeutic level, T, is attained at time $t = dI$ (which is one time interval, I, after the last dose is administered).

We now determine a formula that gives the therapeutic level, T. At time $t = 0$ the patient receives the first P units, so the amount of drug in the body is P at $t = 0$. At time $t = I$ the amount present from the first dose is Pe^{-kI} [by Equation (1)]. In addition, at $t = I$ the second P units are given. Thus the *total* amount of the drug present at $t = I$ is

$$P + Pe^{-kI}$$

At time $t = 2I$, the amount remaining from the first dose is Pe^{-2kI}; from the second dose, which has been in the system for only one time interval, the amount present is Pe^{-kI}. Also, at time $t = 2I$ the third dose of P units is given, so

the total amount of the drug present at $t = 2I$ is

$$P + Pe^{-kI} + Pe^{-2kI}$$

Continuing, the amount of drug present in the system at time $(d-1)I$, the time of the last dose, is

$$P + Pe^{-kI} + Pe^{-2kI} + \cdots + Pe^{-(d-1)\cdot kI}$$

Thus one further time interval later, at time dI, when a dose of P is not administered but the therapeutic level T is reached, we have

$$T = Pe^{-kI} + Pe^{-2kI} + \cdots + Pe^{-dkI} \qquad (3)$$

since each term in the preceding expression decays by a factor of e^{-kI}. This is a good opportunity to use the *summation notation* of Section 1.5 and rewrite Equation (3) as

$$T = P \sum_{i=1}^{d} e^{-ikI} \qquad (4)$$

[15]This discussion is adapted from Gerald M. Armstrong and Calvin P. Midgley, "The Exponential-Decay Law Applied to Medical Dosages," *The Mathematics Teacher*, 80, no. 3 (February 1987), 110–13. By permission of the National Council of Teachers of Mathematics.

The sum is of a special kind that we will study later, but observe that multiplying both sides of Equation (4) by e^{-kI} gives

$$e^{-kI}T = P\sum_{i=2}^{d+1} e^{-ikI} \qquad (5)$$

If we subtract corresponding sides of Equation (5) from those of Equation (4) and common factor, we obtain

$$(1-e^{-kI})T = P\left(\sum_{i=1}^{d} e^{-ikI} - \sum_{i=2}^{d+1} e^{-ikI}\right) = P(e^{-kI} - e^{-(d+1)kI})$$

and you should be sure to check that most of the terms of the two sums *cancel* as indicated. Continuing, we have

$$(1 - e^{-kI})T = Pe^{-kI}(1 - e^{-dkI})$$

$$T = \frac{Pe^{-kI}(1 - e^{-dkI})}{1 - e^{-kI}} \qquad (6)$$

$$T = \frac{P(1 - e^{-dkI})}{e^{kI}(1 - e^{-kI})}$$

$$T = \frac{P(1 - e^{-dkI})}{e^{kI} - 1} \qquad (7)$$

Equation (7) expresses the therapeutic level, T, in terms of the dose, P; the length of the time intervals, I; the number of doses, d; and the half-life, H, of the drug (since $k = (\ln 2)/H$). It is easy to solve Equation (7) for P if T, H, I, and d are known. It is also possible to solve the equation for d if T, H, I, and P are known. (Solving Equation (7) for either H or I in terms of the other quantities can be quite complicated.)

The objective now is to maintain the therapeutic level in the patient's system. To do this, a reduced dose, R, is given at times $t = dI$, $(d + 1)I$, $(d + 2)I$, and so on. We can determine a formula for R in the following way.

At time $t = (d + 1)I$, but before the second reduced dose is given, the amount of drug in the system from the first reduced dose is Re^{-kI} and the amount that remains from the therapeutic level is Te^{-kI}. Suppose you require that the sum of these amounts be the therapeutic level, T; that is,

$$T = Re^{-kI} + Te^{-kI}$$

Solving for R gives

$$Re^{-kI} = T - Te^{-kI}$$

$$R = T(1 - e^{-kI})e^{kI}$$

Replacing T by the right side of Equation (6) gives

$$R = \frac{Pe^{-kI}(1 - e^{-dkI})}{1 - e^{-kI}}(1 - e^{-kI})e^{kI}$$

which simplifies to

$$R = P(1 - e^{-dkI}) \qquad (8)$$

By continuing the reduced doses at time intervals of length I, we are assured that the drug level in the system never falls below T after time $(d + 1)I$. Furthermore, because $-dkI < 0$, we have $0 < e^{-dkI} < 1$. Consequently, the factor $1 - e^{-dkI}$ in Equation (8) is between 0 and 1. This ensures that R is less than P; hence R is indeed a *reduced* dose.

It is interesting to note that Armstrong and Midgley state that "the therapeutic amount T must be chosen from a range of empirically determined values. Medical discretion and experience are needed to select proper intervals and durations of time to administer the drug. Even the half-life of a drug can vary somewhat among different patients." More on medical drugs and their safe use can be found at www.fda.gov/cder.

Problems

1. Solve Equation (7) for (a) P and (b) d.

2. Show that if I is equal to the half-life of the drug, Equation (7) can be written as

$$T = \left(1 - \frac{1}{2^d}\right)P$$

   Note that $0 < 1 - (1/2^d) < 1$ for $d > 0$. Hence, the foregoing equation implies that when doses of P units are administered at time intervals equal to the half-life of the drug, it follows that at one time interval after any dose is given, but before the next dose is given, the total level of the drug in a patient's system is less than P.

3. Theophylline is a drug used to treat bronchial asthma and has a half-life of 8 hours in the system of a relatively healthy nonsmoking patient. Suppose that such a patient achieves the desired therapeutic level of this drug in 12 hours when 100 milligrams is administered every 4 hours. Here $d = 3$. Because of toxicity, the dose must be reduced thereafter. To the nearest milligram, determine (a) the therapeutic level and (b) the reduced dose.

4. Use a graphing calculator to generate a drug concentration graph and verify that Equation (8) correctly gives the maintenance dose. On the calculator, enter $0.5 \rightarrow$ K, $3 \rightarrow$ D, $1 \rightarrow$ I and $1 \rightarrow$ P. Then enter Y1 = P(1 − e^(−D*K*I)) to represent R. Finally, enter Y2 = P e^(−K X) + P e^(−K(X − I))*(X ≥ I) + P e^(−K(X − 2I))*(X ≥ 2I) + Y1 e^(−K(X − 3I))*(X ≥ 3I) + Y1 e^(−K(X − 4I))*(X ≥ 4I). Then select only Y2 to be graphed and graph the function. Experiment with different values for K, D, I, and P. What adjustment is necessary in the expression for Y2 as you change D?

5

MATHEMATICS OF FINANCE

 Treasury Securities

For people who like cars and can afford to buy a nice one, a trip to an auto dealership can be a lot of fun. However, buying a car also has a side that many people find unpleasant: the negotiating. The verbal tug-of-war with the salesperson is especially difficult if the buyer is planning to pay on an installment plan and does not understand the numbers being quoted.

How, for instance, does the fact that the salesperson is offering the car for $12,800 translate into a monthly payment of $281.54? The answer is amortization. The term comes via French from the Latin root *mort-*, meaning "dead"; from this we also get *mortal* and *mortified*. A debt that is gradually paid off is eventually "killed," and the payment plan for doing this is called an amortization schedule. The schedule is determined by a formula you will learn in Section 5.4 and apply in Section 5.5.

Using the formula, we can calculate the monthly payment for the car. If one makes a $900 down payment on a $12,800 car and pays the rest off over four years at 4.8% A.P.R. compounded monthly, the monthly payment for principal and interest only should be $272.97. If the payment is higher than that, it may contain additional charges such as sales tax, registration fees, or insurance premiums, which the buyer should ask about since some of them may be optional. Understanding the mathematics of finance can help consumers make more informed decisions about purchases and investments.

OBJECTIVE

To extend the notion of compound interest to include effective rates and to solve interest problems whose solutions require logarithms.

5.1 Compound Interest

In this chapter we model selected topics in finance that deal with the time value of money, such as investments, loans, and so on. In later chapters, when more mathematics is at our disposal, certain topics will be revisited and expanded.

Let us first review some facts from Section 4.1, where the notion of compound interest was introduced. Under compound interest, at the end of each interest period the interest earned for that period is added to the *principal* (the invested amount) so that it, too, earns interest over the next interest period. The basic formula for the value (or *compound amount*) of an investment after n interest periods under compound interest is as follows:

Compound Interest Formula

For an original principal of P, the formula

$$S = P(1 + r)^n \tag{1}$$

gives the **compound amount** S at the end of n *interest periods* at the *periodic rate* of r.

The compound amount is also called the *accumulated amount,* and the difference between the compound amount and the original principal, $S - P$, is called the *compound interest*.

Recall that an interest rate is usually quoted as an *annual* rate, called the *nominal rate* or the *annual percentage rate* (A.P.R.). The periodic rate (or rate per interest period) is obtained by dividing the nominal rate by the number of interest periods per year.

For example, let us compute the compound amount when $1000 is invested for 5 years at the nominal rate of 8% compounded quarterly. The rate *per period* is 0.08/4, and the number of interest periods is 5×4.

From Equation (1), we have

$$S = 1000 \left(1 + \frac{0.08}{4}\right)^{5 \times 4}$$

$$= 1000(1 + 0.02)^{20} \approx \$1485.95$$

Keep a calculator handy while you read this chapter.

PRINCIPLES IN PRACTICE 1

COMPOUND INTEREST

Suppose you leave an initial amount of $518 in a savings account for three years. If interest is compounded daily (365 times per year), use a graphing calculator to graph the compound amount S as a function of the nominal rate of interest. From the graph, estimate the nominal rate of interest so that there is $600 after three years.

● EXAMPLE 1   **Compound Interest**

Suppose that $500 amounted to $588.38 in a savings account after three years. If interest was compounded semiannually, find the nominal rate of interest, compounded semiannually, that was earned by the money.

Solution: Let r be the semiannual rate. There are $2 \times 3 = 6$ interest periods. From Equation (1),

$$500(1 + r)^6 = 588.38$$

$$(1 + r)^6 = \frac{588.38}{500}$$

$$1 + r = \sqrt[6]{\frac{588.38}{500}}$$

$$r = \sqrt[6]{\frac{588.38}{500}} - 1 \approx 0.0275$$

Thus, the semiannual rate was 2.75%, so the nominal rate was $5\frac{1}{2}$% compounded semiannually.

⬤ EXAMPLE 2   **Doubling Money**

At what nominal rate of interest, compounded yearly, will money double in eight years?

Solution: Let r be the rate at which a principal of P doubles in eight years. Then the compound amount is $2P$. From Equation (1),

$$P(1+r)^8 = 2P$$

$$(1+r)^8 = 2$$

$$1+r = \sqrt[8]{2}$$

$$r = \sqrt[8]{2} - 1 \approx 0.0905$$

Note that the doubling rate is independent of the principal P.

Hence, the desired rate is 9.05%.

◖◖

We can determine how long it takes for a given principal to accumulate to a particular amount by using logarithms, as Example 3 shows.

⬤ EXAMPLE 3   **Compound Interest**

How long will it take for $600 to amount to $900 at an annual rate of 6% compounded quarterly?

Solution: The periodic rate is $r = 0.06/4 = 0.015$. Let n be the number of interest periods it takes for a principal of $P = 600$ to amount to $S = 900$. Then, from Equation (1),

$$900 = 600(1.015)^n \qquad (2)$$

$$(1.015)^n = \frac{900}{600}$$

$$(1.015)^n = 1.5$$

To solve for n, we first take the natural logarithms of both sides:

$$\ln(1.015)^n = \ln 1.5$$

$$n \ln 1.015 = \ln 1.5 \qquad \text{since } \ln m^r = r \ln m$$

$$n = \frac{\ln 1.5}{\ln 1.015} \approx 27.233$$

The number of years that corresponds to 27.233 quarterly interest periods is $27.233/4 \approx 6.8083$, which is about 6 years, $9\frac{1}{2}$ months. Actually, the principal doesn't amount to $900 until 7 years pass, because interest is compounded quarterly.

NOW WORK PROBLEM 20   ◖◖

T E C H N O L O G Y

We can solve Equation (2) in Example 3 by graphing

$$Y_1 = 900$$

$$Y_2 = 600(1.015)^{\wedge}X$$

and finding the intersection. (See Figure 5.1.)

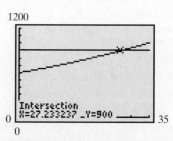

FIGURE 5.1   Solution of Example 3.

Effective Rate

If P dollars are invested at a nominal rate of 10% compounded quarterly for one year, the principal will earn more than 10% that year. In fact, the compound interest is

$$S - P = P\left(1 + \frac{0.10}{4}\right)^4 - P = [(1.025)^4 - 1]P$$
$$\approx 0.103813P$$

which is about 10.38% of P. That is, 10.38% is the approximate rate of interest compounded *annually* that is actually earned, and that rate is called the **effective rate** of interest. The effective rate is independent of P. In general, the effective interest rate is just the rate of *simple* interest earned over a period of one year. Thus, we have shown that the nominal rate of 10% compounded quarterly is equivalent to an effective rate of 10.38%. Following the preceding procedure, we can generalize our result:

Effective Rate

The **effective rate** r_e that is equivalent to a nominal rate of r compounded n times a year is given by

$$r_e = \left(1 + \frac{r}{n}\right)^n - 1 \qquad (3)$$

PRINCIPLES IN PRACTICE 3

EFFECTIVE RATE

An investment is compounded monthly. Use a graphing calculator to graph the effective rate r_e as a function of the nominal rate r. Then use the graph to find the nominal rate that is equivalent to an effective rate of 8%.

●**EXAMPLE 4** **Effective Rate**

What effective rate is equivalent to a nominal rate of 6% compounded (a) semiannually and (b) quarterly?

Solution:

a. From Equation (3), the effective rate is

$$r_e = \left(1 + \frac{0.06}{2}\right)^2 - 1 = (1.03)^2 - 1 = 0.0609 = 6.09\%$$

b. The effective rate is

$$r_e = \left(1 + \frac{0.06}{4}\right)^4 - 1 = (1.015)^4 - 1 \approx 0.061364 = 6.14\%$$

NOW WORK PROBLEM 9 ●●

Example 4 illustrates that, for a given nominal rate r, the effective rate increases as the number of interest periods per year (n) increases. However, in Section 5.3 it is shown that, regardless of how large n is, the maximum effective rate that can be obtained is $e^r - 1$.

●**EXAMPLE 5** **Effective Rate**

To what amount will \$12,000 accumulate in 15 years if it is invested at an effective rate of 5%?

Solution: Since an effective rate is the rate that is compounded annually, we have

$$S = 12,000(1.05)^{15} \approx \$24,947.14$$

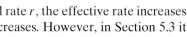

NOW WORK PROBLEM 15 ●●

● EXAMPLE 6 **Doubling Money**

How many years will it take for money to double at the effective rate of r?

Solution: Let n be the number of years it takes for a principal of P to double. Then the compound amount is $2P$. Thus,

$$2P = P(1+r)^n$$

$$2 = (1+r)^n$$

$$\ln 2 = n \ln(1+r) \qquad \text{(taking logarithms of both sides)}$$

Hence,

$$n = \frac{\ln 2}{\ln(1+r)}$$

For example, if $r = 0.06$, the number of years it takes to double a principal is

$$\frac{\ln 2}{\ln 1.06} \approx 11.9 \text{ years}$$

NOW WORK PROBLEM 11 ●○○

We remark that when alternative interest rates are available to an investor, effective rates are used to compare them—that is, to determine which of them is the "best." The next example illustrates.

PRINCIPLES IN PRACTICE 4

COMPARING INTEREST RATES

Suppose you have two investment opportunities. You can invest $10,000 at 11% compounded monthly, or you can invest $9700 at 11.25% compounded quarterly. Which has the better effective rate of interest? Which is the better investment over 20 years?

● EXAMPLE 7 **Comparing Interest Rates**

If an investor has a choice of investing money at 6% compounded daily or $6\frac{1}{8}\%$ compounded quarterly, which is the better choice?

Solution:

Strategy We determine the equivalent effective rate of interest for each nominal rate and then compare our results.

The respective effective rates of interest are

$$r_e = \left(1 + \frac{0.06}{365}\right)^{365} - 1 \approx 6.18\%$$

and

$$r_e = \left(1 + \frac{0.06125}{4}\right)^4 - 1 \approx 6.27\%$$

Since the second choice gives the higher effective rate, it is the better choice (in spite of the fact that daily compounding may be psychologically more appealing).

NOW WORK PROBLEM 21 ●○○

Problems 5.1

In Problems 1 and 2, find (a) the compound amount and (b) the compound interest for the given investment and rate.

1. $6000 for eight years at an effective rate of 8%

2. $750 for 12 months at an effective rate of 7%

In Problems 3–6, find the effective rate that corresponds to the given nominal rate. Round answers to three decimal places.

3. 3% compounded semiannually

4. 5% compounded quarterly

5. 4% compounded daily

6. 6% compounded daily

7. Find the effective rate of interest (rounded to three decimal places) that is equivalent to a nominal rate of 10% compounded

  (a) yearly,   **(b)** semiannually,

  (c) quarterly,   **(d)** monthly,

  (e) daily.

8. Find (i) the compound interest (rounded to two decimal places) and (ii) the effective rate (to three decimal places) if $1000 is invested for five years at an annual rate of 7% compounded

(a) quarterly, (b) monthly,

(c) weekly, (d) daily.

*9. Over a five-year period, an original principal of $2000 accumulated to $2950 in an account in which interest was compounded quarterly. Determine the effective rate of interest, rounded to two decimal places.

10. Suppose that over a seven-year period, $1000 accumulated to $1835 in an investment certificate in which interest was compounded monthly. Find the nominal rate of interest, compounded monthly, that was earned. Round your answer to two decimal places.

In Problems 11 and 12, find how many years it would take to double a principal at the given effective rate. Give your answer to one decimal place.

*11. 9% **12.** 5%

13. A $6000 certificate of deposit is purchased for $6000 and is held for seven years. If the certificate earns an effective rate of 8%, what is it worth at the end of that period?

14. How many years will it take for money to triple at the effective rate of r?

*15. **College Costs** Suppose attending a certain college costs $21,500 in the 2005–2006 school year. This price includes tuition, room, board, books, and other expenses. Assuming an effective 6% inflation rate for these costs, determine what the college costs will be in the 2015–2016 school year.

16. College Costs Repeat Problem 15 for an inflation rate of 2% compounded quarterly.

17. Finance Charge A major credit-card company has a finance charge of $1\frac{1}{2}$% per month on the outstanding indebtedness. (a) What is the nominal rate compounded monthly? (b) What is the effective rate?

18. How long would it take for a principal of P to double if money is worth 12% compounded monthly? Give your answer to the nearest month.

19. To what sum will $2000 amount in eight years if invested at a 6% effective rate for the first four years and at 6% compounded semiannually thereafter?

20. How long will it take for $500 to amount to $700 if invested at 8% compounded quarterly?

*21. An investor has a choice of investing a sum of money at 8% compounded annually or at 7.8% compounded semiannually. Which is the better of the two rates?

22. What nominal rate of interest, compounded monthly, corresponds to an effective rate of 4.5%?

23. Savings Account A bank advertises that it pays interest on savings accounts at the rate of $4\frac{3}{4}$% compounded daily. Find the effective rate if the bank assumes that a year consists of (a) 360 days or (b) 365 days in determining the *daily rate*. Assume that compounding occurs 365 times a year, and round your answer to two decimal places.

24. Savings Account Suppose that $700 amounted to $801.06 in a savings account after two years. If interest was compounded quarterly, find the nominal rate of interest, compounded quarterly, that was earned by the money.

25. Inflation As a hedge against inflation, an investor purchased a painting in 1990 for $100,000. It was sold in 2000 for $300,000. At what effective rate did the painting appreciate in value?

26. Inflation If the rate of inflation for certain goods is $7\frac{1}{4}$% compounded daily, how many years will it take for the average price of such a good to double?

27. Zero-Coupon Bond A *zero-coupon bond* is a bond that is sold for less than its face value (that is, it is *discounted*) and has no periodic interest payments. Instead, the bond is redeemed for its face value at maturity. Thus, in this sense, interest is paid at maturity. Suppose that a zero-coupon bond sells for $420 and can be redeemed in 14 years for its face value of $1000. The bond earns interest at what nominal rate, compounded semiannually?

28. Misplaced Funds Suppose that $1000 is misplaced in a non-interest-bearing checking account and forgotten. Each year, the bank imposes a service charge of 1%. After 20 years, how much remains of the $1000? (*Hint:* Consider Equation (1) with $r = -0.01$.)

OBJECTIVE

To study present value and to solve problems involving the time value of money by using equations of value. To introduce the net present value of cash flows.

5.2 Present Value

Suppose that $100 is deposited in a savings account that pays 6% compounded annually. Then at the end of two years, the account is worth

$$100(1.06)^2 = 112.36$$

To describe this relationship, we say that the compound amount of $112.36 is the *future value* of the $100, and $100 is the *present value* of the $112.36. In general, there are times when we may know the future value of an investment and wish to find the present value. To obtain a formula for doing this, we solve the equation $S = P(1+r)^n$ for P. The result is $P = S/(1+r)^n = S(1+r)^{-n}$.

Present Value

The principal P that must be invested at the periodic rate of r for n interest periods so that the compound amount is S is given by

$$P = S(1 + r)^{-n} \tag{1}$$

and is called the **present value** of S.

EXAMPLE 1 Present Value

Find the present value of $1000 due after three years if the interest rate is 9% compounded monthly.

Solution: We use Equation (1) with $S = 1000$, $r = 0.09/12 = 0.0075$, and $n = 3(12) = 36$:

$$P = 1000(1.0075)^{-36} \approx \$764.15$$

This means that $764.15 must be invested at 9% compounded monthly to have $1000 in three years.

NOW WORK PROBLEM 1

If the interest rate in Example 1 were 10% compounded monthly, the present value would be

$$P = 1000 \left(1 + \frac{0.1}{12} \right)^{-36} \approx \$741.74$$

which is less than before. It is typical that the present value for a given future value decreases as the interest rate per interest period increases.

EXAMPLE 2 Single-Payment Trust Fund

A trust fund for a child's education is being set up by a single payment so that at the end of 15 years there will be $50,000. If the fund earns interest at the rate of 7% compounded semiannually, how much money should be paid into the fund?

Solution: We want the present value of $50,000, due in 15 years. From Equation (1) with $S = 50,000$, $r = 0.07/2 = 0.035$, and $n = 15(2) = 30$, we have

$$P = 50,000(1.035)^{-30} \approx \$17,813.92$$

NOW WORK PROBLEM 13

Equations of Value

Suppose that Mr. Smith owes Mr. Jones two sums of money: $1000, due in two years, and $600, due in five years. If Mr. Smith wishes to pay off the total debt now by a single payment, how much should the payment be? Assume an interest rate of 8% compounded quarterly.

The single payment x due now must be such that it would grow and eventually pay off the debts when they are due. That is, it must equal the sum of the present values of the future payments. As shown in the timeline of Figure 5.2, we have

$$x = 1000(1.02)^{-8} + 600(1.02)^{-20} \tag{2}$$

This equation is called an *equation of value*. We find that

$$x \approx \$1257.27$$

Thus, the single payment now due is $1257.27. Let us analyze the situation in more detail. There are two methods of payment of the debt: a single payment now or two payments in the future. Notice that Equation (2) indicates that the value *now* of all payments under one method must equal the value *now* of all payments under the

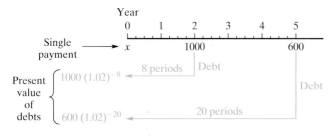

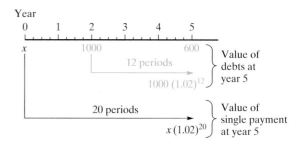

FIGURE 5.2   Replacing two future payments by a single payment now.

FIGURE 5.3   Diagram for equation of value.

Figure 5.2 is a useful tool for visualizing the time value of money. Always draw such a timeline to set up an equation of value.

other method. In general, this is true not just *now*, but at *any time*. For example, if we multiply both sides of Equation (2) by $(1.02)^{20}$, we get the equation of value

$$x(1.02)^{20} = 1000(1.02)^{12} + 600 \qquad (3)$$

The left side of Equation (3) gives the value five years from now of the single payment (see Figure 5.3), while the right side gives the value five years from now of all payments under the other method. Solving Equation (3) for x gives the same result, $x \approx 1257.27$. In general, an **equation of value** illustrates that when one is considering two methods of paying a debt (or of making some other transaction), *at any time* the value of all payments under one method must equal the value of all payments under the other method.

In certain situations, one equation of value may be more convenient to use than another, as Example 3 illustrates.

● EXAMPLE 3   Equation of Value

A debt of $3000 *due six years from now is instead to be paid off by three payments:* $500 *now,* $1500 *in three years, and a final payment at the end of five years. What would this payment be if an interest rate of* 6% *compounded annually is assumed?*

Solution: Let x be the final payment due in five years. For computational convenience, we will set up an equation of value to represent the situation at the end of that time, for in that way the coefficient of x will be 1, as seen in Figure 5.4. Notice that at year 5 we compute the future values of $500 and $1500, and the present value of $3000. The equation of value is

$$500(1.06)^5 + 1500(1.06)^2 + x = 3000(1.06)^{-1}$$

so

$$x = 3000(1.06)^{-1} - 500(1.06)^5 - 1500(1.06)^2$$

$$\approx \$475.68$$

NOW WORK PROBLEM 15 ●●●

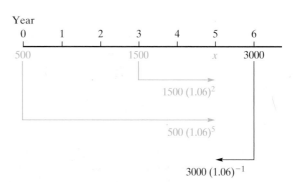

FIGURE 5.4   Time values of payments for Example 3.

When one is considering a choice of two investments, a comparison should be made of the value of each investment at a certain time, as Example 4 shows.

● EXAMPLE 4   Comparing Investments

Suppose that you had the opportunity of investing $5000 in a business such that the value of the investment after five years would be $6300. On the other hand, you could instead put the $5000 in a savings account that pays 6% compounded semiannually. Which investment is better?

Solution: Let us consider the value of each investment at the end of five years. At that time the business investment would have a value of $6300, while the savings account would have a value of $5000(1.03)^{10} \approx \$6719.58$. Clearly, the better choice is putting the money in the savings account.

NOW WORK PROBLEM 21

Net Present Value

If an initial investment will bring in payments at future times, the payments are called **cash flows.** The **net present value,** denoted NPV, of the cash flows is defined to be the sum of the present values of the cash flows, minus the initial investment. If NPV > 0, then the investment is profitable; if NPV < 0, the investment is not profitable.

● EXAMPLE 5   Net Present Value

Suppose that you can invest $20,000 in a business that guarantees you cash flows at the end of years 2, 3, and 5 as indicated in the table to the left. Assume an interest rate of 7% compounded annually, and find the net present value of the cash flows.

Solution: Subtracting the initial investment from the sum of the present values of the cash flows gives

$$\text{NPV} = 10{,}000(1.07)^{-2} + 8000(1.07)^{-3} + 6000(1.07)^{-5} - 20{,}000$$

$$\approx -\$457.31$$

Since NPV < 0, the business venture is not profitable if one considers the time value of money. It would be better to invest the $20,000 in a bank paying 7%, since the venture is equivalent to investing only $20,000 - \$457.31 = \$19{,}542.69$.

NOW WORK PROBLEM 19

Year	Cash Flow
2	$10,000
3	8000
5	6000

Problems 5.2

In Problems 1–10, find the present value of the given future payment at the specified interest rate.

*1. $6000 due in 20 years at 5% compounded annually

2. $3500 due in eight years at 6% effective

3. $4000 due in 12 years at 7% compounded semiannually

4. $1740 due in two years at 18% compounded monthly

5. $9000 due in $5\frac{1}{2}$ years at 8% compounded quarterly

6. $6000 due in $6\frac{1}{2}$ years at 10% compounded semiannually

7. $8000 due in five years at 10% compounded monthly

8. $500 due in three years at $8\frac{3}{4}$% compounded quarterly

9. $10,000 due in four years at $9\frac{1}{2}$% compounded daily

10. $1250 due in $1\frac{1}{2}$ years at $13\frac{1}{2}$% compounded weekly

11. A bank account pays 5.3% annual interest, compounded monthly. How much must be deposited now so that the account contains exactly $12,000 at the end of one year?

12. Repeat Problem 11 for the nominal rate of 7.1% compounded semiannually.

*13. **Trust Fund**   A trust fund for a 10-year-old child is being set up by a single payment so that at age 21 the child will receive $27,000. Find how much the payment is if an interest rate of 6% compounded semiannually is assumed.

14. A debt of $550 due in four years and $550 due in five years is to be repaid by a single payment now. Find how much the payment is if an interest rate of 10% compounded quarterly is assumed.

*15. A debt of $600 due in three years and $800 due in four years is to be repaid by a single payment two years from now. If the interest rate is 8% compounded semiannually, how much is the payment?

16. A debt of $7000 due in five years is to be repaid by a payment of $3000 now and a second payment at the end of five years. How much should the second payment be if the interest rate is 8% compounded monthly?

17. A debt of $5000 due five years from now and $5000 due ten years from now is to be repaid by a payment of $2000 in two years, a payment of $4000 in four years, and a final payment at the end of six years. If the interest rate is 2.5% compounded annually, how much is the final payment?

18. A debt of $3500 due in four years and $5000 due in six years is to be repaid by a single payment of $1500 now and three equal payments that are due each consecutive year from now. If the interest rate is 7% compounded annually, how much are each of the equal payments?

*19. **Cash Flows**   An initial investment of $25,000 in a business guarantees the following cash flows:

Year	Cash Flow
3	$8000
4	$10,000
6	$14,000

Assume an interest rate of 5% compounded semiannually.

(a) Find the net present value of the cash flows.
(b) Is the investment profitable?

20. **Cash Flows**   Repeat Problem 19 for the interest rate of 6% compounded semiannually.

*21. **Decision Making**   Suppose that a person has the following choices of investing $10,000:

(a) placing the money in a savings account paying 6% compounded semiannually;

(b) investing in a business such that the value of the investment after 8 years is $16,000.

Which is the better choice?

22. A owes B two sums of money: $1000 plus interest at 7% compounded annually, which is due in five years, and $2000 plus interest at 8% compounded semiannually, which is due in seven years. If both debts are to be paid off by a single payment at the end of six years, find the amount of the payment if money is worth 6% compounded quarterly.

23. **Purchase Incentive**   A jewelry store advertises that for every $1000 spent on diamond jewelry, the purchaser receives a $1000 bond at absolutely no cost. In reality, the $1000 is the full maturity value of a zero-coupon bond (see problem 27 of Problems 5.1), which the store purchases at a heavily reduced price. If the bond earns interest at the rate of 7.5% compounded quarterly and matures after 20 years, how much does the bond cost the store?

24. Find the present value of $6500 due in four years at a bank rate of 5.8% compounded daily. Assume that the bank uses 360 days in determining the daily rate and that there are 365 days in a year; that is, compounding occurs 365 times in a year.

25. **Promissory Note**   A *(promissory) note* is a written statement agreeing tõ pay a sum of money either on demand or at a definite future time. When a note is purchased for its present value at a given interest rate, the note is said to be *discounted,* and the interest rate is called the *discount rate.* Suppose a $10,000 note due eight years from now is sold to a financial institution for $4700. What is the nominal discount rate with quarterly compounding?

5.3 Interest Compounded Continuously

We have seen that when money is invested at a given annual rate, the interest earned each year depends on how frequently interest is compounded. For example, more interest is earned if it is compounded monthly rather than semiannually. We can successively get still more interest by compounding it weekly, daily, per hour, and so on. However, there is a maximum interest that can be earned, which we now examine.

Suppose a principal of P dollars is invested for t years at an annual rate of r. If interest is compounded k times a year, then the rate per interest period is r/k, and there are kt periods. From Section 4.1, recalled in Section 5.1, the compound amount is given by

$$S = P\left(1 + \frac{r}{k}\right)^{kt}$$

If k, the number of interest periods per year, is increased indefinitely, as we did in the "thought experiment" of Section 4.1 to introduce the number e, then the length of each period approaches 0 and we say that interest is **compounded continuously.** We can make this precise. In fact, with a little algebra we can relate the compound amount to the number e. Let $m = k/r$, so that

$$P\left(1 + \frac{r}{k}\right)^{kt} = P\left(\left(1 + \frac{1}{k/r}\right)^{k/r}\right)^{rt} = P\left(\left(1 + \frac{1}{m}\right)^{m}\right)^{rt} = P\left(\left(\frac{m+1}{m}\right)^{m}\right)^{rt}$$

In Section 4.1 we noted that, for n a positive integer, the numbers $\left(\dfrac{n+1}{n}\right)^n$ increase as n does but they are nevertheless bounded. (For example, it can be shown that all of the numbers $\left(\dfrac{n+1}{n}\right)^n$ are less than 3.) We *defined* e to be the least real number which is greater than all the values $\left(\dfrac{n+1}{n}\right)^n$, where n is a positive integer. It turns out (although it is beyond the scope of this book) that it is not necessary to require that n be an integer. For arbitrary positive m, the numbers $\left(\dfrac{m+1}{m}\right)^m$ increase as m does but they remain bounded and the number e as defined in Section 4.1 is the least real number that is greater than all the values $\left(\dfrac{m+1}{m}\right)^m$.

In the case at hand, for fixed r, the numbers $m = k/r$ increase as k (an integer) does, but the $m = k/r$ are not necessarily integers. However, if one accepts the truth of the preceding paragraph, then it follows that the compound amount $P\left(\left(\dfrac{m+1}{m}\right)^m\right)^{rt}$ approaches the value Pe^{rt} as k, and hence m, is increased indefinitely and we have the following:

Compound Amount under Continuous Interest

The formula

$$S = Pe^{rt} \tag{1}$$

gives the compound amount S of a principal of P dollars after t years at an annual interest rate r compounded continuously.

●EXAMPLE 1   Compound Amount

If $\$100$ is invested at an annual rate of 5% compounded continuously, find the compound amount at the end of

a. *1 year.*

b. *5 years.*

Solution:

a. Here $P = 100$, $r = 0.05$, and $t = 1$, so

$$S = Pe^{rt} = 100e^{(0.05)(1)} \approx \$105.13$$

We can compare this value with the value after one year of a $\$100$ investment at an annual rate of 5% compounded semiannually—namely, $100(1.025)^2 \approx 105.06$.

b. Here $P = 100$, $r = 0.05$, and $t = 5$, so

$$S = 100e^{(0.05)(5)} = 100e^{0.25} \approx \$128.40$$

NOW WORK PROBLEM 1   ●●●

The interest of $5.13 is the maximum amount of compound interest that can be earned at an annual rate of 5%.

We can find an expression that gives the effective rate that corresponds to an annual rate of r compounded continuously. (From Section 5.1, the effective rate is the rate compounded annually that gives rise to the same interest in a year as does the rate and compounding scheme under consideration.) If r_e is the corresponding effective rate, then after one year, a principal P accumulates to $P(1+r_e)$. This must equal the accumulated amount under continuous interest, Pe^r. Thus, $P(1+r_e) = Pe^r$, from which it follows that $1 + r_e = e^r$, so $r_e = e^r - 1$.

Effective Rate under Continuous Interest

The effective rate corresponding to an annual rate of r compounded continuously is

$$r_e = e^r - 1$$

● EXAMPLE 2   **Effective Rate**

Find the effective rate that corresponds to an annual rate of 5% compounded continuously.

Solution: The effective rate is

$$e^r - 1 = e^{0.05} - 1 \approx 0.0513$$

which is 5.13%.

NOW WORK PROBLEM 5 ◖◗●

If we solve $S = Pe^{rt}$ for P, we get $P = S/e^{rt} = Se^{-rt}$. In this formula, P is the principal that must be invested now at an annual rate of r compounded continuously so that at the end of t years the compound amount is S. We call P the **present value** of S.

Present Value under Continuous Interest

The formula

$$P = Se^{-rt}$$

gives the present value P of S dollars due at the end of t years at an annual rate of r compounded continuously.

● EXAMPLE 3   **Trust Fund**

A trust fund is being set up by a single payment so that at the end of 20 years there will be $25,000 in the fund. If interest is compounded continuously at an annual rate of 7%, how much money (to the nearest dollar) should be paid into the fund initially?

Solution: We want the present value of $25,000 due in 20 years. Therefore,

$$P = Se^{-rt} = 25{,}000e^{-(0.07)(20)}$$

$$= 25{,}000e^{-1.4} \approx 6165$$

Thus, $6165 should be paid initially.

NOW WORK PROBLEM 13 ◖◗●

Problems 5.3

In Problems 1 and 2, find the compound amount and compound interest if $4000 is invested for six years and interest is compounded continuously at the given annual rate.

*1. $6\frac{1}{4}$%   2. 9%

In Problems 3 and 4, find the present value of $2500 due eight years from now if interest is compounded continuously at the given annual rate.

3. $6\frac{3}{4}$%   4. 8%

In Problems 5–8, find the effective rate of interest that corresponds to the given annual rate compounded continuously.

*5. 4%   6. 8%

7. 3%   8. 11%

9. **Investment**   If $100 is deposited in a savings account that earns interest at an annual rate of $4\frac{1}{2}$% compounded continuously, what is the value of the account at the end of two years?

10. Investment   If $1000 is invested at an annual rate of 3% compounded continuously, find the compound amount at the end of eight years.

11. Stock Redemption   The board of directors of a corporation agrees to redeem some of its callable preferred stock in five years. At that time, $1,000,000 will be required. If the corporation can invest money at an annual interest rate of 5% compounded continuously, how much should it presently invest so that the future value is sufficient to redeem the shares?

12. Trust Fund   A trust fund is being set up by a single payment so that at the end of 30 years there will be $50,000 in the fund. If interest is compounded continuously at an annual rate of 6%, how much money should be paid into the fund initially?

*13. **Trust Fund**   As a gift for their newly born daughter's 25th birthday, the Smiths want to give her at that time a sum of money which has the same buying power as does $25,000 on the date of her birth. To accomplish this, they will make a single initial payment into a trust fund set up specifically for the purpose.

   (a) Assume that the annual effective rate of inflation is 3.5%. In 25 years, what sum will have the same buying power as does $25,000 at the date of the Smith's daughter's birth? Round your answer to the nearest dollar.

   (b) What should be the amount of the single initial payment into the fund if interest is compounded continuously at an annual rate of 4.5%? Round your answer to the nearest dollar.

14. Investment   Presently, the Smiths have $50,000 to invest for 18 months. They have two options open to them:

   (a) Invest the money in a certificate paying interest at the nominal rate of 5% compounded quarterly;

   (b) Invest the money in a savings account earning interest at the annual rate of 4.5% compounded continuously.

How much money will they have in 18 months with each option?

15. What annual rate compounded continuously is equivalent to an effective rate of 5%?

16. What annual rate r compounded continuously is equivalent to a nominal rate of 6% compounded semiannually? (*Hint:* First show that $r = 2\ln(1.03)$.)

17. If interest is compounded continuously at an annual rate of 0.07, how many years would it take for a principal P to triple? Give your answer to the nearest year.

18. If interest is compounded continuously, at what annual rate will a principal of P quadruple in 30 years? Give your answer to the nearest percent.

19. Savings Options   On July 1, 2001, Mr. Green had $1000 in a savings account at the First National Bank. This account earns interest at an annual rate of 3.5% compounded continuously. A competing bank was attempting to attract new customers by offering to add $20 immediately to any new account opened with a minimum $1000 deposit, and the new account would earn interest at the annual rate of 3.5% compounded semiannually. Mr. Green decided to choose one of the following three options on July 1, 2001:

   (a) Leave the money at the First National Bank;

   (b) Move the money to the competing bank;

   (c) Leave half the money at the First National Bank and move the other half to the competing bank.

For each of these three options, find Mr. Green's accumulated amount on July 1, 2003.

20. Investment

   (a) On November 1, 1996, Ms. Rodgers invested $10,000 in a 10-year certificate of deposit that paid interest at the annual rate of 4% compounded continuously. When the certificate matured on November 1, 2006, she reinvested the entire accumulated amount in corporate bonds, which earn interest at the rate of 5% compounded annually. To the nearest dollar, what will be Ms. Rodgers's accumulated amount on November 1, 2011?

   (b) If Ms. Rodgers had made a single investment of $10,000 in 1996 that matures in 2011 and has an effective rate of interest of 4.5%, would her accumulated amount be more or less than that in part (a) and by how much (to the nearest dollar)?

21. Investment Strategy   Suppose that you have $9000 to invest.

   (a) If you invest it with the First National Bank at the nominal rate of 5% compounded quarterly, find the accumulated amount at the end of one year.

   (b) The First National Bank also offers certificates on which it pays 5.5% compounded continuously. However, a minimum investment of $10,000 is required. Because you have only $9000, the bank is willing to give you a 1-year loan for the extra $1000 that you need. Interest for this loan is at an effective rate of 8%, and both principal and interest are payable at the end of the year. Determine whether or not this strategy of investment is preferable to the strategy in part (a).

OBJECTIVE

To introduce the notions of ordinary annuities and annuities due. To use geometric series to model the present value and future value of an annuity. To determine payments to be placed in a sinking fund.

5.4 Annuities

Sequences and Geometric Series

In mathematics we use the word **sequence** to describe a list of numbers, called *terms*, that are arranged in a definite order. For example, the list

$$2, \ 4, \ 6, \ 8$$

is a (finite) sequence. The first term is 2, the second is 4, and so on. Note that *order and repetitions matter* in a sequence. For example, the sequence above and the following two sequences are all *different:*

$$2, \ 6, \ 4, \ 8$$

$$2, \ 6, \ 6, \ 4, \ 8$$

However, we have equalities of *sets*

$$\{2, 4, 6, 8\} = \{2, 6, 4, 8\} = \{2, 6, 6, 4, 8\}$$

so that the notions of sequence and set are quite distinct.

In the sequence

$$3, \ 6, \ 12, \ 24, \ 48$$

each term after the first can be obtained by multiplying the preceding term by 2:

$$6 = 3(2), \quad 12 = 6(2), \quad \text{and so on}$$

This means that the *ratio* of every two consecutive terms is 2:

$$\frac{6}{3} = 2, \quad \frac{12}{6} = 2, \quad \text{and so on}$$

We call the sequence a *geometric sequence* with *first term* 3 and *common ratio* 2. Note that it can be written as

$$3, \ 3(2), \ 3(2)(2), \ 3(2)(2)(2), \ 3(2)(2)(2)(2)$$

or, more compactly, in the form

$$3, \ 3(2), \ 3(2^2), \ 3(2^3), \ 3(2^4)$$

More generally, if a geometric sequence has n terms such that the first term is a and the common ratio is the constant r, then the sequence has the form

$$a, ar, ar^2, ar^3, \ldots, ar^{n-1}$$

Note that the nth term in the sequence is ar^{n-1}.

DEFINITION

The sequence of n numbers

$$a, ar, ar^2, \ldots, ar^{n-1}, \quad \text{where } a \neq 0 \quad \text{(See footnote 1)}$$

is called a ***geometric sequence*** with ***first term a*** and ***common ratio r.***

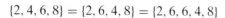

PRINCIPLES IN PRACTICE 1

GEOMETRIC SEQUENCES

A rubber ball always bounces back $\frac{3}{4}$ of its previous height. If the ball is dropped from a height of 64 feet, what are the next five heights of the ball?

CAUTION

Do not confuse sequences and sets.

EXAMPLE 1   Geometric Sequences

a. The geometric sequence with $a = 3$, common ratio $\frac{1}{2}$, and $n = 5$ is

$$3, \ 3\left(\frac{1}{2}\right), \ 3\left(\frac{1}{2}\right)^2, \ 3\left(\frac{1}{2}\right)^3, \ 3\left(\frac{1}{2}\right)^4$$

that is,

$$3, \ \frac{3}{2}, \ \frac{3}{4}, \ \frac{3}{8}, \ \frac{3}{16}$$

b. The numbers

$$1, \ 0.1, \ 0.01, \ 0.001$$

form a geometric sequence with $a = 1$, $r = 0.1$, and $n = 4$.

c. The terms

$$Pe^{-kI}, \ Pe^{-2kI}, \ldots, Pe^{-dkI}$$

---

[1] If $a = 0$, the sequence is $0, 0, 0, \ldots$. We do not consider this constant sequence to be geometric.

form a geometric sequence with $a = Pe^{-kI}$, $r = e^{-kI}$, and $n = d$. See the Mathematical Snapshot of Chapter 4.

NOW WORK PROBLEM 1

● EXAMPLE 2   Geometric Sequence

If $100 is invested at the rate of 6% compounded annually, then the list of compound amounts at the end of each year for eight years is

$$100(1.06), 100(1.06)^2, 100(1.06)^3, \ldots, 100(1.06)^8$$

This is a geometric sequence with common ratio 1.06.

NOW WORK PROBLEM 3 ●●●

The indicated sum of the terms of the geometric sequence $a, ar, ar^2, \ldots, ar^{n-1}$ is called a **geometric series:**

$$a + ar + ar^2 + \cdots + ar^{n-1} \tag{1}$$

For example,

$$1 + \frac{1}{2} + \left(\frac{1}{2}\right)^2 + \cdots + \left(\frac{1}{2}\right)^6$$

is a geometric series with $a = 1$, common ratio $r = \frac{1}{2}$, and $n = 7$.

We compute the sum s of the geometric series in Equation (1), for $r \neq 1$,[2] taking advantage of the opportunity to use and practice the summation notation of Section 1.5:

$$s = \sum_{i=0}^{n-1} ar^i = a \sum_{i=0}^{n-1} r^i \tag{2}$$

Multiplying by r gives

$$rs = ra \sum_{i=0}^{n-1} r^i = a \sum_{i=0}^{n-1} r^{i+1} = a \sum_{i=1}^{n} r^i \tag{3}$$

Subtracting corresponding sides of Equation (3) from Equation (2) yields

$$s - rs = a \sum_{i=0}^{n-1} r^i - a \sum_{i=1}^{n} r^i = a \left(\sum_{i=0}^{n-1} r^i - \sum_{i=1}^{n} r^i \right) = a(1 - r^n)$$

(Be sure to follow the calculation in the parentheses in which subtracting the second sum from the first sum results in cancellation of all terms except the first of the first sum and the last of the second sum.) We now have

$$s(1 - r) = a(1 - r^n)$$

and hence

$$s = \sum_{i=0}^{n-1} ar^i = \frac{a(1 - r^n)}{1 - r} \quad \text{for } r \neq 1$$

It should be noted that this result was derived in the Mathematical Snapshot of Chapter 4 for a rather complicated special case. The general case here is actually easier than the special one.

---

[2]If $r = 1$, then $s = a + \cdots + a = na$.

Sum of Geometric Series

The **sum** of a geometric series of n terms whose first term is a and whose common ratio is $r \neq 1$ is given by

$$\sum_{i=0}^{n-1} ar^i = \frac{a(1-r^n)}{1-r} \qquad (4)$$

PRINCIPLES IN PRACTICE 3

SUM OF GEOMETRIC SERIES

A ball rebounds $\frac{2}{3}$ of its previous height after each bounce. If the ball is tossed up to a height of 6 meters, how far has it traveled in the air when it hits the ground for the 12th time?

◉ EXAMPLE 3   **Sum of Geometric Series**

Find the sum of the geometric series

$$1 + \frac{1}{2} + \left(\frac{1}{2}\right)^2 + \cdots + \left(\frac{1}{2}\right)^6$$

Solution: Here $a = 1, r = \frac{1}{2}$, and $n = 7$ (not 6). From Equation (4), we have

$$s = \frac{a(1-r^n)}{1-r} = \frac{1\left(1 - \left(\frac{1}{2}\right)^7\right)}{1 - \frac{1}{2}} = \frac{\frac{127}{128}}{\frac{1}{2}} = \frac{127}{64}$$

NOW WORK PROBLEM 5 ◐●

PRINCIPLES IN PRACTICE 4

SUM OF GEOMETRIC SERIES

A company earns a profit of $2000 in its first month. Suppose the profit increases by 10% each month for two years. Find the amount of profit the company earns in its first two years.

◉ EXAMPLE 4   **Sum of Geometric Series**

Find the sum of the geometric series

$$3^5 + 3^6 + 3^7 + \cdots + 3^{11}$$

Solution: Here $a = 3^5, r = 3$, and $n = 7$. From Equation (4),

$$s = \frac{3^5(1 - 3^7)}{1 - 3} = \frac{243(1 - 2187)}{-2} = 265{,}599$$

NOW WORK PROBLEM 7 ◐●

Present Value of an Annuity

The notion of a geometric series is the basis of the mathematical model of an *annuity*. An **annuity** is a sequence of payments made at fixed periods of time over a given interval. The fixed period is called the **payment period,** and the given interval is the **term** of the annuity. An example of an annuity is the depositing of $100 in a savings account every three months for a year.

The **present value of an annuity** is the sum of the *present values* of all the payments. It represents the amount that must be invested *now* to purchase the payments due in the future. Unless otherwise specified, we assume that each payment is made at the *end* of a payment period; such an annuity is called an **ordinary annuity.** We also assume that interest is computed at the end of each payment period.

Let us consider an annuity of n payments of R (dollars) each, where the interest rate *per period* is r (see Figure 5.5) and the first payment is due one period from now. The present value of the annuity is given by

$$A = R(1+r)^{-1} + R(1+r)^{-2} + \cdots + R(1+r)^{-n}$$

This is a geometric series of n terms with first term $R(1+r)^{-1}$ and common ratio $(1+r)^{-1}$. Hence, from Equation (4), we obtain the formula

$$A = \frac{R(1+r)^{-1}(1 - (1+r)^{-n})}{1 - (1+r)^{-1}}$$

$$= \frac{R(1 - (1+r)^{-n})}{(1+r)(1 - (1+r)^{-1})} = \frac{R(1 - (1+r)^{-n})}{(1+r) - 1}$$

$$= R \cdot \frac{1 - (1+r)^{-n}}{r}$$

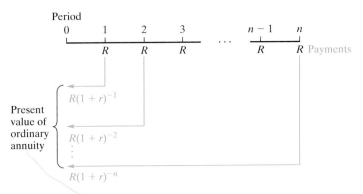

FIGURE 5.5 Present value of ordinary annuity.

Present Value of Annuity

The formula

$$A = R \cdot \frac{1 - (1+r)^{-n}}{r} \qquad (5)$$

gives the **present value** A of an ordinary annuity of R dollars per payment period for n periods at the interest rate of r per period.

In Equation (5), the expression $[1 - (1+r)^{-n}]/r$ is denoted $a_{\overline{n}|r}$ and (letting $R = 1$ in Equation (5)) we see that it represents the present value of an annuity of \$1 per period for n periods at the interest rate of r per period. The symbol $a_{\overline{n}|r}$ is read "a angle n at r". Thus, Equation (5) can be written as

$$A = Ra_{\overline{n}|r} \qquad (6)$$

Students often ask, "What is a in $a_{\overline{n}|r}$?" The simplest answer is that a is the name of a function, of two variables. As such, it is a rule that associates to each *pair* (n, r) of input numbers a unique output number. If standard notation for functions of two variables were used, we would write $a(n, r)$, but the notation $a_{\overline{n}|r}$ is traditional in the mathematics of finance. (Of course $a_{\overline{n}|r}$ is not the first deviation from the standard $f(x)$ nomenclature for functions. We have already seen that $\sqrt{x}$, $|x|$, $n!$, and $\log_2 x$ are other creative notations for particular common functions.)

Selected values of $a_{\overline{n}|r}$ are given in Appendix A. (Most are approximate.)

Whenever a desired value of $a_{\overline{n}|r}$ is not in Appendix A, we will use a calculator to compute it.

⬤ EXAMPLE 5 **Present Value of Annuity**

Find the present value of an annuity of \$100 per month for $3\frac{1}{2}$ years at an interest rate of 6% compounded monthly.

Solution: Substituting in Equation (6), we set $R = 100$, $r = 0.06/12 = 0.005$, and $n = \left(3\frac{1}{2}\right)(12) = 42$. Thus,

$$A = 100a_{\overline{42}|0.005}$$

From Appendix B, $a_{\overline{42}|0.005} \approx 37.798300$. Hence,

$$A \approx 100(37.798300) = \$3779.83$$

NOW WORK PROBLEM 13 ⬤⬤

⬤ EXAMPLE 6 **Present Value of Annuity**

Given an interest rate of 5% compounded annually, find the present value of an annuity of \$2000 due at the end of each year for three years and \$5000 due thereafter at the end of each year for four years. (See Figure 5.6.)

FIGURE 5.6 Annuity of Example 6.

Solution: The present value is obtained by summing the present values of all payments:

$$2000(1.05)^{-1} + 2000(1.05)^{-2} + 2000(1.05)^{-3} + 5000(1.05)^{-4}$$
$$+ 5000(1.05)^{-5} + 5000(1.05)^{-6} + 5000(1.05)^{-7}$$

Rather than evaluating this expression, we can simplify our work by considering the payments to be an annuity of $5000 for seven years, minus an annuity of $3000 for three years, so that the first three payments are $2000 each. Thus, the present value is

$$5000a_{\overline{7}|0.05} - 3000a_{\overline{3}|0.05}$$
$$\approx 5000(5.786373) - 3000(2.723248)$$
$$\approx \$20,762.12$$

NOW WORK PROBLEM 25 ●○●

PRINCIPLES IN PRACTICE 7

PERIODIC PAYMENT OF ANNUITY

Given an annuity with equal payments at the end of each quarter for six years and an interest rate of 4.8% compounded quarterly, use a graphing calculator to graph the present value A as a function of the monthly payment R. Determine the monthly payment if the present value of the annuity is $15,000.

●EXAMPLE 7 **Periodic Payment of Annuity**

If $10,000 is used to purchase an annuity consisting of equal payments at the end of each year for the next four years and the interest rate is 6% compounded annually, find the amount of each payment.

Solution: Here $A = \$10,000$, $n = 4$, $r = 0.06$, and we want to find R. From Equation (6),

$$10,000 = Ra_{\overline{4}|0.06}$$

Solving for R gives

$$R = \frac{10,000}{a_{\overline{4}|0.06}} \approx \frac{10,000}{3.465106} \approx \$2885.91$$

In general, the formula

$$R = \frac{A}{a_{\overline{n}|r}}$$

gives the periodic payment R of an ordinary annuity whose present value is A.

NOW WORK PROBLEM 27 ●○●

PRINCIPLES IN PRACTICE 8

ANNUITY DUE

A man makes house payments of $1200 at the beginning of every month. If the man wishes to pay one year's worth of payments in advance, how much should he pay, provided that the interest rate is 6.8% compounded monthly?

An example of a situation involving an annuity due is an apartment lease for which the first payment is made immediately.

●EXAMPLE 8 **Annuity Due**

The premiums on an insurance policy are $50 per quarter, payable at the beginning of each quarter. If the policyholder wishes to pay one year's premiums in advance, how much should be paid, provided that the interest rate is 4% compounded quarterly?

Solution: We want the present value of an annuity of $50 per period for four periods at a rate of 1% per period. However, each payment is due at the *beginning* of the payment period. Such an annuity is called an **annuity due.** The given annuity can be thought of as an initial payment of $50, followed by an ordinary annuity of $50 for three periods. (See Figure 5.7.) Thus, the present value is

$$50 + 50a_{\overline{3}|0.01} \approx 50 + 50(2.940985) \approx \$197.05$$

We remark that the general formula for the **present value of an annuity due** is $A = R + Ra_{\overline{n-1}|r}$, that is,

$$A = R(1 + a_{\overline{n-1}|r})$$

NOW WORK PROBLEM 17 ●○●

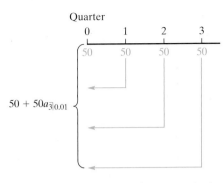

FIGURE 5.7 Annuity due (present value).

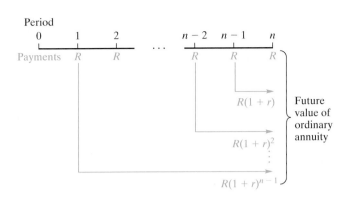

FIGURE 5.8 Future value of ordinary annuity.

Amount of an Annuity

The **amount** (or **future value**) **of an annuity** is the value, at the end of the term, of all payments. That is, it is the sum of the compound amounts of all payments. Let us consider an ordinary annuity of n payments of R (dollars) each, where the interest rate per period is r. The compound amount of the last payment is R, since it occurs at the end of the last interest period and hence does not accrue interest. (See Figure 5.8.) The $(n-1)$th payment earns interest for one period, the $(n-2)$th payment earns interest for two periods, and so on, and the first payment earns interest for $n-1$ periods. Thus, the future value of the annuity is

$$R + R(1+r) + R(1+r)^2 + \cdots + R(1+r)^{n-1}$$

This is a geometric series of n terms with first term R and common ratio $1+r$. Consequently, its sum S is (using Equation (4))

$$S = \frac{R(1-(1+r)^n)}{1-(1+r)} = R \cdot \frac{1-(1+r)^n}{-r} = R \cdot \frac{(1+r)^n-1}{r}$$

Amount of an Annuity

The formula

$$S = R \cdot \frac{(1+r)^n-1}{r} \tag{7}$$

gives the **amount** S of an ordinary annuity of R (dollars) per payment period for n periods at the interest rate of r per period.

The expression $[(1+r)^n-1]/r$ is abbreviated $s_{\overline{n}|r}$, and some approximate values of $s_{\overline{n}|r}$ are given in Appendix A. Thus,

$$S = s_{\overline{n}|r} \tag{8}$$

It follows that $s_{\overline{n}|r}$ is the amount of an ordinary annuity of $1 per payment period for n periods at the interest rate of r per period. Like $a_{\overline{n}|r}$, $s_{\overline{n}|r}$ is also a function of two variables (with a rather unconventional notation).

EXAMPLE 9 Amount of Annuity

Find the amount of an annuity consisting of payments of $50 at the end of every three months for three years at the rate of 6% compounded quarterly. Also, find the compound interest.

Solution: To find the amount of the annuity, we use Equation (8) with $R = 50$, $n = 4(3) = 12$, and $r = 0.06/4 = 0.015$:

$$S = 50s_{\overline{12}|0.015} \approx 50(13.041211) \approx \$652.06$$

The compound interest is the difference between the amount of the annuity and the sum of the payments, namely,

$$652.06 - 12(50) = 652.06 - 600 = \$52.06$$

NOW WORK PROBLEM 19

EXAMPLE 10 Amount of Annuity Due

At the beginning of each quarter, $50 is deposited into a savings account that pays 6% compounded quarterly. Find the balance in the account at the end of three years.

Solution: Since the deposits are made at the beginning of a payment period, we want the amount of an *annuity due*, as defined in Example 8. (See Figure 5.9.) The given annuity can be thought of as an ordinary annuity of $50 for 13 periods, minus the final payment of $50. Thus, the amount is

$$50s_{\overline{13}|0.015} - 50 \approx 50(14.236830) - 50 \approx \$661.84$$

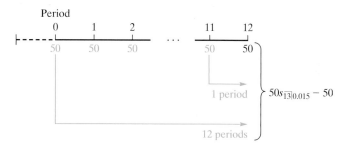

FIGURE 5.9 Future value of annuity due.

The formula for the **future value of an annuity due** is $S = Rs_{\overline{n+1}|r} - R$, which is

$$S = R(s_{\overline{n+1}|r} - 1)$$

NOW WORK PROBLEM 23

Sinking Fund

Our final examples involve the notion of a *sinking fund*.

EXAMPLE 11 Sinking Fund

*A **sinking fund** is a fund into which periodic payments are made in order to satisfy a future obligation. Suppose a machine costing $7000 is to be replaced at the end of eight years, at which time it will have a salvage value of $700. In order to provide money at that time for a new machine costing the same amount, a sinking fund is set up. The amount in the fund at the end of eight years is to be the difference between the replacement cost and the salvage value. If equal payments are placed in the fund at the end of each quarter and the fund earns 8% compounded quarterly, what should each payment be?*

Solution: The amount needed after eight years is $7000 - 700 = \$6300$. Let R be the quarterly payment. The payments into the sinking fund form an annuity with $n = 4(8) = 32$, $r = 0.08/4 = 0.02$, and $S = 6300$. Thus, from Equation (8), we have

$$6300 = Rs_{\overline{32}|0.02}$$

$$R = \frac{6300}{s_{\overline{32}|0.02}} \approx \frac{6300}{44.227030} \approx \$142.45$$

In general, the formula

$$R = \frac{S}{s_{\overline{n}|r}}$$

gives the periodic payment R of an annuity that is to amount to S.

NOW WORK PROBLEM 31

● EXAMPLE 12   **Sinking Fund**

A rental firm estimates that, if purchased, a machine will yield an annual net return of $1000 for six years, after which the machine would be worthless. How much should the firm pay for the machine if it wants to earn 7% annually on its investment and also set up a sinking fund to replace the purchase price? For the fund, assume annual payments and a rate of 5% compounded annually.

Solution: Let x be the purchase price. Each year, the return on the investment is $0.07x$. Since the machine gives a return of $1000 a year, the amount left to be placed into the fund each year is $1000 - 0.07x$. These payments must accumulate to x. Hence,

$$(1000 - 0.07x)s_{\overline{6}|0.05} = x$$

$$1000s_{\overline{6}|0.05} - 0.07xs_{\overline{6}|0.05} = x$$

$$1000s_{\overline{6}|0.05} = x(1 + 0.07s_{\overline{6}|0.05})$$

$$\frac{1000s_{\overline{6}|0.05}}{1 + 0.07s_{\overline{6}|0.05}} = x$$

$$x \approx \frac{1000(6.801913)}{1 + 0.07(6.801913)}$$

$$\approx \$4607.92$$

Another way to look at the problem is as follows: Each year, the $1000 must account for a return of $0.07x$ and also a payment of $\dfrac{x}{s_{\overline{6}|0.05}}$, into the sinking fund. Thus, we have $1000 = 0.07x + \dfrac{x}{s_{\overline{6}|0.05}}$, which, when solved, gives the same result.

NOW WORK PROBLEM 33 ●●●

Problems 5.4

In Problems 1–4, write the geometric sequence satisfying the given conditions. Simplify the terms.

*1. $a = 64, r = \frac{1}{2}, n = 5$

2. $a = 2, r = -3, n = 4$

*3. $a = 100, r = 1.02, n = 3$

4. $a = 81, r = 3^{-1}, n = 4$

In Problems 5–8, find the sum of the given geometric series by using Equation (4) of this section.

*5. $\frac{4}{7} + \left(\frac{4}{7}\right)^2 + \cdots + \left(\frac{4}{7}\right)^5$

6. $1 + \frac{1}{5} + \left(\frac{1}{5}\right)^2 + \cdots + \left(\frac{1}{5}\right)^6$

*7. $1 + 0.1 + (0.1)^2 + \cdots + (0.1)^5$

8. $(1.1)^{-1} + (1.1)^{-2} + \cdots + (1.1)^{-6}$

In Problems 9–12, use Appendix A and find the value of the given expression.

9. $a_{\overline{35}|0.04}$

10. $a_{\overline{15}|0.07}$

11. $s_{\overline{8}|0.0075}$

12. $s_{\overline{11}|0.0125}$

In Problems 13–16, find the present value of the given (ordinary) annuity.

*13. $600 per year for six years at the rate of 6% compounded annually

14. $1000 every six months for four years at the rate of 10% compounded semiannually

15. $2000 per quarter for $4\frac{1}{2}$ years at the rate of 8% compounded quarterly

16. $1500 per month for 15 months at the rate of 9% compounded monthly

In Problems 17 and 18, find the present value of the given annuity due.

*17. $800 paid at the beginning of each six-month period for six years at the rate of 7% compounded semiannually

18. $150 paid at the beginning of each month for five years at the rate of 7% compounded monthly

In Problems 19–22, find the future value of the given (ordinary) annuity.

*19. $2000 per month for three years at the rate of 15% compounded monthly

20. $600 per quarter for four years at the rate of 8% compounded quarterly

21. $5000 per year for 20 years at the rate of 7% compounded annually

22. $2000 every six months for 10 years at the rate of 6% compounded semiannually

In Problems 23 and 24, find the future value of the given annuity due.

*23. $1200 each year for 12 years at the rate of 8% compounded annually

24. $600 every quarter for $7\frac{1}{2}$ years at the rate of 10% compounded quarterly

*25. For an interest rate of 4% compounded monthly, find the present value of an annuity of $150 at the end of each month for eight months and $175 thereafter at the end of each month for a further two years.

26. **Leasing Office Space** A company wishes to lease temporary office space for a period of six months. The rental fee is $1500 a month, payable in advance. Suppose that the company wants to make a lump-sum payment at the beginning of the rental period to cover all rental fees due over the six-month period. If money is worth 9% compounded monthly, how much should the payment be?

*27. An annuity consisting of equal payments at the end of each quarter for three years is to be purchased for $5000. If the interest rate is 6% compounded quarterly, how much is each payment?

28. **Equipment Purchase** A machine is purchased for $3000 down and payments of $250 at the end of every six months for six years. If interest is at 8% compounded semiannually, find the corresponding cash price of the machine.

29. Suppose $50 is placed in a savings account at the end of each month for four years. If no further deposits are made, (a) how much is in the account after six years, and (b) how much of this amount is compound interest? Assume that the savings account pays 6% compounded monthly.

30. **Insurance Settlement Options** The beneficiary of an insurance policy has the option of receiving a lump-sum payment of $275,000 or 10 equal yearly payments, where the first payment is due at once. If interest is at 3.5% compounded annually, find the yearly payment.

*31. **Sinking Find** In 10 years, a $40,000 machine will have a salvage value of $4000. A new machine at that time is expected to sell for $52,000. In order to provide funds for the difference between the replacement cost and the salvage value, a sinking fund is set up into which equal payments are placed at the end of each year. If the fund earns 7% compounded annually, how much should each payment be?

32. **Sinking Fund** A paper company is considering the purchase of a forest that is estimated to yield an annual return of $50,000 for 10 years, after which the forest will have no value. The company wants to earn 8% on its investment and also set up a sinking fund to replace the purchase price. If money is placed in the fund at the end of each year and earns 6% compounded annually, find the price the company should pay for the forest. Round your answer to the nearest hundred dollars.

*33. **Sinking Fund** In order to replace a machine in the future, a company is placing equal payments into a sinking fund at the end of each year so that after 10 years the amount in the fund is $25,000. The fund earns 6% compounded annually. After 6 years, the interest rate increases and the fund pays 7% compounded annually. Because of the higher interest rate, the company decreases the amount of the remaining payments. Find the amount of the new payment. Round your answer to the nearest dollar.

34. A owes B the sum of $5000 and agrees to pay B the sum of $1000 at the end of each year for five years and a final payment at the end of the sixth year. How much should the final payment be if interest is at 8% compounded annually?

In Problems 35–43, use the following formulas:

$$a_{\overline{n}|r} = \frac{1 - (1 + r)^{-n}}{r}$$

$$s_{\overline{n}|r} = \frac{(1 + r)^n - 1}{r}$$

$$R = \frac{A}{a_{\overline{n}|r}} = \frac{Ar}{1 - (1 + r)^{-n}}$$

$$R = \frac{S}{s_{\overline{n}|r}} = \frac{Sr}{(1 + r)^n - 1}$$

35. Find $s_{\overline{60}|0.017}$ to five decimal places.

36. Find $a_{\overline{9}|0.052}$ to five decimal places.

37. Find $750a_{\overline{480}|0.0135}$ to two decimal places.

38. Find $1000s_{\overline{120}|0.01}$ to two decimal places.

39. Equal payments are to be deposited in a savings account at the end of each quarter for five years so that at the end of that time there will be $3000. If interest is at $5\frac{1}{2}\%$ compounded quarterly, find the quarterly payment.

40. **Insurance Proceeds** Suppose that insurance proceeds of $25,000 are used to purchase an annuity of equal payments at the end of each month for five years. If interest is at the rate of 10% compounded monthly, find the amount of each payment.

41. **Lottery** Mary Jones won a state $4,000,000 lottery and will receive a check for $200,000 now and a similar one each year for the next 19 years. To provide these 20 payments, the State Lottery Commission purchased an annuity due at the interest rate of 10% compounded annually. How much did the annuity cost the Commission?

42. **Pension Plan Options** Suppose an employee of a company is retiring and has the choice of two benefit options under the company pension plan. Option A consists of a guaranteed payment of $650 at the end of each month for 15 years. Alternatively, under option B, the employee receives a lump-sum payment equal to the present value of the payments described under option A.

   (a) Find the sum of the payments under option A.
   (b) Find the lump-sum payment under option B if it is determined by using an interest rate of 5.5% compounded monthly. Round your answer to the nearest dollar.

43. **An Early Start to Investing** An insurance agent offers services to clients who are concerned about their personal financial planning for retirement. To emphasize the advantages of an early start to investing, she points out that a 25-year-old person who saves $2000 a year for 10 years (and makes no more contributions after age 34) will earn more than by waiting 10 years and then saving $2000 a year from age 35 until retirement at age 65 (a total of 30 contributions). Find the net earnings (compound amount minus total contributions) at age 65 for both situations. Assume an effective annual rate of 7%, and suppose that deposits are made at the beginning of each year. Round answers to the nearest dollar.

44. **Continuous Annuity** An annuity in which R dollars is paid each year by uniform payments that are payable continuously is called a *continuous annuity*. The present

value of a continuous annuity for t years is

$$R \cdot \frac{1 - e^{-rt}}{r}$$

where r is the annual rate of interest compounded continuously. Find the present value of a continuous annuity of $100 a year for 20 years at 5% compounded continuously.

45. Profit   Suppose a business has an annual profit of $40,000 for the next five years and the profits are earned continuously throughout each year. Then the profits can be thought of as a continuous annuity. (See Problem 44.) If money is worth 4% compounded continuously, find the present value of the profits.

OBJECTIVE

To learn how to amortize a loan and set up an amortization schedule.

5.5 Amortization of Loans

Suppose that a bank lends a borrower $1500 and charges interest at the nominal rate of 12% compounded monthly. The $1500 plus interest is to be repaid by equal payments of R dollars at the end of each month for three months. You could say that by paying the borrower $1500, the bank is purchasing an annuity of three payments of R each. Using the formula from Example 7 of the preceding section, we find that the monthly payment is given by

$$R = \frac{A}{a_{\overline{n}|r}} = \frac{1500}{a_{\overline{3}|0.01}} \approx \frac{1500}{2.940985} \approx \$510.0332$$

We will round the payment to $510.03, which may result in a slightly higher final payment. However, it is not unusual for a bank to round *up* to the nearest cent, in which case the final payment may be less than the other payments.

The bank can consider each payment as consisting of two parts: (1) interest on the outstanding loan and (2) repayment of part of the loan. This is called **amortizing.** A loan is **amortized** when part of each payment is used to pay interest and the remaining part is used to reduce the outstanding principal. Since each payment reduces the outstanding principal, the interest portion of a payment decreases as time goes on. Let us analyze the loan just described.

At the end of the first month, the borrower pays $510.03. The interest on the outstanding principal is $0.01(1500) = \$15$. The balance of the payment, $510.03 - 15 = \$495.03$, is then applied to reduce the principal. Hence, the principal outstanding is now $1500 - 495.03 = \$1004.97$. At the end of the second month, the interest is $0.01(1004.97) \approx \$10.05$. Thus, the amount of the loan repaid is $510.03 - 10.05 = \$499.98$, and the outstanding balance is $1004.97 - 499.98 = \$504.99$. The interest due at the end of the third and final month is $0.01(504.99) \approx \$5.05$, so the amount of the loan repaid is $510.03 - 5.05 = \$504.98$. This would leave an outstanding balance of $504.99 - 504.98 = \$0.01$, so we take the final payment to be $510.04, and the debt is paid off. As we said earlier, the final payment is adjusted to offset rounding errors. An analysis of how each payment in the loan is handled can be given in a table called an **amortization schedule.** (See Table 5.1.) The total interest paid is $30.10, which is often called the **finance charge.**

Many end-of-year mortgage statements are issued in the form of an amortization schedule.

TABLE 5.1 Amortization Schedule

Period	Principal Outstanding at Beginning of Period	Interest for Period	Payment at End of Period	Principal Repaid at End of Period
1	$1500	$15	$510.03	$495.03
2	1004.97	10.05	510.03	499.98
3	504.99	5.05	510.04	504.99
Total		30.10	1530.10	1500.00

When one is amortizing a loan, at the beginning of any period the principal outstanding is the present value of the remaining payments. Using this fact together with our previous development, we obtain the formulas listed in Table 5.2, which

TABLE 5.2 Amortization Formulas

1. Periodic payment: $R = \dfrac{A}{a_{\overline{n}|r}} = A \cdot \dfrac{r}{1 - (1+r)^{-n}}$

2. Principal outstanding at beginning of kth period:

$$R a_{\overline{n-k+1}|r} = R \cdot \frac{1 - (1+r)^{-n+k-1}}{r}$$

3. Interest in kth payment: $R r a_{\overline{n-k+1}|r}$

4. Principal contained in kth payment: $R(1 - r a_{\overline{n-k+1}|r})$

5. Total interest paid: $R(n - a_{\overline{n}|r}) = nR - A$

describe the amortization of an interest-bearing loan of A dollars, at a rate r per period, by n equal payments of R dollars each and such that a payment is made at the end of each period. In particular, notice that, Formula 1 for the periodic payment R involves $a_{\overline{n}|r}$, which, we recall, is defined as $(1 - (1+r)^{-n})/r$.

⬤ **EXAMPLE 1 Amortizing a Loan**

A person amortizes a loan of $170,000 for a new home by obtaining a 20-year mortgage at the rate of 7.5% compounded monthly. Find (a) the monthly payment, (b) the total interest charges, and (c) the principal remaining after five years.

Solution:

a. The number of payment periods is $n = 12(20) = 240$, the interest rate per period is $r = 0.075/12 = 0.00625$, and $A = 170,000$. From Formula 1 in Table 5.2, the monthly payment R is $170,000/a_{\overline{240}|0.00625}$. Since $a_{\overline{240}|0.00625}$ is not in Appendix A, we use the following equivalent formula and a calculator:

$$R = 170,000 \left(\frac{0.00625}{1 - (1.00625)^{-240}} \right)$$

$$\approx \$1369.51$$

b. From Formula 5, the total interest charges are

$$240(1369.51) - 170,000 = 328,682.40 - 170,000$$

$$= \$158,682.40$$

This is almost as much as the loan itself.

c. After five years, we are at the beginning of the 61st period. Using Formula 2 with $n - k + 1 = 240 - 61 + 1 = 180$, we find that the principal remaining is

$$1369.51 \left(\frac{1 - (1.00625)^{-180}}{0.00625} \right) \approx \$147,733.74$$

NOW WORK PROBLEM 1 ⬤⬤

At one time, a very common type of installment loan involved the "add-on method" of determining the finance charge. With that method, the finance charge is found by applying a quoted annual interest rate under simple (that is, noncompounded) interest to the borrowed amount of the loan. The charge is then added to the principal, and the total is divided by the number of *months* of the loan to determine the monthly installment payment. In loans of this type, the borrower may not immediately realize that the true annual rate is significantly higher than the quoted rate, as the following technology example shows.

Problem: A $1000 loan is taken for one year at 9% interest under the add-on method. Estimate the true annual interest rate if monthly compounding is assumed.

Solution: Since the add-on method is used, payments will be made monthly. The finance charge for $1000 at 9% simple interest for one year is $0.09(1000) = \$90$. Adding this to the loan amount gives $1000 + 90 = \$1090$. Thus, the monthly installment payment is $1090/12 \approx \$90.83$. Hence, we have a loan of $1000 with 12 equal payments of $90.83. From Formula 1 in Table 5.2,

$$R = \frac{A}{a_{\overline{n}|r}}$$

$$\frac{1090}{12} = \frac{1000}{a_{\overline{12}|r}}$$

$$a_{\overline{12}|r} = \frac{1000(12)}{1090} \approx 11.009174$$

We now solve $a_{\overline{12}|r} = 11.009174$ for the monthly rate r. We have

$$\frac{1 - (1 + r)^{-12}}{r} = 11.009174$$

Graphing

$$Y_1 = (1 - (1 + X)\wedge -12)/X$$

$$Y_2 = 11.009174$$

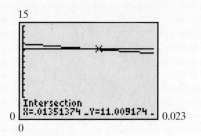

FIGURE 5.10 Solution of $a_{\overline{12}|r} = 11.009174$.

and finding the intersection (see Figure 5.10) gives

$$r \approx 0.01351374$$

which corresponds to an annual rate of

$$12(0.01351374) \approx 0.1622 = 16.22\%$$

Thus, the true annual rate is 16.22%. Federal regulations concerning truth-in-lending laws have made add-on loans virtually obsolete.

The annuity formula

$$A = R \cdot \frac{1 - (1 + r)^{-n}}{r}$$

can be solved for n to give the number of periods of a loan. Multiplying both sides by $\frac{r}{R}$ gives

$$\frac{Ar}{R} = 1 - (1 + r)^{-n}$$

$$(1 + r)^{-n} = 1 - \frac{Ar}{R} = \frac{R - Ar}{R}$$

$$-n \ln(1 + r) = \ln\left(\frac{R - Ar}{R}\right) \qquad \text{(taking logs of both sides)}$$

$$n = -\frac{\ln\left(\dfrac{R - Ar}{R}\right)}{\ln(1 + r)}$$

Using properties of logarithms, we eliminate the minus sign by inverting the quotient in the numerator:

$$n = \frac{\ln\left(\dfrac{R}{R - Ar}\right)}{\ln(1 + r)} \qquad (1)$$

EXAMPLE 2 Periods of a Loan

Muhammar Smith recently purchased a computer for $1500 and agreed to pay it off by making monthly payments of $75. If the store charges interest at the rate of 12% compounded monthly, how many months will it take to pay off the debt?

Solution: From Equation (1),

$$n = \frac{\ln\left(\dfrac{75}{75 - 1500(0.01)}\right)}{\ln(1.01)} \approx 22.4 \text{ months}$$

In reality, there are 23 payments; however, the final payment will be less than $75.

NOW WORK PROBLEM 11

Problems 5.5

*1. A person borrows $8000 from a bank and agrees to pay it off by equal payments at the end of each month for three years. If interest is at 14% compounded monthly, how much is each payment?

2. A person wishes to make a three-year loan and can afford payments of $50 at the end of each month. If interest is at 12% compounded monthly, how much can the person afford to borrow?

3. **Finance Charge**   Determine the finance charge on a 36-month $8000 auto loan with monthly payments if interest is at the rate of 4% compounded monthly.

4. For a one-year loan of $500 at the rate of 15% compounded monthly, find (a) the monthly installment payment and (b) the finance charge.

5. **Car Loan**   A person is amortizing a 36-month car loan of $7500 with interest at the rate of 4% compounded monthly. Find (a) the monthly payment, (b) the interest in the first month, and (c) the principal repaid in the first payment.

6. **Real-Estate Loan**   A person is amortizing a 48-month loan of $35,000 for a house lot. If interest is at the rate of 7.8% compounded monthly, find (a) the monthly payment, (b) the interest in the first payment, and (c) the principal repaid in the first payment.

In Problems 7–10, construct amortization schedules for the indicated debts. Adjust the final payments if necessary.

7. $5000 repaid by four equal yearly payments with interest at 7% compounded annually.

8. $9000 repaid by eight equal semiannual payments with interest at 9.5% compounded semiannually.

9. $900 repaid by five equal quarterly payments with interest at 10% compounded quarterly.

10. $10,000 repaid by five equal monthly payments with interest at 9% compounded monthly.

*11. A loan of $1000 is being paid off by quarterly payments of $100. If interest is at the rate of 8% compounded quarterly, how many *full* payments will be made?

12. A loan of $2000 is being amortized over 48 months at an interest rate of 12% compounded monthly. Find

   (a) the monthly payment;
   (b) the principal outstanding at the beginning of the 36th month;
   (c) the interest in the 36th payment;
   (d) the principal in the 36th payment;
   (e) the total interest paid.

13. A debt of $18,000 is being repaid by 15 equal semiannual payments, with the first payment to be made six months from now. Interest is at the rate of 7% compounded semiannually. However, after two years, the interest rate increases to 8% compounded semiannually. If the debt must be paid off on the original date agreed upon, find the new annual payment. Give your answer to the nearest dollar.

14. A person borrows $2000 and will pay off the loan by equal payments at the end of each month for five years. If interest is at the rate of 16.8% compounded monthly, how much is each payment?

15. **Mortgage**   A $245,000 mortgage for 25 years for a new home is obtained at the rate of 9.2% compounded monthly. Find (a) the monthly payment, (b) the interest in the first payment, (c) the principal repaid in the first payment, and (d) the finance charge.

16. **Auto Loan**   An automobile loan of $8500 is to be amortized over 48 months at an interest rate of 13.2% compounded monthly. Find (a) the monthly payment and (b) the finance charge.

17. **Furniture Loan**   A person purchases furniture for $2000 and agrees to pay off this amount by monthly payments of $100. If interest is charged at the rate of 18% compounded monthly, how many *full* payments will there be?

18. Find the monthly payment of a five-year loan for $9500 if interest is at 9.24% compounded monthly.

19. **Mortgage**   Bob and Mary Rodgers want to purchase a new house and feel that they can afford a mortgage payment of $600 a month. They are able to obtain a 30-year 7.6% mortgage (compounded monthly), but must put down 25% of the cost of the house. Assuming that they have enough savings for the down payment, how expensive a house can they afford? Give your answer to the nearest dollar.

20. **Mortgage**   Suppose you have the choice of taking out a $240,000 mortgage at 6% compounded monthly for either 15 years or 25 years. How much savings is there in the finance charge if you were to choose the 15-year mortgage?

21. On a $25,000 five-year loan, how much less is the monthly payment if the loan were at the rate of 12% compounded monthly rather than at 15% compounded monthly?

22. **Home Loan**   The federal government has a program to aid low-income homeowners in urban areas. This program allows certain qualified homeowners to obtain low-interest home improvement loans. Each loan is processed through a commercial bank. The bank makes home improvement loans at an annual rate of $9\frac{1}{4}\%$ compounded monthly.

However, the government subsidizes the bank so that the loan to the homeowner is at the annual rate of 4% compounded monthly. If the monthly payment at the 4% rate is x dollars (x dollars is the homeowner's monthly payment) and the monthly payment at the $9\frac{1}{4}$% rate is y dollars (y dollars is the monthly payment the bank must receive), then the government makes up the difference $y - x$ to the bank each month. From a practical point of view, the

government does not want to bother with *monthly* payments. Instead, at the beginning of the loan, the government pays the present value of all such monthly differences, at an annual rate of $9\frac{1}{4}$% compounded monthly.

If a qualified homeowner takes out a loan for $5000 for five years, determine the government's payment to the bank at the beginning of the loan.

5.6 Review

Important Terms and Symbols

Examples

Section 5.1	**Compound Interest**	
	effective rate	Ex. 4, p. 199
Section 5.2	**Present Value**	
	present value	Ex. 1, p. 202
	future value equation of value net present value	Ex. 3, p. 203
Section 5.3	**Interest Compounded Continuously**	
	compounded continuously	Ex. 1, p. 206
Section 5.4	**Annuities**	
	geometric sequence geometric series common ratio	Ex. 3, p. 211
	annuity ordinary annuity annuity due	Ex. 8, p. 213
	present value of annuity, $a_{\overline{n}|r}$ amount of annuity, $s_{\overline{n}|r}$	Ex. 9, p. 214
Section 5.5	**Ammortization of Loans**	
	amortizing amortization schedules finance charge	Ex. 1, p. 219

Summary

The concept of compound interest lies at the heart of any discussion dealing with the time value of money—that is, the present value of money due in the future or the future value of money presently invested. Under compound interest, interest is converted into principal and earns interest itself. The basic compound-interest formulas are

$$S = P(1 + r)^n \qquad \text{(future value)}$$

$$P = S(1 + r)^{-n} \qquad \text{(present value)}$$

where S = compound amount (future value)

$\quad P$ = principal (present value)

$\quad r$ = periodic rate

$\quad n$ = number of interest periods

Interest rates are usually quoted as an annual rate called the nominal rate. The periodic rate is obtained by dividing the nominal rate by the number of interest periods each year. The effective rate is the annual simple-interest rate which is equivalent to the nominal rate of r compounded n times a year, and is given by

$$r_e = \left(1 + \frac{r}{n}\right)^n - 1 \qquad \text{(effective rate)}$$

Effective rates are used to compare different interest rates.

If interest is compounded continuously then

$$S = Pe^{rt} \qquad \text{(future value)}$$

$$P = Se^{-rt} \qquad \text{(present value)}$$

where S = compound amount (future value)

$\quad P$ = principal (present value)

$\quad r$ = annual rate

$\quad t$ = number of years

and the effective rate is given by

$$r_e = e^r - 1 \qquad \text{(effective rate)}$$

An annuity is a sequence of payments made at fixed periods of time over some interval. The mathematical basis for formulas dealing with annuities is the notion of the sum of a geometric series—that is,

$$s = \sum_{i=0}^{n-1} ar^i = \frac{a(1 - r^n)}{1 - r} \qquad \text{(sum of geometric series)}$$

where s = sum

$\quad a$ = first term

$\quad r$ = common ratio

$\quad n$ = number of terms

An ordinary annuity is an annuity in which each payment is made at the *end* of a payment period, whereas an annuity due is an annuity in which each payment is made at the *beginning* of a payment period. The basic formulas dealing with ordinary annuities are

$$A = R \cdot \frac{1 - (1 + r)^{-n}}{r} = Ra_{\overline{n}|r} \qquad \text{(present value)}$$

$$S = R \cdot \frac{(1 + r)^n - 1}{r} = Rs_{\overline{n}|r} \qquad \text{(future value)}$$

where A = present value of annuity

S = amount (future value) of annuity

R = amount of each payment

n = number of payment periods

r = periodic rate

For an annuity due, the corresponding formulas are

$$A = R(1 + a_{\overline{n-1}|r}) \qquad \text{(present value)}$$

$$S = R(s_{\overline{n+1}|r} - 1) \qquad \text{(future value)}$$

A loan, such as a mortgage, is amortized when part of each installment payment is used to pay interest and the remaining part is used to reduce the principal. A complete analysis of each payment is given in an amortization schedule. The following formulas deal with amortizing a loan of A dollars, at the periodic rate of r, by n equal payments of R dollars each and such that a payment is made at the end of each period:

Periodic payment:

$$R = \frac{A}{a_{\overline{n}|r}} = A \cdot \frac{r}{1 - (1 + r)^{-n}}$$

Principal outstanding at beginning of kth period:

$$Ra_{\overline{n-k+1}|r} = R \cdot \frac{1 - (1 + r)^{-n+k-1}}{r}$$

Interest in kth payment:

$$Rra_{\overline{n-k+1}|r}$$

Principal contained in kth payment:

$$R(1 - ra_{\overline{n-k+1}|r})$$

Total interest paid:

$$R(n - a_{\overline{n}|r}) = nR - A$$

Review Problems

Problem numbers shown in color indicate problems suggested for use as a practice chapter test.

1. Find the sum of the geometric series

$$3 + 2 + 2 \cdot \frac{2}{3} + \cdots + 3\left(\frac{2}{3}\right)^5$$

2. Find the effective rate that corresponds to a nominal rate of 5% compounded monthly.

3. An investor has a choice of investing a sum of money at either 8.5% compounded annually or 8.2% compounded semiannually. Which is the better choice?

4. Cash Flows   Find the net present value of the following cash flows, which can be purchased by an initial investment of $7000:

Year	Cash Flow
2	$3400
4	3500

Assume that interest is at 7% compounded semiannually.

5. A debt of $1200 due in four years and $1000 due in six years is to be repaid by a payment of $1000 now and a second payment at the end of two years. How much should the second payment be if interest is at 8% compounded semiannually?

6. Find the present value of an annuity of $250 at the end of each month for four years if interest is at 6% compounded monthly.

7. For an annuity of $200 at the end of every six months for $6\frac{1}{2}$ years, find (a) the present value and (b) the future value at an interest rate of 8% compounded semiannually.

8. Find the amount of an annuity due which consists of 13 yearly payments of $150, provided that the interest rate is 4% compounded annually.

9. Suppose $200 is initially placed in a savings account and $200 is deposited at the end of every month for the next

year. If interest is at 8% compounded monthly, how much is in the account at the end of the year?

10. A savings account pays interest at the rate of 5% compounded semiannually. What amount must be deposited now so that $250 can be withdrawn at the end of every six months for the next 10 years?

11. Sinking Fund   A company borrows $5000 on which it will pay interest at the end of each year at the annual rate of 11%. In addition, a sinking fund is set up so that the $5000 can be repaid at the end of five years. Equal payments are placed in the fund at the end of each year, and the fund earns interest at the effective rate of 6%. Find the annual payment in the sinking fund.

12. Car Loan   A debtor is to amortize a $7000 car loan by making equal payments at the end of each month for 36 months. If interest is at 4% compounded monthly, find (a) the amount of each payment and (b) the finance charge.

13. A person has debts of $500 due in three years with interest at 5% compounded annually and $500 due in four years with interest at 6% compounded semiannually. The debtor wants to pay off these debts by making two payments: the first payment now, and the second, which is double the first payment, at the end of the third year. If money is worth 7% compounded annually, how much is the first payment?

14. Construct an amortization schedule for a loan of $3500 repaid by three monthly payments with interest at 16.5% compounded monthly.

15. Construct an amortization schedule for a loan of $15,000 repaid by five monthly payments with interest at 9% compounded monthly.

16. Find the present value of an ordinary annuity of $540 every month for seven years at the rate of 10% compounded monthly.

17. Auto Loan   Determine the finance charge for a 48-month auto loan of $11,000 with monthly payments at the rate of 5.5% compounded monthly.

Mathematical Snapshot

Treasury Securities

The safest single type of investment is in securities issued by the U.S. Treasury. These pay fixed returns on a predetermined schedule, which can span as little as three months or as much as thirty years. The finish date is called the date of maturity.

Although Treasury securities are initially sold by the government, they trade on the open market. Because the prices are free to float up and down, securities' rates of return can change over time. Consider, for example, a six-month Treasury bill, or T-bill, bearing a $10,000 face value and purchased on the date of issue for $9832.84. T-bills pay no interest before maturity but upon maturity are redeemed by the government at face value. This T-bill, if held the full six months, will pay back $\frac{10,000}{9832.84} \approx 101.7\%$ of the original investment, for an annualized effective annual rate of return of $1.017^2 - 1 \approx 3.429\%$. If the same T-bill, however, is sold at midterm for $9913.75, the new owner has a prospective annualized effective rate of return of $\left(\frac{10,000}{9913.75}\right)^4 - 1 \approx$ 3.526% over the remaining three months.

Like T-bills, Treasury notes and bonds are redeemed at face value upon maturity. In addition, however, notes and bonds pay interest twice a year according to a fixed nominal rate.[3] A 6.5%, $20,000, seven-year note pays $0.065(20,000) = \$1300$ every six months. At the end of seven years, the holder receives the final interest payment plus the face value, for a total of $21,300.

Mathematically, it is easier to calculate the present value of a note or bond from an assumed yield than to find the yield given an assumed present value (or price). Notes and bonds differ only in times to maturity: one to ten years for notes, ten to thirty years for bonds. Each note or bond is a guarantee of a lump sum at a future date plus an annuity until then. The present value of the note or bond, then, is the sum of the present value of the future lump sum and the present value of the annuity. We will assume that notes and bonds are evaluated at times when the next interest payment is exactly six months away; that way we can use the formula for the present value of an annuity in Section 5.4.

With semiannual compounding, an annual yield of r corresponds to an interest payment of $\sqrt{1+r} - 1$ every six months. Making the appropriate substitution in the formulas from Sections 5.2 and 5.4, we obtain the following general formula for the present value of a Treasury note or bond:

$$P = S(1 + \sqrt{1+r} - 1)^{-2n} + R \cdot \frac{1 - (1 + \sqrt{1+r} - 1)^{-2n}}{\sqrt{1+r} - 1}$$

which simplifies to

$$P = S(1+r)^{-n} + R \cdot \frac{1 - (1+r)^{-n}}{\sqrt{1+r} - 1}$$

where S is the face value, r is the assumed annual effective rate, and n is the number of years to maturity (so that $2n$ is the number of six-month periods). R is the amount of the semiannual interest payment, which is S times half the bond's nominal rate (for example, $R = 0.03S$ for a 6% bond).

Since we can treat a T-bill as a shorter-term note with a nominal rate of 0%, this formula covers T-bills as well by taking $R = 0$ since there is no annuity component.

To illustrate, if we are seeking a 7.4% effective rate on a new-issue, one-year, $30,000 T-bill (for which $R = 0$), we should be willing to pay

$$30,000(1.074)^{-1} \approx \$27,932.96$$

But if we are seeking a 7.4% effective rate on a 5.5%, $30,000 bond with 17 years left to maturity (here $R = 0.0275 \cdot 30,000 = 825$), we should be willing to pay only

$$30,000(1.074)^{-17} + 825 \cdot \frac{1 - (1.074)^{-17}}{\sqrt{1.074} - 1} \approx 24,870.66$$

Of course, it may happen that our effective rate expectations are unrealistic and that no bonds are for sale at the price we calculate. In that case, we may need to look at market prices and consider whether we can accept the corresponding effective rates of return. But how do we find a security's effective rate of return r from its market price? For T-bills, the second term on the right side of the present value formula drops out, and we can solve the simplified formula for r to obtain

$$r = \left(\frac{S}{P}\right)^{1/n} - 1$$

Calculations for three-month and six-month T-bills use $n = \frac{1}{4}$ and $n = \frac{1}{2}$ (as, for example, in the calculations of the second paragraph of this Snapshot).

Calculating the effective rate of return on a note or bond, on the other hand, involves solving the complete present value equation for r in terms of S, P, and n— and this cannot be done algebraically. However, it can be with a graphing calculator. We set Y_1 equal to the left side of the equation, Y_2 equal to the right side, and find where Y_1 and Y_2 are equal. Suppose, for example, that a 6.8%, $26,000 bond is selling at $26,617.50 eleven years from maturity. Each of the 22 interest payments is $R = 0.034(26,000) = 884$. To find the effective rate, set

$$Y_1 = 26,617.50$$

---

[3] In this context, *nominal rate* does *not* refer to the annual percentage rate. The former is constant, while the latter changes in tandem with the yield.

and

$$Y_2 = 26{,}000(1+X)^{\wedge} - 11$$
$$+ 884(1 - (1+X)^{\wedge} - 11)/(\sqrt{(1+X)} - 1)$$

Then graph Y_1 and Y_2, and find where the two graphs intersect (Figure 5.11).

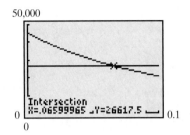

FIGURE 5.11   Finding effective rate.

The graphs intersect for $X \approx 0.0660$, which means that the effective rate is 6.6%.

The graph describing the current effective rates of Treasury securities as a function of time to maturity is called the *yield curve*. Economists keep a daily watch on this curve; you can monitor it yourself on the Web. Most typically, the yield curve looks something like that shown in Figure 5.12 (in which the horizontal time axis has been scaled).

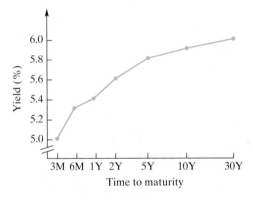

FIGURE 5.12   A typical yield curve.

You can see that the longer the time to maturity, the greater the yield. The usual explanation for this pattern is that having money tied up in a long-term investment means a loss of short-term flexibility—of liquidity, as it is called. To attract buyers, long-term securities must generally be priced for slightly higher effective rates than short-term securities.

Problems

1. Find the present value of an 8.5%, 25-year, $25,000 bond, assuming an annual effective rate of 8.25%.

2. Find the effective rate on a 6.5%, $10,000 note that is selling at $10,389 with seven years left to maturity.

3. In late December of 2000, the yield curve for Treasury securities had the atypical shape shown in Figure 5.13.

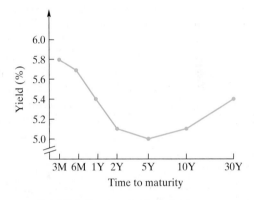

FIGURE 5.13   An atypical yield curve.

T-bills were earning higher yields than five-year notes, contrary to what one would expect. How might investor expectations about future earnings possibilities explain the yield curve?

6

MATRIX ALGEBRA

Mathematical Snapshot   Insulin Requirements as a Linear Process

Matrices, the subject of this chapter, are simply arrays of numbers. Matrices and matrix algebra have potential application whenever numerical information can be meaningfully arranged into rectangular blocks.

One area of application for matrix algebra is computer graphics. An object in a coordinate system can be represented by a matrix that contains the coordinates of each corner. For example, we might set up a connect-the-dots scheme in which the lightning bolt shown is represented by the matrix to its right.

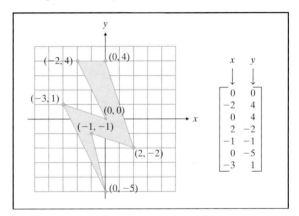

Computer graphics often show objects rotating in space. Computationally, rotation is done by matrix multiplication. The lightning bolt is rotated clockwise 52 degrees about the origin by matrix multiplication, involving a matrix whose entries are functions t_{11}, t_{12}, t_{21}, and t_{22}[1] of the rotation angle:

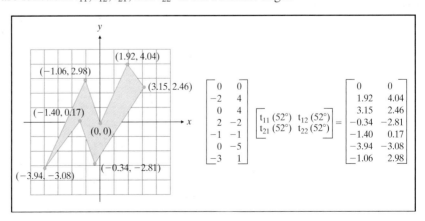

---

[1] Actually, $t_{11} = t_{22}$ and $t_{12} = -t_{21}$, but we will not pursue the details.

To introduce the concept of a matrix and to consider special types of matrices.

6.1 Matrices

Finding ways to describe many situations in mathematics and economics leads to the study of rectangular arrays of numbers. Consider, for example, the system of linear equations

$$\begin{cases} 3x + 4y + 3z = 0 \\ 2x + y - z = 0 \\ 9x - 6y + 2z = 0 \end{cases}$$

If we are organized with our notation, keeping the x's in the first column, the y's in the second column, and so on, then the features that characterize this system are the numerical coefficients in the equations, together with their relative positions. For this reason, the system can be described by the rectangular arrays

$$\begin{bmatrix} 3 & 4 & 3 \\ 2 & 1 & -1 \\ 9 & -6 & 2 \end{bmatrix} \quad \text{and} \quad \begin{bmatrix} 0 \\ 0 \\ 0 \end{bmatrix}$$

CAUTION

Do not use vertical bars, | |, instead of brackets or parentheses, for they have a different meaning.

one for each *side* of the equations, each being called a *matrix* (plural: *matrices*, pronounced may' tri sees). We consider such rectangular arrays to be objects in themselves, and our custom, as just shown, will be to enclose them by brackets. Parentheses are also commonly used. In symbolically representing matrices, we use bold capital letters such as **A**, **B**, **C**, and so on.

In economics it is often convenient to use matrices in formulating problems and displaying data. For example, a manufacturer who produces products A, B, and C could represent the units of labor and material involved in one week's production of these items as in Table 6.1. More simply, the data can be represented by the matrix

TABLE 6.1

	Product		
	A	**B**	**C**
Labor	10	12	16
Material	5	9	7

$$\mathbf{A} = \begin{bmatrix} 10 & 12 & 16 \\ 5 & 9 & 7 \end{bmatrix}$$

The horizontal rows of a matrix are numbered consecutively from top to bottom, and the vertical columns are numbered from left to right. For the foregoing matrix **A**, we have

$$\begin{array}{c} \quad \text{column 1} \quad \text{column 2} \quad \text{column 3} \\ \begin{array}{c} \text{row 1} \\ \text{row 2} \end{array} \begin{bmatrix} 10 & 12 & 16 \\ 5 & 9 & 7 \end{bmatrix} = \mathbf{A} \end{array}$$

Since **A** has two rows and three columns, we say that **A** has *size* 2×3 (read "2 by 3"), where the number of rows is specified first. Similarly, the matrices

$$\mathbf{B} = \begin{bmatrix} 1 & 6 & -2 \\ 5 & 1 & -4 \\ -3 & 5 & 0 \end{bmatrix} \quad \text{and} \quad \mathbf{C} = \begin{bmatrix} 1 & 2 \\ -3 & 4 \\ 5 & 6 \\ 7 & -8 \end{bmatrix}$$

have sizes 3×3 and 4×2, respectively.

The numbers in a matrix are called its **entries.** To denote arbitrary entries in a matrix of size 2×3, say, there are two common methods. First, we may use different letters:

$$\begin{bmatrix} a & b & c \\ d & e & f \end{bmatrix}$$

Second, a single letter may be used, say, a, along with appropriate *double* subscripts to indicate position:

$$\begin{bmatrix} a_{11} & a_{12} & a_{13} \\ a_{21} & a_{22} & a_{23} \end{bmatrix}$$

CAUTION

The row subscript appears to the left of the column subscript. In general, a_{ij} and a_{ji} are quite different.

For the entry a_{12} (read "a sub one-two" or just "a one-two"), the first subscript, 1, specifies the row and the second subscript, 2, the column in which the entry appears. Similarly, the entry a_{23} (read "a two-three") is the entry in the second row and the third column. Generalizing, we say that the symbol a_{ij} denotes the entry in the ith row and jth column.

Our concern in this chapter is the manipulation and application of various types of matrices. For completeness, we now give a formal definition of a matrix.

DEFINITION

A rectangular array of numbers consisting of m horizontal rows and n vertical columns,

$$\begin{bmatrix} a_{11} & a_{12} & \cdots & a_{1n} \\ a_{21} & a_{22} & \cdots & a_{2n} \\ \cdot & \cdot & \cdots & \cdot \\ \cdot & \cdot & \cdots & \cdot \\ \cdot & \cdot & \cdots & \cdot \\ a_{m1} & a_{m2} & \cdots & a_{mn} \end{bmatrix}$$

is called an **$m \times n$ matrix** or a **matrix of size $m \times n$**. For the entry a_{ij}, we call i the row subscript and j the column subscript.

CAUTION

Do not confuse the general entry a_{ij} with the matrix $[a_{ij}]$.

The number of entries in an $m \times n$ matrix is mn. For brevity, an $m \times n$ matrix can be denoted by the symbol $[a_{ij}]_{m \times n}$ or, more simply, $[a_{ij}]$, where the size is understood to be that which is appropriate for the given context. This notation merely indicates what types of symbols we are using to denote the general entry.

A matrix that has exactly one row, such as the 1×4 matrix

$$\mathbf{A} = \begin{bmatrix} 1 & 7 & 12 & 3 \end{bmatrix}$$

is called a **row vector.** A matrix consisting of a single column, such as the 5×1 matrix

$$\begin{bmatrix} 1 \\ -2 \\ 15 \\ 9 \\ 16 \end{bmatrix}$$

is called a **column vector.** Observe that a matrix is 1×1 if and only if it is both a row vector and a column vector. It is safe to treat 1×1 matrices as mere numbers. In other words, we can write $[7] = 7$, and, more generally, $[a] = a$, for any real number a.

PRINCIPLES IN PRACTICE 1

SIZE OF A MATRIX

A manufacturer who uses raw materials A and B is interested in tracking the costs of these materials from three different sources. What is the size of the matrix she would use?

● EXAMPLE 1 **Size of a Matrix**

a. The matrix $\begin{bmatrix} 1 & 2 & 0 \end{bmatrix}$ has size 1×3.

b. The matrix $\begin{bmatrix} 1 & -6 \\ 5 & 1 \\ 9 & 4 \end{bmatrix}$ has size 3×2.

c. The matrix $[7]$ has size 1×1.

d. The matrix $\begin{bmatrix} 1 & 3 & 7 & -2 & 4 \\ 9 & 11 & 5 & 6 & 8 \\ 6 & -2 & -1 & 1 & 1 \end{bmatrix}$ has size 3×5 and $(3)(15) = 15$ entries.

NOW WORK PROBLEM 1a ●●●

CONSTRUCTING MATRICES

An analysis of a workplace uses a 3×5 matrix to describe the time spent on each of three phases of five different projects. Project 1 requires 1 hour for each phase, project 2 requires twice as much time as project 1, project 3 requires twice as much time as project 2, ..., and so on. Construct this time-analysis matrix.

● EXAMPLE 2 **Constructing Matrices**

a. *Construct a three-entry column matrix such that $a_{21} = 6$ and $a_{i1} = 0$ otherwise.*

Solution: Since $a_{11} = a_{31} = 0$, the matrix is

$$\begin{bmatrix} 0 \\ 6 \\ 0 \end{bmatrix}$$

b. *If $\mathbf{A} = [a_{ij}]$ has size 3×4 and $a_{ij} = i + j$, find $\mathbf{A}$.*

Solution: Here $i = 1, 2, 3$ and $j = 1, 2, 3, 4$, and $\mathbf{A}$ has $(3)(4) = 12$ entries. Since $a_{ij} = i + j$, the entry in row i and column j is obtained by adding the numbers i and j. Hence, $a_{11} = 1 + 1 = 2$, $a_{12} = 1 + 2 = 3$, $a_{13} = 1 + 3 = 4$, and so on. Thus,

$$\mathbf{A} = \begin{bmatrix} 1+1 & 1+2 & 1+3 & 1+4 \\ 2+1 & 2+2 & 2+3 & 2+4 \\ 3+1 & 3+2 & 3+3 & 3+4 \end{bmatrix} = \begin{bmatrix} 2 & 3 & 4 & 5 \\ 3 & 4 & 5 & 6 \\ 4 & 5 & 6 & 7 \end{bmatrix}$$

c. *Construct the 3×3 matrix $\mathbf{I}$, given that $a_{11} = a_{22} = a_{33} = 1$ and $a_{ij} = 0$ otherwise.*

Solution: The matrix is given by

$$\mathbf{I} = \begin{bmatrix} 1 & 0 & 0 \\ 0 & 1 & 0 \\ 0 & 0 & 1 \end{bmatrix}$$

NOW WORK PROBLEM 11 ●●

Equality of Matrices

We now define what is meant by saying that two matrices are *equal*.

DEFINITION

Matrices $\mathbf{A} = [a_{ij}]$ and $\mathbf{B} = [b_{ij}]$ are **equal** if and only if they have the same size and $a_{ij} = b_{ij}$ for each i and j (that is, corresponding entries are equal).

Thus,

$$\begin{bmatrix} 1+1 & \frac{2}{2} \\ 2 \cdot 3 & 0 \end{bmatrix} = \begin{bmatrix} 2 & 1 \\ 6 & 0 \end{bmatrix}$$

but

$$\begin{bmatrix} 1 & 1 \end{bmatrix} \neq \begin{bmatrix} 1 \\ 1 \end{bmatrix} \quad \text{and} \quad \begin{bmatrix} 1 & 1 \end{bmatrix} \neq \begin{bmatrix} 1 & 1 & 1 \end{bmatrix} \quad \text{(different sizes)}$$

A matrix equation can define a system of equations. For example, suppose that

$$\begin{bmatrix} x & y+1 \\ 2z & 5w \end{bmatrix} = \begin{bmatrix} 2 & 7 \\ 4 & 2 \end{bmatrix}$$

By equating corresponding entries, we must have

$$\begin{cases} x = 2 \\ y + 1 = 7 \\ 2z = 4 \\ 5w = 2 \end{cases}$$

Solving gives $x = 2$, $y = 6$, $z = 2$, and $w = \frac{2}{5}$.

Transpose of a Matrix

If $\mathbf{A}$ is a matrix, the matrix formed from $\mathbf{A}$ by interchanging its rows with its columns is called the *transpose* of $\mathbf{A}$.

DEFINITION

The **transpose** of an $m \times n$ matrix $\mathbf{A}$, denoted $\mathbf{A}^{\mathrm{T}}$, is the $n \times m$ matrix whose ith row is the ith column of $\mathbf{A}$.

● EXAMPLE 3 **Transpose of a Matrix**

If $\mathbf{A} = \begin{bmatrix} 1 & 2 & 3 \\ 4 & 5 & 6 \end{bmatrix}$, *find* $\mathbf{A}^{\mathrm{T}}$.

Solution: Matrix $\mathbf{A}$ is 2×3, so $\mathbf{A}^{\mathrm{T}}$ is 3×2. Column 1 of $\mathbf{A}$ becomes row 1 of $\mathbf{A}^{\mathrm{T}}$, column 2 becomes row 2, and column 3 becomes row 3. Thus,

$$\mathbf{A}^{\mathrm{T}} = \begin{bmatrix} 1 & 4 \\ 2 & 5 \\ 3 & 6 \end{bmatrix}$$

NOW WORK PROBLEM 19 ●●●

Observe that the columns of $\mathbf{A}^{\mathrm{T}}$ are the rows of $\mathbf{A}$. Also, if we take the transpose of our answer, the original matrix $\mathbf{A}$ is obtained. That is, the transpose operation has the property that

$$(\mathbf{A}^{\mathrm{T}})^{\mathrm{T}} = \mathbf{A}$$

TECHNOLOGY

Graphing calculators have the ability to manipulate matrices. For example, Figure 6.1 shows the result of applying the transpose operation to matrix $\mathbf{A}$.

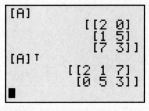

FIGURE 6.1 $\mathbf{A}$ and $\mathbf{A}^{\mathrm{T}}$.

Special Matrices

Certain types of matrices play important roles in matrix theory. We now consider some of these special types.

An $m \times n$ matrix whose entries are all 0 is called the $m \times n$ **zero matrix** and is denoted by $\mathbf{O}_{m \times n}$ or, more simply, by $\mathbf{O}$ if its size is understood. Thus, the 2×3 zero matrix is

$$\mathbf{O} = \begin{bmatrix} 0 & 0 & 0 \\ 0 & 0 & 0 \end{bmatrix}$$

 **CAUTION**

Do not confuse the matrix $\mathbf{O}$ with the real number 0.

and, in general, we have

$$\mathbf{O} = \begin{bmatrix} 0 & 0 & \cdots & 0 \\ 0 & 0 & \cdots & 0 \\ \cdot & \cdot & \cdots & \cdot \\ \cdot & \cdot & \cdots & \cdot \\ \cdot & \cdot & \cdots & \cdot \\ 0 & 0 & \cdots & 0 \end{bmatrix}$$

A matrix having the same number of columns as rows—for example, n rows and n columns—is called a **square matrix** of order n. That is, an $m \times n$ matrix is square if and only if $m = n$. For example, matrices

$$\begin{bmatrix} 2 & 7 & 4 \\ 6 & 2 & 0 \\ 4 & 6 & 1 \end{bmatrix} \quad \text{and} \quad [3]$$

are square with orders 3 and 1, respectively.

In a square matrix of order n, the entries $a_{11}, a_{22}, a_{33}, \ldots, a_{nn}$ lie on the diagonal extending from the upper left corner to the lower right corner of the matrix and are said to constitute the **main diagonal.** Thus, in the matrix

$$\begin{bmatrix} 1 & 2 & 3 \\ 4 & 5 & 6 \\ 7 & 8 & 9 \end{bmatrix}$$

the main diagonal (see the shaded region) consists of $a_{11} = 1, a_{22} = 5$, and $a_{33} = 9$.

A square matrix **A** is called a **diagonal matrix** if all the entries that are off the main diagonal are zero—that is, if $a_{ij} = 0$ for $i \neq j$. Examples of diagonal matrices are

$$\begin{bmatrix} 1 & 0 \\ 0 & 1 \end{bmatrix} \quad \text{and} \quad \begin{bmatrix} 3 & 0 & 0 \\ 0 & 6 & 0 \\ 0 & 0 & 9 \end{bmatrix}$$

A square matrix **A** is said to be an **upper triangular matrix** if all entries *below* the main diagonal are zero—that is, if $a_{ij} = 0$ for $i > j$. Similarly, a matrix **A** is said to be a **lower triangular matrix** if all entries *above* the main diagonal are zero—that is, if $a_{ij} = 0$ for $i < j$. When a matrix is either upper triangular or lower triangular, it is called a **triangular matrix.** Thus, the matrices

> It follows that a matrix is diagonal if and only if it is both upper triangular and lower triangular.

$$\begin{bmatrix} 5 & 1 & 1 \\ 0 & -3 & 7 \\ 0 & 0 & 4 \end{bmatrix} \quad \text{and} \quad \begin{bmatrix} 7 & 0 & 0 & 0 \\ 3 & 2 & 0 & 0 \\ 6 & 5 & -4 & 0 \\ 1 & 6 & 0 & 1 \end{bmatrix}$$

are upper and lower triangular matrices, respectively, and are therefore triangular matrices.

Problems 6.1

*1. Let

$$\mathbf{A} = \begin{bmatrix} 1 & -6 & 2 \\ -4 & 2 & 1 \end{bmatrix} \quad \mathbf{B} = \begin{bmatrix} 1 & 2 & 3 \\ 4 & 5 & 6 \\ 7 & 8 & 9 \end{bmatrix} \quad \mathbf{C} = \begin{bmatrix} 1 & 1 \\ 2 & 2 \\ 3 & 3 \end{bmatrix}$$

$$\mathbf{D} = \begin{bmatrix} 1 & 0 \\ 2 & 3 \end{bmatrix} \quad \mathbf{E} = \begin{bmatrix} 1 & 2 & 3 & 4 \\ 0 & 1 & 6 & 0 \\ 0 & 0 & 2 & 0 \\ 0 & 0 & 6 & 1 \end{bmatrix} \quad \mathbf{F} = \begin{bmatrix} 6 & 2 \end{bmatrix}$$

$$\mathbf{G} = \begin{bmatrix} 5 \\ 6 \\ 1 \end{bmatrix} \quad \mathbf{H} = \begin{bmatrix} 1 & 6 & 2 \\ 0 & 0 & 0 \\ 0 & 0 & 0 \end{bmatrix} \quad \mathbf{J} = [4]$$

(a) State the size of each matrix.
(b) Which matrices are square?
(c) Which matrices are upper triangular? lower triangular?
(d) Which are row vectors?
(e) Which are column vectors?

In Problems 2–9, let

$$\mathbf{A} = [a_{ij}] = \begin{bmatrix} 7 & -2 & 14 & 6 \\ 6 & 2 & 3 & -2 \\ 5 & 4 & 1 & 0 \\ 8 & 0 & 2 & 0 \end{bmatrix}$$

2. What is the order of **A**?

Find the following entries.

3. a_{21} **4.** a_{14} **5.** a_{32}

6. a_{34} **7.** a_{44} **8.** a_{55}

9. What are the main-diagonal entries?

10. Write the upper triangular matrix of order 4, given that all entries which are not required to be 0 are equal to the sum of their suscripts. (For example, $a_{23} = 2 + 3 = 5$.)

*11. Construct the matrix $\mathbf{A} = [a_{ij}]$ if **A** is 3×5 and $a_{ij} = -2i + 3j$.

12. Construct the matrix $\mathbf{B} = [b_{ij}]$ if **B** is 2×2 and $b_{ij} = (-1)^{i+j}(i^2 + j^2)$.

13. If $\mathbf{A} = [a_{ij}]$ is 12×10, how many entries does **A** have? If $a_{ij} = 1$ for $i = j$ and $a_{ij} = 0$ for $i \neq j$, find $a_{33}, a_{52}, a_{10,10}$, and $a_{12,10}$.

14. List the main diagonal of

(a) $\begin{bmatrix} 1 & 4 & -2 & 0 \\ 7 & 0 & 4 & -1 \\ -6 & 6 & -5 & 1 \\ 2 & 1 & 7 & 2 \end{bmatrix}$ (b) $\begin{bmatrix} x & 1 & y \\ 9 & y & 7 \\ y & 0 & z \end{bmatrix}$

15. Write the zero matrix of order (a) 4 and (b) 6.

16. If **A** is a 7×9 matrix, what is the size of $\mathbf{A}^T$?

In Problems 17–20, find $\mathbf{A}^T$.

17. $\mathbf{A} = \begin{bmatrix} 6 & -3 \\ 2 & 4 \end{bmatrix}$ **18.** $\mathbf{A} = \begin{bmatrix} 2 & 4 & 6 & 8 \end{bmatrix}$

*19. $\mathbf{A} = \begin{bmatrix} 1 & 3 & 7 & 3 \\ 3 & 2 & -2 & 0 \\ -4 & 5 & 0 & 1 \end{bmatrix}$ **20.** $\mathbf{A} = \begin{bmatrix} 2 & -1 & 0 \\ -1 & 5 & 1 \\ 0 & 1 & 3 \end{bmatrix}$

21. Let

$$\mathbf{A} = \begin{bmatrix} 7 & 0 \\ 0 & 6 \end{bmatrix} \quad \mathbf{B} = \begin{bmatrix} 1 & 0 & 0 \\ 0 & 2 & 0 \\ 0 & 10 & -3 \end{bmatrix}$$

$$\mathbf{C} = \begin{bmatrix} 0 & 0 & 0 \\ 0 & 0 & 0 \\ 0 & 0 & 0 \end{bmatrix} \quad \mathbf{D} = \begin{bmatrix} 2 & 0 & -1 \\ 0 & 4 & 0 \\ 0 & 0 & 6 \end{bmatrix}$$

(a) Which are diagonal matrices?
(b) Which are triangular matrices?

22. A matrix is *symmetric* if $\mathbf{A}^T = \mathbf{A}$. Is the matrix of Problem 19 symmetric?

23. If

$$\mathbf{A} = \begin{bmatrix} 1 & 0 & -1 \\ 7 & 0 & 9 \end{bmatrix}$$

verify the general property that $(\mathbf{A}^T)^T = \mathbf{A}$ by finding $\mathbf{A}^T$ and then $(\mathbf{A}^T)^T$.

In Problems 24–27, solve the matrix equation.

24. $\begin{bmatrix} 2x & y \\ z & 3w \end{bmatrix} = \begin{bmatrix} 4 & 6 \\ 0 & 7 \end{bmatrix}$

25. $\begin{bmatrix} 6 & 2 \\ x & 7 \\ 3y & 2z \end{bmatrix} = \begin{bmatrix} 6 & 2 \\ 6 & 7 \\ 2 & 7 \end{bmatrix}$

26. $\begin{bmatrix} 4 & 2 & 1 \\ 3x & y & 3z \\ 0 & w & 7 \end{bmatrix} = \begin{bmatrix} 4 & 2 & 1 \\ 6 & 7 & 9 \\ 0 & 9 & 8 \end{bmatrix}$

27. $\begin{bmatrix} 2x & 7 \\ 7 & 2y \end{bmatrix} = \begin{bmatrix} y & 7 \\ 7 & y \end{bmatrix}$

28. Inventory A grocer sold 125 cans of tomato soup, 275 cans of beans and 400 cans of tuna. Write a row vector that gives the number of each item sold. If the items sell for $0.95, $1.03, and $1.25 each, respectively, write this information as a column vector.

29. Sales Analysis The Widget Company has its monthly sales reports given by means of matrices whose rows, in order, represent the number of regular, deluxe, and super-duper models sold, and the columns, in order, give the number of red, white, blue, and purple units sold. The matrices for January (**J**) and February (**F**) are

$$J = \begin{bmatrix} 2 & 6 & 1 & 2 \\ 0 & 1 & 3 & 5 \\ 2 & 7 & 9 & 0 \end{bmatrix} \quad F = \begin{bmatrix} 0 & 2 & 8 & 4 \\ 2 & 3 & 3 & 2 \\ 4 & 0 & 2 & 6 \end{bmatrix}$$

(a) How many white super-duper models were sold in January? (b) How many blue deluxe models were sold in February? (c) In which month were more purple regular models sold? (d) Which model and color sold the same number of units in both months? (e) In which month were more deluxe models sold? (f) In which month were more red widgets sold? (g) How many widgets were sold in January?

30. Input–Output Matrix Input–output matrices, which were developed by W. W. Leontief, indicate the interrelationships that exist among the various sectors of an economy during some period of time. A hypothetical example for a simplified economy is given by matrix **M** at the end of this problem. The consuming sectors are the same as the producing sectors and can be thought of as manufacturers, government, steel, agriculture, households, and so on. Each row shows how the output of a given sector is consumed by the four sectors. For example, of the total output of industry A, 50 went to industry A itself, 70 to B, 200 to C, and 360 to all others. The sum of the entries in row 1—namely, 680—gives the total output of A for a given period. Each column gives the output of each sector that is consumed by a given sector. For example, in producing 680 units, industry A consumed 50 units of A, 90 of B, 120 of C, and 420 from all other producers. For each column, find the sum of the entries. Do the same for each row. What do you observe in comparing these totals? Suppose sector A increases its output by 20%, namely, by 136 units. Assuming that this results in a uniform 20% increase of all its inputs, by how many units will sector B have to increase its output? Answer the same question for C and for all other producers.

	CONSUMERS			
PRODUCERS	Industry A	Industry B	Industry C	All Other Consumers
Industry A	50	70	200	360
Industry B	90	30	270	320
Industry C	120	240	100	1050
All Other Producers	420	370	940	4960

$$M = \begin{bmatrix} 50 & 70 & 200 & 360 \\ 90 & 30 & 270 & 320 \\ 120 & 240 & 100 & 1050 \\ 420 & 370 & 940 & 4960 \end{bmatrix}$$

31. Find all the values of x for which

$$\begin{bmatrix} x^2 + 2000x & \sqrt{x^2} \\ x^2 & \ln(e^x) \end{bmatrix} = \begin{bmatrix} 2001 & -x \\ 2001 - 2000x & x \end{bmatrix}$$

In Problems 32 and 33, find A^T.

32. $A = \begin{bmatrix} 3 & -4 & 5 \\ -2 & 1 & 6 \end{bmatrix}$

33. $A = \begin{bmatrix} 3 & 1 & 4 & 2 \\ 1 & 7 & 3 & 6 \\ 1 & 4 & 1 & 2 \end{bmatrix}$

To define matrix addition and scalar multiplication and to consider properties related to these operations.

6.2 Matrix Addition and Scalar Multiplication

Matrix Addition

Consider a snowmobile dealer who sells two models, Deluxe and Super. Each is available in one of two colors, red and blue. Suppose that the sales for January and February are represented by the matrices

$$\begin{array}{cc} & \text{Deluxe} \quad \text{Super} \\ J = \begin{array}{c} \text{red} \\ \text{blue} \end{array} \begin{bmatrix} 1 & 2 \\ 3 & 5 \end{bmatrix} & F = \begin{bmatrix} 3 & 1 \\ 4 & 2 \end{bmatrix} \end{array}$$

Each row of **J** and **F** gives the number of each model sold for a given color. Each column gives the number of each color sold for a given model. A matrix representing total sales for each model and color over the two months can be obtained by adding the corresponding entries in **J** and **F**:

$$\begin{bmatrix} 4 & 3 \\ 7 & 7 \end{bmatrix}$$

This situation provides some motivation for introducing the operation of matrix addition for two matrices of the same size.

If $\mathbf{A} = [a_{ij}]$ and $\mathbf{B} = [b_{ij}]$ are both $m \times n$ matrices, then the **sum A + B** is the $m \times n$ matrix obtained by adding corresponding entries of $\mathbf{A}$ and $\mathbf{B}$; that is, $\mathbf{A} + \mathbf{B} = [a_{ij} + b_{ij}]$. If the size of $\mathbf{A}$ is different from the size of $\mathbf{B}$, then $\mathbf{A} + \mathbf{B}$ is not defined.

For example, let

$$\mathbf{A} = \begin{bmatrix} 3 & 0 & -2 \\ 2 & -1 & 4 \end{bmatrix} \quad \text{and} \quad \mathbf{B} = \begin{bmatrix} 5 & -3 & 6 \\ 1 & 2 & -5 \end{bmatrix}$$

Since $\mathbf{A}$ and $\mathbf{B}$ are the same size (2×3), their sum is defined. We have

$$\mathbf{A} + \mathbf{B} = \begin{bmatrix} 3+5 & 0+(-3) & -2+6 \\ 2+1 & -1+2 & 4+(-5) \end{bmatrix} = \begin{bmatrix} 8 & -3 & 4 \\ 3 & 1 & -1 \end{bmatrix}$$

MATRIX ADDITION

An office furniture company manufactures desks and tables at two plants, A and B. Matrix **J** represents the production of the two plants in January, and matrix **F** represents the production of the two plants in February. Write a matrix that represents the total production at the two plants for the two months. **J** and **F** are as follows:

$$\begin{array}{cc} & \text{A} \quad\;\; \text{B} \end{array}$$

$$\mathbf{J} = \begin{array}{c} \text{desks} \\ \text{tables} \end{array} \begin{bmatrix} 120 & 80 \\ 105 & 130 \end{bmatrix}$$

$$\mathbf{F} = \begin{array}{c} \text{desks} \\ \text{tables} \end{array} \begin{bmatrix} 110 & 140 \\ 85 & 125 \end{bmatrix}$$

These properties of matrix addition correspond to properties of addition of real numbers.

● EXAMPLE 1   **Matrix Addition**

a. $\begin{bmatrix} 1 & 2 \\ 3 & 4 \\ 5 & 6 \end{bmatrix} + \begin{bmatrix} 7 & -2 \\ -6 & 4 \\ 3 & 0 \end{bmatrix} = \begin{bmatrix} 1+7 & 2-2 \\ 3-6 & 4+4 \\ 5+3 & 6+0 \end{bmatrix} = \begin{bmatrix} 8 & 0 \\ -3 & 8 \\ 8 & 6 \end{bmatrix}$

b. $\begin{bmatrix} 1 & 2 \\ 3 & 4 \end{bmatrix} + \begin{bmatrix} 2 \\ 1 \end{bmatrix}$ is not defined, since the matrices are not the same size.

NOW WORK PROBLEM 7 ●●

If $\mathbf{A}$, $\mathbf{B}$, $\mathbf{C}$, and $\mathbf{O}$ have the same size, then the following properties hold for matrix addition:

Properties of Matrix Addition

1. $\mathbf{A} + \mathbf{B} = \mathbf{B} + \mathbf{A}$	(commutative property)
2. $\mathbf{A} + (\mathbf{B} + \mathbf{C}) = (\mathbf{A} + \mathbf{B}) + \mathbf{C}$	(associative property)
3. $\mathbf{A} + \mathbf{O} = \mathbf{O} + \mathbf{A} = \mathbf{A}$	(identity property)

Property 1 states that matrices can be added in any order, and Property 2 allows matrices to be grouped for the addition operation. Property 3 states that the zero matrix plays the same role in matrix addition as does the number 0 in the addition of real numbers. These properties are illustrated in the next example.

● EXAMPLE 2   **Properties of Matrix Addition**

Let

$$\mathbf{A} = \begin{bmatrix} 1 & 2 & 1 \\ -2 & 0 & 1 \end{bmatrix} \qquad \mathbf{B} = \begin{bmatrix} 0 & 1 & 2 \\ 1 & -3 & 1 \end{bmatrix}$$

$$\mathbf{C} = \begin{bmatrix} -2 & 1 & -1 \\ 0 & -2 & 1 \end{bmatrix} \qquad \mathbf{O} = \begin{bmatrix} 0 & 0 & 0 \\ 0 & 0 & 0 \end{bmatrix}$$

a. *Show that* $\mathbf{A} + \mathbf{B} = \mathbf{B} + \mathbf{A}$.

Solution:

$$\mathbf{A} + \mathbf{B} = \begin{bmatrix} 1 & 3 & 3 \\ -1 & -3 & 2 \end{bmatrix} \qquad \mathbf{B} + \mathbf{A} = \begin{bmatrix} 1 & 3 & 3 \\ -1 & -3 & 2 \end{bmatrix}$$

Thus, $\mathbf{A} + \mathbf{B} = \mathbf{B} + \mathbf{A}$.

b. *Show that* $\mathbf{A} + (\mathbf{B} + \mathbf{C}) = (\mathbf{A} + \mathbf{B}) + \mathbf{C}$.

Solution:

$$\mathbf{A} + (\mathbf{B} + \mathbf{C}) = \mathbf{A} + \begin{bmatrix} -2 & 2 & 1 \\ 1 & -5 & 2 \end{bmatrix} = \begin{bmatrix} -1 & 4 & 2 \\ -1 & -5 & 3 \end{bmatrix}$$

$$(\mathbf{A} + \mathbf{B}) + \mathbf{C} = \begin{bmatrix} 1 & 3 & 3 \\ -1 & -3 & 2 \end{bmatrix} + \mathbf{C} = \begin{bmatrix} -1 & 4 & 2 \\ -1 & -5 & 3 \end{bmatrix}$$

c. *Show that* $\mathbf{A} + \mathbf{O} = \mathbf{A}$.

Solution:

$$\mathbf{A} + \mathbf{O} = \begin{bmatrix} 1 & 2 & 1 \\ -2 & 0 & 1 \end{bmatrix} + \begin{bmatrix} 0 & 0 & 0 \\ 0 & 0 & 0 \end{bmatrix} = \begin{bmatrix} 1 & 2 & 1 \\ -2 & 0 & 1 \end{bmatrix} = \mathbf{A}$$

NOW WORK PROBLEM 1 ⬤⬤⬤

⬤ EXAMPLE 3   **Demand Vectors for an Economy**

Consider a simplified hypothetical economy having three industries, say, coal, electricity, and steel, and three consumers, 1, 2, and 3. Suppose that each consumer may use some of the output of each industry and that each industry uses some of the output of each other industry. Then the needs of each consumer and industry can be represented by a (row) demand vector whose entries, in order, give the amount of coal, electricity, and steel needed by the consumer or industry in some convenient units. For example, the demand vectors for the consumers might be

$$\mathbf{D}_1 = [3 \quad 2 \quad 5] \qquad \mathbf{D}_2 = [0 \quad 17 \quad 1] \qquad \mathbf{D}_3 = [4 \quad 6 \quad 12]$$

and for the industries they might be

$$\mathbf{D}_C = [0 \quad 1 \quad 4] \qquad \mathbf{D}_E = [20 \quad 0 \quad 8] \qquad \mathbf{D}_S = [30 \quad 5 \quad 0]$$

where the subscripts C, E, and S stand for coal, electricity, and steel, respectively. The total demand for these goods by the consumers is given by the sum

$$\mathbf{D}_1 + \mathbf{D}_2 + \mathbf{D}_3 = [3 \quad 2 \quad 5] + [0 \quad 17 \quad 1] + [4 \quad 6 \quad 12] = [7 \quad 25 \quad 18]$$

The total industrial demand is given by the sum

$$\mathbf{D}_C + \mathbf{D}_E + \mathbf{D}_S = [0 \quad 1 \quad 4] + [20 \quad 0 \quad 8] + [30 \quad 5 \quad 0] = [50 \quad 6 \quad 12]$$

Therefore, the total overall demand is given by

$$[7 \quad 25 \quad 18] + [50 \quad 6 \quad 12] = [57 \quad 31 \quad 30]$$

Thus, the coal industry sells a total of 57 units, the total units of electricity sold is 31, and the total units of steel that are sold is 30.[2]

NOW WORK PROBLEM 41 ⬤⬤⬤

Scalar Multiplication

Returning to the snowmobile dealer, recall that February sales were given by the matrix

$$\mathbf{F} = \begin{bmatrix} 3 & 1 \\ 4 & 2 \end{bmatrix}$$

If, in March, the dealer doubles February's sales of each model and color of snowmobile, the sales matrix for March could be obtained by multiplying each entry in $\mathbf{F}$

---

[2]This example, as well as some others in this chapter, is from John G. Kemeny, J. Laurie Snell, and Gerald L. Thompson, *Introduction to Finite Mathematics,* 3d ed. © 1974. Reprinted by permission of Prentice Hall, Inc., Englewood Cliffs, New Jersey.

by 2, yielding

$$\mathbf{M} = \begin{bmatrix} 2(3) & 2(1) \\ 2(4) & 2(2) \end{bmatrix}$$

It seems reasonable to write this operation as

$$\mathbf{M} = 2\mathbf{F} = 2 \begin{bmatrix} 3 & 1 \\ 4 & 2 \end{bmatrix} = \begin{bmatrix} 2 \cdot 3 & 2 \cdot 1 \\ 2 \cdot 4 & 2 \cdot 2 \end{bmatrix} = \begin{bmatrix} 6 & 2 \\ 8 & 4 \end{bmatrix}$$

which is thought of as multiplying a matrix by a real number. In the context of matrices, real numbers are often called *scalars.* Indeed, we have the following definition.

DEFINITION

If **A** is an $m \times n$ matrix and k is a real number then, by $k\mathbf{A}$, we denote the $m \times n$ matrix obtained by multiplying each entry in **A** by k. This operation is called *scalar multiplication,* and $k\mathbf{A}$ is called a *scalar multiple* of **A**.

For example,

$$-3 \begin{bmatrix} 1 & 0 & -2 \\ 2 & -1 & 4 \end{bmatrix} = \begin{bmatrix} -3(1) & -3(0) & -3(-2) \\ -3(2) & -3(-1) & -3(4) \end{bmatrix} = \begin{bmatrix} -3 & 0 & 6 \\ -6 & 3 & -12 \end{bmatrix}$$

● EXAMPLE 4   **Scalar Multiplication**

Let

$$\mathbf{A} = \begin{bmatrix} 1 & 2 \\ 4 & -2 \end{bmatrix} \qquad \mathbf{B} = \begin{bmatrix} 3 & -4 \\ 7 & 1 \end{bmatrix} \qquad \mathbf{O} = \begin{bmatrix} 0 & 0 \\ 0 & 0 \end{bmatrix}$$

Compute the following.

a. $5\mathbf{A}$

Solution:

$$5\mathbf{A} = 5 \begin{bmatrix} 1 & 2 \\ 4 & -2 \end{bmatrix} = \begin{bmatrix} 5(1) & 5(2) \\ 5(4) & 5(-2) \end{bmatrix} = \begin{bmatrix} 5 & 10 \\ 20 & -10 \end{bmatrix}$$

b. $-\dfrac{2}{3}\mathbf{B}$.

Solution:

$$-\frac{2}{3}\mathbf{B} = \begin{bmatrix} -\frac{2}{3}(3) & -\frac{2}{3}(-4) \\ -\frac{2}{3}(7) & -\frac{2}{3}(1) \end{bmatrix} = \begin{bmatrix} -2 & \frac{8}{3} \\ -\frac{14}{3} & -\frac{2}{3} \end{bmatrix}$$

c. $\dfrac{1}{2}\mathbf{A} + 3\mathbf{B}$

Solution:

$$\frac{1}{2}\mathbf{A} + 3\mathbf{B} = \frac{1}{2} \begin{bmatrix} 1 & 2 \\ 4 & -2 \end{bmatrix} + 3 \begin{bmatrix} 3 & -4 \\ 7 & 1 \end{bmatrix}$$

$$= \begin{bmatrix} \frac{1}{2} & 1 \\ 2 & -1 \end{bmatrix} + \begin{bmatrix} 9 & -12 \\ 21 & 3 \end{bmatrix} = \begin{bmatrix} \frac{19}{2} & -11 \\ 23 & 2 \end{bmatrix}$$

d. $0\mathbf{A}$

Solution:

$$0\mathbf{A} = 0 \begin{bmatrix} 1 & 2 \\ 4 & -2 \end{bmatrix} = \begin{bmatrix} 0 & 0 \\ 0 & 0 \end{bmatrix} = \mathbf{O}$$

e. $k\mathbf{O}$

Solution:

$$k\mathbf{O} = k \begin{bmatrix} 0 & 0 \\ 0 & 0 \end{bmatrix} = \begin{bmatrix} 0 & 0 \\ 0 & 0 \end{bmatrix} = \mathbf{O}$$

NOW WORK PROBLEM 5 ●●●

If **A**, **B**, and **O** are the same size, then, for any scalars k, and l, we have the following properties of scalar multiplication:

Properties of Scalar Multiplication

  1. $k(\mathbf{A} + \mathbf{B}) = k\mathbf{A} + k\mathbf{B}$

  2. $(k + l)\mathbf{A} = k\mathbf{A} + l\mathbf{A}$

  3. $k(l\mathbf{A}) = (kl)\mathbf{A}$

  4. $0\mathbf{A} = \mathbf{O}$

  5. $k\mathbf{O} = \mathbf{O}$

Properties 4 and 5 were illustrated in Examples 4(d) and (e); the others will be illustrated in the problems.

We also have the following properties of the transpose operation, where **A** and **B** are of the same size and k is any scalar:

$$(\mathbf{A} + \mathbf{B})^{\mathrm{T}} = \mathbf{A}^{\mathrm{T}} + \mathbf{B}^{\mathrm{T}}$$

$$(k\mathbf{A})^{\mathrm{T}} = k\mathbf{A}^{\mathrm{T}}$$

The first property states that *the transpose of a sum is the sum of the transposes.*

Subtraction of Matrices

If **A** is any matrix, then the scalar multiple $(-1)\mathbf{A}$ is simply written as $-\mathbf{A}$ and is called the **negative of A**:

$$-\mathbf{A} = (-1)\mathbf{A}$$

Thus, if

$$\mathbf{A} = \begin{bmatrix} 3 & 1 \\ -4 & 5 \end{bmatrix}$$

then

$$-\mathbf{A} = (-1)\begin{bmatrix} 3 & 1 \\ -4 & 5 \end{bmatrix} = \begin{bmatrix} -3 & -1 \\ 4 & -5 \end{bmatrix}$$

Note that $-\mathbf{A}$ is the matrix obtained by multiplying each entry of **A** by -1.

Subtraction of matrices is defined in terms of matrix addition:

DEFINITION

More simply, to find **A** − **B**, *we can subtract each entry in* **B** *from the corresponding entry in* **A**.

If **A** and **B** are the same size, then, by $\mathbf{A} - \mathbf{B}$, we mean $\mathbf{A} + (-\mathbf{B})$.

●EXAMPLE 5   **Matrix Subtraction**

a. $\begin{bmatrix} 2 & 6 \\ -4 & 1 \\ 3 & 2 \end{bmatrix} - \begin{bmatrix} 6 & -2 \\ 4 & 1 \\ 0 & 3 \end{bmatrix} = \begin{bmatrix} 2 & 6 \\ -4 & 1 \\ 3 & 2 \end{bmatrix} + (-1)\begin{bmatrix} 6 & -2 \\ 4 & 1 \\ 0 & 3 \end{bmatrix}$

$$= \begin{bmatrix} 2 & 6 \\ -4 & 1 \\ 3 & 2 \end{bmatrix} + \begin{bmatrix} -6 & 2 \\ -4 & -1 \\ 0 & -3 \end{bmatrix}$$

$$= \begin{bmatrix} 2-6 & 6+2 \\ -4-4 & 1-1 \\ 3+0 & 2-3 \end{bmatrix} = \begin{bmatrix} -4 & 8 \\ -8 & 0 \\ 3 & -1 \end{bmatrix}$$

b. If $\mathbf{A} = \begin{bmatrix} 6 & 0 \\ 2 & -1 \end{bmatrix}$ and $\mathbf{B} = \begin{bmatrix} 3 & -3 \\ 1 & 2 \end{bmatrix}$, then

$$\mathbf{A}^T - 2\mathbf{B} = \begin{bmatrix} 6 & 2 \\ 0 & -1 \end{bmatrix} - \begin{bmatrix} 6 & -6 \\ 2 & 4 \end{bmatrix} = \begin{bmatrix} 0 & 8 \\ -2 & -5 \end{bmatrix}$$

NOW WORK PROBLEM 17 ●●

MATRIX EQUATION

A manufacturer of doors, windows, and cabinets writes her yearly profit (in thousands of dollars) for each category in a column vector as $\mathbf{P} = \begin{bmatrix} 248 \\ 319 \\ 532 \end{bmatrix}$.

Her fixed costs of production can be described by the vector $\mathbf{C} = \begin{bmatrix} 40 \\ 30 \\ 60 \end{bmatrix}$. She calculates that, with a new pricing structure that generates an income that is 80% of her competitor's income, she can double her profit, assuming that her fixed costs remain the same. This calculation can be represented by

$$0.8 \begin{bmatrix} x_1 \\ x_2 \\ x_3 \end{bmatrix} - \begin{bmatrix} 40 \\ 30 \\ 60 \end{bmatrix} = 2 \begin{bmatrix} 248 \\ 319 \\ 532 \end{bmatrix}$$

Solve for $x_1, x_2,$ and x_3, which represent her competitor's income from each category.

● **EXAMPLE 6 Matrix Equation**

Solve the equation $2 \begin{bmatrix} x_1 \\ x_2 \end{bmatrix} - \begin{bmatrix} 3 \\ 4 \end{bmatrix} = 5 \begin{bmatrix} 5 \\ -4 \end{bmatrix}$.

Solution:

Strategy We first simplify each side into one matrix. Then, by equality of matrices, we equate corresponding entries.

We have

$$2 \begin{bmatrix} x_1 \\ x_2 \end{bmatrix} - \begin{bmatrix} 3 \\ 4 \end{bmatrix} = 5 \begin{bmatrix} 5 \\ -4 \end{bmatrix}$$

$$\begin{bmatrix} 2x_1 \\ 2x_2 \end{bmatrix} - \begin{bmatrix} 3 \\ 4 \end{bmatrix} = \begin{bmatrix} 25 \\ -20 \end{bmatrix}$$

$$\begin{bmatrix} 2x_1 - 3 \\ 2x_2 - 4 \end{bmatrix} = \begin{bmatrix} 25 \\ -20 \end{bmatrix}$$

By equality of matrices, we must have $2x_1 - 3 = 25$, which gives $x_1 = 14$; from $2x_2 - 4 = -20$, we get $x_2 = -8$.

NOW WORK PROBLEM 35 ●●

T E C H N O L O G Y

The matrix operations of addition, subtraction, and scalar multiplication can be performed on a graphing calculator. For example, Figure 6.2 shows $2\mathbf{A} - 3\mathbf{B}$, where

$$\mathbf{A} = \begin{bmatrix} -2 & 0 \\ 1 & 3 \end{bmatrix} \quad \text{and} \quad \mathbf{B} = \begin{bmatrix} 1 & 2 \\ 4 & 1 \end{bmatrix}$$

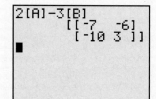

FIGURE 6.2 Matrix operations with graphing calculator.

Problems 6.2

In Problems 1–12, perform the indicated operations.

*1. $\begin{bmatrix} 2 & 0 & -3 \\ -1 & 4 & 0 \\ 1 & -6 & 5 \end{bmatrix} + \begin{bmatrix} 2 & -3 & 4 \\ -1 & 6 & 5 \\ 9 & 11 & -2 \end{bmatrix}$

2. $\begin{bmatrix} 2 & -7 \\ -6 & 4 \end{bmatrix} + \begin{bmatrix} 7 & -4 \\ -2 & 1 \end{bmatrix} + \begin{bmatrix} 2 & 7 \\ 7 & 2 \end{bmatrix}$

3. $\begin{bmatrix} 1 & 4 \\ -2 & 7 \\ 6 & 9 \end{bmatrix} - \begin{bmatrix} 6 & -1 \\ 7 & 2 \\ 1 & 0 \end{bmatrix}$ **4.** $\dfrac{1}{2} \begin{bmatrix} 4 & -2 & 6 \\ 2 & 10 & -12 \\ 0 & 0 & 7 \end{bmatrix}$

*5. $2[2 \quad -1 \quad 3] + 4[-2 \quad 0 \quad 1] - 0[2 \quad 3 \quad 1]$

6. $[7 \quad 7] + 66$ *7. $\begin{bmatrix} 1 & 2 \\ 3 & 4 \end{bmatrix} + \begin{bmatrix} 7 \\ 2 \end{bmatrix}$

8. $\begin{bmatrix} 2 & -1 \\ 7 & 4 \end{bmatrix} + 3 \begin{bmatrix} 0 & 0 \\ 0 & 0 \end{bmatrix}$ **9.** $-6 \begin{bmatrix} 2 & -6 & 7 & 1 \\ 7 & 1 & 6 & -2 \end{bmatrix}$

10. $\begin{bmatrix} 1 & -1 \\ 2 & 0 \\ 3 & -6 \\ 4 & 9 \end{bmatrix} - 3 \begin{bmatrix} -6 & 9 \\ 2 & 6 \\ 1 & -2 \\ 4 & 5 \end{bmatrix}$

11. $\begin{bmatrix} 1 & -5 & 0 \\ -2 & 7 & 0 \\ 4 & 6 & 10 \end{bmatrix} + \dfrac{1}{5} \begin{bmatrix} 10 & 0 & 30 \\ 0 & 5 & 0 \\ 5 & 20 & 25 \end{bmatrix}$

12. $3 \begin{bmatrix} 1 & 0 & 0 \\ 0 & 1 & 0 \\ 0 & 0 & 1 \end{bmatrix} - 3 \left(\begin{bmatrix} 1 & 2 & 0 \\ 0 & -2 & 1 \\ 0 & 0 & 1 \end{bmatrix} - \begin{bmatrix} 4 & -2 & 2 \\ -3 & 21 & -9 \\ 0 & 1 & 0 \end{bmatrix} \right)$

In Problems 13–24, compute the required matrices if

$\mathbf{A} = \begin{bmatrix} 2 & 1 \\ 3 & -3 \end{bmatrix}$ $\mathbf{B} = \begin{bmatrix} -6 & -5 \\ 2 & -3 \end{bmatrix}$ $\mathbf{C} = \begin{bmatrix} -2 & -1 \\ -3 & 3 \end{bmatrix}$ $\mathbf{O} = \begin{bmatrix} 0 & 0 \\ 0 & 0 \end{bmatrix}$

13. $-\mathbf{B}$

14. $-(\mathbf{A} - \mathbf{B})$

15. $2\mathbf{O}$

16. $\mathbf{A} - \mathbf{B} + \mathbf{C}$

*17. $3(2\mathbf{A} - 3\mathbf{B})$

18. $0(\mathbf{A} + \mathbf{B})$

19. $3(\mathbf{A} - \mathbf{C}) + 6$

20. $\mathbf{A} + (\mathbf{C} + \mathbf{B})$

21. $2\mathbf{B} - 3\mathbf{A} + 2\mathbf{C}$

22. $3\mathbf{C} - 2\mathbf{B}$

23. $\frac{1}{2}\mathbf{A} - 2(\mathbf{B} + 2\mathbf{C})$

24. $\frac{1}{2}\mathbf{A} - 5(\mathbf{B} + \mathbf{C})$

In Problems 25–28, verify the equations for the preceding matrices
A, **B**, *and* **C**.

25. $3(\mathbf{A} + \mathbf{B}) = 3\mathbf{A} + 3\mathbf{B}$

26. $(2 + 3)\mathbf{A} = 2\mathbf{A} + 3\mathbf{A}$

27. $k_1(k_2\mathbf{A}) = (k_1 k_2)\mathbf{A}$

28. $k(\mathbf{A} - 2\mathbf{B} + \mathbf{C}) = k\mathbf{A} - 2k\mathbf{B} + k\mathbf{C}$

In Problems 29–34, let

$\mathbf{A} = \begin{bmatrix} 1 & 2 \\ 0 & -1 \\ 7 & 0 \end{bmatrix}$ $\mathbf{B} = \begin{bmatrix} 1 & 3 \\ 4 & -1 \end{bmatrix}$ $\mathbf{C} = \begin{bmatrix} 1 & 0 \\ 1 & 2 \end{bmatrix}$ $\mathbf{D} = \begin{bmatrix} 1 & 2 & -1 \\ 1 & 0 & 2 \end{bmatrix}$

Compute the indicated matrices, if possible.

29. $3\mathbf{A} + \mathbf{D}^{\mathrm{T}}$

30. $(\mathbf{B} - \mathbf{C})^{\mathrm{T}}$

31. $2\mathbf{B}^{\mathrm{T}} - 3\mathbf{C}^{\mathrm{T}}$

32. $2\mathbf{B} + \mathbf{B}^{\mathrm{T}}$

33. $\mathbf{C}^{\mathrm{T}} - \mathbf{D}$

34. $(\mathbf{D} - 2\mathbf{A}^{\mathrm{T}})^{\mathrm{T}}$

*35. Express the matrix equation

$$x \begin{bmatrix} 3 \\ 2 \end{bmatrix} - y \begin{bmatrix} -4 \\ 7 \end{bmatrix} = 3 \begin{bmatrix} 2 \\ 4 \end{bmatrix}$$

as a system of linear equations and solve.

36. In the reverse of the manner used in Problem 35, write the system

$$\begin{cases} 2x - 4y = 16 \\ 5x + 7y = -3 \end{cases}$$

as a matrix equation.

In Problems 37–40, solve the matrix equations.

37. $3 \begin{bmatrix} x \\ y \end{bmatrix} - 3 \begin{bmatrix} -2 \\ 4 \end{bmatrix} = 4 \begin{bmatrix} 6 \\ -2 \end{bmatrix}$

38. $3 \begin{bmatrix} x \\ 2 \end{bmatrix} - 4 \begin{bmatrix} 7 \\ -y \end{bmatrix} = \begin{bmatrix} -x \\ 2y \end{bmatrix}$

39. $\begin{bmatrix} 2 \\ 4 \\ 6 \end{bmatrix} + 2 \begin{bmatrix} x \\ y \\ 4z \end{bmatrix} = \begin{bmatrix} -10 \\ -24 \\ 14 \end{bmatrix}$

40. $x \begin{bmatrix} 2 \\ 0 \\ 2 \end{bmatrix} + 2 \begin{bmatrix} -1 \\ 0 \\ 6 \end{bmatrix} + y \begin{bmatrix} 0 \\ 2 \\ -5 \end{bmatrix} = \begin{bmatrix} 10 \\ 6 \\ 2x + 12 - 5y \end{bmatrix}$

*41. **Production** An auto parts company manufactures distributors, sparkplugs, and magnetos at two plants, I and II. Matrix **X** represents the production of the two plants for retailer X, and matrix **Y** represents the production of the two plants for retailer Y. Write a matrix that represents the total production at the two plants for both retailers. Matrices **X** and **Y** are as follows:

$$\mathbf{X} = \begin{array}{c} \text{DIS} \\ \text{SPG} \\ \text{MAG} \end{array} \overset{\begin{array}{cc} \text{I} & \text{II} \end{array}}{\begin{bmatrix} 30 & 50 \\ 800 & 720 \\ 25 & 30 \end{bmatrix}} \quad \mathbf{Y} = \begin{array}{c} \text{DIS} \\ \text{SPG} \\ \text{MAG} \end{array} \overset{\begin{array}{cc} \text{I} & \text{II} \end{array}}{\begin{bmatrix} 15 & 25 \\ 960 & 800 \\ 10 & 5 \end{bmatrix}}$$

42. **Sales** Let matrix **A** represent the sales (in thousands of dollars) of a toy company in 2003 in three cities, and let **B** represent the sales in the same cities in 2005, where **A** and **B** are given by

$$\mathbf{A} = \begin{array}{c} \text{Action} \\ \text{Educational} \end{array} \begin{bmatrix} 400 & 350 & 150 \\ 450 & 280 & 850 \end{bmatrix}$$

$$\mathbf{B} = \begin{array}{c} \text{Action} \\ \text{Educational} \end{array} \begin{bmatrix} 380 & 330 & 220 \\ 460 & 320 & 750 \end{bmatrix}$$

If the company buys a competitor and doubles its 2005 sales in 2006, what is the change in sales between 2003 and 2006?

43. Suppose the prices of products A, B, and C are given, in that order, by the price row vector

$$\mathbf{P} = [p_1 \quad p_2 \quad p_3]$$

If the prices are to be increased by 10%, the vector for the new prices can be obtained by multiplying **P** by what scalar?

44. Prove that $(\mathbf{A} - \mathbf{B})^{\mathrm{T}} = \mathbf{A}^{\mathrm{T}} - \mathbf{B}^{\mathrm{T}}$. (*Hint:* Use the definition of *subtraction* and properties of the transpose operation.)

In Problems 45–47, compute the given matrices if

$$\mathbf{A} = \begin{bmatrix} 3 & -4 & 5 \\ -2 & 1 & 6 \end{bmatrix} \quad \mathbf{B} = \begin{bmatrix} 1 & 4 & 2 \\ 4 & 1 & 2 \end{bmatrix} \quad \mathbf{C} = \begin{bmatrix} -1 & 1 & 3 \\ 2 & 6 & -6 \end{bmatrix}$$

45. $4\mathbf{A} + 3\mathbf{B}$

46. $-3(\mathbf{A} + 2\mathbf{B}) + \mathbf{C}$

47. $2(3\mathbf{C} - \mathbf{A}) + 2\mathbf{B}$

OBJECTIVE

6.3 Matrix Multiplication

To define multiplication of matrices and to consider associated properties. To express a system as a single matrix equation by using matrix multiplication.

Besides the operations of matrix addition and scalar multiplication, the product **AB** of matrices **A** and **B** can be defined under certain conditions, namely, that *the number of columns of A is equal to the number of rows of B*. Although the following definition of *matrix multiplication* may not appear to you to be a natural one, a more thorough study of matrices would convince you that the definition is a good one and extremely practical for applications.

DEFINITION

Let $\mathbf{A}$ be an $m \times n$ matrix and $\mathbf{B}$ be an $n \times p$ matrix. Then the product $\mathbf{AB}$ is the $m \times p$ matrix $\mathbf{C}$ whose entry c_{ij} (in row i and column j) is given by

$$c_{ij} = \sum_{k=1}^{n} a_{ik}b_{kj} = a_{i1}b_{1j} + a_{i2}b_{2j} + \cdots + a_{in}b_{nj}$$

In words, c_{ij} is obtained by summing the products formed by multiplying, in order, each entry in row i of $\mathbf{A}$ by the corresponding entry in column j of $\mathbf{B}$. If the number of columns of $\mathbf{A}$ is not equal to the number of rows of $\mathbf{B}$, then the product $\mathbf{AB}$ is not defined.

Observe that the definition applies when $\mathbf{A}$ is a row vector with n entries and $\mathbf{B}$ is a column vector with n entries. In this case $\mathbf{A}$ is $1 \times n$, $\mathbf{B}$ is $n \times 1$, and $\mathbf{AB}$ is 1×1. (We noted in Section 6.1 that a $1 \times n$ matrix is just a number.) In fact,

$$\text{if} \quad \mathbf{A} = \begin{bmatrix} a_1 & a_2 & \cdots & a_n \end{bmatrix} \quad \text{and} \quad \mathbf{B} = \begin{bmatrix} b_1 \\ b_2 \\ \vdots \\ b_n \end{bmatrix}$$

$$\text{then} \quad \mathbf{AB} = \sum_{k=1}^{n} a_k b_k = a_1 b_1 + a_2 b_2 + \cdots + a_n b_n$$

Returning to our general definition, it now follows that the *number* c_{ij} is the product of the ith row of $\mathbf{A}$ and the jth column of $\mathbf{B}$. This is very helpful when real computations are performed.

Three points must be completely understood concerning this definition of $\mathbf{AB}$. First, the number of columns of $\mathbf{A}$ must be equal to the number of rows of $\mathbf{B}$. Second, the product $\mathbf{AB}$ has as many rows as $\mathbf{A}$ and as many columns as $\mathbf{B}$.

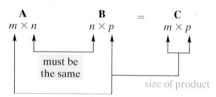

Third, the definition refers to the product $\mathbf{AB}$, *in that order;* $\mathbf{A}$ is the left factor and $\mathbf{B}$ is the right factor. For $\mathbf{AB}$, we say that $\mathbf{B}$ is *premultiplied* by $\mathbf{A}$ or $\mathbf{A}$ is *postmultiplied* by $\mathbf{B}$.

To apply the definition, let us find the product

$$\mathbf{AB} = \begin{bmatrix} 2 & 1 & -6 \\ 1 & -3 & 2 \end{bmatrix} \begin{bmatrix} 1 & 0 & -3 \\ 0 & 4 & 2 \\ -2 & 1 & 1 \end{bmatrix}$$

Matrix $\mathbf{A}$ has size 2×3 ($m \times n$) and matrix $\mathbf{B}$ has size 3×3 ($n \times p$). The number of columns of $\mathbf{A}$ is equal to the number of rows of $\mathbf{B}$ ($n = 3$), so the product $\mathbf{C}$ is defined and will be a 2×3 ($m \times p$) matrix; that is,

$$\mathbf{C} = \begin{bmatrix} c_{11} & c_{12} & c_{13} \\ c_{21} & c_{22} & c_{23} \end{bmatrix}$$

The entry c_{11} is obtained by summing the products of each entry in row 1 of $\mathbf{A}$ by the "corresponding" entry in column 1 of $\mathbf{B}$. Thus,

row 1 entries of $\mathbf{A}$

$$c_{11} = (2)(1) + (1)(0) + (-6)(-2) = 14.$$

column 1 entries of $\mathbf{B}$

At this stage, we have

$$\begin{bmatrix} 2 & 1 & -6 \\ 1 & -3 & 2 \end{bmatrix} \begin{bmatrix} 1 & 0 & -3 \\ 0 & 4 & 2 \\ -2 & 1 & 1 \end{bmatrix} = \begin{bmatrix} 14 & c_{12} & c_{13} \\ c_{21} & c_{22} & c_{23} \end{bmatrix}$$

Here we see that c_{11} is the product of the first row of **A** and the first column of **B**. Similarly, for c_{12}, we use the entries in row 1 of **A** and those in column 2 of **B**:

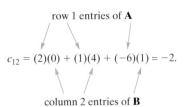

row 1 entries of **A**

$$c_{12} = (2)(0) + (1)(4) + (-6)(1) = -2.$$

column 2 entries of **B**

We now have

$$\begin{bmatrix} 2 & 1 & -6 \\ 1 & -3 & 2 \end{bmatrix} \begin{bmatrix} 1 & 0 & -3 \\ 0 & 4 & 2 \\ -2 & 1 & 1 \end{bmatrix} = \begin{bmatrix} 14 & -2 & c_{13} \\ c_{21} & c_{22} & c_{23} \end{bmatrix}$$

For the remaining entries of **AB**, we obtain

$$c_{13} = (2)(-3) + (1)(2) + (-6)(1) = -10$$

$$c_{21} = (1)(1) + (-3)(0) + (2)(-2) = -3$$

$$c_{22} = (1)(0) + (-3)(4) + (2)(1) = -10$$

$$c_{23} = (1)(-3) + (-3)(2) + (2)(1) = -7$$

Thus,

$$\mathbf{AB} = \begin{bmatrix} 2 & 1 & -6 \\ 1 & -3 & 2 \end{bmatrix} \begin{bmatrix} 1 & 0 & -3 \\ 0 & 4 & 2 \\ -2 & 1 & 1 \end{bmatrix} = \begin{bmatrix} 14 & -2 & -10 \\ -3 & -10 & -7 \end{bmatrix}$$

Note that if we reverse the order of the factors, then the product

$$\mathbf{BA} = \begin{bmatrix} 1 & 0 & -3 \\ 0 & 4 & 2 \\ -2 & 1 & 1 \end{bmatrix} \begin{bmatrix} 2 & 1 & -6 \\ 1 & -3 & 2 \end{bmatrix}$$

Matrix multiplication is not commutative.

is *not* defined, because the number of columns of **B** does *not* equal the number of rows of **A**. This shows that matrix multiplication is not commutative. That is, for any matrices **A** and **B**, it is usually the case that **AB** and **BA** are different, even when both products are defined, so *the order in which the matrices in a product are written is extremely important.*

EXAMPLE 1   Sizes of Matrices and Their Product

Let **A** be a 3×5 matrix and **B** be a 5×3 matrix. Then **AB** is defined and is a 3×3 matrix. Moreover, **BA** is also defined and is a 5×5 matrix.

If **C** is a 3×5 matrix and **D** is a 7×3 matrix, then **CD** is undefined, but **DC** is defined and is a 7×5 matrix.

NOW WORK PROBLEM 7

EXAMPLE 2   Matrix Product

Compute the matrix product

$$\mathbf{AB} = \begin{bmatrix} 2 & -4 & 2 \\ 0 & 1 & -3 \end{bmatrix} \begin{bmatrix} 2 & 1 \\ 0 & 4 \\ 2 & 2 \end{bmatrix}$$

Solution: Since **A** is 2×3 and **B** is 3×2, the product **AB** is defined and will have order 2×2. By simultaneously moving the index finger of your left hand along the rows of **A** and the index finger of your right hand along the columns of **B**, it should not be difficult for you to mentally find the entries of the product. We obtain

$$\begin{bmatrix} 2 & -4 & 2 \\ 0 & 1 & -3 \end{bmatrix} \begin{bmatrix} 2 & 1 \\ 0 & 4 \\ 2 & 2 \end{bmatrix} = \begin{bmatrix} 8 & -10 \\ -6 & -2 \end{bmatrix}$$

NOW WORK PROBLEM 19 ●●

PRINCIPLES IN PRACTICE 1

MATRIX PRODUCTS

A bookstore has 100 dictionaries, 70 cookbooks, and 90 thesauruses in stock. If the value of each dictionary is $28, each cookbook is $22, and each thesaurus is $16, use a matrix product to find the total value of the bookstore's inventory.

● EXAMPLE 3 **Matrix Products**

a. *Compute* $\begin{bmatrix} 1 & 2 & 3 \end{bmatrix} \begin{bmatrix} 4 \\ 5 \\ 6 \end{bmatrix}$.

Solution: The product has order 1×1:

$$\begin{bmatrix} 1 & 2 & 3 \end{bmatrix} \begin{bmatrix} 4 \\ 5 \\ 6 \end{bmatrix} = [32]$$

b. *Compute* $\begin{bmatrix} 1 \\ 2 \\ 3 \end{bmatrix} \begin{bmatrix} 1 & 6 \end{bmatrix}$.

Solution: The product has order 3×2:

$$\begin{bmatrix} 1 \\ 2 \\ 3 \end{bmatrix} \begin{bmatrix} 1 & 6 \end{bmatrix} = \begin{bmatrix} 1 & 6 \\ 2 & 12 \\ 3 & 18 \end{bmatrix}$$

c. $\begin{bmatrix} 1 & 3 & 0 \\ -2 & 2 & 1 \\ 1 & 0 & -4 \end{bmatrix} \begin{bmatrix} 1 & 0 & 2 \\ 5 & -1 & 3 \\ 2 & 1 & -2 \end{bmatrix} = \begin{bmatrix} 16 & -3 & 11 \\ 10 & -1 & 0 \\ -7 & -4 & 10 \end{bmatrix}$

d. $\begin{bmatrix} a_{11} & a_{12} \\ a_{21} & a_{22} \end{bmatrix} \begin{bmatrix} b_{11} & b_{12} \\ b_{21} & b_{22} \end{bmatrix} = \begin{bmatrix} a_{11}b_{11} + a_{12}b_{21} & a_{11}b_{12} + a_{12}b_{22} \\ a_{21}b_{11} + a_{22}b_{21} & a_{21}b_{12} + a_{22}b_{22} \end{bmatrix}$

NOW WORK PROBLEM 25 ●●

 CAUTION

Example 4 shows that even when the matrix products **AB** and **BA** are both defined and the same size, they are not necessarily equal.

● EXAMPLE 4 **Matrix Products**

Compute **AB** *and* **BA** *if*

$$\mathbf{A} = \begin{bmatrix} 2 & -1 \\ 3 & 1 \end{bmatrix} \quad and \quad \mathbf{B} = \begin{bmatrix} -2 & 1 \\ 1 & 4 \end{bmatrix}.$$

Solution: We have

$$\mathbf{AB} = \begin{bmatrix} 2 & -1 \\ 3 & 1 \end{bmatrix} \begin{bmatrix} -2 & 1 \\ 1 & 4 \end{bmatrix} = \begin{bmatrix} -5 & -2 \\ -5 & 7 \end{bmatrix}$$

$$\mathbf{BA} = \begin{bmatrix} -2 & 1 \\ 1 & 4 \end{bmatrix} \begin{bmatrix} 2 & -1 \\ 3 & 1 \end{bmatrix} = \begin{bmatrix} -1 & 3 \\ 14 & 3 \end{bmatrix}$$

Note that although both **AB** and **BA** are defined, and the same size, **AB** and **BA** are not equal.

NOW WORK PROBLEM 37 ●●

T E C H N O L O G Y

Figure 6.3 shows the result of using a graphing calculator to find the product **AB** in Example 4.

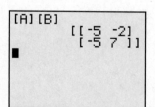

FIGURE 6.3 Calculator solution of matrix product of Example 4.

COST VECTOR

The prices (in dollars per unit) for three textbooks are represented by the price vector $\mathbf{P} = [26.25\ 34.75\ 28.50]$. A university bookstore orders these books in the quantities given by the column vector $\mathbf{Q} = \begin{bmatrix} 250 \\ 325 \\ 175 \end{bmatrix}$. Find the total cost (in dollars) of the purchase.

● **EXAMPLE 5** **Cost Vector**

Suppose that the prices (in dollars per unit) for products A, B, and C are represented by the price vector

Price of
A B C

$$\mathbf{P} = [2 \quad 3 \quad 4]$$

If the quantities (in units) of A, B, and C that are purchased are given by the column vector

$$\mathbf{Q} = \begin{bmatrix} 7 \\ 5 \\ 11 \end{bmatrix} \begin{array}{l} \text{units of A} \\ \text{units of B} \\ \text{units of C} \end{array}$$

then the total cost (in dollars) of the purchases is given by the entry in the cost vector

$$\mathbf{PQ} = [2 \quad 3 \quad 4] \begin{bmatrix} 7 \\ 5 \\ 11 \end{bmatrix} = [(2 \cdot 7) + (3 \cdot 5) + (4 \cdot 11)] = [73]$$

NOW WORK PROBLEM 27 ●●●

● **EXAMPLE 6** **Profit for an Economy**

In Example 3 of Section 6.2, suppose that in the hypothetical economy the price of coal is \$10,000 per unit, the price of electricity is \$20,000 per unit, and the price of steel is \$40,000 per unit. These prices can be represented by the (column) price vector

$$\mathbf{P} = \begin{bmatrix} 10,000 \\ 20,000 \\ 40,000 \end{bmatrix}$$

Consider the steel industry. It sells a total of 30 units of steel at \$40,000 per unit, and its total income is therefore \$1,200,000. Its costs for the various goods are given by the matrix product

$$\mathbf{D_S P} = [30 \quad 5 \quad 0] \begin{bmatrix} 10,000 \\ 20,000 \\ 40,000 \end{bmatrix} = [400,000]$$

Hence, the profit for the steel industry is \$1,200,000 − \$400,000 = \$800,000.

NOW WORK PROBLEM 67 ●●●

Matrix multiplication satisfies the following properties, provided that all sums and products are defined:

Properties of Matrix Multiplication

    1. $\mathbf{A(BC)} = \mathbf{(AB)C}$ (associative property)

    2. $\mathbf{A(B + C)} = \mathbf{AB} + \mathbf{AC}$,   (distributive properties)
        $\mathbf{(A + B)C} = \mathbf{AC} + \mathbf{BC}$

● **EXAMPLE 7** **Associative Property**

If

$$\mathbf{A} = \begin{bmatrix} 1 & -2 \\ -3 & 4 \end{bmatrix} \qquad \mathbf{B} = \begin{bmatrix} 3 & 0 & -1 \\ 1 & 1 & 2 \end{bmatrix} \qquad \mathbf{C} = \begin{bmatrix} 1 & 0 \\ 0 & 2 \\ 1 & 1 \end{bmatrix}$$

compute $\mathbf{ABC}$ *in two ways.*

Solution: Grouping **BC** gives

$$\mathbf{A}(\mathbf{BC}) = \begin{bmatrix} 1 & -2 \\ -3 & 4 \end{bmatrix} \left(\begin{bmatrix} 3 & 0 & -1 \\ 1 & 1 & 2 \end{bmatrix} \begin{bmatrix} 1 & 0 \\ 0 & 2 \\ 1 & 1 \end{bmatrix} \right)$$

$$= \begin{bmatrix} 1 & -2 \\ -3 & 4 \end{bmatrix} \begin{bmatrix} 2 & -1 \\ 3 & 4 \end{bmatrix} = \begin{bmatrix} -4 & -9 \\ 6 & 19 \end{bmatrix}$$

Alternatively, grouping **AB** gives

$$(\mathbf{AB})\mathbf{C} = \left(\begin{bmatrix} 1 & -2 \\ -3 & 4 \end{bmatrix} \begin{bmatrix} 3 & 0 & -1 \\ 1 & 1 & 2 \end{bmatrix} \right) \begin{bmatrix} 1 & 0 \\ 0 & 2 \\ 1 & 1 \end{bmatrix}$$

$$= \begin{bmatrix} 1 & -2 & -5 \\ -5 & 4 & 11 \end{bmatrix} \begin{bmatrix} 1 & 0 \\ 0 & 2 \\ 1 & 1 \end{bmatrix}$$

$$= \begin{bmatrix} -4 & -9 \\ 6 & 19 \end{bmatrix}$$

Note that $\mathbf{A}(\mathbf{BC}) = (\mathbf{AB})\mathbf{C}$.

● EXAMPLE 8 **Distributive Property**

Verify that $\mathbf{A}(\mathbf{B} + \mathbf{C}) = \mathbf{AB} + \mathbf{AC}$ *if*

$$\mathbf{A} = \begin{bmatrix} 1 & 0 \\ 2 & 3 \end{bmatrix} \qquad \mathbf{B} = \begin{bmatrix} -2 & 0 \\ 1 & 3 \end{bmatrix} \qquad \mathbf{C} = \begin{bmatrix} -2 & 1 \\ 0 & 2 \end{bmatrix}$$

Solution: On the left side, we have

$$\mathbf{A}(\mathbf{B} + \mathbf{C}) = \begin{bmatrix} 1 & 0 \\ 2 & 3 \end{bmatrix} \left(\begin{bmatrix} -2 & 0 \\ 1 & 3 \end{bmatrix} + \begin{bmatrix} -2 & 1 \\ 0 & 2 \end{bmatrix} \right)$$

$$= \begin{bmatrix} 1 & 0 \\ 2 & 3 \end{bmatrix} \begin{bmatrix} -4 & 1 \\ 1 & 5 \end{bmatrix} = \begin{bmatrix} -4 & 1 \\ -5 & 17 \end{bmatrix}$$

On the right side,

$$\mathbf{AB} + \mathbf{AC} = \begin{bmatrix} 1 & 0 \\ 2 & 3 \end{bmatrix} \begin{bmatrix} -2 & 0 \\ 1 & 3 \end{bmatrix} + \begin{bmatrix} 1 & 0 \\ 2 & 3 \end{bmatrix} \begin{bmatrix} -2 & 1 \\ 0 & 2 \end{bmatrix}$$

$$= \begin{bmatrix} -2 & 0 \\ -1 & 9 \end{bmatrix} + \begin{bmatrix} -2 & 1 \\ -4 & 8 \end{bmatrix} = \begin{bmatrix} -4 & 1 \\ -5 & 17 \end{bmatrix}$$

Thus, $\mathbf{A}(\mathbf{B} + \mathbf{C}) = \mathbf{AB} + \mathbf{AC}$.

NOW WORK PROBLEM 69 ●●

● EXAMPLE 9 **Raw Materials and Cost**

Suppose that a building contractor has accepted orders for five ranch-style houses, seven Cape Cod–style houses, and 12 colonial-style houses. Then his orders can be represented by the row vector

$$\mathbf{Q} = \begin{bmatrix} 5 & 7 & 12 \end{bmatrix}$$

Furthermore, suppose that the "raw materials" that go into each type of house are steel, wood, glass, paint, and labor. The entries in the following matrix **R** give the number of units of each raw material going into each type of house (the entries are not necessarily realistic, but are chosen for convenience):

	Steel	Wood	Glass	Paint	Labor
Ranch	5	20	16	7	17
Cape Cod	7	18	12	9	21	= **R**
Colonial	6	25	8	5	13

Each row indicates the amount of each raw material needed for a given type of house; each column indicates the amount of a given raw material needed for each type of house. Suppose now that the contractor wishes to compute the amount of each raw material needed to fulfill his orders. Then such information is given by the matrix

$$\mathbf{QR} = \begin{bmatrix} 5 & 7 & 12 \end{bmatrix} \begin{bmatrix} 5 & 20 & 16 & 7 & 17 \\ 7 & 18 & 12 & 9 & 21 \\ 6 & 25 & 8 & 5 & 13 \end{bmatrix}$$

$$= \begin{bmatrix} 146 & 526 & 260 & 158 & 388 \end{bmatrix}$$

Thus, the contractor should order 146 units of steel, 526 units of wood, 260 units of glass, and so on.

The contractor is also interested in the costs he will have to pay for these materials. Suppose steel costs $2500 per unit, wood costs $1200 per unit, and glass, paint, and labor cost $800, $150, and $1500 per unit, respectively. These data can be written as the column cost vector

$$\mathbf{C} = \begin{bmatrix} 2500 \\ 1200 \\ 800 \\ 150 \\ 1500 \end{bmatrix}$$

Then the cost of each type of house is given by the matrix

$$\mathbf{RC} = \begin{bmatrix} 5 & 20 & 16 & 7 & 17 \\ 7 & 18 & 12 & 9 & 21 \\ 6 & 25 & 8 & 5 & 13 \end{bmatrix} \begin{bmatrix} 2500 \\ 1200 \\ 800 \\ 150 \\ 1500 \end{bmatrix} = \begin{bmatrix} 75,850 \\ 81,550 \\ 71,650 \end{bmatrix}$$

Consequently, the cost of materials for the ranch-style house is $75,850, for the Cape Cod house $81,550, and for the colonial house $71,650.

The total cost of raw materials for all the houses is given by

$$\mathbf{QRC} = \mathbf{Q(RC)} = \begin{bmatrix} 5 & 7 & 12 \end{bmatrix} \begin{bmatrix} 75,850 \\ 81,550 \\ 71,650 \end{bmatrix} = \begin{bmatrix} 1,809,900 \end{bmatrix}$$

The total cost is $1,809,900.

NOW WORK PROBLEM 65 ⬤◑◯

Another property of matrices involves scalar and matrix multiplications. If k is a scalar and the product **AB** is defined, then

$$k(\mathbf{AB}) = (k\mathbf{A})\mathbf{B} = \mathbf{A}(k\mathbf{B})$$

The product $k(\mathbf{AB})$ may be written simply as $k\mathbf{AB}$. Thus,

$$k\mathbf{AB} = k(\mathbf{AB}) = (k\mathbf{A})\mathbf{B} = \mathbf{A}(k\mathbf{B})$$

For example,

$$3 \begin{bmatrix} 2 & 1 \\ 0 & -1 \end{bmatrix} \begin{bmatrix} 1 & 3 \\ 2 & 0 \end{bmatrix} = \left(3 \begin{bmatrix} 2 & 1 \\ 0 & -1 \end{bmatrix} \right) \begin{bmatrix} 1 & 3 \\ 2 & 0 \end{bmatrix}$$

$$= \begin{bmatrix} 6 & 3 \\ 0 & -3 \end{bmatrix} \begin{bmatrix} 1 & 3 \\ 2 & 0 \end{bmatrix}$$

$$= \begin{bmatrix} 12 & 18 \\ -6 & 0 \end{bmatrix}$$

There is an interesting property concerning the transpose of a matrix product:

$$(\mathbf{AB})^{\mathrm{T}} = \mathbf{B}^{\mathrm{T}}\mathbf{A}^{\mathrm{T}}$$

In words, the transpose of a product of matrices is equal to the product of their transposes in the *reverse* order.

This property can be extended to the case of more than two factors. For example,

Here we used the fact that $(\mathbf{A}^{\mathrm{T}})^{\mathrm{T}} = \mathbf{A}$.

$$(\mathbf{A}^{\mathrm{T}}\mathbf{BC})^{\mathrm{T}} = \mathbf{C}^{\mathrm{T}}\mathbf{B}^{\mathrm{T}}(\mathbf{A}^{\mathrm{T}})^{\mathrm{T}} = \mathbf{C}^{\mathrm{T}}\mathbf{B}^{\mathrm{T}}\mathbf{A}$$

● EXAMPLE 10   **Transpose of a Product**

Let

$$\mathbf{A} = \begin{bmatrix} 1 & 0 \\ 1 & 2 \end{bmatrix} \quad and \quad \mathbf{B} = \begin{bmatrix} 1 & 2 \\ 1 & 0 \end{bmatrix}$$

Show that $(\mathbf{AB})^{\mathrm{T}} = \mathbf{B}^{\mathrm{T}}\mathbf{A}^{\mathrm{T}}$.

Solution: We have

$$\mathbf{AB} = \begin{bmatrix} 1 & 2 \\ 3 & 2 \end{bmatrix} \quad \text{so} \quad (\mathbf{AB})^{\mathrm{T}} = \begin{bmatrix} 1 & 3 \\ 2 & 2 \end{bmatrix}$$

Now,

$$\mathbf{A}^{\mathrm{T}} = \begin{bmatrix} 1 & 1 \\ 0 & 2 \end{bmatrix} \quad and \quad \mathbf{B}^{\mathrm{T}} = \begin{bmatrix} 1 & 1 \\ 2 & 0 \end{bmatrix}$$

Thus,

$$\mathbf{B}^{\mathrm{T}}\mathbf{A}^{\mathrm{T}} = \begin{bmatrix} 1 & 1 \\ 2 & 0 \end{bmatrix} \begin{bmatrix} 1 & 1 \\ 0 & 2 \end{bmatrix} = \begin{bmatrix} 1 & 3 \\ 2 & 2 \end{bmatrix} = (\mathbf{AB})^{\mathrm{T}}$$

so $(\mathbf{AB})^{\mathrm{T}} = \mathbf{B}^{\mathrm{T}}\mathbf{A}^{\mathrm{T}}$.

Just as the zero matrix plays an important role as the identity in matrix addition, there is a special matrix, called the *identity matrix,* that plays a corresponding role in matrix multiplication:

The $n \times n$ **identity matrix**, denoted $\mathbf{I}_n$, is the diagonal matrix whose main diagonal entries are 1's.

For example, the identity matrices $\mathbf{I}_3$ and $\mathbf{I}_4$ are

$$\mathbf{I}_3 = \begin{bmatrix} 1 & 0 & 0 \\ 0 & 1 & 0 \\ 0 & 0 & 1 \end{bmatrix} \quad and \quad \mathbf{I}_4 = \begin{bmatrix} 1 & 0 & 0 & 0 \\ 0 & 1 & 0 & 0 \\ 0 & 0 & 1 & 0 \\ 0 & 0 & 0 & 1 \end{bmatrix}$$

When the size of an identity matrix is understood, we omit the subscript and simply denote the matrix by $\mathbf{I}$. It should be clear that

$$\mathbf{I}^{\mathrm{T}} = \mathbf{I}$$

The identity matrix plays the same role in matrix multiplication as does the number 1 in the multiplication of real numbers. That is, just as the product of a real number and 1 is the number itself, the product of a matrix and the identity matrix is the matrix itself. For example,

$$\begin{bmatrix} 2 & 4 \\ 1 & 5 \end{bmatrix} \mathbf{I} = \begin{bmatrix} 2 & 4 \\ 1 & 5 \end{bmatrix} \begin{bmatrix} 1 & 0 \\ 0 & 1 \end{bmatrix} = \begin{bmatrix} 2 & 4 \\ 1 & 5 \end{bmatrix}$$

and

$$\mathbf{I}\begin{bmatrix} 2 & 4 \\ 1 & 5 \end{bmatrix} = \begin{bmatrix} 1 & 0 \\ 0 & 1 \end{bmatrix}\begin{bmatrix} 2 & 4 \\ 1 & 5 \end{bmatrix} = \begin{bmatrix} 2 & 4 \\ 1 & 5 \end{bmatrix}$$

In general, if $\mathbf{I}$ is $n \times n$ and $\mathbf{A}$ has n columns, then $\mathbf{AI} = \mathbf{A}$. If $\mathbf{B}$ has n rows, then $\mathbf{IB} = \mathbf{B}$. Moreover, if $\mathbf{A}$ is $n \times n$, then

$$\mathbf{AI} = \mathbf{IA} = \mathbf{A}$$

● EXAMPLE 11 Matrix Operations Involving I and O

If

$$\mathbf{A} = \begin{bmatrix} 3 & 2 \\ 1 & 4 \end{bmatrix} \qquad \mathbf{B} = \begin{bmatrix} \frac{2}{5} & -\frac{1}{5} \\ -\frac{1}{10} & \frac{3}{10} \end{bmatrix}$$

$$\mathbf{I} = \begin{bmatrix} 1 & 0 \\ 0 & 1 \end{bmatrix} \qquad \mathbf{O} = \begin{bmatrix} 0 & 0 \\ 0 & 0 \end{bmatrix}$$

compute each of the following.

a. $\mathbf{I} - \mathbf{A}$.

Solution:

$$\mathbf{I} - \mathbf{A} = \begin{bmatrix} 1 & 0 \\ 0 & 1 \end{bmatrix} - \begin{bmatrix} 3 & 2 \\ 1 & 4 \end{bmatrix} = \begin{bmatrix} -2 & -2 \\ -1 & -3 \end{bmatrix}$$

b. $3(\mathbf{A} - 2\mathbf{I})$.

Solution:

$$3(\mathbf{A} - 2\mathbf{I}) = 3\left(\begin{bmatrix} 3 & 2 \\ 1 & 4 \end{bmatrix} - 2\begin{bmatrix} 1 & 0 \\ 0 & 1 \end{bmatrix}\right)$$

$$= 3\left(\begin{bmatrix} 3 & 2 \\ 1 & 4 \end{bmatrix} - \begin{bmatrix} 2 & 0 \\ 0 & 2 \end{bmatrix}\right)$$

$$= 3\begin{bmatrix} 1 & 2 \\ 1 & 2 \end{bmatrix} = \begin{bmatrix} 3 & 6 \\ 3 & 6 \end{bmatrix}$$

c. $\mathbf{AO}$.

Solution:

$$\mathbf{AO} = \begin{bmatrix} 3 & 2 \\ 1 & 4 \end{bmatrix}\begin{bmatrix} 0 & 0 \\ 0 & 0 \end{bmatrix} = \begin{bmatrix} 0 & 0 \\ 0 & 0 \end{bmatrix} = \mathbf{O}$$

In general, if $\mathbf{AO}$ and $\mathbf{OA}$ are defined, then

$$\mathbf{AO} = \mathbf{OA} = \mathbf{O}$$

d. $\mathbf{AB}$.

Solution:

$$\mathbf{AB} = \begin{bmatrix} 3 & 2 \\ 1 & 4 \end{bmatrix}\begin{bmatrix} \frac{2}{5} & -\frac{1}{5} \\ -\frac{1}{10} & \frac{3}{10} \end{bmatrix} = \begin{bmatrix} 1 & 0 \\ 0 & 1 \end{bmatrix} = \mathbf{I}$$

NOW WORK PROBLEM 55 ●●●

If $\mathbf{A}$ is a square matrix, we can speak of a *power* of $\mathbf{A}$:

If $\mathbf{A}$ is a square matrix and p is a positive integer, then the **pth power** of $\mathbf{A}$, written $\mathbf{A}^p$, is the product of p factors of $\mathbf{A}$:

$$\mathbf{A}^p = \underbrace{\mathbf{A} \cdot \mathbf{A} \cdots \mathbf{A}}_{p \text{ factors}}$$

If $\mathbf{A}$ is $n \times n$, we define $\mathbf{A}^0 = \mathbf{I}_n$.

We remark that $\mathbf{I}^p = \mathbf{I}$.

● EXAMPLE 12   **Power of a Matrix**

If $\mathbf{A} = \begin{bmatrix} 1 & 0 \\ 1 & 2 \end{bmatrix}$, *compute* $\mathbf{A}^3$.

Solution: Since $\mathbf{A}^3 = (\mathbf{A}^2)\mathbf{A}$ and

$$\mathbf{A}^2 = \begin{bmatrix} 1 & 0 \\ 1 & 2 \end{bmatrix} \begin{bmatrix} 1 & 0 \\ 1 & 2 \end{bmatrix} = \begin{bmatrix} 1 & 0 \\ 3 & 4 \end{bmatrix}$$

we have

$$\mathbf{A}^3 = \mathbf{A}^2\mathbf{A} = \begin{bmatrix} 1 & 0 \\ 3 & 4 \end{bmatrix} \begin{bmatrix} 1 & 0 \\ 1 & 2 \end{bmatrix} = \begin{bmatrix} 1 & 0 \\ 7 & 8 \end{bmatrix}$$

NOW WORK PROBLEM 45 ●●●

T E C H N O L O G Y

Use of the graphing calculator to compute $\mathbf{A}^4$, where $\mathbf{A} = \begin{bmatrix} 2 & -3 \\ 1 & 4 \end{bmatrix}$, is shown in Figure 6.4.

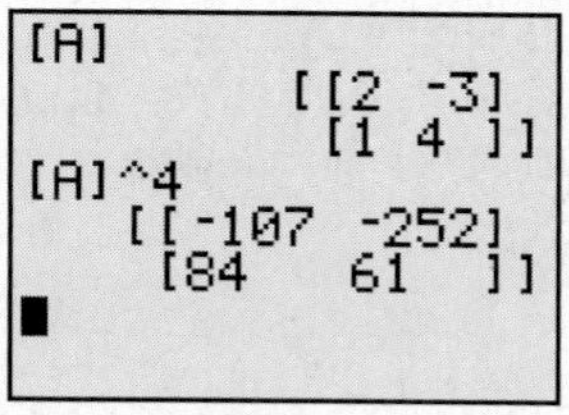

```
[A]
          [[2  -3]
           [1  4 ]]
[A]^4
   [[-107 -252]
    [84    61 ]]
■
```

FIGURE 6.4   Power of a matrix.

Matrix Equations

Systems of linear equations can be represented by using matrix multiplication. For example, consider the matrix equation

$$\begin{bmatrix} 1 & 4 & -2 \\ 2 & -3 & 1 \end{bmatrix} \begin{bmatrix} x_1 \\ x_2 \\ x_3 \end{bmatrix} = \begin{bmatrix} 4 \\ -3 \end{bmatrix} \tag{1}$$

The product on the left side has order 2×1 and hence is a column matrix. Thus,

$$\begin{bmatrix} x_1 + 4x_2 - 2x_3 \\ 2x_1 - 3x_2 + x_3 \end{bmatrix} = \begin{bmatrix} 4 \\ -3 \end{bmatrix}$$

By equality of matrices, corresponding entries must be equal, so we obtain the system

$$\begin{cases} x_1 + 4x_2 - 2x_3 = 4 \\ 2x_1 - 3x_2 + x_3 = -3 \end{cases}$$

Hence, this system of linear equations can be defined by matrix Equation (1). We usually describe Equation (1) by saying that it has the form

$$\mathbf{AX} = \mathbf{B}$$

where $\mathbf{A}$ is the matrix obtained from the coefficients of the variables, $\mathbf{X}$ is a column matrix obtained from the variables, and $\mathbf{B}$ is a column matrix obtained from the constants. Matrix $\mathbf{A}$ is called the *coefficient matrix* for the system.

PRINCIPLES IN PRACTICE 3

MATRIX FORM OF A SYSTEM USING MATRIX MULTIPLICATION

Write the following pair of lines in matrix form, using matrix multiplication.

$$y = -\frac{8}{5}x + \frac{8}{5}, \quad y = -\frac{1}{3}x + \frac{5}{3}$$

● EXAMPLE 13   **Matrix Form of a System Using Matrix Multiplication**

Write the system

$$\begin{cases} 2x_1 + 5x_2 = 4 \\ 8x_1 + 3x_2 = 7 \end{cases}$$

in matrix form by using matrix multiplication.

Solution: If

$$A = \begin{bmatrix} 2 & 5 \\ 8 & 3 \end{bmatrix} \qquad X = \begin{bmatrix} x_1 \\ x_2 \end{bmatrix} \qquad B = \begin{bmatrix} 4 \\ 7 \end{bmatrix}$$

then the given system is equivalent to the single matrix equation

$$AX = B$$

that is,

$$\begin{bmatrix} 2 & 5 \\ 8 & 3 \end{bmatrix} \begin{bmatrix} x_1 \\ x_2 \end{bmatrix} = \begin{bmatrix} 4 \\ 7 \end{bmatrix}$$

NOW WORK PROBLEM 59 ●●

Problems 6.3

If $A = \begin{bmatrix} 1 & 3 & -2 \\ -2 & 1 & -1 \\ 0 & 4 & 3 \end{bmatrix}$, $B = \begin{bmatrix} 0 & -2 & 3 \\ -2 & 4 & -2 \\ 3 & 1 & -1 \end{bmatrix}$, and $AB = C = [c_{ij}]$, find each of the following.

1. c_{11} **2.** c_{23} **3.** c_{32}

4. c_{33} **5.** c_{31} **6.** c_{12}

If A is 2×3, B is 3×1, C is 2×5, D is 4×3, E is 3×2, and F is 2×3, find the size and number of entries of each of the following.

*7. **AE**   8. **DE**   9. **EC**

10. **DB**   11. **FB**   12. **BC**

13. **EE^{T}B**   14. **E(AE)**   15. **E(FB)**

16. **(F + A)B**

Write the identity matrix that has the following order:

17. 4   **18.** 6

In Problems 19–36, perform the indicated operations.

*19. $\begin{bmatrix} 2 & -4 \\ 3 & 2 \end{bmatrix} \begin{bmatrix} 4 & 0 \\ -1 & 3 \end{bmatrix}$ **20.** $\begin{bmatrix} -1 & 1 \\ 0 & 4 \\ 2 & 1 \end{bmatrix} \begin{bmatrix} 1 & -2 \\ 3 & 4 \end{bmatrix}$

21. $\begin{bmatrix} 2 & 0 & 3 \\ -1 & 4 & 5 \end{bmatrix} \begin{bmatrix} 1 \\ 4 \\ 7 \end{bmatrix}$ **22.** $[1 \ \ 0 \ \ 6 \ \ 2] \begin{bmatrix} 0 \\ 1 \\ 2 \\ 3 \end{bmatrix}$

23. $\begin{bmatrix} 1 & 4 & -1 \\ 0 & 0 & 2 \\ -2 & 1 & 1 \end{bmatrix} \begin{bmatrix} 2 & 1 & 0 \\ 0 & -1 & 1 \\ 1 & 1 & 2 \end{bmatrix}$

24. $\begin{bmatrix} 4 & 2 & -2 \\ 3 & 10 & 0 \\ 1 & 0 & 2 \end{bmatrix} \begin{bmatrix} 3 & 1 & 1 & 0 \\ 0 & 0 & 0 & 0 \\ 0 & 1 & 0 & 1 \end{bmatrix}$

*25. $[1 \ \ -2 \ \ 5] \begin{bmatrix} 1 & 5 & -2 & -1 \\ 0 & 0 & 2 & 1 \\ -1 & 0 & 1 & -3 \end{bmatrix}$

26. $[1 \ \ -4] \begin{bmatrix} -2 & 1 \\ 0 & 1 \\ 5 & 0 \end{bmatrix}$ *27. $\begin{bmatrix} 2 \\ 3 \\ -4 \\ 1 \end{bmatrix} [2 \ \ 3 \ \ -2 \ \ 3]$

28. $\begin{bmatrix} 0 & 1 \\ 2 & 3 \end{bmatrix} \left(\begin{bmatrix} 1 & 0 & 1 \\ 1 & 1 & 0 \end{bmatrix} + \begin{bmatrix} 0 & 1 & 0 \\ 0 & 0 & 1 \end{bmatrix} \right)$

29. $3 \left(\begin{bmatrix} -2 & 0 & 2 \\ 3 & -1 & 1 \end{bmatrix} + 2 \begin{bmatrix} -1 & 0 & 2 \\ 1 & 1 & -2 \end{bmatrix} \right) \begin{bmatrix} 1 & 2 \\ 3 & 4 \\ 5 & 6 \end{bmatrix}$

30. $\begin{bmatrix} 1 & -1 \\ 0 & 3 \end{bmatrix} \begin{bmatrix} -1 & 0 & -1 & 0 & 0 \\ 2 & 1 & 2 & 1 & 1 \end{bmatrix}$

31. $\begin{bmatrix} 1 & 2 \\ 3 & 4 \end{bmatrix} \left(\begin{bmatrix} 2 & 0 & 1 \\ 1 & 0 & -2 \end{bmatrix} \begin{bmatrix} 1 & -2 \\ 2 & 1 \\ 3 & 0 \end{bmatrix} \right)$

32. $3 \begin{bmatrix} 1 & 2 \\ -1 & 4 \end{bmatrix} - 4 \left(\begin{bmatrix} 1 & 0 \\ 0 & 1 \end{bmatrix} \begin{bmatrix} -2 & 4 \\ 6 & 1 \end{bmatrix} \right)$

33. $\begin{bmatrix} 0 & 0 & 1 \\ 0 & 1 & 0 \\ 1 & 0 & 0 \end{bmatrix} \begin{bmatrix} x \\ y \\ z \end{bmatrix}$ **34.** $\begin{bmatrix} a_{11} & a_{12} \\ a_{21} & a_{22} \end{bmatrix} \begin{bmatrix} x_1 \\ x_2 \end{bmatrix}$

35. $\begin{bmatrix} 2 & 1 & 3 \\ 4 & 9 & 7 \end{bmatrix} \begin{bmatrix} x_1 \\ x_2 \\ x_3 \end{bmatrix}$ **36.** $\begin{bmatrix} 2 & -3 \\ 0 & 1 \\ 2 & 1 \end{bmatrix} \begin{bmatrix} x_1 \\ x_2 \end{bmatrix}$

In Problems 37–44, compute the required matrices if

$$A = \begin{bmatrix} 1 & -2 \\ 0 & 3 \end{bmatrix} \qquad B = \begin{bmatrix} -2 & 3 & 0 \\ 1 & -4 & 1 \end{bmatrix} \qquad C = \begin{bmatrix} -1 & 1 \\ 0 & 3 \\ 2 & 4 \end{bmatrix}$$

$$D = \begin{bmatrix} 1 & 0 & 0 \\ 0 & 1 & 1 \\ 1 & 2 & 1 \end{bmatrix} \qquad E = \begin{bmatrix} 3 & 0 & 0 \\ 0 & 6 & 0 \\ 0 & 0 & 3 \end{bmatrix} \qquad F = \begin{bmatrix} \frac{1}{3} & 0 & 0 \\ 0 & \frac{1}{6} & 0 \\ 0 & 0 & \frac{1}{3} \end{bmatrix}$$

$$I = \begin{bmatrix} 1 & 0 & 0 \\ 0 & 1 & 0 \\ 0 & 0 & 1 \end{bmatrix}$$

*37. $D - \frac{1}{3}EI$ 38. DD 39. $3A - 2BC$

40. $B(D + E)$ 41. $3I - \frac{2}{3}FE$ 42. $FE(D - I)$

43. $(DC)A$ 44. $A(BC)$

In Problems 45–58, compute the required matrix, if it exists, given that

$$A = \begin{bmatrix} 1 & -1 & 0 \\ 0 & 1 & 1 \end{bmatrix} \qquad B = \begin{bmatrix} 0 & 0 & -1 \\ 2 & -1 & 0 \\ 0 & 0 & 2 \end{bmatrix} \qquad C = \begin{bmatrix} 1 & 0 \\ 2 & -1 \\ 0 & 1 \end{bmatrix}$$

$$I = \begin{bmatrix} 1 & 0 & 0 \\ 0 & 1 & 0 \\ 0 & 0 & 1 \end{bmatrix} \qquad O = \begin{bmatrix} 0 & 0 & 0 \\ 0 & 0 & 0 \\ 0 & 0 & 0 \end{bmatrix}$$

*45. $\mathbf{A}^2$ 46. $\mathbf{A}^T\mathbf{A}$ 47. $\mathbf{B}^3$

48. $\mathbf{A}(\mathbf{B}^T)^2\mathbf{C}$ 49. $(\mathbf{A}\mathbf{I}\mathbf{C})^T$ 50. $\mathbf{A}^T(2\mathbf{C}^T)$

51. $(\mathbf{B}\mathbf{A}^T)^T$ 52. $(2\mathbf{B})^T$ 53. $(2\mathbf{I})^2 - 2\mathbf{I}^2$

54. $(\mathbf{A}^T\mathbf{C}^T\mathbf{B})^0$ *55. $\mathbf{A}(\mathbf{I} - \mathbf{O})$ 56. $\mathbf{I}^T\mathbf{O}$

57. $(\mathbf{A}\mathbf{B})(\mathbf{A}\mathbf{B})^T$ 58. $\mathbf{B}^2 - 3\mathbf{B} + 2\mathbf{I}$

In Problems 59–61, represent the given system by using matrix multiplication.

*59. $\begin{cases} 3x + y = 6 \\ 2x - 9y = 5 \end{cases}$ 60. $\begin{cases} 3x + y + z = 2 \\ x - y + z = 4 \\ 5x - y + 2z = 12 \end{cases}$

61. $\begin{cases} 2r - s + 3t = 9 \\ 5r - s + 2t = 5 \\ 3r - 2s + 2t = 11 \end{cases}$

62. **Secret Messages**   Secret messages can be encoded by using a code and an encoding matrix. Suppose we have the following code:

a	b	c	d	e	f	g	h	i	j	k	l	m
1	2	3	4	5	6	7	8	9	10	11	12	13

n	o	p	q	r	s	t	u	v	w	x	y	z
14	15	16	17	18	19	20	21	22	23	24	25	26

Let the encoding matrix be $\mathbf{E} = \begin{bmatrix} 1 & 3 \\ 2 & 4 \end{bmatrix}$. Then we can encode a message by taking every two letters of the message, converting them to their corresponding numbers, creating a 2×1 matrix, and then multiplying each matrix by $\mathbf{E}$. Use this code and matrix to encode the message "the/falcon/has/landed," leaving the slashes to separate words.

63. **Inventory**   A pet store has 6 kittens, 10 puppies, and 7 parrots in stock. If the value of each kitten is $55, each puppy is $150, and each parrot is $35, find the total value of the pet store's inventory using matrix multiplication.

64. **Stocks**   A stockbroker sold a customer 200 shares of stock A, 300 shares of stock B, 500 shares of stock C, and 250 shares of stock D. The prices per share of A, B, C, and D are $100, $150, $200, and $300, respectively. Write a row vector representing the number of shares of each stock bought. Write a column vector representing the price per share of each stock. Using matrix multiplication, find the total cost of the stocks.

*65. **Construction Cost**   In Example 9, assume that the contractor is to build five ranch-style, two Cape Cod–style, and four colonial-style houses. Using matrix multiplication, compute the total cost of raw materials.

66. **Costs**   In Example 9, assume that the contractor wishes to take into account the cost of transporting raw materials to the building site, as well as the purchasing cost. Suppose the

costs are given in the following matrix:

$$\mathbf{C} = \begin{array}{c} \\ \\ \\ \\ \\ \end{array} \overset{\text{Purchase} \quad \text{Transport}}{\begin{bmatrix} 3500 & 50 \\ 1500 & 50 \\ 1000 & 100 \\ 250 & 10 \\ 3500 & 0 \end{bmatrix}} \begin{array}{l} \text{Steel} \\ \text{Wood} \\ \text{Glass} \\ \text{Paint} \\ \text{Labor} \end{array}$$

(a) By computing $\mathbf{RC}$, find a matrix whose entries give the purchase and transportation costs of the materials for each type of house.

(b) Find the matrix $\mathbf{QRC}$ whose first entry gives the total purchase price and whose second entry gives the total transportation cost.

(c) Let $\mathbf{Z} = \begin{bmatrix} 1 \\ 1 \end{bmatrix}$, and then compute $\mathbf{QRCZ}$, which gives the total cost of materials and transportation for all houses being built.

*67. Perform the following calculations for Example 6.

(a) Compute the amount that each industry and each consumer have to pay for the goods they receive.

(b) Compute the profit earned by each industry.

(c) Find the total amount of money that is paid out by all the industries and consumers.

(d) Find the proportion of the total amount of money found in part (c) paid out by the industries. Find the proportion of the total amount of money found in part (c) that is paid out by the consumers.

68. Prove that if $\mathbf{AB} = \mathbf{BA}$, then $(\mathbf{A} + \mathbf{B})(\mathbf{A} - \mathbf{B}) = \mathbf{A}^2 - \mathbf{B}^2$.

*69. Show that if

$$\mathbf{A} = \begin{bmatrix} 1 & 2 \\ 1 & 2 \end{bmatrix} \quad \text{and} \quad \mathbf{B} = \begin{bmatrix} 2 & -3 \\ -1 & \frac{3}{2} \end{bmatrix}$$

then $\mathbf{AB} = \mathbf{O}$. Observe that since neither $\mathbf{A}$ nor $\mathbf{B}$ is the zero matrix, the algebraic rule for real numbers, "if $ab = 0$, then either a or b is zero," does not hold for matrices. It can also be shown that the cancellation law is not true for matrices; that is, if $\mathbf{AB} = \mathbf{AC}$, then it is not necessarily true that $\mathbf{B} = \mathbf{C}$.

70. Let $\mathbf{D}_1$ and $\mathbf{D}_2$ be two arbitrary 3×3 diagonal matrices. By computing $\mathbf{D}_1\mathbf{D}_2$ and $\mathbf{D}_2\mathbf{D}_1$, show that

(a) Both $\mathbf{D}_1\mathbf{D}_2$ and $\mathbf{D}_2\mathbf{D}_1$ are diagonal matrices.

(b) $\mathbf{D}_1$ and $\mathbf{D}_2$ *commute*, meaning that $\mathbf{D}_1\mathbf{D}_2 = \mathbf{D}_2\mathbf{D}_1$.

In Problems 71–74, compute the required matrices, given that

$$\mathbf{A} = \begin{bmatrix} 3.2 & -4.1 & 5.1 \\ -2.6 & 1.2 & 6.8 \end{bmatrix} \quad \mathbf{B} = \begin{bmatrix} 1.1 & 4.8 \\ -2.3 & 3.2 \\ 4.6 & -1.4 \end{bmatrix}$$

$$\mathbf{C} = \begin{bmatrix} -1.2 & 1.5 \\ 2.4 & 6.2 \end{bmatrix}$$

71. $\mathbf{A}(2\mathbf{B})$ 72. $-3.1(\mathbf{CA})$

73. $3\mathbf{CA}(-\mathbf{B})$ 74. $\mathbf{C}^3$

OBJECTIVE

6.4 Solving Systems by Reducing Matrices

To show how to reduce a matrix and to use matrix reduction to solve a linear system.

In this section we illustrate a method by which matrices can be used to solve a system of linear equations. In developing this *method of reduction*, we will first solve a system by the usual method of elimination. Then we will obtain the same solution by using matrices.

Let us consider the system

$$\begin{cases} 3x - y = 1 & \text{(1)} \\ x + 2y = 5 & \text{(2)} \end{cases}$$

consisting of two linear equations in two unknowns, x and y. Although this system can be solved by various algebraic methods, we willl solve it by a method that is readily adapted to matrices.

For reasons that will be obvious later, we begin by replacing Equation (1) by Equation (2), and Equation (2) by Equation (1), thus obtaining the equivalent system,[3]

$$\begin{cases} x + 2y = 5 & \text{(3)} \\ 3x - y = 1 & \text{(4)} \end{cases}$$

Multiplying both sides of Equation (3) by -3 gives $-3x - 6y = -15$. Adding the left and right sides of this equation to the corresponding sides of Equation (4) produces an equivalent system in which x is eliminated from the second equation:

$$\begin{cases} x + 2y = 5 & \text{(5)} \\ 0x - 7y = -14 & \text{(6)} \end{cases}$$

Now we will eliminate y from the first equation. Multiplying both sides of Equation (6) by $-\frac{1}{7}$ gives the equivalent system,

$$\begin{cases} x + 2y = 5 & \text{(7)} \\ 0x + y = 2 & \text{(8)} \end{cases}$$

From Equation (8), $y = 2$, and, hence, $-2y = -4$. Adding the sides of $-2y = -4$ to the corresponding sides of Equation (7), we get the equivalent system,

$$\begin{cases} x + 0y = 1 \\ 0x + y = 2 \end{cases}$$

Therefore, $x = 1$ and $y = 2$, so the original system is solved.

Note that in solving the original system, we successively replaced it by an equivalent system that was obtained by performing one of the following three operations (called *elementary operations*), which leave the solution unchanged:

1. Interchanging two equations

2. Multiplying one equation by a nonzero constant

3. Adding a constant multiple of the sides of one equation to the corresponding sides of another equation

Before showing a matrix method of solving the original system,

$$\begin{cases} 3x - y = 1 \\ x + 2y = 5 \end{cases}$$

we first need to define some terms. Recall from Section 6.3 that the matrix

$$\begin{bmatrix} 3 & -1 \\ 1 & 2 \end{bmatrix}$$

is the **coefficient matrix** of this system. The entries in the first column correspond to the coefficients of the x's in the equations. For example, the entry in the first row and first column corresponds to the coefficient of x in the first equation; and the entry in the second row and first column corresponds to the coefficient of x in the second equation. Similarly, the entries in the second column correspond to the coefficients of the y's.

Another matrix associated with this system is called the **augmented coefficient matrix** and is given by

$$\begin{bmatrix} 3 & -1 & | & 1 \\ 1 & 2 & | & 5 \end{bmatrix}$$

---

[3] Recall from Section 3.4 that two or more systems are equivalent if they have the same set of solutions.

The first and second columns are the first and second columns, respectively, of the coefficient matrix. The entries in the third column correspond to the constant terms in the system: the entry in the first row of this column is the constant term of the first equation, whereas the entry in the second row is the constant term of the second equation. Although it is not necessary to include the vertical line in the augmented coefficient matrix, it serves to remind us that the 1 and the 5 are the constant terms that appear on the right sides of the equations. The augmented coefficient matrix itself completely describes the system of equations.

The procedure that was used to solve the original system involved a number of equivalent systems. With each of these systems, we can associate its augmented coefficient matrix. Following are the systems that were involved, together with their corresponding augmented coefficient matrices, which we have labeled **A**, **B**, **C**, **D**, and **E**:

$$\begin{cases} 3x - y = 1 \\ x + 2y = 5 \end{cases} \qquad \left[\begin{array}{cc|c} 3 & -1 & 1 \\ 1 & 2 & 5 \end{array}\right] = \mathbf{A}$$

$$\begin{cases} x + 2y = 5 \\ 3x - y = 1 \end{cases} \qquad \left[\begin{array}{cc|c} 1 & 2 & 5 \\ 3 & -1 & 1 \end{array}\right] = \mathbf{B}$$

$$\begin{cases} x + 2y = 5 \\ 0x - 7y = -14 \end{cases} \qquad \left[\begin{array}{cc|c} 1 & 2 & 5 \\ 0 & -7 & -14 \end{array}\right] = \mathbf{C}$$

$$\begin{cases} x + 2y = 5 \\ 0x + y = 2 \end{cases} \qquad \left[\begin{array}{cc|c} 1 & 2 & 5 \\ 0 & 1 & 2 \end{array}\right] = \mathbf{D}$$

$$\begin{cases} x + 0y = 1 \\ 0x + y = 2 \end{cases} \qquad \left[\begin{array}{cc|c} 1 & 0 & 1 \\ 0 & 1 & 2 \end{array}\right] = \mathbf{E}$$

Let us see how these matrices are related.

B can be obtained from **A** by interchanging the first and second rows of **A**. This operation corresponds to interchanging the two equations in the original system.

C can be obtained from **B** by adding to each entry in the second row of **B** −3 times the corresponding entry in the first row of **B**:

$$\mathbf{C} = \left[\begin{array}{cc|c} 1 & 2 & 5 \\ 3 + (-3)(1) & -1 + (-3)(2) & 1 + (-3)(5) \end{array}\right]$$

$$= \left[\begin{array}{cc|c} 1 & 2 & 5 \\ 0 & -7 & -14 \end{array}\right]$$

This operation is described as follows: The addition of −3 times the first row of **B** to the second row of **B**.

D can be obtained from **C** by multiplying each entry in the second row of **C** by $-\frac{1}{7}$. This operation is referred to as multiplying the second row of **C** by $-\frac{1}{7}$. **E** can be obtained from **D** by adding −2 times the second row of **D** to the first row of **D**.

Observe that **E**, which gives the solution, was obtained from **A** by successively performing one of three matrix operations, called **elementary row operations**:

Elementary Row Operations

1. Interchanging two rows of a matrix
2. Multiplying a row of a matrix by a nonzero number
3. Adding a multiple of one row of a matrix to a different row of that matrix

These elementary row operations correspond to the three elementary operations used in the algebraic method of elimination. Whenever a matrix can be obtained from another by one or more elementary row operations, we say that the matrices are **equivalent**. Thus, **A** and **E** are equivalent. (We could also obtain **A** from **E** by performing similar row operations in the reverse order, so the term *equivalent* is appropriate.) When describing particular elementary row operations, we will use the following notation for convenience:

Notation	Corresponding Row Operation
$R_i \leftrightarrow R_j$	Interchange rows R_i and R_j.
kR_i	Multiply row R_i by the nonzero constant k.
$kR_i + R_j$	Add k times row R_i to row R_j (but leave R_i unchanged).

For example, writing

$$\begin{bmatrix} 1 & 0 & -2 \\ 4 & -2 & 1 \\ 5 & 0 & 3 \end{bmatrix} \xrightarrow{-4R_1 + R_2} \begin{bmatrix} 1 & 0 & -2 \\ 0 & -2 & 9 \\ 5 & 0 & 3 \end{bmatrix}$$

means that the second matrix was obtained from the first by adding -4 times row 1 to row 2. Note that we may write $(-k)R_i$ as $-kR_i$.

We are now ready to describe a matrix procedure for solving a system of linear equations. First, we form the augmented coefficient matrix of the system; then, by means of elementary row operations, we determine an equivalent matrix that clearly indicates the solution. Let us be quite specific as to what we mean by a matrix that *clearly indicates the solution.* This is a matrix, called a *reduced matrix,* which will be defined below. It is convenient to define first a **zero-row** of a matrix to be a row that consists *entirely* of zeros. A row that is not a zero-row, meaning that it contains *at least one* nonzero entry, will be called a **nonzero-row.** The first nonzero entry in a nonzero-row is called the **leading entry.**

Reduced Matrix

A matrix is said to be a **reduced matrix** provided that all of the following are true:

1. All zero-rows are at the bottom of the matrix.
2. For each nonzero-row, the leading entry is 1, and all other entries in the *column* in which the leading entry appears are zeros.
3. The leading entry in each row is to the right of the leading entry in any row above it.

It can be shown that each matrix is equivalent to *exactly one* reduced matrix. To solve a system, we find *the* reduced matrix such that the augmented coefficient matrix is equivalent to it. In our previous discussion of elementary row operations, the matrix

$$\mathbf{E} = \begin{bmatrix} 1 & 0 & | & 1 \\ 0 & 1 & | & 2 \end{bmatrix}$$

is a reduced matrix.

EXAMPLE 1 Reduced Matrices

For each of the following matrices, determine whether it is reduced or not reduced.

a. $\begin{bmatrix} 1 & 0 \\ 0 & 3 \end{bmatrix}$ **b.** $\begin{bmatrix} 1 & 0 & 0 \\ 0 & 1 & 0 \end{bmatrix}$ **c.** $\begin{bmatrix} 0 & 1 \\ 1 & 0 \end{bmatrix}$

d. $\begin{bmatrix} 0 & 0 & 0 \\ 0 & 0 & 0 \end{bmatrix}$ **e.** $\begin{bmatrix} 1 & 0 & 0 \\ 0 & 0 & 0 \\ 0 & 1 & 0 \end{bmatrix}$ **f.** $\begin{bmatrix} 0 & 1 & 0 & 3 \\ 0 & 0 & 1 & 2 \\ 0 & 0 & 0 & 0 \end{bmatrix}$

Solution:

a. Not a reduced matrix, because the leading entry in the second row is not 1

b. Reduced matrix

c. Not a reduced matrix, because the leading entry in the second row is not to the right of the leading entry in the first row

d. Reduced matrix

e. Not a reduced matrix, because the second row, which is a zero-row, is not at the bottom of the matrix

f. Reduced matrix

NOW WORK PROBLEM 1

● EXAMPLE 2   Reducing a Matrix

Reduce the matrix

$$\begin{bmatrix} 0 & 0 & 1 & 2 \\ 3 & -6 & -3 & 0 \\ 6 & -12 & 2 & 11 \end{bmatrix}$$

Strategy   To reduce the matrix, we must get the leading entry to be a 1 in the first row, the leading entry a 1 in the second row, and so on, until we arrive at a zero-row, if there are any. Moreover, we must work from left to right, because the leading entry in each row must be to the *left* of all other leading entries in the rows *below* it.

Solution: Since there are no zero-rows to move to the bottom, we proceed to find the first column that contains a nonzero entry; this turns out to be column 1. Accordingly, in the reduced matrix, the leading 1 in the first row must be in column 1. To accomplish this, we begin by interchanging the first two rows so that a nonzero entry is in row 1 of column 1:

$$\begin{bmatrix} 0 & 0 & 1 & 2 \\ 3 & -6 & -3 & 0 \\ 6 & -12 & 2 & 11 \end{bmatrix} \xrightarrow{R_1 \leftrightarrow R_2} \begin{bmatrix} 3 & -6 & -3 & 0 \\ 0 & 0 & 1 & 2 \\ 6 & -12 & 2 & 11 \end{bmatrix}$$

Next, we multiply row 1 by $\frac{1}{3}$ so that the leading entry is a 1:

$$\xrightarrow{\frac{1}{3}R_1} \begin{bmatrix} 1 & -2 & -1 & 0 \\ 0 & 0 & 1 & 2 \\ 6 & -12 & 2 & 11 \end{bmatrix}$$

Now, because we must have zeros below (and above) each leading 1, we add -6 times row 1 to row 3:

$$\xrightarrow{-6R_1 + R_3} \begin{bmatrix} 1 & -2 & -1 & 0 \\ 0 & 0 & 1 & 2 \\ 0 & 0 & 8 & 11 \end{bmatrix}$$

Next, we move to the right of column 1 to find the first column that has a nonzero entry in row 2 or below; this is column 3. Consequently, in the reduced matrix, the leading 1 in the second row must be in column 3. The foregoing matrix already does have a leading 1 there. Thus, all we need do to get zeros below and above the leading 1 is add 1 times row 2 to row 1 and add -8 times row 2 to row 3:

$$\begin{matrix} \xrightarrow{(1)R_2 + R_1} \\ \xrightarrow{-8R_2 + R_3} \end{matrix} \begin{bmatrix} 1 & -2 & 0 & 2 \\ 0 & 0 & 1 & 2 \\ 0 & 0 & 0 & -5 \end{bmatrix}$$

Again, we move to the right to find the first column that has a nonzero entry in row 3; namely, column 4. To make the leading entry a 1, we multiply row 3 by $-\frac{1}{5}$:

$$\xrightarrow{-\frac{1}{5}R_3} \begin{bmatrix} 1 & -2 & 0 & 2 \\ 0 & 0 & 1 & 2 \\ 0 & 0 & 0 & 1 \end{bmatrix}$$

Finally, to get all other entries in column 4 to be zeros, we add -2 times row 3 to both row 1 and row 2:

$$\xrightarrow[\substack{-2R_3+R_2}]{-2R_3+R_1}
\begin{bmatrix}
1 & -2 & 0 & 0 \\
0 & 0 & 1 & 0 \\
0 & 0 & 0 & 1
\end{bmatrix}$$

The last matrix is in reduced form.

NOW WORK PROBLEM 9

T E C H N O L O G Y

Although elementary row operations can be performed on a graphing calculator, the procedure is rather awkward.

The method of reduction described for solving our original system can be generalized to systems consisting of m linear equations in n unknowns. To solve such a system as

$$\begin{cases}
a_{11}x_1 + a_{12}x_2 + \cdots + a_{1n}x_n = c_1 \\
a_{21}x_1 + a_{22}x_2 + \cdots + a_{2n}x_n = c_2 \\
\quad \vdots \qquad\quad \vdots \qquad\qquad\quad \vdots \qquad \vdots \\
a_{m1}x_1 + a_{m2}x_2 + \cdots + a_{mn}x_n = c_m
\end{cases}$$

involves

1. determining the augmented coefficient matrix of the system, which is

$$\begin{bmatrix}
a_{11} & a_{12} & \cdots & a_{1n} & c_1 \\
a_{21} & a_{22} & \cdots & a_{2n} & c_2 \\
\vdots & \vdots & & \vdots & \vdots \\
a_{m1} & a_{m2} & \cdots & a_{mn} & c_m
\end{bmatrix}$$

and

2. determining a reduced matrix such that the augmented coefficient matrix is equivalent to it.

Frequently, step 2 is called *reducing the augmented coefficient matrix.*

PRINCIPLES IN PRACTICE 1

SOLVING A SYSTEM BY REDUCTION

An investment firm offers three stock portfolios: A, B, and C. The number of blocks of each type of stock in each of these portfolios is summarized in the following table:

		Portfolio		
		A	B	C
	High	6	1	3
Risk:	Moderate	3	2	3
	Low	1	5	3

A client wants 35 blocks of high-risk stock, 22 blocks of moderate-risk stock, and 18 blocks of low-risk stock. How many of each portfolio should be suggested?

EXAMPLE 3   Solving a System by Reduction

By using matrix reduction, solve the system

$$\begin{cases}
2x + 3y = -1 \\
2x + \ y = \ 5 \\
\ x + \ y = \ 1
\end{cases}$$

Solution: Reducing the augmented coefficient matrix of the system, we have

$$\begin{bmatrix}
2 & 3 & -1 \\
2 & 1 & 5 \\
1 & 1 & 1
\end{bmatrix}
\xrightarrow{R_1 \leftrightarrow R_3}
\begin{bmatrix}
1 & 1 & 1 \\
2 & 1 & 5 \\
2 & 3 & -1
\end{bmatrix}$$

$$\xrightarrow{-2R_1+R_2}
\begin{bmatrix}
1 & 1 & 1 \\
0 & -1 & 3 \\
2 & 3 & -1
\end{bmatrix}$$

$$\xrightarrow{-2R_1 + R_3} \begin{bmatrix} 1 & 1 & | & 1 \\ 0 & -1 & | & 3 \\ 0 & 1 & | & -3 \end{bmatrix}$$

$$\xrightarrow{(-1)R_2} \begin{bmatrix} 1 & 1 & | & 1 \\ 0 & 1 & | & -3 \\ 0 & 1 & | & -3 \end{bmatrix}$$

$$\xrightarrow{-R_2 + R_1} \begin{bmatrix} 1 & 0 & | & 4 \\ 0 & 1 & | & -3 \\ 0 & 1 & | & -3 \end{bmatrix}$$

$$\xrightarrow{-R_2 + R_3} \begin{bmatrix} 1 & 0 & | & 4 \\ 0 & 1 & | & -3 \\ 0 & 0 & | & 0 \end{bmatrix}$$

The last matrix is reduced and corresponds to the system

$$\begin{cases} x + 0y = 4 \\ 0x + y = -3 \\ 0x + 0y = 0 \end{cases}$$

Since the original system is equivalent to this system, it has a unique solution, namely,

$$x = 4$$

$$y = -3$$

NOW WORK PROBLEM 13

PRINCIPLES IN PRACTICE 2

SOLVING A SYSTEM BY REDUCTION

A health spa customizes the diet and vitamin supplements of each of its clients. The spa offers three different vitamin supplements, each containing different percentages of the recommended daily allowance (RDA) of vitamins A, C, and D. One tablet of supplement X provides 40% of the RDA of A, 20% of the RDA of C, and 10% of the RDA of D. One tablet of supplement Y provides 10% of the RDA of A, 10% of the RDA of C, and 30% of the RDA of D. One tablet of supplement Z provides 10% of the RDA of A, 50% of the RDA of C, and 20% of the RDA of D. The spa staff determines that one client should take 180% of the RDA of vitamin A, 200% of the RDA of vitamin C, and 190% of the RDA of vitamin D each day. How many tablets of each supplement should she take each day?

● EXAMPLE 4 **Solving a System by Reduction**

Using matrix reduction, solve

$$\begin{cases} x + 2y + 4z - 6 = 0 \\ 2z + y - 3 = 0 \\ x + y + 2z - 1 = 0 \end{cases}$$

Solution: Rewriting the system so that the variables are aligned and the constant terms appear on the right sides of the equations, we have

$$\begin{cases} x + 2y + 4z = 6 \\ y + 2z = 3 \\ x + y + 2z = 1 \end{cases}$$

Reducing the augmented coefficient matrix, we obtain

$$\begin{bmatrix} 1 & 2 & 4 & | & 6 \\ 0 & 1 & 2 & | & 3 \\ 1 & 1 & 2 & | & 1 \end{bmatrix} \xrightarrow{-R_1 + R_3} \begin{bmatrix} 1 & 2 & 4 & | & 6 \\ 0 & 1 & 2 & | & 3 \\ 0 & -1 & -2 & | & -5 \end{bmatrix}$$

$$\xrightarrow[\substack{-2R_2 + R_1 \\ (1)R_2 + R_3}]{} \begin{bmatrix} 1 & 0 & 0 & | & 0 \\ 0 & 1 & 2 & | & 3 \\ 0 & 0 & 0 & | & -2 \end{bmatrix}$$

$$\xrightarrow{-\frac{1}{2}R_3} \begin{bmatrix} 1 & 0 & 0 & | & 0 \\ 0 & 1 & 2 & | & 3 \\ 0 & 0 & 0 & | & 1 \end{bmatrix}$$

$$\xrightarrow{-3R_3 + R_2} \begin{bmatrix} 1 & 0 & 0 & | & 0 \\ 0 & 1 & 2 & | & 0 \\ 0 & 0 & 0 & | & 1 \end{bmatrix}$$

The last matrix is reduced and corresponds to

$$\begin{cases} x = 0 \\ y + 2z = 0 \\ 0 = 1 \end{cases}$$

Since $0 \neq 1$, there are no values of x, y, and z for which all equations are satisfied simultaneously. Thus, the original system has no solution.

NOW WORK PROBLEM 15 ●●●

● EXAMPLE 5   **Parametric Form of a Solution**

Using matrix reduction, solve

$$\begin{cases} 2x_1 + 3x_2 + 2x_3 + 6x_4 = 10 \\ x_2 + 2x_3 + x_4 = 2 \\ 3x_1 - 3x_3 + 6x_4 = 9 \end{cases}$$

Solution: Reducing the augmented coefficient matrix, we have

$$\begin{bmatrix} 2 & 3 & 2 & 6 & | & 10 \\ 0 & 1 & 2 & 1 & | & 2 \\ 3 & 0 & -3 & 6 & | & 9 \end{bmatrix} \xrightarrow{\frac{1}{2}R_1} \begin{bmatrix} 1 & \frac{3}{2} & 1 & 3 & | & 5 \\ 0 & 1 & 2 & 1 & | & 2 \\ 3 & 0 & -3 & 6 & | & 9 \end{bmatrix}$$

$$\xrightarrow{-3R_1 + R_3} \begin{bmatrix} 1 & \frac{3}{2} & 1 & 3 & | & 5 \\ 0 & 1 & 2 & 1 & | & 2 \\ 0 & -\frac{9}{2} & -6 & -3 & | & -6 \end{bmatrix}$$

$$\xrightarrow[\frac{9}{2}R_2 + R_3]{-\frac{3}{2}R_2 + R_1} \begin{bmatrix} 1 & 0 & -2 & \frac{3}{2} & | & 2 \\ 0 & 1 & 2 & 1 & | & 2 \\ 0 & 0 & 3 & \frac{3}{2} & | & 3 \end{bmatrix}$$

$$\xrightarrow{\frac{1}{3}R_3} \begin{bmatrix} 1 & 0 & -2 & \frac{3}{2} & | & 2 \\ 0 & 1 & 2 & 1 & | & 2 \\ 0 & 0 & 1 & \frac{1}{2} & | & 1 \end{bmatrix}$$

$$\xrightarrow[-2R_3 + R_2]{2R_3 + R_1} \begin{bmatrix} 1 & 0 & 0 & \frac{5}{2} & | & 4 \\ 0 & 1 & 0 & 0 & | & 0 \\ 0 & 0 & 1 & \frac{1}{2} & | & 1 \end{bmatrix}$$

This matrix is reduced and corresponds to the system

$$\begin{cases} x_1 + \frac{5}{2}x_4 = 4 \\ x_2 = 0 \\ x_3 + \frac{1}{2}x_4 = 1 \end{cases}$$

Thus,

$$x_1 = 4 - \tfrac{5}{2}x_4 \tag{9}$$

$$x_2 = 0 \tag{10}$$

$$x_3 = 1 - \tfrac{1}{2}x_4 \tag{11}$$

The system imposes no restrictions on x_4 so that x_4 may take on *any* real value. If we append

$$x_4 = x_4 \tag{12}$$

to the preceding equations, then we have expressed all four of the unknowns in terms of x_4 and this is the *general* solution of the original system.

For each particular value of x_4 Equations (9)–(12) determine a *particular* solution of the original system. For example, if $x_4 = 0$, then a *particular* solution is

$$x_1 = 4 \qquad x_2 = 0 \qquad x_3 = 1 \qquad x_4 = 0$$

If $x_4 = 2$, then

$$x_1 = -1 \qquad x_2 = 0 \qquad x_3 = 0 \qquad x_4 = 2$$

is another particular solution. Since there are infinitely many possibilities for x_4, there are infinitely many solutions of the original system.

Recall (see Examples 3 and 6 of Section 3.4) that, if we like, we can write $x_4 = r$ and refer to this new variable r as a *parameter*. (However, there is nothing special about the name r so that we could consider x_4 as the parameter on which *all* the original variables depend. Note that we can write $x_2 = 0 + 0x_4$ and $x_4 = 0 + 1x_4$.) Writing r for the parameter, the solution of the original system is given by

$$x_1 = 4 - \tfrac{5}{2}r$$
$$x_2 = 0 + 0r$$
$$x_3 = 1 - \tfrac{1}{2}r$$
$$x_4 = 0 + 1r$$

where r is any real number, and we speak of having a *one-parameter family* of solutions. Now, with matrix addition and scalar multiplication at hand, we can say a little more about such families. Observe that

$$\begin{bmatrix} x_1 \\ x_2 \\ x_3 \\ x_4 \end{bmatrix} = \begin{bmatrix} 4 \\ 0 \\ 1 \\ 0 \end{bmatrix} + r \begin{bmatrix} -\frac{5}{2} \\ 0 \\ -\frac{1}{2} \\ 1 \end{bmatrix}$$

Readers familiar with analytic geometry will see that the solutions form a *line* in $x_1 x_2 x_3 x_4$-space, passing through the *point* $\begin{bmatrix} 4 \\ 0 \\ 1 \\ 0 \end{bmatrix}$ and in the *direction* of the line segment

joining $\begin{bmatrix} 0 \\ 0 \\ 0 \\ 0 \end{bmatrix}$ and $\begin{bmatrix} -\frac{5}{2} \\ 0 \\ -\frac{1}{2} \\ 1 \end{bmatrix}$

NOW WORK PROBLEM 17

Examples 3–5 illustrate the fact that a system of linear equations may have a unique solution, no solution, or infinitely many solutions.

Problems 6.4

In Problems 1–6, determine whether the matrix is reduced or not reduced.

*1. $\begin{bmatrix} 1 & 3 \\ 5 & 0 \end{bmatrix}$

2. $\begin{bmatrix} 1 & 0 & 0 & 3 \\ 0 & 0 & 1 & 2 \end{bmatrix}$

3. $\begin{bmatrix} 1 & 0 & 0 \\ 0 & 1 & 0 \\ 0 & 0 & 1 \end{bmatrix}$

4. $\begin{bmatrix} 1 & 1 \\ 0 & 1 \\ 0 & 0 \\ 0 & 0 \end{bmatrix}$

5. $\begin{bmatrix} 0 & 0 & 0 & 0 \\ 0 & 1 & 0 & 0 \\ 0 & 0 & 1 & 0 \\ 0 & 0 & 0 & 0 \end{bmatrix}$

6. $\begin{bmatrix} 0 & 0 & 1 \\ 1 & 0 & 3 \\ 0 & 1 & 5 \\ 0 & 0 & 0 \end{bmatrix}$

In Problems 7–12, reduce the given matrix.

7. $\begin{bmatrix} 1 & 3 \\ 4 & 0 \end{bmatrix}$

8. $\begin{bmatrix} 0 & -3 & 0 & 2 \\ 1 & 5 & 0 & 2 \end{bmatrix}$

*9. $\begin{bmatrix} 2 & 4 & 6 \\ 1 & 2 & 3 \\ 1 & 2 & 3 \end{bmatrix}$

10. $\begin{bmatrix} 2 & 3 \\ 1 & -6 \\ 4 & 8 \\ 1 & 7 \end{bmatrix}$

11. $\begin{bmatrix} 2 & 0 & 3 & 1 \\ 1 & 4 & 2 & 2 \\ -1 & 3 & 1 & 4 \\ 0 & 2 & 1 & 0 \end{bmatrix}$

12. $\begin{bmatrix} 0 & 0 & 2 \\ 2 & 0 & 3 \\ 0 & -1 & 0 \\ 0 & 4 & 1 \end{bmatrix}$

Solve the systems in Problems 13–26 by the method of reduction.

*13. $\begin{cases} 2x - 7y = 50 \\ x + 3y = 10 \end{cases}$

14. $\begin{cases} x - 3y = -11 \\ 4x + 3y = 9 \end{cases}$

*15. $\begin{cases} 3x + y = 4 \\ 12x + 4y = 2 \end{cases}$

16. $\begin{cases} x + 2y - 3z = 0 \\ -2x - 4y + 6z = 1 \end{cases}$

*17. $\begin{cases} x + 2y + z - 4 = 0 \\ 3x + 2z - 5 = 0 \end{cases}$

18. $\begin{cases} x + 3y + 2z - 1 = 0 \\ x + y + 5z - 10 = 0 \end{cases}$

19. $\begin{cases} x_1 - 3x_2 = 0 \\ 2x_1 + 2x_2 = 3 \\ 5x_1 - x_2 = 1 \end{cases}$

20. $\begin{cases} x_1 + 4x_2 = 9 \\ 3x_1 - x_2 = 6 \\ x_1 - x_2 = 2 \end{cases}$

21. $\begin{cases} x - y - 3z = -5 \\ 2x - y - 4z = -8 \\ x + y - z = -1 \end{cases}$

22. $\begin{cases} x + y - z = 7 \\ 2x - 3y - 2z = 4 \\ x - y - 5z = 23 \end{cases}$

23. $\begin{cases} 2x - 4z = 8 \\ x - 2y - 2z = 14 \\ x + y - 2z = -1 \\ 3x + y + z = 0 \end{cases}$

24. $\begin{cases} x + 3z = -1 \\ 3x + 2y + 11z = 1 \\ x + y + 4z = 1 \\ 2x - 3y + 3z = -8 \end{cases}$

25. $\begin{cases} x_1 - x_2 - x_3 - x_4 - x_5 = 0 \\ x_1 + x_2 - x_3 - x_4 - x_5 = 0 \\ x_1 + x_2 + x_3 - x_4 - x_5 = 0 \\ x_1 + x_2 + x_3 + x_4 - x_5 = 0 \end{cases}$

26. $\begin{cases} x_1 + x_2 + x_3 - x_4 = 0 \\ x_1 - x_2 - x_3 + x_4 = 0 \\ x_1 + x_2 - x_3 - x_4 = 0 \\ x_1 + x_2 - x_3 + x_4 = 0 \end{cases}$

Solve Problems 27–33 by using matrix reduction.

27. Taxes A company has taxable income of $312,000. The federal tax is 25% of that portion that is left after the state tax has been paid. The state tax is 10% of that portion that is left after the federal tax has been paid. Find the company's federal and state taxes.

28. Decision Making A manufacturer produces two products, A and B. For each unit of A sold, the profit is $8, and for each unit of B sold, the profit is $11. From experience, it has been found that 25% more of A can be sold than of B. Next year the manufacturer desires a total profit of $42,000. How many units of each product must be sold?

29. Production Scheduling A manufacturer produces three products: A, B, and C. The profits for each unit of A, B, and C sold are $1, $2, and $3, respectively. Fixed costs are $17,000 per year, and the costs of producing each unit of A, B, and C are $4, $5, and $7, respectively. Next year, a total of 11,000 units of all three products is to be produced and sold, and a total profit of $25,000 is to be realized. If total cost is to be $80,000, how many units of each of the products should be produced next year?

30. Production Allocation National Desk Co. has plants for producing desks on both the east and west coasts. At the East Coast plant, fixed costs are $20,000 per year and the cost of producing each desk is $90. At the West Coast plant, fixed costs are $18,000 per year and the cost of producing each desk is $95. Next year the company wants to produce a total of 800 desks. Determine the production order for each

plant for the forthcoming year if the total cost for each plant is to be the same.

31. Vitamins A person is ordered by a doctor to take 10 units of vitamin A, 9 units of vitamin D, and 19 units of vitamin E each day. The person can choose from three brands of vitamin pills. Brand X contains 2 units of vitamin A, 3 units of vitamin D, and 5 units of vitamin E; brand Y has 1, 3, and 4 units, respectively; and brand Z has 1 unit of vitamin A, none of vitamin D, and 1 of vitamin E.

VITAMIN B

(a) Find all possible combinations of pills that will provide exactly the required amounts of vitamins.
(b) If brand X costs 1 cent a pill, brand Y 6 cents, and brand Z 3 cents, are there any combinations in part (a) costing exactly 15 cents a day?
(c) What is the least expensive combination in part (a)? the most expensive?

32. Production A firm produces three products, A, B, and C, that require processing by three machines, I, II, and III. The time in hours required for processing one unit of each product by the three machines is given by the following table:

	A	B	C
I	3	1	2
II	1	2	1
III	2	4	1

Machine I is available for 490 hours, machine II for 310 hours, and machine III for 560 hours. Find how many units of each product should be produced to make use of all the available time on the machines.

33. Investments An investment company sells three types of pooled funds, Standard (S), Deluxe (D), and Gold Star (G).

Each unit of S contains 12 shares of stock A, 16 of stock B, and 8 of stock C.

Each unit of D contains 20 shares of stock A, 12 of stock B, and 28 of stock C.

Each unit of G contains 32 shares of stock A, 28 of stock B, and 36 of stock C.

Suppose an investor wishes to purchase exactly 220 shares of stock A, 176 shares of stock B, and 264 shares of stock C by buying units of the three funds.

(a) Determine those combinations of units of S, D, and G that will meet the investor's requirements exactly.

(b) Suppose each unit of S (respectively, D, G) costs the investor $300 (respectively, $400, $600). Which of the combinations from part (a) will minimize the total cost to the investor?

6.5 Solving Systems by Reducing Matrices (continued)[4]

OBJECTIVE

To focus our attention on nonhomogeneous systems that involve more than one parameter in their general solution; and to solve, and consider the theory of, homogeneous systems.

As we saw in Section 6.4, a system of linear equations may have a unique solution, no solution, or infinitely many solutions. When there are infinitely many, the general solution is expressed in terms of at least one parameter. For example, the general solution in Example 5 was given in terms of the parameter r:

$$x_1 = 4 - \tfrac{5}{2}r$$

$$x_2 = 0$$

$$x_3 = 1 - \tfrac{1}{2}r$$

$$x_4 = r$$

At times, more than one parameter is necessary,[5] as the next example shows.

● EXAMPLE 1 Two-Parameter Family of Solutions

Using matrix reduction, solve

$$\begin{cases} x_1 + 2x_2 + 5x_3 + 5x_4 = -3 \\ x_1 + x_2 + 3x_3 + 4x_4 = -1 \\ x_1 - x_2 - x_3 + 2x_4 = 3 \end{cases}$$

Solution: The augmented coefficient matrix is

$$\begin{bmatrix} 1 & 2 & 5 & 5 & -3 \\ 1 & 1 & 3 & 4 & -1 \\ 1 & -1 & -1 & 2 & 3 \end{bmatrix}$$

whose reduced form is

$$\begin{bmatrix} 1 & 0 & 1 & 3 & 1 \\ 0 & 1 & 2 & 1 & -2 \\ 0 & 0 & 0 & 0 & 0 \end{bmatrix}$$

Hence,

$$\begin{cases} x_1 + x_3 + 3x_4 = 1 \\ x_2 + 2x_3 + x_4 = -2 \end{cases}$$

from which it follows that

$$x_1 = 1 - x_3 - 3x_4$$

$$x_2 = -2 - 2x_3 - x_4$$

Since no restriction is placed on either x_3 or x_4, they can be arbitrary real numbers, giving us a parametric family of solutions. Setting $x_3 = r$ and $x_4 = s$, we can give the solution of the given system as

$$x_1 = 1 - r - 3s$$

$$x_2 = -2 - 2r - s$$

$$x_3 = r$$

$$x_4 = s$$

---

[4]This section may be omitted.

[5]See Example 7 of Section 3.4.

where the parameters r and s can be any real numbers. By assigning specific values to r and s, we get particular solutions. For example, if $r = 1$ and $s = 2$, then the corresponding particular solution is $x_1 = -6$, $x_2 = -6$, $x_3 = 1$, and $x_4 = 2$. As in the one-parameter case we can now go further and write

$$\begin{bmatrix} x_1 \\ x_2 \\ x_3 \\ x_4 \end{bmatrix} = \begin{bmatrix} 1 \\ -2 \\ 0 \\ 0 \end{bmatrix} + r \begin{bmatrix} -1 \\ -2 \\ 1 \\ 0 \end{bmatrix} + s \begin{bmatrix} -3 \\ -1 \\ 0 \\ 1 \end{bmatrix}$$

which can be shown to exhibit the family of solutions as a *plane* through $\begin{bmatrix} 1 \\ -2 \\ 0 \\ 0 \end{bmatrix}$ in

$x_1 x_2 x_3 x_4$-space.

NOW WORK PROBLEM 1 ◖◖●

It is customary to classify a system of linear equations as being either *homogeneous* or *nonhomogeneous*, depending on whether the constant terms are all zero.

DEFINITION
The system

$$\begin{cases} a_{11}x_1 + a_{12}x_2 + \cdots + a_{1n}x_n = c_1 \\ a_{21}x_1 + a_{22}x_2 + \cdots + a_{2n}x_n = c_2 \\ \quad \vdots \qquad \quad \vdots \qquad \qquad \quad \vdots \\ a_{m1}x_1 + a_{m2}x_2 + \cdots + a_{mn}x_n = c_m \end{cases}$$

is called a **homogeneous system** if $c_1 = c_2 = \cdots = c_m = 0$. The system is a **nonhomogeneous system** if at least one of the c's is not equal to 0.

●EXAMPLE 2   Nonhomogeneous and Homogeneous Systems
The system

$$\begin{cases} 2x + 3y = 4 \\ 3x - 4y = 0 \end{cases}$$

is nonhomogeneous because of the 4 in the top equation. The system

$$\begin{cases} 2x + 3y = 0 \\ 3x - 4y = 0 \end{cases}$$

is homogeneous.

◖●●

If the homogeneous system

$$\begin{cases} 2x + 3y = 0 \\ 3x - 4y = 0 \end{cases}$$

were solved by the method of reduction, first the augmented coefficient matrix would be written

$$\begin{bmatrix} 2 & 3 & | & 0 \\ 3 & -4 & | & 0 \end{bmatrix}$$

Observe that the last column consists entirely of zeros. This is typical of the augmented coefficient matrix of any homogeneous system. We would then reduce this matrix by using elementary row operations:

$$\begin{bmatrix} 2 & 3 & | & 0 \\ 3 & -4 & | & 0 \end{bmatrix} \rightarrow \cdots \rightarrow \begin{bmatrix} 1 & 0 & | & 0 \\ 0 & 1 & | & 0 \end{bmatrix}$$

The last column of the reduced matrix also consists only of zeros. This does not occur by chance. When any elementary row operation is performed on a matrix that has a column consisting entirely of zeros, the corresponding column of the resulting matrix will also be all zeros. For convenience, it will be our custom when solving a homogeneous system by matrix reduction to delete the last column of the matrices involved. That is, we shall reduce only the *coefficient matrix* of the system. For the preceding system, we would have

$$\begin{bmatrix} 2 & 3 \\ 3 & -4 \end{bmatrix} \rightarrow \cdots \rightarrow \begin{bmatrix} 1 & 0 \\ 0 & 1 \end{bmatrix}$$

Here the reduced matrix, called the *reduced coefficient matrix,* corresponds to the system

$$\begin{cases} x + 0y = 0 \\ 0x + y = 0 \end{cases}$$

so the solution is $x = 0$ and $y = 0$.

Let us now consider the number of solutions of the homogeneous system

$$\begin{cases} a_{11}x_1 + a_{12}x_2 + \cdots + a_{1n}x_n = 0 \\ a_{21}x_1 + a_{22}x_2 + \cdots + a_{2n}x_n = 0 \\ \quad \cdot \qquad\quad \cdot \qquad\qquad \cdot \qquad\quad \cdot \\ \quad \cdot \qquad\quad \cdot \qquad\qquad \cdot \qquad\quad \cdot \\ \quad \cdot \qquad\quad \cdot \qquad\qquad \cdot \qquad\quad \cdot \\ a_{m1}x_1 + a_{m2}x_2 + \cdots + a_{mn}x_n = 0 \end{cases}$$

One solution always occurs when $x_1 = 0$, $x_2 = 0$, $\ldots$, and $x_n = 0$, since each equation is satisfied for these values. This solution, called the **trivial solution,** is a solution of *every* homogeneous system.

There is a theorem that allows us to determine whether a homogeneous system has a unique solution (the trivial solution only) or infinitely many solutions. The theorem is based on the number of nonzero-rows that appear in the reduced coefficient matrix of the system. Recall that a *nonzero-row* is a row that does not consist entirely of zeros.

THEOREM

Let **A** be the *reduced* coefficient matrix of a homogeneous system of m linear equations in n unknowns. If **A** has exactly k nonzero-rows, then $k \leq n$. Moreover,

1. if $k < n$, the system has infinitely many solutions, and

2. if $k = n$, the system has a unique solution (the trivial solution).

If a homogeneous system consists of m equations in n unknowns, then the coefficient matrix of the system has size $m \times n$. Thus, if $m < n$ and k is the number of nonzero rows in the reduced coefficient matrix, then $k \leq m$, and hence, $k < n$. By the foregoing theorem, the system must have infinitely many solutions. Consequently, we have the following corollary.

COROLLARY

A homogeneous system of linear equations with fewer equations than unknowns has infinitely many solutions.

CAUTION

The preceding theorem and corollary apply only to **homogeneous** systems of linear equations. For example, consider the system

$$\begin{cases} x + y - 2z = 3 \\ 2x + 2y - 4z = 4 \end{cases}$$

which consists of two linear equations in three unknowns. We **cannot** conclude that this system has infinitely many solutions, since it is not homogeneous. Indeed, you should verify that it has no solution.

PRINCIPLES IN PRACTICE 1

SOLVING HOMOGENEOUS SYSTEMS

A plane in three-dimensional space can be written as $ax + by + cz = d$. We can find the possible intersections of planes in this form by writing them as systems of linear equations and using reduction to solve them. If $d = 0$ in each equation, then we have a homogeneous system with either a unique solution or infinitely many solutions. Determine whether the intersection of the planes

$$5x + 3y + 4z = 0$$

$$6x + 8y + 7z = 0$$

$$3x + 1y + 2z = 0$$

has a unique solution or infinitely many solutions; then solve the system.

● **EXAMPLE 3** **Number of Solutions of a Homogeneous System**

Determine whether the system

$$\begin{cases} x + y - 2z = 0 \\ 2x + 2y - 4z = 0 \end{cases}$$

has a unique solution or infinitely many solutions.

Solution: There are two equations in this homogeneous system, and this number is less than the number of unknowns (three). Thus, by the previous corollary, the system has infinitely many solutions.

NOW WORK PROBLEM 9

● **EXAMPLE 4** **Solving Homogeneous Systems**

Determine whether the following homogeneous systems have a unique solution or infinitely many solutions; then solve the systems.

a. $\begin{cases} x - 2y + z = 0 \\ 2x - y + 5z = 0 \\ x + y + 4z = 0 \end{cases}$

Solution: Reducing the coefficient matrix, we have

$$\begin{bmatrix} 1 & -2 & 1 \\ 2 & -1 & 5 \\ 1 & 1 & 4 \end{bmatrix} \rightarrow \cdots \rightarrow \begin{bmatrix} 1 & 0 & 3 \\ 0 & 1 & 1 \\ 0 & 0 & 0 \end{bmatrix}$$

The number of nonzero rows, 2, in the reduced coefficient matrix is less than the number of unknowns, 3, in the system. By the previous theorem, there are infinitely many solutions.

Since the reduced coefficient matrix corresponds to

$$\begin{cases} x + 3z = 0 \\ y + z = 0 \end{cases}$$

the solution may be given in parametric form by

$$x = -3r$$

$$y = -r$$

$$z = r$$

where r is any real number.

b. $\begin{cases} 3x + 4y = 0 \\ x - 2y = 0 \\ 2x + y = 0 \\ 2x + 3y = 0 \end{cases}$

Solution: Reducing the coefficient matrix, we have

$$\begin{bmatrix} 3 & 4 \\ 1 & -2 \\ 2 & 1 \\ 2 & 3 \end{bmatrix} \rightarrow \cdots \rightarrow \begin{bmatrix} 1 & 0 \\ 0 & 1 \\ 0 & 0 \\ 0 & 0 \end{bmatrix}$$

The number of nonzero rows (2) in the reduced coefficient matrix equals the number of unknowns in the system. By the theorem, the system must have a unique solution, namely, the trivial solution $x = 0$, $y = 0$.

NOW WORK PROBLEM 13

Problems 6.5

In Problems 1–8, solve the systems by using matrix reduction.

*1. $\begin{cases} w + x - y - 9z = -3 \\ 2w + 3x + 2y + 15z = 12 \\ 2w + x + 2y + 5z = 8 \end{cases}$

2. $\begin{cases} 2w + x + 10y + 15z = -5 \\ w - 5x + 2y + 15z = -10 \\ w + x + 6y + 12z = 9 \end{cases}$

3. $\begin{cases} 3w - x - 3y - z = -2 \\ 2w - 2x - 6y - 6z = -4 \\ 2w - x - 3y - 2z = -2 \\ 3w + x + 3y + 7z = 2 \end{cases}$

4. $\begin{cases} w + x + 5z = 1 \\ w + y + 2z = 1 \\ w - 3x + 4y - 7z = 1 \\ x - y + 3z = 0 \end{cases}$

5. $\begin{cases} w + x + 3y - z = 2 \\ 2w + x + 5y - 2z = 0 \\ 2w - x + 3y - 2z = -8 \\ 3w + 2x + 8y - 3z = 2 \\ w + 2y - z = -2 \end{cases}$

6. $\begin{cases} w + x + y + 2z = 4 \\ 2w + x + 2y + 2z = 7 \\ w + 2x + y + 4z = 5 \\ 3w - 2x + 3y - 4z = 7 \\ 4w - 3x + 4y - 6z = 9 \end{cases}$

7. $\begin{cases} 4x_1 - 3x_2 + 5x_3 - 10x_4 + 11x_5 = -8 \\ 2x_1 + x_2 + 5x_3 + 3x_5 = 6 \end{cases}$

8. $\begin{cases} x_1 + 3x_3 + x_4 + 4x_5 = 1 \\ x_2 + x_3 - 2x_4 = 0 \\ 2x_1 - 2x_2 + 3x_3 + 10x_4 + 15x_5 = 10 \\ x_1 + 2x_2 + 3x_3 - 2x_4 + 2x_5 = -2 \end{cases}$

For Problems 9–14, determine whether the system has infinitely many solutions or only the trivial solution. Do not solve the systems.

*9. $\begin{cases} 1.06x + 2.3y - 0.05z = 0 \\ 1.055x - 0.6y + 0.09z = 0 \end{cases}$

10. $\begin{cases} 3w + 5x - 4y + 2z = 0 \\ 7w - 2x + 9y + 3z = 0 \end{cases}$

11. $\begin{cases} 3x - 4y = 0 \\ x + 5y = 0 \\ 4x - y = 0 \end{cases}$

12. $\begin{cases} 2x + 3y + 12z = 0 \\ 3x - 2y + 5z = 0 \\ 4x + y + 14z = 0 \end{cases}$

*13. $\begin{cases} x + y + z = 0 \\ x - z = 0 \\ x - 2y - 5z = 0 \end{cases}$

14. $\begin{cases} 3x + 2y - 2z = 0 \\ 2x + 2y - 2z = 0 \\ -4y + 5z = 0 \end{cases}$

Solve each of the following systems.

15. $\begin{cases} x + y = 0 \\ 3x - 4y = 0 \end{cases}$

16. $\begin{cases} 2x - 5y = 0 \\ 8x - 20y = 0 \end{cases}$

17. $\begin{cases} x + 6y - 2z = 0 \\ 2x - 3y + 4z = 0 \end{cases}$

18. $\begin{cases} 4x + 7y = 0 \\ 2x + 3y = 0 \end{cases}$

19. $\begin{cases} x + y = 0 \\ 3x - 4y = 0 \\ 5x - 8y = 0 \end{cases}$

20. $\begin{cases} 2x - 3y + z = 0 \\ x + 2y - z = 0 \\ x + y + z = 0 \end{cases}$

21. $\begin{cases} x + y + z = 0 \\ -7y - 14z = 0 \\ -2y - 4z = 0 \\ -5y - 10z = 0 \end{cases}$

22. $\begin{cases} x + y + 7z = 0 \\ x - y - z = 0 \\ 2x - 3y - 6z = 0 \\ 3x + y + 13z = 0 \end{cases}$

23. $\begin{cases} w + x + y + 4z = 0 \\ w + x + 5z = 0 \\ 2w + x + 3y + 4z = 0 \\ w - 3x + 2y - 9z = 0 \end{cases}$

24. $\begin{cases} w + x + 2y + 7z = 0 \\ w - 2x - y + z = 0 \\ w + 2x + 3y + 9z = 0 \\ 2w - 3x - y + 4z = 0 \end{cases}$

To determine the inverse of an invertible matrix and to use inverses to solve systems.

6.6 Inverses

We have seen how useful the method of reduction is for solving systems of linear equations. But it is by no means the only method that uses matrices. In this section, we shall discuss a different method that applies to certain systems of n linear equations in n unknowns.

In Section 6.3, we showed how a system of linear equations can be written in matrix form as the single matrix equation $\mathbf{AX} = \mathbf{B}$, where $\mathbf{A}$ is the coefficient matrix. For example, the system

$$\begin{cases} x_1 + 2x_2 = 3 \\ x_1 - x_2 = 1 \end{cases}$$

can be written in the matrix form $\mathbf{AX} = \mathbf{B}$, where

$$\mathbf{A} = \begin{bmatrix} 1 & 2 \\ 1 & -1 \end{bmatrix} \qquad \mathbf{X} = \begin{bmatrix} x_1 \\ x_2 \end{bmatrix} \qquad \mathbf{B} = \begin{bmatrix} 3 \\ 1 \end{bmatrix}$$

If we can determine the values of the entries in the unknown matrix $\mathbf{X}$, we have a solution of the system. Thus, we would like to find a method to solve the matrix equation $\mathbf{AX} = \mathbf{B}$ for $\mathbf{X}$. Some motivation is provided by looking at the procedure for solving the algebraic equation $ax = b$. The latter equation is solved by simply multiplying both sides by the multiplicative inverse of a. (Recall that the multiplicative inverse of a nonzero number a is denoted a^{-1} [which is $1/a$] and has the property that $a^{-1}a = 1$.) For example, if $3x = 11$, then

$$3^{-1}(3x) = 3^{-1}(11) \quad \text{so} \quad x = \frac{11}{3}$$

If we can apply a similar procedure to the *matrix* equation

$$\mathbf{AX} = \mathbf{B} \tag{1}$$

then we need a multiplicative inverse of **A**—that is, a matrix **C** such that $\mathbf{CA} = \mathbf{I}$. Then we can simply multiply both sides of Equation (1) by **C**:

$$\mathbf{C(AX)} = \mathbf{CB}$$

$$\mathbf{(CA)X} = \mathbf{CB}$$

$$\mathbf{IX} = \mathbf{CB}$$

$$\mathbf{X} = \mathbf{CB}$$

Thus, the solution is $\mathbf{X} = \mathbf{CB}$. Of course, this method is based on the existence of a matrix **C** such that $\mathbf{CA} = \mathbf{I}$. When such a matrix does exist, we say that it is an *inverse matrix* (or simply an *inverse*) of **A**.

> ### DEFINITION
> If **A** is a square matrix and there exists a matrix **C** such that $\mathbf{CA} = \mathbf{I}$, then **C** is called an inverse of **A**, and **A** is said to be *invertible.*

PRINCIPLES IN PRACTICE 1

INVERSE OF A MATRIX

Secret messages can be encoded by using a code and an encoding matrix. Suppose we have the following code:

a	b	c	d	e	f	g	h	i	j	k	l	m
1	2	3	4	5	6	7	8	9	10	11	12	13

n	o	p	q	r	s	t	u	v	w	x	y	z
14	15	16	17	18	19	20	21	22	23	24	25	26

Let the encoding matrix be **E**. Then we can encode a message by taking every two letters of the message, converting them to their corresponding numbers, creating a 2×1 matrix, and then multiply each matrix by **E**. The message may be unscrambled with a decoding matrix that is the inverse of the coding matrix—that is, $\mathbf{E}^{-1}$. Determine whether the encoding matrices

$$\begin{bmatrix} 1 & 3 \\ 2 & 4 \end{bmatrix} \quad \text{and} \quad \begin{bmatrix} -2 & 1.5 \\ 1 & -0.5 \end{bmatrix}$$

are inverses of each other.

● **EXAMPLE 1   Inverse of a Matrix**

Let $\mathbf{A} = \begin{bmatrix} 1 & 2 \\ 3 & 7 \end{bmatrix}$ and $\mathbf{C} = \begin{bmatrix} 7 & -2 \\ -3 & 1 \end{bmatrix}$. Since

$$\mathbf{CA} = \begin{bmatrix} 7 & -2 \\ -3 & 1 \end{bmatrix} \begin{bmatrix} 1 & 2 \\ 3 & 7 \end{bmatrix} = \begin{bmatrix} 1 & 0 \\ 0 & 1 \end{bmatrix} = \mathbf{I}$$

matrix **C** is an inverse of **A**.

It can be shown that an invertible matrix has one and only one inverse; that is, an inverse is unique. Thus, in Example 1, matrix **C** is the *only* matrix such that $\mathbf{CA} = \mathbf{I}$. For this reason, we can speak of *the* inverse of an invertible matrix **A**, which we denote by the symbol $\mathbf{A}^{-1}$. Accordingly, $\mathbf{A}^{-1}\mathbf{A} = \mathbf{I}$. Moreover, although matrix multiplication is not generally commutative, it is a fact that $\mathbf{A}^{-1}$ *commutes with* **A**:

$$\mathbf{A}^{-1}\mathbf{A} = \mathbf{AA}^{-1} = \mathbf{I}$$

Returning to the matrix equation $\mathbf{AX} = \mathbf{B}$, Equation (1), we can now state the following:

If **A** is an invertible matrix, then the matrix equation $\mathbf{AX} = \mathbf{B}$ has the unique solution $\mathbf{X} = \mathbf{A}^{-1}\mathbf{B}$.

Probably the idea of an inverse matrix has an air of déjà vu about it. In Section 2.4 we discussed inverse functions, and these can be used to further understand inverse matrices. First, we note that the idea of a function is not limited to the world of numbers. For example, if we have a rule that applies to matrices and gives, for each of certain matrices **X** a unique, well-defined matrix $f(\mathbf{X})$, then we would say that f is a function that takes matrices to matrices. To be a little more specific, suppose that **A** is an $m \times n$ matrix and define $f(\mathbf{X}) = \mathbf{AX}$, for **X** any $n \times 1$ matrix. Then this rule f provides a function from the set of $n \times 1$ matrices to the set of $m \times 1$ matrices. If $m = n$, it can be shown that the function given by $f(\mathbf{X}) = \mathbf{AX}$ has an inverse in the sense of Section 2.4 if and only if **A** has an inverse matrix $\mathbf{A}^{-1}$, in which case

$f^{-1}(\mathbf{X}) = \mathbf{A}^{-1}\mathbf{X}$. There is one caution to be observed here. In general, for a function f to have an inverse, say g, then we require *both $g \circ f = I$ and $f \circ g = I$*, where I is the identity function. It is a rather special fact about matrices that $\mathbf{CA} = \mathbf{I}$ implies also $\mathbf{AC} = \mathbf{I}$.

CAUTION

For general *functions*, if $g \circ f = I$, it does not follow that $f \circ g = I$.

If f is a function, in the most general possible sense, that has an inverse, then any equation of the form $f(x) = b$ has a unique solution, namely $x = f^{-1}(b)$.

PRINCIPLES IN PRACTICE 2

USING THE INVERSE TO SOLVE A SYSTEM

Suppose the encoding matrix $\mathbf{E} = \begin{bmatrix} 1 & 3 \\ 2 & 4 \end{bmatrix}$ was used to encode a message. Use the code from Principles in Practice 1 and the inverse $\mathbf{E}^{-1} = \begin{bmatrix} -2 & 1.5 \\ 1 & -0.5 \end{bmatrix}$ to decode the message, broken into the following pieces:

28, 46, 65, 90

61, 82

59, 88, 57, 86

60, 84, 21, 34, 76, 102

● **EXAMPLE 2**   **Using the Inverse to Solve a System**

Solve the system

$$\begin{cases} x_1 + 2x_2 = 5 \\ 3x_1 + 7x_2 = 18 \end{cases}$$

Solution: In matrix form, we have $\mathbf{AX} = \mathbf{B}$, where

$$\mathbf{A} = \begin{bmatrix} 1 & 2 \\ 3 & 7 \end{bmatrix} \qquad \mathbf{X} = \begin{bmatrix} x_1 \\ x_2 \end{bmatrix} \qquad \mathbf{B} = \begin{bmatrix} 5 \\ 18 \end{bmatrix}$$

In Example 1, we showed that

$$\mathbf{A}^{-1} = \begin{bmatrix} 7 & -2 \\ -3 & 1 \end{bmatrix}$$

Therefore,

$$\mathbf{X} = \mathbf{A}^{-1}\mathbf{B} = \begin{bmatrix} 7 & -2 \\ -3 & 1 \end{bmatrix} \begin{bmatrix} 5 \\ 18 \end{bmatrix} = \begin{bmatrix} -1 \\ 3 \end{bmatrix}$$

so $x_1 = -1$ and $x_2 = 3$.

NOW WORK PROBLEM 19 ●●●

In order that the method of Example 2 apply to a system, two conditions must be met:

1. The system must have the same number of equations as there are unknowns.
2. The coefficient matrix must be invertible.

As far as condition 2 is concerned, we caution that not all square matrices are invertible. For example, if

$$\mathbf{A} = \begin{bmatrix} 0 & 1 \\ 0 & 1 \end{bmatrix}$$

then

$$\begin{bmatrix} a & b \\ c & d \end{bmatrix} \begin{bmatrix} 0 & 1 \\ 0 & 1 \end{bmatrix} = \begin{bmatrix} 0 & a+b \\ 0 & c+d \end{bmatrix} \neq \begin{bmatrix} 1 & 0 \\ 0 & 1 \end{bmatrix}$$

Hence, there is no matrix that when postmultiplied by $\mathbf{A}$, yields the identity matrix. Thus, $\mathbf{A}$ is not invertible.

There is an interesting mechanical procedure that allows us simultaneously to determine whether or not a matrix is invertible *and* find its inverse if it is so. The procedure is based on an observation whose proof would take us too far afield. First, recall that for any matrix $\mathbf{A}$ there is a sequence $E_1, E_2, \ldots, E_k$ of elementary row operations that, when applied to $\mathbf{A}$, produce a reduced matrix. In other words, we have

$$\mathbf{A} \xrightarrow{E_1} \mathbf{A}_1 \xrightarrow{E_2} \mathbf{A}_2 \longrightarrow \cdots \xrightarrow{E_k} \mathbf{A}_k$$

CAUTION

Every identity matrix is a reduced matrix, but not every (square) reduced matix is an identity matrix. For example, any zero matrix $\mathbf{O}$ is reduced

where $\mathbf{A}_k$ is a reduced matrix. We recall, too, that $\mathbf{A}_k$ is unique and determined by $\mathbf{A}$ alone (even though there can be many sequences, of variable lengths, of elementary row operations that accomplish this reduction). If $\mathbf{A}$ is square, say $n \times n$, then we *may* have $\mathbf{A}_k = \mathbf{I}_n$, the $n \times n$ identity matrix.

> **THEOREM**
>
> For square $\mathbf{A}$ and $\mathbf{A}_k$ as previously, $\mathbf{A}$ is invertible if and only if $\mathbf{A}_k = \mathbf{I}$. Moreover, if $E_1, E_2, \ldots, E_k$ is a sequence of elementary row operations that takes $\mathbf{A}$ to $\mathbf{I}$, then the same sequence takes $\mathbf{I}$ to $\mathbf{A}^{-1}$.

● EXAMPLE 3 Determining the Invertibility of a Matrix

Apply the theorem to determine if the matrix

$$\mathbf{A} = \begin{bmatrix} 1 & 0 \\ 2 & 2 \end{bmatrix}$$

is invertible.

Strategy We will *augment* $\mathbf{A}$ with a copy of the (2×2) identity matrix (just as we have often augmented a matrix by a column vector). The result will be 2×4. We will apply elementary row operations to the entire 2×4 matrix until the first n columns form a reduced matrix. If the result is $\mathbf{I}$, then, by the theorem, $\mathbf{A}$ is invertible, but because we have applied the operations to the entire 2×4 matrix, the last n columns will, also by the theorem, be transformed from $\mathbf{I}$ to $\mathbf{A}^{-1}$, if $\mathbf{A}$ is in fact invertible.

Solution: We have

$$[\mathbf{A} \mid \mathbf{I}] = \begin{bmatrix} 1 & 0 & | & 1 & 0 \\ 2 & 2 & | & 0 & 1 \end{bmatrix} \xrightarrow{-2R_1 + R_2} \begin{bmatrix} 1 & 0 & | & 1 & 0 \\ 0 & 2 & | & -2 & 1 \end{bmatrix}$$

$$\xrightarrow{\frac{1}{2}R_2} \begin{bmatrix} 1 & 0 & | & 1 & 0 \\ 0 & 1 & | & -1 & \frac{1}{2} \end{bmatrix} = [\mathbf{I} \mid \mathbf{B}]$$

Since $[\mathbf{A} \mid \mathbf{I}]$ transforms with $\mathbf{I}$ to the left of the augmentation bar, the matrix $\mathbf{A}$ is invertible and the matrix $\mathbf{B}$ to the right of the augmentation bar is $\mathbf{A}^{-1}$. Specifically we conclude that

$$\mathbf{A}^{-1} = \begin{bmatrix} 1 & 0 \\ -1 & \frac{1}{2} \end{bmatrix}$$

NOW WORK PROBLEM 1 ●●●

This procedure is indeed a general one.

For the interested reader, we remark that the matrix $\mathbf{B}$ in the method descibed is in any event invertible and we always have $\mathbf{BA} = \mathbf{R}$.

Method to Find the Inverse of a Matrix

If $\mathbf{A}$ is an $n \times n$ matrix, form the $n \times (2n)$ matrix $[\mathbf{A} \mid \mathbf{I}]$ and perform elementary row operations until the first n columns form a reduced matrix. Assume that the result is $[\mathbf{R} \mid \mathbf{B}]$ so that we have

$$[\mathbf{A} \mid \mathbf{I}] \to \cdots \to [\mathbf{R} \mid \mathbf{B}]$$

If $\mathbf{R} = \mathbf{I}$, then A is invertible and $\mathbf{A}^{-1} = \mathbf{B}$. If $\mathbf{R} \neq \mathbf{I}$ then A is not invertible, meaning that $\mathbf{A}^{-1}$ does not exist (and the matrix $\mathbf{B}$ is of no particular interest to our concerns here).

PRINCIPLES IN PRACTICE 3

FINDING THE INVERSE
OF A MATRIX

We could extend the encoding scheme used in Principles in Practice 1 to a 3×3 matrix, encoding three letters of a message at a time. Find the inverses of the following 3×3 encoding matrices:

$$\mathbf{E} = \begin{bmatrix} 3 & 1 & 2 \\ 2 & 2 & 2 \\ 2 & 1 & 3 \end{bmatrix} \qquad \mathbf{F} = \begin{bmatrix} 2 & 1 & 2 \\ 3 & 2 & 3 \\ 4 & 3 & 4 \end{bmatrix}$$

● EXAMPLE 4 Finding the Inverse of a Matrix

Determine $\mathbf{A}^{-1}$ if $\mathbf{A}$ is invertible.

a. $\mathbf{A} = \begin{bmatrix} 1 & 0 & -2 \\ 4 & -2 & 1 \\ 1 & 2 & -10 \end{bmatrix}$

Solution: Following the foregoing procedure, we have

$$[\mathbf{A} \mid \mathbf{I}] = \begin{bmatrix} 1 & 0 & -2 & 1 & 0 & 0 \\ 4 & -2 & 1 & 0 & 1 & 0 \\ 1 & 2 & -10 & 0 & 0 & 1 \end{bmatrix}$$

$$\xrightarrow[-1R_1 + R_3]{-4R_1 + R_2} \begin{bmatrix} 1 & 0 & -2 & 1 & 0 & 0 \\ 0 & -2 & 9 & -4 & 1 & 0 \\ 0 & 2 & -8 & -1 & 0 & 1 \end{bmatrix}$$

$$\xrightarrow{-\frac{1}{2}R_2} \begin{bmatrix} 1 & 0 & -2 & 1 & 0 & 0 \\ 0 & 1 & -\frac{9}{2} & 2 & -\frac{1}{2} & 0 \\ 0 & 2 & -8 & -1 & 0 & 1 \end{bmatrix}$$

$$\xrightarrow{-2R_2 + R_3} \begin{bmatrix} 1 & 0 & -2 & 1 & 0 & 0 \\ 0 & 1 & -\frac{9}{2} & 2 & -\frac{1}{2} & 0 \\ 0 & 0 & 1 & -5 & 1 & 1 \end{bmatrix}$$

$$\xrightarrow[\frac{9}{2}R_3 + R_2]{2R_3 + R_1} \begin{bmatrix} 1 & 0 & 0 & -9 & 2 & 2 \\ 0 & 1 & 0 & -\frac{41}{2} & 4 & \frac{9}{2} \\ 0 & 0 & 1 & -5 & 1 & 1 \end{bmatrix}$$

The first three columns of the last matrix form **I**. Thus, **A** is invertible and

$$\mathbf{A}^{-1} = \begin{bmatrix} -9 & 2 & 2 \\ -\frac{41}{2} & 4 & \frac{9}{2} \\ -5 & 1 & 1 \end{bmatrix}$$

b. $\mathbf{A} = \begin{bmatrix} 3 & 2 \\ 6 & 4 \end{bmatrix}$

Solution: We have

$$[\mathbf{A} \mid \mathbf{I}] = \begin{bmatrix} 3 & 2 & 1 & 0 \\ 6 & 4 & 0 & 1 \end{bmatrix} \xrightarrow{-2R_1 + R_2} \begin{bmatrix} 3 & 2 & 1 & 0 \\ 0 & 0 & -2 & 1 \end{bmatrix}$$

$$\xrightarrow{\frac{1}{3}R_1} \begin{bmatrix} 1 & \frac{2}{3} & \frac{1}{3} & 0 \\ 0 & 0 & -2 & 1 \end{bmatrix}$$

The first two columns of the last matrix form a reduced matrix different from **I**. Thus, **A** is not invertible.

NOW WORK PROBLEM 7

T E C H N O L O G Y

Finding the inverse of an invertible matrix with a graphing calculator can be a real timesaver. Figure 6.5 shows the inverse of

$$\mathbf{A} = \begin{bmatrix} 3 & 2 \\ 1 & 4 \end{bmatrix}$$

Moreover, on a TI-83 Plus, we can display our answer with fractional entries.

```
[A]⁻¹
    [[.4   -.2]
     [-.1  .3 ]]
Ans▶Frac
    [[2/5   -1/5]
     [-1/10 3/10]]
■
```

FIGURE 6.5 Inverse of **A** with decimal entries and with fractional entries.

Now we will solve a system by using the inverse.

EXAMPLE 5   Using the Inverse to Solve a System

Solve the system

$$\begin{cases} x_1 \quad\quad - 2x_3 = 1 \\ 4x_1 - 2x_2 + x_3 = 2 \\ x_1 + 2x_2 - 10x_3 = -1 \end{cases}$$

by finding the inverse of the coefficient matrix.

Solution: In matrix form the system is $\mathbf{AX} = \mathbf{B}$, where

$$\mathbf{A} = \begin{bmatrix} 1 & 0 & -2 \\ 4 & -2 & 1 \\ 1 & 2 & -10 \end{bmatrix}$$

is the coefficient matrix. From Example 4(a),

$$\mathbf{A}^{-1} = \begin{bmatrix} -9 & 2 & 2 \\ -\frac{41}{2} & 4 & \frac{9}{2} \\ -5 & 1 & 1 \end{bmatrix}$$

The solution is given by $\mathbf{X} = \mathbf{A}^{-1}\mathbf{B}$:

$$\begin{bmatrix} x_1 \\ x_2 \\ x_3 \end{bmatrix} = \begin{bmatrix} -9 & 2 & 2 \\ -\frac{41}{2} & 4 & \frac{9}{2} \\ -5 & 1 & 1 \end{bmatrix} \begin{bmatrix} 1 \\ 2 \\ -1 \end{bmatrix} = \begin{bmatrix} -7 \\ -17 \\ -4 \end{bmatrix}$$

so $x_1 = -7$, $x_2 = -17$, and $x_3 = -4$.

NOW WORK PROBLEM 27

It can be shown that a system of n linear equations in n unknowns has a unique solution if and only if the coefficient matrix is invertible. Indeed, in the previous example the coefficient matrix is invertible, and a unique solution does in fact exist. When the coefficient matrix is not invertible, the system will have either no solution or infinitely many solutions.

While the solution of a system using a matrix inverse is very elegant, we must provide a caution. Given $\mathbf{AX} = \mathbf{B}$, the computational work required to find $\mathbf{A}^{-1}$ is greater than that required to reduce $[\mathbf{A}\,|\,\mathbf{B}]$. If you have several equations to solve, all with the same matrix of coefficients but variable right-hand sides, say $\mathbf{AX} = \mathbf{B_1}$, $\mathbf{AX} = \mathbf{B_2}, \ldots, \mathbf{AX} = \mathbf{B_k}$, then for suitably large k it *might* be faster to compute $\mathbf{A}^{-1}$ than to do k reductions, but a numerical analyst will in most cases convince you of the contrary. Don't forget that even with $\mathbf{A}^{-1}$ in hand you still have to compute $\mathbf{A}^{-1}\mathbf{B}$, and if the order of $\mathbf{A}$ is large, then this too takes considerable time.

⚠ CAUTION

The method of reduction in Sections 6.4 and 6.5 is a faster computation than that of finding a matrix inverse.

EXAMPLE 6   A Coefficient Matrix That Is Not Invertible

Solve the system

$$\begin{cases} x - 2y + z = 0 \\ 2x - y + 5z = 0 \\ x + y + 4z = 0 \end{cases}$$

Solution: The coefficient matrix is

$$\begin{bmatrix} 1 & -2 & 1 \\ 2 & -1 & 5 \\ 1 & 1 & 4 \end{bmatrix}$$

Since

$$\left[\begin{array}{ccc|ccc} 1 & -2 & 1 & 1 & 0 & 0 \\ 2 & -1 & 5 & 0 & 1 & 0 \\ 1 & 1 & 4 & 0 & 0 & 1 \end{array}\right] \rightarrow \cdots \rightarrow \left[\begin{array}{ccc|ccc} 1 & 0 & 3 & -\frac{1}{3} & \frac{2}{3} & 0 \\ 0 & 1 & 1 & -\frac{2}{3} & \frac{1}{3} & 0 \\ 0 & 0 & 0 & 1 & -1 & 1 \end{array}\right]$$

the coefficient matrix is not invertible. Hence, the system *cannot* be solved by inverses. Instead, another method must be used. In Example 4(a) of Section 6.5, the solution was found to be $x = -3r$, $y = -r$, and $z = r$, where r is any real number (thus providing infinitely many solutions).

NOW WORK PROBLEM 31

TECHNOLOGY

To solve the system

$$\begin{cases} 3x + 2y = 6 \\ x + 4y = -8 \end{cases}$$

with a graphing calculator, we enter the coefficient matrix as [A] and the column matrix of constants as [B]. The product $[A]^{-1}[B]$ in Figure 6.6 gives the solution $x = 4$, $y = -3$.

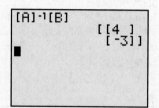

FIGURE 6.6 $[A]^{-1}[B]$ gives solution $x = 4$, $y = -3$ to the system of equations.

Problems 6.6

In Problems 1–18, if the given matrix is invertible, find its inverse.

*1. $\begin{bmatrix} 6 & 1 \\ 7 & 1 \end{bmatrix}$

2. $\begin{bmatrix} 2 & 4 \\ 3 & 6 \end{bmatrix}$

3. $\begin{bmatrix} 2 & 2 \\ 2 & 2 \end{bmatrix}$

4. $\begin{bmatrix} \frac{1}{4} & \frac{3}{8} \\ 0 & -\frac{1}{6} \end{bmatrix}$

5. $\begin{bmatrix} 1 & 0 & 0 \\ 0 & -3 & 0 \\ 0 & 0 & 4 \end{bmatrix}$

6. $\begin{bmatrix} 2 & 0 & 8 \\ -1 & 4 & 0 \\ 2 & 1 & 0 \end{bmatrix}$

*7. $\begin{bmatrix} 1 & 2 & 3 \\ 0 & 0 & 4 \\ 0 & 0 & 5 \end{bmatrix}$

8. $\begin{bmatrix} 2 & 0 & 0 \\ 0 & 0 & 0 \\ 0 & 0 & -4 \end{bmatrix}$

9. $\begin{bmatrix} 1 & 2 \\ 2 & 3 \\ 2 & 5 \end{bmatrix}$

10. $\begin{bmatrix} 0 & 0 & 0 \\ 0 & 0 & 0 \\ 0 & 0 & 0 \end{bmatrix}$

11. $\begin{bmatrix} 1 & 2 & 3 \\ 0 & 1 & 2 \\ 0 & 0 & 1 \end{bmatrix}$

12. $\begin{bmatrix} 1 & 2 & -1 \\ 0 & 1 & 4 \\ 1 & -1 & 2 \end{bmatrix}$

13. $\begin{bmatrix} 7 & 0 & -2 \\ 0 & 1 & 0 \\ -3 & 0 & 1 \end{bmatrix}$

14. $\begin{bmatrix} 2 & 3 & -1 \\ 1 & 2 & 1 \\ -1 & -1 & 3 \end{bmatrix}$

15. $\begin{bmatrix} 2 & 1 & 0 \\ 4 & -1 & 5 \\ 1 & -1 & 2 \end{bmatrix}$

16. $\begin{bmatrix} -1 & 2 & -3 \\ 2 & 1 & 0 \\ 4 & -2 & 5 \end{bmatrix}$

17. $\begin{bmatrix} 1 & 2 & 3 \\ 1 & 3 & 5 \\ 1 & 5 & 12 \end{bmatrix}$

18. $\begin{bmatrix} 2 & -1 & 3 \\ 0 & 2 & 0 \\ 2 & 1 & 1 \end{bmatrix}$

*19. Solve $\mathbf{AX} = \mathbf{B}$ if

$$\mathbf{A}^{-1} = \begin{bmatrix} 1 & 2 \\ 8 & 1 \end{bmatrix} \quad \text{and} \quad \mathbf{B} = \begin{bmatrix} 2 \\ 4 \end{bmatrix}$$

20. Solve $\mathbf{AX} = \mathbf{B}$ if

$$\mathbf{A}^{-1} = \begin{bmatrix} 1 & 0 & 1 \\ 0 & 3 & 0 \\ 2 & 0 & 4 \end{bmatrix} \quad \text{and} \quad \mathbf{B} = \begin{bmatrix} 10 \\ 2 \\ -1 \end{bmatrix}$$

For Problems 21–34, if the coefficient matrix of the system is invertible, solve the system by using the inverse. If not, solve the system by the method of reduction.

21. $\begin{cases} 6x + 5y = 2 \\ x + y = -3 \end{cases}$

22. $\begin{cases} 2x + 4y = 5 \\ -x + 3y = -2 \end{cases}$

23. $\begin{cases} 3x + y = 5 \\ 3x - y = 7 \end{cases}$

24. $\begin{cases} 3x + 2y = 26 \\ 4x + 3y = 37 \end{cases}$

25. $\begin{cases} 2x + 6y = 2 \\ 3x + 9y = 3 \end{cases}$

26. $\begin{cases} 2x + 6y = 8 \\ 3x + 9y = 7 \end{cases}$

*27. $\begin{cases} x + 2y + z = 4 \\ 3x + z = 2 \\ x - y + z = 1 \end{cases}$

28. $\begin{cases} x + y + z = 6 \\ x - y + z = -1 \\ x - y - z = 4 \end{cases}$

29. $\begin{cases} x + y + z = 2 \\ x - y + z = 1 \\ x - y - z = 0 \end{cases}$

30. $\begin{cases} 2x + 8z = 8 \\ -x + 4y = 36 \\ 2x + y = 9 \end{cases}$

*31. $\begin{cases} x + 3y + 3z = 7 \\ 2x + y + z = 4 \\ x + y + z = 4 \end{cases}$

32. $\begin{cases} x + 3y + 3z = 7 \\ 2x + y + z = 4 \\ x + y + z = 3 \end{cases}$

33. $\begin{cases} w \quad\quad + 2y + \ z = 4 \\ w - x \quad\quad + 2z = 12 \\ 2w + x \quad\quad + \ z = 12 \\ w + 2x + \ y + \ z = 12 \end{cases}$ **34.** $\begin{cases} w + x - y \quad\quad = 1 \\ x + y + \ z = 0 \\ -w + x + y \quad\quad = 1 \\ w - x - y + 2z = 1 \end{cases}$

For Problems 35 and 36, find $(\mathbf{I}-\mathbf{A})^{-1}$ for the given matrix $\mathbf{A}$.

35. $\mathbf{A} = \begin{bmatrix} 5 & -2 \\ 1 & 2 \end{bmatrix}$

36. $\mathbf{A} = \begin{bmatrix} -3 & 2 \\ 4 & 3 \end{bmatrix}$

37. Auto Production   Solve the following problems by using the inverse of the matrix involved.

   (a) An automobile factory produces two models, A and B. Model A requires 1 labor hour to paint and $\frac{1}{2}$ labor hour to polish; model B requires 1 labor hour for each process. During each hour that the assembly line is operating, there are 100 labor hours available for painting and 80 labor hours for polishing. How many of each model can be produced each hour if all the labor hours available are to be utilized?

   (b) Suppose each model A requires 10 widgets and 14 shims and each model B requires 7 widgets and 10 shims. The factory can obtain 800 widgets and 1130 shims each hour. How many cars of each model can it produce while using all the parts available?

38. If $\mathbf{A} = \begin{bmatrix} a & 0 & 0 \\ 0 & b & 0 \\ 0 & 0 & c \end{bmatrix}$, where $a, b, c \neq 0$, show that

$$\mathbf{A}^{-1} = \begin{bmatrix} 1/a & 0 & 0 \\ 0 & 1/b & 0 \\ 0 & 0 & 1/c \end{bmatrix}$$

39. (a) If $\mathbf{A}$ and $\mathbf{B}$ are invertible matrices with the same order, show that $(\mathbf{AB})^{-1} = \mathbf{B}^{-1}\mathbf{A}^{-1}$. [*Hint:* Show that

$$(\mathbf{B}^{-1}\mathbf{A}^{-1})(\mathbf{AB}) = \mathbf{I}$$

and use the fact that the inverse is unique.]

   (b) If

$$\mathbf{A}^{-1} = \begin{bmatrix} 1 & 2 \\ 3 & 4 \end{bmatrix} \quad \text{and} \quad \mathbf{B}^{-1} = \begin{bmatrix} 1 & 1 \\ 1 & 2 \end{bmatrix}$$

   find $(\mathbf{AB})^{-1}$.

40. If $\mathbf{A}$ is invertible, it can be shown that $(\mathbf{A}^{\mathrm{T}})^{-1} = (\mathbf{A}^{-1})^{\mathrm{T}}$. Verify this relationship if

$$\mathbf{A} = \begin{bmatrix} 1 & 0 \\ 1 & 2 \end{bmatrix}$$

41. A matrix $\mathbf{P}$ is said to be *orthogonal* if $\mathbf{P}^{-1} = \mathbf{P}^{\mathrm{T}}$. Is the matrix $\mathbf{P} = \dfrac{1}{5}\begin{bmatrix} 3 & -4 \\ 4 & 3 \end{bmatrix}$ orthogonal?

42. Secret Message   A friend has sent you a secret message that consists of three row matrices of numbers as follows:

$$\mathbf{R}_1 = [33 \quad 87 \quad 70] \quad\quad \mathbf{R}_2 = [57 \quad 133 \quad 20]$$

$$\mathbf{R}_3 = [38 \quad 90 \quad 33]$$

Both you and your friend have committed the following matrix to memory (your friend used it to code the message):

$$\mathbf{A} = \begin{bmatrix} 1 & 2 & -1 \\ 2 & 5 & 2 \\ -1 & -2 & 2 \end{bmatrix}$$

Decipher the message by proceeding as follows:

   (a) Calculate the three matrix products $\mathbf{R}_1\mathbf{A}^{-1}$, $\mathbf{R}_2\mathbf{A}^{-1}$, and $\mathbf{R}_3\mathbf{A}^{-1}$.

   (b) Assume that the letters of the alphabet correspond to the numbers 1 through 26, replace the numbers in the preceding three matrices by letters, and determine the message.

43. Investing   A group of investors decides to invest \$500,000 in the stocks of three companies. Company D sells for \$60 a share and has an expected growth of 16% per year. Company E sells for \$80 per share and has an expected growth of 12% per year. Company F sells for \$30 a share and has an expected growth of 9% per year. The group plans to buy four times as many shares of company F as of company E. If the group's goal is 13.68% growth per year, how many shares of each stock should the investors buy?

44. Investing   The investors in Problem 43 decide to try a new investment strategy with the same companies. They wish to buy twice as many shares of company F as of company E, and they have a goal of 14.52% growth per year. How many shares of each stock should they buy?

In Problems 45 and 46, use a graphing calculator to (a) find $\mathbf{A}^{-1}$, and express its entries in decimal form rounded to two decimal places. (b) Express the entries of $\mathbf{A}^{-1}$ in fractional form if your calculator has such capability. [Caution: For part (b), use the calculator matrix $\mathbf{A}^{-1}$ to convert to fractional entries; do not use the matrix of rounded values from part (a).]

45. $\mathbf{A} = \begin{bmatrix} \frac{2}{3} & -\frac{1}{2} \\ -\frac{2}{7} & \frac{4}{5} \end{bmatrix}$

46. $\mathbf{A} = \begin{bmatrix} 2 & 6 & -3 \\ 4 & 8 & 9 \\ -7 & 2 & 5 \end{bmatrix}$

47. If $\mathbf{A} = \begin{bmatrix} 0.4 & -0.6 & -0.3 \\ -0.2 & 0.1 & -0.1 \\ -0.3 & -0.2 & 0.4 \end{bmatrix}$, find $(\mathbf{I} - \mathbf{A})^{-1}$, where $\mathbf{I}$ is the identity matrix of order 3. Round entries to two decimal places.

In Problems 48 and 49, use a graphing calculator to solve the system by using the inverse of the coefficient matrix.

48. $\begin{cases} 0.9x + \ 3y - 4.7z = 13 \\ 2x - 0.4y + \ 2z = 4.7 \\ x - 0.8y - 0.5z = 7.2 \end{cases}$

49. $\begin{cases} \frac{2}{5}w + 4x + \frac{1}{2}y - \frac{3}{7}z = \frac{14}{13} \\ \frac{5}{9}w - \frac{2}{3}x - 4y - \ z = \frac{7}{8} \\ x - \frac{4}{9}y + \frac{5}{6}z = 9 \\ \frac{1}{2}w \quad\quad + 4y - \frac{1}{3}z = \frac{4}{7} \end{cases}$

OBJECTIVE

To use the methods of this chapter to analyze the production of sectors of an economy.

6.7 Leontief's Input–Output Analysis

Input–output matrices, which were developed by Wassily W. Leontief,[6] indicate the supply and demand interrelationships that exist among the various sectors of an economy during some time period. The phrase *input–output* is used because the matrices show the values of outputs of each industry that are sold as inputs to each industry and for final use by consumers.

A hypothetical example for an oversimplified two-industry economy is given by the input–output matrix to be presented next. Before we present and explain the matrix, however, let us say that the *industrial* sectors can be thought of as manufacturing, steel, agriculture, coal, and so on. The *other production factors* sector consists of costs to the respective industries, such as labor, profits, and so on. The *final-demand* sector could be consumption by households, government, and so on. The matrix is as follows:

	Consumers (input)			
Producers (output):	Industry A	Industry B	Final Demand	Totals
Industry A	240	500	460	1200
Industry B	360	200	940	1500
Other Production Factors	600	800	—	
Totals	1200	1500		

Each industry appears in a row and in a column. The row shows the purchases of an industry's output by the industrial sectors and by consumers for final use (hence the term *final demand*). The entries represent the value of the products and might be in units of millions of dollars of product. For example, of the total output of industry A, 240 went as input to industry A itself (for internal use), 500 went to industry B, and 460 went directly to the final-demand sector. The total output of A is the sum of industrial demand and final demand ($240 + 500 + 460 = 1200$).

Each industry column gives the value of what the industry purchased for input from each industry, as well as what it spent for other costs. For example, in order to produce its 1200 units, A purchased 240 units of output from itself, bought 360 of B's output, and had labor and other costs of 600 units.

Note that for each industry, the sum of the entries in its row is equal to the sum of the entries in its column. That is, the value of the total output of A is equal to the value of the total input to A.

Input–output analysis allows us to estimate the total production of each *industrial* sector if there is a change in final demand, *as long as the basic structure of the economy remains the same*. This important assumption means that for each industry, the amount spent on each input for each dollar's worth of output must remain fixed.

For example, in producing 1200 units' worth of product, industry A purchases 240 units' worth from industry A, purchases 360 units' worth from B, and spends 600 units on other costs. Thus, for each dollar's worth of output, industry A spends $\frac{240}{1200} = \frac{1}{5}(= \$0.20)$ on A, $\frac{360}{1200} = \frac{3}{10}(= \$0.30)$ on B, and $\frac{600}{1200} = \frac{1}{2}(= \$0.50)$ on other costs. Combining these fixed ratios of industry A with those of industry B, we can give the input requirements per dollar of output for each industry:

$$
\begin{array}{c}
 & A & B \\
A & \frac{240}{1200} & \frac{500}{1500} \\
B & \frac{360}{1200} & \frac{200}{1500} \\
Other & \frac{600}{1200} & \frac{800}{1500}
\end{array}
=
\begin{array}{c}
 & A & B \\
A & \frac{1}{5} & \frac{1}{3} \\
B & \frac{3}{10} & \frac{2}{15} \\
Other & \frac{1}{2} & \frac{8}{15}
\end{array}
$$

---

[6]Leontief won the 1973 Nobel Prize in economic science for the development of the "input–output" method and its applications to economic problems.

The entries in the matrix are called **input–output coefficients.** The sum of each column is 1.

Now, suppose the value of final demand changes from 460 to 500 for industry *A* and from 940 to 1200 for industry *B*. We would like to estimate the value of *total* output that *A* and *B* must produce for both industry and final demand to meet this goal, provided that the structure in the preceding matrix remains the same.

Let the new values of total outputs for industries *A* and *B* be x_A and x_B, respectively. Now, for *A*,

$$\text{total value of output of } A = \text{value consumed by } A + \text{value consumed by } B + \text{value consumed by final demand}$$

so we have

$$x_A = \frac{1}{5}x_A + \frac{1}{3}x_B + 500$$

Similarly, for *B*,

$$x_B = \frac{3}{10}x_A + \frac{2}{15}x_B + 1200$$

Using matrix notation, we can write

$$\begin{bmatrix} x_A \\ x_B \end{bmatrix} = \begin{bmatrix} \frac{1}{5} & \frac{1}{3} \\ \frac{3}{10} & \frac{2}{15} \end{bmatrix} \begin{bmatrix} x_A \\ x_B \end{bmatrix} + \begin{bmatrix} 500 \\ 1200 \end{bmatrix} \qquad (1)$$

In this matrix equation, let

$$\mathbf{X} = \begin{bmatrix} x_A \\ x_B \end{bmatrix} \qquad \mathbf{A} = \begin{bmatrix} \frac{1}{5} & \frac{1}{3} \\ \frac{3}{10} & \frac{2}{15} \end{bmatrix} \qquad \mathbf{D} = \begin{bmatrix} 500 \\ 1200 \end{bmatrix}$$

We call **X** the **output matrix,** **A** the **coefficient matrix**, and **D** the **final-demand matrix.** From Equation (1),

$$\mathbf{X} = \mathbf{AX} + \mathbf{D}$$

$$\mathbf{X} - \mathbf{AX} = \mathbf{D}$$

Given output **X**, we can show that **AX** is the amount of the output that is required to self-sustain the economy. If we refer to this as the **internal-demand** and observe that the final demand can be construed as an **external-demand** beyond what is necessary for internal production then we arrive at the conceptual equation:

output = internal-demand + external-demand

as a further justification for $\mathbf{X} = \mathbf{AX} + \mathbf{D}$. If **I** is the 2 × 2 identity matrix, then we can derive from the last equation

$$\mathbf{IX} - \mathbf{AX} = \mathbf{D}$$

$$(\mathbf{I} - \mathbf{A})\mathbf{X} = \mathbf{D}$$

which is an ordinary system of linear equations, written in matrix form, with what is usually called the coefficient matrix given by $\mathbf{I} - \mathbf{A}$ (although in Leontief's theory we have used the term *coefficient matrix* for **A**).

If $(\mathbf{I} - \mathbf{A})^{-1}$ exists (and in fact it does when **A** arises as a coefficient matrix via the consideration of this section), then

$$\mathbf{X} = (\mathbf{I} - \mathbf{A})^{-1}\mathbf{D}$$

The matrix $\mathbf{I} - \mathbf{A}$ is called the **Leontief matrix.** In spite of our caution in Section 6.6 about the inefficiency of matrix inverse calculations, we have here one of the best occasions to compute an inverse. Typically, the coefficient matrix **A** remains constant as long as the basic structure of the economy remains the same. Thus the Leontief matrix $\mathbf{I} - \mathbf{A}$ is also a feature of a particular economy. A planner may well want to experiment to find the output **X** necessary to satisfy various final-demand scenarios $\mathbf{D}_1, \mathbf{D}_2, \ldots, \mathbf{D}_k$. If she has at hand the matrix $(\mathbf{I} - \mathbf{A})^{-1}$, then the experiment can be conducted by *k* matrix multiplications: $(\mathbf{I} - \mathbf{A})^{-1}\mathbf{D}_1, (\mathbf{I} - \mathbf{A})^{-1}\mathbf{D}_2, \ldots, (\mathbf{I} - \mathbf{A})^{-1}\mathbf{D}_k$.

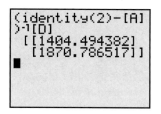

FIGURE 6.7   Evaluating an output matrix.

The issue of computational efficiency can be a serious one. While we treat this topic of input–output analysis with examples of economies divided into two or three sectors, a model with 20 sectors might be more realistic—in which case the Leontief matrix will have 400 entries.

In the case at hand, we enter the matrices $\mathbf{A}$ and $\mathbf{D}$ into a graphing calculator. With a TI-83 Plus, the identity matrix of order 2 is obtained with the command "identity 2." Evaluating $(\mathbf{I} - \mathbf{A})^{-1}\mathbf{D}$ as shown in Figure 6.7 results in the output matrix

$$\mathbf{X} = (\mathbf{I} - \mathbf{A})^{-1}\mathbf{D} = \begin{bmatrix} 1404.49 \\ 1870.79 \end{bmatrix}$$

Here we rounded the entries in $\mathbf{X}$ to two decimal places. Thus, to meet the goal, industry A must produce 1404.49 units of value, and industry B must produce 1870.79. If we were interested in the value of other production factors for A, say, P_A, then

$$P_A = \frac{1}{2}x_A = 702.25$$

● **EXAMPLE 1   Input–Output Analysis**

Given the input–output matrix

		Industry		**Final**
	A	**B**	**C**	**Demand**
Industry: A	240	180	144	36
B	120	36	48	156
C	120	72	48	240
Other	120	72	240	—

suppose final demand changes to 77 for A, 154 for B, and 231 for C. Find the output matrix for the economy. (The entries are in millions of dollars.)

Solution: We separately add the entries in the first three rows. The total values of output for industries A, B, and C are 600, 360, and 480, respectively. To get the coefficient matrix $\mathbf{A}$, we divide the industry entries in each industry column by the total value of output for that industry:

$$\mathbf{A} = \begin{bmatrix} \frac{240}{600} & \frac{180}{360} & \frac{144}{480} \\ \frac{120}{600} & \frac{36}{360} & \frac{48}{480} \\ \frac{120}{600} & \frac{72}{360} & \frac{48}{480} \end{bmatrix}$$

The final-demand matrix is

$$\mathbf{D} = \begin{bmatrix} 77 \\ 154 \\ 231 \end{bmatrix}$$

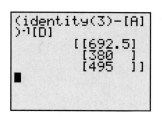

FIGURE 6.8   Evaluating the output matrix of Example 1.

Figure 6.8 shows the result of evaluating $(\mathbf{I} - \mathbf{A})^{-1}\mathbf{D}$. Thus, the output matrix is

$$\mathbf{X} = (\mathbf{I} - \mathbf{A})^{-1}\mathbf{D} = \begin{bmatrix} 692.5 \\ 380 \\ 495 \end{bmatrix}$$

NOW WORK PROBLEM 1 ●●

The data for the coefficient matrix $\mathbf{A}$ (and hence that of the Leontief matrix $\mathbf{I} - \mathbf{A}$) may sometimes be given to you more directly as the next example shows.

● **EXAMPLE 2   Input–Output Analysis**

Suppose that a simple economy consists of three sectors: agriculture (A), manufacturing (M), and transportation (T). Economists have determined that to produce one unit of A requires $\frac{1}{18}$ units of A, $\frac{1}{9}$ units of B, and $\frac{1}{9}$ units of C, while production of one unit of M requires $\frac{3}{16}$ units of A, $\frac{1}{4}$ units of M, and $\frac{3}{16}$ units of T, and production of one unit of T requires $\frac{1}{15}$ units of A, $\frac{1}{3}$ units of M, and $\frac{1}{6}$ units of T. There is an external demand for 40 units of A, 30 units of M, and no units of T. Determine the production levels necessary to meet the external demand.

CAUTION

When input–output data are presented as in this example, take care to enter the coefficient matrix A rather than its transpose.

Strategy   By examining the data in Example 1, we see that to produce 600 units of A required 240 units of A, 120 units of B, and 120 units of C. It follows that to produce *one* unit of A required $\frac{240}{600} = \frac{2}{5}$ units of A, $\frac{120}{600} = \frac{1}{5}$ units of B, and $\frac{120}{600} = \frac{1}{5}$ units of C. The numbers $\frac{240}{600}, \frac{120}{600}$, and $\frac{120}{600}$ constitute, in that order, the first *column* of the coefficient matrix. It follows that the data we have been given describe the coefficient matrix, *column by column*.

Solution: The coefficient matrix is

$$\mathbf{A} = \begin{bmatrix} \frac{1}{18} & \frac{3}{16} & \frac{1}{15} \\ \frac{1}{9} & \frac{1}{4} & \frac{1}{3} \\ \frac{1}{9} & \frac{3}{16} & \frac{1}{6} \end{bmatrix}$$

and the external-demand matrix is

$$\mathbf{D} = \begin{bmatrix} 40 \\ 30 \\ 0 \end{bmatrix}$$

To solve $(\mathbf{I} - \mathbf{A})\mathbf{X} = \mathbf{D}$, we reduce

$$\begin{bmatrix} \frac{17}{18} & -\frac{3}{16} & -\frac{1}{15} & 40 \\ -\frac{1}{9} & \frac{3}{4} & -\frac{1}{3} & 30 \\ -\frac{1}{9} & -\frac{3}{16} & \frac{5}{6} & 0 \end{bmatrix}$$

which we leave as a calculator exercise.

NOW WORK PROBLEM 9

Problems 6.7

*1. Given the input–output matrix

	Industry		**Final**
	Steel	**Coal**	**Demand**
Industry: Steel	200	500	500
Coal	400	200	900
Other	600	800	—

find the output matrix if final demand changes to 600 for steel and 805 for coal. Find the total value of the other production costs that this involves.

2. Given the input–output matrix

	Industry		**Final**
	Education	**Government**	**Demand**
Industry: Education	40	120	40
Government	120	90	90
Other	40	90	—

find the output matrix if final demand changes to (a) 200 for education and 300 for government; (b) 64 for education and 64 for government.

3. Given the input–output matrix

	Industry			**Final**
	Grain	**Fertilizer**	**Cattle**	**Demand**
Industry: Grain	15	30	45	10
Fertilizer	25	30	60	5
Cattle	50	40	60	30
Other	10	20	15	—

find the output matrix (with entries rounded to two decimal places) if final demand changes to (a) 15 for grain, 10 for fertilizer, and 35 for cattle; (b) 10 for grain, 10 for fertilizer, and 10 for cattle.

4. Given the input–output matrix

	Industry			**Final**
	Water	**Electric Power**	**Agriculture**	**Demand**
Industry: Water	100	400	240	260
Electric Power	100	80	480	140
Agriculture	300	160	240	500
Other	500	160	240	—

find the output matrix if final demand changes to 500 for water, 150 for electric power, and 700 for agriculture. Round your entries to two decimal places.

5. Given the input–output matrix

	Industry			Final
	Government	Agriculture	Manufacturing	Demand
Industry: Government	400	200	200	200
Agriculture	200	400	100	300
Manufacturing	200	100	300	400
Other	200	300	400	—

with entries in billions of dollars, find the output matrix for the economy if the final demand changes to 300 for Government, 350 for Agriculture, and 450 for Manufacturing. Round your entries to the nearest billion dollars.

6. Given the input–output matrix in Problem 5, find the output matrix for the economy if the final demand changes to 250 for government, 300 for agriculture, and 350 for manufacturing. Round your entries to the nearest billion dollars.

7. Given the input–output matrix in Problem 5, find the output matrix for the economy if the final demand changes to 300

for government, 400 for agriculture, and 500 for manufacturing. Round your entries to the nearest billion dollars.

8. A very simple economy consists of two sectors: agriculture and milling. To produce one unit of agricultural products requires $\frac{1}{3}$ of a unit of agricultural products and $\frac{1}{4}$ of a unit of milled products. To produce one unit of milled products requires $\frac{3}{4}$ of a unit of agricultural products and no units of milled products. Determine the production levels needed to satisfy an external demand for 300 units of agriculture and 500 units of milled products.

*9. An economy consists of three sectors: coal, steel, and railroads. To produce one unit of coal requires $\frac{1}{10}$ of a unit of coal, $\frac{1}{10}$ of a unit of steel, and $\frac{1}{10}$ of a unit of railroad services. To produce one unit of steel requires $\frac{1}{3}$ of a unit of coal, $\frac{1}{10}$ of a unit of steel, and $\frac{1}{10}$ of a unit of railroad services. To produce one unit of railroad services requires $\frac{1}{4}$ of a unit of coal, $\frac{1}{3}$ of a unit of steel, and $\frac{1}{10}$ of a unit of railroad services. Determine the production levels needed to satisfy an external demand for 300 units of coal, 200 units of steel, and 500 units of railroad services.

6.8 Review

Important Terms and Symbols

		Examples
Section 6.1	**Matrices**	
	matrix size entry, a_{ij} row vector column vector	Ex. 1, p. 228
	equality of matrices transpose of matrix, $\mathbf{A}^T$ zero matrix, $\mathbf{O}$	Ex. 3, p. 230
Section 6.2	**Matrix Addition and Scalar Multiplication**	
	addition and subtraction of matrices scalar multiplication	Ex. 4, p. 235
Section 6.3	**Matrix Multiplication**	
	matrix multiplication identity matrix, $\mathbf{I}$ power of a matrix	Ex. 12, p. 247
	matrix equation, $\mathbf{AX} = \mathbf{B}$	Ex. 13, p. 247
Section 6.4	**Solving Systems by Reducing Matrices**	
	coefficient matrix augmented coefficient matrix	Ex. 3, p. 254
	elementary row operation equivalent matrices reduced matrix	Ex. 4, p. 255
	parameter	Ex. 5, p. 256
Section 6.5	**Solving Systems by Reducing Matrices (continued)**	
	homogeneous system nonhomogeneous system trivial solution	Ex. 4, p. 262
Section 6.6	**Inverses**	
	inverse matrix invertible matrix	Ex. 1, p. 264
Section 6.7	**Leontief's Input–Ouput Analysis**	
	input–output matrix Leontief matrix	Ex. 1, p. 273

Summary

A matrix is a rectangular array of numbers enclosed within brackets. There are a number of special types of matrices, such as zero matrices, $\mathbf{O}$; identity matrices, $\mathbf{I}$; square matrices; and diagonal matrices. Besides the basic operation of scalar multiplication, there are the operations of matrix addition and subtraction, which apply to matrices of the same size. The product $\mathbf{AB}$ is defined when the number of columns of $\mathbf{A}$ is equal to the number of rows of $\mathbf{B}$. Although matrix addition is commutative, matrix multiplication is not. By using matrix multiplication, we can express a system of linear equations as the matrix equation $\mathbf{AX} = \mathbf{B}$.

A system of linear equations may have a unique solution, no solution, or infinitely many solutions. The main method of solving

a system of linear equations using matrices is by applying the three elementary row operations to the augmented coefficient matrix of the system until an equivalent reduced matrix is obtained. The reduced matrix makes any solutions to the system obvious and allows the detection of nonexistence of solutions. If there are infinitely many solutions, the general solution involves at least one parameter.

Occasionally, it is useful to find the inverse of a (square) matrix. The inverse (if it exists) of a square matrix $\mathbf{A}$ is found by augmenting $\mathbf{A}$ with $\mathbf{I}$ and applying elementary row operations to $[\mathbf{A} \mid \mathbf{I}]$ until $\mathbf{A}$ is reduced resulting in $[\mathbf{R} \mid \mathbf{B}]$ (with $\mathbf{R}$ reduced). If $\mathbf{R} = \mathbf{I}$, then $\mathbf{A}$ is invertible and $\mathbf{A}^{-1} = \mathbf{B}$. If $\mathbf{R} \neq \mathbf{I}$, then $\mathbf{A}$ is not invertible, meaning that $\mathbf{A}^{-1}$ does not exist. If the inverse of an $n \times n$ matrix $\mathbf{A}$ exists, then the unique solution to $\mathbf{AX} = \mathbf{B}$ is given by $\mathbf{X} = \mathbf{A}^{-1}\mathbf{B}$. If $\mathbf{A}$ is not invertible, the system has either no solution or infinitely many solutions.

Our final application of matrices dealt with the interrelationships that exist among the various sectors of an economy and is known as Leontief's input–output analysis.

Review Problems

Problem numbers shown in color indicate problems suggested for use as a practice chapter test.

In Problems 1–8, simplify.

1. $2\begin{bmatrix} 3 & 4 \\ -5 & 1 \end{bmatrix} - 3\begin{bmatrix} 1 & 0 \\ 2 & 4 \end{bmatrix}$

2. $8\begin{bmatrix} 1 & 2 \\ 7 & 0 \end{bmatrix} - 2\begin{bmatrix} 1 & 0 \\ 0 & 1 \end{bmatrix}$

3. $\begin{bmatrix} 1 & 7 \\ 2 & -3 \\ 1 & 0 \end{bmatrix}\begin{bmatrix} 1 & 0 & -2 \\ 0 & 6 & 1 \end{bmatrix}$

4. $[2 \quad 3 \quad 7]\begin{bmatrix} 2 & 3 \\ 0 & -1 \\ 5 & 2 \end{bmatrix}$

5. $\begin{bmatrix} 2 & 3 \\ -1 & 3 \end{bmatrix}\left(\begin{bmatrix} 2 & 3 \\ 7 & 6 \end{bmatrix} - \begin{bmatrix} 1 & 8 \\ 4 & 4 \end{bmatrix}\right)$

6. $-\left(\begin{bmatrix} 2 & 0 \\ 7 & 8 \end{bmatrix} + 2\begin{bmatrix} 0 & -5 \\ 6 & -4 \end{bmatrix}\right)$

7. $2\begin{bmatrix} 1 & -2 \\ 3 & 1 \end{bmatrix}^2 [1 \quad -2]^{\mathrm{T}}$

8. $\frac{1}{3}\begin{bmatrix} 3 & 0 \\ 3 & 6 \end{bmatrix}\left(\begin{bmatrix} 1 & 0 \\ 1 & 3 \end{bmatrix}^{\mathrm{T}}\right)^2$

In Problems 9–12, compute the required matrix if

$$\mathbf{A} = \begin{bmatrix} 1 & 1 \\ -1 & 2 \end{bmatrix} \qquad \mathbf{B} = \begin{bmatrix} 1 & 0 \\ 0 & 2 \end{bmatrix}$$

9. $(2\mathbf{A})^{\mathrm{T}} - 3\mathbf{I}^2$

10. $\mathbf{A}(2\mathbf{I}) - \mathbf{AO}^{\mathrm{T}}$

11. $\mathbf{B}^3 + \mathbf{I}^5$

12. $(\mathbf{ABA})^{\mathrm{T}} - \mathbf{A}^{\mathrm{T}}\mathbf{B}^{\mathrm{T}}\mathbf{A}^{\mathrm{T}}$

In Problems 13 and 14, solve for x and y.

13. $\begin{bmatrix} 5 \\ 7 \end{bmatrix}[x] = \begin{bmatrix} 15 \\ y \end{bmatrix}$

14. $\begin{bmatrix} 1 & x \\ 2 & y \end{bmatrix}\begin{bmatrix} 2 & 1 \\ x & 3 \end{bmatrix} = \begin{bmatrix} 3 & 4 \\ 3 & y \end{bmatrix}$

In Problems 15–18, reduce the given matrices.

15. $\begin{bmatrix} 1 & 4 \\ 5 & 8 \end{bmatrix}$

16. $\begin{bmatrix} 0 & 0 & 7 \\ 0 & 5 & 9 \end{bmatrix}$

17. $\begin{bmatrix} 2 & 4 & 7 \\ 1 & 2 & 4 \\ 5 & 8 & 2 \end{bmatrix}$

18. $\begin{bmatrix} 0 & 0 & 0 & 1 \\ 0 & 0 & 0 & 0 \\ 1 & 0 & 0 & 0 \end{bmatrix}$

In Problems 19–22, solve each of the systems by the method of reduction.

19. $\begin{cases} 2x - 5y = 0 \\ 4x + 3y = 0 \end{cases}$

20. $\begin{cases} x - y + 2z = 3 \\ 3x + y + z = 5 \end{cases}$

21. $\begin{cases} x + y + 2z = 1 \\ 3x - 2y - 4z = -7 \\ 2x - y - 2z = 2 \end{cases}$

22. $\begin{cases} x - y - z - 1 = 0 \\ x + y + 2z + 3 = 0 \\ 2x + 2z + 7 = 0 \end{cases}$

In Problems 23–26, find the inverses of the matrices.

23. $\begin{bmatrix} 1 & 5 \\ 3 & 9 \end{bmatrix}$
24. $\begin{bmatrix} 0 & 1 \\ 1 & 0 \end{bmatrix}$

25. $\begin{bmatrix} 1 & 3 & -2 \\ 4 & 1 & 0 \\ 3 & -2 & 2 \end{bmatrix}$

26. $\begin{bmatrix} 5 & 0 & 0 \\ -5 & 2 & 1 \\ -5 & 1 & 3 \end{bmatrix}$

In Problems 27 and 28, solve the given system by using the inverse of the coefficient matrix.

27. $\begin{cases} 3x + y + 4z = 1 \\ x + z = 0 \\ 2y + z = 2 \end{cases}$

28. $\begin{cases} 5x = 3 \\ -5x + 2y + z = 0 \\ -5x + y + 3z = 2 \end{cases}$

29. Let $\mathbf{A} = \begin{bmatrix} 0 & 1 & 1 \\ 0 & 0 & 1 \\ 0 & 0 & 0 \end{bmatrix}$. Find the matrices $\mathbf{A}^2, \mathbf{A}^3, \mathbf{A}^{1000}$, and $\mathbf{A}^{-1}$ (if the inverse exists).

30. $\mathbf{A} = \begin{bmatrix} 2 & 0 \\ 0 & 4 \end{bmatrix}$, show that $(\mathbf{A}^{\mathrm{T}})^{-1} = (\mathbf{A}^{-1})^{\mathrm{T}}$.

31. A consumer wishes to supplement his vitamin intake by *exactly* 13 units of vitamin A, 22 units of vitamin B, and 31 units of vitamin C per week. There are three brands of vitamin capsules available. Brand I contains 1 unit each of vitamins A, B, and C per capsule; brand II contains 1 unit of vitamin A, 2 of B, and 3 of C; and brand III contains 4 units of A, 7 of B, and 10 of C.

(a) What combinations of capsules of brands I, II, and III will produce *exactly* the desired amounts?

(b) If brand I capsules cost 5 cents each, brand II 7 cents each, and brand III 20 cents each, what combination will minimize the consumer's weekly cost?

32. Suppose that $\mathbf{A}$ is an invertible $n \times n$ matrix.

  (a) Prove that $\mathbf{A}^3$ is invertible.

  (b) Prove that if $\mathbf{B}$ and $\mathbf{C}$ are $n \times n$ matrices such that $\mathbf{AB} = \mathbf{AC}$, then $\mathbf{B} = \mathbf{C}$.

  (c) If $\mathbf{A}^2 = \mathbf{A}$ (we say that $\mathbf{A}$ is *idempotent*), find $\mathbf{A}$.

33. If $\mathbf{A} = \begin{bmatrix} 10 & -3 \\ 4 & 7 \end{bmatrix}$ and $\mathbf{B} = \begin{bmatrix} 8 & 6 \\ -7 & -3 \end{bmatrix}$, find $3\mathbf{AB} - 4\mathbf{B}^2$.

34. Solve the system

$$\begin{cases} 7.9x - 4.3y + 2.7z = 11.1 \\ 3.4x + 5.8y - 7.6z = 10.8 \\ 4.5x - 6.2y - 7.4z = 15.9 \end{cases}$$

by using the inverse of the coefficient matrix. Round your answers to two decimal places.

35. Given the input–output matrix

| | Industry | | Final |
	A	B	Demand
Industry: A	10	20	4
B	15	14	10
Other	9	5	—

find the output matrix if final demand changes to 10 for A and 5 for B. (Data are in tens of billions of dollars.)

Mathematical Snapshot

Insulin Requirements as a Linear Process[7]

A vacation lodge in the mountains of Washington State has a well-deserved reputation for attending to the special health needs of its guests. Next week the manager of the lodge is expecting four guests, each of whom has insulin-dependent diabetes. These guests plan to stay at the lodge for 7, 14, 21, and 28 days, respectively.

The lodge is quite a distance from the nearest drugstore, so before the arrival of the guests the manager plans to obtain the total amount of insulin that will be needed. Three different types of insulin are required: semi-lente, lente, and ultra-lente. The manager will store the insulin, and then lodge personnel will administer the daily dose of the three different types of insulin to each of the guests.

The daily requirements of the four guests are as follows:

Guest 1   20 insulin units of semi-lente, 30 units of lente, 10 units of ultra-lente;

Guest 2   40 insulin units of semi-lente, 0 units of lente, 0 units of ultra-lente;

Guest 3   30 insulin units of semi-lente, 10 units of lente, 30 units of ultra-lente;

Guest 4   10 insulin units of semi-lente, 10 units of lente, 50 units of ultra-lente.

This information will be represented by the following "requirement" matrix A:

$A = [a_{ij}]_{3 \times 4}$ where A is given by

	Guest 1	Guest 2	Guest 3	Guest 4
semi-lente insulin	20	40	30	10
lente insulin	30	0	10	10
ultra-lente insulin	10	0	30	50

Recall that Guest 1 will stay for 7 days, Guest 2 for 14 days, Guest 3 for 21 days, and Guest 4 for 28 days. You can let the following column vector T represent the time, in days, that each guest is staying at the lodge:

$$T = \begin{bmatrix} 7 \\ 14 \\ 21 \\ 28 \end{bmatrix}$$

[7]Adapted from Richard F. Baum, "Insulin Requirements as a Linear Process," in R. M. Thrall, J. A. Mortimer, K. R. Rebman, and R. F. Baum, (eds.), *Some Mathematical Models in Biology*, rev. ed. Report 40241-R-7. Prepared at the University of Michigan, 1967.

To determine the total amounts of the different types of insulin needed by the four guests, you compute the matrix product AT.

$$AT = \begin{bmatrix} 20 & 40 & 30 & 10 \\ 30 & 0 & 10 & 10 \\ 10 & 0 & 30 & 50 \end{bmatrix} \begin{bmatrix} 7 \\ 14 \\ 21 \\ 28 \end{bmatrix}$$

$$= 10(7) \begin{bmatrix} 2 & 4 & 3 & 1 \\ 3 & 0 & 1 & 1 \\ 1 & 0 & 3 & 5 \end{bmatrix} \begin{bmatrix} 1 \\ 2 \\ 3 \\ 4 \end{bmatrix}$$

$$= 70 \begin{bmatrix} 23 \\ 10 \\ 30 \end{bmatrix} = \begin{bmatrix} 1610 \\ 700 \\ 2100 \end{bmatrix} = B$$

Vector B (or AT) indicates that a total of 1610 insulin units of semi-lente, 700 insulin units of lente, and 2100 insulin units of ultra-lente are required by the four guests.

Now, change the problem a bit. Suppose that each guest decided to double the original length of stay. The resulting vector that gives the total amount needed of semi-lente, lente, and ultra-lente insulin is

$$A(2T) = 2(AT) = 2B = \begin{bmatrix} 3220 \\ 1400 \\ 4200 \end{bmatrix}$$

In fact, if each guest planned to spend a factor $k(k \geq 0)$ of the original time at the lodge (that is, Guest 1 planned to stay for $k \cdot 7$ days, Guest 2 for $k \cdot 14$ days, and so on), then the insulin requirements would be

$$A(kT) = k(AT) = kB = \begin{bmatrix} k \cdot 1610 \\ k \cdot 700 \\ k \cdot 2100 \end{bmatrix}$$

Similarly, if the guests decided to add 1, 3, 4, and 6 days, respectively, to the times they originally intended to stay, then the amounts of insulin required would be

$$A(T + T_1) = AT + AT_1 \quad \text{where } T_1 = \begin{bmatrix} 1 \\ 3 \\ 4 \\ 6 \end{bmatrix}$$

Based on the results thus far, it is obvious that the following matrix equation generalizes the situation:

$$AX = B$$

that is,

$$\begin{bmatrix} 20 & 40 & 30 & 10 \\ 30 & 0 & 10 & 10 \\ 10 & 0 & 30 & 50 \end{bmatrix} \begin{bmatrix} x_1 \\ x_2 \\ x_3 \\ x_4 \end{bmatrix} = \begin{bmatrix} b_1 \\ b_2 \\ b_3 \end{bmatrix}$$

which represents the linear system

$$\begin{cases} 20x_1 + 40x_2 + 30x_3 + 10x_4 = b_1 \\ 30x_1 + 10x_3 + 10x_4 = b_2 \\ 10x_1 + 30x_3 + 50x_4 = b_3 \end{cases}$$

where x_i is the number of days that Guest i stays at the lodge, and b_1, b_2, b_3 give, respectively, the total number of units of semi-lente, lente, and ultra-lente insulin needed by the four guests for their entire stay at the lodge.

Finally, suppose once again that vector $\mathbf{T}$ represents the number of days that each guest originally planned to stay at the lodge. Furthermore, suppose vector $\mathbf{C}$ gives the cost (in cents) per insulin unit of the three types of insulin, where

$$\mathbf{C} = \begin{bmatrix} 9 \\ 8 \\ 10 \end{bmatrix} = \text{cost matrix}$$

That is, one unit of semi-lente costs 9¢, one unit of lente costs 8¢, and one unit of ultra-lente costs 10¢. Then the total amount paid by the lodge for all the insulin required by the four guests is

$$\mathbf{C}^\mathrm{T}(\mathbf{AT}) = \mathbf{C}^\mathrm{T}\mathbf{B} = \begin{bmatrix} 9 & 8 & 10 \end{bmatrix} \begin{bmatrix} 1610 \\ 700 \\ 2100 \end{bmatrix} = [41{,}090]$$

that is, 41,090 cents, which is $410.90.

Problems

1. Suppose that Guest 1 will stay at the lodge for 7 days, Guest 2 for 10 days, Guest 3 for 7 days, and Guest 4 for 5 days. Assume that the daily requirements of the four guests and the cost matrix are the same as given in the discussion. Find the total amount (in dollars) that the lodge must pay for all the insulin required by the guests.

2. Suppose that the insulin requirements of the four guests add up to 1180 insulin units of semi-lente, 580 of lente, and 1500 of ultra-lente. Assume that same daily requirements for the four guests are as given in the discussion. Using the inverse matrix method on a graphing calculator, find the length of stay for each guest if the total number of days for all four guests is 52.

3. Assume that the daily requirements of the four guests and the cost matrix are the same as given in the discussion. Is it possible, given only the total amount (in dollars) that the lodge must pay for all the required insulin, to determine the length of each guest's stay? Why or why not?

7

LINEAR PROGRAMMING

 Mathematical Snapshot Drug and Radiation Therapies

Linear programming sounds like something involving the writing of computer code. But while linear programming is often done on computers, the "programming" part of the name actually comes from World War II–era military terminology, in which training, supply, and unit-deployment plans were called programs. Each program was a solution to a problem in resource allocation.

For example, suppose that military units in a combat theater need diesel fuel. Each unit has a certain number of tanks, trucks, and other vehicles; each unit uses its vehicles to accomplish an assigned mission; and each unit's mission has some relation to the overall goal of winning the campaign. What fuel distribution program will best contribute to overall victory?

Solving this problem requires quantifying its various elements. Counting gallons of fuel and numbers of each type of vehicle is easy, as is translating gallons of fuel into miles a vehicle can travel. Quantifying the relation between vehicle miles and unit mission accomplishment includes identifying constraints: the maximum gallons per load a tanker truck can carry, the minimum number of miles each unit must travel to reach its combat objective, and so on. Additional quantitative factors include probabilities, such as a unit's chances of winning a key engagement if it maneuvers along one route of travel rather than another.

Quantifying complicated real-world problems in this way is the province of a subject called operations research. Linear programming, one of the oldest and still one of the most important tools of operations research, is used when a problem can be described using equations and inequalities that are all linear.

To represent geometrically the solution of a linear inequality in two variables and to extend this representation to a system of linear inequalities.

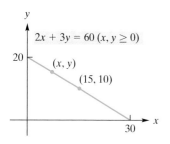

FIGURE 7.1   Budget line.

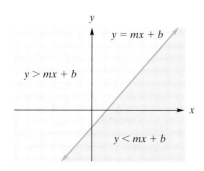

FIGURE 7.2   A nonvertical line determines two half-planes.

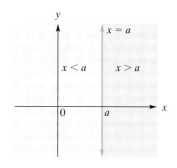

FIGURE 7.3   A vertical line determines two half-planes.

7.1 Linear Inequalities in Two Variables

Suppose a consumer receives a fixed income of $60 per week and uses it *all* to purchase products A and B. If A costs $2 per kilogram and B costs $3 per kilogram, then if our consumer purchases x kilograms of A and y kilograms of B, his cost will be $2x + 3y$. Since he uses all of his $60, x and y must satisfy

$$2x + 3y = 60 \quad \text{where} \quad x, y \geq 0$$

The solutions of this equation, called a *budget equation,* give the possible combinations of A and B that can be purchased for $60. The graph of the equation is the *budget line* in Figure 7.1. Note that $(15, 10)$ lies on the line. This means that if 15 kg of A are purchased, then 10 kg of B must be bought, for a total cost of $60.

On the other hand, suppose the consumer does not necessarily wish to spend all of the $60. In this case, the possible combinations are described by the inequality

$$2x + 3y \leq 60 \quad \text{where} \quad x, y \geq 0 \tag{1}$$

When inequalities in one variable were discussed in Chapter 1, their solutions were represented geometrically by *intervals* on the real-number line. However, for an inequality in two variables, like inequality (1), the solution is usually represented by a *region* in the coordinate plane. We will find the region corresponding to (1) after considering such inequalities in general.

DEFINITION

A ***linear inequality*** in the variables x and y is an inequality that can be written in one of the forms

$$ax + by + c < 0 \quad ax + by + c \leq 0 \quad ax + by + c > 0 \quad ax + by + c \geq 0$$

where a, b, and c are constants and not both a and b are zero.

Geometrically, the solution (or graph) of a linear inequality in x and y consists of all points (x, y) in the plane whose coordinates satisfy the inequality. For example, a solution of $x + 3y < 20$ is the point $(-2, 4)$, because substitution gives

$$-2 + 3(4) < 20,$$

$$10 < 20, \quad \text{which is true}$$

Clearly, there are infinitely many solutions, which is typical of every linear inequality.

To consider linear inequalities in general, we first note that the graph of a nonvertical line $y = mx + b$ separates the plane into three distinct parts (see Figure 7.2):

1. the line itself, consisting of all points (x, y) whose coordinates satisfy the equation $y = mx + b$;

2. the region *above* the line, consisting of all points (x, y) whose coordinates satisfy the inequality $y > mx + b$ (this region is called an *open half-plane*);

3. the open half-plane *below* the line, consisting of all points (x, y) whose coordinates satisfy the inequality $y < mx + b$.

In the situation where the strict inequality "$<$" is replaced by "$\leq$", the solution of $y \leq mx + b$ consists of the line $y = mx + b$, as well as the half-plane below it. In this case, we say that the solution is a *closed half-plane*. A similar statement can be made when "$>$" is replaced by "$\geq$". For a vertical line $x = a$ (see Figure 7.3), we speak of a half-plane to the right ($x > a$) of the line or to its left ($x < a$). Since any linear inequality (in two variables) can be put into one of the forms we have discussed, we can say that *the solution of a linear inequality must be a half-plane.*

To apply these facts, we will solve the linear inequality

$$2x + y < 5$$

From our previous discussion, we know that the solution is a half-plane. To find it, we begin by replacing the inequality symbol by an equality sign and then graphing the

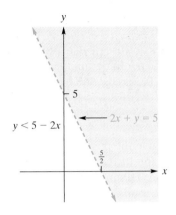

FIGURE 7.4 Graph of $2x + y < 5$.

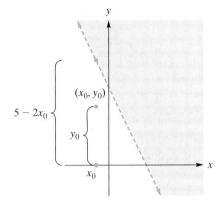

FIGURE 7.5 Analysis of point satisfying $y < 5 - 2x$.

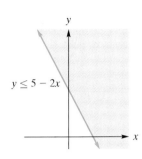

FIGURE 7.6 Graph of $y \leq 5 - 2x$.

Geometrically, the solution of a linear inequality in one variable is an interval on the line, but the solution of a linear inequality in *two* variables is a *region* in the plane.

PRINCIPLES IN PRACTICE 1

SOLVING A LINEAR INEQUALITY

To earn some extra money, you make and sell two types of refrigerator magnets, type A and type B. You have an initial start-up expense of $50. The production cost for type A is $0.90 per magnet, and the production cost for type B is $0.70 per magnet. The price for type A is $2.00 per magnet, and the price for type B is $1.50 per magnet. Let x be the number of type A and y be the number of type B produced and sold. Write an inequality describing revenue greater than cost. Solve the inequality and describe the region. Also, describe what this result means in terms of magnets.

resulting *line*, $2x + y = 5$. This is easily done by choosing two points on the line—for instance, the intercepts $(\frac{5}{2}, 0)$ and $(0, 5)$. (See Figure 7.4.) Because points on the line do not satisfy the "<" inequality, we used a *dashed* line to indicate that the line is not part of the solution. We must now determine whether the solution is the half-plane *above* the line or the one *below* it. This can be done by solving the inequality for y. Once y is isolated, the appropriate half-plane will be apparent. We have

$$y < 5 - 2x$$

From the aforementioned statement 3, we conclude that the solution consists of the half-plane *below* the line. Part of the region that does *not* satisfy this inequality is shaded in Figure 7.4. It will be our custom generally when graphing inequalities to shade the part of the whole plane that does *not* satisfy the condition. Thus, if (x_0, y_0) is *any* point in the unshaded region, then its ordinate y_0 is less than the number $5 - 2x_0$. (See Figure 7.5.) For example, $(-2, -1)$ is in the region, and

$$-1 < 5 - 2(-2)$$
$$-1 < 9$$

If, instead, the original inequality had been $y \leq 5 - 2x$, then the line $y = 5 - 2x$ would have been included in the solution. We would indicate its inclusion by using a solid line rather than a dashed line. This solution, which is a closed half plane, is shown in Figure 7.6. Keep in mind that **a solid line *is* included in the solution, and a dashed line *is not*.**

● EXAMPLE 1 Solving a Linear Inequality

Find the region defined by the inequality $y \leq 5$.

Solution: Since x does not appear, the inequality is assumed to be true for all values of x. The region consists of the line $y = 5$, together with the half-plane below it. (See Figure 7.7, where the solution is the *un*shaded part together with the line itself.)

● EXAMPLE 2 Solving a Linear Inequality

Solve the inequality $2(2x - y) < 2(x + y) - 4$.

Solution: We first solve the inequality for y, so that the appropriate half-plane is obvious. The inequality is equivalent to

$$4x - 2y < 2x + 2y - 4$$
$$4x - 4y < 2x - 4$$
$$-4y < -2x - 4$$
$$y > \frac{x}{2} + 1 \qquad \left(\begin{array}{l}\text{dividing both sides by } -4 \text{ and reversing the sense of the} \\ \text{inequality}\end{array}\right)$$

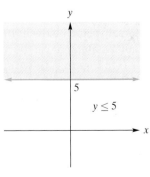

FIGURE 7.7 Graph of $y \leq 5$.

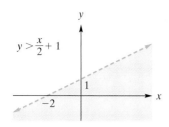

$y > \frac{x}{2} + 1$

FIGURE 7.8   Graph of
$y > \frac{x}{2} + 1$.

   CAUTION

The point where the graphs of
$y = x$ and $y = -2x + 3$ intersect is
not included in the solution. Why?

SOLVING A SYSTEM OF LINEAR INEQUALITIES

A store sells two types of cameras. In order to cover overhead, it must sell at least 50 cameras per week, and in order to satisfy distribution requirements, it must sell at least twice as many of type I as type II. Write a system of inequalities to describe the situation. Let x be the number of type I that the store sells in a week and y be the number of type II that it sells in a week. Find the region described by the system of linear inequalities.

Using a dashed line, we now sketch $y = (x/2) + 1$ by noting that its intercepts are $(0, 1)$ and $(-2, 0)$. Because the inequality symbol is $>$, we shade the half plane below the line. Think of the shading as striking out the points that you do not want. (See Figure 7.8.) Each point in the unshaded region is a solution.

NOW WORK PROBLEM 1

Systems of Inequalities

The solution of a *system* of inequalities consists of all points whose coordinates simultaneously satisfy all of the given inequalities. Geometrically, it is the region that is common to all the regions determined by the given inequalities. For example, let us solve the system

$$\begin{cases} 2x + y > 3 \\ x \geq y \\ 2y - 1 > 0 \end{cases}$$

We first rewrite each inequality so that y is isolated. This gives the equivalent system

$$\begin{cases} y > -2x + 3 \\ y \leq x \\ y > \frac{1}{2} \end{cases}$$

Next, we sketch the corresponding lines $y = -2x + 3$, $y = x$, and $y = \frac{1}{2}$, using dashed lines for the first and third and a solid line for the second. We then shade the region that is below the first line, the region that is above the second line, and the region that is below the third line. The region that is unshaded See (Figure 7.9) together with any solid line boundaries are the points in the solution of the system of inequalities.

EXAMPLE 3   Solving a System of Linear Inequalities

Solve the system

$$\begin{cases} y \geq -2x + 10 \\ y \geq x - 2 \end{cases}$$

Solution: The solution consists of all points that are simultaneously on or above the line $y = -2x + 10$ and on or above the line $y = x - 2$. It is the unshaded region in Figure 7.10.

NOW WORK PROBLEM 9

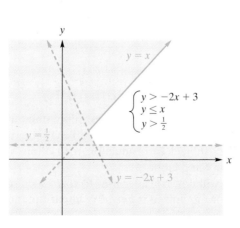

FIGURE 7.9   Solution of a system of linear inequalities.

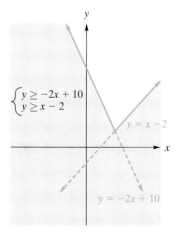

FIGURE 7.10   Solution of a system of linear inequalities.

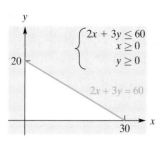

FIGURE 7.11 Solution of a system of linear inequalities.

● EXAMPLE 4 Solving a System of Linear Inequalities

Find the region described by

$$\begin{cases} 2x + 3y \leq 60 \\ x \geq 0 \\ y \geq 0 \end{cases}$$

Solution: This system relates to inequality (1) at the beginning of the section. The first inequality is equivalent to $y \leq -\frac{2}{3}x + 20$. The last two inequalities restrict the solution to points that are both on or to the right of the y-axis *and* also on or above the x-axis. The desired region is unshaded in Figure 7.11.

NOW WORK PROBLEM 17

Problems 7.1

In Problems 1–24, solve the inequalities.

*1. $3x + 4y > 2$

2. $3x - 2y \geq 12$

3. $x + 2y \leq 7$

4. $y > 6 - 2x$

5. $-x \leq 2y - 4$

6. $3x + 5y \geq 12$

7. $3x + y < 0$

8. $2x + 3y < -6$

*9. $\begin{cases} 3x - 2y < 6 \\ x - 3y > 9 \end{cases}$

10. $\begin{cases} 2x + 3y > -6 \\ 3x - y < 6 \end{cases}$

11. $\begin{cases} 2x + 3y \leq 6 \\ x \geq 0 \end{cases}$

12. $\begin{cases} 2y - 3x < 6 \\ x < 0 \end{cases}$

13. $\begin{cases} y - 3x < 5 \\ 2x - 3y > -6 \end{cases}$

14. $\begin{cases} x - y < 1 \\ y - x < 1 \end{cases}$

15. $\begin{cases} 2x - 2 \geq y \\ 2x \leq 3 - 2y \end{cases}$

16. $\begin{cases} 2y < 4x + 2 \\ y < 2x + 1 \end{cases}$

*17. $\begin{cases} x - y > 4 \\ x < 2 \\ y > -5 \end{cases}$

18. $\begin{cases} 5x + 2y < -3 \\ y > -x \\ 3x + 6 < 0 \end{cases}$

19. $\begin{cases} y < 2x + 4 \\ x \geq -2 \\ y < 1 \end{cases}$

20. $\begin{cases} 2x + y \geq 6 \\ x \leq y \\ y \leq 5x + 2 \end{cases}$

21. $\begin{cases} x + y > 1 \\ 3x - 5 \leq y \\ y < 2x \end{cases}$

22. $\begin{cases} 2x - 3y > -12 \\ 3x + y > -6 \\ y > x \end{cases}$

23. $\begin{cases} 3x + y > -6 \\ x - y > -5 \\ x \geq 0 \end{cases}$

24. $\begin{cases} 5y - 2x \leq 10 \\ 4x - 6y \leq 12 \\ y \geq 0 \end{cases}$

If a consumer wants to spend no more than P dollars to purchase quantities x and y of two products having prices of p_1 and p_2

dollars per unit, respectively, then $p_1 x + p_2 y \leq P$, where $x, y \geq 0$. In Problems 25 and 26, find geometrically the possible combinations of purchases by determining the solution of this system for the given values of p_1, p_2, and P.

25. $p_1 = 6$, $p_2 = 4$, $P = 20$

26. $p_1 = 7$, $p_2 = 3$, $P = 25$

27. If a manufacturer wishes to purchase a *total* of no more than 100 lb of product Z from suppliers A and B, set up a system of inequalities that describes the possible combinations of quantities that can be purchased from each supplier. Sketch the solution in the plane.

28. Manufacturing The XYZ Corporation produces two models of home computers: the Alpha model and the Beta model. Let x be the number of Alpha models and y the number of Beta models produced at the San Antonio factory per week. If the factory can produce at most 650 Alpha and Beta models combined in a week, write inequalities to describe this situation.

29. Manufacturing A chair company produces two models of chairs. The Sequoia model takes 3 worker-hours to assemble and $\frac{1}{2}$ worker-hour to paint. The Saratoga model takes 2 worker-hours to assemble and 1 worker-hour to paint. The maximum number of worker-hours available to assemble chairs is 240 per day, and the maximum number of worker-hours available to paint chairs is 80 per day. Write a system of linear inequalities to describe the situation. Let x represent the number of Sequoia models produced in a day and y represent the number of Saratoga models produced in a day. Find the region described by this system of linear inequalities.

OBJECTIVE

To state the nature of a linear programming problem, to introduce terminology associated with it, and to solve it geometrically.

7.2 Linear Programming

Sometimes we want to maximize or minimize a function, subject to certain restrictions (or *constraints*). For example, for a manufacturer we may want to maximize a profit function, subject to production restrictions imposed by limitations on the use of machinery and labor.

We will now consider how to solve such problems when the function to be maximized or minimized is *linear*. A **linear function in x and y** has the form

$$Z = ax + by$$

where a and b are constants. We also require that the corresponding constraints be represented by a system of linear inequalities (involving "$\leq$" or "$\geq$") or linear equations in x and y, and that all variables be nonnegative. A problem involving all of these conditions is called a *linear programming problem*.

Linear programming was developed by George B. Dantzig in the late 1940s and was first used by the U.S. Air Force as an aid in decision making. Today it has wide application in industrial and economic analysis.

In a linear programming problem, the function to be maximized or minimized is called the **objective function.** Although there are usually infinitely many solutions to the system of constraints, which are called **feasible solutions** or **feasible points,** the aim is to find one such solution that is an **optimum solution**—one that maximizes or minimizes the value of the objective function.

We now give a geometrical approach to linear programming. In Section 7.4, a matrix approach will be discussed that will enable us to work with more than two variables and, hence, a wider range of problems.

We consider the following problem. A company produces two types of can openers: manual and electric. Each requires in its manufacture the use of three machines: A, B, and C. Table 7.1 gives data relating to the manufacture of these can openers. Each manual can opener requires the use of machine A for 2 hours, machine B for 1 hour, and machine C for 1 hour. An electric can opener requires 1 hour on A, 2 hours on B, and 1 hour on C. Furthermore, suppose the maximum numbers of hours available per month for the use of machines A, B, and C are 180, 160, and 100, respectively. The profit on a manual can opener is \$4, and on an electric can opener it is \$6. If the company can sell all the can openers it can produce, how many of each type should it make in order to maximize the monthly profit?

CAUTION

A great deal of terminology is used in discussing linear programming, and you are advised to learn this terminology as soon as it is introduced.

TABLE 7.1

	Manual	**Electric**	**Hours Available**
A	2 hr	1 hr	180
B	1 hr	2 hr	160
C	1 hr	1 hr	100
Profit/Unit	\$4	\$6	

To solve the problem, let x and y denote the number of manual and electric can openers, respectively, that are made in a month. Since the number of can openers made is not negative,

$$x \geq 0 \quad \text{and} \quad y \geq 0$$

For machine A, the time needed for working on x manual can openers is $2x$ hours, and the time needed for working on y electric can openers is $1y$ hours. The sum of these times cannot be greater than 180, so

$$2x + y \leq 180$$

Similarly, the restrictions for machines B and C give

$$x + 2y \leq 160 \quad \text{and} \quad x + y \leq 100$$

The profit is a function of x and y and is given by the *profit function*

$$P = 4x + 6y$$

Summarizing, we want to maximize the *objective function*

$$P = 4x + 6y \tag{1}$$

subject to the condition that x and y must be a solution of the system of constraints:

$$\begin{cases} 2x + y \leq 180 & (2) \\ x + 2y \leq 160 & (3) \\ x + y \leq 100 & (4) \\ x \geq 0 & (5) \\ y \geq 0 & (6) \end{cases}$$

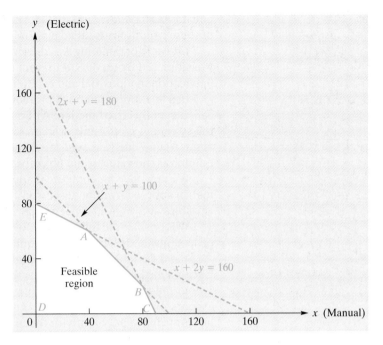

FIGURE 7.12   Feasible region.

Thus, we have a linear programming problem. Constraints (5) and (6) are called **nonnegativity conditions.** The region simultaneously satisfying constraints (2)–(6) is *un*shaded in Figure 7.12. Each point in this region represents a feasible solution, and the region is called the **feasible region.** Although there are infinitely many feasible solutions, we must find one that maximizes the profit function.

Since the objective function, $P = 4x + 6y$, is equivalent to

$$y = -\frac{2}{3}x + \frac{P}{6}$$

it defines a family of parallel lines, one for each possible value of P, each having a slope of $-2/3$ and y-intercept $(0, P/6)$. For example, if $P = 600$, then we obtain the line

$$y = -\frac{2}{3}x + 100$$

shown in Figure 7.13. This line, called an **isoprofit line,** gives all possible combinations of x and y that yield the same profit, \$600. Note that this isoprofit line has no point in common with the feasible region, whereas the isoprofit line for $P = 300$ has infinitely many such points. Let us look for the member of the family that contains a feasible point and whose P-value is maximum. *This will be the line whose y-intercept is farthest from the origin (giving a maximum value of P) and that has at least one point in common with the feasible region.* It is not difficult to observe that such a line

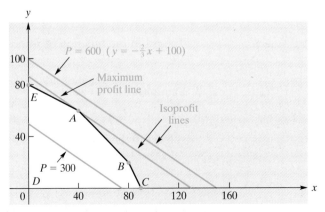

FIGURE 7.13   Isoprofit lines and the feasible region.

will contain the *corner point A.* Any isoprofit line with a greater profit will contain no points of the feasible region.

From Figure 7.12, we see that *A* lies on both the line $x + y = 100$ and the line $x + 2y = 160$. Thus, its coordinates may be found by solving the system

$$\begin{cases} x + \ y = 100 \\ x + 2y = 160 \end{cases}$$

This gives $x = 40$ and $y = 60$. Substituting these values into the equation $P = 4x + 6y$, we find that the maximum profit subject to the constraints is \$520, which is obtained by producing 40 manual can openers and 60 can openers per month.

If a feasible region can be contained within a circle, such as the region in Figure 7.13, it is called a **bounded feasible region.** Otherwise, it is **unbounded.** When a feasible region contains at least one point, it is said to be **nonempty.** Otherwise, it is **empty.** The region in Figure 7.13 is a nonempty bounded feasible region.

It can be shown that

A linear function defined on a nonempty bounded feasible region has a maximum (minimum) value, and this value can be found at a corner point.

This statement gives us a way of finding an optimum solution without drawing isoprofit lines, as we did previously: We simply evaluate the objective function at each of the corner points of the feasible region and then choose a corner point at which the function is optimum.

For example, in Figure 7.13 the corner points are *A, B, C, D,* and *E.* We found *A* before to be (40, 60). To find *B,* we see from Figure 7.12 that we must solve $2x + y = 180$ and $x + y = 100$ simultaneously. This gives the point $B = (80, 20)$. In a similar way, we obtain all the corner points:

$$A = (40, 60) \qquad B = (80, 20) \qquad C = (90, 0)$$

$$D = (0, 0) \qquad E = (0, 80)$$

We now evaluate the objective function $P = 4x + 6y$ at each point:

$$P(A) = 4(40) + 6(60) = 520$$

$$P(B) = 4(80) + 6(20) = 440$$

$$P(C) = 4(90) + 6(0) = 360$$

$$P(D) = 4(0) + 6(0) = 0$$

$$P(E) = 4(0) + 6(80) = 480$$

Thus, *P* has a maximum value of 520 at *A,* where $x = 40$ and $y = 60$.

The optimum solution to a linear programming problem is given by the optimum value of the objective function *and* the point where the optimum value of the objective function occurs.

● EXAMPLE 1 **Solving a Linear Programming Problem**

Maximize the objective function Z = 3x + y subject to the constraints

$$2x + y \le 8$$

$$2x + 3y \le 12$$

$$x \ge 0$$

$$y \ge 0$$

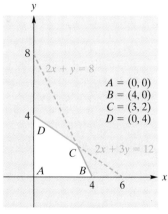

FIGURE 7.14   *A, B, C,* and *D* are corner points of feasible region.

Solution: In Figure 7.14, the feasible region is nonempty and bounded. Thus, *Z* is maximum at one of the four corner points. The coordinates of *A, B,* and *D* are obvious on inspection. To find the coordinates of *C,* we solve the equations $2x + y = 8$ and

$2x + 3y = 12$ simultaneously, which gives $x = 3$, $y = 2$. Thus,

$$A = (0, 0) \quad B = (4, 0) \quad C = (3, 2) \quad D = (0, 4)$$

Evaluating Z at these points, we obtain

$$Z(A) = 3(0) + 0 = 0$$
$$Z(B) = 3(4) + 0 = 12$$
$$Z(C) = 3(3) + 2 = 11$$
$$Z(D) = 3(0) + 4 = 4$$

Hence, the maximum value of Z, subject to the constraints, is 12, and it occurs when $x = 4$ and $y = 0$.

NOW WORK PROBLEM 1

Empty Feasible Region

The next example illustrates a situation where no optimum solution exists.

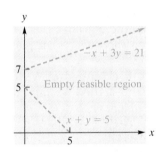

FIGURE 7.15 Empty feasible region.

EXAMPLE 2   Empty Feasible Region

Minimize the objective function $Z = 8x - 3y$, subject to the constraints

$$-x + 3y = 21$$
$$x + y \leq 5$$
$$x \geq 0$$
$$y \geq 0$$

Solution: Notice that the first constraint, $-x + 3y = 21$, is an *equality*. Portions of the lines $-x + 3y = 21$ and $x + y = 5$ for which $x \geq 0$ and $y \geq 0$ are shown in Figure 7.15. A feasible point (x, y) must have $x \geq 0$ and $y \geq 0$, and must lie both on the top line and on or below the bottom line (since $y \leq 5 - x$). However, no such point exists. Hence, the feasible region is *empty*, and the problem has *no* optimum solution.

NOW WORK PROBLEM 5

The situation in Example 2 can be made more general:

Whenever the feasible region of a linear programming problem is empty, no optimum solution exists.

Unbounded Feasible Region

Suppose a feasible region is defined by

$$y = 2$$
$$x \geq 0$$
$$y \geq 0$$

This region is the portion of the horizontal line $y = 2$ indicated in Figure 7.16. Since the region cannot be contained within a circle, it is *unbounded*. Let us consider maximizing

$$Z = x + y$$

subject to the foregoing constraints. Since $y = 2$, $Z = x + 2$. Clearly, as x increases without bound, so does Z. Thus, no feasible point maximizes Z, so no optimum solution exists. In this case, we say that the solution is "unbounded." On the other

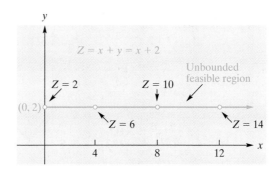

FIGURE 7.16 Unbounded feasible region on which Z has no maximum.

hand, suppose we want to *minimize* $Z = x + y$ over the same region. Since $Z = x + 2$, then Z is minimum when x is as small as possible, namely, when $x = 0$. This gives a minimum value of $Z = x + y = 0 + 2 = 2$, and the optimum solution is the corner point $(0, 2)$.

In general, it can be shown that

If a feasible region is unbounded, and *if* the objective function has a maximum (or minimum) value, then that value occurs at a corner point.

●EXAMPLE 3 **Unbounded Feasible Region**

A produce grower is purchasing fertilizer containing three nutrients, A, B, and C. The minimum needs are 160 units of A, 200 units of B, and 80 units of C. There are two popular brands of fertilizer on the market. Fast Grow, costing $8 a bag, contains 3 units of A, 5 units of B, and 1 unit of C. Easy Grow, costing $6 a bag, contains 2 units of each nutrient. If the grower wishes to minimize cost while still maintaining the nutrients required, how many bags of each brand should be bought? The information is summarized as follows:

	Fast Grow	Easy Grow	Units Required
A	3 units	2 units	160
B	5 units	2 units	200
C	1 units	2 units	80
Cost/Bag	$8	$6	

Solution: Let x be the number of bags of Fast Grow that are bought and y the number of bags of Easy Grow that are bought. Then we wish to *minimize* the cost function

$$C = 8x + 6y \qquad (7)$$

subject to the constraints

$$3x + 2y \geq 160 \qquad (8)$$

$$5x + 2y \geq 200 \qquad (9)$$

$$x + 2y \geq 80 \qquad (10)$$

$$x \geq 0 \qquad (11)$$

$$y \geq 0 \qquad (12)$$

The feasible region satisfying constraints (8)–(12) is unshaded in Figure 7.17, along with *isocost lines* for $C = 400$ and $C = 600$. The feasible region is unbounded.

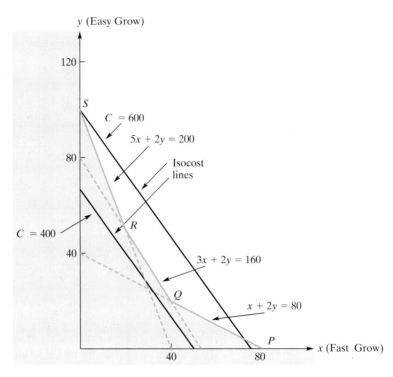

FIGURE 7.17   Minimum cost at corner point Q of unbounded feasible region.

The member of the family of lines $C = 8x + 6y$ that gives a minimum cost, subject to the constraints, intersects the feasible region at the corner point Q. Here we chose the isocost line whose y-intercept was *closest* to the origin and that had at least one point in common with the feasible region. The coordinates of B are found by solving the system

$$\begin{cases} 3x + 2y = 160 \\ x + 2y = 80 \end{cases}$$

Thus, $x = 40$ and $y = 20$, which gives a minimum cost of \$440. The produce grower should buy 40 bags of Fast Grow and 20 bags of Easy Grow.

NOW WORK PROBLEM 15

In Example 3, we found that the function $C = 8x + 6y$ has a minimum value at a corner point of the unbounded feasible region. On the other hand, suppose we want to *maximize* C over that region and take the approach of evaluating C at all corner points. These points are

$$P = (80, 0) \quad Q = (40, 20) \quad R = (20, 50) \quad S = (0, 100)$$

from which we obtain

$$C(P) = 8(80) + 6(0) = 640$$
$$C(Q) = 8(40) + 6(20) = 440$$
$$C(R) = 8(20) + 6(50) = 460$$
$$C(S) = 8(0) + 6(100) = 600$$

CAUTION

When working with an unbounded feasible region, do not simply conclude that an optimum solution exists at a corner point, since there may not be an optimum solution.

A hasty conclusion is that the maximum value of C is 640. This is *false!* There is *no* maximum value, since isocost lines with arbitrarily large values of C intersect the feasible region.

TECHNOLOGY

Problem: Maximize $Z = 4.1x - 3.2y$ subject to the constraints

$$y \le 8 - x \tag{13}$$

$$y \le 6 - 0.2x \tag{14}$$

$$y \ge 2 + 0.3x \tag{15}$$

and

$$x \ge 0 \quad y \ge 0$$

Solution: As shown in Figure 7.18, we enter the objective function as Y_1, where y is entered as "alpha Y." Next, the equations corresponding to constraints (13)–(15) are entered as Y_2, Y_3, and Y_4. To begin, the Y_1 function is "turned off" (that is, the symbol "=" is not highlighted), and we obtain the graphs of Y_2, Y_3, and Y_4. (See Figure 7.19.) It is a good idea to make a pencil sketch of the graphs and label the lines. From the sketch, we determine the feasible region and label any corner points. In Figure 7.19, the feasible region is *shaded* and the corner points are A, B, C, and D. Because the feasible region is nonempty and bounded, the maximum value of Z will occur at one of the corner points.

Point A is the y-intercept of Y_4 and is easily found to be $(0, 2)$. The value of Z at A is found by storing 0 in X, and 2 in Y and then evaluating Y_1. (See Figure 7.20.) Thus, $Z = -6.4$ at this corner point.

Point B is the intersection of Y_2 and Y_4. To find the coordinates of B, we first turn off the Y_3 function and highlight only Y_2 and Y_4. After displaying the graphs of Y_2 and Y_4, we find their intersection point. On the TI-83 Plus it is convenient to use the "intersection" feature. (See Figure 7.21.) The values of X and Y at the intersection are automatically stored in the X and Y registers. Returning to the home screen, we evaluate Y_1 and obtain 8.09 (rounded to two decimal places). Thus, $Z = 8.09$ at corner point B.

Continuing in this fashion, we find the coordinates of C and D and evaluate Y_1 (or Z) there:

$$C = (2.5, 5.5) \quad Z = -7.35$$
$$D = (0, 6) \quad Z = -19.2$$

Hence, the maximum value of Z is 8.09 and occurs at the corner point B, where $x \approx 4.62$ and $y \approx 3.38$.

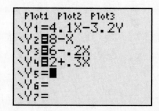

FIGURE 7.18   Entering objective function and equations corresponding to constraints and "turning off" objective function.

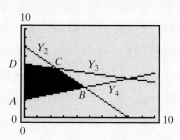

FIGURE 7.19   Determining feasible region and labeling corner points.

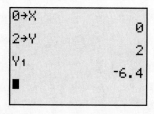

FIGURE 7.20   Evaluating objective function at corner point $A = (0, 2)$.

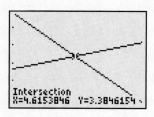

FIGURE 7.21   Determining corner point B.

Problems 7.2

*1. Maximize

$$P = 5x + 7y$$

subject to

$$2x + 3y \le 45$$

$$x - 3y \ge 2$$

$$x, y \ge 0$$

2. Maximize

$$P = 2x + 5y$$

subject to

$$x + y \le 90$$

$$4x + 3y \le 250$$

$$x + 2y \le 225$$

$$x, y \ge 0$$

3. Maximize

$$Z = 4x - 6y$$

subject to

$$y \leq 7$$
$$3x - y \leq 3$$
$$x + y \geq 5$$
$$x, y \geq 0$$

4. Minimize

$$Z = x + y$$

subject to

$$x - y \geq 0$$
$$4x + 3y \geq 12$$
$$9x + 11y \leq 99$$
$$x \leq 8$$
$$x, y \geq 0$$

*5. Maximize

$$Z = 4x - 10y$$

subject to

$$x - 4y \geq 4$$
$$2x - y \leq 2$$
$$x, y \geq 0$$

6. Minimize

$$Z = 20x + 30y$$

subject to

$$2x + y \leq 10$$
$$3x + 4y \leq 24$$
$$8x + 7y \geq 56$$
$$x, y \geq 0$$

7. Minimize

$$Z = 7x + 3y$$

subject to

$$3x - y \geq -2$$
$$x + y \leq 9$$
$$x - y = -1$$
$$x, y \geq 0$$

8. Maximize

$$Z = 0.4x - 0.2y$$

subject to

$$2x - 5y \geq -3$$
$$2x - y \leq 5$$
$$3x + y = 6$$
$$x, y \geq 0$$

9. Minimize

$$C = 3x + 2y$$

subject to

$$2x + y \geq 5$$
$$3x + y \geq 4$$
$$x + 2y \geq 3$$
$$x, y \geq 0$$

10. Minimize

$$C = 2x + 2y$$

subject to

$$x + 2y \geq 80$$
$$3x + 2y \geq 160$$
$$5x + 2y \geq 200$$
$$x, y \geq 0$$

11. Maximize

$$Z = 10x + 2y$$

subject to

$$x + 2y \geq 4$$
$$x - 2y \geq 0$$
$$x, y \geq 0$$

12. Minimize

$$Z = y - x$$

subject to

$$x \geq 3$$
$$x + 3y \geq 6$$
$$x - 3y \geq -6$$
$$x, y \geq 0$$

13. Production for Maximum Profit   A toy manufacturer preparing a production schedule for two new toys, trucks and spinning tops, must use the information concerning their construction times given in the following table:

	Machine A	**Machine B**	**Finishing**
Truck	2 hr	3 hr	5 hr
Spinning Top	1 hr	1 hr	1 hr

For example, each truck requires 2 hours on machine A. The available employee hours per week are as follows: for operating machine A, 80 hours; for B, 50 hours; for finishing, 70 hours. If the profits on each truck and spinning top are \$7 and \$2, respectively, how many of each toy should be made per week in order to maximize profit? What is the maximum profit?

14. Production for Maximum Profit   A manufacturer produces two types of DVD player: Vista and Xtreme. During production, the players require the use of two machines, A and B. The number of hours needed on both machines are indicated in the following table:

	Machine A	Machine B
Vista	1 hr	2 hr
Xtreme	3 hr	2 hr

If each machine can be used 24 hours a day, and the profits on the Vista and Xtreme models are $50 and $80, respectively, how many of each type of player should be made per day to obtain maximum profit? What is the maximum profit?

*15. **Diet Formulation**   A diet is to contain at least 16 units of carbohydrates and 20 units of protein. Food A contains 2 units of carbohydrates and 4 of protein; food B contains 2 units of carbohydrates and 1 of protein. If food A costs $1.20 per unit and food B costs $0.80 per unit, how many units of each food should be purchased in order to minimize cost? What is the minimum cost?

16. Fertilizer Nutrients   A produce grower is purchasing fertilizer containing three nutrients: A, B, and C. The minimum weekly requirements are 80 units of A, 120 of B, and 240 of C. There are two popular blends of fertilizer on the market. Blend I, costing $8 a bag, contains 2 units of A, 6 of B, and 4 of C. Blend II, costing $10 a bag, contains 2 units of A, 2 of B, and 12 of C. How many bags of each blend should the grower buy each week to minimize the cost of meeting the nutrient requirements?

17. Mineral Extraction   A company extracts minerals from ore. The numbers of pounds of minerals A and B that can be extracted from each ton of ores I and II are given in the following table, together with the costs per ton of the ores:

	Ore I	Ore II
Mineral A	100 lb	200 lb
Mineral B	200 lb	50 lb
Cost per ton	$50	$60

If the company must produce at least 3000 lb of A and 2500 lb of B, how many tons of each ore should be processed in order to minimize cost? What is the minimum cost?

18. Production Scheduling   An oil company that has two refineries needs at least 8000, 14,000, and 5000 barrels of low-, medium-, and high-grade oil, respectively. Each day, Refinery I produces 2000 barrels of low-, 3000 barrels of medium-, and 1000 barrels of high-grade oil, whereas Refinery II produces 1000 barrels each of low- and high- and

2000 barrels of medium-grade oil. If it costs $25,000 per day to operate Refinery I and $20,000 per day to operate Refinery II, how many days should each refinery be operated to satisfy the production requirements at minimum cost? What is the minimum cost? (Assume that a minimum cost exists.)

19. Construction Cost   A chemical company is designing a plant for producing two types of polymers, P_1 and P_2. The plant must be capable of producing at least 100 units of P_1 and 420 units of P_2 each day. There are two possible designs for the basic reaction chambers that are to be included in the plant. Each chamber of type A costs $600,000 and is capable of producing 10 units of P_1 and 20 units of P_2 per day; type B is of cheaper design, costing $300,000, and is capable of producing 4 units of P_1 and 30 units of P_2 per day. Because of operating costs, it is necessary to have at least four chambers of each type in the plant. How many chambers of each type should be included to minimize the cost of construction and still meet the required production schedule? (Assume that a minimum cost exists.)

20. Pollution Control   Because of new federal regulations on pollution, a chemical company has introduced into its plant a new, more expensive process to supplement or replace an older process in the production of a particular chemical. The older process discharges 25 grams of carbon dioxide and 50 grams of particulate matter into the atmosphere for each liter of chemical produced. The new process discharges 15 grams of carbon dioxide and 40 grams of particulate matter into the atmosphere for each liter produced. The company makes a profit of 40 cents per liter and 15 cents per liter on the old and new processes, respectively. If the government allows the plant to discharge no more than 12,525 grams of carbon dioxide and no more than 20,000 grams of particulate matter into the atmosphere each day, how many liters of chemical should be produced daily, by each process, to maximize daily profit? What is the maximum daily profit?

21. Construction Discount   The highway department has decided to add exactly 300 km of highway and exactly 200 km of expressway to its road system this year. The standard price for road construction is $2 million per kilometer of highway and $8 million per kilometer of expressway. Only two contractors, company A and company B, can do this kind of construction, so the entire 500 km of road must be built by these two companies. However, company A can construct at most 400 km of roadway (highway and expressway), and company B can construct at most 300 km. For political reasons, each company must be awarded a contract with a standard price of at least $300 million (before discounts). Company A offers a discount of $2000 per kilometer of highway and $6000 per kilometer of expressway; company B offers a discount of $3000 for each kilometer of highway and $5000 for each kilometer of expressway.

(a) Let x and y represent the number of kilometers of highway and expressway, respectively, awarded to company A. Show that the total discount received from both companies is given by

$$D = 1900 - x + y$$

where D is in thousands of dollars.

(b) The highway department wishes to maximize the total discount D. Show that this problem is equivalent to the following linear programming problem, by showing exactly how the first six constraints arise:

Maximize $D = 1900 - x + y$

subject to

$$x + y \leq 400$$
$$x + y \geq 500$$
$$2x + 8y \geq 300$$
$$2x + 8y \leq 1900$$
$$x \leq 300$$
$$y \leq 200$$
$$x, y \geq 0$$

(c) Find the values of x and y that maximize D.

In Problems 22–25, round answers to two decimal places.

22. Maximize

$$Z = 2x + 0.3y$$

subject to

$$y \leq 6 - 4x$$
$$y \geq 2 - 0.5x$$
$$x, y \geq 0$$

23. Maximize

$$Z = 14x - 3y$$

subject to

$$y \geq 12.5 - 4x$$
$$y \leq 9.3 - x$$
$$y \geq 4.7 + 0.8x$$
$$x, y \geq 0$$

24. Minimize

$$Z = 5.1y - 3.5x$$

subject to

$$7.5x + 2y \geq 35$$
$$2.5x + y \leq 7.4$$
$$0.6x - y \geq -0.8$$
$$x, y \geq 0$$

25. Minimize

$$Z = 17.3x - 14.4y$$

subject to

$$0.73x - y \leq -2.4$$
$$1.22x - y \geq -5.1$$
$$0.45x - y \geq -12.4$$
$$x, y \geq 0$$

7.3 Multiple Optimum Solutions[1]

Sometimes an objective function attains its optimum value at more than one feasible point, in which case **multiple optimum solutions** are said to exist. Example 1 will illustrate.

● **EXAMPLE 1** **Multiple Optimum Solutions**

Maximize $Z = 2x + 4y$ subject to the constraints

$$x - 4y \leq -8$$
$$x + 2y \leq 16$$
$$x, y \geq 0$$

Solution: The feasible region appears in Figure 7.22. Since the region is nonempty and bounded, Z has a maximum value at a corner point. The corner points are $A = (0, 2)$, $B = (8, 4)$, and $C = (0, 8)$. Evaluating the objective function at

$$A = (0, 2) \quad B = (8, 4) \quad C = (0, 8)$$

---

[1]This section can be omitted.

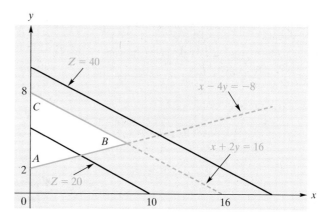

FIGURE 7.22 $Z = 2x + 4y$ is maximized at each point on the line segment $\overline{BC}$.

gives

$$Z(A) = 2(0) + 4(2) = 8$$

$$Z(B) = 2(8) + 4(4) = 32$$

$$Z(C) = 2(0) + 4(8) = 32$$

Thus, the maximum value of Z over the region is 32, and it occurs at *two* corner points, B and C. In fact, this maximum value also occurs at *all* points on the line segment *joining* B and C, for the following reason. Each member of the family of lines $Z = 2x + 4y$ has slope $-\frac{1}{2}$. Moreover, the constraint line $x + 2y = 16$, which contains B and C, also has slope $-\frac{1}{2}$ and hence is parallel to each member of $Z = 2x + 4y$. Figure 7.22 shows lines for $Z = 20$ and $Z = 40$. Note that the member of the family that maximizes Z contains not only B and C, but also all points on the line segment $\overline{BC}$. It thus has infinitely many points in common with the feasible region. Hence, this linear programming problem has infinitely many optimum solutions. In fact, it can be shown that

If (x_1, y_1) and (x_2, y_2) are two corner points at which an objective function is optimum, then the function will also be optimum at all points (x, y) where

$$x = (1 - t)x_1 + tx_2$$

$$y = (1 - t)y_1 + ty_2$$

and

$$0 \leq t \leq 1$$

In our case, if $(x_1, y_1) = B = (8, 4)$ and $(x_2, y_2) = C = (0, 8)$, then Z is maximum at any point (x, y) where

$$x = (1 - t)8 + t \cdot 0 = 8(1 - t)$$

$$y = (1 - t)4 + t \cdot 8 = 4(1 + t)$$

$$\text{for} \quad 0 \leq t \leq 1$$

These equations give the coordinates of any point on the line segment $\overline{BC}$. In particular, if $t = 0$, then $x = 8$ and $y = 4$, which gives the corner point $B = (8, 4)$. If $t = 1$, we get the corner point $C = (0, 8)$. The value $t = \frac{1}{2}$ gives the point $(4, 6)$. Notice that at $(4, 6)$, $Z = 2(4) + 4(6) = 32$, which is the maximum value of Z.

NOW WORK PROBLEM 1 ●●

Problems 7.3

*1. Minimize

$$Z = 3x + 9y$$

subject to

$$y \geq -\tfrac{3}{2}x + 6$$
$$y \geq -\tfrac{1}{3}x + \tfrac{11}{3}$$
$$y \geq x - 3$$
$$x, y \geq 0$$

2. Maximize

$$Z = 2x + 2y$$

subject to

$$2x - y \geq -4$$
$$x - 2y \leq 4$$
$$x + y = 6$$
$$x, y \geq 0$$

3. Maximize

$$Z = 14x + 21y$$

subject to

$$2x + 3y \leq 12$$
$$x + 5y \leq 8$$
$$x, y \geq 0$$

4. **Minimize Cost**   Suppose a car dealer has showrooms in Atherton and Berkeley and warehouses in Concord and Dublin. The cost of delivering a car is $60 from Concord to Atherton, $45 from Concord to Berkeley, $50 from Dublin to Atherton, and $35 from Dublin to Berkeley. Suppose that the showroom in Atherton orders seven cars and the showroom in Berkeley orders four cars. Suppose also that the warehouse in Concord has six cars and the warehouse in Dublin has eight cars available. Find the best way to minimize cost, and find the minimum cost. (*Hint:* Let x be the number of cars delivered from Concord to Atherton and y be the number of cars delivered from Concord to Berkeley. Then $7 - x$ is the number of cars delivered from Dublin to Atherton and $4 - y$ the number of cars delivered from Dublin to Berkeley.)

7.4 The Simplex Method

OBJECTIVE

To show how the simplex method is used to solve a standard linear programming problem. This method enables you to solve problems that cannot be solved geometrically.

Up to now, we have solved linear programming problems by a geometric method. This method is not practical when the number of variables increases to three, and is not possible beyond that. Now we will look at a different technique—the **simplex method,** whose name is linked in more advanced discussions to a geometrical object called a simplex.

The simplex method begins with a feasible solution and tests whether it is optimal. If it is not, the method proceeds to a better solution. We say "better" in the sense that the new solution usually brings you closer to optimization of the objective function.[2] Should this new solution not be optimal, we repeat the procedure. Eventually, the simplex method leads to an optimal solution, if one exists.

Besides being efficient, the simplex method has other advantages. For one, it is completely mechanical. It uses matrices, elementary row operations, and basic arithmetic. Moreover, no graphs need to be drawn; this allows us to solve linear programming problems having any number of constraints and any number of variables.

In this section, we consider only so-called **standard linear programming problems.** These can be put in the following form.

Standard Linear Programming Problem

Maximize the linear function $Z = c_1 x_1 + c_2 x_2 + \cdots + c_n x_n$ subject to the constraints

$$\left. \begin{array}{c} a_{11}x_1 + a_{12}x_2 + \cdots + a_{1n}x_n \leq b_1 \\ a_{21}x_1 + a_{22}x_2 + \cdots + a_{2n}x_n \leq b_2 \\ \cdot \qquad \cdot \qquad\qquad \cdot \qquad \cdot \\ \cdot \qquad \cdot \qquad\qquad \cdot \qquad \cdot \\ \cdot \qquad \cdot \qquad\qquad \cdot \qquad \cdot \\ a_{m1}x_1 + a_{m2}x_2 + \cdots + a_{mn}x_n \leq b_m \end{array} \right\} \tag{1}$$

where $x_1, x_2, \ldots, x_n$ and $b_1, b_2, \ldots, b_m$ are nonnegative.

---

[2]In most cases this is true. In some situations, however, the new solution may be only "just as good" as the previous one. Example 2 will illustrate this.

It is helpful to formulate the problem in matrix notation so as to make its structure more memorable. Let

$$\mathbf{C} = \begin{bmatrix} c_1 & c_2 & \cdots & c_n \end{bmatrix} \quad \text{and} \quad \mathbf{X} = \begin{bmatrix} x_1 \\ x_2 \\ \cdot \\ \cdot \\ \cdot \\ x_n \end{bmatrix}$$

Then the objective function can be written as

$$Z = \mathbf{CX}$$

Now if we write

$$\mathbf{A} = \begin{bmatrix} a_{11} & a_{12} & \cdots & a_{1n} \\ a_{21} & a_{22} & \cdots & a_{2n} \\ \cdot & \cdot & & \cdot \\ \cdot & \cdot & & \cdot \\ \cdot & \cdot & & \cdot \\ a_{m1} & a_{m2} & \cdots & a_{mn} \end{bmatrix} \quad \text{and} \quad \mathbf{B} = \begin{bmatrix} b_1 \\ b_2 \\ \cdot \\ \cdot \\ \cdot \\ b_m \end{bmatrix}$$

then we can say that a standard linear programming problem is one that can be put in the form

CAUTION

Note that $\mathbf{B} \geq \mathbf{0}$ is a condition on the data of the problem and is not a constraint imposed on the variable $\mathbf{X}$.

Maximixe $Z = \mathbf{CX}$

subject to $\begin{cases} \mathbf{AX} \leq \mathbf{B} \\ \mathbf{X} \geq \mathbf{0} \end{cases}$

where $\mathbf{B} \geq \mathbf{0}$

(Matrix inequalities are to be understood like matrix equality. The comparisons refer to matrices of the same size and the inequality is required to hold for all corresponding entries.)

Other types of linear programming problems will be discussed in Sections 7.6 and 7.7.

Note that one feasible solution to a *standard* linear programming problem is always $x_1 = 0, x_2 = 0, \ldots, x_n = 0$ and that at this feasible solution the value of the objective function Z is 0.

The procedure that we follow here will be outlined later in this section.

We now apply the simplex method to the problem in Example 1 of Section 7.2, which can be written

$$\text{maximize } Z = 3x_1 + x_2$$

subject to the constraints

$$2x_1 + x_2 \leq 8 \tag{2}$$

and

$$2x_1 + 3x_2 \leq 12 \tag{3}$$

and

$$x_1 \geq 0, \quad x_2 \geq 0$$

This problem is of standard form. We begin by expressing constraints (2) and (3) as equations. In (2), $2x_1 + x_2$ will *equal* 8 if we add some nonnegative number s_1 to $2x_1 + x_2$, so that

$$2x_1 + x_2 + s_1 = 8 \quad \text{for some } s_1 \geq 0$$

We call s_1 a **slack variable,** since it makes up for the "slack" on the left side of (2) to give us equality. Similarly, inequality (3) can be written as an equation by using the slack variable s_2; we have

$$2x_1 + 3x_2 + s_2 = 12 \quad \text{for some } s_2 \geq 0$$

The variables x_1 and x_2 are called **decision variables.**

Now we can restate the problem in terms of equations:

$$\text{Maximize } Z = 3x_1 + x_2 \tag{4}$$

subject to

$$2x_1 + x_2 + s_1 = 8 \tag{5}$$

and

$$2x_1 + 3x_2 + s_2 = 12 \tag{6}$$

where x_1, x_2, s_1, and s_2 are nonnegative.

From Section 7.2 , we know that the optimum solution occurs at a corner point of the feasible region in Figure 7.23. At each of these points, at least *two* of the variables x_1, x_2, s_1, and s_2 are 0, as the following listing indicates:

1. At A, we have $x_1 = 0$ and $x_2 = 0$.
2. At B, $x_1 = 4$ and $x_2 = 0$. But from Equation (5), $2(4) + 0 + s_1 = 8$. Thus, $s_1 = 0$.
3. At C, $x_1 = 3$ and $x_2 = 2$. But from Equation (5), $2(3) + 2 + s_1 = 8$. Hence, $s_1 = 0$. From Equation (6), $2(3) + 3(2) + s_2 = 12$. Therefore, $s_2 = 0$.
4. At D, $x_1 = 0$ and $x_2 = 4$. From Equation (6), $2(0) + 3(4) + s_2 = 12$. Thus, $s_2 = 0$.

CAUTION

We remarked earlier that there is a great deal of terminology used in the discussion of linear programing. In particular there are many types of *variables*. It is important to understand that the variables called decision variables $x_1, x_2, \ldots, x_n$ remain decision variables throughout the solution of a problem and the same remark applies to the slack variables $s_1, s_2, \ldots, s_m$. In the process of examining the corner points of the feasible region, we find solutions to the system in which at least n of the $n + m$ variables are 0. Precisely n of these are called nonbasic variables and the remaining m are called basic variables. Which m of the $n + m$ variables are *basic* depends on the corner point under consideration. Among other things, the procedure that we are describing provides a mechanical way of keeping track of which variables, at any time, are basic.

It can also be shown that any solution to Equations (5) and (6), such that at least *two* of the four variables x_1, x_2, s_1, and s_2 are zero, corresponds to a corner point. Any such solution where at least two of these variables are zero is called a **basic feasible solution** (abbreviated BFS). This number, 2, is determined by the number n of decision variables, 2 in the present example. For any particular BFS, the variables held at 0 are called **nonbasic variables,** and all the others are called **basic variables** for that BFS. Since there is a total of $n + m$ variables, the number of basic variables in the general system that arises from (1) is m, the number of constraints (other than those expressing nonnegativity). Thus, for the BFS corresponding to item 3 in the

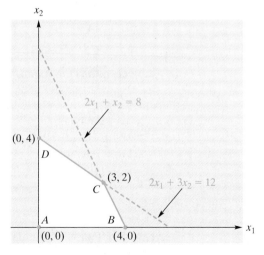

FIGURE 7.23   Optimum solution must occur at corner point of feasible region.

preceding list, s_1 and s_2 are the nonbasic variables x_1 and x_2 are the basic variables, but for the BFS corresponding to item 4, the nonbasic variables are x_1 and s_2 and the basic variables are x_2 and s_1.

We will first find an initial BFS, and hence an initial corner point, and then determine whether the corresponding value of Z can be increased by a different BFS. Since $x_1 = 0$ and $x_2 = 0$ is a feasible solution to this standard linear programming problem, let us initially find the BFS such that the decision variables x_1 and x_2 are nonbasic and hence the slack variables s_1 and s_2 are basic. That is, we choose $x_1 = 0$ and $x_2 = 0$ and find the corresponding values of $s_1, s_2,$ and Z. This can be done most conveniently by matrix techniques, based on the methods developed in Chapter 6.

If we write Equation (4) as $-3x_1 - x_2 + Z = 0$ then Equations (5), (6), and (4) form the linear system

$$\begin{cases} 2x_1 + x_2 + s_1 & = 8 \\ 2x_1 + 3x_2 + s_2 & = 12 \\ -3x_1 - x_2 + Z = 0 \end{cases}$$

in the variables $x_1, x_2, s_1, s_2,$ and Z. Thus, in general, when we add the objective function to the system that provides the constraints, we have $m + 1$ equations in $n + m + 1$ unknowns. In terms of an augmented coefficient matrix, called the **initial simplex table,** we have

$$\begin{array}{c|ccccc|c} B & x_1 & x_2 & s_1 & s_2 & Z & R \\ \hline s_1 & 2 & 1 & 1 & 0 & 0 & 8 \\ s_2 & 2 & 3 & 0 & 1 & 0 & 12 \\ \hline Z & -3 & -1 & 0 & 0 & 1 & 0 \end{array}$$

It is convenient to be generous with labels for matrices that are being used as simplex tables. Thus, the columns in the matrix to the left of the vertical bar are labeled, naturally enough, by the variables to which they correspond. We have chosen R as a label for the column that provides the R)ight sides of the system of equations. We have chosen B to label the list of row labels. The first two rows correspond to the constraints, and the last row, called the **objective row,** corresponds to the objective equation—thus the horizontal separating line. Notice that if $x_1 = 0$ and $x_2 = 0$, then, from rows 1, 2, and 3, we can directly read off the values of $s_1, s_2,$ and Z: $s_1 = 8, s_2 = 12,$ and $Z = 0$. Thus the rows of this *initial* simplex table are labeled to the left by $s_1, s_2,$ and Z. We remind you that s_1 and s_2 are the basic variables. So the column heading B can be understood to stand for B)asic variables. Our initial basic feasible solution is

$$x_1 = 0 \quad x_2 = 0 \quad s_1 = 8 \quad s_2 = 12$$

at which $Z = 0$.

Let us see if we can find a BFS that gives a larger value of Z. The variables x_1 and x_2 are nonbasic in the preceding BFS. We will now look for a BFS in which one of these variables is basic while the other remains nonbasic. Which one should we choose as the basic variable? Let us examine the possibilities. From the Z-row of the preceding matrix, $Z = 3x_1 + x_2$. If x_1 is allowed to become basic, then x_2 remains at 0 and $Z = 3x_1$; thus, for each one-unit increase in x_1, Z increases by three units. On the other hand, if x_2 is allowed to become basic, then x_1 remains at 0 and $Z = x_2$; hence, for each one-unit increase in x_2, Z increases by one unit. Consequently, we get a *greater* increase in the value of Z if x_1, rather than x_2, enters the basic-variable category. In this case, we call x_1 an **entering variable.** Thus, in terms of the simplex table that follows (which is the same as the matrix presented earlier, except for some additional labeling), the entering variable can be found by looking at the "most negative" of the numbers enclosed by the brace in the Z-row. (By *most negative*, we mean the negative indicator having the greatest magnitude.) Since that number is -3

and appears in the x_1-column, x_1 is the entering variable. The numbers in the brace are called **indicators.**

entering
variable
↓

B	x_1	x_2	s_1	s_2	Z	R
s_1	2	1	1	0	0	8
s_2	2	3	0	1	0	12
Z	−3	−1	0	0	1	0

indicators

Let us summarize the information that can be obtained from this table. It gives a BFS where s_1 and s_2 are the basic variables and x_1 and x_2 are nonbasic. The BFS is $s_1 = 8$ (the right-hand side of the s_1-row), $s_2 = 12$ (the right-hand side of the s_2-row), $x_1 = 0$, and $x_2 = 0$. The −3 in the x_1-column of the Z-row indicates that if x_2 remains 0, then Z increases three units for each one-unit increase in x_1. The −1 in the x_2-column of the Z-row indicates that if x_1 remains 0, then Z increases one unit for each one-unit increase in x_2. The column in which the most negative indicator, −3, lies gives the entering variable x_1—that is, the variable that should become basic in the next BFS.

In our new BFS, the larger the increase in x_1 (from $x_1 = 0$), the larger is the increase in Z. Now, by how much can we increase x_1? Since x_2 is still held at 0, from rows 1 and 2 of the simplex table, it follows that

$$s_1 = 8 - 2x_1$$

and

$$s_2 = 12 - 2x_1$$

Since s_1 and s_2 are nonnegative, we have

$$8 - 2x_1 \geq 0$$

and

$$12 - 2x_1 \geq 0$$

From the first inequality, $x_1 \leq \frac{8}{2} = 4$; from the second, $x_1 \leq \frac{12}{2} = 6$. Thus, x_1 must be less than or equal to the smaller of the quotients $\frac{8}{2}$ and $\frac{12}{2}$, which is $\frac{8}{2}$. Hence, x_1 can increase at most by 4. However, in a BFS, two variables must be 0. We already have $x_2 = 0$. Since $s_1 = 8 - 2x_1$, s_1 must be 0 for $x_1 = 4$. Therefore, we have a new BFS with x_1 replacing s_1 as a basic variable. That is, s_1 will *depart* from the list of basic variables in the previous BFS and will be nonbasic in the new BFS. We say that s_1 is the **departing variable** for the previous BFS. In summary, for our new BFS, we want x_1 and s_2 as basic variables with $x_1 = 4$, $s_2 = 12$ (as before) and x_2 and s_1 as nonbasic variables ($x_2 = 0$, $s_1 = 0$).

Before proceeding, let us update our table. To the right of the following table, the quotients $\frac{8}{2}$ and $\frac{12}{2}$ are indicated:

entering variable
(most negative indicator)
↓

B	x_1	x_2	s_1	s_2	Z	R	Quotients
s_1 ← departing variable (smallest quotient)	2	1	1	0	0	8	$8 \div 2 = 4$
s_2	2	3	0	1	0	12	$12 \div 2 = 6$
Z	−3	−1	0	0	1	0	

These quotients are obtained by dividing each entry in the first two rows of the R-column by the entry in the corresponding row of the entering-variable column, that is the x_1-column. Notice that the departing variable is in the same row as the *smaller* quotient, $8 \div 2$.

Since x_1 and s_2 will be basic variables in our new BFS, it would be convenient to change our previous table by elementary row operations into a form in which the values of x_1, s_2, and Z can be read off with ease (just as we were able to do with the solution corresponding to $x_1 = 0$ and $x_2 = 0$). To do this, we want to find a matrix that is equivalent to the preceding table but that has the form

$$
\begin{array}{c}
\begin{array}{ccccccc}
\text{B} & x_1 & x_2 & s_1 & s_2 & Z & \text{R}
\end{array}\\
\begin{array}{c} x_1 \\ s_2 \\ Z \end{array}
\left[
\begin{array}{ccccc|c}
1 & ? & ? & 0 & 0 & ? \\
0 & ? & ? & 1 & 0 & ? \\
0 & ? & ? & 0 & 1 & ?
\end{array}
\right]
\end{array}
$$

where the question marks represent numbers to be determined. Notice here that if $x_2 = 0$ and $s_1 = 0$, then x_1 equals the number in row x_1 of column R, s_2 equals the number in row s_2 of column R, and Z is the number in row Z of column R. Thus, we must transform the table

$$
\begin{array}{c}
\qquad\qquad\qquad\qquad
\begin{array}{c}
\text{entering} \\
\text{variable} \\
\downarrow
\end{array}\\
\begin{array}{ccccccc}
\text{B} & x_1 & x_2 & s_1 & s_2 & Z & \text{R}
\end{array}\\
\begin{array}{rc}
\text{departing} \leftarrow s_1 \\
\text{variable} \quad s_2 \\
Z
\end{array}
\left[
\begin{array}{ccccc|c}
2 & 1 & 1 & 0 & 0 & 8 \\
2 & 3 & 0 & 1 & 0 & 12 \\
-3 & -1 & 0 & 0 & 1 & 0
\end{array}
\right]
\end{array}
\tag{7}
$$

into an equivalent matrix that has a 1 where the "shaded" entry appears and 0's elsewhere in the x_1-column. The shaded entry is called the **pivot entry**—it is in the column of the entering variable (called the *pivot column*) and the row of the departing variable (called the *pivot row*). By elementary row operations, we have

$$
\begin{array}{ccccc|c}
x_1 & x_2 & s_1 & s_2 & Z & \\
\hline
2 & 1 & 1 & 0 & 0 & 8 \\
2 & 3 & 0 & 1 & 0 & 12 \\
-3 & -1 & 0 & 0 & 1 & 0
\end{array}
$$

$$
\xrightarrow{\frac{1}{2}R_1}
\left[
\begin{array}{ccccc|c}
1 & \frac{1}{2} & \frac{1}{2} & 0 & 0 & 4 \\
2 & 3 & 0 & 1 & 0 & 12 \\
-3 & -1 & 0 & 0 & 1 & 0
\end{array}
\right]
$$

$$
\xrightarrow[3R_1 + R_3]{-2R_1 + R_2}
\left[
\begin{array}{ccccc|c}
1 & \frac{1}{2} & \frac{1}{2} & 0 & 0 & 4 \\
0 & 2 & -1 & 1 & 0 & 4 \\
0 & \frac{1}{2} & \frac{3}{2} & 0 & 1 & 12
\end{array}
\right]
$$

Thus, we have a new simplex table:

$$
\begin{array}{c}
\begin{array}{ccccccc}
\text{B} & x_1 & x_2 & s_1 & s_2 & Z & \text{R}
\end{array}\\
\begin{array}{c} x_1 \\ s_2 \\ Z \end{array}
\left[
\begin{array}{ccccc|c}
1 & \frac{1}{2} & \frac{1}{2} & 0 & 0 & 4 \\
0 & 2 & -1 & 1 & 0 & 4 \\
0 & \frac{1}{2} & \frac{3}{2} & 0 & 1 & 12
\end{array}
\right]\\
\qquad\quad \underbrace{\qquad\qquad\qquad}_{\text{indicators}}
\end{array}
\tag{8}
$$

For $x_2 = 0$ and $s_1 = 0$, from the first row, we have $x_1 = 4$; from the second, we obtain $s_2 = 4$. These values give us the new BFS. Note that we replaced the s_1 located to the left of the initial table (7) by x_1 in our new table (8), so that s_1 *departed* and x_1 *entered*. From row 3, for $x_2 = 0$ and $s_1 = 0$, we get $Z = 12$, which is a larger value than we had before. (Before, we had $Z = 0$.)

In our present BFS, x_2 and s_1 are nonbasic variables ($x_2 = 0, s_1 = 0$). Suppose we look for another BFS that gives a larger value of Z and is such that one of x_2 or s_1 is basic. The equation corresponding to the Z-row is given by $\frac{1}{2}x_2 + \frac{3}{2}s_1 + Z = 12$, which can be rewritten as

$$Z = 12 - \frac{1}{2}x_2 - \frac{3}{2}s_1 \tag{9}$$

If x_2 becomes basic and therefore s_1 remains nonbasic, then

$$Z = 12 - \frac{1}{2}x_2 \quad \text{(since } s_1 = 0)$$

Here, each one-unit increase in x_2 *decreases* Z by $\frac{1}{2}$ unit. Thus, any increase in x_2 would make Z smaller than before. On the other hand, if s_1 becomes basic and x_2 remains nonbasic, then, from Equation (9),

$$Z = 12 - \frac{3}{2}s_1 \quad \text{(since } x_2 = 0)$$

Here each one-unit increase in s_1 *decreases* Z by $\frac{3}{2}$ units. Hence, any increase in s_1 would make Z smaller than before. Consequently, we cannot move to a better BFS. In short, no BFS gives a larger value of Z than the BFS $x_1 = 4$, $s_2 = 4$, $x_2 = 0$, and $s_1 = 0$ (which gives $Z = 12$).

In fact, since $x_2 \geq 0$ and $s_1 \geq 0$, and since the coefficients of x_2 and s_1 in Equation (9) are negative, Z is maximum when $x_2 = 0$ and $s_1 = 0$. That is, in (8), *having all nonnegative indicators means that we have an optimum solution.*

In terms of our original problem, if

$$Z = 3x_1 + x_2$$

subject to

$$2x_1 + x_2 \leq 8 \quad 2x_1 + 3x_2 \leq 12 \quad x_1, x_2 \geq 0$$

then Z is maximized when $x_1 = 4$ and $x_2 = 0$, and the maximum value of Z is 12. (This confirms our result in Example 1 of Section 7.2.) Note that the values of s_1 and s_2 do not have to appear here.

Let us outline the simplex method for a standard linear programming problem with three decision variables and four constraints, not counting nonnegativity conditions. The outline suggests how the simplex method works for any number of decision variables and any number of constraints.

Simplex Method

Problem:

$$\text{Maximize } Z = c_1 x_1 + c_2 x_2 + c_3 x_3$$

subject to

$$a_{11}x_1 + a_{12}x_2 + a_{13}x_3 \leq b_1$$
$$a_{21}x_1 + a_{22}x_2 + a_{23}x_3 \leq b_2$$
$$a_{31}x_1 + a_{32}x_2 + a_{33}x_3 \leq b_3$$
$$a_{41}x_1 + a_{42}x_2 + a_{43}x_3 \leq b_4$$

where x_1, x_2, x_3 and b_1, b_2, b_3, b_4 are nonnegative.

Method:

1. Set up the initial simplex table:

B	x_1	x_2	x_3	s_1	s_2	s_3	s_4	Z	R
s_1	a_{11}	a_{12}	a_{13}	1	0	0	0	0	b_1
s_2	a_{21}	a_{22}	a_{23}	0	1	0	0	0	b_2
s_3	a_{31}	a_{32}	a_{33}	0	0	1	0	0	b_3
s_4	a_{41}	a_{42}	a_{43}	0	0	0	1	0	b_4
Z	$-c_1$	$-c_2$	$-c_3$	0	0	0	0	1	0

indicators

There are four slack variables: $s_1, s_2, s_3,$ and s_4—one for each constraint.

2. If all the indicators in the last row are nonnegative, then Z has a maximum with the current list of basic variables and the current value of Z. (In the case of the initial simplex table this gives $x_1 = 0, x_2 = 0,$ and $x_3 = 0$, with maximum value of $Z = 0$.) If there are any negative indicators, locate and mark the column in which the most negative indicator appears. This *pivot column* gives the entering variable. (If more than one column contains the most negative indicator, the choice of pivot column is arbitrary.)

3. Divide each *positive*[3] entry above the objective row in the entering-variable column *into* the corresponding value of column R.

4. Mark the entry in the pivot column that corresponds to the smallest quotient in step 3. This is the pivot entry and the row in which it is located is the *pivot row*. The departing variable is the one that labels the pivot row.

5. Use elementary row operations to transform the table into a new equivalent table that has a 1 where the pivot entry was and 0's elsewhere in that column.

6. In the labels column B, of this table, the entering variable replaces the departing variable.

7. If the indicators of the new table are all nonnegative, we have an optimum solution. The maximum value of Z is the entry in the last row and last column. It occurs when the basic variables as found in the label column, B, are equal to the corresponding entries in column R. All other variables are 0. If at least one of the indicators is negative, repeat the process, beginning with step 2 applied to the new table.

As an aid in understanding the simplex method, we should be able to interpret certain entries in a table. Suppose that we obtain a table in which the last row is as shown in the following array:

B	x_1	x_2	x_3	s_1	s_2	s_3	s_4	Z	R
	.	.	.	.	.	.	.		.
	.	.	.	.	.	.	.		.
	.	.	.	.	.	.	.		.
Z	a	b	c	d	e	f	g	1	h

We can interpret the entry b, for example, as follows: If x_2 is nonbasic and were to become basic, then, for each one-unit increase in x_2,

$$\text{if } b < 0, \quad Z \text{ increases by } |b| \text{ units}$$

$$\text{if } b > 0, \quad Z \text{ decreases by } |b| \text{ units}$$

$$\text{if } b = 0, \quad \text{there is no change in } Z$$

---

[3]This will be discussed after Example 1.

●**EXAMPLE 1**   **The Simplex Method**

Maximize $Z = 5x_1 + 4x_2$ subject to

$$x_1 + x_2 \leq 20$$

$$2x_1 + x_2 \leq 35$$

$$-3x_1 + x_2 \leq 12$$

and $x_1, x_2 \geq 0$.

Solution: This linear programming problem fits the standard form. The initial simplex table is

entering
variable
↓

B	x_1	x_2	s_1	s_2	s_3	Z	R	Quotient
s_1	1	1	1	0	0	0	20	$20 \div 1 = 20$
s_2	2	1	0	1	0	0	35	$35 \div 2 = \frac{35}{2}$
s_3	−3	1	0	0	1	0	12	no quotient, $-3 \not> 0$
Z	−5	−4	0	0	0	1	0	

departing ← s_2
variable

indicators

The most negative indicator, −5, occurs in the x_1-column. Thus, x_1 is the entering variable. The smaller quotient is $\frac{35}{2}$, so s_2 is the departing variable. The pivot entry is 2. Using elementary row operations to get a 1 in the pivot position and 0's elsewhere in its column, we have

$$
\begin{array}{cccccc|c}
x_1 & x_2 & s_1 & s_2 & s_3 & Z & \\
1 & 1 & 1 & 0 & 0 & 0 & 20 \\
2 & 1 & 0 & 1 & 0 & 0 & 35 \\
-3 & 1 & 0 & 0 & 1 & 0 & 12 \\
\hline
-5 & -4 & 0 & 0 & 0 & 1 & 0
\end{array}
$$

$\xrightarrow{\frac{1}{2}R_2}$

$$
\begin{bmatrix}
1 & 1 & 1 & 0 & 0 & 0 & 20 \\
1 & \frac{1}{2} & 0 & \frac{1}{2} & 0 & 0 & \frac{35}{2} \\
-3 & 1 & 0 & 0 & 1 & 0 & 12 \\
\hline
-5 & -4 & 0 & 0 & 0 & 1 & 0
\end{bmatrix}
$$

$\xrightarrow[\substack{3R_2 + R_3 \\ 5R_2 + R_4}]{-1R_2 + R_1}$

$$
\begin{bmatrix}
0 & \frac{1}{2} & 1 & -\frac{1}{2} & 0 & 0 & \frac{5}{2} \\
1 & \frac{1}{2} & 0 & \frac{1}{2} & 0 & 0 & \frac{35}{2} \\
0 & \frac{5}{2} & 0 & \frac{3}{2} & 1 & 0 & \frac{129}{2} \\
\hline
0 & -\frac{3}{2} & 0 & \frac{5}{2} & 0 & 1 & \frac{175}{2}
\end{bmatrix}
$$

Our new table is

entering
variable
↓

B	x_1	x_2	s_1	s_2	s_3	Z	R	Quotients
s_1	0	$\frac{1}{2}$	1	$-\frac{1}{2}$	0	0	$\frac{5}{2}$	$\frac{5}{2} \div \frac{1}{2} = 5$
x_1	1	$\frac{1}{2}$	0	$\frac{1}{2}$	0	0	$\frac{35}{2}$	$\frac{35}{2} \div \frac{1}{2} = 35$
s_3	0	$\frac{5}{2}$	0	$\frac{3}{2}$	1	0	$\frac{129}{2}$	$\frac{129}{2} \div \frac{5}{2} = 25\frac{4}{5}$
Z	0	$-\frac{3}{2}$	0	$\frac{5}{2}$	0	1	$\frac{175}{2}$	

departing ← s_1
variable

indicators

Note that in column B, which keeps track of which variables are basic, x_1 has replaced s_2. Since we still have a negative indicator, $-\frac{3}{2}$, we must continue our process. Evidently, $-\frac{3}{2}$ is the most negative indicator and the entering variable is now x_2. The smallest quotient is 5. Hence, s_1 is the departing variable and $\frac{1}{2}$ is the pivot entry. Using elementary row operations, we have

$$
\begin{array}{ccccccc}
 & x_1 & x_2 & s_1 & s_2 & s_3 & Z & R \\
\left[\begin{array}{cccccc|c}
0 & \frac{1}{2} & 1 & -\frac{1}{2} & 0 & 0 & \frac{5}{2} \\
1 & \frac{1}{2} & 0 & \frac{1}{2} & 0 & 0 & \frac{35}{2} \\
0 & \frac{5}{2} & 0 & \frac{3}{2} & 1 & 0 & \frac{129}{2} \\
\hline
0 & -\frac{3}{2} & 0 & \frac{5}{2} & 0 & 1 & \frac{175}{2}
\end{array}\right]
\end{array}
$$

$$
\xrightarrow[\substack{-1R_1+R_2 \\ -5R_1+R_3 \\ 3R_1+R_4}]{}
\left[\begin{array}{cccccc|c}
0 & \frac{1}{2} & 1 & -\frac{1}{2} & 0 & 0 & \frac{5}{2} \\
1 & 0 & -1 & 1 & 0 & 0 & 15 \\
0 & 0 & -5 & 4 & 1 & 0 & 52 \\
\hline
0 & 0 & 3 & 1 & 0 & 1 & 95
\end{array}\right]
$$

$$
\xrightarrow{2R_1}
\left[\begin{array}{cccccc|c}
0 & 1 & 2 & -1 & 0 & 0 & 5 \\
1 & 0 & -1 & 1 & 0 & 0 & 15 \\
0 & 0 & -5 & 4 & 1 & 0 & 52 \\
\hline
0 & 0 & 3 & 1 & 0 & 1 & 95
\end{array}\right]
$$

Our new table is

$$
\begin{array}{c}
\begin{array}{cccccccc}
B & x_1 & x_2 & s_1 & s_2 & s_3 & Z & R
\end{array} \\
\begin{array}{c}
x_2 \\ x_1 \\ s_3 \\ \\ Z
\end{array}
\left[\begin{array}{cccccc|c}
0 & 1 & 2 & -1 & 0 & 0 & 5 \\
1 & 0 & -1 & 1 & 0 & 0 & 15 \\
0 & 0 & -5 & 4 & 1 & 0 & 52 \\
\hline
0 & 0 & \underbrace{3 \quad 1 \quad 0}_{\text{indicators}} & & & 1 & 95
\end{array}\right]
\end{array}
$$

where x_2 replaced s_1 in column B. Since all indicators are nonnegative, the maximum value of Z is 95 and occurs when $x_2 = 5$ and $x_1 = 15$ (and $s_3 = 52, s_1 = 0$, and $s_2 = 0$).

NOW WORK PROBLEM 1 ●○●

It is interesting to see how the values of Z got progressively "better" in successive tables in Example 1. These are the entries in the last row and last column of each simplex table. In the initial table, we had $Z = 0$. From then on, we obtained $Z = \frac{175}{2} = 87\frac{1}{2}$ and then $Z = 95$, the maximum.

In Example 1, you may wonder why no quotient is considered in the third row of the initial table. The BFS for this table is

$$s_1 = 20, \quad s_2 = 35, \quad s_3 = 12, \quad x_1 = 0, \quad x_2 = 0$$

where x_1 is the entering variable. The quotients 20 and $\frac{35}{2}$ reflect that, for the next BFS, we have $x_1 \leq 20$ and $x_1 \leq \frac{35}{2}$. Since the third row represents the equation $s_3 = 12 + 3x_1 - x_2$, and $x_2 = 0$, it follows that $s_3 = 12 + 3x_1$. But $s_3 \geq 0$, so $12 + 3x_1 \geq 0$, which implies that $x_1 \geq -\frac{12}{3} = -4$. Thus, we have

$$x_1 \leq 20, \quad x_1 \leq \frac{35}{2}, \quad \text{and} \quad x_1 \geq -4$$

Hence, x_1 can increase at most by $\frac{35}{2}$. The condition $x_1 \geq -4$ has no influence in determining the maximum increase in x_1. That is why the quotient $12/(-3) = -4$ is not considered in row 3. In general, *no quotient is considered for a row if the entry in the entering-variable column is negative* (or, of course, 0).

Although the simplex procedure that has been developed in this section applies only to linear programming problems of standard form, other forms may be adapted to fit this form. Suppose that a constraint has the form

$$a_1 x_1 + a_2 x_2 + \cdots + a_n x_n \geq -b$$

where $b > 0$. Here the inequality symbol is "$\geq$", and the constant on the right side is *negative*. Thus, the constraint is not in standard form. However, multiplying both sides by -1 gives

$$-a_1 x_1 - a_2 x_2 - \cdots - a_n x_n \leq b$$

which *does* have the proper form. Accordingly, it may be necessary to rewrite a constraint before proceeding with the simplex method.

In a simplex table, several indicators may "tie" for being most negative. In this case, we choose any one of these indicators to give the column for the entering variable. Likewise, there may be several quotients that "tie" for being the smallest. We may then choose any one of these quotients to determine the departing variable and pivot entry. Example 2 will illustrate this situation. When a tie for the smallest quotient exists, then, along with the nonbasic variables, a BFS will have a basic variable that is 0. In this case we say that the BFS is *degenerate* or that the linear programming problem has a *degeneracy*. More will be said about degeneracies in Section 7.5.

PRINCIPLES IN PRACTICE 1

THE SIMPLEX METHOD

The Toones Company has $30,000 for the purchase of materials to make three types of MP3 players. The company has allocated a total of 1200 hours of assembly time and 180 hours of packaging time for the players. The following table gives the cost per player, the number of hours per player, and the profit per player for each type:

	Type 1	Type 2	Type 3
Cost/ player	$300	$300	$400
Assembly Hours/ player	15	15	10
Packaging Hours/ player	2	2	3
Profit	$150	$250	$200

Find the number of players of each type the company should produce to maximize profit.

● **EXAMPLE 2   The Simplex Method**

Maximize $Z = 3x_1 + 4x_2 + \frac{3}{2}x_3$ *subject to*

$$-x_1 - 2x_2 \qquad\qquad \geq -10 \qquad\qquad (10)$$
$$2x_1 + 2x_2 + x_3 \leq 10$$
$$x_1, x_2, x_3 \geq 0$$

Solution: Constraint (10) does not fit the standard form. However, multiplying both sides of inequality (10) by -1 gives

$$x_1 + 2x_2 \leq 10$$

which *does* have the proper form. Thus, our initial simplex table is table I:

SIMPLEX TABLE I

entering
variable
↓

	B	x_1	x_2	x_3	s_1	s_2	Z	R	Quotients
departing →	s_1	1	2	0	1	0	0	10	$10 \div 2 = 5$
variable	s_2	2	2	1	0	1	0	10	$10 \div 2 = 5$
	Z	-3	-4	$-\frac{3}{2}$	0	0	1	0	

indicators

The entering variable is x_2. Since there is a tie for the smallest quotient, we can choose either s_1 or s_2 as the departing variable. Let us choose s_1. The pivot entry is shaded. Using elementary row operations, we get table II:

SIMPLEX TABLE II

entering
variable
↓

	B	x_1	x_2	x_3	s_1	s_2	Z	R	Quotients
	x_2	$\frac{1}{2}$	1	0	$\frac{1}{2}$	0	0	5	no quotient $0 \not> 0$
departing →	s_2	1	0	1	-1	1	0	0	$0 \div 1 = 0$
variable	Z	-1	0	$-\frac{3}{2}$	2	0	1	20	

indicators

Table II corresponds to a BFS in which a basic variable, s_2, is zero. Thus, the BFS is degenerate. Since there are negative indicators, we continue. The entering variable is now x_3, the departing variable is s_2, and the pivot is shaded. Using elementary row operations, we get table III:

SIMPLEX TABLE III

B	x_1	x_2	x_3	s_1	s_2	Z	R
x_2	$\frac{1}{2}$	1	0	$\frac{1}{2}$	0	0	5
x_3	1	0	1	-1	1	0	0
Z	$\frac{1}{2}$	0	0	$\frac{1}{2}$	$\frac{3}{2}$	1	20

indicators

Since all indicators are nonnegative, Z is maximized when $x_2 = 5$, $x_3 = 0$, and $x_1 = s_1 = s_2 = 0$. The maximum value is $Z = 20$. Note that this value is the same as the value of Z in table II. In degenerate problems, it is possible to arrive at the same value of Z at various stages of the simplex process. In Problem 7 you are asked to solve this example by using s_2 as the departing variable in the initial table.

NOW WORK PROBLEM 7

Because of its mechanical nature, the simplex procedure is readily adaptable to computers to solve linear programming problems involving many variables and constraints.

Problems 7.4

Use the simplex method to solve the following problems.

*1. Maximize

$$Z = x_1 + 2x_2$$

subject to

$$2x_1 + x_2 \leq 8$$
$$2x_1 + 3x_2 \leq 12$$
$$x_1, x_2 \geq 0$$

2. Maximize

$$Z = 2x_1 + x_2$$

subject to

$$-x_1 + x_2 \leq 4$$
$$x_1 + x_2 \leq 6$$
$$x_1, x_2 \geq 0$$

3. Maximize

$$Z = -x_1 + 2x_2$$

subject to

$$3x_1 + 2x_2 \leq 5$$
$$-x_1 + 3x_2 \leq 3$$
$$x_1, x_2 \geq 0$$

4. Maximize

$$Z = 4x_1 + 7x_2$$

subject to

$$2x_1 + 3x_2 \leq 9$$
$$x_1 + 5x_2 \leq 10$$
$$x_1, x_2 \geq 0$$

5. Maximize

$$Z = 8x_1 + 2x_2$$

subject to

$$x_1 - x_2 \leq 1$$
$$x_1 + 2x_2 \leq 8$$
$$x_1 + x_2 \leq 5$$
$$x_1, x_2 \geq 0$$

6. Maximize

$$Z = 2x_1 - 6x_2$$

subject to

$$x_1 - x_2 \leq 4$$
$$-x_1 + x_2 \leq 4$$
$$x_1 + x_2 \leq 6$$
$$x_1, x_2 \geq 0$$

*7. Solve the problem in Example 2 by choosing s_2 as the departing variable in table I.

8. Maximize

$$Z = 2x_1 - x_2 + x_3$$

subject to

$$2x_1 + x_2 - x_3 \leq 4$$

$$x_1 + x_2 + x_3 \leq 2$$

$$x_1, x_2, x_3 \geq 0$$

9. Maximize

$$Z = 2x_1 + x_2 - x_3$$

subject to

$$x_1 + x_2 \leq 1$$

$$x_1 - 2x_2 - x_3 \geq -2$$

$$x_1, x_2, x_3 \geq 0$$

10. Maximize

$$Z = -2x_1 + 3x_2$$

subject to

$$x_1 + x_2 \leq 1$$

$$x_1 - x_2 \leq 2$$

$$x_1 - x_2 \geq -3$$

$$x_1 \leq 5$$

$$x_1, x_2 \geq 0$$

11. Maximize

$$Z = x_1 + x_2$$

subject to

$$2x_1 - x_2 \leq 4$$

$$-x_1 + 2x_2 \leq 6$$

$$5x_1 + 3x_2 \leq 20$$

$$2x_1 + x_2 \leq 10$$

$$x_1, x_2 \geq 0$$

12. Maximize

$$W = 2x_1 + x_2 - 2x_3$$

subject to

$$-2x_1 + x_2 + x_3 \geq -2$$

$$x_1 - x_2 + x_3 \leq 4$$

$$x_1 + x_2 + 2x_3 \leq 6$$

$$x_1, x_2, x_3 \geq 0$$

13. Maximize

$$W = x_1 - 12x_2 + 4x_3$$

subject to

$$4x_1 + 3x_2 - x_3 \leq 1$$

$$x_1 + x_2 - x_3 \geq -2$$

$$-x_1 + x_2 + x_3 \geq -1$$

$$x_1, x_2, x_3 \geq 0$$

14. Maximize

$$W = 4x_1 + 0x_2 - x_3$$

subject to

$$x_1 + x_2 + x_3 \leq 6$$

$$x_1 - x_2 + x_3 \leq 10$$

$$x_1 - x_2 - x_3 \leq 4$$

$$x_1, x_2, x_3 \geq 0$$

15. Maximize

$$Z = 60x_1 + 0x_2 + 90x_3 + 0x_4$$

subject to

$$x_1 - 2x_2 \leq 2$$

$$x_1 + x_2 \leq 5$$

$$x_3 + x_4 \leq 4$$

$$x_3 - 2x_4 \leq 7$$

$$x_1, x_2, x_3, x_4 \geq 0$$

16. Maximize

$$Z = 3x_1 + 2x_2 - 2x_3 - x_4$$

subject to

$$x_1 + x_3 - x_4 \leq 3$$

$$x_1 - x_2 + x_4 \leq 6$$

$$x_1 + x_2 - x_3 + x_4 \leq 5$$

$$x_1, x_2, x_3, x_4 \geq 0$$

17. Freight Shipments A freight company handles shipments by two corporations, A and B, that are located in the same city. Corporation A ships boxes that weigh 3 lb each and have a volume of 2 ft^3; B ships 1 ft^3 boxes that weigh 5 lb each. Both A and B ship to the same destination. The transportation cost for each box from A is $0.75, and from B it is $0.50. The freight company has a truck with 2400 ft^3 of cargo space and a maximum capacity of 36,800 lb. In one haul, how many boxes from each corporation should be transported by this truck so that the freight company receives maximum revenue? What is the maximum revenue?

18. Production A company manufactures three products: X, Y, and Z. Each product requires machine time and finishing time as shown in the following table:

	Machine Time	**Finishing Time**
X	1 hr	4 hr
Y	2 hr	4 hr
Z	3 hr	8 hr

The numbers of hours of machine time and finishing time available per month are 900 and 5000, respectively. The unit profit on X, Y, and Z is $6, $8, and $12, respectively. What is the maximum profit per month that can be obtained?

19. Production A company manufactures three types of patio furniture: chairs, rockers, and chaise lounges. Each requires wood, plastic, and aluminum as shown in the following table:

	Wood	**Plastic**	**Aluminum**
Chair	1 unit	1 unit	2 units
Rocker	1 unit	1 unit	3 units
Chaise lounge	1 unit	2 units	5 units

The company has available 400 units of wood, 500 units of plastic, and 1450 units of aluminum. Each chair, rocker, and chaise lounge sells for $21, $24, and $36, respectively. Assuming that all furniture can be sold, determine a production order so that total revenue will be maximum. What is the maximum revenue?

OBJECTIVE

To consider the simplex method in relation to degeneracy, unbounded solutions, and multiple optimum solutions.

7.5 Degeneracy, Unbounded Solutions, and Multiple Solutions[4]

Degeneracy

In the preceding section, we stated that a BFS is **degenerate** if, along with the nonbasic variables, one of the basic variables is 0. Suppose x_1, x_2, x_3, and x_4 are the variables in a degenerate BFS, where x_1 and x_2 are basic with $x_1 = 0$, x_3 and x_4 are nonbasic, and x_3 is the entering variable. The corresponding simplex table has the form

$$
\begin{array}{c}
\text{entering} \\
\text{variable} \\
\downarrow
\end{array}
$$

$$
\begin{array}{c}
\text{departing} \leftarrow \\
\text{variable}
\end{array}
\quad
\begin{array}{c|cccccc|c}
B & x_1 & x_2 & x_3 & x_4 & Z & & R \\
\hline
x_1 & 1 & 0 & a_{13} & a_{14} & 0 & & 0 \\
x_2 & 0 & 1 & a_{23} & a_{24} & 0 & & a \\
\hline
Z & 0 & 0 & d_1 & d_2 & 1 & & d_3
\end{array}
\quad 0 \div a_{13} = 0
$$

$$\underbrace{}_{\text{indicators}}$$

Thus, the BFS is

$$x_1 = 0, \quad x_2 = a, \quad x_3 = 0, \quad x_4 = 0$$

Suppose $a_{13} > 0$. Then the smaller quotient is 0, and we can choose a_{13} as the pivot entry. Therefore, x_1 is the departing variable. Elementary row operations give the following table, where the question marks represent numbers to be determined:

$$
\begin{array}{c|cccccc|c}
B & x_1 & x_2 & x_3 & x_4 & Z & & R \\
\hline
x_3 & ? & 0 & 1 & ? & 0 & & 0 \\
x_2 & ? & 1 & 0 & ? & 0 & & a \\
\hline
Z & ? & 0 & 0 & ? & 1 & & d_3
\end{array}
$$

For the BFS corresponding to this table, x_3 and x_2 are basic variables, and x_1 and x_4 are nonbasic. The BFS is

$$x_3 = 0 \qquad x_2 = a \qquad x_1 = 0 \qquad x_4 = 0$$

which is the same BFS as before. Actually, these are usually considered different BFS's, with the only distinction being that x_1 is basic in the first BFS, whereas in the second it is nonbasic. The value of Z for both BFSs is the same: d_3. Thus, no "improvement" in Z is obtained.

In a degenerate situation, some problems may develop in the simplex procedure. It is possible to obtain a sequence of tables that correspond to BFSs that give the same Z-value. Moreover, we may eventually return to the first table in the sequence. In Figure 7.24, we arrive at BFS_1, proceed to BFS_2, go on to BFS_3, and finally return to BFS_1. This is called *cycling*. When cycling occurs, we may never obtain the optimum value of Z. This phenomenon is encountered rarely in practical linear programming problems; however, there are techniques (which will not be considered in this text) for resolving such difficulties.

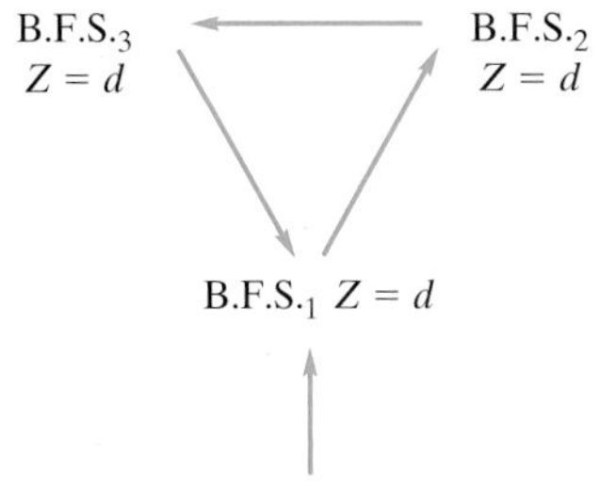

B.F.S.$_3$ $Z = d$

B.F.S.$_2$ $Z = d$

B.F.S.$_1$ $Z = d$

FIGURE 7.24 Cycling.

---

[4]This section may be omitted.

A degenerate BFS will occur when two quotients in a simplex table tie for being the smallest. For example, consider the following (partial) table:

$$\begin{array}{ccccc} \text{B} & x_3 & & \text{R} & \textit{Quotients} \\ x_1 & \begin{bmatrix} q_1 \\ q_2 \end{bmatrix} & \Big| & \begin{bmatrix} p_1 \\ p_2 \end{bmatrix} & \begin{matrix} p_1/q_1 \\ p_2/q_2 \end{matrix} \end{array}$$

Here x_1 and x_2 are basic variables. Suppose x_3 is nonbasic and entering, and $p_1/q_1 = p_2/q_2$ are the smallest quotients. Choosing q_1 as the pivot entry, we obtain, by elementary row operations,

$$\begin{array}{cccc} \text{B} & x_3 & & \text{R} \\ x_3 & \begin{bmatrix} 1 \\ 0 \end{bmatrix} & \Big| & \begin{bmatrix} p_1/q_1 \\ p_2 - q_2\frac{p_1}{q_1} \end{bmatrix} \end{array}$$

Since $p_1/q_1 = p_2/q_2$, we have $p_2 - q_2(p_1/q_1) = 0$. Thus, the BFS corresponding to this table has $x_2 = 0$, which gives a *degenerate* BFS. Although such a BFS may produce cycling, we will not encounter many such situations in this book. However, see problem 11 in Problems 7.5.

Unbounded Solutions

We now turn our attention to "unbounded problems." In Section 7.2, we saw that a linear programming problem may have no maximum value because the feasible region is such that the objective function may become arbitrarily large therein. In this case, the problem is said to have an **unbounded solution.** This is a way of saying specifically that no optimum solution exists. Such a situation occurs when no quotients are possible in a simplex table for an entering variable. For example, consider the following simplex table:

$$\begin{array}{c} \text{entering} \\ \text{variable} \\ \downarrow \end{array}$$

$$\begin{array}{cccccccl} \text{B} & x_1 & x_2 & x_3 & x_4 & Z & & \text{R} \\ x_1 & \begin{bmatrix} 1 & -3 & 0 & 2 & 0 \\ 0 & 0 & 1 & 4 & 0 \\ \hline 0 & -5 & 0 & -2 & 1 \end{bmatrix} & & & & & \Big| & \begin{matrix} 5 \\ 1 \\ 10 \end{matrix} & \begin{matrix} \text{no quotient} \\ \text{no quotient} \end{matrix} \\ Z \end{array}$$

$$\underbrace{\qquad\qquad\qquad}_{\text{indicators}}$$

Here x_2 is the entering variable, and for each one-unit increase in x_2, Z increases by 5. Since there are no positive entries in the first two rows of the x_2 column, no quotients exist. From rows 1 and 2 we get

$$x_1 = 5 + 3x_2 - 2x_4$$

and

$$x_3 = 1 - 4x_4$$

If we try to proceed to the next BFS, what is an upper bound on x_2? In that BFS, x_4 will remain nonbasic ($x_4 = 0$). Thus, $x_1 = 5 + 3x_2$ and $x_3 = 1$. Since $x_1 \geq 0$, $x_2 \geq -\frac{5}{3}$. Therefore, there is no upper bound on x_2. Hence, Z can be arbitrarily large, and we have an unbounded solution. In general,

If no quotients exist in a simplex table, then the linear programming problem has an unbounded solution.

● EXAMPLE 1   Unbounded Solution

Maximize $Z = x_1 + 4x_2 - x_3$ subject to

$$-5x_1 + 6x_2 - 2x_3 \leq 30$$

$$-x_1 + 3x_2 + 6x_3 \leq 12$$

$$x_1, x_2, x_3 \geq 0$$

Solution: The initial simplex table is

entering
variable
↓

B	x_1	x_2	x_3	s_1	s_2	Z	R	Quotients
s_1	-5	6	-2	1	0	0	30	$30 \div 6 = 5$
s_2	-1	3	6	0	1	0	12	$12 \div 3 = 4$
Z	-1	-4	1	0	0	1	0	

departing ← s_2

departing
variable

indicators

The second table is

entering
variable
↓

B	x_1	x_2	x_3	s_1	s_2	Z	R	
s_1	-3	0	-14	1	-2	0	6	no quotient
x_2	$-\frac{1}{3}$	1	2	0	$\frac{1}{3}$	0	4	no quotient
Z	$-\frac{7}{3}$	0	9	0	$\frac{4}{3}$	1	16	

indicators

Here the entering variable is x_1. Since the entries in the first two rows of the x_1-column are negative, no quotients exist. Hence, the problem has an unbounded solution.

NOW WORK PROBLEM 3 ●●

Multiple Optimum Solutions

We conclude this section with a discussion of "multiple optimum solutions." Suppose that

$$x_1 = a_1 \qquad x_2 = a_2 \qquad \ldots \qquad x_n = a_n$$

and

$$x_1 = b_1 \qquad x_2 = b_2 \qquad \ldots \qquad x_n = b_n$$

are two *different* BFSs for which a linear programming problem is optimum. By "different BFSs," we mean that $a_i \neq b_i$ for some i, where $1 \leq i \leq n$. It can be shown that the values

$$\begin{aligned}
x_1 &= (1 - t)a_1 + tb_1 \\
x_2 &= (1 - t)a_2 + tb_2 \\
&\vdots \\
x_n &= (1 - t)a_n + tb_n
\end{aligned}$$

(1)

for any t such that $0 \leq t \leq 1$

also give an optimum solution (although it may not necessarily be a BFS). Thus, there are *multiple (optimum) solutions* to the problem.

We can determine the possibility of multiple optimum solutions from a simplex table that gives an optimum solution, such as the following (partial) simplex table:

B	x_1	x_2	x_3	x_4	Z	R
x_1						p_1
x_2						q_1
Z	0	0	a	0	1	r

indicators

Here a must be nonnegative. The corresponding BFS is

$$x_1 = p_1 \qquad x_2 = q_1 \qquad x_3 = 0 \qquad x_4 = 0$$

and the maximum value of Z is r. If x_4 were to become basic, the indicator 0 in the x_4-column means that for each one-unit increase in x_4, Z does not change. Thus, we can find a BFS in which x_4 is basic and the corresponding Z-value is the same as before. This is done by treating x_4 as an entering variable in the preceding table. If, for instance, x_1 is the departing variable, then the pivot element will be the one in row x_1 and column x_4. The new BFS will have the form

$$x_1 = 0 \quad x_2 = q_2 \quad x_3 = 0 \quad x_4 = p_2$$

where q_2 and p_2 are numbers that result from the pivoting process. If this BFS is different from the previous one, multiple solutions exist. In fact, from Equations (1), an optimum solution is given by any values of x_1, x_2, x_3, and x_4, such that

$$x_1 = (1 - t)p_1 + t \cdot 0 = (1 - t)p_1$$
$$x_2 = (1 - t)q_1 + tq_2$$
$$x_3 = (1 - t) \cdot 0 + t \cdot 0 = 0$$
$$x_4 = (1 - t) \cdot 0 + tp_2 = tp_2$$

where $0 \leq t \leq 1$

Note that when $t = 0$, we get the first optimum BFS; when $t = 1$, we get the second. It may be possible to repeat the procedure by using the table corresponding to the last BFS and obtain more optimum solutions by using Equations (1).

In general,

In a table that gives an optimum solution, a zero indicator for a nonbasic variable suggests the possibility of multiple optimum solutions.

EXAMPLE 2 Multiple Solutions

Maximize $Z = -x_1 + 4x_2 + 6x_3$ *subject to*

$$x_1 + 2x_2 + 3x_3 \leq 6$$
$$-2x_1 - 5x_2 + x_3 \leq 10$$
$$x_1, x_2, x_3 \geq 0$$

Solution: Our initial simplex table is

entering
variable
↓

B	x_1	x_2	x_3	s_1	s_2	Z	R	Quotients
departing ← s_1	1	2	3	1	0	0	6	$6 \div 3 = 2$
s_2	-2	-5	1	0	1	0	10	$10 \div 1 = 10$
Z	1	-4	-6	0	0	1	0	

indicators

Since there is a negative indicator, we continue and obtain

entering
variable
↓

B	x_1	x_2	x_3	s_1	s_2	Z	R	Quotients
departing ← x_3	$\frac{1}{3}$	$\frac{2}{3}$	1	$\frac{1}{3}$	0	0	2	$2 \div \frac{2}{3} = 3$
s_2	$-\frac{7}{3}$	$-\frac{17}{3}$	0	$-\frac{1}{3}$	1	0	8	no quotient
Z	3	0	0	2	0	1	12	

indicators

All indicators are nonnegative: hence, an optimum solution occurs for the BFS

$$x_3 = 2 \quad s_2 = 8 \quad x_1 = 0 \quad x_2 = 0 \quad s_1 = 0$$

and the maximum value of Z is 12. However, since x_2 is a nonbasic variable and its indicator is 0, we will check for multiple solutions. Treating x_2 as an entering variable, we obtain the following table:

B	x_1	x_2	x_3	s_1	s_2	Z	R
x_2	$\frac{1}{2}$	1	$\frac{3}{2}$	$\frac{1}{2}$	0	0	3
s_2	$\frac{1}{2}$	0	$\frac{17}{2}$	$\frac{5}{2}$	1	0	25
Z	3	0	0	2	0	1	12

The BFS here is

$$x_2 = 3, \quad s_2 = 25, \quad x_1 = 0, \quad x_3 = 0, \quad s_1 = 0$$

(for which $Z = 12$, as before) and is different from the previous one. Thus, multiple solutions exist. Since we are concerned only with values of the decision variables, we have an optimum solution,

$$x_1 = (1 - t) \cdot 0 + t \cdot 0 = 0$$

$$x_2 = (1 - t) \cdot 0 + t \cdot 3 = 3t$$

$$x_3 = (1 - t) \cdot 2 + t \cdot 0 = 2(1 - t)$$

for each value of t such that $0 \le t \le 1$. (For example, if $t = \frac{1}{2}$, then $x_1 = 0$, $x_2 = \frac{3}{2}$, and $x_3 = 1$ is an optimum solution.)

In the last BFS, x_3 is nonbasic and its indicator is 0. However, if we repeated the process for determining other optimum solutions, we would return to the second table. Therefore, our procedure gives no other optimum solutions.

NOW WORK PROBLEM 5 ⬤⬤⬤

Problems 7.5

In Problems 1 and 2, does the linear programming problem associated with the given table have a degeneracy? If so, why?

1.

B	x_1	x_2	s_1	s_2	Z	R
x_1	1	2	4	0	0	6
s_2	0	1	1	1	0	3
Z	0	-3	-2	0	1	10

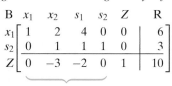
indicators

2.

B	x_1	x_2	x_3	s_1	s_2	Z	R
s_1	2	0	2	1	1	0	4
x_2	3	1	1	0	1	0	0
Z	-5	0	1	0	-3	1	2

indicators

In Problems 3–10, use the simplex method.

*3. Maximize

$$Z = 2x_1 + 7x_2$$

subject to

$$4x_1 - 3x_2 \le 4$$

$$3x_1 - x_2 \le 6$$

$$5x_1 \le 8$$

$$x_1, x_2 \ge 0$$

4. Maximize

$$Z = 2x_1 + x_2$$

subject to

$$x_1 - x_2 \le 7$$

$$-x_1 + x_2 \le 5$$

$$8x_1 + 5x_2 \le 40$$

$$2x_1 + x_2 \le 6$$

$$x_1, x_2 \ge 0$$

*5. Maximize

$$Z = -4x_1 + 8x_2$$

subject to

$$2x_1 - 2x_2 \le 4$$

$$-x_1 + 2x_2 \le 4$$

$$3x_1 + x_2 \le 6$$

$$x_1, x_2 \ge 0$$

6. Maximize

$$Z = 8x_1 + 2x_2 + 4x_3$$

subject to

$$x_1 - x_2 + 4x_3 \le 6$$

$$x_1 - x_2 - x_3 \ge -4$$

$$x_1 - 6x_2 + x_3 \le 8$$

$$x_1, x_2, x_3 \ge 0$$

7. Maximize

$$Z = 5x_1 + 6x_2 + x_3$$

subject to

$$9x_1 + 3x_2 - 2x_3 \le 5$$

$$4x_1 + 2x_2 - x_3 \le 2$$

$$x_1 - 4x_2 + x_3 \le 3$$

$$x_1, x_2, x_3 \ge 0$$

8. Maximize

$$Z = 2x_1 + x_2 - 4x_3$$

subject to

$$6x_1 + 3x_2 - 3x_3 \le 10$$

$$x_1 - x_2 + x_3 \le 1$$

$$2x_1 - x_2 + 2x_3 \le 12$$

$$x_1, x_2, x_3 \ge 0$$

9. Maximize

$$Z = 6x_1 + 2x_2 + x_3$$

subject to

$$2x_1 + x_2 + x_3 \le 7$$

$$-4x_1 - x_2 \ge -6$$

$$x_1, x_2, x_3 \ge 0$$

10. Maximize

$$P = x_1 + 2x_2 + x_3 + 2x_4$$

subject to

$$x_1 - x_2 \le 2$$

$$x_2 - x_3 \le 3$$

$$x_2 - 3x_3 + x_4 \le 4$$

$$x_1, x_2, x_3, x_4 \ge 0$$

11. Production A company manufactures three types of patio furniture: chairs, rockers, and chaise lounges. Each requires wood, plastic, and aluminum as given in the following table:

	Wood	Plastic	Aluminum
Chair	1 unit	1 unit	2 units
Rocker	1 unit	1 unit	3 units
Chaise lounge	1 unit	2 units	5 units

The company has available 400 units of wood, 600 units of plastic, and 1500 units of aluminum. Each chair, rocker, and chaise lounge sells for $24, $32, and $48, respectively. Assuming that all furniture can be sold, what is the maximum total revenue that can be obtained? Determine the possible production orders that will generate this revenue.

OBJECTIVE

To use artificial variables to handle maximization problems that are not of standard form.

7.6 Artificial Variables

To start using the simplex method, a *basic feasible solution, BFS,* is required. (We algebraically start at a *corner point* using the initial simplex table and each subsequent table takes us, algebraically, to another corner point until we reach the one at which an optimum solution is obtaind.) For a *standard* linear programming problem, we begin with the BFS in which all decision variables are zero. However, for a maximization problem that is not of standard form, such a BFS may not exist. In this section, we will learn how the simplex method is used in such situations.

Let us consider the following problem:

$$\text{Maximize } Z = x_1 + 2x_2$$

subject to

$$x_1 + x_2 \le 9 \tag{1}$$

$$x_1 - x_2 \ge 1 \tag{2}$$

$$x_1, x_2 \ge 0$$

Since constraint (2) cannot be written as $a_1x_1 + a_2x_2 \le b$, where b is nonnegative, this problem cannot be put into standard form. Note that $(0, 0)$ is not a feasible point because it does not satisfy constraint (2). (Because $0 - 0 = 0 \ge 1$ is *false!*) To solve the problem, we begin by writing constraints (1) and (2) as equations. Constraint (1) becomes

$$x_1 + x_2 + s_1 = 9 \tag{3}$$

where $s_1 \geq 0$ is a slack variable. For constraint (2), $x_1 - x_2$ will equal 1 if we *subtract* a nonnegative slack variable s_2 from $x_1 - x_2$. That is, by subtracting s_2, we are making up for the "surplus" on the left side of (2), so that we have equality. Thus,

$$x_1 - x_2 - s_2 = 1 \qquad (4)$$

where $s_2 \geq 0$. We can now restate the problem:

$$\text{Maximize } Z = x_1 + 2x_2 \qquad (5)$$

subject to

$$x_1 + x_2 + s_1 = 9 \qquad (6)$$

$$x_1 - x_2 - s_2 = 1 \qquad (7)$$

$$x_1, x_2, s_1, s_2 \geq 0$$

Since $(0, 0)$ is not in the feasible region, we do not have a BFS in which $x_1 = x_2 = 0$. In fact, if $x_1 = 0$ and $x_2 = 0$ are substituted into Eqution (7), then $0 - 0 - s_2 = 1$, which gives $s_2 = -1$, and now the problem is that this contradicts the condition that $s_2 \geq 0$.

To get the simplex method started, we need an initial BFS. Although none is obvious, there is an ingenious method to arrive at one *artificially*. It requires that we consider a related linear programming problem called the *artificial problem*. First, a new equation is formed by adding a nonnegative variable t to the left side of the equation in which the coefficient of the slack variable is -1. The variable t is called an **artificial variable.** In our case, we replace Equation (7) by $x_1 - x_2 - s_2 + t = 1$. Thus, Equations (6) and (7) become

$$x_1 + x_2 + s_1 = 9 \qquad (8)$$

$$x_1 - x_2 - s_2 + t = 1 \qquad (9)$$

$$x_1, x_2, s_1, s_2, t \geq 0$$

An obvious solution to Equations (8) and (9) is found by setting x_1, x_2, and s_2 equal to 0. This gives

$$x_1 = x_2 = s_2 = 0 \quad s_1 = 9 \quad t = 1$$

Note that these values do not satisfy Equation (7). However, it is clear that any solution of Equations (8) and (9) *for which* $t = 0$ will give a solution to Equations (6) and (7), and conversely.

We can *eventually* force t to be 0 if we alter the original objective function. We define the **artificial objective function** to be

$$W = Z - Mt = x_1 + 2x_2 - Mt \qquad (10)$$

where the constant M is a very large positive number. We will not worry about the particular value of M and will proceed to maximize W by the simplex method. Since there are $m = 2$ constraints (excluding the nonnegativity conditions) and $n = 5$ variables in Equations (8) and (9), any BFS must have at least $n - m = 3$ variables equal to zero. We start with the following BFS:

$$x_1 = x_2 = s_2 = 0 \quad s_1 = 9 \quad t = 1 \qquad (11)$$

In this initial BFS, the nonbasic variables are the decision variables and the "surplus" variable s_2. The corresponding value of W is $W = x_1 + 2x_2 - Mt = -M$, which is "extremely" negative since we assume that M is a very large positive number. A significant improvement in W will occur if we can find another BFS for which $t = 0$. Since the simplex method seeks better values of W at each stage, we will apply it until we reach such a BFS, if possible. That solution will be an initial BFS for the original problem.

To apply the simplex method to the artificial problem, we first write Equation (10) as

$$-x_1 - 2x_2 + Mt + W = 0 \qquad (12)$$

The augmented coefficient matrix of Equations (8), (9), and (12) is

$$
\begin{array}{cccccc}
x_1 & x_2 & s_1 & s_2 & t & W \\
\end{array}
$$
$$
\left[\begin{array}{cccccc|c}
1 & 1 & 1 & 0 & 0 & 0 & 9 \\
1 & -1 & 0 & -1 & 1 & 0 & 1 \\
\hline
-1 & -2 & 0 & 0 & M & 1 & 0
\end{array}\right] \qquad (13)
$$

An initial BFS is given by (11). Notice that, from row s_1, when $x_1 = x_2 = s_2 = 0$, we can directly read the value of s_1, namely, $s_1 = 9$. From row 2, we get $t = 1$. From row 3, $Mt + W = 0$. Since $t = 1$, $W = -M$. But in a simplex table we want the value of W to appear in the last row and last column. This is not so in (13); thus, we modify that matrix.

To do this, we transform (13) into an equivalent matrix whose last row has the form

$$
\begin{array}{cccccc}
x_1 & x_2 & s_1 & s_2 & t & W \\
? & ? & 0 & ? & 0 & 1 \mid ?
\end{array}
$$

That is, the M in the t-column is replaced by 0. As a result, if $x_1 = x_2 = s_2 = 0$, then W equals the last entry. Proceeding to obtain such a matrix, we have, by pivotting at the shaded element in column t:

$$
\begin{array}{ccccccc}
x_1 & x_2 & s_1 & s_2 & t & W & R \\
\end{array}
$$
$$
\left[\begin{array}{cccccc|c}
1 & 1 & 1 & 0 & 0 & 0 & 9 \\
1 & -1 & 0 & -1 & 1 & 0 & 1 \\
\hline
-1 & -2 & 0 & 0 & M & 1 & 0
\end{array}\right]
$$

$$
\xrightarrow{-MR_2 + R_3}
\begin{array}{ccccccc}
x_1 & x_2 & s_1 & s_2 & t & W & R \\
\end{array}
$$
$$
\left[\begin{array}{cccccc|c}
1 & 1 & 1 & 0 & 0 & 0 & 9 \\
1 & -1 & 0 & -1 & 1 & 0 & 1 \\
\hline
-1-M & -2+M & 0 & M & 0 & 1 & -M
\end{array}\right]
$$

Let us now check things out. If $x_1 = 0$, $x_2 = 0$, and $s_2 = 0$, then from row 1 we get $s_1 = 9$, from row 2, $t = 1$, and from row 3, $W = -M$. Thus, we now have initial simplex table I:

SIMPLEX TABLE I

entering
variable
↓

B	x_1	x_2	s_1	s_2	t	W	R		Quotients
s_1	1	1	1	0	0	0		9	$9 \div 1 = 9$
departing ← t	1	-1	0	-1	1	0		1	$1 \div 1 = 1$
W	$-1-M$	$-2+M$	0	M	0	1		$-M$	

indicators

From this point, we can use the procedures of Section 7.4. Since M is a large positive number, the most negative indicator is $-1 - M$. Thus, the entering variable is x_1. From the quotients, we get t as the departing variable. The pivot entry is shaded. Using elementary row operations to get 1 in the pivot position and 0's elsewhere in that column, we get simplex table II:

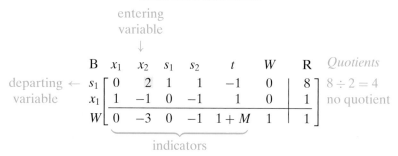

SIMPLEX TABLE II

From table II, we have the following BFS:

$$s_1 = 8, \quad x_1 = 1, \quad x_2 = 0, \quad s_2 = 0, \quad t = 0$$

Since $t = 0$, the values $s_1 = 8$, $x_1 = 1$, $x_2 = 0$, and $s_2 = 0$ form an initial BFS for the *original* problem! The artificial variable has served its purpose. For succeeding tables, we will delete the t-column (since we want to solve the original problem) and change the W's to Z's (since $W = Z$ for $t = 0$). From table II, the entering variable is x_2, the departing variable is s_1, and the pivot entry is shaded. Using elementary row operations (omitting the t-column), we get table III:

SIMPLEX TABLE III

B	x_1	x_2	s_1	s_2	Z	R
x_2	0	1	$\frac{1}{2}$	$\frac{1}{2}$	0	4
x_1	1	0	$\frac{1}{2}$	$-\frac{1}{2}$	0	5
Z	0	0	$\frac{3}{2}$	$\frac{1}{2}$	1	13

indicators

Since all the indicators are nonnegative, the maximum value of Z is 13. It occurs when $x_1 = 5$ and $x_2 = 4$.

It is worthwhile to review the steps we performed to solve our problem:

Here is a summary of the procedure involving artificial variables.

$$\text{Maximize } Z = x_1 + 2x_2$$

subject to

$$x_1 + x_2 \leq 9 \tag{14}$$

$$x_1 - x_2 \geq 1 \tag{15}$$

$$x_1, x_2 \geq 0$$

We write Inequality (14) as

$$x_1 + x_2 + s_1 = 9 \tag{16}$$

Since Inequality (15) involves the symbol $\geq$, and the constant on the right side is nonnegative, we write Inequality (15) in a form having both a surplus variable and an artificial variable:

$$x_1 - x_2 - s_2 + t = 1 \tag{17}$$

The artificial objective equation to consider is $W = x_1 + 2x_2 - Mt$, or, equivalently,

$$-x_1 - 2x_2 + Mt + W = 0 \tag{18}$$

The augmented coefficient matrix of the system formed by Equations (16)–(18) is

$$
\begin{array}{c c}
\text{B} & \begin{array}{c c c c c c c}
x_1 & x_2 & s_1 & s_2 & t & W & R
\end{array} \\
\begin{array}{c}
s_1 \\
t \\
W
\end{array} &
\left[
\begin{array}{c c c c c c | c}
1 & 1 & 1 & 0 & 0 & 0 & 9 \\
1 & -1 & 0 & -1 & 1 & 0 & 1 \\
\hline
-1 & -2 & 0 & 0 & M & 1 & 0
\end{array}
\right]
\end{array}
$$

Next, we remove the M from the artificial variable column and replace it by 0 by using elementary row operations. The resulting simplex table I corresponds to the initial BFS of the artificial problem in which the decision variables, x_1 and x_2, and the surplus variable s_2 are each 0:

SIMPLEX TABLE I

$$
\begin{array}{c c}
\text{B} & \begin{array}{c c c c c c c}
x_1 & x_2 & s_1 & s_2 & t & W & R
\end{array} \\
\begin{array}{c}
s_1 \\
t \\
W
\end{array} &
\left[
\begin{array}{c c c c c c | c}
1 & 1 & 1 & 0 & 0 & 0 & 9 \\
1 & -1 & 0 & -1 & 1 & 0 & 1 \\
\hline
-1-M & -2+M & 0 & M & 0 & 1 & -M
\end{array}
\right]
\end{array}
$$

The basic variables s_1 and t in column B of the table correspond to the nondecision variables in Equations (16) and (17) that have positive coefficients. We now apply the simplex method until we obtain a BFS in which the artificial variable t equals 0. Then we can delete the artificial variable column, change the W's to Z's, and continue the procedure until the maximum value of Z is obtained.

ARTIFICIAL VARIABLES

The GHI Company manufactures two models of snowboards, standard and deluxe, at two different manufacturing plants. The maximum output at plant I is 1200 per month, while the maximum output at plant II is 1000 per month. Due to contractual obligations, the number of deluxe models produced at plant I cannot exceed the number of standard models produced at plant I by more than 200. The profit per standard and deluxe snowboard manufactured at plant I is $40 and $60, respectively, while the profit per standard and deluxe snowboard manufactured at plant II is $45 and $50, respectively. This month, GHI received an order for 1000 standard and 800 deluxe models. Find how many of each model should be produced at each plant to satisfy the order and maximize the profit. (*Hint:* Let x_1 represent the number of standard models and x_2 represent the number of deluxe models manufactured at plant I.)

● EXAMPLE 1   **Artificial Variables**

Use the simplex method to maximize $Z = 2x_1 + x_2$ subject to

$$x_1 + x_2 \le 12 \tag{19}$$

$$x_1 + 2x_2 \le 20 \tag{20}$$

$$-x_1 + x_2 \ge 2 \tag{21}$$

$$x_1, x_2 \ge 0$$

Solution: The equations for (19)–(21) will involve two slack variables, s_1 and s_2, for the two $\le$ constraints, and a surplus variable s_3 and an artificial variable t for the $\ge$ constraint. We thus have

$$x_1 + x_2 + s_1 \qquad\qquad = 12 \tag{22}$$

$$x_1 + 2x_2 \qquad + s_2 \qquad\quad = 20 \tag{23}$$

$$-x_1 + x_2 \qquad\qquad - s_3 + t = 2 \tag{24}$$

We consider $W = Z - Mt = 2x_1 + x_2 - Mt$ as the artificial objective equation, equivalently,

$$-2x_1 - x_2 + Mt + W = 0 \tag{25}$$

where M is a large positive number. Now we construct the augmented coefficient matrix of Equations (22)–(25):

$$
\begin{array}{c c c c c c c}
x_1 & x_2 & s_1 & s_2 & s_3 & t & W
\end{array}
$$
$$
\left[
\begin{array}{c c c c c c c | c}
1 & 1 & 1 & 0 & 0 & 0 & 0 & 12 \\
1 & 2 & 0 & 1 & 0 & 0 & 0 & 20 \\
-1 & 1 & 0 & 0 & -1 & 1 & 0 & 2 \\
\hline
-2 & -1 & 0 & 0 & 0 & M & 1 & 0
\end{array}
\right]
$$

To get simplex table I, we replace the M in the artificial variable column by zero by adding $-M$ times row 3 to row 4:

SIMPLEX TABLE I

entering
variable
↓

	B	x_1	x_2	s_1	s_2	s_3	t	W	R	Quotients
	s_1	1	1	1	0	0	0	0	12	$12 \div 1 = 12$
	s_2	1	2	0	1	0	0	0	20	$20 \div 2 = 10$
departing ←	t	-1	1	0	0	-1	1	0	2	$2 \div 1 = 2$
variable	W	$-2 + M$	$-1 - M$	0	0	M	0	1	$-2M$	

indicators

The variables s_1, s_2, and t in column B—that is, the basic variables—are the nondecision variables with positive coefficients in Equations (22)–(24). Since M is a large positive number, $-1 - M$ is the most negative indicator. The entering variable is x_2, the departing variable is t, and the pivot entry is shaded. Proceeding, we get table II:

SIMPLEX TABLE II

entering
variable
↓

	B	x_1	x_2	s_1	s_2	s_3	t	W	R	Quotients
departing ←	s_1	2	0	1	0	1	-1	0	10	$10 \div 2 = 5$
variable	s_2	3	0	0	1	2	-2	0	16	$16 \div 3 = 5\frac{1}{3}$
	x_2	-1	1	0	0	-1	1	0	2	
	W	-3	0	0	0	-1	$1 + M$	1	2	

indicators

The BFS corresponding to table II has $t = 0$. Thus, we delete the t-column and change W's to Z's in succeeding tables. Continuing, we obtain table III:

SIMPLEX TABLE III

	B	x_1	x_2	s_1	s_2	s_3	Z	R
	x_1	1	0	$\frac{1}{2}$	0	$\frac{1}{2}$	0	5
	s_2	0	0	$-\frac{3}{2}$	1	$\frac{1}{2}$	0	1
	x_2	0	1	$\frac{1}{2}$	0	$-\frac{1}{2}$	0	7
	Z	0	0	$\frac{3}{2}$	0	$\frac{1}{2}$	1	17

indicators

All indicators are nonnegative. Hence, the maximum value of Z is 17. It occurs when $x_1 = 5$ and $x_2 = 7$.

NOW WORK PROBLEM 1

Equality Constraints

When an *equality* constraint of the form

$$a_1 x_1 + a_2 x_2 + \cdots + a_n x_n = b \quad \text{where } b \geq 0$$

occurs in a linear programming problem, artificial variables are used in the simplex method. To illustrate, consider the following problem:

$$\text{Maximize } Z = x_1 + 3x_2 - 2x_3$$

subject to

$$x_1 + x_2 - x_3 = 6 \tag{26}$$

$$x_1, x_2, x_3 \geq 0$$

Constraint (26) is already expressed as an equation, so no slack variable is necessary. Since $x_1 = x_2 = x_3 = 0$ is not a feasible solution, we do not have an obvious starting point for the simplex procedure. Thus, we create an artificial problem by first adding an artificial variable t to the left side of Equation (26):

$$x_1 + x_2 - x_3 + t = 6$$

Here an obvious BFS is $x_1 = x_2 = x_3 = 0, t = 6$. The artificial objective function is

$$W = Z - Mt = x_1 + 3x_2 - 2x_3 - Mt$$

where M is a large positive number. The simplex procedure is applied to this artificial problem until we obtain a BFS in which $t = 0$. This solution will give an initial BFS for the original problem, and we then proceed as before.

In general, the simplex method can be used to

$$\text{maximize } Z = c_1x_1 + c_2x_2 + \cdots + c_nx_n$$

subject to

$$\left.\begin{array}{l} a_{11}x_1 + a_{12}x_2 + \cdots + a_{1n}x_n\{\leq, \geq, =\}\ b_1 \\ a_{21}x_1 + a_{22}x_2 + \cdots + a_{2n}x_n\{\leq, \geq, =\}\ b_2 \\ \quad\vdots \qquad \vdots \qquad\qquad \vdots \qquad \vdots \\ a_{m1}x_1 + a_{m2}x_2 + \cdots + a_{mn}x_n\{\leq, \geq, =\}\ b_m \end{array}\right\} \tag{27}$$

and $x_1 \geq 0$, $x_2 \geq 0$, ..., $x_n \geq 0$. The symbolism $\{\leq, \geq, =\}$ means that one of the relations "$\leq$", "$\geq$", or "$=$" exists for a constraint.

For each $b_i < 0$, multiply the corresponding inequality by -1 (thus changing the sense of the inequality). If, with all $b_i \geq 0$, all constraints involve "$\leq$", the problem is of standard form and the simplex techniques of the previous sections apply directly. If, *with all $b_i \geq 0$*, any constraint involves "$\geq$" or "$=$", we begin with an artificial problem, which is obtained as follows.

Each constraint that contains "$\leq$" is written as an equation involving a slack variable s_i (with coefficient $+1$):

$$a_{i1}x_1 + a_{i2}x_2 + \cdots + a_{in}x_n + s_i = b_i$$

Each constraint that contains "$\geq$" is written as an equation involving a surplus variable s_j (with coefficient -1) and an artificial variable t_j:

$$a_{j1}x_1 + a_{j2}x_2 + \cdots + a_{jn}x_n - s_j + t_j = b_j$$

Each constraint that contains "$=$" is rewritten as an equation with an artificial variable t_k inserted:

$$a_{k1}x_1 + a_{k2}x_2 + \cdots + a_{kn}x_n + t_k = b_k$$

Should the artificial variables involved in this problem be, for example, t_1, t_2, and t_3, then the artificial objective function is

$$W = Z - Mt_1 - Mt_2 - Mt_3$$

where M is a large positive number. An initial BFS occurs when $x_1 = x_2 = \cdots = x_n = 0$ and each *surplus* variable equals 0.

After obtaining an initial simplex table, we apply the simplex procedure until we arrive at a table that corresponds to a BFS in which *all* artificial variables are 0. We then delete the artificial variable columns, change W's to Z's, and continue by using the procedures of the previous sections.

EXAMPLE 2   An Equality Constraint

Use the simplex method to maximize $Z = x_1 + 3x_2 - 2x_3$ subject to

$$-x_1 - 2x_2 - 2x_3 = -6 \tag{28}$$

$$-x_1 - x_2 + x_3 \leq -2 \tag{29}$$

$$x_1, x_2, x_3 \geq 0 \tag{30}$$

Solution: Constraints (28) and (29) will have the forms indicated in (27) (that is, b's positive) if we multiply both sides of each constraint by -1:

$$x_1 + 2x_2 + 2x_3 = 6 \tag{31}$$

$$x_1 + x_2 - x_3 \geq 2 \tag{32}$$

Since constraints (31) and (32) involve "=" and "$\geq$", two artificial variables, t_1 and t_2, will occur. The equations for the artificial problem are

$$x_1 + 2x_2 + 2x_3 \qquad + t_1 \quad = 6 \tag{33}$$

and

$$x_1 + x_2 - x_3 - s_2 \qquad + t_2 = 2 \tag{34}$$

Here the subscript 2 on s_2 reflects the order of the equations. The artificial objective function is $W = Z - Mt_1 - Mt_2$, equivalently,

$$-x_1 - 3x_2 + 2x_3 + Mt_1 + Mt_2 + W = 0 \tag{35}$$

where M is a large positive number. The augmented coefficient matrix of Equations (33)–(35) is

$$
\begin{array}{ccccccc}
x_1 & x_2 & x_3 & s_2 & t_1 & t_2 & W \\
\end{array}
$$
$$
\left[
\begin{array}{ccccccc|c}
1 & 2 & 2 & 0 & 1 & 0 & 0 & 6 \\
1 & 1 & -1 & -1 & 0 & 1 & 0 & 2 \\
\hline
-1 & -3 & 2 & 0 & M & M & 1 & 0
\end{array}
\right]
$$

We now use elementary row operations to remove the bottom-row M's from *all* the artificial variable columns. By adding $-M$ times row 1 to row 3 and adding $-M$ times row 2 to row 3, we get initial simplex table I:

SIMPLEX TABLE I

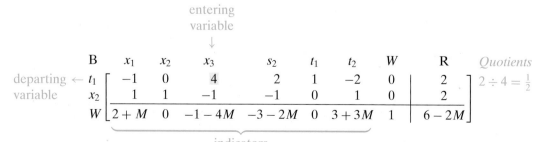

Proceeding, we obtain simplex tables II and III:

SIMPLEX TABLE II

SIMPLEX TABLE III

entering
variable
↓

	B	x_1	x_2	x_3	s_2	t_1	t_2	W	R	Quotients
departing ←	x_3	$-\frac{1}{4}$	0	1	$\frac{1}{2}$	$\frac{1}{4}$	$-\frac{1}{2}$	0	$\frac{1}{2}$	$\frac{1}{2} \div \frac{1}{2} = 1$
variable	x_2	$\frac{3}{4}$	1	0	$-\frac{1}{2}$	$\frac{1}{4}$	$\frac{1}{2}$	0	$\frac{5}{2}$	
	W	$\frac{7}{4}$	0	0	$-\frac{5}{2}$	$\frac{1}{4} + M$	$\frac{5}{2} + M$	1	$\frac{13}{2}$	

indicators

For the BFS corresponding to table III, the artificial variables t_1 and t_2 are both 0. We now can delete the t_1- and t_2-columns and change W's to Z's. Continuing, we obtain simplex table IV:

SIMPLEX TABLE IV

B	x_1	x_2	x_3	s_2	Z	R
s_2	$-\frac{1}{2}$	0	2	1	0	1
x_2	$\frac{1}{2}$	1	1	0	0	3
Z	$\frac{1}{2}$	0	5	0	1	9

indicators

Since all indicators are nonnegative, we have reached the final table. The maximum value of Z is 9, and it occurs when $x_1 = 0$, $x_2 = 3$, and $x_3 = 0$.

NOW WORK PROBLEM 5 ●●●

Empty Feasible Regions

It is possible that the simplex procedure terminates and not all artificial variables are 0. It can be shown that in this situation *the feasible region of the original problem is empty* and, hence, there is *no optimum solution*. The following example will illustrate.

● EXAMPLE 3 **An Empty Feasible Region**

Use the simplex method to maximize $Z = 2x_1 + x_2$ subject to

$$-x_1 + x_2 \geq 2 \tag{36}$$

$$x_1 + x_2 \leq 1,$$

$$x_1, x_2 \geq 0 \tag{37}$$

and $x_1, x_2 \geq 0$.

Solution: Since constraint (36) is of the form $a_{11}x_1 + a_{12}x_2 \geq b_1$, where $b_1 \geq 0$, an artificial variable will occur. The equations to consider are

$$-x_1 + x_2 - s_1 \quad + t_1 = 2 \tag{38}$$

and

$$x_1 + x_2 \quad + s_2 = 1 \tag{39}$$

where s_1 is a surplus variable, s_2 is a slack variable, and t_1 is artificial. The artificial objective function is $W = Z - Mt_1$, equivalently,

$$-2x_1 - x_2 + Mt_1 + W = 0 \tag{40}$$

The augmented coefficient matrix of Equations (38)–(40) is

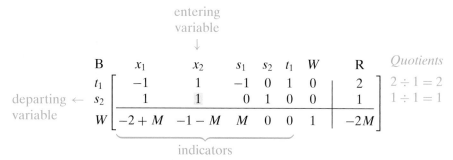

$$\begin{array}{cccccc} x_1 & x_2 & s_1 & s_2 & t_1 & W \\ \end{array}$$
$$\begin{bmatrix} -1 & 1 & -1 & 0 & 1 & 0 & 2 \\ 1 & 1 & 0 & 1 & 0 & 0 & 1 \\ \hline -2 & -1 & 0 & 0 & M & 1 & 0 \end{bmatrix}$$

The simplex tables are as follows:

SIMPLEX TABLE I

entering
variable
↓

B	x_1	x_2	s_1	s_2	t_1	W	R	Quotients
t_1	-1	1	-1	0	1	0	2	$2 \div 1 = 2$
s_2	1	1	0	1	0	0	1	$1 \div 1 = 1$
W	$-2 + M$	$-1 - M$	M	0	0	1	$-2M$	

departing ← s_2
variable

indicators

SIMPLEX TABLE II

B	x_1	x_2	s_1	s_2	t_1	W	R
t_1	-2	0	-1	-1	1	0	1
x_2	1	1	0	1	0	0	1
W	$-1 + 2M$	0	M	$1 + M$	0	1	$1 - M$

indicators

Since M is a large positive number, the indicators in simplex table II are nonnegative, so the simplex procedure terminates. The value of the artificial variable t_1 is 1. Therefore, as previously stated, the feasible region of the original problem is empty and, hence, no solution exists. This result can be obtained geometrically. Figure 7.25 shows the graphs of $-x_1 + x_2 = 2$ and $x_1 + x_2 = 1$ for $x_1, x_2 \geq 0$. Since there is no point (x_1, x_2) that simultaneously lies above $-x_1 + x_2 = 2$ and below $x_1 + x_2 = 1$ such that $x_1, x_2 \geq 0$, the feasible region is empty and, thus, no solution exists.

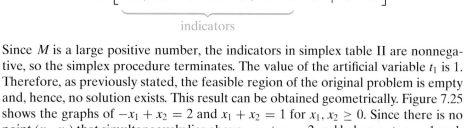

NOW WORK PROBLEM 9

FIGURE 7.25 Empty feasible region (no solution exists).

In the next section we will use the simplex method on minimization problems.

Problems 7.6

Use the simplex method to solve the following problems.

*1. Maximize
$$Z = 2x_1 + x_2$$
subject to
$$x_1 + x_2 \leq 6$$
$$-x_1 + x_2 \geq 4$$
$$x_1, x_2 \geq 0$$

2. Maximize
$$Z = 3x_1 + 4x_2$$
subject to
$$x_1 + 2x_2 \leq 8$$
$$x_1 + 6x_2 \geq 12$$
$$x_1, x_2 \geq 0$$

3. Maximize
$$Z = 2x_1 + x_2 - x_3$$
subject to
$$x_1 + 2x_2 + x_3 \leq 5$$
$$-x_1 + x_2 + x_3 \geq 1$$
$$x_1, x_2, x_3 \geq 0$$

4. Maximize
$$Z = x_1 - x_2 + 4x_3$$
subject to
$$x_1 + x_2 + x_3 \leq 9$$
$$x_1 - 2x_2 + x_3 \geq 6$$
$$x_1, x_2, x_3 \geq 0$$

*5. Maximize

$$Z = 3x_1 + 2x_2 + x_3$$

subject to

$$x_1 + x_2 + x_3 \leq 10$$
$$x_1 - x_2 - x_3 = 6$$
$$x_1, x_2, x_3 \geq 0$$

6. Maximize

$$Z = 2x_1 + x_2 + 3x_3$$

subject to

$$x_2 - 2x_3 \geq 5$$
$$x_1 + x_2 + x_3 = 7$$
$$x_1, x_2, x_3 \geq 0$$

7. Maximize

$$Z = x_1 - 10x_2$$

subject to

$$x_1 - x_2 \leq 1$$
$$x_1 + 2x_2 \leq 8$$
$$x_1 + x_2 \geq 5$$
$$x_1, x_2 \geq 0$$

8. Maximize

$$Z = x_1 + 4x_2 - x_3$$

subject to

$$x_1 + x_2 - x_3 \geq 5$$
$$x_1 + x_2 + x_3 \leq 3$$
$$x_1 - x_2 + x_3 = 7$$
$$x_1, x_2, x_3 \geq 0$$

*9. Maximize

$$Z = 3x_1 - 2x_2 + x_3$$

subject to

$$x_1 + x_2 + x_3 \leq 1$$
$$x_1 - x_2 + x_3 \geq 2$$
$$x_1 - x_2 - x_3 \leq -6$$
$$x_1, x_2, x_3 \geq 0$$

10. Maximize

$$Z = x_1 + 4x_2$$

subject to

$$x_1 + 2x_2 \leq 8$$
$$x_1 + 6x_2 \geq 12$$
$$x_2 \geq 2$$
$$x_1, x_2 \geq 0$$

11. Maximize

$$Z = -3x_1 + 2x_2$$

subject to

$$x_1 - x_2 \leq 4$$
$$-x_1 + x_2 = 4$$
$$x_1 \geq 6$$
$$x_1, x_2 \geq 0$$

12. Maximize

$$Z = 2x_1 - 8x_2$$

subject to

$$x_1 - 2x_2 \geq -12$$
$$-x_1 + x_2 \geq 2$$
$$x_1 + x_2 \geq 10$$
$$x_1, x_2 \geq 0$$

13. **Production**   A company manufactures two types of bookcases: Standard and Executive. Each type requires assembly and finishing times as given in the following table:

	Assembly Time	Finishing Time	Profit per Unit
Standard	2 hr	3 hr	$35
Executive	3 hr	4 hr	$40

The profit on each unit is also indicated. The number of hours available per week in the assembly department is 400, and in the finishing department it is 500. Because of a union contract, the finishing department is guaranteed at least 250 hours of work per week. How many units of each type should the company produce each week to maximize profit?

14. **Production**   A company manufactures three products: X, Y, and Z. Each product requires the use of time on machines A and B as given in the following table:

	Machine A	Machine B
Product X	1 hr	1 hr
Product Y	2 hr	1 hr
Product Z	2 hr	2 hr

The numbers of hours per week that A and B are available for production are 40 and 30, respectively. The profit per unit on X, Y, and Z is $50, $60, and $75, respectively. At least five units of Z must be produced next week. What should be the production order for that period if maximum profit is to be achieved? What is the maximum profit?

15. **Investments**   The prospectus of an investment fund states that all money is invested in bonds that are rated A, AA, and AAA; no more than 30% of the total investment is in A and AA bonds, and at least 50% is in AA and AAA bonds. The A, AA, and AAA bonds, respectively, yield 8%, 7%, and 6% annually. Determine the percentages of the total investment that should be committed to each type of bond so that the fund maximizes its annual yield. What is this yield?

7.7 Minimization

So far we have used the simplex method to *maximize* objective functions. In general, to *minimize* a function it suffices to maximize the negative of the function. To understand why, consider the function $f(x) = x^2 - 4$. In Figure 7.26(a), observe that the minimum value of f is -4, and it occurs when $x = 0$. Figure 7.26(b) shows the graph of $g(x) = -f(x) = -(x^2 - 4)$. This graph is the reflection through the x-axis of the graph of f. Notice that the maximum value of g is 4 and occurs when $x = 0$. Thus the minimum value of $x^2 - 4$ is the negative of the maximum value of $-(x^2 - 4)$. That is,

$$\min f = -\max(-f)$$

Alternatively, think of a point C on the positive half of the number line moving to the left. As it does so, the point $-C$ moves to the right. It is clear that if, for some reason, C stops, then it stops at the minimum value that C encounters. If C stops, then so does $-C$, at the maximum value encountered by $-C$. Since this value of $-C$ is still the negative of the value of C, we see that

$$\min C = -\max(-C)$$

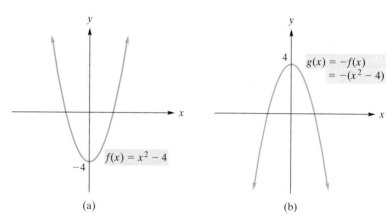

FIGURE 7.26   Minimum value of $f(x)$ is equal to the negative of the maximum value of $-f(x)$.

The problem in Example 1 will be solved more efficiently in Example 4 of Section 7.8.

⬤ **EXAMPLE 1   Minimization**

Use the simplex method to minimize $Z = x_1 + 2x_2$ subject to

$$-2x_1 + x_2 \geq 1 \tag{1}$$

$$-x_1 + x_2 \geq 2 \tag{2}$$

$$x_1, x_2 \geq 0 \tag{3}$$

Solution: To minimize Z, we can maximize $-Z = -x_1 - 2x_2$. Note that constraints (1) and (2) each have the form $a_1x_1 + a_2x_2 \geq b$, where $b \geq 0$. Thus, their equations involve two surplus variables s_1 and s_2, each with coefficient -1, and two artificial variables t_1 and t_2:

$$-2x_1 + x_2 - s_1 + t_1 = 1 \tag{4}$$

$$-x_1 + x_2 - s_2 + t_2 = 2 \tag{5}$$

Since there are *two* artificial variables, we maximize the objective function

$$W = (-Z) - Mt_1 - Mt_2$$

where M is a large positive number. Equivalently,

$$x_1 + 2x_2 + Mt_1 + Mt_2 + W = 0 \tag{6}$$

The augmented coefficient matrix of Equations (4)–(6) is

$$
\begin{array}{ccccccc}
x_1 & x_2 & s_1 & s_2 & t_1 & t_2 & W \\
\end{array}
$$

$$
\left[
\begin{array}{ccccccc|c}
-2 & 1 & -1 & 0 & 1 & 0 & 0 & 1 \\
-1 & 1 & 0 & -1 & 0 & 1 & 0 & 2 \\
\hline
1 & 2 & 0 & 0 & M & M & 1 & 0 \\
\end{array}
\right]
$$

Proceeding, we obtain simplex tables I, II, III:

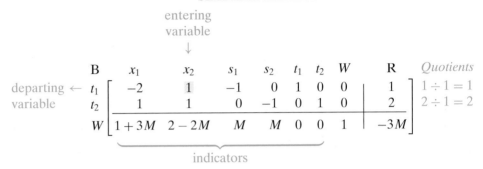

SIMPLEX TABLE I

entering
variable
↓

	B	x_1	x_2	s_1	s_2	t_1	t_2	W	R		Quotients
departing ←	t_1	-2	1	-1	0	1	0	0	1		$1 \div 1 = 1$
variable	t_2	1	1	0	-1	0	1	0	2		$2 \div 1 = 2$
	W	$1+3M$	$2-2M$	M	M	0	0	1	$-3M$		

indicators

SIMPLEX TABLE II

entering
variable
↓

	B	x_1	x_2	s_1	s_2	t_1	t_2	W	R		Quotients
	x_2	-2	1	-1	0	1	0	0	1		
departing ←	t_2	-1	0	1	-1	-1	1	0	1		$1 \div 1 = 1$
variable	W	$5-M$	0	$2-M$	M	$-2+2M$	0	1	$-2-M$		

indicators

SIMPLEX TABLE III

	B	x_1	x_2	s_1	s_2	t_1	t_2	W	R
	x_2	-1	1	0	-1	0	1	0	2
	s_1	-1	0	1	-1	-1	1	0	1
	W	3	0	0	2	M	$-2+M$	1	-4

indicators

The BFS corresponding to table III has both artificial variables equal to 0. Thus, the t_1- and t_2-columns are no longer needed. However, the indicators in the x_1-, x_2-, s_1-, and s_2-columns are nonnegative and, hence, an optimum solution has been reached. Since $W = -Z$ when $t_1 = t_2 = 0$, the maximum value of $-Z$ is -4. Consequently, the *minimum* value of Z is $-(-4) = 4$. It occurs when $x_1 = 0$ and $x_2 = 2$.

NOW WORK PROBLEM 1 ●●●

Here is an interesting example dealing with environmental controls.

●EXAMPLE 2   Reducing Dust Emissions

A cement plant produces 2,500,000 barrels of cement per year. The kilns emit 2 lb of dust for each barrel produced. A governmental agency dealing with environmental protection requires that the plant reduce its dust emissions to no more than 800,000 lb per year. There are two emission control devices available, A and B. Device A reduces emissions to $\frac{1}{2}$ lb per barrel, and its cost is \$0.20 per barrel of cement produced. For device B, emissions are reduced to $\frac{1}{5}$ lb per barrel, and the cost is \$0.25 per barrel of cement produced. Determine the most economical course of action that the plant *should take so that it complies with the agency's requirement and also maintains its annual production of 2,500,000 barrels of cement.*[5]

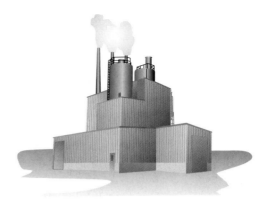

Solution: We must minimize the annual cost of emission control. Let x_1, x_2, and x_3 be the annual numbers of barrels of cement produced in kilns that use device A, device B, and no device, respectively. Then $x_1, x_2, x_3 \geq 0$, and the annual emission control cost (in dollars) is

$$C = \tfrac{1}{5}x_1 + \tfrac{1}{4}x_2 + 0x_3 \tag{7}$$

Since 2,500,000 barrels of cement are produced each year,

$$x_1 + x_2 + x_3 = 2{,}500{,}000 \tag{8}$$

The numbers of pounds of dust emitted annually by the kilns that use device A, device B, and no device are $\tfrac{1}{2}x_1$, $\tfrac{1}{5}x_2$, and $2x_3$, respectively. Since the total number of pounds of dust emission is to be no more than 800,000,

$$\tfrac{1}{2}x_1 + \tfrac{1}{5}x_2 + 2x_3 \leq 800{,}000 \tag{9}$$

To minimize C subject to constraints (8) and (9), where $x_1, x_2, x_3 \geq 0$, we first maximize $-C$ by using the simplex method. The equations to consider are

$$x_1 + x_2 + x_3 + t_1 = 2{,}500{,}000 \tag{10}$$

and

$$\frac{1}{2}x_1 + \frac{1}{5}x_2 + 2x_3 + s_2 = 800{,}000 \tag{11}$$

where t_1 and s_2 are artificial and slack variables, respectively. The artificial objective equation is $W = (-C) - Mt_1$, equivalently,

$$\tfrac{1}{5}x_1 + \tfrac{1}{4}x_2 + 0x_3 + Mt_1 + W = 0 \tag{12}$$

where M is a large positive number. The augmented coefficient matrix of Equations (10)–(12) is

	x_1	x_2	x_3	s_2	t_1	W	
	1	1	1	0	1	0	$2{,}500{,}000$
	$\tfrac{1}{2}$	$\tfrac{1}{5}$	2	1	0	0	$800{,}000$
	$\tfrac{1}{5}$	$\tfrac{1}{4}$	0	0	M	1	0

---

[5]This example is adapted from Robert E. Kohn, "A Mathematical Model for Air Pollution Control," *School Science and Mathematics*, 69 (1969), 487–94.

Once we determine our initial simplex table, we proceed and obtain (after three additional simplex tables) our final table:

$$
\begin{array}{c|ccccc|c}
\text{B} & x_1 & x_2 & x_3 & S_2 & -C & \text{R} \\
\hline
x_2 & 0 & 1 & -5 & -\frac{10}{3} & 0 & 1{,}500{,}000 \\
x_1 & 1 & 0 & 6 & \frac{10}{3} & 0 & 1{,}000{,}000 \\
\hline
-C & 0 & 0 & \frac{1}{20} & \frac{1}{6} & 1 & -575{,}000 \\
\end{array}
$$

indicators

Notice that W is replaced by $-C$ when $t_1 = 0$. The maximum value of $-C$ is $-575{,}000$ and occurs when $x_1 = 1{,}000{,}000$, $x_2 = 1{,}500{,}000$, and $x_3 = 0$. Thus, the *minimum* annual cost of the emission control is $-(-575{,}000) = \$575{,}000$. Device A should be installed on kilns producing 1,000,000 barrels of cement annually, and device B should be installed on kilns producing 1,500,000 barrels annually.

NOW WORK PROBLEM 11

Problems 7.7

Use the simplex method to solve the following problems.

*1. Minimize
$$Z = 2x_1 + 5x_2$$
subject to
$$x_1 - x_2 \geq 7$$
$$2x_1 + x_2 \geq 9$$
$$x_1, x_2 \geq 0$$

2. Minimize
$$Z = 8x_1 + 12x_2$$
subject to
$$2x_1 + 2x_2 \geq 1$$
$$x_1 + 3x_2 \geq 2$$
$$x_1, x_2 \geq 0$$

3. Minimize
$$Z = 12x_1 + 6x_2 + 3x_3$$
subject to
$$x_1 - x_2 - x_3 \geq 18$$
$$x_1, x_2, x_3 \geq 0$$

4. Minimize
$$Z = x_1 + x_2 + 2x_3$$
subject to
$$x_1 + 2x_2 - x_3 \geq 4$$
$$x_1, x_2, x_3 \geq 0$$

5. Minimize
$$Z = 2x_1 + 3x_2 + x_3$$

subject to
$$x_1 + x_2 + x_3 \leq 6$$
$$x_1 - x_3 \leq -4$$
$$x_2 + x_3 \leq 5$$
$$x_1, x_2, x_3 \geq 0$$

6. Minimize
$$Z = 5x_1 + x_2 + 3x_3$$
subject to
$$3x_1 + x_2 - x_3 \leq 4$$
$$2x_1 + 2x_3 \leq 5$$
$$x_1 + x_2 + x_3 \geq 2$$
$$x_1, x_2, x_3 \geq 0$$

7. Minimize
$$Z = x_1 - x_2 - 3x_3$$
subject to
$$x_1 + 2x_2 + x_3 = 4$$
$$x_2 + x_3 = 1$$
$$x_1 + x_2 \leq 6$$
$$x_1, x_2, x_3 \geq 0$$

8. Minimize
$$Z = x_1 - x_2$$
subject to
$$-x_1 + x_2 \geq 4$$
$$x_1 + x_2 = 1$$
$$x_1, x_2 \geq 0$$

9. Minimize

$$Z = x_1 + 8x_2 + 5x_3$$

subject to

$$x_1 + x_2 + x_3 \geq 8$$

$$-x_1 + 2x_2 + x_3 \geq 2$$

$$x_1, x_2, x_3 \geq 0$$

10. Minimize

$$Z = 4x_1 + 4x_2 + 6x_3$$

subject to

$$x_1 - x_2 - x_3 \leq 3$$

$$x_1 - x_2 + x_3 \geq 3$$

$$x_1, x_2, x_3 \geq 0$$

*11. **Emission Control** A cement plant produces 3,300,000 barrels of cement per year. The kilns emit 2 lb of dust for each barrel produced. The plant must reduce its dust emissions to no more than 1,000,000 lb per year. There are two devices available, A and B, that will control emissions. Device A will reduce emissions to $\frac{1}{2}$ lb per barrel, and the cost is \$0.25 per barrel of cement produced. For device B, emissions are reduced to $\frac{1}{4}$ lb per barrel, and the cost is \$0.40 per barrel of cement produced. Determine the most economical course of action the plant should take so that it maintains an annual production of exactly 3,300,000 barrels of cement.

12. **Delivery Truck Scheduling** Because of increased business, a catering service finds that it must rent additional delivery trucks. The minimum needs are 12 units each of refrigerated and nonrefrigerated space. Two standard types of trucks are available in the rental market. Type A has 2 units of refrigerated space and 1 unit of nonrefrigerated space. Type B has 2 units of refrigerated space and 3 units of nonrefrigerated space. The costs per mile are \$0.40 for A and \$0.60 for B. How many of each type of truck should be rented so as to minimize total cost per mile? What is the minimum total cost per mile?

13. **Transportation Costs** A retailer has stores in Columbus and Dayton and has warehouses in Akron and Springfield. Each store requires delivery of exactly 150 DVD players. In the Akron warehouse there are 200 DVD players, and in the Springfield warehouse there are 150.

The transportation costs to ship DVD players from the warehouses to the stores are given in the following table:

	Columbus	Dayton
Akron	\$5	\$7
Springfield	\$3	\$2

For example, the cost to ship a DVD player from Akron to the Columbus store is \$5. How should the retailer order the DVD players so that the requirements of the stores are met and the total transportation costs are minimized? What is the minimum transportation cost?

14. **Parts Purchasing** An auto manufacturer purchases alternators from two suppliers, X and Y. The manufacturer has two plants, A and B, and requires delivery of exactly 7000 alternators to plant A and exactly 5000 to plant B. Supplier X charges \$300 and \$320 per alternator (including transportation cost) to A and B, respectively. For these prices, X requires that the auto manufacturer order at least a total of 3000 alternators. However, X can supply no more than 5000 alternators. Supplier Y charges \$340 and \$280 per alternator to A and B, respectively, and requires a minimum order of 7000 alternators. Determine how the manufacturer should order the necessary alternators so that his total cost is a minimum. What is this minimum cost?

15. **Producing Wrapping Paper** A paper company stocks its holiday wrapping paper in 48-in.-wide rolls, called stock rolls, and cuts such rolls into smaller widths, depending on customers' orders. Suppose that an order for 50 rolls of 15-in.-wide paper and 60 rolls of 10-in.-wide paper is received. From a stock roll, the company can cut three 15-in.-wide rolls and one 3-in.-wide roll. (See Figure 7.27.) Since the 3-in.-wide roll cannot be used in the order, 3 in. is called the trim loss for this roll.

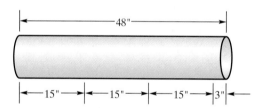

FIGURE 7.27 Diagram for Problem 15.

Similarly, from a stock roll, two 15-in.-wide rolls, one 10-in.-wide roll, and one 8-in.-wide roll could be cut. Here the trim loss would be 8 in. The following table indicates the number of 15-in. and 10-in. rolls, together with trim loss, that can be cut from a stock roll:

Roll width	15 in.	3	2	1	—
	10 in.	0	1	—	—
Trim loss			3	8	—	—

(a) Complete the last two columns of the table. (b) Assume that the company has a sufficient number of stock rolls to fill the order and that *at least* 50 rolls of 15-in.-wide and *at least* 60 rolls of 10-in.-wide wrapping paper will be cut. If x_1, x_2, x_3, and x_4 are the numbers of stock rolls that are cut in a manner described by columns 1–4 of the table, respectively, determine the values of the x's so that the total trim loss is minimized. (c) What is the minimum amount of total trim loss?

OBJECTIVE

To first motivate and then formally define the dual of a linear programming problem.

7.8 The Dual

There is a fundamental principle, called *duality*, that allows us to solve a maximization problem by solving a related minimization problem. Let us illustrate.

TABLE 7.2

	Machine A	**Machine B**	**Profit/Unit**
Manual	1 hr	1 hr	$10
Electric	2 hr	4 hr	$24
Hours available	120	180	

Suppose that a company produces two types of garden shears, manual and electric, and each requires the use of machines A and B in its production. Table 7.2 indicates that manual shears require the use of A for 1 hour and B for 1 hour. Electric shears require A for 2 hours and B for 4 hours. The maximum numbers of hours available per month for machines A and B are 120 and 180, respectively. The profit on manual shears is $10, and on an electric shears it is $24. Assuming that the company can sell all the shears it can produce, we will determine the maximum monthly profit. If x_1 and x_2 are the numbers of manual and electric shears produced per month, respectively, then we want to maximize the monthly profit function

$$P = 10x_1 + 24x_2$$

subject to

$$x_1 + 2x_2 \leq 120 \qquad (1)$$

$$x_1 + 4x_2 \leq 180 \qquad (2)$$

$$x_1, x_2 \geq 0$$

Writing constraints (1) and (2) as equations, we have

$$x_1 + 2x_2 + s_1 = 120 \qquad (3)$$

and

$$x_1 + 4x_2 + s_2 = 180$$

where s_1 and s_2 are slack variables. In Equation (3), $x_1 + 2x_2$ is the number of hours that machine A is used. Since 120 hours on A are available, s_1 is the number of available hours that are *not* used for production. That is, s_1 represents unused capacity (in hours) for A. Similarly, s_2 represents unused capacity for B. Solving this problem by the simplex method, we find that the final table is

$$
\begin{array}{c|ccccc|c}
B & x_1 & x_2 & s_1 & s_2 & P & R \\
\hline
x_1 & 1 & 0 & 2 & -1 & 0 & 60 \\
x_2 & 0 & 1 & -\frac{1}{2} & \frac{1}{2} & 0 & 30 \\
\hline
P & 0 & 0 & 8 & 2 & 1 & 1320
\end{array}
\qquad (4)
$$

indicators

Thus, the maximum profit per month is $1320, which occurs when $x_1 = 60$ and $x_2 = 30$.

Now let us look at the situation from a different point of view. Suppose that the company wishes to rent out machines A and B. What is the minimum monthly rental fee they should charge? Certainly, if the charge is too high, no one would rent the machines. On the other hand, if the charge is too low, it may not pay the company to rent them at all. Obviously, the minimum rent should be $1320. That is, the minimum the company should charge is the profit it could make by using the machines itself. We can arrive at this minimum rental fee directly by solving a linear programming problem.

Let F be the total monthly rental fee. To determine F, suppose the company assigns values or "worths" to each hour of capacity on machines A and B. Let these worths be y_1 and y_2 dollars, respectively, where $y_1, y_2 \geq 0$. Then the monthly worth of machine A is $120y_1$, and for B it is $180y_2$. Thus,

$$F = 120y_1 + 180y_2$$

The total worth of machine time to produce a set of manual shears is $1y_1 + 1y_2$. This should be at least equal to the \$10 profit the company can earn by producing those shears. If not, the company would make more money by using the machine time to produce a set of manual shears. Accordingly,

$$1y_1 + 1y_2 \geq 10$$

Similarly, the total worth of machine time to produce a set of electric shears should be at least \$24:

$$2y_1 + 4y_2 \geq 24$$

Therefore, the company wants to

$$\text{minimize } F = 120y_1 + 180y_2$$

subject to

$$y_1 + y_2 \geq 10 \qquad (5)$$

$$2y_1 + 4y_2 \geq 24 \qquad (6)$$

$$y_1, y_2 \geq 0$$

To minimize F, we maximize $-F$. Since constraints (5) and (6) have the form $a_1y_1 + a_2y_2 \geq b$, where $b \geq 0$, we consider an artificial problem. If r_1 and r_2 are surplus variables and t_1 and t_2 are artificial variables, then we want to maximize

$$W = (-F) - Mt_1 - Mt_2$$

where M is a large positive number, such that

$$y_1 + y_2 - r_1 + t_1 = 10$$

$$2y_1 + 4y_2 - r_2 + t_2 = 24$$

and the y's, r's, and t's are nonnegative. The final simplex table for this problem (with the artificial variable columns deleted and W changed to $-F$) is

B	y_1	y_2	r_1	r_2	$-F$	R
y_1	1	0	-2	$\frac{1}{2}$	0	8
y_2	0	1	1	$-\frac{1}{2}$	0	2
$-F$	0	0	60	30	1	-1320

indicators

Since the maximum value of $-F$ is -1320, the *minimum* value of F is $-(-1320) = \$1320$ (as anticipated). It occurs when $y_1 = 8$ and $y_2 = 2$. We have therefore determined the optimum value of one linear programming problem (maximizing profit) by finding the optimum value of another problem (minimizing rental fee).

The values $y_1 = 8$ and $y_2 = 2$ could have been anticipated from the final table of the maximization problem. In (4), the indicator 8 in the s_1-column means that at the optimum level of production, if s_1 increases by one unit, then the profit P *decreases* by 8. That is, 1 unused hour of capacity on A decreases the maximum profit by \$8. Thus, 1 hour of capacity on A is worth \$8. We say that the **shadow price** of 1 hour of capacity on A is \$8. Now, recall that y_1 in the rental problem is the worth of 1 hour of capacity on A. Therefore, y_1 must equal 8 in the optimum solution for that problem. Similarly, since the indicator in the s_2-column is 2, the shadow price of 1 hour of capacity on B is \$2, which is the value of y_2 in the optimum solution of the rental problem.

Let us now analyze the structure of our two linear programming problems:

Maximize	Minimize
$P = 10x_1 + 24x_2$	$F = 120y_1 + 180y_2$
subject to	subject to
$\left.\begin{array}{l} x_1 + 2x_2 \leq 120 \\ x_1 + 4x_2 \leq 180 \end{array}\right\}$ (7)	$\left.\begin{array}{l} y_1 + y_2 \geq 10 \\ 2y_1 + 4y_2 \geq 24 \end{array}\right\}$ (8)
and $x_1, x_2 \geq 0.$	and $y_1, y_2 \geq 0.$

Note that in (7) the inequalities are all $\leq$, but in (8) they are all $\geq$. The coefficients of the objective function in the minimization problem are the constant terms in (7). The constant terms in (8) are the coefficients of the objective function of the maximization problem. The coefficients of the y_1's in (8) are the coefficients of x_1 and x_2 in the first constraint of (7); the coefficients of the y_2's in (8) are the coefficients of x_1 and x_2 in the second constraint of (7). The minimization problem is called the *dual* of the maximization problem, and vice versa.

In general, with any given linear programming problem, we can associate another linear programming problem called its **dual.** The given problem is called **primal.** If the primal is a maximization problem, then its dual is a minimization problem. Similarly, if the primal involves minimization, then the dual involves maximization.

Any primal maximization problem can be written in the form indicated in Table 7.3. Note that there are no restrictions on the b's.[6] The corresponding dual minimization problem can be written in the form indicated in Table 7.4. Similarly, any primal minimization problem can be put in the form of Table 7.4, and its dual is the maximization problem in Table 7.3.

TABLE 7.3 Primal (Dual)

Maximize $Z = c_1x_1 + c_2x_2 + \cdots + c_nx_n$
subject to

$$\left.\begin{array}{l} a_{11}x_1 + a_{12}x_2 + \cdots + a_{1n}x_n \leq b_1 \\ a_{21}x_1 + a_{22}x_2 + \cdots + a_{2n}x_n \leq b_2 \\ \quad\vdots \qquad\quad \vdots \qquad\qquad\quad \vdots \\ a_{m1}x_1 + a_{m2}x_2 + \cdots + a_{mn}x_n \leq b_m \end{array}\right\} \quad (9)$$

and $x_1, x_2, \ldots, x_n \geq 0$

TABLE 7.4 Dual (Primal)

Minimize $W = b_1y_1 + b_2y_2 + \cdots + b_my_m$
subject to

$$\left.\begin{array}{l} a_{11}y_1 + a_{21}y_2 + \cdots + a_{m1}y_m \geq c_1 \\ a_{12}y_1 + a_{22}y_2 + \cdots + a_{m2}y_m \geq c_2 \\ \quad\vdots \qquad\quad \vdots \qquad\qquad\quad \vdots \\ a_{1n}y_1 + a_{2n}y_2 + \cdots + a_{mn}y_m \geq c_n \end{array}\right\} \quad (10)$$

and $y_1, y_2, \ldots, y_m \geq 0$

---

[6]If an inequality constraint involves $\geq$, multiplying both sides by -1 yields an inequality involving $\leq$. If a constraint is an equality, it can be written in terms of two inequalities, one involving $\leq$ and one involving $\geq$.

Let us compare the primal and its dual in Tables 7.3 and 7.4. For convenience, when we refer to constraints, we will mean those in (9) or (10); we will not include the nonnegativity conditions. Observe that if all the constraints in the primal involve $\leq$ ($\geq$), then all the constraints in its dual involve $\geq$ ($\leq$). The coefficients in the dual's objective function are the constant terms in the primal's constraints. Similarly, the constant terms in the dual's constraints are the coefficients of the primal's objective function. The coefficient matrix of the left sides of the dual's constraints is the *transpose* of the coefficient matrix of the left sides of the primal's constraints. That is,

$$\begin{bmatrix} a_{11} & a_{12} & \cdots & a_{1n} \\ a_{21} & a_{22} & \cdots & a_{2n} \\ \cdot & \cdot & & \cdot \\ \cdot & \cdot & & \cdot \\ \cdot & \cdot & & \cdot \\ a_{m1} & a_{m2} & \cdots & a_{mn} \end{bmatrix}^{\mathrm{T}} = \begin{bmatrix} a_{11} & a_{21} & \cdots & a_{m1} \\ a_{12} & a_{22} & \cdots & a_{m2} \\ \cdot & \cdot & & \cdot \\ \cdot & \cdot & & \cdot \\ \cdot & \cdot & & \cdot \\ a_{1n} & a_{2n} & \cdots & a_{mn} \end{bmatrix}$$

If the primal involves n decision variables and m slack variables, then the dual involves m decision variables and n slack variables. It should be noted that the dual of the *dual* is the primal.

There is an important relationship between the primal and its dual:

If the primal has an optimum solution, then so does the dual, and the optimum value of the primal's objective function is the *same* as that of its dual.

Moreover, suppose that the primal's objective function is

$$Z = c_1 x_1 + c_2 x_2 + \cdots + c_n x_n$$

Then

If s_i is the slack variable associated with the ith constraint in the dual, then the indicator in the s_i-column of the final simplex table of the dual is the value of x_i in the optimum solution of the primal.

Thus, we can solve the primal by merely solving its dual. At times this is more convenient than solving the primal directly. The link between the primal and the dual can be expressed very succinctly using matrix notation. Let

$$\mathbf{C} = \begin{bmatrix} c_1 & c_2 & \cdots & c_n \end{bmatrix} \quad \text{and} \quad \mathbf{X} = \begin{bmatrix} x_1 \\ x_2 \\ \cdot \\ \cdot \\ \cdot \\ x_n \end{bmatrix}$$

Then the objective function of the primal problem can be written as

$$Z = \mathbf{CX}$$

Furthermore, if we write

$$\mathbf{A} = \begin{bmatrix} a_{11} & a_{12} & \cdots & a_{1n} \\ a_{21} & a_{22} & \cdots & a_{2n} \\ \cdot & \cdot & & \cdot \\ \cdot & \cdot & & \cdot \\ \cdot & \cdot & & \cdot \\ a_{m1} & a_{m2} & \cdots & a_{mn} \end{bmatrix} \quad \text{and} \quad \mathbf{B} = \begin{bmatrix} b_1 \\ b_2 \\ \cdot \\ \cdot \\ \cdot \\ b_m \end{bmatrix}$$

then the system of constraints for the primal problem becomes

$$\mathbf{AX} \leq \mathbf{B} \quad \text{and} \quad \mathbf{X} \geq \mathbf{0}$$

where we understand $\leq$ ($\geq$) between matrices of the same size to mean that the inequality holds for each pair of corresponding entries. Now let

$$\mathbf{Y} = \begin{bmatrix} y_1 \\ y_2 \\ \cdot \\ \cdot \\ \cdot \\ y_m \end{bmatrix}$$

The dual problem has objective function given by

$$W = \mathbf{B}^{\mathsf{T}}\mathbf{Y}$$

and its system of constraints is

$$\mathbf{A}^{\mathsf{T}}\mathbf{Y} \geq \mathbf{C}^{\mathsf{T}} \quad \text{and} \quad \mathbf{Y} \geq \mathbf{0}$$

PRINCIPLES IN PRACTICE 1

FINDING THE DUAL OF A
MAXIMIZATION PROBLEM

Find the dual of the following problem: Suppose that the What If Company has $60,000 for the purchase of materials to make three types of gadgets. The company has allocated a total of 2000 hours of assembly time and 120 hours of packaging time for the gadgets. The following table gives the cost per gadget, the number of hours per gadget, and the profit per gadget for each type:

	Type 1	Type 2	Type 3
Cost/ Gadget	$300	$220	$180
Assembly Hours/ Gadget	20	40	20
Packaging Hours/ Gadget	3	1	2
Profit	$300	$200	$200

● EXAMPLE 1 Finding the Dual of a Maximization Problem

Find the dual of the following:

$$\text{Maximize } Z = 3x_1 + 4x_2 + 2x_3$$

subject to

$$x_1 + 2x_2 + 0x_3 \leq 10$$
$$2x_1 + 2x_2 + x_3 \leq 10$$

and $x_1, x_2, x_3 \geq 0$

Solution: The primal is of the form of Table 7.3. Thus, the dual is

$$\text{minimize } W = 10y_1 + 10y_2$$

subject to

$$y_1 + 2y_2 \geq 3$$
$$2y_1 + 2y_2 \geq 4$$
$$0y_1 + y_2 \geq 2$$

and $y_1, y_2 \geq 0$

NOW WORK PROBLEM 1 ●●●

PRINCIPLES IN PRACTICE 2

FINDING THE DUAL OF A
MINIMIZATION PROBLEM

Find the dual of the following problem:
A person decides to take two different dietary supplements. Each supplement contains two essential ingredients, A and B, for which there are minimum daily requirements, and each contains a third ingredient, C, that needs to be minimized.

	Supplement 1	Supplement 2	Daily Requirement
A	20 mg/oz	6 mg/oz	98 mg
B	8 mg/oz	16 mg/oz	80 mg
C	6 mg/oz	2 mg/oz	

● EXAMPLE 2 Finding the Dual of a Minimization Problem

Find the dual of the following:

$$\text{Minimize } Z = 4x_1 + 3x_2$$

subject to

$$3x_1 - x_2 \geq 2 \tag{11}$$
$$x_1 + x_2 \leq 1 \tag{12}$$
$$-4x_1 + x_2 \leq 3 \tag{13}$$

and $x_1, x_2 \geq 0$.

Solution: Since the primal is a minimization problem, we want constraints (12) and (13) to involve $\geq$. (See Table 7.4.) Multiplying both sides of (12) and (13) by -1, we get $-x_1 - x_2 \geq -1$ and $4x_1 - x_2 \geq -3$. Thus, constraints (11)–(13) become

$$3x_1 - x_2 \geq 2$$
$$-x_1 - x_2 \geq -1$$
$$4x_1 - x_2 \geq -3$$

The dual is

$$\text{maximize } W = 2y_1 - y_2 - 3y_3$$

subject to

$$3y_1 - y_2 + 4y_3 \leq 4$$

$$-y_1 - y_2 - \ y_3 \leq 3$$

and $y_1, y_2, y_3 \geq 0$

NOW WORK PROBLEM 3

● EXAMPLE 3 **Applying the Simplex Method to the Dual**

Use the dual and the simplex method to

$$\text{maximize } Z = 4x_1 - x_2 - x_3$$

subject to

$$3x_1 + x_2 - x_3 \leq 4$$

$$x_1 + x_2 + x_3 \leq 2$$

and $x_1, x_2, x_3 \geq 0$.

Solution: The dual is

$$\text{minimize } W = 4y_1 + 2y_2$$

subject to

$$3y_1 + y_2 \geq 4 \tag{14}$$

$$y_1 + y_2 \geq -1 \tag{15}$$

$$-y_1 + y_2 \geq -1 \tag{16}$$

and $y_1, y_2 \geq 0$. To use the simplex method, we must get nonnegative constants in (15) and (16). Multiplying both sides of these equations by -1 gives

$$-y_1 - y_2 \leq 1 \tag{17}$$

$$y_1 - y_2 \leq 1 \tag{18}$$

Since (14) involves $\geq$, an artificial variable is required. The equations corresponding to (14), (17), and (18) are, respectively,

$$3y_1 + y_2 - s_1 + t_1 = 4$$

$$-y_1 - y_2 + s_2 \quad = 1$$

and

$$y_1 - y_2 + s_3 \quad = 1$$

where t_1 is an artificial variable, s_1 is a surplus variable, and s_2 and s_3 are slack variables. To minimize W, we maximize $-W$. The artificial objective function is $U = (-W) - Mt_1$, where M is a large positive number. After computations, we find that the final simplex table is

B	y_1	y_2	s_1	s_2	s_3	$-W$	R
y_2	0	1	$-\frac{1}{4}$	0	$-\frac{3}{4}$	0	$\frac{1}{4}$
s_2	0	0	$-\frac{1}{2}$	1	$-\frac{1}{2}$	0	$\frac{5}{2}$
y_1	1	0	$-\frac{1}{4}$	0	$\frac{1}{4}$	0	$\frac{5}{4}$
$-W$	0	0	$\frac{3}{2}$	0	$\frac{1}{2}$	1	$-\frac{11}{2}$

indicators

The maximum value of $-W$ is $-\frac{11}{2}$, so the *minimum* value of W is $\frac{11}{2}$. Hence, the maximum value of Z is also $\frac{11}{2}$. Note that the indicators in the s_1-, s_2-, and s_3-columns are $\frac{3}{2}$, 0, and $\frac{1}{2}$, respectively. Thus, the maximum value of Z occurs when $x_1 = \frac{3}{2}$, $x_2 = 0$, and $x_3 = \frac{1}{2}$.

NOW WORK PROBLEM 11

In Example 1 of Section 7.7 we used the simplex method to

$$minimize \ Z = x_1 + 2x_2$$

subject to

$$-2x_1 + x_2 \geq 1$$
$$-x_1 + x_2 \geq 2$$

and $x_1, x_2 \geq 0$. The initial simplex table had 24 entries and involved two artificial variables. The table of the dual has only 18 entries and *no artificial variables* and is easier to handle, as Example 4 will show. Thus, there may be a distinct advantage in solving the dual to determine the solution of the primal.

This discussion shows the advantage of solving the dual problem.

● EXAMPLE 4   Using the Dual and the Simplex Method

Use the dual and the simplex method to

$$minimize \ Z = x_1 + 2x_2$$

subject to

$$-2x_1 + x_2 \geq 1$$
$$-x_1 + x_2 \geq 2$$

and $x_1, x_2 \geq 0$.

Solution: The dual is

$$maximize \ W = y_1 + 2y_2$$

subject to

$$-2y_1 - y_2 \leq 1$$
$$y_1 + y_2 \leq 2$$

and $y_1, y_2 \geq 0$. The initial simplex table is table I:

SIMPLEX TABLE I

entering
variable
↓

B	y_1	y_2	s_1	s_2	W	R	Quotients
s_1	-2	-1	1	0	0	1	
s_2	1	1	0	1	0	2	$2 \div 1 = 2$
W	-1	-2	0	0	1	0	

departing ← s_2
variable

indicators

Continuing, we get table II.

SIMPLEX TABLE II

B	y_1	y_2	s_1	s_2	W	R
s_1	-1	0	1	1	0	3
y_2	1	1	0	1	0	2
W	1	0	0	2	1	4

indicators

Since all indicators are nonnegative in table II, the maximum value of W is 4. Hence, the minimum value of Z is also 4. The indicators 0 and 2 in the s_1- and s_2-columns of table II mean that the minimum value of Z occurs when $x_1 = 0$ and $x_2 = 2$.

NOW WORK PROBLEM 9

Problems 7.8

In Problems 1–8, find the duals. Do not solve.

*1. Maximize

$$Z = x_1 + 2x_2$$

subject to

$$x_1 + x_2 \leq 5$$
$$-x_1 + x_2 \leq 3$$
$$x_1, x_2 \geq 0$$

2. Maximize

$$Z = 2x_1 + x_2 - x_3$$

subject to

$$2x_1 + 2x_2 \leq 3$$
$$-x_1 + 4x_2 + 2x_3 \leq 5$$
$$x_1, x_2, x_3 \geq 0$$

*3. Minimize

$$Z = x_1 + 8x_2 + 5x_3$$

subject to

$$x_1 + x_2 + x_3 \geq 8$$
$$-x_1 + 2x_2 + x_3 \geq 2$$
$$x_1, x_2, x_3 \geq 0$$

4. Minimize

$$Z = 8x_1 + 12x_2$$

subject to

$$2x_1 + 2x_2 \geq 1$$
$$x_1 + 3x_2 \geq 2$$
$$x_1, x_2 \geq 0$$

5. Maximize

$$Z = x_1 - x_2$$

subject to

$$-x_1 + 2x_2 \leq 13$$
$$-x_1 + x_2 \geq 3$$
$$x_1 + x_2 \geq 11$$
$$x_1, x_2 \geq 0$$

6. Maximize

$$Z = x_1 - x_2 + 4x_3$$

subject to

$$x_1 + x_2 + x_3 \leq 9$$
$$x_1 - 2x_2 + x_3 \geq 6$$
$$x_1, x_2, x_3 \geq 0$$

7. Minimize

$$Z = 4x_1 + 4x_2 + 6x_3,$$

subject to

$$x_1 - x_2 - x_3 \leq 3,$$
$$x_1 - x_2 + x_3 \geq 3,$$
$$x_1, x_2, x_3 \geq 0.$$

8. Minimize

$$Z = 5x_1 + 4x_2$$

subject to

$$-4x_1 + 3x_2 \geq -10$$
$$8x_1 - 10x_2 \leq 80$$
$$x_1, x_2 \geq 0$$

In Problems 9–14, solve by using duals and the simplex method.

*9. Minimize

$$Z = 2x_1 + 2x_2 + 5x_3$$

subject to

$$x_1 - x_2 + 2x_3 \geq 2$$
$$-x_1 + 2x_2 + x_3 \geq 3$$
$$x_1, x_2, x_3 \geq 0$$

10. Minimize

$$Z = 2x_1 + 2x_2$$

subject to

$$x_1 + 4x_2 \geq 28$$
$$2x_1 - x_2 \geq 2$$
$$-3x_1 + 8x_2 \geq 16$$
$$x_1, x_2 \geq 0$$

*11. Maximize

$$Z = 3x_1 + 8x_2$$

subject to

$$x_1 + 2x_2 \leq 8$$
$$x_1 + 6x_2 \leq 12$$
$$x_1, x_2 \geq 0$$

12. Maximize

$$Z = 2x_1 + 6x_2,$$

subject to

$$3x_1 + x_2 \le 12,$$

$$x_1 + x_2 \le 8,$$

$$x_1, x_2 \ge 0.$$

13. Minimize

$$Z = 6x_1 + 4x_2$$

subject to

$$-x_1 + x_2 \le 1$$

$$x_1 + x_2 \ge 3$$

$$x_1, x_2 \ge 0$$

14. Minimize

$$Z = 2x_1 + x_2 + x_3$$

subject to

$$2x_1 - x_2 - x_3 \le 2$$

$$-x_1 - x_2 + 2x_3 \ge 4$$

$$x_1, x_2, x_3 \ge 0$$

15. Advertising A firm is comparing the costs of advertising in two media—newspaper and radio. For every dollar's worth of advertising, the following table gives the number of people, by income group, reached by these media:

	Under $40,000	Over $40,000
Newspaper	40	100
Radio	50	25

The firm wants to reach at least 80,000 persons earning under $40,000 and at least 60,000 earning over $40,000. Use the dual and the simplex method to find the amounts that the firm should spend on newspaper and radio advertising so as to reach these numbers of people at a minimum total advertising cost. What is the minimum total advertising cost?

16. Use the dual and the simplex method to find the minimum total cost per mile in Problem 12 of Exercise 7.7.

17. Labor Costs A company pays skilled and semiskilled workers in its assembly department $14 and $8 per hour, respectively. In the shipping department, shipping clerks are paid $9 per hour and shipping clerk apprentices are paid $6 per hour. The company requires at least 90 workers in the assembly department and at least 60 in the shipping department. Because of union agreements, at least twice as many semiskilled workers must be employed as skilled workers. Also, at least twice as many shipping clerks must be employed as shipping clerk apprentices. Use the dual and the simplex method to find the number of each type of worker that the company must employ so that the total hourly wage paid to these employees is a minimum. What is the minimum total hourly wage?

7.9 Review

Important Terms and Symbols   Examples

Summary

The solution of a system of linear inequalities consists of all points whose coordinates simultaneously satisfy all of the inequalities. Geometrically, for two variables, it is the region that is common to all of the regions determined by the inequalities.

Linear programming involves maximizing or minimizing a linear function (the objective function) subject to a system of constraints, which are linear inequalities or linear equations. One method for finding an optimum solution for a nonempty feasible region is the corner-point method. The objective function is evaluated at each of the corner points of the feasible region, and we choose a corner point at which the objective function is optimum.

For a problem involving more than two variables, the corner-point method is either impractical or impossible. Instead, we use a matrix method called the simplex method, which is efficient and completely mechanical.

Review Problems

Problem numbers shown in color indicate problems suggested for use as a practice chapter test.

In Problems 1–10, solve the given inequality or system of inequalities.

1. $-3x + 2y > -6$

2. $2x - 3y + 8 \geq 0$

3. $3x \leq -5$

4. $-x < 2$

5. $\begin{cases} y - 3x < 6 \\ x - y \ > -3 \end{cases}$

6. $\begin{cases} x - 2y > 4 \\ x + y \ > 1 \end{cases}$

7. $\begin{cases} x - y < 4 \\ y - x < 4 \end{cases}$

8. $\begin{cases} \quad x > y \\ x + y < 0 \end{cases}$

9. $\begin{cases} 4x + 2y > -6 \\ 3x - 2y > -7 \\ \quad\quad x \geq 0 \end{cases}$

10. $\begin{cases} 2x - y > 5 \\ \quad x < 3 \\ \quad y < 7 \end{cases}$

In Problems 11–18, do not use the simplex method.

11. Maximize

$$Z = x - 2y$$

subject to

$$y - x \leq 2$$
$$x + y \leq 4$$
$$x \leq 3$$
$$x, y \geq 0$$

12. Maximize

$$Z = 4x + 2y$$

subject to

$$x + 2y \leq 10$$
$$x \leq 4$$
$$y \geq 1$$
$$x, y \geq 0$$

13. Minimize

$$Z = 2x - y$$

subject to

$$x - y \geq -2$$
$$x + y \geq 1$$
$$x - 2y \leq 2$$
$$x, y \geq 0$$

14. Minimize

$$Z = x + y$$

subject to

$$x + 2y \leq 12$$
$$4x + 3y \leq 15$$
$$x - 6y \leq 0$$
$$x, y \geq 0$$

15. Minimize

$$Z = 2x + 3y$$

subject to

$$x + y \leq 5$$
$$2x + 5y \leq 10$$
$$5x + 8y \geq 20$$
$$x, y \geq 0$$

[7]16. Minimize

$$Z = 2x + 2y$$

subject to

$$x + y \geq 4$$
$$-x + 3y \leq 18$$
$$x \leq 6$$
$$x, y \geq 0$$

[7]17. Maximize

$$Z = 9x + 6y$$

subject to

$$x + 2y \leq 8$$
$$3x + 2y \leq 12$$
$$x, y \geq 0$$

18. Maximize

$$Z = 4x + y$$

subject to

$$x + 2y \geq 16$$
$$3x + 2y \geq 24$$
$$x, y \geq 0$$

---

[7]Problems 16 and 17 refer to Section 7.3.

In Problems 19–28, use the simplex method.

19. Maximize

$$Z = 4x_1 + 5x_2$$

subject to

$$x_1 + 6x_2 \leq 12$$

$$x_1 + 2x_2 \leq 8$$

$$x_1, x_2 \geq 0$$

20. Maximize

$$Z = 18x_1 + 20x_2$$

subject to

$$2x_1 + 3x_2 \leq 18$$

$$4x_1 + 3x_2 \leq 24$$

$$x_2 \leq 5$$

$$x_1, x_2 \geq 0$$

21. Minimize

$$Z = 3x_1 + 2x_2 + x_3$$

subject to

$$x_1 + 2x_2 + 3x_3 \geq 5$$

$$x_1, x_2, x_3 \geq 0$$

22. Minimize

$$Z = x_1 + 2x_2$$

subject to

$$3x_1 + 5x_2 \geq 20$$

$$x_1 \geq 5$$

$$x_1, x_2 \geq 0$$

23. Maximize

$$Z = x_1 + 2x_2$$

subject to

$$x_1 + x_2 \leq 12$$

$$x_1 + x_2 \geq 5$$

$$x_1 \leq 10$$

$$x_1, x_2 \geq 0$$

24. Minimize

$$Z = 2x_1 + x_2$$

subject to

$$x_1 + 2x_2 \leq 6$$

$$x_1 + x_2 \geq 1$$

$$x_1, x_2 \geq 0$$

25. Minimize

$$Z = x_1 + 2x_2 + x_3$$

subject to

$$x_1 - x_2 - x_3 \leq -1$$

$$6x_1 + 3x_2 + 2x_3 = 12$$

$$x_1, x_2, x_3 \geq 0$$

26. Maximize

$$Z = 2x_1 + 3x_2 + 5x_3$$

subject to

$$x_1 + x_2 + 3x_3 \geq 5$$

$$2x_1 + x_2 + 4x_3 \leq 5$$

$$x_1, x_2, x_3 \geq 0$$

[8]**27.** Maximize

$$Z = x_1 + 4x_2 + 2x_3$$

subject to

$$4x_1 - x_2 \leq 2$$

$$-8x_1 + 2x_2 + 5x_3 \leq 2$$

$$x_1, x_2, x_3 \leq 0$$

[8]**28.** Minimize

$$Z = x_1 + x_2$$

subject to

$$x_1 + x_2 + 2x_3 \leq 4$$

$$x_3 \geq 1$$

$$x_1, x_2, x_3 \geq 0$$

In Problems 29 and 30, solve by using duals and the simplex method.

29. Minimize

$$Z = 2x_1 + 7x_2 + 8x_3$$

subject to

$$x_1 + 2x_2 + 3x_3 \geq 35$$

$$x_1 + x_2 + x_3 \geq 25$$

$$x_1, x_2, x_3 \geq 0$$

30. Maximize

$$Z = x_1 - 2x_2$$

subject to

$$x_1 - x_2 \leq 3$$

$$x_1 + 2x_2 \leq 4$$

$$4x_1 + x_2 \geq 2$$

$$x_1, x_2 \geq 0$$

---

[8]Problems 27 and 28 refer to Section 7.5.

31. Production Order   A company manufactures three products: X, Y, and Z. Each product requires the use of time on machines A and B as given in the following table:

	Machine A	Machine B
Product X	1 hr	1 hr
Product Y	2 hr	1 hr
Product Z	2 hr	2 hr

The numbers of hours per week that A and B are available for production are 40 and 34, respectively. The profit per unit on X, Y, and Z is $10, $15, and $22, respectively. What should be the weekly production order if maximum profit is to be obtained? What is the maximum profit?

32. Repeat Problem 31 if the company must produce at least a total of 24 units per week.

33. Oil Transportation   An oil company has storage facilities for heating fuel in cities A, B, C, and D. Cities C and D are each in need of exactly 500,000 gal of fuel. The company determines that A and B can each sacrifice at most 600,000 gal to satisfy the needs of C and D. The following table gives the costs per gallon to transport fuel between the cities:

	To	
From	**C**	**D**
A	$0.01	$0.02
B	0.02	0.04

How should the company distribute the fuel in order to minimize the total transportation cost? What is the minimum transportation cost?

34. Profit   Jason operates a home business selling two computer games: "Space Traders" and "Green Dwarf." These games are installed for Jason by three friends, Nicole, Hillary, and Katie, each of whom must do some of the work on the installation of each game. The time that each must spend on each game is given in the following table:

	Nicole	Hillary	Katie
Space Traders	30 min	20 min	10 min
Green Dwarf	10 min	10 min	50 min

Jason's friends have other work to do, but they find that each week they can spend up to 300, 200, and 500 minutes, respectively, working on Jason's games. Jason makes a profit of $5 on each sale of Space Traders and $9 on each sale of Green Dwarf. How many of each game should Jason try to sell each week to maximize profit, and what is this maximum profit?

35. Diet Formulation   A technician in a zoo must formulate a diet from two commercial products, food A and food B, for a certain group of animals. In 200 g of food A there are 16 grams of fat, 32 grams of carbohydrate, and 4 grams of protein. In 200 g of food B there are 8 grams of fat, 64 grams of carbohydrate, and 10 grams of protein. The minimum daily requirements are 176 grams of fat, 1024 grams of carbohydrate, and 200 grams of protein. If food A costs 8 cents per 100 grams and food B costs 22 cents per 100 grams, how many grams of each food should be used to meet the minimum daily requirements at the least cost? (Assume that a minimum cost exists.)

In Problems 36 and 37, do not use the simplex method. Round your answers to two decimal places.

36. Minimize

$$Z = 4.2x - 2.1y$$

subject to

$$y \le 3.4 + 1.2x$$
$$y \le -7.6 + 3.5x$$
$$y \le 18.7 - 0.6x$$
$$x, y \ge 0$$

37. Maximize

$$Z = 12.4x + 8.3y$$

subject to

$$1.4x + 1.7y \le 15.9$$
$$-3.6x + 2.6y \le -10.7$$
$$-1.3x + 4.3y \le -5.2$$
$$x, y \ge 0$$

Mathematical Snapshot

Drug and Radiation Therapies[9]

Frequently there are alternative forms of treatment available for a patient diagnosed to have a particular disease complex. With each treatment there may be not only positive effects on the patient but also negative effects, such as toxicity or discomfort. A physician must make the best choice of these treatments or combination of treatments. This choice will depend not only on curative effects but also on the effects of toxicity and discomfort.

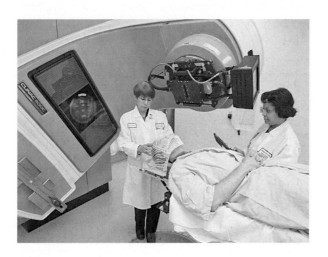

Suppose that you are a physician with a cancer patient under your care and there are two possible treatments available: drug administration and radiation therapy. Let us assume that the efficacies of the treatments are expressed in common units, say, curative units. The drug contains 1000 curative units per ounce, and the radiation gives 1000 curative units per minute. Your analysis indicates that the patient must receive at least 3000 curative units.

However, a degree of toxicity is involved with each treatment. Assume that the toxic effects of each treatment are measured in a common unit of toxicity, say, a toxic unit. The drug contains 400 toxic units per ounce, and the radiation induces 1000 toxic units per minute. Based on your studies, you believe that the patient must receive no more than 2000 toxic units.

In addition, each treatment involves a degree of discomfort to the patient. The drug is three times as noxious per ounce as the radiation per minute.

Table 7.5 summarizes the data. The problem posed to you is to determine the doses of the drug and radiation that

TABLE 7.5

	Curative Units	Toxic Units	Relative Discomfort
Drug (per ounce)	1000	400	3
Radiation (per minute)	1000	1000	1
Requirement	≥ 3000	≤ 2000	

will satisfy the curative and toxicity requirements and, at the same time, minimize the discomfort to the patient.

Let x_1 be the number of ounces of the drug and x_2 be the number of minutes of radiation to be administered. Then you want to minimize the discomfort D given by

$$D = 3x_1 + x_2$$

subject to the curative condition

$$1000x_1 + 1000x_2 \geq 3000$$

and the toxic condition

$$400x_1 + 1000x_2 \leq 2000$$

where $x_1 \geq 0$ and $x_2 \geq 0$. You should recognize that this is a linear programming problem. By graphing, the feasible region shown in Figure 7.28 is obtained. The corner points are $(3, 0)$, $(5, 0)$, and $(\frac{5}{3}, \frac{4}{3})$.

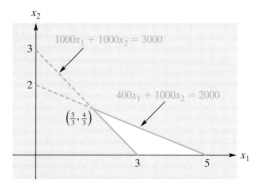

FIGURE 7.28 Feasible region for drug and radiation problem.

---

[9]Adapted from R. S. Ledley and L. B. Lusted, "Medical Diagnosis and Modern Decision Making," *Proceedings of Symposia in Applied Mathematics*, Vol. XIV; *Mathematical Problems in the Biological Sciences* (American Mathematical Society, 1962).

Evaluating D at each corner point gives the following:

$$\text{at } (3, 0) \quad D = 3(3) + 0 = 9$$
$$\text{at } (5, 0) \quad D = 3(5) + 0 = 15$$

and

$$\text{at } \left(\frac{5}{3}, \frac{4}{3}\right) \quad D = 3\left(\frac{5}{3}\right) + \frac{4}{3} = \frac{19}{3} \approx 6.3$$

Because D is minimized at $\left(\frac{5}{3}, \frac{4}{3}\right)$, you should prescribe a treatment of $\frac{5}{3}$ ounces of the drug and $\frac{4}{3}$ minutes of radiation. Thus by solving a linear programming problem, you have determined the "best" treatment for the patient.

The National Institutes of Health maintain a Web site, www.nih.gov/health, that contains current information in a variety of health-related areas.

You may also want to search the Web for online sites that use applets to demonstrate the simplex method. Simply enter "simplex method" and "applet" in any Internet search engine.

Problems

1. Suppose that drug and radiation treatments are available to a patient. Each ounce of the drug contains 500 curative units and 400 toxic units. Each minute of radiation gives 1000 curative units and 600 toxic units. The patient requires at least 2000 curative units and can tolerate no more than 1400 toxic units. If each ounce of the drug is as noxious as each minute of radiation, determine the doses of the drug and radiation so that the discomfort to the patient is minimized. Use the geometric method on a graphing calculator if one is available.

2. Suppose that drug A, drug B, and radiation therapy are treatments available to a patient. Each ounce of drug A contains 600 curative units and 500 toxic units. Each ounce of drug B contains 500 curative units and 100 toxic units. Each minute of radiation gives 1000 curative units and 1000 toxic units. The patient requires at least 3000 curative units and can tolerate no more than 2000 toxic units. If each ounce of drug A and each minute of radiation are equally as noxious, and each ounce of drug B is twice as noxious as each ounce of drug A, determine the doses of the drugs and radiation so that the discomfort to the patient is minimized. Use the simplex method.

3. Which method do you think is easier for doing linear programming, the simplex method or a technology-assisted geometric method? Give reasons for your answer.

8

INTRODUCTION TO PROBABILITY AND STATISTICS

Mathematical Snapshot Probability and Cellular Automata

Modern probability theory began with a very *practical* problem. If a game between two gamblers is interrupted, the player who is ahead surely has a right to more than half the pot of money being contested. But not to all of it! How should the pot be divided? This problem was unsolved in 1654, when a French count shared it with Blaise Pascal.

The solution involves probability. If a player has a 30% chance of winning $150, the current dollar value of the player's situation is 0.30($150) = $45. Why? Because a player who repeatedly finds himself with a 30% chance of winning $150 will in the long run win just as much money by staying in the game as by allowing himself to be bought out for $45 every time. Each player, then, should get a share proportional to his or her probability, just before interruption, of winning if the game had been played to completion.

But how is that probability calculated? Working together, Pascal and Pierre de Fermat established the following result. Suppose a game consists of a series of "rounds" involving chance, such as coin tosses that each has an equal chance of winning, and that the overall winner is the one who first wins a certain number of rounds. If, at interruption, Player 1 need r more rounds to win while Player 2 needs s more rounds to win, then Player 1's probability of winning is

$$\sum_{k=0}^{s-1} \frac{{}_nC_k}{2^n}$$

where $n = r + s - 1$ is the maximum number of rounds the game could still have lasted. Player 2's probability of winning is 1 minus Player 1's probability of winning. The notation ${}_nC_k$, which is read "n choose k," may be unfamiliar to you; you will learn about it in this chapter. (The Σ denotes summation as in Section 1.5.) Once you understand the formula, you will be able to verify that if Player 1 is 1 round away from winning and Player 2 is 3 rounds away from winning, then Player 1 should get $\frac{7}{8}$ of the pot to Player 2's $\frac{1}{8}$. Later in this chapter we will show how Pascal and Fermat arrived at their answer. With practice you will be able to answer questions of your own like this one.

The term *probability* is familiar to most of us. It is not uncommon to hear such phrases as "the probability of precipitation," "the probability of flooding," and "the probability of receiving an A in a course." Loosely speaking, probability refers to a number that indicates the degree of likelihood that some future event will have a particular outcome. For example, before tossing a well-balanced coin, you do not know with certainty whether the outcome will be a head or a tail. However, no doubt you consider these outcomes as being equally likely to occur. This means that if the coin were tossed a large number of times, you would expect that approximately half of the tosses would give heads. Thus, we say that the probability of a head occurring on any toss is $\frac{1}{2}$, or 50%.

The field of probability forms the basis of the study of statistics. In statistics, we are concerned about making an inference—that is, a prediction or decision—about a population (a large set of objects under consideration) by using a sample of data drawn from that population. In other words, in statistics, we make an inference about a population based on a known sample. For example, by drawing a sample of units from an assembly line, we can statistically make an inference about *all* the units in a production run. However, in the study of probability, we work with a known population and consider the likelihood (or probability) of drawing a particular sample from it. For example, if we select a card from a deck, we may be interested in the probability that it will be the ace of hearts.

8.1 Basic Counting Principle and Permutations

OBJECTIVE

To develop and apply a Basic Counting Principle and to extend it to permutations.

Basic Counting Principle

Later on, you will find that computing a probability may require you to calculate the number of elements in a set. Because counting the elements individually may be extremely tedious (or even prohibitive), we spend some time developing efficient counting techniques. We begin by motivating the *Basic Counting Principle,* which is useful in solving a wide variety of problems.

Suppose a manufacturer wants to produce coffee brewers in 2-, 8-, and 10-cup capacities, with each capacity available in colors of white, beige, red, and green. How many types of brewers must the manufacturer produce? To answer the question, it is not necessary that we count the capacity–color pairs one by one (such as 2-white and 8-beige). Since there are three capacities, and for each capacity there are four colors, the number of types is the product $3 \cdot 4 = 12$. We can systematically list the different types by using the **tree diagram** of Figure 8.1. From the starting point, there are three branches that indicate the possible capacities. From each of these branches are four more branches that indicate the possible colors. This tree determines 12 paths, each beginning at the starting point and ending at a tip. Each path determines a different type of coffee brewer. We refer to the diagram as being a *two-level* tree: There is a level for capacity and a level for color.

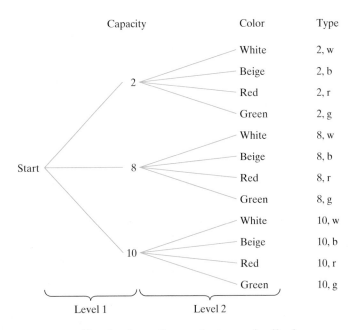

FIGURE 8.1 Two-level tree diagram for types of coffee brewers.

We can consider the listing of the types of coffee brewers as a two-stage procedure. In the first stage we indicate a capacity and in the second a color. The number of types of coffee brewers is the number of ways the first stage can occur (3), times the number of ways the second stage can occur (4), which yields $3 \cdot 4 = 12$. Suppose further that the manufacturer decides to make all of the models available with a timer option that allows the consumer to awake with freshly brewed compnay. Assuming that this really is an option, so that a coffee brewer either comes with a timer or without a timer, counting the number of types of brewers now becomes a three-stage procedure. There are now $3 \cdot 4 \cdot 2 = 24$ types of brewer.

This multiplication procedure can be generalized into a Basic Counting Principle:

Basic Counting Principle

Suppose that a procedure involves a sequence of k stages. Let n_1 be the number of ways the first can occur and n_2 be the number of ways the second can occur. Continuing in this way, let n_k be the number of ways the kth stage can occur. Then the total number of different ways the procedure can occur is

$$n_1 \cdot n_2 \cdots n_k$$

● EXAMPLE 1   Travel Routes

Two roads connect cities A *and* B, *four connect* B *and* C, *and five connect* C *and* D. *(See Figure 8.2.) To drive from* A, *to* B, *to* C, *and then to city* D, *how many different routes are possible?*

Solution: Here we have a three-stage procedure. The first (A → B) has two possibilities, the second (B → C) has four, and the third (C → D) has five. By the Basic Counting Principle, the total number of routes is $2 \cdot 4 \cdot 5 = 40$.

NOW WORK PROBLEM 1

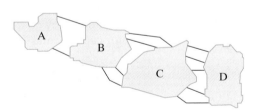

FIGURE 8.2   Roads connecting cities A, B, C, D.

● EXAMPLE 2   Coin Tosses and Roll of a Die

When a coin is tossed, a head (H) *or a tail* (T) *may show. If a die is rolled, a 1, 2, 3, 4, 5, or 6 may show. Suppose a coin is tossed twice and then a die is rolled, and the result is noted (such as* H *on first toss,* T *on second, and 4 on roll of die). How many different results can occur?*

Solution: Tossing a coin twice and then rolling a die can be considered a three-stage procedure. Each of the first two stages (the coin toss) has two possible outcomes. The third stage (rolling the die) has six possible outcomes. By the Basic Counting Principle, the number of different results for the procedure is

$$2 \cdot 2 \cdot 6 = 24$$

NOW WORK PROBLEM 3   ●●

● EXAMPLE 3 **Answering a Quiz**

In how many different ways can a quiz be answered under each of the following conditions?

a. *The quiz consists of three multiple-choice questions with four choices for each.*

Solution: Successively answering the three questions is a three-stage procedure. The first question can be answered in any of four ways. Likewise, each of the other two questions can be answered in four ways. By the Basic Counting Principle, the number of ways to answer the quiz is

$$4 \cdot 4 \cdot 4 = 4^3 = 64$$

b. *The quiz consists of three multiple-choice questions (with four choices for each) and five true–false questions.*

Solution: Answering the quiz can be considered a two-stage procedure. First we can answer the multiple-choice questions (the first stage), and then we can answer the true–false questions (the second stage). From part (a), the three multiple-choice questions can be answered in $4 \cdot 4 \cdot 4$ ways. Each of the true–false questions has two choices ("true" or "false"), so the total number of ways of answering all five of them is $2 \cdot 2 \cdot 2 \cdot 2 \cdot 2$. By the Basic Counting Principle, the number of ways the entire quiz can be answered is

$$\underbrace{(4 \cdot 4 \cdot 4)}_{\substack{\text{multiple} \\ \text{choice}}} \underbrace{(2 \cdot 2 \cdot 2 \cdot 2 \cdot 2)}_{\text{true–false}} = 4^3 \cdot 2^5 = 2048$$

NOW WORK PROBLEM 5

● EXAMPLE 4 **Letter Arrangements**

From the five letters A, B, C, D, *and* E, *how many three-letter horizontal arrangements (called "words") are possible if no letter may be repeated? (A "word" need not make sense.) For example,* BDE *and* DEB *are two acceptable words, but* CAC *is not.*

Solution: To form a word, we must successively fill the positions __ __ __ with different letters. Thus, we have a three-stage procedure. For the first position, we can choose any of the five letters. After filling that position with some letter, we can fill the second position with any of the remaining four letters. After that position is filled, the third position can be filled with any of the three letters that have not yet been used. By the Basic Counting Principle, the total number of three-letter words is

If repetitions are allowed, the number of words is $5 \cdot 5 \cdot 5 = 125$.

$$5 \cdot 4 \cdot 3 = 60$$

NOW WORK PROBLEM 7

Permutations

In Example 4, we selected three different letters from five letters and arranged them in an *order*. Each result is called a *permutation of five letters taken three at a time*. More generally, we have the following definition.

DEFINITION

An ordered selection of r objects, without repetition, taken from n distinct objects is called a *permutation of n objects taken r at a time*. The number of such permutations is denoted $_nP_r$.

Thus, in Example 4, we found that

$$_5P_3 = 5 \cdot 4 \cdot 3 = 60$$

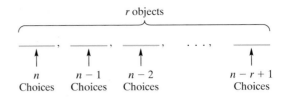

FIGURE 8.3   An ordered arrangement of r objects selected from n objects.

By a similar analysis, we will now find a general formula for $_nP_r$. In making an ordered arrangement of r objects from n objects, for the first position we may choose any one of the n objects. (See Figure. 8.3.) After the first position is filled, there remain $n - 1$ objects that may be chosen for the second position. After that position is filled, there are $n - 2$ objects that may be chosen for the third position. Continuing in this way and using the Basic Counting Principle, we arrive at the following formula:

The number of permutations of n objects taken r at a time is given by

$$_nP_r = \underbrace{n(n-1)(n-2)\cdots(n-r+1)}_{r \text{ factors}} \tag{1}$$

The formula for $_nP_r$ can be expressed in terms of factorials.[1] Multiplying the right side of Equation (1) by

$$\frac{(n-r)(n-r-1)\cdots(2)(1)}{(n-r)(n-r-1)\cdots(2)(1)}$$

gives

$$_nP_r = \frac{n(n-1)(n-2)\cdots(n-r+1)\cdot(n-r)(n-r-1)\cdots(2)(1)}{(n-r)(n-r-1)\cdots(2)(1)}$$

The numerator is simply $n!$, and the denominator is $(n - r)!$. Thus, we have the following result:

The number of permutations of n objects taken r at a time is given by

$$_nP_r = \frac{n!}{(n-r)!} \tag{2}$$

For example, from Equation (2), we have

$$_7P_3 = \frac{7!}{(7-3)!} = \frac{7!}{4!} = \frac{7\cdot6\cdot5\cdot4\cdot3\cdot2\cdot1}{4\cdot3\cdot2\cdot1} = 210$$

Many calculators can directly calculate $_nP_r$.

This calculation can be obtained easily with a calculator by using the factorial key. Alternatively, we can conveniently write

$$\frac{7!}{4!} = \frac{7\cdot6\cdot5\cdot4!}{4!} = 7\cdot6\cdot5 = 210$$

Notice how 7! was written so that the 4!'s would cancel.

● EXAMPLE 5   Club Officers

A club has 20 members. The offices of president, vice president, secretary, and treasurer are to be filled, and no member may serve in more than one office. How many different slates of candidates are possible?

---

[1]Factorials were discussed in Section 2.2.

Solution: We will consider a slate in the order of president, vice president, secretary, and treasurer. Each ordering of four members constitutes a slate, so the number of possible slates is $_{20}P_4$. By Equation (1),

$$_{20}P_4 = 20 \cdot 19 \cdot 18 \cdot 17 = 116{,}280$$

Alternatively, using Equation (2) gives

$$_{20}P_4 = \frac{20!}{(20-4)!} = \frac{20!}{16!} = \frac{20 \cdot 19 \cdot 18 \cdot 17 \cdot 16!}{16!}$$
$$= 20 \cdot 19 \cdot 18 \cdot 17 = 116{,}280$$

Note the large number of slates that are possible!

NOW WORK PROBLEM 11

CAUTION

Calculations with factorials tend to produce very large numbers. To avoid overflow on a calculator, it is frequently important to do some *cancellation* before making entries.

● **EXAMPLE 6 Political Questionnaire**

A politician sends a questionnaire to her constituents to determine their concerns about six important national issues: unemployment, the environment, taxes, interest rates, national defense, and social security. A respondent is to select four issues of personal concern and rank them by placing the number 1, 2, 3, or 4 after each issue to indicate the degree of concern, with 1 indicating the greatest concern and 4 the least. In how many ways can a respondent reply to the questionnaire?

Solution: A respondent is to rank four of the six issues. Thus, we can consider a reply as an ordered arrangement of six items taken four at a time, where the first item is the issue with rank 1, the second is the issue with rank 2, and so on. Hence, we have a permutation problem, and the number of possible replies is

$$_6P_4 = \frac{6!}{(6-4)!} = \frac{6!}{2!} = \frac{6 \cdot 5 \cdot 4 \cdot 3 \cdot 2!}{2!} = 6 \cdot 5 \cdot 4 \cdot 3 = 360$$

NOW WORK PROBLEM 21

In case you want to find the number of permutations of *n* objects taken all at a time, setting $r = n$ in Equation (2) gives

$$_nP_n = \frac{n!}{(n-n)!} = \frac{n!}{0!} = \frac{n!}{1} = n!$$

Each of these permutations is simply called a **permutation of *n* objects.**

The number of permutations of *n* objects is *n*!

For example, the number of permutations of the letters in the word SET is 3!, or 6. These permutations are

SET   STE   EST   ETS   TES   TSE

● **EXAMPLE 7 Name of Legal Firm**

Lawyers Smith, Jones, Jacobs, and Bell want to form a legal firm and will name it by using all four of their last names. How many possible names are there?

Solution: Since order is important, we must find the number of permutations of four names, which is

$$4! = 4 \cdot 3 \cdot 2 \cdot 1 = 24$$

Thus, there are 24 possible names for the firm.

NOW WORK PROBLEM 19

Problems 8.1

*1. **Production Process**   In a production process, a product goes through one of the assembly lines A, B, or C and then goes through one of the finishing lines D or E. Draw a tree diagram that indicates the possible production routes for a unit of the product. How many production routes are possible?

2. **Air Conditioner Models**   A manufacturer produces air conditioners having 6000-, 8000-, and 10,000-BTU capacities. Each capacity is available with one- or two-speed fans. Draw a tree diagram that represents all types of models. How many types are there?

*3. **Dice Rolls**   A red die is rolled and then a green die is rolled. Draw a tree diagram to indicate the possible results. How many results are possible?

4. **Coin Toss**   A coin is tossed four times. Draw a tree diagram to indicate the possible results. How many results are possible?

In Problems 5–10, use the Basic Counting Principle.

*5. **Course Selection**   A student must take a science course and a humanities course. The available science courses are biology, chemistry, physics, computer science, and mathematics. In humanities, the available courses are English, history, speech communications, and classics. How many two-course selections can the student make?

6. **Auto Routes**   A person lives in city A and commutes by automobile to city B. There are five roads connecting A and B. (a) How many routes are possible for a round trip? (b) How many round-trip routes are possible if a different road is to be used for the return trip?

*7. **Dinner Choices**   At a restaurant, a complete dinner consists of an appetizer, an entree, a dessert, and a beverage. The choices for the appetizer are soup and salad; for the entree, the choices are chicken, fish, steak, and lamb; for the dessert, the choices are cherries jubilee, fresh peach cobbler, chocolate truffle cake, and blueberry roly-poly; for the beverage, the choices are coffee, tea, and milk. How many complete dinners are possible?

8. **Multiple-Choice Exam**   In how many ways is it possible to answer a six-question multiple-choice examination if each question has four choices (and one choice is selected for each question)?

9. **True–False Exam**   In how many ways is it possible to answer a 10-question true–false examination?

10. **Canadian Postal Codes**   A Canadian postal code consists of a string of six characters, of which three are letters and three are digits, which begins with a letter and for which each letter is followed by a (single) digit. (For readability, the string is broken into strings of three. For example, B3H 3J5 is a valid postal code.) How many Canadian postal codes are possible?

In Problems 11–16, determine the values.

*11. $_6P_3$

12. $_{95}P_1$

13. $_6P_6$

14. $_9P_4$

15. $_4P_2 \cdot _5P_3$

16. $\dfrac{_{99}P_5}{_{99}P_4}$

17. Compute $1000!/999!$ without using a calculator. Now try it with your calculator, using the factorial feature.

18. Determine $\dfrac{_nP_r}{n!}$.

In Problems 19–42, use any appropriate counting method.

*19. **Name of Firm**   Flynn, Peters, and Walters are forming an advertising firm and agree to name it by their three last names. How many names for the firm are possible?

20. **Softball**   If a softball league has six teams, how many different end-of-the-season rankings are possible? Assume that there are no ties.

*21. **Contest**   In how many ways can a judge award first, second, and third prizes in a contest having eight contestants?

22. **Matching-Type Exam**   On a history exam, each of six items in one column is to be matched with exactly one of eight items in another column. No item in the second column can be selected more than once. In how many ways can the matching be done?

23. **Die Roll**   A die (with six faces) is rolled four times and the outcome of each roll is noted. How many results are possible?

24. **Coin Toss**   A coin is tossed eight times. How many results are possible if the order of the tosses is considered?

25. **Problem Assignment**   In a mathematics class with 12 students, the instructor wants homework problems 1, 3, and 5 put on the board by three different students. In how many ways can the instructor assign the problems?

26. **Combination Lock**   A combination lock has 26 different letters, and a sequence of three different letters must be selected for the lock to open. How many combinations are possible?

27. **Student Questionnaire**   A university issues a questionnaire whereby each student must rank the four items with which he or she is most dissatisfied. The items are

tuition fees	professors
parking fees	cafeteria food
dormitory rooms	class sizes

The ranking is to be indicated by the numbers 1, 2, 3 and 4, where 1 indicates the item involving the greatest dissatisfaction and 4 the least. In how many ways can a student answer the questionnaire?

28. **Die Roll**   A die is rolled three times. How many results are possible if the order of the rolls is considered and the second roll produces a number less than 3?

29. **Letter Arrangements**   How many six-letter words from the letters in the word MEADOW are possible if no letter is repeated?

30. **Letter Arrangements**   Using the letters in the word DISC, how many four-letter words are possible if no letter is repeated?

31. Book Arrangements   In how many ways can five of seven books be arranged on a bookshelf? In how many ways can all seven books be arranged on the shelf?

32. Lecture Hall   A lecture hall has five doors. In how many ways can a student enter the hall by one door and

(a) Exit by a different door?

(b) Exit by any door?

33. Poker Hand   A poker hand consists of 5 cards drawn from a deck of 52 playing cards. The hand is said to be "four of a kind" if four of the cards have the same face value. For example, hands with four 10's or four jacks or four 2's are four-of-a-kind hands. How many such hands are possible?

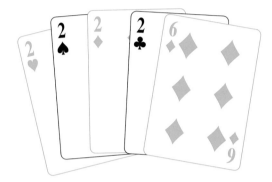

34. Merchandise Choice   In a merchandise catalog, a CD rack is available in the colors of black, red, yellow, gray, and blue. When placing an order for one CD rack, customers must indicate their first and second color choices. In how many ways can this be done?

35. Diner Order   Five students go to a diner and order a hamburger, a cheeseburger, a fish sandwich, a roast beef sandwich, and a grilled cheese sandwhich (one sandwich for each). When the waitress returns with the food, she forgets which student ordered which item and simply places a sandwich before each student. In how many ways can the waitress do this?

36. Group Photograph   In how many ways can three men and two women line up for a group picture? In how many ways can they line up if a woman is to be at each end?

37. Club Officers   A club has 12 members.

(a) In how many ways can the offices of president, vice president, secretary, and treasurer be filled if no member can serve in more than one office?

(b) In how many ways can the four offices be filled if the president and vice president must be different members?

38. Fraternity Names   Suppose a fraternity is named by three Greek letters. (There are 24 letters in the Greek alphabet.)

(a) How many names are possible?

(b) How many names are possible if no letter can be used more than one time?

39. Basketball   In how many ways can a basketball coach assign positions to her five-member team if two of the members are qualified for the center position and all five are qualified for all the other positions?

40. Call Letters   Suppose the call letters of a radio station consist of four letters, of which the first must be a K or a W. How many such identifications not ending in O are possible?

41. Baseball   A baseball manager determines that, of his nine team members, three are strong hitters and six are weak. If the manager wants the strong hitters to be the first three batters in a batting order, how many batting orders are possible?

42. Signal Flags   When at least one of four flags colored red, green, yellow, and blue are arranged vertically on a flagpole, the result indicates a signal (or message). Different arrangements give different signals.

(a) How many different signals are possible if all four flags are used?

(b) How many different signals are possible if at least one flag is used?

OBJECTIVE

To discuss combinations, permutations with repeated objects, and assignments to cells.

8.2 Combinations and Other Counting Principles •••

Combinations

We continue our discussion of counting methods by considering the following. In a 20-member club the offices of president, vice president, secretary, and treasurer are to be filled, and no member may serve in more than one office. If these offices, in the order given, are filled by members A, B, C, and D, respectively, then we can represent this slate by

$$ABCD$$

A different slate is

$$BACD$$

These two slates represent different permutations of 20 members taken four at a time. Now, as a different situation, let us consider four-person *committees* that may be formed from the 20 members. In that case, the two arrangements

$$ABCD \quad \text{and} \quad BACD$$

represent the *same* committee. Here *the order of listing the members is of no concern.* These two arrangements are considered to give the same *combination* of A, B, C, and D.

> **DEFINITION**
>
> A selection of *r* objects, without regard to order and without repetition, selected from *n* distinct objects is called a *combination of n objects taken r at a time*. The number of such combinations is denoted $_nC_r$, which can be read "*n* choose *r*".

The important phrase here is "without regard to order," since order implies a permutation rather than a combination.

● **EXAMPLE 1** **Comparing Combinations and Permutations**

List all combinations and all permutations of the four letters

$$A, \quad B, \quad C, \quad and \quad D$$

when they are taken three at a time.

Solution: The combinations are

$$ABC \quad ABD \quad ACD \quad BCD$$

There are four combinations, so $_4C_3 = 4$. The permutations are

ABC	ABD	ACD	BCD
ACB	ADB	ADC	BDC
BAC	BAD	CAD	CBD
BCA	BDA	CDA	CDB
CAB	DAB	DAC	DBC
CBA	DBA	DCA	DCB

There are 24 permutations.

NOW WORK PROBLEM 1 ●○●

In Example 1, notice that each column consists of all the permutations for the same combination of letters. With this observation, we can determine a formula for $_nC_r$—the number of combinations of *n* objects taken *r* at a time. Suppose one such combination is

$$x_1 x_2 \cdots x_r$$

The number of permutations of these *r* objects is *r*!. If we listed all other such combinations and then listed all permutations of these combinations, we would obtain a complete list of permutations of the *n* objects taken *r* at a time. Thus, by the Basic Counting Principle,

$$_nC_r \cdot r! = {_nP_r}$$

Solving for $_nC_r$ gives

$$_nC_r = \frac{_nP_r}{r!} = \frac{\dfrac{n!}{(n-r)!}}{r!} = \frac{n!}{r!(n-r)!}$$

The number of combinations of *n* objects taken *r* at a time is given by

$$_nC_r = \frac{n!}{r!(n-r)!}$$

Many calculators can directly compute $_nC_r$.

● **EXAMPLE 2** **Committee Selection**

If a club has 20 members, how many different four-member committees are possible?

Solution: Order is not important because, no matter how the members of a committee are arranged, we have the same committee. Thus, we simply have to compute the number of combinations of 20 objects taken four at a time, $_{20}C_4$:

$$_{20}C_4 = \frac{20!}{4!(20-4)!} = \frac{20!}{4!16!}$$

$$= \frac{20 \cdot 19 \cdot 18 \cdot 17 \cdot 16!}{4 \cdot 3 \cdot 2 \cdot 1 \cdot 16!} = 4845$$

Observe how 20! was written so that the 16!'s would cancel.

There are 4845 possible committees.

NOW WORK PROBLEM 9

It is important to remember that if a selection of objects is made and *order is important*, then *permutations* should be considered. If *order is not important*, consider *combinations*. A key aid to memory is that $_nP_r$ is the number of executive slates with r ranks that can be chosen from n people while $_nC_r$ is the number of committees with r members that can be chosen from n people. An executive slate can be thought of as a committee in which every individual has been ranked. There are $r!$ ways to rank the members of a committee with r members. Thus if we think of of forming an executive slate as a two-stage procedure, then using the Basic Counting Principle of Section 8.1 we again get

$$_nP_r = {_nC_r} \cdot r!$$

● EXAMPLE 3 Poker Hand

A **poker hand** consists of 5 cards dealt from an ordinary deck of 52 cards. How many different poker hands are there?

Solution: One possible hand is

2 of hearts,   3 of diamonds,   6 of clubs,
4 of spades,   king of hearts

which we can abbreviate as

2H   3D   6C   4S   KH

The order in which the cards are dealt does not matter, so this hand is the same as

KH   4S   6C   3D   2H

Thus, the number of possible hands is the number of ways that 5 objects can be selected from 52, without regard to order. This is a combination problem. We have

$$_{52}C_5 = \frac{52!}{5!(52-5)!} = \frac{52!}{5!47!}$$

$$= \frac{52 \cdot 51 \cdot 50 \cdot 49 \cdot 48 \cdot 47!}{5 \cdot 4 \cdot 3 \cdot 2 \cdot 1 \cdot 47!}$$

$$= \frac{52 \cdot 51 \cdot 50 \cdot 49 \cdot 48}{5 \cdot 4 \cdot 3 \cdot 2} = 2{,}598{,}960$$

NOW WORK PROBLEM 11

● EXAMPLE 4 Majority Decision and Sum of Combinations

A college promotion committee consists of five members. In how many ways can the committee reach a majority decision in favor of a promotion?

Strategy   A favorable majority decision is reached if, and only if,

exactly three members vote favorably,
or   exactly four members vote favorably,
or   all five members vote favorably

To determine the total number of ways to reach a favorable majority decision, we *add* the number of ways that each of the preceding votes can occur.

Solution: Suppose exactly three members vote favorably. The order of the members is of no concern, and thus we can think of these members as forming a combination. Hence, the number of ways three of the five members can vote favorably is $_5C_3$. Similarly, the number of ways exactly four can vote favorably is $_5C_4$, and the number of ways all five can vote favorably is $_5C_5$ (which, of course, is 1). Thus, the number of ways to reach a majority decision in favor of a promotion is

$$_5C_3 + {_5C_4} + {_5C_5} = \frac{5!}{3!(5-3)!} + \frac{5!}{4!(5-4)!} + \frac{5!}{5!(5-5)!}$$

$$= \frac{5!}{3!2!} + \frac{5!}{4!1!} + \frac{5!}{5!0!}$$

$$= \frac{5 \cdot 4 \cdot 3!}{3! \cdot 2 \cdot 1} + \frac{5 \cdot 4!}{4! \cdot 1} + 1$$

$$= 10 + 5 + 1 = 16$$

NOW WORK PROBLEM 15

Combinations and Sets

The previous example leads rather naturally to some properties of combinations that are quite useful in the study of probability. For example, we will show that

$$_5C_0 + {_5C_1} + {_5C_2} + {_5C_3} + {_5C_4} + {_5C_5} = 2^5$$

and, for any nonnegative iteger n,

$$_nC_0 + {_nC_1} + \cdots + {_nC_{n-1}} + {_nC_n} = 2^n \tag{1}$$

We can build on the last equation of Example 4 to verify the first of these equations:

$$_5C_0 + {_5C_1} + {_5C_2} + {_5C_3} + {_5C_4} + {_5C_5} = {_5C_0} + {_5C_1} + {_5C_2} + 16$$

$$= \frac{5!}{0!(5-0)!} + \frac{5!}{1!(5-1)!} + \frac{5!}{2!(5-2)!} + 16$$

$$= \frac{5!}{0!5!} + \frac{5!}{1!4!} + \frac{5!}{2!3!} + 16$$

$$= 1 + \frac{5 \cdot 4!}{4!} + \frac{5 \cdot 4 \cdot 3!}{2 \cdot 3!} + 16$$

$$= 1 + 5 + 10 + 16$$

$$= 32$$

$$= 2^5$$

However, this calculation is not illuminating and would be impractical if we were to adapt it for values of n much larger than 5.

Thus far we have primarily looked at *sets* in the context of sets of numbers. In examples in the study of probability, we often look at things like sets of playing cards, sets of die rolls, sets of ordered pairs of dice rolls, and the like. Typically these sets are finite. If a set S has n elements, we can, in principle, list its elements. For example, we might write

$$S = \{s_1, s_2, \ldots, s_n\}$$

A *subset E of S* is a set with the property that *every element of E is also an element of S*. When this is the case we write $E \subseteq S$. Formally,

$E \subseteq S$ if and only if, for all x, if x is an element of E then x is an element of S.

For any set S, we always have $\emptyset \subseteq S$ and $S \subseteq S$. If a set S has n elements, then any subset of S has r elements, where $0 \le r \le n$. The empty set, $\emptyset$, is the only subset of

S that has 0 elements. The whole set, S, is the only subset of S that has n elements. What is a general subset of S, containing r elements, where $0 \le r \le n$? According to our previous definition of *combination*, such a subset is exactly a combination of n objects taken r at a time and the number of such combinations is denoted $_nC_r$. Thus we may also think of $_nC_r$ as *the number of r-element subsets of an n-element set*.

For any set S, we can form the set of *all* subsets of S. It is called the *power set* of S and sometimes denoted 2^S. We claim that if S has n elements, then 2^S has 2^n elements. This is quite easy to see. If

$$S = \{s_1, s_2, \cdots, s_n\}$$

then specification of a subset E of S can be thought of as a procedure involving n stages. The first stage is to ask the question "Is s_1 an element of E?", the second stage is to ask, "Is s_2 an element of E?". We continue to ask such questions until we come to the nth stage—the last stage—"Is s_n an element of E?". Observe that each of these questions can be answered in exactly two ways; namely yes or no. According to the Basic Counting Principle of Section 8.1, the total number of ways that specification of a subset of S can occur is

$$\underbrace{2 \cdot 2 \cdot \ldots \cdot 2}_{n \text{ factors}} = 2^n$$

It follows that there are 2^n subsets of an n-element set. It is convenient to write $\#(S)$ for the number of elements of set S. Thus we have

$$\#(2^S) = 2^{\#(S)} \tag{2}$$

If $\#(S) = n$, then for each E in 2^S, we have $\#(E) = r$, for some r satisfying $0 \le r \le n$. For each such r, let us write $\mathcal{S}_r$ for the subset of 2^S consisting of all those elements E with $n(E) = r$. Thus $\mathcal{S}_r$ is the set of all r-element subsets of the n-element set S. From our observations in the last paragraph it follows that

$$\#(\mathcal{S}_r) = {}_nC_r \tag{3}$$

Now we claim that

$$\#(\mathcal{S}_0) + \#(\mathcal{S}_1) + \cdots + \#(\mathcal{S}_{n-1}) + \#(\mathcal{S}_n) = \#(2^S) \tag{4}$$

since every element E of 2^S is in *exactly* one of the sets $\mathcal{S}_r$. Substituting Equation (3), for each $0 \le r \le n$, and Equation (2) in Equation (4), we have Equation (1).

● EXAMPLE 5   A basic combinatorial identity

Establish the identity

$$_nC_r + {}_nC_{r+1} = {}_{n+1}C_{r+1}$$

Solution 1: We can calculate using $_nC_r = \dfrac{n!}{r!(n-r)!}$:

$$
\begin{aligned}
_nC_r + {}_nC_{r+1} &= \frac{n!}{r!(n-r)!} + \frac{n!}{(r+1)!(n-r-1)!} \\
&= \frac{(r+1)n! + (n-r)n!}{(r+1)!(n-r)!} \\
&= \frac{((r+1) + (n-r))n!}{(r+1)!(n-r)!} \\
&= \frac{(n+1)n!}{(r+1)!((n+1) - (r+1))!} \\
&= \frac{(n+1)!}{(r+1)!((n+1) - (r+1))!} \\
&= {}_{n+1}C_{r+1}
\end{aligned}
$$

Solution 2: We can reason using the idea that $_nC_r$ is the number of r-element subsets of an n-element set. Let S be an n-element set that does not contain s_* as an element. Then $S \cup \{s_*\}$ is an $(n+1)$-element set. Now the $(r+1)$-element subsets of $S \cup \{s_*\}$ are disjointly of two kinds:

1. those that contain s_* as an element;
2. those that do not contain s_* as an element.

Let us write $\mathcal{S}_*$ for the $(r+1)$-element subsets of $S \cup \{s_*\}$ that contain s_* and $\mathcal{S}$ for the $(r+1)$-element subsets of $S \cup \{s_*\}$ which do not contain s_*. Then

$$_{n+1}C_{r+1} = \#(\mathcal{S}_*) + \#(\mathcal{S})$$

because every $r+1$-element subset of $S \cup \{s_*\}$ is in exactly one of $\mathcal{S}_*$ or $\mathcal{S}$. Now the $(r+1)$-element subsets of $S \cup \{s_*\}$ that contain s_* are in one-to-one correspondence with the r-element subsets of S so we have

$$\#(\mathcal{S}_*) = {}_nC_r$$

On the other hand, $(r+1)$-element subsets of $S \cup \{s_*\}$ that do not contain s_* are in one-to-one correspondence with the $(r+1)$-element subsets of S so

$$\#(\mathcal{S}) = {}_nC_{r+1}$$

Assembling the last three displayed equations gives

$$_{n+1}C_{r+1} = {}_nC_r + {}_nC_{r+1}$$

as required.

The first solution is good computational practice, but the second solution is illustrative of ideas and arguments that are often useful in the study of probability. The identity we have just established together with

$$_nC_0 = 1 = {}_nC_n$$

for all n, allows us to generate *Pascal's Triangle:*

$$
\begin{array}{ccccccccccc}
 & & & & & 1 & & & & & \\
 & & & & 1 & & 1 & & & & \\
 & & & 1 & & 2 & & 1 & & & \\
 & & 1 & & 3 & & 3 & & 1 & & \\
 & 1 & & 4 & & 6 & & 4 & & 1 & \\
1 & & 5 & & 10 & & 10 & & 5 & & 1 \\
\end{array}
$$

$$\vdots$$

You should convince yourself that the $(r+1)$th entry in the $(n+1)$th row of Pascal's Triangle is $_nC_r$.

Permutations with Repeated Objects

In Section 8.1, we discussed permutations of objects that were all different. Now we examine the case where some of the objects are alike (or *repeated*). For example, consider determining the number of different permutations of the seven letters in the word

SUCCESS

Here the letters C and S are repeated. If the two C's were interchanged, the resulting permutation would be indistinguishable from SUCCESS. Thus, the number of distinct permutations is not 7!, as it would be with 7 different objects. To determine the number of distinct permutations, we use an approach that involves combinations.

S's	U	C's	E	S's	U	C's	E
—, —, —	—	—, —,	—	2, 3, 6	1	5, 7	4

(a)	(b)

FIGURE 8.4 Permutations with repeated objects.

Figure 8.4(a) shows boxes representing the different letters in the word SUCCESS. In these boxes we place the integers from 1 through 7. We place three integers in the S's box (because there are three S's), one in the U box, two in the C's box, and one in the E box. A typical placement is indicated in Figure 8.4(b). That placement can be thought of as indicating a permutation of the seven letters in SUCCESS, namely, the permutation in which (going from left to right) the S's are in the second, third, and sixth positions, the U is in the first position, and so on. Thus, Figure 8.4(b) corresponds to the permutation

USSECSC

To count the number of distinct permutations, it suffices to determine the number of ways the integers from 1 to 7 can be placed in the boxes. Since the order in which they are placed into a box is not important, the S's box can be filled in $_7C_3$ ways. Then the U box can be filled with one of the remaining four integers in $_4C_1$ ways. Then the C's box can be filled with two of the remaining three integers in $_3C_2$ ways. Finally, the E box can be filled with one of the remaining one integers in $_1C_1$ ways. Since we have a four-stage procedure, by the Basic Counting Principle the total number of ways to fill the boxes or, equivalently, the number of distinguishable permutations of the letters in SUCCESS is

$$
_7C_3 \cdot {_4C_1} \cdot {_3C_2} \cdot {_1C_1} = \frac{7!}{3!4!} \cdot \frac{4!}{1!3!} \cdot \frac{3!}{2!1!} \cdot \frac{1!}{1!0!}
$$
$$
= \frac{7!}{3!1!2!1!}
$$
$$
= 420
$$

In summary, the word SUCCESS has four types of letters: S, U, C, and E. There are three S's, one U, two C's, and one E, and the number of distinguishable permutations of the 7 letters is

$$
\frac{7!}{3!1!2!1!}
$$

Observing the forms of the numerator and denominator, we can make the following generalization:

Permutations with Repeated Objects

The number of distinguishable permutations of n objects such that n_1 are of one type, n_2 are of a second type,..., and n_k are of a kth type, where $n_1 + n_2 + \cdots + n_k = n$, is

$$
\frac{n!}{n_1!n_2! \cdots n_k!} \tag{5}
$$

In problems of this kind there are often a number of quite different solutions to the same problem. A solution that seems straightforward to one person may seem complicated to another. Accordingly, we present another solution to the problem of counting the number, N, of different permutations of the letters of

SUCCESS

We will begin by tagging the letters so that they become distinguishable, thus obtaining

$S_1U_1C_1C_2E_1S_2S_3$

Giving a permutation of these seven "different" letters can be described as a multistage procedure. We can begin by permuting as if we can't see the subscripts and by definition there are N ways to accomplish this task. For each of these ways, there are 3! ways to permute the three S's, for each of these, 1! ways to permute the one U, for each of these, 2! ways to permute the two C's, and for each of these, 1! ways to permute the one E. According to the Basic Counting Principle of Section 8.1, there are

$$N \cdot 3! \cdot 1! \cdot 2! \cdot 1!$$

ways to permute the seven "different" letters of $S_1U_1C_1C_2E_1S_2S_3$. On the other hand, we already know that there are

$$7!$$

permutations of seven different letters, so we must have

$$N \cdot 3! \cdot 1! \cdot 2! \cdot 1! = 7!$$

From this we find

$$N = \frac{7!}{3!1!2!1!}$$

in agreement with our earlier finding.

● EXAMPLE 6   Letter Arrangements with and without Repetition

For each of the following words, how many distinguishable permutations of the letters are possible?

a. *APOLLO*

Solution: The word APOLLO has six letters with repetition. We have one A, one P, two O's, and two L's. Using Equation (5), we find that the number of permutations is

$$\frac{6!}{1!1!2!2!} = 180$$

b. *GERM*

Solution: None of the four letters in GERM is repeated, so the number of permutations is

$$_4P_4 = 4! = 24$$

NOW WORK PROBLEM 17 ●●

● EXAMPLE 7   Name of Legal Firm

A group of four lawyers, Smith, Jones, Smith, and Bell (the Smiths are cousins), want to form a legal firm and will name it by using all of their last names. How many possible names exist?

Solution: Each different permutation of the last four names is a name for the firm. There are two Smiths, one Jones, and one Bell. From Equation (5), the number of distinguishable names is

$$\frac{4!}{2!1!1!} = 12$$

NOW WORK PROBLEM 19 ●●

A	B
2, 3, 5	1, 4

FIGURE 8.5
Assignment of people to rooms.

Cells

At times, we want to find the number of ways in which objects can be placed into "compartments," or *cells*. For example, suppose that from a group of five people, three are to be assigned to room A and two to room B. In how many ways can this be done? Figure 8.5 shows one such assignment, where the numbers $1, 2, \ldots, 5$ represent the people. Obviously, the order in which people are placed into the rooms is of no concern. The boxes (or cells) remind us of those in Figure 8.4(b), and, by an analysis similar to the discussion of permutations with repeated objects, the number of ways to assign the people is

$$\frac{5!}{3!2!} = \frac{5 \cdot 4 \cdot 3!}{3!2!} = 10$$

In general, we have the following principle:

Assignment to Cells

Suppose n distinct objects are assigned to k ordered cells with n_i objects in cell $i\,(i = 1, 2, \ldots, k)$ and the order in which the objects are assigned to cell i is of no concern. The number of all such assignment is

$$\frac{n!}{n_1!n_2! \cdots n_k!} \tag{6}$$

where $n_1 + n_2 + \cdots + n_k = n$.

To say it again, slightly differently, there are $_{n_1+n_2+\cdots+n_k}C_{n_1}$ ways to choose n_1 objects to put in the first cell, and for each of these ways there are $_{n_2+n_3+\cdots+n_k}C_{n_2}$ ways to choose n_2 objects to put in the second cell, and so on, giving, by the Basic Counting Principle of Section 8.1,

$$(_{n_1+n_2+\cdots+n_k}C_{n_1})(_{n_2+n_3+\cdots+n_k}C_{n_2}) \ldots (_{n_{k-1}+n_k}C_{n_{k-1}})(_{n_k}C_{n_k})$$

$$= \frac{(n_1 + n_2 + \cdots + n_k)!}{n_1!(n_2 + n_3 + \cdots + n_k)!} \cdot \frac{(n_2 + n_3 + \cdots + n_k)!}{n_2!(n_3 + n_4 + \cdots + n_k)!} \cdots \frac{(n_{k-1} + n_k)!}{n_{k-1}!n_k!} \cdot \frac{n_k!}{n_k!0!}$$

$$= \frac{(n_1 + n_2 + \cdots + n_k)!}{n_1!n_2! \cdots n_k!}$$

which is the number in (6).

● EXAMPLE 8 Assigning Mourners to Limousines

A funeral director must assign 15 mourners to three limousines: 6 in the first limousine, 5 in the second, and 4 in the third. In how many ways can this be done?

Solution: Here 15 people are placed into three cells (limousines): 6 in cell 1, 5 in cell 2, and 4 in cell 3. By Equation (2), the number of ways this can be done is

$$\frac{15!}{6!5!4!} = 630{,}630$$

NOW WORK PROBLEM 23 ●●●

Example 9 will show three different approaches to a counting problem. As we have said, many counting problems have alternative methods of solution.

● EXAMPLE 9 Art Exhibit

An artist has created 20 original paintings, and she will exhibit some of them in three galleries. Four paintings will be sent to gallery A, *four to gallery* B, *and three to gallery* C. *In how many ways can this be done?*

Solution:

Method 1 The artist must send $4 + 4 + 3 = 11$ paintings to the galleries, and the 8 that are not sent can be thought of as staying in her studio. Thus, we can think of this situation as placing 20 paintings into four cells:

4 in gallery A
4 in gallery B
3 in gallery C
9 in the artist's studio

From Equation (6), the number of ways this can be done is

$$\frac{20!}{4!4!3!9!} = 1{,}939{,}938{,}000$$

Method 2   We can handle the problem in terms of a two-stage procedure and use the Basic Counting Principle. First, 11 paintings are selected for exhibit. Then, these are split into three groups (cells) corresponding to the three galleries. We proceed as follows.

Selecting 11 of the 20 paintings for exhibit (order is of no concern) can be done in $_{20}C_{11}$ ways. Once a selection is made, four of the paintings go into one cell (gallery A), four go to a second cell (gallery B), and three go to a third cell (gallery C). By Equation (6), this can be done in $\dfrac{11!}{4!4!3!}$ ways. Applying the Basic Counting Principle gives the number of ways the artist can send the paintings to the galleries:

$$_{20}C_{11} \cdot \frac{11!}{4!4!3!} = \frac{20!}{11!9!} \cdot \frac{11!}{4!4!3!} = 1{,}939{,}938{,}000$$

Method 3   Another approach to this problem is in terms of a three-stage procedure. First, 4 of the 20 paintings are selected for shipment to gallery A. This can be done in $_{20}C_4$ ways. Then, from the remaining 16 paintings, the number of ways 4 can be selected for gallery B is $_{16}C_4$. Finally, the number of ways 3 can be sent to gallery C from the 12 paintings that have not yet been selected is $_{12}C_3$. By the Basic Counting Principle, the entire procedure can be done in

$$_{20}C_4 \cdot _{16}C_4 \cdot _{12}C_3 = \frac{20!}{4!16!} \cdot \frac{16!}{4!12!} \cdot \frac{12!}{3!9!} = \frac{20!}{4!4!3!9!}$$

ways, which gives the previous answer, as expected!

NOW WORK PROBLEM 27

Problems 8.2

In Problems 1–6, determine the values.

*1. $_6C_4$

2. $_6C_2$

3. $_{100}C_{100}$

4. $_{1001}C_1$

5. $_5P_3 \cdot _4C_2$

6. $_4P_2 \cdot _5C_3$

7. Verify that $_nC_r = _nC_{n-r}$.

8. Determine $_nC_n$.

*9. **Committee**   In how many ways can a four-member committee be formed from a group of 17 people?

10. **Horse Race**   In a horse race, a horse is said to *finish in the money* if it finishes in first, second, or third place. For an eight-horse race, in how many ways can the horses finish in the money? Assume no ties.

*11. **Math Exam**   On a 13-question mathematics examination, a student must answer any 9 questions. In how many ways can the 9 questions be chosen (without regard to order)?

12. **Cards**   From a deck of 52 playing cards, how many 4-card hands are there comprised solely of red cards?

13. **Quality Control**   A quality-control technician must select a sample of 10 dresses from a production lot of 74 couture dresses. How many different samples are possible? Express your answer in terms of factorials.

14. **Packaging**   A jelly producer makes seven types of jelly. The producer packages gift boxes containing four jars of jelly, no two of which are of the same type. To reflect the three national chains through which the jelly is distributed, the producer uses three types of boxes. How many different gift boxes are possible?

*15. **Scoring on Exam**   In a 10-question examination, each question is worth 10 points and is graded right or wrong. Considering the individual questions, in how many ways can a student score 80 or better?

16. **Team Results**   A sports team plays 11 games. In how many ways can the outcomes of the games result in four wins, five losses, and two ties?

*17. **Letter Arrangements**   How many distinguishable arrangements of all the letters in the word MISSISSAUGA are possible?

18. **Letter Arrangements**   How many distinguishable arrangements of all the letters in the word STREETSBORO are possible?

*19. **Coin Toss**   If a coin is tossed seven times and the outcome of each toss is noted, in how many ways can four heads and three tails occur?

20. Die Roll A die is rolled six times and the order of the rolls is considered. In how many ways can two 2's, three 3's, and one 4 occur?

21. Repair Scheduling An appliance repairman must go out on six service calls. In how many ways can he arrange his schedule?

22. Baseball A Little League baseball team has 12 members and must play an away game. Three cars will be used for transportation. In how many ways can the manager assign the members to specific cars if each car can accommodate four members?

*23. Project Assignment The director of research and development for a company has nine scientists who are equally qualified to work on projects A, B, and C. In how many ways can the director assign three scientists to each project?

24. Holly Bushes A landscaper plants a row hedge of nine young holly plants. The nursery guarantees that exactly five of them are female (and will thus have the characteristic red berries); however, it is not clear to the landscaper which of the young plants are the females. How many different arrangements of male and female plants may become evident when the females have berries?

25. True–False Exam A biology instructor includes several true–false questions on quizzes. From experience, a student believes that half of the questions are true and half are false. If there are 10 true–false questions on the next quiz, in how many ways can the student answer half of them "true" and the other half "false"?

26. Food Order A waiter takes the following order from a table with seven people: three hamburgers, two cheeseburgers, and two steak sandwiches. Upon returning with the food, he forgets who ordered what item and simply places an item in front of each person. In how many ways can the waiter do this?

*27. Caseworker Assignment A social services office has 15 new clients. The supervisor wants to assign 5 clients to each of three specific caseworkers. In how many ways can this be done?

28. Hockey There are 11 members on a hockey team, and all but one, the goalie, are qualified for the other five positions. In how many ways can the coach form a starting lineup?

29. Flag Signals Colored flags arranged vertically on a flagpole indicate a signal (or message). How many different signals are possible if

(a) two red, three green, and two yellow flags are used?

(b) two red, three green, and three yellow flags are available, and all the red and green flags and at least two yellow flags are used?

30. Hiring A company personnel director must hire six people: four for the assembly department and two for the shipping department. There are 10 applicants who are equally qualified to work in each department. In how many ways can the personnel director fill the positions?

31. Financial Portfolio A financial advisor wants to create a portfolio consisting of eight stocks and four bonds. If twelve stocks and seven bonds are acceptable for the portfolio, in how many ways can the portfolio be created?

32. World Series A baseball team wins the World Series if it is the first team in the series to win four games. Thus, a series could range from four to seven games. For example, a team winning the first four games would be the champion. Likewise, a team losing the first three games and winning the last four would be champion. In how many ways can a team win the World Series?

33. Subcommittee A committee has seven members, three of whom are male and four female. In how many ways can a subcommittee be selected if it is to consist exactly of

(a) three males?

(b) four females?

(c) two males and two females?

34. Subcommittee A committee has four male and four female members. In how many ways can a subcommittee of four be selected if at least two females are to serve on it?

35. Poker Hand A poker hand consists of 5 cards from a deck of 52 playing cards. The hand is a "full house" if there are 3 cards of one denomination and 2 cards of another. For example, three 10's and two jacks form a full house. How many full-house hands are possible?

36. Poker Hand In poker, 2 cards of the same denomination form a "pair." For example, two 8's form a pair. A poker hand (5 cards from a 52-card deck) is said to be a "two-pair" hand if it contains two pairs and there are three different face values involved in the five cards. For example, a pair of 3's, a pair of 8's, and a 10 constitute a two-pair hand. How many two-pair hands are possible?

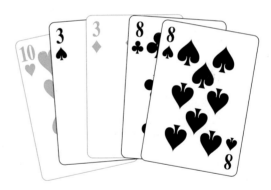

37. Tram Loading At a tourist attraction, two trams carry sightseers up a picturesque mountain. One tram can accommodate seven people and the other eight. A busload of 18 tourists arrives, and both trams are at the bottom of the mountain. Obviously, only 15 tourists can initially go up the mountain. In how many ways can the attendant load 15 tourists onto the two trams?

38. Discussion Groups A history instructor wants to split a class of 10 students into three discussion groups. One group will consist of four students and discuss topic A. The second and third groups will discuss topics B and C, respectively, and consist of three students each.

(a) In how many ways can the instructor form the groups?

(b) If the instructor designates a group leader and a secretary (different students) for each group, in how many ways can the class be split?

OBJECTIVE

8.3 Sample Spaces and Events

To determine a sample space and to consider events associated with it. (These notions involve sets and subsets.) To represent a sample space and events by means of a Venn diagram. To introduce the notions of complement, union, and intersection.

Sample Spaces

Inherent in any discussion of probability is the performance of an experiment (a procedure) in which a particular result, or *outcome*, involves chance. For example, consider the experiment of tossing a coin. There are only two ways the coin can fall, a head (H) or a tail (T), but the actual outcome is determined by chance. (We assume that the coin does not land on its edge.) The set of possible outcomes,

$$\{H, T\}$$

is called a *sample space* for the experiment, and H and T are called *sample points*.

> **DEFINITION**
>
> A *sample space* S for an experiment is the set of all possible outcomes of the experiment. The elements of S are called *sample points*. If there is a finite number of sample points, that number is denoted $\#(S)$, and S is said to be a *finite sample space*.

When determining "possible outcomes" of an experiment, we must be sure that they reflect the situation about which we are concerned. For example, consider the experiment of rolling a die and observing the top face. We could say that a sample space is

$$S_1 = \{1, 2, 3, 4, 5, 6\}$$

The order in which sample points are listed in a sample space is of no concern.

where the possible outcomes are the number of dots on the top face. However, other possible outcomes are

	odd number of dots appear	(odd)
and	even number of dots appear	(even)

Thus, the set

$$S_2 = \{\text{odd, even}\}$$

is also a sample space for the experiment, so you can see that an experiment may have more than one sample space.

If an outcome in S_1 occurred, then we know which outcome in S_2 occurred, but the reverse is not true. To describe this asymmetry, we say that S_1 is a **more primitive** sample space than S_2. Usually, the more primitive a sample space is, the more questions pertinent to the experiment it allows us to answer. For example, with S_1, we can answer such questions as

"Did a 3 occur?"
"Did a number greater than 2 occur?"
"Did a number less than 4 occur?"

But with S_2, we cannot answer these questions. As a rule of thumb, the more primitive a sample space is, the more elements it has and the more detail it indicates. Unless otherwise stated, when an experiment has more than one sample space, it will be our practice to consider only a sample space that gives sufficient detail to answer all pertinent questions relative to the experiment. For example, for the experiment of rolling a die and observing the top face, it will be tacitly understood that we are observing the number of dots. Thus, we will consider the sample space to be

$$S_1 = \{1, 2, 3, 4, 5, 6\}$$

and will refer to it as the *usual* sample space for the experiment.

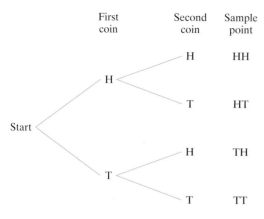

FIGURE 8.6   Tree diagram for toss of two coins.

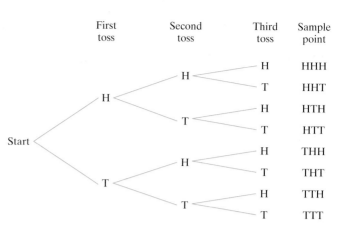

FIGURE 8.7   Tree diagram for three tosses of a coin.

EXAMPLE 1   Sample Space: Toss of Two Coins

Two different coins are tossed, and the result (H or T) for each coin is observed. Determine a sample space.

Solution: One possible outcome is a head on the first coin and a head on the second, which we can indicate by the ordered pair (H, H) or, more simply, HH. Similarly, we indicate a head on the first coin and a tail on the second by HT, and so on. A sample space is

$$S = \{HH, HT, TH, TT\}$$

A tree diagram is given in Figure 8.6 and, in a sense, indicates the sample space. We remark that *S* is also a sample space for the experiment of tossing a single coin twice in succession. In fact, these two experiments can be considered one and the same. Although other sample spaces can be contemplated, we take *S* to be the *usual* sample space for these experiments.

NOW WORK PROBLEM 3

PRINCIPLES IN PRACTICE 1

VIDEO CHOICES

A video store has 400 different movies to rent. A customer wants to rent 3 movies. If she chooses the videos at random, how many 3-movie choices (sample points) does she have?

FIGURE 8.8   Four colored jelly beans in a bag.

EXAMPLE 2   Sample Space: Three Tosses of a Coin

A coin is tossed three times, and the result of each toss is observed. Describe a sample space and determine the number of sample points.

Solution: Because there are three tosses, we choose a sample point to be an ordered *triple*, such as HHT, where each component is either H or T. By the Basic Counting Principle, the total number of sample points is $2 \cdot 2 \cdot 2 = 8$. A sample space (the *usual* one) is

$$S = \{HHH, HHT, HTH, HTT, THH, THT, TTH, TTT\}$$

and a tree diagram appears in Figure 8.7. Note that it is not necessary to list the entire sample space to determine the number of sample points in it.

NOW WORK PROBLEM 9

EXAMPLE 3   Sample Space: Jelly Beans in a Bag

A bag contains four jelly beans: one red, one pink, one black, and one white. (See Figure 8.8.)

a. *A jelly bean is withdrawn at random, its color is noted, and it is put back in the bag. Then a jelly bean is again randomly withdrawn and its color noted. Describe a sample space and determine the number of sample points.*

Solution: In this experiment we say that the two jelly beans are withdrawn **with replacement.** Let R, P, B, and W denote withdrawing a red, pink, black, and white jelly bean, respectively. Then our sample space consists of the sample points RW, PB, RB, WW, and so on, where (for example) RW represents the outcome that the first jelly bean withdrawn is red and the second is white. There are four possibilities for the first withdrawal and, since that jelly bean is placed back in the bag, four possibilities for the second withdrawal. By the Basic Counting Principle, the number of sample points is $4 \cdot 4 = 16$.

b. *Determine the number of sample points in the sample space if two jely beans are selected in succession* **without replacement** *and the colors are noted.*

Solution: The first jelly bean drawn can have any of four colors. Since it is *not* returned to the bag, the second jelly bean drawn can have any of the *three* remaining colors. Thus, the number of sample points is $4 \cdot 3 = 12$. Alternatively, there are $_4P_2 = 12$ sample points.

NOW WORK PROBLEM 7 ◖◗◗

EXAMPLE 4   Sample Space: Poker Hand

From an ordinary deck of 52 playing cards, a poker hand is dealt. Describe a sample space and determine the number of sample points.

Solution: A sample space consists of all combinations of 52 cards taken 5 at a time. From Example 3 of Section 8.2, the number of sample points is $_{52}C_5 = 2{,}598{,}960$.

NOW WORK PROBLEM 13 ◖◗◗

EXAMPLE 5   Sample Space: Roll of Two Dice

A pair of dice is rolled once, and for each die, the number that turns up is observed. Describe a sample space.

Solution: Think of the dice as being distinguishable, as if one were red and the other green. Each die can turn up in six ways, so we can take a sample point to be an ordered pair in which each component is an integer between 1 and 6, inclusive. For example, (4, 6), (3, 2), and (2, 3) are three different sample points. By the Basic Counting Principle, the number of sample points is $6 \cdot 6$, or 36.

NOW WORK PROBLEM 11 ◖◗◗

Events

At times, we are concerned with the outcomes of an experiment that satisfy a particular condition. For example, we may be interested in whether the outcome of rolling a single die is an even number. This condition can be considered as the set of outcomes {2, 4, 6}, which is a subset of the sample space

$$S = \{1, 2, 3, 4, 5, 6\}$$

In general, any subset of a sample space is called an *event* for the experiment. Thus,

$$\{2, 4, 6\}$$

is the event that an even number turns up, which can also be described by

$$\{x \text{ in } S | x \text{ is an even number}\}$$

Note that although an event is a set, it may be possible to describe it verbally as we just did. We often denote an event by E. When several events are involved in a discussion, they may be denoted by E, F, G, H, and so on, or by E_1, E_2, E_3, and so on.

DEFINITION

An *event E* for an experiment is a subset of the sample space for the experiment. If the outcome of the experiment is a sample point in E, then event E is said to *occur*.

In the previous experiment of rolling a die, we saw that $\{2, 4, 6\}$ is an event. Thus, if the outcome is a 2, that event occurs. Some other events are

$$E = \{1, 3, 5\} = \{x \text{ in } S | x \text{ is an odd number}\}$$

$$F = \{3, 4, 5, 6\} = \{x \text{ in } S | x \geq 3\}$$

$$G = \{1\}$$

A sample space is a subset of itself, so it, too, is an event, called the **certain event;** it must occur no matter what the outcome. An event, such as $\{1\}$, that consists of a single sample point is called a **simple event.** We can also consider an event such as $\{x \text{ in } S | x = 7\}$, which can be verbalized as "7 occurs". This event contains no sample points, so it is the empty set $\emptyset$ (the set with no elements in it). In fact, $\emptyset$ is called the **impossible event,** because it can never occur.

● EXAMPLE 6 **Events**

A coin is tossed three times, and the result of each toss is noted. The usual sample space (from Example 2) is

$$\{HHH, HHT, HTH, HTT, THH, THT, TTH, TTT\}$$

Determine the following events.

a. $E = \{\text{one head and two tails}\}$.

Solution:

$$E = \{HTT, THT, TTH\}$$

b. $F = \{\text{at least two heads}\}$.

Solution:

$$F = \{HHH, HHT, HTH, THH\}$$

c. $G = \{\text{all heads}\}$.

Solution:

$$G = \{HHH\}$$

d. $I = \{\text{head on first toss}\}$.

Solution:

$$I = \{HHH, HHT, HTH, HTT\}$$

NOW WORK PROBLEM 15

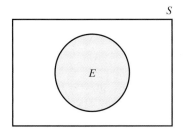

FIGURE 8.9 Venn diagram for sample space S and event E.

Sometimes it is convenient to represent a sample space S and an event E by a *Venn diagram,* as in Figure 8.9. The region inside the rectangle represents the sample points in S. (The sample points are not specifically shown.) The sample points in E are represented by the points inside the circle. Because E is a subset of S, the circular region cannot extend outside the rectangle.

With Venn diagrams, it is easy to see how events for an experiment can be used to form other events. Figure 8.10 shows sample space S and event E. The shaded region inside the rectangle, but outside the circle, represents the set of all sample points in S that are not in E. This set is an event called the *complement of E* and is denoted by E'.

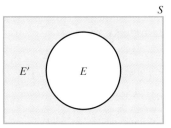

E' is the shaded region

FIGURE 8.10 Venn diagram for the complement of E.

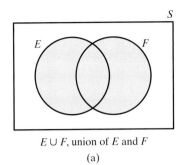
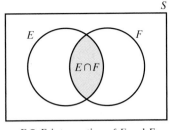

$E \cup F$, union of E and F

(a)

$E \cap F$, intersection of E and F

(b)

FIGURE 8.11   Representation of $E \cup F$ and $E \cap F$.

Figure 8.11(a) shows two events, E and F. The shaded region represents the set of all sample points either in E, or in F, or in both E and F. This set is an event called the *union* of E and F and is denoted by $E \cup F$. The shaded region in Figure 8.11(b) represents the event consisting of all sample points that are common to both E and F. This event is called the *intersection* of E and F and is denoted by $E \cap F$. In summary, we have the following definitions.

DEFINITIONS

Let S be a sample space for an experiment with events E and F. The **complement** of E, denoted by E', is the event consisting of all sample points in S that are not in E. The **union** of E and F, denoted by $E \cup F$, is the event consisting of all sample points that are either in E, or in F, or in both E and F. The **intersection** of E and F, denoted by $E \cap F$, is the event consisting of all sample points that are common to both E and F.

Note that if a sample point is in the event $E \cup F$, then the point is in at least one of the sets E and F. Thus, for the event $E \cup F$ to occur, *at least one* of the events E and F must occur, and conversely. On the other hand, if event $E \cap F$ occurs, then *both* E and F must occur, and conversely. If event E' occurs, then E *does not* occur, and conversely.

● EXAMPLE 7   Complement, Union, Intersection

Given the usual sample space

$$S = \{1, 2, 3, 4, 5, 6\}$$

for the rolling of a die, let E, F, and G be the events

$$E = \{1, 3, 5\} \quad F = \{3, 4, 5, 6\} \quad G = \{1\}$$

Determine each of the following events.

a. E'

Solution: Event E' consists of those sample points in S that are not in E, so

$$E' = \{2, 4, 6\}$$

We note that E' is the event that an even number appears.

b. $E \cup F$

Solution: We want the sample points in E, or F, or both. Thus,

$$E \cup F = \{1, 3, 4, 5, 6\}$$

c. $E \cap F$

Solution: The sample points common to both E and F are 3 and 5, so

$$E \cap F = \{3, 5\}$$

d. $F \cap G$

Solution: Since F and G have no sample point in common,

$$F \cap G = \emptyset$$

e. $E \cup E'$

Solution: Using the result of part (a), we have

$$E \cup E' = \{1, 3, 5\} \cup \{2, 4, 6\} = \{1, 2, 3, 4, 5, 6\} = S$$

f. $E \cap E'$

Solution:

$$E \cap E' = \{1, 3, 5\} \cap \{2, 4, 6\} = \emptyset$$

NOW WORK PROBLEM 17 ◖◗●

The results of Examples 7(e) and 7(f) can be generalized as follows:

If E is any event for an experiment with sample space S, then

$$E \cup E' = S \quad \text{and} \quad E \cap E' = \emptyset$$

Thus, the union of an event and its complement is the sample space; the intersection of an event and its complement is the empty set. These and other properties of events are listed in Table 8.1.

TABLE 8.1 Properties of Events

If E and F are any events for an experiment with sample space S, then

1. $E \cup E = E$	
2. $E \cap E = E$	
3. $(E')' = E$	(the complement of the complement of an event is the event)
4. $E \cup E' = S$	
5. $E \cap E' = \emptyset$	
6. $E \cup S = S$	
7. $E \cap S = E$	
8. $E \cup \emptyset = E$	
9. $E \cap \emptyset = \emptyset$	
10. $E \cup F = F \cup E$	(commutative property of union)
11. $E \cap F = F \cap E$	(commutative property of intersection)
12. $(E \cup F)' = E' \cap F'$	(the complement of a union is the intersection of complements)
13. $(E \cap F)' = E' \cup F'$	(the complement of an intersection is the union of complements)
14. $E \cup (F \cup G) = (E \cup F) \cup G$	(associative property of union)
15. $E \cap (F \cap G) = (E \cap F) \cap G$	(associative property of intersection)
16. $E \cap (F \cup G) = (E \cap F) \cup (E \cap G)$	(distributive property of intersection over union)
17. $E \cup (F \cap G) = (E \cup F) \cap (E \cup G)$	(distributive property of union over intersection)

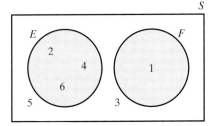

FIGURE 8.12   Mutually exclusive events.

When two events E and F have no sample point in common, that is,

$$E \cap F = \emptyset$$

they are called *mutually exclusive* or *disjoint* events. For example, in the rolling of a die, the events

$$E = \{2, 4, 6\} \quad \text{and} \quad F = \{1\}$$

are mutually exclusive (see Figure 8.12).

DEFINITION

Events E and F are said to be *mutually exclusive* events if and only if $E \cap F = \emptyset$.

When two events are mutually exclusive, the occurrence of one event means that the other event cannot occur; that is, the two events cannot occur simultaneously. An event and its complement are mutually exclusive, since $E \cap E' = \emptyset$.

● EXAMPLE 8   Mutually Exclusive Events

If E, F, and G are events for an experiment and F and G are mutually exclusive, show that events $E \cap F$ and $E \cap G$ are also mutually exclusive.

Solution: Given that $F \cap G = \emptyset$, we must show that the intersection of $E \cap F$ and $E \cap G$ is the empty set. Using the properties in Table 8.1, we have

$$
\begin{aligned}
(E \cap F) \cap (E \cap G) &= (E \cap F \cap E) \cap G && \text{(property 15)} \\
&= (E \cap E \cap F) \cap G && \text{(property 11)} \\
&= (E \cap F) \cap G && \text{(property 2)} \\
&= E \cap (F \cap G) && \text{(property 15)} \\
&= E \cap \emptyset && \text{(given)} \\
&= \emptyset && \text{(property 9)}
\end{aligned}
$$

NOW WORK PROBLEM 31 ●●●

Problems 8.3

In Problems 1–6, determine a sample space for the given experiment.

1. Card Selection   A card is drawn from a four-card deck consisting of the 9 of diamonds, 9 of hearts, 9 of clubs, and 9 of spades.

2. Coin Toss   A coin is tossed four times in succession, and the faces showing are observed.

*3. **Die Roll and Coin Toss**   A die is rolled and then a coin is tossed.

4. Dice Roll   Two dice are rolled, and the sum of the numbers that turn up is observed.

5. Digit Selection   Two different digits are selected, in succession, from those in the number "64901".

6. Genders of Children   The genders of the first, second, third, and fourth children of a four-child family are noted. (Let, for example, BGGB denote that the first, second, third, and fourth children are boy, girl, girl, boy, respectively.)

*7. **Jelly Bean Selection**   A bag contains three colored jelly beans: one red, one white, and one blue. Determine a sample space if (a) two jelly beans are selected with replacement and (b) two jelly beans are selected without replacement.

8. Manufacturing Process   A company makes a product that goes through three processes during its manufacture. The first is an assembly line, the second is a finishing line, and the third is an inspection line. There are three assembly lines (A, B, and C), two finishing lines (D and E), and two inspection lines (F and G). For each process, the company chooses a line at random. Determine a sample space.

In Problems 9–14, describe the nature of a sample space for the given experiment, and determine the number of sample points.

*9. **Coin Toss**   A coin is tossed six times in succession, and the faces showing are observed.

10. Dice Roll   Five dice are rolled, and the numbers that turn up are observed.

*11. **Card and Die**   A card is drawn from an ordinary deck of 52 cards, and then a die is rolled.

12. Rabbit Selection   From a hat containing nine rabbits, four rabbits are pulled successively without replacement.

*13. **Card Deal**   A 10-card hand is dealt from a deck of 52 cards. Do not evaluate your answer.

14. Letter Selection   A four-letter "word" is formed by successively choosing any four letters from the alphabet with replacement.

Suppose that $S = \{1, 2, 3, 4, 5, 6, 7, 8, 9, 10\}$ is the sample space for an experiment with events

$$E = \{1, 3, 5\} \quad F = \{3, 5, 7, 9\} \quad G = \{2, 4, 6, 8\}$$

In Problems 15–22, determine the indicated events.

*15. $E \cup F$ **16.** G'

*17. $E' \cap F$ **18.** $F' \cap G'$

19. F'

20. $(E \cup F)'$

21. $(F \cap G)'$

22. $(E \cup G) \cap F'$

(e) $E_1 \cap E_2$

(f) $(E_1 \cup E_2)'$

(g) $(E_1 \cap E_2)'$

23. Of the following events, which pairs are mutually exclusive?

$$E_1 = \{1, 2, 3\} \quad E_2 = \{3, 4, 5\}$$
$$E_3 = \{1, 2\} \quad E_4 = \{5, 6, 7\}$$

24. Card Selection From a standard deck of 52 playing cards, 2 cards are drawn without replacement. Suppose E_J is the event that both cards are jacks, E_C is the event that both cards are clubs, and E_3 is the event that both cards are 3's. Which pairs of these events are mutually exclusive?

25. Card Selection From a standard deck of 52 playing cards, 1 card is selected. Which pairs of the following events are mutually exclusive?

$$E = \{\text{diamond}\}$$

$$F = \{\text{face card}\}$$

$$G = \{\text{black}\}$$

$$H = \{\text{red}\}$$

$$I = \{\text{ace of diamonds}\}$$

26. Dice A red and a green die are thrown, and the numbers on each are noted. Which pairs of the following events are mutually exclusive?

$$E = \{\text{both are even}\}$$

$$F = \{\text{both are odd}\}$$

$$G = \{\text{sum is 2}\}$$

$$H = \{\text{sum is 4}\}$$

$$I = \{\text{sum is greater than 10}\}$$

27. Coin Toss A coin is tossed three times in succession, and the results are observed. Determine each of the following:

(a) The usual sample space S

(b) The event E_1 that at least one head occurs

(c) The event E_2 that at least one tail occurs

(d) $E_1 \cup E_2$

28. Genders of Children A husband and wife have two children. The outcome of the first child being a boy and the second a girl can be represented by BG. Determine each of the following:

(a) Sample space that describes all the orders of the possible genders of the children

(b) The event that at least one child is a girl

(c) The event that at least one child is a boy

(d) Is the event in part (c) the complement of the event in part (b)?

29. Arrivals Persons A, B, and C enter a building at different times. The outcome of A arriving first, B second, and C third can be indicated by ABC. Determine each of the following:

(a) The sample space involved for the arrivals

(b) The event that A arrives first

(c) The event that A does not arrive first

30. Supplier Selection A grocery store can order fruits and vegetables from suppliers U, V, and W; meat from suppliers U, V, X, and Y; and dry goods from suppliers V, W, X, and Z. The grocery store selects one supplier for each type of item. The outcome of U being selected for fruits and vegetables, V for meat, and W for dry goods can be represented by UVW.

(a) Determine a sample space.

(b) Determine the event E that one supplier supplies all the grocery store's requirements.

(c) Determine E' and give a verbal description of this event.

*31. If E and F are events for an experiment, prove that events $E \cap F$ and $E \cap F'$ are mutually exclusive.

32. If E and F are events for an experiment, show that

$$(E \cap F) \cup (E \cap F') = E$$

Note that from Problem 31, $E \cap F$ and $E \cap F'$ are mutually exclusive events. Thus, the foregoing equation expresses E as a union of mutually exclusive events. (*Hint:* Make use of a distributive property.)

OBJECTIVE

8.4 Probability

To define what is meant by the probability of an event. To develop formulas that are used in computing probabilities. Emphasis is placed on equiprobable spaces.

Equiprobable Spaces

We now introduce the basic concepts underlying the study of probability. Consider tossing a well-balanced die and observing the number that turns up. The usual sample space for the experiment is

$$S = \{1, 2, 3, 4, 5, 6\}$$

Before the experiment is performed, we cannot predict with certainty which of the six possible outcomes (sample points) will occur. But it does seem reasonable that each outcome has the same chance of occurring; that is, the outcomes are *equally likely*. This does not mean that in six tosses each number must turn up once. Rather, it means that if the experiment were performed a large number of times, each outcome would occur about $\frac{1}{6}$ of the time.

To be more specific, let the experiment be performed n times. Each performance of an experiment is called a **trial.** Suppose that we are interested in the event of obtaining a 1 (that is, the simple event consisting of the sample point 1). If a 1 occurs in k of these n trials, then the proportion of times that 1 occurs is k/n. This ratio is called the **relative frequency** of the event. Because getting a 1 is just one of six possible equally likely outcomes, we expect that in the long run a 1 will occur $\frac{1}{6}$ of the time. That is, as n becomes very large, we expect the relative frequency k/n to approach $\frac{1}{6}$. The number $\frac{1}{6}$ is taken to be the probability of getting a 1 on the toss of a well-balanced die, which is denoted $P(1)$. Thus, $P(1) = \frac{1}{6}$. Similarly, $P(2) = \frac{1}{6}$, $P(3) = \frac{1}{6}$, and so on.

In this experiment, all of the simple events in the sample space were understood to be equally likely to occur. To describe this equal likelihood, we say that S is an *equiprobable space*.

DEFINITION

A sample space S is called an *equiprobable space* if and only if all the simple events are equally likely to occur.

We remark that besides the phrase *equally likely*, other words and phrases used in the context of an equiprobable space are *well balanced, fair, unbiased,* and *at random*. For example, we may have a *well-balanced* die (as above), a *fair* coin, or *unbiased* dice, or we may select a jelly bean *at random* from a bag.

We now generalize our discussion of the die experiment to other (finite) equiprobable spaces.

DEFINITION

If S is an equiprobable sample space with N sample points (or outcomes) $s_1, s_2, \ldots, s_N$, then the *probability of the simple event* $\{s_i\}$ is given by

$$P(s_i) = \frac{1}{N}$$

for $i = 1, 2, \ldots, N$. Of course $P(s_i)$ is an abbreviation for $P(\{s_i\})$.

We remark that $P(s_i)$ can be interpreted as the relative frequency of $\{s_i\}$ occurring in the long run.

We can also assign probabilities to events that are not simple. For example, in the die experiment, consider the event E of a 1 or a 2 turning up:

$$E = \{1, 2\}$$

Because the die is well balanced, in n trials (where n is large) we expect that a 1 should turn up approximately $\frac{1}{6}$ of the time and a 2 should turn up approximately $\frac{1}{6}$ of the time. Thus, a 1 or 2 should turn up approximately $\frac{1}{6} + \frac{1}{6}$ of the time, or $\frac{2}{6}$ of the time. Hence, it is reasonable to assume that the long-run relative frequency of E is $\frac{2}{6}$. For this reason, we define $\frac{2}{6}$ to be the probability of E and denote it $P(E)$.

$$P(E) = \frac{1}{6} + \frac{1}{6} = \frac{2}{6}$$

Note that $P(E)$ is simply the sum of the probabilities of the simple events that form E. Equivalently, $P(E)$ is the ratio of the number of sample points in E (two) to the number of sample points in the sample space (six).

DEFINITION

If S is a finite equiprobable space for an experiment and $E = \{s_1, s_2, \ldots, s_j\}$ is an event, then the *probability of E* is given by

$$P(E) = P(s_1) + P(s_2) + \cdots + P(s_j)$$

Equivalently,

$$P(E) = \frac{\#(E)}{\#(S)}$$

where $\#(E)$ is the number of outcomes in E and $\#(S)$ is the number of outcomes in S.

Note that we can think of P as a function that associates with each event E the probability of E, namely, $P(E)$. The probability of E can be interpreted as the relative frequency of E occurring in the long run. Thus, in n trials, we would expect E to occur approximately $n \cdot P(E)$ times, provided that n is large.

EXAMPLE 1 Coin Tossing

Two fair coins are tossed. Determine the probability that

a. *two heads occur*

b. *at least one head occurs*

Solution: The usual sample space is

$$S = \{HH, HT, TH, TT\}$$

Since the four outcomes are equally likely, S is equiprobable and $\#(S) = 4$.

a. If $E = \{HH\}$, then E is a simple event, so

$$P(E) = \frac{\#(E)}{\#(S)} = \frac{1}{4}$$

b. Let $F = \{\text{at least one head}\}$. Then

$$F = \{HH, HT, TH\}$$

which has three outcomes. Thus,

$$P(F) = \frac{\#(F)}{\#(S)} = \frac{3}{4}$$

Alternatively,

$$P(F) = P(HH) + P(HT) + P(TH)$$
$$= \frac{1}{4} + \frac{1}{4} + \frac{1}{4} = \frac{3}{4}$$

Consequently, in 1000 trials of this experiment, we would expect F to occur approximately $1000 \cdot \frac{3}{4} = 750$ times.

NOW WORK PROBLEM 1

EXAMPLE 2 Cards

From an ordinary deck of 52 playing cards, 2 cards are randomly drawn without replacement. If E is the event that one card is a 2 and the other a 3, find P(E).

Solution: We can disregard the order in which the 2 cards are drawn. As our sample space S, we choose the set of all combinations of the 52 cards taken 2 at a time. Thus, S is equiprobable and $\#(S) = {}_{52}C_2$. To find $\#(E)$, we note that since there are four suits, a 2 can be drawn in four ways and a 3 in four ways. Hence, a 2 and a 3 can be drawn in $4 \cdot 4$ ways, so

$$P(E) = \frac{\#(E)}{\#(S)} = \frac{4 \cdot 4}{{}_{52}C_2} = \frac{16}{1326} = \frac{8}{663}$$

NOW WORK PROBLEM 7

EXAMPLE 3 Full House Poker Hand

Find the probability of being dealt a full house in a poker game. A full house is three of one kind and two of another, such as three queens and two 10's. Express your answer in terms of $_nC_r$.

Solution: The set of all combinations of 52 cards taken 5 at a time is an equiprobable sample space. (The order in which the cards are dealt is of no concern.) Thus, $\#(S) = {}_{52}C_5$. We now must find $\#(E)$, where E is the event of being dealt a full house.

Each of the four suits has 13 kinds so three cards of one kind can be dealt in $13 \cdot {}_4C_3$ ways. For each of these there are $12 \cdot {}_4C_2$ ways to be dealt two cards of another kind. Hence, a full house can be dealt in $13 \cdot {}_4C_3 \cdot 12 \cdot {}_4C_2$ ways, and we have

$$P(\text{full house}) = \frac{\#(E)}{\#(S)} = \frac{13 \cdot {}_4C_3 \cdot 12 \cdot {}_4C_2}{{}_{52}C_5}$$

NOW WORK PROBLEM 13 ⬤◯◯

EXAMPLE 4 Selecting a Subcommittee

From a committee of three males and four females, a subcommittee of four is to be randomly selected. Find the probability that it consists of two males and two females.

Solution: Since order of selection is not important, the number of subcommittees of four that can be selected from the seven members is ${}_7C_4$. The two males can be selected in ${}_3C_2$ ways and the two females in ${}_4C_2$ ways. By the Basic Counting Principle, the number of subcommittees of two males and two females is ${}_3C_2 \cdot {}_4C_2$. Thus,

$$P(\text{two males and two females}) = \frac{{}_3C_2 \cdot {}_4C_2}{{}_7C_4}$$

$$= \frac{\dfrac{3!}{2!1!} \cdot \dfrac{4!}{2!2!}}{\dfrac{7!}{4!3!}} = \frac{18}{35}$$

NOW WORK PROBLEM 21 ⬤◯◯

Properties of Probability

We now develop some properties of probability. Let S be an equiprobable sample space with N outcomes; that is, $\#(S) = N$. (We assume a finite sample space throughout this section.) If E is an event, then $0 \le \#(E) \le N$. Dividing each member by $\#(S) = N$ gives

$$0 \le \frac{\#(E)}{\#(S)} \le \frac{N}{N}$$

But $\dfrac{\#(E)}{\#(S)} = P(E)$, so we have the following property:

$$0 \le P(E) \le 1$$

That is, the probability of an event is a number between 0 and 1, inclusive.

Moreover, $P(\emptyset) = \dfrac{\#(\emptyset)}{\#(S)} = \dfrac{0}{N} = 0$. Thus,

$$P(\emptyset) = 0$$

Also, $P(S) = \dfrac{\#(S)}{\#(S)} = \dfrac{N}{N} = 1$, so

$$P(S) = 1$$

Accordingly, the probability of the impossible event is 0 and the probability of the certain event is 1.

Since $P(S)$ is the sum of the probabilities of the outcomes in the sample space, we conclude that the sum of the probabilities of all the simple events for a sample space is 1.

Now let us focus on the probability of the union of two events E and F. The event $E \cup F$ occurs if and only if at least one of the events (E or F) occurs. Thus, $P(E \cup F)$ is the probability that *at least one* of the events E and F occurs. We know that

$$P(E \cup F) = \frac{\#(E \cup F)}{\#(S)}$$

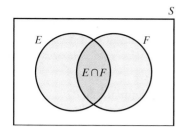

FIGURE 8.13 $E \cap F$ is contained in both E and F.

Note that while we derived Equation 2 for an equiprobable sample space, the result is in fact a general one.

Now

$$\#(E \cup F) = \#(E) + \#(F) - \#(E \cap F) \qquad (1)$$

because $\#(E) + \#(F) = \#(E \cup F) + \#(E \cap F)$. To see the truth of the last statement, look at Figure 8.13 and notice that $E \cap F$ is contained in *both E and F*.

Dividing both sides of Equation 1 for $\#(E \cup F)$ by $\#(S)$ gives the following result:

Probability of a Union of Events

If E and F are events, then

$$P(E \cup F) = P(E) + P(F) - P(E \cap F) \qquad (2)$$

For example, let a fair die be rolled, and let $E = \{1, 3, 5\}$ and $F = \{1, 2, 3\}$. Then $E \cap F = \{1, 3\}$, so

$$P(E \cup F) = P(E) + P(F) - P(E \cap F)$$
$$= \frac{3}{6} + \frac{3}{6} - \frac{2}{6} = \frac{2}{3}$$

Alternatively, $E \cup F = \{1, 2, 3, 5\}$, so $P(E \cup F) = \frac{4}{6} = \frac{2}{3}$.

If E and F are mutually exclusive events, then $E \cap F = \emptyset$ so $P(E \cap F) = P(\emptyset) = 0$. Hence, from Equation (2), we obtain the following law:

Addition Law for Mutually Exclusive Events

If E and F are *mutually exclusive* events, then

$$P(E \cup F) = P(E) + P(F)$$

For example, let a fair die be rolled, and let $E = \{2, 3\}$ and $F = \{1, 5\}$. Then $E \cap F = \emptyset$, so

$$P(E \cup F) = P(E) + P(F) = \frac{2}{6} + \frac{2}{6} = \frac{2}{3}$$

The addition law can be extended to more than two mutually exclusive events.[2] For example, if events E, F, and G are mutually exclusive, then

$$P(E \cup F \cup G) = P(E) + P(F) + P(G)$$

An event and its complement are mutually exclusive, so, by the addition law,

$$P(E \cup E') = P(E) + P(E')$$

But $P(E \cup E') = P(S) = 1$. Thus,

$$1 = P(E) + P(E')$$

so that

$$P(E') = 1 - P(E)$$

equivalently,

$$P(E) = 1 - P(E')$$

In order to find the probability of an event, sometimes it is more convenient first to find the probability of its complement and then subtract the result from 1. See, especially, Example 6(c).

Accordingly, if we know the probability of an event, then we can easily find the probability of its complement, and vice versa. For example, if $P(E) = \frac{1}{4}$, then $P(E') = 1 - \frac{1}{4} = \frac{3}{4}$. $P(E')$ is the probability that E does not occur.

---

[2]Two or more events are **mutually exclusive** if and only if no two of them can occur at the same time. That is, given any two of them, their intersection must be empty. For example, to say the events E, F, and G are mutually exclusive means that

$$E \cap F = E \cap G = F \cap G = \emptyset$$

● EXAMPLE 5 **Quality Control**

From a production run of 5000 *light bulbs,* 2% *of which are defective,* 1 *bulb is selected at random. What is the probability that the bulb is defective? What is the probability that it is not defective?*

Solution: In a sense this is trick question because the statement that "2% are defective" means that "$\frac{2}{100}$ are defective", which in turn means that the chance of getting a defective light bulb is "2 in a hundred", equivalenly, that the probability of getting a defective light bulb is 0.02. However, to reinforce the ideas we have considered so far, let us say that the sample space S consists of the 5000 bulbs. Since a bulb is selected at random, the possible outcomes are equally likely. Let E be the event of selecting a defective bulb. The number of outcomes in E is $0.02 \cdot 5000 = 100$. Thus,

$$P(E) = \frac{\#(E)}{\#(S)} = \frac{100}{5000} = \frac{1}{50} = 0.02$$

Alternatively, since the probability of selecting a particular bulb is $\frac{1}{5000}$ and E contains 100 sample points, by summing probabilities we have

$$P(E) = 100 \cdot \frac{1}{5000} = 0.02$$

The event that the bulb selected is *not* defective is E'. Hence,

$$P(E') = 1 - P(E) = 1 - 0.02 = 0.98$$

NOW WORK PROBLEM 17

● EXAMPLE 6 **Dice**

A pair of well-balanced dice is rolled, and the number on each die is noted. Determine the probability that the sum of the numbers that turn up is (a) 7, *(b)* 7 *or* 11, *and (c) greater than* 3.

Solution: Since each die can turn up in any of six ways, by the Basic Counting Principle the number of possible outcomes is $6 \cdot 6 = 36$. Our sample space consists of the following ordered pairs:

$$\begin{array}{cccccc}
(1,1) & (1,2) & (1,3) & (1,4) & (1,5) & (1,6) \\
(2,1) & (2,2) & (2,3) & (2,4) & (2,5) & (2,6) \\
(3,1) & (3,2) & (3,3) & (3,4) & (3,5) & (3,6) \\
(4,1) & (4,2) & (4,3) & (4,4) & (4,5) & (4,6) \\
(5,1) & (5,2) & (5,3) & (5,4) & (5,5) & (5,6) \\
(6,1) & (6,2) & (6,3) & (6,4) & (6,5) & (6,6)
\end{array}$$

The outcomes are equally likely, so the probability of each outcome is $\frac{1}{36}$. There are a lot of characters present in the preceding list, since each of the 36 ordered pairs involves five (a pair of parentheses, a comma, and 2 digits) for a total of $36 \cdot 5 = 180$ characters. The same information is conveyed by the following coordinatized boxes, requiring just 12 characters and 14 lines.

	1	2	3	4	5	6
1						
2						
3						
4						
5						
6						

(You are encouraged to use abbreviations like this in your written work.)

a. Let E_7 be the event that the sum of the numbers appearing is 7. Then

$$E_7 = \{(1,6), (2,5), (3,4), (4,3), (5,2), (6,1)\}$$

which has six outcomes (and can be seen as the rising diagonal in the coordinatized boxes). Thus,

$$P(E_7) = \frac{6}{36} = \frac{1}{6}$$

b. Let $E_{7 \text{ or } 11}$ be the event that the sum is 7 or 11. If E_{11} is the event that the sum is 11, then

$$E_{11} = \{(5, 6), (6, 5)\}$$

which has two outcomes. Since $E_{7 \text{ or } 11} = E_7 \cup E_{11}$ and E_7 and E_{11} are mutually exclusive, we have

$$P(E_{7 \text{ or } 11}) = P(E_7) + P(E_{11}) = \frac{6}{36} + \frac{2}{36} = \frac{8}{36} = \frac{2}{9}$$

Alternatively, we can determine $P(E_{7 \text{ or } 11})$ by counting the number of outcomes in $E_{7 \text{ or } 11}$. We obtain

$$E_{7 \text{ or } 11} = \{(1, 6), (2, 5), (3, 4), (4, 3), (5, 2), (6, 1), (5, 6), (6, 5)\}$$

which has eight outcomes. Thus,

$$P(E_{7 \text{ or } 11}) = \frac{8}{36} = \frac{2}{9}$$

c. Let E be the event that the sum is greater than 3. The number of outcomes in E is relatively large. Thus, to determine $P(E)$, it is easier to find E', rather than E, and then use the formula $P(E) = 1 - P(E')$. Here E' is the event that the sum is 2 or 3. We have

$$E' = \{(1, 1), (1, 2), (2, 1)\}$$

which has three outcomes. Hence,

$$P(E) = 1 - P(E') = 1 - \frac{3}{36} = \frac{11}{12}$$

NOW WORK PROBLEM 27

● EXAMPLE 7   **Interrupted Gambling**

Obtain Pascal and Fermat's solution to the problem of dividing the pot between two gamblers in an interrupted game of chance, as described in the introduction to this chapter. Recall that when the game was interrupted, Player 1 needed r more "rounds" to win the pot outright and that Player 2 needed s more rounds to win. It is agreed that the pot should be divided so that each player gets the value of the pot multiplied by the probability that he or she would have won an uninterrupted game.

Solution: We need only compute the probability that Player 1 would have won, for if that is p, then the probability that Player 2 would have won is $1 - p$. Now the game can go at most $r + s - 1$ more rounds. To see this, observe that each round produces exactly one winner and let a be the number of the $r + s - 1$ rounds won by Player 1 and let b be the number of the $r + s - 1$ rounds won by Player 2. So $r + s - 1 = a + b$. If neither Player 1 nor Player 2 has won, then $a \leq r - 1$ and $b \leq s - 1$. But in this case we have

$$r + s - 1 = a + b \leq (r - 1) + (s - 1) = r + s - 2$$

which is impossible. It is clear that after $r + s - 2$ there *might* not yet be an overall winner, so we do need to consider a further $r + s - 1$ possible rounds from the time of interruption. Let $n = r + s - 1$. Now Player 1 will win if Player 2 wins k of the n possible further n rounds where $0 \leq k \leq s - 1$. Let E_k be the event that Player 2 wins *exactly* k of the next n rounds. Since the events E_k, for $k = 0, 1, \cdots s - 1$, are mutually exclusive, the probability that Player 1 will win is given by

$$P(E_0 \cup E_1 \cup \cdots \cup E_{s-1}) = P(E_0) + P(E_1) + \cdots + P(E_{s-1}) = \sum_{k=0}^{s-1} P(E_k) \quad (3)$$

It remains to determine $P(E_k)$. We may as well suppose that a round consists of flipping a coin, with outcomes H and T. We further take a single round win for Player 2 to be an outcome of T. Thus Player 2 will win exactly k of the next rounds if exactly k of the next n outcomes are T's. The number of possible outcomes for the next n rounds is of course 2^n, by the multiplication principle. The number of these outcomes which consist of exactly k T's is the number of ways of choosing k from among n. It follows that $P(E_k) = \dfrac{{}_nC_k}{2^n}$, and substituting this value in Equation (3), we obtain

$$\sum_{n=0}^{s-1} \frac{{}_nC_k}{2^n}$$

NOW WORK PROBLEM 29

Probability Functions in General

Many of the properties of equiprobable spaces carry over to sample spaces that are not equiprobable. To illustrate, consider the experiment of tossing two fair coins and observing the number of heads. The coins can fall in one of four ways, namely,

HH   HT   TH   TT

which correspond to two heads, one head, one head, and zero heads, respectively. Because we are interested in the number of heads, we can choose a sample space to be

$$S = \{0, 1, 2\}$$

However, the simple events in S are *not* equally likely to occur because of the four possible ways in which the coins can fall: Two of these ways correspond to the one-head outcome, whereas only one corresponds to the two-head outcome and similarly for the zero-head outcome. In the long run, it is reasonable to expect repeated trials to result in one head about $\frac{2}{4}$ of the time, zero heads about $\frac{1}{4}$ of the time, and two heads about $\frac{1}{4}$ of the time. If we were to assign probabilities to these simple events, it is natural to have

$$P(0) = \frac{1}{4} \quad P(1) = \frac{2}{4} = \frac{1}{2} \quad P(2) = \frac{1}{4}$$

Although S is not equiprobable, these probabilities lie between 0 and 1, inclusive, and their sum is 1. This is consistent with what was stated for an equiprobable space.

Based on our discussion, we can consider a *probability function* that relates to sample spaces in general.

DEFINITION

Let $S = \{s_1, s_2, \ldots, s_N\}$ be a sample space for an experiment. The function P is called a *probability function* if both of the following are true:

1. $0 \le P(s_i) \le 1$ for $i = 1$ to N
2. $P(s_1) + P(s_2) + \cdots + P(s_N) = 1$

If E is an event, then $P(E)$ is the sum of the probabilities of the sample points in E. We define $P(\emptyset)$ to be 0.

From a mathematical point of view, any function P that satisfies conditions 1 and 2 is a probability function for a sample space. For example, consider the sample space for the previous experiment of tossing two fair coins and observing the number of heads:

$$S = \{0, 1, 2\}$$

We could assign probabilities as follows:

$$P(0) = 0.1 \quad P(1) = 0.2 \quad P(2) = 0.7$$

Here P satisfies both conditions 1 and 2 and thus is a legitimate probability function. However, this assignment does not reflect the long-run interpretation of probability and, consequently, would not be acceptable from a practical point of view.

In general, for any probability function defined on a sample space (finite or infinite), the following properties hold:

$$P(E') = 1 - P(E)$$

$$P(S) = 1$$

$$P(E_1 \cup E_2) = P(E_1) + P(E_2) \quad \text{if } E_1 \cap E_2 = \emptyset$$

Empirical Probability

You have seen how easy it is to assign probabilities to simple events when we have an equiprobable sample space. For example, when a fair coin is tossed, we have $S = \{H, T\}$ and $P(H) = P(T) = \frac{1}{2}$. These probabilities are determined by the intrinsic nature of the experiment—namely, that there are two possible outcomes that should have the same probability because the outcomes are equally likely. Such probabilities are called *theoretical* probabilities. However, suppose the coin is not fair. How can probabilities then be assigned? By tossing the coin a number of times, we can determine the relative frequencies of heads and tails occurring. For example, suppose that in 1000 tosses, heads occurs 517 times and tails occurs 483 times. Then the relative frequencies of heads and tails occurring are $\frac{517}{1000}$ and $\frac{483}{1000}$, respectively. In this situation, the assignment $P(H) = 0.517$ and $P(T) = 0.483$ would be quite reasonable. Probabilities assigned in this way are called *empirical probabilities*. In general, probabilities based on sample or historical data are empirical. Now suppose that the coin were tossed 2000 times, and the relative frequencies of heads and tails occurring were $\frac{1023}{2000} = 0.5115$ and $\frac{977}{2000} = 0.4885$, respectively. Then in this case, the assignment $P(H) = 0.5115$ and $P(T) = 0.4885$ would be acceptable. In a certain sense, the latter probabilities may be more indicative of the true nature of the coin than would be the probabilities associated with 1000 tosses.

In the next example, probabilities (empirical) are assigned on the basis of sample data.

● EXAMPLE 8   **Opinion Poll**

An opinion poll of a sample of 150 adult residents of a town was conducted. Each person was asked his or her opinion about floating a bond issue to build a community swimming pool. The results are summarized in Table 8.2.

Suppose an adult resident from the town is randomly selected. Let M be the event "male selected" and F be the event "selected person favors the bond issue." Find each of the following:

a. $P(M)$

b. $P(F)$

c. $P(M \cap F)$

d. $P(M \cup F)$

TABLE 8.2 Opinion Poll

	Favor	Oppose	Total
Male	60	20	80
Female	40	30	70
Total	100	50	150

Strategy We will assume that proportions that apply to the sample also apply to the adult population of the town.

Solution:

a. Of the 150 persons in the sample, 80 are males. Thus, for the adult population of the town (the sample space), we assume that $\frac{80}{150}$ are male. Hence, the (empirical) probability of selecting a male is

$$P(M) = \frac{80}{150} = \frac{8}{15}$$

b. Of the 150 persons in the sample, 100 favor the bond issue. Therefore,

$$P(F) = \frac{100}{150} = \frac{2}{3}$$

c. Table 8.2 indicates that 60 males favor the bond issue. Hence,

$$P(M \cap F) = \frac{60}{150} = \frac{2}{5}$$

d. To find $P(M \cup F)$, we use Equation (1):

$$P(M \cup F) = P(M) + P(F) - P(M \cap F)$$
$$= \frac{80}{150} + \frac{100}{150} - \frac{60}{150} = \frac{120}{150} = \frac{4}{5}$$

NOW WORK PROBLEM 33

Odds

The probability of an event is sometimes expressed in terms of *odds*, especially in gaming situations.

> **DEFINITION**
> The *odds* in favor of event E occurring are the ratio
>
> $$\frac{P(E)}{P(E')}$$
>
> provided that $P(E') \neq 0$. Odds are usually expressed as the ratio $\frac{p}{q}$ (or $p : q$) of two positive integers, which is read "p to q."

● EXAMPLE 9 **Odds for an A in an Exam**

A student believes that the probability of getting an A on the next mathematics exam is 0.2. What are the odds (in favor) of this occurring?

Solution: If $E =$ "gets an A," then $P(E) = 0.2$ and $P(E') = 1 - 0.2 = 0.8$. Hence, the odds of getting an A are

$$\frac{P(E)}{P(E')} = \frac{0.2}{0.8} = \frac{2}{8} = \frac{1}{4} = 1 : 4$$

That is, the odds are 1 to 4. (We remark that the odds *against* getting an A are 4 to 1.)

If the odds that event E occurs are $a : b$, then the probability of E can be easily determined. We are given that

$$\frac{P(E)}{1 - P(E)} = \frac{a}{b}$$

Solving for $P(E)$ gives

$$bP(E) = (1 - P(E))a \qquad \text{(clearing fractions)}$$

$$aP(E) + bP(E) = a$$

$$(a + b)P(E) = a$$

$$P(E) = \frac{a}{a + b}$$

Finding Probability from Odds

If the odds that event E occurs are $a : b$, then

$$P(E) = \frac{a}{a + b}$$

Over the long run, if the odds that E occurs are $a : b$, then, on the average, E should occur a times in every $a + b$ trials of the experiment.

● EXAMPLE 10 **Probability of Winning a Prize**

A $1000 savings bond is one of the prizes listed on a contest brochure received in the mail. The odds in favor of winning the bond are stated to be 1 : 10,000. What is the probability of winning this prize?

Solution: Here $a = 1$ and $b = 10,000$. From the preceding rule,

$$P(\text{winning prize}) = \frac{a}{a + b}$$

$$= \frac{1}{1 + 10,000} = \frac{1}{10,001}$$

NOW WORK PROBLEM 35 ●●

Problems 8.4

*1. In 3000 trials of an experiment, how many times would you expect event E to occur if $P(E) = 0.25$?

2. In 3000 trials of an experiment, how many times would you expect event E to occur if $P(E') = 0.45$?

3. If $P(E) = 0.2$, $P(F) = 0.3$, and $P(E \cap F) = 0.1$, find (a) $P(E')$ and (b) $P(E \cup F)$.

4. If $P(E) = \frac{1}{4}$, $P(F) = \frac{1}{2}$, and $P(E \cap F) = \frac{1}{8}$, find (a) $P(E')$ and (b) $P(E \cup F)$.

5. If $P(E \cap F) = 0.831$, are E and F mutually exclusive?

6. If $P(E) = \frac{1}{2}$, $P(E \cup F) = \frac{13}{20}$, and $P(E \cap F) = \frac{1}{10}$, find $P(F)$.

*7. **Dice**   A pair of well-balanced dice is tossed. Find the probability that the sum of the numbers is (a) 8; (b) 2 or 3; (c) 3, 4, or 5; (d) 12 or 13; (e) even; (f) odd; and (g) less than 10.

8. **Dice**   A pair of fair dice is tossed. Determine the probability that at least one die shows a 2 or a 3.

9. **Card Selection**   A card is randomly selected from a standard deck of 52 playing cards. Determine the probability that the card is (a) the king of hearts, (b) a diamond, (c) a jack, (d) red, (e) a heart or a club, (f) a club and a 4, (g) a club or a 4, (h) red and a king, and (i) a spade and a heart.

10. **Coin and Die**   A fair coin and a fair die are tossed. Find the probability that (a) a head and a 5 show, (b) a head shows, (c) a 3 shows, and (d) a head and an even number show.

11. **Coin, Die, and Card**   A fair coin and a fair die are tossed, and a card is randomly selected from a standard deck of 52 playing cards. Determine the probability that the coin, die, and card, respectively, show (a) a tail, a 3, and the queen of hearts; (b) a tail, a 3, and a queen; (c) a head, a 2 or 3, and a queen; and (d) a head, an even number, and a diamond.

12. **Coins**   Three fair coins are tossed. Find the probability that (a) three heads show, (b) exactly one tail shows, (c) no more than two heads show, and (d) no more than one tail shows.

*13. **Card Selection**   Three cards from a standard deck of 52 playing cards are successively drawn at random without replacement. Find the probability that (a) all three cards are kings and (b) all three cards are hearts.

14. **Card Selection**   Two cards from a standard deck of 52 playing cards are successively drawn at random with replacement. Find the probability that (a) both cards are kings and (b) one card is a king and the other is a heart.

15. **Genders of Children**   Assuming that the gender of a person is determined at random, determine the probability

that a family with three children has (a) three girls, (b) exactly one boy, (c) no girls, and (d) at least one girl.

16. Jelly Bean Selection A jelly bean is randomly taken from a bag that contains seven red, three white, and eight blue jelly beans. Find the probability that the jelly bean is (a) blue, (b) not red, (c) red or white, (d) neither red nor blue, (e) yellow, and (f) red or yellow.

*17. **Stock Selection** A stock is selected at random from a list of 60 utility stocks, 48 of which have an annual dividend yield of 6% or more. Find the probability that the stock pays an annual dividend that yields (a) 6% or more and (b) less than 6%.

18. Inventory A clothing store maintains its inventory of ties so that 40% are 100% pure silk. If a tie is selected at random, what is the probability that it is (a) 100% pure silk and (b) not 100% pure silk?

19. Examination Grades On an examination given to 40 students, 10% received an A, 25% a B, 35% a C, 25% a D, and 5% an F. If a student is selected at random, what is the probability that the student (a) received an A, (b) received an A or a B, (c) received neither a D nor an F, and (d) did not receive an F? (e) Answer questions (a)–(d) if the number of students that were given the examination is unknown.

20. Jelly Bean Selection Two bags contain colored jelly beans. Bag 1 contains three red and two green jelly beans, and bag 2 contains four red and five green jelly beans. A jelly bean is selected at random from each bag. Find the probability that (a) both jelly beans are red and (b) one jelly bean is red and the other is green.

*21. **Committee Selection** From a group of two women and three men, two persons are selected at random to form a committee. Find the probability that the committee consists of women only.

22. Committee Selection For the committee selection in Problem 21, find the probability that the committee consists of a man and a woman.

23. Examination Score A student answers each question on a 10-question true–false examination in a random fashion. If each question is worth 10 points, what is the probability that the student scores (a) 100 points and (b) 90 or more points?

24. Multiple-Choice Examination On an eight-question, multiple-choice examination, there are four choices for each question, only one of which is correct. If a student answers each question in a random fashion, find the probability that the student answers (a) each question correctly and (b) exactly four questions correctly.

25. Poker Hand Find the probability of being dealt four of a kind in a poker game. This simply means four of one kind and one other card, such as four queens and a 10. Express your answer using the symbol $_nC_r$.

26. Suppose $P(E) = \frac{1}{4}$, $P(E \cup F) = \frac{5}{14}$, and $P(E \cap F) = \frac{1}{7}$.

(a) Find $P(F)$ (b) Find $P(E' \cup F)$

[*Hint:*

$$F = (E \cap F) \cup (E' \cap F)$$

where $E \cap F$ and $E' \cap F$ are mutually exclusive.]

*27. **Faculty Committee** The classification of faculty at a college is indicated in Table 8.3. If a committee of three faculty members is selected at random, what is the

probability that it consists of (a) all females; (b) a professor and two associate professors?

TABLE 8.3 Faculty Classification

	Male	Female	Total
Professor	12	3	15
Associate Professor	15	9	24
Assistant Professor	18	8	26
Instructor	20	15	35
Total	65	35	100

28. Biased Die A die is biased such that $P(1) = \frac{3}{10}$, $P(2) = P(5) = \frac{2}{10}$, and $P(3) = P(4) = P(6) = \frac{1}{10}$. If the die is tossed, find P(even number).

*29. **Interrupted Game** A pair of gamblers are tossing a coin and calling so that exactly one of them wins each toss. There is a pot of $25, which they agree will go to the first one to win 10 tosses. Their mothers arrive on the scene and call a halt to the game when Shiloh has won 7 tosses and Caitlin has won 5. Later Shiloh and Caitlin split the money according to the Pascal and Fermat formula. What is Shiloh's share of the pot?

30. Interrupted Game Repeat Problem 30 for Shiloh and Caitlin's next meeting when the police break up their game of 10 tosses for $50, with Shiloh having won 5 tosses and Caitlin only 2.

31. Biased Die When a biased die is tossed, the probabilities of 1, 3, and 5 showing are the same. The probabilities of 2, 4, and 6 showing are also the same, but are twice those of 1, 3, and 5. Determine $P(1)$.

32. For the sample space $\{a, b, c, d, e, f, g\}$, suppose that the probabilities of a, b, c, d, and e are the same and that the probabilities of f and g are the same. Is it possible to determine $P(f)$? If it is also known that $P(\{a, f\}) = \frac{1}{3}$, what more can be said?

*33. **Tax Increase** A legislative body is considering a tax increase to support education. A survey of 100 registered voters was conducted, and the results are indicated in Table 8.4. Assume that the survey reflects the opinions of the voting population. If a person from that population is selected at random, determine each of the following (empirical) probabilities.

(a) P(favors tax increase)
(b) P(opposes tax increase)
(c) P(is a Republican with no opinion)

TABLE 8.4 Tax Increase Survey

	Favor	Oppose	No Opinion	Total
Democrat	32	26	2	60
Republican	15	17	3	35
Other	4	1	0	5
Total	51	44	5	100

34. Digital Camcorder Sales A department store chain has stores in the cities of Exton and Whyton. Each store sells three brands of camcorders, A, B, and C. Over the past year, the average monthly unit sales of the camcorders was determined, and the results are indicated in Table 8.5. Assume that future sales follow the pattern indicated in the table.

(a) Determine the probability that a sale of a camcorder next month is for brand B.

(b) Next month, if a sale occurs at the Exton store, find the probability that it is for brand C.

TABLE 8.5 Unit Sales per Month

	A	B	C
Exton	25	40	30
Whyton	20	25	30

In Problems 35–38, for the given probability, find the odds that E will occur.

*35. $P(E) = \frac{4}{5}$

36. $P(E) = \frac{1}{6}$

37. $P(E) = 0.7$

38. $P(E) = 0.001$

In Problems 37–40, the odds that E will occur are given. Find P(E).

39. $7 : 5$

40. $100 : 1$

41. $4 : 10$

42. $a : a$

43. Weather Forecast A television weather forecaster reported a 75% chance of rain tomorrow. What are the odds that it will rain tomorrow?

44. If the odds of event *E not* occurring are $3 : 5$, what are the odds that *E* does occur? Repeat the question with the odds of event *E not* occurring being $a : b$.

To discuss conditional probability via a reduced sample space as well as the original space. To analyze a stochastic process with the aid of a probability tree. To develop the general multiplication law for $P(E \cap F)$.

8.5 Conditional Probability and Stochastic Processes

Conditional Probability

The probability of an event could be affected when additional related information about the experiment is known. For example, if you guess at the answer to a multiple-choice question having five choices, the probability of getting the correct answer is $\frac{1}{5}$. However, if you know that answers A and B are wrong and thus can be ignored, the probability of guessing the correct answer increases to $\frac{1}{3}$. In this section, we consider similar situations in which we want the probability of an event E when it is known that some other event F has occurred. This is called a **conditional probability** and is denoted by $P(E|F)$, which is read "the conditional probability of E, given F." For instance, in the situation involving the multiple-choice question, we have

$$P(\text{guessing correct answer}|\text{A and B eliminated}) = \frac{1}{3}$$

To investigate the notion of conditional probability, we consider the following situation. A fair die is rolled, and we are interested in the probability of the event

$$E = \{\text{even number shows}\}$$

The usual equiprobable sample space for this experiment is

$$S = \{1, 2, 3, 4, 5, 6\}$$

so

$$E = \{2, 4, 6\}$$

Thus,

$$P(E) = \frac{\#(E)}{\#(S)} = \frac{3}{6} = \frac{1}{2}$$

Now we change the situation a bit. Suppose the die is rolled out of our sight, and then we are told that a number greater than 3 occurred. In light of this additional information, what now is the probability of an even number? To answer that question, we reason as follows. The event F of a number greater than 3 is

$$F = \{4, 5, 6\}$$

Since F already occurred, the set of possible outcomes is no longer S; it is F. That is, F becomes our new sample space, called a **reduced sample space** or a *subspace* of S. The outcomes in F are equally likely, and, of these, only 4 and 6 are favorable to E;

that is,

$$E \cap F = \{4, 6\}$$

Since two of the three outcomes in the reduced sample space are favorable to an even number occurring, we say that $\frac{2}{3}$ is *the conditional probability of an even number, given that a number greater than 3 occurred*:

$$P(E|F) = \frac{\#(E \cap F)}{\#(F)} = \frac{2}{3} \tag{1}$$

The Venn diagram in Figure 8.14 illustrates the situation.

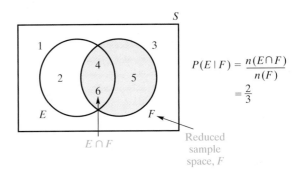

FIGURE 8.14 Venn diagram for conditional probability.

If we compare the conditional probability $P(E|F) = \frac{2}{3}$ with the "unconditional" probability $P(E) = \frac{1}{2}$, we see that $P(E|F) > P(E)$. This means that knowing that a number greater than 3 occurred *increases* the likelihood that an even number occurred. There are situations, however, in which conditional and unconditional probabilities are the same. These are discussed in the next section.

In summary, we have the following generalization of Equation (1):

Formula for a Conditional Probability

If E and F are events associated with an equiprobable sample space and $F \neq \emptyset$, then

$$P(E|F) = \frac{\#(E \cap F)}{\#(F)} \tag{2}$$

Since $E \cap F$ and $E' \cap F$ are disjoint events whose union is F, it is easy to see that

$$P(E|F) + P(E'|F) = 1$$

from which we get

$$P(E'|F) = 1 - P(E|F)$$

● **EXAMPLE 1** **Jelly Beans in a Bag**

A bag contains two blue jelly beans (say, B_1 and B_2) and two white jelly beans (W_1 and W_2). If two jelly beans are randomly taken from the bag, without replacement, find the probability that the second jelly bean taken is white, given that the first one is blue. (See Figure 8.15.)

Solution: For our equiprobable sample space, we take all ordered pairs, such as (B_1, W_2) and (W_2, W_1), whose components indicate the jelly beans selected on the first and on the second draw. Let B and W be the events

$$B = \{\text{blue on first draw}\}$$

$$W = \{\text{white on second draw}\}$$

FIGURE 8.15 Two white and two blue jelly beans in a bag.

We are interested in

$$P(W|B) = \frac{\#(W \cap B)}{\#(B)}$$

The reduced sample space B consists of all outcomes in which a blue jelly bean is drawn first:

$$B = \{(B_1, B_2), (B_1, W_1), (B_1, W_2), (B_2, B_1), (B_2, W_1), (B_2, W_2)\}$$

Event $W \cap B$ consists of the outcomes in B for which the second jelly bean is white:

$$W \cap B = \{(B_1, W_1), (B_1, W_2), (B_2, W_1), (B_2, W_2)\}$$

Since $\#(B) = 6$ and $\#(W \cap B) = 4$, we have

$$P(W|B) = \frac{4}{6} = \frac{2}{3}$$

NOW WORK PROBLEM 1

Example 1 showed how efficient the use of a reduced sample space can be. Note that it was not necessary to list all the outcomes either in the original sample space or in event W. Although we listed the outcomes in B, we could have found $\#(B)$ by using counting methods:

There are two ways in which the first jelly bean can be blue, and three possibilities for the second jelly bean, which can be either the remaining blue jelly bean or one of the two white jelly beans. Thus, $\#(B) = 2 \cdot 3 = 6$.

The number $\#(W \cap B)$ could also be found by means of counting methods.

EXAMPLE 2   Survey

In a survey of 150 people, each person was asked his or her marital status and opinion about floating a bond issue to build a community swimming pool. The results are summarized in Table 8.6. If one of these persons is randomly selected, find each of the following conditional probabilities.

TABLE 8.6 Survey

	Favor (F)	**Oppose (F')**	**Total**
Married (M)	60	20	80
Single (M')	40	30	70
Total	100	50	150

a. *The probability that the person favors the bond issue, given that the person is married.*

Solution: We are interested in $P(F|M)$. The reduced sample space (M) contains 80 married persons, of which 60 favor the bond issue. Thus,

$$P(F|M) = \frac{\#(F \cap M)}{\#(M)} = \frac{60}{80} = \frac{3}{4}$$

b. *The probability that the person is married, given that the person favors the bond issue.*

Solution: We want to find $P(M|F)$. The reduced sample space (F) contains 100 persons who favor the bond issue. Of these, 60 are married. Hence,

$$P(M|F) = \frac{\#(M \cap F)}{\#(F)} = \frac{60}{100} = \frac{3}{5}$$

Note that here $P(M|F) \neq P(F|M)$. Equality is possible precisely if $P(M) = P(F)$, assuming that all of $P(M)$, $P(F)$, and $P(M \cap F)$ are not equal to zero.

Another method of computing a conditional probability is by means of a formula involving *probabilities* with respect to the *original* sample space itself. Before stating the formula, we will provide some motivation so that it seems reasonable to you. (The discussion that follows is oversimplified in the sense that certain assumptions are tacitly made.)

To consider $P(E|F)$, we will assume that event F has probability $P(F)$ and event $E \cap F$ has probability $P(E \cap F)$. Let the experiment associated with this problem be repeated n times, where n is very large. Then the number of trials in which F occurs is approximately $n \cdot P(F)$. Of these, the number in which event E *also* occurs is approximately $n \cdot P(E \cap F)$. For large n, we estimate $P(E|F)$ by the relative frequency of the number of occurrences of $E \cap F$ with respect to the number of occurrences of F, which is approximately

$$\frac{n \cdot P(E \cap F)}{n \cdot P(F)} = \frac{P(E \cap F)}{P(F)}$$

This result strongly suggests the formula that appears in the following formal definition of conditional probability. (The definition applies to any sample space, whether or not it is equiprobable.)

DEFINITION

The *conditional probability* of an event E, given that event F has occurred, is denoted $P(E|F)$ and is defined by

$$P(E|F) = \frac{P(E \cap F)}{P(F)} \qquad \text{if } P(F) \neq 0 \tag{3}$$

Similarly,

$$P(F|E) = \frac{P(F \cap E)}{P(E)} \qquad \text{if } P(E) \neq 0 \tag{4}$$

We emphasize that **the probabilities in Equations (3) and (4) are with respect to the original sample space.** Here we do *not* deal directly with a reduced sample space.

● EXAMPLE 3   **Quality Control**

After the initial production run of a new style of steel desk, a quality control technician found that 40% of the desks had an alignment problem and 10% had both a defective paint job and an alignment problem. If a desk is randomly selected from this run, and it has an alignment problem, what is the probability that it also has a defective paint job?

Solution: Let A and D be the events

$$A = \{\text{alignment problem}\}$$

$$D = \{\text{defective paint job}\}$$

We are interested in $P(D|A)$, the probability of a defective paint job, given an alignment problem. From the given data, we have $P(A) = 0.4$ and $P(D \cap A) = 0.1$. Substituting into Equation (3) gives

$$P(D|A) = \frac{P(D \cap A)}{P(A)} = \frac{0.1}{0.4} = \frac{1}{4}$$

It is convenient to use Equation (3) to solve this problem, because we are given probabilities rather than information about the sample space.

NOW WORK PROBLEM 7

EXAMPLE 4 Genders of Offspring

If a family has two children, find the probability that both are boys, given that one of the children is a boy. Assume that a child of either gender is equally likely and that, for example, having a girl first and a boy second is just as likely as having a boy first and a girl second.

Solution: Let E and F be the events

$$E = \{\text{both children are boys}\}$$

$$F = \{\text{at least one of the children is a boy}\}$$

We are interested in $P(E|F)$. Letting the letter B denote "boy" and G denote "girl," we use the equiprobable sample space

$$S = \{\text{BB, BG, GG, GB}\}$$

where, in each outcome, the order of the letters indicates the order in which the children are born. Thus,

$$E = \{\text{BB}\} \quad F = \{\text{BB, BG, GB}\} \quad \text{and} \quad E \cap F = \{\text{BB}\}$$

From Equation (3),

$$P(E|F) = \frac{P(E \cap F)}{P(F)} = \frac{\frac{1}{4}}{\frac{3}{4}} = \frac{1}{3}$$

Alternatively, this problem can be solved by using the reduced sample space F:

$$P(E|F) = \frac{\#(E \cap F)}{\#(F)} = \frac{1}{3}$$

NOW WORK PROBLEM 9

Equations (3) and (4) can be rewritten in terms of products by clearing fractions. This gives

$$P(E \cap F) = P(F)P(E|F)$$

and

$$P(F \cap E) = P(E)P(F|E)$$

By the commutative law, $P(E \cap F) = P(F \cap E)$, so we can combine the preceding equations to get an important law:

General Multiplication Law

$$P(E \cap F) = P(E)P(F|E) \tag{5}$$

$$= P(F)P(E|F)$$

The general multiplication law states that the probability that two events *both* occur is equal to the probability that one of them occurs, times the conditional probability that the other one occurs, given that the first has occurred.

● EXAMPLE 5   Advertising

A computer hardware company placed an ad for its new modem in a popular computer magazine. The company believes that the ad will be read by 32% of the magazine's readers and that 2% of those who read the ad will buy the modem. Assume that this is true, and find the probability that a reader of the magazine will read the ad and buy the modem.

Solution: Letting R denote the event "read ad" and B denote "buy modem," we are interested in $P(R \cap B)$. We are given that $P(R) = 0.32$. The fact that 2% of the readers of the ad will buy the modem can be written $P(B|R) = 0.02$. By the general multiplication law, Equation (5),

$$P(R \cap B) = P(R)P(B|R) = (0.32)(0.02) = 0.0064$$

NOW WORK PROBLEM 11   ●●●

Stochastic Processes

The general multiplication law is also called the **law of compound probability.** The reason is that it is extremely useful when applied to an experiment that can be expressed as a *sequence* (or a compounding) of two or more other experiments, called **trials** or **stages.** The original experiment is called a **compound experiment,** and the sequence of trials is called a **stochastic process.** The probabilities of the events associated with each trial (beyond the first) could depend on what events occurred in the preceding trials, so they are conditional probabilities.

When we analyze a compound experiment, a tree diagram is extremely useful in keeping track of the possible outcomes at each stage. A complete path from the start to a tip of the tree gives an outcome of the experiment.

The notion of a compound experiment is discussed in detail in the next example. Read it carefully. Although the discussion is lengthy for the sake of developing a new idea, the actual computation takes little time.

● EXAMPLE 6   Cards and Probability Tree

Two cards are drawn without replacement from a standard deck of cards. Find the probability that the second card is red.

Solution: The experiment of drawing two cards without replacement can be thought of as a compound experiment consisting of a sequence of two trials: The first is drawing a card, and the second is drawing a card after the first card has been drawn. The first trial has two possible outcomes:

$$R_1 = \{\text{red card}\} \quad \text{or} \quad B_1 = \{\text{black card}\}$$

(Here the subscript "1" refers to the first trial.) In Figure 8.16, these outcomes are represented by the two branches in the first level of the tree. Keep in mind that these outcomes are mutually exclusive, and they are also *exhaustive* in the sense that there are no other possibilities. Since there are 26 cards of each color, we have

$$P(R_1) = \frac{26}{52} \quad \text{and} \quad P(B_1) = \frac{26}{52}$$

These *unconditional* probabilities are written along the corresponding branches. We appropriately call Figure 8.16 a **probability tree.**

Now, if a red card is obtained in the first trial, then, of the remaining 51 cards, 25 are red and 26 are black. The card drawn in the second trial can be red (R_2) or

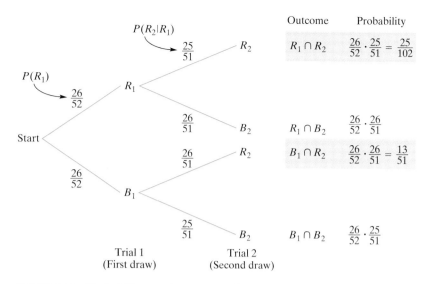

FIGURE 8.16 Probability tree for compound experiment.

black (B_2). Thus, in the tree, the fork at R_1 has two branches: red and black. The *conditional* probabilities $P(R_2|R_1) = \frac{25}{51}$ and $P(B_2|R_1) = \frac{26}{51}$ are placed along these branches. Similarly, if a black card is obtained in the first trial, then, of the remaining 51 cards, 26 are red and 25 are black. Hence, $P(R_2|B_1) = \frac{26}{51}$ and $P(B_2|B_1) = \frac{25}{51}$, as indicated alongside the two branches emanating from B_1. The complete tree has two levels (one for each trial) and four paths (one for each of the four mutually exclusive and exhaustive events of the compound experiment).

Note that the sum of the probabilities along the branches from the vertex "Start" to R_1 and B_1 is 1:

$$\frac{26}{52} + \frac{26}{52} = 1$$

In general, the sum of the probabilities along all the branches emanating from a single vertex to an outcome of that trial must be 1. Thus, for the vertex at R_1,

$$\frac{25}{51} + \frac{26}{51} = 1$$

and for the vertex at B_1,

$$\frac{26}{51} + \frac{25}{51} = 1$$

Now, consider the topmost path. It represents the event "red on first draw and red on second draw." By the general multiplication law,

$$P(R_1 \cap R_2) = P(R_1)P(R_2|R_1) = \frac{26}{52} \cdot \frac{25}{51} = \frac{25}{102}$$

That is, *the probability of an event is obtained by multiplying the probabilities in the branches of the path for that event.* The probabilities for the other three paths are also indicated in the tree.

Returning to the original question, we see that two paths give a red card on the second draw, namely, the paths for $R_1 \cap R_2$ and $B_1 \cap R_2$. Therefore, the event "second card red" is the union of two mutually exclusive events. By the addition law, the probability of the event is the sum of the probabilities for the two paths:

$$P(R_2) = \frac{26}{52} \cdot \frac{25}{51} + \frac{26}{52} \cdot \frac{26}{51} = \frac{25}{102} + \frac{13}{51} = \frac{1}{2}$$

Note how easy it was to find $P(R_2)$ by using a probability tree.

Here is a summary of what we have done:

$$R_2 = (R_1 \cap R_2) \cup (B_1 \cap R_2)$$

$$P(R_2) = P(R_1 \cap R_2) + P(B_1 \cap R_2)$$

$$= P(R_1)P(R_2|R_1) + P(B_1)P(R_2|B_1)$$

$$= \frac{26}{52} \cdot \frac{25}{51} + \frac{26}{52} \cdot \frac{26}{51} = \frac{25}{102} + \frac{13}{51} = \frac{1}{2}$$

NOW WORK PROBLEM 29 ●●●

● EXAMPLE 7 Cards

Two cards are drawn without replacement from a standard deck of cards. Find the probability that both cards are red.

Solution: Refer back to the probability tree in Figure 8.16. Only one path gives a red card on both draws, namely, that for $R_1 \cap R_2$. Thus, multiplying the probabilities along this path gives the desired probability:

$$P(R_1 \cap R_2) = P(R_1)P(R_2|R_1) = \frac{26}{52} \cdot \frac{25}{51} = \frac{25}{102}$$

NOW WORK PROBLEM 33 ●●●

● EXAMPLE 8 Defective Computer Chips

A company uses one computer chip in assembling each unit of a product. The chips are purchased from suppliers A, B, and C and are randomly picked for assembling a unit. Twenty percent come from A, 30% come from B, and the remainder come from C. The company believes that the probability that a chip from A will prove to be defective in the first 24 hours of use is 0.03, and the corresponding probabilities for B and C are 0.04 and 0.01, respectively. If an assembled unit is chosen at random and tested for 24 continuous hours, what is the probability that the chip in it is defective?

Solution: In this problem, there is a sequence of two trials: selecting a chip (A, B, or C) and then testing the selected chip [defective (D) or nondefective (D')]. We are given the unconditional probabilities

$$P(A) = 0.2 \quad \text{and} \quad P(B) = 0.3$$

Since A, B, and C are mutually exclusive and exhaustive,

$$P(C) = 1 - (0.2 + 0.3) = 0.5$$

From the statement of the problem, we also have the conditional probabilities

$$P(D|A) = 0.03 \quad P(D|B) = 0.04 \quad P(D|C) = 0.01$$

We want to find $P(D)$. To begin, we construct the two-level probability tree shown in Figure 8.17. We see that the paths that give a defective chip are those for the events

$$A \cap D \quad B \cap D \quad C \cap D$$

Since these events are mutually exclusive,

$$P(D) = P(A \cap D) + P(B \cap D) + P(C \cap D)$$

$$= P(A)P(D|A) + P(B)P(D|B) + P(C)P(D|C)$$

$$= (0.2)(0.03) + (0.3)(0.04) + (0.5)(0.01) = 0.023$$

NOW WORK PROBLEM 47 ●●●

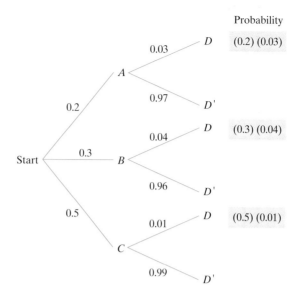

FIGURE 8.17 Probability tree for Example 8.

The general multiplication law can be extended so that it applies to more than two events. For n events, we have

$$P(E_1 \cap E_2 \cap \cdots \cap E_n)$$
$$= P(E_1)P(E_2|E_1)P(E_3|E_1 \cap E_2) \cdots P(E_n|E_1 \cap E_2 \cap \cdots \cap E_{n-1})$$

(We assume that all conditional probabilities are defined.) In words, the probability that two or more events all occur is equal to the probability that one of them occurs, times the conditional probability that a second one occurs given that the first occurred, times the conditional probability that a third occurs given that the first two occurred, and so on. For example, in the manner of Example 7, the probability of drawing three red cards from a deck without replacement is

$$P(R_1 \cap R_2 \cap R_3) = P(R_1)P(R_2|R_1)P(R_3|R_1 \cap R_2) = \frac{26}{52} \cdot \frac{25}{51} \cdot \frac{24}{50}$$

● **EXAMPLE 9   Jelly Beans in a Bag**

Bag I contains one black and two red jelly beans, and Bag II contains one pink jelly bean. (See Figure 8.18.) A bag is selected at random. Then a jelly bean is randomly taken from it and placed in the other bag. A jelly bean is then randomly taken from that bag. Find the probability that this jelly bean is pink.

Solution: This is a compound experiment with three trials:

   a. Selecting a bag
   b. Taking a jelly bean from the bag
   c. Putting it in the other bag and then taking a jelly bean from that bag

We want to find P(pink jelly bean on second draw). We analyze the situation by constructing a three-level probability tree. (See Figure 8.19.) The first trial has two equally likely possible outcomes, "Bag I" or "Bag II, " so each has probability of $\frac{1}{2}$.

If Bag I was selected, the second trial has two possible outcomes, "red" (R) or "black" (B), with conditional probabilities $P(R|\mathrm{I}) = \frac{2}{3}$ and $P(B|\mathrm{I}) = \frac{1}{3}$. If Bag II was selected, there is one possible outcome, "pink" (P), so $P(P|\mathrm{II}) = 1$. Thus, the second level of the tree has three branches.

Now we turn to the third trial. If Bag I was selected and a red jelly bean taken from it and placed in Bag II, then Bag II contains one red and one pink jelly bean. Hence, at the end of the second trial, the fork at vertex R has two branches, R and

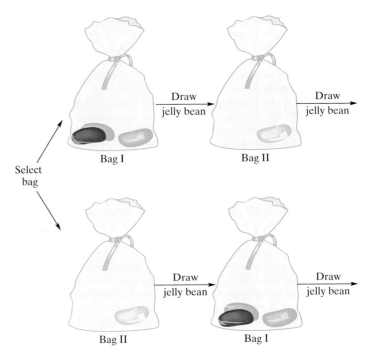

FIGURE 8.18 Jelly bean selections from bags.

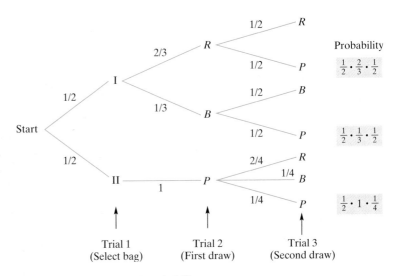

FIGURE 8.19 Three-level probability tree.

P, with conditional probabilities

$$P(R|\text{I} \cap R) = \frac{1}{2} \quad \text{and} \quad P(P|\text{I} \cap R) = \frac{1}{2}$$

Similarly, the tree shows the two possibilities if Bag I was initially selected and a black jelly bean was placed into Bag II. Now, if Bag II was selected in the first trial, then the pink jelly bean in it was taken and placed into Bag I, so Bag I contains two red, one black, and one pink jelly bean. Thus, the fork at P has *three* branches, one with probability $\frac{2}{4}$ and two with probability $\frac{1}{4}$.

We see that three paths give a pink jelly bean on the third trial, so for each, we multiply the probabilities along its branches. For example, the second path from the top represents $\text{I} \rightarrow R \rightarrow P$; the probability of this event is

$$P(\text{I} \cap R \cap P) = P(\text{I}) P(R|\text{I}) P(P|\text{I} \cap R)$$
$$= \frac{1}{2} \cdot \frac{2}{3} \cdot \frac{1}{2}$$

Adding the probabilities for the three paths gives

$$P(\text{pink jelly bean on second draw}) = \frac{1}{2} \cdot \frac{2}{3} \cdot \frac{1}{2} + \frac{1}{2} \cdot \frac{1}{3} \cdot \frac{1}{2} + \frac{1}{2} \cdot 1 \cdot \frac{1}{4}$$

$$= \frac{1}{6} + \frac{1}{12} + \frac{1}{8} = \frac{3}{8}$$

NOW WORK PROBLEM 43

Problems 8.5

*1. Given the equiprobable sample space

$$S = \{1, 2, 3, 4, 5, 6, 7, 8, 9\}$$

and events

$$E = \{1, 3\}$$

$$F = \{1, 2, 4, 5, 6\}$$

$$G = \{5, 6, 7, 8, 9\}$$

find each of the following.

(a) $P(E|F)$ (b) $P(E'|F)$
(c) $P(E|F')$ (d) $P(F|E)$
(e) $P(E|F \cap G)$

2. Given the equiprobable sample space

$$S = \{1, 2, 3, 4, 5\}$$

and events

$$E = \{1, 2\}$$

$$F = \{3, 4\}$$

$$G = \{1, 2, 3\}$$

find each of the following.

(a) $P(E)$ (b) $P(E|F)$
(c) $P(E|G)$ (d) $P(G|E)$
(e) $P(G|F')$ (f) $P(E'|F')$

3. If $P(E) > 0$, find $P(E|E)$.

4. If $P(E) > 0$, find $P(\emptyset|E)$.

5. If $P(E|F) = 0.57$, find $P(E'|F)$.

6. If F and G are mutually exclusive events with positive probabilities, find $P(F|G)$.

*7. If $P(E) = \frac{1}{4}$, $P(F) = \frac{1}{3}$, and $P(E \cap F) = \frac{1}{6}$, find each of the following:

(a) $P(E|F)$ (b) $P(F|E)$

8. If $P(E) = \frac{1}{4}$, $P(F) = \frac{1}{3}$, and $P(E|F) = \frac{3}{4}$, find $P(E \cup F)$. [*Hint:* Use the addition law to find $P(E \cup F)$.]

*9. If $P(E) = \frac{1}{4}$, $P(E \cup F) = \frac{7}{12}$, and $P(E \cap F) = \frac{1}{6}$, find each of the following:

(a) $P(F|E)$
(b) $P(F)$
(c) $P(E|F)$
(d) $P(E|F')$ [*Hint:* Find $P(E \cap F')$ by using the identity $P(E) = P(E \cap F) + P(E \cap F')$.]

10. If $P(E) = \frac{3}{5}$, $P(F) = \frac{3}{10}$, and $P(E \cup F) = \frac{7}{10}$, find $P(E|F)$.

*11. **Gypsy Moth** Because of gypsy moth infestation of three large areas that are densely populated with trees,

consideration is being given to aerial spraying to destroy larvae. A survey was made of the 200 residents of these areas to determine whether or not they favor the spraying. The resulting data are shown in Table 8.7. Suppose that a resident is randomly selected. Let I be the event "the resident is from Area I" and so on. Find each of the following:

(a) $P(F)$ (b) $P(F|II)$
(c) $P(O|I)$ (d) $P(III)$
(e) $P(III|O)$ (f) $P(II|N')$

TABLE 8.7

	Area I	Area II	Area III	Total
Favor (F)	46	35	44	125
Opposed (O)	22	15	10	47
No opinion (N)	10	8	10	28
Total	78	58	64	200

12. **College Selection and Family Income** A survey of 175 students resulted in the data shown in Table 8.8, which shows the type of college the student attends and the income level of the student's family. Suppose a student in the survey is randomly selected.

(a) Find the probability that the student attends a public college, given that the student comes from a middle-income family.
(b) Find the probability that the student is from a high-income family, given that the student attends a private college.
(c) If the student comes from a high-income family, find the probability that the student attends a private college.
(d) Find the probability that the student attends a public college or comes from a low-income family.

TABLE 8.8

Income	College		Total
	Private	Public	
High	14	11	25
Middle	25	55	80
Low	10	60	70
Total	49	126	175

13. **Cola Preference** A survey was taken among cola drinkers to see which of two popular brands people preferred. It was found that 45% liked brand A, 40% liked brand B, and 20%

liked both brands. Suppose that a person in the survey is randomly selected.

(a) Find the probability that the person liked brand A, given that he or she liked brand B.

(b) Find the probability that the person liked brand B, given that he or she liked brand A.

14. **Quality Control** Of the MP3 players produced by a well-known firm, 19% have defective earpieces and 13% have both defective earpieces and scratched screens. If an MP3 player is randomly selected from a shipment and it has defective earpieces, what is the probability that it has a scratched screen?

In Problems 15 and 16, assume that a child of either gender is equally likely and that, for example, having a girl first and a boy second is just as likely as having a boy first and a girl second.

15. **Genders of Offspring** If a family has two children, what is the probability that one child is a girl, given that at least one child is a boy?

16. **Genders of Offspring** If a family has three children, find each of the following.

(a) The probability that it has two girls, given that at least one child is a boy

(b) The probability that it has at least two girls, given that the oldest child is a girl

17. **Coin Toss** If a fair coin is tossed three times in succession, find each of the following.

(a) The probability of getting exactly two tails, given that the second toss is a tail

(b) The probability of getting exactly two tails, given that the second toss is a head

18. **Coin Toss** If a fair coin is tossed four times in succession, find the odds of getting four tails, given that the first toss is a tail.

19. **Die Roll** If a fair die is rolled, find the probability of getting a number less than 4, given that the number is odd.

20. **Cards** If a card is drawn randomly from a standard deck, find the probability of getting a face card, given that the card is red.

21. **Dice Roll** If two fair dice are rolled, find the probability that two 1's occur, given that at least one die shows a 1.

22. **Dice Roll** If a fair red die and a fair green die are rolled, find the probability that the sum is greater than 9, given that a 5 shows on the red die.

23. **Dice Roll** If a fair red die and a fair green die are rolled, find the probability of getting a total of 7, given that the green die shows an even number.

24. **Dice Roll** A fair die is rolled two times in succession.

(a) Find the probability that the sum is 6, given that the second roll is neither a 2 nor a 4.

(b) Find the probability that the sum is 6 and that the second roll is neither a 2 nor a 4.

25. **Die Roll** If a fair die is rolled two times in succession, find the probability of getting a total greater than 7, given that the first roll is greater than 3.

26. **Coin and Die** If a fair coin and a fair die are thrown, find the probability that the coin shows tails, given that the number on the die is odd.

27. **Cards** If a card is randomly drawn from a deck of 52 cards, find the probability that the card is a king, given that it is a heart.

28. **Cards** If a card is randomly drawn from a deck of 52 cards, find the probability that the card is a heart, given that it is a face card (a jack, queen, or king).

*29. **Cards** If two cards are randomly drawn without replacement from a standard deck, find the probability that the second card is not a face card, given that the first card is a face card (a jack, queen, or king).

In Problems 30–35, consider the experiment to be a compound experiment.

30. **Cards** If two cards are randomly drawn from a standard deck, find the probability that both cards are face cards if

(a) the cards are drawn without replacement

(b) the cards are drawn with replacement

31. **Cards** If three cards are randomly drawn without replacement from a standard deck, find the probability of getting a king, a queen, and a jack, in that order.

32. **Cards** If three cards are randomly drawn without replacement from a standard deck, find the probability of getting the ace of spades, the ace of hearts, and the ace of diamonds, in that order.

*33. **Cards** If three cards are randomly drawn without replacement from a standard deck, find the probability that all three cards are jacks.

34. **Cards** If two cards are randomly drawn without replacement from a standard deck of cards, find the probability that the second card is a heart.

35. **Cards** If two cards are randomly drawn without replacement from a standard deck, find the probability of getting two diamonds, given that the first card is red.

36. **Wake-Up Call** Barbara Smith, a sales representative, is staying overnight at a hotel and has a breakfast meeting with an important client the following morning. She asked the desk to give her a 7 A.M. wake-up call that morning in order that she be prompt for the meeting. The probability that she will get the call is 0.9. If she gets the call, the probability that she will be on time is 0.9. If the call is not given, the probability that she will be on time is 0.4. Find the probability that she will be on time for the meeting.

37. **Taxpayer Survey** In a certain school district, a questionnaire was sent to all property-tax payers concerning whether or not a new high school should be built. Of those that responded, 60% favored its construction, 30% opposed it, and 10% had no opinion. Further analysis of the data concerning the area in which the respondents lived gave the results in Table 8.9.

TABLE 8.9

	Urban	Suburban
Favor	45%	55%
Oppose	55%	45%
No opinion	35%	65%

(a) If one of the respondents is selected at random, what is the probability that he or she lives in an urban area?

(b) If a respondent is selected at random, use the result of part (a) to find the probability that he or she favors the construction of the school, given that the person lives in an urban area.

38. Marketing   A travel agency has a computerized telephone that randomly selects telephone numbers for advertising suborbital space trips. The telephone automatically dials the selected number and plays a prerecorded message to the recipient of the call. Experience has shown that 2% of those called show interest and contact the agency. However, of these, only 1.4% actually agree to purchase a trip.

(a) Find the probability that a person called will contact the agency and purchase a trip.

(b) If 100,000 people are called, how many can be expected to contact the agency and purchase a trip?

39. Rabbits in a Tall Hat   A tall hat contains three yellow and two red rabbits.

(a) If two rabbits are randomly pulled from the hat without replacement, find the probability that the second rabbit pulled is yellow, given that the first rabbit pulled is red.

(b) Repeat part (a), but assume that the first rabbit is replaced before the second rabbit is pulled.

40. Jelly Beans in a Bag   Bag 1 contains four green and three red jelly beans, and Bag 2 contains three green, one white, and two red jelly beans. A jelly bean is randomly taken from Bag 1 and placed into Bag 2. If a jelly bean is then randomly taken from Bag 2, find the probability that the jelly bean is green.

41. Balls in a Box   Box 1 contains three red and two white balls. Box 2 contains two red and two white balls. A box is chosen at random and then a ball is chosen at random from it. What is the probability that the ball is white?

42. Balls in a Box   Box 1 contains two red and three white balls. Box 2 contains three red and four white balls. Box 3 contains two red, two white, and two green balls. A box is chosen at random, and then a ball is chosen at random from it.

(a) Find the probability that the ball is white.
(b) Find the probability that the ball is red.
(c) Find the probability that the ball is green.

*43. **Jelly Beans in a Bag**   Bag 1 contains one green and one red jelly bean, and Bag 2 contains one white and one red jelly bean. A bag is selected at random. A jelly bean is randomly taken from it and placed in the other bag. A jelly bean is then randomly drawn from that bag. Find the probability that the jelly bean is white.

44. Dead Batteries   During Hurricane Juan's passage through Halifax, Ms. Wood's lights went out and in the dark she reached in the kitchen drawer for 4 batteries for her flashlight. There were 10 batteries in the drawer, but 5 of them were dead ones (that Mr. Wood should have thrown out). Find the probability that the 4 batteries Ms. Wood took from the drawer were all dead.

45. Quality Control   A soda producer requires the use of a bottle filler on each of its two product lines. The Mountain Spring line produces 20,000 bottles per day, and the Doctor Salt line produces 40,000 bottles per day. Over a period of time, it has been found that the Mountain Spring filler underfills 1% of its bottles, whereas the Doctor Salt filler underfills 3% of its bottles. At the end of a day, a bottle was selected at random from the total production. Find the probability that the bottle was underfilled.

46. Game Show   A TV game show host presents the following situation to a contestant. On a table are three identical boxes. One of them contains two identical envelopes. In one is a check for $5000, and in the other is a check for $1. Another box contains two envelopes with a check for $5000 in each and six envelopes with a check for $1 in each. The remaining box contains one envelope with a check for $5000 inside and five envelopes with a check for $1 inside each. If the contestant must select a box at random and then randomly draw an envelope, find the probability that a check for $5000 is inside.

*47. **Quality Control**   A company uses one computer chip in assembling each unit of a product. The chips are purchased from suppliers A, B, and C and are randomly picked for assembling a unit. Ten percent come from A, 20% come from B, and the remainder come from C. The probability that a chip from A will prove to be defective in the first 24 hours of use is 0.06, and the corresponding probabilities for B and C are 0.04 and 0.05, respectively. If an assembled unit is chosen at random and tested for 24 continuous hours, what is the probability that the chip in it will prove to be defective?

48. Quality Control   A manufacturer of widgets has four assembly lines: A, B, C, and D. The percentages of output produced by the lines are 30%, 20%, 35%, and 15%, respectively, and the percentages of defective units they produce are 6%, 3%, 2%, and 5%. If a widget is randomly selected from stock, what is the probability that it is defective?

49. Voting   In a certain town, 40% of eligible voters are registered Democrats, 35% are Republicans, and the remainder are Independents. In the last primary election, 15% of the Democrats, 20% of the Republicans, and 10% of the Independents voted.

(a) If an eligible voter is chosen at random, what is the probability that he or she is a Democrat who voted?

(b) If an eligible voter is chosen at random, what is the probability that he or she voted?

50. Job Applicants   A restaurant has four openings for waiters. Suppose Allison, Lesley, Alan, Tom, Danica, Bronwyn, Steve, and Richard are the only applicants for these jobs, and all are equally qualified. If four are hired at random, find the probability that Allison, Lesley, Tom, and Bronwyn were chosen, given that Richard was not hired.

51. Committee Selection   Suppose six female and five male students wish to fill three openings on a campus committee on cultural diversity. If three of the students are chosen at random for the committee, find the probability that all three are female, given that at least one is female.

To develop the notion of independent events and apply the special multiplication law.

8.6 Independent Events

In our discussion of conditional probability, you saw that the probability of an event may be affected by the knowledge that another event has occurred. In this section, we consider the situation where the additional information has no effect. That is, the conditional probability $P(E \mid F)$ and the unconditional probability $P(E)$ are the same.

When $P(E \mid F) = P(E)$, we say that E is *independent* of F. If E is independent of F, it follows that F is independent of E (and vice versa). To prove this, assume that $P(E \mid F) = P(E)$ and $P(E) \neq 0$. Then

$$P(F \mid E) = \frac{P(E \cap F)}{P(E)} = \frac{P(F)P(E \mid F)}{P(E)} = \frac{P(F)P(E)}{P(E)} = P(F)$$

which means that F is independent of E. Thus, to prove independence, it suffices to show that either $P(E \mid F) = P(E)$ or $P(F \mid E) = P(F)$, and when one of these is true, we simply say that E and F are *independent events*.

 **CAUTION**

Independence of two events is defined by probabilities, not by a causal relationship.

DEFINITION
Let E and F be events with *positive* probabilities. Then E and F are said to be *independent events* if either

$$P(E \mid F) = P(E) \tag{1}$$

or

$$P(F \mid E) = P(F) \tag{2}$$

If E and F are not independent, they are said to be *dependent events*.

Thus, with dependent events, the occurrence of one of the events *does* affect the probability of the other. If E and F are independent events, it can be shown that the events in each of the following pairs are also independent:

$$E \text{ and } F' \quad E' \text{ and } F \quad E' \text{ and } F'$$

●EXAMPLE 1 Showing That Two Events Are Independent

A fair coin is tossed twice. Let E and F be the events

$$E = \{head \ on \ first \ toss\}$$

$$F = \{head \ on \ second \ toss\}$$

Determine whether or not E and F are independent events.

Solution: We suspect that they are independent, because one coin toss should not influence the outcome of another toss. To confirm our suspicion, we will compare $P(E)$ with $P(E \mid F)$. For the equiprobable sample space $S = \{HH, HT, TH, TT\}$, we have $E = \{HH, HT\}$ and $F = \{HH, TH\}$. Thus,

$$P(E) = \frac{\#(E)}{\#(S)} = \frac{2}{4} = \frac{1}{2}$$

$$P(E \mid F) = \frac{\#(E \cap F)}{\#(F)} = \frac{\#(\{HH\})}{\#(F)} = \frac{1}{2}$$

Since $P(E \mid F) = P(E)$, events E and F are independent.

NOW WORK PROBLEM 7 ●○○

In Example 1 we suspected the result, and certainly there are other situations where we have an intuitive feeling as to whether or not two events are independent. For example, if a red die and green die are tossed, we expect (and it is indeed true) that the events "3 on red die" and "6 on green die" are independent, because the outcome on one die should not be influenced by the outcome on the other die. Similarly, if two cards are drawn *with replacement* from a deck of cards, we would assume that the events "first card is a jack" and "second card is a jack" are independent. However, suppose the cards are drawn *without replacement*. Because the first card drawn is not put back in the deck, it should have an effect on the outcome of the second draw, so we expect the events to be dependent. In many problems, your intuitive notion of independence or the context of the problem may make it clear whether or not independence can be assumed. In spite of one's intuition (which can prove to be wrong!), the only sure way to determine whether events E and F are independent (or dependent) is by showing that Equation (1) or Equation (2) is true (or is not true).

● **EXAMPLE 2**   **Smoking and Sinusitis**

In a study of smoking and sinusitis, 4000 people were studied, with the results as given in Table 8.10. Suppose a person from the study is selected at random. On the basis of the data, determine whether or not the events "having sinusitis" (L) and "smoking" (S) are independent events.

TABLE 8.10 Smoking and Sinusitis

	Smoker	**Nonsmoker**	**Total**
Sinusitis	432	1018	1450
No sinusitis	528	2022	2550
Total	960	3040	4000

Solution: We will compare $P(L)$ with $P(L \mid S)$. The number $P(L)$ is the proportion of the people studied that have sinusitis:

$$P(L) = \frac{1450}{4000} = \frac{29}{80} = 0.3625$$

For $P(L \mid S)$, the sample space is reduced to 960 smokers, of which 432 have sinusitis:

$$P(L \mid S) = \frac{432}{960} = \frac{9}{20} = 0.45$$

Since $P(L \mid S) \neq P(L)$, having sinusitis and smoking are dependent.

NOW WORK PROBLEM 9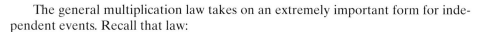

The general multiplication law takes on an extremely important form for independent events. Recall that law:

$$P(E \cap F) = P(E)P(F \mid E)$$

$$= P(F)P(E \mid F)$$

If events E and F are independent, then $P(F \mid E) = P(F)$, so substitution in the first equation gives

$$P(E \cap F) = P(E)P(F)$$

The same result is obtained from the second equation. Thus, we have the following law:

Special Multiplication Law

If E and F are *independent events,* then

$$P(E \cap F) = P(E)P(F) \tag{3}$$

Equation (3) states that if E and F are independent events, then the probability that E and F both occur is the probability that E occurs times the probability that F occurs. Keep in mind that Equation (3) is *not* valid when E and F are dependent.

● EXAMPLE 3 **Survival Rates**

Suppose the probability of the event "Bob lives 20 more years" (B) is 0.8 and the probability of the event "Doris lives 20 more years" (D) is 0.85. Assume that B and D are independent events.

a. *Find the probability that both Bob and Doris live 20 more years.*

Solution: We are interested in $P(B \cap D)$. Since B and D are independent events, the special multiplication law applies:

$$P(B \cap D) = P(B)P(D) = (0.8)(0.85) = 0.68$$

b. *Find the probability that at least one of them lives 20 more years.*

Solution: Here we want $P(B \cup D)$. By the addition law,

$$P(B \cup D) = P(B) + P(D) - P(B \cap D)$$

From part (a), $P(B \cap D) = 0.68$, so

$$P(B \cup D) = 0.8 + 0.85 - 0.68 = 0.97$$

c. *Find the probability that exactly one of them lives 20 more years.*

Solution: We first express the event

$$E = \{\text{exactly one of them lives 20 more years}\}$$

in terms of the given events, B and D. Now, event E can occur in one of two *mutually exclusive* ways: Bob lives 20 more years but Doris does not ($B \cap D'$), or Doris lives 20 more years but Bob does not ($B' \cap D$). Thus,

$$E = (B \cap D') \cup (B' \cap D)$$

By the addition law (for mutually exclusive events),

$$P(E) = P(B \cap D') + P(B' \cap D) \tag{4}$$

To compute $P(B \cap D')$, we note that, since B and D are independent, so are B and D' (from the statement preceding Example 1). Accordingly, we can use the multiplication law and the rule for complements:

$$P(B \cap D') = P(B)P(D')$$
$$= P(B)(1 - P(D)) = (0.8)(0.15) = 0.12$$

Similarly,

$$P(B' \cap D) = P(B')P(D) = (0.2)(0.85) = 0.17$$

Substituting into Equation (4) gives

$$P(E) = 0.12 + 0.17 = 0.29$$

NOW WORK PROBLEM 25 ●●●

In Example 3, it was assumed that events B and D are independent. However, if Bob and Doris are related in some way, it is quite possible that the survival of one of them has a bearing on the survival of the other. In that case the assumption of independence is not justified, and we could not use the special multiplication law, Equation (3).

● EXAMPLE 4   **Cards**

In a math exam, a student was given the following two-part problem. A card is randomly drawn from a deck of 52 cards. Let H, K, and R be the events

$$H = \{heart\ drawn\}$$
$$K = \{king\ drawn\}$$
$$R = \{red\ card\ drawn\}$$

Find $P(H \cap K)$ and $P(H \cap R)$.
For the first part, the student wrote

$$P(H \cap K) = P(H)P(K) = \frac{13}{52} \cdot \frac{4}{52} = \frac{1}{52}$$

and for the second part, she wrote

$$P(H \cap R) = P(H)P(R) = \frac{13}{52} \cdot \frac{26}{52} = \frac{1}{8}$$

The answer was correct for $P(H \cap K)$ but not for $P(H \cap R)$. Why?

Solution: The reason is that the student assumed independence in *both* parts by using the special multiplication law to multiply unconditional probabilities when, in fact, that assumption should *not* have been made. Let us examine the first part of the exam problem for independence. We will see whether $P(H)$ and $P(H \mid K)$ are the same. We have

$$P(H) = \frac{13}{52} = \frac{1}{4}$$

and

$$P(H \mid K) = \frac{1}{4} \qquad \text{(one heart out of four kings)}$$

Since $P(H) = P(H \mid K)$, events H and K are independent, so the student's procedure is valid. (The student was lucky!) For the second part, again we have $P(H) = \frac{1}{4}$, but

$$P(H \mid R) = \frac{13}{26} = \frac{1}{2} \qquad \text{(13 hearts out of 26 red cards)}$$

Since $P(H \mid R) \neq P(H)$, events H and R are dependent, so the student should not have multiplied the unconditional probabilities. (The student was unlucky.) However, the student would have been safe by using the *general* multiplication law, that is,

$$P(H \cap R) = P(H)P(R \mid H) = \frac{13}{52} \cdot 1 = \frac{1}{4}$$

or

$$P(H \cap R) = P(R)P(H \mid R) = \frac{26}{52} \cdot \frac{13}{26} = \frac{1}{4}$$

More simply, observe that $H \cap R = H$, so

$$P(H \cap R) = P(H) = \frac{13}{52} = \frac{1}{4}$$

NOW WORK PROBLEM 33 ◖◗◗

Equation (3) is often used as an alternative means of defining independent events, and we will consider it as such:

Events E and F are independent if and only if

$$P(E \cap F) = P(E)P(F) \tag{3}$$

Putting everything together, we can say that to prove events E and F, with nonzero probabilities, are independent, only one of the following relationships has to be shown:

$$P(E \mid F) = P(E) \tag{1}$$

or

$$P(F \mid E) = P(F) \tag{2}$$

or

$$P(E \cap F) = P(E)P(F) \tag{3}$$

In other words, if any one of these equations is true, then all of them are true; if any is false, then all of them are false, and E and F are dependent.

● EXAMPLE 5   Dice

Two fair dice, one red and the other green, are rolled, and the numbers on the top faces are noted. Let E and F be the events

$$E = \{number\ on\ red\ die\ is\ even\}$$

$$F = \{sum\ is\ 7\}$$

Test whether $P(E \cap F) = P(E)P(F)$ to determine whether E and F are independent.

Solution: Our usual sample space for the roll of two dice has $6 \cdot 6 = 36$ equally likely outcomes. For event E, the red die can fall in any of three ways and the green die any of six ways, so E consists of $3 \cdot 6 = 18$ outcomes. Thus, $P(E) = \frac{18}{36} = \frac{1}{2}$. Event F has six outcomes:

$$F = \{(1, 6), (2, 5), (3, 4), (4, 3), (5, 2), (6, 1)\} \tag{5}$$

where, for example, we take $(1, 6)$ to mean "1" on the red die and "6" on the green die. Therefore, $P(F) = \frac{6}{36} = \frac{1}{6}$, so

$$P(E)P(F) = \frac{1}{2} \cdot \frac{1}{6} = \frac{1}{12}$$

Now, event $E \cap F$ consists of all outcomes in which the red die is even and the sum is 7. Using Equation (5) as an aid, we see that

$$E \cap F = \{(2, 5), (4, 3), (6, 1)\}$$

Thus,

$$P(E \cap F) = \frac{3}{36} = \frac{1}{12}$$

Since $P(E \cap F) = P(E)P(F)$, events E and F are independent. This fact may not have been obvious before the problem was solved.

NOW WORK PROBLEM 17 ●●●

● EXAMPLE 6   Genders of Offspring

For a family with at least two children, let E and F be the events

$$E = \{at\ most\ one\ boy\},$$

$$F = \{at\ least\ one\ child\ of\ each\ gender\}$$

Assume that a child of either gender is equally likely and that, for example, having a girl first and a boy second is just as likely as having a boy first and a girl second. Determine whether E and F are independent in each of the following situations:

a. *The family has exactly two children.*

Solution: We will use the equiprobable sample space

$$S = \{BB, BG, GG, GB\}$$

and test whether $P(E \cap F) = P(E)P(F)$. We have

$$E = \{BG, GB, GG\} \quad F = \{BG, GB\} \quad E \cap F = \{BG, GB\}$$

Thus, $P(E) = \frac{3}{4}$, $P(F) = \frac{2}{4} = \frac{1}{2}$, and $P(E \cap F) = \frac{2}{4} = \frac{1}{2}$. We ask whether

$$P(E \cap F) \overset{?}{=} P(E)P(F)$$

and see that

$$\frac{1}{2} \neq \frac{3}{4} \cdot \frac{1}{2} = \frac{3}{8}$$

so E and F are dependent events.

b. *The family has exactly three children.*

Solution: Based on the result of part (a), you may have an intuitive feeling that E and F are dependent. Nevertheless, we must test this conjecture. For three children, we use the equiprobable sample space

$$S = \{BBB, BBG, BGB, BGG, GBB, GBG, GGB, GGG\}$$

Again we test whether $P(E \cap F) = P(E)P(F)$. We have

$$E = \{BGG, GBG, GGB, GGG\}$$

$$F = \{BBG, BGB, BGG, GBB, GBG, GGB\}$$

$$E \cap F = \{BGG, GBG, GGB\}$$

Hence, $P(E) = \frac{4}{8} = \frac{1}{2}$, $P(F) = \frac{6}{8} = \frac{3}{4}$, and $P(E \cap F) = \frac{3}{8}$, so

$$P(E)P(F) = \frac{1}{2} \cdot \frac{3}{4} = \frac{3}{8} = P(E \cap F)$$

Therefore, we have the *unexpected* result that events E and F are independent. (*Moral:* You cannot always trust your intuition.)

NOW WORK PROBLEM 27

We now generalize our discussion of independence to the case of more than two events.

DEFINITION

The events $E_1, E_2, \ldots, E_n$ are said to be *independent* if and only if for each set of two or more of the events, the probability of the intersection of the events in the set is equal to the product of the probabilities of the events in that set.

For instance, let us apply the definition to the case of three events ($n = 3$). We say that E, F, and G are independent events if the special multiplication law is true for these events, taken two at a time and three at a time. That is, each of the following equations must be true:

$$\left. \begin{array}{l} P(E \cap F) = P(E)P(F) \\ P(E \cap G) = P(E)P(G) \\ P(F \cap G) = P(F)P(G) \end{array} \right\} \text{Two at a time}$$

$$P(E \cap F \cap G) = P(E)P(F)P(G) \} \text{Three at a time}$$

As another example, if events E, F, G, and H are independent, then we can assert such things as

$$P(E \cap F \cap G \cap H) = P(E)P(F)P(G)P(H)$$

$$P(E \cap G \cap H) = P(E)P(G)P(H)$$

and

$$P(F \cap H) = P(F)P(H)$$

Similar conclusions can be made if any of the events are replaced by their complements.

◉ EXAMPLE 7   **Cards**

Four cards are randomly drawn, with replacement, from a deck of 52 cards. Find the probability that the cards chosen, in order, are a king (K), a queen (Q), a jack (J), and a heart (H).

Solution: Since there is replacement, what happens on any draw does not affect the outcome on any other draw, so we can assume independence and multiply the unconditional probabilities. We obtain

$$P(K \cap Q \cap J \cap H) = P(K)P(Q)P(J)P(H)$$

$$= \frac{4}{52} \cdot \frac{4}{52} \cdot \frac{4}{52} \cdot \frac{13}{52} = \frac{1}{8788}$$

NOW WORK PROBLEM 35   ◐◑●

◉ EXAMPLE 8   **Aptitude Test**

Personnel Temps, a temporary-employment agency, requires that each job applicant take the company's aptitude test, which has 80% accuracy.

a. *Find the probability that the test will be accurate for the next three applicants who are tested.*

Solution: Let A, B, and C be the events that the test will be accurate for applicants A, B, and C, respectively. We are interested in

$$P(A \cap B \cap C)$$

Since the accuracy of the test for one applicant should not affect the accuracy for any of the others, it seems reasonable to assume that A, B, and C are independent. Thus, we can multiply probabilities:

$$P(A \cap B \cap C) = P(A)P(B)P(C)$$

$$= (0.8)(0.8)(0.8) = (0.8)^3 = 0.512$$

b. *Find the probability that the test will be accurate for at least two of the next three applicants who are tested.*

Solution: Here, *at least two* means "exactly two or three." In the first case, the possible ways of choosing the two tests that are accurate are

$$A \text{ and } B \quad A \text{ and } C \quad B \text{ and } C$$

In each of these three possibilities, the test for the remaining applicant is not accurate. For example, choosing A and B gives the event $A \cap B \cap C'$, whose probability is

$$P(A)P(B)P(C') = (0.8)(0.8)(0.2) = (0.8)^2(0.2)$$

You should verify that the probability for each of the other two possibilities is also $(0.8)^2(0.2)$. Summing the three probabilities gives

$$P(\text{exactly two accurate}) = 3[(0.8)^2(0.2)] = 0.384$$

Using this result and that of part (a), we obtain

$$P(\text{at least two accurate}) = P(\text{exactly two accurate}) + P(\text{three accurate})$$

$$= 0.384 + 0.512 = 0.896$$

Alternatively, the problem could be solved by computing

$$1 - [P(\text{none accurate}) + P(\text{exactly one accurate})]$$

Why?

NOW WORK PROBLEM 21

We conclude with a note of caution: **Do not confuse independent events with mutually exclusive events.** The concept of independence is defined in terms of probability, whereas mutual exclusiveness is not. When two events are independent, the occurrence of one of them does not affect the probability of the other. However, when two events are mutually exclusive, they cannot occur simultaneously. Although these two concepts are not the same, we can draw some conclusions about their relationship. If E and F are mutually exclusive events *with positive probabilities*, then

$$P(E \cap F) = 0 \neq P(E)P(F) \qquad \text{since } P(E) > 0 \text{ and } P(F) > 0$$

which shows that E and F are dependent. In short, *mutually exclusive events with positive probabilities must be dependent.* Another way of saying this is that *independent events with positive probabilities are not mutually exclusive.*

Problems 8.6

1. If events E and F are independent with $P(E) = \frac{1}{3}$ and $P(F) = \frac{3}{4}$, find each of the following.

  (a) $P(E \cap F)$ **(b)** $P(E \cup F)$

  (c) $P(E \mid F)$ **(d)** $P(E' \mid F)$

  (e) $P(E \cap F')$ **(f)** $P(E \cup F')$

  (g) $P(E \mid F')$

2. If events E, F, and G are independent with $P(E) = 0.1$, $P(F) = 0.3$, and $P(G) = 0.6$, find each of the following.

  (a) $P(E \cap F)$ **(b)** $P(F \cap G)$

  (c) $P(E \cap F \cap G)$ **(d)** $P(E \mid (F \cap G))$

  (e) $P(E' \cap F \cap G')$

3. If events E and F are independent with $P(E) = \frac{2}{7}$ and $P(E \cap F) = \frac{1}{9}$, find $P(F)$.

4. If events E and F are independent with $P(E \mid F) = \frac{1}{3}$, find $P(E')$.

In Problems 5 and 6, events E and F satisfy the given conditions. Determine whether E and F are independent or dependent.

5. $P(E) = \frac{3}{4}$, $P(F) = \frac{8}{9}$, $P(E \cap F) = \frac{2}{3}$

6. $P(E) = 0.28$, $P(F) = 0.15$, $P(E \cap F) = 0.038$

*7. **Stockbrokers** Six hundred investors were surveyed to determine whether a person who uses a full-service stockbroker has better performance in his or her investment portfolio than one who uses a discount broker. In general, discount brokers usually offer no investment advice to their clients, whereas full-service brokers usually offer help in selecting stocks but charge larger fees. The data, based on the last 12 months, are given in Table 8.11. Determine whether the event of having a full-service broker and the

event of having an increase in portfolio value are independent or dependent.

TABLE 8.11 Portfolio Value

	Increase	**Decrease**	**Total**
Full service	320	80	400
Discount	160	40	200
Total	480	120	600

8. Cinema Offenses An observation of 175 patrons in a theater resulted in the data shown in Table 8.12. The table shows three types of cinema offenses committed by male and female patrons. Crunchers include noisy eaters of popcorn and other morsels, as well as cold-drink slurpers. Determine whether the event of being a male and the event of being a cruncher are independent or dependent. (See page 5D of the July 21, 1991, issue of *USA TODAY* for the article "Pests Now Appearing at a Theater Near You.")

TABLE 8.12 Theatre Patrons

	Male	**Female**	**Total**
Talkers	60	10	70
Crunchers	55	25	80
Seat kickers	15	10	25
Total	130	45	175

*9. **Dice** Two fair dice are rolled, one red and one green, and the numbers on the top faces are noted. Let event E be "number on red die is neither 2 nor 3" and event F be "sum is 8." Determine whether E and F are independent or dependent.

10. **Cards**   A card is randomly drawn from an ordinary deck of 52 cards. Let E and F be the events "black card drawn" and "a 2, 3, or 4 drawn," respectively. Determine whether E and F are independent or dependent.

11. **Coins**   If two fair coins are tossed, let E be the event "at most one head" and F be the event "exactly one head." Determine whether E and F are independent or dependent.

12. **Coins**   If three fair coins are tossed, let E be the event "at most one head" and F be the event "at least one head and one tail." Determine whether E and F are independent or dependent.

13. **Chips in a Bowl**   A bowl contains seven chips numbered from 1 to 7. Two chips are randomly withdrawn with replacement. Let E, F, and G be the events

$$E = 3 \text{ on first withdrawal}$$

$$F = 3 \text{ on second withdrawal}$$

$$G = \text{sum is odd}$$

(a) Determine whether E and F are independent or dependent.

(b) Determine whether E and G are independent or dependent.

(c) Determine whether F and G are independent or dependent.

(d) Are E, F, and G independent?

14. **Chips in a Bowl**   A bowl contains six chips numbered from 1 to 6. A chip is randomly withdrawn. Let E be the event of withdrawing a 3 and F be the event of withdrawing a 5.

(a) Are E and F mutually exclusive?

(b) Are E and F independent?

In Problems 15 and 16, events E and F satisfy the given conditions. Determine whether E and F are independent or dependent.

15. $P(E \mid F) = 0.5$, $P(E \cap F) = 0.3$, $P(F \mid E) = 0.4$

16. $P(E \mid F) = \frac{2}{3}$, $P(E \cup F) = \frac{17}{18}$, $P(E \cap F) = \frac{5}{9}$

In Problems 17–37, you may make use of your intuition concerning independent events if nothing to that effect is specified.

*17. **Dice**   Two fair dice are rolled, one red and one green. Find the probability that the red die is a 4 and the green die is a number greater than 4.

18. **Die**   If a fair die is rolled three times, find the probability that a 2 or 3 comes up each time.

19. **Fitness Classes**   At a certain fitness center, the probability that a member regularly attends an aerobics class is $\frac{1}{5}$. If two members are randomly selected, find the probability that both attend the class regularly. Assume independence.

20. **Monopoly**   In the game of Monopoly, a player rolls two fair dice. One special situation that can arise is that the numbers on the top faces of the dice are the same (such as two 3's). This result is called a "double," and when it occurs, the player continues his or her turn and rolls the dice again. The pattern continues, unless the player is unfortunate enough to throw doubles three consecutive times. In that case, the player goes to jail. Find the probability that a player goes to jail in this way.

*21. **Cards**   Three cards are randomly drawn, with replacement, from an ordinary deck of 52 cards. Find the probability that the cards drawn, in order, are an ace, a face card (a jack, queen, or king), and a spade.

22. **Die**   If a fair die is rolled seven times, find each of the following.

(a) The probability of getting a number greater than 4 each time

(b) The probability of getting a number less than 4 each time

23. **Exam Grades**   In a sociology course, the probability that Bill gets an A on the final exam is $\frac{3}{4}$, and for Jim and Linda, the probabilities are $\frac{1}{2}$ and $\frac{4}{5}$, respectively. Assume independence and find each of the following.

(a) The probability that all three of them get an A on the exam

(b) The probability that none of them get an A on the exam

(c) The probability that, of the three, only Linda gets an A

24. **Die**   If a fair die is rolled three times, find the probability of getting at least one 6.

*25. **Survival Rates**   The probability that person A survives 15 more years is $\frac{2}{3}$, and the probability that person B survives 15 more years is $\frac{3}{5}$. Find the probability of each of the following. Assume independence.

(a) A and B both survive 15 years.

(b) B survives 15 years, but A does not.

(c) Exactly one of A and B survives 15 years.

(d) At least one of A and B survives 15 years.

(e) Neither A nor B survives 15 years.

26. **Matching**   In his desk, a secretary has a drawer containing a mixture of two sizes of paper (A and B) and another drawer containing a mixture of envelopes of two corresponding sizes. The percentages of each size of paper and envelope in the drawers are given in Table 8.13. If a piece of paper and an envelope are randomly drawn, find the probability that they are the same size.

TABLE 8.13 Paper and Envelopes

Size	Drawers	
	Paper	Envelopes
A	63%	57%
B	37%	43%

*27. **Jelly Beans in a Bag**   A bag contains five red, seven white, and six green jelly beans. If two jelly beans are randomly taken out with replacement, find each of the following.

(a) The probability that the first jelly bean is white and the second is green.

(b) The probability that one jelly bean is red and the other one is white

28. Dice   Suppose two fair dice are rolled twice. Find the probability of getting a total of 7 on one of the rolls and a total of 12 on the other one.

29. Jelly Beans in a Bag   A bag contains three red, seven white, and nine green jelly beans. If two jelly beans are randomly withdrawn with replacement, find the probability that they have the same color.

30. Die   Find the probability of rolling the same number in three throws of a fair die.

31. Tickets in Hat   Twenty tickets numbered from 1 to 20 are placed in a hat. If two tickets are randomly drawn with replacement, find the probability that the sum is 35.

32. Coins and Dice   Suppose two fair coins are tossed and then two fair dice are rolled. Find each of the following.

(a) The probability that two tails and two 3's occur

(b) The probability that two heads, one 4, and one 6 occur

*33. Carnival Game   In a carnival game, a well-balanced roulette-type wheel has 12 equally spaced slots that are numbered from 1 to 12. The wheel is spun, and a ball travels along the rim of the wheel. When the wheel stops, the number of the slot in which the ball finally rests is considered the result of the spin. If the wheel is spun three times, find each of the following.

(a) The probability that the first number will be 4 and the second and third numbers will be 5

(b) The probability that there will be one even number and two odd numbers

34. Cards   Three cards are randomly drawn, with replacement, from an ordinary deck of 52 cards. Find each of the following.

(a) The probability of drawing, in order, a 10, a spade, and a black jack

(b) The probability of drawing exactly three kings

(c) The probability that one queen, one spade and one black ace are drawn

(d) The probability of drawing exactly one ace

*35. Multiple-Choice Exam   A quiz contains five multiple-choice problems. Each problem has four choices for the answer, but only one of them is correct. Suppose a student randomly guesses the answer to each problem. Find each of the following by assuming that the guesses are independent.

(a) The probability that the student gets exactly four correct answers

(b) The probability that the student gets at least four correct answers

(c) The probability that the student gets three or more correct answers

36. Shooting Gallery   At a shooting gallery, suppose Bill, Jim, and Linda each take one shot at a moving target. The probability that Bill hits the target is 0.5, and for Jim and Linda, the probabilities are 0.4 and 0.7, respectively. Assume independence and find each of the following.

(a) The probability that none of them hit the target

(b) The probability that Linda is the only one of them that hits the target

(c) The probability that exactly one of them hits the target

(d) The probability that exactly two of them hit the target

(e) The probability that all of them hit the target

37. Decision Making[3]   The president of Zeta Construction Company must decide which of two actions to take, namely, to rent or to buy expensive excavating equipment. The probability that the vice president makes a faulty analysis and thus recommends the wrong decision to the president is 0.04. To be thorough, the president hires two consultants, who study the problem independently and make their recommendations. After having observed them at work, the president estimates that the first consultant is likely to recommend the wrong decision with probability 0.05, the other with probability 0.1. He decides to take the action recommended by a majority of the three recommendations he receives. What is the probability that he will make the wrong decision?

OBJECTIVE

To solve a Bayes's problem. To develop Bayess formula.

8.7 Bayes's Formula

In this section, we will be dealing with a two-stage experiment in which we know the outcome of the second stage and are interested in the probability that a particular outcome has occurred in the first stage.

To illustrate, suppose it is believed that of the total population (our sample space), 8% have a particular disease. Imagine also that there is a new blood test for detecting the disease and that researchers have evaluated its effectiveness. Data from extensive testing show that the blood test is not perfect: Not only is it positive for only 95% of those who have the disease, but it is also positive for 3% of those who do not. Suppose a person from the population is selected at random and given

---

[3]Samuel Goldberg, *Probability, an Introduction* (Prentice-Hall, Inc., 1960, Dover Publications, Inc., 1986), p. 113. Adapted by permission of the author.

the blood test. If the result is positive, what is the probability that the person has the disease?

To analyze this problem, we consider the following events:

$$D_1 = \{\text{having the disease}\}$$

$$D_2 = \{\text{not having the disease}\}$$

$$T_1 = \{\text{testing positive}\}$$

$$T_2 = \{\text{testing negative}\}$$

We are given that

$$P(D_1) = 0.08$$

so

$$P(D_2) = 1 - 0.08 = 0.92$$

because D_1 and D_2 are complements. It is reasonable to assume that T_1 and T_2 are also complements; in that case, we have the conditional probabilities

$$P(T_1 \mid D_1) = 0.95 \qquad P(T_2 \mid D_1) = 1 - 0.95 = 0.05$$

$$P(T_1 \mid D_2) = 0.03 \qquad P(T_2 \mid D_2) = 1 - 0.03 = 0.97$$

Figure 8.20 shows a two-stage probability tree that reflects this information. The first stage takes into account either having or not having the disease, and the second stage shows possible test results.

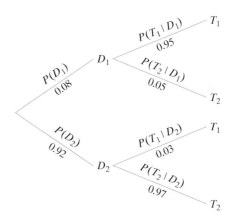

FIGURE 8.20 Two-stage probability tree.

We are interested in the probability that a person who tests positive has the disease. That is, we want to find the conditional probability that D_1 occurred in the first stage, given that T_1 occurred in the second stage:

$$P(D_1 \mid T_1)$$

It is important that you understand the difference between the conditional probabilities $P(D_1 \mid T_1)$ and $P(T_1 \mid D_1)$. The probability $P(T_1 \mid D_1)$, which is *given* to us, is a "typical" conditional probability, in that it deals with the probability of an outcome in the second stage *after* an outcome in the first stage has occurred. However, with $P(D_1 \mid T_1)$, we have a "reverse" situation. Here we must find the probability of an outcome in the *first* stage, given that an outcome in the second stage occurred. In a sense, we have the "cart before the horse" in that this probability does not fit the usual (and more natural) pattern of a typical conditional probability. Fortunately, we have all the tools needed to find $P(D_1 \mid T_1)$. We proceed as follows.

From the definition of conditional probability,

$$P(D_1 \mid T_1) = \frac{P(D_1 \cap T_1)}{P(T_1)} \tag{1}$$

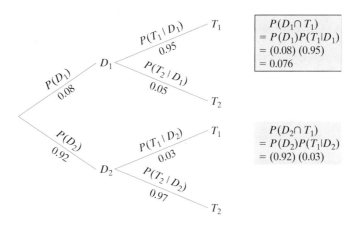

FIGURE 8.21 Probability tree to determine $P(D_1 \mid T_1)$.

Consider the numerator. Applying the general multiplication law gives

$$P(D_1 \cap T_1) = P(D_1)P(T_1 \mid D_1)$$

$$= (0.08)(0.95) = 0.076$$

which is indicated in the path through D_1 and T_1 in Figure 8.21. The denominator, $P(T_1)$, is the sum of the probabilities for all paths of the tree ending in T_1. Thus,

$$P(T_1) = P(D_1 \cap T_1) + P(D_2 \cap T_1)$$

$$= P(D_1)P(T_1 \mid D_1) + P(D_2)P(T_1 \mid D_2)$$

$$= (0.08)(0.95) + (0.92)(0.03) = 0.1036$$

Hence,

$$P(D_1 \mid T_1) = \frac{P(D_1 \cap T_1)}{P(T_1)}$$

$$= \frac{\text{probability of path through } D_1 \text{ and } T_1}{\text{sum of probabilities of all paths to } T_1}$$

$$= \frac{0.076}{0.1036} = \frac{760}{1036} = \frac{190}{259} \approx 0.734$$

So the probability that the person has the disease, given that the test is positive, is approximately 0.734. In other words, about 73.4% of people who test positive actually have the disease. You would probably agree that this probability was relatively easy to find by using basic principles (Equation (1)) and a probability tree (Figure 8.21).

At this point, some terminology should be introduced. The *unconditional* probabilities $P(D_1)$ and $P(D_2)$ are called **prior probabilities,** because they are given *before* we have any knowledge about the outcome of a blood test. The conditional probability $P(D_1 \mid T_1)$ is called a **posterior probability,** because it is found after the outcome (T_1) of the test is known.

From our answer for $P(D_1 \mid T_1)$, we can easily find the posterior probability of not having the disease given a positive test result:

$$P(D_2 \mid T_1) = 1 - P(D_1 \mid T_1) = 1 - \frac{190}{259} = \frac{69}{259} \approx 0.266$$

Of course, this can also be found by using the probability tree:

$$P(D_2 \mid T_1) = \frac{\text{probability of path through } D_2 \text{ and } T_1}{\text{sum of probabilities of all paths to } T_1}$$

$$= \frac{(0.92)(0.03)}{0.1036} = \frac{0.0276}{0.1036} = \frac{276}{1036} = \frac{69}{259} \approx 0.266$$

It is not really necessary to use a probability tree to find $P(D_1 \mid T_1)$. Instead, a formula can be developed. We know that

$$P(D_1 \mid T_1) = \frac{P(D_1 \cap T_1)}{P(T_1)} = \frac{P(D_1)P(T_1 \mid D_1)}{P(T_1)} \tag{2}$$

Although we used a probability tree to express $P(T_1)$ conveniently as a sum of probabilities, the sum can be found another way. Take note that events D_1 and D_2 have two properties: They are mutually exclusive and their union is the sample space S. Such events are collectively called a **partition** of S. Using this partition, we can break up event T_1 into mutually exclusive "pieces":

$$T_1 = T_1 \cap S = T_1 \cap (D_1 \cup D_2)$$

Then, by the distributive and commutative laws,

$$T_1 = (D_1 \cap T_1) \cup (D_2 \cap T_1) \tag{3}$$

Since D_1 and D_2 are mutually exclusive, so are events $D_1 \cap T_1$ and $D_2 \cap T_1$.[4] Thus, T_1 has been expressed as a union of mutually exclusive events. In this form, we can find $P(T_1)$ by adding probabilities. Applying the addition law for mutually exclusive events to Equation (3) gives

$$P(T_1) = P(D_1 \cap T_1) + P(D_2 \cap T_1)$$

$$= P(D_1)P(T_1 \mid D_1) + P(D_2)P(T_1 \mid D_2)$$

Substituting into Equation (2), we obtain

$$P(D_1 \mid T_1) = \frac{P(D_1)P(T_1 \mid D_1)}{P(D_1)P(T_1 \mid D_1) + P(D_2)P(T_1 \mid D_2)} \tag{4}$$

which is a formula for computing $P(D_1 \mid T_1)$.

Equation (4) is a special case (namely, for a partition of S into two events) of the following general formula, called **Bayes's formula,**[5] which has had wide application in decision making:

Bayes's Formula

Suppose $F_1, F_2 \ldots, F_n$ are n events that partition a sample space S. That is, the F_i's are mutually exclusive and their union is S. Furthermore, suppose that E is any event in S, where $P(E) > 0$. Then the conditional probability of F_i given that event E has occurred is expressed by

$$P(F_i \mid E) = \frac{P(F_i)P(E \mid F_i)}{P(F_1)P(E \mid F_1) + P(F_2)P(E \mid F_2) + \cdots + P(F_n)P(E \mid F_n)}$$

for each value of i, where $i = 1, 2, \ldots, n$.

Rather than memorize the formula, a probability tree can be used to obtain $P(F_i \mid E)$. Using the tree in Figure 8.22, we have

$$P(F_i \mid E) = \frac{\text{probability for path through } F_i \text{ and } E}{\text{sum of all probabilities for paths to } E}$$

● EXAMPLE 1 Quality Control

A digital camcorder manufacturer uses one microchip in assembling each camcorder it produces. The microchips are purchased from suppliers A, B, and C and are randomly picked for assembling each camcorder. Twenty percent of the microchips come from A, 35% come from B, and the remainder come from C. Based on past experience, the manufacturer believes that the probability that a microchip from A is defective is

---

[4]See Example 8 of Section 8.3.

[5]After Thomas Bayes (1702–1761), the 18th-century English minister who discovered the formula.

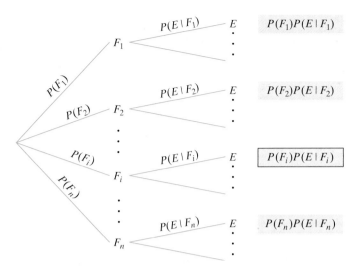

FIGURE 8.22 FIGURE 8.22   Probability tree for $P(F_i \mid E)$.

0.03, *and the corresponding probabilities for* B *and* C *are* 0.02 *and* 0.01, *respectively. A camcorder is selected at random from a day's production, and its microchip is found to be defective. Find the probability that it was supplied (a) from* A, *(b) from* B, *and (c) from* C. *(d) From what supplier was the microchip most likely purchased?*

Solution: We define the following events:

$$S_1 = \{\text{supplier A}\}$$

$$S_2 = \{\text{supplier B}\}$$

$$S_3 = \{\text{supplier C}\}$$

$$D = \{\text{defective microchip}\}$$

We have

$$P(S_1) = 0.2 \quad P(S_2) = 0.35 \quad P(S_3) = 0.45$$

and the conditional probabilities

$$P(D \mid S_1) = 0.03 \quad P(D \mid S_2) = 0.02 \quad P(D \mid S_3) = 0.01$$

which are reflected in the probability tree in Figure 8.23. Note that the figure shows only the portion of the complete probability tree that relates to event D. This is all that actually needs to be drawn, and this abbreviated form is often called a *Bayes's probability tree.*

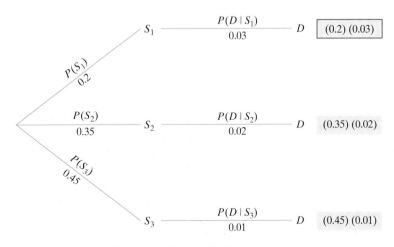

FIGURE 8.23   Bayes's probability tree for Example 1.

For part (a), we want to find the probability of S_1 given that D has occurred. That is,

$$P(S_1 \mid D) = \frac{\text{probability of path through } S_1 \text{ and } D}{\text{sum of probabilities of all paths to } D}$$

$$= \frac{(0.2)(0.03)}{(0.2)(0.03) + (0.35)(0.02) + (0.45)(0.01)}$$

$$= \frac{0.006}{0.006 + 0.007 + 0.0045}$$

$$= \frac{0.006}{0.0175} = \frac{60}{175} = \frac{12}{35}$$

This means that approximately 34.3% of the defective microchips come from supplier A.

For part (b), we have

$$P(S_2 \mid D) = \frac{\text{probability of path through } S_2 \text{ and } D}{\text{sum of probabilities of all paths to } D}$$

$$= \frac{(0.35)(0.02)}{0.0175} = \frac{0.007}{0.0175} = \frac{70}{175} = \frac{14}{35}$$

For part (c),

$$P(S_3 \mid D) = \frac{\text{probability of path through } S_3 \text{ and } D}{\text{sum of probabilities of all paths to } D}$$

$$= \frac{(0.45)(0.01)}{0.0175} = \frac{0.0045}{0.0175} = \frac{45}{175} = \frac{9}{35}$$

For part (d), the greatest of $P(S_1 \mid D)$, $P(S_2 \mid D)$, and $P(S_3 \mid D)$ is $P(S_2 \mid D)$. Thus, the defective microchip was most likely supplied by B.

NOW WORK PROBLEM 9

Bag I   Bag II

FIGURE 8.24 Diagram for Example 2.

● **EXAMPLE 2**   **Jelly Beans in a Bag**

Two identical bags, Bag I and Bag II, are on a table. Bag I contains one red and one black jelly bean; Bag II contains two red jelly beans. (See Figure 8.24.) A bag is selected at random, and then a jelly bean is randomly taken from it. The jelly bean is red. What is the probability that the other jelly bean in the selected bag is red?

Solution: Because the other jelly bean could be red or black, you might hastily conclude that the answer is $\frac{1}{2}$. Sorry, your intuition fails! The question can be restated as follows: Find the probability that the jelly bean came from Bag II, given that the jelly bean is red. We define the events

$$B_1 = \{\text{Bag I selected}\}$$

$$B_2 = \{\text{Bag II selected}\}$$

$$R = \{\text{red jelly bean selected}\}$$

We want to find $P(B_2 \mid R)$. Since a bag is selected at random, $P(B_1) = P(B_2) = \frac{1}{2}$. From Figure 8.24, we conclude that

$$P(R \mid B_1) = \frac{1}{2} \quad \text{and} \quad P(R \mid B_2) = 1$$

We will show two methods of solving this problem, the first with a probability tree and the second with Bayes's formula.

$$B_1 \xrightarrow[\frac{1}{2}]{P(R\,|\,B_1)} R \quad \boxed{\tfrac{1}{2}\cdot\tfrac{1}{2}}$$

$$P(B_1) \atop \frac{1}{2}$$

$$P(B_2) \atop \frac{1}{2}$$

$$B_2 \xrightarrow[1]{P(R\,|\,B_2)} R \quad \boxed{\tfrac{1}{2}\cdot 1}$$

FIGURE 8.25   Bayes's probability tree for Example 2.

Method 1: Probability Tree   Figure 8.25 shows a Bayes's probability tree for our problem. Since all paths end at R,

$$P(B_2 \,|\, R) = \frac{\text{probability for path through } B_2 \text{ and } R}{\text{sum of probabilities of all paths}}$$

$$= \frac{\left(\frac{1}{2}\right)(1)}{\left(\frac{1}{2}\right)\left(\frac{1}{2}\right) + \left(\frac{1}{2}\right)(1)} = \frac{\frac{1}{2}}{\frac{3}{4}} = \frac{2}{3}$$

Note that the unconditional probability of choosing Bag II, namely, $P(B_2) = \frac{1}{2}$, increases to $\frac{2}{3}$, given that a red jelly bean was taken. An increase is reasonable: Since there are only red jelly beans in Bag II, choosing a red jelly bean should make it more likely that it came from Bag II.

Method 2: Bayes's Formula   Because B_1 and B_2 partition the sample space, by Bayes's formula we have

$$P(B_2 \,|\, R) = \frac{P(B_2)\,P(R\,|\,B_2)}{P(B_1)\,P(R\,|\,B_1) + P(B_2)\,P(R\,|\,B_2)}$$

$$= \frac{\left(\frac{1}{2}\right)(1)}{\left(\frac{1}{2}\right)\left(\frac{1}{2}\right) + \left(\frac{1}{2}\right)(1)} = \frac{\frac{1}{2}}{\frac{3}{4}} = \frac{2}{3}$$

NOW WORK PROBLEM 7 ●●

Problems 8.7

1. Suppose events E and F partition a sample space S, where E and F have probabilities

$$P(E) = \frac{2}{5} \quad P(F) = \frac{3}{5}$$

If D is an event such that

$$P(D\,|\,E) = \frac{1}{10} \quad P(D\,|\,F) = \frac{1}{5}$$

find the probabilities $P(E\,|\,D)$ and $P(F\,|\,D')$.

2. A sample space is partitioned by events E_1, E_2, and E_3, whose probabilities are $\frac{1}{5}$, $\frac{3}{10}$, and $\frac{1}{2}$, respectively. Suppose S is an event such that the following conditional probabilities hold:

$$P(S\,|\,E_1) = \frac{2}{5} \quad P(S\,|\,E_2) = \frac{7}{10} \quad P(S\,|\,E_3) = \frac{1}{2}$$

Find the probabilities $P(E_1\,|\,S)$ and $P(E_3\,|\,S')$.

3. Voting   In a certain precinct, 42% of the eligible voters are registered Democrats, 33% are Republicans, and the remainder are Independents. During the last primary election, 25% of the Democrats, 27% of the Republicans,

and 15% of the Independents voted. Find the probability that a person who voted is a Democrat.

4. Imported versus Domestic Tires   Out of 3000 tires in the warehouse of a tire distributor, 2000 tires are domestic and 1000 are imported. Among the domestic tires, 40% are all-season; of the imported tires, 10% are all-season. If a tire is selected at random and it is an all-season, what is the probability that it is imported?

5. **Disease Testing** A new test was developed for detecting Gamma's disease, which is believed to affect 3% of the population. Results of extensive testing indicate that 86% of persons who have this disease will have a positive reaction to the test, whereas 7% of those who do not have the disease will also have a positive reaction.

   (a) What is the probability that a randomly selected person who has a positive reaction will actually have Gamma's disease?

   (b) What is the probability that a randomly selected person who has a negative reaction will actually have Gamma's disease?

6. **Earnings and Dividends** Of the companies in a particular sector of the economy, it is believed that one-third will have an increase in quarterly earnings. Of those that do, the percentage that declare a dividend is 60%. Of those that do not have an increase, the percentage that declare a dividend is 10%. What percentage of companies that declare a dividend will have an increase in quarterly earnings?

*7. **Jelly Beans in a Bag** A bag contains four red and two green jelly beans, and a second bag contains two red and three green jelly beans. A bag is selected at random and a jelly bean is randomly taken from it. The jelly bean is red. What is the probability that it came from the first bag?

8. **Balls in a Bowl** Bowl I contains two red and three white balls. Bowl II contains four red and three white balls. Bowl III contains two red, two white, and two green balls. A bowl is chosen at random, and then a ball is chosen at random from it. The ball is white. Find the probability that it came from Bowl I.

*9. **Quality Control** A manufacturing process requires the use of a robotic welder on each of two assembly lines, A and B, which produce 300 and 500 units of product per day, respectively. Based on experience, it is believed that the welder on A produces 2% defective units, whereas the welder on B produces 5% defective units. At the end of a day, a unit was selected at random from the total production and was found to be defective. What is the probability that it came from line A?

10. **Quality Control** An automobile manufacturer has four plants: A, B, C, and D. The percentages of total daily output that are produced by the four plants are 35%, 20%, 30%, and 15%, respectively. The percentages of defective units produced by the plants are estimated to be 2%, 5%, 3%, and 4%, respectively. Suppose that a car on a dealer's lot is randomly selected and found to be defective. What is the probability that it came from plant (a) A? (b) B? (c) C? (d) D?

11. **Wake-Up Call** Barbara Smith, a sales representative, is staying overnight at a hotel and has a breakfast meeting with an important client the following morning. She asked the front desk to give her a 7 A.M. wake-up call that morning in order that she be prompt for the meeting. The probability that the desk makes the call is 0.95. If the call is made, the probability that she will be on time is 0.9, but if the call is not made, the probability that she will be on time is 0.75. If she is on time for the meeting, what is the probability that the call was made?

12. **Candy Snatcher** On a high shelf are two identical opaque candy jars containing 50 raisin clusters each. The clusters in

one of the jars are made with dark chocolate. In the other jar, 20 are made with dark chocolate and 30 are made with milk chocolate. (They are mixed well, however.) Bob Jones, who has a sudden craving for chocolate, reaches up and randomly takes a raisin cluster from one of the jars. If it is made with dark chocolate, what is the probability that it was taken from the jar containing only dark chocolate?

13. **Physical Fitness Activity** The week of National Employee Health and Fitness Day, the employees of a large company were asked to exercise a minimum of three times that week for at least 20 minutes per session. The purpose was to generate "exercise miles." All participants who completed this requirement received a certificate acknowledging their contribution. The activities reported were walking, bicycling, and running. Of all who participated, $\frac{1}{2}$ reported walking, $\frac{1}{4}$ reported bicycling, and $\frac{1}{4}$ reported running. Suppose that the probability that a participant who walks will complete the requirement is $\frac{9}{10}$, and for bicycling and running it is $\frac{4}{5}$ and $\frac{2}{3}$, respectively. What percentage of persons who completed the requirement do you expect reported walking? (Assume that each participant got his or her exercise from only one activity.)

14. **Battery Reliability** When the weather is extremely frigid, a motorist must charge his car battery during the night in order to improve the likelihood that the car will start early the following morning. If he does not charge it, the probability that the car will not start is $\frac{4}{5}$. If he does charge it, the probability that the car will not start is $\frac{1}{8}$. Past experience shows that the probability that he remembers to charge the battery is $\frac{9}{10}$. One morning, during a cold spell, he cannot start his car. What is the probability that he forgot to charge the battery?

15. **Automobile Satisfaction Survey** In a customer satisfaction survey, $\frac{3}{5}$ of those surveyed had a Japanese-made car, $\frac{1}{10}$ a European-made car, and $\frac{3}{10}$ an American-made car. Of the first group, 85% said they would buy the same make of car again, and for the other two groups the corresponding percentages are 50% and 40%. What is the probability that a person who said they would buy the same make again had a Japanese-made car?

16. **Mineral Test Borings** A geologist believes that the probability that the rare earth mineral dalhousium occurs in a particular region of the country is 0.005. If dalhousium is present in that region, the geologist's test borings will have a positive result 80% of the time. However, if dalhousium is not present, a negative result will occur 85% of the time.

   (a) If a test is positive on a site in the region, find the probability that dalhousium is there.

(b) If a test is negative on such a site, find the probability that dalhousium is there.

17. Physics Exam After a physics exam was given, it turned out that only 75% of the class answered every question. Of those who did, 80% passed, but of those who did not, only 50% passed. If a student passed the exam, what is the probability that the student answered every question? (P.S.: The instructor eventually reached the conclusion that the test was too long and curved the exam grades, to be fair and merciful.)

18. Giving Up Smoking In a 1990 survey of smokers, 75% predicted that they would still be smoking five years later. Five years later, 70% of those who predicted that they would be smoking did not smoke, and of those who predicted that they would not be smoking, 90% did not smoke. What percentage of those who do not smoke predicted that they would be smoking?

19. Alien Communication B. G. Cosmos, a scientist, believes that the probability is $\frac{2}{5}$ that aliens from an advanced civilization on Planet X are trying to communicate with us by sending high-frequency signals to Earth. By using sophisticated equipment, Cosmos hopes to pick up these signals. The manufacturer of the equipment, Trekee, Inc., claims that if aliens are indeed sending signals, the probability that the equipment will detect them is $\frac{3}{5}$. However, if aliens are not sending signals, the probability that the equipment will seem to detect such signals is $\frac{1}{10}$. If the equipment detects signals, what is the probability that aliens are actually sending them?

20. Calculus Grades In an honors Calculus I class, 60% of students had an A average at midterm. Of these, 70% ended up with a course grade of A, and of those who did not have an A average at midterm, 60% ended up with a course grade of A. If one of the students is selected at random and is found to have received an A for the course, what is the probability that the student did not have an A average at midterm?

21. Movie Critique A well-known pair of highly influential movie critics have a popular TV show on which they review new movie releases and recently released videos. Over the past 10 years, they gave a "Two Thumbs Up" to 70% of movies that turned out to be box-office successes; they gave a "Two Thumbs Down" to 80% of movies that proved to be unsuccessful. A new movie, *Math Wizard,* whose release is imminent, is considered favorably by others in the industry who have previewed it; in fact, they give it a prior probability of success of $\frac{8}{10}$. Find the probability that it will be a success, given that the pair of TV critics give it a "Two Thumbs Up" after seeing it. Assume that all films are given either "Two Thumbs Up" or "Two Thumbs Down."

22. Balls in a Bowl Bowl 1 contains five green and four red balls, and Bowl 2 contains three green, one white, and three red balls. A ball is randomly taken from Bowl 1 and placed in Bowl 2. A ball is then randomly taken from Bowl 2. If the ball is green, find the probability that a green ball was taken from Bowl 1.

23. Risky Loan In the loan department of The Bank of Montreal, past experience indicates that 20% of loan requests are considered by bank examiners to fall into the "substandard" class and should not be approved. However, the bank's loan reviewer, M. Blackwell, is lax at times and concludes that a request is not in the substandard class when it is, and vice versa. Suppose that 25% of requests that are actually substandard are not considered substandard by Blackwell and that 15% of requests that are not substandard are considered by Blackwell to be substandard and, hence, not approved.

(a) Find the probability that Blackwell considers that a request is substandard.

(b) Find the probability that a request is substandard, given that Blackwell considers it to be substandard.

(c) Find the probability that Blackwell makes an error in considering a request. (An error occurs when the request is not substandard but is considered substandard, or when the request is substandard but is considered to be not substandard.)

24. Coins in Chests Each of three identical chests has two drawers. The first chest contains a gold coin in each drawer. The second chest contains a silver coin in each drawer, and the third contains a silver coin in one drawer and a gold coin in the other. A chest is chosen at random and a drawer is opened. There is a gold coin in it. What is the probability that the coin in the other drawer of that chest is silver?

25. Product Identification after Flood[6] After a severe flood, a distribution warehouse is stocked with waterproofed boxes of fireworks from which the identification labels have been washed off. There are three types of fireworks: low quality, medium quality, and high quality, each packed in units of 100 in identical boxes. None of the individual fireworks have markings on them, but it is believed that in the entire warehouse, the proportions of boxes with low-, medium-, and high-quality fireworks are 0.25, 0.25, and 0.5, respectively. Because detonating a firework destroys it, extensive testing of the fireworks is impractical. Instead, the distributor decides that two fireworks from each box will be tested. The quality of the fireworks will then be decided on the basis of how many of the two are defective. The manufacturer, on the basis of experience, estimates the conditional probabilities given in Table 8.14. Suppose two fireworks are selected from a box and tested, and both are found to detonate satisfactorily. Let the events L, M, and H be that the box contains low-, medium-, and high-quality fireworks, respectively. Furthermore, let E be the observed event that neither of the fireworks was defective.

TABLE 8.14 Conditional Probabilities of Finding x Defectives, Given That Two Fireworks Were Tested from a Box of Known Quality

Number of Defectives x	Quality of Fireworks		
	Low	Medium	High
0	0.49	0.64	0.81
1	0.44	0.32	0.18
2	0.07	0.04	0.01

---

[6]Samuel Goldberg, *Probability, An Introduction* (Prentice-Hall, Inc., 1960, Dover Publications, Inc., 1986), pp. 97–98. Adapted by permission of the author.

(a) Find $P(L \mid E)$, the probability that the box contains low-quality fireworks, given E.

(b) Find the probability that the box contains medium-quality fireworks, given E.

(c) Find the probability that the box contains high-quality fireworks, given E.

(d) What is the most likely quality of the fireworks in the box, given E?

26. **Product Identification after Flood**

(a) Repeat Problem 25 if E is the observed event that exactly one of the tested fireworks is defective.

(b) Repeat Problem 25 if E is the observed event that both of the tested fireworks are defective.

27. **Weather Forecasting**[7]   J. B. Smith, who has lived in the same city many years, assigns a prior probability of 0.4 that today's weather will be inclement. (He thinks that today will be fair with probability 0.6.) Smith listens to an early morning weather forecast to get information on the day's weather. The forecaster makes one of three predictions: fair weather, inclement weather, or uncertain weather. Smith has made estimates of conditional probabilities of the different predictions, given the day's weather, as shown in Table 8.15. For example, Smith believes that, of the fair days, 70% are correctly forecast, 20% are forecast as inclement, and 10% are forecast as uncertain. Suppose that Smith hears the forecaster predict fair weather. What is the posterior probability of fair weather?

TABLE 8.15 Weather and Forecast

Day's Weather	Forecast		
	Fair	Inclement	Uncertain
Fair	0.7	0.2	0.1
Inclement	0.3	0.6	0.1

8.8 Review

Important Terms and Symbols

Examples

Section 8.1	**Basic Counting Principle and Permuations**	
	tree diagram Basic Counting Principle	Ex. 1, p. 346
	permutation, $_nP_r$	Ex. 5, p. 348
Section 8.2	**Combinations and Other Counting Principles**	
	combination, $_nC_r$	Ex. 2, p. 352
	permutation with repeated objects cells	Ex. 6, p. 358
Section 8.3	**Sample Spaces and Events**	
	sample space sample point finite sample space	Ex. 1, p. 363
	event certain event impossible event simple event	Ex. 6, p. 365
	Venn diagram complement, E' union, $\cup$ intersection, $\cap$	Ex. 7, p. 366
	mutually exclusive events	Ex. 8, p. 368
Section 8.4	**Probability**	
	equally likely outcomes trial relative frequency	Ex. 1, p. 371
	equiprobable space probability of event, $P(E)$	Ex. 2, p. 371
	addition law for mutually exclusive events empirical probability	Ex. 5, p. 374
	odds	Ex. 9, p. 378
Section 8.5	**Conditional Probability and Stochastic Processes**	
	conditional probability, $P(E \mid F)$ reduced sample space	Ex. 1, p. 382
	general multiplication law trial compound experiment	Ex. 5, p. 386
	probability tree	Ex. 6, p. 386
Section 8.6	**Independent Events**	
	independent events dependent events	Ex. 1, p. 394
	special multiplication law	Ex. 3, p. 396
Section 8.7	**Bayes's Fopmula**	
	partition prior probability posterior probability	Ex. 1, p. 406
	Bayes's formula Bayes's probability tree	Ex. 2, p. 408

---

[7]Samuel Goldberg, *Probability, An Introduction* (Prentice-Hall, Inc., 1960, Dover Publications, Inc., 1986), pp. 99–100. Adapted by permission of the author.

Summary

It is important to know the number of ways a procedure can occur. Suppose a procedure involves a sequence of k stages. Let n_1 be the number of ways the first stage can occur, and n_2 the number of ways the second stage can occur, and so on, with n_k the number of ways the kth stage can occur. Then the number of ways the procedure can occur is

$$n_1 \cdot n_2 \cdots n_k$$

This result is called the Basic Counting Principle.

An ordered selection of r objects, without repetition, taken from n distinct objects is called a permutation of the n objects taken r at a time. The number of such permutations is denoted $_nP_r$ and is given by

$$_nP_r = \underbrace{n(n-1)(n-2)\cdots(n-r+1)}_{r \text{ factors}} = \frac{n!}{(n-r)!}$$

If the selection is made without regard to order, then it is simply an r-element subset of an n-element set and is called a combination of n objects taken r at a time. The number of such combinations is denoted $_nC_r$ and is given by

$$_nC_r = \frac{n!}{r!(n-r)!}$$

When some of the objects are repeated, the number of distinguishable permutations of n objects, such that n_1 are of one type, n_2 are of a second type, and so on, and n_k are of a kth type, is

$$\frac{n!}{n_1!n_2!\cdots n_k!} \tag{1}$$

where $n_1 + n_2 + \cdots + n_k = n$.

The expression in Equation (1) can also be used to determine the number of assignments of objects to cells. If n distinct objects are placed into k ordered cells, with n_i objects in cell i, for $i = 1, 2, \ldots, k$, then the number of such assignments is

$$\frac{n!}{n_1!n_2!\cdots n_k!}$$

where $n_1 + n_2 + \cdots + n_k = n$.

A sample space for an experiment is a set S of all possible outcomes of the experiment. These outcomes are called sample points. A subset E of S is called an event. Two special events are the sample space itself, which is a certain event, and the empty set, which is an impossible event. An event consisting of a single sample point is called a simple event. Two events are said to be mutually exclusive when they have no sample point in common. A sample space whose outcomes are equally likely is called an equiprobable space. If E is an event for a finite equiprobable space S, then the probability that E occurs is given by

$$P(E) = \frac{\#(E)}{\#(S)}$$

If F is also an event in S, we have

$$P(E \cup F) = P(E) + P(F) - P(E \cap F)$$

$$P(E \cup F) = P(E) + P(F) \qquad \text{if } E \text{ and } F \text{ are mutually exclusive}$$

$$P(E') = 1 - P(E)$$

$$P(S) = 1$$

$$P(\emptyset) = 0$$

For an event E, the ratio

$$\frac{P(E)}{P(E')} \left(= \frac{P(E)}{1 - P(E)} \right)$$

gives the odds that E occurs. Conversely, if the odds that E occurs are $a : b$, then

$$P(E) = \frac{a}{a + b}$$

The probability that an event E occurs, given that event F has occurred, is called a conditional probability. It is denoted by $P(E \mid F)$ and can be computed either by considering a reduced equiprobable sample space and using the formula

$$P(E \mid F) = \frac{\#(E \cap F)}{\#(F)}$$

or from the formula

$$P(E \mid F) = \frac{P(E \cap F)}{P(F)}$$

which involves probabilities with respect to the original sample space.

To find the probability that two events both occur, we can use the general multiplication law:

$$P(E \cap F) = P(E)P(F \mid E) = P(F)P(E \mid F)$$

Here we multiply the probability that one of the events occurs by the conditional probability that the other one occurs, given that the first has occurred. For more than two events, the corresponding law is

$$P(E_1 \cap E_2 \cap \cdots \cap E_n)$$
$$= P(E_1)P(E_2 \mid E_1)P(E_3 \mid E_1 \cap E_2) \cdots$$
$$P(E_n \mid E_1 \cap E_2 \cap \cdots \cap E_{n-1})$$

The general multiplication law is also called the law of compound probability, because it is useful when applied to a compound experiment—one that can be expressed as a sequence of two or more other experiments, called trials or stages.

When we analyze a compound experiment, a probability tree is extremely useful in keeping track of the possible outcomes for each trial of the experiment. A path is a complete sequence of branches from the start to a tip of the tree. Each path represents an outcome of the compound experiment, and the probability of that path is the product of the probabilities for the branches of the path.

Events E and F are independent when the occurrence of one of them does not affect the probability of the other; that is,

$$P(E\,|\,F) = P(E) \quad \text{or} \quad P(F\,|\,E) = P(F)$$

Events that are not independent are dependent.

If E and F are independent, the general multiplication law simplifies into the special multiplication law:

$$P(E \cap F) = P(E)P(F)$$

Here the probability that E and F both occur is the probability of E times the probability of F. The preceding equation forms the basis of an alternative definition of independence: Events E and F are independent if and only if

$$P(E \cap F) = P(E)P(F)$$

Three or more events are independent if and only if for each set of two or more of the events, the probability of the intersection of the events in that set is equal to the product of the probabilities of those events.

A partition divides a sample space into mutually exclusive events. If E is an event and $F_1, F_2, \ldots, F_n$ is a partition, then, to find the conditional probability of event F_i, given E, when prior and conditional probabilities are known, we can use Bayes's formula:

$$P(F_i\,|\,E)$$
$$= \frac{P(F_i)P(E\,|\,F_i)}{P(F_1)P(E\,|\,F_1) + P(F_2)P(E\,|\,F_2) + \cdots + P(F_n)P(E\,|\,F_n)}$$

A Bayes-type problem can also be solved with the aid of a Bayes probability tree.

Review Problems

Problem numbers shown in color indicate problems suggested for use as a practice chapter test.

In Problems 1–4, determine the values.

1. $_8P_3$

2. $_{20}P_1$

3. $_9C_7$

4. $_{12}C_5$

5. **License Plate**   A six-character license plate consists of three letters followed by three numbers. How many different license plates are possible?

6. **Dinner**   In a restaurant, a complete dinner consists of one appetizer, one entrée, and one dessert. The choices for the appetizer are soup and salad; for the entrée, chicken, steak, lobster, and veal; and for the dessert, ice cream, pie, and pudding. How many complete dinners are possible?

7. **Garage-Door Opener**   The transmitter for an electric garage-door opener transmits a coded signal to a receiver. The code is determined by five switches, each of which is either in an "on" or "off" position. Determine the number of different codes that may be transmitted.

8. **Baseball**   A baseball manager must determine a batting order for his nine-member team. How many batting orders are possible?

9. **Softball**   A softball league has seven teams. In terms of first, second, and third place, in how many ways can the season end? Assume that there are no ties.

10. **Trophies**   In a trophy case, nine different trophies are to be placed—two on the top shelf, three on the middle, and four on the bottom. Considering the order of arrangement on each shelf, in how many ways can the trophies be placed in the case?

11. **Groups**   Eleven stranded wait-listed passengers surge to the counter for boarding passes. But there are only six boarding passes available. How many different groups of passengers can board?

12. **Cards**   From a 52-card deck of playing cards, a five-card hand is dealt. In how many ways can two of the cards be of the same face value, and another two of a different face value and the fifth of yet another face value?

13. **Light Bulbs**   A carton contains 24 light bulbs, one of which is defective. (a) In how many ways can three bulbs be selected? (b) In how many ways can three bulbs be selected if one is defective?

14. **Multiple-Choice Exam**   Each question of a 10-question multiple-choice examination is worth 10 points and has four choices, only one of which is correct. By guessing, in how many ways is it possible to receive a score of 90 or better?

15. **Letter Arrangement**   How many distinguishable horizontal arrangements of the letters in MISSISSIPPI are possible?

16. **Flag Signals**   Colored flags arranged vertically on a flagpole indicate a signal (or message). How many different signals are possible if two red, three green, and four white flags are all used?

17. **Personnel Agency**   A mathematics professor personnel agency provides mathematics professors on a temporary basis to universities that are short of staff. The manager has a pool of nine professors and must send four to Dalhousie University and three to St. Mary's. In how many ways can the manager make assignments?

18. **Tour Operator**   A tour operator has three vans, and each can accommodate seven tourists. Suppose 14 people arrive for a city sightseeing tour and the operator will use only two

vans. In how many ways can the operator assign the people to the vans?

19. Suppose $S = \{1, 2, 3, 4, 5, 6, 7, 8\}$ is the sample space and $E_1 = \{1, 2, 3, 4, 5, 6\}$ and $E_2 = \{4, 5, 6, 7\}$ are events for an experiment. Find (a) $E_1 \cup E_2$, (b) $E_1 \cap E_2$, (c) $E_1' \cup E_2$, (d) $E_1 \cap E_1'$, and (e) $(E_1 \cap E_2')'$. (f) Are E_1 and E_2 mutually exclusive?

20. **Die and Coin** A die is rolled and then a coin is tossed. (a) Determine a sample space for this experiment. Determine the events that (b) a 2 shows and (c) a head and an even number show.

21. **Bags of Jelly Beans** Three bags, labeled 1, 2, and 3, each contain two jelly beans, one red and the other green. A jelly beans is selected at random from each bag. (a) Determine a sample space for this experiment. Determine the events that (b) exactly two jelly beans are red and (c) the jelly beans are the same color.

22. Suppose that E_1 and E_2 are events for an experiment with a finite number of sample points. If $P(E_1) = 0.6$, $P(E_1 \cup E_2) = 0.7$, and $P(E_1 \cap E_2) = 0.2$, find $P(E_2)$.

23. **Quality Control** A manufacturer of computer chips packages 10 chips to a box. For quality control, two chips are selected at random from each box and tested. If any one of the tested chips is defective, the entire box of chips is rejected for sale. For a box that contains exactly one defective chip, what is the probability that the box is rejected?

24. **Drugs** Each of 100 white rats was injected with one of four drugs, A, B, C, or D. Drug A was given to 35%, B to 25%, and C to 15%. If a rat is chosen at random, determine the probability that it was injected with either C or D. If the experiment is repeated on a larger group of 300 rats but with the drugs given in the same proportion, what is the effect on the previous probability?

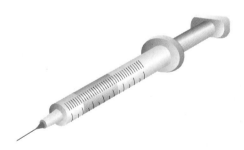

25. **Multiple-Choice Exam** Each question on a five-question multiple-choice examination has four choices, only one of which is correct. If a student answers each question in a random fashion, what is the probability that the student answers exactly two questions incorrectly?

26. **Cola Preference** To determine the national preference of cola drinkers, an advertising agency conducted a survey of 200 of them. Two cola brands, A and B, were involved. The results of the survey are indicated in Table 8.16. If a cola drinker is selected at random, determine the (empirical) probability that the person

(a) Likes both A and B

(b) Likes A, but not B

TABLE 8.16 Cola Preference

Like A only	70
Like B only	80
Like both A and B	35
Like neither A nor B	15
Total	200

27. **Jelly Beans in a Bag** A bag contains four red and six green jelly beans.

(a) If two jelly beans are randomly selected in succession with replacement, determine the probability that both are red.

(b) If the selection is made without replacement, determine the probability that both are red.

28. **Dice** A pair of fair dice is rolled. Determine the probability that the sum of the numbers is (a) 2 or 7, (b) a multiple of 3, and (c) no less than 7.

29. **Cards** Three cards from a standard deck of 52 playing cards are randomly drawn in succession with replacement. Determine the probability that (a) all three cards are black and (b) two cards are black and the other is a diamond.

30. **Cards** Two cards from a standard deck of 52 playing cards are randomly drawn in succession without replacement. Determine the probability that (a) both are hearts and (b) one is an ace and the other is a red king.

In Problems 31 and 32, for the given value of $P(E)$, find the odds that E will occur.

31. $P(E) = \frac{3}{8}$

32. $P(E) = 0.92$

In Problems 33 and 34, the odds that E will occur are given. Find $P(E)$.

33. $6 : 1$

34. $3 : 4$

35. **Cards** If a card is randomly drawn from a fair deck of 52 cards, find the probability that it is not a face card (a jack, queen, or king), given that it is a heart.

36. **Dice** If two fair dice are rolled, find the probability that the sum is less than 7, given that a 6 shows on at least one of the dice.

37. **Novel and TV Movie** The probability that a particular novel will be successful is 0.6, and if it is successful, the probability that the rights will be purchased for a

made-for-TV movie is 0.7. Find the probability that the novel will be successful and made into a TV movie.

38. Cards   Three cards are drawn from a standard deck of cards. Find the probability that the cards are, in order, a queen, a heart, and the ace of clubs if the cards are drawn with replacement.

39. Dice   If two dice are thrown, find each of the following.

(a) The probability of getting a total of 7, given that a 4 occurred on at least one die

(b) The probability of getting a total of 7 and that a 4 occurred on at least one die

40. Die   A fair die is tossed two times in succession. Find the probability that the first toss is less than 4, given that the total is greater than 8.

41. Die   If a fair die is tossed two times in succession, find the probability that the first number is less than or equal to the second number, given that the second number is less than 3.

42. Cards   Three cards are drawn without replacement from a standard deck of cards. Find the probability that the third card is a heart.

43. Seasoning Survey   A survey of 600 adults was made to determine whether or not they liked the taste of a new seasoning. The results are summarized in Table 8.17.

TABLE 8.17 Seasoning Survey

	Like	Dislike	Total
Male	80	40	120
Female	320	160	480
Total	400	200	600

(a) If a person in the survey is selected at random, find the probability that the person dislikes the seasoning (L'), given that the person is a female (F).

(b) Determine whether the events
$L = \{$liking the seasoning$\}$ and $M = \{$being a male$\}$ are independent or dependent.

44. Chips   A bowl contains six chips numbered from 1 to 6. Two chips are randomly withdrawn with replacement. Let E be the event of getting a 4 the first time and F be the event of getting a 4 the second time.

(a) Are E and F mutually exclusive?

(b) Are E and F independent?

45. College and Family Income   A survey of 175 students resulted in the data shown in Table 8.18. The table shows the type of college the student attends and the income level of the student's family. If a student is selected at random, determine whether the event of attending a public college and the event of coming from a middle-class family are independent or dependent.

TABLE 8.18 Student Survey

	College		
Income	Private	Public	Total
High	15	10	25
Middle	25	55	80
Low	10	60	70
Total	50	125	175

46. If $P(E) = \frac{1}{4}$, $P(F) = \frac{1}{3}$, and $P(E \mid F) = \frac{1}{6}$, find $P(E \cup F)$.

47. Shrubs   When a certain type of shrub is planted, the probability that it will take root is 0.7. If four shrubs are planted, find each of the following. Assume independence.

(a) The probability that none of them take root

(b) The probability that exactly two of them take root

(c) The probability that at most two of them take root

48. Antibiotic   A certain antibiotic is effective for 75% of the people who take it. Suppose four persons take this drug. What is the probability that it will be effective for at least three of them? Assume independence.

49. Bags of Jelly Beans   Bag I contains three green and two red jelly beans, and Bag II contains four red, two green, and two white jelly beans. A jelly bean is randomly taken from Bag I and placed in Bag II. If a jelly bean is then randomly taken from Bag II, find the probability that the jelly bean is red.

50. Bags of Jelly Beans   Bag I contains four red and two white jelly beans. Bag II contains two red and three white jelly beans. A bag is chosen at random, and then a jelly bean is randomly taken from it.

(a) What is the probability that the jelly bean is white?

(b) If the jelly bean is white, what is the probability that it was taken from Bag II?

51. Grade Distribution   Last semester, the grade distribution for a certain class taking an upper-level college course was analyzed. It was found that the proportion of students receiving a grade of A was 0.4 and the proportion getting an A and being a graduate student was 0.1. If a student is randomly selected from this class and is found to have received an A, find the probability that the student is a graduate student.

52. Alumni Reunion   At the most recent alumni day at Omega College, 507 persons attended. Of these, 409 lived within the state, and 40% of them were attending for the first time. Among the alumni who lived out of the state, 73% were attending for the first time. That day a raffle was held, and the person who won had also won it the year before. Find the probability that the winner was from out of state.

53. Quality Control   A music company burns CDs on two shifts. The first shift produces 3000 discs per day, and the second produces 5000. From past experience, it is believed that of the output produced by the first and second shifts, 1% and 2% are scratched, respectively. At the end of a day, a disc was selected at random from the total production.

(a) Find the probability that the CD is scratched.

(b) If the CD is scratched, find the probability that it came from the first shift.

54. Aptitude Test   In the past, a company has hired only experienced personnel for its word-processing department. Because of a shortage in this field, the company has decided to hire inexperienced persons and will provide on-the-job training. It has supplied an employment agency with a new aptitude test that has been designed for applicants who desire such a training position. Of those who recently took the test, 35% passed. In order to gauge the effectiveness of the test, everyone who took the test was put in the training program. Of those who passed the test, 80% performed satisfactorily, whereas of those who failed, only 30% did satisfactorily. If one of the new trainees is selected at random and is found to be satisfactory, what is the probability that the person passed the exam?

Mathematical Snapshot

Probability and Cellular Automata[8]

Systems of one, two, or three equations are good for modeling simple processes, like the path of a thrown object or the accrual of interest in a bank account. But how do we model something complicated and irregular, like a bolt of lightning or the spread of a rumor? For this, instead of trying to write and solve a system of equations, we can use a different modeling technique: cellular automata.

Cellular automata represent large, complex phenomena using collections of many small entities each following a few simple rules. The best-known system of cellular automata is the game called LIFE developed by John Conway in the late 1970s. It can be played by hand on a checkerboard, but using a computer program is faster and easier. Downloadable freeware can be found on the Internet. (Go to any search engine and search for "LIFE" and "Conway.")

Intriguing though LIFE is, it is not particularly good for modeling real-life processes. Better for such a task are cellular automata whose rules contain an element of randomness. Here is an example. Let us model the seepage of an oil spill into the soil below. We will model the soil as a pattern of cells layered like bricks (Figure 8.26).

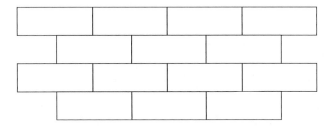

FIGURE 8.26   Cells for an oil spill model.

Each cell represents a pore in the soil, a pocket of space between dirt particles. All cells start in the "empty" state. To simulate an oil spill, we switch the entire top layer of cells (the soil surface) from empty to "filled." Depending on the microstructure of the pore arrangement, oil might flow from a pore either to both pores below, or just to the left one, or just to the right one, or to neither. We will model this by supposing that at every junction between a filled cell and an empty one below it, the filled cell has a probability P of switching the empty cell to filled.

For a TI-83 Plus graphing calculator, the following program models the process:

```
PROGRAM:OIL
:Input "P?", P
:ClrDraw
:AxesOff
:0 → X : 0 → N
:For(Y,0,46)
:Pxl–On(X,2Y + N)
:Pxl–On(X,2Y + N + 1)
:End
:For(X,1,62)
:X–2iPart(X/2) → N
:For(Y,0,46)
:If ((pxl–Test(X–1, 2Y + N) and rand < P) or
    (pxl–Test(X–1,2Y + N + 1) and rand < P))
:Then
:Pxl–On(X,2Y + N)
:Pxl–On(X,2Y + N + 1)
:End:End:End
```

After you have entered this program, set the standard viewing window and run the program. At the P? prompt, enter some value between 0 and 1. Then watch as the screen fills with a simulated oil spill percolating downward through the soil. You should find that for $P < 0.55$, sooner or later the oil stops seeping downward; that for $P > 0.75$, the oil shows no sign of slowdown; and that for $0.55 < P < 0.75$, the seepage exhibits a highly irregular pattern something like the one in Figure 8.27.

FIGURE 8.27   Oil seepage with $P = 0.6$.

The oil spill model is one dimensional in the sense that at any instant, only one row of cells is in transition. Other systems of cellular automata run in two dimensions. We can use a scheme somewhat like the LIFE game to model the spread of a fad—be it mood rings (1975–1977) or collapsible metal scooters (1999–2000). A fad is, by nature, a transient thing: People who pick up a fad soon tire of it and from then

[8]Adapted from L. Charles Biehl, "Forest Fires, Oil Spills, and Fractal Geometry, Part 1: Cellular Automata and Modeling Natural Phenomena," *The Mathematics Teacher*, 91 (November 1998), 682–87. By permission of the National Council of Teachers of Mathematics.

on are "immune." The spread of a fad is somewhat like the spread of a forest fire, where trees catch fire and in turn pass the fire on to others before burning out.

We can model the fad-spreading process on a grid, where each square represents a person in one of three states: prefad, midfad, and postfad. A prefad square that shares a side with a midfad square has a probability P of being "infected" and being turned into a midfad cell. The midfad state only lasts one cycle and is followed by a permanent postfad state.

This model would be hard to implement on a graphing calculator but can readily be done using a computer algebra system, like Maple or *Mathematica*. The results are analogous to the ones for the oil spill: For $P < 0.4$, the fad dies out; for $P > 0.6$, the fad spreads almost uniformly throughout the population; and for $0.4 < P < 0.6$, the fad spreads in an irregular and unpredictable pattern (Figure 8.28).

The interesting result here is that the difference between a fad that spreads widely and one that quickly peters out can be quite small. A bit of fashion that passes from friend to friend with a probability of 0.45 may never really catch on, while a fashion transmitted with a probability of 0.55 has a good chance of becoming a major craze. Of course, the model we used is quite simple. A more sophisticated model might allow for the fact that some people have more friends than others, that some friendships are more fad-transmission prone than others, and so on.

Problems

1. By repeated runs on a graphing calculator, estimate the critical value of P at which the oil spill in the simulation

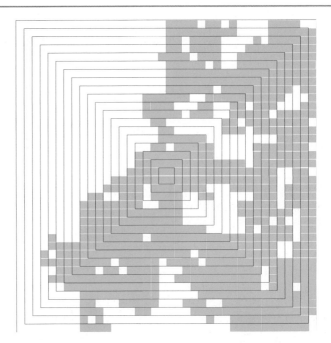

FIGURE 8.28   Fad spreading with $P = 0.5$.

starts to penetrate deeply into the soil. That is, what is the smallest value of P for which the spill does not diminish significantly between the middle and the bottom of the screen?

2. What other phenomena might be modeled using cellular automata? What would the rules look like?

9

ADDITIONAL TOPICS IN PROBABILITY

 Mathematical Snapshot Markov Chains in Game Theory

As we saw at the beginning of Chapter 8, probability can be used to solve the problem of dividing up the pot of money between two gamblers when their game is interrupted. Now we might ask a follow-up question: What are the chances that a game will be interrupted in the first place?

Our answer depends on the details, of course. If the gamblers know in advance that they will play a fixed number of rounds—the "interruption" being scheduled in advance, as it were—then it might be fairly easy to calculate the probability that time will run out before they finish. Or, if the amount of time available is unknown, we might calculate the *expected duration* of a complete game and an *expected time* before the next interruption. Then, if the expected game length came out well under the expected time to the next interruption, we could say that the probability of having to break the game off in midplay was low. But if we wanted to give a more exact, numerical answer we would have to do a more complicated calculation.

The kind of problem encountered here is not unique to gambling. In industry, manufacturers need to know how likely they are to have to interrupt a production cycle due to equipment breakdown. One way they keep this probability low is by logging the usage hours on each machine and replacing it as the hours approach "mean time to failure"—the expected value of the number of usage hours the machine provides in its lifetime. Medical researchers face a related problem when they consider the possibility of having to break off an experiment because too many test subjects drop out. To keep this probability low, researchers often calculate an expected number of drop-outs in advance and include this number, plus a cushion, in the number of people recruited for a study.

The idea of the expected value for a number—the length of time until something happens, or the number of people who drop out of a study—is one of the key concepts of this chapter.

9.1 Discrete Random Variables and Expected Value

With some experiments, we are interested in events associated with numbers. For example, if two coins are tossed, our interest may be in the *number* of heads that occur. Thus, we consider the events

$$\{0\} \quad \{1\} \quad \{2\}$$

If we let X be a variable that represents the number of heads that occur, then the only values that X can assume are 0, 1, and 2. The value of X is determined by the outcome of the experiment, and hence by chance. In general, a variable whose values depend on the outcome of a random process is called a **random variable.** Usually, random variables are denoted by capital letters such as X, Y, or Z, and the values that these variables assume may be denoted by corresponding lowercase letters (x, y, or z). Thus, for the number of heads (X) that occur in the tossing of two coins, we may indicate the possible values by writing

$$X = x \quad \text{where } x = 0, 1, \ 2$$

or, more simply,

$$X = 0, 1, 2$$

● EXAMPLE 1   Random Variables

a. Suppose a die is rolled and X is the number that turns up. Then X is a random variable and $X = 1, 2, 3, 4, 5, 6$.

b. Suppose a coin is successively tossed until a head appears. If Y is the number of such tosses, then Y is a random variable and

$$Y = y \quad \text{where } y = 1, 2, 3, 4, \ldots$$

Note that Y may assume infinitely many values.

c. A student is taking an exam with a one-hour time limit. If X is the number of minutes it takes to complete the exam, then X is a random variable. The values that X may assume form the interval $(0, 60]$. That is, $0 < X \le 60$.

A random variable is called a **discrete random variable** if it may assume only a finite number of values or if its values can be placed in one-to-one correspondence with the positive integers. In Examples 1(a) and 1(b), X and Y are discrete. A random variable is called a **continuous random variable** if it may assume any value in some interval or intervals, such as X does in Example 1(c). In this chapter, we will be concerned with discrete random variables; Chapter 16 deals with continuous random variables.

If X is a random variable, the probability of the event that X assumes the value x is denoted $P(X = x)$. Similarly, we can consider the probabilities of events such as $X \le x$ and $X > x$. If X is discrete, then the function f that assigns the number $P(X = x)$ to each possible value of X is called the **probability function** or the **distribution** of the random variable X. Thus,

$$f(x) = P(X = x)$$

It may be helpful to verbalize this equation as "$f(x)$ is the probability that X assumes the value x".

● EXAMPLE 2   Distribution of a Random Variable

Suppose that X is the number of heads that appear on the toss of two well-balanced coins. Determine the distribution of X.

Solution: We must find the probabilities of the events $X = 0$, $X = 1$, and $X = 2$. The usual equiprobable sample space is

$$S = \{HH, HT, TH, TT\}$$

Probability Table

x	$P(X = x)$
0	1/4
1	2/4
2	1/4

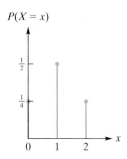

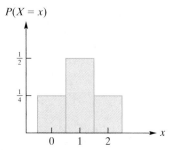

FIGURE 9.1 Graph of the distribution of X.

FIGURE 9.2 Probability histogram for X.

Hence,

$$\text{the event } X = 0 \text{ is } \{TT\}$$

$$\text{the event } X = 1 \text{ is } \{HT, TH\}$$

$$\text{the event } X = 2 \text{ is } \{HH\}$$

The probability for each of these events is given in the **probability table** in the margin. If f is the distribution for X, that is, $f(x) = P(X = x)$, then

$$f(0) = \frac{1}{4} \quad f(1) = \frac{1}{2} \quad f(2) = \frac{1}{4}$$

In Example 2, the distribution f was indicated by the listing

$$f(0) = \frac{1}{4} \quad f(1) = \frac{1}{2} \quad f(2) = \frac{1}{4}$$

However, the probability table for X gives the same information and is an acceptable way of expressing the distribution of X. Another way is by the graph of the distribution, as shown in Figure 9.1. The vertical lines from the x-axis to the points on the graph merely emphasize the heights of the points. Another representation of the distribution of X is the rectangle diagram in Figure 9.2, called the **probability histogram** for X. Here a rectangle is centered over each value of X. The rectangle above x has width 1 and height $P(X = x)$. Thus, its *area* is the probability $1 \cdot P(X = x) = P(X = x)$. This interpretation of probability as an area is important in Chapter 16.

Note in Example 2 that the sum of $f(0)$, $f(1)$, and $f(2)$ is 1:

$$f(0) + f(1) + f(2) = \frac{1}{4} + \frac{1}{2} + \frac{1}{4} = 1$$

This must be the case, because the events $X = 0$, $X = 1$, and $X = 2$ are mutually exclusive and the union of all three is the sample space [and $P(S) = 1$]. We can conveniently indicate the sum $f(0) + f(1) + f(2)$ by the summation notation[1]

$$\sum_x f(x)$$

This usage differs slightly from that in Section 1.5, in that the upper and lower bounds of summation are not given explicitly. Here $\sum_x f(x)$ means that we are to sum all terms of the form $f(x)$, for *all* values of x under consideration (which in this case are 0, 1, and 2). Thus,

$$\sum_x f(x) = f(0) + f(1) + f(2)$$

In general, for any distribution f, we have $0 \le f(x) \le 1$ for all x, and the sum of all function values is 1. Therefore,

$$\sum_x f(x) = 1$$

This means that in any probability histogram, the sum of the areas of the rectangles is 1.

The distribution for a random variable X gives the relative frequencies of the values of X in the long run. However, it is often useful to determine the "average" value of X in the long run. In Example 2, for instance, suppose that the two coins were tossed n times, which resulted in $X = 0$ occurring k_0 times, $X = 1$ occurring k_1 times, and $X = 2$ occurring k_2 times. Then the average value of X for these n tosses is

$$\frac{0 \cdot k_0 + 1 \cdot k_1 + 2 \cdot k_2}{n} = 0 \cdot \frac{k_0}{n} + 1 \cdot \frac{k_1}{n} + 2 \cdot \frac{k_2}{n}$$

---

[1]Summation notation is introduced in Section 1.5.

But the fractions k_0/n, k_1/n and k_2/n are the relative frequencies of the events $X = 0$, $X = 1$, and $X = 2$, respectively, that occur in the n tosses. If n is very large, then these relative frequencies approach the probabilities of the events $X = 0$, $X = 1$, and $X = 2$. Thus, it seems reasonable that the average value of X in the long run is

$$0 \cdot f(0) + 1 \cdot f(1) + 2 \cdot f(2) = 0 \cdot \frac{1}{4} + 1 \cdot \frac{1}{2} + 2 \cdot \frac{1}{4} = 1 \tag{1}$$

This means that if we tossed the coins many times, the average number of heads appearing per toss is very close to 1. We define the sum in Equation (1) to be the *mean* of X. It is also called the *expected value* of X and the *expectation* of X. The mean of X is often denoted by $\mu = \mu(X)$ (μ is the Greek letter "mu") and also by $E(X)$. Note that from Equation (1), μ has the form $\sum_x x f(x)$. In general, we have the following definition.

DEFINITION

If X is a discrete random variable with distribution f, then the **mean** of X is given by

$$\mu = \mu(X) = E(X) = \sum_x x f(x)$$

The mean of X can be interpreted as the average value of X in the long run. In fact, if the values that X takes on are $x_1, x_2, \ldots, x_n$ and these are equiprobable so that we have $f(x_i) = \dfrac{1}{n}$, for $i = 1, 2, \ldots, n$ then

$$\mu = \sum_x x f(x) = \sum_{i=1}^{n} x_i \frac{1}{n} = \frac{\sum_{i=1}^{n} x_i}{n}$$

which is the average *in the usual sense of that word* of the numbers $x_1, x_2, \ldots, x_n$. In the general case, it is useful to think of the mean, μ, as a *weighted average* where the weights are provided by the probabilities, $f(x)$. We emphasisze that the mean does not necessarily have to be an outcome of the experiment. In other words, μ may be different from all the values x that the random variable X actually assumes. The next example will illustrate.

● EXAMPLE 3 Expected Gain

An insurance company offers a $180,000 catastrophic fire insurance policy to home-owners of a certain type of house. The policy provides protection in the event that such a house is totally destroyed by fire in a one-year period. The company has determined that the probability of such an event is 0.002. If the annual policy premium is $379, find the expected gain per policy for the company.

Strategy If an insured house does not suffer a catastrophic fire, the company gains $379. However, if there is such a fire, the company loses $180,000 − $379 (insured value of house minus premium), or $179,621. If X is the gain (in dollars) to the company, then X is a random variable that may assume the values 379 and −179,621. (A loss is considered a negative gain.) The expected gain per policy for the company is the expected value of X.

Solution: If f is the probability function for X, then

$$f(-179{,}621) = P(X = -179{,}621) = 0.002$$

and

$$f(379) = P(X = 379) = 1 - 0.002 = 0.998$$

The expected value of X is given by

$$E(X) = \sum_x x f(x) = -179{,}621\, f(-179{,}621) + 379\, f(379)$$

$$= -179{,}621(0.002) + 379(0.998) = 19$$

Thus, if the company sold many policies, it could expect to gain approximately \$19 per policy, which could be applied to such expenses as advertising, overhead, and profit.

NOW WORK PROBLEM 19 ⬤◐⬤

Since $E(X)$ is the average value of X in the long run, it is a measure of what might be called the *central tendency* of X. However, $E(X)$ does not indicate the *dispersion* or spread of X from the mean in the long run. For example, Figure 9.3 shows the graphs of two distributions, f and g, for the random variables X and Y. It can easily be demonstrated that both X and Y have the same mean: $E(X) = 2$ and $E(Y) = 2$. (You should verify this claim.) But from Figure 9.3, X is more likely to assume the values 1 or 3 than is Y, because $f(1)$ and $f(3)$ are $\frac{2}{5}$, whereas $g(1)$ and $g(3)$ are $\frac{1}{5}$. Thus, X has more likelihood of assuming values away from the mean than does Y, so there is more dispersion for X in the long run.

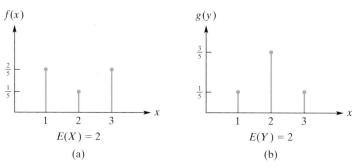

FIGURE 9.3 Probability distributions.

There are various ways to measure dispersion for a random variable X. One way is to determine the long-run average of the absolute values of the deviations from the mean μ—that is, $E(|X - \mu|)$, which is the mean of the derived random variable $|X - \mu|$. In fact, if g is a suitable function and X is a random variable, then $Y = g(X)$ is another random variable. Morever, it can be shown that if $Y = g(X)$, then $E(Y) = \sum_x g(x) f(x)$, where f is the probability function for X. For example, if $Y = |X - \mu|$, then

$$E(|X - \mu|) = \sum_x |x - \mu| f(x)$$

However, while $E(|X - \mu|)$ might appear to be an obvious measure of dispersion, it is not often used.

Many other measures of dispersion can be considered, but two are most widely accepted. One is the *variance,* and the other is the *standard deviation.* The variance of X, denoted by $\mathrm{Var}(X)$, is the long-run average of the *squares* of the deviations of X from μ. In other words, for the variance we consider the random variable $Y = (X - \mu)^2$ and we have

Variance of X

$$\mathrm{Var}(X) = E((X - \mu)^2) = \sum_x (x - \mu)^2 f(x) \tag{2}$$

Since $(X - \mu)^2$ is involved in $\mathrm{Var}(X)$, and both X and μ have the same units of measurement, the units for $\mathrm{Var}(X)$ are those of X^2. For instance, in Example 3, X is in dollars; thus, $\mathrm{Var}(X)$ has units of dollars squared. It is convenient to have a measure of dispersion in the same units as X. Such a measure is $\sqrt{\mathrm{Var}(X)}$, which is called the *standard deviation of X* and is denoted by $\sigma = \sigma(X)$ (σ is the lowercase Greek letter "sigma").

Standard Deviation of X

$$\sigma = \sigma(X) = \sqrt{\text{Var}(X)}$$

Note that σ has the property that

$$\sigma^2 = \text{Var}(X)$$

Both $\text{Var}(X) = \sigma^2$ and σ are measures of the dispersion of X. The greater the value of $\text{Var}(X)$, or σ, the greater is the dispersion. One result of a famous theorem, *Chebyshev's inequality,* is that the probability of X falling within two standard deviations of the mean is at least $\frac{3}{4}$. This means that the probability that X lies in the interval $(\mu - 2\sigma, \mu + 2\sigma)$ is greater than or equal to $\frac{3}{4}$. More generally, for $k > 1$, Chebyshev's inequality tells us that

$$P(X \in (\mu - k\sigma, \mu + k\sigma)) \geq \frac{k^2 - 1}{k^2}$$

To illustrate further, with $k = 4$, this means that, for any probabilistic experiment, at least $\frac{4^2 - 1}{4^2} = \frac{15}{16} = 93.75\%$ of the data values lie in the interval $(\mu - 4\sigma, \mu + 4\sigma)$. To lie in the interval $(\mu - 4\sigma, \mu + 4\sigma)$ is to lie "within four standard deviations of the mean".

We can write the formula for variance in Equation (2) in a different way. It is a good exercise with summation notation.

$$\text{Var}(X) = \sum_x (x - \mu)^2 f(x)$$

$$= \sum_x (x^2 - 2x\mu + \mu^2) f(x)$$

$$= \sum_x (x^2 f(x) - 2x\mu f(x) + \mu^2 f(x))$$

$$= \sum_x x^2 f(x) - 2\mu \sum_x x f(x) + \mu^2 \sum_x f(x)$$

$$= \sum_x x^2 f(x) - 2\mu(\mu) + \mu^2(1) \qquad \left(\text{since } \sum_x x f(x) = \mu \text{ and } \sum_x f(x) = 1\right)$$

Thus we have

$$\text{Var}(X) = \sigma^2 = \sum_x x^2 f(x) - \mu^2 \quad (= E(X^2) - (E(X))^2) \qquad (3)$$

This formula for variance is quite useful, since it usually simplifies computations.

● **EXAMPLE 4   Mean, Variance, and Standard Deviation**

A basket contains 10 balls, each of which shows a number. Five balls show 1, two show 2, and three show 3. A ball is selected at random. If X is the number that shows, determine μ, Var(X), and σ.

Solution: The sample space consists of 10 equally likely outcomes (the balls). The values that X can assume are 1, 2, and 3. The events $X = 1$, $X = 2$, and $X = 3$ contain 5, 2, and 3 sample points, respectively. Thus, if f is the probability function for X,

$$f(1) = P(X = 1) = \frac{5}{10} = \frac{1}{2}$$

$$f(2) = P(X = 2) = \frac{2}{10} = \frac{1}{5}$$

$$f(3) = P(X = 3) = \frac{3}{10}$$

Calculating the mean gives

$$\mu = \sum_x x f(x) = 1 \cdot f(1) + 2 \cdot f(2) + 3 \cdot f(3)$$

$$= 1 \cdot \frac{5}{10} + 2 \cdot \frac{2}{10} + 3 \cdot \frac{3}{10} = \frac{18}{10} = \frac{9}{5}$$

To find $\text{Var}(X)$, either Equation (2) or Equation (3) can be used. Both will be used here so that we can compare the arithmetical computations involved. By Equation (2),

$$\text{Var}(X) = \sum_x (x - \mu)^2 f(x)$$

$$= \left(1 - \frac{9}{5}\right)^2 f(1) + \left(2 - \frac{9}{5}\right)^2 f(2) + \left(3 - \frac{9}{5}\right)^2 f(3)$$

$$= \left(-\frac{4}{5}\right)^2 \cdot \frac{5}{10} + \left(\frac{1}{5}\right)^2 \cdot \frac{2}{10} + \left(\frac{6}{5}\right)^2 \cdot \frac{3}{10}$$

$$= \frac{16}{25} \cdot \frac{5}{10} + \frac{1}{25} \cdot \frac{2}{10} + \frac{36}{25} \cdot \frac{3}{10}$$

$$= \frac{80 + 2 + 108}{250} = \frac{190}{250} = \frac{19}{25}$$

By Equation (3),

$$\text{Var}(X) = \sum_x x^2 f(x) - \mu^2$$

$$= 1^2 \cdot f(1) + 2^2 \cdot f(2) + 3^2 \cdot f(3) - \left(\frac{9}{5}\right)^2$$

$$= 1 \cdot \frac{5}{10} + 4 \cdot \frac{2}{10} + 9 \cdot \frac{3}{10} - \frac{81}{25}$$

$$= \frac{5 + 8 + 27}{10} - \frac{81}{25} = \frac{40}{10} - \frac{81}{25}$$

$$= 4 - \frac{81}{25} = \frac{19}{25}$$

Notice that Equation (2) involves $(x - \mu)^2$, but Equation (3) involves x^2. Because of this, it is often easier to compute variances by Equation (3) than by Equation (2). Since $\sigma^2 = \text{Var}(X) = \frac{19}{25}$, the standard deviation is

$$\sigma = \sqrt{\text{Var}(X)} = \sqrt{\frac{19}{25}} = \frac{\sqrt{19}}{5}$$

NOW WORK PROBLEM 1 ●●●

Problems 9.1

In Problems 1–4, the distribution of the random variable X is given. Determine μ, Var(X), and σ. In Problem 1, construct the probability histogram. In Problem 2, graph the distribution.

*1. $f(0) = 0.1$, $f(1) = 0.4$, $f(2) = 0.2$, $f(3) = 0.3$

2. $f(4) = 0.4$, $f(5) = 0.6$

3. See Figure 9.4.

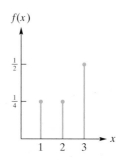

FIGURE 9.4 Distribution for Problem 3.

4. See Figure 9.5.

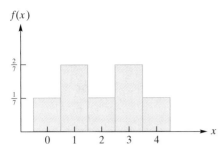

FIGURE 9.5   Distribution for Problem 4.

5. The random variable X has the following distribution:

x	$P(X = x)$
3	
5	0.3
6	0.2
7	0.4

(a) Find $P(X = 3)$.
(b) Find μ.
(c) Find σ^2.

6. The random variable X has the following distribution:

x	$P(X = x)$
2	$6a$
4	$2a$
6	0.2

(a) Find $P(X = 2)$ and $P(X = 4)$.
(b) Find μ.

In Problems 7–10, determine $E(X)$, σ^2, and σ for the random variable X.

7. Coin Toss   Three fair coins are tossed. Let X be the number of heads that occur.

8. Balls in a Basket   A basket contains six balls, each of which shows a number. Four balls show a 1 and two show a 2. A ball is randomly selected and the number that shows, X, is observed.

9. Committee   From a group of two women and three men, two persons are selected at random to form a committee. Let X be the number of men on the committee.

10. Jelly Beans in a Jar   A jar contains two red and three green jelly beans. Two jelly beans are randomly withdrawn in succession with replacement, and the number of red jelly beans, X, is observed.

11. Marbles in a Bag   A bag contains three red and two white marbles. Two marbles are randomly withdrawn in succession without replacement. Let X = the number of red marbles withdrawn. Find the distribution f for X.

12. Subcommittee   From a state government committee consisting of four Whigs and six Tories, a subcommittee of three is to be randomly selected. Let X be the number of

Whigs in the subcommittee. Find a general formula, in terms of combinations, that gives $P(X = x)$, where $x = 0, 1, 2, 3$.

13. Raffle   A charitable organization is having a raffle for a single prize of $5000. Each raffle ticket costs $2, and 8000 tickets have been sold.

(a) Find the expected gain for the purchaser of a single ticket.
(b) Find the expected gain for the purchaser of two tickets.

14. Coin Game   Consider the following game. You are to toss three fair coins. If three heads or three tails turn up, your friend pays you $10. If either one or two heads turn up, you must pay your friend $6. What are your expected winnings or losses per game?

15. Earnings   A landscaper earns $200 per day when working and loses $30 per day when not working. If the probability of working on any day is $\frac{4}{7}$, find the landscaper's expected daily earnings.

16. Fast-Food Restaurant   A fast-food chain estimates that if it opens a restaurant in a shopping center, the probability that the restaurant is successful is 0.65. A successful restaurant earns an annual profit of $75,000; a restaurant that is not successful loses $20,000. What is the expected gain to the chain if it opens a restaurant in a shopping center?

17. Insurance   An insurance company offers a hospitalization policy to individuals in a certain group. For a one-year period, the company will pay $100 per day, up to a maximum of five days, for each day the policyholder is hospitalized. The company estimates that the probability that any person in this group is hospitalized for exactly one day is 0.001; for exactly two days, 0.002; for exactly three days, 0.003; for exactly four days, 0.004; and for five or more days, 0.008. Find the expected gain per policy to the company if the annual premium is $10.

18. Demand   The following table for a small car rental company gives the probability that x cars are rented daily:

x	0	1	2	3	4	5	6	7	8
$P(X = x)$	0.05	0.10	0.15	0.20	0.15	0.15	0.10	0.05	0.05

Determine the expected daily demand for their cars.

*19. Insurance Premium   In Example 3, if the company wants an expected gain of $50 per policy, determine the annual premium.

20. Roulette In the game of roulette, there is a wheel with 37 slots numbered with the integers from 0 to 36, inclusive. A player bets $1 (for example) and chooses a number. The wheel is spun and a ball rolls on the wheel. If the ball lands in the slot showing the chosen number, the player receives the $1 bet plus $35. Otherwise, the player loses the $1 bet. Assume that all numbers are equally likely, and determine the expected gain or loss per play.

21. Coin Game Suppose that you pay $1.25 to play a game in which two fair coins are tossed. You receive the number of dollars equal to the number of heads that occur. What is your expected gain (or loss) on each play? The game is said to be *fair* to you when your expected gain is $0. What should you pay to play if this is to be a fair game?

OBJECTIVE

To develop the binomial distribution and relate it to the binomial theorem.

9.2 The Binomial Distribution

Binomial Theorem

Later in this section you will see that the terms in the expansion of a power of a binomial are useful in describing the distributions of certain random variables. It is worthwhile, therefore, first to discuss the *binomial theorem,* which is a formula for expanding $(a + b)^n$, where n is a positive integer.

Regardless of n, there are patterns in the expansion of $(a + b)^n$. To illustrate, we consider the *cube* of the binomial $a + b$. By successively applying the distributive law, we have

$$
\begin{aligned}
(a + b)^3 &= [(a + b)(a + b)](a + b) \\
&= [a(a + b) + b(a + b)](a + b) \\
&= [aa + ab + ba + bb](a + b) \\
&= aa(a + b) + ab(a + b) + ba(a + b) + bb(a + b) \\
&= aaa + aab + aba + abb + baa + bab + bba + bbb
\end{aligned}
\tag{1}
$$

so that

$$
(a + b)^3 = a^3 + 3a^2b + 3ab^2 + b^3
\tag{2}
$$

Three observations can be made about the right side of Equation (2). First, notice that the number of terms is four, which is one more than the power to which $a + b$ is raised (3). Second, the first and last terms are the *cubes* of a and b; the powers of a *decrease* from left to right (from 3 to 0), and the powers of b *increase* (from 0 to 3). Third, for each term, the sum of the exponents of a and b is 3, which is the power to which $a + b$ is raised.

Let us now focus on the coefficients of the terms in Equation (2). Consider the coefficient of the ab^2-term. It is the number of terms in Equation (1) that involve exactly two b's, namely, 3. But let us see *why* there are three terms that involve two b's. Notice in Equation (1) that each term is the product of three numbers, each of which is either a or b. Because of the distributive law, each of the three $a + b$ factors in $(a + b)^3$ contributes either an a or b to the term. Thus, the number of terms involving one a and two b's is equal to the number of ways of choosing two of the three factors to supply a b, namely, $_3C_2 = \dfrac{3!}{2!1!} = 3$. Similarly,

the coefficient of the a^3-term is $_3C_0$

the coefficient of the a^2b-term is $_3C_1$

and

the coefficient of the b^3-term is $_3C_3$

Generalizing our observations, we obtain a formula for expanding $(a + b)^n$, called the *binomial theorem.*

Binomial Theorem

If n is a positive integer, then

$$(a + b)^n = {}_nC_0 a^n + {}_nC_1 a^{n-1}b + {}_nC_2 a^{n-2}b^2 + \cdots + {}_nC_{n-1}ab^{n-1} + {}_nC_n b^n$$

$$= \sum_{i=0}^{n} {}_nC_i a^{n-i} b^i$$

The numbers ${}_nC_r$ are also called **binomial coefficients** for this reason.

● EXAMPLE 1   **Binomial Theorem**

Use the binomial theorem to expand $(q + p)^4$.

Solution: Here $n = 4$, $a = q$, and $b = p$. Thus,

$$(q + p)^4 = {}_4C_0 q^4 + {}_4C_1 q^3 p + {}_4C_2 q^2 p^2 + {}_4C_3 q p^3 + {}_4C_4 p^4$$

$$= \frac{4!}{0!4!}q^4 + \frac{4!}{1!3!}q^3 p + \frac{4!}{2!2!}q^2 p^2 + \frac{4!}{3!1!}q p^3 + \frac{4!}{4!0!}p^4$$

Recalling that $0! = 1$, we have

$$(q + p)^4 = q^4 + 4q^3 p + 6q^2 p^2 + 4q p^3 + p^4$$

Look back now to the display of *Pascal's Triangle* in Section 8.2, which provides a memorable way to generate the binomial coefficients. For example, the numbers in the $(4 + 1)$th row of Pascal's Triangle, 1 4 6 4 1, are the coefficients found in Example 1.

Binomial Distribution

We now turn our attention to repeated trials of an experiment in which the outcome of any trial does not affect the outcome of any other trial. These are referred to as **independent trials.** For example, when a fair die is rolled five times, the outcome on one roll does not affect the outcome on any other roll. Here we have five independent trials of rolling a die. Together, these five trials can be considered as a five-stage compound experiment involving independent events, so we can use the special multiplication law of Section 8.6 to determine the probability of obtaining specific outcomes of the trials.

To illustrate, let us find the probability of getting exactly two 4's in the five rolls of the die. We will consider getting a 4 as a *success* (S) and getting any of the other five numbers as a *failure* (F). For example, the sequence

SSFFF

denotes getting

4, 4, followed by three other numbers

This sequence can be considered as the intersection of five independent events: success on the first trial, success on the second, failure on the third, and so on. Since the probability of success on any trial is $\frac{1}{6}$ and the probability of failure is $1 - \frac{1}{6} = \frac{5}{6}$, by the special multiplication law for the intersection of independent events, the probability of the sequence SSFFF occurring is

$$\frac{1}{6} \cdot \frac{1}{6} \cdot \frac{5}{6} \cdot \frac{5}{6} \cdot \frac{5}{6} = \left(\frac{1}{6}\right)^2 \left(\frac{5}{6}\right)^3$$

In fact, this is the probability for *any* particular order of the two S's and three F's. Let us determine how many ways a sequence of two S's and three F's can be formed. Out of five trials, the number of ways of choosing the two trials for success is $_5C_2$. Another way to look at this problem is that we are counting "Permutations with Repeated Objects" as in Section 8.2 of the "word" SSFFF. There are $\dfrac{5!}{2! \cdot 3!} = _5C_2$ of these. So the probability of getting exactly two 4's in the five rolls is

$$_5C_2 \left(\frac{1}{6}\right)^2 \left(\frac{5}{6}\right)^3 \tag{3}$$

If we denote the probability of success by p and the probability of failure by $q(= 1 - p)$, then (3) takes the form

$$_5C_2 p^2 q^3$$

which is the term involving p^2 in the expansion of $(q + p)^5$.

More generally, consider the probability of getting exactly x 4's in n rolls of the die. Then $n - x$ of the rolls must be some other number. For a particular order, the probability is

$$p^x q^{n-x}$$

The number of possible orders is $_nC_x$, which again we can see as the question of finding the number of permutations of n symbols, where x of them are S (success) and the remaining $n - x$ are F (failure). According to the result in Section 8.2 on "Permutations with Repeated Objects," there are

$$\frac{n!}{x! \cdot (n-x)!} = _nC_x$$

of these and therefore

$$P(X = x) = _nC_x p^x q^{n-x}$$

which is a general expression for the terms in $(q + p)^n$. In summary, the distribution for X (the number of 4's that occur in n rolls) is given by the terms in $(q + p)^n$.

Whenever we have n independent trials of an experiment in which each trial has only two possible outcomes (success and failure) and the probability of success in each trial remains the same, the trials are called **Bernoulli trials.** Because the distribution of the number of successes corresponds to the expansion of a power of a binomial, the experiment is called a **binomial experiment,** and the distribution of the number of successes is called a **binomial distribution.**

Binomial Distribution

If X is the number of successes in n independent trials of a binomial experiment with probability p of success and q of failure on any trial, then the distribution f for X is given by

$$f(x) = P(X = x) = _nC_x p^x q^{n-x}$$

where x is an integer such that $0 \le x \le n$ and $q = 1 - p$. Any random variable with this distribution is called a **binomial random variable** and is said to have a **binomial distribution**. The mean and standard deviation of X are given, respectively, by

$$\mu = np \quad \sigma = \sqrt{npq}$$

PRINCIPLES IN PRACTICE 1

BINOMIAL DISTRIBUTION

Let X be the number of persons out of four job applicants who are hired. If the probability of any one applicant being hired is 0.3, find the distribution of X.

● EXAMPLE 2 **Binomial Distribution**

Suppose X is a binomial random variable with $n = 4$ and $p = \frac{1}{3}$. Find the distribution for X.

Solution: Here $q = 1 - p = 1 - \frac{1}{3} = \frac{2}{3}$. So we have

$$P(X = x) = _nC_x p^x q^{n-x} \qquad x = 0, 1, 2, 3, 4$$

Thus,

$$P(X=0) = {}_4C_0 \left(\frac{1}{3}\right)^0 \left(\frac{2}{3}\right)^4 = \frac{4!}{0!4!} \cdot 1 \cdot \frac{16}{81} = 1 \cdot 1 \cdot \frac{16}{81} = \frac{16}{81}$$

$$P(X=1) = {}_4C_1 \left(\frac{1}{3}\right)^1 \left(\frac{2}{3}\right)^3 = \frac{4!}{1!3!} \cdot \frac{1}{3} \cdot \frac{8}{27} = 4 \cdot \frac{1}{3} \cdot \frac{8}{27} = \frac{32}{81}$$

$$P(X=2) = {}_4C_2 \left(\frac{1}{3}\right)^2 \left(\frac{2}{3}\right)^2 = \frac{4!}{2!2!} \cdot \frac{1}{9} \cdot \frac{4}{9} = 6 \cdot \frac{1}{9} \cdot \frac{4}{9} = \frac{8}{27}$$

$$P(X=3) = {}_4C_3 \left(\frac{1}{3}\right)^3 \left(\frac{2}{3}\right)^1 = \frac{4!}{3!1!} \cdot \frac{1}{27} \cdot \frac{2}{3} = 4 \cdot \frac{1}{27} \cdot \frac{2}{3} = \frac{8}{81}$$

$$P(X=4) = {}_4C_4 \left(\frac{1}{3}\right)^4 \left(\frac{2}{3}\right)^0 = \frac{4!}{4!0!} \cdot \frac{1}{81} \cdot 1 = 1 \cdot \frac{1}{81} \cdot 1 = \frac{1}{81}$$

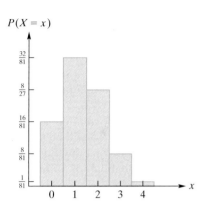

FIGURE 9.6 Binomial distribution, $n = 4$, $p = \frac{1}{3}$.

The probability histogram for X is given in Figure 9.6. Note that the mean μ for X is $np = 4\left(\frac{1}{3}\right) = \frac{4}{3}$, and the standard deviation is

$$\sigma = \sqrt{npq} = \sqrt{4 \cdot \frac{1}{3} \cdot \frac{2}{3}} = \sqrt{\frac{8}{9}} = \frac{2\sqrt{2}}{3}$$

NOW WORK PROBLEM 1 ◖◖◗

● **EXAMPLE 3 At Least Two Heads in Eight Coin Tosses**

A fair coin is tossed eight times. Find the probability of getting at least two heads.

Solution: If X is the number of heads that occur, then X has a binomial distribution with $n = 8$, $p = \frac{1}{2}$, and $q = \frac{1}{2}$. To simplify our work, we use the fact that

$$P(X \geq 2) = 1 - P(X < 2)$$

Now,

$$P(X < 2) = P(X=0) + P(X=1)$$

$$= {}_8C_0 \left(\frac{1}{2}\right)^0 \left(\frac{1}{2}\right)^8 + {}_8C_1 \left(\frac{1}{2}\right)^1 \left(\frac{1}{2}\right)^7$$

$$= 1 \cdot 1 \cdot \frac{1}{256} + 8 \cdot \frac{1}{2} \cdot \frac{1}{128} = \frac{9}{256}$$

Thus,

$$P(X \geq 2) = 1 - \frac{9}{256} = \frac{247}{256}$$

A probability histogram for X is given in Figure 9.7.

NOW WORK PROBLEM 17 ◖●◗

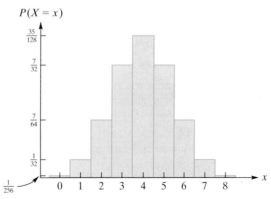

FIGURE 9.7 Binomial distribution, $n = 8$, $p = \frac{1}{2}$.

● EXAMPLE 4 **Income Tax Audit**

For a particular group of individuals, 20% of their income tax returns are audited each year. Of five randomly chosen individuals, what is the probability that exactly two will have their returns audited?

Solution: We will consider this to be a binomial experiment with five trials (selecting an individual). Actually, the experiment is not truly binomial, because selecting an individual from this group affects the probability that another individual's return will be audited. For example, if there are 5000 individuals, then 20%, or 1000, will be audited. The probability that the first individual selected will be audited is $\frac{1000}{5000}$. If that event occurs, the probability that the second individual selected will be audited is $\frac{999}{4999}$. Thus, the trials are not independent. However, we assume that the number of individuals is large, so that for all practical purposes, the probability of auditing an individual remains constant from trial to trial.

For each trial, the two outcomes are *being audited* and *not being audited*. Here we define a success as being audited. Letting X be the number of returns audited, $p = 0.2$, and $q = 1 - 0.2 = 0.8$, we have

$$P(X = 2) = {}_5C_2(0.2)^2(0.8)^3 = \frac{5!}{2!3!}(0.04)(0.512)$$

$$= 10(0.04)(0.512) = 0.2048$$

NOW WORK PROBLEM 15 ●●●

Problems 9.2

In Problems 1–4, determine the distribution f for the binomial random variable X if the number of trials is n and the probability of success on any trial is p. Also, find μ and σ.

*1. $n = 2$, $p = \frac{1}{5}$

2. $n = 3$, $p = \frac{1}{2}$

3. $n = 3$, $p = \frac{2}{3}$

4. $n = 4$, $p = 0.4$

In Problems 5–10, determine the given probability if X is a binomial random variable, n is the number of trials, and p is the probability of success on any trial.

5. $P(X = 5)$; $n = 6$, $p = 0.2$

6. $P(X = 2)$; $n = 5$, $p = \frac{1}{3}$

7. $P(X = 2)$; $n = 4$, $p = \frac{4}{5}$

8. $P(X = 4)$; $n = 7$, $p = 0.2$

9. $P(X < 2)$; $n = 5$, $p = \frac{1}{2}$

10. $P(X \geq 2)$; $n = 6$, $p = \frac{2}{3}$

11. **Coin**   A fair coin is tossed 11 times. What is the probability that exactly eight heads occur?

12. **Multiple-Choice Quiz**   Each question in a six-question multiple-choice quiz has four choices, only one of which is correct. If a student guesses at all six questions, find the probability that exactly three will be correct.

13. **Marbles**   A jar contains five red and seven green marbles. Four marbles are randomly withdrawn in succession with replacement. Determine the probability that exactly two of the marbles withdrawn are green.

14. **Cards**   From a deck of 52 playing cards, 3 cards are randomly selected in succession with replacement. Determine the probability that exactly two cards are aces.

*15. **Quality Control**   A manufacturer produces electrical switches, of which 2% are defective. From a production run

of 50,000 switches, four are randomly selected and each is tested. Determine the probability that the sample contains exactly two defective switches. Round your answer to three decimal places. Assume that the four trials are independent and that the number of defective switches in the sample has a binomial distribution.

16. **Coin**   A coin is biased so that $P(H) = 0.2$ and $P(T) = 0.8$. If X is the number of heads in three tosses, determine a formula for $P(X = x)$.

*17. **Coin**   A biased coin is tossed three times in succession. The probability of heads on any toss is $\frac{1}{4}$. Find the probability that (a) exactly two heads occur and (b) two or three heads occur.

18. **Cards**   From an ordinary deck of 52 playing cards, 7 cards are randomly drawn in succession with replacement. Find the probability that there are (a) exactly four hearts and (b) at least four hearts.

19. **Quality Control**   In a large production lot of electronic devices, it is believed that one fifth are defective. If a sample of six is randomly selected, find the probability that no more than one will be defective.

20. **Computer**   For a certain large population, the probability that a randomly selected person has a computer is 0.7. If five persons are selected at random, find the probability that at least three have a computer.

21. **Baseball**   The probability that a certain baseball player gets a hit is 0.300. Find the probability that if he goes to bat four times, he will get at least one hit.

22. **Stocks**   A financial advisor claims that 60% of the stocks that he recommends for purchase increase in value. From a list of 200 recommended stocks, a client selects 4 at random. Determine the probability, rounded to two decimal places,

that at least 2 of the chosen stocks increase in value. Assume that the selections of the stocks are independent trials and that the number of stocks that increase in value has a binomial distribution.

23. Genders of Children   If a family has five children, find the probability that at least two are girls. (Assume that the probability that a child is a girl is $\frac{1}{2}$.)

24. If X is a binomially distributed random variable with $n = 50$ and $p = \frac{2}{5}$, find σ^2.

25. Suppose X is a binomially distributed random variable such that $\mu = 3$ and $\sigma^2 = 2$. Find $P(X = 2)$.

26. Quality Control   In a production process, the probability of a defective unit is 0.06. Suppose a sample of 15 units is selected at random. Let X be the number of defectives.

  (a) Find the expected number of defective units.

  (b) Find $\text{Var}(X)$.

  (c) Find $P(X \leq 1)$. Round your answer to two decimal places.

OBJECTIVE

To develop the notions of a Markov chain and the associated transition matrix. To find state vectors and the steady-state vector.

9.3 Markov Chains

We conclude this chapter with a discussion of a special type of stochastic process called a *Markov[2] chain*.

Markov Chain

A **Markov chain** is a sequence of trials of an experiment in which the possible outcomes of each trial remain the same from trial to trial, are finite in number, and have probabilities that depend only upon the outcome of the previous trial.

To illustrate a Markov chain, we consider the following situation. Imagine that a small town has only two service stations—say, stations 1 and 2—that handle the servicing needs of the town's automobile owners. (These customers form the population under consideration.) Each time a customer needs car servicing, he or she must make a *choice* of which station to use.

Thus, each customer can be placed into a category according to which of the two stations he or she most recently chose. We can view a customer and the service stations as a *system*. If a customer most recently chose station 1, we will refer to this as *state* 1 of the system. Similarly, if a customer most recently chose station 2, we say that the system is currently in state 2. Hence, at any given time, the system is in one of its two states. Of course, over a period of time, the system may move from one state to the other. For example, the sequence 1, 2, 2, 1 indicates that in four successive car servicings, the system changed from state 1 to state 2, remained at state 2, and then changed to state 1.

This situation can be thought of as a sequence of trials of an experiment (choosing a service station) in which the possible outcomes for each trial are the two states (station 1 and station 2). Each trial involves observing the state of the system at that time.

If we know the current state of the system, we realize that we cannot be sure of its state at the next observation. However, we may know the *likelihood* of its being in a particular state. For example, suppose that if a customer most recently used station 1, then the probability that the customer uses station 1 the next time is 0.7. (This means that, of those customers who used station 1 most recently, 70% continued to use station 1 the next time and 30% changed to station 2.) Assume also that if a customer used station 2 most recently, the probability is 0.8 that the customer also uses station 2 the next time. You may recognize these probabilities as being *conditional* probabilities. That is,

$$P(\text{remaining in state 1} \mid \text{currently in state 1}) = 0.7$$

$$P(\text{changing to state 2} \mid \text{currently in state 1}) = 0.3$$

$$P(\text{remaining in state 2} \mid \text{currently in state 2}) = 0.8$$

$$P(\text{changing to state 1} \mid \text{currently in state 2}) = 0.2$$

---

[2]After the Russian mathematician Andrei Markov (1856–1922).

These four probabilities can be organized in a square matrix $\mathbf{T} = [t_{ij}]$ by letting entry t_{ij} be the probability of a customer being next in state i given that they are currently in state j.

$$\begin{array}{cc} \textit{Next} & \textit{Current State} \\ \textit{State} & \text{State 1} \quad \text{State 2} \end{array}$$

$$\mathbf{T} = \begin{array}{c} \text{State 1} \\ \text{State 2} \end{array} \begin{bmatrix} 0.7 & 0.2 \\ 0.3 & 0.8 \end{bmatrix}$$

For example, the sum of the entries in column 1 of **T** is $0.7 + 0.3 = 1$.

Matrix $\mathbf{T}$ is called a *transition matrix* because it indicates the probabilities of transition from one state to another in *one step*—that is, as we go from one observation period to the next. The entries are called *transition probabilities*. We emphasize that *the transition matrix remains the same at every stage of the sequence of observations.* Note that all entries of the matrix are nonnegative, because they are probabilities. Moreover, the sum of the entries in each column must be 1, because, for each current state, the probabilities account for all possible transitions.

Let us summarize our service station situation up to now. We have a sequence of trials in which the possible outcomes (or states) are the same from trial to trial and are finite in number (two). The probability that the system is in a particular state for a given trial depends only on the state for the preceding trial. Thus, we have a so-called *two-state Markov chain.* A Markov chain determines a square matrix $\mathbf{T}$, called a transition matrix.

Transition Matrix

A **transition matrix** for a k-state Markov chain is a $k \times k$ matrix $\mathbf{T} = [t_{ij}]$ in which the entry t_{ij} is the probability, from one trial to the next, of moving to state i from state j. All entries are nonnegative, and the sum of the entries in each column is 1. We can say

$$t_{ij} = P(\text{next state is } i \mid \text{current state is } j)$$

Suppose that when observations are initially made, 60% of all customers used station 1 most recently and 40% used station 2. This means that, before any additional trials (car servicings) are considered, the probabilities that a customer is in state 1 or 2 are 0.6 and 0.4, respectively. These probabilities are called *initial state probabilities* and are collectively referred to as being the *initial distribution.* They can be represented by a column vector, called an **initial state vector,** which is denoted by $\mathbf{X}_0$. In this case,

A subscript of 0 is used for the initial state vector.

$$\mathbf{X}_0 = \begin{bmatrix} 0.6 \\ 0.4 \end{bmatrix}$$

We would like to find the vector that gives the state probabilities for a customer's *next* visit to a service station. This state vector is denoted by $\mathbf{X}_1$. More generally, a state vector is defined as follows:

State Vector

The **state vector** $\mathbf{X}_n$ for a k-state Markov chain is a k-entry column vector in which the entry x_j is the probability of being in state j after the nth trial.

We can find the entries for $\mathbf{X}_1$ from the probability tree in Figure 9.8. We see that the probability of being in state 1 after the next visit is the sum

$$(0.7)(0.6) + (0.2)(0.4) = 0.5 \tag{1}$$

and the probability of being in state 2 is

$$(0.3)(0.6) + (0.8)(0.4) = 0.5 \tag{2}$$

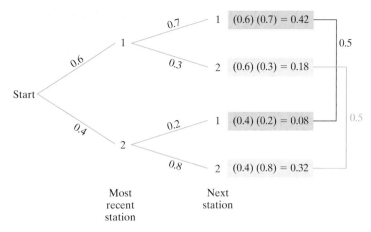

FIGURE 9.8 Probability tree for two-state Markov chain.

Thus,

$$\mathbf{X}_1 = \begin{bmatrix} 0.5 \\ 0.5 \end{bmatrix}$$

The sums of products on the left sides of Equations (1) and (2) remind us of matrix multiplication. In fact, they are the entries in the matrix $\mathbf{TX}_0$ obtained by multiplying the initial state vector on the left by the transition matrix:

$$\mathbf{X}_1 = \mathbf{TX}_0 = \begin{bmatrix} 0.7 & 0.2 \\ 0.3 & 0.8 \end{bmatrix} \begin{bmatrix} 0.6 \\ 0.4 \end{bmatrix} = \begin{bmatrix} 0.5 \\ 0.5 \end{bmatrix}$$

This pattern of taking the product of a state vector and the transition matrix to get the next state vector continues, allowing us to find state probabilities for future observations. For example, to find $\mathbf{X}_2$, the state vector that gives the probabilities for each state after two trials (following the initial observation), we have

$$\mathbf{X}_2 = \mathbf{TX}_1 = \begin{bmatrix} 0.7 & 0.2 \\ 0.3 & 0.8 \end{bmatrix} \begin{bmatrix} 0.5 \\ 0.5 \end{bmatrix} = \begin{bmatrix} 0.45 \\ 0.55 \end{bmatrix}$$

Thus, the probability of being in state 1 after two car servicings is 0.45. Note that, since $\mathbf{X}_1 = \mathbf{TX}_0$, we can write

$$\mathbf{X}_2 = \mathbf{T}(\mathbf{TX}_0)$$

so that

$$\mathbf{X}_2 = \mathbf{T}^2\mathbf{X}_0$$

In general, the nth state vector $\mathbf{X}_n$ can be found by multiplying the previous state vector $\mathbf{X}_{n-1}$ on the left by $\mathbf{T}$.

If $\mathbf{T}$ is the transition matrix for a Markov chain then the state vector $\mathbf{X}_n$ for the nth trial is given by

$$\mathbf{X}_n = \mathbf{TX}_{n-1}$$

Equivalently, we can find $\mathbf{X}_n$ by using only the initial state column vector $\mathbf{X}_0$ and the transition matrix $\mathbf{T}$:

Here we find $\mathbf{X}_n$ by using powers of $\mathbf{T}$.

$$\mathbf{X}_n = \mathbf{T}^n\mathbf{X}_0 \tag{3}$$

Let us now consider the situation in which we know the initial state of the system. For example, take the case of observing initially that a customer has most recently chosen station 1. This means the probability that the system is in state 1 is 1, so the initial state vector must be

$$\mathbf{X}_0 = \begin{bmatrix} 1 \\ 0 \end{bmatrix}$$

Suppose we determine $\mathbf{X}_2$, the state vector that gives the state probabilities after the next two visits. This is given by

$$\mathbf{X}_2 = \mathbf{T}^2 \mathbf{X}_0 = \begin{bmatrix} 0.7 & 0.2 \\ 0.3 & 0.8 \end{bmatrix}^2 \begin{bmatrix} 1 \\ 0 \end{bmatrix}$$

$$= \begin{bmatrix} 0.55 & 0.30 \\ 0.45 & 0.70 \end{bmatrix} \begin{bmatrix} 1 \\ 0 \end{bmatrix} = \begin{bmatrix} 0.55 \\ 0.45 \end{bmatrix}$$

Thus, for this customer, the probabilities of using station 1 or station 2 after two steps are 0.55 and 0.45, respectively. Observe that these probabilities form the *first column* of $\mathbf{T}^2$. On the other hand, if the system were initially in state 2, then the state vector after two steps would be

$$\mathbf{T}^2 \begin{bmatrix} 0 \\ 1 \end{bmatrix} = \begin{bmatrix} 0.55 & 0.30 \\ 0.45 & 0.70 \end{bmatrix} \begin{bmatrix} 0 \\ 1 \end{bmatrix} = \begin{bmatrix} 0.30 \\ 0.70 \end{bmatrix}$$

Hence, for this customer, the probabilities of using station 1 or station 2 after two steps are 0.30 and 0.70, respectively. Observe that these probabilities form the *second column* of $\mathbf{T}^2$. Based on our observations, we now have a way of interpreting $\mathbf{T}^2$: The entries in

$$\mathbf{T}^2 = \begin{matrix} & 1 & 2 \\ 1 \\ 2 \end{matrix} \begin{bmatrix} 0.55 & 0.30 \\ 0.45 & 0.70 \end{bmatrix}$$

give the probabilities of moving to a state from another in *two* steps. In general, we have the following:

This gives the significance of the entries in $\mathbf{T}^n$.

If $\mathbf{T}$ is a transition matrix, then for $\mathbf{T}^n$ the entry in row i and column j gives the probability of being in state i after n steps, starting from state j.

● EXAMPLE 1 Demography

A certain county is divided into three demographic regions. Research indicates that each year 20% of the residents in region 1 move to region 2 and 10% move to region 3. (The others remain in region 1.) Of the residents in region 2, 10% move to region 1 and 10% move to region 3. Of the residents in region 3, 20% move to region 1 and 10% move to region 2.

a. *Find the transition matrix $\mathbf{T}$ for this situation.*

Solution: We have

$$\begin{matrix} & To & & From\ Region \\ & Region & 1 & 2 & 3 \\ & 1 \\ \mathbf{T} = & 2 \\ & 3 \end{matrix} \begin{bmatrix} 0.7 & 0.1 & 0.2 \\ 0.2 & 0.8 & 0.1 \\ 0.1 & 0.1 & 0.7 \end{bmatrix}$$

Note that to find t_{11}, we subtracted the sum of the other two entries in the first column from 1. The entries t_{22} and t_{33} are found similarly.

b. *Find the probability that a resident of region 1 this year is a resident of region 1 next year; in two years.*

Solution: From entry t_{11} in transition matrix $\mathbf{T}$, the probability that a resident of region 1 remains in region 1 after one year is 0.7. The probabilities of moving from one region to another in two steps are given by $\mathbf{T}^2$:

$$\mathbf{T}^2 = \begin{array}{c} \\ 1 \\ 2 \\ 3 \end{array}\begin{array}{c} \begin{array}{ccc} 1 & 2 & 3 \end{array} \\ \begin{bmatrix} 0.53 & 0.17 & 0.29 \\ 0.31 & 0.67 & 0.19 \\ 0.16 & 0.16 & 0.52 \end{bmatrix} \end{array}$$

Powers of $\mathbf{T}$ can be conveniently found with a graphing calculator.

Thus, the probability that a resident of region 1 is in region 1 after two years is 0.53.

c. *This year, suppose* 40% *of county residents live in region* 1, 30% *live in region* 2, *and* 30% *live in region* 3. *Find the probability that a resident of the county lives in region* 2 *after three years.*

Solution: The initial state vector is

$$\mathbf{X}_0 = \begin{bmatrix} 0.40 \\ 0.30 \\ 0.30 \end{bmatrix}$$

The distribution of the population after three years is given by state vector $\mathbf{X}_3$. From Equation (3) with $n = 3$, we have

Of course, $\mathbf{X}_3$ can be easily obtained with a graphing calculator: Enter $\mathbf{X}_0$ and $\mathbf{T}$, and then evaluate $\mathbf{T}^3\mathbf{X}_0$ directly.

$$\begin{aligned} \mathbf{X}_3 &= \mathbf{T}^3\mathbf{X}_0 = \mathbf{T}\mathbf{T}^2\mathbf{X}_0 \\ &= \begin{bmatrix} 0.7 & 0.1 & 0.2 \\ 0.2 & 0.8 & 0.1 \\ 0.1 & 0.1 & 0.7 \end{bmatrix} \begin{bmatrix} 0.53 & 0.17 & 0.29 \\ 0.31 & 0.67 & 0.19 \\ 0.16 & 0.16 & 0.52 \end{bmatrix} \begin{bmatrix} 0.40 \\ 0.30 \\ 0.30 \end{bmatrix} \\ &= \begin{bmatrix} 0.3368 \\ 0.4024 \\ 0.2608 \end{bmatrix} \end{aligned}$$

This result means that in three years, 33.68% of the county residents live in region 1, 40.24% live in region 2, and 26.08% live in region 3. Thus, the probability that a resident lives in region 2 in three years is 0.4024.

Steady-State Vectors

Let us now return to our service station problem. Recall that if the initial state vector is

$$\mathbf{X}_0 = \begin{bmatrix} 0.6 \\ 0.4 \end{bmatrix}$$

then

$$\mathbf{X}_1 = \begin{bmatrix} 0.5 \\ 0.5 \end{bmatrix}$$

$$\mathbf{X}_2 = \begin{bmatrix} 0.45 \\ 0.55 \end{bmatrix}$$

Some state vectors beyond the second are

$$\mathbf{X}_3 = \mathbf{T}\mathbf{X}_2 = \begin{bmatrix} 0.7 & 0.2 \\ 0.3 & 0.8 \end{bmatrix} \begin{bmatrix} 0.45 \\ 0.55 \end{bmatrix} = \begin{bmatrix} 0.425 \\ 0.575 \end{bmatrix}$$

$$\mathbf{X}_4 = \mathbf{T}\mathbf{X}_3 = \begin{bmatrix} 0.7 & 0.2 \\ 0.3 & 0.8 \end{bmatrix} \begin{bmatrix} 0.425 \\ 0.575 \end{bmatrix} = \begin{bmatrix} 0.4125 \\ 0.5875 \end{bmatrix}$$

$$\mathbf{X}_5 = \mathbf{T}\mathbf{X}_4 = \begin{bmatrix} 0.7 & 0.2 \\ 0.3 & 0.8 \end{bmatrix} \begin{bmatrix} 0.4125 \\ 0.5875 \end{bmatrix} = \begin{bmatrix} 0.40625 \\ 0.59375 \end{bmatrix}$$

$$\vdots$$

$$\mathbf{X}_{10} = \mathbf{T}\mathbf{X}_9 \approx \begin{bmatrix} 0.40020 \\ 0.59980 \end{bmatrix}$$

These results strongly suggest, and it is indeed the case, that, as the number of trials increases, the entries in the state vectors tend to get closer and closer to the corresponding entries in the vector

$$\mathbf{Q} = \begin{bmatrix} 0.40 \\ 0.60 \end{bmatrix}$$

(Equivalently, it can be shown that the entries in each column of $\mathbf{T}^n$ approach the corresponding entries in those of $\mathbf{Q}$ as n increases.) Vector $\mathbf{Q}$ has a special property. Observe the result of multiplying $\mathbf{Q}$ on the left by the transition matrix $\mathbf{T}$:

$$\mathbf{TQ} = \begin{bmatrix} 0.7 & 0.2 \\ 0.3 & 0.8 \end{bmatrix} \begin{bmatrix} 0.40 \\ 0.60 \end{bmatrix} = \begin{bmatrix} 0.40 \\ 0.60 \end{bmatrix} = \mathbf{Q}$$

We thus have

$$\mathbf{TQ} = \mathbf{Q}$$

which shows that $\mathbf{Q}$ *remains unchanged from trial to trial.*

In summary, as the number of trials increases, the state vectors get closer and closer to $\mathbf{Q}$, which remains unchanged from trial to trial. The distribution of the population between the service stations stabilizes. That is, in the long run, approximately 40% of the population will have their cars serviced at station 1 and 60% at station 2. To describe this, we say that $\mathbf{Q}$ is the **steady-state vector** of this process. It can be shown that the steady-state vector is unique. (There is only one such vector.) Moreover, $\mathbf{Q}$ does not depend on the initial state vector $\mathbf{X}_0$ but depends only on the transition matrix $\mathbf{T}$. For this reason, we say that $\mathbf{Q}$ is the *steady-state vector for* $\mathbf{T}$.

The steady-state vector is unique and does not depend on the initial distribution.

What we need now is a procedure for finding the steady-state vector $\mathbf{Q}$ without having to compute state vectors for large values of n. Fortunately, the previously stated property that $\mathbf{TQ} = \mathbf{Q}$ can be used to find $\mathbf{Q}$. If we let $\mathbf{Q} = \begin{bmatrix} q_1 \\ q_2 \end{bmatrix}$, we have

$$\mathbf{TQ} = \mathbf{Q} = \mathbf{IQ}$$

$$\mathbf{TQ} - \mathbf{IQ} = \mathbf{O}$$

$$(\mathbf{T} - \mathbf{I})\mathbf{Q} = \mathbf{O}$$

$$\left(\begin{bmatrix} 0.7 & 0.2 \\ 0.3 & 0.8 \end{bmatrix} - \begin{bmatrix} 1 & 0 \\ 0 & 1 \end{bmatrix} \right) \begin{bmatrix} q_1 \\ q_2 \end{bmatrix} = \begin{bmatrix} 0 \\ 0 \end{bmatrix}$$

$$\begin{bmatrix} -0.3 & 0.2 \\ 0.3 & -0.2 \end{bmatrix} \begin{bmatrix} q_1 \\ q_2 \end{bmatrix} = \begin{bmatrix} 0 \\ 0 \end{bmatrix}$$

which suggests that $\mathbf{Q}$ can be found by solving the resulting system of linear equations arising here in matrix form. Using the techniques of Chapter 6, we see immediately that the coefficient matrix of the last equation reduces to

$$\begin{bmatrix} 3 & -2 \\ 0 & 0 \end{bmatrix}$$

which suggests that there are infinitely many possibilities for the steady-state vector $\mathbf{Q}$. However, the entries of a state vector must add up to 1 so that the further equation $q_1 + q_2 = 1$ must be added to the system. We arrive at

$$\begin{bmatrix} 3 & -2 \\ 1 & 1 \end{bmatrix} \begin{bmatrix} q_1 \\ q_2 \end{bmatrix} = \begin{bmatrix} 0 \\ 1 \end{bmatrix}$$

which is easily seen to have the unique solution

$$\mathbf{Q} = \begin{bmatrix} q_1 \\ q_2 \end{bmatrix} = \begin{bmatrix} 0.4 \\ 0.6 \end{bmatrix}$$

which confirms our previous suspicion.

We must point out that for Markov chains in general, the state vectors do not always approach a steady-state vector. However, it can be shown that a steady-state vector for **T** does exist, provided that **T** is *regular:*

A transition matrix **T** is **regular** if there exists an integer power of **T** for which all entries are (strictly) positive.

Only regular transition matrices will be considered in this section. A Markov chain whose transition matrix is regular is called a **regular Markov chain.**

In summary, we have the following:

Suppose **T** is the $k \times k$ transition matrix for a regular Markov chain. Then the steady-state column vector

$$\mathbf{Q} = \begin{bmatrix} q_1 \\ q_2 \\ \vdots \\ q_k \end{bmatrix}$$

is the solution to the matrix equations

$$[1 \quad 1 \quad \cdots \quad 1]\mathbf{Q} = 1 \tag{4}$$

$$(\mathbf{T} - \mathbf{I}_k)\mathbf{Q} = \mathbf{O} \tag{5}$$

where in Equation (4) the (matrix) coefficient of **Q** is the row vector consisting of k entries all of which are 1.

Equations (4) and (5) can always be combined into a single matrix equation:

$$\mathbf{T}^*\mathbf{Q} = \mathbf{O}^*$$

where $\mathbf{T}^*$ is the $(k + 1) \times k$ matrix obtained by pasting the row $[1 \quad 1 \quad \cdots \quad 1]$ to the top of the $k \times k$ matrix $\mathbf{T} - \mathbf{I}_k$ (where $\mathbf{I}_k$ is the $k \times k$ identity matrix) and $\mathbf{O}^*$ is the $k + 1$-column vector obtained by pasting a 1 to the top of the zero k-column vector. We can then find **Q** by reducing the augmented matrix $[\mathbf{T}^* \mid \mathbf{O}^*]$. The next example will illustrate.

● EXAMPLE 2   Steady-State Vector

For the demography problem of Example 1, *in the long run what percentage of county residents will live in each region?*

Solution: The population distribution in the long run is given by the steady-state vector **Q**, which we now proceed to find. The matrix **T** for this example was shown to be

$$\begin{bmatrix} 0.7 & 0.1 & 0.2 \\ 0.2 & 0.8 & 0.1 \\ 0.1 & 0.1 & 0.7 \end{bmatrix}$$

so that $\mathbf{T} - \mathbf{I}$ is

$$\begin{bmatrix} -0.3 & 0.1 & 0.2 \\ 0.2 & -0.2 & 0.1 \\ 0.1 & 0.1 & -0.3 \end{bmatrix}$$

and $[\mathbf{T}^* \mid \mathbf{O}^*]$ is

$$\begin{bmatrix} 1 & 1 & 1 & 1 \\ -0.3 & 0.1 & 0.2 & 0 \\ 0.2 & -0.2 & 0.1 & 0 \\ 0.1 & 0.1 & -0.3 & 0 \end{bmatrix}$$

which reduces to

$$\begin{bmatrix} 1 & 0 & 0 & 5/16 \\ 0 & 1 & 0 & 7/16 \\ 0 & 0 & 1 & 1/4 \\ 0 & 0 & 0 & 0 \end{bmatrix}$$

showing that the steady-state vector $\mathbf{Q} = \begin{bmatrix} 5/16 \\ 7/16 \\ 1/4 \end{bmatrix} = \begin{bmatrix} 0.3125 \\ 0.4375 \\ 0.2500 \end{bmatrix}$. Thus, in the long run, the percentages of county residents living in regions 1, 2, and 3 are 31.25%, 43.75%, and 25%, respectively.

NOW WORK PROBLEM 37 ◖◗

Problems 9.3

In Problems 1–6, can the given matrix be a transition matrix for a Markov chain?

1. $\begin{bmatrix} \frac{1}{2} & \frac{2}{3} \\ -\frac{3}{2} & \frac{1}{3} \end{bmatrix}$

2. $\begin{bmatrix} 0.1 & 1 \\ 0.9 & 0 \end{bmatrix}$

3. $\begin{bmatrix} \frac{1}{2} & \frac{1}{8} & \frac{1}{3} \\ -\frac{1}{4} & \frac{5}{8} & \frac{1}{3} \\ \frac{3}{4} & \frac{1}{4} & \frac{1}{3} \end{bmatrix}$

4. $\begin{bmatrix} 0.2 & 0.6 & 0 \\ 0.7 & 0.2 & 0 \\ 0.1 & 0.2 & 0 \end{bmatrix}$

5. $\begin{bmatrix} 0.4 & 0 & 0.5 \\ 0.2 & 0.1 & 0.3 \\ 0.4 & 0.9 & 0.2 \end{bmatrix}$

6. $\begin{bmatrix} 0.5 & 0.1 & 0.3 \\ 0.4 & 0.3 & 0.3 \\ 0.6 & 0.6 & 0.4 \end{bmatrix}$

In Problems 7–10, a transition matrix for a Markov chain is given. Determine the values of the letter entries.

7. $\begin{bmatrix} \frac{2}{3} & b \\ a & \frac{1}{4} \end{bmatrix}$

8. $\begin{bmatrix} a & b \\ \frac{5}{12} & a \end{bmatrix}$

9. $\begin{bmatrix} 0.4 & a & a \\ a & 0.1 & b \\ 0.3 & b & c \end{bmatrix}$

10. $\begin{bmatrix} a & a & a \\ a & b & b \\ a & \frac{1}{4} & c \end{bmatrix}$

In Problems 11–14, determine whether the given vector could be a state vector for a Markov chain.

11. $\begin{bmatrix} 0.4 \\ 0.6 \end{bmatrix}$

12. $\begin{bmatrix} 1 \\ 0 \end{bmatrix}$

13. $\begin{bmatrix} 0.2 \\ 0.7 \\ 0.5 \end{bmatrix}$

14. $\begin{bmatrix} 0.9 \\ -0.1 \\ 0.2 \end{bmatrix}$

In Problems 15–20, a transition matrix $\mathbf{T}$ and an initial state vector $\mathbf{X}_0$ are given. Compute the state vectors $\mathbf{X}_1$, $\mathbf{X}_2$, and $\mathbf{X}_3$.

15. $\mathbf{T} = \begin{bmatrix} \frac{2}{3} & 1 \\ \frac{1}{3} & 0 \end{bmatrix}$

$\mathbf{X}_0 = \begin{bmatrix} \frac{1}{4} \\ \frac{3}{4} \end{bmatrix}$

16. $\mathbf{T} = \begin{bmatrix} \frac{1}{2} & \frac{1}{4} \\ \frac{1}{2} & \frac{3}{4} \end{bmatrix}$

$\mathbf{X}_0 = \begin{bmatrix} \frac{1}{2} \\ \frac{1}{2} \end{bmatrix}$

17. $\mathbf{T} = \begin{bmatrix} 0.3 & 0.5 \\ 0.7 & 0.5 \end{bmatrix}$

$\mathbf{X}_0 = \begin{bmatrix} 0.4 \\ 0.6 \end{bmatrix}$

18. $\mathbf{T} = \begin{bmatrix} 0.1 & 0.9 \\ 0.9 & 0.1 \end{bmatrix}$

$\mathbf{X}_0 = \begin{bmatrix} 0.2 \\ 0.8 \end{bmatrix}$

19. $\mathbf{T} = \begin{bmatrix} 0.1 & 0 & 0.3 \\ 0.2 & 0.4 & 0.3 \\ 0.7 & 0.6 & 0.4 \end{bmatrix}$

$\mathbf{X}_0 = \begin{bmatrix} 0.2 \\ 0 \\ 0.8 \end{bmatrix}$

20. $\mathbf{T} = \begin{bmatrix} 0.4 & 0.1 & 0.2 & 0.1 \\ 0 & 0.1 & 0.3 & 0.3 \\ 0.4 & 0.7 & 0.4 & 0.4 \\ 0.2 & 0.1 & 0.1 & 0.2 \end{bmatrix}$

$\mathbf{X}_0 = \begin{bmatrix} 0.1 \\ 0.3 \\ 0.4 \\ 0.2 \end{bmatrix}$

In Problems 21–24, a transition matrix $\mathbf{T}$ is given.

(a) Compute $\mathbf{T}^2$ and $\mathbf{T}^3$.

(b) What is the probability of going to state 2 from state 1 after two steps?

(c) What is the probability of going to state 1 from state 2 after three steps?

21. $\begin{bmatrix} \frac{1}{4} & \frac{3}{4} \\ \frac{3}{4} & \frac{1}{4} \end{bmatrix}$

22. $\begin{bmatrix} \frac{1}{3} & \frac{1}{2} \\ \frac{2}{3} & \frac{1}{2} \end{bmatrix}$

23. $\begin{bmatrix} 0 & 0.5 & 0.3 \\ 1 & 0.4 & 0.3 \\ 0 & 0.1 & 0.4 \end{bmatrix}$

24. $\begin{bmatrix} 0.1 & 0.1 & 0.1 \\ 0.2 & 0.1 & 0.1 \\ 0.7 & 0.8 & 0.8 \end{bmatrix}$

In Problems 25–30, find the steady-state vector for the given transition matrix.

25. $\begin{bmatrix} \frac{1}{2} & \frac{2}{3} \\ \frac{1}{2} & \frac{1}{3} \end{bmatrix}$

26. $\begin{bmatrix} \frac{1}{2} & \frac{1}{4} \\ \frac{1}{2} & \frac{3}{4} \end{bmatrix}$

27. $\begin{bmatrix} \frac{1}{5} & \frac{3}{5} \\ \frac{4}{5} & \frac{2}{5} \end{bmatrix}$

28. $\begin{bmatrix} \frac{1}{4} & \frac{1}{3} \\ \frac{3}{4} & \frac{2}{3} \end{bmatrix}$

29. $\begin{bmatrix} 0.4 & 0.6 & 0.6 \\ 0.3 & 0.3 & 0.1 \\ 0.3 & 0.1 & 0.3 \end{bmatrix}$

30. $\begin{bmatrix} 0.1 & 0.4 & 0.3 \\ 0.2 & 0.2 & 0.3 \\ 0.7 & 0.4 & 0.4 \end{bmatrix}$

31. Spread of Flu A flu has attacked a college dorm that has 200 students. Suppose the probability that a student having the flu will still have it 4 days later is 0.1. However, for a student who does not have the flu, the probability of having the flu 4 days later is 0.2.

(a) Find a transition matrix for this situation.

(b) If 120 students now have the flu, how many students (to the nearest integer) can be expected to have the flu 8 days from now? 12 days from now?

32. Physical Fitness A physical-fitness center has found that, of those members who perform high-impact exercising on one visit, 55% will do the same on the next visit and 45% will do low-impact exercising. Of those who perform low-impact exercising on one visit, 75% will do the same on the next visit and 25% will do high-impact exercising. On the last visit, suppose that 65% of members did high-impact exercising and 35% did low-impact exercising. After two more visits, what percentage of members will be performing high-impact exercising?

33. Newspapers In a certain area, two daily newspapers are available. It has been found that if a customer buys newspaper A on one day, then the probability is 0.3 that he or she will change to the other newspaper the next day. If a customer buys newspaper B on one day, then the probability is 0.6 that he or she will buy the same newspaper the next day.

(a) Find the transition matrix for this situation.

(b) Find the probability that a person who buys A on Monday will buy A on Wednesday.

34. Video Rentals A video rental store has three locations in a city. A video can be rented from any of the three locations and returned to any of them. Studies show that videos are rented from one location and returned to a location according to the probabilities given by the following matrix:

Returned to	Rented from 1	2	3
1	0.7	0.2	0.2
2	0.1	0.8	0.2
3	0.2	0	0.6

Suppose that 20% of the videos are initially rented from location 1, 50% from 2, and 30% from 3. Find the percentages of videos that can be expected to be returned to each location:

(a) After this rental

(b) After the next rental

35. Voting In a certain region, voter registration was analyzed according to party affiliation: Democratic, Republican, and other. It was found that on a year-to-year basis, the probability that a voter switches registration to Republican from Democratic is 0.1; to other from Democratic 0.1; to Democratic from Republican, 0.1; to other from Republican, 0.1; to Democratic from other, 0.3; and to Republican from other, 0.2.

(a) Find a transition matrix for this situation.

(b) What is the probability that a presently registered Republican voter will be registered Democratic two years from now?

(c) If 40% of the present voters are Democratic and 40% are Republican, what percentage can be expected to be Republican one year from now?

36. Demography The residents of a certain region are classified as urban (U), suburban (S), or rural (R). A marketing firm has found that over successive 5-year periods, residents shift from one classification to another according to the probabilities given by the following matrix:

	U	S	R
U	0.7	0.1	0.1
S	0.1	0.8	0.1
R	0.2	0.1	0.8

(a) Find the probability that a suburban resident will be a rural resident in 15 years.

(b) Suppose the initial population of the region is 50% urban, 25% suburban, and 25% rural. Determine the expected population distribution in 15 years.

*37. Long-Distance Telephone Service A major long-distance telephone company (company A) has studied the tendency of telephone users to switch from one carrier to another. The company believes that over successive six-month periods, the probability that a customer who uses A's service will switch to a competing service is 0.2 and the probability that a customer of any competing service will switch to A is 0.3.

(a) Find a transition matrix for this situation.

(b) If A presently controls 70% of the market, what percentage can it expect to control six months from now?

(c) What percentage of the market can A expect to control in the long run?

38. Automobile Purchases   In a certain region, a study of car ownership was made. It was determined that if a person presently owns a Ford, then the probability that the next car the person buys is also a Ford is 0.75. If a person does not presently own a Ford, then the probability that the person will buy a Ford on the next car purchase is 0.35.

(a) Find the transition matrix for this situation.

(b) In the long run, what proportion of car purchases in the region can be expected to be Fords?

39. Laboratory Mice   Suppose 100 mice are in a two-compartment cage and are free to move between the compartments. At regular time intervals, the number of mice in each compartment is observed. It has been found that if a mouse is in compartment 1 at one observation, then the probability that the mouse will be in compartment 1 at the next observation is $\frac{5}{7}$. If a mouse is in compartment 2 at one observation, then the probability that the mouse will be in compartment 2 at the next observation is $\frac{4}{7}$. Initially, suppose that 50 mice are placed into each compartment.

(a) Find the transition matrix for this situation.

(b) After two observations, what percentage of mice (rounded to two decimal places) can be expected to be in each compartment?

(c) In the long run, what percentage of mice can be expected in each compartment?

40. Vending Machines   A typical cry of college students is, "Don't put your money into that soda machine; I tried it and the machine didn't work!" Suppose that if a vending machine is working properly one time, then the probability that it will work properly the next time is 0.8. On the other hand, suppose that if the machine is not working properly one time, then the probability that it will not work properly the next time is 0.9.

(a) Find a transition matrix for this situation.

(b) Suppose that four people line up at such a soda machine. If the first person receives a soda, what is the probability that the fourth person receives a soda? (Assume nobody makes more than one attempt.)

(c) If there are 42 such vending machines on a college campus, in the long run how many machines do you expect to work properly?

41. Advertising   A supermarket chain sells bread from bakeries A and B. Presently, A accounts for 50% of the

chain's daily bread sales. To increase sales, A launches a new advertising campaign. The bakery believes that the change in bread sales at the chain will be based on the following transition matrix:

$$
\begin{array}{cc}
 & \text{A} \quad \text{B} \\
\begin{array}{c} \text{A} \\ \text{B} \end{array} &
\left[\begin{array}{cc} \frac{3}{4} & \frac{1}{2} \\ \frac{1}{4} & \frac{1}{2} \end{array} \right]
\end{array}
$$

(a) Find the steady-state vector.

(b) In the long run, by what percentage can A expect to increase present sales at the chain? Assume that the total daily sales of bread at the chain remain the same.

42. Bank Branches   A bank with three branches, A, B, and C, finds that customers usually return to the same branch for their banking needs. However, at times a customer may go to a different branch because of a changed circumstance. For example, a person who usually goes to branch A may sometimes deviate and go to branch B because the person has business to conduct in the vicinity of branch B. For customers of branch A, suppose that 80% return to A on their next visit, 10% go to B, and 10% go to C. For customers of branch B, suppose that 70% return to B on their next visit, 20% go to A, and 10% go to C. For customers of branch C, suppose that 70% return to C on their next visit, 20% go to A, and 10% go to B.

(a) Find a transition matrix for this situation.

(b) If a customer most recently went to branch B, what is the probability that the customer returns to B on the second bank visit from now?

(c) Initially, suppose 200 customers go to A, 200 go to B, and 100 go to C. On their next visit, how many can be expected to go to A? To B? To C?

(d) Of the initial 500 customers, in the long run how many can be expected to go to A? To B? To C?

43. Show that the transition matrix $\mathbf{T} = \begin{bmatrix} \frac{1}{2} & 1 \\ \frac{1}{2} & 0 \end{bmatrix}$ is regular. (*Hint:* Examine the entries in $\mathbf{T}^2$.)

44. Show that the transition matrix $\begin{bmatrix} 0 & 1 \\ 1 & 0 \end{bmatrix}$ is not regular.

9.4 Review

Important Terms and Symbols   Examples

Section 9.1	**Discrete Random Variables and Expected Value**	
	discrete random variable probability function histogram	Ex. 2, p. 421
	mean, μ expected value, $E(X)$	Ex. 3, p. 423
	variance, $\text{Var}(X)$ standard deviation, σ	Ex. 4, p. 425

Summary

If X is a discrete random variable and f is the function such that $f(x) = P(X = x)$, then f is called the probability function, or distribution, of X. In general,

$$\sum_x f(x) = 1$$

The mean, or expected value, of X is the long-run average of X and is denoted μ or $E(X)$:

$$\mu = E(X) = \sum_x x f(x)$$

The mean can be interpreted as a measure of the central tendency of X in the long run. A measure of the dispersion of X is the variance, denoted $\text{Var}(X)$ and is given by

$$\text{Var}(X) = \sum_x (x - \mu)^2 f(x)$$

equivalently, by

$$\text{Var}(X) = \sum_x x^2 f(x) - \mu^2$$

Another measure of dispersion of X is the standard deviation σ:

$$\sigma = \sqrt{\text{Var}(X)}$$

If an experiment is repeated several times, then each performance of the experiment is called a trial. The trials are independent when the outcome of any single trial does not affect the outcome of any other. If there are only two possible outcomes (success and failure) for each independent trial, and the probabilities of success and failure do not change from trial to trial, then the experiment is called a binomial experiment. For such an experiment, if X is the number of successes in n trials, then the distribution f of X is called a binomial distribution, and

$$f(x) = P(X = x) = {}_nC_x p^x q^{n-x}$$

where p is the probability of success on any trial and $q = 1 - p$ is the probability of failure. The mean μ and standard deviation σ of X are given by

$$\mu = np \quad \text{and} \quad \sigma = \sqrt{npq}$$

A binomial distribution is intimately connected with the binomial theorem, which is a formula for expanding the nth power of a binomial, namely,

$$(a + b)^n = \sum_{i=0}^{n} {}_nC_i a^{n-i} b^i$$

where n is a positive integer.

A Markov chain is a sequence of trials of an experiment in which the possible outcomes of each trial, which are called states, remain the same from trial to trial, are finite in number, and have probabilities that depend only upon the outcome of the previous trial. For a k-state Markov chain, if the probability of moving to state i from state j from one trial to the next is represented by t_{ij}, then the $k \times k$ matrix $\mathbf{T} = [t_{ij}]$ is called the transition matrix for the chain. The entries in the nth power of $\mathbf{T}$ also represent probabilities; the entry in the ith row and jth column of $\mathbf{T}^n$ gives the probability of moving to state i from state j in n steps. A k-entry column vector in which the entry x_j is the probability of being in state j after the nth trial is called a state vector and is denoted $\mathbf{X}_n$. The initial state probabilities are represented by the initial state vector $\mathbf{X}_0$. The state vector $\mathbf{X}_n$ can be found by multiplying the previous state vector $\mathbf{X}_{n-1}$ on the left by the transition matrix $\mathbf{T}$:

$$\mathbf{X}_n = \mathbf{T}\mathbf{X}_{n-1}$$

Alternatively, $\mathbf{X}_n$ can be found by multiplying the initial state vector $\mathbf{X}_0$ by $\mathbf{T}^n$:

$$\mathbf{X}_n = \mathbf{T}^n \mathbf{X}_0$$

If the transition matrix $\mathbf{T}$ is regular (that is, if there is a positive integer power of $\mathbf{T}$ for which all entries are positive), then, as the number n of trials increases, $\mathbf{X}_n$ gets closer and closer to the vector $\mathbf{Q}$, called the steady-state vector of $\mathbf{T}$. If

$$\mathbf{Q} = \begin{bmatrix} q_1 \\ q_2 \\ \vdots \\ q_k \end{bmatrix}$$

then the entries of $\mathbf{Q}$ indicate the long-run probability distribution of the states. The vector $\mathbf{Q}$ can be found by solving the matrix equation

$$\mathbf{T}^* \mathbf{Q} = \mathbf{O}^*$$

where $\mathbf{T}^*$ is the $(k + 1) \times k$ matrix obtained by pasting the row $[1 \;\; 1 \;\; \cdots \;\; 1]$ to the top of the $k \times k$ matrix $\mathbf{T} - \mathbf{I}_k$ (where $\mathbf{I}_k$ is the $k \times k$ identity matrix) and $\mathbf{O}^*$ is the $k + 1$-column vector obtained by pasting a 1 to the top of the zero k-column vector. Thus we construct and reduce

$$\left[\begin{array}{c|c} 1 \cdots 1 & 1 \\ \mathbf{T} - \mathbf{I} & \mathbf{O} \end{array} \right]$$

which, if $\mathbf{T}$ is regular, will result in

$$\left[\begin{array}{c|c} \mathbf{I} & \mathbf{Q} \\ \mathbf{O} & 0 \end{array} \right]$$

Review Problems

Problem numbers shown in color indicate problems suggested for use as a practice chapter test.

In Problems 1 and 2, the distribution for the random variable X is given. Construct the probability histogram and determine μ, Var(X), and σ.

1. $f(1) = 0.7$, $f(2) = 0.1$, $f(3) = 0.2$

2. $f(0) = \frac{1}{6}$, $f(1) = \frac{1}{2}$, $f(2) = \frac{1}{3}$

3. **Coin and Die** A fair coin and a fair die are tossed. Let X be the number of dots that show minus the number of heads. Determine (a) the distribution f for X and (b) $E(X)$.

4. **Cards** Two cards from a standard deck of 52 playing cards are randomly drawn in succession without replacement, and the number of aces, X, is observed. Determine (a) the distribution f for X and (b) $E(X)$.

5. **Card Game** In a game, a player pays $0.25 to randomly draw 2 cards, with replacement, from a standard deck of 52 playing cards. For each 10 that appears, the player receives $1. What is the player's expected gain or loss? Give your answer to the nearest cent.

6. **Gas Station Profits** An oil company determines that the probability that a gas station located along an interstate highway is successful is 0.45. A successful station earns an annual profit of $40,000; a station that is not successful loses $10,000 annually. What is the expected gain to the company if it locates a station along an interstate highway?

7. **Mail-Order Computers** A mail-order computer company offers a 30-day money-back guarantee to any customer who is not completely satisfied with its product. The company realizes a profit of $200 for each computer sold, but assumes a loss of $100 for shipping and handling for each unit returned. The probability that a unit is returned is 0.08.

   (a) What is the expected gain for each unit shipped?

   (b) If the distributor ships 4000 units per year, what is the expected annual profit?

8. **Lottery** In a lottery, you pay $1.00 to choose one of 41 million number combinations. If that combination is drawn, you win $15 million. What is your expected gain (or loss) per play?

In Problems 9 and 10, determine the distribution f for the binomial random variable X if the number of trials is n and the probability of success on any trial is p. Also, find μ and σ.

9. $n = 4$, $p = 0.15$

10. $n = 5$, $p = \frac{1}{3}$

In Problems 11 and 12, determine the given probability if X is a binomial random variable, n is the number of trials, and p is the probability of success on any trial.

11. $P(X \le 1)$; $n = 5$, $p = \frac{3}{4}$

12. $P(X > 2)$; $n = 6$, $p = \frac{2}{3}$

13. **Die** A fair die is rolled four times. Find the probability that exactly three of the rolls result in a 2 or a 3.

14. **Planting Success** The probability that a certain type of bush survives planting is 0.9. If four bushes are planted, what is the probability that all of them die?

15. **Coin** A biased coin is tossed five times. The probability that a head occurs on any toss is $\frac{2}{5}$. Find the probability that at least two heads occur.

16. **Jelly Beans** A bag contains two red, three green, and five black jelly beans. Five jelly beans are randomly withdrawn in succession with replacement. Find the probability that at most two of the jelly beans are red.

In Problems 17 and 18, a transition matrix for a Markov chain is given. Determine the values of a, b, and c.

17. $\begin{bmatrix} 0.1 & 2a & a \\ a & b & b \\ 0.6 & b & c \end{bmatrix}$

18. $\begin{bmatrix} a & a & b \\ a & b & b \\ 0.2 & c & a \end{bmatrix}$

In Problems 19 and 20, a transition matrix $\mathbf{T}$ and an initial state vector $\mathbf{X}_0$ for a Markov chain are given. Compute the state vectors $\mathbf{X}_1$, $\mathbf{X}_2$, and $\mathbf{X}_3$.

19. $\mathbf{T} = \begin{bmatrix} 0.1 & 0.3 & 0.1 \\ 0.2 & 0.4 & 0.1 \\ 0.7 & 0.3 & 0.8 \end{bmatrix}$

$\mathbf{X}_0 = \begin{bmatrix} 0.5 \\ 0 \\ 0.5 \end{bmatrix}$

20. $\mathbf{T} = \begin{bmatrix} 0.4 & 0.1 & 0.1 \\ 0.2 & 0.6 & 0.5 \\ 0.4 & 0.3 & 0.4 \end{bmatrix}$

$\mathbf{X}_0 = \begin{bmatrix} 0.1 \\ 0.3 \\ 0.6 \end{bmatrix}$

In Problems 21 and 22, a transition matrix $\mathbf{T}$ for a Markov chain is given.

   (a) Compute $\mathbf{T}^2$ and $\mathbf{T}^3$.

   (b) What is the probability of going to state 1 from state 2 after two steps?

   (c) What is the probability of going to state 2 from state 1 after three steps?

21. $\begin{bmatrix} \frac{1}{7} & \frac{3}{7} \\ \frac{6}{7} & \frac{4}{7} \end{bmatrix}$

22. $\begin{bmatrix} 0 & 0.4 & 0.3 \\ 0 & 0.3 & 0.5 \\ 1 & 0.3 & 0.2 \end{bmatrix}$

In Problems 23 and 24, find the steady-state vector for the given transition matrix for a Markov chain.

23. $\begin{bmatrix} \frac{1}{3} & \frac{2}{3} \\ \frac{2}{3} & \frac{1}{3} \end{bmatrix}$

24. $\begin{bmatrix} 0.4 & 0.4 & 0.3 \\ 0.3 & 0.2 & 0.3 \\ 0.3 & 0.4 & 0.4 \end{bmatrix}$

25. **Automobile Market** For a particular segment of the automobile market, the results of a survey indicate that 80% of people who own a Japanese car would buy a Japanese car the next time and 20% would buy a non-Japanese car. Of owners of non-Japanese cars, 40% would buy a non-Japanese car the next time and 60% would buy a Japanese car.

(a) Of those who currently own a Japanese car, what percentage will buy a Japanese car two cars later?

(b) If 60% of this segment currently own Japanese cars and 40% own non-Japanese cars, what will be the distribution for this segment of the market two cars from now?

(c) How will this segment be distributed in the long run?

26. **Voting**   Suppose that the probabilities of voting for particular parties in a future election depend on the voting patterns in the previous election. For a certain region where there is a three-party political system, assume that these probabilities are contained in the matrix

$$\mathbf{T} = [t_{ij}] = \begin{bmatrix} 0.7 & 0.4 & 0.1 \\ 0.2 & 0.5 & 0.1 \\ 0.1 & 0.1 & 0.8 \end{bmatrix}$$

where t_{ij} represents the probability that a voter will vote for party i in the next election if he or she voted for party j in the last election.

(a) At the last election, 50% of the electorate voted for party 1, 30% for party 2, and 20% for party 3. What is the expected percentage distribution of votes for the next election?

(b) In the long run, what is the percentage distribution of votes? Give your answers to the nearest percent.

Mathematical Snapshot

Markov Chains in Game Theory

Game theory is the mathematical study of how people behave in competitive and cooperative situations. Each situation is represented by a table showing the players' options and their payoffs for various outcomes. Figure 9.9 shows a famous game, called the Prisoner's Dilemma. Each player chooses between "Cooperate" and "Defect." The players' subsequent payoffs are in the cell to the right of Player 1's choice and below Player 2's. Player 1 gets the payoff in the lower left corner of the cell, Player 2 the payoff in the upper right corner.

Player 2

	Cooperate	Defect
Cooperate	4 4	5 1
Defect	1 5	2 2

FIGURE 9.9 The Prisoner's Dilemma.

Most people who have studied the game think that the rational choice for a player trying to maximize his or her payoff is to defect—a move that, if played by both players, leads to a payoff of 2 for each. The paradox is that both players would be better off (payoff of 4) with the supposedly irrational choice, Cooperate. You should study the game for a few minutes and see what you think.

The game gets its name from a fable about two prisoners, but the same puzzle arises whenever two people agree to do each other a favor but have no way of checking up on each other. Both benefit if both cooperate, but each is tempted to cheat. On a larger scale, voting and voluntary garbage recycling are cooperative behaviors in a many-person Prisoner's Dilemma where the players are all the members of a community.

Game theorists have long been troubled by the seeming irrationality of Cooperate. Trying to understand how

it might make sense after all, the theorists have studied scenarios in which the Prisoner's Dilemma is played not just once, but repeatedly. In this Iterated Prisoner's Dilemma, more than just two strategies are possible, because a player's play in a given round can be based on what happened in the previous round.

One much-discussed strategy, called Tit-for-Tat, is to cooperate initially and from then on to do whatever the other player did in the preceding round. If the other player defected in round n, Tit-for-Tat defects in round $n+1$; if the other player cooperated in round n, Tit-for-Tat cooperates in round $n+1$. Of course, "Always Cooperate" and "Always Defect" are also options. "Tit-for-Tat," however, models a more sophisticated way of thinking: It is prepared to cooperate but does not allow itself to be repeatedly exploited by the other's defections.

The succession of rounds in the Iterated Prisoner's Dilemma is a Markov chain with four states, where each state represents a combination of players' choices. Let us number the states as follows: Both cooperate = State 1; Player 1 cooperates/Player 2 defects = State 2; Player 1 defects/Player 2 cooperates = State 3; Both defect = State 4. The transitions between states are fully determined by the players' choice of strategies. If both are playing Tit-for-Tat, then State 1 leads to State 1, State 2 leads to State 3, State 3 leads to State 2, and State 4 leads to State 4. You should take a moment to convince yourself of this. The transition matrix looks like this:

$$
\mathbf{T} = \begin{matrix} & \begin{matrix} 1 & 2 & 3 & 4 \end{matrix} \\ \begin{matrix} 1 \\ 2 \\ 3 \\ 4 \end{matrix} & \begin{bmatrix} 1 & 0 & 0 & 0 \\ 0 & 0 & 1 & 0 \\ 0 & 1 & 0 & 0 \\ 0 & 0 & 0 & 1 \end{bmatrix} \end{matrix}
$$

With this matrix, the course of play crucially depends on what choice the two players open with. If Player 1 cooperates and Player 2 defects, initially, then the successive state vectors look like this:

$$
\mathbf{X}_0 = \begin{bmatrix} 0 \\ 1 \\ 0 \\ 0 \end{bmatrix}, \ \mathbf{T}\mathbf{X}_0 = \begin{bmatrix} 0 \\ 0 \\ 1 \\ 0 \end{bmatrix}, \ \mathbf{T}^2\mathbf{X}_0 = \begin{bmatrix} 0 \\ 1 \\ 0 \\ 0 \end{bmatrix}, \ \mathbf{T}^3\mathbf{X}_0 = \begin{bmatrix} 0 \\ 0 \\ 1 \\ 0 \end{bmatrix} \dots
$$

The players oscillate back and forth between States 2 and 3. On the other hand, States 1 and 4 are both "traps": If two Tit-for-Tat players start off both cooperating, they will cooperate all down the line, but if they both start with Defect, they will continue in that pattern.

It is regrettable that two Tit-for-Tat players, both ready to cooperate, may get locked into the cycle of defection instead. What if we modify Tit-for-Tat slightly, so that it occasionally forgives defection and cooperates in the next round, as a sort of peace offering? We can model this by giving each side, let us say, a 10% chance of cooperating after the other has defected. The transition matrix then ends up looking like this:

$$\mathbf{T} = \begin{array}{c} \\ 1 \\ 2 \\ 3 \\ 4 \end{array} \begin{array}{cccc} 1 & 2 & 3 & 4 \\ \begin{bmatrix} 1 & 0.1 & 0.1 & 0.01 \\ 0 & 0 & 0.9 & 0.09 \\ 0 & 0.9 & 0 & 0.09 \\ 0 & 0 & 0 & 0.81 \end{bmatrix} \end{array}$$

If both defect (State 4), there is a $0.9 \times 0.9 = 0.81$ chance that they will both defect in the next round (State 4 again), a $0.9 \times 0.1 = 0.09$ that Player 1 will cooperate and Player 2 will defect (State 2), and so on. Make sure you understand how the probabilities in the matrix are obtained.

Now two Tit-for-Tat-ers can escape State 4 and settle permanently into State 1. How long will that take? In principle the game could stay in State 4 for a long time. But sooner or later, it will find its way to State 1. How many iterations are necessary before there is at least a 0.50 probability of the players having made it to mutual cooperation?

Formally, if $\mathbf{X}_0 = \begin{bmatrix} 0 \\ 0 \\ 0 \\ 1 \end{bmatrix}$, what value of n puts the first entry of the state vector $\mathbf{X}_n = \mathbf{T}^n\mathbf{X}_0$ at greater than or equal to 0.5? The answer is 12:

$$\mathbf{T}^{12}\mathbf{X}_0 = \begin{bmatrix} 0.5149 \\ 0.2027 \\ 0.2027 \\ 0.0798 \end{bmatrix}$$

So if a game between modified Tit-for-Tat-ers runs to 100 rounds, it is highly likely that most of those rounds will have cooperative outcomes, even if the game started out with mutual distrust. In computer simulations involving interaction among many different strategies, Tit-for-Tat generally emerges as quite profitable—more profitable, it should be noted, than Always Defect. Cooperation, it is comforting to discover, need not be irrational after all.

To learn more on the Internet, visit a search engine and type in "Iterated Prisoner's Dilemma."

Problems

1. Using a graphing calculator, model play between two modified Tit-for-Tat-ers where one starts with Cooperate and the other with Defect. How many rounds are needed to reach at least 0.50 probability of mutual cooperation?

2. Prove that $\begin{bmatrix} 1 \\ 0 \\ 0 \\ 0 \end{bmatrix}$ is the steady-state vector for a game between two modified Tit-for-Tat-ers.

3. Assume that all states are initially possible, but beginning with the second round, Player 2 adopts the strategy of Always Defect, Always Cooperate, or regular Tit-for-Tat. What is the transition matrix for each of these three games if Player 1 is using modified Tit-for-Tat in each case?

4. Use the results from Problem 3 to describe what will happen in the long run with each game.

10

LIMITS AND CONTINUITY

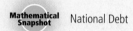 National Debt

The philosopher Zeno of Elea was fond of paradoxes about motion. His most famous one goes something like this. The warrior Achilles agrees to run a race against a tortoise. Achilles can run 10 meters per second and the tortoise only 1 meter per second, so the tortoise gets a 10-meter head start. Since Achilles is so much faster, he should still win. But by the time he has covered his first 10 meters and reached the place where the tortoise started, the tortoise has advanced 1 meter and is still ahead. And after Achilles has covered that 1 meter, the tortoise has advanced another 0.1 meter and is still ahead. And after Achilles has covered that 0.1 meter, the tortoise has advanced another 0.01 meter and is still ahead. And so on. Therefore, Achilles gets closer and closer to the tortoise but can never catch up.

Zeno's audience knew, of course, that the argument was fishy. The position of Achilles at time t after the race has begun is $(10 \text{ m/s})t$. The position of the tortoise at the same time t is $(1 \text{ m/s})t + 10 \text{ m}$. When these are equal, Achilles and the tortoise are side by side. To solve the resulting equation

$$(10 \text{ m/s})t = (1 \text{ m/s})t + 10 \text{ m}$$

for t is to find the time at which Achilles pulls even with the tortoise.

The solution is $t = 1\frac{1}{9}$ seconds, at which time Achilles will have run $\left(1\frac{1}{9}\text{s}\right)(10 \text{ m/s}) = 11\frac{1}{9}$ meters.

What puzzled Zeno and his listeners is how it could be that

$$10 + 1 + \frac{1}{10} + \frac{1}{100} + \cdots = 11\frac{1}{9}$$

where the left side represents an *infinite sum* and the right side is a finite result. The modern solution to this problem is the concept of a limit, which is the key topic of this chapter. The left side of the equation is an infinite geometric series. Using limit notation, summation notation, and the formula from Section 5.4 for the sum of a geometric series, we write

$$\lim_{k \to \infty} \sum_{n=0}^{k} 10^{1-n} = \lim_{k \to \infty} \frac{10\left(1 - \left(\frac{1}{10}\right)^{k+1}\right)}{1 - \frac{1}{10}} = \frac{100}{9} = 11\frac{1}{9}$$

10.1 Limits

Perhaps you have been in a parking-lot situation in which you must "inch up" to the car in front, but yet you do not want to bump or touch it. This notion of getting closer and closer to something, but yet not touching it, is very important in mathematics and is involved in the concept of *limits*, which lies at the foundation of calculus. Basically, we will let a variable "inch up" to a particular value and examine the effect it has on the values of a function.

For example, consider the function

$$f(x) = \frac{x^3 - 1}{x - 1}$$

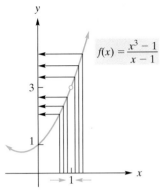

FIGURE 10.1 $\displaystyle\lim_{x \to 1} \frac{x^3 - 1}{x - 1} = 3.$

Although this function is not defined at $x = 1$, we may be curious about the behavior of the function values as x gets very close to 1. Table 10.1 gives some values of x that are slightly less than 1 and some that are slightly greater, as well as their corresponding function values. Notice that as x takes on values closer and closer to 1, regardless of whether x approaches it *from the left* ($x < 1$) or *from the right* ($x > 1$), the corresponding values of $f(x)$ get closer and closer to one and only one number, namely 3. This is also clear from the graph of f in Figure 10.1. Notice there that even though the function is not defined at $x = 1$ (as indicated by the hollow dot), the function values get closer and closer to 3 as x gets closer and closer to 1. To express this, we say that the **limit** of $f(x)$ as x approaches 1 is 3 and write

$$\lim_{x \to 1} \frac{x^3 - 1}{x - 1} = 3$$

We can make $f(x)$ as close to 3 as we wish, and keep it that close, by taking x sufficiently close to, but not equal to, 1. The limit exists at 1, even though 1 is not in the domain of f.

TABLE 10.1

	$x < 1$		$x > 1$	
x	$f(x)$	x	$f(x)$
0.8	2.44	1.2	3.64
0.9	2.71	1.1	3.31
0.95	2.8525	1.05	3.1525
0.99	2.9701	1.01	3.0301
0.995	2.985025	1.005	3.015025
0.999	2.997001	1.001	3.003001

We can also consider the limit of a function as x approaches a number that is in the domain. Let us examine the limit of $f(x) = x + 3$ as x approaches 2:

$$\lim_{x \to 2} (x + 3)$$

Obviously, if x is close to 2 (but not equal to 2), then $x + 3$ is close to 5. This is also apparent from the table and graph in Figure 10.2. Thus,

$$\lim_{x \to 2} (x + 3) = 5$$

Given a function f and a number a, there *may* be two ways of associa~~ ~~ to the pair (f, a). One such number is the *evaluation of f at a*, ~~ ~~

precisely when a is in the domain of f. For example, if f~~ ~~

example, then $f(1)$ does not *exist*. Another way of associating

x < 2		x > 2	
x	f(x)	x	f(x)
1.5	4.5	2.5	5.5
1.9	4.9	2.1	5.1
1.95	4.95	2.05	5.05
1.99	4.99	2.01	5.01
1.999	4.999	2.001	5.001

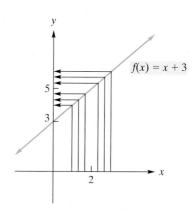

FIGURE 10.2 $\lim_{x \to 2}(x + 3) = 5$.

(f, a) is *the limit of $f(x)$ as x approaches a,* which is denoted $\lim_{x \to a} f(x)$. We have given two examples. Here is the general case.

DEFINITION

The limit of $f(x)$ as x approaches a is the number L, written

$$\lim_{x \to a} f(x) = L$$

provided that $f(x)$ is arbitrarily close to L for all x sufficiently close to, but not equal to, a. If there is no such number, we say that the limit *does not exist.*

We emphasize that, when finding a limit, we are concerned not with what happens to $f(x)$ when x equals a, but only with what happens to $f(x)$ when x is close to a. In fact, even if $f(a)$ *exists,* the preceding definition explicitly rules out consideration of it. In our second example, $f(x) = x + 3$, we have $f(2) = 5$ and also $\lim_{x \to 2}(x + 3) = 5$, but it is quite possible to have a function f and a number a for which both $f(a)$ and $\lim_{x \to a} f(x)$ exist and are different, Moreover, a limit must be independent of the way in which x *approaches* a, meaning the way in which x gets close to a. That is, the limit must be the same whether x approaches a from the left or from the right (for $x < a$ or $x > a$, respectively).

● EXAMPLE 1   Estimating a Limit from a Graph

a. *Estimate* $\lim_{x \to 1} f(x)$, *where the graph of f is given in Figure* 10.3(a).

Solution: If we look at the graph for values of x near 1, we see that $f(x)$ is near 2. Moreover, as x gets closer and closer to 1, $f(x)$ appears to get closer and closer to 2. Thus, we estimate that

$$\lim_{x \to 1} f(x) \text{ is } 2$$

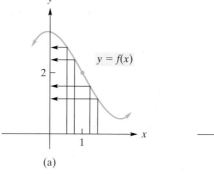

(a)

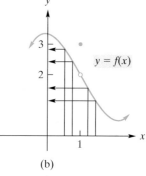

(b)

FIGURE 10.3 Investigation of $\lim_{x \to 1} f(x)$.

b. *Estimate* $\lim_{x \to 1} f(x)$, *where the graph of f is given in Figure* 10.3(b).

Solution: Although $f(1) = 3$, this fact has no bearing whatsoever on the limit of $f(x)$ as x approaches 1. We see that as x gets closer and closer to 1, $f(x)$ appears to get closer and closer to 2. Thus, we estimate that

$$\lim_{x \to 1} f(x) \text{ is } 2$$

NOW WORK PROBLEM 1

Up to now, all of the limits that we have considered did indeed exist. Next we look at some situations in which a limit does not exist.

PRINCIPLES IN PRACTICE 1

LIMITS THAT DO NOT EXIST

The greatest integer function, denoted, $f(x) = \lfloor x \rfloor$, is used every day by cashiers making change for customers. This function tells the amount of paper money for each amount of change owed. (For example, if a customer is owed \$1.25 in change, he or she would get \$1 in paper money; thus, $\lfloor 1.25 \rfloor = 1$.) Formally, $\lfloor x \rfloor$ is defined as the greatest integer less than or equal to x. Graph f, sometimes called a step function, on your graphing calculator in the standard viewing rectangle. (You will find it in the numbers menu; it's called "integer part.") Explore this graph using TRACE. Determine whether $\lim_{x \to a} f(x)$ exists.

● **EXAMPLE 2** **Limits That Do Not Exist**

a. *Estimate* $\lim_{x \to -2} f(x)$ *if it exists, where the graph of f is given in Figure* 10.4.

Solution: As x approaches -2 from the left ($x < -2$), the values of $f(x)$ appear to get closer to 1. But as x approaches -2 from the right ($x > -2$), $f(x)$ appears to get closer to 3. Hence, as x approaches -2, the function values do not settle down to one and only one number. We conclude that

$$\lim_{x \to -2} f(x) \text{ does not exist}$$

Note that the limit does not exist even though the function is defined at $x = -2$.

b. *Estimate* $\lim_{x \to 0} \dfrac{1}{x^2}$ *if it exists.*

Solution: Let $f(x) = 1/x^2$. The table in Figure 10.5 gives values of $f(x)$ for some values of x near 0. As x gets closer and closer to 0, the values of $f(x)$ get larger and larger without bound. This is also clear from the graph. Since the values of $f(x)$ do not approach a *number* as x approaches 0,

$$\lim_{x \to 0} \frac{1}{x^2} \text{ does not exist}$$

NOW WORK PROBLEM 3

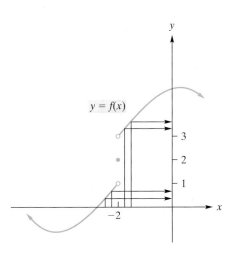

FIGURE 10.4 $\lim_{x \to -2} f(x)$ does not exist.

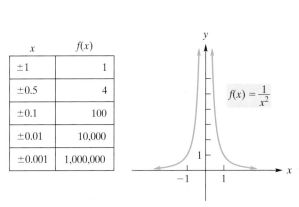

x	$f(x)$
± 1	1
± 0.5	4
± 0.1	100
± 0.01	10,000
± 0.001	1,000,000

$f(x) = \dfrac{1}{x^2}$

FIGURE 10.5 $\lim_{x \to 0} \dfrac{1}{x^2}$ does not exist.

TECHNOLOGY

Problem: Estimate $\lim_{x\to 2} f(x)$ if

$$f(x) = \frac{x^3 + 2.1x^2 - 10.2x + 4}{x^2 + 2.5x - 9}$$

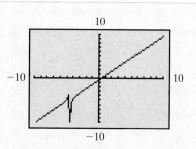

Solution: One method of finding the limit is by constructing a table of function values $f(x)$ when x is close to 2. From Figure 10.6, we estimate the limit to be 1.57. Alternatively, we can estimate the limit from the graph of f. Figure 10.7 shows the graph of f in the standard window of $[-10, 10] \times [-10, 10]$. First we zoom in several times around $x = 2$ and obtain Figure 10.8. After tracing around $x = 2$, we estimate the limit to be 1.57.

FIGURE 10.7 Graph of $f(x)$ in standard window.

X	Y1
1.9	1.4688
1.99	1.5592
1.999	1.5682
1.9999	1.5691
2.01	1.5793
2.001	1.5702
2.0001	1.5693

X=2.0001

FIGURE 10.6 $\lim_{x\to 2} f(x) \approx 1.57$.

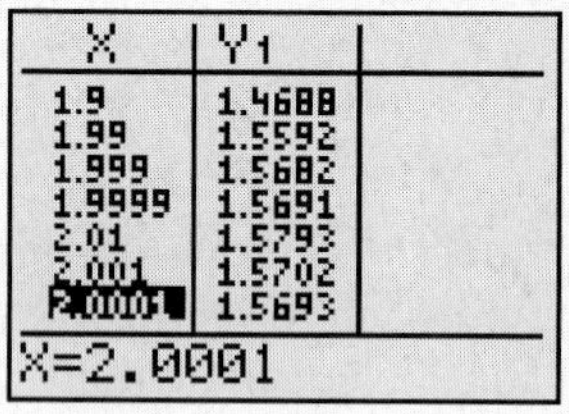

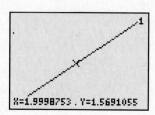

FIGURE 10.8 Zooming and tracing around $x = 2$ gives $\lim_{x\to 2} f(x) \approx 1.57$.

Properties of Limits

To determine limits, we do not always want to compute function values or sketch a graph. Alternatively, there are several properties of limits that we may be able to employ. The following properties may seem reasonable to you:

1. If $f(x) = c$ is a constant function, then

$$\lim_{x\to a} f(x) = \lim_{x\to a} c = c$$

2. $\lim_{x\to a} x^n = a^n$, for any positive integer n

EXAMPLE 3 Applying Limit Properties 1 and 2

a. $\lim_{x\to 2} 7 = 7$; $\lim_{x\to -5} 7 = 7$

b. $\lim_{x\to 6} x^2 = 6^2 = 36$

c. $\lim_{t\to -2} t^4 = (-2)^4 = 16$

NOW WORK PROBLEM 9

Some other properties of limits are as follows:

If $\lim_{x\to a} f(x)$ and $\lim_{x\to a} g(x)$ exist, then

3.
$$\lim_{x\to a}[f(x) \pm g(x)] = \lim_{x\to a} f(x) \pm \lim_{x\to a} g(x)$$

That is, the limit of a sum or difference is the sum or difference, respectively, of the limits.

4.
$$\lim_{x\to a}[f(x) \cdot g(x)] = \lim_{x\to a} f(x) \cdot \lim_{x\to a} g(x)$$

That is, the limit of a product is the product of the limits.

5.
$$\lim_{x\to a}[cf(x)] = c \cdot \lim_{x\to a} f(x), \text{ where } c \text{ is a constant}$$

That is, the limit of a constant times a function is the constant times the limit of the function.

PRINCIPLES IN PRACTICE 2

APPLYING LIMIT PROPERTIES

The volume of helium in a spherical balloon (in cubic centimeters), as a function of the radius r in centimeters, is given by $V(r) = \frac{4}{3}\pi r^3$. Find $\lim_{r\to 1} V(r)$.

● EXAMPLE 4   **Applying Limit Properties**

a. $\lim_{x\to 2}(x^2 + x) = \lim_{x\to 2} x^2 + \lim_{x\to 2} x$ (Property 3)

$$= 2^2 + 2 = 6 \qquad \text{(Property 2)}$$

b. Property 3 can be extended to the limit of a finite number of sums and differences. For example,

$$\lim_{q\to -1}(q^3 - q + 1) = \lim_{q\to -1} q^3 - \lim_{q\to -1} q + \lim_{q\to -1} 1$$

$$= (-1)^3 - (-1) + 1 = 1$$

c. $\lim_{x\to 2}[(x+1)(x-3)] = \lim_{x\to 2}(x+1) \cdot \lim_{x\to 2}(x-3)$ (Property 4)

$$= \left[\lim_{x\to 2} x + \lim_{x\to 2} 1\right] \cdot \left[\lim_{x\to 2} x - \lim_{x\to 2} 3\right]$$

$$= (2+1) \cdot (2-3) = 3(-1) = -3$$

d. $\lim_{x\to -2} 3x^3 = 3 \cdot \lim_{x\to -2} x^3$ (Property 5)

$$= 3(-2)^3 = -24$$

NOW WORK PROBLEM 11 ●●●

PRINCIPLES IN PRACTICE 3

LIMIT OF A POLYNOMIAL

The revenue function for a certain product is given by $R(x) = 500x - 6x^2$. Find $\lim_{x\to 8} R(x)$.

● EXAMPLE 5   **Limit of a Polynomial Function**

Let $f(x) = c_n x^n + c_{n-1} x^{n-1} + \cdots + c_1 x + c_0$ define a polynomial function. Then

$$\lim_{x\to a} f(x) = \lim_{x\to a}(c_n x^n + c_{n-1} x^{n-1} + \cdots + c_1 x + c_0)$$

$$= c_n \cdot \lim_{x\to a} x^n + c_{n-1} \cdot \lim_{x\to a} x^{n-1} + \cdots + c_1 \cdot \lim_{x\to a} x + \lim_{x\to a} c_0$$

$$= c_n a^n + c_{n-1} a^{n-1} + \cdots + c_1 a + c_0 = f(a)$$

Thus, we have the following property:

If f is a polynomial function, then

$$\lim_{x\to a} f(x) = f(a)$$

In other words, if f is a polynomial and a is any number, then both ways of associating a number to the pair (f, a), namely evaluation and forming the limit, exist and are equal.

NOW WORK PROBLEM 13 ●●●

The result of Example 5 allows us to find many limits simply by evaluation. For example, we can find

$$\lim_{x\to -3}(x^3 + 4x^2 - 7)$$

by substituting -3 for x because $x^3 + 4x^2 - 7$ is a polynomial function:

$$\lim_{x \to -3}(x^3 + 4x^2 - 7) = (-3)^3 + 4(-3)^2 - 7 = 2$$

Similarly,

$$\lim_{h \to 3}[2(h-1)] = 2(3-1) = 4$$

We want to stress that one does not find limits simply by evaluating, "plugging in" as some say, unless there is a rule that covers the situation. We were able to find the previous two limits by evaluation because we have a rule that applies to limits of polynomial functions. However, indiscriminate use of evaluation can lead to errors. To illustrate, in Example 1(b) we have $f(1) = 3$, which is not $\lim_{x \to 1} f(x)$; in Example 2(a), $f(-2) = 2$, which is not $\lim_{x \to -2} f(x)$.

The next two limit properties concern quotients and roots.

If $\lim_{x \to a} f(x)$ and $\lim_{x \to a} g(x)$ exist, then

6.
$$\lim_{x \to a} \frac{f(x)}{g(x)} = \frac{\lim_{x \to a} f(x)}{\lim_{x \to a} g(x)} \quad \text{if} \quad \lim_{x \to a} g(x) \ne 0$$

That is, the limit of a quotient is the quotient of limits, provided that the denominator does not have a limit of 0.

7.
$$\lim_{x \to a} \sqrt[n]{f(x)} = \sqrt[n]{\lim_{x \to a} f(x)} \qquad \text{(See Footnote 1)}$$

CAUTION

Note that in Example 6(a) the numerator and denominator of the function are polynomials. In general, we can determine a limit of a rational function by evaluation, provided that the denominator is not 0 at a.

● **EXAMPLE 6   Applying Limit Properties 6 and 7**

a. $\displaystyle \lim_{x \to 1} \frac{2x^2 + x - 3}{x^3 + 4} = \frac{\lim_{x \to 1}(2x^2 + x - 3)}{\lim_{x \to 1}(x^3 + 4)} = \frac{2 + 1 - 3}{1 + 4} = \frac{0}{5} = 0$

b. $\displaystyle \lim_{t \to 4} \sqrt{t^2 + 1} = \sqrt{\lim_{t \to 4}(t^2 + 1)} = \sqrt{17}$

c. $\displaystyle \lim_{x \to 3} \sqrt[3]{x^2 + 7} = \sqrt[3]{\lim_{x \to 3}(x^2 + 7)} = \sqrt[3]{16} = \sqrt[3]{8 \cdot 2} = 2\sqrt[3]{2}$

NOW WORK PROBLEM 15 ●●●

Limits and Algebraic Manipulation

We now consider limits to which our limit properties do not apply and which cannot be determined by evaluation. A fundamental result is the following:

CAUTION

The condition for equality of the limits does not preclude the possibility that $f(a) = g(a)$. The condition only concerns $x \ne a$.

If f and g are two functions for which $f(x) = g(x)$, for all $x \ne a$, then

$$\lim_{x \to a} f(x) = \lim_{x \to a} g(x)$$

(meaning that if either limit exists, then the other exists and they are equal).

The result follows directly from the definition of limit since the value of $\lim_{x \to a} f(x)$ depends only on those values $f(x)$ for x that are close to a. We repeat: The evaluation of f at a, $f(a)$, or lack of its existence, is irrelevant in the

---

[1]If n is even, we require that $\lim_{x \to a} f(x)$ be positive.

determination of $\lim_{x \to a} f(x)$ unless we have a specific rule that applies, such as in the case when f is a polynomial.

APPLYING LIMIT PROPERTY

The rate of change of productivity p (in number of units produced per hour) increases with time on the job by the function

$$p(t) = \frac{50(t^2 + 4t)}{t^2 + 3t + 20}$$

Find $\lim_{t \to 2} p(t)$.

● **EXAMPLE 7   Finding a Limit**

Find $\lim\limits_{x \to -1} \dfrac{x^2 - 1}{x + 1}$.

Solution: As $x \to -1$, both numerator and denominator approach zero. Because the limit of the denominator is 0, we *cannot* use Property 6. However, since what happens to the quotient when x equals -1 is of no concern, we can assume that $x \ne -1$ and simplify the fraction:

$$\frac{x^2 - 1}{x + 1} = \frac{(x + 1)(x - 1)}{x + 1} = x - 1 \quad \text{for } x \ne -1$$

This algebraic manipulation (factoring and cancellation) of the original function $\dfrac{x^2 - 1}{x + 1}$ yields a new function $x - 1$, which is the same as the original function for $x \ne -1$. Thus the fundamental result displayed on the previous page applies and we have

$$\lim_{x \to -1} \frac{x^2 - 1}{x + 1} = \lim_{x \to -1} (x - 1) = -1 - 1 = -2$$

Notice that, although the original function is not defined at -1, it *does* have a limit as $x \to -1$.

NOW WORK PROBLEM 21

When both $f(x)$ and $g(x)$ approach 0 as $x \to a$, then the limit

$$\lim_{x \to a} \frac{f(x)}{g(x)}$$

is said to have the *form* 0/0. Similarly, we speak of *form* $k/0$, for $k \ne 0$ if $f(x)$ approaches $k \ne 0$ as $x \to a$ but $g(x)$ approaches 0 as $x \to a$.

In Example 7, the method of finding a limit by evaluation does not work. Replacing x by -1 gives 0/0, which has no meaning. When the meaningless form 0/0 arises, algebraic manipulation (as in Example 7) may result in a function that agrees with the original function, except possibly at the limiting value. In Example 7 the new function, $x - 1$, is a polynomial and its limit *can* be found by evaluation.

In the beginning of this section, we found

$$\lim_{x \to 1} \frac{x^3 - 1}{x - 1}$$

by examining a table of function values of $f(x) = (x^3 - 1)/(x - 1)$ and also by considering the graph of f. This limit has the form 0/0. Now we will determine the limit by using the technique used in Example 7.

CAUTION

There is frequently confusion about which principle is being used in this example and in Example 7. It is this:

If $f(x) = g(x)$ for $x \ne a$

then $\lim\limits_{x \to a} f(x) = \lim\limits_{x \to a} g(x)$

● **EXAMPLE 8   Form 0/0**

Find $\lim\limits_{x \to 1} \dfrac{x^3 - 1}{x - 1}$.

Solution: As $x \to 1$, both the numerator and denominator approach 0. Thus, we will try to express the quotient in a different form for $x \ne 1$. By factoring, we have

$$\frac{x^3 - 1}{x - 1} = \frac{(x - 1)(x^2 + x + 1)}{(x - 1)} = x^2 + x + 1 \quad \text{for } x \ne 1$$

(Alternatively, long division would give the same result.) Therefore,

$$\lim_{x \to 1} \frac{x^3 - 1}{x - 1} = \lim_{x \to 1}(x^2 + x + 1) = 1^2 + 1 + 1 = 3$$

as we showed before.

NOW WORK PROBLEM 23 ●●●

FORM $0/0$

The length of a material increases as it is heated up according to the equation $l = 125 + 2x$. The rate at which the length is increasing is given by

$$\lim_{h \to 0} \frac{125 + 2(x + h) - (125 + 2x)}{h}$$

Calculate this limit.

The expression

$$\frac{f(x + h) - f(x)}{h}$$

is called a *difference quotient*. The limit of the difference quotient lies at the heart of differential calculus. You will encounter such limits in Chapter 11.

● **EXAMPLE 9** **Form** $0/0$

If $f(x) = x^2 + 1$, find $\displaystyle\lim_{h \to 0} \frac{f(x + h) - f(x)}{h}$.

Solution:

$$\lim_{h \to 0} \frac{f(x + h) - f(x)}{h} = \lim_{h \to 0} \frac{[(x + h)^2 + 1] - (x^2 + 1)}{h}$$

Here we treat x as a constant because h, not x, is changing. As $h \to 0$, both the numerator and denominator approach 0. Therefore, we will try to express the quotient in a different form, for $h \neq 0$. We have

$$\lim_{h \to 0} \frac{[(x + h)^2 + 1] - (x^2 + 1)}{h} = \lim_{h \to 0} \frac{[x^2 + 2xh + h^2 + 1] - x^2 - 1}{h}$$

$$= \lim_{h \to 0} \frac{2xh + h^2}{h}$$

$$= \lim_{h \to 0} \frac{h(2x + h)}{h}$$

$$= \lim_{h \to 0} (2x + h)$$

$$= 2x$$

Note: It is the fourth equality above, $\displaystyle\lim_{h \to 0} \frac{h(2x + h)}{h} = \lim_{h \to 0} (2x + h)$, that uses the fundamental result. When $\dfrac{h(2x + h)}{h}$ and $2x + h$ are considered as *functions of h*, they are seen to be equal, for all $h \neq 0$. It follows that their limits as h approaches 0 are equal.

NOW WORK PROBLEM 35 ●●●

A Special Limit

We conclude this section with a note concerning a most important limit, namely,

$$\lim_{x \to 0} (1 + x)^{1/x}$$

Figure 10.9 shows the graph of $f(x) = (1 + x)^{1/x}$. Although $f(0)$ does not exist, as $x \to 0$ it is clear that the limit of $(1 + x)^{1/x}$ exists. It is approximately 2.71828 and is denoted by the letter e. This, you may recall, is the base of the system of natural logarithms. The limit

$$\lim_{x \to 0} (1 + x)^{1/x} = e$$

This limit will be used in Chapter 12.

can actually be considered the definition of e. It can be shown that this agrees with the definition of e that we gave in Section 4.1.

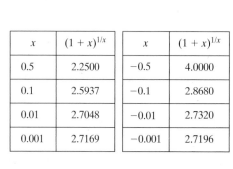

x	$(1 + x)^{1/x}$	x	$(1 + x)^{1/x}$
0.5	2.2500	-0.5	4.0000
0.1	2.5937	-0.1	2.8680
0.01	2.7048	-0.01	2.7320
0.001	2.7169	-0.001	2.7196

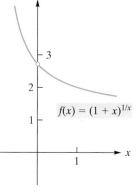

FIGURE 10.9 $\lim_{x \to 0} (1 + x)^{1/x} = e$.

Problems 10.1

In Problems 1–4, use the graph of f to estimate each limit if it exists.

*1. Graph of f appears in Figure 10.10.

   (a) $\lim_{x \to 0} f(x)$ (b) $\lim_{x \to 1} f(x)$ (c) $\lim_{x \to 2} f(x)$

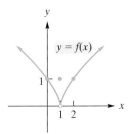

FIGURE 10.10   Diagram for Problem 1.

2. Graph of f appears in Figure 10.11.

   (a) $\lim_{x \to -1} f(x)$ (b) $\lim_{x \to 0} f(x)$ (c) $\lim_{x \to 1} f(x)$

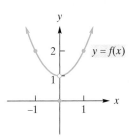

FIGURE 10.11   Diagram for Problem 2.

*3. Graph of f appears in Figure 10.12.

   (a) $\lim_{x \to -1} f(x)$ (b) $\lim_{x \to 1} f(x)$ (c) $\lim_{x \to 2} f(x)$

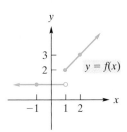

FIGURE 10.12   Diagram for Problem 3.

4. Graph of f appears in Figure 10.13.

   (a) $\lim_{x \to -1} f(x)$ (b) $\lim_{x \to 0} f(x)$ (c) $\lim_{x \to 1} f(x)$

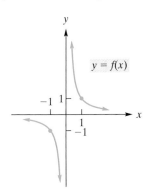

FIGURE 10.13   Diagram for Problem 4.

In Problems 5–8, use your calculator to complete the table, and use your results to estimate the given limit.

5. $\lim_{x \to -1} \dfrac{3x^2 + 2x - 1}{x + 1}$

x	-0.9	-0.99	-0.999	-1.001	-1.01	-1.1
$f(x)$						

6. $\lim_{x \to -3} \dfrac{x^2 - 9}{x + 3}$

x	-3.1	-3.01	-3.001	-2.999	-2.99	-2.9
$f(x)$						

7. $\lim_{x \to 0} \dfrac{e^x - 1}{x}$

x	-0.1	-0.01	-0.001	0.001	0.01	0.1
$f(x)$						

8. $\lim_{h \to 0} \dfrac{\sqrt{1 + h} - 1}{h}$

h	-0.1	-0.01	-0.001	0.001	0.01	0.1
$f(x)$						

In Problems 9–34, find the limits.

*9. $\lim_{x \to 2} 16$

10. $\lim_{x \to 3} 2x$

*11. $\lim_{t \to -5} (t^2 - 5)$

12. $\lim_{t \to 1/3} (5t - 7)$

*13. $\lim_{x \to -2} (3x^3 - 4x^2 + 2x - 3)$

14. $\lim_{r \to 9} \dfrac{4r - 3}{11}$

*15. $\lim_{t \to -3} \dfrac{t - 2}{t + 5}$

16. $\lim_{x \to -6} \dfrac{x^2 + 6}{x - 6}$

17. $\lim_{h \to 0} \dfrac{h}{h^2 - 7h + 1}$

18. $\lim_{z \to 0} \dfrac{z^2 - 5z - 4}{z^2 + 1}$

19. $\lim_{p \to 4} \sqrt{p^2 + p + 5}$

20. $\lim_{y \to 15} \sqrt{y + 3}$

*21. $\lim_{x \to -2} \dfrac{x^2 + 2x}{x + 2}$

22. $\lim_{x \to -1} \dfrac{x + 1}{x + 1}$

*23. $\lim_{x \to 2} \dfrac{x^2 - x - 2}{x - 2}$

24. $\lim_{t \to 0} \dfrac{t^3 + 3t^2}{t^3 - 4t^2}$

25. $\lim_{x \to 3} \dfrac{x^2 - x - 6}{x - 3}$

26. $\lim_{t \to 2} \dfrac{t^2 - 4}{t - 2}$

27. $\lim_{x \to 3} \dfrac{x - 3}{x^2 - 9}$

28. $\lim_{x \to 0} \dfrac{x^2 - 2x}{x}$

29. $\lim_{x \to 4} \dfrac{x^2 - 9x + 20}{x^2 - 3x - 4}$

30. $\lim_{x \to -3} \dfrac{x^4 - 81}{x^2 + 8x + 15}$

31. $\lim_{x \to 2} \dfrac{3x^2 - x - 10}{x^2 + 5x - 14}$

32. $\lim_{x \to -4} \dfrac{x^2 + 2x - 8}{x^2 + 5x + 4}$

33. $\lim_{h \to 0} \dfrac{(2 + h)^2 - 2^2}{h}$

34. $\lim_{x \to 0} \dfrac{(x + 2)^2 - 4}{x}$

*35. Find $\lim_{h \to 0} \dfrac{(x + h)^2 - x^2}{h}$ by treating x as a constant.

36. Find $\lim_{h \to 0} \dfrac{3(x + h)^2 + 7(x + h) - 3x^2 - 7x}{h}$ by treating x as a constant.

In Problems 37–42, find $\lim_{h \to 0} \dfrac{f(x + h) - f(x)}{h}$.

37. $f(x) = 7 - 3x$

38. $f(x) = 2x + 3$

39. $f(x) = x^2 - 3$

40. $f(x) = x^2 + x + 1$

41. $f(x) = x^3 - 4x^2$

42. $f(x) = 3 - x + 4x^2$

43. Find $\lim\limits_{x \to 6} \dfrac{\sqrt{x-2}-2}{x-6}$ (*Hint:* First rationalize the numerator by multiplying both the numerator and denominator by $\sqrt{x-2}+2$.)

44. Find the constant c so that $\lim\limits_{x \to 3} \dfrac{x^2 + x + c}{x^2 - 5x + 6}$ exists. For that value of c, determine the limit. (*Hint:* Find the value of c for which $x - 3$ is a factor of the numerator.)

45. Power Plant The maximum theoretical efficiency of a power plant is given by

$$E = \frac{T_h - T_c}{T_h}$$

where T_h and T_c are the respective absolute temperatures of the hotter and colder reservoirs. Find (a) $\lim\limits_{T_c \to 0} E$ and (b) $\lim\limits_{T_c \to T_h} E$.

46. Satellite When a 3200-lb satellite revolves about the earth in a circular orbit of radius r ft, the total mechanical energy

E of the earth–satellite system is given by

$$E = -\frac{7.0 \times 10^{17}}{r} \text{ ft-lb}$$

Find the limit of E as $r \to 7.5 \times 10^7$ ft.

In Problems 47–50, use a graphing calculator to graph the functions, and then estimate the limits. Round your answers to two decimal places.

47. $\lim\limits_{x \to 2} \dfrac{x^4 + x^3 - 24}{x^2 - 4}$

48. $\lim\limits_{x \to 0} x^x$

49. $\lim\limits_{x \to 9} \dfrac{x - 10\sqrt{x} + 21}{3 - \sqrt{x}}$

50. $\lim\limits_{x \to 1} \dfrac{x^3 + x^2 - 5x + 3}{x^3 + 2x^2 - 7x + 4}$

51. Water Purification The cost of purifying water is given by $C = \dfrac{50{,}000}{p} - 6500$, where p is the percent of impurities remaining after purification. Graph this function on your graphing calculator, and determine $\lim\limits_{p \to 0} C$. Discuss what this means.

52. Profit Function The profit function for a certain business is given by $P(x) = 224x - 3.1x^2 - 800$. Graph this function on your graphing calculator, and use the evaluation function to determine $\lim\limits_{x \to 53.2} P(x)$, utilizing the rule about the limit of a polynomial function.

OBJECTIVE

To study one-sided limits, infinite limits, and limits at infinity.

10.2 Limits (Continued)

One-Sided Limits

Figure 10.14 shows the graph of a function f. Notice that $f(x)$ is not defined when $x = 0$. As x approaches 0 *from the right*, $f(x)$ approaches 1. We write this as

$$\lim_{x \to 0^+} f(x) = 1$$

On the other hand, as x approaches 0 *from the left*, $f(x)$ approaches -1, and we write

$$\lim_{x \to 0^-} f(x) = -1$$

Limits like these are called **one-sided limits.** From the preceding section, we know that the limit of a function as $x \to a$ is independent of the way x approaches a. Thus, the limit will exist if and only if both one-sided limits exist and are equal. We therefore conclude that

$$\lim_{x \to 0} f(x) \text{ does not exist}$$

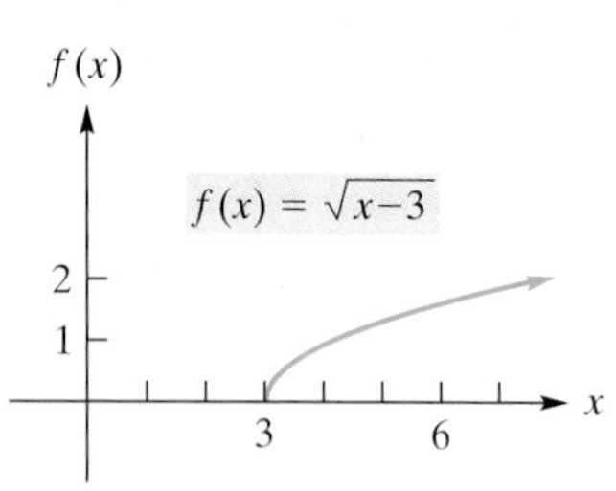

As another example of a one-sided limit, consider $f(x) = \sqrt{x-3}$ as x approaches 3. Since f is defined only when $x \geq 3$, we may speak of the limit as x approaches 3 from the right. If x is slightly greater than 3, then $x - 3$ is a positive number that is close to 0, so $\sqrt{x-3}$ is close to 0. We conclude that

$$\lim_{x \to 3^+} \sqrt{x-3} = 0$$

This limit is also evident from Figure 10.15.

Infinite Limits

In the previous section, we considered limits of the form 0/0—that is, limits where both the numerator and denominator approach 0. Now we will examine limits where the denominator approaches 0, but the numerator approaches a number different from 0. For example, consider

$$\lim_{x \to 0} \frac{1}{x^2}$$

FIGURE 10.14 $\lim_{x \to 0} f(x)$ does not exist.

$y = f(x)$

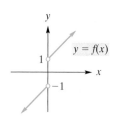

FIGURE 10.15
$\lim_{x \to 3^+} \sqrt{x-3} = 0$.

$f(x) = \sqrt{x-3}$

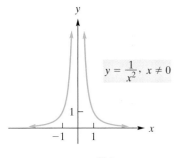

x	$f(x)$
±1	1
±0.5	4
±0.1	100
±0.01	10,000
±0.001	1,000,000

FIGURE 10.16 $\lim\limits_{x \to 0} \dfrac{1}{x^2} = \infty$.

CAUTION

The use of the "equality" sign in this situation does not mean that the limit exists. On the contrary, the symbolism here (∞) is a way of saying specifically that there is no limit, and it indicates why there is no limit.

Here, as x approaches 0, the denominator approaches 0 and the numerator approaches 1. Let us investigate the behavior of $f(x) = 1/x^2$ when x is close to 0. The number x^2 is positive and also close to 0. Thus, dividing 1 by such a number results in a very large number. In fact, the closer x is to 0, the larger the value of $f(x)$. For example, see the table of values in Figure 10.16, which also shows the graph of f. Clearly, as $x \to 0$ both from the left and from the right, $f(x)$ increases without bound. Hence, no limit exists at 0. We say that as $x \to 0$, $f(x)$ becomes positively infinite, and symbolically we express this "infinite limit" by writing

$$\lim_{x \to 0} \frac{1}{x^2} = \infty$$

If $\lim_{x \to a} f(x)$ does not exist, it may be for a reason other than that the values $f(x)$ become arbitrarily large as x gets close to a. For example, look again at the situation in Example 2 of Section 10.1. Here we have

$$\lim_{x \to -2} f(x) \text{ does not exist, but } \lim_{x \to -2} f(x) \neq \infty$$

Consider now the graph of $y = f(x) = 1/x$ for $x \neq 0$. (See Figure 10.17.) As x approaches 0 from the right, $1/x$ becomes positively infinite; as x approaches 0 from the left, $1/x$ becomes negatively infinite. Symbolically, these infinite limits are written

$$\lim_{x \to 0^+} \frac{1}{x} = \infty \quad \text{and} \quad \lim_{x \to 0^-} \frac{1}{x} = -\infty$$

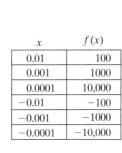

x	$f(x)$
0.01	100
0.001	1000
0.0001	10,000
−0.01	−100
−0.001	−1000
−0.0001	−10,000

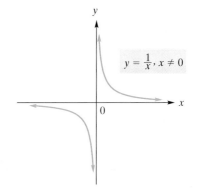

FIGURE 10.17 $\lim\limits_{x \to 0} \dfrac{1}{x}$ does not exist.

Either one of these facts implies that

$$\lim_{x \to 0} \frac{1}{x} \text{ does not exist}$$

● **EXAMPLE 1 Infinite Limits**

Find the limit (if it exists).

a. $\lim\limits_{x \to -1^+} \dfrac{2}{x + 1}$

Solution: As x approaches -1 from the right (think of values of x such as -0.9, -0.99, and so on, as shown in Figure 10.18), $x + 1$ approaches 0, but is always positive. Since we are dividing 2 by positive numbers approaching 0, the results, $2/(x+1)$, are positive numbers that are becoming arbitrarily large. Thus,

$$\lim_{x \to -1^+} \frac{2}{x + 1} = \infty$$

and the limit does not exist. By a similar analysis, you should be able to show that

$$\lim_{x \to -1^-} \frac{2}{x + 1} = -\infty$$

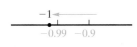

FIGURE 10.18 $x \to -1^+$.

b. $\displaystyle\lim_{x\to 2}\frac{x+2}{x^2-4}$

Solution: As $x \to 2$, the numerator approaches 4 and the denominator approaches 0. Hence, we are dividing numbers near 4 by numbers near 0. The results are numbers that become arbitrarily large in magnitude. At this stage, we can write

$$\lim_{x\to 2}\frac{x+2}{x^2-4} \text{ does not exist}$$

However, let us see if we can use the symbol ∞ or $-\infty$ to be more specific about "does not exist." Notice that

$$\lim_{x\to 2}\frac{x+2}{x^2-4} = \lim_{x\to 2}\frac{x+2}{(x+2)(x-2)} = \lim_{x\to 2}\frac{1}{x-2}$$

Since

$$\lim_{x\to 2^+}\frac{1}{x-2} = \infty \quad \text{and} \quad \lim_{x\to 2^-}\frac{1}{x-2} = -\infty$$

$\displaystyle\lim_{x\to 2}\frac{x+2}{x^2-4}$ is neither ∞ nor $-\infty$.

NOW WORK PROBLEM 31

Example 1 considered limits of the form $k/0$, where $k \neq 0$. It is important that you distinguish the form $k/0$ from the form $0/0$, which was discussed in Section 10.1. These two forms are handled quite differently.

EXAMPLE 2   Finding a Limit

Find $\displaystyle\lim_{t\to 2}\frac{t-2}{t^2-4}$.

Solution: As $t \to 2$, *both* numerator and denominator approach 0 (form 0/0). Thus, we first simplify the fraction, for $t \neq 2$, as we did in Section 10.1, and then take the limit:

$$\lim_{t\to 2}\frac{t-2}{t^2-4} = \lim_{t\to 2}\frac{t-2}{(t+2)(t-2)} = \lim_{t\to 2}\frac{1}{t+2} = \frac{1}{4}$$

NOW WORK PROBLEM 37

Limits at Infinity

Now let us examine the function

$$f(x) = \frac{1}{x}$$

as x becomes infinite, first in a positive sense and then in a negative sense. From Table 10.2, you can see that as x increases without bound through positive values, the values of $f(x)$ approach 0. Likewise, as x decreases without bound through negative values, the values of $f(x)$ also approach 0. These observations are also apparent from the graph in Figure 10.17. There, as you move to the right along the curve through positive x-values, the corresponding y-values approach 0. Similarly, as you move

You should be able to obtain

$$\lim_{x\to\infty}\frac{1}{x} \quad \text{and} \quad \lim_{x\to-\infty}\frac{1}{x}$$

without the benefit of a graph or a table. Dividing 1 by a large positive number results in a small positive number, and as the divisors get arbitrarily large, the quotients get arbitrarily small. A similar argument can be made for the limit as $x \to -\infty$.

TABLE 10.2 Behavior of $f(x)$ as $x \to \pm\infty$

x	$f(x)$	x	$f(x)$
1000	0.001	-1000	-0.001
10,000	0.0001	$-10,000$	-0.0001
100,000	0.00001	$-100,000$	-0.00001
1,000,000	0.000001	$-1,000,000$	-0.000001

to the left along the curve through negative x-values, the corresponding y-values approach 0. Symbolically, we write

$$\lim_{x \to \infty} \frac{1}{x} = 0 \quad \text{and} \quad \lim_{x \to -\infty} \frac{1}{x} = 0$$

Both of these limits are called *limits at infinity*.

● EXAMPLE 3   **Limits at Infinity**

Find the limit (if it exists).

a. $\displaystyle\lim_{x \to \infty} \frac{4}{(x-5)^3}$

Solution: As x becomes very large, so does $x - 5$. Since the cube of a large number is also large, $(x - 5)^3 \to \infty$. Dividing 4 by very large numbers results in numbers near 0. Thus,

$$\lim_{x \to \infty} \frac{4}{(x-5)^3} = 0$$

b. $\displaystyle\lim_{x \to -\infty} \sqrt{4 - x}$

Solution: As x gets negatively infinite, $4-x$ becomes positively infinite. Because square roots of large numbers are large numbers, we conclude that

$$\lim_{x \to -\infty} \sqrt{4 - x} = \infty$$

In our next discussion we will need a certain limit, namely, $\lim_{x \to \infty} 1/x^p$, where $p > 0$. As x becomes very large, so does x^p. Dividing 1 by very large numbers results in numbers near 0. Thus, $\lim_{x \to \infty} 1/x^p = 0$. In general,

$$\lim_{x \to \infty} \frac{1}{x^p} = 0 \quad \text{and} \quad \lim_{x \to -\infty} \frac{1}{x^p} = 0$$

where $p > 0$.[2] For example,

$$\lim_{x \to \infty} \frac{1}{\sqrt[3]{x}} = \lim_{x \to \infty} \frac{1}{x^{1/3}} = 0$$

Let us now find the limit of the rational function

$$f(x) = \frac{4x^2 + 5}{2x^2 + 1}$$

as $x \to \infty$. (Recall from Section 2.2 that a rational function is a quotient of polynomials.) As x gets larger and larger, *both* the numerator and denominator of $f(x)$ become infinite. However, the form of the quotient can be changed, so that we can draw a conclusion as to whether or not it has a limit. To do this, we divide both the numerator and denominator by the greatest power of x that occurs in the denominator. Here it is x^2. This gives

$$\lim_{x \to \infty} \frac{4x^2 + 5}{2x^2 + 1} = \lim_{x \to \infty} \frac{\dfrac{4x^2 + 5}{x^2}}{\dfrac{2x^2 + 1}{x^2}} = \lim_{x \to \infty} \frac{\dfrac{4x^2}{x^2} + \dfrac{5}{x^2}}{\dfrac{2x^2}{x^2} + \dfrac{1}{x^2}}$$

$$= \lim_{x \to \infty} \frac{4 + \dfrac{5}{x^2}}{2 + \dfrac{1}{x^2}} = \frac{\displaystyle\lim_{x \to \infty} 4 + 5 \cdot \lim_{x \to \infty} \frac{1}{x^2}}{\displaystyle\lim_{x \to \infty} 2 + \lim_{x \to \infty} \frac{1}{x^2}}$$

---

[2]For $\lim_{x \to -\infty} 1/x^p$, we assume that p is such that $1/x^p$ is defined for $x < 0$.

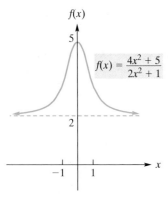

$f(x)$

$$f(x) = \frac{4x^2 + 5}{2x^2 + 1}$$

FIGURE 10.19
$\lim_{x \to \infty} f(x) = 2$ and
$\lim_{x \to -\infty} f(x) = 2$.

Since $\lim_{x \to \infty} 1/x^p = 0$ for $p > 0$,

$$\lim_{x \to \infty} \frac{4x^2 + 5}{2x^2 + 1} = \frac{4 + 5(0)}{2 + 0} = \frac{4}{2} = 2$$

Similarly, the limit as $x \to -\infty$ is 2. These limits are clear from the graph of f in Figure 10.19.

For the preceding function, there is an easier way to find $\lim_{x \to \infty} f(x)$. For *large* values of x, in the numerator the term involving the greatest power of x, namely, $4x^2$, dominates the sum $4x^2 + 5$, and the dominant term in the denominator, $2x^2 + 1$, is $2x^2$. Thus, as $x \to \infty$, $f(x)$ can be approximated by $(4x^2)/(2x^2)$. As a result, to determine the limit of $f(x)$, it suffices to determine the limit of $(4x^2)/(2x^2)$. That is,

$$\lim_{x \to \infty} \frac{4x^2 + 5}{2x^2 + 1} = \lim_{x \to \infty} \frac{4x^2}{2x^2} = \lim_{x \to \infty} 2 = 2$$

as we saw before. In general, we have the following rule:

Limits at Infinity for Rational Functions

If $f(x)$ is a *rational function* and $a_n x^n$ and $b_m x^m$ are the terms in the numerator and denominator, respectively, with the greatest powers of x, then

$$\lim_{x \to \infty} f(x) = \lim_{x \to \infty} \frac{a_n x^n}{b_m x^m}$$

and

$$\lim_{x \to -\infty} f(x) = \lim_{x \to -\infty} \frac{a_n x^n}{b_m x^m}$$

Let us apply this rule to the situation where the degree of the numerator is greater than the degree of the denominator. For example,

$$\lim_{x \to -\infty} \frac{x^4 - 3x}{5 - 2x} = \lim_{x \to -\infty} \frac{x^4}{-2x} = \lim_{x \to -\infty} \left(-\frac{1}{2}x^3\right) = \infty$$

(Note that in the next-to-last step, as x becomes very negative, so does x^3; moreover, $-\frac{1}{2}$ times a very negative number is very positive.) Similarly,

$$\lim_{x \to \infty} \frac{x^4 - 3x}{5 - 2x} = \lim_{x \to \infty} \left(-\frac{1}{2}x^3\right) = -\infty$$

From this illustration, we make the following conclusion:

If the degree of the numerator of a *rational function* is greater than the degree of the denominator, then the function has no limit as $x \to \infty$ and no limit as $x \to -\infty$.

LIMITS AT INFINITY FOR RATIONAL FUNCTIONS

The yearly amount of sales y of a certain company (in thousands of dollars) is related to the amount the company spends on advertising, x (in thousands of dollars), according to the equation $y(x) = \frac{500x}{x + 20}$. Graph this function on your graphing calculator in the window $[0, 1000] \times [0, 550]$. Use TRACE to explore $\lim_{x \to \infty} y(x)$, and determine what this means to the company.

● **EXAMPLE 4   Limits at Infinity for Rational Functions**

Find the limit (if it exists).

a. $\lim_{x \to \infty} \dfrac{x^2 - 1}{7 - 2x + 8x^2}$

Solution:

$$\lim_{x \to \infty} \frac{x^2 - 1}{7 - 2x + 8x^2} = \lim_{x \to \infty} \frac{x^2}{8x^2} = \lim_{x \to \infty} \frac{1}{8} = \frac{1}{8}$$

b. $\lim_{x \to -\infty} \dfrac{x}{(3x - 1)^2}$

Solution:

$$\lim_{x \to -\infty} \frac{x}{(3x - 1)^2} = \lim_{x \to -\infty} \frac{x}{9x^2 - 6x + 1} = \lim_{x \to -\infty} \frac{x}{9x^2}$$

$$= \lim_{x \to -\infty} \frac{1}{9x} = \frac{1}{9} \cdot \lim_{x \to -\infty} \frac{1}{x} = \frac{1}{9}(0) = 0$$

c. $\lim\limits_{x\to\infty} \dfrac{x^5 - x^4}{x^4 - x^3 + 2}$

Solution: Since the degree of the numerator is greater than that of the denominator, there is no limit. More precisely,

$$\lim_{x\to\infty} \frac{x^5 - x^4}{x^4 - x^3 + 2} = \lim_{x\to\infty} \frac{x^5}{x^4} = \lim_{x\to\infty} x = \infty$$

NOW WORK PROBLEM 21 ●●●

CAUTION

The preceding technique applies only to limits of rational functions at *infinity*.

To find $\lim\limits_{x\to 0} \dfrac{x^2 - 1}{7 - 2x + 8x^2}$, we cannot simply determine the limit of $\dfrac{x^2}{8x^2}$. That simplification applies only in case $x \to \infty$ or $x \to -\infty$. Instead, we have

$$\lim_{x\to 0} \frac{x^2 - 1}{7 - 2x + 8x^2} = \frac{0 - 1}{7 - 0 + 0} = -\frac{1}{7}$$

Let us now consider the limit of the polynomial function $f(x) = 8x^2 - 2x$ as $x \to \infty$:

$$\lim_{x\to\infty} (8x^2 - 2x)$$

Because a polynomial is a rational function with denominator 1, we have

$$\lim_{x\to\infty} (8x^2 - 2x) = \lim_{x\to\infty} \frac{8x^2 - 2x}{1} = \lim_{x\to\infty} \frac{8x^2}{1} = \lim_{x\to\infty} 8x^2$$

That is, the limit of $8x^2 - 2x$ as $x \to \infty$ is the same as the limit of the term involving the greatest power of x, namely, $8x^2$. As x becomes very large, so does $8x^2$. Thus,

$$\lim_{x\to\infty} (8x^2 - 2x) = \lim_{x\to\infty} 8x^2 = \infty$$

In general, we have the following:

As $x \to \infty$ (or $x \to -\infty$), the limit of a *polynomial function* is the same as the limit of its term that involves the greatest power of x.

LIMITS AT INFINITY FOR POLYNOMIAL FUNCTIONS

The cost C of producing x units of a certain product is given by $C(x) = 50{,}000 + 200x + 0.3x^2$. Use your graphing calculator to explore $\lim\limits_{x\to\infty} C(x)$ and determine what this means.

● **EXAMPLE 5   Limits at Infinity for Polynomial Functions**

a. $\lim\limits_{x\to-\infty}(x^3 - x^2 + x - 2) = \lim\limits_{x\to-\infty} x^3$. As x becomes very negative, so does x^3. Thus,

$$\lim_{x\to-\infty} (x^3 - x^2 + x - 2) = \lim_{x\to-\infty} x^3 = -\infty$$

b. $\lim\limits_{x\to-\infty}(-2x^3 + 9x) = \lim\limits_{x\to-\infty} -2x^3 = \infty$, because -2 times a very negative number is very positive.

NOW WORK PROBLEM 9 ●●●

The technique of focusing on dominant terms to find limits as $x \to \infty$ or $x \to -\infty$ is valid for *rational functions*, but it is not necessarily valid for other types of functions. For example, consider

Do not use dominant terms when a function is not rational.

$$\lim_{x\to\infty} \left(\sqrt{x^2 + x} - x \right) \tag{1}$$

Notice that $\sqrt{x^2 + x} - x$ is not a rational function. It is *incorrect* to infer that because x^2 dominates in $x^2 + x$, the limit in (1) is the same as

$$\lim_{x\to\infty} \left(\sqrt{x^2} - x \right) = \lim_{x\to\infty} (x - x) = \lim_{x\to\infty} 0 = 0$$

It can be shown (see Problem 62) that the limit in (1) is not 0, but is $\frac{1}{2}$.

The ideas discussed in this section will now be applied to a case-defined function.

PRINCIPLES IN PRACTICE 4

LIMITS FOR A CASE-DEFINED FUNCTION

A plumber charges $100 for the first hour of work at your house and $75 for every hour (or fraction thereof) afterward. The function for what an x-hour visit will cost you is

$$f(x) = \begin{cases} \$100 & \text{if } 0 < x \le 1 \\ \$175 & \text{if } 1 < x \le 2 \\ \$250 & \text{if } 2 < x \le 3 \\ \$325 & \text{if } 3 < x \le 4 \end{cases}$$

Find $\lim_{x \to 1} f(x)$ and $\lim_{x \to 2.5} f(x)$.

● EXAMPLE 6   Limits for a Case-Defined Function

If $f(x) = \begin{cases} x^2 + 1 & \text{if } x \ge 1 \\ 3 & \text{if } x < 1 \end{cases}$, *find the limit (if it exists).*

a. $\lim_{x \to 1^+} f(x)$

Solution: Here x gets close to 1 from the right. For $x > 1$, we have $f(x) = x^2 + 1$. Thus,

$$\lim_{x \to 1^+} f(x) = \lim_{x \to 1^+} (x^2 + 1)$$

If x is greater than 1, but close to 1, then $x^2 + 1$ is close to 2. Therefore,

$$\lim_{x \to 1^+} f(x) = \lim_{x \to 1^+} (x^2 + 1) = 2$$

b. $\lim_{x \to 1^-} f(x)$

Solution: Here x gets close to 1 from the left. For $x < 1$, $f(x) = 3$. Hence,

$$\lim_{x \to 1^-} f(x) = \lim_{x \to 1^-} 3 = 3$$

c. $\lim_{x \to 1} f(x)$

Solution: We want the limit as x approaches 1. However, the rule of the function depends on whether $x \ge 1$ or $x < 1$. Thus, we must consider one-sided limits. The limit as x approaches 1 will exist if and only if both one-sided limits exist and are the same. From parts (a) and (b),

$$\lim_{x \to 1^+} f(x) \ne \lim_{x \to 1^-} f(x) \qquad \text{since } 2 \ne 3$$

Therefore,

$$\lim_{x \to 1} f(x) \qquad \text{does not exist}$$

d. $\lim_{x \to \infty} f(x)$

Solution: For very large values of x, we have $x \ge 1$, so $f(x) = x^2 + 1$. Thus,

$$\lim_{x \to \infty} f(x) = \lim_{x \to \infty} (x^2 + 1) = \lim_{x \to \infty} x^2 = \infty$$

e. $\lim_{x \to -\infty} f(x)$

Solution: For very negative values of x, we have $x < 1$, so $f(x) = 3$. Hence,

$$\lim_{x \to -\infty} f(x) = \lim_{x \to -\infty} 3 = 3$$

All the limits in parts (a) through (c) should be obvious from the graph of f in Figure 10.20.

NOW WORK PROBLEM 57 ●●●

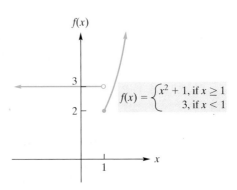

FIGURE 10.20   Graph of case-defined function.

Problems 10.2

1. For the function f given in Figure 10.21, find the following limits. If the limit does not exist, so state that, or use the symbol ∞ or $-\infty$ where appropriate.

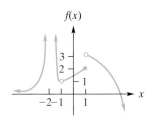

FIGURE 10.21 Diagram for Problem 1.

(a) $\lim_{x \to 1^-} f(x)$ **(b)** $\lim_{x \to 1^+} f(x)$

(c) $\lim_{x \to 1} f(x)$ **(d)** $\lim_{x \to \infty} f(x)$

(e) $\lim_{x \to -2^-} f(x)$ **(f)** $\lim_{x \to -2^+} f(x)$

(g) $\lim_{x \to -2} f(x)$ **(h)** $\lim_{x \to -\infty} f(x)$

(i) $\lim_{x \to -1^-} f(x)$ **(j)** $\lim_{x \to -1^+} f(x)$

(k) $\lim_{x \to -1} f(x)$

2. For the function f given in Figure 10.22, find the following limits. If the limit does not exist, so state that, or use the symbol ∞ or $-\infty$ where appropriate.

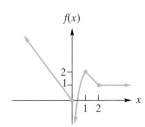

FIGURE 10.22 Diagram for Problem 2.

(a) $\lim_{x \to 0^-} f(x)$ **(b)** $\lim_{x \to 0^+} f(x)$

(c) $\lim_{x \to 0} f(x)$ **(d)** $\lim_{x \to -\infty} f(x)$

(e) $\lim_{x \to 1} f(x)$ **(f)** $\lim_{x \to \infty} f(x)$

(g) $\lim_{x \to 2^+} f(x)$

In each of Problems 3–54, find the limit. If the limit does not exist, so state that, or use the symbol ∞ or $-\infty$ where appropriate.

3. $\lim_{x \to 3^+} (x - 2)$ **4.** $\lim_{x \to -1^+} (1 - x^2)$

5. $\lim_{x \to -\infty} 5x$ **6.** $\lim_{x \to -\infty} 19$

7. $\lim_{x \to 0^-} \dfrac{6x}{x^4}$ **8.** $\lim_{x \to 2} \dfrac{7}{x - 1}$

*9. $\lim_{x \to -\infty} x^2$ **10.** $\lim_{t \to \infty} (t - 1)^3$

11. $\lim_{h \to 0^+} \sqrt{h}$ **12.** $\lim_{h \to 5^-} \sqrt{5 - h}$

13. $\lim_{x \to -2^-} \dfrac{-3}{x + 2}$ **14.** $\lim_{x \to 0^-} 2^{1/2}$

15. $\lim_{x \to 1^+} \left(4\sqrt{x - 1}\right)$ **16.** $\lim_{x \to 2^+} \left(x\sqrt{x^2 - 4}\right)$

17. $\lim_{x \to \infty} \sqrt{x + 10}$ **18.** $\lim_{x \to -\infty} -\sqrt{1 - 10x}$

19. $\lim_{x \to \infty} \dfrac{3}{\sqrt{x}}$ **20.** $\lim_{x \to \infty} \dfrac{-6}{5x\sqrt[3]{x}}$

*21. $\lim_{x \to \infty} \dfrac{x + 8}{x - 3}$ **22.** $\lim_{x \to \infty} \dfrac{2x - 4}{3 - 2x}$

23. $\lim_{x \to -\infty} \dfrac{x^2 - 1}{x^3 + 4x - 3}$ **24.** $\lim_{r \to \infty} \dfrac{r^3}{r^2 + 1}$

25. $\lim_{t \to \infty} \dfrac{3t^3 + 2t^2 + 9t - 1}{5t^2 - 5}$ **26.** $\lim_{x \to \infty} \dfrac{5x}{3x^7 - x^3 + 4}$

27. $\lim_{x \to \infty} \dfrac{7}{2x + 1}$ **28.** $\lim_{x \to -\infty} \dfrac{2}{(4x - 1)^3}$

29. $\lim_{x \to \infty} \dfrac{3 - 4x - 2x^3}{5x^3 - 8x + 1}$ **30.** $\lim_{x \to -\infty} \dfrac{3 - 2x - 2x^3}{7 - 5x^3 + 2x^2}$

*31. $\lim_{x \to 3^-} \dfrac{x + 3}{x^2 - 9}$ **32.** $\lim_{x \to -3^-} \dfrac{3x}{9 - x^2}$

33. $\lim_{w \to \infty} \dfrac{2w^2 - 3w + 4}{5w^2 + 7w - 1}$ **34.** $\lim_{x \to \infty} \dfrac{4 - 3x^3}{x^3 - 1}$

35. $\lim_{x \to \infty} \dfrac{6 - 4x^2 + x^3}{4 + 5x - 7x^2}$ **36.** $\lim_{x \to -\infty} \dfrac{3x - x^3}{x^3 + x + 1}$

*37. $\lim_{x \to -3^-} \dfrac{5x^2 + 14x - 3}{x^2 + 3x}$ **38.** $\lim_{t \to 3} \dfrac{t^2 - 4t + 3}{t^2 - 2t - 3}$

39. $\lim_{x \to 1} \dfrac{x^2 - 3x + 1}{x^2 + 1}$ **40.** $\lim_{x \to -1} \dfrac{3x^3 - x^2}{2x + 1}$

41. $\lim_{x \to 1^+} \left(1 + \dfrac{1}{x - 1}\right)$ **42.** $\lim_{x \to -\infty} -\dfrac{x^5 + 2x^3 - 1}{x^5 - 4x^2}$

43. $\lim_{x \to -7^-} \dfrac{x^2 + 1}{\sqrt{x^2 - 49}}$ **44.** $\lim_{x \to -2^+} \dfrac{x}{\sqrt{16 - x^4}}$

45. $\lim_{x \to 0^+} \dfrac{5}{x + x^2}$ **46.** $\lim_{x \to \infty} \left(x + \dfrac{1}{x}\right)$

47. $\lim_{x \to 1} x(x - 1)^{-1}$ **48.** $\lim_{x \to 1/2} \dfrac{1}{2x - 1}$

49. $\lim_{x \to 1^+} \left(\dfrac{-5}{1 - x}\right)$ **50.** $\lim_{x \to 3} \left(-\dfrac{7}{x - 3}\right)$

51. $\lim_{x \to 0} |x|$ **52.** $\lim_{x \to 0} \left|\dfrac{1}{x}\right|$

53. $\lim_{x \to -\infty} \dfrac{x + 1}{x}$ **54.** $\lim_{x \to \infty} \left(\dfrac{3}{x} - \dfrac{2x^2}{x^2 + 1}\right)$

In Problems 55–58, find the indicated limits. If the limit does not exist, so state that, or use the symbol ∞ or $-\infty$ where appropriate.

55. $f(x) = \begin{cases} 2 & \text{if } x \le 2 \\ 1 & \text{if } x > 2 \end{cases}$

(a) $\lim_{x \to 2^+} f(x)$ **(b)** $\lim_{x \to 2^-} f(x)$

(c) $\lim_{x \to 2} f(x)$ **(d)** $\lim_{x \to \infty} f(x)$

(e) $\lim_{x \to -\infty} f(x)$

56. $f(x) = \begin{cases} x & \text{if } x \le 2 \\ -2 + 4x - x^2 & \text{if } x > 2 \end{cases}$

(a) $\lim_{x \to 2^+} f(x)$ **(b)** $\lim_{x \to 2^-} f(x)$

(c) $\lim_{x \to 2} f(x)$ **(d)** $\lim_{x \to \infty} f(x)$

(e) $\lim_{x \to -\infty} f(x)$

*57. $g(x) = \begin{cases} x & \text{if } x < 0 \\ -x & \text{if } x > 0 \end{cases}$

(a) $\lim_{x \to 0^+} g(x)$ **(b)** $\lim_{x \to 0^-} g(x)$

(c) $\lim_{x \to 0} g(x)$ **(d)** $\lim_{x \to \infty} g(x)$

(e) $\lim_{x \to -\infty} g(x)$

58. $g(x) = \begin{cases} x^2 & \text{if } x < 0 \\ -x & \text{if } x > 0 \end{cases}$

(a) $\lim_{x \to 0^+} g(x)$ **(b)** $\lim_{x \to 0^-} g(x)$

(c) $\lim_{x \to 0} g(x)$ **(d)** $\lim_{x \to \infty} g(x)$

(e) $\lim_{x \to -\infty} g(x)$

59. Average Cost If c is the total cost in dollars to produce q units of a product, then the average cost per unit for an output of q units is given by $\bar{c} = c/q$. Thus, if the total cost equation is $c = 5000 + 6q$, then

$$\bar{c} = \frac{5000}{q} + 6$$

For example, the total cost of an output of 5 units is $5030, and the average cost per unit at this level of production is $1006. By finding $\lim_{q \to \infty} \bar{c}$, show that the average cost approaches a level of stability if the producer continually increases output. What is the limiting value of the average cost? Sketch the graph of the average-cost function.

60. Average Cost Repeat Problem 59, given that the fixed cost is $12,000 and the variable cost is given by the function $c_v = 7q$.

61. Population The population of a certain small city t years from now is predicted to be

$$N = 50{,}000 - \frac{2000}{t+1}$$

Find the population in the long run; that is, find $\lim_{t \to \infty} N$.

62. Show that

$$\lim_{x \to \infty} \left(\sqrt{x^2 + x} - x \right) = \frac{1}{2}$$

(*Hint:* Rationalize the numerator by multiplying the expression $\sqrt{x^2 + x} - x$ by

$$\frac{\sqrt{x^2 + x} + x}{\sqrt{x^2 + x} + x}$$

Then express the denominator in a form such that x is a factor.)

63. Host–Parasite Relationship For a particular host–parasite relationship, it was determined that when the host density (number of hosts per unit of area) is x, the number of hosts parasitized over a period of time is

$$y = \frac{900x}{10 + 45x}$$

If the host density were to increase without bound, what value would y approach?

64. If $f(x) = \begin{cases} \sqrt{2-x} & \text{if } x < 2 \\ x^3 + k(x+1) & \text{if } x \geq 2 \end{cases}$, determine the value of the constant k for which $\lim_{x \to 2} f(x)$ exists.

In Problems 65 and 66, use a calculator to evaluate the given function when $x = 1, 0.5, 0.2, 0.1, 0.01, 0.001,$ and 0.0001. From your results, speculate about $\lim_{x \to 0^+} f(x)$.

65. $f(x) = x^{2x}$ **66.** $f(x) = e^{-1/x}$

67. Graph $f(x) = \sqrt{4x^2 - 1}$. Use the graph to estimate $\lim_{x \to 1/2^+} f(x)$.

68. Graph $f(x) = \dfrac{\sqrt{x^2 - 9}}{x + 3}$. Use the graph to estimate $\lim_{x \to -3^-} f(x)$ if it exists. Use the symbol ∞ or $-\infty$ if appropriate.

69. Graph $f(x) = \begin{cases} 2x^2 + 3 & \text{if } x < 2 \\ 2x + 5 & \text{if } x \geq 2 \end{cases}$. Use the graph to estimate each of the following limits if it exists:

(a) $\lim_{x \to 2^-} f(x)$ **(b)** $\lim_{x \to 2^+} f(x)$ **(c)** $\lim_{x \to 2} f(x)$

OBJECTIVE

To study continuity and to find points of discontinuity for a function.

10.3 Continuity

Many functions have the property that there is no "break" in their graphs. For example, compare the functions

$$f(x) = x \quad \text{and} \quad g(x) = \begin{cases} x & \text{if } x \neq 1 \\ 2 & \text{if } x = 1 \end{cases}$$

whose graphs appear in Figures 10.23 and 10.24, respectively. The graph of f is unbroken, but the graph of g has a break at $x = 1$. Stated another way, if you were to trace both graphs with a pencil, you would have to lift the pencil off the graph of g when $x = 1$, but you would not have to lift it off the graph of f. These situations can be expressed by limits. As x approaches 1, compare the limit of each function with the value of the function at $x = 1$:

$$\lim_{x \to 1} f(x) = 1 = f(1)$$

whereas

$$\lim_{x \to 1} g(x) = 1 \neq 2 = g(1)$$

FIGURE 10.23
Continuous at 1.

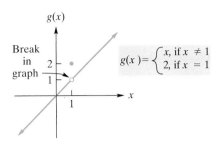

FIGURE 10.24   Discontinuous at 1.

In Section 10.1 we stressed that given a function f and a number a, there are two important ways to associate a number to the pair (f, a). One is simple evaluation, $f(a)$, which *exists* precisely if a is in the domain of f. The other is $\lim_{x \to a} f(x)$, whose existence and determination can be more challenging. For the functions f and g above, the limit of f as $x \to 1$ is the same as $f(1)$, but the limit of g as $x \to 1$ is *not* the same as $g(1)$. For these reasons, we say that f is *continuous* at 1 and g is *discontinuous* at 1.

DEFINITION

A function f is ***continuous*** at a if and only if the following three conditions are met:

1. $f(a)$ exists
2. $\lim_{x \to a} f(x)$ exists
3. $\lim_{x \to a} f(x) = f(a)$

If f is not continuous at a, then f is said to be ***discontinuous*** at a, and a is called a ***point of discontinuity*** of f.

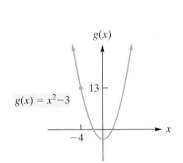

FIGURE 10.25 f is continuous at 7.

● EXAMPLE 1   Applying the Definition of Continuity

a. *Show that $f(x) = 5$ is continuous at 7.*

Solution: We must verify that the preceding three conditions are met. First, $f(7) = 5$, so f is defined at $x = 7$. Second,

$$\lim_{x \to 7} f(x) = \lim_{x \to 7} 5 = 5$$

Thus, f has a limit as $x \to 7$. Third,

$$\lim_{x \to 7} f(x) = 5 = f(7)$$

Therefore, f is continuous at 7. (See Figure 10.25.)

b. *Show that $g(x) = x^2 - 3$ is continuous at -4.*

Solution: The function g is defined at $x = -4 : g(-4) = 13$. Also,

$$\lim_{x \to -4} g(x) = \lim_{x \to -4} (x^2 - 3) = 13 = g(-4)$$

Therefore, g is continuous at -4. (See Figure 10.26.)

NOW WORK PROBLEM 1 ●●

FIGURE 10.26 g is continuous at -4.

We say that a function is *continuous on an interval* if it is continuous at each point there. In this situation, the graph of the function is connected over the interval. For example, $f(x) = x^2$ is continuous on the interval $[2, 5]$. In fact, in Example 5 of Section 10.1, we showed that, for *any* polynomial function f, for any number a, $\lim_{x \to a} f(x) = f(a)$. This means that

A polynomial function is continuous at every point.

It follows that such a function is continuous on every interval. We say that a function is **continuous on its domain** if it is continuous at each point in its domain. If the domain of such a function is the set of all real numbers, we may simply say that the function is continuous.

● EXAMPLE 2   **Continuity of Polynomial Functions**

The functions $f(x) = 7$ and $g(x) = x^2 - 9x + 3$ are polynomial functions. Therefore, they are continuous on their domains. For example, they are continuous at 3.

NOW WORK PROBLEM 13

When is a function discontinuous? We can say that a function f defined on an open interval containing a is discontinuous at a if

1. f has no limit as $x \to a$

   or

2. as $x \to a$, f has a limit that is different from $f(a)$

If f is not defined at a, we will say also, in that case, that f is discontinuous at a. In Figure 10.27, we can find points of discontinuity by inspection.

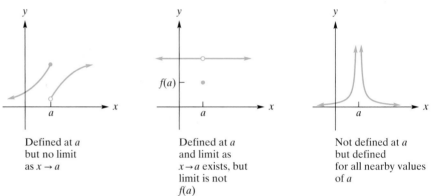

FIGURE 10.27   Discontinuities at a.

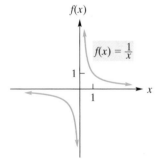

FIGURE 10.28   Infinite discontinuity at 0.

● EXAMPLE 3   **Discontinuities**

a. Let $f(x) = 1/x$. (See Figure 10.28.) Note that f is not defined at $x = 0$, but it is defined for all other x nearby. Thus, f is discontinuous at 0. Moreover, $\lim_{x \to 0^+} f(x) = \infty$ and $\lim_{x \to 0^-} f(x) = -\infty$. A function is said to have an **infinite discontinuity** at a when at least one of the one-sided limits is either ∞ or $-\infty$ as $x \to a$. Hence, f has an *infinite discontinuity* at $x = 0$.

b. Let $f(x) = \begin{cases} 1 & \text{if } x > 0 \\ 0 & \text{if } x = 0 \\ -1 & \text{if } x < 0 \end{cases}$.

(See Figure 10.29.) Although f is defined at $x = 0$, $\lim_{x \to 0} f(x)$ does not exist. Thus, f is discontinuous at 0.

NOW WORK PROBLEM 29

The following property indicates where the discontinuities of a rational function occur:

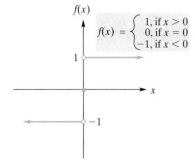

FIGURE 10.29   Discontinuous case-defined function.

Discontinuities of a Rational Function

A rational function is discontinuous at points where the denominator is 0 and is continuous otherwise. Thus, a rational function is continuous on its domain.

● **EXAMPLE 4 Locating Discontinuities in Rational Functions**

For each of the following functions, find all points of discontinuity.

a. $f(x) = \dfrac{x^2 - 3}{x^2 + 2x - 8}$

Solution: This rational function has denominator

$$x^2 + 2x - 8 = (x + 4)(x - 2)$$

which is 0 when $x = -4$ or $x = 2$. Thus, f is discontinuous only at -4 and 2.

b. $h(x) = \dfrac{x + 4}{x^2 + 4}$

Solution: For this rational function, the denominator is never 0. (It is always positive.) Therefore, h has no discontinuity.

NOW WORK PROBLEM 19 ●●●

The rational function $f(x) = \dfrac{x+1}{x+1}$ is continuous on its domain but it is not defined at -1. It is discontinuous at -1. The graph of f is a horizontal straight line with a "hole" in it at -1.

● **EXAMPLE 5 Locating Discontinuities in Case-Defined Functions**

For each of the following functions, find all points of discontinuity.

a. $f(x) = \begin{cases} x + 6 & \text{if } x \geq 3 \\ x^2 & \text{if } x < 3 \end{cases}$

Solution: The cases defining the function are given by polynomials, which are continuous, so the only possible place for a discontinuity is at $x = 3$, where the separation of cases occurs. We know that $f(3) = 3 + 6 = 9$. So because

$$\lim_{x \to 3^+} f(x) = \lim_{x \to 3^+} (x + 6) = 9$$

and

$$\lim_{x \to 3^-} f(x) = \lim_{x \to 3^-} x^2 = 9$$

we can conclude that $\lim_{x \to 3} f(x) = 9 = f(3)$ and the function has no points of discontinuity. We can reach the same conclusion by inspecting the graph of f in Figure 10.30.

b. $f(x) = \begin{cases} x + 2 & \text{if } x > 2 \\ x^2 & \text{if } x < 2 \end{cases}$

Solution: Since f is not defined at $x = 2$, it is discontinuous at 2. Note, however, that

$$\lim_{x \to 2^-} f(x) = \lim_{x \to 2^-} x^2 = 4 = \lim_{x \to 2^+} x + 2 = \lim_{x \to 2^+} f(x)$$

shows that $\lim_{x \to 2} f(x)$ exists. (See Figure 10.31.)

NOW WORK PROBLEM 31 ●●●

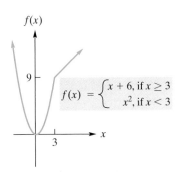

FIGURE 10.30 Continuous case-defined function.

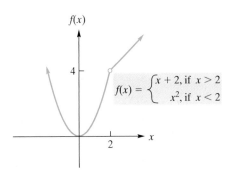

FIGURE 10.31 Discontinuous at 2.

● EXAMPLE 6 "Post-Office" Function

The post-office function

$$c = f(x) = \begin{cases} 39 & \text{if } 0 < x \le 1 \\ 63 & \text{if } 1 < x \le 2 \\ 87 & \text{if } 2 < x \le 3 \\ 111 & \text{if } 3 < x \le 4 \end{cases}$$

gives the cost c (in cents) of mailing, first-class, an item of weight x (ounces), for $0 < x \le 4$, in July 2006. It is clear from its graph in Figure 10.32 that f has discontinuities at 1, 2, and 3 and is constant for values of x between successive discontinuities. Such a function is called a *step function* because of the appearance of its graph.

NOW WORK PROBLEM 35 ●●

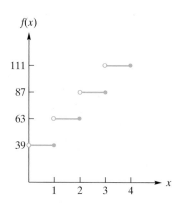

FIGURE 10.32 Post-office function.

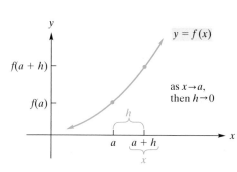

FIGURE 10.33 Diagram for continuity at a.

There is another way to express continuity besides that given in the definition. If we take the statement

$$\lim_{x \to a} f(x) = f(a)$$

and replace x by $a+h$, then as $x \to a$, we have $h \to 0$; and as $h \to 0$ we have $x \to a$. It follows that $\lim_{x \to a} f(x) = \lim_{h \to 0} f(a + h)$, provided the limits exist (Figure 10.33). Thus, the statement

$$\lim_{h \to 0} f(a + h) = f(a)$$

This method of expressing continuity at a is used frequently in mathematical proofs.

assuming both sides exist, also defines continuity at a.

T E C H N O L O G Y

By observing the graph of a function, we may be able to determine where a discontinuity occurs. However, we can be fooled. For example, the function

$$f(x) = \frac{x - 1}{x^2 - 1}$$

is discontinuous at ± 1, but the discontinuity at 1 is not obvious from the graph of f in Figure 10.34. On the other hand, the discontinuity at -1 is obvious. Note that f is defined neither at -1 nor at 1.

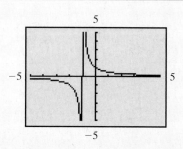

FIGURE 10.34 Discontinuity at 1 is not apparent from graph of $f(x) = \dfrac{x - 1}{x^2 - 1}$.

TABLE 10.3 Demand Schedule

Price/Unit, p	Quantity/Week, q
$20	0
10	5
5	15
4	20
2	45
1	95

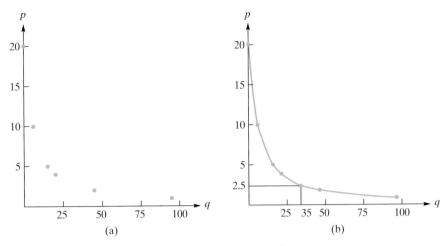

FIGURE 10.35 Viewing data via a continuous function.

Often, it is helpful to describe a situation by a continuous function. For example, the demand schedule in Table 10.3 indicates the number of units of a particular product that consumers will demand per week at various prices. This information can be given graphically, as in Figure 10.35(a), by plotting each quantity–price pair as a point. Clearly, the graph does not represent a continuous function. Furthermore, it gives us no information as to the price at which, say, 35 units would be demanded. However, if we connect the points in Figure 10.35(a) by a smooth curve [see Figure 10.35(b)], we get a so-called demand curve. From it, we could guess that at about $2.50 per unit, 35 units would be demanded.

Frequently, it is possible and useful to describe a graph, as in Figure 10.35(b), by means of an equation that defines a continuous function f. Such a function not only gives us a demand equation, $p = f(q)$, which allows us to anticipate corresponding prices and quantities demanded, but also permits a convenient mathematical analysis of the nature and basic properties of demand. Of course, some care must be used in working with equations such as $p = f(q)$. Mathematically, f may be defined when $q = \sqrt{37}$, but from a practical standpoint, a demand of $\sqrt{37}$ units could be meaningless to our particular situation. For example, if a unit is an egg, then a demand of $\sqrt{37}$ eggs make no sense.

We remark that exponential functions and logarithmic functions are continuous on their domains. Thus, exponential functions have no discontinuities while a logarithmic function has only a discontinuity at 0 (which is an infinite discontinuity).

Problems 10.3

In Problems 1–6, use the definition of continuity to show that the given function is continuous at the indicated point.

*1. $f(x) = x^3 - 5x; x = 2$

2. $f(x) = \dfrac{x-3}{5x}; x = -3$

3. $g(x) = \sqrt{2 - 3x}; x = 0$

4. $f(x) = \dfrac{x}{8}; x = 2$

5. $h(x) = \dfrac{x-4}{x+4}; x = 4$

6. $f(x) = \sqrt[3]{x}; x = -1$

In Problems 7–12, determine whether the function is continuous at the given points.

7. $f(x) = \dfrac{x+4}{x-2}; -2, 0$

8. $f(x) = \dfrac{x^2 - 4x + 4}{6}; 2, -2$

9. $g(x) = \dfrac{x-3}{x^2 - 9}; 3, -3$

10. $h(x) = \dfrac{3}{x^2 + 4}; 2, -2$

11. $f(x) = \begin{cases} x + 2 & \text{if } x \geq 2 \\ x^2 & \text{if } x < 2 \end{cases}; 2, 0$

12. $f(x) = \begin{cases} \dfrac{1}{x} & \text{if } x \neq 0 \\ 0 & \text{if } x = 0 \end{cases}; 0, -1$

In Problems 13–16, give a reason why the function is continuous on its domain.

*13. $f(x) = 2x^2 - 3$

14. $f(x) = \dfrac{2 + 3x - x^2}{5}$

15. $f(x) = \dfrac{x-1}{x^2 + 4}$

16. $f(x) = x(1 - x)$

In Problems 17–34, find all points of discontinuity.

17. $f(x) = 3x^2 - 3$

18. $h(x) = x - 2$

*19. $f(x) = \dfrac{3}{x+4}$

20. $f(x) = \dfrac{x^2 + 3x - 4}{x^2 - 4}$

21. $g(x) = \dfrac{(2x^2 - 3)^3}{15}$

22. $f(x) = -1$

23. $f(x) = \dfrac{x^2 + 6x + 9}{x^2 + 2x - 15}$ **24.** $g(x) = \dfrac{x - 3}{x^2 + x}$

25. $h(x) = \dfrac{x - 7}{x^3 - x}$ **26.** $f(x) = \dfrac{2x - 3}{3 - 2x}$

27. $p(x) = \dfrac{x}{x^2 + 1}$ **28.** $f(x) = \dfrac{x^4}{x^4 - 1}$

*29. $f(x) = \begin{cases} 1 & \text{if } x \geq 0 \\ -1 & \text{if } x < 0 \end{cases}$ **30.** $f(x) = \begin{cases} 2x + 1 & \text{if } x \geq -1 \\ 1 & \text{if } x < -1 \end{cases}$

*31. $f(x) = \begin{cases} 0 & \text{if } x \leq 1 \\ x - 1 & \text{if } x > 1 \end{cases}$ **32.** $f(x) = \begin{cases} x - 3 & \text{if } x > 2 \\ 3 - 2x & \text{if } x < 2 \end{cases}$

33. $f(x) = \begin{cases} x^2 + 1 & \text{if } x > 2 \\ 8x & \text{if } x < 2 \end{cases}$ **34.** $f(x) = \begin{cases} \dfrac{16}{x^2} & \text{if } x \geq 2 \\ 3x - 2 & \text{if } x < 2 \end{cases}$

*35. Telephone Rates   Suppose the long-distance rate for a telephone call from Hazleton, Pennsylvania, to Los Angeles, California, is \$0.10 for the first minute and \$0.06 for each additional minute or fraction thereof. If $y = f(t)$ is a function that indicates the total charge y for a call of t minutes' duration, sketch the graph of f for $0 < t \leq 4\frac{1}{2}$. Use

your graph to determine the values of t, where $0 < t \leq 4\frac{1}{2}$, at which discontinuities occur.

36. The *greatest integer function,* $f(x) = \lfloor x \rfloor$, is defined to be the greatest integer less than or equal to x, where x is any real number. For example, $\lfloor 3 \rfloor = 3$, $\lfloor 1.999 \rfloor = 1$, $\lfloor \frac{1}{4} \rfloor = 0$, and $\lfloor -4.5 \rfloor = -5$. Sketch the graph of this function for $-3.5 \leq x \leq 3.5$. Use your sketch to determine the values of x at which discontinuities occur.

37. Inventory   Sketch the graph of

$$y = f(x) = \begin{cases} -100x + 600 & \text{if } 0 \leq x < 5 \\ -100x + 1100 & \text{if } 5 \leq x < 10 \\ -100x + 1600 & \text{if } 10 \leq x < 15 \end{cases}$$

A function such as this might describe the inventory y of a company at time x. Is f continuous at 2? At 5? At 10?

38. Graph $g(x) = e^{-1/x^2}$. Because g is not defined at $x = 0$, g is discontinuous at 0. Based on the graph of g, is

$$f(x) = \begin{cases} e^{-1/x^2} & \text{if } x \neq 0 \\ 0 & \text{if } x = 0 \end{cases}$$

continuous at 0?

10.4 Continuity Applied to Inequalities

In Section 1.2, we solved linear inequalities. We now turn our attention to showing how the notion of continuity can be applied to solving a nonlinear inequality such as $x^2 + 3x - 4 < 0$. The ability to do this will be important in our study of calculus.

Recall (from Section 2.5) that the x-intercepts of the graph of a function g (that is, the points where the graph meets the x-axis) have x-coordinates that are the roots of the equation $g(x) = 0$, and these are called zeros of g. Conversely, any zero of g gives rise to an x-intercept of the graph of the function. Hence, from the graph of $y = g(x)$ in Figure 10.36, we conclude that $r_1, r_2,$ and r_3 are zeros of g and any other zeros of g will give rise to x-intercepts (beyond what is actually shown of the graph). Assume that in fact all the zeros of g, and hence all the x-intercepts, are shown. Note further from Figure 10.36 that the three zeros determine four open intervals on the x-axis:

$$(-\infty, r_1) \quad (r_1, r_2) \quad (r_2, r_3) \quad (r_3, \infty)$$

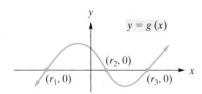

FIGURE 10.36 $r_1, r_2,$ and r_3 are zeros of g.

To solve $x^2 + 3x - 4 > 0$, we let

$$f(x) = x^2 + 3x - 4 = (x + 4)(x - 1)$$

Because f is a polynomial function, it is continuous. The zeros of f are -4 and 1; hence, the graph of f has x-intercepts $(-4, 0)$ and $(1, 0)$. (See Figure 10.37.) The zeros determine three intervals on the x-axis:

$$(-\infty, -4) \quad (-4, 1) \quad (1, \infty)$$

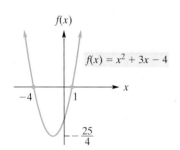

FIGURE 10.37 -4 and 1 are zeros of f.

Consider the interval $(-\infty, -4)$. Since f is continuous on this interval, we claim that either $f(x) > 0$ or $f(x) < 0$ *throughout* the interval. If this were not the case, then $f(x)$ would indeed change sign on the interval. By the continuity of f, there would be a point where the graph intersects the x-axis—for example, at $(x_0, 0)$. (See Figure 10.38.) But then x_0 would be a zero of f. However, this cannot be, because there is no zero of f that is less than -4. Hence, $f(x)$ must be strictly positive or strictly negative on $(-\infty, -4)$. A similar argument can be made for each of the other intervals.

To determine the sign of $f(x)$ on any one of the three intervals, it suffices to determine its sign at any point in the interval. For instance, -5 is in $(-\infty, -4)$ and

$$f(-5) = 6 > 0 \qquad \text{Thus, } f(x) > 0 \text{ on } (-\infty, -4)$$

Similarly, 0 is in $(-4, 1)$, and

$$f(0) = -4 < 0 \qquad \text{Thus, } f(x) < 0 \text{ on } (-4, 1)$$

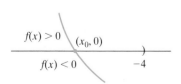

FIGURE 10.38   Change of sign for a continuous function.

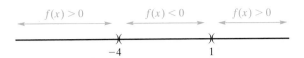

FIGURE 10.39   Sign chart for $x^2 + 3x - 4$.

Finally, 3 is in $(1, \infty)$, and

$$f(3) = 14 > 0 \qquad \text{Thus, } f(x) > 0 \text{ on } (1, \infty)$$

(See the "sign chart" in Figure 10.39.) Therefore,

$$x^2 + 3x - 4 > 0 \quad \text{on} \quad (-\infty, -4) \text{ and } (1, \infty)$$

so we have solved the inequality. These results are obvious from the graph in Figure 10.37. The graph lies above the x-axis (that is, $f(x) > 0$) on $(-\infty, -4)$ and on $(1, \infty)$.

● EXAMPLE 1   Solving a Quadratic Inequality

Solve $x^2 - 3x - 10 > 0$.

Solution: If $f(x) = x^2 - 3x - 10$, then f is a polynomial (quadratic) function and thus is continuous everywhere. To find the real zeros of f, we have

$$x^2 - 3x - 10 = 0$$

$$(x + 2)(x - 5) = 0$$

$$x = -2, 5$$

The zeros -2 and 5 are shown in Figure 10.40. These zeros determine three intervals:

$$(-\infty, -2) \quad (-2, 5) \quad (5, \infty)$$

To determine the sign of $f(x)$ on $(-\infty, -2)$, we choose a point in that interval, say, -3. The sign of $f(x)$ throughout $(-\infty, -2)$ is the same as that of $f(-3)$. Because

$$f(x) = (x + 2)(x - 5) \qquad \text{[factored form of } f(x)\text{]}$$

we have

$$f(-3) = (-3 + 2)(-3 - 5)$$

and hence

$$\text{sign}(f(-3)) = \text{sign}(-3 + 2)(-3 - 5) = \text{sign}(-3 + 2)\text{sign}(-3 - 5) = (-)(-) = +$$

We found the sign of $f(-3)$ by using the signs of the factors of $f(x)$ evaluated at -3. Thus, $f(x) > 0$ on $(-\infty, -2)$. To test the other two intervals, we will choose the points 0 and 6. We find that

$$\text{sign}(f(0)) = \text{sign}(0 + 2)\text{sign}(0 - 5) = (+)(-) = -$$

so $f(x) < 0$ on $(-2, 5)$, and

$$\text{sign}(f(6)) = \text{sign}(6 + 2)\text{sign}(6 - 5) = (+)(+) = +$$

so $f(x) > 0$ on $(5, \infty)$. A summary of our results is in the sign chart in Figure 10.41. Therefore, $x^2 - 3x - 10 > 0$ on $(-\infty, -2) \cup (5, \infty)$.

NOW WORK PROBLEM 1 ●●

FIGURE 10.40   Zeros of $x^2 - 3x - 10$.

FIGURE 10.41   Sign chart for $(x + 2)(x - 5)$.

An open box is formed by cutting a square piece out of each corner of an 8-inch-by-10-inch piece of metal. If each side of the cutout squares is x inches long, the volume of the box is given by $V(x) = x(8 - 2x)(10 - 2x)$. This problem makes sense only when this volume is positive. Find the values of x for which the volume is positive.

● EXAMPLE 2 Solving a Polynomial Inequality

Solve $x(x - 1)(x + 4) \leq 0$.

Solution: If $f(x) = x(x-1)(x+4)$, then f is a polynomial function and is continuous everywhere. The zeros of f are 0, 1, and -4, which are shown in Figure 10.42. These zeros determine four intervals:

$$(-\infty, -4) \quad (-4, 0) \quad (0, 1) \quad (1, \infty)$$

Now, at a test point in each interval, we find the sign of $f(x)$:

$$\text{sign}(f(-5)) = (-)(-)(-) = - \qquad \text{so } f(x) < 0 \text{ on } (-\infty, -4)$$
$$\text{sign}(f(-2)) = (-)(-)(+) = + \qquad \text{so } f(x) > 0 \text{ on } (-4, 0)$$
$$\text{sign}\left(f\left(\frac{1}{2}\right)\right) = (+)(-)(+) = - \qquad \text{so } f(x) < 0 \text{ on } (0, 1)$$
$$\text{sign}(f(2)) = (+)(+)(+) = + \qquad \text{so } f(x) > 0 \text{ on } (1, \infty)$$

Figure 10.43 shows the sign chart for $f(x)$. Thus, $x(x - 1)(x + 4) \leq 0$ on $(-\infty, -4]$ and $[0, 1]$. Note that -4, 0, and 1 are included in the solution because these values satisfy the equality $(=)$ part of the inequality $(\leq)$.

NOW WORK PROBLEM 11

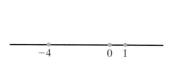

FIGURE 10.42 Zeros of $x(x - 1)(x + 4)$.

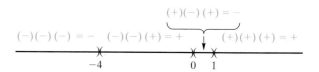

FIGURE 10.43 Sign chart for $x(x - 1)(x + 4)$.

● EXAMPLE 3 Solving a Rational Function Inequality

Solve $\dfrac{x^2 - 6x + 5}{x} \geq 0$.

Solution: Let $f(x) = \dfrac{x^2 - 6x + 5}{x} = \dfrac{(x - 1)(x - 5)}{x}$. For a rational function f, we solve the inequality by considering the intervals determined by the zeros of f and the points where f is discontinuous, for it is around such points that the sign of $f(x)$ may change. Here the zeros are 1 and 5. The function is discontinuous at 0 and continuous otherwise. In Figure 10.44, we have placed a hollow dot at 0 to indicate that f is not defined there. We thus consider the intervals

$$(-\infty, 0) \quad (0, 1) \quad (1, 5) \quad (5, \infty)$$

Determining the sign of $f(x)$ at a test point in each interval, we find that

$$\text{sign}(f(-1)) = \frac{(-)(-)}{(-)} = - \qquad \text{so } f(x) < 0 \text{ on } (-\infty, 0)$$
$$\text{sign}\left(f\left(\frac{1}{2}\right)\right) = \frac{(-)(-)}{(+)} = + \qquad \text{so } f(x) > 0 \text{ on } (0, 1)$$
$$\text{sign}(f(2)) = \frac{(+)(-)}{(+)} = - \qquad \text{so } f(x) < 0 \text{ on } (1, 5)$$
$$\text{sign}(f(6)) = \frac{(+)(+)}{(+)} = + \qquad \text{so } f(x) > 0 \text{ on } (5, \infty)$$

The sign chart for $f(x)$ is shown in Figure 10.45. Therefore, $f(x) \geq 0$ on $(0, 1]$ and $[5, \infty)$. (Why are 1 and 5 included, but 0 is excluded?) Figure 10.46 shows the graph of f. Notice that the solution of $f(x) \geq 0$ consists of all x-values for which the graph lies on or above the x-axis.

NOW WORK PROBLEM 17

FIGURE 10.44 Zeros and points of discontinuity for $\dfrac{(x - 1)(x - 5)}{x}$.

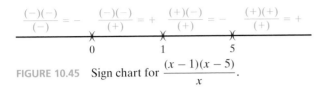

FIGURE 10.46   Graph of $f(x) = \dfrac{x^2 - 6x + 5}{x}$.

FIGURE 10.45   Sign chart for $\dfrac{(x-1)(x-5)}{x}$.

● EXAMPLE 4   Solving Nonlinear Inequalities

a. *Solve $x^2 + 1 > 0$.*

Solution: The equation $x^2 + 1 = 0$ has no real roots. Thus, $f(x) = x^2 + 1$ has no real zero. Also, f is continuous. Therefore, $f(x)$ is always positive or is always negative. But x^2 is always positive or zero, so $x^2 + 1$ is always positive. Hence, the solution of $x^2 + 1 > 0$ is $(-\infty, \infty)$.

b. *Solve $x^2 + 1 < 0$.*

Solution: From part (a), $x^2 + 1$ is always positive, so $x^2 + 1 < 0$ has no solution.

NOW WORK PROBLEM 7   ●●●

Problems 10.4

In Problems 1–26, solve the inequalities by the technique discussed in this section.

*1. $x^2 - 3x - 4 > 0$

2. $x^2 - 8x + 15 > 0$

3. $x^2 - 3x - 10 \le 0$

4. $14 - 5x - x^2 \le 0$

5. $2x^2 + 11x + 14 < 0$

6. $x^2 - 4 < 0$

*7. $x^2 + 4 < 0$

8. $2x^2 - x - 2 \le 0$

9. $(x+2)(x-3)(x+6) \le 0$

10. $(x+5)(x+2)(x-7) \le 0$

*11. $-x(x-5)(x+4) > 0$

12. $(x+2)^2 > 0$

13. $x^3 + 4x \ge 0$

14. $(x+2)^2(x^2 - 1) < 0$

15. $x^3 + 8x^2 + 15x \le 0$

16. $x^3 + 6x^2 + 9x < 0$

*17. $\dfrac{x}{x^2 - 9} < 0$

18. $\dfrac{x^2 - 1}{x} < 0$

19. $\dfrac{4}{x - 1} \ge 0$

20. $\dfrac{3}{x^2 - 5x + 6} > 0$

21. $\dfrac{x^2 - x - 6}{x^2 + 4x - 5} \ge 0$

22. $\dfrac{x^2 + 4x - 5}{x^2 + 3x + 2} \le 0$

23. $\dfrac{3}{x^2 + 6x + 5} \le 0$

24. $\dfrac{2x + 1}{x^2} \le 0$

25. $x^2 + 2x \ge 2$

26. $x^4 - 16 \ge 0$

27. Revenue   Suppose that consumers will purchase q units of a product when the price of *each* unit is $28 - 0.2q$ dollars. How many units must be sold in order that sales revenue will at least \$750?

28. Forest Management   A lumber company owns a forest that is of rectangular shape, 1 mi × 2 mi. The company wants to cut a uniform strip of trees along the outer edges of the forest. At most, how wide can the strip be if the company wants at least $1\frac{5}{16}$ mi² of forest to remain?

29. Container Design   A container manufacturer wishes to make an open box by cutting a 4-in. square from each corner of a square sheet of aluminum and then turning up the sides. The box is to contain at least 324 in.³. Find the dimensions of the smallest sheet of aluminum that can be used.

30. Workshop Participation   Imperial Education Services (I.E.S.) is offering a workshop in data processing to key personnel at Zeta Corporation. The price per person is \$50, and Zeta Corporation guarantees that at least 50 people will attend. Suppose I.E.S. offers to reduce the charge for *everybody* by \$0.50 for each person over the 50 who attends. How should I.E.S. limit the size of the group so that the total revenue it receives will never be less than that received for 50 persons?

31. Graph $f(x) = x^3 + 7x^2 - 5x + 4$. Use the graph to determine the solution of
$$x^3 + 7x^2 - 5x + 4 \le 0$$

32. Graph $f(x) = \dfrac{3x^2 - 0.5x + 2}{6.2 - 4.1x}$. Use the graph to determine the solution of
$$\frac{3x^2 - 0.5x + 2}{6.2 - 4.1x} > 0$$

A novel way of solving a nonlinear inequality like $f(x) > 0$ is by examining the graph of $g(x) = f(x)/|f(x)|$, whose range consists only of 1 and -1:
$$g(x) = \frac{f(x)}{|f(x)|} = \begin{cases} 1 & \text{if } f(x) > 0 \\ -1 & \text{if } f(x) < 0 \end{cases}$$

The solution of $f(x) > 0$ consists of all intervals for which $g(x) = 1$. Using this technique, solve the inequalities in Problems 33 and 34.

33. $6x^2 - x - 2 > 0$

34. $\dfrac{x^2 + x - 1}{x^2 + x - 2} < 0$

10.5 Review

Important Terms and Symbols	**Examples**

Summary

The notion of a limit lies at the foundation of calculus. To say that $\lim_{x \to a} f(x) = L$ means that the values of $f(x)$ can be made as close to the number L as we like by taking x sufficiently close to, but different from, a. If $\lim_{x \to a} f(x)$ and $\lim_{x \to a} g(x)$ exist and c is a constant, then

1. $\lim_{x \to a} c = c$

2. $\lim_{x \to a} x^n = a^n$

3. $\lim_{x \to a}[f(x) \pm g(x)] = \lim_{x \to a} f(x) \pm \lim_{x \to a} g(x)$

4. $\lim_{x \to a}[f(x) \cdot g(x)] = \lim_{x \to a} f(x) \cdot \lim_{x \to a} g(x)$

5. $\lim_{x \to a}[cf(x)] = c \cdot \lim_{x \to a} f(x)$

6. $\lim_{x \to a} \dfrac{f(x)}{g(x)} = \dfrac{\lim_{x \to a} f(x)}{\lim_{x \to a} g(x)}$ if $\lim_{x \to a} g(x) \neq 0,$

7. $\lim_{x \to a} \sqrt[n]{f(x)} = \sqrt[n]{\lim_{x \to a} f(x)}$

8. If f is a polynomial function, then $\lim_{x \to a} f(x) = f(a)$

Property 8 implies that the limit of a polynomial function as $x \to a$ can be found by simply evaluating the polynomial at a. However, with other functions, f, evaluation at a may lead to the meaningless form $0/0$. In such cases, algebraic manipulation such as factoring and cancellation may yield a function g that agrees with f, for $x \neq a$, and for which the limit can be determined.

If $f(x)$ approaches L as x approaches a from the right, then we write $\lim_{x \to a^+} f(x) = L$. If $f(x)$ approaches L as x approaches a from the left, we write $\lim_{x \to a^-} f(x) = L$. These limits are called one-sided limits.

The infinity symbol ∞, which does not represent a number, is used in describing limits. The statement

$$\lim_{x \to \infty} f(x) = L$$

means that as x increases without bound, the values of $f(x)$ approach the number L. A similar statement applies for the situation when $x \to -\infty$, which means that x is decreasing without bound. In general, if $p > 0$, then

$$\lim_{x \to \infty} \frac{1}{x^p} = 0 \quad \text{and} \quad \lim_{x \to -\infty} \frac{1}{x^p} = 0$$

If $f(x)$ increases without bound as $x \to a$, then we write $\lim_{x \to a} f(x) = \infty$. Similarly, if $f(x)$ decreases without bound, we have $\lim_{x \to a} f(x) = -\infty$. To say that the limit of a function is ∞ (or $-\infty$) does not mean that the limit exists. Rather, it is a way of saying that the limit does not exist and tells *why* there is no limit.

There is a rule for evaluating the limit of a rational function (quotient of polynomials) as $x \to \infty$ or $-\infty$. If $f(x)$ is a rational function and $a_n x^n$ and $b_m x^m$ are the terms in the numerator and denominator, respectively, with the greatest powers of x, then

$$\lim_{x \to \infty} f(x) = \lim_{x \to \infty} \frac{a_n x^n}{b_m x^m}$$

and

$$\lim_{x \to -\infty} f(x) = \lim_{x \to -\infty} \frac{a_n x^n}{b_m x^m}$$

In particular, as $x \to \infty$ or $-\infty$, the limit of a polynomial is the same as the limit of the term that involves the greatest power of x. This means that, for a nonconstant polynomial, the limit as $x \to \infty$ or $-\infty$ is either ∞ or $-\infty$.

A function f is continuous at a if and only if

1. $f(a)$ exists

2. $\lim_{x \to a} f(x)$ exists

3. $\lim_{x \to a} f(x) = f(a)$

Geometrically this means that the graph of f has no break when $x = a$. If a function is not continuous at a, then the function is said to be discontinuous at a. Polynomial functions and rational functions are continuous on their domains. Thus polynomial functions have no discontinuities and a rational function is discontinuous only at points where its denominator is zero.

To solve the inequality $f(x) > 0$ (or $f(x) < 0$), we first find the real zeros of f and the values of x for which f is discontinuous. These values determine intervals, and on each interval, $f(x)$ is either always positive or always negative. To find the sign on any one of these intervals, it suffices to find the sign of $f(x)$ at any point there. After the signs are determined for all intervals and assembled on a sign chart, it is easy to give the solution of $f(x) > 0$ (or $f(x) < 0$).

Review Problems

Problem numbers shown in color indicate problems suggested for use as a practice chapter test.

In Problems 1–28, find the limits if they exist. If the limit does not exist, so state that, or use the symbol ∞ or $-\infty$ where appropriate.

1. $\lim\limits_{x \to -1} (2x^2 + 6x - 1)$

2. $\lim\limits_{x \to 0} \dfrac{2x^2 - 3x + 1}{2x^2 - 2}$

3. $\lim\limits_{x \to 3} \dfrac{x^2 - 9}{x^2 - 3x}$

4. $\lim\limits_{x \to -4} \dfrac{2x + 3}{x^2 - 4}$

5. $\lim\limits_{h \to 0} (x + h)$

6. $\lim\limits_{x \to 2} \dfrac{x^2 - 4}{x^2 - 3x + 2}$

7. $\lim\limits_{x \to -4} \dfrac{x^3 + 4x^2}{x^2 + 2x - 8}$

8. $\lim\limits_{x \to 1} \dfrac{x^2 + x - 2}{x^2 + 4x - 5}$

9. $\lim\limits_{x \to \infty} \dfrac{2}{x + 1}$

10. $\lim\limits_{x \to \infty} \dfrac{x^2 + 1}{2x^2}$

11. $\lim\limits_{x \to \infty} \dfrac{2x + 5}{7x - 4}$

12. $\lim\limits_{x \to -\infty} \dfrac{1}{x^4}$

13. $\lim\limits_{t \to 3} \dfrac{2t - 3}{t - 3}$

14. $\lim\limits_{x \to -\infty} \dfrac{x^6}{x^5}$

15. $\lim\limits_{x \to -\infty} \dfrac{x + 3}{1 - x}$

16. $\lim\limits_{x \to 4} \sqrt[3]{64}$

17. $\lim\limits_{x \to \infty} \dfrac{x^2 - 1}{(3x + 2)^2}$

18. $\lim\limits_{x \to 1} \dfrac{x^2 + x - 2}{x - 1}$

19. $\lim\limits_{x \to 3^-} \dfrac{x + 3}{x^2 - 9}$

20. $\lim\limits_{x \to 2} \dfrac{2 - x}{x - 2}$

21. $\lim\limits_{x \to \infty} \sqrt{3x}$

22. $\lim\limits_{y \to 5^+} \sqrt{y - 5}$

23. $\lim\limits_{x \to \infty} \dfrac{x^{100} + (1/x^3)}{\pi - x^{97}}$

24. $\lim\limits_{x \to -\infty} \dfrac{ex^2 - x^4}{31x - 2x^3}$

25. $\lim\limits_{x \to 1} f(x)$ if $f(x) = \begin{cases} x^2 & \text{if } 0 \le x < 1 \\ x & \text{if } x > 1 \end{cases}$

26. $\lim\limits_{x \to 3} f(x)$ if $f(x) = \begin{cases} x + 5 & \text{if } x < 3 \\ 6 & \text{if } x \ge 3 \end{cases}$

27. $\lim\limits_{x \to 4^+} \dfrac{\sqrt{x^2 - 16}}{4 - x}$ *(Hint: For $x > 4$,*

$\sqrt{x^2 - 16} = \sqrt{x - 4}\sqrt{x + 4}$.)

28. $\lim\limits_{x \to 5^+} \dfrac{x^2 - 3x - 10}{\sqrt{x - 5}}$ *(Hint: For $x > 5$, $\dfrac{x - 5}{\sqrt{x - 5}} = \sqrt{x - 5}$.)*

29. If $f(x) = 8x - 2$, find $\lim\limits_{h \to 0} \dfrac{f(x + h) - f(x)}{h}$.

30. If $f(x) = 2x^2 - 3$, find $\lim\limits_{h \to 0} \dfrac{f(x + h) - f(x)}{h}$.

31. Host–Parasite Relationship   For a particular host–parasite relationship, it was determined that when the host density (number of hosts per unit of area) is x, then the number of hosts parasitized over a certain period of time is

$$y = 23\left(1 - \dfrac{1}{1 + 2x}\right)$$

If the host density were to increase without bound, what value would y approach?

32. Predator–Prey Relationship   For a particular predator–prey relationship, it was determined that the number y of prey consumed by an individual predator over a period of time was a function of the prey density x (the number of prey per unit of area). Suppose

$$y = f(x) = \dfrac{10x}{1 + 0.1x}$$

If the prey density were to increase without bound, what value would y approach?

33. Using the definition of continuity, show that the function $f(x) = x + 5$ is continuous at $x = 7$.

34. Using the definition of continuity, show that the function $f(x) = \dfrac{x - 5}{x^2 + 2}$ is continuous at $x = 5$.

35. State whether $f(x) = x^2/5$ is continuous at each real number. Give a reason for your answer.

36. State whether $f(x) = x^2 - 2$ is continuous everywhere. Give a reason for your answer.

In Problems 37–44, find the points of discontinuity (if any) for each function.

37. $f(x) = \dfrac{x^2}{x + 3}$

38. $f(x) = \dfrac{0}{x^3}$

39. $f(x) = \dfrac{x - 1}{2x^2 + 3}$

40. $f(x) = (2 - 3x)^3$

41. $f(x) = \dfrac{4 - x^2}{x^2 + 3x - 4}$

42. $f(x) = \dfrac{2x + 6}{x^3 + x}$

43. $f(x) = \begin{cases} x + 4 & \text{if } x > -2 \\ 3x + 6 & \text{if } x \le -2 \end{cases}$

44. $f(x) = \begin{cases} 1/x & \text{if } x < 1 \\ 1 & \text{if } x \ge 1 \end{cases}$

In Problems 45–52, solve the given inequalities.

45. $x^2 + 4x - 12 > 0$

46. $3x^2 - 3x - 6 \le 0$

47. $x^5 \le 7x^4$

48. $x^3 + 8x^2 + 15x \ge 0$

49. $\dfrac{x + 5}{x^2 - 1} < 0$

50. $\dfrac{x(x + 5)(x + 8)}{3} < 0$

51. $\dfrac{x^2 + 3x}{x^2 + 2x - 8} \ge 0$

52. $\dfrac{x^2 - 9}{x^2 - 16} \le 0$

53. Graph $f(x) = \dfrac{x^3 + 3x^2 - 19x + 18}{x^3 - 2x^2 + x - 2}$. Use the graph to estimate $\lim_{x \to 2} f(x)$.

54. Graph $f(x) = \dfrac{\sqrt{x + 3} - 2}{x - 1}$. From the graph, estimate $\lim_{x \to 1} f(x)$.

55. Graph $f(x) = x \ln x$. From the graph, estimate the one-sided limit $\lim_{x \to 0^+} f(x)$.

56. Graph $f(x) = \dfrac{e^x - 1}{(e^x + 1)(e^{2x} - e^x)}$. Use the graph to estimate $\lim_{x \to 0} f(x)$.

57. Graph $f(x) = x^3 - x^2 + x - 6$. Use the graph to determine the solution of

$$x^3 - x^2 + x - 6 \ge 0$$

58. Graph $f(x) = \dfrac{x^5 - 4}{x^3 + 1}$. Use the graph to determine the solution of

$$\dfrac{x^5 - 4}{x^3 + 1} \le 0$$

Mathematical Snapshot

National Debt

The size of the U.S. national debt is of great concern to many people and is frequently a topic in the news. The magnitude of the debt affects the confidence in the U.S. economy of both domestic and foreign investors, corporate officials, and political leaders. There are those who believe that to reduce the debt there must be cuts in government spending, which could affect government programs, or there must be an increase in revenues, possibly through tax increases.

Suppose that it is possible for the debt to be reduced continuously at an annual fixed rate. This is similar to compounding interest continuously, as studied in Chapter 5, except that instead of adding interest to an amount at each instant of time, you would be subtracting from the debt at each instant. Let us see how you could model this situation.

Suppose the debt D_0 at time $t = 0$ is reduced at an annual rate r. Furthermore, assume that there are k time periods of equal length in a year. At the end of the first period, the original debt is reduced by $D_0\left(\dfrac{r}{k}\right)$, so the new debt is

$$D_0 - D_0\left(\frac{r}{k}\right) = D_0\left(1 - \frac{r}{k}\right)$$

At the end of the second period, this debt is reduced by $D_0\left(1 - \dfrac{r}{k}\right)\dfrac{r}{k}$, so the new debt is

$$D_0\left(1 - \frac{r}{k}\right) - D_0\left(1 - \frac{r}{k}\right)\frac{r}{k}$$

$$= D_0\left(1 - \frac{r}{k}\right)\left(1 - \frac{r}{k}\right)$$

$$= D_0\left(1 - \frac{r}{k}\right)^2$$

The pattern continues. At the end of the third period the debt is $D_0\left(1 - \dfrac{r}{k}\right)^3$, and so on. At the end of t years the number of periods is kt and the debt is $D_0\left(1 - \dfrac{r}{k}\right)^{kt}$. If the debt is to be reduced at each instant of time, then $k \to \infty$. Thus you want to find

$$\lim_{k \to \infty} D_0\left(1 - \frac{r}{k}\right)^{kt}$$

which can be rewritten as

$$D_0\left[\lim_{k \to \infty}\left(1 - \frac{r}{k}\right)^{-k/r}\right]^{-rt}$$

If you let $x = -r/k$, then the condition $k \to \infty$ implies that $x \to 0$. Hence the limit inside the brackets has the form $\lim_{x \to 0}(1 + x)^{1/x}$, which we pointed out in Section 10.1 is e. Therefore, if the debt D_0 at time $t = 0$ is reduced

continuously at an annual rate r, then t years later the debt D is given by

$$D = D_0 e^{-rt}$$

For example, assume the U.S. national debt of \$8432 billion (rounded to the nearest billion) in July 2006 and a continuous reduction rate of 6% annually. Then the debt t years from now is given by

$$D = 8432 e^{-0.06t}$$

where D is in billions of dollars. This means that in 10 years, the debt will be $8432 e^{-0.6} \approx \$4628$ billion. Figure 10.47 shows the graph of $D = 8432 e^{-rt}$ for various rates r. Of course, the greater the value of r, the faster the debt

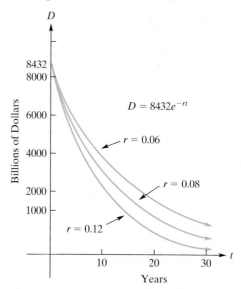

FIGURE 10.47 Budget debt is reduced continuously.

reduction. Notice that for $r = 0.06$, the debt at the end of 30 years is still considerable (approximately $1394 billion).

It is interesting to note that decaying radioactive elements also follow the model of continuous debt reduction, $D = D_0 e^{-rt}$.

To find out where the U.S. national debt currently stands, visit one of the national debt clocks on the Internet. You can find them by looking for "national debt clock" using any search engine.

Problems

In the following problems, assume a current national debt of $8432 billion.

1. If the debt were reduced to $8000 billion a year from now, what annual rate of continuous debt reduction would be involved? Give your answer to the nearest percent.

2. For a continuous debt reduction at an annual rate of 6%, determine the number of years from now required for the debt to be reduced by one-half. Give your answer to the nearest year.

3. What assumptions underlie a model of debt reduction that uses an exponential function? What are the limitations of this approach?

DIFFERENTIATION

Mathematical Snapshot   Marginal Propensity to Consume

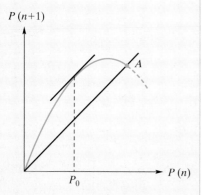

$P(n+1)$

P_0 $P(n)$

Government regulations generally limit the number of fish taken from a given fishing ground by commercial fishing boats in a season. This prevents over-fishing, which depletes the fish population and leaves, in the long run, fewer fish to catch.

From a strictly commercial perspective, the ideal regulations would maximize the number of fish available for the year-to-year fish harvest. The key to finding those ideal regulations is a mathematical function called the reproduction curve. For a given fish habitat, this function estimates the a fish population a year from now, $P(n+1)$, based on the population now, $P(n)$, assuming no external intervention (i.e., no fishing, no influx of predators, and so on).

The figure to the bottom left shows a typical reproduction curve. Also graphed is the line $P(n + 1) = P(n)$, the line along which the populations $P(n + 1)$ and $P(n)$ would be equal. Notice the intersection of the curve and the straight line at point A. This is where, because of habitat crowding, the population has reached its maximum sustainable size. A population that is this size one year will be the same size the next year.

For any point on the horizontal axis, the distance between the reproduction curve and the line $P(n + 1) = P(n)$ represents the sustainable harvest: the number of fish that could be caught, after the spawn have grown to maturity, so that in the end the population is back at the same size it was a year ago.

Commercially speaking, the optimal population size is the one where the distance between the reproduction curve and the line $P(n + 1) = P(n)$ is the greatest. This condition is met where the slopes of the reproduction curve and the line $P(n + 1) = P(n)$ are equal. (The slope of $P(n + 1) = P(n)$ is of course 1.) Thus, for a maximum fish harvest year after year, regulations should aim to keep the fish population fairly close to P_0.

A central idea here is that of the slope of a curve at a given point. That idea is the cornerstone concept of this chapter.

Now we begin our study of calculus. The ideas involved in calculus are completely different from those of algebra and geometry. The power and importance of these ideas and their applications will be clear to you later in the book. In this chapter we introduce the *derivative* of a function, and you will learn important rules for finding derivatives. You will also see how the derivative is used to analyze the rate of change of a quantity, such as the rate at which the position of a body is changing.

OBJECTIVE

To develop the idea of a tangent line to a curve, to define the slope of a curve, and to define a derivative and give it a geometric interpretation. To compute derivatives by using the limit definition.

11.1 The Derivative

• • •

One of the main problems with which calculus deals is finding the slope of the *tangent line* at a point on a curve. In geometry you probably thought of a tangent line, or *tangent*, to a circle as a line that meets the circle at exactly one point (Figure 11.1). However, this idea of a tangent is not very useful for other kinds of curves. For example, in Figure 11.2(a), the lines L_1 and L_2 intersect the curve at exactly one point P. Although we would not think of L_2 as the tangent at this point, it seems natural that L_1 is. In Figure 11.2(b) we intuitively would consider L_3 to be the tangent at point P, even though L_3 intersects the curve at other points.

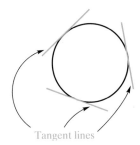

FIGURE 11.1   Tangent lines to a circle.

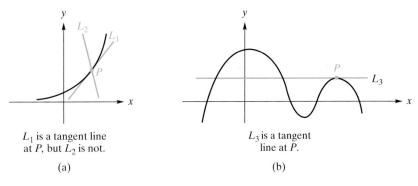

L_1 is a tangent line at P, but L_2 is not.

(a)

L_3 is a tangent line at P.

(b)

FIGURE 11.2   Tangent line at a point.

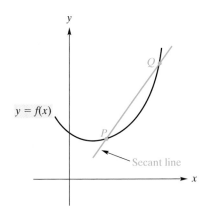

FIGURE 11.3   Secant line PQ.

From the previous examples, you can see that we must drop the idea that a tangent is simply a line that intersects a curve at only one point. To obtain a suitable definition of tangent line, we use the limit concept and the geometric notion of a *secant line*. A **secant line** is a line that intersects a curve at two or more points.

Look at the graph of the function $y = f(x)$ in Figure 11.3. We wish to define the tangent line at point P. If Q is a different point on the curve, the line PQ is a secant line. If Q moves along the curve and approaches P from the right (see Figure 11.4), typical secant lines are PQ', PQ'', and so on. As Q approaches P from the left, they are PQ_1, PQ_2, and so on. *In both cases, the secant lines approach the same limiting position.* This common limiting position of the secant lines is defined to be the **tangent line** to the curve at P. This definition seems reasonable and applies to curves in general, not just circles.

A curve does not necessarily have a tangent line at each of its points. For example, the curve $y = |x|$ does not have a tangent at $(0, 0)$. As you can see in Figure 11.5, a secant line through $(0, 0)$ and a nearby point to its right on the curve must always be

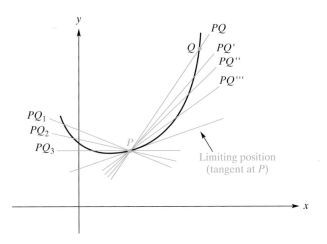

FIGURE 11.4   The tangent line is a limiting position of secant lines.

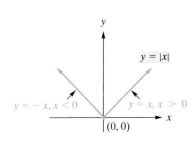

FIGURE 11.5   No tangent line to graph of $y = |x|$ at $(0, 0)$.

the line $y = x$. Thus the limiting position of such secant lines is also the line $y = x$. However, a secant line through $(0, 0)$ and a nearby point to its left on the curve must always be the line $y = -x$. Hence, the limiting position of such secant lines is also the line $y = -x$. Since there is no common limiting position, there is no tangent line at $(0, 0)$.

Now that we have a suitable definition of a tangent to a curve at a point, we can define the *slope of a curve* at a point.

DEFINITION

The **slope of a curve** at a point P is the slope, if it exists, of the tangent line at P.

Since the tangent at P is a limiting position of secant lines PQ, we consider the slope of the tangent to be the limiting value of the slopes of the secant lines as Q approaches P. For example, let us consider the curve $f(x) = x^2$ and the slopes of some secant lines PQ, where $P = (1, 1)$. For the point $Q = (2.5, 6.25)$, the slope of PQ (see Figure 11.6) is

$$m_{PQ} = \frac{\text{rise}}{\text{run}} = \frac{6.25 - 1}{2.5 - 1} = 3.5$$

Table 11.1 includes other points Q on the curve, as well as the corresponding slopes of PQ. Notice that as Q approaches P, the slopes of the secant lines seem to approach 2. Thus, we expect the slope of the indicated tangent line at $(1, 1)$ to be 2. This will be confirmed later, in Example 1. But first, we wish to generalize our procedure.

For the curve $y = f(x)$ in Figure 11.7, we will find an expression for the slope at the point $P = (a, f(a))$. If $Q = (z, f(z))$, the slope of the secant line PQ is

$$m_{PQ} = \frac{f(z) - f(a)}{z - a}$$

If the difference $z - a$ is called h, then we can write z as $a + h$. Here we must have $h \neq 0$, for if $h = 0$, then $z = a$, and no secant line exists. Accordingly,

$$m_{PQ} = \frac{f(z) - f(a)}{z - a} = \frac{f(a + h) - f(a)}{h}$$

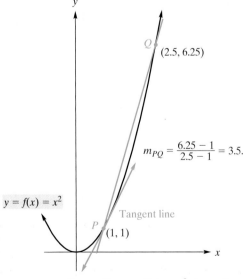

FIGURE 11.6 Secant line to $f(x) = x^2$ through $(1, 1)$ and $(2.5, 6.25)$.

TABLE 11.1 Slopes of Secant Lines to the Curve $f(x) = x^2$ at $P = (1, 1)$

Q	Slope of PQ
$(2.5, 6.25)$	$(6.25 - 1)/(2.5 - 1) = 3.5$
$(2, 4)$	$(4 - 1)/(2 - 1) = 3$
$(1.5, 2.25)$	$(2.25 - 1)/(1.5 - 1) = 2.5$
$(1.25, 1.5625)$	$(1.5625 - 1)/(1.25 - 1) = 2.25$
$(1.1, 1.21)$	$(1.21 - 1)/(1.1 - 1) = 2.1$
$(1.01, 1.0201)$	$(1.021 - 1)/(1.01 - 1) = 2.01$

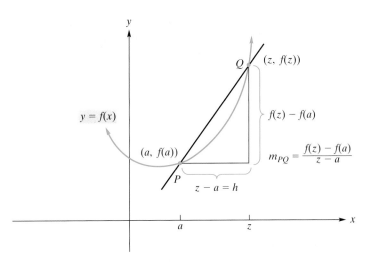

FIGURE 11.7 Secant line through P and Q.

Which of these two forms for m_{PQ} is most convenient depends on the nature of the function f. As Q moves along the curve toward P, z approaches a. This means that h approaches zero. The limiting value of the slopes of the secant lines—which is the slope of the tangent line at $(a, f(a))$—is

$$m_{\tan} = \lim_{z \to a} \frac{f(z) - f(a)}{z - a} = \lim_{h \to 0} \frac{f(a+h) - f(a)}{h} \tag{1}$$

Again, which of these two forms is most convenient—which limit is easiest to determine—depends on the nature of the function f. In Example 1, we will use this limit to confirm our previous expectation that the slope of the tangent line to the curve $f(x) = x^2$ at $(1, 1)$ is 2.

● EXAMPLE 1   Finding the Slope of a Tangent Line

Find the slope of the tangent line to the curve $y = f(x) = x^2$ at the point $(1, 1)$.

Solution: The slope is the limit in Equation (1) with $f(x) = x^2$ and $a = 1$:

$$\lim_{h \to 0} \frac{f(1+h) - f(1)}{h} = \lim_{h \to 0} \frac{(1+h)^2 - (1)^2}{h}$$

$$= \lim_{h \to 0} \frac{1 + 2h + h^2 - 1}{h} = \lim_{h \to 0} \frac{2h + h^2}{h}$$

$$= \lim_{h \to 0} \frac{h(2 + h)}{h} = \lim_{h \to 0} (2 + h) = 2$$

Therefore, the tangent line to $y = x^2$ at $(1, 1)$ has slope 2. (Refer to Figure 11.6.)

NOW WORK PROBLEM 1 ●●●

We can generalize Equation (1) so that it applies to any point $(x, f(x))$ on a curve. Replacing a by x gives a function, called the *derivative* of f, whose input is x and whose output is the slope of the tangent line to the curve at $(x, f(x))$, provided that the tangent line *exists* and *has* a slope. (If the tangent line exists but is *vertical* then it has no slope.) We thus have the following definition, which forms the basis of differential calculus:

DEFINITION

The **derivative** of a function f is the function denoted f' (read "f prime") and defined by

$$f'(x) = \lim_{z \to x} \frac{f(z) - f(x)}{z - x} = \lim_{h \to 0} \frac{f(x + h) - f(x)}{h} \qquad (2)$$

provided that this limit exists. If $f'(a)$ can be found (while perhaps not all $f'(x)$ can be found) f is said to be **differentiable** at a, and $f'(a)$ is called the derivative of f at a or the derivative of f with respect to x at a. The process of finding the derivative is called **differentiation.**

In the definition of the derivative, the expression

$$\frac{f(z) - f(x)}{z - x} = \frac{f(x + h) - f(x)}{h}$$

where $z = x + h$, is called a **difference quotient.** Thus $f'(x)$ is the limit of a difference quotient.

● **EXAMPLE 2   Using the Definition to Find the Derivative**

If $f(x) = x^2$, find the derivative of f.

Solution: Applying the definition of a derivative gives

> Don't be sloppy when applying the limit definition of a derivative. Write $\lim_{h \to 0}$ at each step before the limit is actually taken. Unfortunately, some students neglect to take the final limit, and h appears in their answer. This is a quick way to lose points on an examination.

$$f'(x) = \lim_{h \to 0} \frac{f(x + h) - f(x)}{h}$$

$$= \lim_{h \to 0} \frac{(x + h)^2 - x^2}{h} = \lim_{h \to 0} \frac{x^2 + 2xh + h^2 - x^2}{h}$$

$$= \lim_{h \to 0} \frac{2xh + h^2}{h} = \lim_{h \to 0} \frac{h(2x + h)}{h} = \lim_{h \to 0}(2x + h) = 2x$$

Observe that, in taking the limit, we treated x as a constant, because it was h, not x, that was changing. Also, note that $f'(x) = 2x$ defines a function of x, which we can interpret as giving the slope of the tangent line to the graph of f at $(x, f(x))$. For example, if $x = 1$, then the slope is $f'(1) = 2(1) = 2$, which confirms the result in Example 1.

NOW WORK PROBLEM 3 ●●●

Besides the notation $f'(x)$, other common ways to denote the derivative of $y = f(x)$ at x are

CAUTION

The notation $\dfrac{dy}{dx}$, which is called *Leibniz notation*, should **not** be thought of as a fraction, although it looks like one. It is a single symbol for a derivative. We have not yet attached any meaning to individual symbols such as dy and dx.

$$\frac{dy}{dx} \qquad \text{(pronounced "dee } y, \text{ dee } x\text{" or "dee } y \text{ by dee } x\text{")}$$

$$\frac{d}{dx}(f(x)) \qquad \text{("dee } f(x), \text{ dee } x\text{" or "dee by dee } x \text{ of } f(x)\text{")}$$

$$y' \qquad \text{("} y \text{ prime")}$$

$$D_x y \qquad \text{("dee } x \text{ of } y\text{")}$$

$$D_x(f(x)) \qquad \text{("dee } x \text{ of } f(x)\text{")}$$

Because the derivative gives the slope of the tangent line, $f'(a)$ is the slope of the line tangent to the graph of $y = f(x)$ at $(a, f(a))$.

Two other notations for the derivative of f at a are

$$\frac{dy}{dx}\Big|_{x=a} \qquad \text{and} \qquad y'(a)$$

● EXAMPLE 3   Finding an Equation of a Tangent Line

If $f(x) = 2x^2 + 2x + 3$, find an equation of the tangent line to the graph of f at $(1, 7)$.

Solution:

> **Strategy**   We will first determine the slope of the tangent line by computing the derivative and evaluating it at $x = 1$. Using this result and the point $(1, 7)$ in a point–slope form gives an equation of the tangent line.

We have

$$f'(x) = \lim_{h \to 0} \frac{f(x + h) - f(x)}{h}$$

$$= \lim_{h \to 0} \frac{(2(x + h)^2 + 2(x + h) + 3) - (2x^2 + 2x + 3)}{h}$$

$$= \lim_{h \to 0} \frac{2x^2 + 4xh + 2h^2 + 2x + 2h + 3 - 2x^2 - 2x - 3}{h}$$

$$= \lim_{h \to 0} \frac{4xh + 2h^2 + 2h}{h} = \lim_{h \to 0} (4x + 2h + 2)$$

So

$$f'(x) = 4x + 2$$

and

$$f'(1) = 4(1) + 2 = 6$$

In Example 3 it is *not* correct to say that, since the derivative is $4x + 2$, the tangent line at $(1, 7)$ is $y - 7 = (4x + 2)(x - 1)$. The derivative must be **evaluated** at the point of tangency to determine the slope of the tangent line.

Thus, the tangent line to the graph at $(1, 7)$ has slope 6. A point–slope form of this tangent is

$$y - 7 = 6(x - 1)$$

which in slope–intercept form is

$$y = 6x + 1$$

NOW WORK PROBLEM 25   ●●●

● EXAMPLE 4   Finding the Slope of a Curve at a Point

Find the slope of the curve $y = 2x + 3$ at the point where $x = 6$.

Solution: The slope of the curve is the slope of the tangent line. Letting $y = f(x) = 2x + 3$, we have

$$\frac{dy}{dx} = \lim_{h \to 0} \frac{f(x + h) - f(x)}{h} = \lim_{h \to 0} \frac{(2(x + h) + 3) - (2x + 3)}{h}$$

$$= \lim_{h \to 0} \frac{2h}{h} = \lim_{h \to 0} 2 = 2$$

Since $dy/dx = 2$, the slope when $x = 6$, or in fact at any point, is 2. Note that the curve is a straight line and thus has the same slope at each point.

NOW WORK PROBLEM 19   ●●●

● EXAMPLE 5   A Function with a Vertical Tangent Line

Find $\dfrac{d}{dx}(\sqrt{x})$.

Solution: Letting $f(x) = \sqrt{x}$, we have

$$\frac{d}{dx}(\sqrt{x}) = \lim_{h \to 0} \frac{f(x + h) - f(x)}{h} = \lim_{h \to 0} \frac{\sqrt{x + h} - \sqrt{x}}{h}$$

You should become familiar with the procedure of rationalizing the *numerator*.

As $h \to 0$, both the numerator and denominator approach zero. This can be avoided by rationalizing the *numerator*:

$$\frac{\sqrt{x+h} - \sqrt{x}}{h} = \frac{\sqrt{x+h} - \sqrt{x}}{h} \cdot \frac{\sqrt{x+h} + \sqrt{x}}{\sqrt{x+h} + \sqrt{x}}$$

$$= \frac{(x+h) - x}{h(\sqrt{x+h} + \sqrt{x})} = \frac{h}{h(\sqrt{x+h} + \sqrt{x})}$$

Therefore,

$$\frac{d}{dx}(\sqrt{x}) = \lim_{h \to 0} \frac{h}{h(\sqrt{x+h} + \sqrt{x})} = \lim \frac{1}{\sqrt{x+h} + \sqrt{x}} = \frac{1}{\sqrt{x} + \sqrt{x}} = \frac{1}{2\sqrt{x}}$$

Note that the original function, $\sqrt{x}$, is defined for $x \geq 0$, but its derivative, $1/(2\sqrt{x})$, is defined only when $x > 0$. The reason for this is clear from the graph of $y = \sqrt{x}$ in Figure 11.8. When $x = 0$, the tangent is a vertical line, so its slope is not defined.

NOW WORK PROBLEM 17

In Example 5 we saw that the function $y = \sqrt{x}$ is not differentiable when $x = 0$, because the tangent line is vertical at that point. It is worthwhile mentioning that $y = |x|$ also is not differentiable when $x = 0$, but for a different reason: There is *no* tangent line at all at that point. (Refer back to Figure 11.5.) Both examples show that the domain of f' may be strictly contained in the domain of f.

To indicate a derivative, Leibniz notation is often useful because it makes it convenient to emphasize the independent and dependent variables involved. For example, if the variable p is a function of the variable q, we speak of the derivative of p with respect to q, written dp/dq.

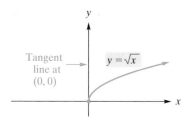

FIGURE 11.8 Vertical tangent line at $(0, 0)$.

It is wise for you to see variables other than x and y involved in a problem. Example 6 illustrates the use of other variables.

● **EXAMPLE 6** **Finding the Derivative of p with Respect to q**

If $p = f(q) = \dfrac{1}{2q}$, find $\dfrac{dp}{dq}$.

Solution: We will do this problem first using the $h \to 0$ limit (the only one we have used so far) and then using $r \to q$ to illustrate the other variant of the limit.

$$\frac{dp}{dq} = \frac{d}{dq}\left(\frac{1}{2q}\right) = \lim_{h \to 0} \frac{f(q+h) - f(q)}{h}$$

$$= \lim_{h \to 0} \frac{\dfrac{1}{2(q+h)} - \dfrac{1}{2q}}{h} = \lim_{h \to 0} \frac{\dfrac{q - (q+h)}{2q(q+h)}}{h}$$

$$= \lim_{h \to 0} \frac{q - (q+h)}{h(2q(q+h))} = \lim_{h \to 0} \frac{-h}{h(2q(q+h))}$$

$$= \lim_{h \to 0} \frac{-1}{2q(q+h)} = -\frac{1}{2q^2}$$

We also have

$$\frac{dp}{dq} = \lim_{r \to q} \frac{f(r) - f(q)}{r - q}$$

$$= \lim_{r \to q} \frac{\dfrac{1}{2r} - \dfrac{1}{2q}}{r - q} = \lim_{r \to q} \frac{\dfrac{q - r}{2rq}}{r - q}$$

$$= \lim_{r \to q} \frac{-1}{2rq} = \frac{-1}{2q^2}$$

We leave it you to decide which form leads to the simpler limit calculation in this case.

Note that when $q = 0$, neither the function nor its derivative exists.

NOW WORK PROBLEM 15

Keep in mind that the derivative of $y = f(x)$ at x is nothing more than a limit, namely

$$\lim_{h \to 0} \frac{f(x + h) - f(x)}{h}$$

equivalently

$$\lim_{z \to x} \frac{f(z) - f(x)}{z - x}$$

whose use we have just illustrated. Although we can interpret the derivative as a function that gives the slope of the tangent line to the curve $y = f(x)$ at the point $(x, f(x))$, this interpretation is simply a geometric convenience that assists our understanding. The preceding limit may exist, aside from any geometric considerations at all. As you will see later, there are other useful interpretations of the derivative.

In Section 11.4, we will make technical use of the following relationship between differentiability and continuity. However, it is of fundamental importnace and needs to be understood from the outset.

If f is differentiable at a, then f is continuous at a.

To establish this result, we will assume that f is differentiable at a. Then $f'(a)$ exists, and

$$\lim_{h \to 0} \frac{f(a + h) - f(a)}{h} = f'(a)$$

Consider the numerator $f(a + h) - f(a)$ as $h \to 0$. We have

$$\lim_{h \to 0} (f(a + h) - f(a)) = \lim_{h \to 0} \left(\frac{f(a + h) - f(a)}{h} \cdot h \right)$$

$$= \lim_{h \to 0} \frac{f(a + h) - f(a)}{h} \cdot \lim_{h \to 0} h$$

$$= f'(a) \cdot 0 = 0$$

Thus, $\lim_{h \to 0} (f(a + h) - f(a)) = 0$. This means that $f(a + h) - f(a)$ approaches 0 as $h \to 0$. Consequently,

$$\lim_{h \to 0} f(a + h) = f(a)$$

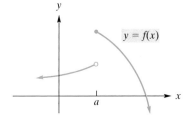

FIGURE 11.9 f is not continuous at a, so f is not differentiable at a.

As stated in Section 10.3, this condition means that f is continuous at a. The foregoing, then, proves that f is continuous at a when f is differentiable there. More simply, we say that **differentiability at a point implies continuity at that point.**

If a function is not continuous at a point, then it cannot have a derivative there. For example, the function in Figure 11.9 is discontinuous at a. The curve has no tangent at that point, so the function is not differentiable there.

● EXAMPLE 7   Continuity and Differentiability

a. Let $f(x) = x^2$. The derivative, $2x$, is defined for all values of x, so $f(x) = x^2$ must be continuous for all values of x.

b. The function $f(p) = \dfrac{1}{2p}$ is not continuous at $p = 0$ because f is not defined there. Thus, the derivative does not exist at $p = 0$.

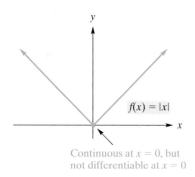

Continuous at $x = 0$, but
not differentiable at $x = 0$

FIGURE 11.10   Continuity does not
imply differentiability.

The converse of the statement that differentiability implies continuity is *false*. That is, it is false that continuity implies differentiability. In Example 8, you will see a function that is continuous at a point, but not differentiable there.

● **EXAMPLE 8   Continuity Does Not Imply Differentiability**

The function $y = f(x) = |x|$ is continuous at $x = 0$. (See Figure 11.10.) As we mentioned earlier, there is no tangent line at $x = 0$. Thus, the derivative does not exist there. This shows that continuity does *not* imply differentiability.

● ● ●

TECHNOLOGY

Many graphing calculators have a numerical derivative feature that estimates the derivative of a function at a point. With the TI-83 Plus we would use the "nDeriv" command, in which we must enter the function, the variable, and the value of the variable (separated by commas) in the format

nDeriv(function, variable, value of variable)

For example, the derivative of $f(x) = \sqrt{x^3 + 2}$ at $x = 1$ is estimated in Figure 11.11. Thus, $f'(1) \approx 0.866$.

On the other hand, we can take the "limit of a difference quotient" approach to estimate this derivative. To make use of the table feature of a graphing calculator, we can enter $f(x)$ as Y_1. Then, for Y_2, we enter the following form of the difference quotient:

$$(Y_1(1 + X) - Y_1(1))/X$$

(Here, x plays the role of h.) Figure 11.12 shows a table for Y_2 as x approaches 0 from both the left and right. This table strongly suggests that $f'(1) \approx 0.866$.

```
nDeriv(√(X^3+2),
X,1)
        .8660253677
■
```

FIGURE 11.11   Numerical derivative.

X	Y₂
.01	.87252
.001	.86667
1E⁻4	.86609
1E⁻5	.86603
⁻.01	.85953
⁻.001	.86538
⁻1E⁻4	.86596

X=⁻1E⁻4

FIGURE 11.12   Limit of a difference quotient as $x \to 0$.

Problems 11.1

In Problems 1 and 2, a function f and a point P on its graph are given.

(a) *Find the slope of the secant line PQ for each point $Q = (x, f(x))$ whose x-value is given in the table. Round your answers to four decimal places.*

(b) *Use your results from part (a) to estimate the slope of the tangent line at P.*

*1. $f(x) = x^3 + 3$, $P = (-2, -5)$

x-value of Q	−3	−2.5	−2.2	−2.1	−2.01	−2.001
m_{PQ}						

2. $f(x) = e^{2x}$, $P = (0, 1)$

x-value of Q	1	0.5	0.2	0.1	0.01	0.001
m_{PQ}						

In Problems 3–18, use the definition of the derivative to find each of the following.

*3. $f'(x)$ if $f(x) = x$

4. $f'(x)$ if $f(x) = 4x - 1$

5. $\dfrac{dy}{dx}$ if $y = 3x + 5$

6. $\dfrac{dy}{dx}$ if $y = -5x$

7. $\dfrac{d}{dx}(5 - 4x)$

8. $\dfrac{d}{dx}\left(1 - \dfrac{x}{2}\right)$

9. $f'(x)$ if $f(x) = 3$

10. $f'(x)$ if $f(x) = 7.01$

11. $\dfrac{d}{dx}(x^2 + 4x - 8)$

12. y' if $y = x^2 + 5x + 1$

13. $\dfrac{dp}{dq}$ if $p = 3q^2 + 2q + 1$

14. $\dfrac{d}{dx}(x^2 - x - 3)$

*15. y' if $y = \dfrac{6}{x}$

16. $\dfrac{dC}{dq}$ if $C = 7 + 2q - 3q^2$

*17. $f'(x)$ if $f(x) = \sqrt{x + 2}$

18. $H'(x)$ if $H(x) = \dfrac{3}{x - 2}$

*19. Find the slope of the curve $y = x^2 + 4$ at the point $(-2, 8)$.

20. Find the slope of the curve $y = 1 - x^2$ at the point $(1, 0)$.

21. Find the slope of the curve $y = 4x^2 - 5$ when $x = 0$.

22. Find the slope of the curve $y = \sqrt{x}$ when $x = 1$.

In Problems 23–28, find an equation of the tangent line to the curve at the given point.

23. $y = x + 4$; $(3, 7)$

24. $y = 3x^2 - 4$; $(1, -1)$

*****25.** $y = x^2 + 2x + 3$; $(1, 6)$

26. $y = (x - 7)^2$; $(6, 1)$

27. $y = \dfrac{3}{x - 1}$; $(2, 3)$

28. $y = \dfrac{5}{1 - 3x}$; $(2, -1)$

29. Banking Equations may involve derivatives of functions. In an article on interest rate deregulation, Christofi and Agapos[1] solve the equation

$$r = \left(\frac{\eta}{1 + \eta}\right)\left(r_L - \frac{dC}{dD}\right)$$

for η (the Greek letter "eta"). Here r is the deposit rate paid by commercial banks, r_L is the rate earned by commercial banks, C is the administrative cost of transforming deposits into return-earning assets, D is the savings deposits level, and η is the deposit elasticity with respect to the deposit rate. Find η.

In Problems 30 and 31, use the numerical derivative feature of your graphing calculator to estimate the derivatives of the functions at the indicated values. Round your answers to three decimal places.

30. $f(x) = \sqrt{2x^2 + 3x}$; $x = 1$, $x = 2$

31. $f(x) = e^x(4x - 7)$; $x = 0$, $x = 1.5$

In Problems 32 and 33, use the "limit of a difference quotient" approach to estimate $f'(x)$ at the indicated values of x. Round your answers to three decimal places.

32. $f(x) = \dfrac{e^x}{x + 1}$; $x = 1$, $x = 10$

33. $f(x) = \dfrac{x^2 + 4x + 2}{x^3 - 3}$; $x = 2$, $x = -4$

34. Find an equation of the tangent line to the curve $f(x) = x^2 + x$ at the point $(-2, 2)$. Graph both the curve and the tangent line. Notice that the tangent line is a good approximation to the curve near the point of tangency.

35. The derivative of $f(x) = x^3 - x + 2$ is $f'(x) = 3x^2 - 1$. Graph both the function f and its derivative f'. Observe that there are two points on the graph of f where the tangent line is horizontal. For the x-values of these points, what are the corresponding values of $f'(x)$? Why are these results expected? Observe the intervals where $f'(x)$ is positive. Notice that tangent lines to the graph of f have positive slopes over these intervals. Observe the interval where $f'(x)$ is negative. Notice that tangent lines to the graph of f have negative slopes over this interval.

In Problems 36 and 37, verify the identity
$(z - x)\left(\sum_{i=0}^{n-1} x^i z^{n-1-i}\right) = z^n - x^n$ *for the indicated values of n and calculate the derivative using the $z \to x$ form of the definition of the derivative in Equation (2).*

36. $n = 4, n = 3, n = 2$; $f'(x)$ if $f(x) = 2x^4 + x^3 - 3x^2$

37. $n = 5, n = 3$; $f'(x)$ if $f(x) = 4x^5 - 3x^3$

OBJECTIVE

To develop basic differentiation rules, namely, formulas for the derivative of a constant, of x^n, of a constant times a function, and of sums and differences of functions.

11.2 Rules for Differentiation

You would probably agree that differentiating a function by direct use of the definition of a derivative can be tedious. Fortunately, there are rules that give us completely mechanical and efficient procedures for differentiation. They also avoid the direct use of limits. We will look at some of these rules in this section.

We begin by showing that the derivative of a constant function is zero. Recall that the graph of the constant function $f(x) = c$ is a horizontal line (see Figure 11.13), which has a slope of zero at each point. This means that $f'(x) = 0$ regardless of x. As a formal proof of this result, we apply the definition of the derivative to $f(x) = c$:

$$f'(x) = \lim_{h \to 0} \frac{f(x + h) - f(x)}{h} = \lim_{h \to 0} \frac{c - c}{h}$$

$$= \lim_{h \to 0} \frac{0}{h} = \lim_{h \to 0} 0 = 0$$

Thus, we have our first rule:

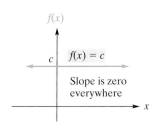

FIGURE 11.13 The slope of a constant function is 0.

RULE 1   Derivative of a Constant

If c is a constant, then

$$\frac{d}{dx}(c) = 0$$

That is, the derivative of a constant function is zero.

---

[1] A. Christofi and A. Agapos, "Interest Rate Deregulation: An Empirical Justification," *Review of Business and Economic Research*, XX, no. 1 (1984), 39–49.

●**EXAMPLE 1** **Derivatives of Constant Functions**

a. $\dfrac{d}{dx}(3) = 0$ because 3 is a constant function.

b. If $g(x) = \sqrt{5}$, then $g'(x) = 0$ because g is a constant function. For example, the derivative of g when $x = 4$ is $g'(4) = 0$.

c. If $s(t) = (1,938,623)^{807.4}$, then $ds/dt = 0$.

NOW WORK PROBLEM 1

The next rule gives a formula for the derivative of "x raised to a constant power"—that is, the derivative of $f(x) = x^n$, where n is an arbitrary real number. A function of this form is called a **power function.** For example, $f(x) = x^2$ is a power function. While the rule we record is valid for all real n, we will establish it only in the case where n is a positive integer. The rule is so central to differential calculus that it warrants a detailed calculation—if only in the case where n is a positive integer. Whether we use the $h \to 0$ form of the definition of derivative or the $z \to x$ form, the calculation of $\dfrac{dx^n}{dx}$ is instructive and provides good practice with summation notation, whose use is more essential in later chapters. We provide a calculation for each possibility. As you will see, we must either expand $(x + h)^n$, to use the $h \to 0$ form of Equation (2) from Section 11.1, or factor $z^n - x^n$, to use the $z \to x$ form.

For the first of these we recall the *binomial theorem* of Section 9.2:

$$(x + h)^n = \sum_{i=0}^{n} {}_nC_i x^{n-i} h^i$$

where the ${}_nC_i$ are the binomial coefficients, whose precise descriptions, except for ${}_nC_0 = 1$ and ${}_nC_1 = n$, are not necessary here (but are given in Section 8.2). For the second we have

$$(z - x)\left(\sum_{i=0}^{n-1} x^i z^{n-1-i}\right) = z^n - x^n$$

which is easily verified by carrying out the multiplication using the rules for manipulating summations given in Section 1.5. In fact we have

$$(z - x)\left(\sum_{i=0}^{n-1} x^i z^{n-1-i}\right) = z\sum_{i=0}^{n-1} x^i z^{n-1-i} - x\sum_{i=0}^{n-1} x^i z^{n-1-i}$$

$$= \sum_{i=0}^{n-1} x^i z^{n-i} - \sum_{i=0}^{n-1} x^{i+1} z^{n-1-i}$$

$$= \left(z^n + \sum_{i=1}^{n-1} x^i z^{n-i}\right) - \left(\sum_{i=0}^{n-2} x^{i+1} z^{n-1-i} + x^n\right)$$

$$= z^n - x^n$$

where you should check that the two summations in the second to last line really do cancel as shown.

RULE 2 **Derivative of x^n**

If n is any real number, then

$$\dfrac{d}{dx}(x^n) = nx^{n-1}$$

That is, the derivative of a constant power of x is the exponent times x raised to a power one less than the given power.

For n a positive integer, if $f(x) = x^n$, the definition of the derivative gives

$$f'(x) = \lim_{h \to 0} \frac{f(x+h) - f(x)}{h} = \lim_{h \to 0} \frac{(x+h)^n - x^n}{h}$$

By our previous discussion on expanding $(x+h)^n$,

$$f'(x) = \lim_{h \to 0} \frac{\sum_{i=0}^{n} {}_nC_i x^{n-i} h^i - x^n}{h}$$

$$\overset{(1)}{=} \lim_{h \to 0} \frac{\sum_{i=1}^{n} {}_nC_i x^{n-i} h^i}{h}$$

$$\overset{(2)}{=} \lim_{h \to 0} \frac{h \sum_{i=1}^{n} {}_nC_i x^{n-i} h^{i-1}}{h}$$

$$\overset{(3)}{=} \lim_{h \to 0} \sum_{i=1}^{n} {}_nC_i x^{n-i} h^{i-1}$$

$$\overset{(4)}{=} \lim_{h \to 0} \left(nx^{n-1} + \sum_{i=2}^{n} {}_nC_i x^{n-i} h^{i-1} \right)$$

$$\overset{(5)}{=} nx^{n-1}$$

where we justify the further steps as follows:

(1) The $i = 0$ term in the summation is ${}_nC_0 x^n h^0 = x^n$ so it cancels with the separate, last, term: $-x^n$.

(2) We are able to extract a common factor of h from each term in the sum.

(3) This is the crucial step. The expressions separated by the equal sign are limits as $h \to 0$ of functions of h that are equal for $h \neq 0$.

(4) The $i = 1$ term in the summation is ${}_nC_1 x^{n-1} h^0 = nx^{n-1}$. It is the only one that does not contain a factor of h, and we separated it from the other terms.

(5) Finally, in determining the limit we made use of the fact that the isolated term is independent of h; while all the others contain h as a factor and so have limit 0 as $h \to 0$.

Now, using the $z \to x$ limit for the definition of the derivative and $f(x) = x^n$, we have

$$f'(x) = \lim_{z \to x} \frac{f(z) - f(x)}{z - x} = \lim_{h \to 0} \frac{z^n - x^n}{z - x}$$

By our previous discussion on factoring $z^n - x^n$ we have

$$f'(x) = \lim_{z \to x} \frac{(z - x)\left(\sum_{i=0}^{n-1} x^i z^{n-1-i} \right)}{z - x}$$

$$\overset{(1)}{=} \lim_{z \to x} \sum_{i=0}^{n-1} x^i z^{n-1-i}$$

$$\overset{(2)}{=} \sum_{i=0}^{n-1} x^i x^{n-1-i}$$

$$\overset{(3)}{=} \sum_{i=0}^{n-1} x^{n-1}$$

$$\overset{(4)}{=} nx^{n-1}$$

where this time we justify the further steps as follows:

(1) Here the crucial step comes first. The expressions separated by the equal sign are limits as $z \to x$ of functions of z that are equal for $z \neq x$.

(2) The limit is given by evaluation because the expression is a polynomial in the variable z.

(3) An obvious rule for exponents is used.

(4) Each term in the sum is x^{n-1}, independent of i, and there are n such terms.

● EXAMPLE 2   Derivatives of Powers of x

a. By Rule 2, $\dfrac{d}{dx}(x^2) = 2x^{2-1} = 2x$.

b. If $F(x) = x = x^1$, then $F'(x) = 1 \cdot x^{1-1} = 1 \cdot x^0 = 1$. Thus, the derivative of x with respect to x is 1.

c. If $f(x) = x^{-10}$, then $f'(x) = -10x^{-10-1} = -10x^{-11}$.

NOW WORK PROBLEM 3 ●●●

When we apply a differentiation rule to a function, sometimes the function must first be rewritten so that it has the proper form for that rule. For example, to differentiate $f(x) = \dfrac{1}{x^{10}}$ we would first rewrite f as $f(x) = x^{-10}$ and then proceed as in Example 2(c).

● EXAMPLE 3   Rewriting Functions in the Form x^n

a. To differentiate $y = \sqrt{x}$, we rewrite $\sqrt{x}$ as $x^{1/2}$ so that it has the form x^n. Thus,

$$\frac{dy}{dx} = \frac{1}{2}x^{(1/2)-1} = \frac{1}{2}x^{-1/2} = \frac{1}{2\sqrt{x}}$$

which agrees with our limit calculation in Example 5 of Section 11.1.

b. Let $h(x) = \dfrac{1}{x\sqrt{x}}$. To apply Rule 2, we must rewrite $h(x)$ as $h(x) = x^{-3/2}$ so that it has the form x^n. We have

$$h'(x) = \frac{d}{dx}(x^{-3/2}) = -\frac{3}{2}x^{(-3/2)-1} = -\frac{3}{2}x^{-5/2}$$

NOW WORK PROBLEM 39 ●●●

CAUTION

In Example 3(b), do not rewrite $\dfrac{1}{x\sqrt{x}}$ as $\dfrac{1}{x^{3/2}}$ and then merely differentiate the denominator.

Now that we can say immediately that the derivative of x^3 is $3x^2$, the question arises as to what we could say about the derivative of a *multiple* of x^3, such as $5x^3$. Our next rule will handle this situation of differentiating a constant times a function.

RULE 3   Constant Factor Rule

If f is a differentiable function and c is a constant, then $cf(x)$ is differentiable, and

$$\frac{d}{dx}(cf(x)) = cf'(x)$$

That is, the derivative of a constant times a function is the constant times the derivative of the function.

Proof. If $g(x) = cf(x)$, applying the definition of the derivative of g gives

$$g'(x) = \lim_{h \to 0} \frac{g(x+h) - g(x)}{h} = \lim_{h \to 0} \frac{cf(x+h) - cf(x)}{h}$$

$$= \lim_{h \to 0} \left(c \cdot \frac{f(x+h) - f(x)}{h} \right) = c \cdot \lim_{h \to 0} \frac{f(x+h) - f(x)}{h}$$

But $\displaystyle\lim_{h \to 0} \frac{f(x+h) - f(x)}{h}$ is $f'(x)$; so $g'(x) = cf'(x)$.

● **EXAMPLE 4   Differentiating a Constant Times a Function**

Differentiate the following functions.

a. $g(x) = 5x^3$

Solution: Here g is a constant (5) times a function (x^3). So

$$\frac{d}{dx}(5x^3) = 5\frac{d}{dx}(x^3) \qquad \text{(Rule 3)}$$

$$= 5(3x^{3-1}) = 15x^2 \qquad \text{(Rule 2)}$$

b. $f(q) = \dfrac{13q}{5}$

Solution:

Strategy   We first rewrite f as a constant times a function and then apply Rule 2.

Because $\dfrac{13q}{5} = \dfrac{13}{5}q$, f is the constant $\dfrac{13}{5}$ times the function q. Thus,

$$f'(q) = \frac{13}{5}\frac{d}{dq}(q) \qquad \text{(Rule 3)}$$

$$= \frac{13}{5}\cdot 1 = \frac{13}{5} \qquad \text{(Rule 2)}$$

c. $y = \dfrac{0.25}{\sqrt[5]{x^2}}$

Solution: We can express y as a constant times a function:

$$y = 0.25 \cdot \frac{1}{\sqrt[5]{x^2}} = 0.25x^{-2/5}$$

Hence,

$$y' = 0.25\frac{d}{dx}(x^{-2/5}) \qquad \text{(Rule 3)}$$

$$= 0.25\left(-\frac{2}{5}x^{-7/5}\right) = -0.1x^{-7/5} \qquad \text{(Rule 2)}$$

NOW WORK PROBLEM 7 ●●

The next rule involves derivatives of sums and differences of functions.

CAUTION

To differentiate $f(x) = (4x)^3$, you may be tempted to write $f'(x) = 3(4x)^2$. **This is incorrect!** Do you see why? The reason is that Rule 2 applies to a power of the variable x, **not** a power of an expression involving x, such as $4x$. To apply our rules, we must get a suitable form for $f(x)$. We can rewrite $(4x)^3$ as $4^3x^3 = 64x^3$. Thus,

$$f'(x) = 64\frac{d}{dx}(x^3) = 64(3x^2) = 192x^2$$

RULE 4   Sum or Difference Rule

If f and g are differentiable functions, then $f + g$ and $f - g$ are differentiable, and

$$\frac{d}{dx}(f(x) + g(x)) = f'(x) + g'(x)$$

and

$$\frac{d}{dx}(f(x) - g(x)) = f'(x) - g'(x)$$

That is, the derivative of the sum (difference) of two functions is the sum (difference) of their derivatives.

Proof. For the case of a sum, if $F(x) = f(x) + g(x)$, applying the definition of the derivative of F gives

$$F'(x) = \lim_{h \to 0} \frac{F(x+h) - F(x)}{h}$$

$$= \lim_{h \to 0} \frac{(f(x+h) + g(x+h)) - (f(x) + g(x))}{h}$$

$$= \lim_{h \to 0} \frac{(f(x+h) - f(x)) + (g(x+h) - g(x))}{h} \qquad \text{(regrouping)}$$

$$= \lim_{h \to 0} \left(\frac{f(x+h) - f(x)}{h} + \frac{g(x+h) - g(x)}{h} \right)$$

Because the limit of a sum is the sum of the limits,

$$F'(x) = \lim_{h \to 0} \frac{f(x+h) - f(x)}{h} + \lim_{h \to 0} \frac{g(x+h) - g(x)}{h}$$

But these two limits are $f'(x)$ and $g'(x)$. Thus,

$$F'(x) = f'(x) + g'(x)$$

The proof for the derivative of a difference of two functions is similar.

Rule 4 can be extended to the derivative of any number of sums and differences of functions. For example,

$$\frac{d}{dx}[f(x) - g(x) + h(x) + k(x)] = f'(x) - g'(x) + h'(x) + k'(x)$$

● EXAMPLE 5   Differentiating Sums and Differences of Functions

Differentiate the following functions.

a. $F(x) = 3x^5 + \sqrt{x}$

Solution: Here F is the sum of two functions, $3x^5$ and $\sqrt{x}$. Therefore,

$$F'(x) = \frac{d}{dx}(3x^5) + \frac{d}{dx}(x^{1/2}) \qquad \text{(Rule 4)}$$

$$= 3\frac{d}{dx}(x^5) + \frac{d}{dx}(x^{1/2}) \qquad \text{(Rule 3)}$$

$$= 3(5x^4) + \frac{1}{2}x^{-1/2} = 15x^4 + \frac{1}{2\sqrt{x}} \qquad \text{(Rule 2)}$$

b. $f(z) = \dfrac{z^4}{4} - \dfrac{5}{z^{1/3}}$

Solution: To apply our rules, we will rewrite $f(z)$ in the form $f(z) = \frac{1}{4}z^4 - 5z^{-1/3}$. Since f is the difference of two functions,

$$f'(z) = \frac{d}{dz}\left(\frac{1}{4}z^4\right) - \frac{d}{dz}(5z^{-1/3}) \qquad \text{(Rule 4)}$$

$$= \frac{1}{4}\frac{d}{dz}(z^4) - 5\frac{d}{dz}(z^{-1/3}) \qquad \text{(Rule 3)}$$

$$= \frac{1}{4}(4z^3) - 5\left(-\frac{1}{3}z^{-4/3}\right) \qquad \text{(Rule 2)}$$

$$= z^3 + \frac{5}{3}z^{-4/3}$$

c. $y = 6x^3 - 2x^2 + 7x - 8$

Solution:

$$\frac{dy}{dx} = \frac{d}{dx}(6x^3) - \frac{d}{dx}(2x^2) + \frac{d}{dx}(7x) - \frac{d}{dx}(8)$$

$$= 6\frac{d}{dx}(x^3) - 2\frac{d}{dx}(x^2) + 7\frac{d}{dx}(x) - \frac{d}{dx}(8)$$

$$= 6(3x^2) - 2(2x) + 7(1) - 0$$
$$= 18x^2 - 4x + 7$$

NOW WORK PROBLEM 47

In Examples 6 and 7, we need to rewrite the given function in a form to which our rules apply.

● EXAMPLE 6   Finding a Derivative

Find the derivative of $f(x) = 2x(x^2 - 5x + 2)$ *when* $x = 2$.

Solution: We multiply and then differentiate each term:

$$f(x) = 2x^3 - 10x^2 + 4x$$
$$f'(x) = 2(3x^2) - 10(2x) + 4(1)$$
$$= 6x^2 - 20x + 4$$
$$f'(2) = 6(2)^2 - 20(2) + 4 = -12$$

NOW WORK PROBLEM 75

● EXAMPLE 7   Finding an Equation of a Tangent Line

Find an equation of the tangent line to the curve

$$y = \frac{3x^2 - 2}{x}$$

when $x = 1$.

Solution:

Strategy   First we find $\dfrac{dy}{dx}$, which gives the slope of the tangent line at any point. Evaluating $\dfrac{dy}{dx}$ when $x = 1$ gives the slope of the required tangent line. We then determine the y-coordinate of the point on the curve when $x = 1$. Finally, we substitute the slope and both of the coordinates of the point in point–slope form to obtain an equation of the tangent line.

Rewriting y as a difference of two functions, we have

$$y = \frac{3x^2}{x} - \frac{2}{x} = 3x - 2x^{-1}$$

Thus,

$$\frac{dy}{dx} = 3(1) - 2((-1)x^{-2}) = 3 + \frac{2}{x^2}$$

The slope of the tangent line to the curve when $x = 1$ is

$$\left.\frac{dy}{dx}\right|_{x=1} = 3 + \frac{2}{1^2} = 5$$

To find the y-coordinate of the point on the curve where $x = 1$, we evaluate $y = \dfrac{3x^2 - 2}{x}$ at $x = 1$. This gives

$$y = \frac{3(1)^2 - 2}{1} = 1$$

CAUTION

To obtain the y-value of the point on the curve when $x = 1$, evaluate the original function, not the derived function.

Hence, the point $(1, 1)$ lies on both the curve and the tangent line. Therefore, an equation of the tangent line is

$$y - 1 = 5(x - 1)$$

In slope–intercept form, we have

$$y = 5x - 4$$

Problems 11.2

In Problems 1–74, differentiate the functions.

*1. $f(x) = 5$

2. $f(x) = \left(\frac{6}{7}\right)^{2/3}$

*3. $y = x^6$

4. $f(x) = x^{21}$

5. $y = x^{80}$

6. $y = x^{5.3}$

*7. $f(x) = 9x^2$

8. $y = 4x^3$

9. $g(w) = 8w^7$

10. $v(x) = x^e$

11. $y = \frac{2}{3}x^4$

12. $f(p) = \sqrt{3}p^4$

13. $f(t) = \dfrac{t^7}{25}$

14. $y = \dfrac{x^7}{7}$

15. $f(x) = x + 3$

16. $f(x) = 3x - 2$

17. $f(x) = 4x^2 - 2x + 3$

18. $F(x) = 5x^2 - 9x$

19. $g(p) = p^4 - 3p^3 - 1$

20. $f(t) = -13t^2 + 14t + 1$

21. $y = x^3 - \sqrt{x}$

22. $y = -8x^4 + \ln 2$

23. $y = -13x^3 + 14x^2 - 2x + 3$

24. $V(r) = r^8 - 7r^6 + 3r^2 + 1$

25. $f(x) = 2(13 - x^4)$

26. $\phi(t) = 5(t^3 - 3^2)$

27. $g(x) = \dfrac{13 - x^4}{3}$

28. $f(x) = \dfrac{5(x^4 - 6)}{2}$

29. $h(x) = 4x^4 + x^3 - \dfrac{9x^2}{2} + 8x$

30. $k(x) = -2x^2 + \dfrac{5}{3}x + 11$

31. $f(x) = \dfrac{3x^4}{10} + \dfrac{7}{3}x^3$

32. $p(x) = \dfrac{x^7}{7} + \dfrac{2x}{3}$

33. $f(x) = x^{3/5}$

34. $f(x) = 2x^{-14/5}$

35. $y = x^{3/4} + 2x^{5/3}$

36. $y = 5x^3 - x^{-2/5}$

37. $y = 11\sqrt{x}$

38. $y = \sqrt{x^7}$

*39. $f(r) = 6\sqrt[3]{r}$

40. $y = 4\sqrt[8]{x^2}$

41. $f(x) = x^{-4}$

42. $f(s) = 2s^{-3}$

43. $f(x) = x^{-3} + x^{-5} - 2x^{-6}$

44. $f(x) = 100x^{-3} + 10x^{1/2}$

45. $y = \dfrac{1}{x}$

46. $f(x) = \dfrac{2}{x^3}$

*47. $y = \dfrac{8}{x^5}$

48. $y = \dfrac{1}{4x^5}$

49. $g(x) = \dfrac{4}{3x^3}$

50. $y = \dfrac{1}{x^2}$

51. $f(t) = \dfrac{1}{2t}$

52. $g(x) = \dfrac{7}{9x}$

53. $f(x) = \dfrac{x}{7} + \dfrac{7}{x}$

54. $\Phi(x) = \dfrac{x^3}{3} - \dfrac{3}{x^3}$

55. $f(x) = -9x^{1/3} + 5x^{-2/5}$

56. $f(z) = 3z^{1/4} - 12^2 - 8z^{-3/4}$

57. $q(x) = \dfrac{1}{\sqrt[3]{8x^2}}$

58. $f(x) = \dfrac{3}{\sqrt[4]{x^3}}$

59. $y = \dfrac{2}{\sqrt{x}}$

60. $y = \dfrac{1}{2\sqrt{x}}$

61. $y = x^2\sqrt{x}$

62. $f(x) = (2x^3)(4x^2)$

63. $f(x) = x(3x^2 - 10x + 7)$

64. $f(x) = x^3(3x^6 - 5x^2 + 4)$

65. $f(x) = x^3(3x)^2$

66. $s(x) = \sqrt[3]{x}(\sqrt[4]{x} - 6x + 3)$

67. $v(x) = x^{-2/3}(x + 5)$

68. $f(x) = x^{3/5}(x^2 + 7x + 11)$

69. $f(q) = \dfrac{3q^2 + 4q - 2}{q}$

70. $f(w) = \dfrac{w - 5}{w^5}$

71. $f(x) = (x + 1)(x + 3)$

72. $f(x) = x^2(x - 2)(x + 4)$

73. $w(x) = \dfrac{x^2 + x^3}{x^2}$

74. $f(x) = \dfrac{7x^3 + x}{6\sqrt{x}}$

For each curve in Problems 75–78, find the slopes at the indicated points.

*75. $y = 3x^2 + 4x - 8$; $(0, -8)$, $(2, 12)$, $(-3, 7)$

76. $y = 5 - 6x - 2x^3$; $(0, 5)$, $\left(\frac{3}{2}, -\frac{43}{4}\right)$, $(-3, 77)$

77. $y = 4$; when $x = -4$, $x = 7$, $x = 22$

78. $y = 3x - 4\sqrt{x}$; when $x = 4$, $x = 9$, $x = 25$

In Problems 79–82, find an equation of the tangent line to the curve at the indicated point.

79. $y = 4x^2 + 5x + 6$; $(1, 15)$

80. $y = \dfrac{1 - x^2}{5}$; $(4, -3)$

81. $y = \dfrac{1}{x^3}$; $\left(2, \frac{1}{8}\right)$

82. $y = -\sqrt[3]{x}$; $(8, -2)$

83. Find an equation of the tangent line to the curve

$$y = 3 + x - 5x^2 + x^4$$

when $x = 0$.

84. Repeat Problem 83 for the curve

$$y = \frac{\sqrt{x}(2 - x^2)}{x}$$

when $x = 4$.

85. Find all points on the curve

$$y = \frac{5}{2}x^2 - x^3$$

where the tangent line is horizontal.

86. Repeat Problem 85 for the curve

$$y = \frac{x^5}{5} - x + 1$$

87. Find all points on the curve

$$y = x^2 - 5x + 3$$

where the slope is 1.

88. Repeat Problem 87 for the curve

$$y = x^4 - 31x + 11$$

89. If $f(x) = \sqrt{x} + \dfrac{1}{\sqrt{x}}$, evaluate the expression

$$\frac{x - 1}{2x\sqrt{x}} - f'(x)$$

90. **Economics**   Eswaran and Kotwal[2] consider agrarian economies in which there are two types of workers, permanent and casual. Permanent workers are employed on long-term contracts and may receive benefits such as holiday gifts and emergency aid. Casual workers are hired on a daily basis and perform routine and menial tasks such as weeding, harvesting, and threshing. The difference z in the present-value cost of hiring a permanent worker over

_____

[2]M. Eswaran and A. Kotwal, "A Theory of Two-Tier Labor Markets in Agrarian Economies," *The American Economic Review*, 75, no. 1 (1985), 162–77.

that of hiring a casual worker is given by

$$z = (1 + b)w_p - bw_c$$

where w_p and w_c are wage rates for permanent labor and casual labor, respectively, b is a constant, and w_p is a function of w_c. Eswaran and Kotwal claim that

$$\frac{dz}{dw_c} = (1 + b)\left[\frac{dw_p}{dw_c} - \frac{b}{1 + b}\right]$$

Verify this.

91. Find an equation of the tangent line to the graph of $y = x^3 - 3x$ at the point $(2, 2)$. Graph both the function and the tangent line on the same screen. Notice that the line passes through $(2, 2)$ and the line appears to be tangent to the curve.

92. Find an equation of the tangent line to the graph of $y = \sqrt[3]{x}$, at the point $(-8, -2)$. Graph both the function and the tangent line on the same screen. Notice that the line passes through $(-8, -2)$ and the line appears to be tangent to the curve.

11.3 The Derivative as a Rate of Change

OBJECTIVE

To motivate the instantaneous rate of change of a function by means of velocity and to interpret the derivative as an instantaneous rate of change. To develop the "marginal" concept, which is frequently used in business and economics.

FIGURE 11.14 Motion along a number line.

We have given a geometric interpretation of the derivative as being the slope of the tangent line to a curve at a point. Historically, an important application of the derivative involves the motion of an object traveling in a straight line. This gives us a convenient way to interpret the derivative as a *rate of change.*

To denote the change in a variable such as x, the symbol Δx (read "delta x") is commonly used. For example, if x changes from 1 to 3, then the change in x is $\Delta x = 3 - 1 = 2$. The new value of $x(= 3)$ is the old value plus the change, or $1 + \Delta x$. Similarly, if t increases by Δt, the new value is $t + \Delta t$. We will use Δ-notation in the discussion that follows.

Suppose an object moves along the number line in Figure 11.14 according to the equation

$$s = f(t) = t^2$$

where s is the position of the object at time t. This equation is called an *equation of motion,* and f is called a **position function.** Assume that t is in seconds and s is in meters. At $t = 1$ the position is $s = f(1) = 1^2 = 1$, and at $t = 3$ the position is $s = f(3) = 3^2 = 9$. Over this two-second time interval, the object has a change in position, or a *displacement,* of $9 - 1 = 8$ m, and the *average velocity* of the object is defined as

$$v_{\text{ave}} = \frac{\text{displacement}}{\text{length of time interval}} \tag{1}$$

$$= \frac{8}{2} = 4 \text{ m/s}$$

To say that the average velocity is 4 m/s from $t = 1$ to $t = 3$ means that, *on the average,* the position of the object changed by 4 m to the right each second during that time interval. Let us denote the changes in s-values and t-values by Δs and Δt, respectively. Then the average velocity is given by

$$v_{\text{ave}} = \frac{\Delta s}{\Delta t} = 4 \text{ m/s} \quad \text{(for the interval } t = 1 \text{ to } t = 3)$$

The ratio $\Delta s/\Delta t$ is also called the **average rate of change of s with respect to** t over the interval from $t = 1$ to $t = 3$.

Now, let the time interval be only 1 second long (that is, $\Delta t = 1$). Then, for the *shorter* interval from $t = 1$ to $t = 1 + \Delta t = 2$, we have $f(2) = 2^2 = 4$, so

$$v_{\text{ave}} = \frac{\Delta s}{\Delta t} = \frac{f(2) - f(1)}{\Delta t} = \frac{4 - 1}{1} = 3 \text{ m/s}$$

More generally, over the time interval from $t = 1$ to $t = 1 + \Delta t$, the object moves from position $f(1)$ to position $f(1 + \Delta t)$. Thus, its displacement is

$$\Delta s = f(1 + \Delta t) - f(1)$$

Since the time interval has length Δt, the object's average velocity is given by

$$v_{\text{ave}} = \frac{\Delta s}{\Delta t} = \frac{f(1 + \Delta t) - f(1)}{\Delta t}$$

TABLE 11.2

Length of Time Interval Δt	Time Interval $t = 1$ to $t = 1 + \Delta t$	Average Velocity, $\dfrac{\Delta s}{\Delta t} = \dfrac{f(1 + \Delta t) - f(1)}{\Delta t}$
0.1	$t = 1$ to $t = 1.1$	2.1 m/s
0.07	$t = 1$ to $t = 1.07$	2.07 m/s
0.05	$t = 1$ to $t = 1.05$	2.05 m/s
0.03	$t = 1$ to $t = 1.03$	2.03 m/s
0.01	$t = 1$ to $t = 1.01$	2.01 m/s
0.001	$t = 1$ to $t = 1.001$	2.001 m/s

If Δt were to become smaller and smaller, the average velocity over the interval from $t = 1$ to $t = 1 + \Delta t$ would be close to what we might call the *instantaneous velocity* at time $t = 1$, that is, the velocity at a *point* in time ($t = 1$), as opposed to the velocity over an *interval* of time. For some typical values of Δt between 0.1 and 0.001, we get the average velocities in Table 11.2, which you can verify.

The table suggests that as the length of the time interval approaches zero, the average velocity approaches the value 2 m/s. In other words, as Δt approaches 0, $\Delta s / \Delta t$ approaches 2 m/s. We define the limit of the average velocity as $\Delta t \to 0$ to be the **instantaneous velocity** (or simply the **velocity**), v, at time $t = 1$. This limit is also called the **instantaneous rate of change** of s with respect to t at $t = 1$:

$$v = \lim_{\Delta t \to 0} v_{\text{ave}} = \lim_{\Delta t \to 0} \frac{\Delta s}{\Delta t} = \lim_{\Delta t \to 0} \frac{f(1 + \Delta t) - f(1)}{\Delta t}$$

If we think of Δt as h, then the limit on the right is simply the derivative of s with respect to t at $t = 1$. Thus, the instantaneous velocity of the object at $t = 1$ is just ds/dt at $t = 1$. Because $s = t^2$ and

$$\frac{ds}{dt} = 2t$$

the velocity at $t = 1$ is

$$v = \left. \frac{ds}{dt} \right|_{t=1} = 2(1) = 2 \text{ m/s}$$

which confirms our previous conclusion.

In summary, if $s = f(t)$ is the position function of an object moving in a straight line, then the average velocity of the object over the time interval $[t, t + \Delta t]$ is given by

$$v_{\text{ave}} = \frac{\Delta s}{\Delta t} = \frac{f(t + \Delta t) - f(t)}{\Delta t}$$

and the velocity at time t is given by

$$v = \lim_{\Delta t \to 0} \frac{f(t + \Delta t) - f(t)}{\Delta t} = \frac{ds}{dt}$$

Selectively combining equations for v, we have

$$\frac{ds}{dt} = \lim_{\Delta t \to 0} \frac{\Delta s}{\Delta t}$$

which provides motivation for the otherwise bizarre Leibniz notation. (After all, Δ is the [uppercase] Greek letter corresponding to d.)

● EXAMPLE 1   **Finding Average Velocity and Velocity**

Suppose the position function of an object moving along a number line is given by $s = f(t) = 3t^2 + 5$, where t is in seconds and s is in meters.

a. Find the average velocity over the interval $[10, 10.1]$.

b. Find the velocity when $t = 10$.

Solution:

a. Here $t = 10$ and $\Delta t = 10.1 - 10 = 0.1$. So we have

$$v_{\text{ave}} = \frac{\Delta s}{\Delta t} = \frac{f(t + \Delta t) - f(t)}{\Delta t}$$

$$= \frac{f(10 + 0.1) - f(10)}{0.1}$$

$$= \frac{f(10.1) - f(10)}{0.1}$$

$$= \frac{311.03 - 305}{0.1} = \frac{6.03}{0.1} = 60.3 \text{ m/s}$$

b. The velocity at time t is given by

$$v = \frac{ds}{dt} = 6t$$

When $t = 10$, the velocity is

$$\left. \frac{ds}{dt} \right|_{t=10} = 6(10) = 60 \text{ m/s}$$

Notice that the average velocity over the interval $[10, 10.1]$ is close to the velocity at $t = 10$. This is to be expected because the length of the interval is small.

NOW WORK PROBLEM 1 ●●

Our discussion of the rate of change of s with respect to t applies equally well to *any* function $y = f(x)$. This means that we have the following:

If $y = f(x)$, then

$$\frac{\Delta y}{\Delta x} = \frac{f(x + \Delta x) - f(x)}{\Delta x} = \begin{cases} \text{average rate of change} \\ \text{of } y \text{ with respect to } x \\ \text{over the interval from} \\ x \text{ to } x + \Delta x \end{cases}$$

and

$$\frac{dy}{dx} = \lim_{\Delta x \to 0} \frac{\Delta y}{\Delta x} = \begin{cases} \text{instantaneous rate of change} \\ \text{of } y \text{ with respect to } x \end{cases} \tag{2}$$

Because the instantaneous rate of change of $y = f(x)$ at a point is a derivative, it is also the *slope of the tangent line* to the graph of $y = f(x)$ at that point. For convenience, we usually refer to the instantaneous rate of change simply as the **rate of change.** The interpretation of a derivative as a rate of change is extremely important.

Let us now consider the significance of the rate of change of y with respect to x. From Equation (2), if Δx (a change in x) is close to 0, then $\Delta y / \Delta x$ is close to dy/dx. That is,

$$\frac{\Delta y}{\Delta x} \approx \frac{dy}{dx}$$

Therefore,

$$\Delta y \approx \frac{dy}{dx} \Delta x \tag{3}$$

That is, if x changes by Δx, then the change in y, Δy, is approximately dy/dx times the change in x. In particular,

if x changes by 1, an estimate of the change in y is $\dfrac{dy}{dx}$

Suppose that the profit P made by selling a certain product at a price of p per unit is given by $P = f(p)$ and the rate of change of that profit with respect to change in price is $\dfrac{dP}{dp} = 5$ at $p = 25$. Estimate the change in the profit P if the price changes from 25 to 25.5.

The position of an object thrown upward at a speed of 16 feet/s from a height of 0 feet is given by $y(t) = 16t - 16t^2$. Find the rate of change of y with respect to t, and evaluate it when $t = 0.5$. Use your graphing calculator to graph $y(t)$. Use the graph to interpret the behavior of the object when $t = 0.5$.

● EXAMPLE 2   **Estimating Δy by Using dy/dx**

Suppose that $y = f(x)$ and $\dfrac{dy}{dx} = 8$ when $x = 3$. Estimate the change in y if x changes from 3 to 3.5.

Solution: We have $dy/dx = 8$ and $\Delta x = 3.5 - 3 = 0.5$. The change in y is given by Δy, and, from Equation (3),

$$\Delta y \approx \frac{dy}{dx} \Delta x = 8(0.5) = 4$$

We remark that, since $\Delta y = f(3.5) - f(3)$, we have $f(3.5) = f(3) + \Delta y$. For example, if $f(3) = 5$, then $f(3.5)$ can be estimated by $5 + 4 = 9$.

● EXAMPLE 3   **Finding a Rate of Change**

Find the rate of change of $y = x^4$ with respect to x, and evaluate it when $x = 2$ and when $x = -1$. Interpret your results.

Solution: The rate of change is

$$\frac{dy}{dx} = 4x^3$$

When $x = 2$, $dy/dx = 4(2)^3 = 32$. This means that if x increases, from 2, by a small amount, then y increases approximately 32 times as much. More simply, we say that, when $x = 2$, y is increasing 32 times as fast as x does. When $x = -1$, $dy/dx = 4(-1)^3 = -4$. The significance of the minus sign on -4 is that, when $x = -1$, y is *decreasing* 4 times as fast as x increases.

NOW WORK PROBLEM 11 ●●

● EXAMPLE 4   **Rate of Change of Price with Respect to Quantity**

Let $p = 100 - q^2$ be the demand function for a manufacturer's product. Find the rate of change of price p per unit with respect to quantity q. How fast is the price changing with respect to q when $q = 5$? Assume that p is in dollars.

Solution: The rate of change of p with respect to q is

$$\frac{dp}{dq} = \frac{d}{dq}(100 - q^2) = -2q$$

Thus,

$$\left.\frac{dp}{dq}\right|_{q=5} = -2(5) = -10$$

This means that when five units are demanded, an *increase* of one extra unit demanded corresponds to a decrease of approximately \$10 in the price per unit that consumers are willing to pay.

● EXAMPLE 5   **Rate of Change of Volume**

A spherical balloon is being filled with air. Find the rate of change of the volume of air in the balloon with respect to its radius. Evaluate this rate of change when the radius is 2 ft.

Solution: The formula for the volume V of a ball of radius r is $V = \frac{4}{3}\pi r^3$. The rate of change of V with respect to r is

$$\frac{dV}{dr} = \frac{4}{3}\pi(3r^2) = 4\pi r^2$$

When $r = 2$ ft, the rate of change is

$$\left.\frac{dV}{dr}\right|_{r=2} = 4\pi(2)^2 = 16\pi \, \frac{\text{ft}^3}{\text{ft}}$$

This means that when the radius is 2 ft, changing the radius by 1 ft will change the volume by approximately 16π ft^3.

⬤⬤⬤

⬤ **EXAMPLE 6   Rate of Change of Enrollment**

A sociologist is studying various suggested programs that can aid in the education of preschool-age children in a certain city. The sociologist believes that x years after the beginning of a particular program, f(x) thousand preschoolers will be enrolled, where

$$f(x) = \frac{10}{9}(12x - x^2) \quad 0 \le x \le 12$$

At what rate would enrollment change (a) after three years from the start of this program and (b) after nine years?

Solution: The rate of change of $f(x)$ is

$$f'(x) = \frac{10}{9}(12 - 2x)$$

a. After three years, the rate of change is

$$f'(3) = \frac{10}{9}(12 - 2(3)) = \frac{10}{9} \cdot 6 = \frac{20}{3} = 6\frac{2}{3}$$

Thus, enrollment would be increasing at the rate of $6\frac{2}{3}$ thousand preschoolers per year.

b. After nine years, the rate is

$$f'(9) = \frac{10}{9}(12 - 2(9)) = \frac{10}{9}(-6) = -\frac{20}{3} = -6\frac{2}{3}$$

Thus, enrollment would be *decreasing* at the rate of $6\frac{2}{3}$ thousand preschoolers per year.

NOW WORK PROBLEM 9 ⬤⬤⬤

Applications of Rate of Change to Economics

A manufacturer's **total-cost function,** $c = f(q)$, gives the total cost c of producing and marketing q units of a product. The rate of change of c with respect to q is called the **marginal cost.** Thus,

$$\text{marginal cost} = \frac{dc}{dq}$$

For example, suppose $c = f(q) = 0.1q^2 + 3$ is a cost function, where c is in dollars and q is in pounds. Then

$$\frac{dc}{dq} = 0.2q$$

The marginal cost when 4 lb are produced is dc/dq, evaluated when $q = 4$:

$$\left.\frac{dc}{dq}\right|_{q=4} = 0.2(4) = 0.80$$

This means that if production is increased by 1 lb, from 4 lb to 5 lb, then the change in cost is approximately $0.80. That is, the additional pound costs about $0.80. In general, *we interpret marginal cost as the approximate cost of one additional unit of output.* After all, the difference $f(q + 1) - f(q)$ can be seen as a difference quotient

$$\frac{f(q + 1) - f(q)}{1}$$

(the case where $h = 1$). Any difference quotient can be regarded as an approximation of the corresponding derivative and, conversely, any derivative can be regarded as an approximation of any of its corresponding difference quotients. Thus, for any function f of q we can always regard $f'(q)$ and $f(q + 1) - f(q)$ as approximations of each other. In economics, the latter can usually be regarded as the exact value of the cost, or profit depending upon the function, of the $(q + 1)$th item when q are produced. The derivative is often easier to compute than the exact value. (In the case at hand, the actual cost of producing one more pound beyond 4 lb is $f(5) - f(4) = 5.5 - 4.6 = \$0.90$.)

If c is the total cost of producing q units of a product, then the **average cost per unit,** $\bar{c}$, is

$$\bar{c} = \frac{c}{q} \tag{4}$$

For example, if the total cost of 20 units is \$100, then the average cost per unit is $\bar{c} = 100/20 = \$5$. By multiplying both sides of Equation (4) by q, we have

$$c = q\bar{c}$$

That is, total cost is the product of the number of units produced and the average cost per unit.

● EXAMPLE 7 Marginal Cost

If a manufacturer's average-cost equation is

$$\bar{c} = 0.0001q^2 - 0.02q + 5 + \frac{5000}{q}$$

find the marginal-cost function. What is the marginal cost when 50 units are produced?

Solution:

> **Strategy**   The marginal-cost function is the derivative of the total-cost function c. Thus, we first find c by multiplying $\bar{c}$ by q. We have

$$c = q\bar{c}$$

$$= q \left(0.0001q^2 - 0.02q + 5 + \frac{5000}{q} \right)$$

$$c = 0.0001q^3 - 0.02q^2 + 5q + 5000$$

Differentiating c, we have the marginal-cost function:

$$\frac{dc}{dq} = 0.0001(3q^2) - 0.02(2q) + 5(1) + 0$$

$$= 0.0003q^2 - 0.04q + 5$$

The marginal cost when 50 units are produced is

$$\left.\frac{dc}{dq}\right|_{q=50} = 0.0003(50)^2 - 0.04(50) + 5 = 3.75$$

If c is in dollars and production is increased by one unit, from $q = 50$ to $q = 51$, then the cost of the additional unit is approximately \$3.75. If production is increased by $\frac{1}{3}$ unit, from $q = 50$, then the cost of the additional output is approximately $\left(\frac{1}{3}\right)(3.75) = \$1.25$.

NOW WORK PROBLEM 21 ●●

Suppose $r = f(q)$ is the **total-revenue function** for a manufacturer. The equation $r = f(q)$ states that the total dollar value received for selling q units of a product is r. The **marginal revenue** is defined as the rate of change of the total dollar value received with respect to the total number of units sold. Hence, marginal revenue is merely the derivative of r with respect to q:

$$\text{marginal revenue} = \frac{dr}{dq}$$

Marginal revenue indicates the rate at which revenue changes with respect to units sold. We interpret it as *the approximate revenue received from selling one additional unit of output.*

● EXAMPLE 8   Marginal Revenue

Suppose a manufacturer sells a product at \$2 per unit. If q units are sold, the total revenue is given by

$$r = 2q$$

The marginal-revenue function is

$$\frac{dr}{dq} = \frac{d}{dq}(2q) = 2$$

which is a constant function. Thus, the marginal revenue is 2 regardless of the number of units sold. This is what we would expect, because the manufacturer receives \$2 for each unit sold.

NOW WORK PROBLEM 23

Relative and Percentage Rates of Change

For the total-revenue function in Example 8, namely, $r = f(q) = 2q$, we have

$$\frac{dr}{dq} = 2$$

This means that revenue is changing at the rate of \$2 per unit, regardless of the number of units sold. Although this is valuable information, it may be more significant when compared to r itself. For example, if $q = 50$, then $r = 2(50) = 100$. Thus, the rate of change of revenue is $2/100 = 0.02$ *of r*. On the other hand, if $q = 5000$, then $r = 2(5000) = \$10,000$, so the rate of change of r is $2/10,000 = 0.0002$ *of r*. Although r changes at the same rate at each level, compared to r itself, this rate is relatively smaller when $r = 10,000$ than when $r = 100$. By considering the ratio

$$\frac{dr/dq}{r}$$

we have a means of comparing the rate of change of r with r itself. This ratio is called the *relative rate of change* of r. We have shown that the relative rate of change when $q = 50$ is

$$\frac{dr/dq}{r} = \frac{2}{100} = 0.02$$

and when $q = 5000$, it is

$$\frac{dr/dq}{r} = \frac{2}{10,000} = 0.0002$$

By multiplying relative rates by 100%, we obtain the so-called *percentage rates of change*. The percentage rate of change when $q = 50$ is $(0.02)(100\%) = 2\%$; when $q = 5000$, it is $(0.0002)(100\%) = 0.02\%$. For example, if an additional unit beyond 50 is sold, then revenue increases by approximately 2%.

In general, for any function f, we have the following definition:

DEFINITION

The **relative rate of change** of $f(x)$ is

$$\frac{f'(x)}{f(x)}$$

The **percentage rate of change** of $f(x)$ is

$$\frac{f'(x)}{f(x)} \cdot 100\%$$

PRINCIPLES IN PRACTICE 3

RELATIVE AND PERCENTAGE RATES OF CHANGE

The volume V enclosed by a capsule-shaped container with a cylindrical height of 4 feet and radius r is given by

$$V(r) = \frac{4}{3}\pi r^3 + 4\pi r^2$$

Determine the relative and percentage rates of change of volume with respect to the radius when the radius is 2 feet.

● **EXAMPLE 9** **Relative and Percentage Rates of Change**

Determine the relative and percentage rates of change of

$$y = f(x) = 3x^2 - 5x + 25$$

when $x = 5$.

Solution: Here

$$f'(x) = 6x - 5$$

Since $f'(5) = 6(5) - 5 = 25$ and $f(5) = 3(5)^2 - 5(5) + 25 = 75$, the relative rate of change of y when $x = 5$ is

$$\frac{f'(5)}{f(5)} = \frac{25}{75} \approx 0.333$$

Multiplying 0.333 by 100% gives the percentage rate of change: $(0.333)(100) = 33.3\%$.

NOW WORK PROBLEM 35 ●●

Problems 11.3

*1. Suppose that the position function of an object moving along a straight line is $s = f(t) = 2t^2 + 3t$, where t is in seconds and s is in meters. Find the average velocity $\Delta s/\Delta t$ over the interval $[1, 1 + \Delta t]$, where Δt is given in the following table:

Δt	1	0.5	0.2	0.1	0.01	0.001
$\Delta s/\Delta t$						

From your results, estimate the velocity when $t = 1$. Verify your estimate by using differentiation.

2. If $y = f(x) = \sqrt{2x + 5}$, find the average rate of change of y with respect to x over the interval $[3, 3 + \Delta x]$, where Δx is given in the following table:

Δx	1	0.5	0.2	0.1	0.01	0.001
$\Delta y/\Delta x$						

From your result, estimate the rate of change of y with respect to x when $x = 3$.

In each of Problems 3–8, a position function is given, where t is in seconds and s is in meters.

(a) *Find the position at the given t-value.*
(b) *Find the average velocity over the given interval.*
(c) *Find the velocity at the given t-value.*

3. $s = 2t^2 - 4t$; $[7, 7.5]$; $t = 7$

4. $s = \frac{1}{2}t + 1$; $[2, 2.1]$; $t = 2$

5. $s = 2t^3 + 6$; $[1, 1.02]$; $t = 1$

6. $s = -3t^2 + 2t + 1$; $[1, 1.25]$; $t = 1$

7. $s = t^4 - 2t^3 + t$; $[2, 2.1]$; $t = 2$

8. $s = 3t^4 - t^{7/2}$; $[0, \frac{1}{4}]$; $t = 0$

*9. **Income–Education** Sociologists studied the relation between income and number of years of education for members of a particular urban group. They found that a person with x years of education before seeking regular employment can expect to receive an average yearly income of y dollars per year, where

$$y = 5x^{5/2} + 5900 \qquad 4 \le x \le 16$$

Find the rate of change of income with respect to number of years of education. Evaluation the expression when $x = 9$.

10. Find the rate of change of the area A of a disc, with respect to its radius r, when $r = 3$ m. The area A of a disc as a function of its radius r is given by

$$A = A(r) = \pi r^2$$

*11. **Skin Temperature** The approximate temperature T of the skin in terms of the temperature T_e of the environment is given by

$$T = 32.8 + 0.27(T_e - 20)$$

where T and T_e are in degrees Celsius.[3] Find the rate of change of T with respect to T_e.

12. Biology   The volume V of a spherical cell is given by $V = \frac{4}{3}\pi r^3$, where r is the radius. Find the rate of change of volume with respect to the radius when $r = 6.3 \times 10^{-4}$ cm.

In Problems 13–18, cost functions are given, where c is the cost of producing q units of a product. In each case, find the marginal-cost function. What is the marginal cost at the given value(s) of q?

13. $c = 500 + 10q; q = 100$

14. $c = 5000 + 6q; q = 36$

15. $c = 0.1q^2 + 3q + 450; q = 5$

16. $c = 0.1q^2 + 3q + 2; q = 3$

17. $c = q^2 + 50q + 1000; q = 15, q = 16, q = 17$

18. $c = 0.04q^3 - 0.5q^2 + 4.4q + 7500; q = 5, q = 25, q = 1000$

In Problems 19–22, $\bar{c}$ represents average cost per unit, which is a function of the number q of units produced. Find the marginal-cost function and the marginal cost for the indicated values of q.

19. $\bar{c} = 0.01q + 5 + \dfrac{500}{q}; q = 50, q = 100$

20. $\bar{c} = 2 + \dfrac{1000}{q}; q = 25, q = 235$

***21.** $\bar{c} = 0.00002q^2 - 0.01q + 6 + \dfrac{20,000}{q}; q = 100, q = 500$

22. $\bar{c} = 0.002q^2 - 0.5q + 60 + \dfrac{7000}{q}; q = 15, q = 25$

In Problems 23–26, r represents total revenue and is a function of the number q of units sold. Find the marginal-revenue function and the marginal revenue for the indicated values of q.

***23.** $r = 0.8q; q = 9, q = 300, q = 500$

24. $r = q\left(15 - \frac{1}{30}q\right); q = 5, q = 15, q = 150$

25. $r = 250q + 45q^2 - q^3; q = 5, q = 10, q = 25$

26. $r = 2q(30 - 0.1q); q = 10, q = 20$

27. Hosiery Mill   The total-cost function for a hosiery mill is estimated by Dean[4] to be

$$c = -10,484.69 + 6.750q - 0.000328q^2$$

where q is output in dozens of pairs and c is total cost in dollars. Find the marginal-cost function and the average cost function and evaluate each when $q = 2000$.

28. Light and Power Plant   The total-cost function for an electric light and power plant is estimated by Nordin[5] to be

$$c = 32.07 - 0.79q + 0.02142q^2 - 0.0001q^3 \quad 20 \le q \le 90$$

where q is the eight-hour total output (as a percentage of capacity) and c is the total fuel cost in dollars. Find the marginal-cost function and evaluate it when $q = 70$.

29. Urban Concentration   Suppose the 100 largest cities in the United States in 1920 are ranked according to magnitude (areas of cities). From Lotka,[6] the following relation holds approximately:

$$PR^{0.93} = 5,000,000$$

Here, P is the population of the city having respective rank R. This relation is called the *law of urban concentration* for 1920. Solve for P in terms of R, and then find how fast the population is changing with respect to rank.

30. Depreciation   Under the straight-line method of depreciation, the value v of a certain machine after t years have elapsed is given by

$$v = 85,000 - 10,500t$$

where $0 \le t \le 9$. How fast is v changing with respect to t when $t = 2$? $t = 3$? at any time?

31. Winter Moth   A study of the winter moth was made in Nova Scotia (adapted from Embree).[7] The prepupae of the moth fall onto the ground from host trees. At a distance of x ft from the base of a host tree, the prepupal density (number of prepupae per square foot of soil) was y, where

$$y = 59.3 - 1.5x - 0.5x^2 \quad 1 \le x \le 9$$

(a) At what rate is the prepupal density changing with respect to distance from the base of the tree when $x = 6$?

(b) For what value of x is the prepupal density decreasing at the rate of 6 prepupae per square foot per foot?

32. Cost Function   For the cost function

$$c = 0.4q^2 + 4q + 5$$

find the rate of change of c with respect to q when $q = 2$. Also, what is $\Delta c/\Delta q$ over the interval $[2, 3]$?

In Problems 33–38, find (a) the rate of change of y with respect to x and (b) the relative rate of change of y. At the given value of x, find (c) the rate of change of y, (d) the relative rate of change of y, and (e) the percentage rate of change of y.

33. $y = f(x) = x + 4; x = 5$

34. $y = f(x) = 7 - 3x; x = 6$

***35.** $y = 3x^2 + 7; x = 2$

36. $y = 5 - 3x^3; x = 1$

37. $y = 8 - x^3; x = 1$

38. $y = x^2 + 3x - 4; x = -1$

39. Cost Function   For the cost function

$$c = 0.3q^2 + 3.5q + 9$$

how fast does c change with respect to q when $q = 10$? Determine the percentage rate of change of c with respect to q when $q = 10$.

[3]R. W. Stacy et al., *Essentials of Biological and Medical Physics* (New York: McGraw-Hill Book Company, 1955).

[4]J. Dean, "Statistical Cost Functions of a Hosiery Mill," *Studies in Business Administration*, XI, no. 4 (Chicago: University of Chicago Press, 1941).

[5]J. A. Nordin, "Note on a Light Plant's Cost Curves," *Econometrica*, 15 (1947), 231–35.

[6]A. J. Lotka, *Elements of Mathematical Biology* (New York: Dover Publications, Inc., 1956).

[7]D. G. Embree, "The Population Dynamics of the Winter Moth in Nova Scotia, 1954–1962," *Memoirs of the Entomological Society of Canada*, no. 46 (1965).

40. Organic Matters/Species Diversity In a discussion of contemporary waters of shallows seas, Odum[8] claims that in such waters the total organic matter y (in milligrams per liter) is a function of species diversity x (in number of species per thousand individuals). If $y = 100/x$, at what rate is the total organic matter changing with respect to species diversity when $x = 10$? What is the percentage rate of change when $x = 10$?

41. Revenue For a certain manufacturer, the revenue obtained from the sale of q units of a product is given by

$$r = 30q - 0.3q^2$$

(a) How fast does r change with respect to q? When $q = 10$, (b) find the relative rate of change of r, and (c) to the nearest percent, find the percentage rate of change of r.

42. Revenue Repeat Problem 43 for the revenue function given by $r = 10q - 0.2q^2$ and $q = 25$.

43. Weight of Limb The weight of a limb of a tree is given by $W = 2t^{0.432}$, where t is time. Find the relative rate of change of W with respect to t.

44. Response to Shock A psychological experiment[9] was conducted to analyze human responses to electrical shocks (stimuli). The subjects received shocks of various intensities. The response R to a shock of intensity I (in microamperes) was to be a number that indicated the perceived magnitude relative to that of a "standard" shock. The standard shock was assigned a magnitude of 10. Two groups of subjects were tested under slightly different conditions. The responses R_1 and R_2 of the first and second groups to a shock of intensity I were given by

$$R_1 = \frac{I^{1.3}}{1855.24} \quad 800 \le I \le 3500$$

and

$$R_2 = \frac{I^{1.3}}{1101.29} \quad 800 \le I \le 3500$$

(a) For each group, determine the relative rate of change of response with respect to intensity.

(b) How do these changes compare with each other?

(c) In general, if $f(x) = C_1 x^n$ and $g(x) = C_2 x^n$, where C_1 and C_2 are constants, how do the relative rates of change of f and g compare?

45. Cost A manufacturer of mountain bikes has found that when 20 bikes are produced per day, the average cost is $150 and the marginal cost is $125. Based on that information, approximate the total cost of producing 21 bikes per day.

46. Marginal and Average Costs Suppose that the cost function for a certain product is $c = f(q)$. If the relative rate of change of c (with respect to q) is $\dfrac{1}{q}$, prove that the marginal-cost function and the average-cost function are equal.

In Problems 47 and 48, use the numerical derivative feature of your graphing calculator.

47. If the total-cost function for a manufacturer is given by

$$c = \frac{5q^2}{\sqrt{q^2 + 3}} + 5000$$

where c is in dollars, find the marginal cost when 10 units are produced. Round your answer to the nearest cent.

48. The population of a city t years from now is given by

$$P = 250{,}000e^{0.04t}$$

Find the rate of change of population with respect to time t three years from now. Round your answer to the nearest integer.

OBJECTIVE

To find derivatives by applying the product and quotient rules, and to develop the concepts of marginal propensity to consume and marginal propensity to save.

11.4 The Product Rule and the Quotient Rule

The equation $F(x) = (x^2 + 3x)(4x + 5)$ expresses $F(x)$ as a product of two functions: $x^2 + 3x$ and $4x + 5$. To find $F'(x)$ by using only our previous rules, we first multiply the functions. Then we differentiate the result, term by term:

$$F(x) = (x^2 + 3x)(4x + 5) = 4x^3 + 17x^2 + 15x$$

$$F'(x) = 12x^2 + 34x + 15 \tag{1}$$

However, in many problems that involve differentiating a product of functions, the multiplication is not as simple as it is here. At times, it is not even practical to attempt it. Fortunately, there is a rule for differentiating a product, and the rule avoids such multiplications. Since the derivative of a sum of functions is the sum of their derivatives, you might expect a similar rule for products. However, the situation is rather subtle.

---

[8]H. T. Odum, "Biological Circuits and the Marine Systems of Texas," in *Pollution and Marine Biology*, ed. T. A. Olsen and F. J. Burgess (New York: Interscience Publishers, 1967).

[9]H. Babkoff, "Magnitude Estimation of Short Electrocutaneous Pulses," *Psychological Research*, 39, no. 1 (1976), 39–49.

RULE 1 The Product Rule

If f and g are differentiable functions, then the product fg is differentiable, and

$$\frac{d}{dx}(f(x)g(x)) = f'(x)g(x) + f(x)g'(x)$$

That is, the derivative of the product of two functions is the derivative of the first function times the second, plus the first function times the derivative of the second.

$$\frac{d}{dx}(\text{product}) = \left(\begin{array}{c}\text{derivative}\\\text{of first}\end{array}\right)(\text{second}) + (\text{first})\left(\begin{array}{c}\text{derivative}\\\text{of second}\end{array}\right)$$

Proof. If $F(x) = f(x)g(x)$, then, by the definition of the derivative of F,

$$F'(x) = \lim_{h \to 0} \frac{F(x+h) - F(x)}{h}$$

$$= \lim_{h \to 0} \frac{f(x+h)g(x+h) - f(x)g(x)}{h}$$

Now we use a "trick." Adding and subtracting $f(x)g(x+h)$ in the numerator, we have

$$F'(x) = \lim_{h \to 0} \frac{f(x+h)g(x+h) - f(x)g(x) + f(x)g(x+h) - f(x)g(x+h)}{h}$$

Regrouping gives

$$F'(x) = \lim_{h \to 0} \frac{(f(x+h)g(x+h) - f(x)g(x+h)) + (f(x)g(x+h) - f(x)g(x))}{h}$$

$$= \lim_{h \to 0} \frac{(f(x+h) - f(x))g(x+h) + f(x)(g(x+h) - g(x))}{h}$$

$$= \lim_{h \to 0} \frac{(f(x+h) - f(x))g(x+h)}{h} + \lim_{h \to 0} \frac{f(x)(g(x+h) - g(x))}{h}$$

$$= \lim_{h \to 0} \frac{f(x+h) - f(x)}{h} \cdot \lim_{h \to 0} g(x+h) + \lim_{h \to 0} f(x) \cdot \lim_{h \to 0} \frac{g(x+h) - g(x)}{h}$$

Since we assumed that f and g are differentiable,

$$\lim_{h \to 0} \frac{f(x+h) - f(x)}{h} = f'(x)$$

and

$$\lim_{h \to 0} \frac{g(x+h) - g(x)}{h} = g'(x)$$

The differentiability of g implies that g is continuous, so from Section 10.3,

$$\lim_{h \to 0} g(x+h) = g(x)$$

Thus,

$$F'(x) = f'(x)g(x) + f(x)g'(x)$$

● **EXAMPLE 1 Applying the Product Rule**

If $F(x) = (x^2 + 3x)(4x + 5)$, find $F'(x)$.

Solution: We will consider F as a product of two functions:

$$F(x) = \underbrace{(x^2 + 3x)}_{f(x)}\underbrace{(4x + 5)}_{g(x)}$$

CAUTION

It is worthwhile to repeat that the derivative of the product of two functions is somewhat subtle. Do not be tempted to make up a simpler rule.

Therefore, we can apply the product rule:

$$F'(x) = f'(x)g(x) + f(x)g'(x)$$

$$= \underbrace{\frac{d}{dx}(x^2 + 3x)}_{\substack{\text{Derivative} \\ \text{of first}}} \underbrace{(4x + 5)}_{\text{Second}} + \underbrace{(x^2 + 3x)}_{\text{First}} \underbrace{\frac{d}{dx}(4x + 5)}_{\substack{\text{Derivative} \\ \text{of second}}}$$

$$= (2x + 3)(4x + 5) + (x^2 + 3x)(4)$$

$$= 12x^2 + 34x + 15 \qquad \text{(simplifying)}$$

This agrees with our previous result. (See Equation (1).) Although there doesn't seem to be much advantage to using the product rule here, you will see that there are times when it is impractical to avoid it.

 NOW WORK PROBLEM 1

● **EXAMPLE 2   Applying the Product Rule**

If $y = (x^{2/3} + 3)(x^{-1/3} + 5x)$, find dy/dx.

Solution: Applying the product rule gives

$$\frac{dy}{dx} = \frac{d}{dx}(x^{2/3} + 3)(x^{-1/3} + 5x) + (x^{2/3} + 3)\frac{d}{dx}(x^{-1/3} + 5x)$$

$$= \left(\frac{2}{3}x^{-1/3}\right)(x^{-1/3} + 5x) + (x^{2/3} + 3)\left(\frac{-1}{3}x^{-4/3} + 5\right)$$

$$= \frac{25}{3}x^{2/3} + \frac{1}{3}x^{-2/3} - x^{-4/3} + 15$$

Alternatively, we could have found the derivative without the product rule by first finding the product $(x^{2/3} + 3)(x^{-1/3} + 5x)$ and then differentiating the result, term by term.

 NOW WORK PROBLEM 15

● **EXAMPLE 3   Differentiating a Product of Three Factors**

If $y = (x + 2)(x + 3)(x + 4)$, find y'.

Solution:

Strategy   We would like to use the product rule, but as given it applies only to *two* factors. By treating the first two factors as a single factor, we can consider y to be a product of two functions:

$$y = [(x + 2)(x + 3)](x + 4)$$

The product rule gives

$$y' = \frac{d}{dx}[(x + 2)(x + 3)](x + 4) + [(x + 2)(x + 3)]\frac{d}{dx}(x + 4)$$

$$= \frac{d}{dx}[(x + 2)(x + 3)](x + 4) + [(x + 2)(x + 3)](1)$$

Applying the product rule again, we have

$$y' = \left(\frac{d}{dx}(x + 2)(x + 3) + (x + 2)\frac{d}{dx}(x + 3)\right)(x + 4) + (x + 2)(x + 3)$$

$$= [(1)(x + 3) + (x + 2)(1)](x + 4) + (x + 2)(x + 3)$$

After simplifying, we obtain

$$y' = 3x^2 + 18x + 26$$

Two other ways of finding the derivative are as follows:

1. Multiply the first two factors of y to obtain

$$y = (x^2 + 5x + 6)(x + 4)$$

and then apply the product rule.

2. Multiply all three factors to obtain

$$y = x^3 + 9x^2 + 26x + 24$$

and then differentiate term by term.

NOW WORK PROBLEM 19

It is sometimes helpful to remember differentiation rules in more streamlined notation. For example,

$$(fg)' = f'g + fg'$$

is a correct equality of functions that expresses the product rule. We can then calculate

$$(fgh)' = ((fg)h)'$$
$$= (fg)'h + (fg)h'$$
$$= (f'g + fg')h + (fg)h'$$
$$= f'gh + fg'h + fgh'$$

It is not suggested that you try to commit to memory derived rules like

$$(fgh)' = f'gh + fg'h + fgh'$$

Because $f'g + fg' = gf' + fg'$, using commutativity of the product of functions, we can express the product rule with the derivatives as second factors:

$$(fg)' = gf' + fg'$$

and using commutativity of addition

$$(fg)' = fg' + gf'$$

Some people prefer these forms.

PRINCIPLES IN PRACTICE 2

DERIVATIVE OF A PRODUCT WITHOUT THE PRODUCT RULE

One hour after x milligrams of a particular drug are given to a person, the change in body temperature $T(x)$, in degrees Fahrenheit, is given approximately by $T(x) = x^2\left(1 - \frac{x}{3}\right)$. The rate at which T changes with respect to the size of the dosage x, $T'(x)$, is called the *sensitivity* of the body to the dosage. Find the sensitivity when the dosage is 1 milligram. Do not use the product rule.

● **EXAMPLE 4** **Using the Product Rule to Find Slope**

Find the slope of the graph of $f(x) = (7x^3 - 5x + 2)(2x^4 + 7)$ when $x = 1$.

Solution:

Strategy We find the slope by evaluating the derivative when $x = 1$. Because f is a product of two functions, we can find the derivative by using the product rule.

We have

$$f'(x) = (7x^3 - 5x + 2)\frac{d}{dx}(2x^4 + 7) + (2x^4 + 7)\frac{d}{dx}(7x^3 - 5x + 2)$$
$$= (7x^3 - 5x + 2)(8x^3) + (2x^4 + 7)(21x^2 - 5)$$

Since we must compute $f'(x)$ when $x = 1$, *there is no need to simplify $f'(x)$ before evaluating it.* Substituting into $f'(x)$, we obtain

$$f'(1) = 4(8) + 9(16) = 176$$

NOW WORK PROBLEM 49

Usually, we do not use the product rule when simpler ways are obvious. For example, if $f(x) = 2x(x+3)$, then it is quicker to write $f(x) = 2x^2 + 6x$, from which $f'(x) = 4x + 6$. Similarly, we do not usually use the product rule to differentiate $y = 4(x^2 - 3)$. Since the 4 is a constant factor, by the constant-factor rule we have $y' = 4(2x) = 8x$.

The next rule is used for differentiating a *quotient* of two functions.

The product rule (and quotient rule that follows) should not be applied when a more direct and efficient method is available.

RULE 2   The Quotient Rule

If f and g are differentiable functions and $g(x) \neq 0$, then the quotient f/g is also differentiable, and

$$\frac{d}{dx}\left(\frac{f(x)}{g(x)}\right) = \frac{g(x)f'(x) - f(x)g'(x)}{(g(x))^2}$$

With the understanding about the denominator not being zero, we can write

$$\left(\frac{f}{g}\right)' = \frac{gf' - fg'}{g^2}$$

That is, the derivative of the quotient of two functions is the denominator times the derivative of the numerator, minus the numerator times the derivative of the denominator, all divided by the square of the denominator.

$$\frac{d}{dx}(\text{quotient})$$

$$= \frac{(\text{denominator})\left(\begin{array}{c}\text{derivative}\\\text{of numerator}\end{array}\right) - (\text{numerator})\left(\begin{array}{c}\text{derivative}\\\text{of denominator}\end{array}\right)}{(\text{denominator})^2}$$

Proof. If $F(x) = \dfrac{f(x)}{g(x)}$, then

$$F(x)g(x) = f(x)$$

By the product rule,

$$F(x)g'(x) + g(x)F'(x) = f'(x)$$

Solving for $F'(x)$, we have

$$F'(x) = \frac{f'(x) - F(x)g'(x)}{g(x)}$$

But $F(x) = f(x)/g(x)$. Thus,

$$F'(x) = \frac{f'(x) - \dfrac{f(x)g'(x)}{g(x)}}{g(x)}$$

Simplifying gives[10]

$$F'(x) = \frac{g(x)f'(x) - f(x)g'(x)}{(g(x))^2}$$

CAUTION

The derivative of the quotient of two functions is trickier still than the product rule. One must remember where the minus sign goes!

●EXAMPLE 5   Applying the Quotient Rule

If $F(x) = \dfrac{4x^2 + 3}{2x - 1}$, *find* $F'(x)$.

Solution:

Strategy   We recognize F as a quotient, so we can apply the quotient rule.

---

[10] You may have observed that this proof assumes the existence of $F'(x)$. However, the rule can be proven without this assumption.

Let $f(x) = 4x^2 + 3$ and $g(x) = 2x - 1$. Then

$$F'(x) = \frac{g(x)f'(x) - f(x)g'(x)}{(g(x))^2}$$

$$= \frac{\overbrace{(2x - 1)}^{\text{Denominator}} \overbrace{\frac{d}{dx}(4x^2 + 3)}^{\substack{\text{Derivative} \\ \text{of numerator}}} - \overbrace{(4x^2 + 3)}^{\text{Numerator}} \overbrace{\frac{d}{dx}(2x - 1)}^{\substack{\text{Derivative of} \\ \text{numerator}}}}{\underbrace{(2x - 1)^2}_{\substack{\text{Square of} \\ \text{denominator}}}}$$

$$= \frac{(2x - 1)(8x) - (4x^2 + 3)(2)}{(2x - 1)^2}$$

$$= \frac{8x^2 - 8x - 6}{(2x - 1)^2} = \frac{2(2x + 1)(2x - 3)}{(2x - 1)^2}$$

NOW WORK PROBLEM 21

● EXAMPLE 6   **Rewriting before Differentiating**

Differentiate $y = \dfrac{1}{x + \dfrac{1}{x + 1}}$.

Solution:

Strategy   To simplify the differentiation, we will rewrite the function so that no fraction appears in the denominator.

We have

$$y = \frac{1}{x + \dfrac{1}{x + 1}} = \frac{1}{\dfrac{x(x + 1) + 1}{x + 1}} = \frac{x + 1}{x^2 + x + 1}$$

$$\frac{dy}{dx} = \frac{(x^2 + x + 1)(1) - (x + 1)(2x + 1)}{(x^2 + x + 1)^2} \qquad \text{(quotient rule)}$$

$$= \frac{(x^2 + x + 1) - (2x^2 + 3x + 1)}{(x^2 + x + 1)^2}$$

$$= \frac{-x^2 - 2x}{(x^2 + x + 1)^2} = -\frac{x^2 + 2x}{(x^2 + x + 1)^2}$$

NOW WORK PROBLEM 45

Although a function may have the form of a quotient, this does not necessarily mean that the quotient rule must be used to find the derivative. The next example illustrates some typical situations in which, although the quotient rule can be used, a simpler and more efficient method is available.

● EXAMPLE 7   **Differentiating Quotients without Using**
the Quotient Rule

Differentiate the following functions.

a. $f(x) = \dfrac{2x^3}{5}$

Solution: Rewriting, we have $f(x) = \frac{2}{5}x^3$. By the constant-factor rule,

$$f'(x) = \frac{2}{5}(3x^2) = \frac{6x^2}{5}$$

b. $f(x) = \dfrac{4}{7x^3}$

Solution: Rewriting, we have $f(x) = \frac{4}{7}(x^{-3})$. Thus,

$$f'(x) = \frac{4}{7}(-3x^{-4}) = -\frac{12}{7x^4}$$

c. $f(x) = \dfrac{5x^2 - 3x}{4x}$

Solution: Rewriting, we have $f(x) = \dfrac{1}{4}\left(\dfrac{5x^2 - 3x}{x}\right) = \dfrac{1}{4}(5x - 3)$ for $x \neq 0$. Thus,

$$f'(x) = \frac{1}{4}(5) = \frac{5}{4} \qquad \text{for } x \neq 0$$

Since the function f is not defined for $x = 0$, f' is not defined for $x = 0$ either.

NOW WORK PROBLEM 17 ●●

CAUTION

To differentiate $f(x) = \dfrac{1}{x^2 - 2}$, you might be tempted first to rewrite the quotient as $(x^2 - 2)^{-1}$. It would be a mistake to do this because we presently have no rule for differentiating that form. In short, we have no choice but to use the quotient rule. However, in the next section we will develop a rule that allows us to differentiate $(x^2 - 2)^{-1}$ in a direct and efficient way.

●**EXAMPLE 8   Marginal Revenue**

If the demand equation for a manufacturer's product is

$$p = \frac{1000}{q + 5}$$

where p is in dollars, find the marginal-revenue function and evaluate it when $q = 45$.

Solution:

Strategy   First we must find the revenue function. The revenue r received for selling q units when the price per unit is p is given by

$$\textbf{revenue} = (\textbf{price})(\textbf{quantity}), \quad \text{that is,} \quad r = pq$$

Using the demand equation, we will express r in terms of q only. Then we will differentiate to find the marginal-revenue function, dr/dq.

The revenue function is

$$r = \left(\frac{1000}{q + 5}\right)q = \frac{1000q}{q + 5}$$

Thus, the marginal-revenue function is given by

$$\frac{dr}{dq} = \frac{(q + 5)\dfrac{d}{dq}(1000q) - (1000q)\dfrac{d}{dq}(q + 5)}{(q + 5)^2}$$

$$= \frac{(q + 5)(1000) - (1000q)(1)}{(q + 5)^2} = \frac{5000}{(q + 5)^2}$$

and

$$\left.\frac{dr}{dq}\right|_{q=45} = \frac{5000}{(45 + 5)^2} = \frac{5000}{2500} = 2$$

This means that selling one additional unit beyond 45 results in approximately $2 more in revenue.

NOW WORK PROBLEM 59 ●●

Consumption Function

A function that plays an important role in economic analysis is the **consumption function.** The consumption function $C = f(I)$ expresses a relationship between the total national income I and the total national consumption C. Usually, both I and C are expressed in billions of dollars and I is restricted to some interval. The *marginal propensity to consume* is defined as the rate of change of consumption with respect to income. It is merely the derivative of C with respect to I:

$$\text{Marginal propensity to consume} = \frac{dC}{dI}$$

If we assume that the difference between income I and consumption C is savings S, then

$$S = I - C$$

Differentiating both sides with respect to I gives

$$\frac{dS}{dI} = \frac{d}{dI}(I) - \frac{d}{dI}(C) = 1 - \frac{dC}{dI}$$

We define dS/dI as the **marginal propensity to save.** Thus, the marginal propensity to save indicates how fast savings change with respect to income, and

$$\begin{matrix} \text{Marginal propensity} \\ \text{to save} \end{matrix} = 1 - \begin{matrix} \text{Marginal propensity} \\ \text{to consume} \end{matrix}$$

● EXAMPLE 9   **Finding Marginal Propensities to Consume and to Save**

If the consumption function is given by

$$C = \frac{5(2\sqrt{I^3} + 3)}{I + 10}$$

determine the marginal propensity to consume and the marginal propensity to save when $I = 100$.

Solution:

$$\frac{dC}{dI} = 5\left(\frac{(I+10)\dfrac{d}{dI}(2I^{3/2}+3) - (2\sqrt{I^3}+3)\dfrac{d}{dI}(I+10)}{(I+10)^2}\right)$$

$$= 5\left(\frac{(I+10)(3I^{1/2}) - (2\sqrt{I^3}+3)(1)}{(I+10)^2}\right)$$

When $I = 100$, the marginal propensity to consume is

$$\left.\frac{dC}{dI}\right|_{I=100} = 5\left(\frac{1297}{12,100}\right) \approx 0.536$$

The marginal propensity to save when $I = 100$ is $1 - 0.536 = 0.464$. This means that if a current income of \$100 billion increases by \$1 billion, the nation consumes approximately 53.6% (536/1000) and saves 46.4% (464/1000) of that increase.

NOW WORK PROBLEM 69 ●●●

Problems 11.4

In Problems 1–48, differentiate the functions.

*1. $f(x) = (4x + 1)(6x + 3)$

2. $f(x) = (3x - 1)(7x + 2)$

3. $s(t) = (5 - 3t)(t^3 - 2t^2)$

4. $Q(x) = (3 + x)(5x^2 - 2)$

5. $f(r) = (3r^2 - 4)(r^2 - 5r + 1)$

6. $C(I) = (2I^2 - 3)(3I^2 - 4I + 1)$

7. $f(x) = x^2(2x^2 - 5)$

8. $f(x) = 3x^3(x^2 - 2x + 2)$

9. $y = (x^2 + 3x - 2)(2x^2 - x - 3)$

10. $\phi(x) = (3 - 5x + 2x^2)(2 + x - 4x^2)$

11. $f(w) = (w^2 + 3w - 7)(2w^3 - 4)$

12. $f(x) = (3x - x^2)(3 - x - x^2)$

13. $y = (x^2 - 1)(3x^3 - 6x + 5) - 4(4x^2 + 2x + 1)$

14. $h(x) = 4(x^5 - 3) + 3(8x^2 - 5)(2x + 2)$

*15. $F(p) = \frac{3}{2}(5\sqrt{p} - 2)(3p - 1)$

16. $g(x) = (\sqrt{x} + 5x - 2)(\sqrt[3]{x} - 3\sqrt{x})$

*17. $y = 7 \cdot \frac{2}{3}$

18. $y = (x - 1)(x - 2)(x - 3)$

*19. $y = (2x - 1)(3x + 4)(x + 7)$

20. $y = \frac{2x - 3}{4x + 1}$

*21. $f(x) = \frac{5x}{x - 1}$

22. $H(x) = \frac{-5x}{5 - x}$

23. $f(x) = \frac{-13}{3x^5}$

24. $f(x) = \frac{5(x^2 - 2)}{7}$

25. $y = \frac{x + 2}{x - 1}$

26. $h(w) = \frac{3w^2 + 5w - 1}{w - 3}$

27. $h(z) = \frac{6 - 2z}{z^2 - 4}$

28. $z = \frac{2x^2 + 5x - 2}{3x^2 + 5x + 3}$

29. $y = \frac{8x^2 - 2x + 1}{x^2 - 5x}$

30. $f(x) = \frac{x^3 - x^2 + 1}{x^2 + 1}$

31. $y = \frac{x^2 - 4x + 3}{2x^2 - 3x + 2}$

32. $F(z) = \frac{z^4 + 4}{3z}$

33. $g(x) = \frac{1}{x^{100} + 7}$

34. $y = \frac{-9}{2x^5}$

35. $u(v) = \frac{v^3 - 8}{v}$

36. $y = \frac{x - 5}{8\sqrt{x}}$

37. $y = \frac{3x^2 - x - 1}{\sqrt[3]{x}}$

38. $y = \frac{x^{0.3} - 2}{2x^{2.1} + 1}$

39. $y = 7 - \frac{4}{x - 8} + \frac{2x}{3x + 1}$

40. $q(x) = 2x^3 + \frac{5x + 1}{3x - 5} - \frac{2}{x^3}$

41. $y = \frac{x - 5}{(x + 2)(x - 4)}$

42. $y = \frac{(9x - 1)(3x + 2)}{4 - 5x}$

43. $s(t) = \frac{t^2 + 3t}{(t^2 - 1)(t^3 + 7)}$

44. $f(s) = \frac{17}{s(5s^2 - 10s + 4)}$

*45. $y = 3x - \frac{\frac{2}{x} - \frac{3}{x - 1}}{x - 2}$

46. $y = 3 - 12x^3 + \frac{1 - \frac{5}{x^2 + 2}}{x^2 + 5}$

47. $f(x) = \frac{a + x}{a - x}$, where a is a constant

48. $f(x) = \frac{x^{-1} + a^{-1}}{x^{-1} - a^{-1}}$, where a is a constant

*49. Find the slope of the curve $y = (4x^2 + 2x - 5)(x^3 + 7x + 4)$ at $(-1, 12)$.

50. Find the slope of the curve $y = \frac{x^3}{x^4 + 1}$ at $(-1, -\frac{1}{2})$.

In Problems 51–54, find an equation of the tangent line to the curve at the given point.

51. $y = \frac{6}{x - 1}$; $(3, 3)$

52. $y = \frac{x + 5}{x^2}$; $(1, 6)$

53. $y = (2x + 3)[2(x^4 - 5x^2 + 4)]$; $(0, 24)$

54. $y = \frac{x + 1}{x^2(x - 4)}$; $\left(2, -\frac{3}{8}\right)$

In Problems 55 and 56, determine the relative rate of change of y with respect to x for the given value of x.

55. $y = \frac{x}{2x - 6}$; $x = 1$

56. $y = \frac{1 - x}{1 + x}$; $x = 5$

57. **Motion** The position function for an object moving in a straight line is

$$s = \frac{2}{t^3 + 1}$$

where t is in seconds and s is in meters. Find the position and velocity of the object at $t = 1$.

58. **Motion** The position function for an object moving in a straight-line path is

$$s = \frac{t + 3}{t^2 + 7}$$

where t is in seconds and s is in meters. Find the positive value(s) of t for which the velocity of the object is 0.

In Problems 59–62, each equation represents a demand function for a certain product, where p denotes the price per unit for q units. Find the marginal-revenue function in each case. Recall that revenue = pq.

*59. $p = 50 - 0.01q$

60. $p = 500/q$

61. $p = \frac{108}{q + 2} - 3$

62. $p = \frac{q + 750}{q + 50}$

63. **Consumption Function** For the United States (1922–1942), the consumption function is estimated by[11]

$$C = 0.672I + 113.1$$

Find the marginal propensity to consume.

64. **Consumption Function** Repeat Problem 63 if $C = 0.712I + 95.05$ for the United States for 1929–1941.[12]

In Problems 65–68, each equation represents a consumption function. Find the marginal propensity to consume and the marginal propensity to save for the given value of I.

65. $C = 3 + \sqrt{I} + 2\sqrt[3]{I}$; $I = 1$

66. $C = 6 + \frac{3I}{4} - \frac{\sqrt{I}}{3}$; $I = 25$

67. $C = \frac{16\sqrt{I} + 0.8\sqrt{I^3} - 0.2I}{\sqrt{I} + 4}$; $I = 36$

68. $C = \frac{20\sqrt{I} + 0.5\sqrt{I^3} - 0.4I}{\sqrt{I} + 5}$; $I = 100$

*69. **Consumption Function** Suppose that a country's consumption function is given by

$$C = \frac{10\sqrt{I} + 0.7\sqrt{I^3} - 0.2I}{\sqrt{I}}$$

where C and I are expressed in billions of dollars.

(a) Find the marginal propensity to save when income is 25 billion dollars.

(b) Determine the relative rate of change of C with respect to I when income is 25 billion dollars.

---

[11]T. Haavelmo, "Methods of Measuring the Marginal Propensity to Consume," *Journal of the American Statistical Association*, XLII (1947), 105–22.

[12]Ibid.

70. Marginal Propensities to Consume and to Save   Suppose that the savings function of a country is

$$S = \frac{I - 2\sqrt{I} - 8}{\sqrt{I} + 2}$$

where the national income (I) and the national savings (S) are measured in billions of dollars. Find the country's marginal propensity to consume and its marginal propensity to save when the national income is \$150 billion. (*Hint:* It may be helpful to first factor the numerator.)

71. Marginal Cost   If the total-cost function for a manufacturer is given by

$$c = \frac{6q^2}{q + 2} + 6000$$

find the marginal-cost function.

72. Marginal and Average Costs   Given the cost function $c = f(q)$, show that if $\frac{d}{dq}(\bar{c}) = 0$, then the marginal-cost function and average-cost function are equal.

73. Host–Parasite Relation   For a particular host–parasite relationship, it is determined that when the host density (number of hosts per unit of area) is x, the number of hosts that are parasitized is y, where

$$y = \frac{900x}{10 + 45x}$$

At what rate is the number of hosts parasitized changing with respect to host density when $x = 2$?

74. Acoustics   The persistence of sound in a room after the source of the sound is turned off is called *reverberation*. The *reverberation time* RT of the room is the time it takes for the intensity level of the sound to fall 60 decibels. In the acoustical design of an auditorium, the following formula may be used to compute the RT of the room:[13]

$$\text{RT} = \frac{0.05V}{A + xV}$$

Here V is the room volume, A is the total room absorption, and x is the air absorption coefficient. Assuming that A and x are positive constants, show that the rate of change of RT with respect to V is always positive. If the total room volume increases by one unit, does the reverberation time increase or decrease?

75. Predator–Prey   In a predator-prey experiment,[14] it was statistically determined that the number of prey consumed, y, by an individual predator was a function of the prey density x (the number of prey per unit of area), where

$$y = \frac{0.7355x}{1 + 0.02744x}$$

Determine the rate of change of prey consumed with respect to prey density.

76. Social Security Benefits   In a discussion of social security benefits, Feldstein[15] differentiates a function of the form

$$f(x) = \frac{a(1 + x) - b(2 + n)x}{a(2 + n)(1 + x) - b(2 + n)x}$$

where a, b, and n are constants. He determines that

$$f'(x) = \frac{-1(1 + n)ab}{(a(1 + x) - bx)^2(2 + n)}$$

Verify this. (*Hint:* For convenience, let $2 + n = c$.) Next observe that Feldstein's function f is of the form

$$g(x) = \frac{A + Bx}{C + Dx} \quad \text{where } A, B, C, \text{ and } D \text{ are constants}$$

Show that $g'(x)$ is a constant divided by a nonnegative function of x. What does this mean?

77. Business   The manufacturer of a product has found that when 20 units are produced per day, the average cost is \$150 and the marginal cost is \$125. What is the relative rate of change of average cost with respect to quantity when $q = 20$?

78. Use the result $(fgh)' = f'gh + fg'h + fgh'$ to find dy/dx if

$$y = (3x + 1)(2x - 1)(x - 4)$$

OBJECTIVE

To introduce and apply the chain rule, to derive the power rule as a special case of the chain rule, and to develop the concept of the marginal-revenue product as an application of the chain rule.

11.5 The Chain Rule and the Power Rule

Our next rule, the *chain rule*, is ultimately the most important rule for finding derivatives. It involves a situation in which y is a function of the variable u, but u is a function of x, and we want to find the derivative of y with respect to x. For example, the equations

$$y = u^2 \quad \text{and} \quad u = 2x + 1$$

define y as a function of u and u as a function of x. If we substitute $2x + 1$ for u in the first equation, we can consider y to be a function of x:

$$y = (2x + 1)^2$$

To find dy/dx, we first expand $(2x + 1)^2$:

$$y = 4x^2 + 4x + 1$$

Then

$$\frac{dy}{dx} = 8x + 4$$

---

[13]L. L. Doelle, *Environmental Acoustics* (New York: McGraw-Hill Book Company, 1972).

[14]C. S. Holling, "Some Characteristics of Simple Types of Predation and Parasitism," *The Canadian Entomologist*, XCI, no. 7 (1959), 385–98.

[15]M. Feldstein, "The Optimal Level of Social Security Benefits," *The Quarterly Journal of Economics*, C, no. 2 (1985), 303–20.

From this example, you can see that finding dy/dx by first performing a substitution *could* be quite involved. For instance, if originally we had been given $y = u^{100}$ instead of $y = u^2$, we wouldn't even want to try substituting. Fortunately, the chain rule will allow us to handle such situations with ease.

RULE 1 The Chain Rule

If y is a differentiable function of u and u is a differentiable function of x, then y is a differentiable function of x and

$$\frac{dy}{dx} = \frac{dy}{du} \cdot \frac{du}{dx}$$

We can show you why the chain rule is reasonable by considering rates of change. Suppose

$$y = 8u + 5 \quad \text{and} \quad u = 2x - 3$$

Let x change by one unit. How does u change? To answer this question, we differentiate and find $du/dx = 2$. But for *each* one-unit change in u, there is a change in y of $dy/du = 8$. Therefore, what is the change in y if x changes by one unit; that is, what is dy/dx? The answer is $8 \cdot 2$, which is $\dfrac{dy}{du} \cdot \dfrac{du}{dx}$. Thus, $\dfrac{dy}{dx} = \dfrac{dy}{du} \cdot \dfrac{du}{dx}$.

We will now use the chain rule to redo the problem at the beginning of this section. If

$$y = u^2 \quad \text{and} \quad u = 2x + 1$$

then

$$\frac{dy}{dx} = \frac{dy}{du} \cdot \frac{du}{dx} = \frac{d}{du}(u^2) \cdot \frac{d}{dx}(2x + 1)$$
$$= (2u)2 = 4u$$

Replacing u by $2x + 1$ gives

$$\frac{dy}{dx} = 4(2x + 1) = 8x + 4$$

which agrees with our previous result.

PRINCIPLES IN PRACTICE 1

USING THE CHAIN RULE

If an object moves horizontally according to $x = 6t$, where t is in seconds, and vertically according to $y = 4x^2$, find its vertical velocity $\dfrac{dy}{dt}$.

● EXAMPLE 1 Using the Chain Rule

a. If $y = 2u^2 - 3u - 2$ *and* $u = x^2 + 4$, *find* dy/dx.

Solution: By the chain rule,

$$\frac{dy}{dx} = \frac{dy}{du} \cdot \frac{du}{dx} = \frac{d}{du}(2u^2 - 3u - 2) \cdot \frac{d}{dx}(x^2 + 4)$$
$$= (4u - 3)(2x)$$

We can write our answer in terms of x alone by replacing u by $x^2 + 4$.

$$\frac{dy}{dx} = [4(x^2 + 4) - 3](2x) = [4x^2 + 13](2x) = 8x^3 + 26x$$

b. If $y = \sqrt{w}$ and $w = 7 - t^3$, *find* dy/dt.

Solution: Here, y is a function of w and w is a function of t, so we can view y as a function of t. By the chain rule,

$$\frac{dy}{dt} = \frac{dy}{dw} \cdot \frac{dw}{dt} = \frac{d}{dw}(\sqrt{w}) \cdot \frac{d}{dt}(7 - t^3)$$
$$= \left(\frac{1}{2}w^{-1/2}\right)(-3t^2) = \frac{1}{2\sqrt{w}}(-3t^2)$$
$$= -\frac{3t^2}{2\sqrt{w}} = -\frac{3t^2}{2\sqrt{7 - t^3}}$$

NOW WORK PROBLEM 1 ●●●

● EXAMPLE 2   **Using the Chain Rule**

If $y = 4u^3 + 10u^2 - 3u - 7$ *and* $u = 4/(3x - 5)$, *find* dy/dx *when* $x = 1$.

Solution: By the chain rule,

$$\frac{dy}{dx} = \frac{dy}{du} \cdot \frac{du}{dx} = \frac{d}{du}(4u^3 + 10u^2 - 3u - 7) \cdot \frac{d}{dx}\left(\frac{4}{3x - 5}\right)$$

$$= (12u^2 + 20u - 3) \cdot \frac{(3x - 5)\dfrac{d}{dx}(4) - 4\dfrac{d}{dx}(3x - 5)}{(3x - 5)^2}$$

$$= (12u^2 + 20u - 3) \cdot \frac{-12}{(3x - 5)^2}$$

Do not simply replace x by 1 and leave your answer in terms of u.

Even though dy/dx is in terms of x's and u's, we can evaluate it when $x = 1$ if we determine the corresponding value of u. When $x = 1$,

$$u = \frac{4}{3(1) - 5} = -2$$

Thus,

$$\left.\frac{dy}{dx}\right|_{x=1} = [12(-2)^2 + 20(-2) - 3] \cdot \frac{-12}{[3(1) - 5]^2}$$

$$= 5 \cdot (-3) = -15$$

NOW WORK PROBLEM 5 ●●

The chain rules states that if $y = f(u)$ and $u = g(x)$, then

$$\frac{dy}{dx} = \frac{dy}{du} \cdot \frac{du}{dx}$$

Actually, the chain rule applies to a composite function, because

$$y = f(u) = f(g(x)) = (f \circ g)(x)$$

Thus y, as a function of x, is $f \circ g$. This means that we can use the chain rule to differentiate a function when we recognize the function as a composition. However, we must first break down the function into composite parts.

For example, to differentiate

$$y = (x^3 - x^2 + 6)^{100}$$

we think of the function as a composition. Let

$$y = f(u) = u^{100} \quad \text{and} \quad u = g(x) = x^3 - x^2 + 6$$

Then $y = (x^3 - x^2 + 6)^{100} = (g(x))^{100} = f(g(x))$. Now that we have a composite, we differentiate. Since $y = u^{100}$ and $u = x^3 - x^2 + 6$, by the chain rule we have

$$\frac{dy}{dx} = \frac{dy}{du} \cdot \frac{du}{dx}$$

$$= (100u^{99})(3x^2 - 2x)$$

$$= 100(x^3 - x^2 + 6)^{99}(3x^2 - 2x)$$

We have just used the chain rule to differentiate $y = (x^3 - x^2 + 6)^{100}$, which is a power of a *function* of x, not simply a power of x. The following rule, called the *power rule,* generalizes our result and is a special case of the chain rule:

RULE 2   **The Power Rule**

If u is a differentiable function of x and n is any real number, then

$$\frac{d}{dx}(u^n) = nu^{n-1}\frac{du}{dx}$$

Proof. Let $y = u^n$. Since y is a differentiable function of u and u is a differentiable function of x, the chain rule gives

$$\frac{dy}{dx} = \frac{dy}{du} \cdot \frac{du}{dx}$$

But $dy/du = nu^{n-1}$. Thus,

$$\frac{dy}{dx} = nu^{n-1}\frac{du}{dx}$$

which is the power rule.

Another way of writing the power-rule formula is

$$\frac{d}{dx}((u(x))^n) = n(u(x))^{n-1}u'(x)$$

●EXAMPLE 3   Using the Power Rule

If $y = (x^3 - 1)^7$, find y'.

Solution: Since y is a power of a *function* of x, the power rule applies. Letting $u(x) = x^3 - 1$ and $n = 7$, we have

$$y' = n[u(x)]^{n-1}u'(x)$$

$$= 7(x^3 - 1)^{7-1}\frac{d}{dx}(x^3 - 1)$$

$$= 7(x^3 - 1)^6(3x^2) = 21x^2(x^3 - 1)^6$$

NOW WORK PROBLEM 9 ◖◗◗

●EXAMPLE 4   Using the Power Rule

If $y = \sqrt[3]{(4x^2 + 3x - 2)^2}$, find dy/dx when $x = -2$.

Solution: Since $y = (4x^2 + 3x - 2)^{2/3}$, we use the power rule with

$$u = 4x^2 + 3x - 2$$

and $n = \frac{2}{3}$. We have

$$\frac{dy}{dx} = \frac{2}{3}(4x^2 + 3x - 2)^{(2/3)-1}\frac{d}{dx}(4x^2 + 3x - 2)$$

$$= \frac{2}{3}(4x^2 + 3x - 2)^{-1/3}(8x + 3)$$

$$= \frac{2(8x + 3)}{3\sqrt[3]{4x^2 + 3x - 2}}$$

Thus,

$$\frac{dy}{dx}\bigg|_{x=-2} = \frac{2(-13)}{3\sqrt[3]{8}} = -\frac{13}{3}$$

NOW WORK PROBLEM 19 ◖◗◗

●EXAMPLE 5   Using the Power Rule

The technique used in Example 5 is frequently used when the numerator of a quotient is a constant and the denominator is not.

If $y = \dfrac{1}{x^2 - 2}$, find $\dfrac{dy}{dx}$.

Solution: Although the quotient rule can be used here, a more efficient approach is to treat the right side as the power $(x^2 - 2)^{-1}$ and use the power rule. Let $u = x^2 - 2$.

Then $y = u^{-1}$, and

$$\frac{dy}{dx} = (-1)(x^2 - 2)^{-1-1}\frac{d}{dx}(x^2 - 2)$$

$$= (-1)(x^2 - 2)^{-2}(2x)$$

$$= -\frac{2x}{(x^2 - 2)^2}$$

NOW WORK PROBLEM 27 ◖●●

● EXAMPLE 6   Differentiating a Power of a Quotient

If $z = \left(\dfrac{2s + 5}{s^2 + 1}\right)^4$, *find* $\dfrac{dz}{ds}$.

The problem here is to recognize the basic form of the function to be differentiated. In this case it is a power, not a quotient.

Solution: Since z is a power of a function, we first use the power rule:

$$\frac{dz}{ds} = 4\left(\frac{2s + 5}{s^2 + 1}\right)^{4-1}\frac{d}{ds}\left(\frac{2s + 5}{s^2 + 1}\right)$$

Now we use the quotient rule:

$$\frac{dz}{ds} = 4\left(\frac{2s + 5}{s^2 + 1}\right)^3\left(\frac{(s^2 + 1)(2) - (2s + 5)(2s)}{(s^2 + 1)^2}\right)$$

Simplifying, we have

$$\frac{dz}{ds} = 4 \cdot \frac{(2s + 5)^3}{(s^2 + 1)^3}\left(\frac{-2s^2 - 10s + 2}{(s^2 + 1)^2}\right)$$

$$= -\frac{8(s^2 + 5s - 1)(2s + 5)^3}{(s^2 + 1)^5}$$

NOW WORK PROBLEM 41 ◖●●

● EXAMPLE 7   Differentiating a Product of Powers

If $y = (x^2 - 4)^5(3x + 5)^4$, *find* y'.

Solution: Since y is a product, we first apply the product rule:

$$y' = (x^2 - 4)^5\frac{d}{dx}((3x + 5)^4) + (3x + 5)^4\frac{d}{dx}((x^2 - 4)^5)$$

Now we use the power rule:

$$y' = (x^2 - 4)^5(4(3x + 5)^3(3)) + (3x + 5)^4(5(x^2 - 4)^4(2x))$$

$$= 12(x^2 - 4)^5(3x + 5)^3 + 10x(3x + 5)^4(x^2 - 4)^4$$

In differentiating a product in which at least one factor is a power, simplifying the derivative usually involves factoring.

To simplify, we first remove common factors:

$$y' = 2(x^2 - 4)^4(3x + 5)^3[6(x^2 - 4) + 5x(3x + 5)]$$

$$= 2(x^2 - 4)^4(3x + 5)^3(21x^2 + 25x - 24)$$

NOW WORK PROBLEM 39 ◖●●

Usually, the power rule should be used to differentiate $y = [u(x)]^n$. Although a function such as $y = (x^2+2)^2$ can be written $y = x^4+4x^2+4$ and differentiated easily, this method is impractical for a function such as $y = (x^2+2)^{1000}$. Since $y = (x^2+2)^{1000}$ is of the form $y = [u(x)]^n$, we have

$$y' = 1000(x^2 + 2)^{999}(2x)$$

Marginal-Revenue Product

Let us now use our knowledge of calculus to develop a concept relevant to economic studies. Suppose a manufacturer hires m employees who produce a total of q units of a product per day. We can think of q as a function of m. If r is the total revenue the manufacturer receives for selling these units, then r can also be considered a function of m. Thus, we can look at dr/dm, the rate of change of revenue with respect to the number of employees. The derivative dr/dm is called the **marginal-revenue product.** It approximates the change in revenue that results when a manufacturer hires an extra employee.

● EXAMPLE 8   Marginal-Revenue Product

A manufacturer determines that m employees will produce a total of q units of a product per day, where

$$q = \frac{10m^2}{\sqrt{m^2 + 19}} \tag{1}$$

If the demand equation for the product is $p = 900/(q + 9)$, determine the marginal-revenue product when $m = 9$.

Solution: We must find dr/dm, where r is revenue. Note that, by the chain rule,

$$\frac{dr}{dm} = \frac{dr}{dq} \cdot \frac{dq}{dm}$$

Thus, we must find both dr/dq and dq/dm when $m = 9$. We begin with dr/dq. The revenue function is given by

$$r = pq = \left(\frac{900}{q + 9}\right) q = \frac{900q}{q + 9} \tag{2}$$

so, by the quotient rule,

$$\frac{dr}{dq} = \frac{(q + 9)(900) - 900q(1)}{(q + 9)^2} = \frac{8100}{(q + 9)^2}$$

In order to evaluate this expression when $m = 9$, we first use the given equation $q = 10m^2/\sqrt{m^2 + 19}$ to find the corresponding value of q:

$$q = \frac{10(9)^2}{\sqrt{9^2 + 19}} = 81$$

Hence,

$$\left.\frac{dr}{dq}\right|_{m=9} = \left.\frac{dr}{dq}\right|_{q=81} = \frac{8100}{(81 + 9)^2} = 1$$

Now we turn to dq/dm. From the quotient and power rules, we have

$$\frac{dq}{dm} = \frac{d}{dm}\left(\frac{10m^2}{\sqrt{m^2 + 19}}\right)$$

$$= \frac{(m^2 + 19)^{1/2}\dfrac{d}{dm}(10m^2) - (10m^2)\dfrac{d}{dm}[(m^2 + 19)^{1/2}]}{[(m^2 + 19)^{1/2}]^2}$$

$$= \frac{(m^2 + 19)^{1/2}(20m) - (10m^2)[\frac{1}{2}(m^2 + 19)^{-1/2}(2m)]}{m^2 + 19}$$

so

$$\left.\frac{dq}{dm}\right|_{m=9} = \frac{(81 + 19)^{1/2}(20 \cdot 9) - (10 \cdot 81)[\frac{1}{2}(81 + 19)^{-1/2}(2 \cdot 9)]}{81 + 19}$$

$$= 10.71$$

A direct formula for the marginal-revenue product is

$$\frac{dr}{dm} = \frac{dq}{dm}\left(p + q\frac{dp}{dq}\right)$$

Therefore, from the chain rule,

$$\left.\frac{dr}{dm}\right|_{m=9} = (1)(10.71) = 10.71$$

This means that if a tenth employee is hired, revenue will increase by approximately $10.71 per day.

NOW WORK PROBLEM 80 ●●●

Problems 11.5

In Problems 1–8, use the chain rule.

*1. If $y = u^2 - 2u$ and $u = x^2 - x$, find dy/dx.

2. If $y = 2u^3 - 8u$ and $u = 7x - x^3$, find dy/dx.

3. If $y = \dfrac{1}{w^2}$ and $w = 2 - x$, find dy/dx.

4. If $y = \sqrt[3]{z}$ and $z = x^5 - x^4 + 3$, find dy/dx.

*5. If $w = u^3$ and $u = \dfrac{t-1}{t+1}$, find dw/dt when $t = 1$.

6. If $z = u^2 + \sqrt{u} + 9$ and $u = 2s^2 - 1$, find dz/ds when $s = -1$.

7. If $y = 3w^2 - 8w + 4$ and $w = 2x^2 + 1$, find dy/dx when $x = 0$.

8. If $y = 3u^3 - u^2 + 7u - 2$ and $u = 5x - 2$, find dy/dx when $x = 1$.

In Problems 9–52, find y'.

*9. $y = (3x + 2)^6$

10. $y = (x^2 - 4)^4$

11. $y = (3 + 2x^3)^5$

12. $y = (x^2 + x)^4$

13. $y = 2(x^3 - 8x^2 + x)^{100}$

14. $y = \dfrac{(2x^2 + 1)^4}{2}$

15. $y = (x^2 - 2)^{-3}$

16. $y = (2x^3 - 8x)^{-12}$

17. $y = 2(x^2 + 5x - 2)^{-5/7}$

18. $y = 4(7x - x^4)^{-3/2}$

*19. $y = \sqrt{5x^2 - x}$

20. $y = \sqrt{3x^2 - 7}$

21. $y = \sqrt[4]{2x - 1}$

22. $y = \sqrt[3]{8x^2 - 1}$

23. $y = 2\sqrt[5]{(x^3 + 1)^2}$

24. $y = 7\sqrt[3]{(x^5 - 3)^5}$

25. $y = \dfrac{6}{2x^2 - x + 1}$

26. $y = \dfrac{3}{x^4 + 2}$

*27. $y = \dfrac{1}{(x^2 - 3x)^2}$

28. $y = \dfrac{1}{(2 + x)^4}$

29. $y = \dfrac{4}{\sqrt{9x^2 + 1}}$

30. $y = \dfrac{3}{(3x^2 - x)^{2/3}}$

31. $y = \sqrt[3]{7x} + \sqrt[3]{7x}$

32. $y = \sqrt{2x} + \dfrac{1}{\sqrt{2x}}$

33. $y = x^2(x - 4)^5$

34. $y = x(x + 4)^4$

35. $y = 4x^2\sqrt{5x + 1}$

36. $y = 4x^3\sqrt{1 - x^2}$

37. $y = (x^2 + 2x - 1)^3(5x)$

38. $y = x^2(x^3 - 1)^4$

*39. $y = (8x - 1)^3(2x + 1)^4$

40. $y = (3x + 2)^5(4x - 5)^2$

*41. $y = \left(\dfrac{x - 3}{x + 2}\right)^{12}$

42. $y = \left(\dfrac{2x}{x + 2}\right)^4$

43. $y = \sqrt{\dfrac{x - 2}{x + 3}}$

44. $y = \sqrt[3]{\dfrac{8x^2 - 3}{x^2 + 2}}$

45. $y = \dfrac{2x - 5}{(x^2 + 4)^3}$

46. $y = \dfrac{(4x - 2)^4}{3x^2 + 7}$

47. $y = \dfrac{(8x - 1)^5}{(3x - 1)^3}$

48. $y = \sqrt[3]{(x - 2)^2(x + 2)}$

49. $y = 6(5x^2 + 2)\sqrt{x^4 + 5}$

50. $y = 6 + 3x - 4x(7x + 1)^2$

51. $y = 8t + \dfrac{t - 1}{t + 4} - \left(\dfrac{8t - 7}{4}\right)^2$

52. $y = \dfrac{(2x^3 + 6)(7x - 5)}{(2x + 4)^2}$

In Problems 53 and 54, use the quotient rule and power rule to find y'. Do not simplify your answer.

53. $y = \dfrac{(2x + 1)^3(x + 3)^2}{(x^3 - 5)^5}$

54. $y = \dfrac{\sqrt{x + 2}(4x^2 - 1)^2}{9x - 3}$

55. If $y = (5u + 6)^3$ and $u = (x^2 + 1)^4$, find dy/dx when $x = 0$.

56. If $z = 2y^2 - 4y + 5$, $y = 6x - 5$, and $x = 2t$, find dz/dt when $t = 1$.

57. Find the slope of the curve $y = (x^2 - 7x - 8)^3$ at the point $(8, 0)$.

58. Find the slope of the curve $y = \sqrt{x + 1}$ at the point $(8, 3)$.

In Problems 59–62, find an equation of the tangent line to the curve at the given point.

59. $y = \sqrt[3]{(x^2 - 8)^2}$; $(3, 1)$

60. $y = (x + 3)^3$; $(-1, 8)$

61. $y = \dfrac{\sqrt{7x + 2}}{x + 1}$; $\left(1, \dfrac{3}{2}\right)$

62. $y = \dfrac{-3}{(3x^2 + 1)^3}$; $(0, -3)$

In Problems 63 and 64, determine the percentage rate of change of y with respect to x for the given value of x.

63. $y = (x^2 + 9)^3$; $x = 4$

64. $y = \dfrac{1}{(x^2 - 1)^3}$; $x = 2$

In Problems 65–68, q is the total number of units produced per day by m employees of a manufacturer, and p is the price per unit at which the q units are sold. In each case, find the marginal-revenue product for the given value of m.

65. $q = 5m$, $p = -0.4q + 50$; $m = 6$

66. $q = (200m - m^2)/20$, $p = -0.1q + 70$; $m = 40$

67. $q = 10m^2/\sqrt{m^2 + 9}$, $p = 525/(q + 3)$; $m = 4$

68. $q = 100m/\sqrt{m^2 + 19}$, $p = 4500/(q + 10)$; $m = 9$

69. Demand Equation   Suppose $p = 100 - \sqrt{q^2 + 20}$ is a demand equation for a manufacturer's product.

(a) Find the rate of change of p with respect to q.
(b) Find the relative rate of change of p with respect to q.
(c) Find the marginal-revenue function.

70. Marginal-Revenue Product   If $p = k/q$, where k is a constant, is the demand equation for a manufacturer's product and $q = f(m)$ defines a function that gives the total number of units produced per day by m employees, show that the marginal-revenue product is always zero.

71. Cost Function   The cost c of producing q units of a product is given by

$$c = 5500 + 12q + 0.2q^2$$

If the price per unit p is given by the equation

$$q = 900 - 1.5p$$

use the chain rule to find the rate of change of cost with respect to price per unit when $p = 85$.

72. Hospital Discharges   A governmental health agency examined the records of a group of individuals who were hospitalized with a particular illness. It was found that the total proportion that had been discharged at the end of t days of hospitalization was given by

$$f(t) = 1 - \left(\frac{250}{250 + t}\right)^3$$

Find $f'(100)$ and interpret your answer.

73. Marginal Cost   If the total-cost function for a manufacturer is given by

$$c = \frac{5q^2}{\sqrt{q^2 + 3}} + 5000$$

find the marginal-cost function.

74. Salary/Education   For a certain population, if E is the number of years of a person's education and S represents average annual salary in dollars, then for $E \geq 7$,

$$S = 340E^2 - 4360E + 42{,}800$$

(a) How fast is salary changing with respect to education when $E = 16$?
(b) At what level of education does the rate of change of salary equal $5000 per year of education?

75. Biology   The volume of a spherical cell is given by $V = \frac{4}{3}\pi r^3$, where r is the radius. At time t seconds, the radius (in centimeters) is given by

$$r = 10^{-8}t^2 + 10^{-7}t$$

Use the chain rule to find dV/dt when $t = 10$.

76. Pressure in Body Tissue   Under certain conditions, the pressure p developed in body tissue by ultrasonic beams is given as a function of the beam's intensity via the equation[16]

$$p = (2\rho V I)^{1/2}$$

where ρ (a Greek letter read "rho") is density of the affected tissue and V is the velocity of propagation of the

beam. Here ρ and V are constants. (a) Find the rate of change of p with respect to I. (b) Find the relative rate of change of p with respect to I.

77. Demography   Suppose that, for a certain group of 20,000 births, the number of people surviving to age x years is

$$l_x = -0.000354x^4 + 0.00452x^3 + 0.848x^2 - 34.9x + 20{,}000$$
$$0 \leq x \leq 95.2$$

(a) Find the rate of change of l_x with respect to x, and evaluate your answer for $x = 65$.
(b) Find the relative rate of change and the percentage rate of change of l_x when $x = 65$. Round your answers to three decimal places.

78. Muscle Contraction   A muscle has the ability to shorten when a load, such as a weight, is imposed on it. The equation

$$(P + a)(v + b) = k$$

is called the "fundamental equation of muscle contraction."[17] Here P is the load imposed on the muscle, v is the velocity of the shortening of the muscle fibers, and a, b, and k are positive constants. Express v as a function of P. Use your result to find dv/dP.

79. Economics   Suppose $pq = 100$ is the demand equation for a manufacturer's product. Let c be the total cost, and assume that the marginal cost is 0.01 when $q = 200$. Use the chain rule to find dc/dp when $q = 200$.

***80. Marginal-Revenue Product**   A monopolist who employs m workers finds that they produce

$$q = 2m(2m + 1)^{3/2}$$

units of product per day. The total revenue r (in dollars) is given by

$$r = \frac{50q}{\sqrt{1000 + 3q}}$$

(a) What is the price per unit (to the nearest cent) when there are 12 workers?
(b) Determine the marginal revenue when there are 12 workers.
(c) Determine the marginal-revenue product when $m = 12$.

81. Suppose $y = f(x)$, where $x = g(t)$. Given that $g(2) = 3$, $g'(2) = 4$, $f(2) = 5$, $f'(2) = 6$, $g(3) = 7$, $g'(3) = 8$, $f(3) = 9$, and $f'(3) = 10$, determine the value of $\left.\frac{dy}{dt}\right|_{t=2}$.

82. Business   A manufacturer has determined that, for his product, the daily average cost (in hundreds of dollars) is given by

$$\bar{c} = \frac{324}{\sqrt{q^2 + 35}} + \frac{5}{q} + \frac{19}{18}$$

(a) As daily production increases, the average cost approaches a constant dollar amount. What is this amount?
(b) Determine the manufacturer's marginal cost when 17 units are produced per day.
(c) The manufacturer determines that if production (and sales) were increased to 18 units per day, revenue would increase by $275. Should this move be made? Why?

---

[16]R. W. Stacy et al., *Essentials of Biological and Medical Physics* (New York: McGraw-Hill Book Company, 1955).

[17]Ibid.

83. If

$$y = (u + 1)\sqrt{u + 5}$$

and

$$u = x(x^2 + 5)^5$$

find dy/dx when $x = 0.1$. Round your answer to two decimal places.

84. If

$$y = \frac{2u + 3}{u^3 - 2}$$

and

$$u = \frac{x + 4}{(2x + 3)^3}$$

find dy/dx when $x = -1$. Round your answer to two decimal places.

11.6 Review

Important Terms and Symbols   Examples

Section 11.1   The Derivative

secant line   tangent line   slope of a curve   Ex. 1, p. 483

derivative $\displaystyle\lim_{h \to 0} \frac{f(x + h) - f(x)}{h}$ $\displaystyle\lim_{z \to x} \frac{f(z) - f(x)}{z - x}$ Ex. 2, p. 484

difference quotient $f'(x)$ y' $\dfrac{d}{dx}(f(x))$ $\dfrac{dy}{dx}$ Ex. 4, p. 485

Section 11.2   Rules for Differentiation

power function   constant factor rule   sum or difference rule   Ex. 5, p. 494

Section 11.3   The Derivative as a Rate of Change

position function Δx velocity   rate of change   Ex. 1, p. 498
total-cost function   marginal cost   average cost   Ex. 7, p. 502
total-revenue function   marginal revenue   Ex. 8, p. 503
relative rate of change   percentage rate of change   Ex. 9, p. 504

Section 11.4   The Product Rule and the Quotient Rule

product rule   quotient rule   Ex. 5, p. 510
consumption function   marginal propensity to consume and to save   Ex. 9, p. 513

Section 11.5   The Chain Rule and the Power Rule

chain rule   power rule   marginal-revenue product   Ex. 8, p. 520

Summary

The tangent line (or tangent) to a curve at point P is the limiting position of secant lines PQ as Q approaches P along the curve. The slope of the tangent at P is called the slope of the curve at P.

If $y = f(x)$, the derivative of f at x is the function $f'(x)$ defined by the limit in the equation

$$f'(x) = \lim_{h \to 0} \frac{f(x + h) - f(x)}{h}$$

Geometrically, the derivative gives the slope of the curve $y = f(x)$ at the point $(x, f(x))$. An equation of the tangent line at a particular point $(a, f(a))$ is obtained by evaluating $f'(a)$, which is the slope of the tangent line, and using the point–slope form of a line: $y - f(a) = f'(a)(x - a)$. Any function that is differentiable at a point must also be continuous there.

The basic rules for finding derivatives are as follows, where we assume that all functions are differentiable:

$$\frac{d}{dx}(c) = 0, \text{ where } c \text{ is any constant}$$

$$\frac{d}{dx}(x^n) = nx^{n-1}, \text{ where } n \text{ is any real number}$$

$$\frac{d}{dx}(cf(x)) = cf'(x), \text{ where } c \text{ is a constant}$$

$$\frac{d}{dx}(f(x) + g(x)) = f'(x) + g'(x)$$

$$\frac{d}{dx}(f(x) - g(x)) = f'(x) - g'(x)$$

$$\frac{d}{dx}(f(x)g(x)) = f(x)g'(x) + g(x)f'(x)$$

$$\frac{d}{dx}\left(\frac{f(x)}{g(x)}\right) = \frac{g(x)f'(x) - f(x)g'(x)}{(g(x))^2}$$

$$\frac{dy}{dx} = \frac{dy}{du} \cdot \frac{du}{dx}, \text{ where } y \text{ is a function of } u \text{ and } u \text{ is a function of } x$$

$$\frac{d}{dx}(u^n) = nu^{n-1}\frac{du}{dx}, \text{ where } u \text{ is a function of } x \text{ and } n \text{ is any real number}$$

The derivative dy/dx can also be interpreted as giving the (instantaneous) rate of change of y with respect to x:

$$\frac{dy}{dx} = \lim_{\Delta x \to 0} \frac{\Delta y}{\Delta x} = \lim_{\Delta x \to 0} \frac{\text{change in } y}{\text{change in } x}$$

In particular, if $s = f(t)$ is a position function, where s is position at time t, then

$$\frac{ds}{dt} = \text{velocity at time } t$$

In economics, the term *marginal* is used to describe derivatives of specific types of functions. If $c = f(q)$ is a total-cost function (c is the total cost of q units of a product), then the rate of change

$$\frac{dc}{dq} \text{ is called marginal cost}$$

We interpret marginal cost as the approximate cost of one additional unit of output. (Average cost per unit, $\bar{c}$, is related to total cost c by $\bar{c} = c/q$, or $c = \bar{c}q$.)

A total-revenue function $r = f(q)$ gives a manufacturer's revenue r for selling q units of product. (Revenue r and price p are related by $r = pq$.) The rate of change

$$\frac{dr}{dq} \text{ is called marginal revenue}$$

which is interpreted as the approximate revenue obtained from selling one additional unit of output.

If r is the revenue that a manufacturer receives when the total output of m employees is sold, then the derivative dr/dm is called the marginal-revenue product and gives the approximate change in revenue that results when the manufacturer hires an extra employee.

If $C = f(I)$ is a consumption function, where I is national income and C is national consumption, then

$$\frac{dC}{dI} \text{ is marginal propensity to consume}$$

and

$$1 - \frac{dC}{dI} \text{ is marginal prospensity to save}$$

For any function, the relative rate of change of $f(x)$ is

$$\frac{f'(x)}{f(x)}$$

which compares the rate of change of $f(x)$ with $f(x)$ itself. The percentage rate of change is

$$\frac{f'(x)}{f(x)} \cdot 100\%$$

Review Problems

Problem numbers shown in color indicate problems suggested for use as a practice chapter test.

In Problems 1–4, use the definition of the derivative to find $f'(x)$.

1. $f(x) = 2 - x^2$

2. $f(x) = 2x^2 - 3x + 1$

3. $f(x) = \sqrt{3x}$

4. $f(x) = \dfrac{2}{1 + 4x}$

In Problems 5–38, differentiate.

5. $y = 7^4$

6. $y = ex$

7. $y = 7x^4 - 6x^3 + 5x^2 + 1$

8. $y = 4(x^2 + 5) - 7x$

9. $f(s) = s^2(s^2 + 2)$

10. $y = \sqrt{x + 3}$

11. $y = \dfrac{x^2 + 1}{5}$

12. $y = -\dfrac{2}{2x^2}$

13. $y = (x^3 + 7x^2)(x^3 - x^2 + 5)$

14. $y = (x^2 + 1)^{100}(x - 6)$

15. $f(x) = (2x^2 + 4x)^{100}$

16. $f(w) = w\sqrt{w} + w^2$

17. $y = \dfrac{3}{2x + 1}$

18. $y = \dfrac{5x^2 - 8x}{2x}$

19. $y = (8 + 2x)(x^2 + 1)^4$

20. $g(z) = (2z)^{3/5} + 5$

21. $f(z) = \dfrac{z^2 - 1}{z^2 + 4}$

22. $y = \dfrac{x - 5}{(x + 2)^2}$

23. $y = \sqrt[3]{4x - 1}$

24. $f(x) = (1 + 2^3)^{12}$

25. $y = \dfrac{1}{\sqrt{1 - x^2}}$

26. $y = \dfrac{x(x + 1)}{2x^2 + 3}$

27. $h(x) = (x - 6)^4(x + 5)^3$

28. $y = \dfrac{(x + 3)^5}{x}$

29. $y = \dfrac{5x - 4}{x + 6}$

30. $f(x) = 5x^3\sqrt{3 + 2x^4}$

31. $y = 2x^{-3/8} + (2x)^{-3/8}$

32. $y = \sqrt{\dfrac{x}{2}} + \sqrt{\dfrac{2}{x}}$

33. $y = \dfrac{x^2 + 6}{\sqrt{x^2 + 5}}$

34. $y = \sqrt[3]{(7 - 3x^2)^2}$

35. $y = (x^3 + 6x^2 + 9)^{3/5}$

36. $z = 0.4x^2(x + 1)^{-3} + 0.5$

37. $g(z) = \dfrac{-z}{(z - 1)^{-2}}$

38. $g(z) = \dfrac{-3}{4(z^5 + 2z - 5)^4}$

In Problems 39–42, find an equation of the tangent line to the curve at the point corresponding to the given value of x.

39. $y = x^2 - 6x + 4, x = 1$

40. $y = -2x^3 + 6x + 1, x = 2$

41. $y = \sqrt[3]{x}, x = 8$

42. $y = \dfrac{x^2}{x - 12}, x = 13$

43. If $f(x) = 4x^2 + 2x + 8$ find the relative and percentage rates of change of $f(x)$ when $x = 1$.

44. If $f(x) = x/(x + 4)$, find the relative and percentage rates of change of $f(x)$ when $x = 1$.

45. Marginal Revenue If $r = q(20 - 0.1q)$ is a total-revenue function, find the marginal-revenue function.

46. Marginal Cost If

$$c = 0.0001q^3 - 0.02q^2 + 3q + 6000$$

is a total-cost function, find the marginal cost when $q = 100$.

47. Consumption Function If

$$C = 7 + 0.6I - 0.25\sqrt{I}$$

is a consumption function, find the marginal propensity to consume and the marginal propensity to save when $I = 16$.

48. Demand Equation If $p = \dfrac{q + 12}{q + 5}$ is a demand equation, find the rate of change of price p with respect to quantity q.

49. Demand Equation If $p = -0.1q + 500$ is a demand equation, find the marginal-revenue function.

50. Average Cost If $\bar{c} = 0.03q + 1.2 + \dfrac{3}{q}$ is an average-cost function, find the marginal cost when $q = 100$.

51. Power-Plant Cost Function The total-cost function of an electric light and power plant is estimated by[18]

$$c = 16.68 + 0.125q + 0.00439q^2 \qquad 20 \le q \le 90$$

[18]J. A. Nordin, "Note on a Light Plant's Cost Curves," *Econometrica, 15* (1947), 231–55.

where q is the eight-hour total output (as a percentage of capacity) and c is the total fuel cost in dollars. Find the marginal-cost function and evaluate it when $q = 70$.

52. Marginal-Revenue Product A manufacturer has determined that m employees will produce a total of q units of product per day, where

$$q = m(50 - m)$$

If the demand function is given by

$$p = -0.01q + 9$$

find the marginal-revenue product when $m = 10$.

53. Winter Moth In a study of the winter moth in Nova Scotia,[19] it was determined that the average number of eggs, y, in a female moth was a function of the female's abdominal width x (in millimeters), where

$$y = f(x) = 14x^3 - 17x^2 - 16x + 34$$

and $1.5 \le x \le 3.5$. At what rate does the number of eggs change with respect to abdominal width when $x = 2$?

54. Host–Parasite Relation For a particular host–parasite relationship, it is found that when the host density (number of hosts per unit of area) is x, the number of hosts that are parasitized is

$$y = 12\left(1 - \frac{1}{1 + 3x}\right) \qquad x \ge 0$$

For what value of x does dy/dx equal $\frac{1}{3}$?

55. Bacteria Growth Bacteria are growing in a culture. The time t (in hours) for the number of bacteria to double in number (the generation time) is a function of the temperature T (in degrees Celsius) of the culture and is given by

$$t = f(T) = \begin{cases} \frac{1}{24}T + \frac{11}{4} & \text{if } 30 \le T \le 36 \\ \frac{4}{3}T - \frac{175}{4} & \text{if } 36 < T \le 39 \end{cases}$$

Find dt/dT when (a) $T = 38$ and (b) $T = 35$.

56. Motion The position function of a particle moving in a straight line is

$$s = \frac{9}{2t^2 + 3}$$

where t is in seconds and s is in meters. Find the velocity of the particle at $t = 1$.

57. Rate of Change The volume of a sphere is given by $V = \frac{1}{6}\pi d^3$, where d is the diameter. Find the rate of change of V with respect to d when $d = 4$ ft.

58. Motion The position function for a ball thrown vertically upward from the ground is

$$s = 218t - 16t^2$$

where s is the height in feet above the ground after t seconds. For what value(s) of t is the velocity 64 ft/s?

59. Find the marginal-cost function if the average-cost function is

$$\bar{c} = 2q + \frac{10{,}000}{q^2}$$

60. Find an equation of the tangent line to the curve

$$y = \frac{(x^3 + 2)\sqrt{x + 1}}{x^4 + 2x}$$

at the point on the curve where $x = 1$.

61. A manufacturer has found that when m employees are working, the number of units of product produced per day is

$$q = 10\sqrt{m^2 + 4900} - 700$$

The demand equation for the product is

$$8q + p^2 - 19{,}300 = 0$$

where p is the selling price when the demand for the product is q units per day.

(a) Determine the manufacturer's marginal-revenue product when $m = 240$.

(b) Find the relative rate of change of revenue with respect to the number of employees when $m = 240$.

(c) Suppose it would cost the manufacturer $400 more per day to hire an additional employee. Would you advise the manufacturer to hire the 241st employee? Why?

62. If $f(x) = x^2 \ln x$, use the "limit of a difference quotient" approach to estimate $f'(5)$. Round your answer to three decimal places.

63. If $f(x) = \sqrt[3]{x^2 + 3x - 4}$, use the numerical derivative feature of your graphing calculator to estimate the derivative when $x = 10$. Round your answer to three decimal places.

64. The total-cost function for a manufacturer is given by

$$c = \frac{5q^2 + 4}{\sqrt{q^2 + 6}} + 2500$$

where c is in dollars. Use the numerical derivative feature of your graphing calculator to estimate the marginal cost when 15 units are produced. Round your answer to the nearest cent.

65. If

$$y = (u + 3)\sqrt{u + 6}$$

and

$$u = \frac{x + 4}{x + 3}$$

find dy/dx when $x = 0.3$. Round your answer to two decimal places.

[19]D. G. Embree, "The Population Dynamics of the Winter Moth in Nova Scotia, 1954–1962," *Memoirs of the Entomological Society of Canada*, no. 46 (1965).

Mathematical Snapshot

Marginal Propensity to Consume

A consumption function can be defined either for a nation, as in Section 11.4, or for an individual family. In either case, the function relates total consumption to total income. A savings function, similarly, relates total savings to total income, either at the national or at the family level.

Data about income, consumption, and savings for the United States as a whole can be found in the National Income and Product Accounts (NIPA) tables compiled by the Bureau of Economic Analysis, a division of the U.S. Department of Commerce. The tables are downloadable at www.bea.gov. For the years 1959–1999, the national consumption function is indicated by the scatterplot in Figure 11.15.

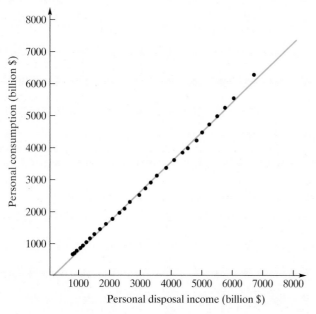

FIGURE 11.15   U.S. national consumption function.

Notice that the points lie more or less along a straight line. A linear regression gives the equation for this as $y = 0.9314x - 99.1936$.

The marginal propensity to consume derived from this graph is simply the slope of the line, that is, about 0.931 or 93.1%. At the national level, then, an increase of $1 billion in total disposable income produces an increase of $931 million in consumption. And if we assume that the rest is saved, there is an increase of $69 million in total savings.[20]

Perhaps somewhat easier to relate to, because of the smaller numbers involved, is the consumption function for

an individual household. This function is documented in Consumer Expenditure Surveys conducted by the Bureau of Labor Statistics, which is part of the U.S. Department of Labor. The survey results for each year can be downloaded at www.bls.gov/cex/.

Each year's survey gives information for five quintiles, as they are called, where a quintile represents one-fifth of American households. The quintiles are ordered by income, so that the bottom quintile represents the poorest 20% of Americans and the top quintile represents the richest 20%.

TABLE 11.3 U.S. Family Income and Expenses, 1999

After-Tax Income	Total Expenses
$7101	$16,766
$17,576	$24,850
$30,186	$33,078
$48,607	$46,015
$98,214	$75,080

For the year 1999, income and consumption are as shown in Table 11.3. The numbers are average values within each quintile. If these data values are plotted using a graphing calculator, the points lie in a pattern that could be reasonably well approximated by a straight line but could be even better approximated by a curve—a curve shaped, qualitatively, like a square root function (Figure 11.16).

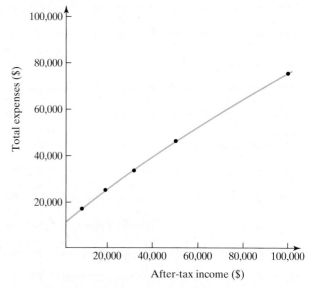

FIGURE 11.16   U.S. family consumption function.

---

[20]In reality, we must also account for interest payments and other outlays not counted as consumption. But we will ignore this complication from now on.

Most graphing calculators do not have a regression function for a square root–type function. They do, however, have a quadratic regression function—and the inverse of a quadratic function is a square root–type function. (Inverse functions were defined in Section 2.4.) So, we proceed as follows. First, using the statistics capabilities of a graphics calculator, enter the numbers in the *second* column in Table 11.3 as x-values and those in the *first* column as y-values. Second, perform a quadratic regression. The function obtained is given by

$$y = (4.4627 \times 10^{-6})x^2 + 1.1517x - 13{,}461$$

Third, swap the lists of x- and y-values in preparation for plotting. Fourth, replace y with x and x with y in the quadratic regression equation and solve the result for y (using the quadratic formula) to obtain the equation

$$y = \frac{-1.1517 \pm \sqrt{1.1517^2 - 4(4.4627 \times 10^{-6})(-13{,}461 - x)}}{2(4.4627 \times 10^{-6})}$$

or, more simply,

$$y = -129{,}036 \pm \sqrt{1.9667 \times 10^{10} + 224{,}080x}$$

Finally, enter the upper half of the curve (corresponding to the $+$ part of the $\pm$ sign) as a function to be graphed; then display it together with a plot of the data. The result looks as shown in Figure 11.17.

To find the marginal consumption for a given income level, we now use the dy/dx function. To find the marginal consumption at \$50,000, for instance, we select dy/dx, then enter 50000. The calculator returns the value 0.637675, which represents a marginal consumption of about 63.8%. In other words, a family earning \$50,000 per year will, if given an extra \$1000, spend \$638 of it and save the rest.

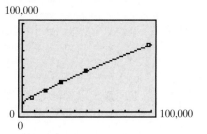

FIGURE 11.17   Graph of regression curve.

Problems

1. Compare the consumption function for Figure 11.15 with the consumption functions in Problems 63 and 64 of Section 11.5. Give two ways that these consumption functions differ significantly and interpret the differences qualitatively.

2. The first row in Table 11.3 has \$7101 in the first column and \$16,766 in the second column. What does this mean?

3. Suppose a family earning \$25,000 per year in 1999 received an unexpected bonus check for \$1000. How much of that check would you expect the family to spend? How much to save?

4. Suppose a family earning \$90,000 per year in 1999 received an unexpected \$1000 bonus check. How much would it spend?

5. What are the likely real-life reasons for the different answers in Problems 3 and 4?

![12](image of chapter number 12 in a circle)

ADDITIONAL DIFFERENTIATION TOPICS

 Mathematical Snapshot Economic Order Quantity

A fter an uncomfortable trip in a vehicle, passengers sometimes describe the ride as "jerky." But what is jerkiness, exactly? What does it mean for, say, an engineer designing a new transportation system?

Travel in a straight line at a constant speed is called *uniform motion,* and there is nothing jerky about it. But if either the path or the speed changes, the ride may become jerky. Change in velocity over time is, formally, the derivative of velocity. Called acceleration, the change in velocity is the *second derivative* of position with respect to time—the derivative of the derivative of position. One of the important concepts covered in this chapter is that of a higher-order derivative, of which acceleration is an example.

But is acceleration responsible for jerkiness? The feeling of being jerked back and forth on a roller coaster is certainly related to acceleration. On the other hand, automotive magazines often praise a car for having *smooth* acceleration. So apparently acceleration has something to do with jerkiness but is not itself the cause.

The derivative of acceleration is the *third* derivative of position with respect to time. When this third derivative is large, the acceleration is changing rapidly. A roller coaster in a steady turn to the left is undergoing steady leftward acceleration. But when the coaster changes abruptly from a hard left turn to a hard right turn, the acceleration changes directions—and the riders experience a jerk. The third derivative of position is, in fact, so apt a measure of jerkiness that it is customarily called the *jerk,* just as the second derivative is called the acceleration.

Jerk has implications not only for passenger comfort in vehicles but also for equipment reliability. Engineers designing equipment for spacecraft, for instance, follow guidelines about the maximum jerk the equipment must be able to survive without damage to its internal components.

To develop a differentiation formula for $y = \ln u$, to apply the formula, and to use it to differentiate a logarithmic function to a base other than e.

12.1 Derivatives of Logarithmic Functions

In this section, we develop formulas for differentiating logarithmic functions. We begin with the derivative of $f(x) = \ln x$, where $x > 0$. By the definition of the derivative,

$$\frac{d}{dx}(\ln x) = \lim_{h \to 0} \frac{f(x+h) - f(x)}{h} = \lim_{h \to 0} \frac{\ln(x+h) - \ln x}{h}$$

Using the property of logarithms that $\ln m - \ln n = \ln(m/n)$, we have

$$\frac{d}{dx}(\ln x) = \lim_{h \to 0} \frac{\ln\left(\dfrac{x+h}{x}\right)}{h}$$

$$= \lim_{h \to 0} \left(\frac{1}{h} \ln\left(\frac{x+h}{x}\right)\right) = \lim_{h \to 0} \left(\frac{1}{h} \ln\left(1 + \frac{h}{x}\right)\right)$$

Writing $\dfrac{1}{h}$ as $\dfrac{1}{x} \cdot \dfrac{x}{h}$ gives

$$\frac{d}{dx}(\ln x) = \lim_{h \to 0} \left(\frac{1}{x} \cdot \frac{x}{h} \ln\left(1 + \frac{h}{x}\right)\right)$$

$$= \lim_{h \to 0} \left(\frac{1}{x} \ln\left(1 + \frac{h}{x}\right)^{x/h}\right) \qquad \text{(since } r \ln m = \ln m^r\text{)}$$

$$= \frac{1}{x} \cdot \lim_{h \to 0} \left(\ln\left(1 + \frac{h}{x}\right)^{x/h}\right)$$

Recall from Section 4.2 the graph of the function $f(x) = \ln x$. It is manifestly unbroken on its domain, which is the set of all positive real numbers. It can be *proved* that the logarithm function is continuous, as suggested by the unbroken graph, and from continuity we have that the limit of a logarithm is the logarithm of the limit ($\lim \ln u = \ln \lim u$). Consequently, we have

$$\frac{d}{dx}(\ln x) = \frac{1}{x} \ln\left(\lim_{h \to 0}\left(1 + \frac{h}{x}\right)^{x/h}\right) \tag{1}$$

To evaluate $\lim\limits_{h \to 0}\left(1 + \dfrac{h}{x}\right)^{x/h}$, first note that if $h \to 0$, then, for fixed x, $\dfrac{h}{x} \to 0$ and, conversely, if $\dfrac{h}{x} \to 0$, then $h \to 0$. Thus, if we replace $\dfrac{h}{x}$ by k, the limit becomes

$$\lim_{k \to 0}(1 + k)^{1/k} = e$$

as in Section 10.1. Hence, Equation (1) becomes

$$\frac{d}{dx}(\ln x) = \frac{1}{x} \ln e = \frac{1}{x}(1) = \frac{1}{x} \quad \text{for } x > 0$$

Some care is required with this equation because while the left-hand side is defined only for $x > 0$, the right-hand side is defined for all $x \neq 0$. For $x < 0$, $\ln(-x)$ is defined and by the chain rule we have

$$\frac{d}{dx}(\ln(-x)) = \frac{1}{-x}\frac{d}{dx}(-x) = \frac{-1}{-x} = \frac{1}{x} \quad \text{for } x < 0$$

We can combine the last two equations by using the absolute function to get

$$\frac{d}{dx}(\ln|x|) = \frac{1}{x} \quad \text{for } x \neq 0 \tag{2}$$

EXAMPLE 1   **Differentiating Functions Involving ln x**

a. *Differentiate* $f(x) = 5 \ln x$.

Solution: Here f is a constant (5) times a function $(\ln x)$, so by Equation (2), we have

$$f'(x) = 5 \frac{d}{dx}(\ln x) = 5 \cdot \frac{1}{x} = \frac{5}{x} \quad \text{for } x > 0$$

b. *Differentiate* $y = \dfrac{\ln x}{x^2}$.

Solution: By the quotient rule and Equation (2),

$$y' = \frac{x^2 \dfrac{d}{dx}(\ln x) - (\ln x)\dfrac{d}{dx}(x^2)}{(x^2)^2}$$

$$= \frac{x^2 \left(\dfrac{1}{x}\right) - (\ln x)(2x)}{x^4} = \frac{x - 2x \ln x}{x^4} = \frac{1 - 2 \ln x}{x^3} \quad \text{for } x > 0$$

NOW WORK PROBLEM 1 ⬤◯⬤

We will now extend Equation (2) to cover a broader class of functions. Let $y = \ln |u|$, where u is a differentiable function of x. By the chain rule,

The chain rule is used to develop the differentiation formula for ln |u|.

$$\frac{d}{dx}(\ln |u|) = \frac{dy}{du} \cdot \frac{du}{dx} = \frac{d}{du}(\ln |u|) \cdot \frac{du}{dx} = \frac{1}{u} \cdot \frac{du}{dx} \quad \text{for } u \neq 0$$

Thus,

$$\frac{d}{du}(\ln |u|) = \frac{1}{u} \cdot \frac{du}{dx} \quad \text{for } u \neq 0 \tag{3}$$

Of course, Equation (3) gives us $\dfrac{d}{du}(\ln u) = \dfrac{1}{u} \cdot \dfrac{du}{dx}$ for $u > 0$.

PRINCIPLES IN PRACTICE 1

DIFFERENTIATING FUNCTIONS INVOLVING ln u

The supply of q units of a product at a price of p dollars per unit is given by $q(p) = 25 + 2\ln(3p^2 + 4)$. Find the rate of change of supply with respect to price, $\dfrac{dq}{dp}$.

EXAMPLE 2   **Differentiating Functions Involving ln u**

a. *Differentiate* $y = \ln(x^2 + 1)$.

Solution: This function has the form $\ln u$ with $u = x^2 + 1$ and since $x^2 + 1 > 0$, for all x, $y = \ln(x^2 + 1)$ is defined for all x. Using Equation (3), we have

$$\frac{dy}{dx} = \frac{1}{x^2 + 1} \frac{d}{dx}(x^2 + 1) = \frac{1}{x^2 + 1}(2x) = \frac{2x}{x^2 + 1}$$

b. *Differentiate* $y = x^2 \ln(4x + 2)$.

Solution: Using the product rule gives

$$\frac{dy}{dx} = x^2 \frac{d}{dx}(\ln(4x + 2)) + (\ln(4x + 2))\frac{d}{dx}(x^2)$$

By Equation (3) with $u = 4x + 2$,

$$\frac{dy}{dx} = x^2 \left(\frac{1}{4x + 2}\right)(4) + (\ln(4x + 2))(2x)$$

$$= \frac{2x^2}{2x + 1} + 2x \ln(4x + 2) \quad \text{for } 4x + 2 > 0$$

Since $4x + 2 > 0$ exactly when $x > -1/2$, we have

$$\frac{d}{dx}(x^2 \ln(4x + 2)) = \frac{2x^2}{2x + 1} + 2x \ln(4x + 2) \quad \text{for } x > -1/2$$

c. *Differentiate $y = \ln|\ln|x||$.*

Solution: This has the form $y = \ln|u|$ with $u = \ln|x|$. Using Equation (3), we obtain

$$y' = \frac{1}{\ln|x|}\frac{d}{dx}(\ln|x|) = \frac{1}{\ln|x|}\left(\frac{1}{x}\right) = \frac{1}{x\ln|x|} \quad \text{for } x, u \neq 0$$

Since $\ln|x| = 0$ when $x = -1, 1$, we have

$$\frac{d}{dx}(\ln|\ln|x||) = \frac{1}{x\ln|x|} \quad \text{for } x \neq -1, 0, 1$$

NOW WORK PROBLEM 9

Frequently, we can reduce the work involved in differentiating the logarithm of a product, quotient, or power by using properties of logarithms to rewrite the logarithm *before* differentiating. The next example will illustrate.

● **EXAMPLE 3   Rewriting Logarithmic Functions before Differentiating**

a. *Find $\dfrac{dy}{dx}$ if $y = \ln(2x + 5)^3$.*

Solution: Here we have the logarithm of a power. First we simplify the right side by using properties of logarithms. Then we differentiate. We have

$$y = \ln(2x + 5)^3 = 3\ln(2x + 5) \quad \text{for } 2x + 5 > 0$$

$$\frac{dy}{dx} = 3\left(\frac{1}{2x + 5}\right)(2) = \frac{6}{2x + 5} \quad \text{for } x > -5/2$$

Comparing both methods, we note that the easier one is to simplify first and then differentiate.

Alternatively, if the simplification were not performed first, we would write

$$\frac{dy}{dx} = \frac{1}{(2x + 5)^3}\frac{d}{dx}((2x + 5)^3)$$

$$= \frac{1}{(2x + 5)^3}(3)(2x + 5)^2(2) = \frac{6}{2x + 5}$$

b. *Find $f'(p)$ if $f(p) = \ln((p + 1)^2(p + 2)^3(p + 3)^4)$.*

Solution: We simplify the right side and then differentiate:

$$f(p) = 2\ln(p + 1) + 3\ln(p + 2) + 4\ln(p + 3)$$

$$f'(p) = 2\left(\frac{1}{p + 1}\right)(1) + 3\left(\frac{1}{p + 2}\right)(1) + 4\left(\frac{1}{p + 3}\right)(1)$$

$$= \frac{2}{p + 1} + \frac{3}{p + 2} + \frac{4}{p + 3}$$

NOW WORK PROBLEM 5

● **EXAMPLE 4   Differentiating Functions Involving Logarithms**

a. *Find $f'(w)$ if $f(w) = \ln\sqrt{\dfrac{1 + w^2}{w^2 - 1}}$.*

Solution: We simplify by using properties of logarithms and then differentiate:

$$f(w) = \frac{1}{2}(\ln(1 + w^2) - \ln(w^2 - 1))$$

$$f'(w) = \frac{1}{2}\left(\frac{1}{1 + w^2}(2w) - \frac{1}{w^2 - 1}(2w)\right)$$

$$= \frac{w}{1 + w^2} - \frac{w}{w^2 - 1} = -\frac{2w}{w^4 - 1}$$

b. *Find $f'(x)$ if $f(x) = \ln^3(2x + 5)$.*

Solution: The exponent 3 refers to the cubing of $\ln(2x + 5)$. That is,

$$f(x) = \ln^3(2x + 5) = [\ln(2x + 5)]^3$$

By the power rule,

$$f'(x) = 3(\ln(2x + 5))^2 \frac{d}{dx}(\ln(2x + 5))$$

$$= 3(\ln(2x + 5))^2 \left(\frac{1}{2x + 5}(2) \right)$$

$$= \frac{6}{2x + 5}(\ln(2x + 5))^2$$

NOW WORK PROBLEM 39

CAUTION

Do not confuse $\ln^3(2x + 5)$ with $\ln(2x + 5)^3$, which occurred in Example 3(a). It is advisable to write $\ln^3(2x + 5)$ explicitly as $[\ln(2x + 5)]^3$ and avoid $\ln^3(2x + 5)$.

Derivatives of Logarithmic Functions to the Base b

To differentiate a logarithmic function to a base different from e, we can first convert the logarithm to natural logarithms via the change-of-base formula and then differentiate the resulting expression. For example, consider $y = \log_b u$, where u is a differentiable function of x. By the change-of-base formula,

$$y = \log_b u = \frac{\ln u}{\ln b} \quad \text{for } u > 0$$

Differentiating, we have

$$\frac{d}{dx}(\log_b u) = \frac{d}{dx}\left(\frac{\ln u}{\ln b} \right) = \frac{1}{\ln b}\frac{d}{dx}(\ln u) = \frac{1}{\ln b} \cdot \frac{1}{u}\frac{du}{dx}$$

Summarizing,

$$\frac{d}{dx}(\log_b u) = \frac{1}{(\ln b)u} \cdot \frac{du}{dx} \quad \text{for } u > 0$$

CAUTION

Note that $\ln b$ is just a constant!

Rather than memorize this rule, we suggest that you remember the procedure used to obtain it.

> **Procedure to Differentiate $\log_b u$**
>
> Convert $\log_b u$ to natural logarithms to obtain $\dfrac{\ln u}{\ln b}$, and then differentiate.

EXAMPLE 5   Differentiating a Logarithmic Function to the Base 2

Differentiate $y = \log_2 x$.

Solution: Following the foregoing procedure, we have

$$\frac{d}{dx}(\log_2 x) = \frac{d}{dx}\left(\frac{\ln x}{\ln 2} \right) = \frac{1}{\ln 2}\frac{d}{dx}(\ln x) = \frac{1}{(\ln 2)x}$$

It is worth mentioning that we can write our answer in terms of the original base. Because

$$\frac{1}{\ln b} = \frac{1}{\dfrac{\log_b b}{\log_b e}} = \frac{\log_b e}{1} = \log_b e$$

we can express $\dfrac{1}{(\ln 2)x}$ as $\dfrac{\log_2 e}{x}$. More generally, $\dfrac{d}{dx}(\log_b u) = \dfrac{\log_b e}{u} \cdot \dfrac{du}{dx}$.

NOW WORK PROBLEM 15

PRINCIPLES IN PRACTICE 2

DIFFERENTIATING A LOGARITHMIC FUNCTION TO THE BASE 10

The intensity of an earthquake is measured on the Richter scale. The reading is given by $R = \log \dfrac{I}{I_0}$, where I is the intensity and I_0 is a standard minimum intensity. If $I_0 = 1$, find $\dfrac{dR}{dI}$, the rate of change of the Richter-scale reading with respect to the intensity.

● EXAMPLE 6   **Differentiating a Logarithmic Function to the Base 10**

If $y = \log(2x + 1)$, *find the rate of change of y with respect to x.*

Solution: The rate of change is dy/dx, and the base involved is 10. Therefore, we have

$$\frac{dy}{dx} = \frac{d}{dx}(\log(2x + 1)) = \frac{d}{dx}\left(\frac{\ln(2x + 1)}{\ln 10}\right)$$

$$= \frac{1}{\ln 10} \cdot \frac{1}{2x + 1}(2) = \frac{2}{\ln 10(2x + 1)}$$

Problems 12.1

In Problems 1–44, differentiate the functions. If possible, first use properties of logarithms to simplify the given function.

*1. $y = 4 \ln x$

2. $y = \dfrac{5 \ln x}{9}$

3. $y = \ln(3x - 7)$

4. $y = \ln(5x - 6)$

*5. $y = \ln x^2$

6. $y = \ln(3x^2 + 2x + 1)$

7. $y = \ln(1 - x^2)$

8. $y = \ln(-x^2 + 6x)$

*9. $f(X) = \ln(4X^6 + 2X^3)$

10. $f(r) = \ln(2r^4 - 3r^2 + 2r + 1)$

11. $f(t) = t \ln t$

12. $y = x^2 \ln x$

13. $y = x^3 \ln(2x + 5)$

14. $y = (ax + b)^3 \ln(ax + b)$

*15. $y = \log_3(8x - 1)$

16. $f(w) = \log(w^2 + w)$

17. $y = x^2 + \log_2(x^2 + 4)$

18. $y = x^2 \log_2 x$

19. $f(z) = \dfrac{\ln z}{z}$

20. $y = \dfrac{x^2}{\ln x}$

21. $y = \dfrac{x^2 + 3}{(\ln x)^2}$

22. $y = \ln x^{100}$

23. $y = \ln(x^2 + 4x + 5)^3$

24. $y = 6 \ln \sqrt[3]{x}$

25. $y = 9 \ln \sqrt{1 + x^2}$

26. $f(t) = \ln\left(\dfrac{t^5}{1 + 3t^2 + t^4}\right)$

27. $f(l) = \ln\left(\dfrac{1 + l}{1 - l}\right)$

28. $y = \ln\left(\dfrac{2x + 3}{3x - 4}\right)$

29. $y = \ln \sqrt[4]{\dfrac{1 + x^2}{1 - x^2}}$

30. $y = \ln \sqrt[3]{\dfrac{x^3 - 1}{x^3 + 1}}$

31. $y = \ln[(x^2 + 2)^2(x^3 + x - 1)]$

32. $y = \ln[(5x + 2)^4(8x - 3)^6]$

33. $y = 13 \ln\left(x^2 \sqrt[3]{5x + 2}\right)$

34. $y = 6 \ln \dfrac{x}{\sqrt{2x + 1}}$

35. $y = (x^2 + 1) \ln(2x + 1)$

36. $y = (ax + b) \ln(ax)$

37. $y = \ln x^3 + \ln^3 x$

38. $y = x^{\ln 2}$

*39. $y = \ln^4(ax)$

40. $y = \ln^2(2x + 11)$

41. $y = x \ln \sqrt{x - 1}$

42. $y = \ln\left(x^3 \sqrt[4]{2x + 1}\right)$

43. $y = \sqrt{4 + 3 \ln x}$

44. $y = \ln\left(x + \sqrt{1 + x^2}\right)$

45. Find an equation of the tangent line to the curve
$$y = \ln(x^2 - 3x - 3)$$
when $x = 4$.

46. Find an equation of the tangent line to the curve
$$y = x[\ln(x) - 1]$$
at the point where $x = e$.

47. Find the slope of the curve $y = \dfrac{x}{\ln x}$ when $x = 3$.

48. **Marginal Revenue**   Find the marginal-revenue function if the demand function is $p = 25/\ln(q + 2)$.

49. **Marginal Cost**   A total-cost function is given by
$$c = 25 \ln(q + 1) + 12$$
Find the marginal cost when $q = 6$.

50. **Marginal Cost**   A manufacturer's average-cost function, in dollars, is given by
$$\bar{c} = \frac{500}{\ln(q + 20)}$$
Find the marginal cost (rounded to two decimal places) when $q = 50$.

51. **Supply Change**   The supply of q units of a product at a price of p dollars per unit is given by $q(p) = 25 + 10 \ln(2p + 1)$. Find the rate of change of supply with respect to price, $\dfrac{dq}{dp}$.

52. **Sound Perception**   The loudness of sound (L, measured in decibels) perceived by the human ear depends upon intensity levels (I) according to $L = 10 \log \dfrac{I}{I_0}$, where I_0 is the standard threshold of audibility. If $I_0 = 17$, find $\dfrac{dL}{dI}$, the rate of change of the loudness with respect to the intensity.

53. **Biology**   In a certain experiment with bacteria, it is observed that the relative activeness of a given bacteria colony is described by
$$A = 6 \ln\left(\frac{T}{a - T} - a\right)$$
where a is a constant and T is the surrounding temperature. Find the rate of change of A with respect to T.

54. Show that the relative rate of change of $y = f(x)$ with respect to x is equal to the derivative of $y = \ln f(x)$.

55. Show that $\dfrac{d}{dx}(\log_b u) = \dfrac{1}{u}(\log_b e)\dfrac{du}{dx}$.

In Problems 56 and 57, use differentiation rules to find $f'(x)$. Then use your graphing calculator to find all zeros of $f'(x)$. Round your answers to two decimal places.

📱 56. $f(x) = x^3 \ln x$

📱 57. $f(x) = \dfrac{\ln(x^2)}{x^2}$

12.2 Derivatives of Exponential Functions

We now obtain a formula for the derivative of the exponential function

$$y = e^u$$

where u is a differentiable function of x. In logarithmic form, we have

$$u = \ln y$$

Differentiating both sides with respect to x gives

$$\frac{d}{dx}(u) = \frac{d}{dx}(\ln y)$$

$$\frac{du}{dx} = \frac{1}{y}\frac{dy}{dx}$$

Solving for dy/dx and then replacing y by e^u gives

$$\frac{dy}{dx} = y\frac{du}{dx} = e^u\frac{du}{dx}$$

Thus,

$$\frac{d}{dx}(e^u) = e^u\frac{du}{dx} \qquad (1)$$

As a special case, let $u = x$. Then $du/dx = 1$, and

$$\frac{d}{dx}(e^x) = e^x \qquad (2)$$

Note that the function and its derivative are the same.

CAUTION

The power rule does not apply to e^x and other exponential functions, b^x. The power rule applies to power functions, x^a. Note the location of the variable.

● **EXAMPLE 1 Differentiating Functions Involving e^x**

a. *Find* $\dfrac{d}{dx}(3e^x)$. Since 3 is a constant factor,

$$\frac{d}{dx}(3e^x) = 3\frac{d}{dx}(e^x)$$

$$= 3e^x \qquad \text{(by Equation (2))}$$

If a quotient can be easily rewritten as a product, then you can use the somewhat simpler product rule rather than the quotient rule.

b. *If* $y = \dfrac{x}{e^x}$, *find* $\dfrac{dy}{dx}$.

Solution: We could use first the quotient rule and then Equation (2), but it is a little easier to first rewrite the function as $y = xe^{-x}$ and use the product rule and Equation (1):

$$\frac{dy}{dx} = e^{-x}\frac{d}{dx}(x) + x\frac{d}{dx}(e^{-x}) = e^{-x}(1) + x(e^{-x})(-1) = e^{-x}(1-x) = \frac{1-x}{e^x}$$

c. *If* $y = e^2 + e^x + \ln 3$, *find* y'.

Solution: Since e^2 and $\ln 3$ are constants, $y' = 0 + e^x + 0 = e^x$.

NOW WORK PROBLEM 1 ●●●

PRINCIPLES IN PRACTICE 1

DIFFERENTIATING FUNCTIONS INVOLVING e^u

When an object is moved from one environment to another, the change in temperature of the object is given by $T = Ce^{kt}$, where C is the temperature difference between the two environments, t is the time in the new environment, and k is a constant. Find the rate of change of temperature with respect to time.

● **EXAMPLE 2 Differentiating Functions Involving e^u**

a. *Find* $\dfrac{d}{dx}\left(e^{x^3+3x}\right)$.

Solution: The function has the form e^u with $u = x^3 + 3x$. From Equation (1),

$$\frac{d}{dx}\left(e^{x^3+3x}\right) = e^{x^3+3x}\frac{d}{dx}(x^3 + 3x) = e^{x^3+3x}(3x^2 + 3)$$

$$= 3(x^2 + 1)e^{x^3+3x}$$

 $\dfrac{d}{dx}(e^u) = e^u\dfrac{du}{dx}$. Don't forget the $\dfrac{du}{dx}$.

b. *Find* $\dfrac{d}{dx}(e^{x+1}\ln(x^2+1))$.

Solution: By the product rule,

$$\frac{d}{dx}(e^{x+1}\ln(x^2+1)) = e^{x+1}\frac{d}{dx}(\ln(x^2+1)) + (\ln(x^2+1))\frac{d}{dx}(e^{x+1})$$

$$= e^{x+1}\left(\frac{1}{x^2+1}\right)(2x) + (\ln(x^2+1))e^{x+1}(1)$$

$$= e^{x+1}\left(\frac{2x}{x^2+1} + \ln(x^2+1)\right)$$

NOW WORK PROBLEM 3 ◖◗●

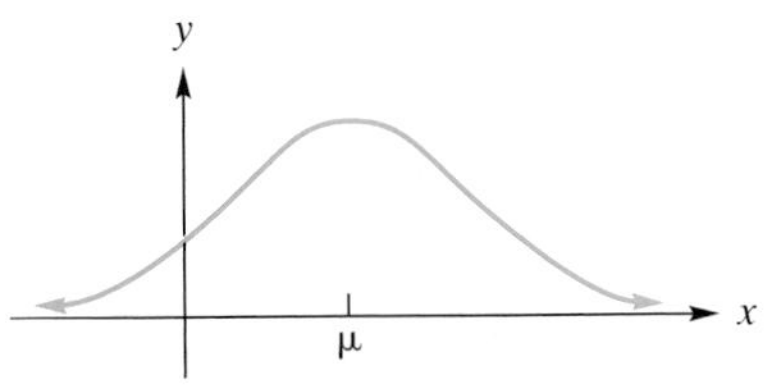

FIGURE 12.1 The normal-distribution density function.

EXAMPLE 3 The Normal-Distribution Density Function

*An important function used in the social sciences is the **normal-distribution density** function*

$$y = f(x) = \frac{1}{\sigma\sqrt{2\pi}}e^{-(1/2)((x-\mu)/\sigma)^2}$$

where σ (a Greek letter read "sigma") and μ (a Greek letter read "mu") are constants. The graph of this function, called the normal curve, is bell shaped. (See Figure 12.1.) Determine the rate of change of y with respect to x when $x = \mu + \sigma$.

Solution: The rate of change of y with respect to x is dy/dx. We note that the factor $\dfrac{1}{\sigma\sqrt{2\pi}}$ is a constant and the second factor has the form e^u, where

$$u = -\frac{1}{2}\left(\frac{x-\mu}{\sigma}\right)^2$$

Thus,

$$\frac{dy}{dx} = \frac{1}{\sigma\sqrt{2\pi}}\left(e^{-(1/2)((x-\mu)/\sigma)^2}\right)\left(-\frac{1}{2}(2)\left(\frac{x-\mu}{\sigma}\right)\left(\frac{1}{\sigma}\right)\right)$$

Evaluating dy/dx when $x = \mu + \sigma$, we obtain

$$\frac{dy}{dx}\bigg|_{x=\mu+\sigma} = \frac{1}{\sigma\sqrt{2\pi}}\left(e^{-(1/2)((\mu+\sigma-\mu)/\sigma)^2}\right)\left(-\frac{1}{2}(2)\left(\frac{\mu+\sigma-\mu}{\sigma}\right)\left(\frac{1}{\sigma}\right)\right)$$

$$= \frac{1}{\sigma\sqrt{2\pi}}\left(e^{-(1/2)}\right)\left(-\frac{1}{\sigma}\right)$$

$$= \frac{-e^{-(1/2)}}{\sigma^2\sqrt{2\pi}} = \frac{-1}{\sigma^2\sqrt{2\pi e}}$$

◖◗●

Differentiating Exponential Functions to the Base *b*

Now that we are familiar with the derivative of e^u, we consider the derivative of the more general exponential function b^u. Because $b = e^{\ln b}$, we can express b^u as an exponential function with the base e, a form we can differentiate. We have

$$\frac{d}{dx}(b^u) = \frac{d}{dx}((e^{\ln b})^u) = \frac{d}{dx}(e^{(\ln b)u})$$

$$= e^{(\ln b)u}\frac{d}{dx}((\ln b)u)$$

$$= e^{(\ln b)u}(\ln b)\left(\frac{du}{dx}\right)$$

$$= b^u(\ln b)\frac{du}{dx} \qquad (\text{since } e^{(\ln b)u} = b^u)$$

Summarizing,

$$\frac{d}{dx}(b^u) = b^u(\ln b)\frac{du}{dx} \tag{3}$$

Note that if $b = e$, then the factor $\ln b$ in Equation (3) is 1. Thus, if exponential functions to the base e are used, we have a simpler differentiation formula with which to work. This is the reason natural exponential functions are used extensively in calculus. Rather than memorizing Equation (3), we suggest that you remember the procedure for obtaining it.

Procedure to Differentiate b^u

Convert b^u to a natural exponential function by using the property that $b = e^{\ln b}$, and then differentiate.

The next example will illustrate this procedure.

EXAMPLE 4   Differentiating an Exponential Function to the Base 4

Find $\dfrac{d}{dx}(4^x)$.

Solution: Using the preceding procedure, we have

$$\frac{d}{dx}(4^x) = \frac{d}{dx}((e^{\ln 4})^x)$$

$$= \frac{d}{dx}(e^{(\ln 4)x}) \qquad \left(\text{form}: \frac{d}{dx}(e^u)\right)$$

$$= e^{(\ln 4)x}(\ln 4) \qquad \text{(by Equation (1))}$$

$$= 4^x(\ln 4)$$

Verify the result by using Equation (3) directly.

NOW WORK PROBLEM 15

EXAMPLE 5   Differentiating Different Forms

Find $\dfrac{d}{dx}\left(e^2 + x^e + 2^{\sqrt{x}}\right)$.

Solution: Here we must differentiate three different forms; do not confuse them! The first (e^2) is a constant base to a constant power, so it is a constant itself. Thus, its derivative is zero. The second (x^e) is a variable base to a constant power, so the power rule applies. The third $(2^{\sqrt{x}})$ is a constant base to a variable power, so we must differentiate an exponential function. Taken all together, we have

$$\frac{d}{dx}\left(e^2 + x^e + 2^{\sqrt{x}}\right) = 0 + ex^{e-1} + \frac{d}{dx}\left[e^{(\ln 2)\sqrt{x}}\right]$$

$$= ex^{e-1} + \left[e^{(\ln 2)\sqrt{x}}\right](\ln 2)\left(\frac{1}{2\sqrt{x}}\right)$$

$$= ex^{e-1} + \frac{2^{\sqrt{x}}\ln 2}{2\sqrt{x}}$$

NOW WORK PROBLEM 17

EXAMPLE 6   Differentiating Inverse Functions

We can apply the technique of this section to differentiate any inverse function f^{-1} once we know the derivative of f. Suppose that

$$y = f^{-1}(u)$$

where, as usual, u is a differentiable function of x and we wish to find dy/dx. In terms of f, we have

$$f(y) = u$$

and differentiating both sides with respect to x, we obtain

$$f'(y)\frac{dy}{dx} = \frac{du}{dx}$$

using the chain rule for the left side. Solving for dy/dx and replacing y by $f^{-1}(u)$ gives

$$\frac{dy}{dx} = \frac{1}{f'(f^{-1}(u))}\frac{du}{dx}$$

so that

$$\frac{d}{dx}(f^{-1}(u)) = \frac{1}{f'(f^{-1}(u))}\frac{du}{dx} \quad \text{for } f'(f^{-1}(u)) \neq 0 \tag{4}$$

and, in particular,

$$\frac{d}{dx}(f^{-1}(x)) = \frac{1}{f'(f^{-1}(x))} \quad \text{for } f'(f^{-1}(x)) \neq 0 \tag{5}$$

The reader may have noticed here, and in our treatment of the derivative of the exponential function, that we have assumed that f^{-1} is differentiable. The assumption can be avoided since it follows from the differentiability of f. A geometric plausibility argument can be given as follows. At each point $(x, f(x))$ on the graph of f there is a nonvertical tangent line whose slope is $f'(x)$. Since the graph of f^{-1} is obtained from the graph of f by reflection in the line $y = x$, it seems clear that at each point $(x, f^{-1}(x)$ on the graph of f^{-1} there will also be a tangent line—whose slope is $(f^{-1})'(x)$. Of course, if the tangent at $(f^{-1}(x), f(f^{-1}(x))) = (f^{-1}(x), x)$ is horizontal, then the tangent line at the reflected point $(x, f^{-1}(x))$ is vertical so that there the derivative of f^{-1} does not exist. Observe that our algebraic derivation accounts for those exceptions!

As with the chain rule, Leibniz notation is well suited for inverse functions. Indeed, if $y = f^{-1}(x)$, then $\frac{d}{dx}(f^{-1}(x)) = \frac{dy}{dx}$. Now since $f(y) = x$, $\frac{dx}{dy} = f'(y)$ and we see that Equation (5) can be rewritten as

$$\frac{dy}{dx} = \frac{1}{\dfrac{dx}{dy}} \tag{6}$$

● **EXAMPLE 7** **Differentiating Power Functions Again**

We have often used the rule $d/dx(x^a) = ax^{a-1}$, but we have only *proved* it for a a positive integer and a few other special cases. At least for $x > 0$, we can now improve our understanding of power functions, using Equation (1).

For $x > 0$, we can write $x^a = e^{a \ln x}$. So we have

$$\frac{d}{dx}(x^a) = \frac{d}{dx}e^{a \ln x} = e^{a \ln x}\frac{d}{dx}(a \ln x) = x^a(ax^{-1}) = ax^{a-1}$$

Problems 12.2

In Problems 1–28, differentiate the functions.

*1. $y = 5e^x$

2. $y = \dfrac{2e^x}{5}$

*3. $y = e^{2x^2+3}$

4. $y = e^{2x^2+5}$

5. $y = e^{9-5x}$

6. $f(q) = e^{-q^3+6q-1}$

7. $f(r) = e^{3r^2+4r+4}$

8. $y = e^{x^2+6x^3+1}$

9. $y = xe^x$

10. $y = 3x^4e^{-x}$

11. $y = x^2e^{-x^2}$

12. $y = xe^{3x}$

13. $y = \dfrac{e^x + e^{-x}}{3}$

14. $y = \dfrac{e^x - e^{-x}}{e^x + e^{-x}}$

*15. $y = 5^{2x^3}$

16. $y = 2^x x^2$

*17. $f(w) = \dfrac{e^{2w}}{w^2}$

18. $y = e^{x-\sqrt{x}}$

19. $y = e^{1+\sqrt{x}}$

20. $y = (e^{2x} + 1)^3$

21. $y = x^5 - 5^x$

22. $f(z) = e^{-1/z^2}$

23. $y = \dfrac{e^x - 1}{e^x + 1}$

24. $y = e^{2x}(x + 6)$

25. $y = \ln e^x$.

26. $y = e^{-x} \ln x$

27. $y = e^{x^2 \ln x^2}$

28. $y = \ln e^{4x+1}$

29. If $f(x) = ee^x e^{x^2}$, find $f'(-1)$.

30. If $f(x) = 5^{x^2 \ln x}$, find $f'(1)$.

31. Find an equation of the tangent line to the curve $y = e^x$ when $x = -2$.

32. Find an equation of the tangent line to the curve $y = e^x$ at the point $(1, e)$.

For each of the demand equations in Problems 33 and 34, find the rate of change of price p with respect to quantity q. What is the rate of change for the indicated value of q?

33. $p = 15e^{-0.001q}$; $q = 500$

34. $p = 9e^{-5q/750}$; $q = 300$

In Problems 35 and 36, $\bar{c}$ is the average cost of producing q units of a product. Find the marginal-cost function and the marginal cost for the given values of q.

35. $\bar{c} = \dfrac{7000e^{q/700}}{q}$; $q = 350, q = 700$

36. $\bar{c} = \dfrac{850}{q} + 4000\dfrac{e^{(2q+6)/800}}{q}$; $q = 97, q = 197$

37. If $w = e^{x^3 - 4x} + x \ln(x-1)$ and $x = \dfrac{t+1}{t-1}$, find $\dfrac{dw}{dt}$ when $t = 3$.

38. If $f'(x) = x^3$ and $u = e^x$, show that

$$\frac{d}{dx}[f(u)] = e^{4x}$$

39. Determine the value of the positive constant c if

$$\frac{d}{dx}(c^x - x^c)\bigg|_{x=1} = 0$$

40. Calculate the relative rate of change of

$$f(x) = 10^{-x} + \ln(8 + x) + 0.01e^{x-2}$$

when $x = 2$. Round your answer to four decimal places.

41. Production Run   For a firm, the daily output on the tth day of a production run is given by

$$q = 500(1 - e^{-0.2t})$$

Find the rate of change of output q with respect to t on the tenth day.

42. Normal-Density Function   For the normal-density function

$$f(x) = \frac{1}{\sqrt{2\pi}}e^{-x^2/2}$$

find $f'(1)$.

43. Population   The population, in millions, of the greater Seattle area t years from 1970 is estimated by $P = 1.92e^{0.0176t}$. Show that $dP/dt = kP$, where k is a constant. This means that the rate of change of population at any time is proportional to the population at that time.

44. Market Penetration   In a discussion of diffusion of a new process into a market, Hurter and Rubenstein[1] refer to an equation of the form

$$Y = k\alpha^{\beta^t}$$

where Y is the cumulative level of diffusion of the new process at time t and k, α, and β are positive constants. Verify their claim that

$$\frac{dY}{dt} = k\alpha^{\beta^t}(\beta^t \ln \alpha) \ln \beta$$

45. Finance   After t years, the value S of a principal of P dollars invested at the annual rate of r compounded continuously is given by $S = Pe^{rt}$. Show that the relative rate of change of S with respect to t is r.

46. Predator–Prey Relationship   In an article concerning predators and prey, Holling[2] refers to an equation of the form

$$y = K(1 - e^{-ax})$$

where x is the prey density, y is the number of prey attacked, and K and a are constants. Verify his statement that

$$\frac{dy}{dx} = a(K - y)$$

47. Earthquakes   According to Richter,[3] the number of earthquakes of magnitude M or greater per unit of time is given by $N = 10^A 10^{-bM}$, where A and b are constants. Find dN/dM.

48. Psychology   Short-term retention was studied by Peterson and Peterson.[4] The two researchers analyzed a procedure in which an experimenter verbally gave a subject a three-letter consonant syllable, such as CHJ, followed by a three-digit number, such as 309. The subject then repeated the number and counted backward by 3's, such as 309, 306, 303, ... After a period of time, the subject was signaled by a light to recite the three-letter consonant syllable. The time between the experimenter's completion of the last consonant to the onset of the light was called the *recall interval*. The time between the onset of the light and the completion of a response was referred to as *latency*. After many trials, it was determined that, for a recall interval of t seconds, the approximate proportion of correct recalls with latency below 2.83 seconds was

$$p = 0.89[0.01 + 0.99(0.85)^t]$$

(a) Find dp/dt and interpret your result.

(b) Evaluate dp/dt when $t = 2$. Round your answer to two decimal places.

49. Medicine   Suppose a tracer, such as a colored dye, is injected instantly into the heart at time $t = 0$ and mixes uniformly with blood inside the heart. Let the initial concentration of the tracer in the heart be C_0, and assume that the heart has constant volume V. Also assume that, as

[1] A. P. Hurter, Jr., A. H. Rubenstein, et al., "Market Penetration by New Innovations: The Technological Literature," *Technological Forecasting and Social Change*, 11 (1978), 197–221.

[2] C. S. Holling, "Some Characteristics of Simple Types of Predation and Parasitism," *The Canadian Entomologist*, XCI, no. 7 (1959), 385–98.

[3] C. F. Richter, *Elementary Seismology* (San Francisco: W. H. Freeman and Company, Publishers, 1958).

[4] L. R. Peterson and M. J. Peterson, "Short-Term Retention of Individual Verbal Items," *Journal of Experimental Psychology*, 58 (1959), 193–98.

fresh blood flows into the heart, the diluted mixture of blood and tracer flows out at the constant positive rate r. Then the concentration of the tracer in the heart at time t is given by

$$C(t) = C_0 e^{-(r/V)t}$$

Show that $dC/dt = (-r/V)C(t)$.

50. Medicine   In Problem 49, suppose the tracer is injected at a constant rate R. Then the concentration at time t is

$$C(t) = \frac{R}{r}\left[1 - e^{-(r/V)t}\right]$$

(a) Find $C(0)$.
(b) Show that $\dfrac{dC}{dt} = \dfrac{R}{V} - \dfrac{r}{V}C(t)$.

51. Schizophrenia   Several models have been used to analyze the length of stay in a hospital. For a particular group of schizophrenics, one such model is[5]

$$f(t) = 1 - e^{-0.008t}$$

where $f(t)$ is the proportion of the group that was discharged at the end of t days of hospitalization. Find the rate of discharge (the proportion discharged per day) at the end of 100 days. Round your answer to four decimal places.

52. Savings and Consumption   A country's savings S (in billions of dollars) is related to its national income I (in billions of dollars) by the equation

$$S = \ln \frac{5}{3 + e^{-I}}$$

(a) Find the marginal propensity to consume as a function of income.
(b) To the nearest million dollars, what is the national income when the marginal propensity to save is $\dfrac{1}{8}$? (*Hint*: 1 billion = 1000 million).

In Problems 53 and 54, use differentiation rules to find $f'(x)$. Then use your graphing calculator to find all real zeros of $f'(x)$. Round your answers to two decimal places.

53. $f(x) = e^{2x^3 + x^2 - 3x}$

54. $f(x) = x + e^{-x}$

---

[5]W. W. Eaton and G. A. Whitmore, "Length of Stay as a Stochastic Process: A General Approach and Application to Hospitalization for Schizophrenia," *Journal of Mathematical Sociology*, 5 (1977), 273–92.

OBJECTIVE

To give a mathematical analysis of the economic concept of elasticity.

12.3 Elasticity of Demand

Elasticity of demand is a means by which economists measure how a change in the price of a product will affect the quantity demanded. That is, it refers to consumer response to price changes. Loosely speaking, elasticity of demand is the ratio of the resulting percentage change in quantity demanded to a given percentage change in price:

$$\frac{\text{percentage change in quantity}}{\text{percentage change in price}}$$

For example, if, for a price increase of 5%, quantity demanded were to decrease by 2%, we would loosely say that elasticity of demand is $-2/5$.

To be more general, suppose $p = f(q)$ is the demand function for a product. Consumers will demand q units at a price of $f(q)$ per unit and will demand $q + h$ units at a price of $f(q + h)$ per unit (Figure 12.2). The *percentage* change in quantity demanded from q to $q + h$ is

$$\frac{(q + h) - q}{q} \cdot 100\% = \frac{h}{q} \cdot 100\%$$

The corresponding percentage change in price per unit is

$$\frac{f(q + h) - f(q)}{f(q)} \cdot 100\%$$

The ratio of these percentage changes is

$$\frac{\dfrac{h}{q} \cdot 100\%}{\dfrac{f(q + h) - f(q)}{f(q)} \cdot 100\%} = \frac{h}{q} \cdot \frac{f(q)}{f(q + h) - f(q)}$$

$$= \frac{f(q)}{q} \cdot \frac{h}{f(q + h) - f(q)}$$

$$= \frac{\dfrac{f(q)}{q}}{\dfrac{f(q + h) - f(q)}{h}} \qquad (1)$$

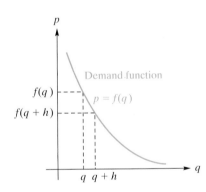

FIGURE 12.2   Change in demand.

If f is differentiable, then as $h \to 0$, the limit of $[f(q+h) - f(q)]/h$ is $f'(q) = dp/dq$. Thus, the limit of (1) is

$$\frac{\dfrac{f(q)}{q}}{f'(q)} = \frac{\dfrac{p}{q}}{\dfrac{dp}{dq}} \qquad (\text{since } p = f(q))$$

which is called the *point elasticity of demand*.

> **DEFINITION**
>
> If $p = f(q)$ is a differentiable demand function, the ***point elasticity of demand,*** denoted by the Greek letter η (eta), at (q, p) is given by
>
> $$\eta = \eta(q) = \frac{\dfrac{p}{q}}{\dfrac{dp}{dq}}$$

 CAUTION

Since p is a function of q, dp/dq is a function of q and thus the ratio that defines η is a function of q. That is why we write $\eta = \eta(q)$.

To illustrate, let us find the point elasticity of demand for the demand function $p = 1200 - q^2$. We have

$$\eta = \frac{\dfrac{p}{q}}{\dfrac{dp}{dq}} = \frac{\dfrac{1200 - q^2}{q}}{-2q} = -\frac{1200 - q^2}{2q^2} = -\left(\frac{600}{q^2} - \frac{1}{2}\right) \qquad (2)$$

For example, if $q = 10$, then $\eta = -\left((600/10^2) - \frac{1}{2}\right) = -5\frac{1}{2}$. Since

$$\eta \approx \frac{\% \text{ change in demand}}{\% \text{ change in price}}$$

we have

$$(\% \text{ change in price})(\eta) \approx \% \text{ change in demand}$$

Thus, if price were increased by 1% when $q = 10$, then quantity demanded would change by approximately

$$(1\%)\left(-5\frac{1}{2}\right) = -5\frac{1}{2}\%$$

That is, demand would decrease $5\frac{1}{2}\%$. Similarly, decreasing price by $\frac{1}{2}\%$ when $q = 10$ results in a change in demand of approximately

$$\left(-\frac{1}{2}\%\right)\left(-5\frac{1}{2}\right) = 2\frac{3}{4}\%$$

Hence, demand increases by $2\frac{3}{4}\%$.

Note that when elasticity is evaluated, no units are attached to it—it is nothing more than a real number. In fact, the 100%'s arising from the word *percentage* cancel, so that elasticity is really an approximation of the ratio

$$\frac{\text{relative change in quantity}}{\text{relative change in price}}$$

and each of the relative changes is no more than a real number. For usual behavior of demand, an increase (decrease) in price corresponds to a decrease (increase) in quantity. This means that if price is plotted as a function of quantity then the graph will have a negative slope at each point. Thus, dp/dq will typically be negative, and since p and q are positive, η will typically be negative too. Some economists disregard the minus sign; in the preceding situation, they would consider the elasticity to be $5\frac{1}{2}$. We will not adopt this practice.

There are three categories of elasticity:

1. When $|\eta| > 1$, demand is *elastic*.
2. When $|\eta| = 1$, demand has *unit elasticity*.
3. When $|\eta| < 1$, demand is *inelastic*.

For example, in Equation (2), since $|\eta| = 5\frac{1}{2}$ when $q = 10$, demand is elastic. If $q = 20$, then $|\eta| = \left| -\left[(600/20^2) - \frac{1}{2} \right] \right| = 1$ so demand has unit elasticity. If $q = 25$, then $|\eta| = \left| -\frac{23}{50} \right|$, and demand is inelastic.

Loosely speaking, for a given percentage change in price, there is a greater percentage change in quantity demanded if demand is elastic, a smaller percentage change if demand is inelastic, and an equal percentage change if demand has unit elasticity. To better understand elasticity, it is helpful to think of typical examples. Demand for an essential utilty such as electricity tends to be inelastic through a wide range of prices. If electricity prices are increased by 10%, consumers can be expected to reduce their consumption somewhat but a full 10% decrease may not be possible if most of their electricity usage is for essentials of life such as heating and food preparation. On the other hand, demand for luxury goods tends to be quite elastic. A 10% increase in the price of jewelry, for example, may result in a 50% decrease in demand.

● EXAMPLE 1   Finding Point Elasticity of Demand

Determine the point elasticity of the demand equation

$$p = \frac{k}{q} \quad \text{where } k > 0 \text{ and } q > 0$$

Solution: From the definition, we have

$$\eta = \frac{\dfrac{p}{q}}{\dfrac{dp}{dq}} = \frac{\dfrac{k}{q^2}}{\dfrac{-k}{q^2}} = -1$$

Thus, the demand has unit elasticity for all $q > 0$. The graph of $p = k/q$ is called an *equilateral hyperbola* and is often found in economics texts in discussions of elasticity. (See Figure 2.14 for a graph of such a curve.)

NOW WORK PROBLEM 1 ●●

If we are given $p = f(q)$ for our demand equation, as in our discussion thus far, then it is usually straightforward to calculate $dp/dq = f'(q)$. However, if instead we are given q as a function of p, then we will have $q = f^{-1}(p)$ and from Example 6 in Section 12.2,

$$\frac{dp}{dq} = \frac{1}{\dfrac{dq}{dp}}$$

It follows that

$$\eta = \frac{\dfrac{p}{q}}{\dfrac{dp}{dq}} = \frac{p}{q} \cdot \frac{dq}{dp} \tag{3}$$

provides another useful expression for η. Notice too that if $q = g(p)$, then

$$\eta = \eta(p) = \frac{p}{q} \cdot \frac{dq}{dp} = \frac{p}{g(p)} \cdot g'(p) = p \cdot \frac{g'(p)}{g(p)}$$

and thus

elasticity $=$ price $\cdot$ relative rate of change of quantity as a function of price   (4)

●EXAMPLE 2 **Finding Point Elasticity of Demand**

Determine the point elasticity of the demand equation

$$q = p^2 - 40p + 400 \quad \text{where } q > 0$$

Solution: Here we have q given as a function of p and it is easy to see that $dq/dp = 2p - 40$. Thus,

$$\eta(p) = \frac{p}{q} \cdot \frac{dq}{dp} = \frac{p}{q(p)}(2p - 40)$$

For example, if $p = 15$, then $q = q(15) = 25$; hence, $\eta(15) = (15(-10))/25 = -6$, so demand is elastic for $p = 15$.

NOW WORK PROBLEM 13 ●●●

Here we analyze elasticity for linear demand.

Point elasticity for a *linear* demand equation is quite interesting. Suppose the equation has the form

$$p = mq + b \quad \text{where } m < 0 \text{ and } b > 0$$

(See Figure 12.3.) We assume that $q > 0$; thus, $p < b$. The point elasticity of demand is

$$\eta = \frac{\dfrac{p}{q}}{\dfrac{dp}{dq}} = \frac{\dfrac{p}{q}}{m} = \frac{p}{mq} = \frac{p}{p - b}$$

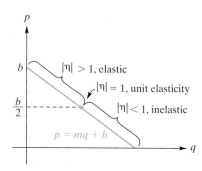

By considering $d\eta/dp$, we will show that η is a decreasing function of p. By the quotient rule,

$$\frac{d\eta}{dp} = \frac{(p - b) - p}{(p - b)^2} = -\frac{b}{(p - b)^2}$$

FIGURE 12.3 Elasticity for linear demand.

Since $b > 0$ and $(p - b)^2 > 0$, it follows that $d\eta/dp < 0$, meaning that the graph of $\eta = \eta(p)$ has a negative slope. This means that as price p increases, elasticity η decreases. However, p ranges between 0 and b, and at the midpoint of this range, $b/2$,

$$\eta = \eta(b) = \frac{\dfrac{b}{2}}{\dfrac{b}{2} - b} = \frac{\dfrac{b}{2}}{-\dfrac{b}{2}} = -1$$

Therefore, if $p < b/2$, then $\eta > -1$; if $p > b/2$, then $\eta < -1$. Because we typically have $\eta < 0$, we can state these facts another way: When $p < b/2$, $|\eta| < 1$, and demand is inelastic; when $p = b/2$, $|\eta| = 1$, and demand has unit elasticity; when $p > b/2$, $|\eta| > 1$ and demand is elastic. This shows that the slope of a demand curve is not a measure of elasticity. The slope of the line in Figure 12.3 is m everywhere, but elasticity varies with the point on the line. Of course, this is in accord with Equation (4).

Elasticity and Revenue

Here we analyze the relationship between elasticity and the rate of change of revenue.

Turning to a different situation, we can relate how elasticity of demand affects changes in revenue (marginal revenue). If $p = f(q)$ is a manufacturer's demand function, the total revenue is given by

$$r = pq$$

To find the marginal revenue, dr/dq, we differentiate r by using the product rule:

$$\frac{dr}{dq} = p + q\frac{dp}{dq}. \tag{5}$$

Factoring the right side of Equation (5), we have

$$\frac{dr}{dq} = p\left(1 + \frac{q}{p}\frac{dp}{dq}\right)$$

But

$$\frac{q}{p}\frac{dp}{dq} = \frac{\dfrac{dp}{dq}}{\dfrac{p}{q}} = \frac{1}{\eta}$$

Thus,

$$\frac{dr}{dq} = p\left(1 + \frac{1}{\eta}\right) \tag{6}$$

If demand is elastic, then $\eta < -1$, so $1 + \dfrac{1}{\eta} > 0$. If demand is inelastic, then $\eta > -1$, so $1 + \dfrac{1}{\eta} < 0$. We can assume that $p > 0$. From Equation (6) we can conclude that $dr/dq > 0$ on intervals for which demand is elastic. As we will soon see, a function is increasing on intervals for which its derivative is positive and a function is decreasing on intervals for which its derivative is negative. Hence, total revenue r is increasing on intervals for which demand is elastic and total revenue is decreasing on intervals for which demand is inelastic.

Thus, we conclude from the preceding argument that as more units are sold, a manufacturer's total revenue increases if demand is elastic, but decreases if demand is inelastic. That is, if demand is elastic, a lower price will increase revenue. This means that a lower price will cause a large enough increase in demand to actually increase revenue. If demand is inelastic, a lower price will decrease revenue. For unit elasticity, a lower price leaves total revenue unchanged.

Problems 12.3

In Problems 1–14, find the point elasticity of the demand equations for the indicated values of q or p, and determine whether demand is elastic, is inelastic, or has unit elasticity.

*1. $p = 40 - 2q$; $q = 5$

2. $p = 10 - 0.04q$; $q = 100$

3. $p = \dfrac{3500}{q}$; $q = 288$

4. $p = \dfrac{500}{q^2}$; $q = 52$

5. $p = \dfrac{500}{q+2}$; $q = 104$

6. $p = \dfrac{800}{2q+1}$; $q = 24$

7. $p = 150 - e^{q/100}$; $q = 100$

8. $p = 100e^{-q/200}$; $q = 200$

9. $q = 1200 - 150p$; $p = 4$

10. $q = 100 - p$; $p = 50$

11. $q = \sqrt{500 - p}$; $p = 400$

12. $q = \sqrt{2500 - p^2}$; $p = 20$

*13. $q = \dfrac{(p - 100)^2}{2}$; $p = 20$

14. $q = p^2 - 50p + 850$; $p = 20$

15. For the linear demand equation $p = 13 - 0.05q$, verify that demand is elastic when $p = 10$, is inelastic when $p = 3$, and has unit elasticity when $p = 6.50$.

16. For what value (or values) of q do the following demand equations have unit elasticity?

   (a) $p = 36 - 0.25q$
   (b) $p = 300 - q^2$

17. The demand equation for a product is

$$q = 500 - 40p + p^2$$

where p is the price per unit (in dollars) and q is the quantity of units demanded (in thousands). Find the point elasticity of demand when $p = 15$. If this price of 15 is increased by $\frac{1}{2}\%$, what is the approximate change in demand?

18. The demand equation for a certain product is

$$q = \sqrt{2500 - p^2}$$

where p is in dollars. Find the point elasticity of demand when $p = 30$, and use this value to compute the approximate percentage change in demand if the price of \$30 is decreased to \$28.50.

19. For the demand equation $p = 500 - 2q$, verify that demand is elastic and total revenue is increasing for $0 < q < 125$. Verify that demand is inelastic and total revenue is decreasing for $125 < q < 250$.

20. Verify that $\dfrac{dr}{dq} = p\left(1 + \dfrac{1}{\eta}\right)$ if $p = 50 - 3q$.

21. Repeat Problem 20 for $p = \dfrac{1000}{q^2}$.

22. Suppose $p = mq + b$ is a linear demand equation, where $m \neq 0$ and $b > 0$.

   (a) Show that $\lim_{p \to b^-} \eta = -\infty$.
   (b) Show that $\eta = 0$ when $p = 0$.

23. The demand equation for a manufacturer's product has the form

$$p = \frac{a}{\sqrt{b + cq^2}}$$

where a, b, and c are constants, with $c \neq 0$.

   (a) Show that elasticity does not depend on a.
   (b) Show that if b and c are positive, then demand is elastic for all $q > 0$.
   (c) For which value or values of the constants is unit elasticity possible?

24. Given the demand equation $q^2(1 + p)^2 = p$, determine the point elasticity of demand when $p = 9$.

25. The demand equation for a product is

$$q = \frac{60}{p} + \ln(65 - p^3)$$

(a) Determine the point elasticity of demand when $p = 4$, and classify the demand as elastic, inelastic, or of unit elasticity at this price level.

(b) If the price is lowered by 2% (from $4.00 to $3.92), use the answer to part (a) to estimate the corresponding percentage change in quantity sold.

(c) Will the changes in part (b) result in an increase or decrease in revenue? Explain.

26. The demand equation for a manufacturer's product is

$$p = 50(151 - q)^{0.02\sqrt{q+19}}$$

(a) Find the value of dp/dq when 150 units are demanded.

(b) Using the result in part (a), determine the point elasticity of demand when 150 units are demanded. At this level, is demand elastic, inelastic, or of unit elasticity?

(c) Use the result in part (b) to approximate the price per unit if demand decreases from 150 to 140 units.

(d) If the current demand is 150 units, should the manufacturer increase or decrease price in order to increase revenue? (Justify your answer.)

27. A manufacturer of aluminum doors currently is able to sell 500 doors per week at a price of $80 each. If the price were lowered to $75 each, an additional 50 doors per week could be sold. Estimate the current elasticity of demand for the doors, and also estimate the current value of the manufacturer's marginal-revenue function.

28. Given the demand equation

$$p = 2000 - q^2$$

where $5 \le q \le 40$, for what value of q is $|\eta|$ a maximum? For what value is it a minimum?

29. Repeat Problem 28 for

$$p = \frac{200}{q+5}$$

such that $5 \le q \le 95$.

OBJECTIVE

To discuss the notion of a function defined implicitly and to determine derivatives by means of implicit differentiation.

12.4 Implicit Differentiation

Implicit differentiation is a technique for differentiating functions that are not given in the usual form $y = f(x)$ (nor in the form $x = g(y)$). To introduce this technique, we will find the slope of a tangent line to a circle. Let us take the circle of radius 2 whose center is at the origin (Figure 12.4). Its equation is

$$x^2 + y^2 = 4$$

$$x^2 + y^2 - 4 = 0 \tag{1}$$

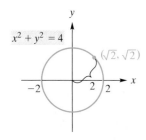

FIGURE 12.4   The circle $x^2 + y^2 = 4$.

The point $(\sqrt{2}, \sqrt{2})$ lies on the circle. To find the slope at this point, we need to find dy/dx there. Until now, we have always had y given explicitly (directly) in terms of x before determining y'; that is, our equation was always in the form $y = f(x)$ (or in the form $x = g(y)$). In Equation (1), this is not so. We say that Equation (1) has the form $F(x, y) = 0$, where $F(x, y)$ denotes a function of two variables. The obvious thing to do is solve Equation (1) for y in terms of x:

$$x^2 + y^2 - 4 = 0$$

$$y^2 = 4 - x^2$$

$$y = \pm\sqrt{4 - x^2} \tag{2}$$

A problem now occurs: Equation (2) may give two values of y for a value of x. It does not define y explicitly as a function of x. We can, however, suppose that Equation (1) defines y as one of two different functions of x,

$$y = +\sqrt{4 - x^2} \quad \text{and} \quad y = -\sqrt{4 - x^2}$$

whose graphs are given in Figure 12.5. Since the point $(\sqrt{2}, \sqrt{2})$ lies on the graph of $y = \sqrt{4 - x^2}$, we should differentiate that function:

$$y = \sqrt{4 - x^2}$$

$$\frac{dy}{dx} = \frac{1}{2}(4 - x^2)^{-1/2}(-2x)$$

$$= -\frac{x}{\sqrt{4 - x^2}}$$

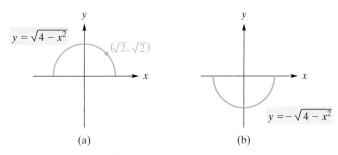

FIGURE 12.5 $x^2 + y^2 = 4$ gives rise to two different functions of x.

So

$$\frac{dy}{dx}\bigg|_{x=\sqrt{2}} = -\frac{\sqrt{2}}{\sqrt{4-2}} = -1$$

Thus, the slope of the circle $x^2 + y^2 - 4 = 0$ at the point $(\sqrt{2}, \sqrt{2})$ is -1.

Let us summarize the difficulties we had. First, y was not originally given explicitly in terms of x. Second, after we tried to find such a relation, we ended up with more than one function of x. In fact, depending on the equation given, it may be very complicated or even impossible to find an explicit expression for y. For example, it would be difficult to solve $ye^x + \ln(x + y) = 0$ for y. We will now consider a method that avoids such difficulties.

An equation of the form $F(x, y) = 0$, such as we had originally, is said to express y *implicitly* as a function of x. The word *implicitly* is used, since y is not given explicitly as a function of x. However, it is assumed or *implied* that the equation defines y as at least one differentiable function of x. Thus, we assume that Equation (1), $x^2 + y^2 - 4 = 0$, defines some differentiable function of x, say, $y = f(x)$. Next, we treat y as a function of x and differentiate both sides of Equation (1) with respect to x. Finally, we solve the result for dy/dx. Applying this procedure, we obtain

$$\frac{d}{dx}(x^2 + y^2 - 4) = \frac{d}{dx}(0)$$

$$\frac{d}{dx}(x^2) + \frac{d}{dx}(y^2) - \frac{d}{dx}(4) = \frac{d}{dx}(0) \tag{3}$$

We know that $\dfrac{d}{dx}(x^2) = 2x$ and that both $\dfrac{d}{dx}(4)$ and $\dfrac{d}{dx}(0)$ are 0. But $\dfrac{d}{dx}(y^2)$ is **not** $2y$, because we are differentiating with respect to x, not y. That is, y is not the independent variable. Since y is assumed to be a function of x, y^2 has the form u^n, where y plays the role of u. Just as the power rule states that $\dfrac{d}{dx}(u^2) = 2u\dfrac{du}{dx}$, we have $\dfrac{d}{dx}(y^2) = 2y\dfrac{dy}{dx}$. Hence, Equation (3) becomes

$$2x + 2y\frac{dy}{dx} = 0$$

Solving for dy/dx gives

$$2y\frac{dy}{dx} = -2x$$

$$\frac{dy}{dx} = -\frac{x}{y} \quad \text{for } y \neq 0 \tag{4}$$

Notice that the expression for dy/dx involves the variable y as well as x. This means that to find dy/dx at a point, both coordinates of the point must be substituted into dy/dx. Thus,

$$\frac{dy}{dx}\bigg|_{(\sqrt{2}, \sqrt{2})} = -\frac{\sqrt{2}}{\sqrt{2}} = -1$$

as before. This method of finding dy/dx is called **implicit differentiation.** We note that Equation (4) is not defined when $y = 0$. Geometrically, this is clear, since the tangent line to the circle at either $(2, 0)$ or $(-2, 0)$ is vertical, and the slope is not defined.

Here are the steps to follow when differentiating implicitly:

Implicit Differentiation Procedure

For an equation that we assume defines y implicitly as a differentiable function of x, the derivative $\dfrac{dy}{dx}$ can be found as follows:

1. Differentiate both sides of the equation with respect to x.
2. Collect all terms involving $\dfrac{dy}{dx}$ on one side of the equation, and collect all other terms on the other side.
3. Factor $\dfrac{dy}{dx}$ from the side involving the $\dfrac{dy}{dx}$ terms.
4. Solve for $\dfrac{dy}{dx}$, noting any restrictions.

CAUTION

The derivative of y^3 with respect to x is $3y^2 \dfrac{dy}{dx}$, not $3y^2$.

● EXAMPLE 1 Implicit Differentiation

Find $\dfrac{dy}{dx}$ by implicit differentiation if $y + y^3 - x = 7$.

Solution: Here y is not given as an explicit function of x [that is, not in the form $y = f(x)$]. Thus, we assume that y is an implicit (differentiable) function of x and apply the preceding four-step procedure:

1. Differentiating both sides with respect to x, we have

$$\frac{d}{dx}(y + y^3 - x) = \frac{d}{dx}(7)$$

$$\frac{d}{dx}(y) + \frac{d}{dx}(y^3) - \frac{d}{dx}(x) = \frac{d}{dx}(7)$$

Now, $\dfrac{d}{dx}(y)$ can be written $\dfrac{dy}{dx}$, and $\dfrac{d}{dx}(x) = 1$. By the power rule,

$$\frac{d}{dx}(y^3) = 3y^2 \frac{dy}{dx}$$

Hence, we obtain

$$\frac{dy}{dx} + 3y^2 \frac{dy}{dx} - 1 = 0$$

2. Collecting all $\dfrac{dy}{dx}$ terms on the left side and all other terms on the right side gives

$$\frac{dy}{dx} + 3y^2 \frac{dy}{dx} = 1$$

3. Factoring $\dfrac{dy}{dx}$ from the left side, we have

$$\frac{dy}{dx}(1 + 3y^2) = 1$$

In an implicit-differentiation problem, we are able to find the derivative of a function without knowing the function.

4. We solve for $\dfrac{dy}{dx}$ by dividing both sides by $1 + 3y^2$:

$$\frac{dy}{dx} = \frac{1}{1 + 3y^2}$$

Because step 4 of the process often involves division by an expression involving the variables, the answer obtained must often be restricted to exclude those values of the variables that would make the denominator zero. Here the denominator is always greater than or equal to 1, so there is no restriction.

NOW WORK PROBLEM 3 ●●●

Suppose that P, the proportion of people affected by a certain disease, is described by $\ln\left(\dfrac{P}{1-P}\right) = 0.5t$, where t is the time in months. Find $\dfrac{dP}{dt}$, the rate at which P grows with respect to time.

● EXAMPLE 2 **Implicit Differentiation**

Find $\dfrac{dy}{dx}$ *if* $x^3 + 4xy^2 - 27 = y^4$.

Solution: Since y is not given explicitly in terms of x, we will use the method of implicit differentiation:

1. Assuming that y is a function of x and differentiating both sides with respect to x, we get

$$\frac{d}{dx}(x^3 + 4xy^2 - 27) = \frac{d}{dx}(y^4)$$

$$\frac{d}{dx}(x^3) + 4\frac{d}{dx}(xy^2) - \frac{d}{dx}(27) = \frac{d}{dx}(y^4)$$

To find $\dfrac{d}{dx}(xy^2)$, we use the product rule:

$$3x^2 + 4\left[x\frac{d}{dx}(y^2) + y^2\frac{d}{dx}(x)\right] - 0 = 4y^3\frac{dy}{dx}$$

$$3x^2 + 4\left[x\left(2y\frac{dy}{dx}\right) + y^2(1)\right] = 4y^3\frac{dy}{dx}$$

$$3x^2 + 8xy\frac{dy}{dx} + 4y^2 = 4y^3\frac{dy}{dx}$$

2. Collecting $\dfrac{dy}{dx}$ terms on the left side and other terms on the right gives

$$8xy\frac{dy}{dx} - 4y^3\frac{dy}{dx} = -3x^2 - 4y^2$$

3. Factoring $\dfrac{dy}{dx}$ from the left side yields

$$\frac{dy}{dx}(8xy - 4y^3) = -3x^2 - 4y^2$$

4. Solving for $\dfrac{dy}{dx}$, we have

$$\frac{dy}{dx} = \frac{-3x^2 - 4y^2}{8xy - 4y^3} = \frac{3x^2 + 4y^2}{4y^3 - 8xy}$$

which gives the value of dy/dx at points (x, y) for which $4y^3 - 8xy \neq 0$.

NOW WORK PROBLEM 11 ●●

The volume V enclosed by a spherical balloon of radius r is given by the equation $V = \dfrac{4}{3}\pi r^3$. If the radius is increasing at a rate of 5 inches/minute $\left(\text{that is, } \dfrac{dr}{dt} = 5\right)$, then find $\left.\dfrac{dV}{dt}\right|_{r=12}$, the rate of increase of the volume, when the radius is 12 inches.

● EXAMPLE 3 **Implicit Differentiation**

Find the slope of the curve $x^3 = (y - x^2)^2$ *at* $(1, 2)$.

Solution: The slope at $(1, 2)$ is the value of dy/dx at that point. Finding dy/dx by implicit differentiation, we have

$$\frac{d}{dx}(x^3) = \frac{d}{dx}[(y - x^2)^2]$$

$$3x^2 = 2(y - x^2)\left(\frac{dy}{dx} - 2x\right)$$

$$3x^2 = 2\left(y\frac{dy}{dx} - 2xy - x^2\frac{dy}{dx} + 2x^3\right)$$

$$3x^2 = 2y\frac{dy}{dx} - 4xy - 2x^2\frac{dy}{dx} + 4x^3$$

$$3x^2 + 4xy - 4x^3 = 2y\frac{dy}{dx} - 2x^2\frac{dy}{dx}$$

$$3x^2 + 4xy - 4x^3 = 2\frac{dy}{dx}(y - x^2)$$

$$\frac{dy}{dx} = \frac{3x^2 + 4xy - 4x^3}{2(y - x^2)} \qquad \text{for } y - x^2 \neq 0$$

For the point $(1, 2)$, $y - x^2 = 2 - 1^2 = 1 \neq 0$. Thus, the slope of the curve at $(1, 2)$ is

$$\frac{dy}{dx}\bigg|_{(1,2)} = \frac{3(1)^2 + 4(1)(2) - 4(1)^3}{2(2 - (1)^2)} = \frac{7}{2}$$

NOW WORK PROBLEM 25 ●●●

PRINCIPLES IN PRACTICE 3

IMPLICIT DIFFERENTIATION

A 10-foot ladder is placed against a vertical wall. Suppose the bottom of the ladder slides away from the wall at a constant rate of 3 ft/s. $\left(\text{That is, } \dfrac{dx}{dt} = 3.\right)$ How fast is the top of the ladder sliding down the wall when the top of the ladder is 8 feet from the ground (that is, when $y = 8$)? (That is, what is $\dfrac{dy}{dt}$?) (Use the Pythagorean theorem for right triangles, $x^2 + y^2 = z^2$, where x and y are the legs of the triangle and z is the hypotenuse.)

●**EXAMPLE 4   Implicit Differentiation**

If $q - p = \ln q + \ln p$, find dq/dp.

Solution: We assume that q is a function of p and differentiate both sides with respect to p:

$$\frac{d}{dp}(q) - \frac{d}{dp}(p) = \frac{d}{dp}(\ln q) + \frac{d}{dp}(\ln p)$$

$$\frac{dq}{dp} - 1 = \frac{1}{q}\frac{dq}{dp} + \frac{1}{p}$$

$$\frac{dq}{dp} - \frac{1}{q}\frac{dq}{dp} = \frac{1}{p} + 1$$

$$\frac{dq}{dp}\left(1 - \frac{1}{q}\right) = \frac{1}{p} + 1$$

$$\frac{dq}{dp}\left(\frac{q - 1}{q}\right) = \frac{1 + p}{p}$$

$$\frac{dq}{dp} = \frac{(1 + p)q}{p(q - 1)} \qquad \text{for } p(q - 1) \neq 0$$

NOW WORK PROBLEM 19 ●●●

Problems 12.4

In Problems 1–24, find dy/dx by implicit differentiation.

1. $x^2 + 4y^2 = 4$

2. $3x^2 + 6y^2 = 1$

***3.** $2y^3 - 7x^2 = 5$

4. $2x^2 - 3y^2 = 4$

5. $\sqrt[3]{x} + \sqrt[3]{y} = 3$

6. $x^{1/5} + y^{1/5} = 4$

7. $x^{3/4} + y^{3/4} = 5$

8. $y^3 = 4x$

9. $xy = 4$

10. $x^2 + xy - 2y^2 = 0$

***11.** $xy - y - 11x = 5$

12. $x^3 - y^3 = 3x^2y - 3xy^2$

13. $2x^3 + y^3 - 12xy = 0$

14. $2x^3 + 3xy + y^3 = 0$

15. $x = \sqrt{y} + \sqrt[4]{y}$

16. $x^3y^3 + x = 9$

17. $5x^3y^4 - x + y^2 = 25$

18. $y^2 + y = \ln x$

***19.** $y \ln x = xe^y$

20. $\ln(xy) + x = 4$

21. $xe^y + y = 13$

22. $4x^2 + 9y^2 = 16$

23. $(1 + e^{3x})^2 = 3 + \ln(x + y)$ **24.** $e^{x+y} = \ln(x + y)$

***25.** If $x + xy + y^2 = 7$, find dy/dx at $(1, 2)$.

26. If $x\sqrt{y + 1} = y\sqrt{x + 1}$, find dy/dx at $(3, 3)$.

27. Find the slope of the curve $4x^2 + 9y^2 = 1$ at the point $\left(0, \frac{1}{3}\right)$; at the point (x_0, y_0).

28. Find the slope of the curve $(x^2 + y^2)^2 = 4y^2$ at the point $(0, 2)$.

29. Find an equation of the tangent line to the curve of

$$x^3 + xy + y^2 = -1$$

at the point $(-1, 1)$.

30. Repeat Problem 29 for the curve

$$y^2 + xy - x^2 = 5$$

at the point $(4, 3)$.

For the demand equations in Problems 31–34, find the rate of change of q with respect to p.

31. $p = 100 - q^2$

32. $p = 400 - \sqrt{q}$

33. $p = \dfrac{20}{(q + 5)^2}$

34. $p = \dfrac{10}{q^2 + 3}$

35. Radioactivity   The relative activity I/I_0 of a radioactive element varies with elapsed time according to the equation

$$\ln\left(\frac{I}{I_0}\right) = -\lambda\, t$$

where λ (a Greek letter read "lambda") is the disintegration constant and I_0 is the initial intensity (a constant). Find the rate of change of the intensity I with respect to the elapsed time t.

36. Earthquakes   The magnitude M of an earthquake and its energy E are related by the equation[6]

$$1.5M = \log\left(\frac{E}{2.5 \times 10^{11}}\right)$$

Here M is given in terms of Richter's preferred scale of 1958 and E is in ergs. Determine the rate of change of energy with respect to magnitude and the rate of change of magnitude with respect to energy.

37. Physical Scale   The relationship between the speed (v), frequency (f), and wavelength (λ) of any wave is given by

$$v = f\lambda$$

Find $df/d\lambda$ by differentiating implicitly. (Treat v as a constant.) Then show that the same result is obtained if you first solve the equation for f and then differentiate with respect to λ.

38. Biology   The equation $(P + a)(v + b) = k$ is called the "fundamental equation of muscle contraction."[7] Here P is the load imposed on the muscle, v is the velocity of the shortening of the muscle fibers, and a, b, and k are positive constants. Use implicit differentiation to show that, in terms of P,

$$\frac{dv}{dP} = -\frac{k}{(P + a)^2}$$

39. Marginal Propensity to Consume   A country's savings S is defined implicitly in terms of its national income I by the equation

$$S^2 + \frac{1}{4}I^2 = SI + I$$

where both S and I are in billions of dollars. Find the marginal propensity to consume when $I = 16$ and $S = 12$.

40. Technological Substitution   New products or technologies often tend to replace old ones. For example, today most commercial airlines use jet engines rather than prop engines. In discussing the forecasting of technological substitution, Hurter and Rubenstein[8] refer to the equation

$$\ln\frac{f(t)}{1 - f(t)} + \sigma\frac{1}{1 - f(t)} = C_1 + C_2 t$$

where $f(t)$ is the market share of the substitute over time t and C_1, C_2, and σ (a Greek letter read "sigma") are constants. Verify their claim that the rate of substitution is

$$f'(t) = \frac{C_2 f(t)[1 - f(t)]^2}{\sigma f(t) + [1 - f(t)]}$$

OBJECTIVE

To describe the method of logarithmic differentiation and to show how to differentiate a function of the form u^v.

12.5 Logarithmic Differentiation

A technique called **logarithmic differentiation** often simplifies the differentiation of $y = f(x)$ when $f(x)$ involves products, quotients, or powers. The procedure is as follows:

Logarithmic Differentiation

To differentiate $y = f(x)$,

1. Take the natural logarithm of both sides. This results in

$$\ln y = \ln(f(x))$$

2. Simplify $\ln(f(x))$ by using properties of logarithms.

3. Differentiate both sides with respect to x.

4. Solve for $\dfrac{dy}{dx}$.

5. Express the answer in terms of x only. This requires substituting $f(x)$ for y.

There are a couple of points worth noting. First, irrespective of any simplification, the procedure produces

$$\frac{y'}{y} = \frac{d}{dx}(\ln(f(x)))$$

---

[6] K. E. Bullen, *An Introduction to the Theory of Seismology* (Cambridge, U.K.: Cambridge at the University Press, 1963).

[7] R. W. Stacy et al., *Essentials of Biological and Medical Physics* (New York: McGraw-Hill Book Company, 1955).

[8] A. P. Hurter, Jr., A. H. Rubenstein et al., "Market Penetration by New Innovations: The Technological Literature," *Technological Forecasting and Social Change*, 11 (1978), 197–221.

so that

$$\frac{dy}{dx} = y\frac{d}{dx}(\ln(f(x)))$$

is a formula that you can memorize, if you prefer. Second, the quantity $\dfrac{f'(x)}{f(x)}$, which results from differentiating $\ln(f(x))$, is what was called the *relative rate of change of* $f(x)$ in Section 11.3.

The next example illustrates the procedure.

● EXAMPLE 1   **Logarithmic Differentiation**

Find y' if $y = \dfrac{(2x-5)^3}{x^2\sqrt[4]{x^2+1}}$.

Solution: Differentiating this function in the usual way is messy because it involves the quotient, power, and product rules. Logarithmic differentiation makes the work less of a chore.

1. We take the natural logarithm of both sides:

$$\ln y = \ln\frac{(2x-5)^3}{x^2\sqrt[4]{x^2+1}}$$

2. Simplifying by using properties of logarithms, we have

$$\ln y = \ln(2x-5)^3 - \ln\left(x^2\sqrt[4]{x^2+1}\right)$$
$$= 3\ln(2x-5) - (\ln x^2 + \ln(x^2+1)^{1/4})$$
$$= 3\ln(2x-5) - 2\ln x - \frac{1}{4}\ln(x^2+1)$$

3. Differentiating with respect to x gives

$$\frac{y'}{y} = 3\left(\frac{1}{2x-5}\right)(2) - 2\left(\frac{1}{x}\right) - \frac{1}{4}\left(\frac{1}{x^2+1}\right)(2x)$$
$$= \frac{6}{2x-5} - \frac{2}{x} - \frac{x}{2(x^2+1)}$$

4. Solving for y' yields

$$y' = y\left(\frac{6}{2x-5} - \frac{2}{x} - \frac{x}{2(x^2+1)}\right)$$

5. Substituting the original expression for y gives y' in terms of x only:

$$y' = \frac{(2x-5)^3}{x^2\sqrt[4]{x^2+1}}\left[\frac{6}{2x-5} - \frac{2}{x} - \frac{x}{2(x^2+1)}\right]$$

NOW WORK PROBLEM 1 ◖●●

CAUTION

Since y is a function of x, differentiating $\ln y$ with respect to x gives $\dfrac{y'}{y}$.

Logarithmic differentiation can also be used to differentiate a function of the form $y = u^v$, where both u and v are differentiable functions of x. Because neither the base nor the exponent is necessarily a constant, the differentiation techniques for u^n and a^u do not apply here.

● EXAMPLE 2   **Differentiating the Form u^v**

Differentiate $y = x^x$ by using logarithmic differentiation.

Solution: This example is a good candidate for the *formula* approach to logarithmic differentiation.

$$y' = y\frac{d}{dx}(\ln x^x) = x^x\frac{d}{dx}(x\ln x) = x^x\left((1)(\ln x) + (x)\left(\frac{1}{x}\right)\right) = x^x(\ln x + 1)$$

It is worthwhile mentioning that an alternative technique for differentiating a function of the form $y = u^v$ is to convert it to an exponential function to the base e.

To illustrate, for the function in this example, we have

$$y = x^x = (e^{\ln x})^x = e^{x \ln x}$$

$$y' = e^{x \ln x} \left(1 \ln x + x \frac{1}{x} \right) = x^x (\ln x + 1)$$

NOW WORK PROBLEM 15

● EXAMPLE 3   Relative Rate of Change of a Product

Show that the relative rate of change of a product is the sum of the relative rates of change of its factors. Use this result to express the percentage rate of change in revenue in terms of the percentage rate of change in price.

Solution: Recall that the relative rate of change of a function r is $\dfrac{r'}{r}$. We are to show that if $r = pq$, then $\dfrac{r'}{r} = \dfrac{p'}{p} + \dfrac{q'}{q}$. From $r = pq$ we have $\ln r = \ln p + \ln q$, which, when both sides are differentiated, gives

$$\frac{r'}{r} = \frac{p'}{p} + \frac{q'}{q}$$

as required. Multiplying both sides by 100% gives an expression for the percentage rate of change of r in terms of those of p and q:

$$\frac{r'}{r} 100\% = \frac{p'}{p} 100\% + \frac{q'}{q} 100\%$$

If p is *price* per item and q is *quantity* sold, then $r = pq$ is total *revenue*. In this case we take differentiation to be with respect to p and note that now $\dfrac{q'}{q} = \eta \dfrac{p'}{p}$, where η is the elasticity of demand as in Section 12.3. It follows that in this case we have

$$\frac{r'}{r} 100\% = (1 + \eta) \frac{p'}{p} 100\%$$

expressing the percentage rate of change in revenue in terms of the percentage rate of change in price. For example, if at a given price and quantity, $\eta = -5$, then a 1% increase in price will result in a $(1 - 5)\% = -4\%$ increase in revenue, which is to say a 4% *decrease* in revenue, while a 3% decrease in price—that is, a -3% *increase* in price—will result in a $(1 - 5)(-3)\% = 12\%$ increase in revenue. It is also clear that at points at which there is unit elasticity ($\eta = -1$), any percentage change in price produces no percentage change in revenue.

NOW WORK PROBLEM 29

● EXAMPLE 4   Differentiating the Form u^v

Find the derivative of $y = (1 + e^x)^{\ln x}$.

Solution: This has the form $y = u^v$, where $u = 1 + e^x$ and $v = \ln x$. Using logarithmic differentiation, we have

$$\ln y = \ln((1 + e^x)^{\ln x})$$

$$\ln y = (\ln x) \ln(1 + e^x)$$

$$\frac{y'}{y} = \left(\frac{1}{x} \right) (\ln(1 + e^x)) + (\ln x) \left(\frac{1}{1 + e^x} \cdot e^x \right)$$

$$\frac{y'}{y} = \frac{\ln(1 + e^x)}{x} + \frac{e^x \ln x}{1 + e^x}$$

$$y' = y \left(\frac{\ln(1 + e^x)}{x} + \frac{e^x \ln x}{1 + e^x} \right)$$

$$y' = (1 + e^x)^{\ln x} \left(\frac{\ln(1 + e^x)}{x} + \frac{e^x \ln x}{1 + e^x} \right)$$

NOW WORK PROBLEM 17

Alternatively, we can differentiate even a general function of the form $y = u(x)^{v(x)}$ with $u(x) > 0$ by using the equation

$$u^v = e^{v \ln u}$$

Indeed, if $y = u(x)^{v(x)} = e^{v(x) \ln u(x)}$ for $u(x) > 0$, then

$$\frac{dy}{dx} = \frac{d}{dx} \left(e^{v(x) \ln u(x)} \right) = e^{v(x) \ln u(x)} \frac{d}{dx} (v(x) \ln u(x)) = u^v \left(v'(x) \ln u(x) + v(x) \frac{u'(x)}{u(x)} \right)$$

which could be summarized as

$$(u^v)' = u^v \left(v' \ln u + v \frac{u'}{u} \right)$$

As is often the case, there is no suggestion that you should memorize the preceding formula. The point here is that we have shown *any* function of the form u^v can be differentiated using the equation $u^v = e^{v \ln u}$. The same result will be obtained from logarithmic differentiation:

$$\ln y = \ln(u^v)$$

$$\ln y = v \ln u$$

$$\frac{y'}{y} = v' \ln u + v \frac{u'}{u}$$

$$y' = y \left(v' \ln u + v \frac{u'}{u} \right)$$

$$(u^v)' = u^v \left(v' \ln u + v \frac{u'}{u} \right)$$

After completing this section, you should understand how to differentiate each of the following forms:

$$y = \begin{cases} (f(x))^a & \text{(a)} \\ b^{f(x)} & \text{(b)} \\ (f(x))^{g(x)} & \text{(c)} \end{cases}$$

For type (a), use the power rule. For type (b), use the differentiation formula for exponential functions. [If $b \neq e$, you can first convert $b^{f(x)}$ to an e^u function.] For type (c), use logarithmic differentiation or first convert to an e^u function. Do not apply a rule in a situation where the rule does not apply. For example, the power rule does not apply to x^x.

Problems 12.5

In Problems 1–12, find y′ by using logarithmic differentiation.

*1. $y = (x + 1)^2(x - 2)(x^2 + 3)$

2. $y = (3x + 4)(8x - 1)^2(3x^2 + 1)^4$

3. $y = (3x^3 - 1)^2(2x + 5)^3$

4. $y = (2x^2 + 1)\sqrt{8x^2 - 1}$

5. $y = \sqrt{x + 1}\sqrt{x^2 - 2}\sqrt{x + 4}$

6. $y = (2x + 1)\sqrt{x^3 + 2}\sqrt[3]{2x + 5}$

7. $y = \dfrac{\sqrt{1 - x^2}}{1 - 2x}$

8. $y = \sqrt{\dfrac{x^2 + 5}{x + 9}}$

9. $y = \dfrac{(2x^2 + 2)^2}{(x + 1)^2(3x + 2)}$

10. $y = \dfrac{x(1 + x^2)^2}{\sqrt{2 + x^2}}$

11. $y = \sqrt{\dfrac{(x + 3)(x - 2)}{2x - 1}}$

12. $y = \sqrt[3]{\dfrac{6(x^3 + 1)^2}{x^6 e^{-4x}}}$

In Problems 13–20, find y′.

13. $y = x^{x^2 + 1}$

14. $y = (2x)^{\sqrt{x}}$

*15. $y = x^{1/x}$

16. $y = \left(\dfrac{3}{x^2} \right)^x$

*17. $y = (3x + 1)^{2x}$

18. $y = (x^2 + 1)^{x+1}$

19. $y = 4e^x x^{3x}$

20. $y = (\ln x)^{e^x}$

21. If $y = (4x - 3)^{2x+1}$, find dy/dx when $x = 1$.

22. If $y = (\ln x)^{\ln x}$, find dy/dx when $x = e$.

23. Find an equation of the tangent line to
$$y = (x + 1)(x + 2)^2(x + 3)^2$$
at the point where $x = 0$.

24. Find an equation of the tangent line to the graph of
$$y = x^x$$
at the point where $x = 1$.

25. Find an equation of the tangent line to the graph of

$$y = e^x(x^2 + 1)^x$$

at the point where $x = 1$.

26. If $y = x^x$, find the relative rate of change of y with respect to x when $x = 1$.

27. If $y = (3x)^{-2x}$, find the value of x for which the *percentage* rate of change of y with respect to x is 60.

28. Suppose $f(x)$ is a positive differentiable function and g is a differentiable function and $y = (f(x))^{g(x)}$. Use logarithmic

differentiation to show that

$$\frac{dy}{dx} = (f(x))^{g(x)}\left(f'(x)\frac{g(x)}{f(x)} + g'(x)\ln(f(x))\right)$$

*29. The demand equation for a compact disc is

$$q = 500 - 40p + p^2$$

If the price of \$15 is increased by 1/2%, find the corresponding percentage change in revenue.

30. Repeat Problem 29 with the same information except for a 10% *decrease* in price.

OBJECTIVE

To approximate real roots of an equation by using calculus. The method shown is suitable for calculators.

12.6 Newton's Method

It is quite easy to solve equations of the form $f(x) = 0$ when f is a linear or quadratic function. For example, we can solve $x^2 + 3x - 2 = 0$ by the quadratic formula. However, if $f(x)$ has degree greater than 2 (or is not a polynomial), it may be difficult, or even impossible, to find solutions (or roots) of $f(x) = 0$ by the methods to which you are accustomed. For this reason, we may settle for approximate solutions, which can be obtained in a variety of efficient ways. For example, a graphing calculator may be used to estimate the real roots of $f(x) = 0$. In this section, you will learn how the derivative may be so used (provided that f is differentiable). The procedure we will develop, called *Newton's method,* is well suited to a calculator or computer.

Newton's method requires an initial estimate for a root of $f(x) = 0$. One way of obtaining this estimate is by making a rough sketch of the graph of $y = f(x)$ and estimating the root from the graph. A point on the graph where $y = 0$ is an x-intercept, and the x-value of this point is a root of $f(x) = 0$. Another way of locating a root is based on the following fact:

> If f is continuous on the interval $[a, b]$ and $f(a)$ and $f(b)$ have opposite signs, then the equation $f(x) = 0$ has at least one real root between a and b.

Figure 12.6 depicts this situation. The x-intercept between a and b corresponds to a root of $f(x) = 0$, and we can use either a or b to approximate this root.

Assuming that we have an estimated (but incorrect) value for a root, we turn to a way of getting a better approximation. In Figure 12.7, you can see that $f(r) = 0$, so r is a root of the equation $f(x) = 0$. Suppose x_1 is an initial approximation to r (and one that is close to r). Observe that the tangent line to the curve at $(x_1, f(x_1))$ intersects the x-axis at the point $(x_2, 0)$, and x_2 is a better approximation to r than is x_1.

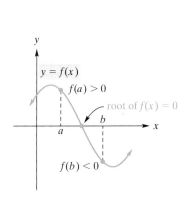

FIGURE 12.6   Root of $f(x) = 0$ between a and b, where $f(a)$ and $f(b)$ have opposite signs.

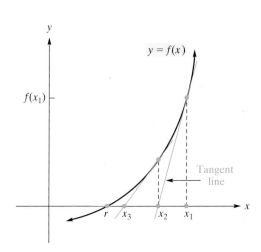

FIGURE 12.7   Improving approximation of root via tangent line.

We can find x_2 from the equation of the tangent line. The slope of the tangent line is $f'(x_1)$, so a point–slope form for this line is

$$y - f(x_1) = f'(x_1)(x - x_1) \tag{1}$$

Since $(x_2, 0)$ is on the tangent line, its coordinates must satisfy Equation (1). This gives

$$0 - f(x_1) = f'(x_1)(x_2 - x_1)$$

$$-\frac{f(x_1)}{f'(x_1)} = x_2 - x_1 \qquad [\text{if } f'(x_1) \neq 0]$$

Thus,

$$x_2 = x_1 - \frac{f(x_1)}{f'(x_1)} \tag{2}$$

To get a better approximation to r, we again perform the procedure described, but this time we use x_2 as our starting point. This gives the approximation

$$x_3 = x_2 - \frac{f(x_2)}{f'(x_2)} \tag{3}$$

Repeating (or *iterating*) this computation over and over, we hope to obtain better approximations, in the sense that the sequence of values

$$x_1, x_2, x_3, \ldots$$

will approach r. In practice, we terminate the process when we have reached a desired degree of accuracy.

If you analyze Equations (2) and (3), you can see how x_2 is obtained from x_1 and how x_3 is obtained from x_2. In general, x_{n+1} is obtained from x_n by means of the following general formula, called **Newton's method:**

Newton's Method

$$x_{n+1} = x_n - \frac{f(x_n)}{f'(x_n)} \quad n = 1, 2, 3, \ldots \tag{4}$$

A formula, like Equation (4), that indicates how one number in a sequence is obtained from the preceding one is called a **recursion formula,** or an *iteration equation.*

PRINCIPLES IN PRACTICE 1

APPROXIMATING A ROOT BY NEWTON'S METHOD

If the total profit (in dollars) from the sale of x televisions is $P(x) = 20x - 0.01x^2 - 850 + 3\ln(x)$, use Newton's method to approximate the break-even quantities. (*Note:* There are two break-even quantities; one is between 10 and 50, and the other is between 1900 and 2000.) Give the x-value to the nearest integer.

In the event that a root lies between a and b, and $f(a)$ and $f(b)$ are equally close to 0, choose either a or b as the first approximation.

● EXAMPLE 1 **Approximating a Root by Newton's Method**

Approximate the root of $x^4 - 4x + 1 = 0$ that lies between 0 and 1. Continue the approximation procedure until two successive approximations differ by less than 0.0001.

Solution: Letting $f(x) = x^4 - 4x + 1$, we have

$$f(0) = 0 - 0 + 1 = 1$$

and

$$f(1) = 1 - 4 + 1 = -2$$

(Note the change in sign.) Since $f(0)$ is closer to 0 than is $f(1)$, we choose 0 to be our first approximation, x_1. Now,

$$f'(x) = 4x^3 - 4$$

so

$$f(x_n) = x_n^4 - 4x_n + 1 \quad \text{and} \quad f'(x_n) = 4x_n^3 - 4$$

Substituting into Equation (4) gives the recursion formula

$$x_{n+1} = x_n - \frac{f(x_n)}{f'(x_n)} = x_n - \frac{x_n^4 - 4x_n + 1}{4x_n^3 - 4}$$

$$= \frac{4x_n^4 - 4x_n - x_n^4 + 4x_n - 1}{4x_n^3 - 4}$$

so

$$x_{n+1} = \frac{3x_n^4 - 1}{4x_n^3 - 4} \tag{5}$$

Since $x_1 = 0$, letting $n = 1$ in Equation (5) gives

$$x_2 = \frac{3x_1^4 - 1}{4x_1^3 - 4} = \frac{3(0)^4 - 1}{4(0)^3 - 4} = 0.25$$

Letting $n = 2$ in Equation (5) gives

$$x_3 = \frac{3x_2^4 - 1}{4x_2^3 - 4} = \frac{3(0.25)^4 - 1}{4(0.25)^3 - 4} \approx 0.25099$$

Letting $n = 3$ in Equation (5) gives

$$x_4 = \frac{3x_3^4 - 1}{4x_3^3 - 4} = \frac{3(0.25099)^4 - 1}{4(0.25099)^3 - 4} \approx 0.25099$$

The data obtained thus far are displayed in Table 12.1. Since the values of x_3 and x_4 differ by less than 0.0001, we take the root to be 0.25099 (that is, x_4).

NOW WORK PROBLEM 1 ⬤⬤

TABLE 12.1

n	x_n	x_{n+1}
1	0.00000	0.25000
2	0.25000	0.25099
3	0.25099	0.25099

⬤ **EXAMPLE 2   Approximating a Root by Newton's Method**

Approximate the root of $x^3 = 3x - 1$ that lies between -1 and -2. Continue the approximation procedure until two successive approximations differ by less than 0.0001.

Solution: Letting $f(x) = x^3 - 3x + 1$ (we need the form $f(x) = 0$), we find that

$$f(-1) = (-1)^3 - 3(-1) + 1 = 3$$

and

$$f(-2) = (-2)^3 - 3(-2) + 1 = -1$$

(Note the change in sign.) Since $f(-2)$ is closer to 0 than is $f(-1)$, we choose -2 to be our first approximation, x_1. Now,

$$f'(x) = 3x^2 - 3$$

so

$$f(x_n) = x_n^3 - 3x_n + 1 \quad \text{and} \quad f'(x_n) = 3x_n^2 - 3$$

Substituting into Equation (4) gives the recursion formula

$$x_{n+1} = x_n - \frac{f(x_n)}{f'(x_n)} = x_n - \frac{x_n^3 - 3x_n + 1}{3x_n^2 - 3}$$

so

$$x_{n+1} = \frac{2x_n^3 - 1}{3x_n^2 - 3} \tag{6}$$

TABLE 12.2

n	x_n	$x_n + 1$
1	−2.00000	−1.88889
2	−1.88889	−1.87945
3	−1.87945	−1.87939

Since $x_1 = -2$, letting $n = 1$ in Equation (6) gives

$$x_2 = \frac{2x_1^3 - 1}{3x_1^2 - 3} = \frac{2(-2)^3 - 1}{3(-2)^2 - 3} \approx -1.88889$$

Continuing in this way, we obtain Table 12.2. Because the values of x_3 and x_4 differ by 0.00006, which is less than 0.0001, we take the root to be −1.87939 (that is, x_4).

NOW WORK PROBLEM 3 ⬤⬤

The situation where x_1 leads to a derivative of 0 occurs in Problems 2 and 8 of Problems 12.6. In case your choice for the initial approximation, x_1, gives the derivative a value of zero, choose a different number that is close to the desired root. A graph of f could be helpful in this situation. Finally, we remark that there are times when the sequence of approximations does not approach the root. A discussion of such situations is beyond the scope of this book.

TECHNOLOGY

Figure 12.8 gives a short TI-83 Plus program for Newton's method. Before the program is executed, the first approximation to the root of $f(x) = 0$ is stored as X, and $f(x)$ and $f'(x)$ are stored as Y_1 and Y_2, respectively.

```
PROGRAM:NEWTON
:Lbl A
:X-Y₁(X)/Y₂(X)
:Ans→X
:Disp X
:Pause
:Goto A█
```

FIGURE 12.8 Calculator program for Newton's method.

When executed, the program computes the first iteration and pauses. Successive iterations are obtained by successively pressing ENTER. Figure 12.9 shows the iterations for the problem in Example 2.

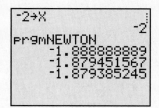

FIGURE 12.9 Iterations for problem in Example 2.

Problems 12.6

In Problems 1–10, use Newton's method to approximate the indicated root of the given equation. Continue the approximation procedure until the difference of two successive approximations is less than 0.0001.

*1. $x^3 - 4x + 1 = 0$; root between 0 and 1

2. $x^3 + 2x^2 - 1 = 0$; root between 0 and 1

*3. $x^3 - x - 1 = 0$; root between 1 and 2

4. $x^3 - 9x + 6 = 0$; root between 2 and 3

5. $x^3 + x + 1 = 0$; root between −1 and 0

6. $x^3 = 2x + 5$; root between 2 and 3

7. $x^4 = 3x - 1$; root between 0 and 1

8. $x^4 + 4x - 1 = 0$; root between −2 and −1

9. $x^4 - 2x^3 + x^2 - 3 = 0$; root between 1 and 2

10. $x^4 - x^3 + x - 2 = 0$; root between 1 and 2

11. Estimate, to three-decimal-place accuracy, the cube root of 71. (*Hint:* Show that the problem is equivalent to finding a root of $f(x) = x^3 - 71 = 0$.) Choose 4 as the initial estimate. Continue the iteration until two successive approximations, rounded to three decimal places, are the same.

12. Estimate $\sqrt[4]{19}$, to two-decimal-place accuracy. Use 2 as your initial estimate.

13. Find, to two-decimal-place accuracy, all real solutions of the equation $e^x = x + 5$. (*Hint:* A rough sketch of the graphs of $y = e^x$ and $y = x + 5$ makes it clear how many solutions there are. Use nearby integer values for your initial estimates.)

14. Find, to three-decimal-place accuracy, all real solutions of the equation $\ln x = 5 - x$.

15. **Break-Even Quantity** The cost of manufacturing q tons of a certain product is given by

$$c = 250 + 2q - 0.1q^3$$

and the revenue obtained by selling the q tons is given by

$$r = 3q$$

Approximate, to two-decimal-place accuracy, the break-even quantity. (*Hint:* Approximate a root of $r - c = 0$ by choosing 13 as your initial estimate.)

16. **Break-Even Quantity** The total cost of manufacturing q hundred pencils is c dollars, where

$$c = 40 + 3q + \frac{q^2}{1000} + \frac{1}{q}$$

Pencils are sold for $7 per hundred.

(a) Show that the break-even quantity is a solution of the equation

$$f(q) = \frac{q^3}{1000} - 4q^2 + 40q + 1 = 0$$

(b) Use Newton's method to approximate the solution of $f(q) = 0$, where $f(q)$ is given in part (a). Use 10 as your initial approximation, and give your answer to two-decimal-place accuracy.

17. **Equilibrium** Given the supply equation $p = 2q + 5$ and the demand equation $p = \dfrac{100}{q^2 + 1}$, use Newton's method to

estimate the market equilibrium quantity. Give your answer to three-decimal-place accuracy.

18. **Equilibrium**   Given the supply equation

$$p = 0.2q^3 + 0.5q + 2$$

and the demand equation $p = 10 - q$, use Newton's method to estimate the market equilibrium quantity, and find the corresponding equilibrium price. Use 5 as an initial estimate

for the required value of q, and give your answer to two-decimal-place accuracy.

19. Use Newton's method to approximate (to two-decimal-place accuracy) a critical value of the function

$$f(x) = \frac{x^3}{3} - x^2 - 5x + 1$$

on the interval $[3, 4]$.

12.7 Higher-Order Derivatives

To find higher-order derivatives both directly and implicitly.

We know that the derivative of a function $y = f(x)$ is itself a function, $f'(x)$. If we differentiate $f'(x)$, the resulting function is called the **second derivative** of f at x. It is denoted $f''(x)$, which is read "f double prime of x." Similarly, the derivative of the second derivative is called the **third derivative,** written $f'''(x)$. Continuing in this way, we get *higher-order derivatives*. Some notations for higher-order derivatives are given in Table 12.3. To avoid clumsy notation, primes are not used beyond the third derivative.

CAUTION

The symbol d^2y/dx^2 represents the second derivative of y. It is not the same as $(dy/dx)^2$, the square of the first derivative of y.

TABLE 12.3

First derivative:	y'	$f'(x)$	$\dfrac{dy}{dx}$	$\dfrac{d}{dx}(f(x))$	$D_x y$
Second derivative:	y''	$f''(x)$	$\dfrac{d^2y}{dx^2}$	$\dfrac{d^2}{dx^2}(f(x))$	$D_x^2 y$
Third derivative:	y'''	$f'''(x)$	$\dfrac{d^3y}{dx^3}$	$\dfrac{d^3}{dx^3}(f(x))$	$D_x^3 y$
Fourth derivative:	$y^{(4)}$	$f^{(4)}(x)$	$\dfrac{d^4y}{dx^4}$	$\dfrac{d^4}{dx^4}(f(x))$	$D_x^4 y$

● **EXAMPLE 1   Finding Higher-Order Derivatives**

a. If $f(x) = 6x^3 - 12x^2 + 6x - 2$, find all *higher-order derivatives.*

Solution: Differentiating $f(x)$ gives

$$f'(x) = 18x^2 - 24x + 6$$

Differentiating $f'(x)$ yields

$$f''(x) = 36x - 24$$

Similarly,

$$f'''(x) = 36$$
$$f^{(4)}(x) = 0$$

All successive derivatives are also 0: $f^{(5)}(x) = 0$, and so on.

b. If $f(x) = 7$, find $f''(x)$.

Solution:

$$f'(x) = 0$$
$$f''(x) = 0$$

NOW WORK PROBLEM 1 ●●●

PRINCIPLES IN PRACTICE 1

FINDING A SECOND-ORDER DERIVATIVE

The height $h(t)$ of a rock dropped off of a 200-foot building is given by $h(t) = 200 - 16t^2$, where t is the time measured in seconds. Find $\dfrac{d^2h}{dt^2}$, the acceleration of the rock at time t.

● **EXAMPLE 2   Finding a Second-Order Derivative**

If $y = e^{x^2}$, find $\dfrac{d^2y}{dx^2}$.

Solution:

$$\frac{dy}{dx} = e^{x^2}(2x) = 2xe^{x^2}$$

By the product rule,

$$\frac{d^2y}{dx^2} = 2[x(e^{x^2})(2x) + e^{x^2}(1)] = 2e^{x^2}(2x^2 + 1)$$

NOW WORK PROBLEM 5 ◖●●

●EXAMPLE 3   **Evaluating a Second-Order Derivative**

If $y = f(x) = \dfrac{16}{x+4}$, *find* $\dfrac{d^2y}{dx^2}$ *and evaluate it when* $x = 4$.

Solution: Since $y = 16(x+4)^{-1}$, the power rule gives

$$\frac{dy}{dx} = -16(x+4)^{-2}$$

$$\frac{d^2y}{dx^2} = 32(x+4)^{-3} = \frac{32}{(x+4)^3}$$

Evaluating when $x = 4$, we obtain

$$\left.\frac{d^2y}{dx^2}\right|_{x=4} = \frac{32}{8^3} = \frac{1}{16}$$

The second derivative evaluated at $x = 4$ is also denoted $f''(4)$ or $y''(4)$.

NOW WORK PROBLEM 21 ◖●●

●EXAMPLE 4   **Finding the Rate of Change of $f''(x)$**

If $f(x) = x \ln x$, *find the rate of change of* $f''(x)$.

Solution: To find the rate of change of any function, we must find its derivative. Thus, we want the derivative of $f''(x)$, which is $f'''(x)$. Accordingly,

$$f'(x) = x\left(\frac{1}{x}\right) + (\ln x)(1) = 1 + \ln x$$

$$f''(x) = 0 + \frac{1}{x} = \frac{1}{x}$$

$$f'''(x) = \frac{d}{dx}(x^{-1}) = (-1)x^{-2} = -\frac{1}{x^2}$$

NOW WORK PROBLEM 17 ◖●●

Higher-Order Implicit Differentiation

We will now find a higher-order derivative by means of implicit differentiation. Keep in mind that we will assume y to be a function of x.

●EXAMPLE 5   **Higher-Order Implicit Differentiation**

Find $\dfrac{d^2y}{dx^2}$ *if* $x^2 + 4y^2 = 4$.

Solution: Differentiating both sides with respect to x, we obtain

$$2x + 8y\frac{dy}{dx} = 0$$

$$\frac{dy}{dx} = \frac{-x}{4y} \tag{1}$$

$$\frac{d^2y}{dx^2} = \frac{4y\dfrac{d}{dx}(-x) - (-x)\dfrac{d}{dx}(4y)}{(4y)^2}$$

$$= \frac{4y(-1) - (-x)\left(4\dfrac{dy}{dx}\right)}{16y^2}$$

$$= \frac{-4y + 4x\dfrac{dy}{dx}}{16y^2}$$

$$\frac{d^2y}{dx^2} = \frac{-y + x\dfrac{dy}{dx}}{4y^2} \tag{2}$$

Although we have found an expression for d^2y/dx^2, our answer involves the derivative dy/dx. It is customary to express the answer without the derivative—that is, in terms of x and y only. This is easy to do. From Equation (1), $\dfrac{dy}{dx} = \dfrac{-x}{4y}$, so by substituting into Equation (2), we have

$$\frac{d^2y}{dx^2} = \frac{-y + x\left(\dfrac{-x}{4y}\right)}{4y^2} = \frac{-4y^2 - x^2}{16y^3} = -\frac{4y^2 + x^2}{16y^3}$$

We can further simplify the answer. Since $x^2 + 4y^2 = 4$ (the original equation),

In Example 5, the simplification of d^2y/dx^2 by making use of the original equation is not unusual.

$$\frac{d^2y}{dx^2} = -\frac{4}{16y^3} = -\frac{1}{4y^3}$$

NOW WORK PROBLEM 23

● EXAMPLE 6 Higher-Order Implicit Differentiation

Find $\dfrac{d^2y}{dx^2}$ *if* $y^2 = e^{x+y}$.

Solution: Differentiating both sides with respect to x gives

$$2y\frac{dy}{dx} = e^{x+y}\left(1 + \frac{dy}{dx}\right)$$

Solving for dy/dx, we obtain

$$2y\frac{dy}{dx} = e^{x+y} + e^{x+y}\frac{dy}{dx}$$

$$2y\frac{dy}{dx} - e^{x+y}\frac{dy}{dx} = e^{x+y}$$

$$(2y - e^{x+y})\frac{dy}{dx} = e^{x+y}$$

$$\frac{dy}{dx} = \frac{e^{x+y}}{2y - e^{x+y}}$$

Since $y^2 = e^{x+y}$ (the original equation),

$$\frac{dy}{dx} = \frac{y^2}{2y - y^2} = \frac{y}{2 - y}$$

$$\frac{d^2y}{dx^2} = \frac{(2 - y)\dfrac{dy}{dx} - y\left(-\dfrac{dy}{dx}\right)}{(2 - y)^2} = \frac{2\dfrac{dy}{dx}}{(2 - y)^2}$$

Now we express our answer without dy/dx. Since $\dfrac{dy}{dx} = \dfrac{y}{2 - y}$,

$$\frac{d^2y}{dx^2} = \frac{2\left(\dfrac{y}{2 - y}\right)}{(2 - y)^2} = \frac{2y}{(2 - y)^3}$$

NOW WORK PROBLEM 31

Problems 12.7

In Problems 1–20, find the indicated derivatives.

*1. $y = 4x^3 - 12x^2 + 6x + 2$, y'''

2. $y = x^5 - 2x^4 + 7x^2 - 2$, y'''

3. $y = 8 - x$, $\dfrac{d^2 y}{dx^2}$

4. $y = -x - x^2$, $\dfrac{d^2 y}{dx^2}$

*5. $y = x^3 + e^x$, $y^{(4)}$

6. $F(q) = \ln(q + 1)$, $\dfrac{d^3 F}{dq^3}$

7. $f(x) = x^2 \ln x$, $f''(x)$

8. $y = \dfrac{1}{x}$, y'''

9. $f(q) = \dfrac{1}{2q^4}$, $f'''(q)$

10. $f(x) = \sqrt{x}$, $f''(x)$

11. $f(r) = \sqrt{9 - r}$, $f''(r)$

12. $y = e^{-4x^2}$, y''

13. $y = \dfrac{1}{2x + 3}$, $\dfrac{d^2 y}{dx^2}$

14. $y = (3x + 7)^5$, y''

15. $y = \dfrac{x + 1}{x - 1}$, y''

16. $y = 2x^{1/2} + (2x)^{1/2}$, y''

*17. $y = \ln[x(x + 6)]$, y''

18. $y = \ln \dfrac{(2x + 5)(5x - 2)}{x + 1}$, y''

19. $f(z) = z^2 e^z$, $f''(z)$

20. $y = \dfrac{x}{e^x}$, $\dfrac{d^2 y}{dx^2}$

*21. If $y = e^{2x} + e^{3x}$, find $\left.\dfrac{d^5 y}{dx^5}\right|_{x=0}$.

22. If $y = e^{2 \ln(x^3 + 1)}$, find y'' when $x = 1$.

In Problems 23–32, find y''.

*23. $x^2 + 4y^2 - 16 = 0$

24. $x^2 - y^2 = 16$

25. $y^2 = 4x$

26. $9x^2 + 16y^2 = 25$

27. $\sqrt{x} + 4\sqrt{y} = 4$

28. $y^2 - 6xy = 4$

29. $xy + y - x = 4$

30. $x^2 + 2xy + y^2 = 1$

*31. $y = e^{x+y}$

32. $e^x - e^y = x^2 + y^2$

33. If $x^2 + 3x + y^2 = 4y$, find $d^2 y/dx^2$ when $x = 0$ and $y = 0$.

34. Show that the equation

$$f''(x) + 4f'(x) + 4f(x) = 0$$

is satisfied if $f(x) = (3x - 5)e^{-2x}$.

35. Find the rate of change of $f'(x)$ if $f(x) = (5x - 3)^4$.

36. Find the rate of change of $f''(x)$ if

$$f(x) = 6\sqrt{x} + \frac{1}{6\sqrt{x}}$$

37. **Marginal Cost** If $c = 0.3q^2 + 2q + 850$ is a cost function, how fast is marginal cost changing when $q = 100$?

38. **Marginal Revenue** If $p = 400 - 40q - q^2$ is a demand equation, how fast is marginal revenue changing when $q = 4$?

39. If $f(x) = x^4 - 6x^2 + 5x - 6$, determine the values of x for which $f''(x) = 0$.

40. Suppose that $e^y = y^2 e^x$. (a) Determine dy/dx, and express your answer in terms of y only. (b) Determine $d^2 y/dx^2$, and express your answer in terms of y only.

In Problems 41 and 42, determine $f''(x)$. Then use your graphing calculator to find all real zeros of $f''(x)$. Round your answers to two decimal places.

41. $f(x) = 6e^x - x^3 - 15x^2$

42. $f(x) = \dfrac{x^5}{20} + \dfrac{x^4}{12} + \dfrac{5x^3}{6} + \dfrac{x^2}{2}$

12.8 Review

Important Terms and Symbols — Examples

Summary

The derivative formulas for natural logarithmic and exponential functions are

$$\frac{d}{dx}(\ln u) = \frac{1}{u}\frac{du}{dx}$$

and

$$\frac{d}{dx}(e^u) = e^u\frac{du}{dx}$$

To differentiate logarithmic and exponential functions in bases other than e, you can first transform the function to base e and then differentiate the result. Alternatively, differentiation formulas can be applied:

$$\frac{d}{dx}(\log_b u) = \frac{1}{(\ln b)u}\cdot\frac{du}{dx}$$

$$\frac{d}{dx}(b^u) = b^u(\ln b)\cdot\frac{du}{dx}$$

Point elasticity of demand is a function that measures how consumer demand is affected by a change in price. It is given by

$$\eta = \frac{p}{q}\frac{dq}{dp}$$

where p is the price per unit at which q units are demanded. The three categories of elasticity are as follows:

$	\eta(p)	> 1$	demand is elastic
$	\eta(p)	= 1$	unit elasticity
$	\eta(p)	< 1$	demand is inelastic

For a given percentage change in price, if there is a greater (respectively lesser) percentage change in quantity demanded, then demand is elastic (respectively, inelastic) and conversely.

The relationship between elasticity and the rate of change of revenue is given by

$$\frac{dr}{dq} = p\left(1 + \frac{1}{\eta}\right)$$

If an equation implicitly defines y as a function of x (rather than defining it explicitly in the form $y = f(x)$), then dy/dx can be found by implicit differentiation. With this method, we treat y as a function of x and differentiate both sides of the equation with respect to x. When doing this, remember that

$$\frac{d}{dx}(y^n) = ny^{n-1}\frac{dy}{dx}$$

and, more generally, that

$$\frac{d}{dx}(f(y)) = f'(y)\frac{dy}{dx}$$

Finally, we solve the resulting equation for dy/dx.

Suppose that $f(x)$ consists of products, quotients, or powers. To differentiate $y = \log_b(f(x))$, it may be helpful to use properties of logarithms to rewrite $\log_b(f(x))$ in terms of simpler logarithms and then differentiate that form. To differentiate $y = f(x)$, where $f(x)$ consists of products, quotients, or powers, the method of logarithmic differentiation may be used. In that method, we take the natural logarithm of both sides of $y = f(x)$ to obtain $\ln y = \ln(f(x))$. After simplifying $\ln(f(x))$ by using properties of logarithms, we differentiate both sides of $\ln y = \ln(f(x))$ with respect to x and then solve for y'. Logarithmic differentiation can also be used to differentiate $y = u^v$, where both u and v are functions of x.

Newton's method is the name given to the following formula, which is used to approximate the roots of the equation $f(x) = 0$, provided that f is differentiable:

$$x_{n+1} = x_n - \frac{f(x_n)}{f'(x_n)}, \quad n = 1, 2, 3, \ldots$$

In most cases you might encounter, the approximation improves as n increases.

Because the derivative $f'(x)$ of a function $y = f(x)$ is itself a function, it can be successively differentiated to obtain the second derivative $f''(x)$, the third derivative $f'''(x)$, and other higher-order derivatives.

Review Problems

Problem numbers shown in color indicate problems suggested for use as a practice chapter test.

In Problems 1–30, differentiate.

1. $y = 3e^x + e^2 + e^{x^2} + x^{e^2}$

2. $f(w) = we^w + w^2$

3. $f(r) = \ln(3r^2 + 7r + 1)$

4. $y = e^{\ln x}$

5. $y = e^{x^2+4x+5}$

6. $f(t) = \log_6\sqrt{t^2 + 1}$

7. $y = e^x(x^2 + 2)$

8. $y = 3^{5x^3}$

9. $y = \sqrt{(x - 6)(x + 5)(9 - x)}$

10. $f(t) = e^{1/t}$

11. $y = \dfrac{\ln x}{e^x}$

12. $y = \dfrac{e^x + e^{-x}}{x^2}$

13. $f(q) = \ln[(q + 1)^2(q + 2)^3]$

14. $y = (x + 2)^3(x + 1)^4(x - 2)^2$

15. $y = 2^{2x^2+2x-5}$

16. $y = (e + e^2)^0$

17. $y = \dfrac{4e^{3x}}{xe^{x-1}}$

18. $y = \dfrac{e^x}{\ln x}$

19. $y = \log_2(8x + 5)^2$

20. $y = \ln\left(\dfrac{5}{x^2}\right)$

21. $f(l) = \ln(1 + l + l^2 + l^3)$

22. $y = (x^2)^{x^2}$

23. $y = (x + 1)^{x+1}$

24. $y = \dfrac{1 + e^x}{1 - e^x}$

25. $\phi(t) = \ln\left(t\sqrt{4 - t^2}\right)$

26. $y = (x + 3)^{\ln x}$

27. $y = \dfrac{(x^2 + 1)^{1/2}(x^2 + 2)^{1/3}}{(2x^3 + 6x)^{2/5}}$

28. $y = \dfrac{\ln x}{\sqrt{x}}$

29. $y = (x^x)^x$

30. $y = x^{(x^x)}$

In Problems 31–34, evaluate y' at the given value of x.

31. $y = (x + 1)\ln x^2, x = 1$

32. $y = \dfrac{e^{x^2+1}}{\sqrt{x^2 + 1}}, x = 1$

33. $y = e^{e + x\ln(1/x)}, x = e$

34. $y = \left[\dfrac{2^{5x}(x^2 - 3x + 5)^{1/3}}{(x^2 - 3x + 7)^3}\right]^{-1}, x = 0$

In Problems 35 and 36, find an equation of the tangent line to the curve at the point corresponding to the given value of x.

35. $y = 3e^x, x = \ln 2$

36. $y = x + x^2 \ln x, x = 1$

37. Find the y-intercept of the tangent line to the graph of $y = x(2^{2-x^2})$ at the point where $x = 1$.

38. If $w = 2^{x+1} + \ln(1 + x^2)$ and

$$x = \log_2(t^2 + 1) - e^{(t-1)^2}$$

find w and dw/dt when $t = 1$.

In Problems 39–42, find the indicated derivative at the given point.

39. $y = e^{x^2-2x+1}, y'', (1, 1)$

40. $y = x^2 e^x, y''', (1, e)$

41. $y = \ln(2x), y''', (1, \ln 2)$

42. $y = x \ln x, y'', (1, 0)$

In Problems 43–46, find dy/dx.

43. $2xy + y^2 = 10$

44. $x^3 y^3 = 3$

45. $\ln(xy^2) = xy$

46. $y^2 e^{y \ln x} = e^2$

In Problems 47 and 48, find d^2y/dx^2 at the given point.

47. $x + xy + y = 5, (2, 1)$

48. $xy + y^2 = 2, (-1, 2)$

49. If y is defined implicitly by $e^y = (y + 1)e^x$, determine both dy/dx and d^2y/dx^2 as explicit functions of y only.

50. If $\sqrt{x} + \sqrt{y} = 1$, find $\dfrac{d^2y}{dx^2}$.

51. Schizophrenia Several models have been used to analyze the length of stay in a hospital. For a particular group of schizophrenics, one such model is[9]

$$f(t) = 1 - (0.8e^{-0.01t} + 0.2e^{-0.0002t})$$

where $f(t)$ is the proportion of the group that was discharged at the end of t days of hospitalization. Determine the discharge rate (proportion discharged per day) at the end of t days.

52. Earthquakes According to Richter,[10] the number N of earthquakes of magnitude M or greater per unit of time is given by $\log N = A - bM$, where A and b are constants. He claims that

$$\log\left(-\dfrac{dN}{dM}\right) = A + \log\left(\dfrac{b}{q}\right) - bM$$

where $q = \log e$. Verify this statement.

53. If $f(x) = e^{3x^4 + 2x^3 - 25x}$, find all real zeros of $f'(x)$. Round your answers to two decimal places.

54. If $f(x) = \dfrac{x^5}{10} + \dfrac{x^4}{6} + \dfrac{2x^3}{3} + x^2 + 1$, find all zeros of $f''(x)$. Round your answers to two decimal places.

For the demand equations in Problems 55–57, determine whether demand is elastic, is inelastic, or has unit elasticity for the indicated value of q.

55. $p = \dfrac{500}{q}; \quad q = 200$

56. $p = 900 - q^2; \quad q = 10$

57. $p = 18 - 0.02q; \quad q = 600$

58. The demand equation for a product is

$$p = 20 - 2\sqrt{q}$$

(a) Find the point elasticity of demand when $p = 8$.

(b) Find all values of p for which demand is elastic.

---

[9]Adapted from W. W. Eaton and G. A. Whitmore, "Length of Stay as a Stochastic Process: A General Approach and Application to Hospitalization for Schizophrenia," *Journal of Mathematical Sociology*, 5 (1977) 273–92.

[10]C. F. Richter, *Elementary Seismology* (San Francisco: W. H. Freeman and Company, Publishers, 1958).

59. The demand equation of a product is

$$q = \sqrt{2500 - p^2}$$

Find the point elasticity of demand when $p = 30$. If the price of 30 decreases $\frac{2}{3}\%$, what is the approximate change in demand?

60. The demand equation for a product is

$$q = \sqrt{100 - p}, \quad \text{where } 0 < p < 100$$

(a) Find all prices that correspond to elastic demand.

(b) Compute the point elasticity of demand when $p = 40$. Use your answer to estimate the percentage increase or decrease in demand when price is increased by 5% to $p = 42$.

61. The equation $x^3 - 2x - 2 = 0$ has a root between 1 and 2. Use Newton's method to estimate the root. Continue the approximation procedure until the difference of two successive approximations is less than 0.0001. Round your answer to four decimal places.

62. Find, to an accuracy of three decimal places, all real solutions of the equation $e^x = 3x$.

Mathematical Snapshot

Economic Order Quantity

In inventory management, the economic order quantity is the most cost-efficient size for resupply orders. To find this optimum size, we need to have an idea of how stock depletion and resupply take place, and of how costs accrue.

Here are the classic assumptions:

1. Inventory is depleted through purchases at a constant rate D, which is measured in units per year.
2. Resupply orders are all the same size, and each arrives in a single lump shipment just as stock is running out.
3. Besides the cost per item, each order also involves a fixed cost per order, F.
4. Each unit in stock has a constant value, V, measured in dollars.
5. The cost of inventory storage is a fixed fraction, R, of total current inventory value. This carrying cost factor is measured in dollars per dollar per year.

Assumptions 1 and 2 entail a graph of inventory over time that looks like Figure 12.10.

We now wish to minimize the cost, in dollars per year, of managing an inventory in the way Figure 12.10 depicts. If resupply is ordered in lots of q units each, then there are $\dfrac{D}{q}$ orders per year, for an annual ordering cost of $\dfrac{FD}{q}$. (The yearly expense due to the per-item ordering cost cannot be adjusted by changing the order size, so this cost is ignored in our calculations.) With an average inventory level of $\dfrac{q}{2}$, the annual carrying cost is $\dfrac{RVq}{2}$. The annual inventory-related cost, C, is then the sum of the ordering cost and the carrying cost:

$$C = C(q) = \frac{FD}{q} + \frac{RVq}{2}$$

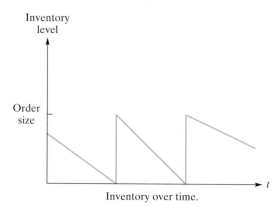

FIGURE 12.10 Inventory over time.

This function C clearly takes on large values both when q gets large and when q approaches zero. It follows from arguments that we will study in detail in the next chapter that if there is a unique value of q where $\dfrac{dC}{dq}$ equals zero then this value of q will provide a minimum value of C. Let us try to find such a q.

$$\frac{dC}{dq} = \frac{-FD}{q^2} + \frac{RV}{2} = 0$$

$$q^2 = \frac{2FD}{RV}$$

$$q = \sqrt{\frac{2FD}{RV}}$$

This formula is called the Wilson lot size formula, after an industry consultant who popularized its use. If we substitute $F = \$10$ per order, $D = 1500$ units per year, $R = \$0.10$ dollars per dollar per year, and $V = \$10$, then q comes out as

$$q = \sqrt{\frac{2(10)(1500)}{(0.10)(10)}} \approx 173.2$$

The most cost-efficient order size is 173 units.

Variations of the Wilson formula relax one or more of the five assumptions on which it is based. One assumption that can be relaxed is assumption 5. Suppose that carrying cost as a percentage of inventory value rises when inventory is low. (Think of a large warehouse sitting nearly empty.) We will model this by replacing R with $R(1+ke^{-sq})$. R is the per-dollar annual carrying cost for large inventory levels, and the term ke^{-sq} ($k, s > 0$) raises the cost for low inventory levels. The total annual inventory cost now becomes

$$C = \frac{FD}{q} + \frac{RVq(1 + ke^{-sq})}{2}$$

Again, we wish to minimize this quantity, and again C gets large both as q gets large and as q approaches zero. The minimum is where

$$\frac{dC}{dq} = \frac{-FD}{q^2} + \frac{RV(1 + ke^{-sq} - ksqe^{-sq})}{2} = 0$$

Suppose $k = 1, s = \frac{\ln 2}{1000} \approx 0.000693$. Then the per-dollar carrying cost is twice as great as for a small inventory as for a very large one and is midway between those two costs at an inventory level of 1000. If we keep F, D, R, and V the same as before and then use a graphing calculator or other numeric solution technique, we find that $\frac{dC}{dq} = 0$ when $q \approx 127.9$. The optimum order size is 128 units. Note that even though the assumptions now include economies of scale, the carrying cost is greater at all inventory levels and has led to a lower economic order quantity.

Problems

1. Use the Wilson lot size formula to calculate the economic order quantity for an item that is worth $36.50, costs 5% of its value annually to store, sells at a rate of 3400 units per year, and is purchased from a supplier that charges a flat $25 processing fee for every order.

2. Suppose that assumptions 1 and 3–5 are kept but 2 is modified: A manager never allows inventory to drop to zero but instead maintains a safety margin of a certain number of units. What difference does this make to the calculation of the economic order quantity?

3. What other assumptions, besides assumptions 2 and 5, might realistically be relaxed? Explain.

13

CURVE SKETCHING

Mathematical Snapshot Population Change over Time

I n the mid-1970s, economist Arthur Laffer was explaining his views on taxes to a politician—as the story goes, it was either presidential aspirant Ronald Reagan or Ford administration staff member Richard Cheney (later Vice President under George W. Bush). To illustrate his argument, Laffer grabbed a paper napkin and sketched the graph that now bears his name: the Laffer curve.[1]

The Laffer curve describes total government tax revenue as a function of the tax rate. Obviously, if the tax rate is zero, the government gets nothing. But if the tax rate is 100%, revenue would again equal zero, because there is no incentive to earn money if it will all be taken away. Since tax rates between 0% and 100% do generate revenue, Laffer reasoned, the curve relating revenue to tax rate must look, qualitatively, more or less as shown in the figure below.

Laffer's argument was not meant to show that the optimal tax rate was 50%. It was meant to show that under some circumstances, namely when the tax rate is to the right of the peak of the curve, it is possible to *raise government revenue by lowering taxes*. This was a key argument made for the tax cuts passed by Congress during the first term of the Reagan presidency.

Because the Laffer curve is only a qualitative picture, it does not actually give an optimal tax rate. Revenue-based arguments for tax cuts involve the claim that the point of peak revenue lies to the left of the current taxation scheme on the horizontal axis. By the same token, those who urge raising taxes to raise government income are assuming either a different relationship between rates and revenues or a different location of the curve's peak.

By itself, then, the Laffer curve is too abstract to be of much help in determining the optimal tax rate. But even very simple sketched curves, like supply and demand curves and the Laffer curve, can help economists describe the causal factors that drive an economy. In this chapter, we will discuss techniques for sketching and interpreting curves.

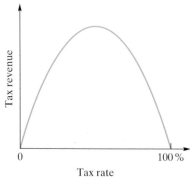

---

[1] For one version of the story, see Jude Wanniski, *The Way the World Works,* 3rd ed. (Morristown, NJ: Polyconomics, 1989), p. 299.

To find when a function is increasing or decreasing, to find critical values, to locate relative maxima and relative minima, and to state the first-derivative test. Also, to sketch the graph of a function by using the information obtained from the first derivative.

13.1 Relative Extrema

Increasing or Decreasing Nature of a Function

Examining the graphical behavior of functions is a basic part of mathematics and has applications to many areas of study. When we sketch a curve, just plotting points may not give enough information about its shape. For example, the points $(-1, 0)$, $(0, -1)$, and $(1, 0)$ satisfy the equation given by $y = (x + 1)^3(x - 1)$. On the basis of these points, you might hastily conclude that the graph should appear as in Figure 13.1(a), but in fact the true shape is given in Figure 13.1(b). In this chapter we will explore the powerful role that differentiation plays in analyzing a function, so that we can determine the true shape and behavior of its graph.

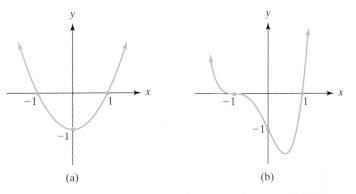

(a)   (b)

FIGURE 13.1   Curves passing through $(-1, 0)$, $(0, -1)$, and $(1, 0)$.

We begin by analyzing the graph of the function $y = f(x)$ in Figure 13.2. Notice that as x increases (goes from left to right) on the interval I_1, between a and b, the values of $f(x)$ increase and the curve is rising. Mathematically, this observation means that if x_1 and x_2 are any two points in I_1 such that $x_1 < x_2$, then $f(x_1) < f(x_2)$. Here f is said to be an *increasing function* on I_1. On the other hand, as x increases on the interval I_2 between c and d, the curve is falling. On this interval, $x_3 < x_4$ implies that $f(x_3) > f(x_4)$, and f is said to be a *decreasing function* on I_2. We summarize these observations in the following definition.

DEFINITION

A function f is said to be ***increasing*** on an interval I when, for any two numbers x_1, x_2 in I, if $x_1 < x_2$, then $f(x_1) < f(x_2)$. A function f is ***decreasing*** on an interval I when, for any two numbers x_1, x_2 in I, if $x_1 < x_2$, then $f(x_1) > f(x_2)$.

In terms of the graph of the function, f is increasing on I if the curve rises to the right and f is decreasing on I if the curve falls to the right. Recall that a straight line

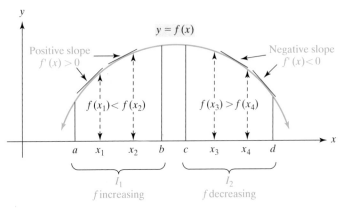

FIGURE 13.2   Increasing or decreasing nature of function.

with positive slope rises to the right while a straight line with negative slope falls to the right.

Turning again to Figure 13.2, we note that over the interval I_1, tangent lines to the curve have positive slopes, so $f'(x)$ must be positive for all x in I_1. A positive derivative implies that the curve is rising. Over the interval I_2, the tangent lines have negative slopes, so $f'(x) < 0$ for all x in I_2. The curve is falling where the derivative is negative. We thus have the following rule, which allows us to use the derivative to determine when a function is increasing or decreasing:

RULE 1 Criteria for Increasing or Decreasing Function

Let f be differentiable on the interval (a, b). If $f'(x) > 0$ for all x in (a, b), then f is increasing on (a, b). If $f'(x) < 0$ for all x in (a, b), then f is decreasing on (a, b).

To illustrate these ideas, we will use Rule 1 to find the intervals on which $y = 18x - \frac{2}{3}x^3$ is increasing and the intervals on which y is decreasing. Letting $y = f(x)$, we must determine when $f'(x)$ is positive and when $f'(x)$ is negative. We have

$$f'(x) = 18 - 2x^2 = 2(9 - x^2) = 2(3 + x)(3 - x)$$

Using the technique of Section 10.4, we can find the sign of $f'(x)$ by testing the intervals determined by the roots of $2(3 + x)(3 - x) = 0$, namely, -3 and 3. These should be arranged in increasing order on the top of a sign chart for f' so as to divide the domain of f into intervals. (See Figure 13.3.) In each interval, the sign of $f'(x)$ is determined by the signs of its factors:

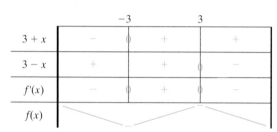

FIGURE 13.3   Sign chart for $f'(x) = 18 - 9x^2$ and its interpretation for $f(x)$.

If $x < -3$,   then sign($f'(x)$) = 2(−)(+) = −, so f is *decreasing*.

If $-3 < x < 3$,   then sign($f'(x)$) = 2(+)(+) = +, so f is *increasing*.

If $x > 3$,   then sign($f'(x)$) = 2(+)(−) = −, so f is *decreasing*.

These results are indicated in the sign chart given by Figure 13.3, where the bottom line is a schematic version of what the signs of f' say about f itself. Notice that the horizontal line segments in the bottom row indicate horizontal tangents for f at -3 and at 3. Thus, f is decreasing on $(-\infty, -3)$ and $(3, \infty)$, and is increasing on $(-3, 3)$. This corresponds to the rising and falling nature of the graph of f shown in Figure 13.4. Indeed, the point of a well-constructed sign chart is to provide a schematic for subsequent construction of the graph itself.

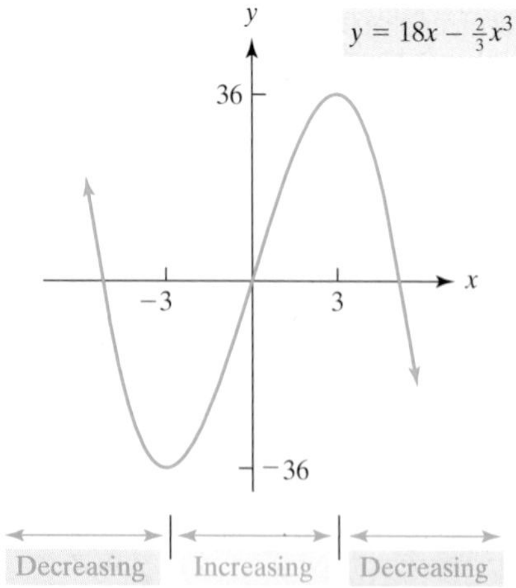

Extrema

Look now at the graph of $y = f(x)$ in Figure 13.5. Some observations can be made. First, there is something special about the points P, Q, and R. Notice that P is *higher* than any other "nearby" point on the curve—and likewise for R. The point Q is *lower* than any other "nearby" point on the curve. Since P, Q, and R may not necessarily be the highest or lowest points on the *entire* curve, we say that the graph of f has *relative maxima at a and at c; and has a relative minimum at b.* The function f has

y

36

$y = 18x - \frac{2}{3}x^3$

-3 3 x

-36

Decreasing   Increasing   Decreasing

FIGURE 13.4   Increasing/ decreasing for $y = 18x - \frac{2}{3}x^3$.

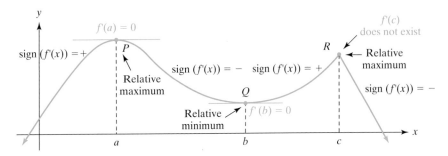

FIGURE 13.5 Relative maxima and relative minima.

CAUTION

Be sure to note the difference between relative extreme *values* and *where* they occur.

relative maximum values of $f(a)$ at a and $f(c)$ at c; and has a relative minimum value of $f(b)$ at b. We also say that $(a, f(a))$ and $(c, f(c))$ are *relative maximum points* and $(b, f(b))$ is a *relative minimum point* on the graph of f.

Turning back to the graph, we see that there is an *absolute maximum* (highest point on the entire curve) at a, but there is no *absolute minimum* (lowest point on the entire curve) because the curve is assumed to extend downward indefinitely. More precisely, we define these new terms as follows:

DEFINITION

A function f has a **relative maximum** at a if there is an open interval containing a on which $f(a) \geq f(x)$ for all x in the interval. The relative maximum value is $f(a)$. A function f has a **relative minimum** at a if there is an open interval containing a on which $f(a) \leq f(x)$ for all x in the interval. The relative minimum value is $f(a)$.

DEFINITION

A function f has an **absolute maximum** at a if $f(a) \geq f(x)$ for all x in the domain of f. The absolute maximum value is $f(a)$. A function f has an **absolute minimum** at a if $f(a) \leq f(x)$ for all x in the domain of f. The absolute minimum value is $f(a)$.

If it exists, an absolute maximum value is unique; however, it may occur at more than one value of x. A similar statement is true for an absolute minimum.

We refer to either a relative maximum or a relative minimum as a **relative extremum** (plural: *relative extrema*). Similarly, we speak of **absolute extrema.**

When dealing with relative extrema, we compare the function value at a point with values of nearby points; however, when dealing with absolute extrema, we compare the function value at a point with all other values determined by the domain. Thus, relative extrema are *local* in nature, whereas absolute extrema are *global* in nature.

Referring back to Figure 13.5, we notice that at a relative extremum the derivative may not be defined (as when $x = c$). But whenever it is defined at a relative extremum, it is 0 (as when $x = a$ and when $x = b$), and hence the tangent line is horizontal. We can state the following:

RULE 2   A Necessary Condition for Relative Extrema

If f has a relative extremum at a, then $f'(a) = 0$ or $f'(a)$ does not exist.

The implication in Rule 2 goes in only one direction:

$$\left. \begin{array}{c} \text{relative extremum} \\ \text{at } a \end{array} \right\} \quad \underset{\Longrightarrow}{\text{implies}} \quad \left\{ \begin{array}{c} f'(a) = 0 \\ \text{or} \\ f'(a) \text{ does not exist} \end{array} \right.$$

Rule 2 does *not* say that if $f'(a)$ is 0 or $f'(a)$ does not exist, then there must be a relative extremum at a. In fact, there may not be one at all. For example, in Figure 13.6(a), $f'(a)$ is 0 because the tangent line is horizontal at a, but there is no relative extremum

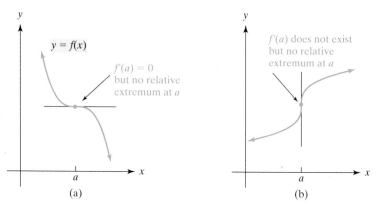

FIGURE 13.6   No relative extremum at a.

there. In Figure 13.6(b), $f'(a)$ does not exist because the tangent line is vertical at a, but again there is no relative extremum there.

But if we want to find all relative extrema of a function—and this is an important task—what Rule 2 *does* tell us is that we can limit our search to those values of x in the domain of f for which *either* $f'(x) = 0$ *or* $f'(x)$ does not exist. Typically, in applications, this cuts down our search for relative extrema from the infinitely many x for which f is defined to a small finite number of *possibilities*. Because these values of x are so important for locating the relative extrema of f, they are called the *critical values* for f, and if a is a critical value for f then we also say that $(a, f(a))$ is a *critical point* on the graph of f. Thus, in Figure 13.5, the numbers a, b, and c are critical values, and P, Q, and R are critical points.

DEFINITION

For a in the domain of f, if either $f'(a) = 0$ or $f'(a)$ does not exsit, then a is called a **critical value** for f. If a is a critical value, then the point $(a, f(a))$ is called a **critical point** for f.

At a critical point, there may be a relative maximum, a relative minimum, or neither. Moreover, from Figure 13.5, we observe that each relative extremum occurs at a point around which the sign of $f'(x)$ is changing. For the relative maximum at a, the sign of $f'(x)$ goes from $+$ for $x < a$ to $-$ for $x > a$, *as long as x is near a*. For the relative minimum at b, the sign of $f'(x)$ goes from $-$ to $+$, and for the relative maximum at c, it again goes from $+$ to $-$. Thus, *around relative maxima, f is increasing and then decreasing, and the reverse holds for relative minima.* More precisely, we have the following rule:

RULE 3   Criteria for Relative Extrema

Suppose f is continuous on an open interval I that contains the critical value a and f is differentiable on I, except possibly at a.

1. If $f'(x)$ changes from positive to negative as x increases through a, then f has a relative maximum at a.

2. If $f'(x)$ changes from negative to positive as x increases through a, then f has a relative minimum at a.

To illustrate Rule 3 with a concrete example, refer again to Figure 13.3, the sign chart for $f'(x) = 18 - 2x^2$. The row labeled by $f'(x)$ shows clearly that $f(x) = 18x - \frac{2}{3}x^2$ has a relative minimum at -3 and a relative maximum at 3. The row providing the interpretation of the chart for f, labeled $f(x)$, is immediately deduced from the row above it. The significance of the $f(x)$ row is that it provides

We point out again that not every critical value corresponds to a relative extremum. For example, if $y = f(x) = x^3$, then $f'(x) = 3x^2$. Since $f'(0) = 0$, 0 is a critical value. But if $x < 0$, then $3x^2 > 0$ and if $x > 0$, then $3x^2 > 0$. Since $f'(x)$ does not change sign at 0, there is no relative extremum at 0. Indeed, since $f'(x) \geq 0$ for all x, the graph of f never falls, and f is said to be *nondecreasing*. (See Figure 13.8.)

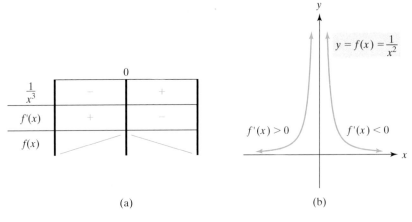

FIGURE 13.7 $f'(0)$ is not defined, but 0 is not a critical value because 0 is not in the domain of f.

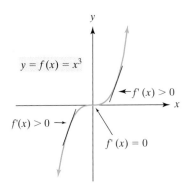

FIGURE 13.8   Zero is a critical value, but does not give a relative extremum.

an intermediate step in actually sketching the graph of f. In this row it stands out, visually, that f has a relative minimum at -3 and a relative maximum at 3.

When searching for extrema of a function f, care must be paid to those a that are not in the domain of f but that are near values in the domain of f. Consider the following example. If

$$y = f(x) = \frac{1}{x^2} \quad \text{then} \quad f'(x) = -\frac{2}{x^3}$$

Although $f'(x)$ does not exist at 0, 0 is not a critical value, because 0 is not in the domain of f. Thus, a relative extremum cannot occur at 0. Nevertheless, the derivative may change sign around any x-value where $f'(x)$ is not defined, so such values are important in determining intervals over which f is increasing or decreasing. In particular, such values should be included in a sign chart for f'. See Figure 13.7(a) and the accompanying graph Figure 13.7(b).

Observe that the thick vertical rule at 0 on the chart serves to indicate that 0 is not in the domain of f. Here there are no extrema of any kind.

In Rule 3 the hypotheses must be satisfied, or the conclusion need not hold. For example, consider the case-defined function

$$f(x) = \begin{cases} \dfrac{1}{x^2} & \text{if } x \neq 0 \\ 0 & \text{if } x = 0 \end{cases}$$

Here, 0 is explicitly in the domain of f but f is not continuous at 0. We recall from Section 11.1 that if a function f is not continuous at a, then f is not differentiable at a, meaning that $f'(a)$ does not exist. Thus $f'(0)$ does not exist and 0 is a critical value that must be included in the sign chart for f' shown in Figure 13.9(a). We extend

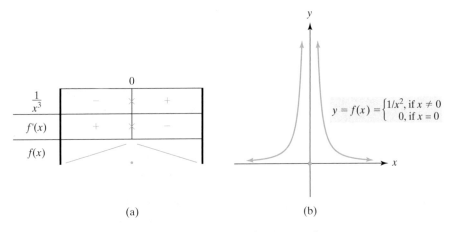

FIGURE 13.9   Zero is a critical value but Rule 3 does not apply.

our sign chart conventions by indicating with a $\times$ symbol those values for which f' does not exist. We see in this example that $f'(x)$ changes from positive to negative as x increases through 0 but f does *not* have a relative maximum at 0. Here Rule 3 does not apply because its continuity hypothesis is not met. In Figure 13.9(b), 0 is displayed in the domain of f. It is clear that f has an absolute *minimum* at 0 because $f(0) = 0$ and, for all $x \neq 0$, $f(x) > 0$.

Summarizing the results of this section, we have the *first-derivative test* for the relative extrema of $y = f(x)$:

First-Derivative Test for Relative Extrema

Step 1. Find $f'(x)$.

Step 2. Determine all critical values of f (those a where $f'(a) = 0$ or $f'(a)$ does not exist) and any a that are not in the domain of f but that are near values in the domain of f, and construct a sign chart that shows for each of the intervals determined by these values whether f is increasing ($f'(x) > 0$) or decreasing ($f'(x) < 0$).

Step 3. For each critical value a at which f is continuous, determine whether $f'(x)$ changes sign as x increases through a. There is a relative maximum at a if $f'(x)$ changes from $+$ to $-$ going from left to right and a relative minimum if $f'(x)$ changes from $-$ to $+$ going from left to right. If $f'(x)$ does not change sign, there is no relative extremum at a.

Step 4. For critical values a at which f is not continuous, analyze the situation by using the definitions of extrema directly.

PRINCIPLES IN PRACTICE 1

FIRST-DERIVATIVE TEST

The cost equation for a hot dog stand is given by $c(q) = 2q^3 - 21q^2 + 60q + 500$, where q is the number of hot dogs sold, and $c(q)$ is the cost in dollars. Use the first-derivative test to find where relative extrema occur.

● **EXAMPLE 1 First-Derivative Test**

If $y = f(x) = x + \dfrac{4}{x+1}$, for $x \neq 1$ use the first-derivative test to find where relative extrema occur.

Solution:

Step 1. $f(x) = x + 4(x+1)^{-1}$, so

$$f'(x) = 1 + 4(-1)(x+1)^{-2} = 1 - \frac{4}{(x+1)^2}$$

$$= \frac{(x+1)^2 - 4}{(x+1)^2} = \frac{x^2 + 2x - 3}{(x+1)^2}$$

$$= \frac{(x+3)(x-1)}{(x+1)^2} \quad \text{for } x \neq -1$$

Note that we expressed $f'(x)$ as a quotient with numerator and denominator fully factored. This enables us in Step 2 to determine easily where $f'(x)$ is 0 or does not exist and the signs of f'.

Step 2. Setting $f'(x) = 0$ gives $x = -3, 1$. The denominator of $f'(x)$ is 0 when x is -1. We note that -1 is not in the domain of f but all values near -1 are in the domain of f. We construct a sign chart, headed by the values -3, -1, and 1 (which we have placed in increasing order). See Figure 13.10.

The three values lead us to test four intervals as shown in our sign chart. On each of these intervals, f is differentiable and is not zero. We determine the sign of f' on each interval by first determining the sign of each of its factors on each iterval. For example, considering first the interval $(-\infty, -3)$, it is not easy to see immediately that $f'(x) > 0$ there; but it is easy to see that $x+3 < 0$ for $x < -3$, while $(x+1)^{-2} > 0$ for all $x \neq -1$, and $x-1 < 0$ for $x < 1$. These observations account for the signs of the factors in the $(-\infty, -3)$ column of the chart. The sign of $f'(x)$ in that column is obtained by "multiplying signs" (downward): $(-)(+)(-) = +$. We repeat these considerations for the other three intervals. Note that the thick vertical line at -1 in the chart indicates that -1 is not in the domain of f and hence cannot give rise to any extrema.

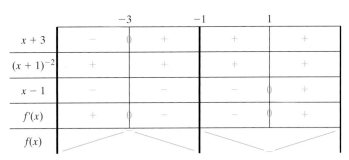

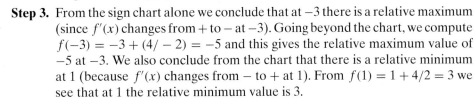

FIGURE 13.10 Sign chart for $f'(x) = \dfrac{(x+3)(x-1)}{(x+1)^2}$.

In the bottom row of the sign chart we record, graphically, the nature of tangent lines to $f(x)$ in each interval and at the values where f' is 0.

Step 3. From the sign chart alone we conclude that at -3 there is a relative maximum (since $f'(x)$ changes from $+$ to $-$ at -3). Going beyond the chart, we compute $f(-3) = -3 + (4/-2) = -5$ and this gives the relative maximum value of -5 at -3. We also conclude from the chart that there is a relative minimum at 1 (because $f'(x)$ changes from $-$ to $+$ at 1). From $f(1) = 1 + 4/2 = 3$ we see that at 1 the relative minimum value is 3.

Step 4. There are no critical values at which f is not continuous, so our considerations above provide the whole story about the relative extrema of $f(x)$, whose graph is given in Figure 13.11. Note that the general shape of the graph was indeed forecast by the bottom row of the sign chart (Figure 13.10).

NOW WORK PROBLEM 37

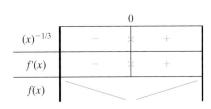

FIGURE 13.11 Graph of $y = x + \dfrac{4}{x+1}$.

● EXAMPLE 2 A Relative Extremum where $f'(x)$ Does Not Exist

Test $y = f(x) = x^{2/3}$ *for relative extrema.*

Solution: We have

$$f'(x) = \frac{2}{3}x^{-1/3}$$

$$= \frac{2}{3\sqrt[3]{x}}$$

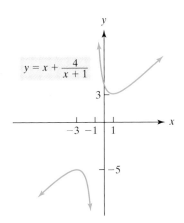

FIGURE 13.12 Sign chart for $f'(x) = \dfrac{2}{3\sqrt[3]{x}}$.

and the sign chart is which requires little comment, except to note that again we use the symbol $\times$ on the vertical line at 0 to indicate that $f'(0)$ does not exist. Since f is continuous at 0, we conclude from Rule 3 that f has a relative minimum at 0 of $f(0) = 0$, and there are no other relative extrema. We note further, by inspection, that f has an *absolute* minimum at 0. The graph of f follows as Figure 13.13. Note that we could have predicted its shape from the the bottom line of line of the sign chart Figure 13.12, which shows there can be no tangent with a slope at 0. (Of course, the tangent does exist at 0 but it is a vertical line.)

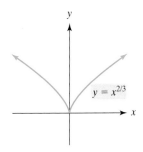

FIGURE 13.13
Derivative does not exist
at 0 and there is a
minimum at 0.

NOW WORK PROBLEM 41

FINDING RELATIVE EXTREMA

A drug is injected into a patient's bloodstream. The concentration of the drug in the bloodstream t hours after the injection is approximated by $C(t) = \dfrac{0.14t}{t^2 + 4t + 4}$. Find the relative extrema for $t > 0$, and use them to determine when the drug is at its greatest concentration.

● EXAMPLE 3   Finding Relative Extrema

Test $y = f(x) = x^2 e^x$ for relative extrema.

Solution: By the product rule,

$$f'(x) = x^2 e^x + e^x(2x) = xe^x(x + 2)$$

Noting that e^x is always positive, we obtain the critical values 0 and -2. From the sign chart of $f'(x)$ given in Figure 13.14, we conclude that there is a relative maximum when $x = -2$ and a relative minimum when $x = 0$.

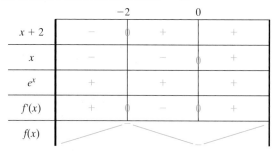

FIGURE 13.14   Sign chart for $f'(x) = x(x + 2)e^x$.

NOW WORK PROBLEM 49 ●●●

Curve Sketching

In the next example we show how the first-derivative test, in conjunction with the notions of intercepts and symmetry, can be used as an aid in sketching the graph of a function.

● EXAMPLE 4   Curve Sketching

Sketch the graph of $y = f(x) = 2x^2 - x^4$ with the aid of intercepts, symmetry, and the first-derivative test.

Solution:

Intercepts   If $x = 0$, then $f(x) = 0$ so that the y-intercept is $(0, 0)$. Next note that

$$f(x) = 2x^2 - x^4 = x^2(2 - x^2) = x^2\left(\sqrt{2} + x\right)\left(\sqrt{2} - x\right)$$

So if $y = 0$, then $x = 0, \pm\sqrt{2}$ and the x-intercepts are $(-\sqrt{2}, 0)$, $(0, 0)$, and $(\sqrt{2}, 0)$. We have the sign chart *for f itself* (Figure 13.15), which shows the intervals over which the graph of $y = f(x)$ is above the x-axis ($+$) and the intervals over which the graph of $y = f(x)$ is below the x-axis ($-$).

	$-\sqrt{2}$		0		$\sqrt{2}$	
$\sqrt{2} + x$	$-$	0	$+$		$+$		$+$
x^2	$+$		$+$	0	$+$		$+$
$\sqrt{2} - x$	$+$		$+$		$+$	0	$-$
$f(x)$	$-$	0	$+$	0	$+$	0	$-$

FIGURE 13.15   Sign chart for $f(x) = (\sqrt{2} + x)x^2(\sqrt{2} - x)$.

Symmetry   Testing for y-axis symmetry, we have

$$f(-x) = 2(-x)^2 - (-x)^4 = 2x^2 - x^4 = f(x)$$

So the graph is symmetric with respect to the y-axis. Because y is a function (and not the zero function), there is no x-axis symmetry and hence no symmetry about the origin.

First-Derivative Test

Step 1. $y' = 4x - 4x^3 = 4x(1 - x^2) = 4x(1 + x)(1 - x)$.

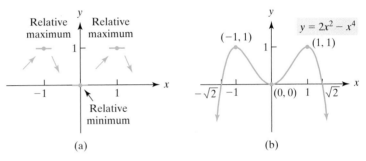

FIGURE 13.16 Sign chart of $y' = (1 + x)4x(1 - x)$.

Step 2. Setting $y' = 0$ gives the critical values $x = 0, \pm 1$. Since f is a polynomial, it is defined and differentiable for all x. Thus the only values to head the sign chart for f' are $-1, 0, 1$ (in increasing order) and the sign chart is given in Figure 13.16. Since we are interested in the graph, the critical *points* are important to us. By substituting the critical values into the *original* equation, $y = 2x^2 - x^4$, we obtain the y-coordinates of these points. We find the critical points to be $(-1, 1)$, $(0, 0)$, and $(1, 1)$.

Step 3. From the sign chart and evaluations in step 2, it is clear that f has relative maxima $(-1, 1)$ and $(1, 1)$ and relative minimum $(0, 0)$. (Step 4 does not apply here.)

FIGURE 13.17 Putting together the graph of $y = 2x^2 - x^4$.

Discussion In Figure 13.17(a), we have indicated the horizontal tangents at the relative maximum and minimum points. We know the curve rises from the left, has a relative maximum, then falls, has a relative minimum, then rises to a relative maximum, and falls thereafter. By symmetry, it suffices to sketch on one side of the y-axis and construct a mirror image on the other side. We also know, from the sign chart for f, where the graph crosses and touches the x-axis, and this adds further precision to our sketch, which is shown in Figure 13.17(b).

As a passing comment, we note that *absolute* maxima occur at $x = \pm 1$. See Figure 13.17(b). There is no absolute minimum.

NOW WORK PROBLEM 59

T E C H N O L O G Y

A graphing calculator is a powerful tool for investigating relative extrema. For example, consider the function

$$f(x) = 3x^4 - 4x^3 + 4$$

whose graph is shown in Figure 13.18. It appears that there is a relative minimum near $x = 1$. We can locate this minimum by either using "trace and zoom" or (on a TI-83 Plus) using the "minimum" feature. Figure 13.19 shows the latter approach. The relative minimum point is estimated to be $(1.00, 3)$.

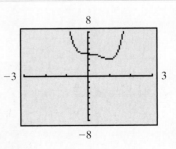

FIGURE 13.18 Graph of $f(x) = 3x^4 - 4x^3 + 4$.

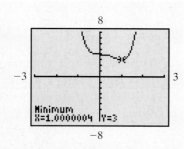

FIGURE 13.19 Relative minimum at (1.00, 3).

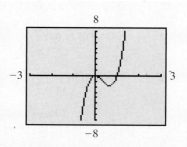

FIGURE 13.20 Graph of $f'(x) = 12x^3 - 12x^2$.

Now let us see how the graph of f' indicates when extrema occur. We have

$$f'(x) = 12x^3 - 12x^2$$

whose graph is shown in Figure 13.20. It appears that $f'(x)$ is 0 at two points. Using "trace and zoom" or the "zero" feature, we estimate the zeros of f' (the critical values of f) to be 1 and 0. Around $x = 1$, we see that $f'(x)$ goes from negative values to positive values. (That is, the graph of f' goes from below the x-axis to above it.) Thus, we conclude that f has a relative minimum at $x = 1$, which confirms our previous result.

Around the critical value $x = 0$, the values of $f'(x)$ are negative. Since $f'(x)$ does not change sign, we conclude that there is no relative extremum at $x = 0$. This is also apparent from the graph in Figure 13.18.

It is worthwhile to note that we can approximate the graph of f' without determining $f'(x)$ itself. We make use of the "nDeriv" feature. First we enter the function f as Y_1. Then we set

$$Y_2 = \text{nDeriv}(Y_1, X, X)$$

The graph of Y_2 approximates the graph of $f'(x)$.

Problems 13.1

In Problems 1–4, the graph of a function is given (Figures 13.21–13.24). Find the open intervals on which the function is increasing or decreasing, and find the coordinates of all relative extrema.

1.

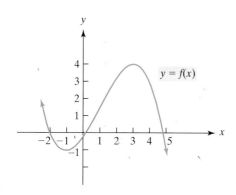

FIGURE 13.21 Graph for Problem 1.

2.

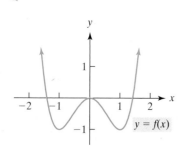

FIGURE 13.22 Graph for Problem 2.

3.

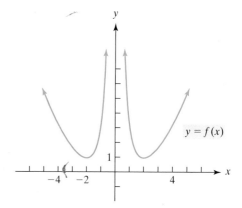

FIGURE 13.23 Graph for Problem 3.

4.

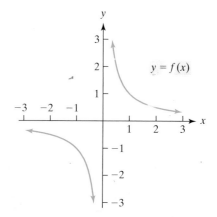

FIGURE 13.24 Graph for Problem 4.

In Problems 5–8, the derivative of a continuous function f is given. Find the open intervals on which f is increasing or decreasing, and find the x-values of all relative extrema.

5. $f'(x) = (x + 3)(x - 1)(x - 2)$

6. $f'(x) = 2x(x - 1)^3$

7. $f'(x) = (x + 1)(x - 3)^2$ **8.** $f'(x) = \dfrac{x(x + 2)}{x^2 + 1}$

In Problems 9–52, determine where the function is increasing or decreasing, and determine where relative maxima and minima occur. Do not sketch the graph.

9. $y = 2x^3 + 1$ **10.** $y = x^2 + 4x + 3$

11. $y = x - x^2 + 2$ **12.** $y = x^3 - \dfrac{5}{2}x^2 - 2x + 6$

13. $y = -\dfrac{x^3}{3} - 2x^2 + 5x - 2$ **14.** $y = \dfrac{x^4}{4} + x^3$

15. $y = x^4 - 2x^2$ **16.** $y = -3 + 12x - x^3$

17. $y = x^3 - \dfrac{7}{2}x^2 + 2x - 5$ **18.** $y = x^3 - 6x^2 + 12x - 6$

19. $y = 2x^3 - \dfrac{11}{2}x^2 - 10x + 2$ **20.** $y = -5x^3 + x^2 + x - 7$

21. $y = \dfrac{x^3}{3} - 5x^2 + 22x + 1$ **22.** $y = \dfrac{9}{5}x^5 - \dfrac{47}{3}x^3 + 10x$

23. $y = 3x^5 - 5x^3$ **24.** $y = 3x - \dfrac{x^6}{2}$

25. $y = -x^5 - 5x^4 + 200$ **26.** $y = \dfrac{3x^4}{2} - 4x^3 + 17$

27. $y = 8x^4 - x^8$ **28.** $y = \dfrac{4}{5}x^5 - \dfrac{13}{3}x^3 + 3x + 4$

29. $y = (x^2 - 1)^4$ **30.** $y = \sqrt[3]{x}(x - 2)$

31. $y = \dfrac{5}{x - 1}$ **32.** $y = \dfrac{3}{x}$

33. $y = \dfrac{10}{\sqrt{x}}$ **34.** $y = \dfrac{3x}{2x + 5}$

35. $y = \dfrac{x^2}{2 - x}$ **36.** $y = 4x^2 + \dfrac{1}{x}$

37. $y = \dfrac{x^2 - 3}{x + 2}$ **38.** $y = \dfrac{2x^2}{4x^2 - 25}$

39. $y = \dfrac{5x + 2}{x^2 + 1}$ **40.** $y = \sqrt[3]{x^3 - 9x}$

41. $y = (x - 1)^{2/3}$ **42.** $y = x^2(x + 3)^4$

43. $y = x^3(x - 6)^4$ **44.** $y = x(1 - x)^{2/5}$

45. $y = e^{-\pi x} + \pi$ **46.** $y = x \ln x$

47. $y = x^2 - 9 \ln x$ **48.** $y = x^{-1}e^x$

49. $y = e^x + e^{-x}$ **50.** $y = e^{-x^2/2}$

51. $y = x \ln x - x$ **52.** $y = (x^2 + 1)e^{-x}$

In Problems 53–64, determine intervals on which the function is increasing or decreasing, relative maxima and minima, symmetry, and those intercepts that can be obtained conveniently. Then sketch the graph.

53. $y = x^2 - 3x - 10$ **54.** $y = 2x^2 - 5x - 12$

55. $y = 3x - x^3$ **56.** $y = x^4 - 16$

57. $y = 2x^3 - 9x^2 + 12x$ **58.** $y = 2x^3 - x^2 - 4x + 4$

59. $y = x^4 + 4x^3 + 4x^2$ **60.** $y = x^6 - \dfrac{6}{5}x^5$

61. $y = (x - 1)^2(x + 2)^2$ **62.** $y = \sqrt{x}(x^2 - x - 2)$

63. $y = 2\sqrt{x} - x$ **64.** $y = x^{5/3} + 5x^{2/3}$

65. Sketch the graph of a continuous function f such that $f(2) = 2$, $f(4) = 6$, $f'(2) = f'(4) = 0$, $f'(x) < 0$ for $x < 2$, $f'(x) > 0$ for $2 < x < 4$, f has a relative maximum at 4, and $\lim_{x \to \infty} f(x) = 0$.

66. Sketch the graph of a continuous function f such that $f(1) = 2$, $f(4) = 5$, $f'(1) = 0$, $f'(x) \geq 0$ for $x < 4$, f has a relative maximum when $x = 4$, and there is a vertical tangent line when $x = 4$.

67. Average Cost   If $c_f = 25{,}000$ is a fixed-cost function, show that the average fixed-cost function $\bar{c}_f = c_f/q$ is a decreasing function for $q > 0$. Thus, as output q increases, each unit's portion of fixed cost declines.

68. Marginal Cost   If $c = 3q - 3q^2 + q^3$ is a cost function, when is marginal cost increasing?

69. Marginal Revenue   Given the demand function

$$p = 400 - 2q$$

find when marginal revenue is increasing.

70. Cost Function   For the cost function $c = \sqrt{q}$, show that marginal and average costs are always decreasing for $q > 0$.

71. Revenue   For a manufacturer's product, the revenue function is given by $r = 240q + 57q^2 - q^3$. Determine the output for maximum revenue.

72. Labor Markets   Eswaran and Kotwal[2] consider agrarian economies in which there are two types of workers, permanent and casual. Permanent workers are employed on long-term contracts and may receive benefits such as holiday gifts and emergency aid. Casual workers are hired on a daily basis and perform routine and menial tasks such as weeding, harvesting, and threshing. The difference z in the present-value cost of hiring a permanent worker over that of hiring a casual worker is given by

$$z = (1 + b)w_p - bw_c$$

where w_p and w_c are wage rates for permanent labor and casual labor, respectively, b is a positive constant, and w_p is a function of w_c.

(a) Show that

$$\frac{dz}{dw_c} = (1 + b)\left[\frac{dw_p}{dw_c} - \frac{b}{1 + b}\right]$$

(b) If $dw_p/dw_c < b/(1 + b)$, show that z is a decreasing function of w_c.

73. Thermal Pollution   In Shonle's discussion of thermal pollution,[3] the efficiency of a power plant is given by

$$E = 0.71\left(1 - \frac{T_c}{T_h}\right)$$

where T_h and T_c are the respective absolute temperatures of the hotter and colder reservoirs. Assume that T_c is a positive constant and that T_h is positive. Using calculus, show that as T_h increases, the efficiency increases.

[2]M. Eswaran and A. Kotwal, "A Theory of Two-Tier Labor Markets in Agrarian Economics," *The American Economic Review*, 75, no. 1 (1985), 162–77.

[3]J. I. Shonle, *Environmental Applications of General Physics* (Reading, MA: Addison-Wesley Publishing Company, Inc., 1975).

74. Telephone Service   In a discussion of the pricing of local telephone service, Renshaw[4] determines that total revenue r is given by

$$r = 2F + \left(1 - \frac{a}{b}\right)p - p^2 + \frac{a^2}{b}$$

where p is an indexed price per call, and a, b, and F are constants. Determine the value of p that maximizes revenue.

75. Storage and Shipping Costs   In his model for storage and shipping costs of materials for a manufacturing process, Lancaster[5] derives the cost function

$$C(k) = 100\left(100 + 9k + \frac{144}{k}\right) \quad 1 \le k \le 100$$

where $C(k)$ is the total cost (in dollars) of storage and transportation for 100 days of operation if a load of k tons of material is moved every k days.

(a) Find $C(1)$.

(b) For what value of k does $C(k)$ have a minimum?

(c) What is the minimum value?

76. Physiology—The Bends   When a deep-sea diver undergoes decompression or a pilot climbs to a high altitude, nitrogen may bubble out of the blood, causing what is commonly called *the bends*. Suppose the percentage P of people who suffer effects of the bends at an altitude of h thousand feet is given by[6]

$$P = \frac{100}{1 + 100{,}000e^{-0.36h}}$$

Is P an increasing function of h?

In Problems 77–80, from the graph of the function, find the coordinates of all relative extrema. Round your answers to two decimal places.

77. $y = 0.3x^2 + 2.3x + 5.1$

78. $y = 3x^4 - 4x^3 - 5x + 1$

79. $y = \dfrac{8.2x}{0.4x^2 + 3}$

80. $y = \dfrac{e^x(3 - x)}{7x^2 + 1}$

81. Graph the function

$$f(x) = [x(x - 2)(2x - 3)]^2$$

in the window $-1 \le x \le 3$, $-1 \le y \le 3$. Upon first glance, it may appear that this function has two relative minimum points and one relative maximum point. However, in reality, it has three relative minimum points and two relative maximum points. Determine the x-values of all these points. Round answers to two decimal places.

82. If $f(x) = 3x^3 - 7x^2 + 4x + 2$, display the graphs of f and f' on the same screen. Notice that $f'(x) = 0$ where relative extrema of f occur.

83. Let $f(x) = 6 + 4x - 3x^2 - x^3$. (a) Find $f'(x)$. (b) Graph $f'(x)$. (c) Observe where $f'(x)$ is positive and where it is negative. Give the intervals (rounded to two decimal places) where f is increasing and where f is decreasing. (d) Graph f and f' on the same screen, and verify your results to part (c).

84. If $f(x) = x^4 - x^2 - (x + 2)^2$, find $f'(x)$. Determine the critical values of f. Round your answers to two decimal places.

OBJECTIVE

To find extreme values on a closed interval.

13.2 Absolute Extrema on a Closed Interval

If a function f is *continuous* on a *closed* interval $[a, b]$, it can be shown that of *all* the function values $f(x)$ for x in $[a, b]$, there must be an (absolute) maximum value and an (absolute) minimum value. These two values are called **extreme values** of f on that interval. This important property of continuous functions is called the *extreme-value theorem*.

Extreme-Value Theorem

If a function is continuous on a closed interval, then the function has *both* a maximum value *and* a minimum value on that interval.

For example, each function in Figure 13.25 is continuous on the closed interval $[1, 3]$. Geometrically, the extreme-value theorem assures us that over this interval each graph has a highest point and a lowest point.

In the extreme-value theorem, you must realize that we are dealing with

1. a closed interval and

2. a function continuous on that interval

---

[4]E. Renshaw, "A Note on Equity and Efficiency in the Pricing of Local Telephone Services," *The American Economic Review*, 75, no. 3 (1985), 515–18.

[5]P. Lancaster, *Mathematics: Models of the Real World* (Englewood Cliffs, NJ: Prentice-Hall, Inc., 1976).

[6]Adapted from G. E. Folk, Jr., *Textbook of Environmental Physiology*, 2nd ed. (Philadelphia: Lea & Febiger, 1974).

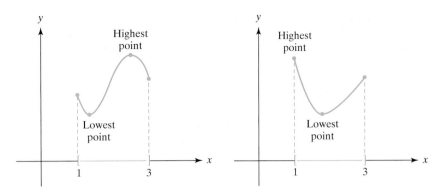

FIGURE 13.25   Illustrating the extreme-value theorem.

If either condition (1) or condition (2) is not met, then extreme values are not guaranteed. For example, Figure 13.26(a) shows the graph of the continuous function $f(x) = x^2$ on the *open* interval $(-1, 1)$. You can see that f has no maximum value on the interval (although f has a minimum value there). Now consider the function $f(x) = 1/x^2$ on the closed interval $[-1, 1]$. Here f is *not continuous* at 0. From the graph of f in Figure 13.26(b), you can see that f has no maximum value (although there is a minimum value).

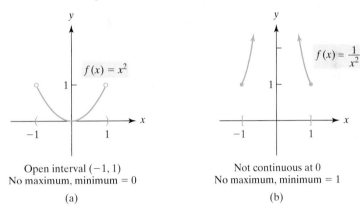

FIGURE 13.26   Extreme-value theorem does not apply.

In the previous section, our emphasis was on relative extrema. Now we will focus our attention on absolute extrema and make use of the extreme-value theorem where possible. If the domain of a function is a closed interval, to determine *absolute* extrema we must examine the function not only at critical values, but also at the endpoints. For example, Figure 13.27 shows the graph of the continuous function $y = f(x)$ over $[a, b]$. The extreme-value theorem guarantees absolute extrema over the interval. Clearly, the important points on the graph occur at $x = a, b, c,$ and $d,$

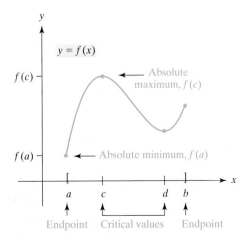

FIGURE 13.27   Absolute extrema.

which correspond to endpoints or critical values. Notice that the absolute maximum occurs at the critical value c and the absolute minimum occurs at the endpoint a. These results suggest the following procedure:

Procedure to Find Absolute Extrema for a Function f That Is Continuous on $[a, b]$

Step 1. Find the critical values of f.

Step 2. Evaluate $f(x)$ at the endpoints a and b and at the critical values in (a, b).

Step 3. The maximum value of f is the greatest of the values found in step 2. The minimum value of f is the least of the values found in step 2.

●EXAMPLE 1   Finding Extreme Values on a Closed Interval

Find absolute extrema for $f(x) = x^2 - 4x + 5$ over the closed interval $[1, 4]$.

Solution: Since f is continuous on $[1, 4]$, the foregoing procedure applies.

Step 1. To find the critical values of f, we first find f':

$$f'(x) = 2x - 4 = 2(x - 2)$$

This gives the critical value $x = 2$.

Step 2. Evaluating $f(x)$ at the endpoints 1 and 4 and at the critical value 2, we have

$$\begin{aligned} f(1) &= 2 \\ f(4) &= 5 \end{aligned} \quad \text{values of } f \text{ at endpoints}$$

and

$$f(2) = 1 \quad \text{value of } f \text{ at critical value 2 in } (1, 4)$$

Step 3. From the function values in step 2, we conclude that the maximum is $f(4) = 5$ and the minimum is $f(2) = 1$. (See Figure 13.28.)

NOW WORK PROBLEM 1 ●●●

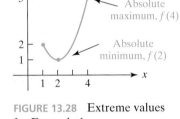

FIGURE 13.28   Extreme values for Example 1.

Problems 13.2

In Problems 1–14, find the absolute extrema of the given function on the given interval.

*1. $f(x) = x^2 - 2x + 3, [0, 3]$

2. $f(x) = -2x^2 - 6x + 5, [-3, 2]$

3. $f(x) = \frac{1}{3}x^3 + \frac{1}{2}x^2 - 2x + 1, [-1, 0]$

4. $f(x) = \frac{1}{4}x^4 - \frac{3}{2}x^2, [0, 1]$

5. $f(x) = 4x^3 + 3x^2 - 18x + 3, [\frac{1}{2}, 3]$

6. $f(x) = x^{2/3}, [-8, 8]$

7. $f(x) = -3x^5 + 5x^3, [-2, 0]$

8. $f(x) = \frac{7}{3}x^3 + 2x^2 - 3x + 1, [0, 3]$

9. $f(x) = 3x^4 - x^6, [-1, 2]$

10. $f(x) = \frac{1}{4}x^4 - \frac{1}{2}x^2 + 3, [-2, 3]$

11. $f(x) = x^4 - 9x^2 + 2, [-1, 3]$

12. $f(x) = \dfrac{x}{x^2 + 1}, [0, 2]$

13. $f(x) = (x - 1)^{2/3}, [-26, 28]$

14. $f(x) = 0.2x^3 - 3.6x^2 + 2x + 1, [-1, 2]$

15. Consider the function

$$f(x) = x^4 + 8x^3 + 21x^2 + 20x + 9$$

over the interval $[-4, 9]$.

   (a) Determine the value(s) (rounded to two decimal places) of x at which f attains a minimum value.
   (b) What is the minimum value (rounded to two decimal places) of f?
   (c) Determine the value(s) of x at which f attains a maximum value.
   (d) What is the maximum value of f?

OBJECTIVE

To test a function for concavity and inflection points. Also, to sketch curves with the aid of the information obtained from the first and second derivatives.

13.3 Concavity

You have seen that the first derivative provides much information for sketching curves. It is used to determine when a function is increasing or decreasing and to locate relative maxima and minima. However, to be sure we know the true shape of a curve, we may need more information. For example, consider the curve $y = f(x) = x^2$. Since $f'(x) = 2x$, $x = 0$ is a critical value. If $x < 0$, then $f'(x) < 0$, and f is decreasing; if $x > 0$, then $f'(x) > 0$, and f is increasing. Thus, there is a relative minimum when $x = 0$. In Figure 13.29, both curves meet the preceding conditions. But which one

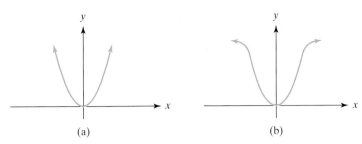

FIGURE 13.29   Two functions with $f'(x) < 0$ for $x < 0$ and $f'(x) > 0$ for $x > 0$.

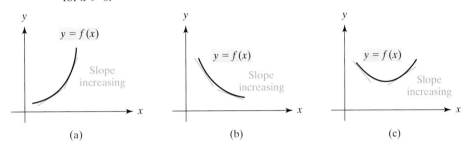

FIGURE 13.30   Each curve is concave up.

truly describes the curve $y = x^2$? This question will be settled easily by using the second derivative and the notion of *concavity*.

In Figure 13.30, note that each curve $y = f(x)$ "bends" (or opens) upward. This means that if tangent lines are drawn to each curve, the curves lie *above* them. Moreover, the slopes of the tangent lines *increase* in value as x increases: In part (a), the slopes go from small positive values to larger values; in part (b), they are negative and approaching zero (and thus increasing); in part (c), they pass from negative values to positive values. Since $f'(x)$ gives the slope at a point, an increasing slope means that f' must be an increasing function. To describe this property, each curve (or function f) in Figure 13.30 is said to be *concave up*.

In Figure 13.31, it can be seen that each curve lies *below* the tangent lines and the curves are bending downward. As x increases, the slopes of the tangent lines are *decreasing*. Thus, f' must be a decreasing function here, and we say that f is *concave down*.

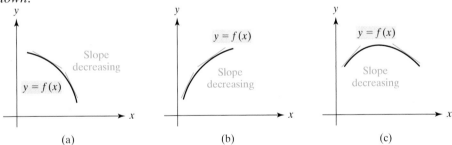

FIGURE 13.31   Each curve is concave down.

![CAUTION icon] **C A U T I O N**

Concavity relates to whether f', not f, is increasing or decreasing. In Figure 13.30(b), note that f is concave up and decreasing; however, in Figure 13.31(a), f is concave down and decreasing.

DEFINITION

Let f be differentiable on the interval (a, b). Then f is said to be ***concave up*** [***concave down***] on (a, b) if f' is increasing [decreasing] on (a, b).

Remember: If f is concave up on an interval, then, geometrically, its graph is bending upward there. If f is concave down, then its graph is bending downward.

Since f' is increasing when its derivative $f''(x)$ is positive, and f' is decreasing when $f''(x)$ is negative, we can state the following rule:

RULE 1   Criteria for Concavity

Let f' be differentiable on the interval (a, b). If $f''(x) > 0$ for all x in (a, b), then f is concave up on (a, b). If $f''(x) < 0$ for all x in (a, b), then f is concave down on (a, b).

A function f is also said to be concave up at a point c if there exists an open interval around c on which f is concave up. In fact, for the functions that we will consider, if $f''(c) > 0$, then f is concave up at c. Similarly, f is concave down at c if $f''(c) < 0$.

● EXAMPLE 1 Testing for Concavity

Determine where the given function is concave up and where it is concave down.

a. $y = f(x) = (x - 1)^3 + 1$.

Solution: To apply Rule 1, we must examine the signs of y''. Now, $y' = 3(x - 1)^2$, so

$$y'' = 6(x - 1)$$

Thus, f is concave up when $6(x - 1) > 0$, that is, when $x > 1$. And f is concave down when $6(x - 1) < 0$, that is, when $x < 1$. We now use a sign chart for f'' (together with an interpretation line for f) to organize our findings. (See Figure 13.32.)

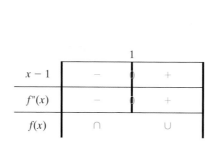
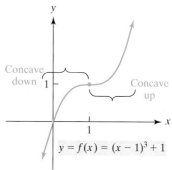

FIGURE 13.32 Sign chart for f'' and concavity for $f(x) = (x - 1)^3 + 1$.

b. $y = x^2$.

Solution: We have $y' = 2x$ and $y'' = 2$. Because y'' is always positive, the graph of $y = x^2$ must always be concave up, as in Figure 13.29(a). The graph cannot appear as in Figure 13.29(b), for that curve is sometimes concave down.

NOW WORK PROBLEM 1 ●●

A point on a graph where concavity changes from concave down to concave up, or vice versa, such as $(1, 1)$ in Figure 13.32, is called an *inflection point* or a *point of inflection*. Around such a point, the sign of $f''(x)$ goes from $-$ to $+$ or from $+$ to $-$. More precisely, we have the following definition:

DEFINITION

> The definition of an inflection point implies that a is in the domain of f.

A function f has an ***inflection point*** at a if and only if f is continuous at a and f changes concavity at a.

To test a function for concavity and inflection points, first find the values of x where $f''(x)$ is 0 or not defined. These values of x determine intervals. On each interval, determine whether $f''(x) > 0$ (f is concave up) or $f''(x) < 0$ (f is concave down). If concavity changes around one of these x-values and f is continuous there, then f has an inflection point at this x-value. The continuity requirement implies that the x-value must be in the domain of the function. In brief, a *candidate* for an inflection point must satisfy two conditions:

1. f'' must be 0 or fail to exist at that point.

2. f must be continuous at that point.

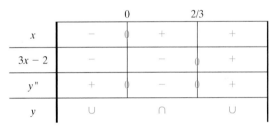

FIGURE 13.33 Inflection point for $f(x) = x^{1/3}$.

The candidate *will be* an inflection point if concavity changes around it. For example, if $f(x) = x^{1/3}$, then $f'(x) = \frac{1}{3}x^{-2/3}$ and

$$f''(x) = -\frac{2}{9}x^{-5/3} = -\frac{2}{9x^{5/3}}$$

Because f'' does not exist at 0, but f is continuous at 0, there is a candidate for an inflection point at 0. If $x > 0$, then $f''(x) < 0$, so f is concave down for $x > 0$; if $x < 0$, then $f''(x) > 0$, so f is concave up for $x < 0$. Because concavity changes at 0, there is an inflection point there. (See Figure 13.33.)

● EXAMPLE 2 Concavity and Inflection Points

Test $y = 6x^4 - 8x^3 + 1$ for concavity and inflection points.

Solution: We have

$$y' = 24x^3 - 24x^2$$

$$y'' = 72x^2 - 48x = 24x(3x - 2)$$

		0		2/3	
x	$-$	0	$+$		$+$
$3x - 2$	$-$		$-$	0	$+$
y''	$+$	0	$-$	0	$+$
y	$\cup$		$\cap$		$\cup$

FIGURE 13.34 Sign chart of $y'' = 24x(3x - 2)$ for $y = 6x^4 - 8x^3 + 1$.

To find where $y'' = 0$, we set each factor in y'' equal to 0. This gives $x = 0, \frac{2}{3}$. We also note that y'' is never undefined. Thus, there are three intervals to consider, as recorded on the top of the sign chart in Figure 13.34. Since y is continuous at 0 and $\frac{2}{3}$, these points are candidates for inflection points. Having completed the sign chart, we see that concavity changes at 0 and at $\frac{2}{3}$. Thus these candidates are indeed inflection points. (See Figure 13.35.) In summary, the curve is concave up on $(-\infty, 0)$ and $(\frac{2}{3}, \infty)$ and is concave down on $(0, \frac{2}{3})$. Inflection points occur at 0 and at $\frac{2}{3}$. These points are $(0, y(0)) = (0, 1)$ and $(\frac{2}{3}, y(\frac{2}{3})) = (\frac{2}{3}, -\frac{5}{27})$.

NOW WORK PROBLEM 13 ●●●

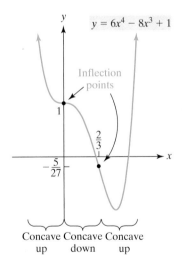

FIGURE 13.35 Graph of $y = 6x^4 - 8x^3 + 1$.

As we did in the analysis of increasing and decreasing, so we must in concavity analysis consider also those points a that are not in the domain of f but that are near points in the domain of f. The next example will illustrate.

● EXAMPLE 3 A Change in Concavity with No Inflection Point

Discuss concavity and find all inflection points for $f(x) = \dfrac{1}{x}$.

FIGURE 13.36 Sign chart for $f''(x)$.

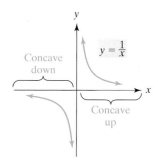

FIGURE 13.37 Graph of $y = \dfrac{1}{x}$.

Solution: Since $f(x) = x^{-1}$ for $x \neq 0$,

$$f'(x) = -x^{-2} \quad \text{for } x \neq 0$$

$$f''(x) = 2x^{-3} = \frac{2}{x^3} \quad \text{for } x \neq 0$$

We see that $f''(x)$ is never 0 but it is not defined when $x = 0$. Since f is not continuous at 0, we conclude that 0 is not a candidate for an inflection point. Thus, the given function has no inflection point. However, 0 must be considered in an analysis of concavity. See the sign chart in Figure 13.36; note that we have a thick verical line at 0 to indicate that 0 is not in the domain of f and cannot correspond to an inflection point. If $x > 0$, then $f''(x) > 0$; if $x < 0$, then $f''(x) < 0$. Hence, f is concave up on $(0, \infty)$ and concave down on $(-\infty, 0)$. (See Figure 13.37.) Although concavity changes around $x = 0$, there is no inflection point there because f is not continuous at 0 (nor is it even defined there).

NOW WORK PROBLEM 23 ◖◗◗

 CAUTION

A candidate for an inflection point may not necessarily be an inflection point. For example, if $f(x) = x^4$, then $f''(x) = 12x^2$ and $f''(0) = 0$. But $f''(x) > 0$ both when $x < 0$ and when $x > 0$. Thus, concavity does not change, and there are no inflection points. (See Figure 13.38.)

Curve Sketching

◖ **EXAMPLE 4   Curve Sketching**

Sketch the graph of $y = 2x^3 - 9x^2 + 12x$.

Solution:

Intercepts   If $x = 0$, then $y = 0$. Setting $y = 0$ gives $0 = x(2x^2 - 9x + 12)$. Clearly, $x = 0$ is a solution, and using the quadratic formula on $2x^2 - 9x + 12 = 0$ gives no real roots. Thus, the only intercept is $(0, 0)$. In fact, since $2x^2 - 9x + 12$ is a continuous function whose value at 0 is $2 \cdot 0^2 - 9 \cdot 0 + 12 = 12 > 0$, we conclude that $2x^2 - 9x + 12 > 0$ for all x, which gives the sign chart Figure 13.39 for y.

Note that this chart tells us the graph of $y = 2x^3 - 9x^2 + 12x$ is confined to the third and first quadrants of the xy-plane.

Symmetry   None.

Maxima and Minima   We have

$$y' = 6x^2 - 18x + 12 = 6(x^2 - 3x + 2) = 6(x - 1)(x - 2)$$

The critical values are $x = 1, 2$, so these and the factors $x - 1$ and $x - 2$ determine the sign chart of y' (Figure 13.40).

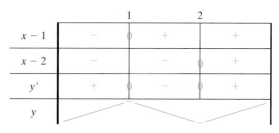

FIGURE 13.38 Graph of $f(x) = x^4$.

FIGURE 13.39 Sign chart for y.

FIGURE 13.40 Sign chart of $y' = 6(x - 1)(x - 2)$.

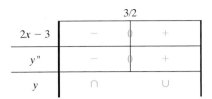

FIGURE 13.41 Sign chart of y''.

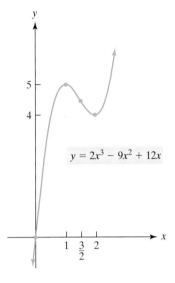

$y = 2x^3 - 9x^2 + 12x$

FIGURE 13.42 Graph of
$y = 2x^3 - 9x^2 + 12x$.

From the sign chart for y' we see that there is a relative maximum at 1 and a relative minimum at 2. Note too that the bottom line of Figure 13.40, together with that of Figure 13.39, comes close to determining a precise graph of $y = 2x^3 - 9x^2 + 12x$. Of course, it will help to know the relative maximum $y(1) = 5$, which occurs at 1, and the relative minimum $y(2) = 4$, which occurs at 2, so that in addition to the intercept $(0, 0)$ we will actually plot also $(1, 5)$ and $(2, 4)$.

Concavity

$$y'' = 12x - 18 = 6(2x - 3)$$

Setting $y'' = 0$ gives a possible inflection point at $x = \frac{3}{2}$ from which we construct the simple sign chart for y'' in Figure 13.41.

Since concavity changes at $x = \frac{3}{2}$, at which point f is certainly continuous, there is an inflection point at $\frac{3}{2}$.

Discussion We know the coordinates of three of the important points on the graph. The only other important point from our perspective is the inflection point, and since $y(3/2) = 2(3/2)^3 - 9(3/2)^2 + 12(3/2) = 9/2$ the inflection point is $(3/2, 9/2)$.

We plot the four points noted above and observe from all three sign charts jointly that the curve increases through the third quadrant and passes through $(0, 0)$, all the while concave down until a relative maximum is attained at $(1, 5)$. The curve then falls until it reaches a relative minimum at $(2, 4)$. However, along the way the concavity changes at $(3/2, 9/2)$ from concave down to concave up and remains so for the rest of the curve. After $(2, 4)$ the curve increases through the first quadrant. The curve is shown in Figure 13.42.

NOW WORK PROBLEM 39 ◖◗

TECHNOLOGY

Suppose that you need to find the inflection points for

$$f(x) = \frac{1}{20}x^5 - \frac{17}{16}x^4 + \frac{273}{32}x^3 - \frac{4225}{128}x^2 + \frac{750}{4}$$

The second derivative of f is given by

$$f''(x) = x^3 - \frac{51}{4}x^2 + \frac{819}{16}x - \frac{4225}{64}$$

Here the zeros of f'' are not obvious. Thus, we will graph f'' using a graphing calculator. (See Figure 13.43.) We find that the zeros of f'' are 3.25 and 6.25. Around $x = 6.25$, $f''(x)$ goes from negative to positive values. Therefore, at $x = 6.25$, there is an inflection point. Around $x = 3.25$, $f''(x)$ does not change sign, so no inflection point exists at $x = 3.25$. Comparing our results with the graph of f in Figure 13.44, we see that everything checks out.

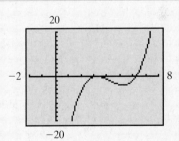

FIGURE 13.43 Graph of f''; zeros of f'' are 3.25 and 6.25.

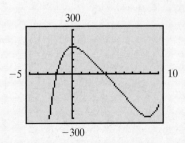

FIGURE 13.44 Graph of f; inflection point at $x = 6.25$, but not at $x = 3.25$.

Problems 13.3

In Problems 1–6, a function and its second derivative are given. Determine the concavity of f and x-values where points of inflection occur.

*1. $f(x) = 2x^4 + 3x^3 + 2x - 3;\ f''(x) = 6x(4x + 3)$

2. $f(x) = \dfrac{x^5}{20} + \dfrac{x^4}{4} - 2x^2;\ f''(x) = (x - 1)(x + 2)^2$

3. $f(x) = \dfrac{2 + x - x^2}{x^2 - 2x + 1};\ f''(x) = \dfrac{2(7 - x)}{(x - 1)^4}$

4. $f(x) = \dfrac{x^2}{(x - 1)^2};\ f''(x) = \dfrac{2(2x + 1)}{(x - 1)^4}$

5. $f(x) = \dfrac{x^2 + 1}{x^2 - 2};\ f''(x) = \dfrac{6(3x^2 + 2)}{(x^2 - 2)^3}$

6. $f(x) = x\sqrt{4 - x^2};\ f''(x) = \dfrac{2x(x^2 - 6)}{(4 - x^2)^{3/2}}$

In Problems 7–34, determine concavity and the x-values where points of inflection occur. Do not sketch the graphs.

7. $y = -2x^2 + 4x$

8. $y = -74x^2 + 19x - 37$

9. $y = 4x^3 + 12x^2 - 12x$

10. $y = x^3 - 6x^2 + 9x + 1$

11. $y = 2x^3 - 5x^2 + 5x - 2$

12. $y = x^4 - 8x^2 - 6$

*13. $y = 2x^4 - 48x^2 + 7x + 3$

14. $y = -\dfrac{x^4}{4} + \dfrac{9x^2}{2} + 2x$

15. $y = 2x^{1/5}$

16. $y = \dfrac{7}{x^3}$

17. $y = \dfrac{x^4}{2} + \dfrac{19x^3}{6} - \dfrac{7x^2}{2} + x + 5$

18. $y = -\dfrac{5}{2}x^4 - \dfrac{1}{6}x^3 + \dfrac{1}{2}x^2 + \dfrac{1}{3}x - \dfrac{2}{5}$

19. $y = \dfrac{1}{20}x^5 - \dfrac{1}{4}x^4 + \dfrac{1}{6}x^3 - \dfrac{1}{2}x - \dfrac{2}{3}$

20. $y = \dfrac{1}{10}x^5 - 3x^3 + 17x + 43$

21. $y = \dfrac{1}{30}x^6 - \dfrac{7}{12}x^4 + 5x^2 + 2x - 1$

22. $y = x^6 - 3x^4$

*23. $y = \dfrac{x + 1}{x - 1}$

24. $y = 1 - \dfrac{1}{x^2}$

25. $y = \dfrac{x^2}{x^2 + 1}$

26. $y = \dfrac{4x^2}{x + 3}$

27. $y = \dfrac{21x + 40}{6(x + 3)^2}$

28. $y = 3(x^2 - 2)^2$

29. $y = 5e^x$

30. $y = e^x - e^{-x}$

31. $y = 3xe^x$

32. $y = xe^{x^2}$

33. $y = \dfrac{\ln x}{2x}$

34. $y = \dfrac{x^2 + 1}{3e^x}$

In Problems 35–62, determine intervals on which the function is increasing, decreasing, concave up, and concave down; relative maxima and minima; inflection points; symmetry; and those intercepts that can be obtained conveniently. Then sketch the graph.

35. $y = x^2 - x - 6$

36. $y = x^2 + 2$

37. $y = 5x - 2x^2$

38. $y = x - x^2 + 2$

*39. $y = x^3 - 9x^2 + 24x - 19$

40. $y = x^3 - 25x^2$

41. $y = \dfrac{x^3}{3} - 4x$

42. $y = x^3 - 6x^2 + 9x$

43. $y = x^3 - 3x^2 + 3x - 3$

44. $y = 2x^3 + \dfrac{5}{2}x^2 + 2x$

45. $y = 4x^3 - 3x^4$

46. $y = -x^3 + 2x^2 - x + 4$

47. $y = -2 + 12x - x^3$

48. $y = (3 + 2x)^3$

49. $y = 2x^3 - 6x^2 + 6x - 2$

50. $y = \dfrac{x^5}{100} - \dfrac{x^4}{20}$

51. $y = 5x - x^5$

52. $y = x^2(x - 1)^2$

53. $y = 3x^4 - 4x^3 + 1$

54. $y = 3x^5 - 5x^3$

55. $y = 4x^2 - x^4$

56. $y = x^4 - x^2$

57. $y = x^{1/3}(x - 8)$

58. $y = (x - 1)^2(x + 2)^2$

59. $y = 4x^{1/3} + x^{4/3}$

60. $y = (x + 1)\sqrt{x + 4}$

61. $y = 6x^{2/3} - \dfrac{x}{2}$

62. $y = 5x^{2/3} - x^{5/3}$

63. Sketch the graph of a continuous function f such that $f(2) = 4$, $f'(2) = 0$, $f'(x) < 0$ if $x < 2$, and $f''(x) > 0$ if $x > 2$.

64. Sketch the graph of a continuous function f such that $f(4) = 4$, $f'(4) = 0$, $f''(x) < 0$ for $x < 4$, and $f''(x) > 0$ for $x > 4$.

65. Sketch the graph of a continuous function f such that $f(1) = 1$, $f'(1) = 0$, and $f''(x) < 0$ for all x.

66. Sketch the graph of a continuous function f such that $f(3) = 4$, both $f'(x) > 0$ and $f''(x) > 0$ for $x < 3$, and both $f'(x) < 0$ and $f''(x) > 0$ for $x > 3$.

67. **Demand Equation** Show that the graph of the demand equation $p = \dfrac{100}{q + 2}$ is decreasing and concave up for $q > 0$.

68. **Average Cost** For the cost function

$$c = q^2 + 2q + 1$$

show that the graph of the average-cost function $\bar{c}$ is always concave up for $q > 0$.

69. **Species of Plants** The number of species of plants on a plot may depend on the size of the plot. For example, in Figure 13.45, we see that on 1-m^2 plots there are three species (A, B, and C on the left plot, A, B, and D on the right plot), and on a 2-m^2 plot there are four species (A, B, C, and D).

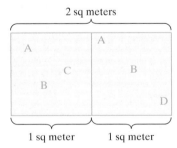

FIGURE 13.45 Species of plants.

In a study of rooted plants in a certain geographic region,[7] it was determined that the average number of species, S, occurring on plots of size A (in square meters) is given by

$$S = f(A) = 12\sqrt[4]{A} \quad 0 \le A \le 625$$

---

[7]Adapted from R. W. Poole, *An Introduction to Quantitative Ecology* (New York: McGraw-Hill Book Company, 1974).

Sketch the graph of f. (*Note:* Your graph should be rising and concave down. Thus, the number of species is increasing with respect to area, but at a decreasing rate.)

70. Inferior Good   In a discussion of an inferior good, Persky[8] considers a function of the form

$$g(x) = e^{(U_0/A)} e^{-x^2/(2A)}$$

where x is a quantity of a good, U_0 is a constant that represents utility, and A is a positive constant. Persky claims that the graph of g is concave down for $x < \sqrt{A}$ and concave up for $x > \sqrt{A}$. Verify this.

71. Psychology   In a psychological experiment involving conditioned response,[9] subjects listened to four tones, denoted 0, 1, 2, and 3. Initially, the subjects were conditioned to tone 0 by receiving a shock whenever this tone was heard. Later, when each of the four tones (stimuli) were heard without shocks, the subjects' responses were recorded by means of a tracking device that measures galvanic skin reaction. The average response to each stimulus (without shock) was determined, and the results were plotted on a coordinate plane where the x- and y-axes represent the stimuli (0, 1, 2, 3) and the average galvanic responses, respectively. It was determined that the points fit a curve that is approximated by the graph of

$$y = 12.5 + 5.8(0.42)^x$$

Show that this function is decreasing and concave up.

72. Entomology   In a study of the effects of food deprivation on hunger,[10] an insect was fed until its appetite was completely satisfied. Then it was deprived of food for t hours (the deprivation period). At the end of this period, the insect was re-fed until its appetite was again completely satisfied. The weight H (in grams) of the food that was consumed at this time was statistically found to be a function of t, where

$$H = 1.00[1 - e^{-(0.0464t + 0.0670)}]$$

Here H is a measure of hunger. Show that H is increasing with respect to t and is concave down.

73. Insect Dispersal   In an experiment on the dispersal of a particular insect,[11] a large number of insects are placed at a release point in an open field. Surrounding this point are traps that are placed in a concentric circular arrangement at a distance of 1 m, 2 m, 3 m, and so on from the release point. Twenty-four hours after the insects are released, the number of insects in each trap is counted. It is determined that at a distance of r meters from the release point, the average number of insects contained in a trap is

$$n = f(r) = 0.1 \ln(r) + \frac{7}{r} - 0.8 \quad 1 \le r \le 10$$

(a) Show that the graph of f is always falling and concave up. (b) Sketch the graph of f. (c) When $r = 5$, at what rate is the average number of insects in a trap decreasing with respect to distance?

74. Graph $y = -0.35x^3 + 4.1x^2 + 8.3x - 7.4$, and from the graph determine the number of (a) relative maximum points, (b) relative minimum points, and (c) inflection points.

75. Graph $y = x^5(x - 2.3)$, and from the graph determine the number of inflection points. Now, prove that for any $a \ne 0$, the curve $y = x^5(x - a)$ has two points of inflection.

76. Graph $y = 1 - 2^{-x^2}$, and from the graph determine the number of inflection points.

77. Graph the curve $y = x^3 - 2x^2 + x + 3$, and also graph the tangent line to the curve at $x = 2$. Around $x = 2$, does the curve lie above or below the tangent line? From your observation determine the concavity at $x = 2$.

78. If $f(x) = 2x^3 + 3x^2 - 6x + 1$, find $f'(x)$ and $f''(x)$. Note that where f' has a relative minimum, f changes its direction of bending. Why?

79. If $f(x) = x^6 + 3x^5 - 4x^4 + 2x^2 + 1$, find the x-values (rounded to two decimal places) of the inflection points of f.

80. If $f(x) = \dfrac{x+1}{x^2+1}$, find the x-values (rounded to two decimal places) of the inflection points of f.

OBJECTIVE

To locate relative extrema by applying the second-derivative test.

13.4 The Second-Derivative Test

The second derivative can be used to test certain critical values for relative extrema. Observe in Figure 13.46 that at a, there is a horizontal tangent; that is, $f'(a) = 0$. Furthermore, around a, the function is concave up (that is, $f''(a) > 0$). This leads us to conclude that there is a relative minimum at a. On the other hand, around b, the function is concave down (that is, $f''(b) < 0$). Because the tangent line is horizontal at b, we conclude that a relative maximum exists there. This technique of examining the second derivative at points where the first derivative is 0 is called the *second-derivative test* for relative extrema.

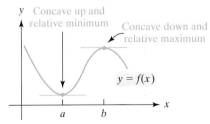

FIGURE 13.46   Relating concavity to relative extrema.

---

[8]A. L. Persky, "An Inferior Good and a Novel Indifference Map," *The American Economist* XXIX, no. 1 (1985), 67–69.

[9]Adapted from C. I. Hovland, "The Generalization of Conditioned Responses: I. The Sensory Generalization of Conditioned Responses with Varying Frequencies of Tone," *Journal of General Psychology*, 17 (1937), 125–48.

[10]C. S. Holling, "The Functional Response of Invertebrate Predators to Prey Density," *Memoirs of the Entomological Society of Canada*, no. 48 (1966).

[11]Adapted from Poole, op. cit.

Second-Derivative Test for Relative Extrema

Suppose $f'(a) = 0$.

If $f''(a) < 0$, then f has a relative maximum at a.

If $f''(a) > 0$, then f has a relative minimum at a.

We want to emphasize that *the second-derivative test does* not *apply when* $f''(a) = 0$. If both $f'(a) = 0$ and $f''(a) = 0$, then there may be a relative maximum, a relative minimum, or neither, at a. In such cases, the first-derivative test should be used to analyze what is happening at a. (Also, the second-derivative test does not apply when $f'(a)$ does not exist.)

● EXAMPLE 1 Second-Derivative Test

Test the following for relative maxima and minima. Use the second-derivative test, if possible.

a. $y = 18x - \frac{2}{3}x^3$.

Solution:

$$y' = 18 - 2x^2 = 2(9 - x^2) = 2(3 + x)(3 - x)$$

$$y'' = -4x \qquad \left(\text{taking } \frac{d}{dx} \text{ of } 18 - 2x^2 \right)$$

Solving $y' = 0$ gives the critical values $x = \pm 3$.

$$\text{If } x = 3, \quad \text{then } y'' = -4(3) = -12 < 0.$$

There is a relative maximum when $x = 3$.

$$\text{If } x = -3, \quad \text{then } y'' = -4(-3) = 12 > 0.$$

There is a relative minimum when $x = -3$. (Refer back to Figure 13.4.)

b. $y = 6x^4 - 8x^3 + 1$.

Solution:

$$y' = 24x^3 - 24x^2 = 24x^2(x - 1)$$
$$y'' = 72x^2 - 48x$$

Solving $y' = 0$ gives the critical values $x = 0, 1$. We see that

$$\text{if } x = 0, \quad \text{then } y'' = 0$$

and

$$\text{if } x = 1, \quad \text{then } y'' > 0$$

By the second-derivative test, there is a relative minimum when $x = 1$. We cannot apply the test when $x = 0$ because $y'' = 0$ there. To analyze what is happening at 0, we turn to the first-derivative test:

$$\text{If } x < 0, \quad \text{then } y' < 0.$$

$$\text{If } 0 < x < 1, \quad \text{then } y' < 0.$$

Thus, no maximum or minimum exists when $x = 0$. (Refer back to Figure 13.35.)

NOW WORK PROBLEM 5 ●●●

CAUTION

Although the second-derivative test can be very useful, do not depend entirely on it. Not only may the test fail to apply, but also it may be awkward to find the second derivative.

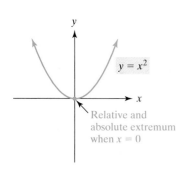

FIGURE 13.47 Exactly one relative extremum implies an absolute extremum.

If a continuous function has *exactly one* relative extremum on an interval, it can be shown that the relative extremum must also be an *absolute* extremum on the interval. To illustrate, in Figure 13.47 the function $y = x^2$ has a relative minimum when $x = 0$, and there are no other relative extrema. Since $y = x^2$ is continuous, this relative minimum is also an absolute minimum for the function.

EXAMPLE 2   Absolute Extrema

If $y = f(x) = x^3 - 3x^2 - 9x + 5$, determine when absolute extrema occur on the interval $(0, \infty)$.

Solution: We have

$$f'(x) = 3x^2 - 6x - 9 = 3(x^2 - 2x - 3)$$
$$= 3(x + 1)(x - 3)$$

The only critical value on the interval $(0, \infty)$ is 3. Applying the second-derivative test at this point gives

$$f''(x) = 6x - 6$$
$$f''(3) = 6(3) - 6 = 12 > 0$$

Thus, there is a relative minimum at 3. Since this is the only relative extremum on $(0, \infty)$ and f is continuous there, we conclude by our previous discussion that there is an *absolute* minimum value at 3; this value is $f(3) = -22$. (See Figure 13.48.)

NOW WORK PROBLEM 3 ◖●●

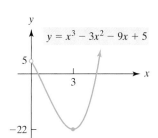

FIGURE 13.48   On $(0, \infty)$, there is an absolute minimum at 3.

Problems 13.4

In Problems 1–14, test for relative maxima and minima. Use the second-derivative test, if possible. In Problems 1–4, state whether the relative extrema are also absolute extrema.

1. $y = x^2 - 5x + 6$

*3. $y = -4x^2 + 2x - 8$

*5. $y = \frac{1}{3}x^3 + 2x^2 - 5x + 1$

2. $y = 5x^2 + 20x + 2$

4. $y = 3x^2 - 5x + 6$

6. $y = x^3 - 12x + 1$

7. $y = -x^3 + 3x^2 + 1$

9. $y = 7 - 2x^4$

11. $y = 81x^5 - 5x$

13. $y = (x^2 + 7x + 10)^2$

8. $y = x^4 - 2x^2 + 4$

10. $y = -2x^7$

12. $y = \frac{55}{3}x^3 - x^2 - 21x - 3$

14. $y = -x^3 + 3x^2 + 9x - 2$

OBJECTIVE

To determine horizontal and vertical asymptotes for a curve and to sketch the graphs of functions having asymptotes.

13.5 Asymptotes

Vertical Asymptotes

In this section, we conclude our discussion of curve-sketching techniques by investigating functions having *asymptotes*. An asymptote is a line that a curve approaches arbitrarily closely. For example, in each part of Figure 13.49, the dashed line $x = a$ is an asymptote. But to be precise about it, we need to make use of infinite limits. In Figure 13.49(a), notice that as $x \to a^+$, $f(x)$ becomes positively infinite:

$$\lim_{x \to a^+} f(x) = \infty$$

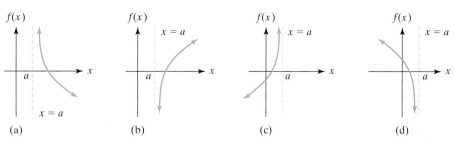

(a)   (b)   (c)   (d)

FIGURE 13.49   Vertical asymptotes $x = a$.

In Figure 13.49(b), as $x \to a^+$, $f(x)$ becomes negatively infinite:

$$\lim_{x \to a^+} f(x) = -\infty$$

In Figure. 13.49(c) and (d), we have

$$\lim_{x \to a^-} f(x) = \infty \quad \text{and} \quad \lim_{x \to a^-} f(x) = -\infty$$

respectively.

Loosely speaking, we can say that each graph in Figure 13.49 "blows up" around the dashed vertical line $x = a$, in the sense that a one-sided limit of $f(x)$ at a is either ∞ or $-\infty$. The line $x = a$ is called a *vertical asymptote* for the graph. A vertical asymptote is not part of the graph but is a useful aid in sketching it because part of the graph approaches the asymptote. Because of the explosion around $x = a$, the function is *not* continuous at a.

DEFINITION

The line $x = a$ is a **vertical asymptote** for the graph of the function f if and only if at least one of the following is true:

$$\lim_{x \to a^+} f(x) = \pm\infty$$

or

$$\lim_{x \to a^-} f(x) = \pm\infty$$

To determine vertical asymptotes, we must find values of x around which $f(x)$ increases or decreases without bound. For a rational function (a quotient of two polynomials) *expressed in lowest terms* these x-values are precisely those for which the denominator is zero but the numerator is not zero. For example, consider the rational function

$$f(x) = \frac{3x - 5}{x - 2}$$

When x is 2, the denominator is 0, but the numerator is not. If x is slightly larger than 2, then $x - 2$ is both close to 0 and positive, and $3x - 5$ is close to 1. Thus, $(3x - 5)/(x - 2)$ is very large, so

$$\lim_{x \to 2^+} \frac{3x - 5}{x - 2} = \infty$$

This limit is sufficient to conclude that the line $x = 2$ is a vertical asymptote. Because we are ultimately interested in the behavior of a function around a vertical asymptote, it is worthwhile to examine what happens to this function as x approaches 2 from the left. If x is slightly less than 2, then $x - 2$ is very close to 0 but negative, and $3x - 5$ is close to 1. Hence, $(3x - 5)/(x - 2)$ is "very negative," so

$$\lim_{x \to 2^-} \frac{3x - 5}{x - 2} = -\infty$$

We conclude that the function increases without bound as $x \to 2^+$ and decreases without bound as $x \to 2^-$. The graph appears in Figure 13.50.

In summary, we have a rule for vertical asymptotes.

Vertical-Asymptote Rule for Rational Functions

Suppose that

$$f(x) = \frac{P(x)}{Q(x)}$$

where P and Q are polynomial functions and the quotient is in lowest terms. The line $x = a$ is a vertical asymptote for the graph of f if and only if $Q(a) = 0$ and $P(a) \neq 0$.

 CAUTION

To see that the proviso about *lowest terms* is necessary, observe that

$$f(x) = \frac{3x - 5}{x - 2} = \frac{(3x - 5)(x - 2)}{(x - 2)^2} \text{ so}$$

that $x = 2$ is a vertical asymptote of $\frac{(3x - 5)(x - 2)}{(x - 2)^2}$, and here 2 makes both the denominator *and* the numerator 0.

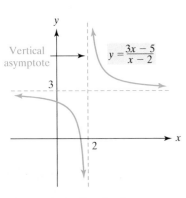

FIGURE 13.50 Graph of $y = \dfrac{3x - 5}{x - 2}$.

(It might be thought here that "lowest terms" rules out the possibility of a value *a* making *both* denominator *and* numerator 0, but consider the rational function $\frac{(3x-5)(x-2)}{(x-2)}$. Here we cannot divide numerator and denominator by $x-2$, to obtain the polynomial $3x-5$, because the domain of the latter is not equal to the domain of the former.)

● **EXAMPLE 1   Finding Vertical Asymptotes**

Determine vertical asymptotes for the graph of

$$f(x) = \frac{x^2 - 4x}{x^2 - 4x + 3}$$

Solution: Since f is a rational function, the vertical-asymptote rule applies. Writing

$$f(x) = \frac{x(x-4)}{(x-3)(x-1)} \quad \text{(factoring)}$$

makes it clear that the denominator is 0 when x is 3 or 1. Neither of these values makes the numerator 0. Thus, the lines $x = 3$ and $x = 1$ are vertical asymptotes. (See Figure 13.51.)

NOW WORK PROBLEM 1 ●●●

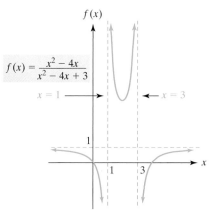

FIGURE 13.51   Graph of $f(x) = \frac{x^2 - 4x}{x^2 - 4x + 3}$.

Although the vertical-asymptote rule guarantees that the lines $x = 3$ and $x = 1$ are vertical asymptotes, it does not indicate the precise nature of the ``blow-up'' around these lines. A precise analysis requires the use of one-sided limits.

Horizontal and Oblique Asymptotes

A curve $y = f(x)$ may have other kinds of asymptote. In Figure 13.52(a), as x increases without bound ($x \to \infty$), the graph approaches the horizontal line $y = b$. That is,

$$\lim_{x \to \infty} f(x) = b$$

In Figure 13.52(b), as x becomes negatively infinite, the graph approaches the horizontal line $y = b$. That is,

$$\lim_{x \to -\infty} f(x) = b$$

In each case, the dashed line $y = b$ is called a *horizontal asymptote* for the graph. It is a horizontal line around which the graph "settles" either as $x \to \infty$ or as $x \to -\infty$.

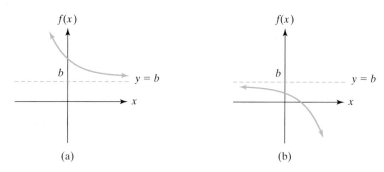

FIGURE 13.52   Horizontal asymptotes $y = b$.

In summary, we have the following definition:

DEFINITION

Let f be a function. The line $y = b$ is a **horizontal asymptote** for the graph of f if and only if at least one of the following is true:

$$\lim_{x \to \infty} f(x) = b \quad \text{or} \quad \lim_{x \to -\infty} f(x) = b$$

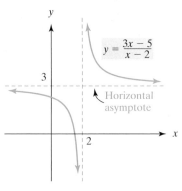

FIGURE 13.53 Graph of $f(x) = \dfrac{3x-5}{x-2}$.

To test for horizontal asymptotes, we must find the limits of $f(x)$ as $x \to \infty$ and as $x \to -\infty$. To illustrate, we again consider

$$f(x) = \frac{3x-5}{x-2}$$

Since this is a rational function, we can use the procedures of Section 10.2 to find the limits. Because the dominant term in the numerator is $3x$ and the dominant term in the denominator is x, we have

$$\lim_{x \to \infty} \frac{3x-5}{x-2} = \lim_{x \to \infty} \frac{3x}{x} = \lim_{x \to \infty} 3 = 3$$

Thus, the line $y = 3$ is a horizontal asymptote. (See Figure 13.53.) Also,

$$\lim_{x \to -\infty} \frac{3x-5}{x-2} = \lim_{x \to -\infty} \frac{3x}{x} = \lim_{x \to -\infty} 3 = 3$$

Hence, the graph settles down near the horizontal line $y = 3$ both as $x \to \infty$ and as $x \to -\infty$.

● EXAMPLE 2 **Finding Horizontal Asymptotes**

Find horizontal asymptotes for the graph of

$$f(x) = \frac{x^2 - 4x}{x^2 - 4x + 3}$$

Solution: We have

$$\lim_{x \to \infty} \frac{x^2 - 4x}{x^2 - 4x + 3} = \lim_{x \to \infty} \frac{x^2}{x^2} = \lim_{x \to \infty} 1 = 1$$

Therefore, the line $y = 1$ is a horizontal asymptote. The same result is obtained for $x \to -\infty$. (Refer back to Figure 13.51.)

NOW WORK PROBLEM 11 ●●●

Horizontal asymptotes arising from limits such as $\lim_{t \to \infty} f(t) = b$, where t is thought of as *time,* can be important in business applications as expressions of long-term behavior. For example, in Section 9.3 we discussed long-term market share.

If we rewrite $\lim_{x \to \infty} f(x) = b$ as $\lim_{x \to \infty}(f(x) - b) = 0$, then another possibility is suggested. For it might be that the long-term behavior of f, while not constant, is linear. This leads us to the following:

DEFINITION

Let f be a function. The line $y = mx + b$ is a ***nonvertical asymptote*** for the graph of f if and only if at least one of the following is true:

$$\lim_{x \to \infty} (f(x) - (mx + b)) = 0 \quad \text{or} \quad \lim_{x \to -\infty} (f(x) - (mx + b)) = 0$$

Of course, if $m = 0$, then we have just repeated the definition of horizontal asymptote. But if $m \neq 0$ then $y = mx + b$ is the equation of a nonhorizontal (and nonvertical) line with slope m that is sometimes described as *oblique.* Thus to say that $\lim_{x \to \infty}(f(x) - (mx + b)) = 0$ is to say that for large values of x, the graph settles down near the line $y = mx + b$, often called an *oblique asympote* for the graph.

If $f(x) = \dfrac{P(x)}{Q(x)}$, where the degree of P is one more than the degree of Q, then long divison allows us to write $\dfrac{P(x)}{Q(x)} = (mx + b) + \dfrac{R(x)}{Q(x)}$, where $m \neq 0$ and where either $R(x)$ is the zero polynomial or the degree of R is strictly less than the degree of Q. In this case, $y = mx + b$ will be an oblique asymptote for the graph of f. The next example will illustrate.

● EXAMPLE 3 **Finding an Oblique Asymptote**

Find the oblique asyptote for the graph of the rational function

$$y = f(x) = \frac{10x^2 + 9x + 5}{5x + 2}$$

Solution: Since the degree of the numerator is 2, one greater than the degree of the denominator, we use long division to express

$$f(x) = \frac{10x^2 + 9x + 5}{5x + 2} = 2x + 1 + \frac{3}{5x + 2}$$

Thus

$$\lim_{x \to \pm\infty} (f(x) - (2x + 1)) = \lim_{x \to \pm\infty} \frac{3}{5x + 2} = 0$$

which shows that $y = 2x + 1$ is an oblique asymptote, in fact the only nonvertical asymptote, as we explain below. On the other hand, it is clear that $x = -\frac{2}{5}$ is a vertical asymptote—and the only one. (See Figure 13.54.)

NOW WORK PROBLEM 35 ●●●

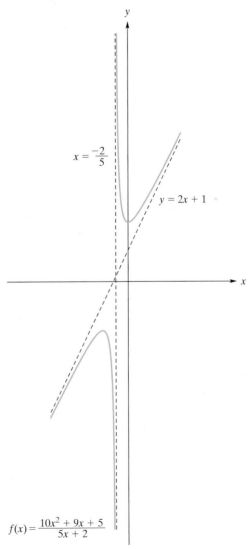

FIGURE 13.54 Graph of $f(x) = \dfrac{10x^2 + 9x + 5}{5x + 2}$
has an oblique asymptote.

A few remarks about asymptotes are appropriate now. With vertical asymptotes, we are examining the behavior of a graph around specific x-values. However, with nonvertical asymptotes we are examining the graph as x becomes unbounded. Although a graph may have numerous vertical asymptotes, it can have at most two different nonvertical asymptotes—possibly one for $x \to \infty$ and possibly one for $x \to -\infty$. If, for example, the graph has two horizontal asymptotes, then there can be no oblique asymptotes.

From Section 10.2, when the numerator of a rational function has degree greater than that of the denominator, no limit exists as $x \to \infty$ or $x \to -\infty$. From this observation, we conclude that *whenever the degree of the numerator of a rational function is greater than the degree of the denominator, the graph of the function cannot have a horizontal asymptote.* Similarly, it can be shown that if the degree of the numerator of a rational function is more than one greater than the degree of the denominator, the function cannot have an oblique asymptote.

EXAMPLE 4 Finding Horizontal and Vertical Asymptotes

Find vertical and horizontal asymptotes for the graph of the polynomial function

$$y = f(x) = x^3 + 2x$$

Solution: We begin with vertical asymptotes. This is a rational function with denominator 1, which is never zero. By the vertical-asymptote rule, there are no vertical asymptotes. Because the degree of the numerator (3) is greater than the degree of the denominator (0), there are no horizontal asymptotes. However, let us examine the behavior of the graph of f as $x \to \infty$ and $x \to -\infty$. We have

$$\lim_{x \to \infty} (x^3 + 2x) = \lim_{x \to \infty} x^3 = \infty$$

and

$$\lim_{x \to -\infty} (x^3 + 2x) = \lim_{x \to -\infty} x^3 = -\infty$$

Thus, as $x \to \infty$, the graph must extend indefinitely upward, and as $x \to -\infty$, the graph must extend indefinitely downward. (See Figure 13.55.)

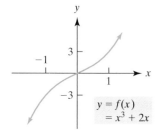

FIGURE 13.55 Graph of $y = x^3 + 2x$ has neither horizontal nor vertical asymptotes.

NOW WORK PROBLEM 9

The results in Example 3 can be generalized to any polynomial function:

A polynomial function of degree greater than 1 has no asymptotes.

EXAMPLE 5 Finding Horizontal and Vertical Asymptotes

Find horizontal and vertical asymptotes for the graph of $y = e^x - 1$.

Solution: Testing for horizontal asymptotes, we let $x \to \infty$. Then e^x increases without bound, so

$$\lim_{x \to \infty} (e^x - 1) = \infty$$

Thus, the graph does not settle down as $x \to \infty$. However, as $x \to -\infty$, we have $e^x \to 0$, so

$$\lim_{x \to -\infty} (e^x - 1) = \lim_{x \to -\infty} e^x - \lim_{x \to -\infty} 1 = 0 - 1 = -1$$

Therefore, the line $y = -1$ is a horizontal asymptote. The graph has no vertical asymptotes because $e^x - 1$ neither increases nor decreases without bound around any fixed value of x. (See Figure 13.56.)

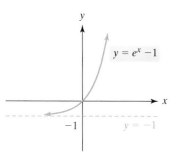

FIGURE 13.56 Graph of $y = e^x - 1$ has a horizontal asymptote.

NOW WORK PROBLEM 23

Curve Sketching

In this section we show how to graph a function by making use of all the curve-sketching tools that we have developed.

● EXAMPLE 6   **Curve Sketching**

Sketch the graph of $y = \dfrac{1}{4 - x^2}$.

Solution:

Intercepts   When $x = 0$, $y = \frac{1}{4}$. If $y = 0$, then $0 = 1/(4 - x^2)$, which has no solution. Thus $(0, \frac{1}{4})$ is the only intercept. However, the factorization

$$y = \frac{1}{4 - x^2} = \frac{1}{(2 + x)(2 - x)}$$

allows us to construct the following sign chart for y, showing where the graph lies below the x-axis $(-)$ and where it lies above the the x-axis $(+)$.

		-2		2	
$\dfrac{1}{2 + x}$		$-$		$+$	$+$
$\dfrac{1}{2 - x}$		$+$		$+$	$-$
y		$-$		$+$	$-$

FIGURE 13.57   Sign chart for $y = \dfrac{1}{4 - x^2}$.

Symmetry   There is symmetry about the y-axis:

$$y(-x) = \frac{1}{4 - (-x)^2} = \frac{1}{4 - x^2} = y(x)$$

Since y is a function of x (and not the constant function 0), there can be no symmetry about the x-axis and hence no symmetry about the origin. Since x is not a function of y (and y is a function of x), there can be no symmetry about $y = x$ either.

Asymptotes   From the factorization of y above, we see that $x = -2$ and $x = 2$ are vertical asymptotes. Testing for horizontal asymptotes, we have

$$\lim_{x \to \pm\infty} \frac{1}{4 - x^2} = \lim_{x \to \pm\infty} \frac{1}{-x^2} = -\lim_{x \to \pm\infty} \frac{1}{x^2} = 0$$

Thus, $y = 0$ (the x-axis) is the only nonvertical asymptote.

Maxima and Minima   Since $y = (4 - x^2)^{-1}$,

$$y' = -1(4 - x^2)^{-2}(-2x) = \frac{2x}{(4 - x^2)^2}$$

We see that y' is 0 when $x = 0$ and y' is undefined when $x = \pm 2$. However, only 0 is a critical value, because y is not defined at ± 2. The sign chart for y' follows. (See Figure 13.58.)

The sign chart shows clearly that the function is decreasing on $(-\infty, -2)$ and $(-2, 0)$, increasing on $(0, 2)$ and $(2, \infty)$, and that there is a relative minimum at 0.

		-2		0		2	
$2x$	$-$		$-$	0	$+$		$+$
$\dfrac{1}{(4 - x^2)^2}$	$+$		$+$		$+$		$+$
y'	$-$		$-$	0	$+$		$+$
y							

FIGURE 13.58   Sign chart for $y' = \dfrac{2x}{(4 - x^2)^2}$

Concavity

$$y'' = \frac{(4-x^2)^2(2) - (2x)2(4-x^2)(-2x)}{(4-x^2)^4}$$

$$= \frac{2(4-x^2)[(4-x^2) - (2x)(-2x)]}{(4-x^2)^4} = \frac{2(4+3x^2)}{(4-x^2)^3}$$

Setting $y'' = 0$, we get no real roots. However, y'' is undefined when $x = \pm 2$. Although concavity may change around these values of x, the values cannot correspond to inflection points because they are not in the domain of the function. There are three intervals to test for concavity. (See the sign chart Figure 13.59.)

The sign chart shows that the graph is concave up on $(-2, 2)$ and concave down on $(-\infty, -2)$ and $(2, \infty)$.

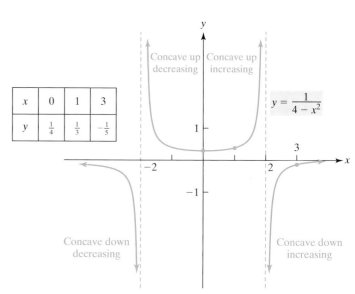

x	0	1	3
y	$\frac{1}{4}$	$\frac{1}{3}$	$-\frac{1}{5}$

		-2		2	
$4 + 3x^2$	$+$		$+$		$+$
$\frac{1}{(4-x^2)^3}$	$-$		$+$		$-$
y''	$-$		$+$		$-$
y	$\cap$		$\cup$		$\cap$

FIGURE 13.59 Concavity analysis.

FIGURE 13.60 Graph of $y = \dfrac{1}{4-x^2}$.

Discussion Only one point on the curve, $(0, 1/4)$, has arisen as a special point that must be plotted (both because it is an intercept and a local minimum). We might wish to plot a few more points as in the table in Figure 13.60, but note that any such extra points are only of value if they are on the same side of the y-axis (because of symmetry). Taking account of all the information gathered, we obtain the graph in Figure 13.60.

 NOW WORK PROBLEM 31

EXAMPLE 7 Curve Sketching

Sketch the graph of $y = \dfrac{4x}{x^2 + 1}$.

Solution:

Intercepts When $x = 0$, $y = 0$; when $y = 0$, $x = 0$. Thus, $(0, 0)$ is the only intercept. Since the denominator of y is always positive, we see that the sign of y is that of x. Here we dispense with a sign chart for y. From the observations so far it follows that the graph proceeds from the third quadrant (negative x and negative y), through $(0, 0)$ to the positive quadrant (positive x and positive y).

Symmetry There is symmetry about the origin:

$$y(-x) = \frac{4(-x)}{(-x)^2 + 1} = \frac{-4x}{x^2 + 1} = -y(x)$$

No other symmetry exists.

Asymptotes   The denominator of this rational function is never 0, so there are no vertical asymptotes. Testing for horizontal asymptotes, we have

$$\lim_{x \to \pm\infty} \frac{4x}{x^2 + 1} = \lim_{x \to \pm\infty} \frac{4x}{x^2} = \lim_{x \to \pm\infty} \frac{4}{x} = 0$$

Thus, $y = 0$ (the x-axis) is a horizontal asymptote and the only nonvertical asymptote.

Maxima and Minima   We have

$$y' = \frac{(x^2 + 1)(4) - 4x(2x)}{(x^2 + 1)^2} = \frac{4 - 4x^2}{(x^2 + 1)^2} = \frac{4(1 + x)(1 - x)}{(x^2 + 1)^2}$$

The critical values are $x = \pm 1$, so there are three intervals to consider in the sign chart. (See Figure 13.61.)

We see that y is decreasing on $(-\infty, -1)$ and on $(1, \infty)$, increasing on $(-1, 1)$, with relative minimum at -1 and relative maximum at 1. The relative minimum is $(-1, y(-1)) = (-1, -2)$; the relative maximum is $(1, y(1)) = (1, 2)$.

	-1		1	
$1 + x$	$-$	0	$+$	$+$
$1 - x$	$+$		$+$	0	$-$
$\dfrac{1}{(x^2 + 1)^2}$	$+$		$+$		$+$
y'	$-$	0	$+$	0
y				

FIGURE 13.61   Sign chart for y'.

Concavity   Since $y' = \dfrac{4 - 4x^2}{(x^2 + 1)^2}$,

$$y'' = \frac{(x^2 + 1)^2(-8x) - (4 - 4x^2)(2)(x^2 + 1)(2x)}{(x^2 + 1)^4}$$

$$= \frac{8x(x^2 + 1)(x^2 - 3)}{(x^2 + 1)^4} = \frac{8x(x + \sqrt{3})(x - \sqrt{3})}{(x^2 + 1)^3}$$

Setting $y'' = 0$, we conclude that the possible points of inflection are when $x = \pm\sqrt{3}, 0$. There are four intervals to consider in the sign chart. (See Figure 13.62.)

	$-\sqrt{3}$		0		$\sqrt{3}$	
$x + \sqrt{3}$	$-$	0	$+$		$+$		$+$
x	$-$		$-$	0	$+$		$+$
$x - \sqrt{3}$	$-$		$-$		$-$	0	$+$
$\dfrac{1}{(x^2 + 1)^3}$	$+$		$+$		$+$		$+$
y''	$-$	0	$+$	0	$-$	0	$+$
y	$\cap$		$\cup$		$\cap$		$\cup$

FIGURE 13.62   Concavity analysis for $y = \dfrac{4x}{x^2 + 1}$.

Inflection points occur at $x = 0$ and $\pm\sqrt{3}$. The inflection points are

$$(-\sqrt{3}, y(\sqrt{3})) = (-\sqrt{3}, -\sqrt{3}) \quad (0, y(0)) = (0, 0) \quad (\sqrt{3}, y(\sqrt{3})) = (\sqrt{3}, \sqrt{3})$$

x	0	1	$\sqrt{3}$	-1	$-\sqrt{3}$
y	0	2	$\sqrt{3}$	-2	$-\sqrt{3}$

FIGURE 13.63   Graph of $y = \dfrac{4x}{x^2 + 1}$.

Discussion   After consideration of all of the preceding information, the graph of $y = 4x/(x^2 + 1)$ is given in Figure 13.63, together with a table of important points.

NOW WORK PROBLEM 39 ◉◉◉

Problems 13.5

In Problems 1–24, find the vertical asymptotes and the nonvertical asymptotes for the graphs of the functions. Do not sketch the graphs.

*1. $y = \dfrac{x}{x - 1}$

2. $y = \dfrac{x + 1}{x}$

3. $f(x) = \dfrac{x + 2}{3x - 5}$

4. $y = \dfrac{2x + 1}{2x + 1}$

5. $y = \dfrac{4}{x}$

6. $y = 1 - \dfrac{2}{x^2}$

7. $y = \dfrac{1}{x^2 - 1}$

8. $y = \dfrac{x}{x^2 - 4}$

*9. $y = x^2 - 5x + 5$

10. $y = \dfrac{x^4}{x^3 - 4}$

*11. $f(x) = \dfrac{2x^2}{x^2 + x - 6}$

12. $f(x) = \dfrac{x^3}{5}$

13. $y = \dfrac{2x^2 + 3x + 1}{x^2 - 5}$

14. $y = \dfrac{2x^3 + 1}{3x(2x - 1)(4x - 3)}$

15. $y = \dfrac{2}{x - 3} + 5$

16. $f(x) = \dfrac{x^2 - 1}{2x^2 - 9x + 4}$

17. $f(x) = \dfrac{3 - x^4}{x^3 + x^2}$

18. $y = \dfrac{x^2 + 4x^3 + 6x^4}{3x^2}$

19. $y = \dfrac{x^2 - 3x - 4}{1 + 4x + 4x^2}$

20. $y = \dfrac{x^4 + 1}{1 - x^4}$

21. $y = \dfrac{9x^2 - 16}{2(3x + 4)^2}$

22. $y = \dfrac{2}{5} + \dfrac{2x}{12x^2 + 5x - 2}$

*23. $y = 2e^{x+2} + 4$

24. $f(x) = 12e^{-x}$

In Problems 25–46, determine intervals on which the function is increasing, decreasing, concave up, and concave down; relative maxima and minima; inflection points; symmetry; vertical and nonvertical asymptotes; and those intercepts that can be obtained conveniently. Then sketch the curve.

25. $y = \dfrac{3}{x}$

26. $y = \dfrac{2}{2x - 3}$

27. $y = \dfrac{x}{x - 1}$

28. $y = \dfrac{10}{\sqrt{x}}$

29. $y = x^2 + \dfrac{1}{x^2}$

30. $y = \dfrac{3x^2 - 5x - 1}{x - 2}$

*31. $y = \dfrac{1}{x^2 - 1}$

32. $y = \dfrac{1}{x^2 + 1}$

33. $y = \dfrac{1 + x}{1 - x}$

34. $y = \dfrac{1 + x}{x^2}$

*35. $y = \dfrac{x^2}{7x + 4}$

36. $y = \dfrac{x^3 + 1}{x}$

37. $y = \dfrac{9}{9x^2 - 6x - 8}$

38. $y = \dfrac{8x^2 + 3x + 1}{2x^2}$

*39. $y = \dfrac{3x + 1}{(3x - 2)^2}$

40. $y = \dfrac{3x + 1}{(6x + 5)^2}$

41. $y = \dfrac{x^2 - 1}{x^3}$

42. $y = \dfrac{3x}{(x - 2)^2}$

43. $y = x + \dfrac{1}{x + 1}$

44. $y = \dfrac{3x^4 + 1}{x^3}$

45. $y = \dfrac{-3x^2 + 2x - 5}{3x^2 - 2x - 1}$

46. $y = 3x + 2 + \dfrac{1}{3x + 2}$

47. Sketch the graph of a function f such that $f(0) = 0$, there is a horizontal asymptote $y = 1$ for $x \to \pm\infty$, there is a vertical asymptote $x = 2$, both $f'(x) < 0$ and $f''(x) < 0$ for $x < 2$, and both $f'(x) < 0$ and $f''(x) > 0$ for $x > 2$.

48. Sketch the graph of a function f such that $f(0) = 0$, there is a horizontal asymptote $y = 2$ for $x \to \pm\infty$, there is a vertical asymptote $x = -1$, both $f'(x) > 0$ and $f''(x) > 0$ for $x < -1$, and both $f'(x) > 0$ and $f''(x) < 0$ for $x > -1$.

49. Sketch the graph of a function f such that $f(0) = 0$, there is a horizontal asymptote $y = 0$ for $x \to \pm\infty$, there are vertical asymptotes $x = -1$ and $x = 2$, $f'(x) < 0$ for $x < -1$ and $-1 < x < 2$, and $f''(x) < 0$ for $x > 2$.

50. Sketch the graph of a function f such that $f(-2) = 2$, $f(0) = 0$, $f(2) = 0$, there is a horizontal asymptote $y = 1$ for $x \to \pm\infty$, there are vertical asymptotes $x = -1$ and $x = 1$, $f''(x) > 0$ for $x < -1$, and $f'(x) < 0$ for $-1 < x < 1$ and $f''(x) < 0$ for $1 < x$.

51. Purchasing Power In discussing the time pattern of purchasing, Mantell and Sing[12] use the curve

$$y = \frac{x}{a + bx}$$

as a mathematical model. Find the asymptotes for their model.

52. Sketch the graphs of $y = 6 - 3e^{-x}$ and $y = 6 + 3e^{-x}$. Show that they are asymptotic to the same line. What is the equation of this line?

53. Market for Product For a new product, the yearly number of thousand packages sold, y, t years after its introduction is predicted to be given by

$$y = f(t) = 150 - 76e^{-t}$$

Show that $y = 150$ is a horizontal asymptote for the graph. This reveals that after the product is established with consumers, the market tends to be constant.

54. Graph $y = \dfrac{x^2 - 2}{x^3 + \frac{7}{2}x^2 + 12x + 1}$. From the graph, locate any horizontal or vertical asymptotes.

55. Graph $y = \dfrac{6x^3 - 2x^2 + 6x - 1}{3x^3 - 2x^2 - 18x + 12}$. From the graph, locate any horizontal or vertical asymptotes.

56. Graph $y = \dfrac{\ln(x + 4)}{x^2 - 8x + 5}$ in the standard window. The graph suggests that there are two vertical asymptotes of the form $x = k$, where $k > 0$. Also, it appears that the graph "begins" near $x = -4$. As $x \to -4^+$, $\ln(x + 4) \to -\infty$ and $x^2 - 8x + 5 \to 53$. Thus, $\lim_{x \to 4^+} y = -\infty$. This gives the vertical asymptote $x = -4$. So, in reality, there are *three* vertical asymptotes. Use the zoom feature to make the asymptote $x = -4$ apparent from the display.

57. Graph $y = \dfrac{0.34e^{0.7x}}{4.2 + 0.71e^{0.7x}}$, where $x > 0$. From the graph, determine an equation of the horizontal asymptote by examining the y-values as $x \to \infty$. To confirm this equation algebraically, find $\lim_{x \to \infty} y$ by first dividing both the numerator and denominator by $e^{0.7x}$.

OBJECTIVE

To model situations involving maximizing or minimizing a quantity.

13.6 Applied Maxima and Minima

By using techniques from this chapter, we can solve problems that involve maximizing or minimizing a quantity. For example, we might want to maximize profit or minimize cost. The crucial part is expressing the quantity to be maximized or minimized as a function of some variable in the problem. Then we differentiate and test the resulting critical values. For this, the first-derivative test or the second-derivative test can be used, although it may be obvious from the nature of the problem whether or not a critical value represents an appropriate answer. Because our interest is in *absolute* maxima and minima, sometimes we must examine endpoints of the domain of the function. (Very often the function used to model the situation of a problem will be the restriction to a closed interval of a function that has a large natural domain. Such *real-world* limitations tend to generate endpoints.)

The aim of this example is to set up a cost function from which cost is minimized.

● **EXAMPLE 1** **Minimizing the Cost of a Fence**

For insurance purposes, a manufacturer plans to fence in a 10,800-ft² rectangular storage area adjacent to a building by using the building as one side of the enclosed area. The fencing parallel to the building faces a highway and will cost $3 per foot installed, whereas the fencing for the other two sides costs $2 per foot installed. Find the amount of each type of fence so that the total cost of the fence will be a minimum. What is the minimum cost?

Solution: As a first step in a problem like this, it is a good idea to draw a diagram that reflects the situation. In Figure 13.64, we have labeled the length of the side parallel to the building as x and the lengths of the other two sides as y, where x and y are in feet.

Since we want to minimize cost, our next step is to determine a function that gives cost. The cost obviously depends on how much fencing is along the highway and how much is along the other two sides. Along the highway the cost per foot is 3 (dollars), so the total cost of that fencing is $3x$. Similarly, along *each* of the other two sides, the cost is $2y$. Thus, the total cost of the fencing is given by the cost function

$$C = 3x + 2y + 2y$$

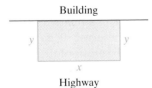

Building

Highway

FIGURE 13.64 Fencing problem of Example 1.

---

[12]L. H. Mantell and F. P. Sing, *Economics for Business Decisions* (New York: McGraw-Hill Book Company, 1972), p. 107.

that is,

$$C = 3x + 4y \tag{1}$$

We need to find the absolute minimum value of C. To do this, we use the techniques discussed in this chapter; that is, we examine C at critical values (and any endpoints) in the domain. But in order to differentiate, we need to first express C as a function of one variable only. [Equation (1) gives C as a function of *two* variables, x and y.] We can accomplish this by first finding a relationship between x and y. From the statement of the problem, we are told that the storage area, which is xy, must be 10,800:

$$xy = 10,800 \tag{2}$$

With this equation, we can express one variable (say, y) in terms of the other (x). Then, substitution into Equation (1) will give C as a function of one variable only. Solving Equation (2) for y gives

$$y = \frac{10,800}{x} \tag{3}$$

Substituting into Equation (1), we have

$$C = C(x) = 3x + 4 \left(\frac{10,800}{x} \right)$$

$$C(x) = 3x + \frac{43,200}{x} \tag{4}$$

From the physical nature of the problem, the domain of C is $x > 0$.

We now find dC/dx, set it equal to 0, and solve for x. We have

$$\frac{dC}{dx} = 3 - \frac{43,200}{x^2} \qquad \left(\frac{d}{dx}(43,200x^{-1}) = -43,200x^{-2} \right)$$

$$3 - \frac{43,200}{x^2} = 0$$

$$3 = \frac{43,200}{x^2}$$

from which it follows that

$$x^2 = \frac{43,200}{3} = 14,400$$

$$x = 120 \qquad\qquad \text{(since } x > 0\text{)}$$

Thus, 120 is the *only* critical value, and there are no endpoints to consider. To test this value, we will use the second-derivative test.

$$\frac{d^2C}{dx^2} = \frac{86,400}{x^3}$$

When $x = 120$, $d^2C/dx^2 > 0$, so we conclude that $x = 120$ gives a relative minimum. However, since 120 is the only critical value on the open interval $(0, \infty)$ and C is continuous on that interval, this relative minimum must also be an absolute minimum.

We are not done yet! The questions posed in the problem must be answered. For minimum cost, the number of feet of fencing along the highway is 120. When $x = 120$, we have, from Equation (3), $y = 10,800/120 = 90$. Therefore, the number of feet of fencing for the other two sides is $2y = 180$. It follows that 120 ft of the \$3 fencing and 180 ft of the \$2 fencing are needed. The minimum cost can be obtained from the cost function, Equation (4), and is

$$C(120) = 3x + \left. \frac{43,200}{x} \right|_{x=120} = 3(120) + \frac{43,200}{120} = 720$$

NOW WORK PROBLEM 3 ●○○

Based on Example 1, the following guide may be helpful in solving an applied maximum or minimum problem:

Guide for Solving Applied Max–Min Problems

Step 1. When appropriate, draw a diagram that reflects the information in the problem.

Step 2. Set up a function for the quantity that you want to maximize or minimize.

Step 3. Express the function in step 2 as a function of one variable only, and note the domain of this function. The domain may be implied by the nature of the problem itself.

Step 4. Find the critical values of the function. After testing each critical value, determine which one gives the absolute extreme value you are seeking. If the domain of the function includes endpoints, be sure to also examine function values at these endpoints.

Step 5. Based on the results of step 4, answer the question(s) posed in the problem.

● EXAMPLE 2 Maximizing Revenue

The demand equation for a manufacturer's product is

$$p = \frac{80 - q}{4} \quad 0 \le q \le 80$$

This example involves maximizing revenue when a demand equation is known.

where q is the number of units and p is the price per unit. At what value of q will there be maximum revenue? What is the maximum revenue?

Solution: Let r represent total revenue, which is the quantity to be maximized. Since

$$\text{revenue} = (\text{price})(\text{quantity})$$

we have

$$r = pq = \frac{80 - q}{4} \cdot q = \frac{80q - q^2}{4} = r(q)$$

where $0 \le q \le 80$. Setting $dr/dq = 0$, we obtain

$$\frac{dr}{dq} = \frac{80 - 2q}{4} = 0$$

$$80 - 2q = 0$$

$$q = 40$$

Thus, 40 is the only critical value. Now we see whether this gives a maximum. Examining the first derivative for $0 \le q < 40$, we have $dr/dq > 0$, so r is increasing. If $q > 40$, then $dr/dq < 0$, so r is decreasing. Because to the left of 40 we have r increasing, and to the right r is decreasing, we conclude that $q = 40$ gives the *absolute* maximum revenue, namely

$$r(40) = (80(40) - (40)^2)/4 = 400$$

NOW WORK PROBLEM 7 ●●

● EXAMPLE 3 Minimizing Average Cost

A manufacturer's total-cost function is given by

This example involves minimizing average cost when the cost function is known.

$$c = c(q) = \frac{q^2}{4} + 3q + 400$$

where c is the total cost of producing q units. At what level of output will average cost per unit be a minimum? What is this minimum?

Solution: The quantity to be minimized is the average cost $\bar{c}$. The average-cost function is

$$\bar{c} = \bar{c}(q) = \frac{c}{q} = \frac{\frac{q^2}{4} + 3q + 400}{q} = \frac{q}{4} + 3 + \frac{400}{q} \tag{5}$$

Here q must be positive. To minimize $\bar{c}$, we differentiate:

$$\frac{d\bar{c}}{dq} = \frac{1}{4} - \frac{400}{q^2} = \frac{q^2 - 1600}{4q^2}$$

To get the critical values, we solve $d\bar{c}/dq = 0$:

$$q^2 - 1600 = 0$$

$$(q - 40)(q + 40) = 0$$

$$q = 40 \qquad \text{(since } q > 0\text{)}$$

To determine whether this level of output gives a relative minimum, we will use the second-derivative test. We have

$$\frac{d^2\bar{c}}{dq^2} = \frac{800}{q^3}$$

which is positive for $q = 40$. Thus, $\bar{c}$ has a relative minimum when $q = 40$. We note that $\bar{c}$ is continuous for $q > 0$. Since $q = 40$ is the only relative extremum, we conclude that this relative minimum is indeed an absolute minimum. Substituting $q = 40$ in Equation (5) gives the minimum average cost $\bar{c}(40) = \frac{40}{4} + 3 + \frac{400}{40} = 23$.

NOW WORK PROBLEM 5 ●●

● EXAMPLE 4 **Maximization Applied to Enzymes**

This example is a biological application involving maximizing the rate at which an enzyme is formed. The equation involved is a literal equation.

An enzyme is a protein that acts as a catalyst for increasing the rate of a chemical reaction that occurs in cells. In a certain reaction, an enzyme is converted to another enzyme called the product. The product acts as a catalyst for its own formation. The rate R at which the product is formed (with respect to time) is given by

$$R = kp(l - p)$$

where l is the total initial amount of both enzymes, p is the amount of the product enzyme, and k is a positive constant. For what value of p will R be a maximum?

Solution: We can write $R = k(pl - p^2)$. Setting $dR/dp = 0$ and solving for p gives

$$\frac{dR}{dp} = k(l - 2p) = 0$$

$$p = \frac{l}{2}$$

Now, $d^2R/dp^2 = -2k$. Since $k > 0$, the second derivative is always negative. Hence, $p = l/2$ gives a relative maximum. Moreover, since R is a continuous function of p, we conclude that we indeed have an absolute maximum when $p = l/2$.

●●

Calculus can be applied to inventory decisions, as the following example shows.

● EXAMPLE 5 **Economic Lot Size**

This example involves determining the number of units in a production run in order to minimize certain costs.

*A company annually produces and sells 10,000 units of a product. Sales are uniformly distributed throughout the year. The company wishes to determine the number of units to be manufactured in each production run in order to minimize total annual setup costs and carrying costs. The same number of units is produced in each run. This number is referred to as the **economic lot size or economic order quantity**. The production cost* *of each unit is $20, and carrying costs (insurance, interest, storage, etc.) are estimated to be 10% of the value of the average inventory. Setup costs per production run are $40. Find the economic lot size.*

Solution: Let q be the number of units in a production run. Since sales are distributed at a uniform rate, we will assume that inventory varies uniformly from q to 0 between production runs. Thus, we take the average inventory to be $q/2$ units. The production costs are $20 per unit, so the value of the average inventory is $20(q/2)$. Carrying costs are 10% of this value:

$$0.10(20)\left(\frac{q}{2}\right)$$

The number of production runs per year is $10,000/q$. Hence, the total setup costs are

$$40\left(\frac{10,000}{q}\right)$$

Therefore, the total of the annual carrying costs and setup costs is given by

$$C = 0.10(20)\left(\frac{q}{2}\right) + 40\left(\frac{10,000}{q}\right)$$

$$= q + \frac{400,000}{q} \qquad\qquad (q > 0)$$

$$\frac{dC}{dq} = 1 - \frac{400,000}{q^2} = \frac{q^2 - 400,000}{q^2}$$

Setting $dC/dq = 0$, we get

$$q^2 = 400,000$$

Since $q > 0$,

$$q = \sqrt{400,000} = 200\sqrt{10} \approx 632.5$$

To determine whether this value of q minimizes C, we will examine the first derivative. If $0 < q < \sqrt{400,000}$, then $dC/dq < 0$. If $q > \sqrt{400,000}$, then $dC/dq > 0$. We conclude that there is an *absolute* minimum at $q = 632.5$. The number of production runs is $10,000/632.5 \approx 15.8$. For practical purposes, there would be 16 lots, each having the economic lot size of 625 units.

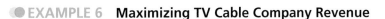

NOW WORK PROBLEM 29

● EXAMPLE 6 Maximizing TV Cable Company Revenue

The Vista TV Cable Co. currently has 100,000 subscribers who are each paying a monthly rate of $40. A survey reveals that there will be 1000 more subscribers for each $0.25 decrease in the rate. At what rate will maximum revenue be obtained, and how many subscribers will there be at this rate?

The aim of this example is to set up a revenue function from which revenue is maximized over a closed interval.

Solution: Let x be the number of $0.25 decreases. The monthly rate is then $40 - 0.25x$, where $0 \leq x \leq 160$ (the rate cannot be negative), and the number of *new* subscribers is $1000x$. Thus, the total number of subscribers is $100,000 + 1000x$. We want to maximize the revenue, which is given by

$$r = (\text{number of subscribers})(\text{rate per subscriber})$$

$$= (100,000 + 1000x)(40 - 0.25x)$$

$$= 1000(100 + x)(40 - 0.25x)$$

$$= 1000(4000 + 15x - 0.25x^2)$$

Setting $r' = 0$ and solving for x, we have

$$r' = 1000(15 - 0.5x) = 0$$

$$x = 30$$

Since the domain of r is the closed interval $[0, 160]$, the absolute maximum value of r must occur at $x = 30$ or at one of the endpoints of the interval. We now compute r at these three points:

$$r(0) = 1000(4000 + 15(0) - 0.25(0)^{2)} = 4,000,000$$

$$r(30) = 1000(4000 + 15(30) - 0.25(30)^2) = 4,225,000$$

$$r(160) = 1000(4000 + 15(160) - 0.25(160)^2 = 0$$

Accordingly, the maximum revenue occurs when $x = 30$. This corresponds to thirty $0.25 decreases, for a total decrease of $7.50; that is, the monthly rate is $40 - $7.50 = $32.50. The number of subscribers at that rate is $100,000 + 30(1000) = 130,000$.

NOW WORK PROBLEM 19

EXAMPLE 7 Maximizing the Number of Recipients of Health-Care Benefits

An article in a sociology journal stated that if a particular health-care program for the elderly were initiated, then t years after its start, n thousand elderly people would receive direct benefits, where

$$n = \frac{t^3}{3} - 6t^2 + 32t \quad 0 \le t \le 12$$

For what value of t does the maximum number receive benefits?

Solution: Setting $dn/dt = 0$, we have

$$\frac{dn}{dt} = t^2 - 12t + 32 = 0$$

$$(t - 4)(t - 8) = 0$$

$$t = 4 \quad \text{or} \quad t = 8$$

Since the domain of n is the closed interval $[0, 12]$, the absolute maximum value of n must occur at $t = 0, 4, 8,$ or 12:

$$n(0) = \frac{0^3}{3} - 6(0^2)0 + 32(0) = 0$$

$$n(4) = \frac{4^3}{3} - 6(4^2) + 32(4) = \frac{160}{3}$$

$$n(8) = \frac{8^3}{3} - 6(8^2) + 32(8) = \frac{128}{3}$$

$$n(12) = \frac{12^3}{3} - 6(12)^2 + 32(12) = \frac{288}{3} = 96$$

Thus, an absolute maximum occurs when $t = 12$. A graph of the function is given in Figure 13.65.

NOW WORK PROBLEM 15

In the next example, we use the word *monopolist*. Under a situation of monopoly, there is only one seller of a product for which there are no similar substitutes, and the seller—that is, the monopolist—controls the market. By considering the demand equation for the product, the monopolist may set the price (or volume of output) so that maximum profit will be obtained.

EXAMPLE 8 Profit Maximization

Suppose that the demand equation for a monopolist's product is $p = 400 - 2q$ and the average-cost function is $\bar{c} = 0.2q + 4 + (400/q)$, where q is number of units, and both p and $\bar{c}$ are expressed in dollars per unit.

Here we maximize a function over a closed interval.

FIGURE 13.65 Graph of $n = \dfrac{t^3}{3} - 6t^2 + 32t$ on $[0, 12]$.

CAUTION

The preceding example illustrates that you should not ignore endpoints when finding absolute extrema on a closed interval.

This example involves maximizing profit when the demand and average-cost functions are known. In the last part, a tax is imposed on the monopolist, and a new profit function is analyzed.

a. *Determine the level of output at which profit is maximized.*

b. *Determine the price at which maximum profit occurs.*

c. *Determine the maximum profit.*

d. *If, as a regulatory device, the government imposes a tax of $22 per unit on the monopolist, what is the new price for profit maximization?*

Solution: We know that

$$\text{profit} = \text{total revenue} - \text{total cost}$$

Since total revenue r and total cost c are given by

$$r = pq = 400q - 2q^2$$

and

$$c = q\bar{c} = 0.2q^2 + 4q + 400$$

the profit is

$$P = r - c = 400q - 2q^2 - (0.2q^2 + 4q + 400)$$

so that

$$P(q) = 396q - 2.2q^2 - 400 \quad \text{for } q > 0 \tag{6}$$

a. To maximize profit, we set $dP/dq = 0$:

$$\frac{dP}{dq} = 396 - 4.4q = 0$$

$$q = 90$$

Now, $d^2P/dq^2 = -4.4$ is always negative, so it is negative at the critical value $q = 90$. By the second-derivative test, then, there is a relative maximum there. Since $q = 90$ is the only critical value on $(0, \infty)$, we must have an absolute maximum there.

b. The price at which maximum profit occurs is obtained by setting $q = 90$ in the demand equation:

$$p = 400 - 2(90) = 220$$

c. The maximum profit is obtained by evaluating $P(90)$. We have

$$P(90) = 396(90) - 2.2(90)^2 - 400 = 17{,}420$$

d. The tax of $22 per unit means that for q units, the total cost increases by $22q$. The new cost function is $c_1 = 0.2q^2 + 4q + 400 + 22q$, and the new profit is given by

$$P_1 = 400q - 2q^2 - (0.2q^2 + 4q + 400 + 22q)$$

$$= 374q - 2.2q^2 - 400$$

Setting $dP_1/dq = 0$ gives

$$\frac{dP_1}{dq} = 374 - 4.4q = 0$$

$$q = 85$$

Since $d^2P_1/dq^2 = -4.4 < 0$, we conclude that, to maximize profit, the monopolist must restrict output to 85 units at a higher price of $p_1 = 400 - 2(85) = \$230$. Since this price is only $10 more than before, only part of the tax has been shifted to the consumer, and the monopolist must bear the cost of the balance. The profit now is $15,495, which is less than the former profit.

NOW WORK PROBLEM 13

This discussion leads to the economic principle that when profit is maximum, marginal revenue is equal to marginal cost.

We conclude this section by using calculus to develop an important principle in economics. Suppose $p = f(q)$ is the demand function for a firm's product, where p is price per unit and q is the number of units produced and sold. Then the total revenue is given by $r = qp = qf(q)$, which is a function of q. Let the total cost of producing q units be given by the cost function $c = g(q)$. Thus, the total profit, which is total revenue minus total cost, is also a function of q, namely,

$$P(q) = r - c = qf(q) - g(q)$$

Let us consider the most profitable output for the firm. Ignoring special cases, we know that profit is maximized when $dP/dq = 0$ and $d^2P/dq^2 < 0$. We have

$$\frac{dP}{dq} = \frac{d}{dq}(r - c) = \frac{dr}{dq} - \frac{dc}{dq}$$

Consequently, $dP/dq = 0$ when

$$\frac{dr}{dq} = \frac{dc}{dq}$$

That is, at the level of maximum profit, the slope of the tangent to the total-revenue curve must equal the slope of the tangent to the total-cost curve (Figure 13.66). But dr/dq is the marginal revenue MR, and dc/dq is the marginal cost MC. Thus, under typical conditions, to maximize profit, it is necessary that

$$MR = MC$$

For this to indeed correspond to a maximum, it is necessary that $d^2P/dq^2 < 0$:

$$\frac{d^2P}{dq^2} = \frac{d^2}{dq^2}(r - c) = \frac{d^2r}{dq^2} - \frac{d^2c}{dq^2} < 0 \quad \text{equivalently} \quad \frac{d^2r}{dq^2} < \frac{d^2c}{dq^2}$$

That is, when $MR = MC$, in order to ensure maximum profit, the slope of the marginal-revenue curve must be less than the slope of the marginal-cost curve.

The condition that $d^2P/dq^2 < 0$ when $dP/dq = 0$ can be viewed another way. Equivalently, to have $MR = MC$ correspond to a maximum, dP/dq must go from $+$ to $-$; that is, it must go from $dr/dq - dc/dq > 0$ to $dr/dq - dc/dq < 0$. Hence, as output increases, we must have $MR > MC$ and then $MR < MC$. This means that at the point q_1 of maximum profit, *the marginal-cost curve must cut the marginal-revenue curve from below* (Figure 13.67). For production up to q_1, the revenue from additional output would be greater than the cost of such output, and the total profit would increase. For output beyond q_1, $MC > MR$, and each unit of output would add more to total costs than to total revenue. Hence, total profits would decline.

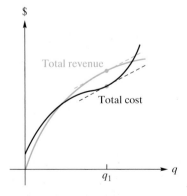

FIGURE 13.66 At maximum profit, marginal revenue equals marginal cost.

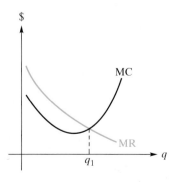

FIGURE 13.67 At maximum profit, the marginal-cost curve cuts the marginal-revenue curve from below.

Problems 13.6

In this set of problems, unless otherwise specified, p is price per unit (in dollars) and q is output per unit of time. Fixed costs refer to costs that remain constant at all levels of production during a given time period. (An example is rent.)

1. Find two numbers whose sum is 82 and whose product is as big as possible.

2. Find two nonnegative numbers whose sum is 20 and for which the product of twice one number and the square of the other number will be a maximum.

*3.** **Fencing**   A company has set aside $9000 to fence in a rectangular portion of land adjacent to a stream by using the stream for one side of the enclosed area. The cost of the fencing parallel to the stream is $15 per foot installed, and the fencing for the remaining two sides costs $9 per foot installed. Find the dimensions of the maximum enclosed area.

4. **Fencing**   The owner of the Laurel Nursery Garden Center wants to fence in 1200 ft^2 of land in a rectangular plot to be used for different types of shrubs. The plot is to be divided into five equal plots with four fences parallel to the same pair of sides, as shown in Figure 13.68. What is the least number of feet of fence needed?

FIGURE 13.68   Diagram for Problem 4.

*5.** **Average Cost**   A manufacturer finds that the total cost c of producing a product is given by the cost function

$$c = 0.05q^2 + 5q + 500$$

At what level of output will average cost per unit be a minimum?

6. **Automobile Expense**   The cost per hour (in dollars) of operating an automobile is given by

$$C = 0.12s - 0.0012s^2 + 0.08 \qquad 0 \le s \le 60$$

where s is the speed in miles per hour. At what speed is the cost per hour a minimum?

*7.** **Revenue**   The demand equation for a monopolist's product is

$$p = -5q + 30$$

At what price will revenue be maximized?

8. **Revenue**   Suppose that the demand function for a monopolist's product is of the form

$$q = Ae^{-Bp}$$

for positive constants A and B. In terms of A and B, find the value of p for which maximum revenue is obtained. Can you explain why your answer does not depend on A?

9. **Weight Gain**   A group of biologists studied the nutritional effects on rats that were fed a diet containing 10% protein.[13] The protein consisted of yeast and cottonseed flour. By varying the percent p of yeast in the protein mix, the group found that the (average) weight gain (in grams) of a rat over a period of time was

$$f(p) = 160 - p - \frac{900}{p + 10} \qquad 0 \le p \le 100$$

Find (a) the maximum weight gain and (b) the minimum weight gain.

10. **Drug Dose**   The severity of the reaction of the human body to an initial dose D of a drug is given by[14]

$$R = f(D) = D^2 \left(\frac{C}{2} - \frac{D}{3} \right)$$

where the constant C denotes the maximum amount of the drug that may be given. Show that R has a maximum *rate of change* when $D = C/2$.

11. **Profit**   For a monopolist's product, the demand function is

$$p = 85 - 0.05q$$

and the cost function is

$$c = 600 + 35q$$

At what level of output will profit be maximized? At what price does this occur, and what is the profit?

12. **Profit**   For a monopolist, the cost per unit of producing a product is $3, and the demand equation is

$$p = \frac{10}{\sqrt{q}}$$

What price will give the greatest profit?

---

[13]Adapted from R. Bressani, "The Use of Yeast in Human Foods," in *Single-Cell Protein*, ed. R. I. Mateles and S. R. Tannenbaum (Cambridge, MA: MIT Press, 1968).

[14]R. M. Thrall, J. A. Mortimer. K. R. Rebman, and R. F. Baum, eds., *Some Mathematical Models in Biology*, rev. ed., Report No. 40241-R-7. Prepared at University of Michigan, 1967.

*13. **Profit**   For a monopolist's product, the demand equation is

$$p = 42 - 4q$$

and the average-cost function is

$$\bar{c} = 2 + \frac{80}{q}$$

Find the profit-maximizing price.

14. **Profit**   For a monopolist's product, the demand function is

$$p = \frac{40}{\sqrt{q}}$$

and the average-cost function is

$$\bar{c} = \frac{1}{3} + \frac{2000}{q}$$

Find the profit-maximizing price and output. At this level, show that marginal revenue is equal to marginal cost.

*15. **Profit**   A manufacturer can produce at most 120 units of a certain product each year. The demand equation for the product is

$$p = q^2 - 100q + 3200$$

and the manufacturer's average-cost function is

$$\bar{c} = \frac{2}{3}q^2 - 40q + \frac{10,000}{q}$$

Determine the profit-maximizing output q and the corresponding maximum profit.

16. **Cost**   A manufacturer has determined that, for a certain product, the average cost (in dollars per unit) is given by

$$\bar{c} = 2q^2 - 42q + 228 + \frac{210}{q}$$

where $3 \leq q \leq 12$.

(a) At what level within the interval [3, 12] should production be fixed in order to minimize total cost? What is the minimum total cost?

(b) If production were required to lie within the interval [7, 12], what value of q would minimize total cost?

17. **Profit**   For XYZ Manufacturing Co., total fixed costs are $1200, material and labor costs combined are $2 per unit, and the demand equation is

$$p = \frac{100}{\sqrt{q}}$$

What level of output will maximize profit? Show that this occurs when marginal revenue is equal to marginal cost. What is the price at profit maximization?

18. **Revenue**   A real-estate firm owns 100 garden-type apartments. At $400 per month, each apartment can be rented. However, for each $10-per-month increase, there will be two vacancies with no possibility of filling them. What rent per apartment will maximize monthly revenue?

*19. **Revenue**   A TV cable company has 4800 subscribers who are each paying $18 per month. It can get 150 more subscribers for each $0.50 decrease in the monthly fee. What rate will yield maximum revenue, and what will this revenue be?

20. **Profit**   A manufacturer of a product finds that, for the first 600 units that are produced and sold, the profit is $40 per unit. The profit on each of the units beyond 600 is decreased by $0.05 times the number of additional units produced. For example, the total profit when 602 units are produced and sold is $600(40) + 2(39.90)$. What level of output will maximize profit?

21. **Container Design**   A container manufacturer is designing a rectangular box, open at the top and with a square base, that is to have a volume of 32 ft³. If the box is to require the least amount of material, what must be its dimensions? (See Figure 13.69.)

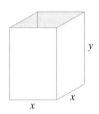

FIGURE 13.69
Open-top box
for Problems 21
and 22.

22. **Container Design**   An open-top box with a square base is to be constructed from 192 ft² of material. What should be the dimensions of the box if the volume is to be a maximum? What is the maximum volume? (See Figure 13.69.)

23. **Container Design**   An open box is to be made by cutting equal squares from each corner of a L-inch-square piece of cardboard and then folding up the sides. Find the length of the side of the square (in terms of L) that must be cut out if the volume of the box is to be maximized. What is the maximum volume? (See Figure 13.70.)

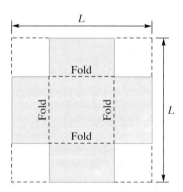

FIGURE 13.70   Box for
Problem 23.

24. **Poster Design**   A rectangular cardboard poster is to have 240 in² for printed matter. It is to have a 5-in. margin at the top and bottom and a 3-in. margin on each side. Find the dimensions of the poster so that the amount of cardboard used is minimized. (See Figure 13.71.)

(*Hint:* First find the values of x and y in Figure 13.71 that minimize the amount of cardboard.)

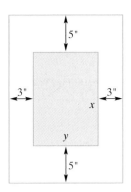

FIGURE 13.71 Poster for Problem 24.

25. Container Design A cylindrical can, open at the top, is to have a fixed volume of K. Show that if the least amount of material is to be used, then both the radius and height are equal to $\sqrt[3]{K/\pi}$. (See Figure 13.72.)

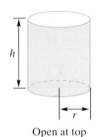

$$\text{Volume} = \pi r^2 h$$
$$\text{Surface area} = 2\pi rh + \pi r^2$$

Open at top

FIGURE 13.72 Can for Problems 25 and 26.

26. Container Design A cylindrical can, open at the top, is to be made from a fixed amount of material, K. If the volume is to be a maximum, show that both the radius and height are equal to $\sqrt{K/(3\pi)}$. (See Figure 13.72.)

27. Profit The demand equation for a monopolist's product is

$$p = 600 - 2q$$

and the total-cost function is

$$c = 0.2q^2 + 28q + 200$$

Find the profit-maximizing output and price, and determine the corresponding profit. If the government were to impose a tax of $22 per unit on the manufacturer, what would be the new profit-maximizing output and price? What is the profit now?

28. Profit Use the *original* data in Problem 27, and assume that the government imposes a license fee of $1000 on the manufacturer. This is a lump-sum amount without regard to output. Show that the profit-maximizing price and output remain the same. Show, however, that there will be less profit.

*29. Economic Lot Size A manufacturer has to produce 1000 units annually of a product that is sold at a uniform rate during the year. The production cost of each unit is $10, and carrying costs (insurance, interest, storage, etc.) are estimated to be 12.8% of the value of average inventory. Setup costs per production run are $40. Find the economic lot size.

30. Profit For a monopolist's product, the cost function is

$$c = 0.004q^3 + 20q + 5000$$

and the demand function is

$$p = 450 - 4q$$

Find the profit-maximizing output.

31. Workshop Attendance Imperial Educational Services (I.E.S.) is considering offering a workshop in resource allocation to key personnel at Acme Corp. To make the offering economically feasible, I.E.S. feels that at least 30 persons must attend at a cost of $50 each. Moreover, I.E.S. will agree to reduce the charge for *everybody* by $1.25 for each person over the 30 who attends. How many people should be in the group for I.E.S. to maximize revenue? Assume that the maximum allowable number in the group is 40.

32. Cost of Leasing Motor The Kiddie Toy Company plans to lease an electric motor that will be used 80,000 horsepower-hours per year in manufacturing. One horsepower-hour is the work done in 1 hour by a 1-horsepower motor. The annual cost to lease a suitable motor is $200, plus $0.40 per horsepower. The cost per horsepower-hour of operating the motor is $0.008/N$, where N is the horsepower. What size motor, in horsepower, should be leased in order to minimize cost?

33. Transportation Cost The cost of operating a truck on a thruway (excluding the salary of the driver) is

$$0.165 + \frac{s}{200}$$

dollars per mile, where s is the (steady) speed of the truck in miles per hour. The truck driver's salary is $18 per hour. At what speed should the truck driver operate the truck to make a 700-mile trip most economical?

34. Cost For a manufacturer, the cost of making a part is $30 per unit for labor and $10 per unit for materials; overhead is fixed at $20,000 per week. If more than 5000 units are made each week, labor is $45 per unit for those units in excess of 5000. At what level of production will average cost per unit be a minimum?

35. Profit Ms. Jones owns a small insurance agency that sells policies for a large insurance company. For each policy sold, Ms. Jones, who does not sell policies herself, is paid a commission of $50 by the insurance company. From previous experience, Ms. Jones has determined that, when she employs m salespeople,

$$q = m^3 - 15m^2 + 92m$$

policies can be sold per week. She pays each of the m salespeople a salary of $1000 per week, and her weekly fixed cost is $3000. Current office facilities can accommodate at most eight salespeople. Determine the number of salespeople that Ms. Jones should hire to maximize her weekly profit. What is the corresponding maximum profit?

Insurance Policy

36. Profit A manufacturing company sells high-quality jackets through a chain of specialty shops. The demand equation for these jackets is

$$p = 400 - 50q$$

where p is the selling price (in dollars per jacket) and q is the demand (in thousands of jackets). If this company's marginal-cost function is given by

$$\frac{dc}{dq} = \frac{800}{q+5}$$

show that there is a maximum profit, and determine the number of jackets that must be sold to obtain this maximum profit.

37. Chemical Production Each day, a firm makes x tons of chemical A ($x \le 4$) and

$$y = \frac{24 - 6x}{5 - x}$$

tons of chemical B. The profit on chemical A is $2000 per ton, and on B it is $1000 per ton. How much of chemical A should be produced per day to maximize profit? Answer the same question if the profit on A is P per ton and that on B is $P/2$ per ton.

38. Rate of Return To erect an office building, fixed costs are $1.44 million and include land, architect's fees, a basement, a foundation, and so on. If x floors are constructed, the cost (excluding fixed costs) is

$$c = 10x[120,000 + 3000(x - 1)]$$

The revenue per month is $60,000 per floor. How many floors will yield a maximum rate of return on investment? (Rate of return = total revenue/total cost.)

39. Gait and Power Output of an Animal In a model by Smith,[15] the power output of an animal at a given speed as a

function of its movement or *gait j*, is found to be

$$P(j) = Aj\frac{L^4}{V} + B\frac{V^3 L^2}{1 + j}$$

where A and B are constants, j is a measure of the "jumpiness" of the gait, L is a constant representing linear dimension, and V is a constant forward speed.

Assume that P is a minimum when $dP/dj = 0$. Show that when this occurs,

$$(1 + j)^2 = \frac{BV^4}{AL^2}$$

As a passing comment, Smith indicates that "at top speed, j is zero for an elephant, 0.3 for a horse, and 1 for a greyhound, approximately."

40. Traffic Flow In a model of traffic flow on a lane of a freeway, the number of cars the lane can carry per unit time is given by[16]

$$N = \frac{-2a}{-2at_r + v - \dfrac{2al}{v}}$$

where a is the acceleration of a car when stopping ($a < 0$), t_r is the reaction time to begin braking, v is the average speed of the cars, and l is the length of a car. Assume that a, t_r, and l are constant. To find how many cars a lane can carry at most, we want to find the speed v that maximizes N. To maximize N, it suffices to minimize the denominator

$$-2at_r + v - \frac{2al}{v}$$

(a) Find the value of v that minimizes the denominator.

(b) Evaluate your answer in part (a) when $a = -19.6$ (ft/s²), $l = 20$ (ft), and $t_r = 0.5$ (s). Your answer will be in feet per second.

(c) Find the corresponding value of N to one decimal place. Your answer will be in cars per second. Convert your answer to cars per hour.

(d) Find the relative change in N that results when l is reduced from 20 ft to 15 ft, for the maximizing value of v.

41. Average Cost During the Christmas season, a promotional company purchases cheap red felt stockings, glues fake white fur and sequins onto them, and packages them for distribution. The total cost of producing q cases of stockings is given by

$$c = 3q^2 + 50q - 18q \ln q + 120$$

Find the number of cases that should be processed in order to minimize the average cost per case. Determine (to two decimal places) this minimum average cost.

[15]J. M. Smith, *Mathematical Ideas in Biology* (London: Cambridge University Press, 1968).

[16]J. I. Shonle, *Environmental Applications of General Physics* (Reading, MA: Addison-Wesley Publishing Co., 1975).

42. Profit A monopolist's demand equation is given by

$$p = q^2 - 20q + 160$$

where p is the selling price (in thousands of dollars) per ton when q tons of product are sold. Suppose that fixed cost is $50,000 and that each ton costs $30,000 to produce. If current equipment has a maximum production capacity of 12 tons, use the graph of the profit function to determine at what production level the maximum profit occurs. Find the corresponding maximum profit and selling price per ton.

13.7 Review

Important Terms and Symbols   Examples

Summary

Calculus is a great aid in sketching the graph of a function. The first derivative is used to determine when a function is increasing or decreasing and to locate relative maxima and minima. If $f'(x)$ is positive throughout an interval, then over that interval, f is increasing and its graph rises (from left to right). If $f'(x)$ is negative throughout an interval, then over that interval, f is decreasing and its graph is falling.

A point $(a, f(a))$ on the graph at which $f'(a)$ is 0 or is not defined is a candidate for a relative extremum, and a is called a critical value. For a relative extremum to occur at a, the first derivative must change sign around a. The following procedure is the first-derivative test for the relative extrema of $y = f(x)$:

First-Derivative Test for Relative Extrema

Step 1. Find $f'(x)$.

Step 2. Determine all values a where $f'(a) = 0$ or $f'(a)$ is not defined.

Step 3. On the intervals defined by the values in step 2, determine whether f is increasing ($f'(x) > 0$) or decreasing ($f'(x) < 0$).

Step 4. For each critical value a at which f is continuous, determine whether $f'(x)$ changes sign as x increases through a. There is a relative maximum at a if $f'(x)$ changes from $+$ to $-$, and a relative minimum if $f'(x)$ changes from $-$ to $+$. If $f'(x)$ does not change sign, there is no relative extremum at a.

Under certain conditions, a function is guaranteed to have absolute extrema. The extreme-value theorem states that if f is continuous on a closed interval, then f has an absolute maximum value and an absolute minimum value over the interval. To locate absolute extrema, the following procedure can be used:

Procedure to Find Absolute Extrema for a Function f Continuous on $[a, b]$

Step 1. Find the critical values of f.

Step 2. Evaluate $f(x)$ at the endpoints a and b and at the critical values in (a, b).

Step 3. The maximum value of f is the greatest of the values found in step 2. The minimum value of f is the least of the values found in step 2.

The second derivative is used to determine concavity and points of inflection. If $f''(x) > 0$ throughout an interval, then f is concave up over that interval, and its graph bends upward. If $f''(x) < 0$ over an interval, then f is concave down throughout that interval, and its graph bends downward. A point on the graph where f is continuous and its concavity changes is an inflection point. The point $(a, f(a))$ on the graph is a possible point of inflection if either $f''(a) = 0$ or $f''(a)$ is not defined and f is continuous at a.

The second derivative also provides a means for testing certain critical values for relative extrema:

Second-Derivative Test for Relative Extrema

Suppose $f'(a) = 0$. Then

If $f''(a) < 0$, then f has a relative maximum at a.

If $f''(a) > 0$, then f has a relative minimum at a.

Asymptotes are also aids in curve sketching. Graphs "blow up" near vertical asymptotes, and they "settle" near horizontal asymptotes and oblique asymptotes. The line $x = a$ is a vertical asymptote for the graph of a function f if $\lim f(x) = \infty$ or $-\infty$ as x approaches a from the right $(x \to a^+)$ or the left $(x \to a^-)$. For the case of a rational function, $f(x) = P(x)/Q(x)$ in lowest terms, we can find vertical asymptotes without evaluating limits. If $Q(a) = 0$ but $P(a) \neq 0$, then the line $x = a$ is a vertical asymptote.

The line $y = b$ is a horizontal asymptote for the graph of a function f if at least one of the following is true:

$$\lim_{x \to \infty} f(x) = b \quad \text{or} \quad \lim_{x \to -\infty} f(x) = b$$

The line $y = mx + b$ is an oblique asymptote for the graph of a function f if at least one of the following is true:

$$\lim_{x \to \infty} (f(x) - (mx + b)) = 0 \quad \text{or} \quad \lim_{x \to -\infty} (f(x) - (mx + b)) = 0$$

In particular, a polynomial function of degree greater than 1 has no asymptotes. Moreover, a rational function whose numerator has degree greater than that of the denominator does not have a horizontal asymptote and a rational function whose numerator has degree more than one greater than that of the denominator does not have an oblique asymptote.

Applied Maxima and Minima

In applied work the importance of calculus in maximization and minimization problems can hardly be overstated. For example, in the area of economics, we can maximize profit or minimize cost. Some important relationships that are used in economics problems are the following:

$$\bar{c} = \frac{c}{q} \quad \text{average cost per unit} = \frac{\text{total cost}}{\text{quantity}}$$

$$r = pq \quad \text{revenue} = (\text{price})(\text{quantity})$$

$$P = r - c \quad \text{profit} = \text{total revenue} - \text{total cost}$$

Review Problems

Problem numbers shown in color indicate problems suggested for use as a practice chapter test.

In Problems 1–4, find horizontal and vertical asymptotes.

1. $y = \dfrac{3x^2}{x^2 - 16}$

2. $y = \dfrac{x + 3}{9x - 3x^2}$

3. $y = \dfrac{5x^2 - 3}{(3x + 2)^2}$

4. $y = \dfrac{4x + 1}{3x - 5} - \dfrac{3x + 1}{2x - 11}$

In Problems 5–8, find critical values.

5. $f(x) = \dfrac{5x^2}{3 - x^2}$

6. $f(x) = 8(x - 1)^2(x + 6)^4$

7. $f(x) = \dfrac{\sqrt[3]{x + 1}}{3 - 4x}$

8. $f(x) = \dfrac{13xe^{-5x/6}}{6x + 5}$

In Problems 9–12, find intervals on which the function is increasing or decreasing.

9. $f(x) = -\frac{5}{3}x^3 + 15x^2 + 35x + 10$

10. $f(x) = \dfrac{2x^2}{(x + 1)^2}$

11. $f(x) = \dfrac{6x^4}{x^2 - 3}$

12. $f(x) = 4\sqrt[3]{5x^3 - 7x}$

In Problems 13–18, find intervals on which the function is concave up or concave down.

13. $f(x) = x^4 - x^3 - 14$

14. $f(x) = \dfrac{x - 2}{x + 2}$

15. $f(x) = \dfrac{1}{2x - 1}$

16. $f(x) = x^3 + 2x^2 - 5x + 2$

17. $f(x) = (2x + 1)^3(3x + 2)$

18. $f(x) = (x^2 - x - 1)^2$

In Problems 19–24, test for relative extrema.

19. $f(x) = 2x^3 - 9x^2 + 12x + 7$

20. $f(x) = \dfrac{2x + 1}{x^2}$

21. $f(x) = \dfrac{x^{10}}{10} + \dfrac{x^5}{5}$

22. $f(x) = \dfrac{x^2}{x^2 - 4}$

23. $f(x) = x^{2/3}(x + 1)$

24. $f(x) = x^3(x - 2)^4$

In Problems 25–30, find the x-values where inflection points occur.

25. $y = x^5 - 5x^4 + 3x$

26. $y = \dfrac{x^2 + 2}{5x}$

27. $y = 4(3x - 5)(x^4 + 2)$

28. $y = x^2 + 2\ln(-x)$

29. $y = \dfrac{x^3}{e^x}$

30. $y = 6(x^2 - 4)^3$

In Problems 31–34, test for absolute extrema on the given interval.

31. $f(x) = 3x^4 - 4x^3, [0, 2]$

32. $f(x) = 2x^3 - 15x^2 + 36x, [0, 3]$

33. $f(x) = \dfrac{x}{(5x - 6)^2}, [-2, 0]$

34. $f(x) = (x+1)^2(x-1)^{2/3}$, [2, 3]

35. Let $f(x) = (x^2+1)e^{-x}$.

(a) Determine the values of x at which relative maxima and relative minima, if any, occur.

(b) Determine the interval(s) on which the graph of f is concave down, and find the coordinates of all points of inflection.

36. Let $f(x) = \dfrac{x}{x^2-1}$.

(a) Determine whether the graph of f is symmetric about the x-axis, y-axis, or origin.

(b) Find the interval(s) on which f is increasing.

(c) Find the coordinates of all relative extrema of f.

(d) Determine $\lim_{x \to -\infty} f(x)$ and $\lim_{x \to \infty} f(x)$.

(e) Sketch the graph of f.

(f) State the absolute minimum and absolute maximum values of $f(x)$ (if they exist).

In Problems 37–48, indicate intervals on which the function is increasing, decreasing, concave up, or concave down; indicate relative maximum points, relative minimum points, points of inflection, horizontal asymptotes, vertical asymptotes, symmetry, and those intercepts that can be obtained conveniently. Then sketch the graph.

37. $y = x^2 - 2x - 24$

38. $y = 2x^3 + 15x^2 + 36x + 9$

39. $y = x^3 - 12x + 20$

40. $y = x^4 - 4x^3 - 20x^2 + 150$

41. $y = x^3 - x$

42. $y = \dfrac{x+2}{x-3}$

43. $f(x) = \dfrac{100(x+5)}{x^2}$

44. $y = \dfrac{x^2-4}{x^2-1}$

45. $y = \dfrac{2x}{(3x-1)^3}$

46. $y = 6x^{1/3}(2x-1)$

47. $f(x) = \dfrac{e^x + e^{-x}}{2}$

48. $f(x) = 1 - \ln(x^3)$

49. Are the following statements true or false?

(a) If $f'(x_0) = 0$, then f must have a relative extremum at x_0.

(b) Since the function $f(x) = 1/x$ is decreasing on the intervals $(-\infty, 0)$ and $(0, \infty)$, it is impossible to find x_1 and x_2 in the domain of f such that $x_1 < x_2$ and $f(x_1) < f(x_2)$.

(c) On the interval $(-1, 1]$, the function $f(x) = x^4$ has an absolute maximum and an absolute minimum.

(d) If $f''(x_0) = 0$, then $(x_0, f(x_0))$ must be a point of inflection.

(e) A function f defined on the interval $(-2, 2)$ with exactly one relative maximum must have an absolute maximum.

50. An important function in probability theory is the standard normal-density function

$$f(x) = \frac{1}{\sqrt{2\pi}} e^{-x^2/2}$$

(a) Determine whether the graph of f is symmetric about the x-axis, y-axis, or origin.

(b) Find the intervals on which f is increasing and those on which it is decreasing.

(c) Find the coordinates of all relative extrema of f.

(d) Find $\lim_{x \to -\infty} f(x)$ and $\lim_{x \to \infty} f(x)$.

(e) Find the intervals on which the graph of f is concave up and those on which it is concave down.

(f) Find the coordinates of all points of inflection.

(g) Sketch the graph of f.

(h) Find all absolute extrema.

51. Marginal Cost   If $c = q^3 - 6q^2 + 12q + 18$ is a total-cost function, for what values of q is marginal cost increasing?

52. Marginal Revenue   If $r = 320q^{3/2} - 2q^2$ is the revenue function for a manufacturer's product, determine the intervals on which the marginal-revenue function is increasing.

53. Revenue Function   The demand equation for a manufacturer's product is

$$p = 200 - \frac{\sqrt{q}}{5} \quad \text{where } q > 0$$

Show that the graph of the revenue function is concave down wherever it is defined.

54. Contraception   In a model of the effect of contraception on birthrate,[17] the equation

$$R = f(x) = \frac{x}{4.4 - 3.4x} \quad 0 \le x \le 1$$

gives the proportional reduction R in the birthrate as a function of the efficiency x of a contraception method. An efficiency of 0.2 (or 20%) means that the probability of becoming pregnant is 80% of the probability of becoming pregnant without the contraceptive. Find the reduction (as a percentage) when efficiency is (a) 0, (b) 0.5, and (c) 1. Find dR/dx and d^2R/dx^2, and sketch the graph of the equation.

55. Learning and Memory   If you were to recite members of a category, such as four-legged animals, the words that you utter would probably occur in "chunks," with distinct pauses between such chunks. For example, you might say the following for the category of four-legged animals:

> dog, cat, mouse, rat,
> (pause)
> horse, donkey, mule,
> (pause)
> cow, pig, goat, lamb,
> etc.

The pauses may occur because you must mentally search for subcategories (animals around the house, beasts of burden, farm animals, etc.).

The elapsed time between onsets of successive words is called *interresponse time*. A function has been used to analyze the length of time for pauses and the chunk size

---

[17]R. K. Leik and B. F. Meeker, *Mathematical Sociology* (Englewood Cliffs, NJ: Prentice-Hall, Inc., 1975).

(number of words in a chunk).[18] This function f is such that

$$f(t) = \begin{cases} \text{the average number of words} \\ \text{that occur in succession with} \\ \text{interresponse times less than } t \end{cases}$$

The graph of f has a shape similar to that in Figure 13.73 and is best fit by a third-degree polynomial, such as

$$f(t) = At^3 + Bt^2 + Ct + D$$

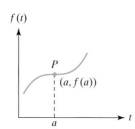

FIGURE 13.73 Diagram for Problem 55.

The point P has special meaning. It is such that the value a separates interresponse times *within* chunks from those *between* chunks. Mathematically, P is a critical point that is also a point of inflection. Assume these two conditions, and show that (a) $a = -B/(3A)$ and (b) $B^2 = 3AC$.

56. Market Penetration   In a model for the market penetration of a new product, sales S of the product at time t are given by[19]

$$S = g(t) = \frac{m(p+q)^2}{p} \left[\frac{e^{-(p+q)t}}{\left(\frac{q}{p} e^{-(p+q)t} + 1 \right)^2} \right]$$

where p, q, and m are nonzero constants.

(a) Show that

$$\frac{dS}{dt} = \frac{\frac{m}{p}(p+q)^3 e^{-(p+q)t} \left[\frac{q}{p} e^{-(p+q)t} - 1 \right]}{\left(\frac{q}{p} e^{-(p+q)t} + 1 \right)^3}$$

(b) Determine the value of t for which maximum sales occur. You may assume that S attains a maximum when $dS/dt = 0$.

In Problems 57–60, where appropriate, round your answers to two decimal places.

57. From the graph of $y = 4x^3 + 5.3x^2 - 7x + 3$, find the coordinates of all relative extrema.

58. From the graph of $f(x) = x^4 - 2x^3 + 3x - 1$, determine the absolute extrema of f over the interval $[-1, 1]$.

59. The graph of a function f has exactly one inflection point. If

$$f''(x) = \frac{x^3 + 3x + 2}{5x^2 - 2x + 4}$$

use the graph of f'' to determine the x-value of the inflection point of f.

60. Graph $y = \dfrac{3x - 6x^2}{x^3 + 4x + 1}$. From the graph, locate any horizontal or vertical asymptotes.

61. Maximization of Production   A manufacturer determined that m employees on a certain production line will produce q units per month, where

$$q = 80m^2 - 0.1m^4$$

To obtain maximum monthly production, how many employees should be assigned to the production line?

62. Revenue   The demand function for a manufacturer's product is given by $p = 100e^{-0.1q}$. For what value of q does the manufacturer maximize total revenue?

63. Revenue   The demand function for a monopolist's product is

$$p = \sqrt{500 - q}$$

If the monopolist wants to produce at least 100 units, but not more than 200 units, how many units should be produced to maximize total revenue?

64. Average Cost   If $c = 0.01q^2 + 5q + 100$ is a cost function, find the average-cost function. At what level of production q is there a minimum average cost?

65. Profit   The demand function for a monopolist's product is

$$p = 500 - 3q$$

and the average cost per unit for producing q units is

$$\bar{c} = q + 200 + \frac{1000}{q}$$

where p and $\bar{c}$ are in dollars per unit. Find the maximum profit that the monopolist can achieve.

66. Container Design   A rectangular box is to be made by cutting out equal squares from each corner of a piece of cardboard 10 in. by 16 in. and then folding up the sides. What must be the length of the side of the square cut out if the volume of the box is to be maximum?

67. Fencing   A rectangular portion of a field is to be enclosed by a fence and divided equally into three parts by two fences parallel to one pair of the sides. If a total of 800 ft of fencing is to be used, find the dimensions that will maximize the fenced area.

68. Poster Design   A rectangular poster having an area of 500 in^2 is to have a 4-in. margin at each side and at the bottom and a 6-in. margin at the top. The remainder of the poster is for printed matter. Find the dimensions of the poster so that the area for the printed matter is maximized.

69. Cost   A furniture company makes personal-computer stands. For a certain model, the total cost (in thousands of dollars) when q *hundred* stands are produced is given by

$$c = 2q^3 - 9q^2 + 12q + 20$$

[18]A. Graesser and G. Mandler, "Limited Processing Capacity Constrains the Storage of Unrelated Sets of Words and Retrieval from Natural Categories," *Human Learning and Memory*, 4, no. 1 (1978), 86–100.

[19]A. P. Hurter, Jr., A. H. Rubenstein et al., "Market Penetration by New Innovations: The Technological Literature," *Technological Forecasting and Social Change*, vol. 11 (1978), 197–221.

(a) The company is currently capable of manufacturing between 75 and 600 stands (inclusive) per week. Determine the number of stands that should be produced per week to minimize the total cost, and find the corresponding average cost per stand.

(b) Suppose that between 300 and 600 stands must be produced. How many should the company now produce in order to minimize total cost?

70. Bacteria   In a laboratory, an experimental antibacterial agent is applied to a population of 100 bacteria. Data indicate that the number of bacteria t hours after the agent is introduced is given by

$$N = \frac{12{,}100 + 110t + 100t^2}{121 + t^2}$$

For what value of t does the maximum number of bacteria in the population occur? What is this maximum number?

Mathematical Snapshot

Population Change Over Time

Now that we know how to find the derivative of a function, we might ask whether there is a way to run the process in reverse: to find a function, given its derivative. Ultimately, this is what integration (Chapters 14–15) is all about. Meanwhile, however, we can use the derivative of a function to find the function *approximately* even without knowing how to do integration.

To illustrate, suppose we wish to describe the population, over time, of a small town in a frontier area. Let use imagine that the things we know about the town are all facts about how its population, P, changes over time, t, with population measured in people and time in years:

1. Births exceed deaths, so that over the course of a year there is a 25% increase in the population before other factors are accounted for. Thus, the annual change due to the birth/death difference is $0.25P$.

2. Every year, of the travelers passing through, ten decide to stop and settle down. This contributes a constant 10 to the annual change.

3. Loneliness causes some people to leave when the town is too small for them. At the extreme, 99% of people will leave over the course of a year if they are all alone (population $= 1$). When the population is 100, 10% of the residents leave per year due to loneliness.

Assuming an exponential relationship, we write the likelihood that a given person leaves within a year due to loneliness as Ae^{-kP}, where A and k are positive constants. The numbers tell us that $Ae^{-k\cdot1} = 0.99$ and $Ae^{-k\cdot100} = 0.10$. Solving this pair of equations for A and k yields

$$k = \frac{\ln 9.9}{99} \approx 0.02316$$

and

$$A = 0.99e^{(\ln 9.9)/99} \approx 1.01319$$

And if Ae^{-kP} is the likelihood of a single person's leaving, the population change per year due to loneliness is $-P$ times that, namely $-1.01319Pe^{-0.02316P}$. (The negative sign is due to the fact that the change is downward.)

4. Crowding causes some people to leave when the town is too large for them. Nobody has a crowding problem when they are all alone (population $= 1$), but when the population is 100, 10% of the residents leave per year due to crowding.

Again assuming an exponential relationship, we write the likelihood that a given person leaves within a year due to crowding as $1 - Ae^{-kP}$. This time, the numbers tell us that $1 - Ae^{-k\cdot1} = 0$ and $1 - Ae^{-k\cdot100} = 0.10$. Solving this pair of equations for A and k yields

$$k = -\frac{\ln 0.9}{99} \approx 0.001064$$

and

$$A = e^{-(\ln 0.9)/99} \approx 1.001065$$

If $1 - Ae^{-kP}$ is the likelihood of a single person's leaving, the population change per year due to crowding is $-P$ times that, namely $-P(1 - 1.001065e^{-0.001064P})$.

The overall rate of change in the population now is the net effect of all these factors added together. In equation form,

$$\frac{dP}{dt} = 0.25P + 10 - 1.01319Pe^{-0.02316P}$$
$$-P(1 - 1.001065e^{-0.001064P})$$

Before we try to reconstruct the function $P(t)$, let us graph the derivative. On a graphing calculator, it looks as shown in Figure 13.74. Note that $\dfrac{dP}{dt}$ is depicted as a

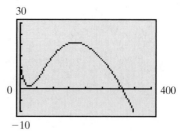

FIGURE 13.74 $\dfrac{dP}{dt}$ as a function of P.

function of P. This is a different graph from what we would get if we knew P as a function of t, found its derivative, and graphed that in the standard manner, namely as a function of t. Nonetheless, this graph reveals some significant facts. First, the derivative is positive from $P = 0$ to $P = 311$; this means that the population will have positive growth in that entire range and thus can be expected to grow from nothing into a substantial community.

Growth does fall to near zero at around $P = 30$. Apparently departures due to loneliness nearly bring growth to a halt when the population is still small. But once the town has grown through that phase, its size increases steadily, at one point (around $P = 170$) adding 21 people per year.

Eventually, departures due to crowding start to take a toll. Above 312 the derivative is negative. This means that if the population ever fluctuated above 312, population losses would shrink it back down to that level. In short, the population of this town stabilizes at 311 or 312—not exactly a city, but this is, after all, a frontier environment.

If we now wish to graph the town's population as a function of time, here is how we do this: We approximate the graph with a string of line segments, each of which has a slope given by the expression we obtained for dP/dt. We begin with a known time and population and calculate the initial slope. Let us grow the town from nothing, setting $t = 0$ at $P = 0$. Then $\dfrac{dP}{dt} = 10$. Now we advance the clock forward by a convenient time interval—let us choose 1 year—and, since the slope at $(0, 0)$ equals 10, we increase the population from 0 to 10. The new values for t and P are 1 and 10, respectively, so we draw a line segment from $(0, 0)$ to $(1, 10)$. Now, with $t = 1$ and $P = 10$, we recalculate the slope and go through the same steps again, and we repeat this process until we have drawn as much of the curve as we want to see.

Obviously, this would be extremely tedious to do by hand. On a graphing calculator, however, we can use the programming and line-drawing features. For a TI-83 Plus, the following program does the job nicely, after the expression for $\dfrac{dP}{dt}$ is entered as Y_1 (keeping P as the variable):

```
PROGRAM:POPLTN
:Input "P?",P
:Input "T?", T
:ClrDraw
:T → S
:For(I, S + 1, S + 55)
:Line(T,P,I,P + Y₁)
:I → T
:(P + Y₁) → P
:End
```

Deselect the function Y_1. Set the graphing window to display the coordinate plane from 0 to 55 horizontally and from 0 to 350 vertically. Then run the program and, at the prompt, give initial values for P and t. The program will draw 55 line segments, enough to take the population to its final size from $P = 0, t = 0$. The result is shown in Figure 13.75.

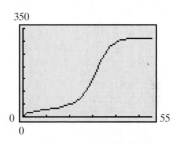

350

0

0

55

FIGURE 13.75 P as a function of t.

Problems

1. What information does Figure 13.75 give that is not evident from Figure 13.74?

2. What happens when an initial value of 450 is selected for P? (The display should be adjusted to run from 0 to 500 vertically.) Does this seem right?

3. Why is this procedure for obtaining a graph of $P(t)$ only approximate? How could the approximation be improved?

INTEGRATION

Mathematical Snapshot   Delivered Price

Anyone who runs a business knows the need for accurate cost estimates. When jobs are individually contracted, determining how much a job will cost is generally the first step in deciding how much to bid.

For example, a painter must determine how much paint a job will take. Since a gallon of paint will cover a certain number of square feet, the key is to determine the area of the surfaces to be painted. Normally, even this requires only simple arithmetic—walls and ceilings are rectangular, and so total area is a sum of products of base and height.

But not all area calculations are as simple. Suppose, for instance, that the bridge shown below must be sandblasted to remove accumulated soot. How would the contractor who charges for sandblasting by the square foot calculate the area of the vertical face on each side of the bridge?

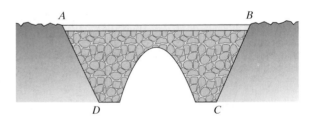

The area could be estimated as perhaps three-quarters of the area of the trapezoid formed by points *A*, *B*, *C*, and *D*. But a more accurate calculation—which might be desirable if the bid were for dozens of bridges of the same dimensions (as along a stretch of railroad)—would require a more refined approach.

If the shape of the bridge's arch can be described mathematically by a function, the contractor could use the method introduced in this chapter: integration. Integration has many applications, the simplest of which is finding areas of regions bounded by curves. Other applications include calculating the total deflection of a beam due to bending stress, calculating the distance traveled underwater by a submarine, and calculating the electricity bill for a company that consumes power at differing rates over the course of a month. Chapters 11–13 dealt with differential calculus. We differentiated a function and obtained another function, its derivative. *Integral calculus* is concerned with the reverse process: We are given the derivative of a function and must find the original function. The need for doing this arises in a natural way. For example, we might have a marginal-revenue function and want to find the revenue function from it. Integral calculus also involves a concept that allows us to take the limit of a special kind of sum as the number of terms in the sum becomes infinite. This is the real power of integral calculus! With such a notion, we can find the area of a region that cannot be found by any other convenient method.

To define the differential, interpret it geometrically, and use it in approximations. Also, to restate the reciprocal relationship between dx/dy and dy/dx.

14.1 Differentials

We will soon give you a reason for using the symbol dy/dx to denote the derivative of y with respect to x. To do this, we introduce the notion of the *differential* of a function.

DEFINITION

Let $y = f(x)$ be a differentiable function of x, and let Δx denote a change in x, where Δx can be any real number. Then the **differential** of y, denoted dy or $d(f(x))$, is given by

$$dy = f'(x)\,\Delta x$$

Note that dy depends on two variables, namely, x and Δx. In fact, dy is a function of two variables.

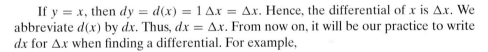 **EXAMPLE 1   Computing a Differential**

Find the differential of $y = x^3 - 2x^2 + 3x - 4$, and evaluate it when $x = 1$ and $\Delta x = 0.04$.

Solution: The differential is

$$dy = \frac{d}{dx}(x^3 - 2x^2 + 3x - 4)\,\Delta x$$

$$= (3x^2 - 4x + 3)\,\Delta x$$

When $x = 1$ and $\Delta x = 0.04$,

$$dy = [3(1)^2 - 4(1) + 3](0.04) = 0.08$$

NOW WORK PROBLEM 1

If $y = x$, then $dy = d(x) = 1\,\Delta x = \Delta x$. Hence, the differential of x is Δx. We abbreviate $d(x)$ by dx. Thus, $dx = \Delta x$. From now on, it will be our practice to write dx for Δx when finding a differential. For example,

$$d(x^2 + 5) = \frac{d}{dx}(x^2 + 5)\,dx = 2x\,dx$$

Summarizing, we say that if $y = f(x)$ defines a differentiable function of x, then

$$dy = f'(x)\,dx$$

where dx is any real number. Provided that $dx \neq 0$, we can divide both sides by dx:

$$\frac{dy}{dx} = f'(x)$$

That is, dy/dx can be viewed either as the quotient of two differentials, namely, dy divided by dx, or as one symbol for the derivative of f at x. It is for this reason that we introduced the symbol dy/dx to denote the derivative.

EXAMPLE 2   Finding a Differential in Terms of dx

a. If $f(x) = \sqrt{x}$, then

$$d(\sqrt{x}) = \frac{d}{dx}(\sqrt{x})\,dx = \frac{1}{2}x^{-1/2}dx = \frac{1}{2\sqrt{x}}dx$$

b. If $u = (x^2 + 3)^5$, then $du = 5(x^2 + 3)^4(2x)\,dx = 10x(x^2 + 3)^4 dx$.

NOW WORK PROBLEM 3

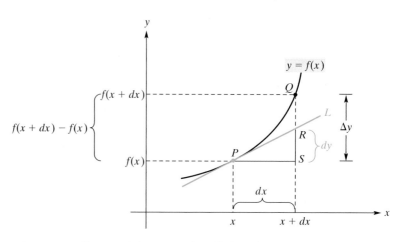

FIGURE 14.1   Geometric interpretation of dy and Δx.

The differential can be interpreted geometrically. In Figure 14.1, the point $P(x, f(x))$ is on the curve $y = f(x)$. Suppose x changes by dx, a real number, to the new value $x + dx$. Then the new function value is $f(x + dx)$, and the corresponding point on the curve is $Q(x + dx, f(x + dx))$. Passing through P and Q are horizontal and vertical lines, respectively, that intersect at S. A line L tangent to the curve at P intersects segment QS at R, forming the right triangle PRS. Observe that the graph of f near P is approximated by the tangent line at P. The slope of L is $f'(x)$ but it is also given by $\overline{SR}/\overline{PS}$ so that

$$f'(x) = \frac{\overline{SR}}{\overline{PS}}$$

Since $dy = f'(x)\,dx$ and $dx = \overline{PS}$,

$$dy = f'(x)\,dx = \frac{\overline{SR}}{\overline{PS}} \cdot \overline{PS} = \overline{SR}$$

Thus, if dx is a change in x at P, then dy is the corresponding vertical change along the **tangent line** at P. Note that for the same dx, the vertical change along the **curve** is $\Delta y = \overline{SQ} = f(x + dx) - f(x)$. Do not confuse Δy with dy. However, from Figure 14.1, the following is apparent:

When dx is close to 0, dy is an approximation to Δy. Therefore,

$$\Delta y \approx dy$$

This fact is useful in estimating Δy, a change in y, as Example 3 shows.

● EXAMPLE 3   Using the Differential to Estimate a Change in a Quantity

A governmental health agency examined the records of a group of individuals who were hospitalized with a particular illness. It was found that the total proportion P that are discharged at the end of t days of hospitalization is given by

$$P = P(t) = 1 - \left(\frac{300}{300 + t}\right)^3$$

Use differentials to approximate the change in the proportion discharged if t changes from 300 to 305.

Solution: The change in t from 300 to 305 is $\Delta t = dt = 305 - 300 = 5$. The change in P is $\Delta P = P(305) - P(300)$. We approximate ΔP by dP:

$$\Delta P \approx dP = P'(t)\,dt = -3\left(\frac{300}{300 + t}\right)^2 \left(-\frac{300}{(300 + t)^2}\right) dt = 3\frac{300^3}{(300 + t)^4}\,dt$$

When $t = 300$ and $dt = 5$,

$$dP = 3\frac{300^3}{600^4}5 = \frac{15}{2^3 600} = \frac{1}{2^3 40} = \frac{1}{320} \approx 0.0031$$

For a comparison, the true value of ΔP is

$$P(305) - P(300) = 0.87807 - 0.87500 = 0.00307$$

(to five decimal places).

NOW WORK PROBLEM 11

We said that if $y = f(x)$, then $\Delta y \approx dy$ if dx is close to zero. Thus,

$$\Delta y = f(x + dx) - f(x) \approx dy$$

so that

Formula (1) is used to approximate a function value, whereas the formula $\Delta y \approx dy$ is used to approximate a change in function values.

$$f(x + dx) \approx f(x) + dy \qquad \textbf{(1)}$$

This formula gives us a way of estimating a function value $f(x + dx)$. For example, suppose we estimate $\ln(1.06)$. Letting $y = f(x) = \ln x$, we need to estimate $f(1.06)$. Since $d(\ln x) = (1/x)\, dx$, we have, from Formula (1),

$$f(x + dx) \approx f(x) + dy$$

$$\ln(x + dx) \approx \ln x + \frac{1}{x}\, dx$$

We know the exact value of $\ln 1$, so we will let $x = 1$ and $dx = 0.06$. Then $x + dx = 1.06$, and dx is close to zero. Therefore,

$$\ln(1 + 0.06) \approx \ln(1) + \frac{1}{1}(0.06)$$

$$\ln(1.06) \approx 0 + 0.06 = 0.06$$

The true value of $\ln(1.06)$ to five decimal places is 0.05827.

● EXAMPLE 4 Using the Differential to Estimate a Function Value

The demand function for a product is given by

$$p = f(q) = 20 - \sqrt{q}$$

where p is the price per unit in dollars for q units. By using differentials, approximate the price when 99 units are demanded.

Solution: We want to approximate $f(99)$. By Formula (1),

$$f(q + dq) \approx f(q) + dp$$

where

$$dp = -\frac{1}{2\sqrt{q}}\, dq \qquad \left(\frac{dp}{dq} = -\frac{1}{2}q^{-1/2}\right)$$

We choose $q = 100$ and $dq = -1$ because $q + dq = 99$, dq is small, and it is easy to compute $f(100) = 20 - \sqrt{100} = 10$. We thus have

$$f(99) = f[100 + (-1)] \approx f(100) - \frac{1}{2\sqrt{100}}(-1)$$

$$f(99) \approx 10 + 0.05 = 10.05$$

Hence, the price per unit when 99 units are demanded is approximately $10.05.

NOW WORK PROBLEM 17

The equation $y = x^3 + 4x + 5$ defines y as a function of x. We could write $f(x) = x^3 + 4x + 5$. However, the equation also defines x implicitly as a function of y. In fact, if we restrict the domain of f to some set of real numbers x so that

$y = f(x)$ is a one-to-one function, then in principle we could solve for x in terms of y and get $x = f^{-1}(y)$. (Actually, no restriction of the domain is necessary here. Since $f'(x) = 3x^2 + 4 > 0$, for all x, we see that f is strictly increasing on $(-\infty, \infty)$ and is thus one-to-one on $(-\infty, \infty)$.) As we did in Example 6 of Section 12.2, we can look at the derivative of x with respect to y, dx/dy and we have seen that it is given by

$$\frac{dx}{dy} = \frac{1}{\dfrac{dy}{dx}} \qquad \text{provided that } dy/dx \neq 0$$

Since dx/dy can be considered a quotient of differentials, we now see that it is the reciprocal of the quotient of differentials dy/dx. Thus

$$\frac{dx}{dy} = \frac{1}{3x^2 + 4}$$

It is important to understand that it is not necessary to be able to solve $y = x^3 + 4x + 5$ for x in terms of y and the equation $\dfrac{dx}{dy} = \dfrac{1}{3x^2 + 4}$ holds for all x.

● EXAMPLE 5 **Finding dp/dq from dq/dp**

Find $\dfrac{dp}{dq}$ *if* $q = \sqrt{2500 - p^2}$.

Solution:

Strategy There are a number of ways to find dp/dq. One approach is to solve the given equation for p explicitly in terms of q and then differentiate directly. Another approach to find dp/dq is to use implicit differentiation. However, since q is given explicitly as a function of p, we can easily find dq/dp and then use the preceding reciprocal relation to find dp/dq. We will take this approach.

We have

$$\frac{dq}{dp} = \frac{1}{2}(2500 - p^2)^{-1/2}(-2p) = -\frac{p}{\sqrt{2500 - p^2}}$$

Hence,

$$\frac{dp}{dq} = \frac{1}{\dfrac{dq}{dp}} = -\frac{\sqrt{2500 - p^2}}{p}$$

NOW WORK PROBLEM 27 ●●●

Problems 14.1

In Problems 1–10, find the differential of the function in terms of x and dx.

*1. $y = 5x - 7$

2. $y = 2$

*3. $f(x) = \sqrt{x^4 - 9}$

4. $f(x) = (4x^2 - 5x + 2)^3$

5. $u = \dfrac{1}{x^2}$

6. $u = \dfrac{1}{\sqrt{x}}$

7. $p = \ln(x^2 + 7)$

8. $p = e^{x^3 + 2x - 5}$

9. $y = (9x + 3)e^{2x^2 + 3}$

10. $y = \ln\sqrt{x^2 + 12}$

In Problems 11–16, find Δy and dy for the given values of x and dx.

*11. $y = 4 - 7x$; $x = 3$, $dx = 0.02$

12. $y = 5x^2$; $x = -1$, $dx = -0.02$

13. $y = 2x^2 + 5x - 7$; $x = -2$, $dx = 0.1$

14. $y = (3x + 2)^2$; $x = -1$, $dx = -0.03$

15. $y = \sqrt{32 - x^2}$; $x = 4$, $dx = -0.05$ Round your answer to three decimal places.

16. $y = \ln(-x)$; $x = -5$, $dx = 0.1$

*17. Let $f(x) = \dfrac{x + 5}{x + 1}$

(a) Evaluate $f'(1)$.

(b) Use differentials to estimate the value of $f(1.1)$.

18. Let $f(x) = x^{3x}$

(a) Evaluate $f'(1)$.

(b) Use differentials to estimate the value of $f(0.98)$.

In Problems 19–26, approximate each expression by using differentials.

19. $\sqrt{288}$ (*Hint:* $17^2 = 289$.)   **20.** $\sqrt{122}$

21. $\sqrt[3]{65.5}$ **22.** $\sqrt[4]{16.3}$

23. $\ln 0.97$ **24.** $\ln 1.01$

25. $e^{0.001}$ **26.** $e^{-0.01}$

In Problems 27–32, find dx/dy or dp/dq.

*27. $y = 2x - 1$ **28.** $y = 5x^2 + 3x + 2$

29. $q = (p^2 + 5)^3$ **30.** $q = \sqrt{p + 5}$

31. $q = \dfrac{1}{p}$ **32.** $q = e^{4 - 2p}$

33. If $y = 7x^2 - 6x + 3$, find the value of dx/dy when $x = 3$.

34. If $y = \ln x^2$, find the value of dx/dy when $x = 3$.

In Problems 35 and 36, find the rate of change of q with respect to p for the indicated value of q.

35. $p = \dfrac{500}{q + 2}$; $q = 18$ **36.** $p = 50 - \sqrt{q}$; $q = 100$

37. Profit   Suppose that the profit (in dollars) of producing q units of a product is

$$P = 397q - 2.3q^2 - 400$$

Using differentials, find the approximate change in profit if the level of production changes from $q = 90$ to $q = 91$. Find the true change.

38. Revenue   Given the revenue function

$$r = 250q + 45q^2 - q^3$$

use differentials to find the approximate change in revenue if the number of units increases from $q = 40$ to $q = 41$. Find the true change.

39. Demand   The demand equation for a product is

$$p = \frac{10}{\sqrt{q}}$$

Using differentials, approximate the price when 24 units are demanded.

40. Demand   Given the demand function

$$p = \frac{200}{\sqrt{q + 8}}$$

use differentials to estimate the price per unit when 40 units are demanded.

41. If $y = f(x)$, then the *proportional change in y* is defined to be $\Delta y/y$, which can be approximated with differentials

by dy/y. Use this last form to approximate the proportional change in the cost function

$$c = f(q) = \frac{q^4}{2} + 3q + 400$$

when $q = 10$ and $dq = 2$. Round your answer to one decimal place.

42. Status/Income   Suppose that S is a numerical value of status based on a person's annual income I (in thousands of dollars). For a certain population, suppose $S = 20\sqrt{I}$. Use differentials to approximate the change in S if annual income decreases from \$45,000 to \$44,500.

43. Biology   The volume of a spherical cell is given by $V = \frac{4}{3}\pi r^3$, where r is the radius. Estimate the change in volume when the radius changes from 6.5×10^{-4} cm to 6.6×10^{-4} cm.

44. Muscle Contraction   The equation

$$(P + a)(v + b) = k$$

is called the "fundamental equation of muscle contraction."[1] Here P is the load imposed on the muscle, v is the velocity of the shortening of the muscle fibers, and a, b, and k are positive constants. Find P in terms of v, and then use the differential to approximate the change in P due to a small change in v.

45. Demand   The demand, q, for a monopolist's product is related to the price per unit, p, according to the equation

$$2 + \frac{q^2}{200} = \frac{4000}{p^2}$$

(a) Verify that 40 units will be demanded when the price per unit is \$20.

(b) Show that $\dfrac{dq}{dp} = -2.5$ when the price per unit is \$20.

(c) Use differentials and the results of parts (a) and (b) to approximate the number of units that will be demanded if the price per unit is reduced to \$19.20.

46. Profit   The demand equation for a monopolist's product is

$$p = \frac{1}{2}q^2 - 66q + 7000$$

and the average-cost function is

$$\bar{c} = 500 - q + \frac{80,000}{2q}$$

(a) Find the profit when 100 units are demanded.

(b) Use differentials and the result of part (a) to estimate the profit when 98 units are demanded.

OBJECTIVE

To define the antiderivative and the indefinite integral and to apply basic integration formulas.

14.2 The Indefinite Integral

Given a function f, if F is a function such that

$$F'(x) = f(x) \qquad (1)$$

then F is called an *antiderivative* of f. Thus,

> An antiderivative of f is simply a function whose derivative is f.

---

[1] R. W. Stacy et al., *Essentials of Biological and Medical Physics* (New York: McGraw-Hill, 1955).

Multiplying both sides of Equation (1) by the differential dx gives $F'(x)\,dx = f(x)\,dx$. However, because $F'(x)\,dx$ is the differential of F, we have $dF = f(x)\,dx$. Hence, we can think of an antiderivative of f as a function whose differential is $f(x)\,dx$.

DEFINITION

An ***antiderivative*** of a function f is a function F such that
$$F'(x) = f(x)$$

Equivalently, in differential notation,
$$dF = f(x)\,dx$$

For example, because the derivative of x^2 is $2x$, x^2 is an antiderivative of $2x$. However, it is not the only antiderivative of $2x$: Since
$$\frac{d}{dx}(x^2 + 1) = 2x \qquad \text{and} \qquad \frac{d}{dx}(x^2 - 5) = 2x$$

both $x^2 + 1$ and $x^2 - 5$ are also antiderivatives of $2x$. In fact, it is obvious that because the derivative of a constant is zero, $x^2 + C$ is also an antiderivative of $2x$ for *any* constant C. Thus, $2x$ has infinitely many antiderivatives. More importantly, *all* antiderivatives of $2x$ must be functions of the form $x^2 + C$, because of the following fact:

Any two antiderivatives of a function differ only by a constant.

Since $x^2 + C$ describes all antiderivatives of $2x$, we can refer to it as being the *most general antiderivative* of $2x$, denoted by $\int 2x\,dx$, which is read "the *indefinite integral* of $2x$ with respect to x." Thus, we write
$$\int 2x\,dx = x^2 + C$$

The symbol $\int$ is called the **integral sign,** $2x$ is the **integrand,** and C is the **constant of integration.** The dx is part of the integral notation and indicates the variable involved. Here x is the **variable of integration.**

More generally, the **indefinite integral** of any function f with respect to x is written $\int f(x)\,dx$ and denotes the most general antiderivative of f. Since all antiderivatives of f differ only by a constant, if F is any antiderivative of f, then
$$\int f(x)\,dx = F(x) + C \quad \text{where } C \text{ is a constant}$$

To *integrate* f means to find $\int f(x)\,dx$. In summary,
$$\int f(x)\,dx = F(x) + C \qquad \text{if and only if} \qquad F'(x) = f(x)$$

Thus we have
$$\frac{d}{dx}\left(\int f(x)\,dx\right) = f(x) \quad \text{and} \quad \int \frac{d}{dx}(F(x))\,dx = F(x) + C$$

which shows the extent to which differentiation and indefinite integration are inverse procedures.

CAUTION

A common mistake is to omit C, the constant of integration.

● EXAMPLE 1   **Finding an Indefinite Integral**

Find $\int 5\, dx$.

Solution:

Strategy   First we must find (perhaps better words are "guess at") a function whose derivative is 5. Then we add the constant of integration.

Since we know that the derivative of $5x$ is 5, $5x$ is an antiderivative of 5. Therefore,

$$\int 5\, dx = 5x + C$$

NOW WORK PROBLEM 1

TABLE 14.1 Basic Integration Formulas

1. $\displaystyle\int k\, dx = kx + C$ k is a constant

2. $\displaystyle\int x^n\, dx = \frac{x^{n+1}}{n+1} + C$ $n \neq -1$

3. $\displaystyle\int x^{-1}\, dx = \int \frac{1}{x}\, dx = \int \frac{dx}{x} = \ln x + C$ for $x > 0$

4. $\displaystyle\int e^x\, dx = e^x + C$

5. $\displaystyle\int k f(x)\, dx = k \int f(x)\, dx$ k is a constant

6. $\displaystyle\int (f(x) \pm g(x))\, dx = \int f(x)\, dx \pm \int g(x)\, dx$

Using differentiation formulas from Chapters 11 and 12, we have compiled a list of basic integration formulas in Table 14.1. These formulas are easily verified. For example, Formula 2 is true because the derivative of $x^{n+1}/(n+1)$ is x^n for $n \neq -1$. (We must have $n \neq -1$ because the denominator is 0 when $n = -1$.) Formula 2 states that the indefinite integral of a power of x (except x^{-1}) is obtained by increasing the exponent of x by 1, dividing by the new exponent, and adding on the constant of integration. The indefinite integral of x^{-1} will be discussed in Section 14.4.

To verify Formula 5, we must show that the derivative of $k \int f(x)\, dx$ is $kf(x)$. Since the derivative of $k \int f(x)\, dx$ is simply k times the derivative of $\int f(x)\, dx$, and the derivative of $\int f(x)\, dx$ is $f(x)$, Formula 5 is verified. You should verify the other formulas. Formula 6 can be extended to any number of sums or differences.

● EXAMPLE 2   **Indefinite Integrals of a Constant and of a Power of x**

a. *Find* $\int 1\, dx$.

Solution: By Formula 1 with $k = 1$

$$\int 1\, dx = 1x + C = x + C$$

Usually, we write $\int 1\, dx$ as $\int dx$. Thus, $\int dx = x + C$.

b. Find $\int x^5\, dx$.

Solution: By Formula 2 with $n = 5$,

$$\int x^5\, dx = \frac{x^{5+1}}{5+1} + C = \frac{x^6}{6} + C$$

NOW WORK PROBLEM 3

PRINCIPLES IN PRACTICE 2

INDEFINITE INTEGRAL OF A
CONSTANT TIMES A FUNCTION

If the rate of change of a company's revenues can be modeled by $\dfrac{dR}{dt} = 0.12t^2$, then find $\int 0.12t^2\, dt$, which gives the form of the company's revenue function.

 CAUTION

Only a *constant* factor of the integrand can "pass through" an integral sign. Because x is not a constant, one cannot find $\int 7x\, dx$ as $7x$ times $\int dx$.

● EXAMPLE 3   Indefinite Integral of a Constant Times a Function

Find $\int 7x\, dx$.

Solution: By Formula 5 with $k = 7$ and $f(x) = x$,

$$\int 7x\, dx = 7\int x\, dx$$

Since x is x^1, by Formula 2 we have

$$\int x^1\, dx = \frac{x^{1+1}}{1+1} + C_1 = \frac{x^2}{2} + C_1$$

where C_1 is the constant of integration. Therefore,

$$\int 7x\, dx = 7\int x\, dx = 7\left(\frac{x^2}{2} + C_1\right) = \frac{7}{2}x^2 + 7C_1$$

Since $7C_1$ is just an arbitrary constant, we will replace it by C for simplicity. Thus,

$$\int 7x\, dx = \frac{7}{2}x^2 + C$$

It is not necessary to write all intermediate steps when integrating. More simply, we write

$$\int 7x\, dx = (7)\frac{x^2}{2} + C = \frac{7}{2}x^2 + C$$

NOW WORK PROBLEM 5

● EXAMPLE 4   Indefinite Integral of a Constant Times a Function

Find $\int -\frac{3}{5}e^x\, dx$.

Solution:

$$\int -\frac{3}{5}e^x\, dx = -\frac{3}{5}\int e^x\, dx \qquad \text{(Formula 5)}$$

$$= -\frac{3}{5}e^x + C \qquad \text{(Formula 4)}$$

NOW WORK PROBLEM 21

PRINCIPLES IN PRACTICE 3

FINDING INDEFINITE INTEGRALS

Due to new competition, the number of subscriptions to a certain magazine is declining at a rate of $\dfrac{dS}{dt} = -\dfrac{480}{t^3}$ subscriptions per month, where t is the number of months since the competition entered the market. Find the form of the equation for the number of subscribers to the magazine.

● EXAMPLE 5   Finding Indefinite Integrals

a. Find $\int \dfrac{1}{\sqrt{t}}\, dt$.

Solution: Here t is the variable of integration. We rewrite the integrand so that a basic formula can be used. Since $1/\sqrt{t} = t^{-1/2}$, applying Formula 2 gives

$$\int \frac{1}{\sqrt{t}}\, dt = \int t^{-1/2}\, dt = \frac{t^{(-1/2)+1}}{-\dfrac{1}{2}+1} + C = \frac{t^{1/2}}{\dfrac{1}{2}} + C = 2\sqrt{t} + C$$

b. *Find* $\int \dfrac{1}{6x^3}\, dx.$

Solution:

$$\int \frac{1}{6x^3}\, dx = \frac{1}{6}\int x^{-3}\, dx = \left(\frac{1}{6}\right)\frac{x^{-3+1}}{-3+1} + C$$

$$= -\frac{x^{-2}}{12} + C = -\frac{1}{12x^2} + C$$

NOW WORK PROBLEM 9 ●●

INDEFINITE INTEGRAL OF A SUM

The rate of growth of the population of a new city is estimated by $\dfrac{dN}{dt} = 500 + 300\sqrt{t}$, where t is in years. Find

$$\int (500 + 300\sqrt{t})\, dt$$

When integrating an expression involving more than one term, only one constant of integration is needed.

●EXAMPLE 6 **Indefinite Integral of a Sum**

Find $\int (x^2 + 2x)\, dx.$

Solution: By Formula 6,

$$\int (x^2 + 2x)\, dx = \int x^2\, dx + \int 2x\, dx$$

Now,

$$\int x^2\, dx = \frac{x^{2+1}}{2+1} + C_1 = \frac{x^3}{3} + C_1$$

and

$$\int 2x\, dx = 2\int x\, dx = (2)\frac{x^{1+1}}{1+1} + C_2 = x^2 + C_2$$

Thus,

$$\int (x^2 + 2x)\, dx = \frac{x^3}{3} + x^2 + C_1 + C_2$$

For convenience, we will replace the constant $C_1 + C_2$ by C. We then have

$$\int (x^2 + 2x)\, dx = \frac{x^3}{3} + x^2 + C$$

Omitting intermediate steps, we simply integrate term by term and write

$$\int (x^2 + 2x)\, dx = \frac{x^3}{3} + (2)\frac{x^2}{2} + C = \frac{x^3}{3} + x^2 + C$$

NOW WORK PROBLEM 11 ●●

INDEFINITE INTEGRAL OF A SUM AND DIFFERENCE

Suppose the rate of savings in the United States is given by $\dfrac{dS}{dt} = 2.1t^2 - 65.4t + 491.6$, where t is the time in years and S is the amount of money saved in billions of dollars. Find the form of the equation for the amount of money saved.

●EXAMPLE 7 **Indefinite Integral of a Sum and Difference**

Find $\int (2\sqrt[5]{x^4} - 7x^3 + 10e^x - 1)\, dx.$

Solution:

$$\int (2\sqrt[5]{x^4} - 7x^3 + 10e^x - 1)\, dx$$

$$= 2\int x^{4/5}\, dx - 7\int x^3\, dx + 10\int e^x\, dx - \int 1\, dx \qquad \text{(Formulas 5 \& 6)}$$

$$= (2)\frac{x^{9/5}}{\frac{9}{5}} - (7)\frac{x^4}{4} + 10e^x - x + C \qquad \text{(Formulas 1, 2, \& 4)}$$

$$= \frac{10}{9}x^{9/5} - \frac{7}{4}x^4 + 10e^x - x + C$$

NOW WORK PROBLEM 15 ●●

Sometimes, in order to apply the basic integration formulas, it is necessary first to perform algebraic manipulations on the integrand, as Example 8 shows.

EXAMPLE 8 Using Algebraic Manipulation to Find an Indefinite Integral

Find $\int y^2 \left(y + \dfrac{2}{3} \right) dy$.

Solution: The integrand does not fit a familiar integration form. However, by multiplying the integrand we get

$$\int y^2 \left(y + \frac{2}{3} \right) dy = \int \left(y^3 + \frac{2}{3} y^2 \right) dy$$

$$= \frac{y^4}{4} + \left(\frac{2}{3} \right) \frac{y^3}{3} + C = \frac{y^4}{4} + \frac{2y^3}{9} + C$$

NOW WORK PROBLEM 41

CAUTION

In Example 8, we first multiplied the factors in the integrand. The answer could not have been found simply in terms of $\int y^2 \, dy$ and $\int (y + \frac{2}{3}) \, dy$. There is not a formula for the integral of a general product of functions.

EXAMPLE 9 Using Algebraic Manipulation to Find an Indefinite Integral

a. Find $\int \dfrac{(2x - 1)(x + 3)}{6} \, dx$.

Solution: By factoring out the constant $\frac{1}{6}$ and multiplying the binomials, we get

$$\int \frac{(2x - 1)(x + 3)}{6} \, dx = \frac{1}{6} \int (2x^2 + 5x - 3) \, dx$$

$$= \frac{1}{6} \left((2)\frac{x^3}{3} + (5)\frac{x^2}{2} - 3x \right) + C$$

$$= \frac{x^3}{9} + \frac{5x^2}{12} - \frac{x}{2} + C$$

Another algebraic approach to part (b) is

$$\int \frac{x^3 - 1}{x^2} \, dx = \int (x^3 - 1)x^{-2} \, dx$$

$$= \int (x - x^{-2}) \, dx$$

and so on.

b. Find $\int \dfrac{x^3 - 1}{x^2} \, dx$.

Solution: We can break up the integrand into fractions by dividing each term in the numerator by the denominator:

$$\int \frac{x^3 - 1}{x^2} \, dx = \int \left(\frac{x^3}{x^2} - \frac{1}{x^2} \right) dx = \int (x - x^{-2}) \, dx$$

$$= \frac{x^2}{2} - \frac{x^{-1}}{-1} + C = \frac{x^2}{2} + \frac{1}{x} + C$$

NOW WORK PROBLEM 49

Problems 14.2

In Problems 1–52, find the indefinite integrals.

*1. $\displaystyle\int 7 \, dx$

2. $\displaystyle\int \frac{1}{2x} \, dx$

*3. $\displaystyle\int x^8 \, dx$

4. $\displaystyle\int 5x^{24} \, dx$

*5. $\displaystyle\int 5x^{-7} \, dx$

6. $\displaystyle\int \frac{z^{-3}}{3} \, dz$

7. $\displaystyle\int \frac{2}{x^{10}} \, dx$

8. $\displaystyle\int \frac{7}{x^4 \, dx}$

*9. $\displaystyle\int \frac{1}{t^{7/4}} \, dt$

10. $\displaystyle\int \frac{7}{2x^{9/4}} \, dx$

*11. $\displaystyle\int (4 + t) \, dt$

12. $\displaystyle\int (r^3 + 2r) \, dr$

13. $\displaystyle\int (y^5 - 5y) \, dy$

14. $\displaystyle\int (5 - 2w - 6w^2) \, dw$

*15. $\displaystyle\int (3t^2 - 4t + 5) \, dt$

16. $\displaystyle\int (1 + t^2 + t^4 + t^6) \, dt$

17. $\int (7 + e)\, dx$

18. $\int (5 - 2^{-1})\, dx$

19. $\int \left(\frac{x}{7} - \frac{3}{4}x^4 \right) dx$

20. $\int \left(\frac{2x^2}{7} - \frac{8}{3}x^4 \right) dx$

*21. $\int \pi e^x\, dx$

22. $\int \left(\frac{e^x}{3} + 2x \right) dx$

23. $\int (x^{8.3} - 9x^6 + 3x^{-4} + x^{-3})\, dx$

24. $\int (0.7y^3 + 10 + 2y^{-3})\, dy$

25. $\int \frac{-2\sqrt{x}}{3}\, dx$

26. $\int dz$

27. $\int \frac{1}{4\sqrt[8]{x^2}}\, dx$

28. $\int \frac{-4}{(3x)^3}\, dx$

29. $\int \left(\frac{x^3}{3} - \frac{3}{x^3} \right) dx$

30. $\int \left(\frac{1}{2x^3} - \frac{1}{x^4} \right) dx$

31. $\int \left(\frac{3w^2}{2} - \frac{2}{3w^2} \right) dw$

32. $\int \frac{4}{e^{-s}}\, ds$

33. $\int \frac{3u - 4}{5}\, du$

34. $\int \frac{1}{12} \left(\frac{1}{3} e^x \right) dx$

35. $\int (u^e + e^u)\, du$

36. $\int \left(3y^3 - 2y^2 + \frac{e^y}{6} \right) dy$

37. $\int (2\sqrt{x} - 3\sqrt[4]{x})\, dx$

38. $\int 0\, dt$

39. $\int \left(-\frac{\sqrt[3]{x^2}}{5} - \frac{7}{2\sqrt{x}} + 6x \right) dx$

40. $\int \left(\sqrt[3]{u} + \frac{1}{\sqrt{u}} \right) du$

*41. $\int (x^2 + 5)(x - 3)\, dx$

42. $\int x^4(x^3 + 8x^2 + 7)\, dx$

43. $\int \sqrt{x}(x + 3)\, dx$

44. $\int (z + 2)^2\, dz$

45. $\int (3u + 2)^3\, du$

46. $\int \left(\frac{2}{\sqrt[5]{x}} - 1 \right)^2 dx$

47. $\int v^{-2}(2v^4 + 3v^2 - 2v^{-3})\, dv$

48. $\int (6e^u - u^3(\sqrt{u} + 1))\, du$

*49. $\int \frac{z^4 + 10z^3}{2z^2}\, dz$

50. $\int \frac{x^4 - 5x^2 + 2x}{5x^2}\, dx$

51. $\int \frac{e^x + e^{2x}}{e^x}\, dx$

52. $\int \frac{(x^3 + 1)^2}{x^2}\, dx$

53. If $F(x)$ and $G(x)$ are such that $F'(x) = G'(x)$, is it true that $F(x) - G(x)$ must be zero?

54. **(a)** Find a function F such that $\int F(x)\, dx = xe^x + C$.
   (b) Is there only one function F satisfying the equation given in part (a), or are there many such functions?

55. Find $\int \frac{d}{dx} \left(\frac{1}{\sqrt{x^2 + 1}} \right) dx$.

14.3 Integration with Initial Conditions

If we know the rate of change, f', of the function f, then the function f itself is an antiderivative of f' (since the derivative of f is f'). Of course, there are many antiderivatives of f', and the most general one is denoted by the indefinite integral. For example, if

$$f'(x) = 2x$$

then

$$f(x) = \int f'(x)\, dx = \int 2x\, dx = x^2 + C. \qquad (1)$$

That is, *any* function of the form $f(x) = x^2 + C$ has its derivative equal to $2x$. Because of the constant of integration, notice that we do not know $f(x)$ specifically. However, if f must assume a certain function value for a particular value of x, then we can determine the value of C and thus determine $f(x)$ specifically. For instance, if $f(1) = 4$, then from Equation (1),

$$f(1) = 1^2 + C$$
$$4 = 1 + C$$
$$C = 3$$

Thus,

$$f(x) = x^2 + 3$$

That is, we now know the particular function $f(x)$ for which $f'(x) = 2x$ and $f(1) = 4$. The condition $f(1) = 4$, which gives a function value of f for a specific value of x, is called an *initial condition*.

EXAMPLE 1 Initial-Condition Problem

If y is a function of x such that $y' = 8x - 4$ and $y(2) = 5$, find y. [Note: $y(2) = 5$ means that $y = 5$ when $x = 2$.] Also, find $y(4)$.

Solution: Here $y(2) = 5$ is the initial condition. Since $y' = 8x - 4$, y is an antiderivative of $8x - 4$:

$$y = \int (8x - 4)\, dx = 8 \cdot \frac{x^2}{2} - 4x + C = 4x^2 - 4x + C \tag{2}$$

We can determine the value of C by using the initial condition. Because $y = 5$ when $x = 2$, from Equation (2), we have

$$5 = 4(2)^2 - 4(2) + C$$
$$5 = 16 - 8 + C$$
$$C = -3$$

Replacing C by -3 in Equation (2) gives the function that we seek:

$$y = 4x^2 - 4x - 3 \tag{3}$$

To find $y(4)$, we let $x = 4$ in Equation (3):

$$y(4) = 4(4)^2 - 4(4) - 3 = 64 - 16 - 3 = 45$$

NOW WORK PROBLEM 1

EXAMPLE 2 Initial-Condition Problem Involving y''

Given that $y'' = x^2 - 6$, $y'(0) = 2$, and $y(1) = -1$, find y.

Solution:

Strategy To go from y'' to y, two integrations are needed: the first to take us from y'' to y' and the other to take us from y' to y. Hence, there will be two constants of integration, which we will denote by C_1 and C_2.

Since $y'' = \dfrac{d}{dx}(y') = x^2 - 6$, y' is an antiderivative of $x^2 - 6$. Thus,

$$y' = \int (x^2 - 6)\, dx = \frac{x^3}{3} - 6x + C_1 \tag{4}$$

Now, $y'(0) = 2$ means that $y' = 2$ when $x = 0$; therefore, from Equation (4), we have

$$2 = \frac{0^3}{3} - 6(0) + C_1$$

Hence, $C_1 = 2$, so

$$y' = \frac{x^3}{3} - 6x + 2$$

By integration, we can find y:

$$y = \int \left(\frac{x^3}{3} - 6x + 2 \right) dx$$

$$= \left(\frac{1}{3} \right) \frac{x^4}{4} - (6)\frac{x^2}{2} + 2x + C_2$$

so

$$y = \frac{x^4}{12} - 3x^2 + 2x + C_2 \tag{5}$$

Now, since $y = -1$ when $x = 1$, we have, from Equation (5),

$$-1 = \frac{1^4}{12} - 3(1)^2 + 2(1) + C_2$$

Thus, $C_2 = -\frac{1}{12}$, so

$$y = \frac{x^4}{12} - 3x^2 + 2x - \frac{1}{12}$$

NOW WORK PROBLEM 5 ●○●

Integration with initial conditions is applicable to many applied situations, as the next three examples illustrate.

● **EXAMPLE 3 Income and Education**

For a particular urban group, sociologists studied the current average yearly income y (in dollars) that a person can expect to receive with x years of education before seeking regular employment. They estimated that the rate at which income changes with respect to education is given by

$$\frac{dy}{dx} = 100x^{3/2} \quad 4 \le x \le 16$$

where $y = 28{,}720$ when $x = 9$. Find y.

Solution: Here y is an antiderivative of $100x^{3/2}$. Thus,

$$y = \int 100x^{3/2}\, dx = 100 \int x^{3/2}\, dx$$

$$= (100)\frac{x^{5/2}}{\dfrac{5}{2}} + C$$

$$y = 40x^{5/2} + C \tag{6}$$

The initial condition is that $y = 28{,}720$ when $x = 9$. By putting these values into Equation (6), we can determine the value of C:

$$28{,}720 = 40(9)^{5/2} + C$$

$$= 40(243) + C$$

$$28{,}720 = 9720 + C$$

Therefore, $C = 19{,}000$, and

$$y = 40x^{5/2} + 19{,}000$$

NOW WORK PROBLEM 17 ●○●

● **EXAMPLE 4 Finding the Demand Function from Marginal Revenue**

If the marginal-revenue function for a manufacturer's product is

$$\frac{dr}{dq} = 2000 - 20q - 3q^2$$

find the demand function.

Solution:

Strategy By integrating dr/dq and using an initial condition, we can find the revenue function r. But revenue is also given by the general relationship $r = pq$, where p is the price per unit. Thus, $p = r/q$. Replacing r in this equation by the revenue function yields the demand function.

Since dr/dq is the derivative of total revenue r,

$$r = \int (2000 - 20q - 3q^2)\, dq$$

$$= 2000q - (20)\frac{q^2}{2} - (3)\frac{q^3}{3} + C$$

so that

$$r = 2000q - 10q^2 - q^3 + C \tag{7}$$

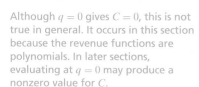
Revenue is 0 when q is 0.

We assume that **when no units are sold, there is no revenue;** that is, $r = 0$ when $q = 0$. This is our initial condition. Putting these values into Equation (7) gives

$$0 = 2000(0) - 10(0)^2 - 0^3 + C$$

Although $q = 0$ gives $C = 0$, this is not true in general. It occurs in this section because the revenue functions are polynomials. In later sections, evaluating at $q = 0$ may produce a nonzero value for C.

Hence, $C = 0$, and

$$r = 2000q - 10q^2 - q^3$$

To find the demand function, we use the fact that $p = r/q$ and substitute for r:

$$p = \frac{r}{q} = \frac{2000q - 10q^2 - q^3}{q}$$

$$p = 2000 - 10q - q^2$$

NOW WORK PROBLEM 11 ●●●

● EXAMPLE 5   Finding Cost from Marginal Cost

In the manufacture of a product, fixed costs per week are $4000. (Fixed costs are costs, such as rent and insurance, that remain constant at all levels of production during a given time period.) If the marginal-cost function is

$$\frac{dc}{dq} = 0.000001(0.002q^2 - 25q) + 0.2$$

where c is the total cost (in dollars) of producing q pounds of product per week, find the cost of producing 10,000 lb in 1 week.

Solution: Since dc/dq is the derivative of the total cost c,

$$c(q) = \int [0.000001(0.002q^2 - 25q) + 0.2]\, dq$$

$$= 0.000001 \int (0.002q^2 - 25q)\, dq + \int 0.2\, dq$$

$$c(q) = 0.000001 \left(\frac{0.002q^3}{3} - \frac{25q^2}{2} \right) + 0.2q + C$$

When q is 0, total cost is equal to fixed cost.

Fixed costs are constant regardless of output. Therefore, when $q = 0, c = 4000$, which is our initial condition. Putting $c(0) = 4000$ in the last equation, we find that $C = 4000$, so

Although $q = 0$ gives C a value equal to fixed costs, this is not true in general. It occurs in this section because the cost functions are polynomials. In later sections, evaluating at $q = 0$ may produce a value for C that is different from fixed cost.

$$c(q) = 0.000001 \left(\frac{0.002q^3}{3} - \frac{25q^2}{2} \right) + 0.2q + 4000 \tag{8}$$

From Equation (8), we have $c(10,000) = 5416\frac{2}{3}$. Thus, the total cost for producing 10,000 pounds of product in 1 week is $5416.67.

NOW WORK PROBLEM 15 ●●●

Problems 14.3

In Problems 1 and 2, find y subject to the given conditions.

*1. $dy/dx = 3x - 4$; $y(-1) = \frac{13}{2}$

2. $dy/dx = x^2 - x$; $y(3) = \frac{19}{2}$

In Problems 3 and 4, if y satisfies the given conditions, find y(x) for the given value of x.

3. $y' = 5/\sqrt{x}$, $y(9) = 50$; $x = 16$

4. $y' = -x^2 + 2x$, $y(2) = 1$; $x = 1$

In Problems 5–8, find y subject to the given conditions.

*5. $y'' = -3x^2 + 4x$; $y'(1) = 2$, $y(1) = 3$

6. $y'' = x + 1$; $y'(0) = 0$, $y(0) = 5$

7. $y''' = 2x$; $y''(-1) = 3$, $y'(3) = 10$, $y(0) = 13$

8. $y''' = e^x + 1$; $y''(0) = 1$, $y'(0) = 2$, $y(0) = 3$

In Problems 9–12, dr/dq is a marginal-revenue function. Find the demand function.

9. $dr/dq = 0.7$

10. $dr/dq = 10 - \frac{1}{16}q$

*11. $dr/dq = 275 - q - 0.3q^2$

12. $dr/dq = 5{,}000 - 3(2q + 2q^3)$

In Problems 13–16, dc/dq is a marginal-cost function and fixed costs are indicated in braces. For Problems 13 and 14, find the total-cost function. For Problems 15 and 16, find the total cost for the indicated value of q.

13. $dc/dq = 1.35$;   {200}

14. $dc/dq = 2q + 75$;   {2000}

*15. $dc/dq = 0.08q^2 - 1.6q + 6.5$;   {8000}; $q = 25$

16. $dc/dq = 0.000204q^2 - 0.046q + 6$;   {15,000}; $q = 200$

*17. **Diet for Rats**   A group of biologists studied the nutritional effects on rats that were fed a diet containing 10% protein.[2] The protein consisted of yeast and corn flour.

Over a period of time, the group found that the (approximate) rate of change of the average weight gain G (in grams) of a rat with respect to the percentage P of yeast in the protein mix was

$$\frac{dG}{dP} = -\frac{P}{25} + 2 \qquad 0 \le P \le 100$$

If $G = 38$ when $P = 10$, find G.

18. Winter Moth   A study of the winter moth was made in Nova Scotia.[3] The prepupae of the moth fall onto the ground from host trees. It was found that the (approximate) rate at which prepupal density y (the number of prepupae per square foot of soil) changes with respect to distance x (in feet) from the base of a host tree is

$$\frac{dy}{dx} = -1.5 - x \quad 1 \le x \le 9$$

If $y = 57.3$ when $x = 1$, find y.

19. Fluid Flow   In the study of the flow of fluid in a tube of constant radius R, such as blood flow in portions of the body, one can think of the tube as consisting of concentric tubes of radius r, where $0 \le r \le R$. The velocity v of the fluid is a function of r and is given by[4]

$$v = \int -\frac{(P_1 - P_2)r}{2l\eta}\, dr$$

where P_1 and P_2 are pressures at the ends of the tube, η (a Greek letter read "eta") is fluid viscosity, and l is the length of the tube. If $v = 0$ when $r = R$, show that

$$v = \frac{(P_1 - P_2)(R^2 - r^2)}{4l\eta}$$

20. Elasticity of Demand   The sole producer of a product has determined that the marginal-revenue function is

$$\frac{dr}{dq} = 100 - 3q^2$$

Determine the point elasticity of demand for the product when $q = 5$. (*Hint:* First find the demand function.)

21. Average Cost   A manufacturer has determined that the marginal-cost function is

$$\frac{dc}{dq} = 0.003q^2 - 0.4q + 40$$

where q is the number of units produced. If marginal cost is $27.50 when $q = 50$ and fixed costs are $5000, what is the *average* cost of producing 100 units?

22. If $f''(x) = 30x^4 + 12x$ and $f'(1) = 10$, evaluate

$$f(965.335245) - f(-965.335245)$$

OBJECTIVE

To learn and apply the formulas for $\int u^n\, du$, $\int e^u\, du$, and $\int \frac{1}{u}\, du$.

14.4 More Integration Formulas

Power Rule for Integration

The formula

$$\int x^n\, dx = \frac{x^{n+1}}{n+1} + C \qquad \text{if } n \ne -1$$

---

[2] Adapted from R. Bressani, "The Use of Yeast in Human Foods," in *Single-Cell Protein*, ed. R. I. Mateles and S. R. Tannenbaum (Cambridge, MA: MIT Press, 1968).

[3] Adapted from D. G. Embree, "The Population Dynamics of the Winter Moth in Nova Scotia, 1954–1962," *Memoirs of the Entomological Society of Canada*, no. 46 (1965).

[4] R. W. Stacy et al., *Essentials of Biological and Medical Physics* (New York: McGraw-Hill, 1955).

which applies to a power of x, can be generalized to handle a power of a *function* of x. Let u be a differentiable function of x. By the power rule for differentiation, if $n \neq -1$, then

$$\frac{d}{dx}\left(\frac{(u(x))^{n+1}}{n+1}\right) = \frac{(n+1)(u(x))^n \cdot u'(x)}{n+1} = (u(x))^n \cdot u'(x)$$

Thus,

$$\int (u(x))^n \cdot u'(x)\, dx = \frac{(u(x))^{n+1}}{n+1} + C \quad n \neq -1$$

We call this the *power rule for integration*. Note that $u'(x)dx$ is the differential of u, namely du. In mathematical shorthand, we can replace $u(x)$ by u and $u'(x)\, dx$ by du:

Power Rule for Integration

If u is differentiable, then

$$\int u^n\, du = \frac{u^{n+1}}{n+1} + C \qquad \text{if } n \neq -1 \tag{1}$$

It is essential that you realize the difference between the power rule for integration and the formula for $\int x^n\, dx$. In the power rule, u represents a function, whereas in $\int x^n\, dx$, x is a variable.

● EXAMPLE 1   Applying the Power Rule for Integration

a. *Find* $\int (x+1)^{20}\, dx$.

Solution: Since the integrand is a power of the function $x + 1$, we will set $u = x + 1$. Then $du = dx$, and $\int (x+1)^{20}\, dx$ has the form $\int u^{20}\, du$. By the power rule for integration,

$$\int (x+1)^{20}\, dx = \int u^{20}\, du = \frac{u^{21}}{21} + C = \frac{(x+1)^{21}}{21} + C$$

Note that we give our answer not in terms of u, but explicitly in terms of x.

b. *Find* $\int 3x^2(x^3+7)^3\, dx$.

Solution: We observe that the integrand contains a power of the function $x^3 + 7$. Let $u = x^3 + 7$. Then $du = 3x^2\, dx$. Fortunately, $3x^2$ appears as a factor in the integrand and we have

$$\int 3x^2(x^3+7)^3\, dx = \int (x^3+7)^3[3x^2\, dx] = \int u^3\, du$$

$$= \frac{u^4}{4} + C = \frac{(x^3+7)^4}{4} + C$$

NOW WORK PROBLEM 3 ●○○

After integrating, you may wonder what happened to $3x^2$. We note again that $du = 3x^2\, dx$.

In order to apply the power rule for integration, sometimes an adjustment must be made to obtain du in the integrand, as Example 2 illustrates.

● EXAMPLE 2   Adjusting for *du*

Find $\int x\sqrt{x^2+5}\, dx$.

Solution: We can write this as $\int x(x^2+5)^{1/2}\, dx$. Notice that the integrand contains a power of the function $x^2 + 5$. If $u = x^2 + 5$, then $du = 2x\, dx$. Since the *constant* factor 2 in du does *not* appear in the integrand, this integral does not have the form $\int u^n\, du$. However, from $du = 2x\, dx$ we can write $x\, dx = \dfrac{du}{2}$ so that the integral

becomes

$$\int x(x^2+5)^{1/2}\,dx = \int (x^2+5)^{1/2}[x\,dx] = \int u^{1/2}\frac{du}{2}$$

Moving the *constant* factor $\frac{1}{2}$ in front of the integral sign, we have

$$\int x(x^2+5)^{1/2}\,dx = \frac{1}{2}\int u^{1/2}\,du = \frac{1}{2}\left(\frac{u^{3/2}}{\frac{3}{2}}\right)+C = \frac{1}{3}u^{3/2}+C$$

which in terms of x (as is required) gives

$$\int x\sqrt{x^2+5}\,dx = \frac{(x^2+5)^{3/2}}{3}+C$$

 **CAUTION**

The answer to an integration problem must be expressed in terms of the original variable.

NOW WORK PROBLEM 15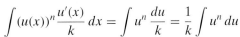

In Example 2, the integrand $x\sqrt{x^2+5}$ missed being of the form $(u(x))^{1/2}u'(x)$ by the *constant factor* of 2. In general, if we have $\displaystyle\int (u(x))^n \frac{u'(x)}{k}\,dx$, for k a nonzero constant, then we can write

 CAUTION

We can adjust for constant factors, but not variable factors.

$$\int (u(x))^n \frac{u'(x)}{k}\,dx = \int u^n \frac{du}{k} = \frac{1}{k}\int u^n\,du$$

to simplify the integral, but such *adjustments* of the integrand are *not possible for variable factors*.

When using the form $\int u^n\,du$, do not neglect du. For example,

$$\int (4x+1)^2\,dx \neq \frac{(4x+1)^3}{3}+C$$

The correct way to do this problem is as follows. Let $u = 4x+1$, from which it follows that $du = 4\,dx$. Thus $dx = \dfrac{du}{4}$ and

$$\int (4x+1)^2\,dx = \int u^2\left[\frac{du}{4}\right] = \frac{1}{4}\int u^2\,du = \frac{1}{4}\cdot\frac{u^3}{3}+C = \frac{(4x+1)^3}{12}+C$$

● **EXAMPLE 3 Applying the Power Rule for Integration**

a. *Find* $\displaystyle\int \sqrt[3]{6y}\,dy.$

Solution: The integrand is $(6y)^{1/3}$, a power of a function. However, in this case the obvious substitution $u = 6y$ can be avoided. More simply, we have

$$\int \sqrt[3]{6y}\,dy = \int 6^{1/3}y^{1/3}\,dy = \sqrt[3]{6}\int y^{1/3}\,dy = \sqrt[3]{6}\frac{y^{4/3}}{\frac{4}{3}}+C = \frac{3\sqrt[3]{6}}{4}y^{4/3}+C$$

b. *Find* $\displaystyle\int \frac{2x^3+3x}{(x^4+3x^2+7)^4}\,dx.$

Solution: We can write this as $\int (x^4+3x^2+7)^{-4}(2x^3+3x)\,dx$. Let us try to use the power rule for integration. If $u = x^4+3x^2+7$, then $du = (4x^3+6x)\,dx$, which is two times the quantity $(2x^3+3x)\,dx$ in the integral. Thus $(2x^3+3x)\,dx = \dfrac{du}{2}$ and we again illustrate the *adjustment* technique:

$$\int (x^4+3x^2+7)^{-4}[(2x^3+3x)\,dx] = \int u^{-4}\left[\frac{du}{2}\right] = \frac{1}{2}\int u^{-4}\,du$$

$$= \frac{1}{2}\cdot\frac{u^{-3}}{-3}+C = -\frac{1}{6u^3}+C = -\frac{1}{6(x^4+3x^2+7)^3}+C$$

NOW WORK PROBLEM 5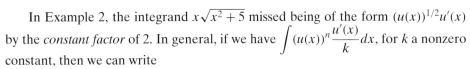

In using the power rule for integration, take care when making your choice for u. In Example 3(b), you would *not* be able to proceed very far if, for instance, you let $u = 2x^3 + 3x$. At times you may find it necessary to try many different choices. So don't just sit and look at the integral. Try something even if it is wrong, because it may give you a hint as to what might work. **Skill at integration comes only after many hours of practice and conscientious study.**

EXAMPLE 4   An Integral to Which the Power Rule Does Not Apply

Find $\displaystyle\int 4x^2(x^4 + 1)^2\, dx.$

Solution: If we set $u = x^4 + 1$, then $du = 4x^3\, dx$. To get du in the integral, we need an additional factor of the *variable* x. However, we can adjust only for **constant** factors. Thus, we cannot use the power rule. Instead, to find the integral, we will first expand $(x^4 + 1)^2$:

$$\int 4x^2(x^4 + 1)^2\, dx = 4\int x^2(x^8 + 2x^4 + 1)\, dx$$

$$= 4\int (x^{10} + 2x^6 + x^2)\, dx$$

$$= 4\left(\frac{x^{11}}{11} + \frac{2x^7}{7} + \frac{x^3}{3}\right) + C$$

NOW WORK PROBLEM 67

Integrating Natural Exponential Functions

We now turn our attention to integrating exponential functions. If u is a differentiable function of x, then

$$\frac{d}{dx}(e^u) = e^u\frac{du}{dx}$$

Corresponding to this differentiation formula is the integration formula

$$\int e^u\frac{du}{dx}\, dx = e^u + C$$

But $\dfrac{du}{dx}\, dx$ is the differential of u, namely, du. Thus,

$$\int e^u\, du = e^u + C \tag{2}$$

CAUTION

Do not apply the power-rule formula for $\int u^n\, du$ to $\int e^u\, du$.

PRINCIPLES IN PRACTICE 1

INTEGRALS INVOLVING EXPONENTIAL FUNCTIONS

When an object is moved from one environment to another, its temperature T changes at a rate given by $\dfrac{dT}{dt} = kCe^{kt}$, where t is the time (in hours) after changing environments, C is the temperature difference (original minus new) between the environments, and k is a constant. If the original environment is $70°$, the new environment is $60°$, and $k = -0.5$, find the general form of $T(t)$.

EXAMPLE 5   Integrals Involving Exponential Functions

a. *Find* $\displaystyle\int 2xe^{x^2}\, dx.$

Solution: Let $u = x^2$. Then $du = 2x\, dx$, and by Equation (2),

$$\int 2xe^{x^2}\, dx = \int e^{x^2}[2x\, dx] = \int e^u\, du$$

$$= e^u + C = e^{x^2} + C$$

b. *Find* $\int (x^2 + 1)e^{x^3+3x}\, dx$.

Solution: If $u = x^3 + 3x$, then $du = (3x^2 + 3)\, dx = 3(x^2 + 1)\, dx$. If the integrand contained a factor of 3, the integral would have the form $\int e^u\, du$. Thus, we write

$$\int (x^2 + 1)e^{x^3+3x}\, dx = \int e^{x^3+3x}[(x^2 + 1)\, dx]$$

$$= \frac{1}{3} \int e^u\, du = \frac{1}{3}e^u + C$$

$$= \frac{1}{3}e^{x^3+3x} + C$$

where in the second step we replaced $(x^2 + 1)\, dx$ by $\frac{1}{3}\, du$ but wrote $\frac{1}{3}$ outside the integral.

NOW WORK PROBLEM 41

Integrals Involving Logarithmic Functions

As you know, the power-rule formula $\int u^n\, du = u^{n+1}/(n + 1) + C$ does not apply when $n = -1$. To handle that situation, namely, $\int u^{-1}\, du = \int \frac{1}{u}\, du$, we first recall from Section 12.1 that

$$\frac{d}{dx}(\ln |u|) = \frac{1}{u}\frac{du}{dx} \quad \text{for } u \neq 0$$

which gives us the integration formula

$$\int \frac{1}{u}\, du = \ln |u| + C \quad \text{for } u \neq 0 \qquad (3)$$

In particular, if $u = x$, then $du = dx$, and

$$\int \frac{1}{x}\, dx = \ln |x| + C \quad \text{for } x \neq 0 \qquad (4)$$

INTEGRALS INVOLVING $\frac{1}{u}\, du$

If the rate of vocabulary memorization of the average student in a foreign language is given by $\frac{dv}{dt} = \frac{35}{t + 1}$, where v is the number of vocabulary words memorized in t hours of study, find the general form of $v(t)$.

● **EXAMPLE 6** **Integrals Involving** $\frac{1}{u}\, du$

a. Find $\int \frac{7}{x}\, dx$.

Solution: From Equation (4),

$$\int \frac{7}{x}\, dx = 7 \int \frac{1}{x}\, dx = 7 \ln |x| + C$$

Using properties of logarithms, we can write this answer another way:

$$\int \frac{7}{x}\, dx = \ln |x^7| + C$$

b. *Find* $\int \frac{2x}{x^2 + 5}\, dx$.

Solution: Let $u = x^2 + 5$. Then $du = 2x\, dx$. From Equation (3),

$$\int \frac{2x}{x^2 + 5}\, dx = \int \frac{1}{x^2 + 5}[2x\, dx] = \int \frac{1}{u}\, du$$

$$= \ln |u| + C = \ln |x^2 + 5| + C$$

Since $x^2 + 5$ is always positive, we can omit the absolute-value bars:

$$\int \frac{2x}{x^2 + 5}\, dx = \ln(x^2 + 5) + C$$

NOW WORK PROBLEM 31

●EXAMPLE 7 **An Integral Involving** $\dfrac{1}{u} du$

Find $\displaystyle\int \frac{(2x^3 + 3x)\,dx}{x^4 + 3x^2 + 7}.$

Solution: If $u = x^4 + 3x^2 + 7$, then $du = (4x^3 + 6x)\,dx$, which is two times the numerator giving $(2x^3 + 3x)\,dx = \dfrac{du}{2}$. To apply Equation (3), we write

$$\int \frac{2x^3 + 3x}{x^4 + 3x^2 + 7}\,dx = \frac{1}{2}\int \frac{1}{u}\,du$$

$$= \frac{1}{2}\ln|u| + C$$

$$= \frac{1}{2}\ln|x^4 + 3x^2 + 7| + C \qquad \text{(Rewrite } u \text{ in terms of } x.)$$

$$= \frac{1}{2}\ln(x^4 + 3x^2 + 7) + C \qquad (x^4 + 3x^2 + 7 > 0 \quad \text{for all } x)$$

NOW WORK PROBLEM 51 ●●

●EXAMPLE 8 **An Integral Involving Two Forms**

Find $\displaystyle\int \left(\frac{1}{(1-w)^2} + \frac{1}{w-1}\right) dw.$

Solution:

$$\int \left(\frac{1}{(1-w)^2} + \frac{1}{w-1}\right) dw = \int (1-w)^{-2}dw + \int \frac{1}{w-1}dw$$

$$= -1\int (1-w)^{-2}[-dw] + \int \frac{1}{w-1}dw$$

The first integral has the form $\int u^{-2}\,du$, and the second has the form $\displaystyle\int \frac{1}{v}\,dv$. Thus,

$$\int \left(\frac{1}{(1-w)^2} + \frac{1}{w-1}\right) dw = -\frac{(1-w)^{-1}}{-1} + \ln|w-1| + C$$

$$= \frac{1}{1-w} + \ln|w-1| + C$$

●●

For your convenience, we list in Table 14.2 the basic integration formulas so far discussed. We assume that u is a function of x.

TABLE 14.2 Basic Integration Formulas

1. $\displaystyle\int k\,du = ku + C \qquad k$ a constant

2. $\displaystyle\int u^n\,du = \frac{u^{n+1}}{n+1} + C \qquad n \neq -1$

3. $\displaystyle\int \frac{1}{u}\,du = \ln|u| + C \qquad u \neq 0$

4. $\displaystyle\int e^u\,du = e^u + C$

5. $\displaystyle\int kf(x)\,dx = k\int f(x)\,dx \qquad k$ a constant

6. $\displaystyle\int [f(x) \pm g(x)]\,dx = \int f(x)\,dx \pm \int g(x)\,dx$

Problems 14.4

In Problems 1–80, find the indefinite integrals.

1. $\int (x+5)^7 \, dx$

2. $\int 15(x+2)^4 \, dx$

*3. $\int 2x(x^2+3)^5 \, dx$

4. $\int (3x^2+10x)(x^3+5x^2+6) \, dx$

*5. $\int (3y^2+6y)(y^3+3y^2+1)^{2/3} \, dy$

6. $\int (15t^2-6t+1)(5t^3-3t^2+t)^{17} \, dt$

7. $\int \frac{5}{(3x-1)^3} \, dx$ **8.** $\int \frac{4x}{(2x^2-7)^{10}} \, dx$

9. $\int \sqrt{2x-1} \, dx$ **10.** $\int \frac{1}{\sqrt{x-5}} \, dx$

11. $\int (7x-6)^4 \, dx$ **12.** $\int x^2(3x^3+7)^3 \, dx$

13. $\int u(5u^2-9)^{14} \, du$ **14.** $\int 9x\sqrt{1+2x^2} \, dx$

*15. $\int 4x^4(27+x^5)^{1/3} \, dx$ **16.** $\int (4-5x)^9 \, dx$

17. $\int 3e^{3x} \, dx$ **18.** $\int 5e^{3t+7} \, dt$

19. $\int (2t+1)e^{t^2+t} \, dt$ **20.** $\int -3w^2 e^{-w^3} \, dw$

21. $\int xe^{7x^2} \, dx$ **22.** $\int x^3 e^{4x^4} \, dx$

23. $\int 4e^{-3x} \, dx$ **24.** $\int x^4 e^{-6x^5} \, dx$

25. $\int \frac{1}{x+5} \, dx$ **26.** $\int \frac{12x^2+4x+2}{x+x^2+2x^3} \, dx$

27. $\int \frac{3x^2+4x^3}{x^3+x^4} \, dx$ **28.** $\int \frac{6x^2-6x}{1-3x^2+2x^3} \, dx$

29. $\int \frac{6z}{(z^2-6)^5} \, dx$ **30.** $\int \frac{3}{(5v-1)^4} \, dv$

*31. $\int \frac{4}{x} \, dx$ **32.** $\int \frac{3}{1+2y} \, dy$

33. $\int \frac{s^2}{s^3+5} \, ds$ **34.** $\int \frac{2x^2}{3-4x^3} \, dx$

35. $\int \frac{5}{4-2x} \, dx$ **36.** $\int \frac{7t}{5t^2-6} \, dt$

37. $\int \sqrt{5x} \, dx$ **38.** $\int \frac{1}{(3x)^6} \, dx$

39. $\int \frac{x}{\sqrt{x^2-4}} \, dx$ **40.** $\int \frac{9}{1-3x} \, dx$

*41. $\int 2y^3 e^{y^4+1} \, dx$ **42.** $\int 2\sqrt{2x-1} \, dx$

43. $\int v^2 e^{-2v^3+1} \, dv$ **44.** $\int \frac{x^2}{\sqrt[3]{2x^3+9}} \, dx$

45. $\int (e^{-5x}+2e^x) \, dx$ **46.** $\int 4\sqrt[3]{y+1} \, dy$

47. $\int (8x+10)(7-2x^2-5x)^3 \, dx$

48. $\int 2ye^{3y^2} \, dy$ **49.** $\int \frac{x^2+2}{x^3+6x} \, dx$

50. $\int (e^x+2e^{-3x}-e^{5x}) \, dx$ ***51.** $\int \frac{16s-4}{3-2s+4s^2} \, ds$

52. $\int (6t^2+4t)(t^3+t^2+1)^6 \, dt$

53. $\int x(2x^2+1)^{-1} \, dx$

54. $\int (8w^5+w^2-2)(6w-w^3-4w^6)^{-4} \, dw$

55. $\int -(x^2-2x^5)(x^3-x^6)^{-10} \, dx$

56. $\int \frac{3}{5}(v-2)e^{2-4v+v^2} \, dv$ **57.** $\int (2x^3+x)(x^4+x^2) \, dx$

58. $\int (e^{3.1})^2 \, dx$ **59.** $\int \frac{7+14x}{(4-x-x^2)^5} \, dx$

60. $\int (e^x-e^{-x})^2 \, dx$ **61.** $\int x(2x+1)e^{4x^3+3x^2-4} \, dx$

62. $\int (u^3-ue^{6-3u^2}) \, du$ **63.** $\int x\sqrt{(8-5x^2)^3} \, dx$

64. $\int e^{-x/7} \, dx$ **65.** $\int \left(\sqrt{2x}-\frac{1}{\sqrt{2x}}\right) \, dx$

66. $\int 3\frac{x^4}{e^{x^5}} \, dx$ ***67.** $\int (x^2+1)^2 \, dx$

68. $\int \left[x(x^2-16)^2-\frac{1}{2x+5}\right] \, dx$

69. $\int \left[\frac{x}{x^2+1}+\frac{x^5}{(x^6+1)^2}\right] \, dx$

70. $\int \left[\frac{3}{x-1}+\frac{1}{(x-1)^2}\right] \, dx$

71. $\int \left[\frac{2}{4x+1}-(4x^2-8x^5)(x^3-x^6)^{-8}\right] \, dx$

72. $\int (r^3+5)^2 \, dr$ **73.** $\int \left[\sqrt{3x+1}-\frac{x}{x^2+3}\right] \, dx$

74. $\int \left[\frac{x}{3x^2+5}-\frac{x^2}{(x^3+1)^3}\right] \, dx$

75. $\int \frac{e^{\sqrt{x}}}{\sqrt{x}} \, dx$ **76.** $\int (e^5-3^e) \, dx$

77. $\int \frac{1+e^{2x}}{4e^x} \, dx$ **78.** $\int \frac{2}{t^2}\sqrt{\frac{1}{t}+9} \, dt$

79. $\int \frac{x+1}{x^2+2x} \ln(x^2+2x) \, dx$ **80.** $\int \sqrt[3]{xe^{\sqrt[3]{8x^4}}} \, dx$

In Problems 81–84, find y subject to the given conditions.

81. $y'=(3-2x)^2$; $y(0)=1$

82. $y'=\frac{x}{x^2+6}$; $y(1)=0$

83. $y'' = \dfrac{1}{x^2}; \quad y'(-2) = 3, y(1) = 2$

84. $y'' = (x+1)^{3/2}; \quad y'(3) = 0, y(3) = 0$

85. Real Estate The rate of change of the value of a house that cost \$350,000 to build can be modeled by $\dfrac{dV}{dt} = 8e^{0.05t}$, where t is the time in years since the house was built and V is the value (in thousands of dollars) of the house. Find $V(t)$.

86. Life Span If the rate of change of the expected life span l at birth of people born in the United States can be modeled by $\dfrac{dl}{dt} = \dfrac{12}{2t+50}$, where t is the number of years after 1940 and the expected life span was 63 years in 1940, find the expected life span for people born in 1998.

87. Oxygen in Capillary In a discussion of the diffusion of oxygen from capillaries,[5] concentric cylinders of radius r are used as a model for a capillary. The concentration C of oxygen in the capillary is given by

$$C = \int \left(\frac{Rr}{2K} + \frac{B_1}{r} \right) dr$$

where R is the constant rate at which oxygen diffuses from the capillary, and K and B_1 are constants. Find C. (Write the constant of integration as B_2.)

88. Find $f(2)$ if $f\left(\frac{1}{3}\right) = 2$ and $f'(x) = e^{3x+2} - 3x$.

14.5 Techniques of Integration

We turn now to some more difficult integration problems.

When you are integrating fractions, sometimes a preliminary division is needed to get familiar integration forms, as the next example shows.

● EXAMPLE 1 **Preliminary Division before Integration**

a. *Find* $\displaystyle\int \frac{x^3 + x}{x^2}\, dx$.

Solution: A familiar integration form is not apparent. However, we can break up the integrand into two fractions by dividing each term in the numerator by the denominator. We then have

$$\int \frac{x^3 + x}{x^2}\, dx = \int \left(\frac{x^3}{x^2} + \frac{x}{x^2} \right) dx = \int \left(x + \frac{1}{x} \right) dx$$

$$= \frac{x^2}{2} + \ln |x| + C$$

Here we split up the integrand.

b. Find $\displaystyle\int \frac{2x^3 + 3x^2 + x + 1}{2x + 1}\, dx$.

Solution: Here the integrand is a quotient of polynomials in which the degree of the numerator is greater than or equal to that of the denominator. In such a situation we first use long division. Recall that if f and g are polynomials, with the degree of f greater than or equal to the degree of g, then long division allows us to find (uniquely) polynomials q and r, where either r is the zero polynomial or the degree of r is strictly less than the degree of g, satisfying

$$\frac{f}{g} = q + \frac{r}{g}$$

Using an obvious, abbreviated notation, we see that

$$\int \frac{f}{g} = \int \left(q + \frac{r}{g} \right) = \int q + \int \frac{r}{g}$$

Since integrating a polynomial is easy, we see that integrating rational functions reduces to the task of integrating *proper rational functions*—those for which the degree of the numerator is strictly less than the degree of the denominator. In

---

[5]W. Simon, *Mathematical Techniques for Physiology and Medicine* (New York: Academic Press, Inc., 1972).

this case we obtain

$$\int \frac{2x^3 + 3x^2 + x + 1}{2x + 1} \, dx = \int \left(x^2 + x + \frac{1}{2x + 1} \right) dx$$

$$= \frac{x^3}{3} + \frac{x^2}{2} + \int \frac{1}{2x + 1} \, dx$$

$$= \frac{x^3}{3} + \frac{x^2}{2} + \frac{1}{2} \int \frac{1}{2x + 1} \, d(2x + 1)$$

$$= \frac{x^3}{3} + \frac{x^2}{2} + \frac{1}{2} \ln |2x + 1| + C$$

NOW WORK PROBLEM 1

● EXAMPLE 2   **Indefinite Integrals**

a. *Find* $\int \dfrac{1}{\sqrt{x}(\sqrt{x} - 2)^3} \, dx.$

Solution: We can write this integral as $\int \dfrac{(\sqrt{x} - 2)^{-3}}{\sqrt{x}} \, dx.$ Let us try the power

rule for integration with $u = \sqrt{x} - 2.$ Then $du = \dfrac{1}{2\sqrt{x}} \, dx,$ so that $\dfrac{dx}{\sqrt{x}} = 2du,$

and

$$\int \frac{(\sqrt{x} - 2)^{-3}}{\sqrt{x}} \, dx = \int (\sqrt{x} - 2)^{-3} \left[\frac{dx}{\sqrt{x}} \right]$$

$$= 2 \int u^{-3} \, du = 2 \left(\frac{u^{-2}}{-2} \right) + C$$

$$= -\frac{1}{u^2} + C = -\frac{1}{(\sqrt{x} - 2)^2} + C$$

b. *Find* $\int \dfrac{1}{x \ln x} \, dx.$

Solution: If $u = \ln x,$ then $du = \dfrac{1}{x} \, dx,$ and

$$\int \frac{1}{x \ln x} \, dx = \int \frac{1}{\ln x} \left(\frac{1}{x} \, dx \right) = \int \frac{1}{u} \, du$$

$$= \ln |u| + C = \ln |\ln x| + C$$

c. *Find* $\int \dfrac{5}{w(\ln w)^{3/2}} \, dw.$

Solution: If $u = \ln w,$ then $du = \dfrac{1}{w} dw.$ Applying the power rule for integration, we have

$$\int \frac{5}{w(\ln w)^{3/2}} dw = 5 \int (\ln w)^{-3/2} \left[\frac{1}{w} dw \right]$$

$$= 5 \int u^{-3/2} \, du = 5 \cdot \frac{u^{-1/2}}{-\frac{1}{2}} + C$$

$$= \frac{-10}{u^{1/2}} + C = -\frac{10}{(\ln w)^{1/2}} + C$$

NOW WORK PROBLEM 23

Integrating b^u

In Section 14.4, we integrated an exponential function to the base e:

$$\int e^u \, du = e^u + C$$

Now let us consider the integral of an exponential function with an arbitrary base, b.

$$\int b^u \, du$$

To find this integral, we first convert to base e using

$$b^u = e^{(\ln b)u} \tag{1}$$

(as we did in many differentiation examples too). Example 3 will illustrate.

● **EXAMPLE 3**   **An Integral Involving b^u**

Find $\displaystyle\int 2^{3-x} \, dx$.

Solution:

Strategy   We want to integrate an exponential function to the base 2. To do this, we will first convert from base 2 to base e by using Equation (1).

$$\int 2^{3-x} \, dx = \int e^{(\ln 2)(3-x)} \, dx$$

The integrand of the second integral is of the form e^u, where $u = (\ln 2)(3 - x)$. Since $du = -\ln 2 \, dx$, we can solve for dx and write

$$\int e^{(\ln 2)(3-x)} \, dx = -\frac{1}{\ln 2} \int e^u \, du$$

$$= -\frac{1}{\ln 2} e^u + C = -\frac{1}{\ln 2} e^{(\ln 2)(3-x)} + C = -\frac{1}{\ln 2} 2^{3-x} + C$$

Thus,

$$\int 2^{3-x} \, dx = -\frac{1}{\ln 2} 2^{3-x} + C$$

Notice that we expressed our answer in terms of an exponential function to the base 2, the base of the original integrand.

NOW WORK PROBLEM 27

Generalizing the procedure described in Example 3, we can obtain a formula for integrating b^u:

$$\int b^u \, du = \int e^{(\ln b)u} \, du$$

$$= \frac{1}{\ln b} \int e^{(\ln b)u} \, d((\ln b)u) \qquad (\ln b \text{ is a constant})$$

$$= \frac{1}{\ln b} e^{(\ln b)u} + C$$

$$= \frac{1}{\ln b} b^u + C$$

Hence, we have

$$\int b^u \, du = \frac{1}{\ln b} b^u + C$$

Applying this formula to the integral in Example 3 gives

$$\int 2^{3-x} \, dx \qquad (b = 2, u = 3 - x)$$

$$= -\int 2^{3-x} \, d(3 - x) \qquad (-d(3 - x) = dx)$$

$$= -\frac{1}{\ln 2} 2^{3-x} + C$$

which is the same result that we obtained before.

Application of Integration

We will now consider an application of integration that relates a consumption function to the marginal propensity to consume.

● EXAMPLE 4 **Finding a Consumption Function from Marginal Propensity to Consume**

For a certain country, the marginal propensity to consume is given by

$$\frac{dC}{dI} = \frac{3}{4} - \frac{1}{2\sqrt{3I}}$$

where consumption C is a function of national income I. Here I is expressed in large denominations of money. Determine the consumption function for the country if it is known that consumption is 10 (C = 10) when I = 12.

Solution: Since the marginal propensity to consume is the derivative of C, we have

$$C = C(I) = \int \left(\frac{3}{4} - \frac{1}{2\sqrt{3I}} \right) dI = \int \frac{3}{4} dI - \frac{1}{2} \int (3I)^{-1/2} dI$$

$$= \frac{3}{4}I - \frac{1}{2} \int (3I)^{-1/2} dI$$

If we let $u = 3I$, then $du = 3\,dI = d(3I)$, and

$$C = \frac{3}{4}I - \left(\frac{1}{2} \right) \frac{1}{3} \int (3I)^{-1/2} d(3I)$$

$$= \frac{3}{4}I - \frac{1}{6} \frac{(3I)^{1/2}}{\frac{1}{2}} + K$$

$$C = \frac{3}{4}I - \frac{\sqrt{3I}}{3} + K$$

This is an example of an initial-value problem.

When $I = 12$, $C = 10$, so

$$10 = \frac{3}{4}(12) - \frac{\sqrt{3(12)}}{3} + K$$

$$10 = 9 - 2 + K$$

Thus, $K = 3$, and the consumption function is

$$C = \frac{3}{4}I - \frac{\sqrt{3I}}{3} + 3$$

NOW WORK PROBLEM 61 ●●

Problems 14.5

In Problems 1–56, determine the indefinite integrals.

*1. $\displaystyle\int \frac{2x^6 + 8x^4 - 4x}{2x^2} \, dx$

2. $\displaystyle\int \frac{9x^2 + 5}{3x} \, dx$

3. $\displaystyle\int (3x^2 + 2)\sqrt{2x^3 + 4x + 1} \, dx$

4. $\displaystyle\int \frac{x}{\sqrt[4]{x^2 + 1}} \, dx$

5. $\displaystyle\int \frac{9}{\sqrt{2 - 3x}} \, dx$

6. $\displaystyle\int \frac{2xe^{x^2} \, dx}{e^{x^2} - 2}$

7. $\displaystyle\int 4^{7x} \, dx$

8. $\displaystyle\int 5^t \, dt$

9. $\displaystyle\int 2x(7 - e^{x^2/4}) \, dx$

10. $\displaystyle\int \left(e^x + x^e + ex + \frac{e}{x} \right) dx$

11. $\displaystyle\int \frac{6x^2 - 11x + 5}{3x - 1} \, dx$

12. $\displaystyle\int \frac{(3x + 2)(x - 4)}{x - 3} \, dx$

13. $\displaystyle\int \frac{5e^{2x}}{7e^{2x} + 4} \, dx$

14. $\displaystyle\int 6(e^{4 - 3x})^2 \, dx$

15. $\displaystyle\int \frac{e^{7/x}}{x^2} \, dx$

16. $\displaystyle\int \frac{2x^4 - 6x^3 + x - 2}{x - 2} \, dx$

17. $\displaystyle\int \frac{5x^3}{x^2 + 9} \, dx$

18. $\displaystyle\int \frac{5 - 4x^2}{3 + 2x} \, dx$

19. $\displaystyle\int \frac{(\sqrt{x} + 2)^2}{3\sqrt{x}} \, dx$

20. $\int \dfrac{5e^s}{1 + 3e^s}\, ds$

21. $\int \dfrac{5(x^{1/3} + 2)^4}{\sqrt[3]{x^2}}\, dx$

22. $\int \dfrac{\sqrt{1 + \sqrt{x}}}{\sqrt{x}}\, dx$

*23. $\int \dfrac{\ln x}{x}\, dx$

24. $\int \sqrt{t}\,(3 - t\sqrt{t})^{0.6}\, dt$

25. $\int \dfrac{\ln^2(r + 1)}{r + 1}\, dr$

26. $\int \dfrac{9x^5 - 6x^4 - ex^3}{7x^2}\, dx$

*27. $\int \dfrac{3^{\ln x}}{x}\, dx$

28. $\int \dfrac{4}{x \ln(2x^2)}\, dx$

29. $\int x^2 \sqrt{e^{x^3+1}}\, dx$

30. $\int \dfrac{x + 3}{x + 6}\, dx$

31. $\int \dfrac{8}{(x + 3)\ln(x + 3)}\, dx$

32. $\int (e^{e^2} + x^e - 2x)\, dx$

33. $\int \dfrac{x^3 + x^2 - x - 3}{x^2 - 3}\, dx$

34. $\int \dfrac{4x \ln \sqrt{1 + x^2}}{1 + x^2}\, dx$

35. $\int \dfrac{6x^2 \sqrt{\ln(x^3 + 1)^2}}{x^3 + 1}\, dx$

36. $\int 3(x^2 + 2)^{-1/2} x e^{\sqrt{x^2+2}}\, dx$

37. $\int \left(\dfrac{x^3 - 1}{\sqrt{x^4 - 4x}} - \ln 7 \right) dx$

38. $\int \dfrac{x - x^{-2}}{x^2 + 2x^{-1}}\, dx$

39. $\int \dfrac{2x^4 - 8x^3 - 6x^2 + 4}{x^3}\, dx$

40. $\int \dfrac{e^x + e^{-x}}{e^x - e^{-x}}\, dx$

41. $\int \dfrac{x}{x + 1}\, dx$

42. $\int \dfrac{2x}{(x^2 + 1)\ln(x^2 + 1)}\, dx$

43. $\int \dfrac{xe^{x^2}}{\sqrt{e^{x^2} + 2}}\, dx$

44. $\int \dfrac{5}{(3x + 1)[1 + \ln(3x + 1)]^2}\, dx$

45. $\int \dfrac{(e^{-x} + 6)^2}{e^x}\, dx$

46. $\int \left[\dfrac{1}{8x + 1} - \dfrac{1}{e^x(8 + e^{-x})^2} \right] dx$

47. $\int (x^3 + ex)\sqrt{x^2 + e}\, dx$

48. $\int 3^{x \ln x}(1 + \ln x)\, dx$ $\left(\textit{Hint: } \dfrac{d}{dx}(x \ln x) = 1 + \ln x \right)$

49. $\int \sqrt{x}\sqrt{(8x)^{3/2} + 3}\, dx$

50. $\int \dfrac{2}{x(\ln x)^{2/3}}\, dx$

51. $\int \dfrac{\sqrt{s}}{e^{\sqrt{s^3}}}\, ds$

52. $\int \dfrac{\ln^3 x}{3x}\, dx$

53. $\int e^{\ln(x^2+1)}\, dx$

54. $\int dx$

55. $\int \dfrac{\ln(xe^x)}{x}\, dx$

56. $\int e^{f(x) + \ln(f'(x))}\, dx$ assuming $f' > 0$

In Problems 57 and 58, dr/dq is a marginal-revenue function. Find the demand function.

57. $\dfrac{dr}{dq} = \dfrac{200}{(q + 2)^2}$

58. $\dfrac{dr}{dq} = \dfrac{900}{(2q + 3)^3}$

In Problems 59 and 60, dc/dq is a marginal-cost function. Find the total-cost function if fixed costs in each case are 2000.

59. $\dfrac{dc}{dq} = \dfrac{20}{q + 5}$

60. $\dfrac{dc}{dq} = 3e^{0.002q}$

In Problems 61–63, dC/dI represents the marginal propensity to consume. Find the consumption function subject to the given condition.

*61. $\dfrac{dC}{dI} = \dfrac{1}{\sqrt{I}}$; $C(9) = 8$

62. $\dfrac{dC}{dI} = \dfrac{1}{2} - \dfrac{1}{2\sqrt{2I}}$; $C(2) = \dfrac{3}{4}$

63. $\dfrac{dC}{dI} = \dfrac{3}{4} - \dfrac{1}{6\sqrt{I}}$; $C(25) = 23$.

64. **Cost Function** The marginal-cost function for a manufacturer's product is given by

$$\dfrac{dc}{dq} = 10 - \dfrac{100}{q + 10}$$

where c is the total cost in dollars when q units are produced. When 100 units are produced, the average cost is $50 per unit. To the nearest dollar, determine the manufacturer's fixed cost.

65. **Cost Function** Suppose the marginal-cost function for a manufacturer's product is given by

$$\dfrac{dc}{dq} = \dfrac{100q^2 - 3998q + 60}{q^2 - 40q + 1}$$

where c is the total cost in dollars when q units are produced.

(a) Determine the marginal cost when 40 units are produced.

(b) If fixed costs are $10,000, find the total cost of producing 40 units.

(c) Use the results of parts (a) and (b) and differentials to approximate the total cost of producing 42 units.

66. **Cost Function** The marginal-cost function for a manufacturer's product is given by

$$\dfrac{dc}{dq} = \dfrac{9}{10}\sqrt{q}\sqrt{0.04q^{3/4} + 4}$$

where c is the total cost in dollars when q units are produced. Fixed costs are $360.

(a) Determine the marginal cost when 25 units are produced.

(b) Find the total cost of producing 25 units.

(c) Use the results of parts (a) and (b) and differentials to approximate the total cost of producing 23 units.

67. **Value of Land** It is estimated that t years from now the value V (in dollars) of an acre of land near the ghost town of Cherokee, California, will be increasing at the rate of

$$\dfrac{8t^3}{\sqrt{0.2t^4 + 8000}}$$ dollars per year. If the land is currently worth $500 per acre, how much will it be worth in 10 years? Express your answer to the nearest dollar.

68. **Revenue Function** The marginal-revenue function for a manufacturer's product is of the form

$$\dfrac{dr}{dq} = \dfrac{a}{e^q + b}$$

for constants a and b, where r is the total revenue received (in dollars) when q units are produced and sold. Find the demand function, and express it in the form $p = f(q)$. (*Hint: Rewrite dr/dq by multiplying both numerator and denominator by e^{-q}.*)

69. Savings A certain country's marginal propensity to save is given by

$$\frac{dS}{dI} = \frac{5}{(I+2)^2}$$

where S and I represent total national savings and income, respectively, and are measured in billions of dollars. If total national consumption is $7.5 billion when total national income is $8 billion, for what value(s) of I is total national savings equal to zero?

70. Consumption Function A certain country's marginal propensity to save is given by

$$\frac{dS}{dI} = \frac{1}{2} - \frac{1.8}{\sqrt[3]{3I^2}}$$

where S and I represent total national savings and income, respectively, and are measured in billions of dollars.

(a) Determine the marginal propensity to consume when total national income is $81 billion.

(b) Determine the consumption function, given that savings are $3 billion when total national income is $24 billion.

(c) Use the result in part (b) to show that consumption is $54.9 billion when total national income is $81 billion.

(d) Use differentials and the results in parts (a) and (c) to approximate consumption when total national income is $78 billion.

OBJECTIVE

To motivate, by means of the concept of area, the definite integral as a limit of a special sum; to evaluate simple definite integrals by using a limiting process.

14.6 The Definite Integral

Figure 14.2 shows the region R bounded by the lines $y = f(x) = 2x$, $y = 0$ (the x-axis), and $x = 1$. The region is simply a right triangle. If b and h are the lengths of the base and the height, respectively, then, from geometry, the area of the triangle is $A = \frac{1}{2}bh = \frac{1}{2}(1)(2) = 1$ square unit. (Henceforth, we will treat areas as pure numbers and write *square unit* only if it seems necessary for emphasis.) We will now find this area by another method, which, as you will see later, applies to more complex regions. This method involves the summation of areas of rectangles.

Let us divide the interval $[0, 1]$ on the x-axis into four subintervals of equal length by means of the equally spaced points $x_0 = 0$, $x_1 = \frac{1}{4}$, $x_2 = \frac{2}{4}$, $x_3 = \frac{3}{4}$, and $x_4 = \frac{4}{4} = 1$. (See Figure 14.3.) Each subinterval has length $\Delta x = \frac{1}{4}$. These subintervals determine four subregions of R: R_1, R_2, R_3, and R_4, as indicated.

With each subregion, we can associate a *circumscribed* rectangle (Figure 14.4)—that is, a rectangle whose base is the corresponding subinterval and whose height is the *maximum* value of $f(x)$ on that subinterval. Since f is an increasing function, the maximum value of $f(x)$ on each subinterval occurs when x is the right-hand endpoint. Thus, the areas of the circumscribed rectangles associated with regions R_1, R_2, R_3, and R_4 are $\frac{1}{4}f(\frac{1}{4})$, $\frac{1}{4}f(\frac{2}{4})$, $\frac{1}{4}f(\frac{3}{4})$, and $\frac{1}{4}f(\frac{4}{4})$, respectively. The area of each rectangle is an approximation to the area of its corresponding subregion. Hence, the sum of the areas of these rectangles, denoted by $\overline{S}_4$ (read "S upper bar sub 4" or "the fourth

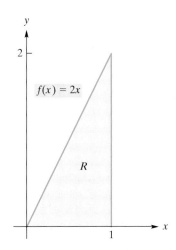

FIGURE 14.2 Region bounded by $f(x) = 2x$, $y = 0$, and $x = 1$.

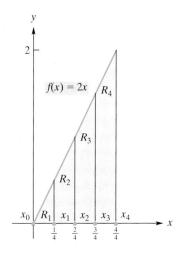

FIGURE 14.3 Four subregions of R.

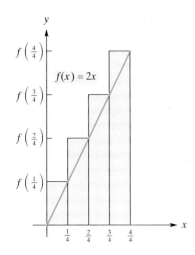

FIGURE 14.4   Four circumscribed rectangles.

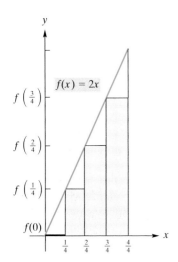

FIGURE 14.5   Four inscribed rectangles.

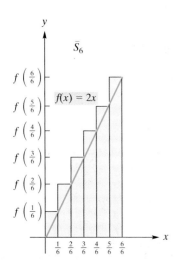

FIGURE 14.6   Six circumscribed rectangles.

upper sum"), approximates the area A of the triangle. We have

$$\overline{S}_4 = \tfrac{1}{4} f\left(\tfrac{1}{4}\right) + \tfrac{1}{4} f\left(\tfrac{2}{4}\right) + \tfrac{1}{4} f\left(\tfrac{3}{4}\right) + \tfrac{1}{4} f\left(\tfrac{4}{4}\right)$$

$$= \tfrac{1}{4}\left(2\left(\tfrac{1}{4}\right) + 2\left(\tfrac{2}{4}\right) + 2\left(\tfrac{3}{4}\right) + 2\left(\tfrac{4}{4}\right)\right) = \tfrac{5}{4}$$

You can verify that $\overline{S}_4 = \sum_{i=1}^{4} f(x_i)\Delta x$. The fact that $\overline{S}_4$ is greater than the actual area of the triangle might have been expected, since $\overline{S}_4$ includes areas of shaded regions that are not in the triangle. (See Figure 14.4.)

On the other hand, with each subregion we can also associate an *inscribed* rectangle (Figure 14.5)—that is, a rectangle whose base is the corresponding subinterval, but whose height is the *minimum* value of $f(x)$ on that subinterval. Since f is an increasing function, the minimum value of $f(x)$ on each subinterval will occur when x is the left-hand endpoint. Thus, the areas of the four inscribed rectangles associated with R_1, R_2, R_3, and R_4 are $\tfrac{1}{4} f(0)$, $\tfrac{1}{4} f\left(\tfrac{1}{4}\right)$, $\tfrac{1}{4} f\left(\tfrac{2}{4}\right)$, and $\tfrac{1}{4} f\left(\tfrac{3}{4}\right)$, respectively. Their sum, denoted $\underline{S}_4$ (read "S lower bar sub 4" or "the fourth lower sum"), is also an approximation to the area A of the triangle. We have

$$\underline{S}_4 = \tfrac{1}{4} f(0) + \tfrac{1}{4} f\left(\tfrac{1}{4}\right) + \tfrac{1}{4} f\left(\tfrac{2}{4}\right) + \tfrac{1}{4} f\left(\tfrac{3}{4}\right)$$

$$= \tfrac{1}{4}\left(2(0) + 2\left(\tfrac{1}{4}\right) + 2\left(\tfrac{2}{4}\right) + 2\left(\tfrac{3}{4}\right)\right) = \tfrac{3}{4}$$

Using summation notation, we can write $\underline{S}_4 = \sum_{i=0}^{3} f(x_i)\Delta x$. Note that $\underline{S}_4$ is less than the area of the triangle, because the rectangles do not account for the portion of the triangle that is not shaded in Figure 14.5.

Since

$$\frac{3}{4} = \underline{S}_4 \le A \le \overline{S}_4 = \frac{5}{4}$$

we say that $\underline{S}_4$ is an approximation to A from *below* and $\overline{S}_4$ is an approximation to A from *above*.

If $[0, 1]$ is divided into more subintervals, we expect that better approximations to A will occur. To test this, let us use six subintervals of equal length $\Delta x = \tfrac{1}{6}$. Then $\overline{S}_6$, the total area of six circumscribed rectangles (see Figure 14.6), and $\underline{S}_6$, the total area of six inscribed rectangles (see Figure 14.7), are

$$\overline{S}_6 = \tfrac{1}{6} f\left(\tfrac{1}{6}\right) + \tfrac{1}{6} f\left(\tfrac{2}{6}\right) + \tfrac{1}{6} f\left(\tfrac{3}{6}\right) + \tfrac{1}{6} f\left(\tfrac{4}{6}\right) + \tfrac{1}{6} f\left(\tfrac{5}{6}\right) + \tfrac{1}{6} f\left(\tfrac{6}{6}\right)$$

$$= \tfrac{1}{6}\left(2\left(\tfrac{1}{6}\right) + 2\left(\tfrac{2}{6}\right) + 2\left(\tfrac{3}{6}\right) + 2\left(\tfrac{4}{6}\right) + 2\left(\tfrac{5}{6}\right) + 2\left(\tfrac{6}{6}\right)\right) = \tfrac{7}{6}$$

and

$$\underline{S}_6 = \tfrac{1}{6} f(0) + \tfrac{1}{6} f\left(\tfrac{1}{6}\right) + \tfrac{1}{6} f\left(\tfrac{2}{6}\right) + \tfrac{1}{6} f\left(\tfrac{3}{6}\right) + \tfrac{1}{6} f\left(\tfrac{4}{6}\right) + \tfrac{1}{6} f\left(\tfrac{5}{6}\right)$$

$$= \tfrac{1}{6}\left(2(0) + 2\left(\tfrac{1}{6}\right) + 2\left(\tfrac{2}{6}\right) + 2\left(\tfrac{3}{6}\right) + 2\left(\tfrac{4}{6}\right) + 2\left(\tfrac{5}{6}\right)\right) = \tfrac{5}{6}$$

Note that $\underline{S}_6 \le A \le \overline{S}_6$, and, with appropriate labeling, both $\overline{S}_6$ and $\underline{S}_6$ will be of the form $\Sigma f(x)\,\Delta x$. Clearly, using six subintervals gives better approximations to the area than does four subintervals, as expected.

More generally, if we divide $[0, 1]$ into n subintervals of equal length Δx, then $\Delta x = 1/n$, and the endpoints of the subintervals are $x = 0, 1/n, 2/n, \ldots,$ $(n - 1)/n$, and $n/n = 1$. (See Figure 14.8.) The endpoints of the kth subinterval, for $k = 1, \ldots n$, are $(k - 1)/n$ and k/n and the maximum value of f occurs at the right-hand endpoint k/n. It follows the area of the kth circumscribed rectangle is $1/n \cdot f(k/n) = 1/n \cdot 2(k/n) = 2k/n^2$, for $k = 1, \ldots n$. The total area of *all n circumscribed* rectangles is

$$\overline{S}_n = \sum_{k=1}^{n} f(k/n)\Delta x = \sum_{k=1}^{n} \frac{2k}{n^2} \tag{1}$$

$$= \frac{2}{n^2} \sum_{k=1}^{n} k \qquad \left(\text{by factoring } \frac{2}{n^2} \text{ from each term}\right)$$

$$= \frac{2}{n^2} \cdot \frac{n(n + 1)}{2} \qquad \text{(from Section 1.5)}$$

$$= \frac{n + 1}{n}$$

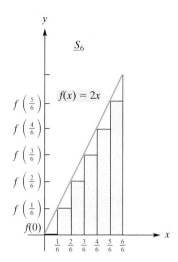

FIGURE 14.7   Six inscribed rectangles.

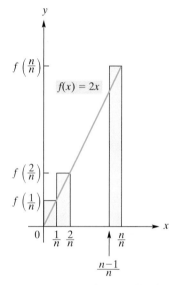

FIGURE 14.8 n circumscribed rectangles.

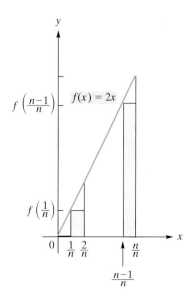

FIGURE 14.9 n inscribed rectangles.

(We recall that $\sum_{k=1}^{n} k = 1 + 2 + \cdots + n$ is the sum of the first n positive integers and the formula used above was derived in Section 1.5 in anticipation of its application here.)

For *inscribed* rectangles, we note that the minimum value of f occurs at the left-hand endpoint, $(k-1)/n$, of $[(k-1)/n, k/n]$ so that the area of the kth inscribed rectangle is $1/n \cdot f(k-1/n) = 1/n \cdot 2((k-1)/n) = 2(k-1)/n^2$, for $k = 1, \ldots n$. The total area determined of *all n inscribed* rectangles (see Figure 14.9) is

$$\underline{S}_n = \sum_{k=1}^{n} f((k-1)/n)\,\Delta x = \sum_{k=1}^{n} \frac{2(k-1)}{n^2} \tag{2}$$

$$= \frac{2}{n^2} \sum_{k=1}^{n} k - 1 \qquad \left(\text{by factoring } \frac{2}{n^2} \text{ from each term} \right)$$

$$= \frac{2}{n^2} \sum_{k=0}^{n-1} k \qquad \text{(adjusting the summation)}$$

$$= \frac{2}{n^2} \cdot \frac{(n-1)n}{2} \qquad \text{(adapted from Section 1.5)}$$

$$= \frac{n-1}{n}$$

From Equations (1) and (2), we again see that both $\overline{S}_n$ and $\underline{S}_n$ are sums of the form $\sum f(x)\Delta x$, namely, $\overline{S}_n = \sum_{k=1}^{n} f\left(\frac{k}{n}\right) \Delta x$ and $\underline{S}_n = \sum_{k=1}^{n} f\left(\frac{k-1}{n}\right) \Delta x$.

From the nature of $\overline{S}_n$ and $\underline{S}_n$, it seems reasonable—and it is indeed true—that

$$\underline{S}_n \leq A \leq \overline{S}_n$$

As n becomes larger, $\underline{S}_n$ and $\overline{S}_n$ become better approximations to A. In fact, let us take the limits of $\underline{S}_n$ and $\overline{S}_n$ as n approaches ∞ through positive integral values:

$$\lim_{n \to \infty} \underline{S}_n = \lim_{n \to \infty} \frac{n-1}{n} = \lim_{n \to \infty} \left(1 - \frac{1}{n} \right) = 1$$

$$\lim_{n \to \infty} \overline{S}_n = \lim_{n \to \infty} \frac{n+1}{n} = \lim_{n \to \infty} \left(1 + \frac{1}{n} \right) = 1$$

Since $\overline{S}_n$ and $\underline{S}_n$ have the same limit, namely,

$$\lim_{n \to \infty} \overline{S}_n = \lim_{n \to \infty} \underline{S}_n = 1 \tag{3}$$

and since

$$\underline{S}_n \leq A \leq \overline{S}_n$$

we will take this limit to be the area of the triangle. Thus $A = 1$, which agrees with our prior finding. It is important to understand that here we developed a *definition of the notion of area* that is applicable to many different regions.

We call the common limit of $\overline{S}_n$ and $\underline{S}_n$, namely, 1, the *definite integral* of $f(x) = 2x$ on the interval from $x = 0$ to $x = 1$, and we denote this quantity by writing

$$\int_0^1 2x \, dx = 1 \tag{4}$$

The reason for using the term *definite integral* and the symbolism in Equation (4) will become apparent in the next section. The numbers 0 and 1 appearing with the integral sign $\int$ in Equation (4) are called the *limits of integration;* 0 is the *lower limit* and 1 is the *upper limit.*

In general, for a function f defined on the interval from $x = a$ to $x = b$, where $a < b$, we can form the sums $\overline{S}_n$ and $\underline{S}_n$, which are obtained by considering the maximum and minimum values, respectively, on each of n subintervals of equal length Δx.[6] We can now state the following:

The common limit of $\overline{S}_n$ and $\underline{S}_n$ as $n \to \infty$, if it exists, is called the **definite integral** of f over $[a, b]$ and is written

$$\int_a^b f(x)\, dx$$

The numbers a and b are called **limits of integration;** a is the **lower limit** and b is the **upper limit.** The symbol x is called the **variable of integration** and $f(x)$ is the **integrand.**

In terms of a limiting process, we have

$$\sum f(x)\, \Delta x \to \int_a^b f(x)\, dx$$

The definite integral is the limit of sums of the form $\sum f(x)\, \Delta x$. This interpretation will be useful in later sections.

Two points must be made about the definite integral. First, the definite integral is the limit of a sum of the form $\sum f(x)\, \Delta x$. In fact, we can think of the integral sign as an elongated "S," the first letter of "Summation." Second, for an arbitrary function f defined on an interval, we may be able to calculate the sums $\overline{S}_n$ and $\underline{S}_n$ and determine their common limit if it exists. However, some terms in the sums may be negative if $f(x)$ is negative at points in the interval. These terms are not areas of rectangles (an area is never negative), so the common limit may not represent an area. Thus, **the definite integral is nothing more than a real number; it may or may not represent an area.**

As you saw in Equation (3), $\lim_{n \to \infty} \underline{S}_n$ is equal to $\lim_{n \to \infty} \overline{S}_n$. For an arbitrary function, this is not always true. However, for the functions that we will consider, these limits will be equal, and the definite integral will always exist. To save time, we will just use the **right-hand endpoint** of each subinterval in computing a sum. For the functions in this section, this sum will be denoted S_n.

PRINCIPLES IN PRACTICE 1

COMPUTING AN AREA BY USING RIGHT-HAND ENDPOINTS

A company has determined that its marginal-revenue function is given by $R'(x) = 600 - 0.5x$, where R is the revenue (in dollars) received when x units are sold. Find the total revenue received for selling 10 units by finding the area in the first quadrant bounded by $y = R'(x) = 600 - 0.5x$ and the lines $y = 0$, $x = 0$, and $x = 10$.

In general, over $[a, b]$, we have

$$\Delta x = \frac{b - a}{n}$$

● **EXAMPLE 1** **Computing an Area by Using Right-Hand Endpoints**

Find the area of the region in the first quadrant bounded by $f(x) = 4 - x^2$ and the lines $x = 0$ and $y = 0$.

Solution: A sketch of the region appears in Figure 14.10. The interval over which x varies in this region is seen to be $[0, 2]$, which we divide into n subintervals of equal length Δx. Since the length of $[0, 2]$ is 2, we take $\Delta x = 2/n$. The endpoints of the subintervals are $x = 0, 2/n, 2(2/n), \ldots, (n-1)(2/n)$, and $n(2/n) = 2$, which are shown in Figure 14.11. The diagram also shows the corresponding rectangles obtained by using the right-hand endpoint of each subinterval. The area of the kth rectangle, for $k = 1, \ldots n$, is the product of its width, $2/n$, and its height, $f(k(2/n)) = 4 - (2k/n)^2$, which is the function value at the right-hand endpoint of its base. Summing these areas, we get

$$S_n = \sum_{k=1}^{n} f\left(k \cdot \left(\frac{2}{n}\right)\right) \Delta x = \sum_{k=1}^{n} \left(4 - \left(\frac{2k}{n}\right)^2\right) \frac{2}{n}$$

$$= \sum_{k=1}^{n} \left(\frac{8}{n} - \frac{8k^2}{n^3}\right) = \sum_{k=1}^{n} \frac{8}{n} - \sum_{k=1}^{n} \frac{8k^2}{n^3} = \frac{8}{n} \sum_{k=1}^{n} 1 - \frac{8}{n^3} \sum_{k=1}^{n} k^2$$

$$= \frac{8}{n} n - \frac{8}{n^3} \frac{n(n+1)(2n+1)}{6}$$

$$= 8 - \frac{4}{3} \left(\frac{(n+1)(2n+1)}{n^2}\right)$$

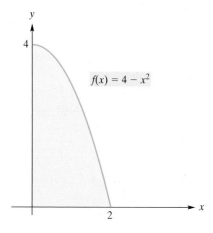

FIGURE 14.10 Region of Example 1.

$f(x) = 4 - x^2$

_____

[6]Here we assume that the maximum and minimum values exist.

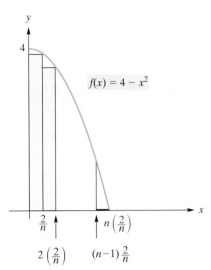

$$f(x) = 4 - x^2$$

$$\frac{2}{n} \qquad n\left(\frac{2}{n}\right)$$

$$2\left(\frac{2}{n}\right) \qquad (n-1)\frac{2}{n}$$

FIGURE 14.11 *n* subintervals and corresponding rectangles for Example 1.

The second line of the preceding computations uses basic summation manipulations as discussed in Section 1.5. The third line uses two specific summation formulas, also from Section 1.5: The sum of *n* copies of 1 is *n* and the sum of the first *n* squares is $\dfrac{n(n+1)(2n+1)}{6}$.

Finally, we take the limit of the S_n as $n \to \infty$:

$$
\begin{aligned}
\lim_{n\to\infty} S_n &= \lim_{n\to\infty}\left(8 - \frac{4}{3}\left(\frac{(n+1)(2n+1)}{n^2}\right)\right) \\
&= 8 - \frac{4}{3}\lim_{n\to\infty}\left(\frac{2n^2 + 3n + 1}{n^2}\right) \\
&= 8 - \frac{4}{3}\lim_{n\to\infty}\left(2 + \frac{3}{n} + \frac{1}{n^2}\right) \\
&= 8 - \frac{8}{3} = \frac{16}{3}
\end{aligned}
$$

Hence, the area of the region is $\dfrac{16}{3}$.

NOW WORK PROBLEM 7 ●●●

● EXAMPLE 2   Evaluating a Definite Integral

Evaluate $\displaystyle\int_0^2 (4 - x^2)\,dx$.

Solution: We want to find the definite integral of $f(x) = 4 - x^2$ over the interval $[0, 2]$. Thus, we must compute $\lim_{n\to\infty} S_n$. But this limit is precisely the limit $\dfrac{16}{3}$ found in Example 1, so we conclude that

$$\int_0^2 (4 - x^2)\,dx = \frac{16}{3}$$

No units are attached to the answer, since a definite integral is simply a number.

NOW WORK PROBLEM 19 ●●●

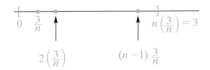

$$0 \quad \frac{3}{n} \qquad\qquad\qquad n\left(\frac{3}{n}\right) = 3$$

$$2\left(\frac{3}{n}\right) \qquad (n-1)\frac{3}{n}$$

FIGURE 14.12   Dividing $[0, 3]$ into *n* subintervals.

● EXAMPLE 3   Integrating a Function over an Interval

Integrate $f(x) = x - 5$ *from* $x = 0$ *to* $x = 3$; *that is, evaluate* $\int_0^3 (x - 5)\,dx$.

Solution: We first divide $[0, 3]$ into *n* subintervals of equal length $\Delta x = 3/n$. The endpoints are $0, 3/n, 2(3/n), \ldots, (n-1)(3/n), n(3/n) = 3$. (See Figure 14.12.) Using right-hand endpoints, we form the sum and simplify

$$
\begin{aligned}
S_n &= \sum_{k=1}^{n} f\left(k\frac{3}{n}\right)\frac{3}{n} \\
&= \sum_{k=1}^{n}\left(\left(k\frac{3}{n} - 5\right)\frac{3}{n}\right) = \sum_{k=1}^{n}\left(\frac{9}{n^2}k - \frac{15}{n}\right) = \frac{9}{n^2}\sum_{k=1}^{n}k - \frac{15}{n}\sum_{k=1}^{n}1 \\
&= \frac{9}{n^2}\left(\frac{n(n+1)}{2}\right) - \frac{15}{n}(n) \\
&= \frac{9}{2}\frac{n+1}{n} - 15 = \frac{9}{2}\left(1 + \frac{1}{n}\right) - 15
\end{aligned}
$$

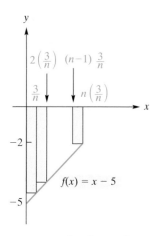

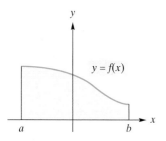

FIGURE 14.13 $f(x)$ is negative at each right-hand endpoint.

FIGURE 14.14 If f is continuous and $f(x) \geq 0$ on $[a, b]$, then $\int_a^b f(x)\,dx$ represents the area under the curve.

Taking the limit, we obtain

$$\lim_{n \to \infty} S_n = \lim_{n \to \infty} \left(\frac{9}{2} \left(1 + \frac{1}{n} \right) - 15 \right) = \frac{9}{2} - 15 = -\frac{21}{2}$$

Thus,

$$\int_0^3 (x - 5)\,dx = -\frac{21}{2}$$

Note that the definite integral here is a *negative* number. The reason is clear from the graph of $f(x) = x - 5$ over the interval $[0, 3]$. (See Figure 14.13.) Since the value of $f(x)$ is negative at each right-hand endpoint, each term in S_n must also be negative. Hence, $\lim_{n \to \infty} S_n$, which is the definite integral, is negative.

Geometrically, each term in S_n is the negative of the area of a rectangle. (Refer again to Figure 14.13.) Although the definite integral is simply a number, here we can interpret it as representing the negative of the area of the region bounded by $f(x) = x - 5$, $x = 0$, $x = 3$, and the x-axis ($y = 0$).

NOW WORK PROBLEM 17

In Example 3, it was shown that *the definite integral does not have to represent an area*. In fact, there the definite integral was negative. However, if f is continuous and $f(x) \geq 0$ on $[a, b]$, then $S_n \geq 0$ for all values of n. Therefore, $\lim_{n \to \infty} S_n \geq 0$, so $\int_a^b f(x)\,dx \geq 0$. Furthermore, this definite integral gives the area of the region bounded by $y = f(x)$, $y = 0$, $x = a$, and $x = b$. (See Figure 14.14.)

Although the approach that we took to discuss the definite integral is sufficient for our purposes, it is by no means rigorous. **The important thing to remember about the definite integral is that it is the limit of a special sum.**

TECHNOLOGY

Here is a program for the TI-83 Plus graphing calculator that will estimate the limit of S_n as $n \to \infty$ for a function f defined on $[a, b]$.

PROGRAM:RIGHTSUM
Lbl 1
Input "SUBINTV",N
(B − A)/N → H
Ø → S
A + H → X
1 → I
Lbl 2
Y$_1$ + S → S
X + H → X
I + 1 → I
If I ≤ N
Goto 2
H*S → S
Disp S
Pause
Goto 1

```
prgmRIGHTSUM
SUBINTV100
              -10.455
SUBINTV1000
              -10.4955
SUBINTV2000
              -10.49775
■
```

FIGURE 14.15 Values of S_n for $f(x) = x - 5$ on $[0, 3]$.

value of S_n. Each time ENTER is pressed, the program repeats. In this way, a display of values of S_n for various numbers of subintervals may be obtained. Figure 14.15 shows values of $S_n (n = 100, 1000, \text{and } 2000)$ for the function $f(x) = x - 5$ on the interval $[0, 3]$. As $n \to \infty$, it appears that $S_n \to -10.5$. Thus, we estimate that

$$\lim_{n \to \infty} S_n \approx -10.5$$

Equivalently,

$$\int_0^3 (x - 5)\,dx \approx -10.5$$

which agrees with our result in Example 3.

It is interesting to note that the time required for an older calculator to compute S_{2000} in Figure 14.15 was in excess of 1.5 minutes. The time required on a TI-84 Plus is less than 1 minute.

RIGHTSUM will compute S_n for a given number n of subintervals. Before executing the program, store $f(x)$, a, and b as Y_1, A, and B, respectively. Upon execution of the program, you will be prompted to enter the number of subintervals. Then the program proceeds to display the

Problems 14.6

In Problems 1–4, sketch the region in the first quadrant that is bounded by the given curves. Approximate the area of the region by the indicated sum. Use the right-hand endpoint of each subinterval.

1. $f(x) = x, y = 0, x = 1$; S_3

2. $f(x) = 3x, y = 0, x = 1$; S_5

3. $f(x) = x^2, y = 0, x = 1$; S_4

4. $f(x) = x^2 + 1, y = 0, x = 0, x = 1$; S_2

In Problems 5 and 6, by dividing the indicated interval into n subintervals of equal length, find S_n for the given function. Use the right-hand endpoint of each subinterval. Do not find $\lim_{n\to\infty} S_n$.

5. $f(x) = 4x$; $[0, 1]$ **6.** $f(x) = 3x + 2$; $[0, 3]$

In Problems 7 and 8, (a) simplify S_n and (b) find $\lim_{n\to\infty} S_n$.

*7. $S_n = \dfrac{1}{n}\left[\left(\dfrac{1}{n}+1\right) + \left(\dfrac{2}{n}+1\right) + \cdots + \left(\dfrac{n}{n}+1\right)\right]$

8. $S_n = \dfrac{2}{n}\left[\left(\dfrac{2}{n}\right)^2 + \left(2 \cdot \dfrac{2}{n}\right)^2 + \cdots + \left(n \cdot \dfrac{2}{n}\right)^2\right]$

In Problems 9–14, sketch the region in the first quadrant that is bounded by the given curves. Determine the exact area of the region by considering the limit of S_n as $n \to \infty$. Use the right-hand endpoint of each subinterval.

9. Region as described in Problem 1

10. Region as described in Problem 2

11. Region as described in Problem 3

12. $y = x^2, y = 0, x = 1, x = 2$

13. $f(x) = 3x^2, y = 0, x = 1$

14. $f(x) = 9 - x^2, y = 0, x = 0$

In Problems 15–20, evaluate the given definite integral by taking the limit of S_n. Use the right-hand endpoint of each subinterval. Sketch the graph, over the given interval, of the function to be integrated.

15. $\displaystyle\int_1^3 5x\,dx$

16. $\displaystyle\int_0^4 9\,dx$

*17. $\displaystyle\int_0^3 -4x\,dx$

18. $\displaystyle\int_1^4 (2x+1)\,dx$

*19. $\displaystyle\int_0^1 (x^2+x)\,dx$

20. $\displaystyle\int_1^2 (x+2)\,dx$

21. Find $D_x\left[\displaystyle\int_2^3 \sqrt{x^2+1}\,dx\right]$ without the use of limits.

22. Find $\displaystyle\int_0^3 f(x)\,dx$ without the use of limits, where

$$f(x) = \begin{cases} 2 & \text{if } 0 \le x < 1 \\ 4 - 2x & \text{if } 1 \le x < 2 \\ 5x - 10 & \text{if } 2 \le x \le 3 \end{cases}$$

23. Find $\displaystyle\int_{-1}^3 f(x)\,dx$ without the use of limits, where

$$f(x) = \begin{cases} 1 & \text{if } x \le 1 \\ 2 - x & \text{if } 1 \le x \le 2 \\ -1 + \dfrac{x}{2} & \text{if } x > 2 \end{cases}$$

*In each of Problems 24–26, use a program, such as **RIGHTSUM**, to estimate the area of the region in the first quadrant bounded by the given curves. Round your answer to one decimal place.*

24. $f(x) = x^3 + 1, y = 0, x = 2, x = 3.7$

25. $f(x) = 4 - \sqrt{x}, y = 0, x = 1, x = 9$

26. $f(x) = e^x, y = 0, x = 0, x = 1$

*In each of Problems 27–30, use a program, such as **RIGHTSUM**, to estimate the value of the definite integral. Round your answer to one decimal place.*

27. $\displaystyle\int_2^5 \dfrac{x+1}{x+2}\,dx$

28. $\displaystyle\int_{-3}^{-1} \dfrac{1}{x^2}\,dx$

29. $\displaystyle\int_{-1}^2 (4x^2 + x - 13)\,dx$

30. $\displaystyle\int_1^2 \ln x\,dx$

OBJECTIVE

To informally develop the Fundamental Theorem of Integral Calculus and to use it to compute definite integrals.

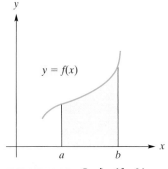

FIGURE 14.16 On $[a, b]$, f is continuous and $f(x) \ge 0$.

14.7 The Fundamental Theorem of Integral Calculus

The Fundamental Theorem

Thus far, the limiting processes of both the derivative and definite integral have been considered separately. We will now bring these fundamental ideas together and develop the important relationship that exists between them. As a result, we will be able to evaluate definite integrals more efficiently.

The graph of a function f is given in Figure 14.16. Assume that f is continuous on the interval $[a, b]$ and that its graph does not fall below the x-axis. That is, $f(x) \ge 0$. From the preceding section, the area of the region below the graph and above the x-axis from $x = a$ to $x = b$ is given by $\int_a^b f(x)\,dx$. We will now consider another way to determine this area.

Suppose that there is a function $A = A(x)$, which we will refer to as an area function, that gives the area of the region below the graph of f and above the x-axis from a to x, where $a \le x \le b$. This region is shaded in Figure 14.17. Do not confuse $A(x)$, which is an area, with $f(x)$, which is the height of the graph at x.

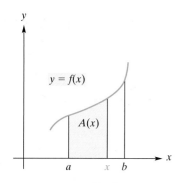

FIGURE 14.17 $A(x)$ is an area function.

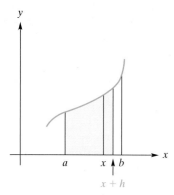

FIGURE 14.18 $A(x + h)$ gives the area of the shaded region.

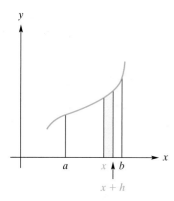

FIGURE 14.19 Area of shaded region is $A(x + h) - A(x)$.

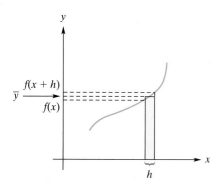

FIGURE 14.20 Area of rectangle is the same as area of shaded region in Figure 14.19.

From its definition, we can state two properties of A immediately:

1. $A(a) = 0$, since there is "no area" from a to a
2. $A(b)$ is the area from a to b; that is,

$$A(b) = \int_a^b f(x)\, dx$$

If x is increased by h units, then $A(x + h)$ is the area of the shaded region in Figure 14.18. Hence, $A(x + h) - A(x)$ is the difference of the areas in Figure. 14.18 and 14.17, namely, the area of the shaded region in Figure 14.19. For h sufficiently close to zero, the area of this region is the same as the area of a rectangle (Figure 14.20) whose base is h and whose height is some value $\overline{y}$ between $f(x)$ and $f(x + h)$. Here $\overline{y}$ is a function of h. Thus, on the one hand, the area of the rectangle is $A(x + h) - A(x)$, and, on the other hand, it is $h\overline{y}$, so

$$A(x + h) - A(x) = h\overline{y}$$

Equivalently,

$$\frac{A(x + h) - A(x)}{h} = \overline{y} \qquad \text{(dividing by } h\text{)}$$

Since $\overline{y}$ is between $f(x)$ and $f(x + h)$, it follows that as $h \to 0$, $\overline{y}$ approaches $f(x)$, so

$$\lim_{h \to 0} \frac{A(x + h) - A(x)}{h} = f(x) \tag{1}$$

But the left side is merely the derivative of A. Thus, Equation (1) becomes

$$A'(x) = f(x)$$

We conclude that the area function A has the additional property that its derivative A' is f. That is, A is an antiderivative of f. Now, suppose that F is *any* antiderivative of f. Then, since both A and F are antiderivatives of the same function, they differ at most by a constant C:

$$A(x) = F(x) + C. \tag{2}$$

Recall that $A(a) = 0$. So, evaluating both sides of Equation (2) when $x = a$ gives

$$0 = F(a) + C$$

so that

$$C = -F(a)$$

Thus, Equation (2) becomes

$$A(x) = F(x) - F(a) \tag{3}$$

If $x = b$, then, from Equation (3),

$$A(b) = F(b) - F(a) \tag{4}$$

But recall that

$$A(b) = \int_a^b f(x)\, dx \tag{5}$$

From Equations (4) and (5), we get

$$\int_a^b f(x)\, dx = F(b) - F(a)$$

A relationship between a definite integral and antidifferentiation has now become clear. To find $\int_a^b f(x)\, dx$, it suffices to find an antiderivative of f, say, F, and subtract the value of F at the lower limit a from its value at the upper limit b. We assumed here that f was continuous and $f(x) \geq 0$ so that we could appeal to the concept of an area. However, our result is true for any continuous function[7] and is known as the *Fundamental Theorem of Integral Calculus*.

Fundamental Theorem of Integral Calculus

If f is continuous on the interval $[a, b]$ and F is any antiderivative of f on $[a, b]$, then

$$\int_a^b f(x) \, dx = F(b) - F(a)$$

It is important that you understand the difference between a definite integral and an indefinite integral. The **definite integral** $\int_a^b f(x) \, dx$ is a **number** defined to be the limit of a sum. The Fundamental Theorem states that the **indefinite integral** $\int f(x) \, dx$ (the most general antiderivative of f), which is a **function** of x related to the differentiation process, can be used to determine this limit.

The definite integral is a number, and an indefinite integral is a function.

Suppose we apply the Fundamental Theorem to evaluate $\int_0^2 (4 - x^2) \, dx$. Here $f(x) = 4 - x^2$, $a = 0$, and $b = 2$. Since an antiderivative of $4 - x^2$ is $F(x) = 4x - (x^3/3)$, it follows that

$$\int_0^2 (4 - x^2) \, dx = F(2) - F(0) = \left(8 - \frac{8}{3}\right) - (0) = \frac{16}{3}$$

This confirms our result in Example 2 of Section 14.6. If we had chosen $F(x)$ to be $4x - (x^3/3) + C$, then we would have

$$F(2) - F(0) = \left[\left(8 - \frac{8}{3}\right) + C\right] - [0 + C] = \frac{16}{3}$$

as before. Since the choice of the value of C is immaterial, for convenience we will always choose it to be 0, as originally done. Usually, $F(b) - F(a)$ is abbreviated by writing

$$F(b) - F(a) = F(x) \Big|_a^b$$

Since F in the Fundamental Theorem of Calculus is *any* antiderivative of f and $\int f(x) \, dx$ is the most general antiderivative of f, it showcases the notation to write

$$\int_a^b f(x) \, dx = \left(\int f(x) \, dx\right) \Big|_a^b$$

Using the $\Big|_a^b$ notation, we have

$$\int_0^2 (4 - x^2) \, dx = \left(4x - \frac{x^3}{3}\right) \Big|_0^2 = \left(8 - \frac{8}{3}\right) - 0 = \frac{16}{3}$$

PRINCIPLES IN PRACTICE 1

APPLYING THE FUNDAMENTAL THEOREM

The income (in dollars) from a fast-food chain is increasing at a rate of $f(t) = 10{,}000e^{0.02t}$, where t is in years. Find $\int_3^6 10{,}000e^{0.02t} \, dt$, which gives the total income for the chain between the third and sixth years.

● EXAMPLE 1   **Applying the Fundamental Theorem**

Find $\displaystyle\int_{-1}^3 (3x^2 - x + 6) \, dx.$

Solution: An antiderivative of $3x^2 - x + 6$ is

$$x^3 - \frac{x^2}{2} + 6x$$

_____

[7]If f is continuous on $[a, b]$, it can be shown that $\int_a^b f(x) \, dx$ does indeed exist.

Thus,

$$\int_{-1}^{3} (3x^2 - x + 6)\, dx$$

$$= \left(x^3 - \frac{x^2}{2} + 6x\right)\Big|_{-1}^{3}$$

$$= \left[3^3 - \frac{3^2}{2} + 6(3)\right] - \left[(-1)^3 - \frac{(-1)^2}{2} + 6(-1)\right]$$

$$= \left(\frac{81}{2}\right) - \left(-\frac{15}{2}\right) = 48$$

NOW WORK PROBLEM 1 ◖●●

Properties of the Definite Integral

For $\int_{a}^{b} f(x)\, dx$, we have assumed that $a < b$. We now define the cases in which $a > b$ or $a = b$. First,

$$\text{If } a > b, \quad \text{then} \quad \int_{a}^{b} f(x)\, dx = -\int_{b}^{a} f(x)\, dx.$$

That is, interchanging the limits of integration changes the integral's sign. For example,

$$\int_{2}^{0} (4 - x^2)\, dx = -\int_{0}^{2} (4 - x^2)\, dx$$

If the limits of integration are equal, we have

$$\int_{a}^{a} f(x)\, dx = 0$$

Some properties of the definite integral deserve mention. The first of the properties that follow restates more formally our comment from the preceding section concerning area.

Properties of the Definite Integral

1. If f is continuous and $f(x) \geq 0$ on $[a, b]$, then $\int_{a}^{b} f(x)\, dx$ can be interpreted as the area of the region bounded by the curve $y = f(x)$, the x-axis, and the lines $x = a$ and $x = b$.
2. $\int_{a}^{b} kf(x)\, dx = k\int_{a}^{b} f(x)\, dx$ where k is a constant
3. $\int_{a}^{b} [f(x) \pm g(x)]\, dx = \int_{a}^{b} f(x)\, dx \pm \int_{a}^{b} g(x)\, dx$

Properties 2 and 3 are similar to rules for indefinite integrals because a definite integral may be evaluated by the Fundamental Theorem in terms of an antiderivative. Two more properties of definite integrals are as follows.

4. $\int_{a}^{b} f(x)\, dx = \int_{a}^{b} f(t)\, dt$

The variable of integration is a "dummy variable" in the sense that any other variable produces the same result—that is, the same number.

To illustrate property 4, you can verify, for example, that

$$\int_{0}^{2} x^2\, dx = \int_{0}^{2} t^2\, dt$$

5. If f is continuous on an interval I and a, b, and c are in I, then

$$\int_a^c f(x)\,dx = \int_a^b f(x)\,dx + \int_b^c f(x)\,dx$$

Property 5 means that the definite integral over an interval can be expressed in terms of definite integrals over subintervals. Thus,

$$\int_0^2 (4 - x^2)\,dx = \int_0^1 (4 - x^2)\,dx + \int_1^2 (4 - x^2)\,dx$$

We will look at some examples of definite integration now and compute some areas in Section 14.9.

● EXAMPLE 2   Using the Fundamental Theorem

Find $\displaystyle\int_0^1 \frac{x^3}{\sqrt{1 + x^4}}\,dx.$

Solution: To find an antiderivative of the integrand, we will apply the power rule for integration:

$$\int_0^1 \frac{x^3}{\sqrt{1 + x^4}}\,dx = \int_0^1 x^3 (1 + x^4)^{-1/2}\,dx$$

$$= \frac{1}{4} \int_0^1 (1 + x^4)^{-1/2}\,d(1 + x^4) = \left(\frac{1}{4}\right) \left.\frac{(1 + x^4)^{1/2}}{\frac{1}{2}}\right|_0^1$$

$$= \frac{1}{2}(1 + x^4)^{1/2}\Big|_0^1 = \frac{1}{2}\left((2)^{1/2} - (1)^{1/2}\right)$$

$$= \frac{1}{2}(\sqrt{2} - 1)$$

CAUTION

In Example 2, the value of the antiderivative $\frac{1}{2}(1 + x^4)^{1/2}$ at the lower limit 0 is $\frac{1}{2}(1)^{1/2}$. **Do not** assume that an evaluation at the limit zero will yield 0.

NOW WORK PROBLEM 13 ●●

● EXAMPLE 3   Evaluating Definite Integrals

a. *Find* $\displaystyle\int_1^2 \left[4t^{1/3} + t(t^2 + 1)^3\right]dt.$

Solution:

$$\int_1^2 \left[4t^{1/3} + t(t^2 + 1)^3\right]dt = 4\int_1^2 t^{1/3}\,dt + \frac{1}{2}\int_1^2 (t^2 + 1)^3\,d(t^2 + 1)$$

$$= (4)\left.\frac{t^{4/3}}{\frac{4}{3}}\right|_1^2 + \left(\frac{1}{2}\right)\left.\frac{(t^2 + 1)^4}{4}\right|_1^2$$

$$= 3(2^{4/3} - 1) + \frac{1}{8}(5^4 - 2^4)$$

$$= 3 \cdot 2^{4/3} - 3 + \frac{609}{8}$$

$$= 6\sqrt[3]{2} + \frac{585}{8}$$

b. *Find* $\displaystyle\int_0^1 e^{3t}\,dt.$

Solution:

$$\int_0^1 e^{3t}\,dt = \frac{1}{3}\int_0^1 e^{3t}\,d(3t)$$

$$= \left(\frac{1}{3}\right) e^{3t}\Big|_0^1 = \frac{1}{3}(e^3 - e^0) = \frac{1}{3}(e^3 - 1)$$

NOW WORK PROBLEM 15 ●●

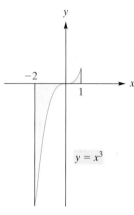

FIGURE 14.21 Graph of $y = x^3$ on the interval $[-2, 1]$.

 C A U T I O N

Remember that $\int_a^b f(x)\,dx$ is a limit of a sum. In some cases this limit represents an area. In others it does not. When $f(x) \geq 0$ on $[a, b]$ the integral represents the area between the graph of f and the x-axis from $x = a$ to $x = b$.

● EXAMPLE 4   **Finding and Interpreting a Definite Integral**

Evaluate $\int_{-2}^{1} x^3\,dx$.

Solution:

$$\int_{-2}^{1} x^3\,dx = \frac{x^4}{4}\bigg|_{-2}^{1} = \frac{1^4}{4} - \frac{(-2)^4}{4} = \frac{1}{4} - \frac{16}{4} = -\frac{15}{4}$$

The reason the result is negative is clear from the graph of $y = x^3$ on the interval $[-2, 1]$. (See Figure 14.21.) For $-2 \leq x < 0$, $f(x)$ is negative. Since a definite integral is a limit of a sum of the form $\Sigma f(x)\,\Delta x$, it follows that $\int_{-2}^{0} x^3\,dx$ is not only a negative number, but also the negative of the area of the shaded region in the third quadrant. On the other hand, $\int_0^1 x^3\,dx$ is the area of the shaded region in the first quadrant, since $f(x) \geq 0$ on $[0, 1]$. The definite integral over the entire interval $[-2, 1]$ is the *algebraic* sum of these numbers, because, from property 5,

$$\int_{-2}^{1} x^3\,dx = \int_{-2}^{0} x^3\,dx + \int_{0}^{1} x^3\,dx$$

Thus, $\int_{-2}^{1} x^3\,dx$ does not represent the area between the curve and the x-axis. However, if area is desired, it can be given by

$$\left| \int_{-2}^{0} x^3\,dx \right| + \int_{0}^{1} x^3\,dx$$

NOW WORK PROBLEM 25 ●●●

The Definite Integral of a Derivative

Since a function f is an antiderivative of f', by the Fundamental Theorem we have

$$\int_a^b f'(x)\,dx = f(b) - f(a) \tag{6}$$

But $f'(x)$ is the rate of change of f with respect to x. Hence, if we know the rate of change of f and want to find the difference in function values $f(b) - f(a)$, it suffices to evaluate $\int_a^b f'(x)\,dx$.

CHANGE IN FUNCTION VALUES

A managerial service determines that the rate of increase in maintenance costs (in dollars per year) for a particular apartment complex is given by $M'(x) = 90x^2 + 5000$, where x is the age of the apartment complex in years and $M(x)$ is the total (accumulated) cost of maintenance for x years. Find the total cost for the first five years.

● EXAMPLE 5   **Finding a Change in Function Values by Definite Integration**

A manufacturer's marginal-cost function is

$$\frac{dc}{dq} = 0.6q + 2$$

If production is presently set at $q = 80$ units per week, how much more would it cost to increase production to 100 units per week?

Solution: The total-cost function is $c = c(q)$, and we want to find the difference $c(100) - c(80)$. The rate of change of c is dc/dq, so by Equation (6),

$$c(100) - c(80) = \int_{80}^{100} \frac{dc}{dq}\,dq = \int_{80}^{100}(0.6q + 2)\,dq$$

$$= \left[\frac{0.6q^2}{2} + 2q \right]\bigg|_{80}^{100} = [0.3q^2 + 2q]\bigg|_{80}^{100}$$

$$= [0.3(100)^2 + 2(100)] - [0.3(80)^2 + 2(80)]$$

$$= 3200 - 2080 = 1120$$

If c is in dollars, then the cost of increasing production from 80 units to 100 units is \$1120.

NOW WORK PROBLEM 59 ●●●

TECHNOLOGY

Many graphing calculators have the capability to estimate the value of a definite integral. On a TI-83 Plus, to estimate

$$\int_{80}^{100} (0.6q + 2)\, dq$$

we use the "fnInt(" command, as indicated in Figure 14.22. The four parameters that must be entered with this command are

<table>
<tr><td>function to</td><td>variable of</td><td>lower</td><td>upper</td></tr>
<tr><td>be integrated</td><td>integration</td><td>limit</td><td>limit</td></tr>
</table>

We see that the value of the definite integral is approximately 1120, which agrees with the result in Example 5.
   Similarly, to estimate

$$\int_{-2}^{1} x^3\, dx$$

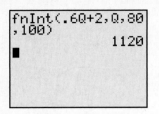

FIGURE 14.22 Estimating $\int_{80}^{100}(0.6q+2)\,dq$.

we enter

$$\text{fnInt}(X^3, X, -2, 1)$$

or, alternatively, if we first store x^3 as Y_1, we can enter

$$\text{fnInt}(Y_1, X, -2, 1)$$

In each case we obtain -3.75, which agrees with the result in Example 4.

Problems 14.7

In Problems 1–43, evaluate the definite integral.

*1. $\displaystyle\int_0^3 5\,dx$

2. $\displaystyle\int_2^4 (1 - e)\,dx$

3. $\displaystyle\int_1^2 5x\,dx$

4. $\displaystyle\int_2^8 -5x\,dx$

5. $\displaystyle\int_{-3}^1 (2x - 3)\,dx$

6. $\displaystyle\int_{-1}^1 (4 - 9y)\,dy$

7. $\displaystyle\int_2^3 (y^2 - 2y + 1)\,dy$

8. $\displaystyle\int_4^1 (2t - 3t^2)\,dt$

9. $\displaystyle\int_{-2}^{-1} (3w^2 - w - 1)\,dw$

10. $\displaystyle\int_8^9 dt$

11. $\displaystyle\int_1^3 3t^{-3}\,dt$

12. $\displaystyle\int_1^2 \frac{x^{-2}}{2}\,dx$

*13. $\displaystyle\int_{-8}^8 \sqrt[3]{x^4}\,dx$

14. $\displaystyle\int_{1/2}^{3/2} (x^2 + x + 1)\,dx$

*15. $\displaystyle\int_{1/2}^3 \frac{1}{x^2}\,dx$

16. $\displaystyle\int_9^{36} (\sqrt{x} - 2)\,dx$

17. $\displaystyle\int_{-1}^1 (z + 1)^5\,dz$

18. $\displaystyle\int_1^8 (x^{1/3} - x^{-1/3})\,dx$

19. $\displaystyle\int_0^1 2x^2(x^3 - 1)^3\,dx$

20. $\displaystyle\int_2^3 (x + 2)^3\,dx$

21. $\displaystyle\int_1^8 \frac{4}{y}\,dy$

22. $\displaystyle\int_{-(e^e)}^{-1} \frac{6}{x}\,dx$

23. $\displaystyle\int_0^1 e^5\,dx$

24. $\displaystyle\int_2^{e+1} \frac{1}{x - 1}\,dx$

*25. $\displaystyle\int_0^1 5x^2 e^{x^3}\,dx$

26. $\displaystyle\int_0^1 (3x^2 + 4x)(x^3 + 2x^2)^4\,dx$

27. $\displaystyle\int_4^5 \frac{2}{(x - 3)^3}\,dx$

28. $\displaystyle\int_{-1/3}^{20/3} \sqrt{3x + 5}\,dx$

29. $\displaystyle\int_{1/3}^2 \sqrt{10 - 3p}\,dp$

30. $\displaystyle\int_{-1}^1 q\sqrt{q^2 + 3}\,dq$

31. $\displaystyle\int_0^1 x^2 \sqrt[3]{7x^3 + 1}\,dx$

32. $\displaystyle\int_0^{\sqrt{7}} \left(3x - \frac{x}{(x^2 + 2)^{4/3}}\right) dx$

33. $\displaystyle\int_0^1 \frac{2x^3 + x}{x^2 + x^4 + 1}\,dx$

34. $\displaystyle\int_a^b (m + ny)\,dy$

35. $\displaystyle\int_0^1 \frac{e^x - e^{-x}}{2}\,dx$

36. $\displaystyle\int_{-2}^1 8|x|\,dx$

37. $\displaystyle\int_\pi^e 3(x^{-2} + x^{-3} - x^{-4})\,dx$

38. $\displaystyle\int_1^2 \left(6\sqrt{x} - \frac{1}{\sqrt{2x}}\right) dx$

39. $\displaystyle\int_1^3 (x + 1)e^{x^2 + 2x}\,dx$

40. $\displaystyle\int_1^{95} \frac{x}{\ln e^x}\,dx$

41. $\displaystyle\int_0^2 \frac{x^6 + 6x^4 + x^3 + 8x^2 + x + 5}{x^3 + 5x + 1}\,dx$

42. $\displaystyle\int_{-1}^1 \frac{2}{1 + e^x}\,dx$ (*Hint:* Multiply the integrand by $\frac{e^{-x}}{e^{-x}}$.)

43. $\displaystyle\int_0^2 f(x)\,dx$ where $f(x) = \begin{cases} 4x^2 & \text{if } 0 \le x < \frac{1}{2} \\ 2x & \text{if } \frac{1}{2} \le x \le 2 \end{cases}$

44. Evaluate $\left(\displaystyle\int_1^3 x\,dx\right)^3 - \displaystyle\int_1^3 x^3\,dx$.

45. Suppose $f(x) = \displaystyle\int_1^x 3\frac{1}{t^2}\,dt$. Evaluate $\displaystyle\int_e^1 f(x)\,dx$.

46. Evaluate $\int_7^7 e^{x^2}\,dx + \int_0^{\sqrt{2}} \dfrac{1}{3\sqrt{2}}\,dx$

47. If $\int_1^3 f(x)\,dx = 4$ and $\int_3^2 f(x)\,dx = 3$, find $\int_1^2 f(x)\,dx$.

48. If $\int_1^4 f(x)\,dx = 6$, $\int_2^4 f(x)\,dx = 5$, and $\int_1^3 f(x)\,dx = 2$, find $\int_2^3 f(x)\,dx$.

49. Evaluate $\int_2^3 \left(\dfrac{d}{dx} \int_2^3 e^{x^3}\,dx \right) dx$

(*Hint:* It is not necessary to find $\int_2^3 e^{x^3}\,dx$.)

50. Suppose that $f(x) = \int_e^x \dfrac{e^t - e^{-t}}{e^t + e^{-t}}\,dt$ where $x > e$. Find $f'(x)$.

51. Severity Index   In discussing traffic safety, Shonle[8] considers how much acceleration a person can tolerate in a crash so that there is no major injury. The *severity index* is defined as

$$\text{S.I.} = \int_0^T \alpha^{5/2}\,dt$$

where α (a Greek letter read "alpha") is considered a constant involved with a weighted average acceleration, and T is the duration of the crash. Find the severity index.

52. Statistics   In statistics, the mean μ (a Greek letter read "mu") of the continuous probability density function f defined on the interval $[a, b]$ is given by

$$\mu = \int_a^b [x \cdot f(x)]\,dx$$

and the variance σ^2 (σ is a Greek letter read "sigma") is given by

$$\sigma^2 = \int_a^b (x - \mu)^2 f(x)\,dx$$

Compute μ and then σ^2 if $a = 0$, $b = 1$, and $f(x) = 1$.

53. Distribution of Incomes   The economist Pareto[9] has stated an empirical law of distribution of higher incomes that gives the number N of persons receiving x or more dollars. If

$$\dfrac{dN}{dx} = -Ax^{-B}$$

where A and B are constants, set up a definite integral that gives the total number of persons with incomes between a and b, where $a < b$.

54. Biology   In a discussion of gene mutation,[10] the following integral occurs:

$$\int_0^{10^{-4}} x^{-1/2}\,dx$$

Evaluate this integral.

55. Continuous Income Flow   The present value (in dollars) of a continuous flow of income of \$2000 a year for five years at 6% compounded continuously is given by

$$\int_0^5 2000e^{-0.06t}\,dt$$

Evaluate the present value to the nearest dollar.

56. Biology   In biology, problems frequently arise involving the transfer of a substance between compartments. An example is a transfer from the bloodstream to tissue. Evaluate the following integral, which occurs in a two-compartment diffusion problem:[11]

$$\int_0^t (e^{-a\tau} - e^{-b\tau})\,d\tau$$

Here, τ (read "tau") is a Greek letter; a and b are constants.

57. Demography   For a certain population, suppose l is a function such that $l(x)$ is the number of persons who reach the age of x in any year of time. This function is called a *life table function*. Under appropriate conditions, the integral

$$\int_x^{x+n} l(t)\,dt$$

gives the expected number of people in the population between the exact ages of x and $x + n$, inclusive. If

$$l(x) = 10{,}000\sqrt{100 - x}$$

determine the number of people between the exact ages of 36 and 64, inclusive. Give your answer to the nearest integer, since fractional answers make no sense.

58. Mineral Consumption   If C is the yearly consumption of a mineral at time $t = 0$, then, under continuous consumption, the total amount of the mineral used in the interval $[0, t]$ is

$$\int_0^t Ce^{k\tau}\,d\tau$$

where k is the rate of consumption. For a rare-earth mineral, it has been determined that $C = 3000$ units and $k = 0.05$. Evaluate the integral for these data.

*59. Marginal Cost   A manufacturer's marginal-cost function is

$$\dfrac{dc}{dq} = 0.2q + 8$$

If c is in dollars, determine the cost involved to increase production from 65 to 75 units.

60. Marginal Cost   Repeat Problem 59 if

$$\dfrac{dc}{dq} = 0.004q^2 - 0.5q + 50$$

and production increases from 90 to 180 units.

---

[8]J. I. Shonle, *Environmental Applications of General Physics* (Reading, MA: Addison-Wesley Publishing Company, Inc., 1975).

[9]G. Tintner, *Methodology of Mathematical Economics and Econometrics* (Chicago: University of Chicago Press, 1967), p. 16.

[10]W. J. Ewens, *Population Genetics* (London: Methuen & Company Ltd., 1969).

[11]W. Simon, *Mathematical Techniques for Physiology and Medicine* (New York: Academic Press, Inc., 1972).

61. Marginal Revenue A manufacturer's marginal-revenue function is

$$\frac{dr}{dq} = \frac{2000}{\sqrt{300q}}$$

If r is in dollars, find the change in the manufacturer's total revenue if production is increased from 500 to 800 units.

62. Marginal Revenue Repeat Problem 61 if

$$\frac{dr}{dq} = 250 + 90q - 3q^2$$

and production is increased from 10 to 20 units.

63. Crime Rate A sociologist is studying the crime rate in a certain city. She estimates that t months after the beginning of next year, the total number of crimes committed will increase at the rate of $8t + 10$ crimes per month. Determine the total number of crimes that can be expected to be committed next year. How many crimes can be expected to be committed during the last six months of that year?

64. Hospital Discharges For a group of hospitalized individuals, suppose the discharge rate is given by

$$f(t) = \frac{81 \times 10^6}{(300 + t)^4}$$

where $f(t)$ is the proportion of the group discharged per day at the end of t days. What proportion has been discharged by the end of 700 days?

65. Production Imagine a one-dimensional country of length $2R$. (See Figure 14.23.[12]) Suppose the production of goods for this country is continuously distributed from border to border. If the amount produced each year per unit of distance is $f(x)$, then the country's total yearly production is given by

$$G = \int_{-R}^{R} f(x)\, dx$$

Evaluate G if $f(x) = i$, where i is constant.

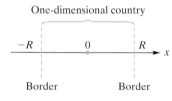

One-dimensional country

FIGURE 14.23 Diagram for Problem 65.

66. Exports For the one-dimensional country of Problem 65, under certain conditions the amount of the country's exports is given by

$$E = \int_{-R}^{R} \frac{i}{2}[e^{-k(R-x)} + e^{-k(R+x)}]\, dx$$

where i and k are constants ($k \neq 0$). Evaluate E.

67. Average Delivered Price In a discussion of a delivered price of a good from a mill to a customer, DeCanio[13] claims that the average delivered price paid by consumers is given by

$$A = \frac{\displaystyle\int_0^R (m + x)[1 - (m + x)]\, dx}{\displaystyle\int_0^R [1 - (m + x)]\, dx}$$

where m is mill price, and x is the maximum distance to the point of sale. DeCanio determines that

$$A = \frac{m + \dfrac{R}{2} - m^2 - mR - \dfrac{R^2}{3}}{1 - m - \dfrac{R}{2}}$$

Verify this.

In Problems 68–70, use the Fundamental Theorem of Integral Calculus to determine the value of the definite integral. Confirm your result with your calculator.

68. $\displaystyle\int_{2.5}^{3.5} (1 + 2x + 3x^2)\, dx$

69. $\displaystyle\int_0^4 \frac{1}{(4x + 4)^2}\, dx$

70. $\displaystyle\int_0^1 e^{3t}\, dt$ Round your answer to two decimal places.

In Problems 71–74, estimate the value of the definite integral. Round your answer to two decimal places.

71. $\displaystyle\int_{-1}^{5} \frac{x^2 + 1}{x^2 + 4}\, dx$ **72.** $\displaystyle\int_3^4 \frac{1}{x \ln x}\, dx$

73. $\displaystyle\int_0^3 2\sqrt{t^2 + 3}\, dt$

74. $\displaystyle\int_{-1}^{1} \frac{6\sqrt{q + 1}}{q + 3}\, dq$

OBJECTIVE

To estimate the value of a definite integral by using either the trapezoidal rule or Simpson's rule.

14.8 Approximate Integration

Trapezoidal Rule

Any function f constructed from polynomials, exponentials, and logarithms using algebraic operations and composition can be differentiated and the resulting function f' is again of the same kind—one that can be constructed from polynomials, exponentials, and logarithms using algebraic operations and composition. Let us call such

---

[12]R. Taagepera, "Why the Trade/GNP Ratio Decreases with Country Size," *Social Science Research*, 5 (1976), 385–404.

[13]S. J. DeCanio, "Delivered Pricing and Multiple Basing Point Equationilibria: A Reevaluation," *The Quarterly Journal of Economics*, XCIX, no. 2 (1984), 329–49.

functions *elementary* (although the term usually has a slightly different meaning). In this terminology, the derivative of an elementary function is also elementary. Integration is more complicated. If an elementary function f has F as an antiderivative, then F may fail to be elementary. Said otherwise, even for a fairly simple-looking function f it is sometimes impossible to find $\int f(x)\,dx$ in terms of the functions that we consider in this book. For example, there is no elementary function whose derivative is e^{x^2} so that you cannot expect to "do" the integral $\int e^{x^2}\,dx$.

On the other hand, consider a function f that is continuous on a closed interval $[a, b]$ with $f(x) \geq 0$ for all x in $[a, b]$. Then $\int_a^b f(x)\,dx$ is simply the *number* that gives the area of the region bounded by the curves $y = f(x)$, $y = 0$, $x = a$, and $x = b$. It is unsatisfying, and perhaps impractical, to not say anything about the number $\int_a^b f(x)\,dx$ because of an inability to "do" the integral $\int f(x)\,dx$. This also applies when the integral $\int f(x)\,dx$ is merely too difficult for the person who needs to find the number $\int_a^b f(x)\,dx$.

Since $\int_a^b f(x)\,dx$ is defined as a limit of sums of the form $\sum f(x)\,\Delta x$, any particular well-formed sum of the form $\sum f(x)\,\Delta x$ can be regarded as an approximation of $\int_a^b f(x)\,dx$. At least for nonnegative f such sums can be regarded as sums of areas of thin rectagles. Consider for example Figure 14.11 in Section 14.6, in which two rectangles are explicitly shown. It is clear that the error that arises from such rectangles is associated with the small side at the top. The error would be reduced if we replaced the rectangles by shapes that have a top side that is closer to the shape of the curve. We will consider two possibilities: using thin trapezoids rather than thin rectangles, the *trapezoidal rule;* and using thin regions surmounted by parabolic arcs, *Simpson's rule.* In each case only a finite number of numerical values of $f(x)$ need be known and the calculations involved are especially suitable for computers or calculators. In both cases, we assume that f is continuous on $[a, b]$.

In developing the trapezoidal rule, for convenience we will also assume that $f(x) \geq 0$ on $[a, b]$, so that we can think in terms of area. Basically, this rule involves approximating the graph of f by straight-line segments.

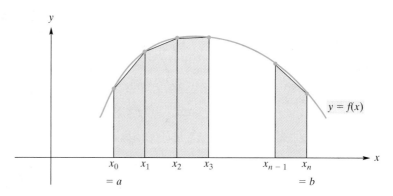

FIGURE 14.24   Approximating an area by using trapezoids.

In Figure 14.24, the interval $[a, b]$ is divided into n subintervals of equal length by the points $a = x_0, x_1, x_2, \ldots,$ and $x_n = b$. Since the length of $[a, b]$ is $b - a$, the length of each subinterval is $(b - a)/n$, which we will call h.

Clearly,

$$x_1 = a + h, \; x_2 = a + 2h, \ldots, x_n = a + nh = b$$

With each subinterval, we can associate a trapezoid (a four-sided figure with two parallel sides). The area A of the region bounded by the curve, the x-axis, and the lines $x = a$ and $x = b$ is $\int_a^b f(x)\,dx$ and can be approximated by the sum of the areas of the trapezoids determined by the subintervals.

Consider the first trapezoid, which is redrawn in Figure 14.25. Since the area of a trapezoid is equal to one-half the base times the sum of the lengths of the parallel sides, this trapezoid has area

$$\tfrac{1}{2}h[\,f(a) + f(a + h)]$$

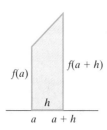

FIGURE 14.25   First trapezoid.

Similarly, the second trapezoid has area

$$\tfrac{1}{2}h[f(a+h)+f(a+2h)]$$

The area A under the curve is approximated by the sum of the areas of n trapezoids:

$$A \approx \tfrac{1}{2}h[f(a)+f(a+h)]+\tfrac{1}{2}h[f(a+h)+f(a+2h)]$$
$$+\tfrac{1}{2}h[f(a+2h)+f(a+3h)]+\cdots+\tfrac{1}{2}h[f(a+(n-1)h)+f(b)]$$

Since $A = \int_a^b f(x)\,dx$, by simplifying the preceding formula we have the trapezoidal rule:

The Trapezoidal Rule

$$\int_a^b f(x)\,dx \approx \frac{h}{2}[f(a)+2f(a+h)+2f(a+2h)$$
$$+\cdots+2f(a+(n-1)h)+f(b)]$$

where $h = (b-a)/n$.

The pattern of the coefficients inside the braces is 1, 2, 2, ..., 2, 1. Usually, the more subintervals, the better is the approximation. In our development, we assumed for convenience that $f(x) \geq 0$ on $[a, b]$. However, the trapezoidal rule is valid without this restriction.

TRAPEZOIDAL RULE

An oil tanker is losing oil at a rate of $R'(t) = \dfrac{60}{\sqrt{t^2+9}}$, where t is the time in minutes and $R(t)$ is the radius of the oil slick in feet. Use the trapezoidal rule with $n=5$ to approximate $\displaystyle\int_0^5 \frac{60}{\sqrt{t^2+9}}\,dt$, the size of the radius after five seconds.

● EXAMPLE 1 Trapezoidal Rule

Use the trapezoidal rule to estimate the value of

$$\int_0^1 \frac{1}{1+x^2}\,dx$$

for $n = 5$. Compute each term to four decimal places, and round the answer to three decimal places.

Solution: Here $f(x) = 1/(1+x^2)$, $n = 5$, $a = 0$, and $b = 1$. Thus,

$$h = \frac{b-a}{n} = \frac{1-0}{5} = \frac{1}{5} = 0.2$$

The terms to be added are

$$
\begin{aligned}
f(a) &= f(0) &= 1.0000 \\
2f(a+h) &= 2f(0.2) &= 1.9231 \\
2f(a+2h) &= 2f(0.4) &= 1.7241 \\
2f(a+3h) &= 2f(0.6) &= 1.4706 \\
2f(a+4h) &= 2f(0.8) &= 1.2195 \\
f(b) &= f(1) &= \underline{0.5000} \quad (a+nh=b)\\
& & 7.8373 = \text{sum}
\end{aligned}
$$

Hence, our estimate for the integral is

$$\int_0^1 \frac{1}{1+x^2}\,dx \approx \frac{0.2}{2}(7.8373) \approx 0.784$$

The actual value of the integral is approximately 0.785.

NOW WORK PROBLEM 1 ●●

Simpson's Rule

Another method for estimating $\int_a^b f(x)\,dx$ is given by Simpson's rule, which involves approximating the graph of f by parabolic segments. We will omit the derivation.

Simpson's Rule

$$\int_a^b f(x)\,dx \approx \frac{h}{3}[f(a) + 4f(a+h) + 2f(a+2h)$$
$$+ \cdots + 4f(a + (n-1)h) + f(b)]$$

where $h = (b-a)/n$ and n is even.

The pattern of coefficients inside the braces is $1, 4, 2, 4, 2, \ldots, 2, 4, 1$, which requires that **$n$ be even.** Let us use this rule for the integral in Example 1.

SIMPSON'S RULE

A yeast culture is growing at the rate of $A'(t) = 0.3e^{0.2t^2}$, where t is the time in hours and $A(t)$ is the amount in grams. Use Simpson's rule with $n = 8$ to approximate $\int_0^4 0.3e^{0.2t^2}\,dt$, the amount the culture grew over the first four hours.

● **EXAMPLE 2 Simpson's Rule**

Use Simpson's rule to estimate the value of $\int_0^1 \frac{1}{1+x^2}\,dx$ *for* $n = 4$. *Compute each term to four decimal places, and round the answer to three decimal places.*

Solution: Here $f(x) = 1/(1+x^2)$, $n = 4$, $a = 0$, and $b = 1$. Thus, $h = (b-a)/n = 1/4 = 0.25$. The terms to be added are

$$\begin{aligned}
f(a) = \quad f(0) \quad &= 1.0000 \\
4f(a+h) = 4f(0.25) &= 3.7647 \\
2f(a+2h) = 2f(0.5) &= 1.6000 \\
4f(a+3h) = 4f(0.75) &= 2.5600 \\
f(b) = \quad f(1) \quad &= \underline{0.5000} \\
9.4247 &= \text{sum}
\end{aligned}$$

Therefore, by Simpson's rule,

$$\int_0^1 \frac{1}{1+x^2}\,dx \approx \frac{0.25}{3}(9.4247) \approx 0.785$$

This is a better approximation than that which we obtained in Example 1 by using the trapezoidal rule.

NOW WORK PROBLEM 5 ●◖◗

Both Simpson's rule and the trapezoidal rule can be used if we know only $f(a)$, $f(a+h)$, and so on; we do not need to know $f(x)$ for all x in $[a, b]$. Example 3 will illustrate.

In Example 3, a definite integral is estimated from data points; the function itself is not known.

● **EXAMPLE 3 Demography**

*A function often used in demography (the study of births, marriages, mortality, etc., in a population) is the **life-table function,** denoted l. In a population having 100,000 births in any year of time, l(x) represents the number of persons who reach the age of x in any year of time. For example, if l(20) = 98,857, then the number of persons who attain age 20 in any year of time is 98,857. Suppose that the function l applies to all people born over an extended period of time. It can be shown that, at any time, the expected number of persons in the population between the exact ages of x and x + m, inclusive, is given by*

$$\int_x^{x+m} l(t)\,dt$$

The following table gives values of l(x) for males and females in the United States[14] *Approximate the number of women in the 20–35 age group by using the trapezoidal rule with n = 3.*

---

[14]*National Vital Statistics Report*, Vol. 48, No. 18, February 7, 2001.

Life Table

Age = x	l(x) Males	l(x) Females	Age = x	l(x) Males	l(x) Females
0	100,000	100,000	45	93,717	96,582
5	99,066	99,220	50	91,616	95,392
10	98,967	99,144	55	88,646	93,562
15	98,834	99,059	60	84,188	90,700
20	98,346	98,857	65	77,547	86,288
25	97,648	98,627	70	68,375	79,926
30	96,970	98,350	75	56,288	70,761
35	96,184	97,964	80	42,127	58,573
40	95,163	97,398			

Solution: We want to estimate

$$\int_{20}^{35} l(t)\, dt$$

We have $h = \dfrac{b-a}{n} = \dfrac{35-20}{3} = 5$. The terms to be added under the trapezoidal rule are

$$l(20) = 98,857$$

$$2l(25) = 2(98,627) = 197,254$$

$$2l(30) = 2(98,350) = 196,700$$

$$l(35) = \underline{97,964}$$

$$590,775 = \text{sum}$$

By the trapezoidal rule,

$$\int_{20}^{35} l(t)\, dt \approx \frac{5}{2}(590,775) = 1,476,937.5$$

NOW WORK PROBLEM 17

Formulas used to determine the accuracy of answers obtained with the trapezoidal or Simpson's rule can be found in standard texts on numerical analysis.

Problems 14.8

In Problems 1 and 2, use the trapezoidal rule or Simpson's rule (as indicated) and the given value of n to estimate the integral.

*1. $\displaystyle\int_{-2}^{4} \frac{170}{1+x^2}\, dx$; trapezoidal rule, $n = 6$

2. $\displaystyle\int_{-2}^{4} \frac{170}{1+x^2}\, dx$; Simpson's rule, $n = 6$

In Problems 3–8, use the trapezoidal rule or Simpson's rule (as indicated) and the given value of n to estimate the integral. Compute each term to four decimal places, and round the answer to three decimal places. In Problems 3–6, also evaluate the integral by antidifferentiation (the Fundamental Theorem of Integral Calculus).

3. $\displaystyle\int_{0}^{1} x^2\, dx$; trapezoidal rule, $n = 5$

4. $\displaystyle\int_{0}^{1} x^2\, dx$; Simpson's rule, $n = 4$

*5. $\displaystyle\int_{1}^{4} \frac{dx}{x^2}$; Simpson's rule, $n = 4$

6. $\displaystyle\int_{1}^{4} \frac{dx}{x}$; trapezoidal rule, $n = 6$

7. $\displaystyle\int_{0}^{2} \frac{x\, dx}{x+1}$; trapezoidal rule, $n = 4$

8. $\displaystyle\int_{2}^{4} \frac{dx}{x+x^2}$; Simpson's rule, $n = 4$

In Problems 9 and 10, use the life table in Example 3 to estimate the given integrals by the trapezoidal rule.

9. $\int_{45}^{70} l(t)\,dt$, males, $n = 5$

10. $\int_{35}^{55} l(t)\,dt$, females, $n = 4$

In Problems 11 and 12, suppose the graph of a continuous function f, where $f(x) \geq 0$, contains the given points. Use Simpson's rule and all of the points to approximate the area between the graph and the x-axis on the given interval. Round the answer to one decimal place.

11. $(1, 0.4), (2, 0.6), (3, 1.2), (4, 0.8), (5, 0.5);$ $[1,5]$

12. $(2, 0), (2.5, 6), (3, 10), (3.5, 11), (4, 14), (4.5, 15), (5, 16);$ $[2,5]$

13. Using all the information given in Figure 14.26, estimate $\int_1^3 f(x)\,dx$ by using Simpson's rule. Give your answer in fractional form.

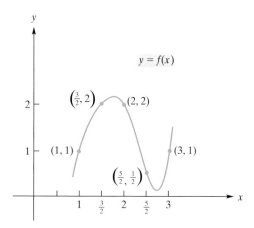

FIGURE 14.26 Graph of f for Problem 13.

In Problems 14 and 15, use Simpson's rule and the given value of n to estimate the integral. Compute each term to four decimal places, and round the answer to three decimal places.

14. $\int_1^3 \dfrac{2}{\sqrt{1+x}}\,dx; n = 4$ Also, evaluate the integral by the Fundamental Theorem of Integral Calculus.

15. $\int_0^1 \sqrt{1 - x^2}\,dx; n = 4$

16. Revenue Use Simpson's rule to approximate the total revenue received from the production and sale of 80 units of a product if the values of the marginal-revenue function dr/dq are as follows:

q (units)	0	10	20	30	40	50	60	70	80
$\dfrac{dr}{dq}$ ($ per unit)	10	9	8.5	8	8.5	7.5	7	6.5	7

*17. **Area of Lake** A straight stretch of highway runs alongside a lake. A surveyor who wishes to know the approximate area of the lake measures the distance from various points along the road to the near and far shores of the lake according to the following table:

Distance along highway (km)	0.0	0.5	1.0	1.5	2.0	2.5	3.0	3.5	4.0
Distance to near shore (km)	0.5	0.3	0.7	1.0	0.5	0.2	0.5	0.8	1.0
Distance to far shore (km)	0.5	2.3	2.2	3.0	2.5	2.2	1.5	1.3	1.0

Draw a rough sketch of the geographical situation. Then use Simpson's rule to give the best estimate of the lake's area. Give your answer in fractional form.

18. Manufacturing A manufacturer estimated both marginal cost (MC) and marginal revenue (MR) at various levels of output (q). These estimates are given in the following table:

q (units)	0	20	40	60	80	100
MC ($ per unit)	260	250	240	200	240	250
MR ($ per unit)	410	350	300	250	270	250

(a) Using the trapezoidal rule, estimate the total variable costs of production for 100 units.

(b) Using the trapezoidal rule, estimate the total revenue from the sale of 100 units.

(c) If we assume that maximum profit occurs when MR = MC (that is, when $q = 100$), estimate the maximum profit if fixed costs are $2000.

OBJECTIVE

To use vertical strips and the definite integral to find the area of the region between a curve and the x-axis.

14.9 Area

In Section 14.7 we saw that the area of a region could be found by evaluating the limit of a sum of the form $\Sigma f(x)\,\Delta x$, where $f(x)\,\Delta x$ represents the area of a rectangle. This limit is a special case of a definite integral, so it can easily be found by using the Fundamental Theorem.

When using the definite integral to determine area, you should make a rough sketch of the region involved. Let us consider the area of the region bounded by $y = f(x)$ and the x-axis from $x = a$ to $x = b$, where $f(x) \geq 0$ on $[a, b]$. (See Figure 14.27.) To set up the integral, a sample rectangle should be included in the sketch because the area of the region is a limit of sums of areas of rectangles. This will not only help you understand the integration process, but it will also help you find areas of more complicated regions. Such a rectangle (see Figure 14.27) is called a

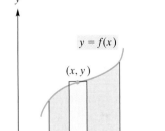

FIGURE 14.27   Region with vertical element.

vertical element of area (or a **vertical strip**). In the diagram, the width of the vertical element is Δx. The height is the y-value of the curve. Hence, the rectangle has area $y \, \Delta x$ or $f(x) \, \Delta x$. The area of the entire region is found by summing the areas of all such elements between $x = a$ and $x = b$ and finding the limit of this sum, which is the definite integral. Symbolically, we have

$$\Sigma f(x) \, \Delta x \rightarrow \int_a^b f(x) \, dx = \text{area}$$

Example 1 will illustrate.

● EXAMPLE 1   **Using the Definite Integral to Find Area**

Find the area of the region bounded by the curve

$$y = 6 - x - x^2$$

and the x-axis.

Solution: First we must sketch the curve so that we can visualize the region. Since

$$y = -(x^2 + x - 6) = -(x - 2)(x + 3)$$

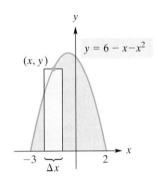

FIGURE 14.28   Region of Example 1 with vertical element.

the x-intercepts are $(2, 0)$ and $(-3, 0)$. Using techniques of graphing that were previously discussed, we obtain the graph and region shown in Figure 14.28. *With this region, it is crucial that the x-intercepts of the curve be found, because they determine the interval over which the areas of the elements must be summed.* That is, *these x-values are the limits of integration.* The vertical element shown has width Δx and height y. Hence, the area of the element is $y \, \Delta x$. Summing the areas of all such elements from $x = -3$ to $x = 2$ and taking the limit via the definite integral gives the area:

$$\sum y \, \Delta x \rightarrow \int_{-3}^{2} y \, dx = \text{area}$$

To evaluate the integral, we must express the integrand in terms of the variable of integration, x. Since $y = 6 - x - x^2$,

$$\text{area} = \int_{-3}^{2} (6 - x - x^2) \, dx = \left(6x - \frac{x^2}{2} - \frac{x^3}{3} \right) \Bigg|_{-3}^{2}$$

$$= \left(12 - \frac{4}{2} - \frac{8}{3} \right) - \left(-18 - \frac{9}{2} - \frac{-27}{3} \right) = \frac{125}{6}$$

NOW WORK PROBLEM 1   ◖●●

● EXAMPLE 2   **Finding the Area of a Region**

Find the area of the region bounded by the curve $y = x^2 + 2x + 2$, the x-axis, and the lines $x = -2$ and $x = 1$.

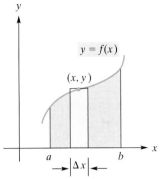

FIGURE 14.29   Diagram for Example 2.

Solution: A sketch of the region is given in Figure 14.29. We have

$$\text{area} = \int_{-2}^{1} y \, dx = \int_{-2}^{1} (x^2 + 2x + 2) \, dx$$

$$= \left(\frac{x^3}{3} + x^2 + 2x \right) \Bigg|_{-2}^{1} = \left(\frac{1}{3} + 1 + 2 \right) - \left(-\frac{8}{3} + 4 - 4 \right)$$

$$= 6$$

NOW WORK PROBLEM 7   ◖●●

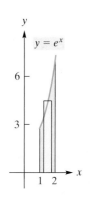

FIGURE 14.30
Diagram for
Example 3.

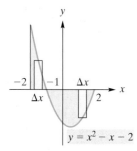

FIGURE 14.31 Diagram for
Example 4.

 CAUTION

It is wrong to write hastily that the area is $\int_{-2}^{2} y\, dx$, for the following reason: For the left rectangle, the height is y. However, for the rectangle on the right, y is negative, so its height is the positive number $-y$. This points out the importance of sketching the region.

● **EXAMPLE 3   Finding the Area of a Region**

Find the area of the region between the curve $y = e^x$ and the x-axis from $x = 1$ to $x = 2$.

Solution: A sketch of the region is given in Figure 14.30. We write

$$\text{area} = \int_{1}^{2} y\, dx = \int_{1}^{2} e^x dx = e^x \Big|_{1}^{2}$$

$$= e^2 - e = e(e - 1)$$

NOW WORK PROBLEM 27 ●●●

● **EXAMPLE 4   An Area Requiring Two Definite Integrals**

Find the area of the region bounded by the curve

$$y = x^2 - x - 2$$

and the line $y = 0$ (the x-axis) from $x = -2$ to $x = 2$.

Solution: A sketch of the region is given in Figure 14.31. Notice that the x-intercepts are $(-1, 0)$ and $(2, 0)$.

On the interval $[-2, -1]$, the area of the element is

$$y\, \Delta x = (x^2 - x - 2)\, \Delta x$$

On $[-1, 2]$ the area is

$$-y\, \Delta x = -(x^2 - x - 2)\, \Delta x$$

Thus,

$$\text{area} = \int_{-2}^{-1} (x^2 - x - 2)\, dx + \int_{-1}^{2} -(x^2 - x - 2)\, dx$$

$$= \left(\frac{x^3}{3} - \frac{x^2}{2} - 2x \right) \Big|_{-2}^{-1} - \left(\frac{x^3}{3} - \frac{x^2}{2} - 2x \right) \Big|_{-1}^{2}$$

$$= \left[\left(-\frac{1}{3} - \frac{1}{2} + 2 \right) - \left(-\frac{8}{3} - \frac{4}{2} + 4 \right) \right]$$

$$- \left[\left(\frac{8}{3} - \frac{4}{2} - 4 \right) - \left(-\frac{1}{3} - \frac{1}{2} + 2 \right) \right]$$

$$= \frac{19}{3}$$

NOW WORK PROBLEM 31 ●●●

The next example shows the use of area as a probability in statistics.

● **EXAMPLE 5   Statistics Application**

*In statistics, a (probability) **density function** f of a variable x, where x assumes all values in the interval $[a, b]$, has the following properties:*

(i) $f(x) \geq 0$

(ii) $\int_{a}^{b} f(x)\, dx = 1$

The probability that x assumes a value between c and d, which is written $P(c \leq x \leq d)$, where $a \leq c \leq d \leq b$, is represented by the area of the region bounded by the graph of f and the x-axis between $x = c$ and $x = d$. Hence (see Figure 14.32),

$$P(c \leq x \leq d) = \int_{c}^{d} f(x)\, dx$$

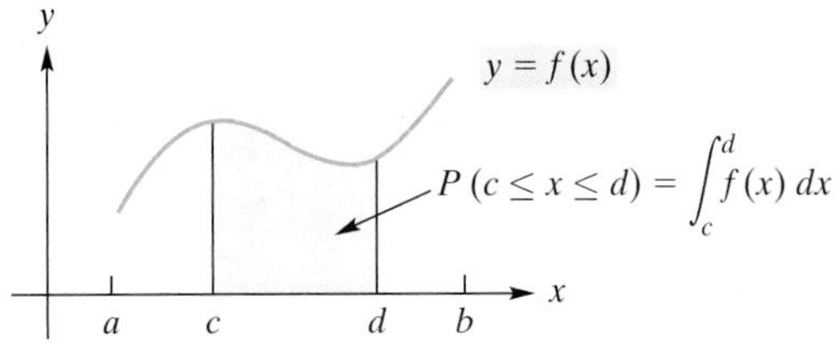

FIGURE 14.32   Probability as an area.

For the density function $f(x) = 6(x - x^2)$, *where* $0 \le x \le 1$, *find each of the following probabilities.*

a. $P(0 \le x \le \frac{1}{4})$

Solution: Here $[a, b]$ is $[0, 1]$, c is 0, and d is $\frac{1}{4}$. We have

$$P\left(0 \le x \le \tfrac{1}{4}\right) = \int_0^{1/4} 6(x - x^2)\, dx = 6\int_0^{1/4} (x - x^2)\, dx$$

$$= 6\left(\frac{x^2}{2} - \frac{x^3}{3}\right)\Bigg|_0^{1/4} = (3x^2 - 2x^3)\Bigg|_0^{1/4}$$

$$= \left(3\left(\frac{1}{4}\right)^2 - 2\left(\frac{1}{4}\right)^3\right) - 0 = \frac{5}{32}$$

b. $P\left(x \ge \frac{1}{2}\right)$

Solution: Since the domain of f is $0 \le x \le 1$, to say that $x \ge \frac{1}{2}$ means that $\frac{1}{2} \le x \le 1$. Thus,

$$P\left(x \ge \frac{1}{2}\right) = \int_{1/2}^1 6(x - x^2)\, dx = 6\int_{1/2}^1 (x - x^2)\, dx$$

$$= 6\left(\frac{x^2}{2} - \frac{x^3}{3}\right)\Bigg|_{1/2}^1 = (3x^2 - 2x^3)\Bigg|_{1/2}^1 = \frac{1}{2}$$

NOW WORK PROBLEM 37 ◖◕◕

Problems 14.9

In Problems 1–34, use a definite integral to find the area of the region bounded by the given curve, the x-axis, and the given lines. In each case, first sketch the region. Watch out for areas of regions that are below the x-axis.

*1. $y = 4x$, $x = 2$

2. $y = \dfrac{3}{4}x + 1$, $x = 0$, $x = 16$

3. $y = 5x + 2$, $x = 1$, $x = 4$

4. $y = x + 5$, $x = 2$, $x = 4$

5. $y = x - 1$, $x = 5$

6. $y = 3x^2$, $x = 1$, $x = 3$

*7. $y = x^2$, $x = 2$, $x = 3$

8. $y = 2x^2 - x$, $x = -2$, $x = -1$

9. $y = x^2 + 2$, $x = -1$, $x = 2$

10. $y = x + x^2 + x^3$, $x = 1$

11. $y = x^2 - 2x$, $x = -3$, $x = -1$

12. $y = 3x^2 - 4x$, $x = -2$, $x = -1$

13. $y = 2 - x - x^2$

14. $y = \dfrac{4}{x}$, $x = 1$, $x = 2$

15. $y = 2 - x - x^3$, $x = -3$, $x = 0$

16. $y = e^x$, $x = 1$, $x = 3$

17. $y = 3 + 2x - x^2$

18. $y = \dfrac{1}{(x-1)^2}$, $x = 2$, $x = 3$

19. $y = \dfrac{1}{x}$, $x = 1$, $x = e$

20. $y = \dfrac{1}{x}$, $x = 1$, $x = e^2$

21. $y = \sqrt{x + 9}$, $x = -9$, $x = 0$

22. $y = x^2 - 4x$, $x = 2$, $x = 6$

23. $y = \sqrt{2x - 1}$, $x = 1$, $x = 5$

24. $y = x^3 + 3x^2$, $x = -2$, $x = 2$

25. $y = \sqrt[3]{x}$, $x = 2$

26. $y = x^2 + 4x - 5$, $x = -5$, $x = 1$

*27. $y = e^x + 1$, $x = 0$, $x = 1$

28. $y = |x|$, $x = -2$, $x = 2$

29. $y = x + \dfrac{2}{x}$, $x = 1$, $x = 2$

30. $y = 4 + 3x - x^2$

*31. $y = x^3$, $x = -2$, $x = 4$

32. $y = \sqrt{x - 2}$, $x = 2$, $x = 6$

33. $y = 2x - x^2$, $x = 1$, $x = 3$

34. $y = x^2 + 1$, $x = 0$, $x = 4$

35. Given that

$$f(x) = \begin{cases} 3x^2 & \text{if } 0 \le x < 2 \\ 16 - 2x & \text{if } x \ge 2 \end{cases}$$

determine the area of the region bounded by the graph of $y = f(x)$, the x-axis, and the line $x = 3$. Include a sketch of the region.

36. Under conditions of a continuous uniform distribution (a topic in statistics) the proportion of persons with incomes between a and t, where $a \le t \le b$, is the area of the region between the curve $y = 1/(b - a)$ and the x-axis from $x = a$ to $x = t$. Sketch the graph of the curve and determine the area of the given region.

*37. Suppose $f(x) = x/8$, where $0 \le x \le 4$. If f is a density function (refer to Example 5), find each of the following.

(a) $P(0 \le x \le 1)$

(b) $P(2 \le x \le 4)$

(c) $P(x \ge 3)$

38. Suppose $f(x) = \frac{1}{3}(1 - x)^2$, where $0 \le x \le 3$. If f is a density function (refer to Example 5), find each of the following.

(a) $P(1 \le x \le 2)$

(b) $P\left(1 \le x \le \frac{5}{2}\right)$

(c) $P(x \le 1)$

(d) $P(x \ge 1)$ using your result from part (c)

39. Suppose $f(x) = 1/x$, where $e \le x \le e^2$. If f is a density function (refer to Example 5), find each of the following.

(a) $P(3 \le x \le 7)$

(b) $P(x \le 5)$

(c) $P(x \ge 4)$

(d) Verify that $P(e \le x \le e^2) = 1$

40. (a) Let r be a real number, where $r > 1$. Evaluate

$$\int_1^r \frac{1}{x^2}\, dx$$

(b) Your answer to part (a) can be interpreted as the area of a certain region of the plane. Sketch this region.

(c) Evaluate $\lim_{r \to \infty} \left(\int_1^r \frac{1}{x^2}\, dx \right)$.

(d) Your answer to part (c) can be interpreted as the area of a certain region of the plane. Sketch this region.

In Problems 41–44, use definite integration to estimate the area of the region bounded by the given curve, the x-axis, and the given lines. Round your answer to two decimal places.

41. $y = \dfrac{1}{x^2 + 1}$, $x = -2$, $x = 1$

42. $y = \dfrac{x}{\sqrt{x + 5}}$, $x = 2$, $x = 7$

43. $y = x^4 - 2x^3 - 2$, $x = 1$, $x = 3$

44. $y = 1 + 3x - x^4$

OBJECTIVE

To find the area of a region bounded by two or more curves by using either vertical or horizontal strips.

14.10 Area between Curves

Vertical Elements

We will now find the area of a region enclosed by several curves. As before, our procedure will be to draw a sample element of area and use the definite integral to "add together" the areas of all such elements.

For example, consider the area of the region in Figure 14.33 that is bounded on the top and bottom by the curves $y = f(x)$ and $y = g(x)$ and on the sides by the lines $x = a$ and $x = b$. The width of the indicated vertical element is Δx, and the height is the y-value of the upper curve minus the y-value of the lower curve, which we will write as $y_{upper} - y_{lower}$. Thus, the area of the element is

$$[y_{upper} - y_{lower}]\, \Delta x$$

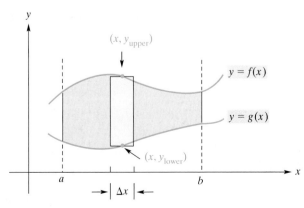

FIGURE 14.33 Region between curves.

which is

$$[f(x) - g(x)]\,\Delta x$$

Summing the areas of all such elements from $x = a$ to $x = b$ by the definite integral gives the area of the region:

$$\sum [f(x) - g(x)]\,\Delta x \to \int_a^b [f(x) - g(x)]\,dx = \text{area}$$

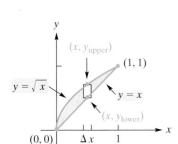

FIGURE 14.34 Diagram for Example 1.

It should be obvious to you that knowing the points of intersection is important in determining the limits of integration.

 **EXAMPLE 1 Finding an Area between Two Curves**

Find the area of the region bounded by the curves $y = \sqrt{x}$ and $y = x$.

Solution: A sketch of the region appears in Figure 14.34. To determine where the curves intersect, we solve the system formed by the equations $y = \sqrt{x}$ and $y = x$. Eliminating y by substitution, we obtain

$$\sqrt{x} = x$$
$$x = x^2 \qquad\qquad \text{(squaring both sides)}$$
$$0 = x^2 - x = x(x - 1)$$
$$x = 0 \quad \text{or} \quad x = 1$$

Since we squared both sides, we must check the solutions found with respect to the *original* equation. It is easily determined that both $x = 0$ and $x = 1$ are solutions of $\sqrt{x} = x$. If $x = 0$, then $y = 0$; if $x = 1$, then $y = 1$. Thus, the curves intersect at $(0, 0)$ and $(1, 1)$. The width of the indicated element of area is Δx. The height is the y-value on the upper curve minus the y-value on the lower curve:

$$y_{\text{upper}} - y_{\text{lower}} = \sqrt{x} - x$$

Hence, the area of the element is $(\sqrt{x} - x)\,\Delta x$. Summing the areas of all such elements from $x = 0$ to $x = 1$ by the definite integral, we get the area of the entire region:

$$\text{area} = \int_0^1 (\sqrt{x} - x)\,dx$$

$$= \int_0^1 (x^{1/2} - x)\,dx = \left(\frac{x^{3/2}}{\frac{3}{2}} - \frac{x^2}{2} \right)\Bigg|_0^1$$

$$= \left(\frac{2}{3} - \frac{1}{2} \right) - (0 - 0) = \frac{1}{6}$$

NOW WORK PROBLEM 9

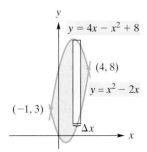

FIGURE 14.35 Diagram for Example 2.

 **EXAMPLE 2 Finding an Area between Two Curves**

Find the area of the region bounded by the curves $y = 4x - x^2 + 8$ and $y = x^2 - 2x$.

Solution: A sketch of the region appears in Figure 14.35. To find where the curves intersect, we solve the system of equations $y = 4x - x^2 + 8$ and $y = x^2 - 2x$:

$$4x - x^2 + 8 = x^2 - 2x,$$
$$-2x^2 + 6x + 8 = 0,$$
$$x^2 - 3x - 4 = 0,$$
$$(x + 1)(x - 4) = 0 \qquad \text{(factoring)}$$
$$x = -1 \quad \text{or} \quad x = 4$$

When $x = -1$, then $y = 3$; when $x = 4$, then $y = 8$. Thus, the curves intersect at $(-1, 3)$ and $(4, 8)$. The width of the indicated element is Δx. The height is the y-value on the upper curve minus the y-value on the lower curve:

$$y_{\text{upper}} - y_{\text{lower}} = (4x - x^2 + 8) - (x^2 - 2x)$$

Therefore, the area of the element is

$$[(4x - x^2 + 8) - (x^2 - 2x)]\, \Delta x = (-2x^2 + 6x + 8)\, \Delta x$$

Summing all such areas from $x = -1$ to $x = 4$, we have

$$\text{area} = \int_{-1}^{4} (-2x^2 + 6x + 8)\, dx = 41\tfrac{2}{3}$$

NOW WORK PROBLEM 25

EXAMPLE 3 Area of a Region Having Two Different Upper Curves

Find the area of the region between the curves $y = 9 - x^2$ and $y = x^2 + 1$ from $x = 0$ to $x = 3$.

Solution: The region is sketched in Figure 14.36. The curves intersect when

$$9 - x^2 = x^2 + 1$$
$$8 = 2x^2$$
$$4 = x^2$$
$$x = \pm 2 \qquad \text{(two solutions)}$$

When $x = \pm 2$, then $y = 5$, so the points of intersection are $(\pm 2, 5)$. Because we are interested in the region from $x = 0$ to $x = 3$, the intersection point that is of concern to us is $(2, 5)$. Notice in Figure 14.36 that in the region to the *left* of the intersection point $(2, 5)$, an element has

$$y_{\text{upper}} = 9 - x^2 \quad \text{and} \quad y_{\text{lower}} = x^2 + 1$$

but for an element to the *right* of $(2, 5)$ the reverse is true, namely,

$$y_{\text{upper}} = x^2 + 1 \quad \text{and} \quad y_{\text{lower}} = 9 - x^2$$

Thus, from $x = 0$ to $x = 2$, the area of an element is

$$(y_{\text{upper}} - y_{\text{lower}})\, \Delta x = [(9 - x^2) - (x^2 + 1]\, \Delta x$$
$$= (8 - 2x^2)\, \Delta x$$

but from $x = 2$ to $x = 3$, it is

$$(y_{\text{upper}} - y_{\text{lower}})\, \Delta x = [(x^2 + 1) - (9 - x^2)]\, \Delta x$$
$$= (2x^2 - 8)\, \Delta x$$

Therefore, to find the area of the entire region, we need *two* integrals:

$$\text{area} = \int_{0}^{2} (8 - 2x^2)\, dx + \int_{2}^{3} (2x^2 - 8)\, dx$$
$$= \left(8x - \frac{2x^3}{3} \right)\Big|_{0}^{2} + \left(\frac{2x^3}{3} - 8x \right)\Big|_{2}^{3}$$
$$= \left[\left(16 - \frac{16}{3} \right) - 0 \right] + \left[(18 - 24) - \left(\frac{16}{3} - 16 \right) \right]$$
$$= \frac{46}{3}$$

NOW WORK PROBLEM 33

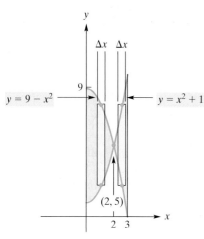

FIGURE 14.36 y_{upper} is $9 - x^2$ on $[0, 2]$ and is $x^2 + 1$ on $[2, 3]$.

Horizontal Elements

Sometimes area can more easily be determined by summing areas of horizontal elements rather than vertical elements. In the following example, an area will be found by both methods. In each case, the element of area determines the form of the integral.

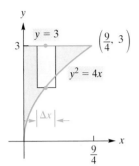

FIGURE 14.37 Vertical element of area.

 C A U T I O N

With horizontal elements, the width is Δy, not Δx.

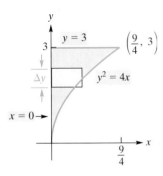

FIGURE 14.38 Horizontal element of area.

● EXAMPLE 4 **The Methods of Vertical Elements and Horizontal Elements**

Find the area of the region bounded by the curve $y^2 = 4x$ and the lines $y = 3$ and $x = 0$ (the y-axis).

Solution: The region is sketched in Figure 14.37. When the curves $y = 3$ and $y^2 = 4x$ intersect, $9 = 4x$, so $x = \frac{9}{4}$. Thus, the intersection point is $(\frac{9}{4}, 3)$. Since the width of the vertical strip is Δx, we integrate with respect to the variable x. Accordingly, y_{upper} and y_{lower} must be expressed as functions of x. For the lower curve, $y^2 = 4x$, we have $y = \pm 2\sqrt{x}$. But $y \geq 0$ for the portion of this curve that bounds the region, so we use $y = 2\sqrt{x}$. The upper curve is $y = 3$. Hence, the height of the strip is

$$y_{upper} - y_{lower} = 3 - 2\sqrt{x}$$

Therefore, the strip has an area of $(3 - 2\sqrt{x})\,\Delta x$, and we wish to sum all such areas from $x = 0$ to $x = \frac{9}{4}$. We have

$$\text{area} = \int_0^{9/4} (3 - 2\sqrt{x})\,dx = \left(3x - \frac{4x^{3/2}}{3} \right)\Bigg|_0^{9/4}$$

$$= \left[3\left(\frac{9}{4}\right) - \frac{4}{3}\left(\frac{9}{4}\right)^{3/2} \right] - (0)$$

$$= \frac{27}{4} - \frac{4}{3}\left[\left(\frac{9}{4}\right)^{1/2}\right]^3 = \frac{27}{4} - \frac{4}{3}\left(\frac{3}{2}\right)^3 = \frac{9}{4}$$

Let us now approach this problem from the point of view of a **horizontal element of area** (or **horizontal strip**) as shown in Figure 14.38. The width of the element is Δy. The length of the element is the *x-value on the rightmost curve minus the x-value on the leftmost curve*. Thus, the area of the element is

$$(x_{right} - x_{left})\,\Delta y$$

We wish to sum all such areas from $y = 0$ to $y = 3$:

$$\sum (x_{right} - x_{left})\,\Delta y \to \int_0^3 (x_{right} - x_{left})\,dy$$

Since the variable of integration is y, we must express x_{right} and x_{left} as functions of y. The rightmost curve is $y^2 = 4x$ so that $x = y^2/4$. The left curve is $x = 0$. Thus,

$$\text{area} = \int_0^3 (x_{right} - x_{left})\,dy$$

$$= \int_0^3 \left(\frac{y^2}{4} - 0\right)dy = \frac{y^3}{12}\Bigg|_0^3 = \frac{9}{4}$$

Note that for this region, horizontal strips make the definite integral easier to evaluate (and set up) than an integral with vertical strips. In any case, remember that **the limits of integration are those limits for the variable of integration.**

NOW WORK PROBLEM 23 ●◖●

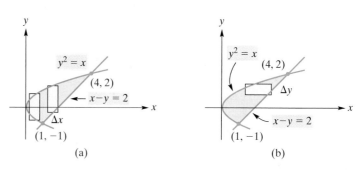

FIGURE 14.39 Region of Example 5 with vertical and horizontal elements.

●EXAMPLE 5 Advantage of Horizontal Elements

Find the area of the region bounded by the graphs of $y^2 = x$ and $x - y = 2$.

Solution: The region is sketched in Figure 14.39. The curves intersect when $y^2 - y = 2$. Thus, $y^2 - y - 2 = 0$; equivalently, $(y + 1)(y - 2) = 0$, from which it follows that $y = -1$ or $y = 2$. This gives the intersection points $(1, -1)$ and $(4, 2)$. Let us try vertical elements of area. (See Figure 14.39(a).) Solving $y^2 = x$ for y gives $y = \pm\sqrt{x}$. As seen in Figure 14.39(a), to the *left* of $x = 1$, the upper end of the element lies on $y = \sqrt{x}$ and the lower end lies on $y = -\sqrt{x}$. To the *right* of $x = 1$, the upper curve is $y = \sqrt{x}$ and the lower curve is $x - y = 2$ (or $y = x - 2$). Thus, with vertical strips, *two* integrals are needed to evaluate the area:

$$\text{area} = \int_0^1 (\sqrt{x} - (-\sqrt{x}))\, dx + \int_1^4 (\sqrt{x} - (x - 2))\, dx$$

Perhaps the use of horizontal strips can simplify our work. In Figure 14.39(b), the width of the strip is Δy. The rightmost curve is *always* $x - y = 2$ (or $x = y + 2$), and the leftmost curve is always $y^2 = x$ (or $x = y^2$). Therefore, the area of the horizontal strip is $[(y + 2) - y^2]\, \Delta y$, so the total area is

$$\text{area} = \int_{-1}^2 (y + 2 - y^2)\, dy = \frac{9}{2}$$

Clearly, the use of horizontal strips is the most desirable approach to solving the problem. Only a single integral is needed, and it is much simpler to compute.

NOW WORK PROBLEM 19 ●●●

₌TECHNOLOGY

Problem: Estimate the area of the region bounded by the graphs of

$$y = x^4 - 2x^3 - 2 \quad \text{and} \quad y = 1 + 2x - 2x^2$$

Solution: With a TI-83 Plus, we enter $x^4 - 2x^3 - 2$ as Y_1 and $1 + 2x - 2x^2$ as Y_2 and display their graphs. The region in question is shaded in Figure 14.40; y_{upper} corresponds to Y_2 and y_{lower} corresponds to Y_1. Using vertical strips, we have

$$\text{area} = \int_A^B (Y_2 - Y_1)\, dx$$

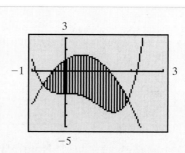

FIGURE 14.40 Graphs of $Y_1(y_{\text{lower}})$ and $Y_2(y_{\text{upper}})$.

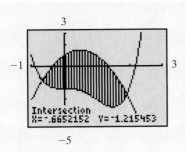

FIGURE 14.41   Intersection point in Quadrant III.

Figure 14.42.) Similarly, we find the *x*-value of the intersection point in Quadrant IV, which we store as B. Using the "fnInt(" command (Figure 14.42), we estimate the area of the region to be 7.54 square units.

where A and B are the *x*-values of the intersection points in Quadrants III and IV, respectively. With the intersection feature we find A, as indicated in Figure 14.41. This value of *x* is then stored from the home screen as A. (See

FIGURE 14.42   Storing *x*-values of intersection points and estimating area.

Problems 14.10

In Problems 1–6, express the area of the shaded region in terms of an integral (or integrals). Do not evaluate your expression.

1. See Figure 14.43.

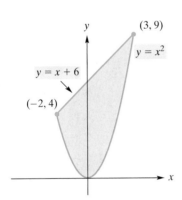

FIGURE 14.43   Region for Problem 1.

2. See Figure 14.44.

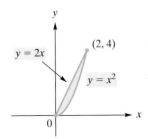

FIGURE 14.44   Region for Problem 2.

3. See Figure 14.45.

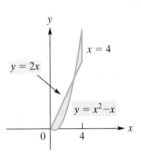

FIGURE 14.45   Region for Problem 3.

4. See Figure 14.46.

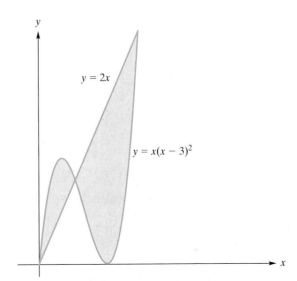

FIGURE 14.46   Region for Problem 4.

5. See Figure 14.47.

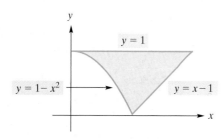

FIGURE 14.47 Region for Problem 5.

6. See Figure 14.48.

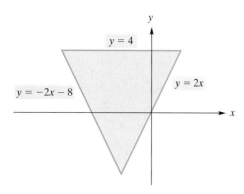

FIGURE 14.48 Region for Problem 6.

7. Express, in terms of a single integral, the total area of the region to the right of the line $x = 1$ that is between the curves $y = x^2 - 5$ and $y = 7 - 2x^2$. Do *not* evaluate the integral.

8. Express, in terms of a single integral, the total area of the region in the first quadrant bounded by the x-axis and the graphs of $y^2 = x$ and $2y = 3 - x$. Do *not* evaluate the integral.

In Problems 9–32, find the area of the region bounded by the graphs of the given equations. Be sure to find any needed points of intersection. Consider whether the use of horizontal strips makes the integral simpler than when vertical strips are used.

*9. $y = x^2$, $y = 2x$

10. $y = x$, $y = -x + 3$, $y = 0$

11. $y = x^2 + 1$, $x \geq 0$, $x = 0$, $y = 3$

12. $y = x^2 + 1$, $y = x + 3$

13. $y = 10 - x^2$, $y = 4$

14. $y^2 = x + 1$, $x = 1$

15. $x = 8 + 2y$, $x = 0$, $y = -1$, $y = 3$

16. $y = x - 6$, $y^2 = x$

17. $y = 4 - x^2$, $y = -3x$

18. $x = y^2 + 2$, $x = 6$

*19. $y^2 = 4x$, $y = 2x - 4$

20. $y = x^3$, $y = x + 6$, $x = 0$.
(*Hint:* The only real root of $x^3 - x - 6 = 0$ is 2.)

21. $2y = 4x - x^2$, $2y = x - 4$

22. $y = \sqrt{x}$, $y = x^2$

*23. $y^2 = 3x$, $3x - 2y = 15$

24. $y = 2 - x^2$, $y = x$

*25. $y = 8 - x^2$, $y = x^2$, $x = -1$, $x = 1$

26. $y^2 = 6 - x$, $3y = x + 12$

27. $y = x^2$, $y = 2$, $y = 5$

28. $y = x^3 + x$, $y = 0$, $x = -1$, $x = 2$

29. $y = x^3 - 1$, $y = x - 1$

30. $y = x^3$, $y = \sqrt{x}$

31. $4x + 4y + 17 = 0$, $y = \dfrac{1}{x}$

32. $y^2 = -x - 2$, $x - y = 5$, $y = -1$, $y = 1$

*33. Find the area of the region that is between the curves

$$y = x - 1 \quad \text{and} \quad y = 5 - 2x$$

from $x = 0$ to $x = 4$.

34. Find the area of the region that is between the curves

$$y = x^2 - 4x + 4 \quad \text{and} \quad y = 10 - x^2$$

from $x = 2$ to $x = 4$.

35. Lorenz Curve A *Lorenz curve* is used in studying income distributions. If x is the cumulative percentage of income recipients, ranked from poorest to richest, and y is the cumulative percentage of income, then equality of income distribution is given by the line $y = x$ in Figure 14.49, where x and y are expressed as decimals. For example, 10% of the people receive 10% of total income, 20% of the people receive 20% of the income, and so on. Suppose the actual distribution is given by the Lorenz curve defined by

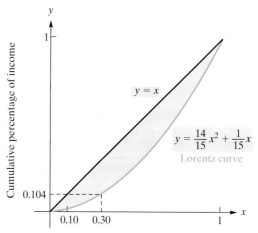

FIGURE 14.49 Diagram for Problem 35.

$$y = \frac{14}{15}x^2 + \frac{1}{15}x$$

Note, for example, that 30% of the people receive only 10.4% of total income. The degree of deviation from equality is measured by the *coefficient of inequality*[15] for a Lorenz curve. This coefficient is defined to be the area

---

[15]G. Stigler, *The Theory of Price*, 3rd ed. (New York: The Macmillan Company, 1966), pp. 293–94.

between the curve and the diagonal, divided by the area under the diagonal:

$$\frac{\text{area between curve and diagonal}}{\text{area under diagonal}}$$

For example, when all incomes are equal, the coefficient of inequality is zero. Find the coefficient of inequality for the Lorenz curve just defined.

36. Lorenz curve Find the coefficient of inequality as in Problem 35 for the Lorenz curve defined by $y = \frac{11}{12}x^2 + \frac{1}{12}x$.

37. Find the area of the region bounded by the graphs of the equations $y^2 = 3x$ and $y = mx$, where m is a positive constant.

38. (a) Find the area of the region bounded by the graphs of $y = x^2 - 1$ and $y = 2x + 2$.

(b) What percentage of the area in part (a) lies above the x-axis?

39. The region bounded by the curve $y = x^2$ and the line $y = 4$ is divided into two parts of equal area by the line $y = k$, where k is a constant. Find the value of k.

In Problems 40–44, estimate the area of the region bounded by the graphs of the given equations. Round your answer to two decimal places.

40. $y = x^2 - 4x + 1, \quad y = -\dfrac{6}{x}$

41. $y = \sqrt{25 - x^2}, \quad y = 7 - 2x - x^4$

42. $y = x^3 - 8x + 1, \quad y = x^2 - 5$

43. $y = x^5 - 3x^3 + 2x, \quad y = 3x^2 - 4$

44. $y = x^4 - 3x^3 - 15x^2 + 19x + 30, \quad y = x^3 + x^2 - 20x$

14.11 Consumers' and Producers' Surplus

OBJECTIVE

To develop the economic concepts of consumers' surplus and producers' surplus, which are represented by areas.

Determining the area of a region has applications in economics. Figure 14.50 shows a supply curve for a product. The curve indicates the price p per unit at which the manufacturer will sell (or supply) q units. The diagram also shows a demand curve for the product. This curve indicates the price p per unit at which consumers will purchase (or demand) q units. The point (q_0, p_0) where the two curves intersect is called the *point of equilibrium*. Here p_0 is the price per unit at which consumers will purchase the same quantity q_0 of a product that producers wish to sell at that price. In short, p_0 is the price at which stability in the producer–consumer relationship occurs.

Let us assume that the market is at equilibrium and the price per unit of the product is p_0. According to the demand curve, there are consumers who would be willing to pay *more* than p_0. For example, at the price per unit of p_1, consumers would buy q_1 units. These consumers are benefiting from the lower equilibrium price p_0.

The vertical strip in Figure 14.50 has area $p \, \Delta q$. This expression can also be thought of as the total amount of money that consumers would spend by buying Δq units of the product if the price per unit were p. Since the price is actually p_0, these consumers spend only $p_0 \, \Delta q$ for the Δq units and thus benefit by the amount $p\Delta q - p_0 \, \Delta q$. This expression can be written $(p - p_0) \, \Delta q$, which is the area of a rectangle of width Δq and height $p - p_0$. (See Figure 14.51.) Summing the areas of all such rectangles from $q = 0$ to $q = q_0$ by definite integration, we have

$$\int_0^{q_0} (p - p_0) \, dq$$

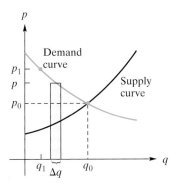

FIGURE 14.50 Supply and demand curves.

This integral, under certain conditions, represents the total gain to consumers who are willing to pay more than the equilibrium price. This total gain is called **consumers' surplus**, abbreviated CS. If the demand function is given by $p = f(q)$, then

$$\text{CS} = \int_0^{q_0} [f(q) - p_0] \, dq$$

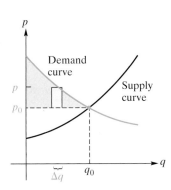

FIGURE 14.51 Benefit to consumers for Δq units.

Geometrically (see Figure 14.52), consumers' surplus is represented by the area between the line $p = p_0$ and the demand curve $p = f(q)$ from $q = 0$ to $q = q_0$.

Some of the producers also benefit from the equilibrium price, since they are willing to supply the product at prices *less* than p_0. Under certain conditions, the

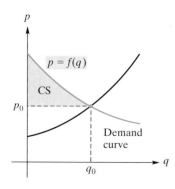

FIGURE 14.52   Consumers' surplus.

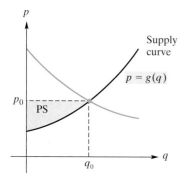

FIGURE 14.53   Producers' surplus.

total gain to the producers is represented geometrically in Figure 14.53 by the area between the line $p = p_0$ and the supply curve $p = g(q)$ from $q = 0$ to $q = q_0$. This gain, called **producers' surplus** and abbreviated PS, is given by

$$\text{PS} = \int_0^{q_0} [p_0 - g(q)]\,dq$$

●EXAMPLE 1   **Finding Consumers' Surplus and Producers' Surplus**

The demand function for a product is

$$p = f(q) = 100 - 0.05q$$

where p is the price per unit (in dollars) for q units. The supply function is

$$p = g(q) = 10 + 0.1q$$

Determine consumers' surplus and producers' surplus under market equilibrium.

Solution: First we must find the equilibrium point (p_0, q_0) by solving the system formed by the functions $p = 100 - 0.05q$ and $p = 10 + 0.1q$. We thus equate the two expressions for p and solve:

$$10 + 0.1q = 100 - 0.05q$$
$$0.15q = 90$$
$$q = 600$$

When $q = 600$ then $p = 10 + 0.1(600) = 70$. Hence, $q_0 = 600$ and $p_0 = 70$. Consumers' surplus is

$$\text{CS} = \int_0^{q_0} [f(q) - p_0]\,dq = \int_0^{600} (100 - 0.05q - 70)\,dq$$

$$= \left(30q - 0.05\frac{q^2}{2}\right)\Bigg|_0^{600} = 9000$$

Producers' surplus is

$$\text{PS} = \int_0^{q_0} [p_0 - g(q)]\,dq = \int_0^{600} [70 - (10 + 0.1q)]\,dq$$

$$= \left(60q - 0.1\frac{q^2}{2}\right)\Bigg|_0^{600} = 18,000$$

Therefore, consumers' surplus is \$9000 and producers' surplus is \$18,000.

NOW WORK PROBLEM 1 ●●●

●EXAMPLE 2   **Using Horizontal Strips to Find Consumers' Surplus and Producers' Surplus**

The demand equation for a product is

$$q = f(p) = \frac{90}{p} - 2$$

and the supply equation is $q = g(p) = p - 1$. Determine consumers' surplus and producers' surplus when market equilibrium has been established.

Solution: Determining the equilibrium point, we have

$$p - 1 = \frac{90}{p} - 2$$
$$p^2 + p - 90 = 0$$
$$(p + 10)(p - 9) = 0$$

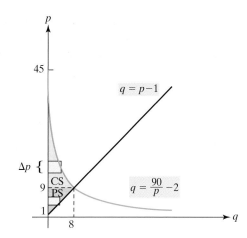

FIGURE 14.54 Diagram for Example 2.

Thus, $p_0 = 9$, so $q_0 = 9 - 1 = 8$. (See Figure 14.54.) Note that the demand equation expresses q as a function of p. Since consumers' surplus can be considered an area, this area can be determined by means of horizontal strips of width Δp and length $q = f(p)$. The areas of these strips are summed from $p = 9$ to $p = 45$ by integrating with respect to p:

$$\text{CS} = \int_9^{45} \left(\frac{90}{p} - 2 \right) dp = (90 \ln |p| - 2p) \Big|_9^{45}$$

$$= 90 \ln 5 - 72 \approx 72.85$$

Using horizontal strips for producers' surplus, we have

$$\text{PS} = \int_1^9 (p - 1) \, dp = \frac{(p-1)^2}{2} \Big|_1^9 = 32$$

NOW WORK PROBLEM 5

Problems 14.11

In Problems 1–6, the first equation is a demand equation and the second is a supply equation of a product. In each case, determine consumers' surplus and producers' surplus under market equilibrium.

*1. $p = 22 - 0.8q$
$p = 6 + 1.2q$

2. $p = 2200 - q^2$
$p = 400 + q^2$

3. $p = \dfrac{50}{q + 5}$
$p = \dfrac{q}{10} + 4.5$

4. $p = 400 - q^2$
$p = 20q + 100$

*5. $q = 100(10 - 2p)$
$q = 50(2p - 1)$

6. $q = \sqrt{100 - p}$
$q = \dfrac{p}{2} - 10$

7. The demand equation for a product is

$$q = 10\sqrt{100 - p}$$

Calculate consumers' surplus under market equilibrium, which occurs at a price of $84.

8. The demand equation for a product is

$$q = 400 - p^2$$

and the supply equation is

$$p = \frac{q}{60} + 5$$

Find producers' surplus and consumers' surplus under market equilibrium.

9. The demand equation for a product is $p = 2^{11-q}$, and the supply equation is $p = 2^{q+1}$, where p is the price per unit (in hundreds of dollars) when q units are demanded or supplied. Determine, to the nearest thousand dollars, consumers' surplus under market equilibrium.

10. The demand equation for a product is

$$(p + 10)(q + 20) = 1000$$

and the supply equation is

$$q - 4p + 10 = 0$$

(a) Verify, by substitution, that market equilibrium occurs when $p = 10$ and $q = 30$.
(b) Determine consumers' surplus under market equilibrium.

11. The demand equation for a product is

$$p = 60 - \frac{50q}{\sqrt{q^2 + 3600}}$$

and the supply equation is

$$p = 10\ln(q + 20) - 26$$

Determine consumers' surplus and producers' surplus under market equilibrium. Round your answers to the nearest integer.

12. Producers' Surplus The supply function for a product is given by the following table, where p is the price per unit (in dollars) at which q units are supplied to the market:

q	0	10	20	30	40	50
p	25	49	59	71	80	94

Use the trapezoidal rule to estimate the producers' surplus if the selling price is $80.

14.12 Review

Important Terms and Symbols   Examples

Summary

If $y = f(x)$ is a differentiable function of x, we define the differential dy by

$$dy = f'(x)\,dx$$

where $dx = \Delta x$ is a change in x and can be any real number. (Thus dy is a function of two variables, namely x and dx.) If dx is close to zero, then dy is an approximation to $\Delta y = f(x + \Delta x) - f(x)$.

$$\Delta y \approx dy$$

Moreover, dy can be used to approximate a function value using

$$f(x + \Delta x) \approx f(x) + dy$$

An antiderivative of a function f is a function F such that $F'(x) = f(x)$. Any two antiderivatives of f differ at most by a constant. The most general antiderivative of f is called the indefinite integral of f and is denoted $\int f(x)\,dx$. Thus,

$$\int f(x)\,dx = F(x) + C$$

where C is called the constant of integration, if and only if $F' = f$.

Some basic integration formulas are as follows:

$$\int k \, dx = kx + C \qquad k \text{ a constant}$$

$$\int x^n \, dx = \frac{x^{n+1}}{n+1} + C \qquad n \neq -1$$

$$\int \frac{1}{x} \, dx = \ln x + C \qquad \text{for } x > 0$$

$$\int e^x \, dx = e^x + C$$

$$\int k f(x) \, dx = k \int f(x) \, dx \qquad k \text{ a constant}$$

and $$\int [f(x) \pm g(x)] \, dx = \int f(x) \, dx \pm \int g(x) \, dx$$

Another formula is the power rule for integration:

$$\int u^n \, du = \frac{u^{n+1}}{n+1} + C, \quad \text{if } n \neq -1$$

Here u represents a differentiable function of x, and du is its differential. In applying the power rule to a given integral, it is important that the integral be written in a form that precisely matches the power rule. Other integration formulas are

$$\int e^u \, du = e^u + C$$

and $$\int \frac{1}{u} \, du = \ln |u| + C \qquad u \neq 0$$

If the rate of change of a function f is known—that is, if f' is known—then f is an antiderivative of f'. In addition, if we know that f satisfies an initial condition, then we can find the particular antiderivative. For example, if a marginal-cost function dc/dq is given to us, then by integration, we can find the most general form of c. That form involves a constant of integration. However, if we are also given fixed costs (that is, costs involved when $q = 0$), then we can determine the value of the constant of integration and thus find the particular cost function c. Similarly, if we are given a marginal-revenue function dr/dq, then by integration and by using the fact that $r = 0$ when $q = 0$, we can determine the particular revenue function r. Once r is known, the corresponding demand equation can be found by using the equation $p = r/q$.

It is helpful at this point to review summation notation from Section 1.5. This notation is especially useful in determining areas. To find the area of the region bounded by $y = f(x)$ [where $f(x) \geq 0$ and f is continuous] and the x-axis from $x = a$ to $x = b$, we divide the interval $[a, b]$ into n subintervals of equal length Δx. If x_i is the right-hand endpoint of an arbitrary subinterval, then the product $f(x_i) \Delta x$ is the area of a rectangle. Denoting the sum of all such areas of rectangles for the n subintervals by S_n, we define the limit of S_n as $n \to \infty$ as the area of the entire region:

$$\lim_{n \to \infty} S_n = \lim_{n \to \infty} \sum_{i=1}^{n} f(x_i) \Delta x = \text{area}$$

If the restriction that $f(x) \geq 0$ is omitted, this limit is defined as the definite integral of f over $[a,b]$:

$$\lim_{n \to \infty} \sum_{i=1}^{n} f(x_i) \Delta x = \int_a^b f(x) \, dx$$

Instead of evaluating definite integrals by using limits, we may be able to employ the Fundamental Theorem of Integral Calculus. Mathematically,

$$\int_a^b f(x) \, dx = F(x) \Big|_a^b = F(b) - F(a)$$

where F is any antiderivative of f.

Some properties of the definite integral are

$$\int_a^b k f(x) \, dx = k \int_a^b f(x) \, dx \qquad k \text{ a constant}$$

$$\int_a^b [f(x) \pm g(x)] \, dx = \int_a^b f(x) \, dx \pm \int_a^b g(x) \, dx$$

and

$$\int_a^c f(x) \, dx = \int_a^b f(x) \, dx + \int_b^c f(x) \, dx$$

If $f(x) \geq 0$ and is continuous on $[a, b]$, then the definite integral can be used to find the area of the region bounded by $y = f(x)$, the x-axis, $x = a$, and $x = b$. The definite integral can also be used to find areas of more complicated regions. In these situations, an element of area should be drawn in the region. This will allow you to set up the proper definite integral. In some situations vertical elements should be considered, whereas in others horizontal elements are more advantageous.

One application of finding areas involves consumers' surplus and producers' surplus. Suppose the market for a product is at equilibrium and (q_0, p_0) is the equilibrium point (the point of intersection of the supply and demand curves for the product). Then consumers' surplus, CS, corresponds to the area from $q = 0$ to $q = q_0$, bounded above by the demand curve and below by the line $p = p_0$. Thus,

$$\text{CS} = \int_0^{q_0} [f(q) - p_0] \, dq$$

where f is the demand function. Producers' surplus, PS, corresponds to the area from $q = 0$ to $q = q_0$, bounded above by the line $p = p_0$ and below by the supply curve. Therefore,

$$\text{PS} = \int_0^{q_0} [p_0 - g(q)] \, dq$$

where g is the supply function.

Review Problems

Problem numbers shown in color indicate problems suggested for use as a practice chapter test.

In Problems 1–40, determine the integrals.

1. $\int (x^3 + 2x - 7)\, dx$

2. $\int dx$

3. $\int_0^8 (\sqrt{2x} + 2x)\, dx$

4. $\int \dfrac{4}{5 - 3x}\, dx$

5. $\int \dfrac{6}{(x + 5)^3}\, dx$

6. $\int_3^9 (y - 6)^{301}\, dy$

7. $\int \dfrac{6x^2 - 12}{x^3 - 6x + 1}\, dx$

8. $\int_0^2 xe^{4 - x^2}\, dx$

9. $\int_0^1 \sqrt[3]{3t + 8}\, dt$

10. $\int \dfrac{4 - 2x}{7}\, dx$

11. $\int y(y + 1)^2\, dy$

12. $\int_0^1 10^{-8}\, dx$

13. $\int \dfrac{\sqrt[5]{t} - \sqrt[3]{t}}{\sqrt{t}}\, dt$

14. $\int \dfrac{(0.5x - 0.1)^4}{0.4}\, dx$

15. $\int_1^3 \dfrac{2t^2}{3 + 2t^3}\, dt$

16. $\int \dfrac{4x^2 - x}{x}\, dx$

17. $\int x^2 \sqrt{3x^3 + 2}\, dx$

18. $\int (8x^3 + 4x)(x^4 + x^2)^{5/2}\, dx$

19. $\int (e^{2y} - e^{-2y})\, dy$

20. $\int \dfrac{8x}{3\sqrt[3]{7 - 2x^2}}\, dx$

21. $\int \left(\dfrac{1}{x} + \dfrac{2}{x^2} \right) dx$

22. $\int_0^2 \dfrac{3e^{3x}}{1 + e^{3x}}\, dx$

23. $\int_{-2}^1 10(y^4 - y + 1)\, dy$

24. $\int_7^{70} dx$

25. $\int_1^2 5x\sqrt{5 - x^2}\, dx$

26. $\int_0^1 (2x + 1)(x^2 + x)^4\, dx$

27. $\int_0^1 \left[2x - \dfrac{1}{(x + 1)^{2/3}} \right] dx$

28. $\int_3^{27} 3(\sqrt{3x} - 2x + 1)\, dx$

29. $\int \dfrac{\sqrt{t} - 3}{t^2}\, dt$

30. $\int \dfrac{3z^3}{z - 1}\, dz$

31. $\int_{-1}^0 \dfrac{x^2 + 4x - 1}{x + 2}\, dx$

32. $\int \dfrac{(x^2 + 4)^2}{x^2}\, dx$

33. $\int 9\sqrt{x}\sqrt{x^{3/2} + 1}\, dx$

34. $\int \dfrac{e^{\sqrt{5x}}}{\sqrt{3x}}\, dx$

35. $\int_1^e \dfrac{e^{\ln x}}{x^2}\, dx$

36. $\int \dfrac{6x^2 + 4}{e^{x^3 + 2x}}\, dx$

37. $\int \dfrac{(1 + e^{2x})^3}{e^{-2x}}\, dx$

38. $\int \dfrac{3}{e^{3x}(6 + e^{-3x})^2}\, dx$

39. $\int 3\sqrt{10^{3x}}\, dx$

40. $\int \dfrac{5x^3 + 15x^2 + 37x + 3}{x^2 + 3x + 7}\, dx$

In Problems 41 and 42, find y, subject to the given condition.

41. $y' = e^{2x} + 3, \quad y(0) = -\dfrac{1}{2}$

42. $y' = \dfrac{x + 5}{x}, \quad y(1) = 3$

In Problems 43–50, determine the area of the region bounded by the given curve, the x-axis, and the given lines.

43. $y = x^2 - 1, \quad x = 2$

44. $y = 4e^x, \quad x = 0, x = 3$

45. $y = \sqrt{x + 4}, \quad x = 0$

46. $y = x^2 - x - 6, \quad x = -4, \quad x = 3$

47. $y = 5x - x^2$

48. $y = \sqrt[4]{x}, \quad x = 1, \quad x = 16$

49. $y = \dfrac{1}{x} + 2, \quad x = 1, \quad x = 4$

50. $y = x^3 - 1, \quad x = -1$

In Problems 51–58, find the area of the region bounded by the given curves.

51. $y^2 = 4x, \quad x = 0, \quad y = 2$

52. $y = 3x^2 - 5, \quad x = 0, \quad y = 4$

53. $y = x^2 + 4x - 5, \quad y = 0$

54. $y = 2x^2, \quad y = x^2 + 9$

55. $y = x^2 - x, \quad y = 10 - x^2$

56. $y = \sqrt{x}, \quad x = 0, \quad y = 3$

57. $y = \ln x, \quad x = 0, \quad y = 0, \quad y = 1$

58. $y = 2 - x, \quad y = x - 3, \quad y = 0, \quad y = 2$

59. **Marginal Revenue**   If marginal revenue is given by

$$\dfrac{dr}{dq} = 100 - \dfrac{3}{2}\sqrt{2q}$$

determine the corresponding demand equation.

60. **Marginal Cost**   If marginal cost is given by

$$\dfrac{dc}{dq} = q^2 + 7q + 6$$

and fixed costs are 2500, determine the total cost of producing six units. Assume that costs are in dollars.

61. **Marginal Revenue**   A manufacturer's marginal-revenue function is

$$\dfrac{dr}{dq} = 250 - q - 0.2q^2$$

If r is in dollars, find the increase in the manufacturer's total revenue if production is increased from 15 to 25 units.

62. **Marginal Cost**   A manufacturer's marginal-cost function is

$$\dfrac{dc}{dq} = \dfrac{1000}{\sqrt{3q + 70}}$$

If c is in dollars, determine the cost involved to increase production from 10 to 33 units.

63. **Hospital Discharges**   For a group of hospitalized individuals, suppose the discharge rate is given by

$$f(t) = 0.008e^{-0.008t}$$

where $f(t)$ is the proportion discharged per day at the end of t days of hospitalization. What proportion of the group is discharged at the end of 100 days?

64. Business Expenses   The total expenditures (in dollars) of a business over the next five years are given by

$$\int_0^5 4000 e^{0.05t}\, dt$$

Evaluate the expenditures.

65. Find the area of the region between the curves $y = 9 - 2x$ and $y = x$ from $x = 0$ to $x = 4$.

66. Find the area of the region between the curves $y = 2x^2$ and $y = 2 - 5x$ from $x = -1$ to $x = \frac{1}{3}$.

67. Consumers' and Producers' Surplus   The demand equation for a product is

$$p = 0.01q^2 - 1.1q + 30$$

and the supply equation is

$$p = 0.01q^2 + 8$$

Determine consumers' surplus and producers' surplus when market equilibrium has been established.

68. Consumers' Surplus   The demand equation for a product is

$$p = (q - 5)^2$$

and the supply equation is

$$p = q^2 + q + 3$$

where p (in thousands of dollars) is the price per 100 units when q hundred units are demanded or supplied. Determine consumers' surplus under market equilibrium.

69. Biology   In a discussion of gene mutation,[16] the equation

$$\int_{q_0}^{q_n} \frac{dq}{q - \widehat{q}} = -(u + v) \int_0^n dt$$

occurs, where u and v are gene mutation rates, the q's are gene frequencies, and n is the number of generations. Assume that all letters represent constants, except q and t. Integrate both sides and then use your result to show that

$$n = \frac{1}{u + v} \ln \left| \frac{q_0 - \widehat{q}}{q_n - \widehat{q}} \right|$$

70. Fluid Flow   In studying the flow of a fluid in a tube of constant radius R, such as blood flow in portions of the body, we can think of the tube as consisting of concentric tubes of radius r, where $0 \le r \le R$. The velocity v of the fluid is a function of r and is given by[17]

$$v = \frac{(P_1 - P_2)(R^2 - r^2)}{4\eta l}$$

where P_1 and P_2 are pressures at the ends of the tube, η (a Greek letter read "eta") is the fluid viscosity, and l is the length of the tube. The volume rate of flow through the tube, Q, is given by

$$Q = \int_0^R 2\pi r v\, dr$$

Show that $Q = \dfrac{\pi R^4 (P_1 - P_2)}{8\eta l}$. Note that R occurs as a factor to the fourth power. Thus, doubling the radius of the tube has the effect of increasing the flow by a factor of 16. The formula that you derived for the volume rate of flow is called *Poiseuille's law*, after the French physiologist Jean Poiseuille.

71. Inventory   In a discussion of inventory, Barbosa and Friedman[18] refer to the function

$$g(x) = \frac{1}{k} \int_1^{1/x} k u^r\, du$$

where k and r are constants, $k > 0$ and $r > -2$, and $x > 0$. Verify the claim that

$$g'(x) = -\frac{1}{x^{r+2}}$$

(*Hint:* Consider two cases: when $r \ne -1$ and when $r = -1$.)

In Problems 72–74, estimate the area of the region bounded by the given curves. Round your answer to two decimal places.

72. $y = x^3 + 9x^2 + 14x - 24$, $y = 0$

73. $y = x^3 - 3x - 2$, $y = 3 + x - 2x^2$

74. $y = x^3 + x^2 - 5x - 3$, $y = x^2 + 2x + 3$

75. The demand equation for a product is

$$p = \frac{200}{\sqrt{q + 20}}$$

and the supply equation is

$$p = 2\ln(q + 10) + 5$$

Determine consumers' surplus and producers' surplus under market equilibrium. Round your answers to the nearest integer.

---

[16]W. B. Mather, *Principles of Quantitative Genetics* (Minneapolis: Burgess Publishing Company, 1964).

[17]R. W. Stacy et al., *Essentials of Biological and Medical Physics* (New York: McGraw-Hill, 1955).

[18]L. C. Barbosa and M. Friedman, "Deterministic Inventory Lot Size Models—a General Root Law," *Management Science,* 24, no. 8 (1978), 819–26.

Mathematical Snapshot

Delivered Price

Suppose that you are a manufacturer of a product whose sales occur within R miles of your mill. Assume that you charge customers for shipping at the rate s, in dollars per mile, for each unit of product sold. If m is the unit price (in dollars) at the mill, then the delivered unit price p to a customer x miles from the mill is the mill price plus the shipping charge sx:

$$p = m + sx \qquad 0 \le x \le R \qquad (1)$$

The problem is to determine the average delivered price of the units sold.

Suppose that there is a function f such that $f(t) \ge 0$ on the interval $[0, R]$ and such that the area under the graph of f and above the t-axis from $t = 0$ to $t = x$ represents the total number of units Q sold to customers within x miles of the mill. [See Figure 14.55(a).] You can refer to f as the distribution of demand. Because Q is a function of x and is represented by area,

$$Q(x) = \int_0^x f(t)\, dt$$

In particular, the total number of units sold within the market area is

$$Q(R) = \int_0^R f(t)\, dt$$

[see Figure 14.55(b)]. For example, if $f(t) = 10$ and $R = 100$, then the total number of units sold within the market area is

$$Q(100) = \int_0^{100} 10\, dt = 10t \Big|_0^{100} = 1000 - 0 = 1000$$

The average delivered price A is given by

$$A = \frac{\text{total revenue}}{\text{total number of units sold}}$$

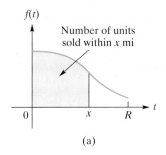

(a)

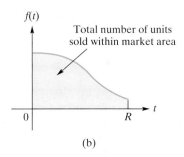

(b)

FIGURE 14.55 Number of units sold as an area.

Because the denominator is $Q(R)$, A can be determined once the total revenue is found.

To find the total revenue, first consider the number of units sold over an interval. If $t_1 < t_2$ [see Figure 14.56(a)], then the area under the graph of f and above the t-axis from $t = 0$ to $t = t_1$ represents the number of units sold within t_1 miles of the mill. Similarly, the area under the graph of f and above the t-axis from $t = 0$ to $t = t_2$ represents the number of units sold within t_2 miles of the mill. Thus the difference in these areas is geometrically the area of the shaded region in Figure 14.56(a) and represents the number of units sold between t_1 and t_2 miles of the mill, which is $Q(t_2) - Q(t_1)$. Thus

$$Q(t_2) - Q(t_1) = \int_{t_1}^{t_2} f(t)\, dt$$

For example, if $f(t) = 10$, then the number of units sold to customers located between 4 and 6 miles of the mill is

$$Q(6) - Q(4) = \int_4^6 10\, dt = 10t \Big|_4^6 = 60 - 40 = 20$$

The area of the shaded region in Figure 14.56(a) can be approximated by the area of a rectangle [see Figure 14.56(b)] whose height is $f(t)$ and whose width is Δt, where $\Delta t = t_2 - t_1$. Thus the number of units sold over the interval

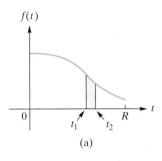

(a)

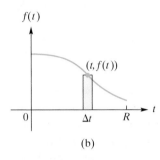

(b)

FIGURE 14.56   Number of units sold over an interval.

of length Δt is approximately $f(t)\,\Delta t$. Because the price of each of these units is [from Equation (1)] approximately $m + st$, the revenue received is approximately

$$(m + st) f(t) \, \Delta t$$

The sum of all such products from $t = 0$ to $t = R$ approximates the total revenue. Definite integration gives

$$\sum (m + st) f(t) \, \Delta t \to \int_0^R (m + st) f(t) \, dt$$

Thus,

$$\text{total revenue} = \int_0^R (m + st) f(t) \, dt$$

Consequently, the average delivered price A is given by

$$A = \frac{\displaystyle\int_0^R (m + st) f(t) \, dt}{Q(R)}$$

Equivalently,

$$A = \frac{\displaystyle\int_0^R (m + st) f(t) \, dt}{\displaystyle\int_0^R f(t) \, dt}$$

For example, if $f(t) = 10,\, m = 200,\, s = 0.25$, and $R = 100$, then

$$\int_0^R (m + st) f(t) \, dt = \int_0^{100} (200 + 0.25t) \cdot 10 \, dt$$

$$= 10 \int_0^{100} (200 + 0.25t) \, dt$$

$$= 10 \left(200t + \frac{t^2}{8} \right) \Big|_0^{100}$$

$$= 10 \left[\left(20{,}000 + \frac{10{,}000}{8} \right) - 0 \right]$$

$$= 212{,}500$$

From before,

$$\int_0^R f(t) \, dt = \int_0^{100} 10 \, dt = 1000$$

Thus, the average delivered price is $212{,}500/1000 = \$212.50$.

Problems

1. If $f(t) = 100 - 2t$, determine the number of units sold to customers located (a) within 5 miles of the mill, and (b) between 20 and 25 miles of the mill.

2. If $f(t) = 40 - 0.5t,\, m = 50,\, s = 0.20$, and $R = 80$, determine (a) the total revenue, (b) the total number of units sold, and (c) the average delivered price.

3. If $f(t) = 900 - t^2,\, m = 100,\, s = 1$, and $R = 30$, determine (a) the total revenue, (b) the total number of units sold, and (c) the average delivered price. Use a graphing calculator if you like.

4. How do real-world sellers of such things as books and clothing generally determine shipping charges for an order? (Visit an online retailer to find out.) How would you calculate average delivered price for their products? Is the procedure fundamentally different from the one in the Snapshot?

METHODS AND APPLICATIONS
OF INTEGRATION

 Mathematical Snapshot   Dieting

We now know how to find the derivative of a function, and in some cases we know how to find a function from its derivative through integration. However, the integration process is not always straightforward.

Suppose we model the gradual disappearance of a chemical substance using the equations $M' = -0.004t$ and $M(0) = 3000$, where the amount M, in grams, is a function of time t in days. This initial-condition problem is easily solved by integration with respect to t: $M = -0.002t^2 + 3000$. But what if, instead, the disappearance of the substance were modeled by the equations $M' = -0.004M$ and $M(0) = 3000$? The simple replacement of t in the first equation with M changes the character of the problem. We have not yet learned how to find a function when its derivative is described in terms of the function itself.

If you worked through the Mathematical Snapshot in Chapter 13, you may remember a similar situation, involving an equation with P on one side and the derivative of P on the other. There, we used an approximation to solve the problem. In this chapter, we will learn a method that yields an exact solution for many problems of this type.

Equations of the form $y' = ky$, where k is a constant, are especially common. When y represents the amount of a radioactive substance, $y' = ky$ can represent the rate of its disappearance through radioactive decay. And if y is the temperature of a chicken just taken out of the oven or just put into a freezer, then a related formula, called Newton's law of cooling, can be used to describe the change in the chicken's internal temperature over time. Newton's law, which is discussed in this chapter, might be used to write procedures for a restaurant kitchen, so that food prone to contamination through bacterial growth does not spend too much time in the temperature danger zone (40°F to 140°F). (Bacterial growth, for that matter, also follows a $y' = ky$ law!)

15.1 Integration by Parts[1]

Many integrals cannot be found by our previous methods. However, there are ways of changing certain integrals to forms that are easier to integrate. Of these methods, we will discuss two: *integration by parts* and (in Section 15.2) *integration using partial fractions*.

If u and v are differentiable functions of x, we have, by the product rule,

$$(uv)' = uv' + vu'$$

Rearranging gives

$$uv' = (uv)' - vu'$$

Integrating both sides with respect to x, we get

$$\int uv' \, dx = \int (uv)' \, dx - \int vu' \, dx \qquad (1)$$

For $\int (uv)' \, dx$, we must find a function whose derivative with respect to x is $(uv)'$. Clearly, uv is such a function. Hence $\int (uv)' \, dx = uv + C_1$, and Equation (1) becomes

$$\int uv' \, dx = uv + C_1 - \int vu' \, dx$$

Absorbing C_1 into the constant of integration for $\int vu' \, dx$ and replacing $v' \, dx$ by dv and $u' \, dx$ by du, we have the *formula for integration by parts:*

Formula for Integration by Parts

$$\int u \, dv = uv - \int v \, du \qquad (2)$$

This formula expresses an integral, $\int u \, dv$, in terms of another integral, $\int v \, du$, that may be easier to find.

To apply the formula to a given integral $\int f(x) \, dx$, we must write $f(x) \, dx$ as the product of two factors (or *parts*) by choosing a function u and a differential dv such that $f(x) \, dx = u \, dv$. However, for the formula to be useful, we must be able to integrate the part chosen for dv. To illustrate, consider

$$\int xe^x \, dx$$

This integral cannot be determined by previous integration formulas. One way to write $xe^x \, dx$ in the form $u \, dv$ is by letting

$$u = x \quad \text{and} \quad dv = e^x \, dx$$

To apply the formula for integration by parts, we must find du and v:

$$du = dx \quad \text{and} \quad v = \int e^x \, dx = e^x + C_1$$

Thus,

$$\int xe^x \, dx = \int u \, dv$$

$$= uv - \int v \, du$$

$$= x(e^x + C_1) - \int (e^x + C_1) \, dx$$

$$= xe^x + C_1 x - e^x - C_1 x + C$$

$$= xe^x - e^x + C$$

$$= e^x(x - 1) + C$$

---

[1]This section may be omitted without loss of continuity.

The first constant, C_1, does not appear in the final answer. It is easy to prove that the constant involved in finding v from dv will always drop out, so from now on we will not write it when we find v.

When you are using the formula for integration by parts, sometimes the "best choice" for u and dv may not be obvious. In some cases, one choice may be as good as another; in other cases, only one choice may be suitable. Insight into making a good choice (if any exists) will come only with practice and, of course, trial and error.

● EXAMPLE 1 **Integration by Parts**

Find $\displaystyle\int \frac{\ln x}{\sqrt{x}}\, dx$ *by integration by parts.*

Solution: We try

$$u = \ln x \quad \text{and} \quad dv = \frac{1}{\sqrt{x}}\, dx$$

Then

$$du = \frac{1}{x}\, dx \quad \text{and} \quad v = \int x^{-1/2}\, dx = 2x^{1/2}$$

Thus,

$$\int \ln x \left(\frac{1}{\sqrt{x}}\, dx \right) = \int u\, dv = uv - \int v\, du$$

$$= (\ln x)(2\sqrt{x}) - \int (2x^{1/2}) \left(\frac{1}{x}\, dx \right)$$

$$= 2\sqrt{x}\ln x - 2 \int x^{-1/2}\, dx$$

$$= 2\sqrt{x}\ln x - 2(2\sqrt{x}) + C \qquad (x^{1/2} = \sqrt{x})$$

$$= 2\sqrt{x}[\ln(x) - 2] + C$$

NOW WORK PROBLEM 3 ◖◗◗

● EXAMPLE 2 **Integration by Parts**

Evaluate $\displaystyle\int_1^2 x \ln x\, dx.$

Solution: Since the integral does not fit a familiar form, we will try integration by parts. Let $u = x$ and $dv = \ln x\, dx$. Then $du = dx$, but $v = \int \ln x\, dx$ is not apparent by inspection. So we will make a different choice for u and dv. Let

$$u = \ln x \quad \text{and} \quad dv = x\, dx$$

Then

$$du = \frac{1}{x}\, dx \quad \text{and} \quad v = \int x\, dx = \frac{x^2}{2}$$

Therefore,

$$\int_1^2 x \ln x\, dx = (\ln x) \left(\frac{x^2}{2} \right) \Big|_1^2 - \int_1^2 \left(\frac{x^2}{2} \right) \frac{1}{x}\, dx$$

$$= (\ln x) \left(\frac{x^2}{2} \right) \Big|_1^2 - \frac{1}{2} \int_1^2 x\, dx$$

$$= \frac{x^2 \ln x}{2} \Big|_1^2 - \frac{1}{2} \left(\frac{x^2}{2} \right) \Big|_1^2$$

$$= (2 \ln 2 - 0) - \left(1 - \tfrac{1}{4} \right) = 2 \ln 2 - \tfrac{3}{4}$$

NOW WORK PROBLEM 5 ◖◗◗

EXAMPLE 3   Integration by Parts where u Is the Entire Integrand

Determine $\displaystyle\int \ln y \, dy$.

Solution: We cannot integrate $\ln y$ by previous methods, so we will try integration by parts. Let $u = \ln y$ and $dv = dy$. Then $du = (1/y) \, dy$ and $v = y$. So we have

$$\int \ln y \, dy = (\ln y)(y) - \int y \left(\frac{1}{y} \, dy\right)$$

$$= y \ln y - \int dy = y \ln y - y + C$$

$$= y[\ln(y) - 1] + C$$

NOW WORK PROBLEM 37

Before trying integration by parts, see whether the technique is really needed. Sometimes the integral can be handled by a basic technique, as Example 4 shows.

EXAMPLE 4   Basic Integration Form

Determine $\displaystyle\int x e^{x^2} \, dx$.

CAUTION

Do not forget about basic integration forms. Integration by parts is not needed here!

Solution: This integral can be fit to the form $\displaystyle\int e^u \, du$.

$$\int x e^{x^2} \, dx = \frac{1}{2} \int e^{x^2} (2x \, dx)$$

$$= \frac{1}{2} \int e^u \, du \qquad (\text{where } u = x^2)$$

$$= \frac{1}{2} e^u + C = \frac{1}{2} e^{x^2} + C$$

NOW WORK PROBLEM 17

Sometimes integration by parts must be used more than once, as shown in the following example.

PRINCIPLES IN PRACTICE 2

APPLYING INTEGRATION BY PARTS TWICE

Suppose a population of bacteria grows at a rate of

$$P'(t) = 0.1t(\ln t)^2$$

Find the general form of $P(t)$.

EXAMPLE 5   Applying Integration by Parts Twice

Determine $\displaystyle\int x^2 e^{2x+1} \, dx$.

Solution: Let $u = x^2$ and $dv = e^{2x+1} \, dx$. Then $du = 2x \, dx$ and $v = e^{2x+1}/2$.

$$\int x^2 e^{2x+1} \, dx = \frac{x^2 e^{2x+1}}{2} - \int \frac{e^{2x+1}}{2} (2x \, dx)$$

$$= \frac{x^2 e^{2x+1}}{2} - \int x e^{2x+1} \, dx$$

To find $\int x e^{2x+1} \, dx$, we will again use integration by parts. Here, let $u = x$ and $dv = e^{2x+1} \, dx$. Then $du = dx$ and $v = e^{2x+1}/2$, and we have

$$\int x e^{2x+1} \, dx = \frac{x e^{2x+1}}{2} - \int \frac{e^{2x+1}}{2} \, dx$$

$$= \frac{x e^{2x+1}}{2} - \frac{e^{2x+1}}{4} + C_1$$

Thus,

$$\int x^2 e^{2x+1}\, dx = \frac{x^2 e^{2x+1}}{2} - \frac{xe^{2x+1}}{2} + \frac{e^{2x+1}}{4} + C \qquad \text{(where } C = -C_1)$$

$$= \frac{e^{2x+1}}{2}\left(x^2 - x + \frac{1}{2}\right) + C$$

NOW WORK PROBLEM 23 ●◐

Problems 15.1

1. In applying integration by parts to

$$\int f(x)\, dx$$

a student found that $u = x$, $du = dx$, $dv = (x+5)^{1/2}$, and $v = \frac{2}{3}(x+5)^{3/2}$. Use this information to find $\int f(x)\, dx$.

2. Use integration by parts to find

$$\int xe^{3x+1}\, dx$$

by choosing $u = x$ and $dv = e^{3x+1}\, dx$.

In Problems 3–29, find the integrals.

*3. $\displaystyle\int xe^{-x}\, dx$

4. $\displaystyle\int xe^{-5x}\, dx$

*5. $\displaystyle\int y^3 \ln y\, dy$

6. $\displaystyle\int x^2 \ln x\, dx$

7. $\displaystyle\int \ln(4x)\, dx$

8. $\displaystyle\int \frac{t}{e^t}\, dt$

9. $\displaystyle\int 3x\sqrt{2x+3}\, dx$

10. $\displaystyle\int \frac{12x}{\sqrt{1+4x}}\, dx$

11. $\displaystyle\int \frac{x}{(5x+2)^3}\, dx$

12. $\displaystyle\int \frac{\ln(x+1)}{2(x+1)}\, dx$

13. $\displaystyle\int \frac{\ln x}{x^2}\, dx$

14. $\displaystyle\int \frac{3x+5}{e^{2x}}\, dx$

15. $\displaystyle\int_1^2 4xe^{2x}\, dx$

16. $\displaystyle\int_1^2 2xe^{-3x}\, dx$

*17. $\displaystyle\int_0^1 xe^{-x^2}\, dx$

18. $\displaystyle\int \frac{3x^3}{\sqrt{4-x^2}}\, dx$

19. $\displaystyle\int_1^2 \frac{3x}{\sqrt{4-x}}\, dx$

20. $\displaystyle\int (\ln x)^2\, dx$

21. $\displaystyle\int 3(2x-2)\ln(x-2)\, dx$

22. $\displaystyle\int \frac{xe^x}{(x+1)^2}\, dx$

*23. $\displaystyle\int x^2 e^x\, dx$

24. $\displaystyle\int_e^3 \sqrt[3]{x}\ln(x^5)\, dx$

25. $\displaystyle\int (x - e^{-x})^2\, dx$

26. $\displaystyle\int x^2 e^{3x}\, dx$

27. $\displaystyle\int x^3 e^{x^2}\, dx$

28. $\displaystyle\int x^5 e^{x^2}\, dx$

29. $\displaystyle\int (2^x + x)^2\, dx$

30. Find $\int \ln(x + \sqrt{x^2+1})\, dx$. *Hint:* Show that

$$\frac{d}{dx}[\ln(x + \sqrt{x^2+1})] = \frac{1}{\sqrt{x^2+1}}$$

31. Find the area of the region bounded by the x-axis, the curve $y = \ln x$, and the line $x = e^3$.

32. Find the area of the region bounded by the x-axis and the curve $y = x^2 e^x$ between $x = 0$ and $x = 1$.

33. Find the area of the region bounded by the x-axis and the curve $y = x^2 \ln x$ between $x = 1$ and $x = 2$.

34. **Consumers' Surplus**   Suppose the demand equation for a manufacturer's product is given by

$$p = 10(q + 10)e^{-(0.1q+1)}$$

where p is the price per unit (in dollars) when q units are demanded. Assume that market equilibrium occurs when $q = 20$. Determine consumers' surplus at market equilibrium.

35. **Revenue**   Suppose total revenue r and price per unit p are differentiable functions of output q.

   (a) Use integration by parts to show that

$$\int p\, dq = r - \int q\frac{dp}{dq}\, dq$$

   (b) Using part (a), show that

$$r = \int \left(p + q\frac{dp}{dq}\right) dq$$

   (c) Using part (b), prove that

$$r(q_0) = \int_0^{q_0} \left(p + q\frac{dp}{dq}\right) dq$$

   (*Hint:* Refer to Section 14.7.)

36. Suppose f is a differentiable function. Apply integration by parts to $\int f(x)e^x\, dx$ to prove that

$$\int f(x)e^x\, dx + \int f'(x)e^x\, dx = f(x)e^x + C$$

$$\left(\text{Hence, } \int [f(x) + f'(x)]e^x\, dx = f(x)e^x + C\right)$$

*37. Suppose that f has an inverse and that $F' = f$. Use integration by parts to develop a useful formula for $\int f^{-1}(x)\, dx$ in terms of F and f^{-1}. (*Hint:* Review Example 3. It used the idea required here, for the special case of $f(x) = e^x$.)

To show how to integrate a proper rational
function by first expressing it as a sum of its
partial fractions.

15.2 Integration by Partial Fractions[2]

Recall that a *rational function* is a quotient of polynomials $N(x)/D(x)$ and that it is *proper* if N and D have no common polynomial factor and the degree of the numerator N is less than the degree of the denominator D. If N/D is not proper, then we can use long division to divide $N(x)$ by $D(x)$:

$$D(x)\overline{)N(x)}\,\dfrac{Q(x)}{}\quad \text{thus}\quad \dfrac{N(x)}{D(x)} = Q(x) + \dfrac{R(x)}{D(x)}$$

$$\vdots$$
$$\overline{R(x)}$$

Here the quotient $Q(x)$ and the remainder $R(x)$ are also polynomials and either $R(x)$ is the constant 0-polynomial or the degree of $R(x)$ is less than that of $D(x)$. Thus R/D is a proper rational function. Since

$$\int \dfrac{N(x)}{D(x)}\,dx = \int \left(Q(x) + \dfrac{R(x)}{D(x)} \right) dx = \int Q(x)\,dx + \int \dfrac{R(x)}{D(x)}\,dx$$

and we already know how to integrate a polynomial, it follows that the task of integrating rational functions reduces to that of integrating *proper* rational functions. We emphasize that the technique we are about to explain requires that a rational function be proper so that the long divison step is not optional. For example,

$$\int \dfrac{2x^4 - 3x^3 - 4x^2 - 17x - 6}{x^3 - 2x^2 - 3x}\,dx = \int \left(2x + 1 + \dfrac{4x^2 - 14x - 6}{x^3 - 2x^2 - 3x} \right) dx$$

$$= x^2 + x + \int \dfrac{4x^2 - 14x - 6}{x^3 - 2x^2 - 3x}\,dx$$

Distinct Linear Factors

Therefore, we will consider

$$\int \dfrac{4x^2 - 14x - 6}{x^3 - 2x^2 - 3x}\,dx$$

It is essential that the denominator be expressed in factored form:

$$\int \dfrac{4x^2 - 14x - 6}{x(x + 1)(x - 3)}\,dx$$

Observe that in this example the denominator consists only of **linear factors** and that each factor occurs exactly once. It can be shown that, to each such factor $x - a$, there corresponds a *partial fraction* of the form

$$\dfrac{A}{x - a} \qquad (A \text{ a constant})$$

such that the integrand is the sum of the partial fractions. If there are n such *distinct* linear factors, there will be n such partial fractions, each of which is easily integrated. Applying these facts, we can write

$$\dfrac{4x^2 - 14x - 6}{x(x + 1)(x - 3)} = \dfrac{A}{x} + \dfrac{B}{x + 1} + \dfrac{C}{x - 3} \tag{1}$$

To determine the constants A, B, and C, we first combine the terms on the right side:

$$\dfrac{4x^2 - 14x - 6}{x(x + 1)(x - 3)} = \dfrac{A(x + 1)(x - 3) + Bx(x - 3) + Cx(x + 1)}{x(x + 1)(x - 3)}$$

[2]This section may be omitted without loss of continuity.

Since the denominators of both sides are equal, we can equate their numerators:

$$4x^2 - 14x - 6 = A(x+1)(x-3) + Bx(x-3) + Cx(x+1) \qquad (2)$$

Although Equation (1) is not defined for $x = 0$, $x = -1$, and $x = 3$, we want to find values for A, B, and C that will make Equation (2) true for all values of x, so that the two sides of the equality provide equal functions. By successively setting x in Equation (2) equal to any three different numbers, we can obtain a system of equations that can be solved for A, B, and C. In particular, the work can be simplified by letting x be the roots of $D(x) = 0$; in our case, $x = 0$, $x = -1$, and $x = 3$. Using Equation (2), we have, for $x = 0$,

$$-6 = A(1)(-3) + B(0) + C(0) = -3A, \quad \text{so } A = 2$$

If $x = -1$,

$$12 = A(0) + B(-1)(-4) + C(0) = 4B, \quad \text{so } B = 3$$

If $x = 3$,

$$-12 = A(0) + B(0) + C(3)(4) = 12C, \quad \text{so } C = -1$$

Thus Equation (1) becomes

$$\frac{4x^2 - 14x - 6}{x(x+1)(x-3)} = \frac{2}{x} + \frac{3}{x+1} - \frac{1}{x-3}$$

Hence,

$$\int \frac{4x^2 - 14x - 6}{x(x+1)(x-3)}\,dx$$

$$= \int \left(\frac{2}{x} + \frac{3}{x+1} - \frac{1}{x-3} \right) dx$$

$$= 2\int \frac{dx}{x} + 3\int \frac{dx}{x+1} - \int \frac{dx}{x-3}$$

$$= 2\ln|x| + 3\ln|x+1| - \ln|x-3| + C$$

For the *original* integral, we can now state that

$$\int \frac{2x^4 - 3x^3 - 4x^2 - 17x - 6}{x^3 - 2x^2 - 3x}\,dx = x^2 + x + 2\ln|x| + 3\ln|x+1| - \ln|x-3| + C$$

An alternative method of determining A, B, and C involves expanding the right side of Equation (2) and combining like terms:

$$4x^2 - 14x - 6 = A(x^2 - 2x - 3) + B(x^2 - 3x) + C(x^2 + x)$$

$$= Ax^2 - 2Ax - 3A + Bx^2 - 3Bx + Cx^2 + Cx$$

$$4x^2 - 14x - 6 = (A + B + C)x^2 + (-2A - 3B + C)x + (-3A)$$

For this last equation to express an equality of functions, the coefficients of corresponding powers of x on the left and right sides must be equal:

$$\begin{cases} 4 = A + B + C \\ -14 = -2A - 3B + C \\ -6 = -3A \end{cases}$$

Solving gives $A = 2$, $B = 3$, and $C = -1$ as before.

The marginal revenue for a company manufacturing q radios per week is given by $r'(q) = \dfrac{5(q+4)}{q^2+4q+3}$, where $r(q)$ is the revenue in thousands of dollars. Find the equation for $r(q)$.

● EXAMPLE 1 Distinct Linear Factors

Determine $\displaystyle\int \frac{2x+1}{3x^2-27}\,dx$ *by using partial fractions.*

Solution: Since the degree of the numerator is less than the degree of the denominator, no long division is necessary. The integral can be written as

$$\frac{1}{3}\int \frac{2x+1}{x^2-9}\,dx$$

Expressing $(2x+1)/(x^2-9)$ as a sum of partial fractions, we have

$$\frac{2x+1}{x^2-9} = \frac{2x+1}{(x+3)(x-3)} = \frac{A}{x+3} + \frac{B}{x-3}$$

Combining terms and equating numerators gives

$$2x+1 = A(x-3) + B(x+3)$$

If $x = 3$, then

$$7 = 6B, \quad \text{so } B = \frac{7}{6}$$

If $x = -3$, then

$$-5 = -6A, \quad \text{so } A = \frac{5}{6}$$

Thus,

$$\int \frac{2x+1}{3x^2-27}\,dx = \frac{1}{3}\left(\int \frac{\frac{5}{6}\,dx}{x+3} + \int \frac{\frac{7}{6}\,dx}{x-3}\right)$$

$$= \frac{1}{3}\left(\frac{5}{6}\ln|x+3| + \frac{7}{6}\ln|x-3|\right) + C$$

NOW WORK PROBLEM 1 ●●●

Repeated Linear Factors

If the denominator of $N(x)/D(x)$ contains only linear factors, some of which are repeated, then, for each factor $(x-a)^k$, where k is the maximum number of times $x-a$ occurs as a factor, there will correspond the sum of k partial fractions:

$$\frac{A}{x-a} + \frac{B}{(x-a)^2} + \cdots + \frac{K}{(x-a)^k}$$

● EXAMPLE 2 Repeated Linear Factors

Determine $\displaystyle\int \frac{6x^2+13x+6}{(x+2)(x+1)^2}\,dx$ *by using partial fractions.*

Solution: Since the degree of the numerator, namely 2, is less than that of the denominator, namely 3, no long division is necessary. In the denominator, the linear factor $x+2$ occurs once and the linear factor $x+1$ occurs twice. There will thus be three partial fractions and three constants to determine, and we have

$$\frac{6x^2+13x+6}{(x+2)(x+1)^2} = \frac{A}{x+2} + \frac{B}{x+1} + \frac{C}{(x+1)^2}$$

$$6x^2+13x+6 = A(x+1)^2 + B(x+2)(x+1) + C(x+2)$$

Let us choose $x = -2$, $x = -1$, and, for convenience, $x = 0$. For $x = -2$, we have

$$4 = A$$

If $x = -1$, then

$$-1 = C$$

If $x = 0$, then

$$6 = A + 2B + 2C = 4 + 2B - 2 = 2 + 2B$$

$$4 = 2B$$

$$2 = B$$

Therefore,

$$\int \frac{6x^2 + 13x + 6}{(x + 2)(x + 1)^2} \, dx = 4 \int \frac{dx}{x + 2} + 2 \int \frac{dx}{x + 1} - \int \frac{dx}{(x + 1)^2}$$

$$= 4 \ln |x + 2| + 2 \ln |x + 1| + \frac{1}{x + 1} + C$$

$$= \ln[(x + 2)^4 (x + 1)^2] + \frac{1}{x + 1} + C$$

The last line above is somewhat optional (depending on what you need the integral for). It merely illustrates that in problems of this kind the logarithms that arise can often be combined.

NOW WORK PROBLEM 5

Distinct Irreducible Quadratic Factors

Suppose a quadratic factor $x^2 + bx + c$ occurs in $D(x)$ and it cannot be expressed as a product of two linear factors with real coefficients. Such a factor is said to be an *irreducible quadratic factor over the real numbers*. To each distinct irreducible quadratic factor that occurs exactly once in $D(x)$, there will correspond a partial fraction of the form

$$\frac{Ax + B}{x^2 + bx + c}$$

Note that even after you have expressed a rational function in terms of partial fractions, you may still find it impossible to integrate using only the calculus you have been taught so far. For example, a very simple irreducible quadratic factor is $x^2 + 1$ and yet

$$\int \frac{1}{x^2 + 1} \, dx = \int \frac{dx}{x^2 + 1} = \tan^{-1} x + C$$

where $\tan^{-1}$ is the inverse of the trigonometric function $\tan$ when $\tan$ is restricted to $(-\pi/2, \pi/2)$. We do not discuss trigonometric functions in this book, but note that any good calculator has a $\tan^{-1}$ key.

EXAMPLE 3 An Integral with a Distinct Irreducible Quadratic Factor

Determine $\int \dfrac{-2x - 4}{x^3 + x^2 + x} \, dx$ *by using partial fractions.*

Solution: Since $x^3 + x^2 + x = x(x^2 + x + 1)$, we have the linear factor x and the quadratic factor $x^2 + x + 1$, which does not seem factorable on inspection. If it were factorable as $(x - r_1)(x - r_2)$, with r_1 and r_2 real, then r_1 and r_2 would be roots of the equation $x^2 + x + 1 = 0$. By the quadratic formula, the roots are

$$x = \frac{-1 \pm \sqrt{1 - 4}}{2}$$

Since there are no real roots, we conclude that $x^2 + x + 1$ is irreducible. Thus there will be two partial fractions and *three* constants to determine. We have

$$\frac{-2x - 4}{x(x^2 + x + 1)} = \frac{A}{x} + \frac{Bx + C}{x^2 + x + 1}$$

$$-2x - 4 = A(x^2 + x + 1) + (Bx + C)x$$

$$= Ax^2 + Ax + A + Bx^2 + Cx$$

$$0x^2 - 2x - 4 = (A + B)x^2 + (A + C)x + A$$

Equating coefficients of like powers of x, we obtain

$$\begin{cases} 0 = A + B \\ -2 = A + C \\ -4 = A \end{cases}$$

Solving gives $A = -4$, $B = 4$, and $C = 2$. Hence,

$$\int \frac{-2x - 4}{x(x^2 + x + 1)}\, dx = \int \left(\frac{-4}{x} + \frac{4x + 2}{x^2 + x + 1} \right) dx$$

$$= -4 \int \frac{dx}{x} + 2 \int \frac{2x + 1}{x^2 + x + 1}\, dx$$

Both integrals have the form $\displaystyle\int \frac{du}{u}$, so

$$\int \frac{-2x - 4}{x(x^2 + x + 1)}\, dx = -4 \ln |x| + 2 \ln |x^2 + x + 1| + C$$

$$= \ln \left[\frac{(x^2 + x + 1)^2}{x^4} \right] + C$$

NOW WORK PROBLEM 7 ●●

Repeated Irreducible Quadratic Factors

Suppose $D(x)$ contains factors of the form $(x^2 + bx + c)^k$, where k is the maximum number of times the irreducible factor $x^2 + bx + c$ occurs. Then, to each such factor there will correspond a sum of k partial fractions of the form

$$\frac{A + Bx}{x^2 + bx + c} + \frac{C + Dx}{(x^2 + bx + c)^2} + \cdots + \frac{M + Nx}{(x^2 + bx + c)^k}$$

● EXAMPLE 4   **Repeated Irreducible Quadratic Factors**

Determine $\displaystyle\int \frac{x^5}{(x^2 + 4)^2}\, dx$ *by using partial fractions.*

Solution: Since the numerator has degree 5 and the denominator has degree 4, we first use long division, which gives

$$\frac{x^5}{x^4 + 8x^2 + 16} = x - \frac{8x^3 + 16x}{(x^2 + 4)^2}$$

The quadratic factor $x^2 + 4$ in the denominator of $(8x^3 + 16x)/(x^2 + 4)^2$ is irreducible and occurs as a factor twice. Thus, to $(x^2 + 4)^2$ there correspond two partial fractions and *four* coefficients to be determined. Accordingly, we set

$$\frac{8x^3 + 16x}{(x^2 + 4)^2} = \frac{Ax + B}{x^2 + 4} + \frac{Cx + D}{(x^2 + 4)^2}$$

and obtain

$$8x^3 + 16x = (Ax + B)(x^2 + 4) + Cx + D$$

$$8x^3 + 0x^2 + 16x + 0 = Ax^3 + Bx^2 + (4A + C)x + 4B + D$$

Equating like powers of x yields

$$\begin{cases} 8 = A \\ 0 = B \\ 16 = 4A + C \\ 0 = 4B + D \end{cases}$$

Solving gives $A = 8$, $B = 0$, $C = -16$, and $D = 0$. Therefore,

$$\int \frac{x^5}{(x^2 + 4)^2}\, dx = \int \left(x - \left(\frac{8x}{x^2 + 4} - \frac{16x}{(x^2 + 4)^2} \right) \right) dx$$

$$= \int x\, dx - 4 \int \frac{2x}{x^2 + 4}\, dx + 8 \int \frac{2x}{(x^2 + 4)^2}\, dx$$

The second integral on the preceding line has the form $\int \dfrac{du}{u}$, and the third integral has the form $\int \dfrac{du}{u^2}$. So

$$\int \frac{x^5}{(x^2 + 4)^2} = \frac{x^2}{2} - 4\ln(x^2 + 4) - \frac{8}{x^2 + 4} + C$$

NOW WORK PROBLEM 27 ●●●

From our examples, you may have deduced that the number of constants needed to express $N(x)/D(x)$ by partial fractions is equal to the degree of $D(x)$, if it is assumed that $N(x)/D(x)$ defines a proper rational function. This is indeed the case. Note also that the representation of a proper rational function by partial fractions is unique; that is, there is only one choice of constants that can be made. Furthermore, regardless of the complexity of the polynomial $D(x)$, it can always (theoretically) be expressed as a product of linear and irreducible quadratic factors with real coefficients.

CAUTION

Do not forget about basic integration forms.

●**EXAMPLE 5   An Integral Not Requiring Partial Fractions**

Find $\displaystyle \int \frac{2x + 3}{x^2 + 3x + 1}\, dx$.

Solution: This integral has the form $\displaystyle \int \frac{1}{u}\, du$. Thus,

$$\int \frac{2x + 3}{x^2 + 3x + 1}\, dx = \ln |x^2 + 3x + 1| + C$$

NOW WORK PROBLEM 17 ●●●

Problems 15.2

In Problems 1–8, express the given rational function in terms of partial fractions. Watch out for any preliminary divisions.

*1. $f(x) = \dfrac{10x}{x^2 + 7x + 6}$

2. $f(x) = \dfrac{x + 5}{x^2 - 1}$

3. $f(x) = \dfrac{2x^2}{x^2 + 5x + 6}$

4. $f(x) = \dfrac{2x^2 - 15}{x^2 + 5x}$

*5. $f(x) = \dfrac{x + 4}{x^2 + 4x + 4}$

6. $f(x) = \dfrac{2x + 3}{x^2(x - 1)}$

*7. $f(x) = \dfrac{x^2 + 3}{x^3 + x}$

8. $f(x) = \dfrac{3x^2 + 5}{(x^2 + 4)^2}$

In Problems 9–30, determine the integrals.

9. $\displaystyle \int \frac{5x - 2}{x^2 - x}\, dx$

10. $\displaystyle \int \frac{7x + 6}{x^2 + 3x}\, dx$

11. $\displaystyle \int \frac{x + 10}{x^2 - x - 2}\, dx$

12. $\displaystyle \int \frac{2x - 1}{x^2 - x - 12}\, dx$

13. $\displaystyle \int \frac{3x^3 - 3x + 4}{4x^2 - 4}\, dx$

14. $\displaystyle \int \frac{7(4 - x^2)}{(x - 4)(x - 2)(x + 3)}\, dx$

15. $\displaystyle \int \frac{3x - 4}{x^3 - x^2 - 2x}\, dx$

16. $\displaystyle \int \frac{4 - x}{x^4 - x^2}\, dx$

*17. $\displaystyle \int \frac{2(3x^5 + 4x^3 - x)}{x^6 + 2x^4 - x^2 - 2}\, dx$

18. $\displaystyle \int \frac{x^4 - 2x^3 + 6x^2 - 11x + 2}{x^3 - 3x^2 + 2x}\, dx$

19. $\displaystyle \int \frac{2x^2 - 5x - 2}{(x - 2)^2(x - 1)}\, dx$

20. $\displaystyle \int \frac{-3x^3 + 2x - 3}{x^2(x^2 - 1)}\, dx$

21. $\displaystyle \int \frac{2(x^2 + 8)}{x^3 + 4x}\, dx$

22. $\displaystyle \int \frac{4x^3 - 3x^2 + 2x - 3}{(x^2 + 3)(x + 1)(x - 2)}\, dx$

23. $\displaystyle \int \frac{-x^3 + 8x^2 - 9x + 2}{(x^2 + 1)(x - 3)^2}\, dx$

24. $\displaystyle \int \frac{5x^4 + 9x^2 + 3}{x(x^2 + 1)^2}\, dx$

25. $\displaystyle \int \frac{14x^3 + 24x}{(x^2 + 1)(x^2 + 2)}\, dx$

26. $\displaystyle \int \frac{12x^3 + 20x^2 + 28x + 4}{3(x^2 + 2x + 3)(x^2 + 1)}\, dx$

*27. $\displaystyle \int \frac{3x^3 + 8x}{(x^2 + 2)^2}\, dx$

28. $\displaystyle \int \frac{3x^2 - 8x + 4}{x^3 - 4x^2 + 4x - 6}\, dx$

29. $\displaystyle\int_0^1 \frac{2-2x}{x^2+7x+12}\,dx$

30. $\displaystyle\int_1^2 \frac{3x^2+15x+13}{x^2+4x+3}\,dx$

31. Find the area of the region bounded by the graph of

$$y = \frac{6(x^2+1)}{(x+2)^2}$$

and the x-axis from $x=0$ to $x=1$.

32. Consumers' Surplus Suppose the demand equation for a manufacturer's product is given by

$$p = \frac{200(q+3)}{q^2+7q+6}$$

where p is the price per unit (in dollars) when q units are demanded. Assume that market equilibrium occurs at the point $(q, p) = (10, 325/22)$. Determine consumers' surplus at market equilibrium.

OBJECTIVE

To illustrate the use of the table of integrals in Appendix B.

15.3 Integration by Tables

Certain forms of integrals that occur frequently may be found in standard tables of integration formulas.[3] A short table appears in Appendix B, and its use will be illustrated in this section.

A given integral may have to be replaced by an equivalent form before it will fit a formula in the table. The equivalent form must match the formula *exactly*. Consequently, the steps that you perform should *not* be done mentally. *Write them down!* Failure to do this can easily lead to incorrect results. Before proceeding with the exercises, be sure you understand the examples *thoroughly*.

In the following examples, the formula numbers refer to the Table of Selected Integrals given in Appendix B.

● **EXAMPLE 1 Integration by Tables**

Find $\displaystyle\int \frac{x\,dx}{(2+3x)^2}$.

Solution: Scanning the table, we identify the integrand with Formula 7:

$$\int \frac{u\,du}{(a+bu)^2} = \frac{1}{b^2}\left(\ln|a+bu| + \frac{a}{a+bu}\right) + C$$

Now we see if we can exactly match the given integrand with that in the formula. If we replace x by u, 2 by a, and 3 by b, then $du = dx$, and by substitution we have

$$\int \frac{x\,dx}{(2+3x)^2} = \int \frac{u\,du}{(a+bu)^2} = \frac{1}{b^2}\left(\ln|a+bu| + \frac{a}{a+bu}\right) + C$$

Returning to the variable x and replacing a by 2 and b by 3, we obtain

$$\int \frac{x\,dx}{(2+3x)^2} = \frac{1}{9}\left(\ln|2+3x| + \frac{2}{2+3x}\right) + C$$

Note that the answer must be given in terms of x, the *original* variable of integration.

NOW WORK PROBLEM 5 ●●●

● **EXAMPLE 2 Integration by Tables**

Find $\displaystyle\int x^2\sqrt{x^2-1}\,dx$.

Solution: This integral is identified with Formula 24:

$$\int u^2\sqrt{u^2\pm a^2}\,du = \frac{u}{8}(2u^2\pm a^2)\sqrt{u^2\pm a^2} - \frac{a^4}{8}\ln|u+\sqrt{u^2\pm a^2}| + C$$

In the preceding formula, if the bottommost sign in the dual symbol "$\pm$" on the left side is used, then the bottommost sign in the dual symbols on the right side must also be used. In the original integral, we let $u = x$ and $a = 1$. Then $du = dx$, and by

[3]See, for example, W. H. Beyer (ed.), *CRC Standard Mathematical Tables and Formulae,* 30th ed. (Boca Raton, FL: CRC Press, 1996).

substitution the integral becomes

$$\int x^2 \sqrt{x^2 - 1} \, dx = \int u^2 \sqrt{u^2 - a^2} \, du$$

$$= \frac{u}{8}(2u^2 - a^2)\sqrt{u^2 - a^2} - \frac{a^4}{8} \ln |u + \sqrt{u^2 - a^2}| + C$$

Since $u = x$ and $a = 1$,

$$\int x^2 \sqrt{x^2 - 1} \, dx = \frac{x}{8}(2x^2 - 1)\sqrt{x^2 - 1} - \frac{1}{8} \ln |x + \sqrt{x^2 - 1}| + C$$

NOW WORK PROBLEM 17 ●●●

This example, as well as Examples 4, 5, and 7, shows how to adjust an integral so that it conforms to one in the table.

● **EXAMPLE 3   Integration by Tables**

Find $\displaystyle\int \frac{dx}{x\sqrt{16x^2 + 3}}$.

Solution: The integrand can be identified with Formula 28:

$$\int \frac{du}{u\sqrt{u^2 + a^2}} = \frac{1}{a} \ln \left| \frac{\sqrt{u^2 + a^2} - a}{u} \right| + C$$

If we let $u = 4x$ and $a = \sqrt{3}$, then $du = 4 \, dx$. Watch closely how, by inserting 4's in the numerator and denominator, we transform the given integral into an equivalent form that matches Formula 28:

$$\int \frac{dx}{x\sqrt{16x^2 + 3}} = \int \frac{(4 \, dx)}{(4x)\sqrt{(4x)^2 + (\sqrt{3})^2}} = \int \frac{du}{u\sqrt{u^2 + a^2}}$$

$$= \frac{1}{a} \ln \left| \frac{\sqrt{u^2 + a^2} - a}{u} \right| + C$$

$$= \frac{1}{\sqrt{3}} \ln \left| \frac{\sqrt{16x^2 + 3} - \sqrt{3}}{4x} \right| + C$$

NOW WORK PROBLEM 7 ●●●

● **EXAMPLE 4   Integration by Tables**

Find $\displaystyle\int \frac{dx}{x^2(2 - 3x^2)^{1/2}}$.

Solution: The integrand is identified with Formula 21:

$$\int \frac{du}{u^2\sqrt{a^2 - u^2}} = -\frac{\sqrt{a^2 - u^2}}{a^2 u} + C$$

Letting $u = \sqrt{3}x$ and $a^2 = 2$, we have $du = \sqrt{3} \, dx$. Hence, by inserting two factors of $\sqrt{3}$ in both the numerator and denominator of the original integral, we have

$$\int \frac{dx}{x^2(2 - 3x^2)^{1/2}} = \sqrt{3} \int \frac{(\sqrt{3} \, dx)}{(\sqrt{3}x)^2[2 - (\sqrt{3}x)^2]^{1/2}} = \sqrt{3} \int \frac{du}{u^2(a^2 - u^2)^{1/2}}$$

$$= \sqrt{3} \left[-\frac{\sqrt{a^2 - u^2}}{a^2 u} \right] + C = \sqrt{3} \left[-\frac{\sqrt{2 - 3x^2}}{2(\sqrt{3}x)} \right] + C$$

$$= -\frac{\sqrt{2 - 3x^2}}{2x} + C$$

NOW WORK PROBLEM 35 ●●●

● EXAMPLE 5   **Integration by Tables**

Find $\displaystyle\int 7x^2 \ln(4x)\, dx$.

Solution: This is similar to Formula 42 with $n = 2$:

$$\int u^n \ln u\, du = \frac{u^{n+1} \ln u}{n+1} - \frac{u^{n+1}}{(n+1)^2} + C$$

If we let $u = 4x$, then $du = 4\, dx$. Hence,

$$\int 7x^2 \ln(4x)\, dx = \frac{7}{4^3} \int (4x)^2 \ln(4x)(4\, dx)$$

$$= \frac{7}{64} \int u^2 \ln u\, du = \frac{7}{64}\left(\frac{u^3 \ln u}{3} - \frac{u^3}{9}\right) + C$$

$$= \frac{7}{64}\left(\frac{(4x)^3 \ln(4x)}{3} - \frac{(4x)^3}{9}\right) + C$$

$$= 7x^3\left(\frac{\ln(4x)}{3} - \frac{1}{9}\right) + C$$

$$= \frac{7x^3}{9}(3\ln(4x) - 1) + C$$

NOW WORK PROBLEM 45 ◖◗◗

● EXAMPLE 6   **Integral Table Not Needed**

Find $\displaystyle\int \frac{e^{2x}\, dx}{7 + e^{2x}}$.

Solution: At first glance, we do not identify the integrand with any form in the table. Perhaps rewriting the integral will help. Let $u = 7 + e^{2x}$, then $du = 2e^{2x}\, dx$. So

$$\int \frac{e^{2x}\, dx}{7 + e^{2x}} = \frac{1}{2} \int \frac{(2e^{2x}\, dx)}{7 + e^{2x}} = \frac{1}{2} \int \frac{du}{u} = \frac{1}{2} \ln|u| + C$$

$$= \frac{1}{2} \ln|7 + e^{2x}| + C = \frac{1}{2} \ln(7 + e^{2x}) + C$$

Thus, we had only to use our knowledge of basic integration forms. Actually, this form appears as Formula 2 in the table.

NOW WORK PROBLEM 39 ◖◗◗

● EXAMPLE 7   **Finding a Definite Integral by Using Tables**

Evaluate $\displaystyle\int_1^4 \frac{dx}{(4x^2 + 2)^{3/2}}$.

Solution: We will use Formula 32 to get the indefinite integral first:

$$\int \frac{du}{(u^2 \pm a^2)^{3/2}} = \frac{\pm u}{a^2 \sqrt{u^2 \pm a^2}} + C$$

Letting $u = 2x$ and $a^2 = 2$, we have $du = 2\, dx$. Thus,

$$\int \frac{dx}{(4x^2 + 2)^{3/2}} = \frac{1}{2} \int \frac{(2\, dx)}{((2x)^2 + 2)^{3/2}} = \frac{1}{2} \int \frac{du}{(u^2 + 2)^{3/2}}$$

$$= \frac{1}{2}\left(\frac{u}{2\sqrt{u^2 + 2}}\right) + C$$

Instead of substituting back to x and evaluating from $x = 1$ to $x = 4$, we can determine the corresponding limits of integration with respect to u and then evaluate the last expression between those limits. Since $u = 2x$, when $x = 1$ we have $u = 2$; when

CAUTION

When changing the variable of integration x to the variable of integration u, be sure to change the limits of integration so that they agree with u.

$x = 4$ we have $u = 8$. Hence,

$$\int_1^4 \frac{dx}{(4x^2 + 2)^{3/2}} = \frac{1}{2} \int_2^8 \frac{du}{(u^2 + 2)^{3/2}}$$

$$= \frac{1}{2} \left(\frac{u}{2\sqrt{u^2 + 2}} \right)\Bigg|_2^8 = \frac{2}{\sqrt{66}} - \frac{1}{2\sqrt{6}}$$

NOW WORK PROBLEM 15 ●●

Integration Applied to Annuities

Tables of integrals are useful when we deal with integrals associated with annuities. Suppose that you must pay out $100 at the end of each year for the next two years. Recall from Chapter 5 that a series of payments over a period of time, such as this, is called an *annuity*. If you were to pay off the debt now instead, you would pay the present value of the $100 that is due at the end of the first year, plus the present value of the $100 that is due at the end of the second year. The sum of these present values is the present value of the annuity. (The present value of an annuity is discussed in Section 5.4.) We will now consider the present value of payments made continuously over the time interval from $t = 0$ to $t = T$, with t in years, when interest is compounded continuously at an annual rate of r.

Suppose a payment is made at time t such that on an annual basis this payment is $f(t)$. If we divide the interval $[0, T]$ into subintervals $[t_{i-1}, t_i]$ of length Δt (where Δt is small), then the total amount of all payments over such a subinterval is approximately $f(t_i)\,\Delta t$. (For example, if $f(t) = 2000$ and Δt were one day, the total amount of the payments would be $2000(\frac{1}{365})$.) The present value of these payments is approximately $e^{-rt_i} f(t_i)\,\Delta t$. (See Section 5.3.) Over the interval $[0, T]$, the total of all such present values is

$$\sum e^{-rt_i} f(t_i)\,\Delta t$$

This sum approximates the present value A of the annuity. The smaller Δt is, the better the approximation. That is, as $\Delta t \to 0$, the limit of the sum is the present value. However, this limit is also a definite integral. That is,

$$A = \int_0^T f(t)e^{-rt}\,dt \tag{1}$$

where A is the **present value of a continuous annuity** at an annual rate r (compounded continuously) for T years if a payment at time t is at the rate of $f(t)$ per year.

We say that Equation (1) gives the **present value of a continuous income stream.** Equation (1) can also be used to find the present value of future profits of a business. In this situation, $f(t)$ is the annual rate of profit at time t.

We can also consider the *future* value of an annuity rather than its present value. If a payment is made at time t, then it has a certain value at the *end* of the period of the annuity—that is, $T - t$ years later. This value is

$$\begin{pmatrix} \text{amount of} \\ \text{payment} \end{pmatrix} + \begin{pmatrix} \text{interest on this} \\ \text{payment for } T - t \text{ years} \end{pmatrix}$$

If S is the total of such values for all payments, then S is called the *accumulated amount of a continuous annuity* and is given by the formula

$$S = \int_0^T f(t)e^{r(T-t)}\,dt$$

where S is the **accumulated amount of a continuous annuity** at the end of T years at an annual rate r (compounded continuously) when a payment at time t is at the rate of $f(t)$ per year.

● EXAMPLE 8 **Present Value of a Continuous Annuity**

Find the present value (to the nearest dollar) of a continuous annuity at an annual rate of 8% for 10 years if the payment at time t is at the rate of t^2 dollars per year.

Solution: The present value is given by

$$A = \int_0^T f(t)e^{-rt}\, dt = \int_0^{10} t^2 e^{-0.08t}\, dt$$

We will use Formula 39,

$$\int u^n e^{au}\, du = \frac{u^n e^{au}}{a} - \frac{n}{a}\int u^{n-1} e^{au}\, du$$

This is called a *reduction formula,* since it reduces one integral to an expression that involves another integral that is easier to determine. If $u = t$, $n = 2$, and $a = -0.08$, then $du = dt$, and we have

$$A = \frac{t^2 e^{-0.08t}}{-0.08}\bigg|_0^{10} - \frac{2}{-0.08}\int_0^{10} t e^{-0.08t}\, dt$$

In the new integral, the exponent of t has been reduced to 1. We can match this integral with Formula 38,

$$\int u e^{au}\, du = \frac{e^{au}}{a^2}(au - 1) + C$$

by letting $u = t$ and $a = -0.08$. Then $du = dt$, and

$$A = \int_0^{10} t^2 e^{-0.08t}\, dt = \frac{t^2 e^{-0.08t}}{-0.08}\bigg|_0^{10} - \frac{2}{-0.08}\left(\frac{e^{-0.08t}}{(-0.08)^2}(-0.08t - 1)\right)\bigg|_0^{10}$$

$$= \frac{100 e^{-0.8}}{-0.08} - \frac{2}{-0.08}\left(\frac{e^{-0.8}}{(-0.08)^2}(-0.8 - 1) - \frac{1}{(-0.08)^2}(-1)\right)$$

$$\approx 185$$

The present value is $185.

NOW WORK PROBLEM 59 ●●

Problems 15.3

In Problems 1 and 2, use Formula 19 in Appendix B to determine the integrals.

1. $\displaystyle\int \frac{dx}{(9 - x^2)^{3/2}}$

2. $\displaystyle\int \frac{dx}{(25 - 4x^2)^{3/2}}$

In Problems 3 and 4, use Formula 30 in Appendix B to determine the integrals.

3. $\displaystyle\int \frac{dx}{x^2\sqrt{16x^2 + 3}}$

4. $\displaystyle\int \frac{3\,dx}{x^3\sqrt{x^4 - 9}}$

In Problems 5–38, find the integrals by using the table in Appendix B

*5. $\displaystyle\int \frac{dx}{x(6 + 7x)}$

6. $\displaystyle\int \frac{3x^2\,dx}{(2 + 5x)^2}$

*7. $\displaystyle\int \frac{dx}{x\sqrt{x^2 + 9}}$

8. $\displaystyle\int \frac{dx}{(x^2 + 7)^{3/2}}$

9. $\displaystyle\int \frac{x\,dx}{(2 + 3x)(4 + 5x)}$

10. $\displaystyle\int 2^{5x}\, dx$

11. $\displaystyle\int \frac{dx}{5 + 2e^{3x}}$

12. $\displaystyle\int x^2\sqrt{1 + x}\, dx$

13. $\displaystyle\int \frac{7\,dx}{x(5 + 2x)^2}$

14. $\displaystyle\int \frac{dx}{x\sqrt{5 - 11x^2}}$

*15. $\displaystyle\int_0^1 \frac{x\,dx}{2 + x}$

16. $\displaystyle\int \frac{2x^2\,dx}{3 + 7x}$

*17. $\displaystyle\int \sqrt{x^2 - 3}\, dx$

18. $\displaystyle\int \frac{dx}{(1 + 5x)(2x + 3)}$

19. $\displaystyle\int_0^{1/12} x e^{12x}\, dx$

20. $\displaystyle\int \sqrt{\frac{2 + 3x}{5 + 3x}}\, dx$

21. $\displaystyle\int x^2 e^x\, dx$

22. $\displaystyle\int_1^2 \frac{4\,dx}{x^2(1 + x)}$

23. $\displaystyle\int \frac{\sqrt{5x^2 + 1}}{2x^2}\, dx$

24. $\displaystyle\int \frac{dx}{x\sqrt{2 - x}}$

25. $\displaystyle\int \frac{x\,dx}{(1 + 3x)^2}$

26. $\displaystyle\int \frac{3\,dx}{\sqrt{(5 + 3x)(6 + 3x)}}$

27. $\displaystyle\int \frac{dx}{7 - 5x^2}$

28. $\displaystyle\int 7x^2\sqrt{3x^2 - 6}\, dx$

29. $\displaystyle\int 36x^5 \ln(3x)\, dx$

30. $\displaystyle\int \frac{5\,dx}{x^2(3 + 2x)^2}$

31. $\displaystyle\int 270x\sqrt{1+3x}\,dx$ **32.** $\displaystyle\int 9x^2 \ln x\,dx$

33. $\displaystyle\int \frac{dx}{\sqrt{4x^2-13}}$ **34.** $\displaystyle\int \frac{dx}{x\ln(2x)}$

*35. $\displaystyle\int \frac{2\,dx}{x^2\sqrt{16-9x^2}}$ **36.** $\displaystyle\int \frac{\sqrt{2-3x^2}}{x}\,dx$

37. $\displaystyle\int \frac{dx}{\sqrt{x}(\pi+7e^{4\sqrt{x}})}$ **38.** $\displaystyle\int_0^1 \frac{3x^2\,dx}{1+2x^3}$

In Problems 39–56, find the integrals by any method.

*39. $\displaystyle\int \frac{x\,dx}{x^2+1}$ **40.** $\displaystyle\int 3x\sqrt{x}e^{x^{5/2}}\,dx$

41. $\displaystyle\int 6x\sqrt{2x^2+1}\,dx$ **42.** $\displaystyle\int \frac{5x^3-\sqrt{x}}{2x}\,dx$

43. $\displaystyle\int \frac{dx}{x^2-5x+6}$ **44.** $\displaystyle\int \frac{e^{2x}}{\sqrt{e^{2x}+3}}\,dx$

*45. $\displaystyle\int x^3 \ln x\,dx$ **46.** $\displaystyle\int_0^3 xe^{-x}\,dx$

47. $\displaystyle\int 4x^3 e^{3x^2}\,dx$ **48.** $\displaystyle\int_1^2 35x^2\sqrt{3+2x}\,dx$

49. $\displaystyle\int \ln^2 x\,dx$ **50.** $\displaystyle\int_1^e 3x\ln x^2\,dx$

51. $\displaystyle\int_1^2 \frac{x\,dx}{\sqrt{4-x}}$ **52.** $\displaystyle\int_2^3 x\sqrt{2+3x}\,dx$

53. $\displaystyle\int_0^1 \frac{2x\,dx}{\sqrt{8-x^2}}$ **54.** $\displaystyle\int_0^{\ln 2} x^2 e^{3x}\,dx$

55. $\displaystyle\int_1^2 x\ln(2x)\,dx$ **56.** $\displaystyle\int_1^2 dx$

57. Biology In a discussion about gene frequency,[4] the integral

$$\int_{q_0}^{q_n} \frac{dq}{q(1-q)}$$

occurs, where the q's represent gene frequencies. Evaluate this integral.

58. Biology Under certain conditions, the number n of generations required to change the frequency of a gene from 0.3 to 0.1 is given by[5]

$$n = -\frac{1}{0.4}\int_{0.3}^{0.1} \frac{dq}{q^2(1-q)}$$

Find n (to the nearest integer).

*59. Continuous Annuity Find the present value, to the nearest dollar, of a continuous annuity at an annual rate of r for T years if the payment at time t is at the annual rate of $f(t)$ dollars, given that

  (a) $r=0.04$ $T=9$ $f(t)=1000$
  (b) $r=0.06$ $T=10$ $f(t)=500t$

60. If $f(t)=k$, where k is a positive constant, show that the value of the integral in Equation (1) of this section is

$$k\left(\frac{1-e^{-rT}}{r}\right)$$

61. Continuous Annuity Find the accumulated amount, to the nearest dollar, of a continuous annuity at an annual rate of r for T years if the payment at time t is at an annual rate of $f(t)$ dollars, given that

  (a) $r=0.06$ $T=10$ $f(t)=400$
  (b) $r=0.04$ $T=5$ $f(t)=40t$

62. Value of Business Over the next five years, the profits of a business at time t are estimated to be $50{,}000t$ dollars per year. The business is to be sold at a price equal to the present value of these future profits. To the nearest 10 dollars, at what price should the business be sold if interest is compounded continuously at the annual rate of 7%?

OBJECTIVE

To develop the concept of the average value of a function.

15.4 Average Value of a Function

If we are given the three numbers 1, 2, and 9, then their average value, or *mean*, is their sum divided by 3. Denoting this average by $\overline{y}$, we have

$$\overline{y} = \frac{1+2+9}{3} = 4$$

Similarly, suppose we are given a function f defined on the interval $[a, b]$, and the points $x_1, x_2, \ldots, x_n$ are in the interval. Then the average value of the n corresponding function values $f(x_1), f(x_2), \ldots, f(x_n)$ is

$$\overline{y} = \frac{f(x_1)+f(x_2)+\cdots+f(x_n)}{n} = \frac{\displaystyle\sum_{i=1}^{n} f(x_i)}{n} \tag{1}$$

We can go a step further. Let us divide the interval $[a, b]$ into n subintervals of equal length. We will choose x_i to be the right-hand endpoint of the ith subinterval. Because

---

[4]W. B. Mather, *Principles of Quantitative Genetics* (Minneapolis: Burgess Publishing Company, 1964).

[5]E. O. Wilson and W. H. Bossert, *A Primer of Population Biology* (Stamford, CT: Sinauer Associates, Inc., 1971).

$[a, b]$ has length $b - a$, each subinterval has length $\dfrac{b-a}{n}$, which we will call Δx. Thus, Equation (1) can be written

$$\overline{y} = \frac{\displaystyle\sum_{i=1}^{n} f(x_i)\left(\dfrac{\Delta x}{\Delta x}\right)}{n} = \frac{\dfrac{1}{\Delta x}\displaystyle\sum_{i=1}^{n} f(x_i)\,\Delta x}{n} = \frac{1}{n\,\Delta x}\sum_{i=1}^{n} f(x_i)\,\Delta x \qquad (2)$$

Since $\Delta x = \dfrac{b-a}{n}$, it follows that $n\,\Delta x = b - a$. So the expression $\dfrac{1}{n\,\Delta x}$ in Equation (2) can be replaced by $\dfrac{1}{b-a}$. Moreover, as $n \to \infty$, the number of function values used in computing $\overline{y}$ increases, and we get the so-called *average value of the function f*, denoted by $\overline{f}$:

$$\overline{f} = \lim_{n\to\infty}\left[\frac{1}{b-a}\sum_{i=1}^{n} f(x_i)\,\Delta x\right] = \frac{1}{b-a}\lim_{n\to\infty}\sum_{i=1}^{n} f(x_i)\,\Delta x$$

But the limit on the right is just the definite integral $\int_a^b f(x)\,dx$. This motivates the following definition:

DEFINITION
The ***average value of a function*** $f(x)$ over the interval $[a, b]$ is denoted $\overline{f}$ and is given by

$$\overline{f} = \frac{1}{b-a}\int_a^b f(x)\,dx$$

EXAMPLE 1   Average Value of a Function

Find the average value of the function $f(x) = x^2$ over the interval $[1, 2]$.

Solution:

$$\overline{f} = \frac{1}{b-a}\int_a^b f(x)\,dx$$

$$= \frac{1}{2-1}\int_1^2 x^2\,dx = \frac{x^3}{3}\bigg|_1^2 = \frac{7}{3}$$

NOW WORK PROBLEM 1

In Example 1, we found that the average value of $y = f(x) = x^2$ over the interval $[1, 2]$ is $\frac{7}{3}$. We can interpret this value geometrically. Since

$$\frac{1}{2-1}\int_1^2 x^2\,dx = \frac{7}{3}$$

by solving for the integral we have

$$\int_1^2 x^2\,dx = \frac{7}{3}(2-1)$$

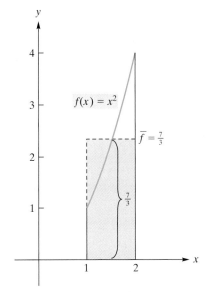

FIGURE 15.1 Geometric interpretation of the average value of a function.

However, this integral gives the area of the region bounded by $f(x) = x^2$ and the x-axis from $x = 1$ to $x = 2$. (See Figure 15.1.) From the preceding equation, this area is $\left(\frac{7}{3}\right)(2-1)$, which is the area of a rectangle whose height is the average value $\overline{f} = \frac{7}{3}$ and whose width is $b - a = 2 - 1 = 1$.

EXAMPLE 2   Average Flow of Blood

Suppose the flow of blood at time t in a system is given by

$$F(t) = \frac{F_1}{(1+\alpha t)^2} \qquad 0 \le t \le T$$

where F_1 and α (a Greek letter read "alpha") are constants.[6] Find the average flow $\overline{F}$ on the interval $[0, T]$.

Solution:

$$\overline{F} = \frac{1}{T-0} \int_0^T F(t)\,dt$$

$$= \frac{1}{T} \int_0^T \frac{F_1}{(1+\alpha t)^2}\,dt = \frac{F_1}{\alpha T} \int_0^T (1+\alpha t)^{-2}(\alpha\,dt)$$

$$= \frac{F_1}{\alpha T}\left(\frac{(1+\alpha t)^{-1}}{-1}\right)\Big|_0^T = \frac{F_1}{\alpha T}\left(-\frac{1}{1+\alpha T}+1\right)$$

$$= \frac{F_1}{\alpha T}\left(\frac{-1+1+\alpha T}{1+\alpha T}\right) = \frac{F_1}{\alpha T}\left(\frac{\alpha T}{1+\alpha T}\right) = \frac{F_1}{1+\alpha T}$$

NOW WORK PROBLEM 11 ●●●

Problems 15.4

In Problems 1–8, find the average value of the function over the given interval.

*1. $f(x) = x^2$; $[-1, 3]$

2. $f(x) = 3x - 1$; $[1, 2]$

3. $f(x) = 2 - 3x^2$; $[-1, 2]$

4. $f(x) = x^2 + x + 1$; $[1, 3]$

5. $f(t) = 2t^5$; $[-3, 3]$

6. $f(t) = t\sqrt{t^2 + 9}$; $[0, 4]$

7. $f(x) = 6\sqrt{x}$; $[1, 9]$

8. $f(x) = 5/x^2$; $[1, 3]$

9. **Profit** The profit (in dollars) of a business is given by

$$P = P(q) = 369q - 2.1q^2 - 400$$

where q is the number of units of the product sold. Find the average profit on the interval from $q = 0$ to $q = 100$.

10. **Cost** Suppose the cost (in dollars) of producing q units of a product is given by

$$c = 4000 + 10q + 0.1q^2$$

Find the average cost on the interval from $q = 100$ to $q = 500$.

*11. **Investment** An investment of $3000 earns interest at an annual rate of 5% compounded continuously. After t years,

its value S (in dollars) is given by $S = 3000e^{0.05t}$. Find the average value of a two-year investment.

12. **Medicine** Suppose that colored dye is injected into the bloodstream at a constant rate R. At time t, let

$$C(t) = \frac{R}{F(t)}$$

be the concentration of dye at a location distant (distal) from the point of injection, where $F(t)$ is as given in Example 2. Show that the average concentration on $[0, T]$ is

$$\overline{C} = \frac{R\left(1 + \alpha T + \frac{1}{3}\alpha^2 T^2\right)}{F_1}$$

13. **Revenue** Suppose a manufacturer receives revenue r from the sale of q units of a product. Show that the average value of the marginal-revenue function over the interval $[0, q_0]$ is the price per unit when q_0 units are sold.

14. Find the average value of $f(x) = \dfrac{1}{x^2 - 4x + 5}$ over the interval $[0, 1]$ using an approximate integration tewchnique. Round your answer to two decimal places.

OBJECTIVE

To solve a differential equation by using the method of separation of variables. To discuss particular solutions and general solutions. To develop interest compounded continuously in terms of a differential equation. To discuss exponential growth and decay.

15.5 Differential Equations

Occasionally, you may have to solve an equation that involves the derivative of an unknown function. Such an equation is called a **differential equation.** An example is

$$y' = xy^2 \tag{1}$$

More precisely, Equation (1) is called a **first-order differential equation,** since it involves a derivative of the first order and none of higher order. A solution of Equation (1) is any function $y = f(x)$ that is defined on an interval and satisfies the equation for all x in the interval.

---

[6]W. Simon, *Mathematical Techniques for Physiology and Medicine* (New York: Academic Press, Inc., 1972).

To solve $y' = xy^2$, equivalently,

$$\frac{dy}{dx} = xy^2 \tag{2}$$

we think of dy/dx as a quotient of differentials and algebraically "separate variables" by rewriting the equation so that each side contains only one variable and a differential is not in a denominator:

$$\frac{dy}{y^2} = x\,dx$$

Integrating both sides and combining the constants of integration, we obtain

$$\int \frac{1}{y^2}\,dy = \int x\,dx$$

$$-\frac{1}{y} = \frac{x^2}{2} + C_1$$

$$-\frac{1}{y} = \frac{x^2 + 2C_1}{2}$$

Since $2C_1$ is an arbitrary constant, we can replace it by C.

$$-\frac{1}{y} = \frac{x^2 + C}{2} \tag{3}$$

Solving Equation (3) for y, we have

$$y = -\frac{2}{x^2 + C} \tag{4}$$

We can verify that y is a solution to the differential equation (2):
    For if y is given by Equation (4), then

$$\frac{dy}{dx} = \frac{4x}{(x^2 + C)^2}$$

while also

$$xy^2 = x\left[-\frac{2}{x^2 + C}\right]^2 = \frac{4x}{(x^2 + C)^2}$$

showing that our y satisfies (2). Note in Equation (4) that, for *each* value of C, a different solution is obtained. We call Equation (4) the **general solution** of the differential equation. The method that we used to find it is called **separation of variables.**

    In the foregoing example, suppose we are given the condition that $y = -\frac{2}{3}$ when $x = 1$, that is, $y(1) = -\frac{2}{3}$. Then the *particular* function that satisfies both Equation (2) and this condition can be found by substituting the values $x = 1$ and $y = -\frac{2}{3}$ into Equation (4) and solving for C:

$$-\frac{2}{3} = -\frac{2}{1^2 + C}$$

$$C = 2$$

Therefore, the solution of $dy/dx = xy^2$ such that $y(1) = -\frac{2}{3}$ is

$$y = -\frac{2}{x^2 + 2} \tag{5}$$

We call Equation (5) a **particular solution** of the differential equation.

● EXAMPLE 1   **Separation of Variables**

Solve $y' = -\dfrac{y}{x}$ *if* $x, y > 0$.

Solution: Writing y' as dy/dx, separating variables, and integrating, we have

$$\frac{dy}{dx} = -\frac{y}{x}$$

$$\frac{dy}{y} = -\frac{dx}{x}$$

$$\int \frac{1}{y}\, dy = -\int \frac{1}{x}\, dx$$

$$\ln |y| = C_1 - \ln |x|$$

Since $x, y > 0$, we can omit the absolute-value bars:

$$\ln y = C_1 - \ln x \qquad (6)$$

To solve for y, we convert Equation (6) to exponential form:

$$y = e^{C_1 - \ln x}$$

So

$$y = e^{C_1} e^{-\ln x} = \frac{e^{C_1}}{e^{\ln x}}$$

Replacing e^{C_1} by C, where $C > 0$, and rewriting $e^{\ln x}$ as x gives

$$y = \frac{C}{x} \qquad C, x > 0$$

NOW WORK PROBLEM 1 ●◗◗

In Example 1, note that Equation (6) expresses the solution implicitly, whereas the final equation ($y = C/x$) states the solution y explicitly in terms of x. You will find that the solutions of certain differential equations are often expressed in implicit form for convenience (or necessity because of the difficulty involved in obtaining an explicit form).

Exponential Growth and Decay

In Section 5.3, the notion of interest compounded continuously was developed. Let us now take a different approach to this topic that involves a differential equation. Suppose P dollars are invested at an annual rate r compounded n times a year. Let the function $S = S(t)$ give the compound amount S (or total amount present) after t years from the date of the initial investment. Then the initial principal is $S(0) = P$. Furthermore, since there are n interest periods per year, each period has length $1/n$ years, which we will denote by Δt. At the end of the first period, the accrued interest for that period is added to the principal, and the sum acts as the principal for the second period, and so on. Hence, if the beginning of an interest period occurs at time t, then the increase in the amount present at the end of a period Δt is $S(t + \Delta t) - S(t)$, which we write as ΔS. This increase, ΔS, is also the interest earned for the period. Equivalently, the interest earned is principal times rate times time:

$$\Delta S = S \cdot r \cdot \Delta t$$

Dividing both sides by Δt, we obtain

$$\frac{\Delta S}{\Delta t} = rS \qquad (7)$$

As $\Delta t \to 0$, then $n = \dfrac{1}{\Delta t} \to \infty$, and consequently interest is being *compounded continuously*; that is, the principal is subject to continuous growth at every instant. However, as $\Delta t \to 0$, then $\Delta S/\Delta t \to dS/dt$, and Equation (7) takes the form

$$\frac{dS}{dt} = rS \qquad (8)$$

This differential equation means that *when interest is compounded continuously, the rate of change of the amount of money present at time t is proportional to the amount present at time t.* The constant of proportionality is r.

To determine the actual function S, we solve the differential equation (8) by the method of separation of variables:

$$\frac{dS}{dt} = rS$$

$$\frac{dS}{S} = r \, dt$$

$$\int \frac{1}{S} \, dS = \int r \, dt$$

$$\ln |S| = rt + C_1$$

We assume that $S > 0$, so $\ln |S| = \ln S$. Thus,

$$\ln S = rt + C_1$$

To get an explicit form, we can solve for S by converting to exponential form.

$$S = e^{rt + C_1} = e^{C_1} e^{rt}$$

For simplicity, e^{C_1} can be replaced by C to obtain the general solution

$$S = C e^{rt}$$

The condition $S(0) = P$ allows us to find the value of C:

$$P = C e^{r(0)} = C \cdot 1$$

Hence $C = P$, so

$$S = P e^{rt} \qquad (9)$$

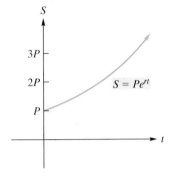

FIGURE 15.2 Compounding continuously.

Equation (9) gives the total value after t years of an initial investment of P dollars compounded continuously at an annual rate r. (See Figure 15.2.)

In our discussion of compound interest, we saw from Equation (8) that the rate of change in the amount present was proportional to the amount present. There are many natural quantities, such as population, whose rate of growth or decay at any time is considered proportional to the amount of that quantity present. If N denotes the amount of such a quantity at time t, then this rate of growth means that

$$\frac{dN}{dt} = kN$$

where k is a constant. If we separate variables and solve for N as we did for Equation (8), we get

$$N = N_0 e^{kt} \qquad (10)$$

where N_0 is a constant. In particular, if $t = 0$, then $N = N_0 e^0 = N_0 \cdot 1 = N_0$. Thus, the constant N_0 is simply $N(0)$. Due to the form of Equation (10), we say that the quantity follows an **exponential law of growth** if k is positive and an **exponential law of decay** if k is negative.

● EXAMPLE 2 Population Growth

In a certain city, the rate at which the population grows at any time is proportional to the size of the population. If the population was 125,000 in 1970 and 140,000 in 1990, what is the expected population in 2010?

Solution: Let N be the size of the population at time t. Since the exponential law of growth applies,

$$N = N_0 e^{kt}$$

To find the population in 2010, we must first find the particular law of growth involved by determining the values of N_0 and k. Let the year 1970 correspond to $t = 0$. Then $t = 20$ in 1990 and $t = 40$ in 2010. We have

$$N_0 = N(0) = 125,000$$

Thus,

$$N = 125,000e^{kt}$$

To find k, we use the fact that $N = 140,000$ when $t = 20$:

$$140,000 = 125,000e^{20k}$$

Hence,

$$e^{20k} = \frac{140,000}{125,000} = 1.12$$

Therefore, the law of growth is

$$N = 125,0000e^{kt}$$
$$= 125,000(e^{20k})^{t/20}$$
$$= 125,000(1.12)^{t/20} \qquad (11)$$

Setting $t = 40$ gives the expected population in 2010:

$$N = N(40) = 125,000(1.12)^2 = 156,800$$

We remark that from $e^{20k} = 1.12$ we have $20k = \ln(1.12)$ and hence $k = \frac{\ln(1.12)}{20} \approx 0.0057$, which can be placed in $N = 125,000e^{kt}$ to give

$$N \approx 125,000e^{0.0057t} \qquad (12)$$

NOW WORK PROBLEM 23 ◖●●

In Chapter 4, radioactive decay was discussed. Here we will consider this topic from the perspective of a differential equation. The rate at which a radioactive element decays at any time is found to be proportional to the amount of that element present. If N is the amount of a radioactive substance at time t, then the rate of decay is given by

$$\frac{dN}{dt} = -\lambda N. \qquad (13)$$

The positive constant λ (a Greek letter read "lambda") is called the **decay constant,** and the minus sign indicates that N is decreasing as t increases. Thus, we have exponential decay. From Equation (10), the solution of this differential equation is

$$N = N_0 e^{-\lambda t} \qquad (14)$$

If $t = 0$, then $N = N_0 \cdot 1 = N_0$, so N_0 represents the amount of the radioactive substance present when $t = 0$.

The time for one-half of the substance to decay is called the **half-life** of the substance. In Section 4.2, it was shown that the half-life is given by

$$\text{half-life} = \frac{\ln 2}{\lambda} \approx \frac{0.69315}{\lambda} \qquad (15)$$

Note that the half-life depends on λ. In Chapter 4, Figure 4.13 shows the graph of radioactive decay.

◖ **EXAMPLE 3**   **Finding the Decay Constant and Half-Life**

If 60% of a radioactive substance remains after 50 days, find the decay constant and the half-life of the element.

Solution: From Equation (14),

$$N = N_0 e^{-\lambda t}$$

where N_0 is the amount of the element present at $t = 0$. When $t = 50$, then $N = 0.6N_0$, and we have

$$0.6N_0 = N_0 e^{-50\lambda}$$

$$0.6 = e^{-50\lambda}$$

$$-50\lambda = \ln(0.6) \qquad \text{(logarithmic form)}$$

$$\lambda = -\frac{\ln(0.6)}{50} \approx 0.01022$$

Thus, $N \approx N_0 e^{-0.01022t}$. The half-life, from Equation (15), is

$$\frac{\ln 2}{\lambda} \approx 67.82 \text{ days}$$

NOW WORK PROBLEM 27

Radioactivity is useful in dating such things as fossil plant remains and archaeological remains made from organic material. Plants and other living organisms contain a small amount of radioactive carbon $14(^{14}C)$ in addition to ordinary carbon (^{12}C). The ^{12}C atoms are stable, but the ^{14}C atoms are decaying exponentially. However, ^{14}C is formed in the atmosphere due to the effect of cosmic rays. This ^{14}C is taken up by plants during photosynthesis and replaces what has decayed. As a result, the ratio of ^{14}C atoms to ^{12}C atoms is considered constant in living tissues over a long period of time. When a plant dies, it stops absorbing ^{14}C, and the remaining ^{14}C atoms decay. By comparing the proportion of ^{14}C to ^{12}C in a fossil plant to that of plants found today, we can estimate the age of the fossil. The half-life of ^{14}C is approximately 5730 years. Thus, if a fossil is found to have a ^{14}C-to-^{12}C ratio that is half that of a similar substance found today, we would estimate the fossil to be 5730 years old.

EXAMPLE 4 Estimating the Age of an Ancient Tool

A wood tool found in a Middle East excavation site is found to have a ^{14}C-to-^{12}C ratio that is 0.6 of the corresponding ratio in a present-day tree. Estimate the age of the tool to the nearest hundred years.

Solution: Let N be the amount of ^{14}C present in the wood t years after the tool was made. Then $N = N_0 e^{-\lambda t}$, where N_0 is the amount of ^{14}C when $t = 0$. Since the ratio of ^{14}C to ^{12}C is 0.6 of the corresponding ratio in a present-day tree, this means that we want to find the value of t for which $N = 0.6N_0$. Thus, we have

$$0.6N_0 = N_0 e^{-\lambda t}$$

$$0.6 = e^{-\lambda t}$$

$$-\lambda t = \ln(0.6) \qquad \text{(logarithmic form)}$$

$$t = -\frac{1}{\lambda} \ln(0.6)$$

From Equation (15), the half-life is $(\ln 2)/\lambda$, which equals 5730, so $\lambda = (\ln 2)/5730$. Consequently,

$$t = -\frac{1}{(\ln 2)/5730} \ln(0.6)$$

$$= -\frac{5730 \ln(0.6)}{\ln 2}$$

$$\approx 4223 \text{ years}$$

NOW WORK PROBLEM 29

Problems 15.5

In Problems 1–8, solve the differential equations.

*1. $y' = 2xy^2$

2. $y' = x^2 y^2$

3. $\dfrac{dy}{dx} - 3x\sqrt{x^2+1} = 0$

4. $\dfrac{dy}{dx} = \dfrac{x}{y}$

5. $\dfrac{dy}{dx} = y, \; y > 0$

6. $y' = e^x y^3$

7. $y' = \dfrac{y}{x}, \; x, \, y > 0$

8. $\dfrac{dy}{dx} + xe^x = 0$

In Problems 9–18, solve each of the differential equations, subject to the given conditions.

9. $y' = \dfrac{1}{y^2}; \; y(1) = 1$

10. $y' = e^{x-y}; \; y(0) = 0$ (*Hint:* $e^{x-y} = e^x/e^y$.)

11. $e^y y' - x^2 = 0; \quad y = 0$ when $x = 0$

12. $x^2 y' + \dfrac{1}{y^2} = 0; \quad y(1) = 2$

13. $(3x^2 + 2)^3 y' - xy^2 = 0; \quad y(0) = \dfrac{3}{2}$

14. $y' + x^3 y = 0; \quad y = e$ when $x = 0$

15. $\dfrac{dy}{dx} = \dfrac{3x\sqrt{1+y^2}}{y}; \quad y > 0, \, y(1) = \sqrt{8}$

16. $2y(x^3 + 2x + 1)\dfrac{dy}{dx} = \dfrac{3x^2 + 2}{\sqrt{y^2+9}}; \quad y(0) = 0$

17. $2\dfrac{dy}{dx} = \dfrac{xe^{-y}}{\sqrt{x^2+3}}; \quad y(1) = 0$

18. $x(y^2 + 1)^{3/2}\, dx = e^{x^2} y\, dy; \quad y(0) = 0$

19. **Cost** Find the manufacturer's cost function $c = f(q)$ given that

$$(q+1)^2 \dfrac{dc}{dq} = cq$$

and fixed cost is e.

20. Find $f(2)$, given that $f(1) = 0$ and that $y = f(x)$ satisfies the differential equation

$$\dfrac{dy}{dx} = xe^{x-y}$$

21. **Circulation of Money** A country has 1.00 billion dollars of paper money in circulation. Each week 25 million dollars is brought into the banks for deposit, and the same amount is paid out. The government decides to issue new paper money; whenever the old money comes into the banks, it is destroyed and replaced by new money. Let y be the amount of old money (in millions of dollars) in circulation at time t (in weeks). Then y satisfies the differential equation

$$\dfrac{dy}{dt} = -0.025y$$

How long will it take for 95% of the paper money in circulation to be new? Round your answer to the nearest week. (*Hint:* If money is 95% new, then y is 5% of 1000.)

22. **Marginal Revenue and Demand** Suppose that a monopolist's marginal-revenue function is given by the differential equation

$$\dfrac{dr}{dq} = (50 - 4q)e^{-r/5}$$

Find the demand equation for the monopolist's product.

*23. **Population Growth** In a certain town, the population at any time changes at a rate proportional to the population. If the population in 1985 was 40,000 and in 1995 was 48,000, find an equation for the population at time t, where t is the number of years past 1985. Write your answer in two forms, one involving e. You may assume that $\ln 1.2 = 0.18$. What is the expected population in 2005?

24. **Population Growth** The population of a town increases by natural growth at a rate proportional to the number N of persons present. If the population at time $t = 0$ is 50,000, find two expressions for the population N, t years later, if the population doubles in 50 years. Assume that $\ln 2 = 0.69$. Also, find N for $t = 100$.

25. **Population Growth** Suppose that the population of the world in 1930 was 2 billion and in 1960 was 3 billion. If the exponential law of growth is assumed, what is the expected population in 2015? Give your answer in terms of e.

26. **Population Growth** If exponential growth is assumed, in approximately how many years will a population double if it triples in 100 years? (*Hint:* Let the population at $t = 0$ be N_0.)

*27. **Radioactivity** If 30% of the initial amount of a radioactive sample remains after 100 seconds, find the decay constant and the half-life of the element.

28. **Radioactivity** If 30% of the initial amount of a radioactive sample has *decayed* after 100 seconds, find the decay constant and the half-life of the element.

*29. **Carbon Dating** An Egyptian scroll was found to have a ^{14}C-to-^{12}C ratio 0.7 of the corresponding ratio in similar present-day material. Estimate the age of the scroll, to the nearest hundred years.

30. **Carbon Dating** A recently discovered archaeological specimen has a ^{14}C-to-^{12}C ratio 0.1 of the corresponding ratio found in present-day organic material. Estimate the age of the specimen, to the nearest hundred years.

31. **Population Growth** Suppose that a population follows exponential growth given by $dN/dt = kN$ for $t \geq t_0$. Suppose also that $N = N_0$ when $t = t_0$. Find N, the population size at time t.

32. **Radioactivity** Polonium-210 has a half-life of about 140 days. (a) Find the decay constant in terms of $\ln 2$. (b) What fraction of the original amount of a sample of polonium-210 remains after one year?

33. **Radioactivity** Radioactive isotopes are used in medical diagnoses as tracers to determine abnormalities that may exist in an organ. For example, if radioactive iodine is swallowed, after some time it is taken up by the thyroid gland. With the use of a detector, the rate at which it is taken up can be measured, and a determination can be made as to whether the uptake is normal. Suppose radioactive technetium-99m, which has a half-life of six hours, is to be used in a brain scan two hours from now. What should be its activity now if the activity when it is used is to be 10 units? Give your answer to one decimal place. (*Hint:* In Equation (14), let $N =$ activity t hours from now and $N_0 =$ activity now.)

34. Radioactivity   A radioactive substance that has a half-life of eight days is to be temporarily implanted in a hospital patient until three-fifths of the amount originally present remains. How long should the implant remain in the patient?

35. Ecology   In a forest, natural litter occurs, such as fallen leaves and branches, dead animals, and so on.[7] Let $A = A(t)$ denote the amount of litter present at time t, where $A(t)$ is expressed in grams per square meter and t is in years. Suppose that there is no litter at $t = 0$. Thus, $A(0) = 0$. Assume that

(a) Litter falls to the ground continuously at a constant rate of 200 grams per square meter per year.

(b) The accumulated litter decomposes continuously at the rate of 50% of the amount present per year (which is $0.50A$).

The difference of the two rates is the rate of change of the amount of litter present with respect to time:

$$\begin{pmatrix} \text{rate of change} \\ \text{of litter present} \end{pmatrix} = \begin{pmatrix} \text{rate of falling} \\ \text{to ground} \end{pmatrix} - \begin{pmatrix} \text{rate of} \\ \text{decomposition} \end{pmatrix}$$

Therefore,

$$\frac{dA}{dt} = 200 - 0.50A$$

Solve for A. To the nearest gram, determine the amount of litter per square meter after one year.

36. Profit and Advertising   A certain company determines that the rate of change of monthly net profit P, as a function of monthly advertising expenditure x, is proportional to the difference between a fixed amount, \$150,000, and $2P$; that is, dP/dx is proportional to \$150,000 $- 2P$. Furthermore, if no money is spent on monthly advertising, the monthly net profit is \$15,000; if \$1000 is spent on monthly advertising, the monthly net profit is \$70,000. What would the monthly net profit be if \$2000 were spent on advertising each month?

37. Value of a Machine   The value of a certain machine depreciates 25% in the first year after the machine is purchased. The rate at which the machine subsequently depreciates is proportional to its value. Suppose that such a machine was purchased new on July 1, 1995, for \$80,000 and was valued at \$38,900 on January 1, 2006.

(a) Determine a formula that expresses the value V of the machine in terms of t, the number of years after July 1, 1996.

(b) Use the formula in part (a) to determine the year and month in which the machine has a value of exactly \$14,000.

OBJECTIVE

To develop the logistic function as a solution of a differential equation. To model the spread of a rumor. To discuss and apply Newton's law of cooling.

15.6 More Applications of Differential Equations

Logistic Growth

In the previous section, we found that if the number N of individuals in a population at time t follows an exponential law of growth, then $N = N_0 e^{kt}$, where $k > 0$ and N_0 is the population when $t = 0$. This law assumes that at time t the rate of growth, dN/dt, of the population is proportional to the number of individuals in the population. That is, $dN/dt = kN$.

Under exponential growth, a population would get infinitely large as time goes on. In reality, however, when the population gets large enough, environmental factors slow down the rate of growth. Examples are food supply, predators, overcrowding, and so on. These factors cause dN/dt to decrease eventually. It is reasonable to assume that the size of a population is limited to some maximum number M, where $0 < N < M$, and as $N \to M$, $dN/dt \to 0$, and the population size tends to be stable.

In summary, we want a population model that has exponential growth initially but that also includes the effects of environmental resistance to large population growth. Such a model is obtained by multiplying the right side of $dN/dt = kN$ by the factor $(M - N)/M$:

$$\frac{dN}{dt} = kN\left(\frac{M - N}{M}\right)$$

Notice that if N is small, then $(M - N)/M$ is close to 1, and we have growth that is approximately exponential. As $N \to M$, then $M - N \to 0$ and $dN/dt \to 0$, as we wanted in our model. Because k/M is a constant, we can replace it by K. Thus,

$$\frac{dN}{dt} = KN(M - N) \tag{1}$$

This states that the rate of growth is proportional to the product of the size of the population and the difference between the maximum size and the actual size of the population. We can solve for N in the differential equation (1) by the method of

---

[7]R. W. Poole, *An Introduction to Quantitative Ecology* (New York: McGraw-Hill Book Company, 1974).

separation of variables:

$$\frac{dN}{N(M - N)} = K \, dt$$

$$\int \frac{1}{N(M - N)} \, dN = \int K \, dt \tag{2}$$

The integral on the left side can be found by using Formula 5 in the table of integrals in Appendix B. Thus, Equation (2) becomes

$$\frac{1}{M} \ln \left| \frac{N}{M - N} \right| = Kt + C$$

so

$$\ln \left| \frac{N}{M - N} \right| = MKt + MC$$

Since $N > 0$ and $M - N > 0$, we can write

$$\ln \frac{N}{M - N} = MKt + MC$$

In exponential form, we have

$$\frac{N}{M - N} = e^{MKt + MC} = e^{MKt} e^{MC}$$

Replacing the positive constant e^{MC} by A and solving for N gives

$$\frac{N}{M - N} = Ae^{MKt}$$

$$N = (M - N) Ae^{MKt}$$

$$N = MAe^{MKt} - NAe^{MKt}$$

$$NAe^{MKt} + N = MAe^{MKt}$$

$$N(Ae^{MKt} + 1) = MAe^{MKt},$$

$$N = \frac{MAe^{MKt}}{Ae^{MKt} + 1}$$

Dividing numerator and denominator by Ae^{MKt}, we have

$$N = \frac{M}{1 + \dfrac{1}{Ae^{MKt}}} = \frac{M}{1 + \dfrac{1}{A} e^{-MKt}}$$

Replacing $1/A$ by b and MK by c yields the so-called *logistic function:*

Logistic Function

The function defined by

$$N = \frac{M}{1 + be^{-ct}} \tag{3}$$

is called the **logistic function** or the **Verhulst–Pearl logistic function.**

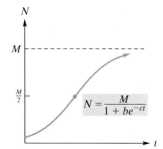

FIGURE 15.3 Logistic curve.

The graph of Equation (3), called a *logistic curve,* is S-shaped and appears in Figure 15.3. Notice that the line $N = M$ is a horizontal asymptote; that is,

$$\lim_{t \to \infty} \frac{M}{1 + be^{-ct}} = \frac{M}{1 + b(0)} = M$$

Moreover, from Equation (1), the rate of growth is

$$KN(M - N)$$

which can be considered a function of N. To find when the maximum rate of growth occurs, we solve $\frac{d}{dN}[KN(M-N)] = 0$ for N:

$$\frac{d}{dN}[KN(M-N)] = \frac{d}{dN}[K(MN - N^2)]$$
$$= K[M - 2N] = 0$$

Thus, $N = M/2$. In other words, the rate of growth increases until the population size is $M/2$ and decreases thereafter. The maximum rate of growth occurs when $N = M/2$ and corresponds to a point of inflection in the graph of N. To find the value of t for which this occurs, we substitute $M/2$ for N in Equation (3) and solve for t:

$$\frac{M}{2} = \frac{M}{1 + be^{-ct}}$$
$$1 + be^{-ct} = 2$$
$$e^{-ct} = \frac{1}{b}$$
$$e^{ct} = b$$
$$ct = \ln b \qquad \text{(logarithmic form)}$$
$$t = \frac{\ln b}{c}$$

Therefore, the maximum rate of growth occurs at the point $((\ln b)/c, M/2)$.

We remark that in Equation (3) we can replace e^{-c} by C, and then the logistic function has the following form:

Alternative Form of Logistic Function

$$N = \frac{M}{1 + bC^t}$$

● EXAMPLE 1 Logistic Growth of Club Membership

Suppose the membership in a new country club is to be a maximum of 800 persons, due to limitations of the physical plant. One year ago the initial membership was 50 persons, and now there are 200. Provided that enrollment follows a logistic function, how many members will there be three years from now?

Solution: Let N be the number of members enrolled t years after the formation of the club. Then, from Equation (3),

$$N = \frac{M}{1 + be^{-ct}}$$

Here $M = 800$, and when $t = 0$, we have $N = 50$. So

$$50 = \frac{800}{1 + b}$$
$$1 + b = \frac{800}{50} = 16$$
$$b = 15$$

Thus,

$$N = \frac{800}{1 + 15e^{-ct}} \tag{4}$$

When $t = 1$, then $N = 200$, so we have

$$200 = \frac{800}{1 + 15e^{-c}}$$

$$1 + 15e^{-c} = \frac{800}{200} = 4$$

$$e^{-c} = \frac{3}{15} = \frac{1}{5}$$

Hence, $c = -\ln \frac{1}{5} = \ln 5$. Rather than substituting this value of c into Equation (4), it is more convenient to substitute the value of e^{-c} there:

$$N = \frac{800}{1 + 15 \left(\frac{1}{5}\right)^t}$$

Three years from now, t will be 4. Therefore,

$$N = \frac{800}{1 + 15 \left(\frac{1}{5}\right)^4} \approx 781$$

NOW WORK PROBLEM 5 ●●●

Modeling the Spread of a Rumor

Let us now consider a simplified model[8] of how a rumor spreads in a population of size M. A similar situation would be the spread of an epidemic or new fad.

Let $N = N(t)$ be the number of persons who know the rumor at time t. We will assume that those who know the rumor spread it randomly in the population and that those who are told the rumor become spreaders of the rumor. Furthermore, we will assume that each knower tells the rumor to k individuals per unit of time. (Some of these k individuals may already know the rumor.) We want an expression for the rate of increase of the knowers of the rumor. Over a unit of time, each of approximately N persons will tell the rumor to k persons. Thus, the total number of persons who are told the rumor over the unit of time is (approximately) Nk. However, we are interested only in *new* knowers. The proportion of the population that does not know the rumor is $(M - N)/M$. Hence, the total number of new knowers of the rumor is

$$Nk \left(\frac{M - N}{M} \right)$$

which can be written $(k/M)N(M - N)$. Therefore,

$$\frac{dN}{dt} = \frac{k}{M} N(M - N)$$

$$= KN(M - N) \quad \text{where } K = \frac{k}{M}$$

This differential equation has the form of Equation (1), so its solution, from Equation (3), is a *logistic function*:

$$N = \frac{M}{1 + be^{-ct}}$$

●EXAMPLE 2 Campus Rumor

In a large university of 45,000 students, a sociology major is researching the spread of a new campus rumor. When she begins her research, she determines that 300 students know the rumor. After one week, she finds that 900 know it. Estimate the number of students who know it four weeks after the research begins by assuming logistic growth. Give the answer to the nearest thousand.

Solution: Let N be the number of students who know the rumor t weeks after the research begins. Then

$$N = \frac{M}{1 + be^{-ct}}$$

———

[8]More simplified, that is, than the model described in the Mathematical Snapshot for Chapter 8.

Here M, the size of the population, is 45,000, and when $t = 0$, $N = 300$. So we have

$$300 = \frac{45,000}{1 + b}$$

$$1 + b = \frac{45,000}{300} = 150$$

$$b = 149$$

Thus,

$$N = \frac{45,000}{1 + 149e^{-ct}}$$

When $t = 1$, then $N = 900$. Hence,

$$900 = \frac{45,000}{1 + 149e^{-c}}$$

$$1 + 149e^{-c} = \frac{45,000}{900} = 50$$

Therefore, $e^{-c} = \frac{49}{149}$, so

$$N = \frac{45,000}{1 + 149\left(\frac{49}{149}\right)^t}$$

When $t = 4$,

$$N = \frac{45,000}{1 + 149\left(\frac{49}{149}\right)^4} \approx 16,000$$

After four weeks, approximately 16,000 students know the rumor.

NOW WORK PROBLEM 3 ⬤⬤⬤

Newton's Law of Cooling

We conclude this section with an interesting application of a differential equation. If a homicide is committed, the temperature of the victim's body will gradually decrease from 37°C (normal body temperature) to the temperature of the surroundings (ambient temperature). In general, the temperature of the cooling body changes at a rate proportional to the difference between the temperature of the body and the ambient temperature. This statement is known as **Newton's law of cooling.** Thus, if $T(t)$ is the temperature of the body at time t and the ambient temperature is a, then

$$\frac{dT}{dt} = k(T - a)$$

where k is the constant of proportionality. Therefore, Newton's law of cooling is a differential equation. It can be applied to determine the time at which a homicide was committed, as the next example illustrates.

⬤ EXAMPLE 3 **Time of Murder**

A wealthy industrialist was found murdered in his home. Police arrived on the scene at 11:00 P.M. The temperature of the body at that time was 31°C, and one hour later it was 30°C. The temperature of the room in which the body was found was 22°C. Estimate the time at which the murder occurred.

Solution: Let t be the number of hours after the body was discovered and $T(t)$ be the temperature (in degrees Celsius) of the body at time t. We want to find the value of t for which $T = 37$ (normal body temperature). This value of t will, of course, be negative. By Newton's law of cooling,

$$\frac{dT}{dt} = k(T - a)$$

where k is a constant and a (the ambient temperature) is 22. Thus,

$$\frac{dT}{dt} = k(T - 22)$$

Separating variables, we have

$$\frac{dT}{T - 22} = k\,dt$$

$$\int \frac{dT}{T - 22} = \int k\,dt$$

$$\ln|T - 22| = kt + C$$

Because $T - 22 > 0$,

$$\ln(T - 22) = kt + C$$

When $t = 0$, then $T = 31$. Therefore,

$$\ln(31 - 22) = k \cdot 0 + C$$

$$C = \ln 9$$

Hence,

$$\ln(T - 22) = kt + \ln 9$$

$$\ln(T - 22) - \ln 9 = kt$$

$$\ln \frac{T - 22}{9} = kt \qquad \left(\ln a - \ln b = \ln \frac{a}{b}\right)$$

When $t = 1$, then $T = 30$, so

$$\ln \frac{30 - 22}{9} = k \cdot 1$$

$$k = \ln \frac{8}{9}$$

Thus,

$$\ln \frac{T - 22}{9} = t \ln \frac{8}{9}$$

Now we find t when $T = 37$:

$$\ln \frac{37 - 22}{9} = t \ln \frac{8}{9}$$

$$t = \frac{\ln(15/9)}{\ln(8/9)} \approx -4.34$$

Accordingly, the murder occurred about 4.34 hours *before* the time of discovery of the body (11:00 P.M.). Since 4.34 hours is (approximately) 4 hours and 20 minutes, the industrialist was murdered about 6:40 P.M.

NOW WORK PROBLEM 9 ●◖◗

Problems 15.6

1. **Population**   The population of a city follows logistic growth and is limited to 100,000. If the population in 1995 was 50,000 and in 2000 was 60,000, what will the population be in 2005? Give your answer to the nearest hundred.

2. **Production**   A company believes that the production of its product in present facilities will follow logistic growth. Presently, 200 units per day are produced, and production will increase to 300 units per day in one year. If production is limited to 500 units per day, what is the anticipated daily production in two years? Give your answer to the nearest unit.

*3. **Spread of Rumor**   In a university of 40,000 students, the administration holds meetings to discuss the idea of bringing in a major rock band for homecoming weekend. Before the plans are officially announced, students representatives on the administrative council spread information about the event as a rumor. At the end of one week, 100 people know the rumor. Assuming logistic growth, how many people know the rumor after two weeks? Give your answer to the nearest hundred.

4. **Spread of Fad**   A new fad is sweeping a college campus of 30,000 students. The college newspaper feels that its readers would be interested in a series on the fad. It assigns a reporter when the number of faddists is 400. One week later, there are 1200 faddists. Assuming logistic growth, find a formula for the number N of faddists t weeks after the assignment of the reporter.

*5. **Flu Outbreak**   In a city whose population is 100,000, an outbreak of flu occurs. When the city health department begins its record keeping, there are 500 infected persons.

One week later, there are 1000 infected persons. Assuming logistic growth, estimate the number of infected persons two weeks after record keeping begins.

6. **Sigmoid Function**   A very special case of the logistic function defined by Equation (3) is the *sigmoid function,* obtained by taking $M = b = c = 1$ so that we have

$$N(t) = \frac{1}{1 + e^{-t}}$$

(a) Show directly that the sigmoid function is the solution of the differential equation

$$\frac{dN}{dt} = N(1 - N)$$

and the initial condition $N(0) = 1/2$.

(b) Show that $(0, 1/2)$ is an inflection point on the graph of the sigmoid function.

(c) Show that the function

$$f(t) = \frac{1}{1 + e^{-t}} - \frac{1}{2}$$

is symmetric about the origin.

(d) Explain how (c) above shows that the sigmoid function is *symmetric about the point* $(0, 1/2)$, explaining at the same time what this means.

(e) Sketch the graph of the sigmoid function.

7. **Biology**   In an experiment,[9] five *Paramecia* were placed in a test tube containing a nutritive medium. The number N of *Paramecia* in the tube at the end of t days is given approximately by

$$N = \frac{375}{1 + e^{5.2 - 2.3t}}$$

(a) Show that this equation can be written as

$$N = \frac{375}{1 + 181.27e^{-2.3t}}$$

and hence is a logistic function.

(b) Find $\lim_{t \to \infty} N$.

8. **Biology**   In a study of the growth of a colony of unicellular organisms,[10] the equation

$$N = \frac{0.2524}{e^{-2.128x} + 0.005125} \qquad 0 \leq x \leq 5$$

was obtained, where N is the estimated area of the growth in square centimeters and x is the age of the colony in days after being first observed.

(a) Put this equation in the form of a logistic function.

(b) Find the area when the age of the colony is 0.

*9. **Time of Murder**   A waterfront murder was committed, and the victim's body was discovered at 4:15 A.M. by police. At that time, the temperature of the body was 28°C. One hour later, the body temperature was 20°C. After checking with the weather bureau, it was determined that the temperature

at the waterfront was −10°C from 11:00 P.M. to 6:00 A.M. About what time did the murder occur?

10. **Enzyme Formation**   An enzyme is a protein that acts as a catalyst for increasing the rate of a chemical reaction that occurs in cells. In a certain reaction, an enzyme A is converted to another enzyme B. Enzyme B acts as a catalyst for its own formation. Let p be the amount of enzyme B at time t and I be the total amount of both enzymes when $t = 0$. Suppose the rate of formation of B is proportional to $p(I - p)$. Without directly using calculus, find the value of p for which the rate of formation will be a maximum.

11. **Fund-Raising**   A small town decides to conduct a fund-raising drive for a fire engine whose cost is $200,000. The initial amount in the fund is $50,000. On the basis of past drives, it is determined that t months after the beginning of the drive, the rate dx/dt at which money is contributed to such a fund is proportional to the difference between the desired goal of $200,000 and the total amount x in the fund at that time. After one month, a total of $100,000 is in the fund. How much will be in the fund after three months?

12. **Birthrate**   In a discussion of unexpected properties of mathematical models of population, Bailey[11] considers the case in which the birthrate per *individual* is proportional to the population size N at time t. Since the growth rate per individual is $\dfrac{1}{N}\dfrac{dN}{dt}$, this means that

$$\frac{1}{N}\frac{dN}{dt} = kN$$

so that

$$\frac{dN}{dt} = kN^2 \qquad \text{(subject to } N = N_0 \text{ at } t = 0\text{)}$$

where $k > 0$. Show that

$$N = \frac{N_0}{1 - kN_0 t}$$

Use this result to show that

$$\lim N = \infty \quad \text{as} \quad t \to \left(\frac{1}{kN_0}\right)^{-}$$

This means that over a finite interval of time, there is an infinite amount of growth. Such a model might be useful only for rapid growth over a short interval of time.

13. **Population**   Suppose that the rate of growth of a population is proportional to the difference between some maximum size M and the number N of individuals in the population at time t. Suppose that when $t = 0$, the size of the population is N_0. Find a formula for N.

[9]G. F. Gause, *The Struggle for Existence* (New York: Hafner Publishing Co., 1964).

[10]A. J. Lotka, *Elements of Mathematical Biology* (New York: Dover Publications, Inc., 1956).

[11]N. T. J. Bailey, *The Mathematical Approach to Biology and Medicine* (New York: John Wiley & Sons, Inc., 1967).

OBJECTIVE

To define and evaluate improper integrals.

15.7 Improper Integrals[12]

Suppose $f(x)$ is continuous and nonnegative for $a \leq x < \infty$. (See Figure 15.4.) We know that the integral $\int_a^r f(x)\,dx$ is the area of the region between the curve $y = f(x)$ and the x-axis from $x = a$ to $x = r$. As $r \to \infty$, we can think of

$$\lim_{r \to \infty} \int_a^r f(x)\,dx$$

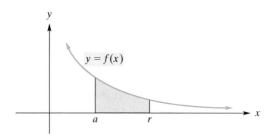

FIGURE 15.4   Area from a to r.

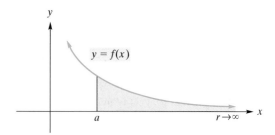

FIGURE 15.5   Area from a to r as $r \to \infty$.

as the area of the unbounded region that is shaded in Figure 15.5. This limit is abbreviated by

$$\int_a^\infty f(x)\,dx \tag{1}$$

and called an **improper integral.** If the limit exists, $\int_a^\infty f(x)\,dx$ is said to be **convergent** and the improper integral *converges* to that limit. In this case the unbounded region is considered to have a finite area, and this area is represented by $\int_a^\infty f(x)\,dx$. If the limit does not exist, the improper integral is said to be **divergent,** and the region does not have a finite area.

We can remove the restriction that $f(x) \geq 0$. In general, the improper integral $\int_a^\infty f(x)\,dx$ is defined by

$$\int_a^\infty f(x)\,dx = \lim_{r \to \infty} \int_a^r f(x)\,dx$$

Other types of improper integrals are

$$\int_{-\infty}^b f(x)\,dx \tag{2}$$

and

$$\int_{-\infty}^\infty f(x)\,dx \tag{3}$$

In each of the three types of improper integrals [(1), (2), and (3)], the interval over which the integral is evaluated has infinite length. The improper integral in (2) is defined by

$$\int_{-\infty}^b f(x)\,dx = \lim_{r \to -\infty} \int_r^b f(x)\,dx$$

If this limit exists, $\int_{-\infty}^b f(x)\,dx$ is said to be convergent. Otherwise, it is divergent. We will define the improper integral in (3) after the following example.

● EXAMPLE 1   **Improper Integrals of the Form $\int_a^\infty f(x)\,dx$ and $\int_{-\infty}^b f(x)\,dx$**

Determine whether the following improper integrals are convergent or divergent. For any convergent integral, determine its value.

PRINCIPLES IN PRACTICE 1

IMPROPER INTEGRALS OF THE FORM $\int_a^\infty f(x)\,dx$ AND $\int_{-\infty}^b f(x)\,dx$

The rate at which the human body eliminates a certain drug from its system may be approximated by $R(t) = 3e^{-0.1t} - 3e^{-0.3t}$, where $R(t)$ is in milliliters per minute and t is the time in minutes since the drug was taken. Find $\int_0^\infty (3e^{-0.1t} - 3e^{-0.3t})\,dt$, the total amount of the drug that is eliminated.

---

[12]This section can be omitted if Chapter 16 will not be covered.

a. $\displaystyle\int_1^\infty \frac{1}{x^3}\,dx$

Solution:

$$\int_1^\infty \frac{1}{x^3}\,dx = \lim_{r\to\infty}\int_1^r x^{-3}\,dx = \lim_{r\to\infty} \left. -\frac{x^{-2}}{2}\right|_1^r$$

$$= \lim_{r\to\infty}\left[-\frac{1}{2r^2}+\frac{1}{2}\right] = -0 + \frac{1}{2} = \frac{1}{2}$$

Therefore, $\displaystyle\int_1^\infty \frac{1}{x^3}\,dx$ converges to $\dfrac{1}{2}$.

b. $\displaystyle\int_{-\infty}^0 e^x\,dx$

Solution:

$$\int_{-\infty}^0 e^x\,dx = \lim_{r\to-\infty}\int_r^0 e^x\,dx = \lim_{r\to-\infty}\left. e^x\right|_r^0$$

$$= \lim_{r\to-\infty}(1-e^r) = 1-0 = 1 \qquad (e^0 = 1)$$

(Here we used the fact that as $r\to-\infty$, the graph of $y=e^r$ approaches the r-axis, so $e^r\to 0$.) Therefore, $\int_{-\infty}^0 e^x\,dx$ converges to 1.

c. $\displaystyle\int_1^\infty \frac{1}{\sqrt{x}}\,dx$

Solution:

$$\int_1^\infty \frac{1}{\sqrt{x}}\,dx = \lim_{r\to\infty}\int_1^r x^{-1/2}\,dx = \lim_{r\to\infty}\left. 2x^{1/2}\right|_1^r$$

$$= \lim_{r\to\infty} 2(\sqrt{r}-1) = \infty$$

Therefore, the improper integral diverges.

NOW WORK PROBLEM 3 ●●●

The improper integral $\int_{-\infty}^\infty f(x)\,dx$ is defined in terms of improper integrals of the forms (1) and (2):

$$\int_{-\infty}^\infty f(x)\,dx = \int_{-\infty}^0 f(x)\,dx + \int_0^\infty f(x)\,dx \qquad (4)$$

If *both* integrals on the right side of Equation (4) are convergent, then $\int_{-\infty}^\infty f(x)\,dx$ is said to be convergent; otherwise, it is divergent.

● **EXAMPLE 2   An Improper Integral of the Form $\int_{-\infty}^\infty f(x)\,dx$**

Determine whether $\displaystyle\int_{-\infty}^\infty e^x\,dx$ *is convergent or divergent.*

Solution:

$$\int_{-\infty}^\infty e^x\,dx = \int_{-\infty}^0 e^x\,dx + \int_0^\infty e^x\,dx$$

By Example 1(b), $\displaystyle\int_{-\infty}^0 e^x\,dx = 1$. On the other hand,

$$\int_0^\infty e^x\,dx = \lim_{r\to\infty}\int_0^r e^x\,dx = \lim_{r\to\infty}\left. e^x\right|_0^r = \lim_{r\to\infty}(e^r - 1) = \infty$$

Since $\int_0^\infty e^x\,dx$ is divergent, $\int_{-\infty}^\infty e^x\,dx$ is also divergent.

NOW WORK PROBLEM 11 ●●●

EXAMPLE 3   Density Function

In statistics, a function f is called a density function if $f(x) \geq 0$ and

$$\int_{-\infty}^{\infty} f(x)\,dx = 1$$

Suppose

$$f(x) = \begin{cases} ke^{-x} & \text{for } x \geq 0 \\ 0 & \text{elsewhere} \end{cases}$$

is a density function. Find k.

Solution: We write the equation $\int_{-\infty}^{\infty} f(x)\,dx = 1$ as

$$\int_{-\infty}^{0} f(x)\,dx + \int_{0}^{\infty} f(x)\,dx = 1$$

Since $f(x) = 0$ for $x < 0$, $\int_{-\infty}^{0} f(x)\,dx = 0$. Thus,

$$\int_{0}^{\infty} ke^{-x}\,dx = 1$$

$$\lim_{r \to \infty} \int_{0}^{r} ke^{-x}\,dx = 1$$

$$\lim_{r \to \infty} -ke^{-x}\Big|_{0}^{r} = 1$$

$$\lim_{r \to \infty} (-ke^{-r} + k) = 1$$

$$0 + k = 1 \qquad (\lim_{r \to \infty} e^{-r} = 0)$$

$$k = 1$$

NOW WORK PROBLEM 13

Problems 15.7

In Problems 1–12, determine the integrals if they exist. Indicate those that are divergent.

1. $\int_{3}^{\infty} \dfrac{1}{x^3}\,dx$

2. $\int_{1}^{\infty} \dfrac{1}{(3x-1)^2}\,dx$

*3. $\int_{1}^{\infty} \dfrac{1}{x}\,dx$

4. $\int_{2}^{\infty} \dfrac{1}{\sqrt[3]{(x+2)^2}}\,dx$

5. $\int_{1}^{\infty} e^{-x}\,dx$

6. $\int_{0}^{\infty} (5 + e^{-x})\,dx$

7. $\int_{1}^{\infty} \dfrac{1}{\sqrt{x}}\,dx$

8. $\int_{4}^{\infty} \dfrac{x\,dx}{\sqrt{(x^2+9)^3}}$

9. $\int_{-\infty}^{-3} \dfrac{1}{(x+1)^2}\,dx$

10. $\int_{-\infty}^{3} \dfrac{1}{\sqrt{7-x}}\,dx$

*11. $\int_{-\infty}^{\infty} 2xe^{-x^2}\,dx$

12. $\int_{-\infty}^{\infty} (5 - 3x)\,dx$

*13. Density Function   The density function for the life x, in hours, of an electronic component in a radiation meter is given by

$$f(x) = \begin{cases} \dfrac{k}{x^2} & \text{for } x \geq 800 \\ 0 & \text{for } x < 800 \end{cases}$$

(a) If k satisfies the condition that $\int_{800}^{\infty} f(x)\,dx = 1$, find k.

(b) The probability that the component will last at least 1200 hours is given by $\int_{1200}^{\infty} f(x)\,dx$. Evaluate this integral.

14. Density Function   Given the density function

$$f(x) = \begin{cases} ke^{-2x} & \text{for } x \geq 1 \\ 0 & \text{elsewhere} \end{cases}$$

find k. (*Hint:* See Example 3.)

15. Future Profits   For a business, the present value of all future profits at an annual interest rate r compounded continuously is given by

$$\int_{0}^{\infty} p(t)e^{-rt}\,dt$$

where $p(t)$ is the profit per year in dollars at time t. If $p(t) = 240{,}000$ and $r = 0.06$, evaluate this integral.

16. Psychology   In a psychological model for signal detection,[13] the probability α (a Greek letter read "alpha") of reporting a signal when no signal is present is given by

$$\alpha = \int_{x_c}^{\infty} e^{-x}\,dx \quad x \geq 0$$

---

[13]D. Laming, *Mathematical Psychology* (New York: Academic Press, Inc., 1973).

The probability β (a Greek letter read "beta") of detecting a signal when it is present is

$$\beta = \int_{x_c}^{\infty} ke^{-kx}\, dx \quad x \geq 0$$

In both integrals, x_c is a constant (called a criterion value in this model). Find α and β if $k = \frac{1}{8}$.

17. Find the area of the region in the third quadrant bounded by the curve $y = e^{3x}$ and the x-axis.

18. Economics In discussing entrance of a firm into an industry, Stigler[14] uses the equation

$$V = \pi_0 \int_0^{\infty} e^{\theta t} e^{-\rho t}\, dt$$

where π_0, θ (a Greek letter read "theta"), and ρ (a Greek letter read "rho") are constants. Show that $V = \pi_0/(\rho - \theta)$ if $\theta < \rho$.

19. Population The predicted rate of growth per year of the population of a certain small city is given by

$$\frac{40{,}000}{(t + 2)^2}$$

where t is the number of years from now. In the long run (that is, as $t \to \infty$), what is the expected change in population from today's level?

15.8 Review

Important Terms and Symbols

Examples

Summary

Sometimes we can easily determine an integral whose form is $\int u\, dv$, where u and v are functions of the same variable, by applying the formula for integration by parts:

$$\int u\, dv = uv - \int v\, du$$

A proper rational function can be integrated by applying the technique of partial fractions (although *some* of the partial fractions that may result have integrals that are beyond the scope of this book). Here we express the rational function as a sum of fractions, each of which is easier to integrate than the original function.

To determine an integral that does not have a familiar form, you may be able to match it with a formula in a table of integrals.

However, it may be necessary to transform the given integral into an equivalent form before the matching can occur.

An annuity is a series of payments over a period of time. Suppose payments are made continuously for T years such that a payment at time t is at the rate of $f(t)$ per year. If the annual rate of interest is r compounded continuously then the present value of the continuous annuity is given by

$$A = \int_0^T f(t)e^{-rt}\, dt$$

and the accumulated amount is given by

$$S = \int_0^T f(t)e^{r(T-t)}\, dt$$

---

[14]G. Stigler, *The Theory of Price*, 3rd ed. (New York: Macmillan Publishing Company, 1966), p. 344.

The average value $\overline{f}$ of a function f over the interval $[a, b]$ is given by

$$\overline{f} = \frac{1}{b - a} \int_a^b f(x)\, dx$$

An equation that involves the derivative of an unknown function is called a differential equation. If the highest-order derivative that occurs is the first, the equation is called a first-order differential equation. Some first-order differential equations can be solved by the method of separation of variables. In that method, by considering the derivative to be a quotient of differentials, we rewrite the equation so that each side contains only one variable and a single differential in the numerator. Integrating both sides of the resulting equation gives the solution. This solution involves a constant of integration and is called the general solution of the differential equation. If the unknown function must satisfy the condition that it has a specific function value for a given value of the independent variable, then a particular solution can be found.

Differential equations arise when we know a relation involving the rate of change of a function. For example, if a quantity N at time t is such that it changes at a rate proportional to the amount present, then

$$\frac{dN}{dt} = kN \qquad \text{where } k \text{ is a constant}$$

The solution of this differential equation is

$$N = N_0 e^{kt}$$

where N_0 is the quantity present at $t = 0$. The value of k may be determined when the value of N is known for a given value of t other than $t = 0$. If k is positive, then N follows an exponential law of growth; if k is negative, N follows an exponential law of decay. If N represents a quantity of a radioactive element, then

$$\frac{dN}{dt} = -\lambda N \qquad \text{where } \lambda \text{ is a positive constant}$$

Thus, N follows an exponential law of decay, and hence

$$N = N_0 e^{-\lambda t}$$

The constant λ is called the decay constant. The time for one half of the element to decay is the half-life of the element:

$$\text{half-life} = \frac{\ln 2}{\lambda} \approx \frac{0.69315}{\lambda}$$

A quantity N may follow a rate of growth given by

$$\frac{dN}{dt} = KN(M - N) \qquad \text{where } K, M \text{ are constants}$$

Solving this differential equation gives a function of the form

$$N = \frac{M}{1 + be^{-ct}} \qquad \text{where } b, c \text{ are constants}$$

which is called a logistic function. Many population sizes can be described by a logistic function. In this case, M represents the limit of the size of the population. A logistic function is also used in analyzing the spread of a rumor.

Newton's law of cooling states that the temperature T of a cooling body at time t changes at a rate proportional to the difference $T - a$, where a is the ambient temperature. Thus,

$$\frac{dT}{dt} = k(T - a) \qquad \text{where } k \text{ is a constant}$$

The solution of this differential equation can be used to determine, for example, the time at which a homicide was committed.

An integral of the form

$$\int_a^\infty f(x)\, dx \qquad \int_{-\infty}^b f(x)\, dx \qquad \text{or} \qquad \int_{-\infty}^\infty f(x)\, dx$$

is called an improper integral. The first two integrals are defined as follows:

$$\int_a^\infty f(x)\, dx = \lim_{r \to \infty} \int_a^r f(x)\, dx$$

and

$$\int_{-\infty}^b f(x)\, dx = \lim_{r \to -\infty} \int_r^b f(x)\, dx$$

If $\int_a^\infty f(x)\, dx$ (or $\int_{-\infty}^b f(x)\, dx$) is a finite number, we say that the integral is convergent; otherwise, it is divergent. The improper integral $\int_{-\infty}^\infty f(x)\, dx$ is defined by

$$\int_{-\infty}^\infty f(x)\, dx = \int_{-\infty}^0 f(x)\, dx + \int_0^\infty f(x)\, dx$$

If both integrals on the right side are convergent, $\int_{-\infty}^\infty f(x)\, dx$ is said to be convergent; otherwise, it is divergent.

Review Problems

Problem numbers shown in color indicate problems suggested for use as a practice chapter test.

In Problems 1–22, determine the integrals.

1. $\displaystyle\int x \ln x\, dx$

2. $\displaystyle\int \frac{1}{\sqrt{4x^2 + 1}}\, dx$

3. $\displaystyle\int_0^2 \sqrt{9x^2 + 16}\, dx$

4. $\displaystyle\int \frac{16x}{3 - 4x}\, dx$

5. $\displaystyle\int \frac{15x - 2}{(3x + 1)(x - 2)}\, dx$

6. $\displaystyle\int_e^{e^2} \frac{1}{x \ln x}\, dx$

7. $\displaystyle\int \frac{dx}{x(x + 2)^2}$

8. $\displaystyle\int \frac{dx}{x^2 - 1}$

9. $\displaystyle\int \frac{dx}{x^2 \sqrt{9 - 16x^2}}$

10. $\displaystyle\int x^3 \ln x^2\, dx$

11. $\displaystyle\int \frac{9\, dx}{x^2 - 9}$

12. $\displaystyle\int \frac{x}{\sqrt{2 + 5x}}\, dx$

13. $\displaystyle\int 49xe^{7x}\, dx$

14. $\displaystyle\int \frac{dx}{2 + 3e^{4x}}$

15. $\displaystyle\int \frac{dx}{2x \ln x^2}$

16. $\displaystyle\int \frac{dx}{x(2 + x)}$

17. $\displaystyle\int \frac{2x}{3 + 2x}\, dx$

18. $\displaystyle\int \frac{dx}{x^2 \sqrt{4x^2 - 9}}$

[15]19. $\displaystyle\int \frac{5x^2 + 2}{x^3 + x}\, dx$

[15]20. $\displaystyle\int \frac{3x^3 + 5x^2 + 4x + 3}{x^4 + x^3 + x^2}\, dx$

[16]21. $\displaystyle\int \frac{\ln(x + 1)}{\sqrt{x + 1}}\, dx$

[16]22. $\displaystyle\int x^2 e^x\, dx$

---

[15]Problems 19 and 20 refer to Section 15.2.

[16]Problems 21 and 22 refer to Section 15.1.

23. Find the average value of $f(x) = 3x^2 + 2x$ over the interval $[2, 4]$.

24. Find the average value of $f(t) = t^2 e^t$ over the interval $[0, 1]$.

In Problems 25 and 26, solve the differential equations.

25. $y' = 3x^2 y + 2xy$ $y > 0$

26. $y' - 2xe^{x^2 - y + 3} = 0$ $y(0) = 3$

In Problems 27–30, determine the improper integrals if they exist.[17] *Indicate those that are divergent.*

27. $\displaystyle\int_1^\infty \frac{1}{x^{2.5}}\, dx$

28. $\displaystyle\int_{-\infty}^0 e^{2x}\, dx$

29. $\displaystyle\int_1^\infty \frac{1}{2x}\, dx$

30. $\displaystyle\int_{-\infty}^\infty xe^{1 - x^2}\, dx$

31. Population The population of a city in 1985 was 100,000 and in 2000 was 120,000. Assuming exponential growth, project the population in 2015.

32. Population The population of a city doubles every 10 years due to exponential growth. At a certain time, the population is 40,000. Find an expression for the number of people N at time t years later. Give your answer in terms of $\ln 2$.

33. Radioactive If 95% of a radioactive substance remains after 100 years, find the decay constant, and, to the nearest percent, give the percentage of the original amount present after 200 years.

34. Medicine Suppose q is the amount of penicillin in the body at time t, and let q_0 be the amount at $t = 0$. Assume that the rate of change of q with respect to t is proportional to q and that q decreases as t increases. Then we have $dq/dt = -kq$, where $k > 0$. Solve for q. What percentage of the original amount present is there when $t = 7/k$?

35. Biology Two organisms are initially placed in a medium and begin to multiply. The number N of organisms that are present after t days is recorded on a graph with the horizontal axis labeled t and the vertical axis labeled N. It is observed that the points lie on a logistic curve. The number of organisms present after 6 days is 300, and beyond 10 days the number approaches a limit of 450. Find the logistic equation.

36. College Enrollment A college believes that enrollment follows logistic growth. Last year enrollment was 1000, and this year it is 1100. If the college can accommodate a maximum of 2000 students, what is the anticipated enrollment next year? Give your answer to the nearest hundred.

37. Time of Murder A coroner is called in on a murder case. He arrives at 6:00 P.M. and finds that the victim's temperature is 35°C. One hour later the body temperature is 34°C. The temperature of the room is 25°C. About what time was the murder committed? (Assume that normal body temperature is 37°C.)

38. Annuity Find the present value, to the nearest dollar, of a continuous annuity at an annual rate of 6% for 12 years if the payment at time t is at the annual rate of $f(t) = 10t$ dollars.

[18]**39. Hospital Discharges** For a group of hospitalized individuals, suppose the proportion that has been discharged at the end of t days is given by

$$\int_0^t f(x)\, dx$$

where $f(x) = 0.007e^{-0.01x} + 0.00005e^{-0.0002x}$. Evaluate

$$\int_0^\infty f(x)\, dx$$

[18]**40. Product Consumption** Suppose that $A(t)$ is the amount of a product that is consumed at time t and that A follows an exponential law of growth. If $t_1 < t_2$ and at time t_2 the amount consumed, $A(t_2)$, is double the amount consumed at time t_1, $A(t_1)$, then $t_2 - t_1$ is called a doubling period. In a discussion of exponential growth, Shonle[19] states that under exponential growth, "the amount of a product consumed during one doubling period is equal to the total used for all time up to the beginning of the doubling period in question." To justify this statement, reproduce his argument as follows. The amount of the product used up to time t_1 is given by

$$\int_{-\infty}^{t_1} A_0 e^{kt}\, dt \quad k > 0$$

where A_0 is the amount when $t = 0$. Show that this is equal to $(A_0/k)e^{kt_1}$. Next, the amount used during the time interval from t_1 to t_2 is

$$\int_{t_1}^{t_2} A_0 e^{kt}\, dt$$

Show that this is equal to

$$\frac{A_0}{k} e^{kt_1}[e^{k(t_2 - t_1)} - 1] \qquad (1)$$

If the interval $[t_1, t_2]$ is a doubling period, then

$$A_0 e^{kt_2} = 2 A_0 e^{kt_1}$$

Show that this relationship implies that $e^{k(t_2 - t_1)} = 2$. Substitute this value into Equation (1); your result should be the same as the total used during all time up to t_1, namely, $(A_0/k)e^{kt_1}$.

41. Revenue, Cost, and Profit The following table gives values of a company's marginal-revenue (MR) and marginal-cost (MC) functions:

q	0	3	6	9	12	15	18
MR	25	22	18	13	7	3	0
MC	15	14	12	10	7	4	2

The company's fixed cost is 25. Assume that profit is a maximum when MR = MC and that this occurs when $q = 12$. Moreover, assume that the output of the company is chosen to maximize the profit. Use the trapezoidal rule and Simpson's rule for each of the following parts.

(a) Estimate the total revenue by using as many data values as possible.

(b) Estimate the total cost by using as few data values as possible.

(c) Determine how the maximum profit is related to the area enclosed by the line $q = 0$ and the MR and MC curves, and use this relation to estimate the maximum profit as accurately as possible.

---

[17]Problems 27–30 refer to Section 15.7.

[18]Problems 39 and 40 refer to Section 15.7.

[19]J. I. Shonle, *Environmental Applications of General Physics* (Reading, MA: Addison-Wesley Publishing Company, Inc., 1975).

Mathematical Snapshot

Dieting

Today there is great concern about diet and weight loss. Some people want to lose weight in order to "look good." Others lose weight for physical fitness or for health reasons. In fact, some lose weight because of peer pressure. Advertisements for weight control programs frequently appear on television and in newspapers and magazines. In many bookstores, entire sections are devoted to diet and weight control.

Suppose you want to determine a mathematical model of the weight of a person on a restricted caloric diet.[20] A person's weight depends both on the daily rate of energy intake, say C calories per day, and on the daily rate of energy consumption, which is typically between 15 and 20 calories per day for each pound of body weight. Consumption depends on age, sex, metabolic rate, and so on. For an average

value of 17.5 calories per pound per day, a person weighing w pounds expends $17.5w$ calories per day. If $C = 17.5w$, then his or her weight remains constant; otherwise weight gain or loss occurs according to whether C is greater or less than $17.5w$.

How fast will weight gain or loss occur? The most plausible physiological assumption is that dw/dt is proportional to the net excess (or deficit) $C - 17.5w$ in the number of calories per day. That is,

$$\frac{dw}{dt} = K(C - 17.5w) \tag{1}$$

where K is a constant. The left side of the equation has units of pounds per day, and $C - 17.5w$ has units of calories per day. Hence the units of K are pounds per calorie. Therefore, you need to know how many pounds each excess or deficit calorie puts on or takes off. The commonly used dietetic conversion factor is that 3500 calories is equivalent to one pound. Thus $K = 1/3500$ lb per calorie.

Now, the differential equation modeling weight gain or loss is

$$\frac{dw}{dt} = \frac{1}{3500}(C - 17.5w) \tag{2}$$

If C is constant, the equation is separable and its solution is

$$w(t) = \frac{C}{17.5} + \left(w_0 - \frac{C}{17.5}\right)e^{-0.005t} \tag{3}$$

where w_0 is the initial weight and t is in days. In the long run, note that the equilibrium weight (that is, the weight as $t \to \infty$) is $w_{eq} = C/17.5$.

For example, if someone initially weighing 180 lb adopts a diet of 2500 calories per day, then we have $w_{eq} = 2500/17.5 \approx 143$ lb and the weight function is

$$w(t) \approx 143 + (180 - 143)e^{-0.005t}$$
$$= 143 + 37e^{-0.005t}$$

Figure 15.6 shows the graph of $w(t)$. Notice how long it takes to get close to the equilibrium weight of 143 lb. The half-life for the process is $(\ln 2)/0.005 \approx 138.6$ days, about 20 weeks. (It would take about 584 days, or 83 weeks, to get to 145 lb.) This may be why so many dieters give up in frustration.

[20]Adapted from A. C. Segal, "A Linear Diet Model," *The College Mathematics Journal,* 18, no. 1 (1987), 44–45. By permission of the Mathematical Association of America.

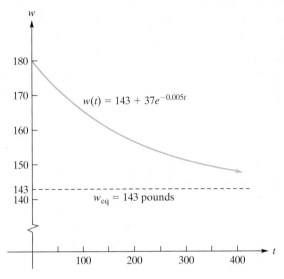

$$w(t) = 143 + 37e^{-0.005t}$$

$$w_{eq} = 143 \text{ pounds}$$

Weight as a function of time.

Problems

1. If a person weighing 200 lb adopts a 2000-calorie-per-day diet, determine, to the nearest pound, the equilibrium weight w_{eq}. To the nearest day, after how many days will the person reach a weight of 175 lb? Obtain the answer either algebraically or using a graphing calculator.

2. Show that the solution of Equation (2) is given by Equation (3).

3. The weight of a person on a restricted caloric diet at time t is given by $w(t)$. [See Equation (3).] The difference between this weight and the equilibrium weight w_{eq} is $w(t) - w_{eq}$. Suppose it takes d days for the person to lose half of the weight difference. Then

$$w(t + d) = w(t) - \tfrac{1}{2}[w(t) - w_{eq}]$$

By solving this equation for d, show that $d = \dfrac{\ln 2}{0.005}$.

4. Ideally, weight loss goals should be set in consultation with a physician. In general, however, one's ideal weight is related to one's height by the body mass index (BMI), which equals weight in kilograms divided by height in meters squared. The optimal BMI range is 18.5 to 24.9.

   How many pounds would a 5′8″-tall, 190-pound woman need to lose to be within the ideal BMI range? (Be mindful of units as you calculate the answer.) To the nearest day, how long would it take for her to lose this much weight on a 2200-calorie-per-day diet?

   Further information on weight and dieting can be found at
   www.consumer.gov/weightloss/setgoals.htm.

5. What are the pros and cons of a "crash" diet, one based on drastic changes in eating habits to achieve rapid weight loss?

16

CONTINUOUS RANDOM VARIABLES

 Mathematical Snapshot Cumulative Distribution from Data

Suppose you are designing a cellular phone network for a large urban area. Ideally, the system would always have enough capacity to meet demand. However, you know that demand fluctuates. Some periods of increased demand can be foreseen, such as holidays, when many people call their families. But other times are not predictable, such as after an earthquake or some other natural disaster, when many people may be calling emergency services or checking in with friends and relatives. Building and operating a phone system with enough capacity to handle any sudden rise in demand, no matter how great, would be hugely expensive. How do you strike a balance between the goal of serving customers and the need to limit costs?

A sensible approach would be to design a system capable of handling the load of phone traffic under normally busy conditions, and to accept the fact that on rare occasions, heavy traffic will lead to overloads. You cannot always predict when overloads will occur, since disasters such as earthquakes are unforeseen occurrences. But some good *probabilistic* predictions of future traffic volume would suffice. You could build a system that would meet demand 99.4% of the time, for example. The remaining 0.6% of the time, customers would simply have to put up with intermittent delays in service.

A probabilistic description of traffic on a phone network is an example of a probability density function. Such functions are the focus of this chapter. They have a wide range of applications—not only calculating how often a system will be overloaded, for example, but also calculating its average load. This allows prediction of such things as average power consumption and average volume of system maintenance activity.

16.1 Continuous Random Variables

Density Functions

In Chapter 9, the random variables that we considered were primarily discrete. Now we will concern ourselves with *continuous* **random variables.** A random variable is continuous if it can assume any value in some interval or intervals. A continuous random variable usually represents data that are *measured*, such as heights, weights, distances, and periods of time. By contrast, the discrete random variables of Chapter 9 usually represent data that are *counted*.

For example, the number of hours of life of a calculator battery is a continuous random variable X. If the maximum possible life is 1000 hours, then X can assume any value in the interval [0, 1000]. In a practical sense, the likelihood that X will assume a single specified value, such as 764.1238, is extremely remote. It is more meaningful to consider the likelihood of X lying within an *interval,* such as that between 764 and 765. Thus, $764 < X < 765$. (For that matter, the nature of measurement of physical quantities, like time, tells us that a statement such as $X = 764.1238$ is really one of the form $764.123750 < X < 764.123849$.) In general, *with a continuous random variable, our concern is the likelihood that it falls within an interval and not that it assumes a particular value.*

As another example, consider an experiment in which a number X is randomly selected from the interval [0, 2]. Then X is a continuous random variable. What is the probability that X lies in the interval [0, 1]? Because we can loosely think of [0, 1] as being "half" the interval [0, 2], a reasonable (and correct) answer is $\frac{1}{2}$. Similarly, if we think of the interval $[0, \frac{1}{2}]$ as being one-fourth of [0, 2], then $P(0 \le X \le \frac{1}{2}) = \frac{1}{4}$. Actually, each one of these probabilities is simply the length of the given interval divided by the length of [0, 2]. For example,

$$P\left(0 \le X \le \frac{1}{2}\right) = \frac{\text{length of } [0, \frac{1}{2}]}{\text{length of } [0, 2]} = \frac{\frac{1}{2}}{2} = \frac{1}{4}$$

Let us now consider a similar experiment in which X denotes a number chosen at random from the interval [0, 1]. As you might expect, the probability that X will assume a value in any given interval within [0, 1] is equal to the length of the given interval divided by the length of [0, 1]. Because [0, 1] has length 1, we can simply say that the probability of X falling in an interval is the length of the interval. For example,

$$P(0.2 \le X \le 0.5) = 0.5 - 0.2 = 0.3$$

and $P(0.2 \le X \le 0.2001) = 0.0001$. Clearly, as the length of an interval approaches 0, the probability that X assumes a value in that interval approaches 0. Keeping this in mind, we can think of a single number such as 0.2 as the limiting case of an interval as the length of the interval approaches 0. (Think of $[0.2, 0.2 + x]$ as $x \to 0$.) Thus, $P(X = 0.2) = 0$. In general, *the probability that a continuous random variable X assumes a particular value is* 0. As a result, **the probability that X lies in some interval is not affected by whether or not either of the endpoints of the interval is included or excluded.** For example,

$$P(X \le 0.4) = P(X < 0.4) + P(X = 0.4)$$
$$= P(X < 0.4) + 0$$
$$= P(X < 0.4)$$

Similarly, $P(0.2 \le X \le 0.5) = P(0.2 < X < 0.5)$.

We can geometrically represent the probabilities associated with a continuous random variable X. This is done by means of the graph of a function $y = f(x)$ such that the area under this graph (and above the x-axis) between the lines $x = a$

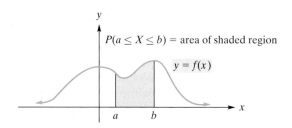

FIGURE 16.1   Probability density function.

and $x = b$ represents the probability that X assumes a value between a and b. (See Figure 16.1.) Since this area is given by the definite integral $\int_a^b f(x)\,dx$, we have

$$P(a \le X \le b) = \int_a^b f(x)\,dx$$

We call the function f the *probability density function* for X (or simply the *density function* for X) and say that it defines the *distribution of X*. Because probabilities are always nonnegative, it is always true that $f(x) \ge 0$. Also, because the event $-\infty < X < \infty$ must occur, the total area under the density function curve must be 1. That is, $\int_{-\infty}^{\infty} f(x)\,dx = 1$. In summary, we have the following definition.

DEFINITION

If X is a continuous random variable, then a function $y = f(x)$ is called a **(probability) density function** for X if and only if it has the following properties:

1. $f(x) \ge 0$
2. $\int_{-\infty}^{\infty} f(x)\,dx = 1$

We then define

3. $P(a \le X \le b) = \int_a^b f(x)\,dx$

To illustrate a density function, we return to the previous experiment in which a number X is chosen at random from the interval $[0, 1]$. Recall that

$$P(a \le X \le b) = \text{length of } [a, b] = b - a \tag{1}$$

where a and b are in $[0, 1]$. We will show that the function

$$f(x) = \begin{cases} 1 & \text{if } 0 \le x \le 1 \\ 0 & \text{otherwise} \end{cases} \tag{2}$$

whose graph appears in Figure 16.2(a), is a density function for X. To do this, we must verify that $f(x)$ satisfies the three conditions stated in the definition of a density

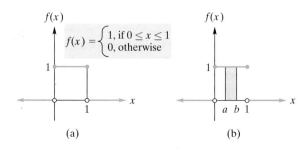

FIGURE 16.2   Probability density function.

function. First, $f(x)$ is either 0 or 1, so $f(x) \geq 0$. Next, since $f(x) = 0$ for x outside $[0, 1]$,

$$\int_{-\infty}^{\infty} f(x)\, dx = \int_0^1 1\, dx = x \Big|_0^1 = 1$$

Finally, to verify that $P(a \leq X \leq b) = \int_a^b f(x)\, dx$, we compute the area under the graph between $x = a$ and $x = b$ [Figure 16.2(b)]. We have

$$\int_a^b f(x)\, dx = \int_a^b 1\, dx = x \Big|_a^b = b - a$$

which, as stated in Equation (1), is $P(a \leq X \leq b)$.

The function in Equation (2) is called the **uniform density function** over $[0, 1]$, and X is said to have a **uniform distribution.** The word *uniform* is meaningful in the sense that the graph of the density function is horizontal, or "flat," over $[0, 1]$. As a result, X is just as likely to assume a value in one interval within $[0, 1]$ as in another of equal length. A more general uniform distribution is given in Example 1.

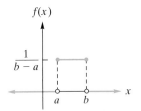

FIGURE 16.3 Uniform density function over $[a, b]$.

EXAMPLE 1 Uniform Density Function

The uniform density function over $[a, b]$ for the random variable X is given by

$$f(x) = \begin{cases} \dfrac{1}{b-a} & \text{if } a \leq x \leq b \\ 0 & \text{otherwise} \end{cases}$$

(See Figure 16.3.) Note that over $[a, b]$, the region under the graph is a rectangle with height $1/(b-a)$ and width $b-a$. Thus, its area is given by $(1/(b-a))(b-a) = 1$ so $\int_{-\infty}^{\infty} f(x)\, dx = 1$, as must be the case for a density function. If $[c, d]$ is any interval within $[a, b]$ then

$$P(c \leq X \leq d) = \int_c^d f(x)\, dx = \int_c^d \frac{1}{b-a}\, dx$$

$$= \frac{x}{b-a} \Big|_c^d = \frac{d-c}{b-a}$$

For example, suppose X is uniformly distributed over the interval $[1, 4]$ and we need to find $P(2 < X < 3)$. Then $a = 1$, $b = 4$, $c = 2$, and $d = 3$. Therefore,

$$P(2 < X < 3) = \frac{3-2}{4-1} = \frac{1}{3}$$

NOW WORK PROBLEM 3(a)–(g)

PRINCIPLES IN PRACTICE 1

DENSITY FUNCTION

Suppose the time (in minutes) passengers must wait for an airplane is uniformly distributed with density function $f(x) = \frac{1}{60}$, where $0 \leq x \leq 60$, and $f(x) = 0$ elsewhere. What is the probability that a passenger must wait between 25 and 45 minutes?

EXAMPLE 2 Density Function

The density function for a random variable X is given by

$$f(x) = \begin{cases} kx & \text{if } 0 \leq x \leq 2 \\ 0 & \text{otherwise} \end{cases}$$

where k is a constant.

a. *Find k.*

Solution: Since $\int_{-\infty}^{\infty} f(x)\, dx$ must be 1 and $f(x) = 0$ outside $[0, 2]$, we have

$$\int_{-\infty}^{\infty} f(x)\, dx = \int_0^2 kx\, dx = \frac{kx^2}{2} \Big|_0^2 = 2k = 1$$

Thus, $k = \frac{1}{2}$, so $f(x) = \frac{1}{2}x$ on $[0, 2]$.

b. *Find* $P(\frac{1}{2} < X < 1)$.

Solution:

$$P\left(\frac{1}{2} < X < 1\right) = \int_{1/2}^{1} \frac{1}{2} x \, dx = \frac{x^2}{4}\bigg|_{1/2}^{1} = \frac{1}{4} - \frac{1}{16} = \frac{3}{16}$$

c. *Find* $P(X < 1)$.

Solution: Since $f(x) = 0$ for $x < 0$, we need only compute the area under the density function between 0 and 1. Thus,

$$P(x < 1) = \int_{0}^{1} \frac{1}{2} x \, dx = \frac{x^2}{4}\bigg|_{0}^{1} = \frac{1}{4}$$

NOW WORK PROBLEM 9(a)–(d), (g), (h) ●●●

●**EXAMPLE 3   Exponential Density Function**

*The **exponential density function** is defined by*

$$f(x) = \begin{cases} ke^{-kx} & \text{if } x \geq 0 \\ 0 & \text{if } x < 0 \end{cases}$$

*where k is a positive constant, called a **parameter**, whose value depends on the experiment under consideration. If X is a random variable with this density function, then X is said to have an **exponential distribution**. Let $k = 1$. Then $f(x) = e^{-x}$ for $x \geq 0$, and $f(x) = 0$ for $x < 0$ (Figure 16.4).*

a. *Find* $P(2 < X < 3)$.

Solution:

$$P(2 < X < 3) = \int_{2}^{3} e^{-x} \, dx = -e^{-x}\bigg|_{2}^{3}$$

$$= -e^{-3} - (-e^{-2}) = e^{-2} - e^{-3} \approx 0.086$$

b. *Find* $P(X > 4)$.

Solution:

$$P(X > 4) = \int_{4}^{\infty} e^{-x} \, dx = \lim_{r \to \infty} \int_{4}^{r} e^{-x} \, dx$$

$$= \lim_{r \to \infty} -e^{-x}\bigg|_{4}^{r} = \lim_{r \to \infty} (-e^{-r} + e^{-4})$$

$$= \lim_{r \to \infty} \left(-\frac{1}{e^r} + e^{-4}\right) = 0 + e^{-4}$$

$$\approx 0.018$$

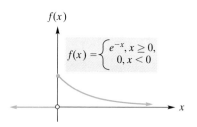

$$f(x) = \begin{cases} e^{-x}, & x \geq 0, \\ 0, & x < 0 \end{cases}$$

FIGURE 16.4   Exponential density function.

Alternatively, we can avoid an improper integral because

$$P(X > 4) = 1 - P(X \leq 4) = 1 - \int_{0}^{4} e^{-x} \, dx$$

NOW WORK PROBLEM 7(a)–(c), (e) ●●●

The **cumulative distribution function** F for the continuous random variable X with density function f is defined by

$$F(x) = P(X \leq x) = \int_{-\infty}^{x} f(t) \, dt$$

For example, $F(2)$ represents the entire area under the density curve that is to the left of the line $x = 2$ (Figure 16.5). Where $f(x)$ is continuous, it can be shown that

$$F'(x) = f(x)$$

That is, the derivative of the cumulative distribution function is the density function. Thus, F is an antiderivative of f, and by the Fundamental Theorem of Integral

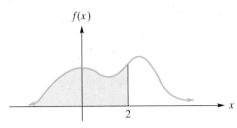

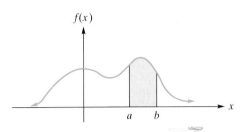

FIGURE 16.5 $F(2) = P(X \le 2) =$ area of shaded region.

FIGURE 16.6 $P(a < X < b)$.

Calculus,

$$P(a < X < b) = \int_a^b f(x)\,dx = F(b) - F(a) \tag{3}$$

This means that the area under the density curve between a and b (Figure 16.6) is simply the area to the left of b minus the area to the left of a.

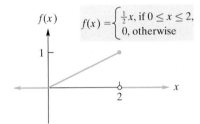

FIGURE 16.7   Density function for Example 4.

● EXAMPLE 4   Finding and Applying the Cumulative Distribution Function

Suppose X is a random variable with density function given by

$$f(x) = \begin{cases} \frac{1}{2}x & \text{if } 0 \le x \le 2 \\ 0 & \text{otherwise} \end{cases}$$

as shown in Figure 16.7.

a. *Find and sketch the cumulative distribution function.*

Solution: Because $f(x) = 0$ if $x < 0$, the area under the density curve to the left of $x = 0$ is 0. Hence, $F(x) = 0$ if $x < 0$. If $0 \le x \le 2$, then

$$F(x) = \int_{-\infty}^x f(t)\,dt = \int_0^x \frac{1}{2}t\,dt = \frac{t^2}{4}\Big|_0^x = \frac{x^2}{4}$$

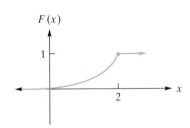

FIGURE 16.8   Cumulative distribution function for Example 4.

Since f is a density function and $f(x) = 0$ for $x < 0$ and also for $x > 2$, the area under the density curve from $x = 0$ to $x = 2$ is 1. Thus, if $x > 2$, the area to the left of x is 1, so $F(x) = 1$. Hence, the cumulative distribution function is

$$F(x) = \begin{cases} 0 & \text{if } x < 0 \\ \dfrac{x^2}{4} & \text{if } 0 \le x \le 2 \\ 1 & \text{if } x > 2 \end{cases}$$

which is shown in Figure 16.8.

b. *Find $P(X < 1)$ and $P(1 < X < 1.1)$.*

Solution: Using the results of part (a), we have

$$P(X < 1) = F(1) = \frac{1^2}{4} = \frac{1}{4}$$

From Equation (3),

$$P(1 < X < 1.1) = F(1.1) - F(1) = \frac{1.1^2}{4} - \frac{1}{4} = 0.0525$$

NOW WORK PROBLEM 1   ●○○

Mean, Variance, and Standard Deviation

For a random variable X with density function f, the **mean** μ (also called the **expectation** of X, $E(X)$) is given by

$$\mu = E(X) = \int_{-\infty}^{\infty} x f(x)\, dx$$

if the integral is convergent, and can be thought of as the average value of X in the long run. The **variance** σ^2 (also written $\text{Var}(X)$) is given by

$$\sigma^2 = \text{Var}(X) = \int_{-\infty}^{\infty} (x - \mu)^2 f(x)\, dx$$

if the integral is convergent. You may have noticed that these formulas are similar to the corresponding ones in Chapter 9 for a discrete random variable. It is easy to show that an alternative formula for the variance is

$$\sigma^2 = \text{Var}(X) = \int_{-\infty}^{\infty} x^2 f(x)\, dx - \mu^2$$

The **standard deviation** is

$$\sigma = \sqrt{\text{Var}(X)}$$

For example, it can be shown that if X is exponentially distributed (see Example 3), then $\mu = 1/k$ and $\sigma = 1/k$. As with a discrete random variable, the standard deviation of a continuous random variable X is small if X is likely to assume values close to the mean but unlikely to assume values far from the mean. The standard deviation is large if the opposite is true.

PRINCIPLES IN PRACTICE 3

FINDING THE MEAN AND STANDARD DEVIATION

The life expectancy (in years) of patients after they have contracted a certain disease is exponentially distributed with $k = 0.2$. Use the information in the paragraph that precedes Example 5 to find the mean life expectancy and the standard deviation.

● EXAMPLE 5 Finding the Mean and Standard Deviation

If X is a random variable with density function given by

$$f(x) = \begin{cases} \frac{1}{2}x & \text{if } 0 \le x \le 2 \\ 0 & \text{otherwise} \end{cases}$$

find its mean and standard deviation.

Solution: The mean is simply given by

$$\mu = \int_{-\infty}^{\infty} x f(x)\, dx = \int_{0}^{2} x \cdot \frac{1}{2}x\, dx = \frac{x^3}{6}\Big|_{0}^{2} = \frac{4}{3}$$

By the alternative formula for variance, we have

$$\sigma^2 = \int_{-\infty}^{\infty} x^2 f(x)\, dx - \mu^2 = \int_{0}^{2} x^2 \cdot \frac{1}{2}x\, dx - \left(\frac{4}{3}\right)^2$$

$$= \frac{x^4}{8}\Big|_{0}^{2} - \frac{16}{9} = 2 - \frac{16}{9} = \frac{2}{9}$$

Thus, the standard deviation is

$$\sigma = \sqrt{\frac{2}{9}} = \frac{\sqrt{2}}{3}$$

NOW WORK PROBLEM 5 ●●●

We conclude this section by emphasizing that a density function for a continuous random variable must not be confused with a probability distribution function for a discrete random variable. Evaluating such a probability distribution function at a *point* gives a probability. But evaluating a density function at a point does not. Instead, the *area* under the density function curve over an *interval* is interpreted as a probability. That is, probabilities associated with a continuous random variable are given by integrals.

Problems 16.1

*1. Suppose X is a continuous random variable with density function given by

$$f(x) = \begin{cases} \frac{1}{6}(x+1) & \text{if } 1 < x < 3 \\ 0 & \text{otherwise} \end{cases}$$

(a) Find $P(1 < X < 2)$.   (b) Find $P(X < 2.5)$.
(c) Find $P(X \geq \frac{3}{2})$.
(d) Find c such that $P(X < c) = \frac{1}{2}$. Give your answer in radical form.

2. Suppose X is a continuous random variable with density function given by

$$f(x) = \begin{cases} \dfrac{1000}{x^2} & \text{if } x > 1000 \\ 0 & \text{otherwise} \end{cases}$$

(a) Find $P(3000 < X < 4000)$.
(b) Find $P(X > 2000)$.

*3. Suppose X is a continuous random variable that is uniformly distributed on $[1, 4]$.

(a) What is the formula of the density function for X? Sketch its graph.
(b) Find $P\left(\frac{3}{2} < X < \frac{7}{2}\right)$.
(c) Find $P(0 < X < 1)$.
(d) Find $P(X \leq 3.5)$.
(e) Find $P(X > 3)$.
(f) Find $P(X = 2)$.
(g) Find $P(X < 5)$.
(h) Find μ.
(i) Find σ.
(j) Find the cumulative distribution function F and sketch its graph. Use F to find $P(X < 2)$ and $P(1 < X < 3)$.

4. Suppose X is a continuous random variable that is uniformly distributed on $[0, 5]$.

(a) What is the formula of the density function for X? Sketch its graph.
(b) Find $P(1 < X < 3)$.
(c) Find $P(4.5 \leq X < 5)$.
(d) Find $P(X = 4)$.
(e) Find $P(X > 2)$.
(f) Find $P(X < 5)$.
(g) Find $P(X > 5)$.
(h) Find μ.
(i) Find σ.
(j) Find the cumulative distribution function F and sketch its graph. Use F to find $P(1 < X < 3.5)$.

*5. Suppose X is uniformly distributed on $[a, b]$.

(a) What is the density function for X?
(b) Find μ.
(c) Find σ^2 and σ.

6. Suppose X is a continuous random variable with density function given by

$$f(x) = \begin{cases} k & \text{if } a \leq x \leq b \\ 0 & \text{otherwise} \end{cases}$$

(a) Show that $k = \dfrac{1}{b-a}$ and thus X is uniformly distributed.
(b) Find the cumulative distribution function F.

*7. Suppose the random variable X is exponentially distributed with $k = 3$.

(a) Find $P(1 < X < 4)$.
(b) Find $P(X < 4)$.
(c) Find $P(X > 6)$.
(d) Find $P(\mu - 2\sigma < X < \mu + 2\sigma)$.
(e) Verify that the density function in question satisfies the requirement that the area under the curve is 1.
(f) Find the cumulative distribution function F.

8. Suppose the random variable X is exponentially distributed with $k = 0.5$.

(a) Find $P(X > 4)$.
(b) Find $P(0.5 < X < 2.6)$.
(c) Find $P(X < 5)$.
(d) Find $P(X = 4)$.
(e) Find c such that $P(0 < X < c) = \frac{1}{2}$.

*9. The density function for a random variable X is given by

$$f(x) = \begin{cases} kx & \text{if } 0 \leq x \leq 4 \\ 0 & \text{otherwise} \end{cases}$$

(a) Find k.
(b) Find $P(2 < X < 3)$.
(c) Find $P(X > 2.5)$.
(d) Find $P(X > 0)$.
(e) Find μ.
(f) Find σ.
(g) Find c such that $P(X < c) = \frac{1}{2}$.
(h) Find $P(3 < X < 5)$.

10. The density function for a random variable X is given by

$$f(x) = \begin{cases} \frac{1}{2}x + k & \text{if } 2 \leq x \leq 4 \\ 0 & \text{otherwise} \end{cases}$$

(a) Find k.   (b) Find $P(X \geq 2.5)$.
(c) Find μ.   (d) Find $P(2 < X < \mu)$.

11. **Waiting Time** At a bus stop, the time X (in minutes) that a randomly arriving person must wait for a bus is uniformly distributed with density function $f(x) = \frac{1}{10}$, where $0 \leq x \leq 10$ and $f(x) = 0$ otherwise. What is the probability that a person must wait at most seven minutes? What is the average time that a person must wait?

12. **Soft-Drink Dispensing** An automatic soft-drink dispenser at a fast-food restaurant dispenses X ounces of cola in a 12-ounce drink. If X is uniformly distributed over $[11.93, 12.07]$, what is the probability that less than 12 ounces will be dispensed? What is the probability that exactly 12 ounces will be dispensed? What is the average amount dispensed?

13. **Emergency Room Arrivals** At a particular hospital, the length of time X (in hours) between successive arrivals at the emergency room is exponentially distributed with $k = 3$. What is the probability that more than one hour passes without an arrival?

14. **Electronic Component Life** The length of life X (in years) of a computer component has an exponential distribution with $k = \frac{2}{5}$. What is the probability that such a component will fail within three years of use? What is the probability that it will last more than five years?

OBJECTIVE

To discuss the normal distribution, standard units, and the table of areas under the standard normal curve (Appendix C).

16.2 The Normal Distribution

Quite often, measured data in nature—such as heights of individuals in a population—are represented by a random variable whose density function may be approximated by the bell-shaped curve in Figure 16.9. The curve extends indefinitely to the right and left and never touches the x-axis. This curve, called the **normal curve**, is the graph of the most important of all density functions, the *normal density function*.

> **DEFINITION**
>
> A continuous random variable X is a **normal random variable**, equivalently has a **normal** (also called Gaussian[1]) **distribution**, if its density function is given by
>
> $$f(x) = \frac{1}{\sigma\sqrt{2\pi}}e^{-(1/2)[(x-\mu)/\sigma]^2} \qquad -\infty < x < \infty$$
>
> called the **normal density function.** The parameters μ and σ are the mean and standard deviation of X, respectively.

Observe in Figure 16.9 that $f(x) \to 0$ as $x \to \pm\infty$. That is, the normal curve has the x-axis as a horizontal asymptote. Also note that the normal curve is symmetric about the vertical line $x = \mu$. That is, the height of a point on the curve d units to the right of $x = \mu$ is the same as the height of the point on the curve that is d units to the left of $x = \mu$. Because of this symmetry and the fact that the area under the normal curve is 1, the area to the right (or left) of the mean must be $\frac{1}{2}$.

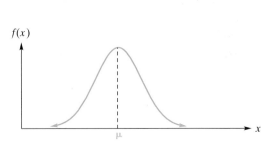

FIGURE 16.9   Normal curve.

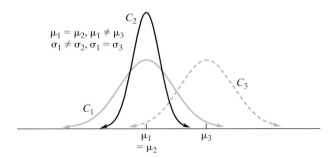

FIGURE 16.10   Normal curves.

Each choice of values for μ and σ determines a different normal curve. The value of μ determines where the curve is "centered," and σ determines how "spread out" the curve is. The smaller the value of σ, the less spread out is the area near μ. For example, Figure 16.10 shows normal curves C_1, C_2, and C_3, where C_1 has mean μ_1 and standard deviation σ_1, C_2 has mean μ_2, and so on. Here C_1 and C_2 have the same mean but different standard deviations: $\sigma_1 > \sigma_2$. C_1 and C_3 have the same standard deviation but different means: $\mu_1 < \mu_3$. Curves C_2 and C_3 have different means and different standard deviations.

The standard deviation plays a significant role in describing probabilities associated with a normal random variable X. More precisely, the probability that X will lie within one standard deviation of the mean is approximately 0.68:

$$P(\mu - \sigma < X < \mu + \sigma) = 0.68$$

In other words, approximately 68% of the area under a normal curve is within one standard deviation of the mean (Figure 16.11). Between $\mu \pm 2\sigma$ is about 95% of the area, and between $\mu \pm 3\sigma$ is about 99.7%:

$$P(\mu - 2\sigma < X < \mu + 2\sigma) = 0.95$$
$$P(\mu - 3\sigma < X < \mu + 3\sigma) = 0.997$$

You are encouraged to become familiar with the percentages in Figure 16.11.

Thus, it is highly likely that X will lie within three standard deviations of the mean.

---

[1] After the German mathematician Carl Friedrich Gauss (1777–1855).

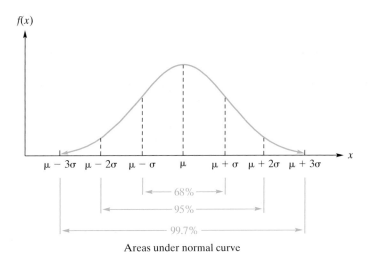

Areas under normal curve

FIGURE 16.11   Probability and number of standard deviations from μ.

● EXAMPLE 1   **Analysis of Test Scores**

Let X be a random variable whose values are the scores obtained on a nationwide test given to high school seniors. Suppose, for modeling purposes, that X is normally distributed with mean 600 and standard deviation 90. Then the probability that X lies within $2\sigma = 2(90) = 180$ points of 600 is 0.95. In other words, 95% of the scores lie between 420 and 780. Similarly, 99.7% of the scores are within $3\sigma = 3(90) = 270$ points of 600—that is, between 330 and 870.

NOW WORK PROBLEM 17   ●●●

If Z is a normally distributed random variable with $\mu = 0$ and $\sigma = 1$, we obtain the normal curve of Figure 16.12, called the **standard normal curve.**

DEFINITION

A continuous random variable Z is a ***standard normal random variable*** (or has a ***standard normal distribution***) if its density function is given by

$$f(z) = \frac{1}{\sqrt{2\pi}}e^{-z^2/2}$$

called the ***standard normal density function.*** The variable Z has mean 0 and standard deviation 1.

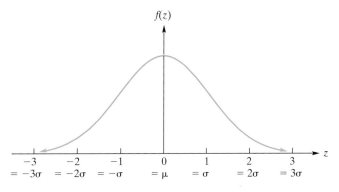

FIGURE 16.12   Standard normal curve: $\mu = 0, \sigma = 1$.

Because a standard normal random variable Z has mean 0 and standard deviation 1, its values are in units of standard deviations from the mean, which are called **standard units.** For example, if $0 < Z < 2.54$, then Z lies within 2.54 standard

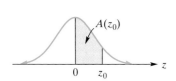

FIGURE 16.13
$A(z_0) = P(0 < Z < z_0)$.

deviations to the right of 0, the mean. That is, $0 < Z < 2.54\sigma$. To find the probability $P(0 < Z < 2.54)$, we have

$$P(0 < Z < 2.54) = \int_0^{2.54} \frac{e^{-z^2/2}}{\sqrt{2\pi}}\, dz$$

The integral on the right cannot be evaluated by elementary functions. However, values for integrals of this kind have been approximated and put in tabular form.

One such table is given in Appendix C. The table there gives the area under a standard normal curve between $z = 0$ and $z = z_0$, where $z_0 \geq 0$. This area is shaded in Figure 16.13 and is denoted by $A(z_0)$. In the left-hand columns of the table are z-values to the nearest tenth. The numbers across the top are the hundredths' values. For example, the entry in the row for 2.5 and column under 0.04 corresponds to $z = 2.54$ and is 0.4945. Thus, the area under a standard normal curve between $z = 0$ and $z = 2.54$ is (approximately) 0.4945:

$$P(0 < Z < 2.54) = A(2.54) \approx 0.4945$$

The numbers in the table are necessarily approximate, but for the balance of this chapter we will write $A(2.54) = 0.4945$ and the like in the interests of improved readability. Similarly, you should verify that $A(2) = 0.4772$ and $A(0.33) = 0.1293$.

Using symmetry, we compute an area to the left of $z = 0$ by computing the corresponding area to the right of $z = 0$. For example,

$$P(-z_0 < Z < 0) = P(0 < Z < z_0) = A(z_0)$$

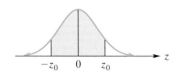

FIGURE 16.14
$P(-z_0 < Z < 0) =$
$P(0 < Z < z_0)$.

as shown in Figure 16.14. Hence, $P(-2.54 < Z < 0) = A(2.54) = 0.4945$.

When computing probabilities for a standard normal variable, you may have to add or subtract areas. A useful aid for doing this properly is a rough sketch of a standard normal curve in which you have shaded the entire area that you want to find, as Example 2 shows.

● **EXAMPLE 2   Probabilities for Standard Normal Variable Z**

a. *Find $P(Z > 1.5)$.*

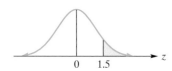

FIGURE 16.15 $P(Z > 1.5)$.

Solution: This probability is the area to the right of $z = 1.5$ (Figure 16.15). That area is equal to the difference between the total area to the right of $z = 0$, which is 0.5, and the area between $z = 0$ and $z = 1.5$, which is $A(1.5)$. Thus,

$$P(Z > 1.5) = 0.5 - A(1.5)$$
$$= 0.5 - 0.4332 = 0.0668 \qquad \text{(from Appendix C)}$$

b. *Find $P(0.5 < Z < 2)$.*

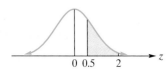

FIGURE 16.16
$P(0.5 < Z < 2)$.

Solution: This probability is the area between $z = 0.5$ and $z = 2$ (Figure 16.16). That area is the difference of two areas. It is the area between $z = 0$ and $z = 2$, which is $A(2)$, minus the area between $z = 0$ and $z = 0.5$, which is $A(0.5)$. Thus,

$$P(0.5 < Z < 2) = A(2) - A(0.5)$$
$$= 0.4772 - 0.1915 = 0.2857$$

c. *Find $P(Z \leq 2)$.*

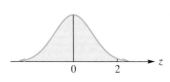

FIGURE 16.17 $P(Z \leq 2)$.

Solution: This probability is the area to the left of $z = 2$ (Figure 16.17). That area is equal to the sum of the area to the left of $z = 0$, which is 0.5, and the area between $z = 0$ and $z = 2$, which is $A(2)$. Thus,

$$P(Z \leq 2) = 0.5 + A(2)$$
$$= 0.5 + 0.4772 = 0.9772$$

NOW WORK PROBLEM 1 ◖◉●

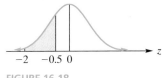

FIGURE 16.18
$P(-2 < Z < -0.5)$.

● EXAMPLE 3   Probabilities for Standard Normal Variable Z

a. *Find $P(-2 < Z < -0.5)$.*

Solution: This probability is the area between $z = -2$ and $z = -0.5$ (Figure 16.18). By symmetry, that area is equal to the area between $z = 0.5$ and $z = 2$, which was computed in Example 2(b). We have

$$P(-2 < Z < -0.5) = P(0.5 < Z < 2)$$

$$= A(2) - A(0.5) = 0.2857$$

b. *Find z_0 such that $P(-z_0 < Z < z_0) = 0.9642$.*

Solution: Figure 16.19 shows the corresponding area. Because the total area is 0.9642, by symmetry the area between $z = 0$ and $z = z_0$ is $\frac{1}{2}(0.9642) = 0.4821$, which is $A(z_0)$. Looking at the body of the table in Appendix C, we see that 0.4821 corresponds to a Z-value of 2.1. Thus, $z_0 = 2.1$.

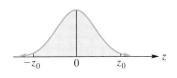

FIGURE 16.19
$P(-z_0 < Z < z_0) = 0.9642$.

NOW WORK PROBLEM 3   ●●●

Transforming to a Standard Normal Variable Z

If X is normally distributed with mean μ and standard deviation σ, you might think that a table of areas is needed for each pair of values of μ and σ. Fortunately, this is not the case. Appendix C is still used. But you must first express the area of a given region as an equal area under a standard normal curve. This involves transforming X into a standard variable Z (with mean 0 and standard deviation 1) by using the following change-of-variable formula:

$$Z = \frac{X - \mu}{\sigma} \tag{1}$$

Here we convert a normal variable to a standard normal variable.

On the right side, subtracting μ from X gives the distance from μ to X. Dividing by σ expresses this distance in terms of units of standard deviation. Thus, Z is the number of standard deviations that X is from μ. That is, formula (1) converts units of X into standard units (Z-values). For example, if $X = \mu$, then using formula (1) gives $Z = 0$. Hence, μ is zero standard deviations from μ.

Suppose X is normally distributed with $\mu = 4$ and $\sigma = 2$. Then to find—for example—$P(0 < X < 6)$, we first use formula (1) to convert the X-values 0 and 6 to Z-values (standard units):

$$z_1 = \frac{x_1 - \mu}{\sigma} = \frac{0 - 4}{2} = -2$$

$$z_2 = \frac{x_2 - \mu}{\sigma} = \frac{6 - 4}{2} = 1$$

It can be shown that

$$P(0 < X < 6) = P(-2 < Z < 1)$$

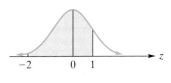

FIGURE 16.20 $P(-2 < Z < 1)$.

This means that the area under a normal curve with $\mu = 4$ and $\sigma = 2$ between $x = 0$ and $x = 6$ is equal to the area under a standard normal curve between $z = -2$ and $z = 1$ (Figure 16.20). This area is the sum of the area A_1 between $z = -2$ and $z = 0$ and the area A_2 between $z = 0$ and $z = 1$. Using symmetry for A_1, we have

$$P(-2 < Z < 1) = A_1 + A_2 = A(2) + A(1)$$

$$= 0.4772 + 0.3413 = 0.8185$$

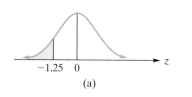
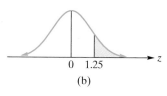

FIGURE 16.21 Diagram for Example 4.

EXAMPLE 4 Employees' Salaries

The weekly salaries of 5000 employees of a large corporation are assumed to be normally distributed with mean $640 and standard deviation $56. How many employees earn less than $570 per week?

Solution: Converting to standard units, we have

$$P(X < 570) = P\left(Z < \frac{570 - 640}{56}\right) = P(Z < -1.25)$$

This probability is the area shown in Figure 16.21(a). By symmetry, that area is equal to the area in Figure 16.21(b) that corresponds to $P(Z > 1.25)$. This area is the difference between the total area to the right of $z = 0$, which is 0.5, and the area between $z = 0$ and $z = 1.25$, which is $A(1.25)$. Thus,

$$P(X < 570) = P(Z < -1.25) = P(Z > 1.25)$$

$$= 0.5 - A(1.25) = 0.5 - 0.3944 = 0.1056$$

That is, 10.56% of the employees have salaries less than $570. This corresponds to $0.1056(5000) = 528$ employees.

NOW WORK PROBLEM 21

Problems 16.2

*1. If Z is a standard normal random variable, find each of the following probabilities.
  (a) $P(0 < Z < 1.7)$ (b) $P(0.43 < Z < 2.89)$
  (c) $P(Z > -1.23)$ (d) $P(Z \le 2.91)$
  (e) $P(-2.51 < Z \le 1.3)$ (f) $P(Z > 0.03)$

2. If Z is a standard normal random variable, find each of the following.
  (a) $P(-1.96 < Z < 1.96)$ (b) $P(-2.11 < Z < -1.35)$
  (c) $P(Z < -1.05)$ (d) $P(Z > 3\sigma)$
  (e) $P(|Z| > 2)$ (f) $P(|Z| < \frac{1}{2})$

In Problems 3–8, find z_0 such that the given statement is true. Assume that Z is a standard normal random variable.

*3. $P(Z < z_0) = 0.5517$ 4. $P(Z < z_0) = 0.0668$
5. $P(Z > z_0) = 0.8599$ 6. $P(Z > z_0) = 0.4129$
7. $P(-z_0 < Z < z_0) = 0.2662$ 8. $P(|Z| > z_0) = 0.3174$

9. If X is normally distributed with $\mu = 16$ and $\sigma = 4$, find each of the following probabilities.
  (a) $P(X < 27)$ (b) $P(X < 10)$
  (c) $P(10.8 < X < 12.4)$

10. If X is normally distributed with $\mu = 200$ and $\sigma = 40$, find each of the following probabilities.
  (a) $P(X > 150)$ (b) $P(210 < X < 250)$

11. If X is normally distributed with $\mu = -3$ and $\sigma = 2$, find $P(X > -2)$.

12. If X is normally distributed with $\mu = 0$ and $\sigma = 1.5$, find $P(X < 3)$.

13. If X is normally distributed with $\mu = 65$ and $\sigma^2 = 100$, find $P(35 < X \le 95)$.

14. If X is normally distributed with $\mu = 8$ and $\sigma = 1$, find $P(X > \mu - \sigma)$.

15. If X is normally distributed such that $\mu = 40$ and $P(X > 54) = 0.0401$, find σ.

16. If X is normally distributed with $\mu = 16$ and $\sigma = 2.25$, find x_0 such that the probability that X is between x_0 and 16 is 0.4641.

*17. **Test Scores** The scores on a national achievement test are normally distributed with mean 500 and standard deviation 100. What percentage of those who took the test had a score between 300 and 700?

18. **Test Scores** In a test given to a large group of people, the scores were normally distributed with mean 65 and standard deviation 10. What is the least whole-number score that a person could get and yet score in about the top 20%?

19. **Adult Heights** The heights (in inches) of adults in a large population are normally distributed with $\mu = 68$ and $\sigma = 3$. What percentage of the group is under 6 feet tall?

20. **Income** The yearly income for a group of 10,000 professional people is normally distributed with $\mu = \$60,000$ and $\sigma = \$5000$.
  (a) What is the probability that a person from this group has a yearly income less than $46,000?
  (b) How many of these people have yearly incomes over $75,000?

*21. **IQ**   The IQs of a large population of children are normally distributed with mean 100.4 and standard deviation 11.6.

  (a) What percentage of the children have IQs greater than 125?

(b) About 90% of the children have IQs greater than what value?

22. Suppose X is a random variable with $\mu = 10$ and $\sigma = 2$. If $P(4 < X < 16) = 0.25$, can X be normally distributed?

16.3 The Normal Approximation to the Binomial Distribution

OBJECTIVE

To show the technique of estimating the binomial distribution by using the normal distribution.

We conclude this chapter by bringing together the notions of a discrete random variable and a continuous random variable. Recall from Chapter 9 that if X is a binomial random variable (which is discrete), and if the probability of success on any trial is p, then for n independent trials, the probability of x successes is given by

$$P(X = x) = {}_nC_x p^x q^{n-x}$$

where $q = 1 - p$. You would no doubt agree that calculating probabilities for a binomial random variable can be quite tedious when the number of trials is large. For example, just imagine trying to compute ${}_{100}C_{40}(0.3)^{40}(0.7)^{60}$. To handle expressions like this, we can approximate a binomial distribution by a normal distribution and then use a table of areas.

To show how this is done, let us take a simple example. Figure 16.22 gives a probability histogram for a binomial experiment with $n = 10$ and $p = 0.5$. The rectangles centered at $x = 0$ and $x = 10$ are not shown because their heights are very close to 0. Superimposed on the histogram is a normal curve, which approximates it. The approximation would be even better if n were larger. That is, as n gets larger, the width of each unit interval appears to get smaller, and the outline of the histogram tends to take on the appearance of a smooth curve. In fact, *it is not unusual to think of a density curve as the limiting case of a probability histogram*. In spite of the fact that in our case n is only 10, the approximation shown does not seem too bad. The question that now arises is, "Which normal distribution approximates the binomial distribution?" Since the mean and standard deviation are measures of central tendency and dispersion of a random variable, we choose the approximating normal distribution to have the same mean and standard deviation as that of the binomial distribution. For this choice, we can estimate the areas of rectangles in the histogram (that is, the binomial probabilities) by finding the corresponding area under the normal curve. In summary, we have the following:

If X is a binomial random variable and n is sufficiently large, then the distribution of X can be approximated by a normal random variable whose mean and standard deviation are the same as for X, which are np and $\sqrt{npq}$, respectively.

Perhaps a word of explanation is appropriate concerning the phrase "n is sufficiently large." Generally speaking, a normal approximation to a binomial distribution is not good if n is small and p is near 0 or 1, because much of the area in the binomial histogram would be concentrated at one end of the distribution (that is, at 0 or n). Thus, the distribution would not be fairly symmetric, and a normal curve would not "fit" well. A general rule you can follow is that the normal approximation to the binomial distribution is reasonable if np and nq are at least 5. This is the case in our example: $np = 10(0.5) = 5$ and $nq = 10(0.5) = 5$.

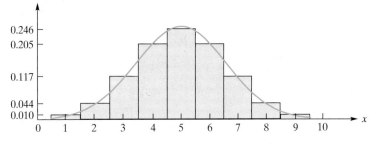

FIGURE 16.22   Normal approximation to binomial distribution.

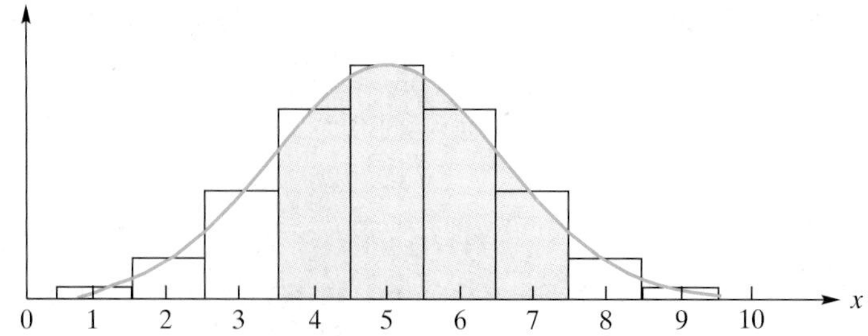

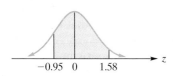

FIGURE 16.23 Normal approximation to $P(4 \le X \le 7)$.

Let us now use the normal approximation to estimate a binomial probability for $n = 10$ and $p = 0.5$. If X denotes the number of successes, then its mean is

$$np = 10(0.5) = 5$$

and its standard deviation is

$$\sqrt{npq} = \sqrt{10(0.5)(0.5)} = \sqrt{2.5} \approx 1.58$$

The probability function for X is given by

$$f(x) = {}_{10}C_x(0.5)^x(0.5)^{10-x}$$

We approximate this distribution by the normal distribution with $\mu = 5$ and $\sigma = \sqrt{2.5}$.

Suppose we estimate the probability that there are between 4 and 7 successes, inclusive, which is given by

$$P(4 \le X \le 7) = P(X = 4) + P(X = 5) + P(X = 6) + P(X = 7)$$

$$= \sum_{x=4}^{7} {}_{10}C_x(0.5)^x(0.5)^{10-x}$$

This probability is the sum of the areas of the *rectangles* for $X = 4, 5, 6$, and 7 in Figure 16.23. Under the normal curve, we have shaded the corresponding area that we will compute as an approximation to this probability. Note that the shading extends not from 4 to 7, but from $4 - \frac{1}{2}$ to $7 + \frac{1}{2}$, that is, from 3.5 to 7.5. This "continuity correction" of 0.5 on each end of the interval allows most of the area in the appropriate rectangles to be included in the approximation, and *such a correction must always be made*. The phrase *continuity correction* is used because X is treated as though it were a continuous random variable. We now convert the X-values 3.5 and 7.5 to Z-values:

$$z_1 = \frac{3.5 - 5}{\sqrt{2.5}} \approx -0.95$$

$$z_2 = \frac{7.5 - 5}{\sqrt{2.5}} \approx 1.58$$

Thus,

$$P(4 \le X \le 7) \approx P(-0.95 \le Z \le 1.58)$$

FIGURE 16.24
$P(-0.95 \le Z \le 1.58)$.

which corresponds to the area under a standard normal curve between $z = -0.95$ and $z = 1.58$ (Figure 16.24). This area is the sum of the area between $z = -0.95$ and $z = 0$, which, by symmetry, is $A(0.95)$, and the area between $z = 0$ and $z = 1.58$, which is $A(1.58)$. Hence,

$$P(4 \le X \le 7) \approx P(-0.95 \le Z \le 1.58)$$

$$= A(0.95) + A(1.58)$$

$$= 0.3289 + 0.4429 = 0.7718$$

This result is close to the true value, 0.7734 (to four decimal places).

PRINCIPLES IN PRACTICE 1

NORMAL APPROXIMATION TO A BINOMIAL DISTRIBUTION

On a game show, the grand prize is hidden behind one of four doors. Assume that the probability of selecting the grand prize is $p = \frac{1}{4}$. There were 20 winners among the last 60 contestants. Suppose that X is the number of contestants that win the grand prize, and X is binomial with $n = 60$. Approximate $P(X = 20)$ by using the normal approximation.

CAUTION

Do not ignore the continuity correction.

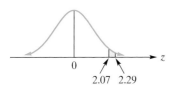

FIGURE 16.25
$P(2.07 \leq Z \leq 2.29)$.

EXAMPLE 1   Normal Approximation to a Binomial Distribution

Suppose X is a binomial random variable with $n = 100$ and $p = 0.3$. Estimate $P(X = 40)$ by using the normal approximation.

Solution: We have

$$P(X = 40) = {}_{100}C_{40}(0.3)^{40}(0.7)^{60}$$

using the formula that was mentioned at the beginning of this section. We use a normal distribution with

$$\mu = np = 100(0.3) = 30$$

and

$$\sigma = \sqrt{npq} = \sqrt{100(0.3)(0.7)} = \sqrt{21} \approx 4.58$$

Converting the corrected X-values 39.5 and 40.5 to Z-values gives

$$z_1 = \frac{39.5 - 30}{\sqrt{21}} \approx 2.07$$

$$z_2 = \frac{40.5 - 30}{\sqrt{21}} \approx 2.29$$

Therefore,

$$P(X = 40) \approx P(2.07 \leq Z \leq 2.29)$$

This probability is the area under a standard normal curve between $z = 2.07$ and $z = 2.29$ (Figure 16.25). That area is the difference of the area between $z = 0$ and $z = 2.29$, which is $A(2.29)$, and the area between $z = 0$ and $z = 2.07$, which is $A(2.07)$. Thus,

$$P(X = 40) \approx P(2.07 \leq Z \leq 2.29)$$

$$= A(2.29) - A(2.07)$$

$$= 0.4890 - 0.4808 = 0.0082 \qquad \text{(from Appendix C)}$$

NOW WORK PROBLEM 3

EXAMPLE 2   Quality Control

In a quality-control experiment, a sample of 500 items is taken from an assembly line. Customarily, 8% of the items produced are defective. What is the probability that more than 50 defective items appear in the sample?

Solution: If X is the number of defective items in the sample, then we will consider X to be binomial with $n = 500$ and $p = 0.08$. To find $P(X \geq 51)$, we use the normal approximation to the binomial distribution with

$$\mu = np = 500(0.08) = 40$$

and

$$\sigma = \sqrt{npq} = \sqrt{500(0.08)(0.92)} = \sqrt{36.8} \approx 6.07$$

Converting the corrected value 50.5 to a Z-value gives

$$z = \frac{50.5 - 40}{\sqrt{36.8}} \approx 1.73$$

Thus,

$$P(X \geq 51) \approx P(Z \geq 1.73)$$

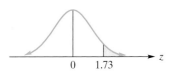

FIGURE 16.26 $P(Z \geq 1.73)$.

This probability is the area under a standard normal curve to the right of $z = 1.73$ (Figure 16.26). That area is the difference of the area to the right of $z = 0$, which is 0.5, and the area between $z = 0$ and $z = 1.73$, which is $A(1.73)$. Hence,

$$P(X \geq 51) \approx P(Z \geq 1.73)$$

$$= 0.5 - A(1.73) = 0.5 - 0.4582 = 0.0418$$

NOW WORK PROBLEM 7

Problems 16.3

In Problems 1–4, X is a binomial random variable with the given values of n and p. Calculate the indicated probabilities by using the normal approximation.

1. $n = 150$, $p = 0.4$; $P(X \geq 52)$, $P(X \geq 74)$

2. $n = 50$, $p = 0.3$; $P(X = 19)$, $P(X \leq 18)$

*3. $n = 200$, $p = 0.6$; $P(X = 125)$, $P(110 \leq X \leq 135)$

4. $n = 25$, $p = 0.25$; $P(X \geq 7)$

5. **Die Tossing**   Suppose a fair die is tossed 300 times. What is the probability that a 5 turns up between 45 and 60 times, inclusive?

6. **Coin Tossing**   For a biased coin, $P(H) = 0.4$ and $P(T) = 0.6$. If the coin is tossed 200 times, what is the probability of getting between 90 and 100 heads, inclusive?

*7. **Truck Breakdown**   A delivery service has a fleet of 60 trucks. At any given time, the probability of a truck being out of use due to factors such as breakdowns and maintenance is 0.1. What is the probability that 7 or more trucks are out of service at any time?

8. **Quality Control**   In a manufacturing plant, a sample of 200 items is taken from the assembly line. For each item in the sample, the probability of being defective is 0.05. What is the probability that there are 7 or more defective items in the sample?

9. **True–False Exam**   In a true–false exam with 25 questions, what is the probability of getting at least 13 correct answers by just guessing on all the questions? If there are 100 questions instead of 25, what is the probability of getting at least 60 correct answers by just guessing?

10. **Multiple-Choice Exam**   In a multiple-choice test with 50 questions, each question has four answers, only one of which is correct. If a student guesses on the last 20 questions, what is the probability of getting at least half of them correct?

11. **Poker**   In a poker game, the probability of being dealt a hand consisting of three cards of one suit and two cards of another suit (in any order) is about 0.1. In 100 dealt hands, what is the probability that 16 or more of them will be as just described?

12. **Taste Test**   A major cola company sponsors a national taste test, in which subjects sample its cola as well as the best-selling brand. Neither cola is identified by brand. The subjects are then asked to choose the cola that tastes better. If each of the 35 subjects in a supermarket actually have no preference and arbitrarily choose one of the colas, what is the probability that 25 or more of them choose the cola from the sponsoring company?

16.4 Review

Important Terms and Symbols   Examples

Section 16.1	**Continuous Random Variables**	
	continuous random variable uniform density function	Ex. 1, p. 727
	exponential density function exponential distribution	Ex. 3, p. 728
	cumulative distribution function	Ex. 4, p. 729
	mean, μ variance, σ^2 standard deviation, σ	Ex. 5, p. 730
Section 16.2	**The Normal Distribution**	
	normal distribution normal density function	Ex. 1, p. 733
	standard normal curve standard normal random variable	Ex. 2, p. 734
	standard normal distribution standard normal density function	Ex. 4, p. 736
Section 16.3	**The Normal Approximation to the Binomial Distribution**	
	continuity correction	Ex. 1, p. 739

Summary

A continuous random variable X can assume any value in an interval or intervals. A density function for X is a function that has the following properties:

1. $f(x) \geq 0$ 2. $\displaystyle\int_{-\infty}^{\infty} f(x)\, dx = 1$

3. $P(a \leq X \leq b) = \displaystyle\int_{a}^{b} f(x)\, dx$

Property 3 means that the area under the graph of f and above the x-axis from $x = a$ to $x = b$ is $P(a \leq X \leq b)$. The probability that X assumes a particular value is 0.

The continuous random variable X has a uniform distribution over $[a, b]$ if its density function is given by

$$f(x) = \begin{cases} \dfrac{1}{b-a} & \text{if } a \leq x \leq b \\ 0 & \text{otherwise} \end{cases}$$

X has an exponential density function f if

$$f(x) = \begin{cases} ke^{-kx} & \text{if } x \geq 0 \\ 0 & \text{if } x < 0 \end{cases}$$

where k is a positive constant.

The cumulative distribution function F for the continuous random variable X with density function f is given by

$$F(x) = P(X \leq x) = \int_{-\infty}^{x} f(t)\, dt$$

Geometrically, $F(x)$ represents the area under the density curve to the left of x. By using F, we are able to find $P(a \leq x \leq b)$:

$$P(a \leq x \leq b) = F(b) - F(a)$$

The mean μ of X (also called expectation of X, $E(X)$) is given by

$$\mu = E(X) = \int_{-\infty}^{\infty} x f(x)\, dx$$

provided that the integral is convergent. The variance is given by

$$\sigma^2 = \mathrm{Var}(X) = \int_{-\infty}^{\infty} (x - \mu)^2 f(x)\, dx$$

$$= \int_{-\infty}^{\infty} x^2 f(x)\, dx - \mu^2$$

provided that the integral is convergent. The standard deviation is given by

$$\sigma = \sqrt{\mathrm{Var}(X)}$$

The graph of the normal density function

$$f(x) = \frac{1}{\sigma \sqrt{2\pi}}\, e^{-(1/2)((x-\mu)/\sigma)^2}$$

is called a normal curve and is bell shaped. If X has a normal distribution, then the probability that X lies within one standard deviation of the mean μ is (approximately) 0.68; within two standard deviations, the probability is 0.95; and within three standard deviations, it is 0.997. If Z is a normal random variable with $\mu = 0$ and $\sigma = 1$, then Z is called a standard normal random variable. The probability $P(0 < Z < z_0)$ is the area under the graph of the standard normal curve from $z = 0$ to $z = z_0$ and is denoted $A(z_0)$. Values of $A(z_0)$ appear in Appendix C.

If X is normally distributed with mean μ and standard deviation σ, then X can be transformed into a standard normal random variable by the change-of-variable formula

$$Z = \frac{X - \mu}{\sigma}$$

With this formula, probabilities for X can be found by using areas under the standard normal curve.

If X is a binomial random variable and the number n of independent trials is large, then the distribution of X can be approximated by using a normal random variable with mean np and standard deviation $\sqrt{npq}$, where p is the probability of success on any trial and $q = 1 - p$. It is important that continuity corrections be considered when we estimate binomial probabilities by a normal random variable.

Review Problems

Problem numbers shown in color indicate problems suggested for use as a practice chapter test.

1. Suppose X is a continuous random variable with density function given by

$$f(x) = \begin{cases} \frac{1}{3} + kx^2 & \text{if } 0 \leq x \leq 1 \\ 0 & \text{otherwise} \end{cases}$$

   (a) Find k.

   (b) Find $P(\frac{1}{2} < X < \frac{3}{4})$.

   (c) Find $P(X \geq \frac{1}{2})$.

   (d) Find the cumulative distribution function.

2. Suppose X is exponentially distributed with $k = \frac{1}{3}$. Find $P(X > 2)$.

3. Suppose X is a random variable with density function given by

$$f(x) = \begin{cases} \frac{2}{25}x & \text{if } 0 \leq x \leq 5 \\ 0 & \text{otherwise} \end{cases}$$

   (a) Find μ.

   (b) Find σ.

4. Let X be uniformly distributed over the interval $[2, 6]$. Find $P(X < 5)$.

Let X be normally distributed with mean 20 and standard deviation 4. In Problems 5–10, determine the given probabilities.

5. $P(X > 22)$

6. $P(X < 21)$

7. $P(14 < X < 18)$

8. $P(X > 10)$

9. $P(X < 23)$

10. $P(23 < X < 33)$

In Problems 11 and 12, X is a binomial random variable with $n = 100$ and $p = 0.35$. Find the given probabilities by using the normal approximation.

11. $P(25 \leq X \leq 47)$

12. $P(X = 48)$

13. **Heights of Individuals**   The heights (in inches) of individuals in a certain group are normally distributed with mean 68 and standard deviation 2. Find the probability that an individual from this group is taller than 6 ft.

14. **Coin Tossing**   If a fair coin is tossed 500 times, use the normal approximation to the binomial distribution to estimate the probability that a head comes up at least 215 times.

Cumulative Distribution from Data

W hat Section 16.3 said about histograms of discrete random variables is even more directly true for continuous random variables: The probability density curve can be thought of as the limiting case of a probability histogram. Indeed, this fact is often used to explain the idea of a probability density function.

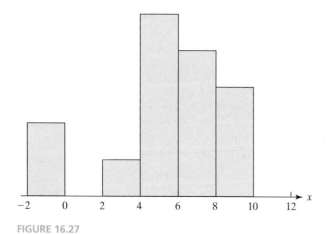

With continuous variables, a histogram divides the range of possible values into a series of intervals, called bins. Above each bin is a bar whose height indicates how much of the data set lies in that bin. Figure 16.27 illustrates this. The rightmost bin is the interval from 8 to 10. Because one-fifth of the data values lie in that bin, the bar covers one-fifth of the area of all the bars put together.

FIGURE 16.27

A probability density curve is the limit of a histogram's outline as the data set gets very large and the bin size gets very small. Figure 16.28 illustrates a larger data set and smaller bin size.

Unfortunately, dozens of data values are normally needed before a histogram's contours begin to smooth out. As a practical matter, we might want to "cheat" and get an

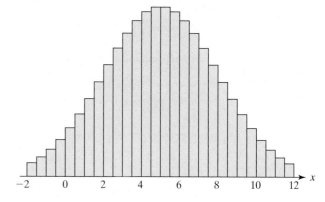

FIGURE 16.28

idea of the shape of the probability density function using fewer data values—and without having to draw a histogram.

Here is a way to do that: First use the data values to plot points that reveal the *cumulative distribution* curve, and then use this curve to infer the shape of the probability density curve. The steps are as follows:

Step 1: Determine n, the sample size.

Step 2: Arrange the data values in order, least to greatest. The smallest data value is $v(1)$, the next-smallest is $v(2)$, and so on.

Step 3: On a set of coordinate axes, plot n points with coordinates $\left(v(i), \dfrac{i - \frac{1}{2}}{n} \right)$ for i ranging from 1 to n.

Step 4: Determine what cumulative distribution function the plot suggests.

Step 5: Find the probability density function as the derivative of the cumulative distribution function.

In step 3, we are plotting each data value $v(i)$ against the experimental probability, based on the data, that a new value would be less than $v(i)$. This probability is calculated by taking the number of values below $v(i)$, namely $i - 1$; adding a term of $\frac{1}{2}$ to split the ith data point in two and count half of it as "below $v(i)$"; and dividing the result by n, the total number of values.

Let us see how this works with data values generated using a known probability density function. On a graphing calculator, we use the *rand* command or its equivalent to generate 15 values (step 1) using the uniform density function that equals 1 on the interval from 0 to 1 and equals 0 elsewhere. One such run produces the following values (ordered from lowest to highest and rounded to three decimal places): 0.043, 0.074, 0.093, 0.198, 0.293, 0.311, 0.399, 0.400, 0.409, 0.566, 0.654, 0.665, 0.760, 0.919, 0.967 (step 2).

When the corresponding probabilities are plotted (step 3), the result is as shown in Figure 16.29.

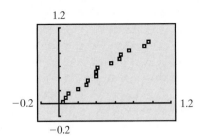

FIGURE 16.29

Even if we did not know how the points had been generated, we could still notice that all data values are between 0 and 1 and that the plotted points lie fairly close to the line $y = x$. This suggests (step 4) the cumulative distribution function

$$F(x) = \begin{cases} 0 & x < 0 \\ x & 0 \le x \le 1 \\ 1 & x > 1 \end{cases}$$

The derivative of this function with respect to x (step 5) is

$$P(x) = \begin{cases} 0 & x < 0 \\ 1 & 0 < x < 1 \\ 0 & x > 1 \end{cases}$$

which is exactly the function that was used to generate the data values.[2] The method works.

A word of caution, however. By nature, cumulative distribution functions are increasing from left to right and therefore all have a broadly similar shape. The plot in Figure 16.29, for example, might instead be interpreted as reflecting the cumulative distribution function

$$F_2(x) = \begin{cases} 0 & x < 0 \\ 1 - e^{-2x} & 0 \le x \end{cases}$$

The fit would not be quite so good as for $F(x)$, but it would not look out of the question either. So the method described here for identifying a possible cumulative distribution function and its derivative, the probability density function, is best supplemented by other information, such as knowledge of the process producing the data.

Problems

1. Generate your own run of 15 values using a uniform density function, and plot the results to obtain a picture of the cumulative distribution function. How well do your results fit with the known distribution function?

2. Repeat Problem 1 for a normal density function. (On the TI-83 Plus, use the *randNorm(* command.) In step 4, perform a logistic regression. Although the cumulative distribution function for a normal density function is *not* actually a logistic function, it has a very similar shape, so that a logistic function can be used as an approximation. Then, if your calculator has the capability, plot the derivative of the logistic function. Otherwise, note where the logistic function's slope is greatest and what happens to the slope at the ends of the curve. How does the behavior of the derivative compare with the behavior of the normal density function used to generate the data values?

3. Visit a Web site with a table of data values and see if you can determine the nature of the probability density function. You might try the near-real-time earthquake list at the U.S. Geological Survey (wwwneic.cr.usgs.gov) or one of the many data sets published by the U.S. Census (www.census.org). Why do you think the data are distributed as they are?

[2]Notice that $P(i)$ is undefined for $x = 0$ and $x = 1$. This does not matter.

MULTIVARIABLE CALCULUS

 Mathematical Snapshot   Data Analysis to Model Cooling

We know (from Chapter 13) how to maximize a company's profit when both revenue and cost are written as functions of a single quantity, namely the number of units produced. But of course the production level is itself determined by other factors—and, in general, no single variable can represent them.

The amount of oil pumped from an oil field each week, for example, depends on both the number of pumps and the number of hours that the pumps are operated. The number of pumps in the field will depend on the amount of capital originally available to build the pumps as well as the size and shape of the field. The number of hours that the pumps can be operated depends on the labor available to run and maintain the pumps. In addition, the amount of oil that the owner will be willing to have pumped from the oil field will depend on the current demand for oil—which is related to the price of the oil.

Maximizing the weekly profit from an oil field will require a balance between the number of pumps and the amount of time each pump can be operated. The maximum profit will not be achieved by building more pumps than can be operated or by running a few pumps full-time.

This is an example of the general problem of maximizing profit when production depends on several factors. The solution involves an analysis of the production function, which relates production output to resources allocated for production. Because, in general, several variables are needed to describe the resource allocation, the most profitable allocation cannot be found by differentiation with respect to a single variable, as in preceding chapters. The more advanced techniques necessary to do the job will be covered in this chapter.

OBJECTIVE

To discuss functions of several variables and to compute function values. To discuss three-dimensional coordinates and sketch simple surfaces.

17.1 Functions of Several Variables

Suppose a manufacturer produces two products, X and Y. Then the total cost depends on the levels of production of *both* X and Y. Table 17.1 is a schedule that indicates total cost at various levels. For example, when 5 units of X and 6 units of Y are produced, the total cost c is 17. In this situation, it seems natural to associate the number 17 with the *ordered pair* (5, 6):

$$(5, 6) \mapsto 17$$

The first element of the ordered pair, 5, represents the number of units of X produced, while the second element, 6, represents the number of units of Y produced. Corresponding to the other production situations, we have

$$(5, 7) \mapsto 19$$

$$(6, 6) \mapsto 18$$

and

$$(6, 7) \mapsto 20$$

TABLE 17.1

No. of Units of X Produced, x	No. of Units of Y Produced, y	Total Cost of Production, c
5	6	17
5	7	19
6	6	18
6	7	20

This correspondence can be considered an input–output relation where the inputs are ordered pairs. With each input, we associate exactly one output. Thus, the correspondence defines a function f whose domain consists of (5, 6), (5, 7), (6, 6), (6, 7) and whose range consists of 17, 19, 18, 20. In function notation,

$$f(5, 6) = 17 \quad f(5, 7) = 19$$

$$f(6, 6) = 18 \quad f(6, 7) = 20$$

We say that the total-cost schedule can be described by $c = f(x, y)$, a function of the two independent variables x and y. The letter c is the dependent variable.

Turning to another function of two variables, we see that the equation

$$z = \frac{2}{x^2 + y^2}$$

defines z as a function of x and y:

$$z = f(x, y) = \frac{2}{x^2 + y^2}$$

The domain of f is all ordered pairs of real numbers (x, y) for which the equation has meaning when the first and second elements of (x, y) are substituted for x and y, respectively, in the equation. Thus, the domain of f is all ordered pairs except $(0, 0)$. For example, to find $f(2, 3)$, we substitute $x = 2$ and $y = 3$ into the expression $2/(x^2 + y^2)$. We obtain $f(2, 3) = 2/(2^2 + 3^2) = 2/13$.

We do not have to look as far as the previous f for functions of several variables. Ordinary addition of real numbers defines the function

$$z = f(x, y) = x + y$$

which generalizes to the linear functions studied in Chapter 7. We can just as easily consider functions where the inputs are ordered triples of real numbers. In Example 2 of Section 7.4 we had

$$Z = 3x_1 + 4x_2 + \frac{3}{2}x_3$$

and we could certainly write $Z = f(x_1, x_2, x_3)$. There is no reason why a function of two variables must be defined for pairs of arbitrary *real* numbers. In Section 5.4 we studied $a_{\overline{n}|r}$, the present value of n one-dollar payments at the interest rate of r per period. The notation is strange, but if we write

$$A = a(n, r) = a_{\overline{n}|r}$$

we see that for input pairs (n, r), where n is a positive integer and r is a (rational) number in the the interval $(0, 1]$, the function a so defined provides real numbers as given, approximately, by the partial listing of Appendix A. For still another example, this time from Section 8.2, consider

$$C(n, r) = {}_nC_r$$

the number of combinations of n objects taken r at a time.

PRINCIPLES IN PRACTICE 1

FUNCTIONS OF TWO VARIABLES

The cost per day for manufacturing both 12-ounce and 20-ounce coffee mugs is given by $c = 160 + 2x + 3y$, where x is the number of 12-ounce mugs and y is the number of 20-ounce mugs. What is the cost per day of manufacturing

1. 500 12-ounce and 700 20-ounce mugs?
2. 1000 12-ounce and 750 20-ounce mugs?

EXAMPLE 1 Functions of Two Variables

a. $f(x, y) = \dfrac{x + 3}{y - 2}$ is a function of two variables. Because the denominator is zero when $y = 2$, the domain of f is all (x, y) such that $y \neq 2$. Some function values are

$$f(0, 3) = \frac{0 + 3}{3 - 2} = 3$$

$$f(3, 0) = \frac{3 + 3}{0 - 2} = -3$$

Note that $f(0, 3) \neq f(3, 0)$.

b. $h(x, y) = 4x$ defines h as a function of x and y. The domain is all ordered pairs of real numbers. Some function values are

$$h(2, 5) = 4(2) = 8$$

$$h(2, 6) = 4(2) = 8$$

Note that the function values are independent of the choice of y.

c. If $z^2 = x^2 + y^2$ and $x = 3$ and $y = 4$, then $z^2 = 3^2 + 4^2 = 25$. Consequently, $z = \pm 5$. Thus, with the ordered pair $(3, 4)$, we *cannot* associate exactly one output number. Hence $z^2 = x^2 + y^2$ does not define z as a function of x and y.

NOW WORK PROBLEM 1

EXAMPLE 2 Temperature–Humidity Index

On hot and humid days, many people tend to feel uncomfortable. The degree of discomfort is numerically given by the temperature–humidity index, THI, *which is a function of two variables,* t_d *and* t_w:

$$\text{THI} = f(t_d, t_w) = 15 + 0.4(t_d + t_w)$$

where t_d *is the dry-bulb temperature (in degrees Fahrenheit) and* t_w *is the wet-bulb temperature (in degrees Fahrenheit) of the air. Evaluate the* THI *when* $t_d = 90$ *and* $t_w = 80$.

Solution: We want to find $f(90, 80)$:

$$f(90, 80) = 15 + 0.4(90 + 80) = 15 + 68 = 83$$

When the THI is greater than 75, most people are uncomfortable. In fact, the THI was once called the "discomfort index." Many electric utilities closely follow this index so that they can anticipate the demand that air-conditioning places on their systems.

NOW WORK PROBLEM 3

If $y = f(x)$ is a function of one variable, the domain of f can be geometrically represented by points on the real-number line. The function itself can be represented by its graph in a coordinate plane, sometimes called a two-dimensional coordinate system. However, for a function of two variables, $z = f(x, y)$, the domain (consisting of ordered pairs of real numbers) can be geometrically represented by a *region* in the plane. The function itself can be geometrically represented in a ***three*-dimensional rectangular coordinate system.** Such a system is formed when three mutually perpendicular real-number lines in space intersect at the origin of each line, as in Figure 17.1. The three number lines are called the x-, y-, and z-axes, and their point of intersection is called the origin of the system. The arrows indicate the positive directions of the axes, and the negative portions of the axes are shown as dashed lines.

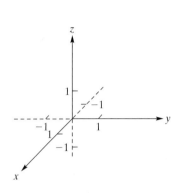

FIGURE 17.1
Three-dimensional rectangular coordinate system.

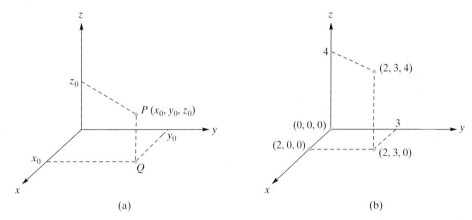

(a)   (b)

FIGURE 17.2 Points in space.

To each point P in space, we can assign a unique ordered triple of numbers, called the *coordinates* of P. To do this [see Figure 17.2(a)], from P, we construct a line perpendicular to the x,y-plane—that is, the plane determined by the x- and y-axes. Let Q be the point where the line intersects this plane. From Q, we construct perpendiculars to the x- and y-axes. These lines intersect the x- and y-axes at x_0 and y_0, respectively. From P, a perpendicular to the z-axis is constructed that intersects the axis at z_0. Thus, we assign to P the ordered triple (x_0, y_0, z_0). It should also be evident that with each ordered triple of numbers we can assign a unique point in space. Due to this one-to-one correspondence between points in space and ordered triples, an ordered triple can be called a point. In Figure 17.2(b), points $(2, 0, 0)$, $(2, 3, 0)$, and $(2, 3, 4)$ are shown. Note that the origin corresponds to $(0, 0, 0)$. Typically, the negative portions of the axes are not shown unless required.

We can represent a function of two variables, $z = f(x, y)$, geometrically. To each ordered pair (x, y) in the domain of f, we assign the point $(x, y, f(x, y))$. The set of all such points is called the *graph* of f. Such a graph appears in Figure 17.3. You can consider $z = f(x, y)$ as representing a surface in space.[1]

In Chapter 10, the continuity of a function of one variable was discussed. If $y = f(x)$, then to say that f is continuous at $x = a$ is to say that we can make the values $f(x)$ a close as we like to $f(a)$ by taking x sufficiently close to, but different from, a. This concept extends to a function of two variables. We say that the function $z = f(x, y)$ is continuous at (a, b) if we can make the values $f(x, y)$ a close as we like to $f(a, b)$ by taking (x, y) sufficiently close to, but different from, (a, b). Loosely interpreting this concept, and without delving into it in great depth, we can say that a function of two variables is continuous on its domain (that is, continuous at each point in its domain) if its graph is an unbroken surface.

In general, a **function of n variables** is a function whose domain consists of ordered n-tuples $(x_1, x_2, \ldots, x_n)$. For example, $f(x, y, z) = 2x + 3y + 4z$ is a function of three variables with a domain consisting of all ordered triples. The function

---

[1] We will freely use the term *surface* in an intuitive sense.

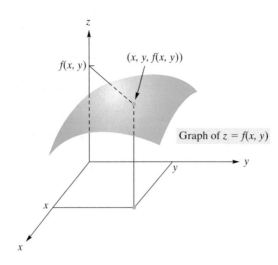

FIGURE 17.3 Graph of a function of two variables.

$g(x_1, x_2, x_3, x_4) = x_1x_2x_3x_4$ is a function of four variables with a domain consisting of all ordered 4-tuples. Although functions of several variables are extremely important and useful, we cannot visualize the graphs of functions of more than two variables.

We now give a brief discussion of sketching surfaces in space. We begin with planes that are parallel to a coordinate plane. By a "coordinate plane" we mean a plane containing two coordinate axes. For example, the plane determined by the x- and y-axes is the x,y-**plane.** Similarly, we speak of the x,z-**plane** and the y,z-**plane.** The coordinate planes divide space into eight parts, called *octants*. In particular, the part containing all points (x, y, z) such that x, y, and z are positive is called the **first octant.**

Names are not usually assigned to the remaining seven octants.

Suppose S is a plane that is parallel to the x,y-plane and passes through the point $(0, 0, 5)$. [See Figure 17.4(a).] Then the point (x, y, z) will lie on S if and only if $z = 5$; that is, x and y can be any real numbers, but z must equal 5. For this reason, we say that $z = 5$ is an equation of S. Similarly, an equation of the plane parallel to the x,z-plane and passing through the point $(0, 2, 0)$ is $y = 2$ [Figure 17.4(b)]. The equation $x = 3$ is an equation of the plane passing through $(3, 0, 0)$ and parallel to the y,z-plane [Figure 17.4(c)]. Now let us look at planes in general.

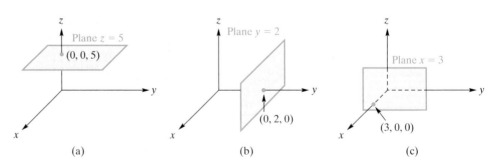

FIGURE 17.4 Planes parallel to coordinate planes.

In space, the graph of an equation of the form

$$Ax + By + Cz + D = 0$$

where D is a constant and A, B, and C are constants that are not all zero, is a plane. Since three distinct points (not lying on the same line) determine a plane, a convenient way to sketch a plane is to first determine the points, if any, where the plane intersects the x-, y-, and z-axes. These points are called *intercepts*.

● EXAMPLE 3   **Graphing a Plane**

Sketch the plane $2x + 3y + z = 6$.

Solution: The plane intersects the x-axis when $y = 0$ and $z = 0$. Thus, $2x = 6$, which gives $x = 3$. Similarly, if $x = z = 0$ then $y = 2$; if $x = y = 0$, then $z = 6$. Therefore, the intercepts are $(3, 0, 0)$, $(0, 2, 0)$ and $(0, 0, 6)$. After these points are plotted, a plane is passed through them. The portion of the plane in the first octant is shown in Figure 17.5(a); however, you should realize that the plane extends indefinitely into space.

NOW WORK PROBLEM 19

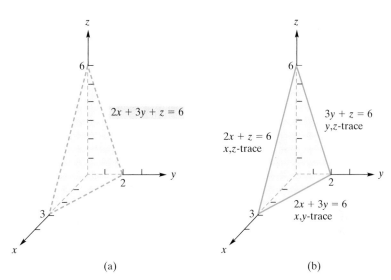

(a)   (b)

FIGURE 17.5   The plane $2x + 3y + z = 6$ and its traces.

A surface can be sketched with the aid of its **traces.** These are the intersections of the surface with the coordinate planes. To illustrate, for the plane $2x + 3y + z = 6$ in Example 3, the trace in the x,y-plane is obtained by setting $z = 0$. This gives $2x + 3y = 6$, which is an equation of a *line* in the x,y-plane. Similarly, setting $x = 0$ gives the trace in the y,z-plane: the line $3y + z = 6$. The x,z-trace is the line $2x + z = 6$. [See Figure 17.5(b).]

● EXAMPLE 4   **Sketching a Surface**

Sketch the surface $2x + z = 4$.

Solution: This equation has the form of a plane. The x- and z-intercepts are $(2, 0, 0)$ and $(0, 0, 4)$, and there is no y-intercept, since x and z cannot both be zero. Setting $y = 0$ gives the x,z-trace $2x + z = 4$, which is a line in the x,z-plane. In fact, the intersection of the surface with *any* plane $y = k$ is also $2x + z = 4$. Hence, the plane appears as in Figure 17.6.

NOW WORK PROBLEM 21

Our final examples deal with surfaces that are not planes but whose graphs can be easily obtained.

● EXAMPLE 5   **Sketching a Surface**

Sketch the surface $z = x^2$.

Solution: The x,z-trace is the curve $z = x^2$, which is parabola. In fact, for *any* fixed value of y, we get $z = x^2$. Thus, the graph appears as in Figure 17.7.

NOW WORK PROBLEM 25

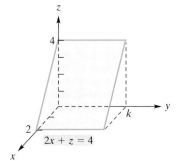

FIGURE 17.6   The plane $2x + z = 4$.

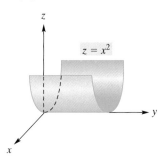

FIGURE 17.7   The surface $z = x^2$.

Note that this equation places no restriction on y.

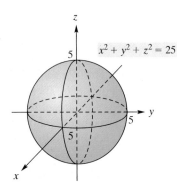

FIGURE 17.8 The surface $x^2 + y^2 + z^2 = 25$.

● EXAMPLE 6 **Sketching a Surface**

Sketch the surface $x^2 + y^2 + z^2 = 25$.

Solution: Setting $z = 0$ gives the x,y-trace $x^2 + y^2 = 25$, which is a circle of radius 5. Similarly, the y,z- and x,z-traces are the circles $y^2 + z^2 = 25$ and $x^2 + z^2 = 25$, respectively. Note also that since $x^2 + y^2 = 25 - z^2$, the intersection of the surface with the plane $z = k$, where $-5 \le k \le 5$, is a circle. For example, if $z = 3$, the intersection is the circle $x^2 + y^2 = 16$. If $z = 4$, the intersection is $x^2 + y^2 = 9$. That is, cross sections of the surface that are parallel to the x,y-plane are circles. The surface appears in Figure 17.8; it is a sphere.

NOW WORK PROBLEM 27

Problems 17.1

In Problems 1–12, determine the indicated function values for the given functions.

*1. $f(x, y) = 4x - y^2 + 3$; $f(1, 2)$

2. $f(x, y) = 3x^2 y - 4y$; $f(2, -1)$

*3. $g(x, y, z) = e^{2x}(3y + z)$; $g(0, 3, -1)$

4. $g(x, y, z) = x^2 yz + xy^2 z + xyz^2$; $g(3, 1, -2)$

5. $h(r, s, t, u) = \dfrac{rs}{t^2 - u^2}$; $h(-3, 3, 5, 4)$

6. $h(r, s, t, u) = \ln(ru)$; $h(1, 5, 3, 1)$

7. $g(p_A, p_B) = 2p_A(p_A^2 - 5)$; $g(4, 8)$

8. $g(p_A, p_B) = p_A \sqrt{p_B} + 10$; $g(8, 4)$

9. $F(x, y, z) = 3$; $F(2, 0, -1)$

10. $F(x, y, z) = \dfrac{2x}{(y + 1)z}$; $F(1, 0, 3)$

11. $f(x, y) = e^{x+y}$; $f(x_0 + h, y_0)$

12. $f(x, y) = x^2 y - 3y^3$; $f(r + t, r)$

13. **Ecology** A method of ecological sampling to determine animal populations in a given area involves first marking all the animals obtained in a sample of R animals from the area and then releasing them so that they can mix with unmarked animals. At a later date a second sample is taken of M animals and the number of these that are marked, S, is noted. Based on R, M, and S, an estimate of the total population of animals in the sample area is given by

$$N = f(R, M, S) = \frac{RM}{S}$$

Find $f(400, 400, 80)$. This method is called the *mark-and-recapture procedure*.[2]

14. **Genetics** Under certain conditions, if two brown-eyed parents have exactly k children, the probability that there will be exactly r blue-eyed children is given by

$$P(r, k) = \frac{k! \left(\frac{1}{4}\right)^r \left(\frac{3}{4}\right)^{k-r}}{r!(k - r)!} \quad r = 0, 1, 2, \ldots, k$$

Find the probability that, out of a total of four children, exactly three will be blue-eyed.

In Problems 15–18, find equations of the planes that satisfy the given conditions.

15. Parallel to the x,z-plane and also passes through the point $(0, 2, 0)$

16. Parallel to the y,z-plane and also passes through the point $(-2, 0, 0)$

17. Parallel to the x,y-plane and also passes through the point $(2, 7, 6)$

18. Parallel to the y,z-plane and also passes through the point $(-4, -2, 7)$

In Problems 19–28, sketch the given surfaces.

*19. $x + y + z = 1$ 20. $2x + y + 2z = 6$

*21. $3x + 6y + 2z = 12$ 22. $2x + 3y + 5z = 1$

23. $x + 2y = 2$ 24. $y = 3z + 2$

*25. $z = 4 - x^2$ 26. $y = z^2$

*27. $x^2 + y^2 + z^2 = 9$ 28. $3x^2 + 2y^2 = 1$

OBJECTIVE

To compute partial derivatives.

17.2 Partial Derivatives

Figure 17.9 shows the surface $z = f(x, y)$ and a plane that is parallel to the x,z-plane and that passes through the point $(a, b, f(a, b))$ on the surface. The equation of this plane is $y = b$. Hence, any point on the curve that is the intersection of the surface $z = f(x, y)$ with the plane $y = b$ must have the form $(x, b, f(x, b))$. Thus, the curve can be described by the equation $z = f(x, b)$. Since b is constant, $z = f(x, b)$ can be considered a function of one variable, x. When the derivative of this function is evaluated at a, it gives the slope of the tangent line to this curve at the point

---

[2]E. P. Odum, *Ecology* (New York: Holt, Rinehart and Winston, 1966).

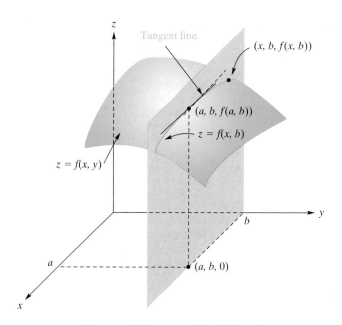

FIGURE 17.9 Geometric interpretation of $f_x(a, b)$.

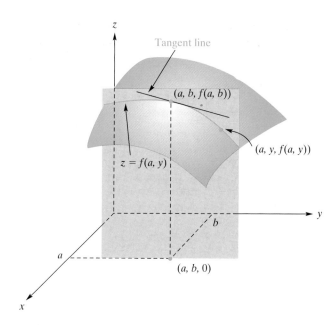

FIGURE 17.10 Geometric interpretation of $f_y(a, b)$.

$(a, b, f(a, b))$. (See Figure 17.9.) This slope is called the *partial derivative of f with respect to x at* (a, b) and is denoted $f_x(a, b)$. In terms of limits,

$$f_x(a, b) = \lim_{h \to 0} \frac{f(a + h, b) - f(a, b)}{h} \tag{1}$$

On the other hand, in Figure 17.10, the plane $x = a$ is parallel to the y, z-plane and cuts the surface $z = f(x, y)$ in a curve given by $z = f(a, y)$, a function of y. When the derivative of this function is evaluated at b, it gives the slope of the tangent line to this curve at the point $(a, b, f(a, b))$. This slope is called the *partial derivative of f with respect to y at* (a, b) and is denoted $f_y(a, b)$. In terms of limits,

$$f_y(a, b) = \lim_{h \to 0} \frac{f(a, b + h) - f(a, b)}{h} \tag{2}$$

> This gives us a geometric interpretation of a partial derivative.

We say that $f_x(a, b)$ is the slope of the tangent line to the graph of f at $(a, b, f(a, b))$ *in the x-direction;* similarly, $f_y(a, b)$ is the slope of the tangent line *in the y-direction.*

For generality, by replacing a and b in Equations (1) and (2) by x and y, respectively, we get the following definition.

DEFINITION

If $z = f(x, y)$, the **partial derivative of f with respect to x**, denoted f_x, is the function given by

$$f_x(x, y) = \lim_{h \to 0} \frac{f(x + h, y) - f(x, y)}{h}$$

provided that the limit exists.

The **partial derivative of f with respect to y**, denoted f_y, is the function given by

$$f_y(x, y) = \lim_{h \to 0} \frac{f(x, y + h) - f(x, y)}{h}$$

provided that the limit exists.

By analyzing the foregoing definition, we can state the following procedure to find f_x and f_y:

This gives us a mechanical way to find partial derivatives.

Procedure to Find $f_x(x, y)$ and $f_y(x, y)$

To find f_x, treat y as a constant, and differentiate f with respect to x in the usual way.

To find f_y, treat x as a constant, and differentiate f with respect to y in the usual way.

● **EXAMPLE 1   Finding Partial Derivatives**

If $f(x, y) = xy^2 + x^2y$, find $f_x(x, y)$ and $f_y(x, y)$. Also, find $f_x(3, 4)$ and $f_y(3, 4)$.

Solution: To find $f_x(x, y)$, we treat y as a constant and differentiate f with respect to x:

$$f_x(x, y) = (1)y^2 + (2x)y = y^2 + 2xy$$

To find $f_y(x, y)$, we treat x as a constant and differentiate with respect to y:

$$f_y(x, y) = x(2y) + x^2(1) = 2xy + x^2$$

Note that $f_x(x, y)$ and $f_y(x, y)$ are each functions of the two variables x and y. To find $f_x(3, 4)$ we evaluate $f_x(x, y)$ when $x = 3$ and $y = 4$:

$$f_x(3, 4) = 4^2 + 2(3)(4) = 40$$

Similarly,

$$f_y(3, 4) = 2(3)(4) + 3^2 = 33$$

NOW WORK PROBLEM 1 ●●●

Notations for partial derivatives of $z = f(x, y)$ are in Table 17.2. Table 17.3 gives notations for partial derivatives evaluated at (a, b). Note that the symbol ∂ (not d) is used to denote a partial derivative. The symbol $\partial z/\partial x$ is read "the partial derivative of z with respect to x."

TABLE 17.2

Partial Derivative of f (or z) with Respect to x	Partial Derivative of f (or z) with Respect to y
$f_x(x, y)$	$f_y(x, y)$
$\dfrac{\partial}{\partial x}(f(x, y))$	$\dfrac{\partial}{\partial y}(f(x, y))$
$\dfrac{\partial z}{\partial x}$	$\dfrac{\partial z}{\partial y}$

TABLE 17.3

Partial Derivative of f (or z) with Respect to x Evaluated at (a, b)	Partial Derivative of f (or z) with Respect to y Evaluated at (a, b)
$f_x(a, b)$	$f_y(a, b)$
$\dfrac{\partial z}{\partial x}\bigg|_{(a,b)}$	$\dfrac{\partial z}{\partial y}\bigg|_{(a,b)}$
$\dfrac{\partial z}{\partial x}\bigg|_{\substack{x=a\\y=b}}$	$\dfrac{\partial z}{\partial y}\bigg|_{\substack{x=a\\y=b}}$

● **EXAMPLE 2   Finding Partial Derivatives**

a. If $z = 3x^3y^3 - 9x^2y + xy^2 + 4y$, find $\dfrac{\partial z}{\partial x}, \dfrac{\partial z}{\partial y}, \dfrac{\partial z}{\partial x}\bigg|_{(1,0)}$ and $\dfrac{\partial z}{\partial y}\bigg|_{(1,0)}$.

Solution: To find $\partial z/\partial x$, we differentiate z with respect to x while treating y as a constant:

$$\frac{\partial z}{\partial x} = 3(3x^2)y^3 - 9(2x)y + (1)y^2 + 0$$

$$= 9x^2y^3 - 18xy + y^2$$

Evaluating the latter equation at $(1, 0)$, we obtain

$$\frac{\partial z}{\partial x}\bigg|_{(1,0)} = 9(1)^2(0)^3 - 18(1)(0) + 0^2 = 0$$

To find $\partial z/\partial y$, we differentiate z with respect to y while treating x as a constant:

$$\frac{\partial z}{\partial y} = 3x^3(3y^2) - 9x^2(1) + x(2y) + 4(1)$$

$$= 9x^3y^2 - 9x^2 + 2xy + 4$$

Thus,

$$\frac{\partial z}{\partial y}\bigg|_{(1,0)} = 9(1)^3(0)^2 - 9(1)^2 + 2(1)(0) + 4 = -5$$

b. *If* $w = x^2 e^{2x+3y}$*, find* $\partial w/\partial x$ *and* $\partial w/\partial y$*.*

Solution: To find $\partial w/\partial x$, we treat y as a constant and differentiate with respect to x. Since $x^2 e^{2x+3y}$ is a product of two functions, each involving x, we use the product rule:

$$\frac{\partial w}{\partial x} = x^2 \frac{\partial}{\partial x}(e^{2x+3y}) + e^{2x+3y} \frac{\partial}{\partial x}(x^2)$$

$$= x^2(2e^{2x+3y}) + e^{2x+3y}(2x)$$

$$= 2x(x + 1)e^{2x+3y}$$

To find $\partial w/\partial y$, we treat x as a constant and differentiate with respect to y:

$$\frac{\partial w}{\partial y} = x^2 \frac{\partial}{\partial y}(e^{2x+3y}) = 3x^2 e^{2x+3y}$$

NOW WORK PROBLEM 27

We have seen that, for a function of two variables, two partial derivatives can be considered. Actually, the concept of partial derivatives can be extended to functions of more than two variables. For example, with $w = f(x, y, z)$ we have three partial derivatives:

the partial with respect to x, denoted $f_x(x, y, z)$, $\partial w/\partial x$, and so on;
the partial with respect to y, denoted $f_y(x, y, z)$, $\partial w/\partial y$, and so on;
and
the partial with respect to z, denoted $f_z(x, y, z)$, $\partial w/\partial z$, and so on

To determine $\partial w/\partial x$, treat y and z as constants, and differentiate w with respect to x. For $\partial w/\partial y$, treat x and z as constants, and differentiate with respect to y. For $\partial w/\partial z$, treat x and y as constants, and differentiate with respect to z. For a function of n variables, we have n partial derivatives, which are determined in an analogous way.

EXAMPLE 3   Partial Derivatives of a Function of Three Variables

If $f(x, y, z) = x^2 + y^2 z + z^3$*, find* $f_x(x, y, z)$*,* $f_y(x, y, z)$*, and* $f_z(x, y, z)$*.*

Solution: To find $f_x(x, y, z)$, we treat y and z as constants and differentiate f with respect to x:

$$f_x(x, y, z) = 2x$$

Treating x and z as constants and differentiating with respect to y, we have

$$f_y(x, y, z) = 2yz$$

Treating x and y as constants and differentiating with respect to z, we have

$$f_z(x, y, z) = y^2 + 3z^2$$

NOW WORK PROBLEM 23

●EXAMPLE 4   **Partial Derivatives of a Function of Four Variables**

If $p = g(r, s, t, u) = \dfrac{rsu}{rt^2 + s^2 t}$, *find* $\dfrac{\partial p}{\partial s}$, $\dfrac{\partial p}{\partial t}$, *and* $\dfrac{\partial p}{\partial t}\Big|_{(0,1,1,1)}$.

Solution: To find $\partial p/\partial s$, first note that p is a quotient of two functions, each involving the variable s. Thus, we use the quotient rule and treat r, t, and u as constants:

$$\frac{\partial p}{\partial s} = \frac{(rt^2 + s^2 t)\dfrac{\partial}{\partial s}(rsu) - rsu\dfrac{\partial}{\partial s}(rt^2 + s^2 t)}{(rt^2 + s^2 t)^2}$$

$$= \frac{(rt^2 + s^2 t)(ru) - (rsu)(2st)}{(rt^2 + s^2 t)^2}$$

Simplification gives

$$\frac{\partial p}{\partial s} = \frac{ru(rt - s^2)}{t(rt + s^2)^2} \qquad \text{(a factor of } t \text{ cancels)}$$

To find $\partial p/\partial t$, we can first write p as

$$p = rsu(rt^2 + s^2 t)^{-1}$$

Next, we use the power rule and treat r, s, and u as constants:

$$\frac{\partial p}{\partial t} = rsu(-1)(rt^2 + s^2 t)^{-2}\frac{\partial}{\partial t}(rt^2 + s^2 t)$$

$$= -rsu(rt^2 + s^2 t)^{-2}(2rt + s^2)$$

so that

$$\frac{\partial p}{\partial s} = -\frac{rsu(2rt + s^2)}{(rt^2 + s^2 t)^2}$$

Letting $r = 0$, $s = 1$, $t = 1$, and $u = 1$ gives

$$\frac{\partial p}{\partial t}\Big|_{(0,1,1,1)} = -\frac{0(1)(1)(2(0)(1) + (1)^2)}{(0(1)^2 + (1)^2(1))^2} = 0$$

NOW WORK PROBLEM 31 ●●●

Problems 17.2

In Problems 1–26, a function of two or more variables is given. Find the partial derivative of the function with respect to each of the variables.

*1. $f(x, y) = 4x^2 + 3y^2 - 7$ 2. $f(x, y) = 2x^2 + 3xy$

3. $f(x, y) = 2y + 1$ 4. $f(x, y) = \ln 2$

5. $g(x, y) = 3x^4 y + 2xy^2 - 5xy + 8x - 9y$

6. $g(x, y) = (x + 1)^2 + (y - 3)^3 + 5xy^3 - 2$

7. $g(p, q) = \sqrt{pq}$ 8. $g(w, z) = \sqrt[3]{w^2 + z^2}$

9. $h(s, t) = \dfrac{s^2 + 4}{t - 3}$ 10. $h(u, v) = \dfrac{8uv^2}{u^2 + v^2}$

11. $u(q_1, q_2) = \frac{1}{2}\ln(q_1 + 2) + \frac{1}{3}\ln(q_2 + 5)$

12. $Q(l, k) = 2l^{0.38}k^{1.79} - 3l^{1.03} + 2k^{0.13}$

13. $h(x, y) = \dfrac{x^2 + 3xy + y^2}{\sqrt{x^2 + y^2}}$ 14. $h(x, y) = \dfrac{\sqrt{x + 9}}{x^2 y + y^2 x}$

15. $z = e^{5xy}$ 16. $z = (x^2 + y^2)e^{2x+3y+1}$

17. $z = 5x\ln(x^2 + y)$ 18. $z = \ln(5x^3 y^2 + 2y^4)^4$

19. $f(r, s) = \sqrt{r + 2s}\,(r^3 - 2rs + s^2)$

20. $f(r, s) = \sqrt{rs}\, e^{2+r}$

21. $f(r, s) = e^{3-r}\ln(7 - s)$

22. $f(r, s) = (5r^2 + 3s^3)(2r - 5s)$

*23. $g(x, y, z) = 2x^3 y^2 + 2xy^3 z + 4z^2$

24. $g(x, y, z) = 2xy^2 z^6 - 4x^2 y^3 z^2 + 3xyz$

25. $g(r, s, t) = e^{s+t}(r^2 + 7s^3)$

26. $g(r, s, t, u) = rs\ln(2t + 5u)$

In Problems 27–34, evaluate the given partial derivatives.

*27. $f(x, y) = x^3 y + 7x^2 y^2$; $f_x(1, -2)$

28. $z = \sqrt{2x^3 + 5xy + 2y^2}$; $\dfrac{\partial z}{\partial x}\Big|_{\substack{x=0 \\ y=1}}$

29. $g(x, y, z) = e^x\sqrt{y + 2z}$; $g_z(0, 6, 4)$

30. $g(x, y, z) = \dfrac{3x^2 y^2 + 2xy + x - y}{xy - yz + xz}$, $g_y(1, 1, 5)$

*31. $h(r, s, t, u) = (s^2 + tu)\ln(2r + 7st)$; $h_s(1, 0, 0, 1)$

32. $h(r, s, t, u) = \dfrac{7r + 3s^2 u^2}{s}; \quad h_t(4, 3, 2, 1)$

33. $f(r, s, t) = rst(r^2 + s^3 + t^4); \quad f_s(1, -1, 2)$

34. $z = \dfrac{x^2 + y^2}{e^{x^2 + y^2}}; \quad \dfrac{\partial z}{\partial x}\Big|_{\substack{x=0 \\ y=0}}, \quad \dfrac{\partial z}{\partial y}\Big|_{\substack{x=1 \\ y=1}}$

35. If $z = xe^{x-y} + ye^{y-x}$, show that

$$\frac{\partial z}{\partial x} + \frac{\partial z}{\partial y} = e^{x-y} + e^{y-x}$$

36. Stock Prices of a Dividend Cycle   In a discussion of stock prices of a dividend cycle, Palmon and Yaari[3] consider the function f given by

$$u = f(t, r, z) = \frac{(1+r)^{1-z} \ln(1+r)}{(1+r)^{1-z} - t}$$

where u is the instantaneous rate of ask-price appreciation, r is an annual opportunity rate of return, z is the fraction of a dividend cycle over which a share of stock is held by a midcycle seller, and t is the effective rate of capital gains tax. They claim that

$$\frac{\partial u}{\partial z} = \frac{t(1+r)^{1-z} \ln^2(1+r)}{[(1+r)^{1-z} - t]^2}$$

Verify this.

37. Money Demand   In a discussion of inventory theory of money demand, Swanson[4] considers the function

$$F(b, C, T, i) = \frac{bT}{C} + \frac{iC}{2}$$

and determines that $\dfrac{\partial F}{\partial C} = -\dfrac{bT}{C^2} + \dfrac{i}{2}$. Verify this partial derivative.

38. Interest Rate Deregulation   In an article on interest rate deregulation, Christofi and Agapos[5] arrive at the equation

$$r_L = r + D\frac{\partial r}{\partial D} + \frac{dC}{dD} \tag{3}$$

where r is the deposit rate paid by commercial banks, r_L is the rate earned by commercial banks, C is the administrative cost of transforming deposits into return-earning assets, and D is the savings deposit level. Christofi and Agapos state that

$$r_L = r\left[\frac{1+\eta}{\eta}\right] + \frac{dC}{dD} \tag{4}$$

where $\eta = \dfrac{r/D}{\partial r/\partial D}$ is the deposit elasticity with respect to the deposit rate. Express Equation (3) in terms of η to verify Equation (4).

39. Advertising and Profitability   In an analysis of advertising and profitability, Swales[6] considers a function f given by

$$R = f(r, a, n) = \frac{r}{1 + a\left(\dfrac{n-1}{2}\right)}$$

where R is the adjusted rate of profit, r is the accounting rate of profit, a is a measure of advertising expenditures, and n is the number of years that advertising fully depreciates. In the analysis, Swales determines $\partial R/\partial n$. Find this partial derivative.

To develop the notions of partial marginal cost, marginal productivity, and competitive and complementary products.

Here we have "rate of change" interpretations of partial derivatives.

17.3 Applications of Partial Derivatives

From Section 17.2, we know that if $z = f(x, y)$, then $\partial z/\partial x$ and $\partial z/\partial y$ can be geometrically interpreted as giving the slopes of the tangent lines to the surface $z = f(x, y)$ in the x- and y-directions, respectively. There are other interpretations: Because $\partial z/\partial x$ is the derivative of z with respect to x when y is held fixed, and because a derivative is a rate of change, we have

$\dfrac{\partial z}{\partial x}$ is the rate of change of z with respect to x when y is held fixed.

Similarly,

$\dfrac{\partial z}{\partial y}$ is the rate of change of z with respect to y when x is held fixed.

We will now look at some applications in which the "rate of change" notion of a partial derivative is very useful.

---

[3]D. Palmon and U. Yaari, "Taxation of Capital Gains and the Behavior of Stock Prices over the Dividend Cycle," *The American Economist*, XXVII, no. 1 (1983), 13–22.

[4]P. E. Swanson, "Integer Constraints on the Inventory Theory of Money Demand," *Quarterly Journal of Business and Economics*, 23, no. 1 (1984), 32–37.

[5]A. Christofi and A. Agapos, "Interest Rate Deregulation: An Empirical Justification," *Review of Business and Economic Research*, XX (1984), 39–49.

[6]J. K. Swales, "Advertising as an Intangible Asset: Profitability and Entry Barriers: A Comment on Reekie and Bhoyrub," *Applied Economics*, 17, no. 4 (1985), 603–17.

Suppose a manufacturer produces x units of product X and y units of product Y. Then the total cost c of these units is a function of x and y and is called a **joint-cost function.** If such a function is $c = f(x, y)$, then $\partial c/\partial x$ is called the **(partial) marginal cost with respect to** x and is the rate of change of c with respect to x when y is held fixed. Similarly, $\partial c/\partial y$ is the **(partial) marginal cost with respect to** y and is the rate of change of c with respect to y when x is held fixed.

For example, if c is expressed in dollars and $\partial c/\partial y = 2$, then the cost of producing an extra unit of Y when the level of production of X is fixed is approximately two dollars.

If a manufacturer produces n products, the joint-cost function is a function of n variables, and there are n (partial) marginal-cost functions.

EXAMPLE 1   Marginal Costs

A company manufactures two types of skis, the Lightning and the Alpine models. Suppose the joint-cost function for producing x pairs of the Lightning model and y pairs of the Alpine model per week is

$$c = f(x, y) = 0.07x^2 + 75x + 85y + 6000$$

where c is expressed in dollars. Determine the marginal costs $\partial c/\partial x$ and $\partial c/\partial y$ when $x = 100$ and $y = 50$, and interpret the results.

Solution: The marginal costs are

$$\frac{\partial c}{\partial x} = 0.14x + 75 \quad \text{and} \quad \frac{\partial c}{\partial y} = 85$$

Thus,

$$\left.\frac{\partial c}{\partial x}\right|_{(100,50)} = 0.14(100) + 75 = 89 \tag{1}$$

and

$$\left.\frac{\partial c}{\partial y}\right|_{(100,50)} = 85 \tag{2}$$

Equation (1) means that increasing the output of the Lightning model from 100 to 101, while maintaining production of the Alpine model at 50, increases costs by approximately \$89. Equation (2) means that increasing the output of the Alpine model from 50 to 51 and holding production of the Lightning model at 100 will increase costs by approximately \$85. In fact, since $\partial c/\partial y$ is a constant function, the marginal cost with respect to y is \$85 at all levels of production.

NOW WORK PROBLEM 1

EXAMPLE 2   Loss of Body Heat

On a cold day, a person may feel colder when the wind is blowing than when the wind is calm because the rate of heat loss is a function of both temperature and wind speed. The equation

$$H = (10.45 + 10\sqrt{w} - w)(33 - t)$$

indicates the rate of heat loss H (in kilocalories per square meter per hour) when the air temperature is t (in degrees Celsius) and the wind speed is w (in meters per second). For $H = 2000$, exposed flesh will freeze in one minute.[7]

a. *Evaluate H when $t = 0$ and $w = 4$.*

Solution: When $t = 0$ and $w = 4$,

$$H = (10.45 + 10\sqrt{4} - 4)(33 - 0) = 872.85$$

---

[7]G. E. Folk, Jr., *Textbook of Environmental Physiology,* 2nd ed. (Philadelphia: Lea & Febiger, 1974).

b. *Evaluate* $\partial H/\partial w$ *and* $\partial H/\partial t$ *when* $t = 0$ *and* $w = 4$, *and interpret the results.*

Solution:

$$\frac{\partial H}{\partial w} = \left(\frac{5}{\sqrt{w}} - 1\right)(33 - t), \quad \frac{\partial H}{\partial w}\bigg|_{\substack{t=0 \\ w=4}} = 49.5$$

$$\frac{\partial H}{\partial t} = (10.45 + 10\sqrt{w} - w)(-1), \quad \frac{\partial H}{\partial t}\bigg|_{\substack{t=0 \\ w=4}} = -26.45$$

These equations mean that when $t = 0$ and $w = 4$, increasing w by a small amount while keeping t fixed will make H increase approximately 49.5 times as much as w increases. Increasing t by a small amount while keeping w fixed will make H *decrease* approximately 26.45 times as much as t increases.

c. *When* $t = 0$ *and* $w = 4$, *which has a greater effect on H: a change in wind speed of 1 m/s or a change in temperature of* $1°C$?

Solution: Since the partial derivative of H with respect to w is greater in magnitude than the partial with respect to t when $t = 0$ and $w = 4$, a change in wind speed of 1 m/s has a greater effect on H.

NOW WORK PROBLEM 13

The output of a product depends on many factors of production. Among these may be labor, capital, land, machinery, and so on. For simplicity, let us suppose that output depends only on labor and capital. If the function $P = f(l, k)$ gives the output P when the producer uses l units of labor and k units of capital, then this function is called a **production function.** We define the **marginal productivity with respect to** l to be $\partial P/\partial l$. This is the rate of change of P with respect to l when k is held fixed. Likewise, the **marginal productivity with respect to** k is $\partial P/\partial k$ and is the rate of change of P with respect to k when l is held fixed.

● EXAMPLE 3 **Marginal Productivity**

A manufacturer of a popular toy has determined that the production function is $P = \sqrt{lk}$, *where* l *is the number of labor-hours per week and* k *is the capital (expressed in hundreds of dollars per week) required for a weekly production of* P *gross of the toy. (One gross is 144 units.) Determine the marginal productivity functions, and evaluate them when* $l = 400$ *and* $k = 16$. *Interpret the results.*

Solution: Since $P = (lk)^{1/2}$,

$$\frac{\partial P}{\partial l} = \frac{1}{2}(lk)^{-1/2}k = \frac{k}{2\sqrt{lk}}$$

and

$$\frac{\partial P}{\partial k} = \frac{1}{2}(lk)^{-1/2}l = \frac{l}{2\sqrt{lk}}$$

Evaluating these equations when $l = 400$ and $k = 16$, we obtain

$$\frac{\partial P}{\partial l}\bigg|_{\substack{l=400 \\ k=16}} = \frac{16}{2\sqrt{400(16)}} = \frac{1}{10}$$

and

$$\frac{\partial P}{\partial k}\bigg|_{\substack{l=400 \\ k=16}} = \frac{400}{2\sqrt{400(16)}} = \frac{5}{2}$$

Thus, if $l = 400$ and $k = 16$, increasing l to 401 and holding k at 16 will increase output by approximately $\frac{1}{10}$ gross. But if k is increased to 17 while l is held at 400, the output increases by approximately $\frac{5}{2}$ gross.

NOW WORK PROBLEM 5

Competitive and Complementary Products

Sometimes two products may be related such that changes in the price of one of them affect the demand for the other. A typical example is that of butter and margarine. If such a relationship exists between products A and B, then the demand for each product is dependent on the prices of both. Suppose q_A and q_B are the quantities demanded for A and B, respectively, and p_A and p_B are their respective prices. Then both q_A and q_B are functions of p_A and p_B:

$$q_A = f(p_A, p_B) \qquad \text{demand function for A}$$

$$q_B = g(p_A, p_B) \qquad \text{demand function for B}$$

We can find four partial derivatives:

$$\frac{\partial q_A}{\partial p_A} \ \textit{the marginal demand for A with respect to } p_A$$

$$\frac{\partial q_A}{\partial p_B} \ \textit{the marginal demand for A with respect to } p_B$$

$$\frac{\partial q_B}{\partial p_A} \ \textit{the marginal demand for B with respect to } p_A$$

$$\frac{\partial q_B}{\partial p_B} \ \textit{the marginal demand for B with respect to } p_B$$

Under typical conditions, if the price of B is fixed and the price of A increases, then the quantity of A demanded will decrease. Thus, $\partial q_A / \partial p_A < 0$. Similarly, $\partial q_B / \partial p_B < 0$. However, $\partial q_A / \partial p_B$ and $\partial q_B / \partial p_A$ may be either positive or negative. If

$$\frac{\partial q_A}{\partial p_B} > 0 \quad \text{and} \quad \frac{\partial q_B}{\partial p_A} > 0$$

then A and B are said to be **competitive products** or **substitutes.** In this situation, an increase in the price of B causes an increase in the demand for A, if it is assumed that the price of A does not change. Similarly, an increase in the price of A causes an increase in the demand for B when the price of B is held fixed. Butter and margarine are examples of substitutes.

Proceeding to a different situation, we say that if

$$\frac{\partial q_A}{\partial p_B} < 0 \quad \text{and} \quad \frac{\partial q_B}{\partial p_A} < 0$$

then A and B are **complementary products.** In this case, an increase in the price of B causes a decrease in the demand for A if the price of A does not change. Similarly, an increase in the price of A causes a decrease in the demand for B when the price of B is held fixed. For example, cars and gasoline are complementary products. An increase in the price of gasoline will make driving more expensive. Hence, the demand for cars will decrease. And an increase in the price of cars will reduce the demand for gasoline.

● EXAMPLE 4   **Determining Whether Products Are Competitive or Complementary**

The demand functions for products A *and* B *are each a function of the prices of* A *and* B *and are given by*

$$q_A = \frac{50\sqrt[3]{p_B}}{\sqrt{p_A}} \quad \textit{and} \quad q_B = \frac{75 p_A}{\sqrt[3]{p_B^2}}$$

respectively. Find the four marginal-demand functions, and determine whether A *and* B *are competitive products, complementary products, or neither.*

Solution: Writing $q_A = 50p_A^{-1/2}p_B^{1/3}$ and $q_B = 75p_A p_B^{-2/3}$, we have

$$\frac{\partial q_A}{\partial p_A} = 50\left(-\frac{1}{2}\right)p_A^{-3/2}p_B^{1/3} = -25p_A^{-3/2}p_B^{1/3}$$

$$\frac{\partial q_A}{\partial p_B} = 50p_A^{-1/2}\left(\frac{1}{3}\right)p_B^{-2/3} = \frac{50}{3}p_A^{-1/2}p_B^{-2/3}$$

$$\frac{\partial q_B}{\partial p_A} = 75(1)p_B^{-2/3} = 75p_B^{-2/3}$$

$$\frac{\partial q_B}{\partial p_B} = 75p_A\left(-\frac{2}{3}\right)p_B^{-5/3} = -50p_A p_B^{-5/3}$$

Since p_A and p_B represent prices, they are both positive. Hence, $\partial q_A/\partial p_B > 0$ and $\partial q_B/\partial p_A > 0$. We conclude that A and B are competitive products.

NOW WORK PROBLEM 19

Problems 17.3

For the joint-cost functions in Problems 1–3, find the indicated marginal cost at the given production level.

*1. $c = 7x + 0.3y^2 + 2y + 900;$ $\dfrac{\partial c}{\partial y}, x = 20, y = 30$

2. $c = x\sqrt{x+y} + 5000;$ $\dfrac{\partial c}{\partial x}, x = 40, y = 60$

3. $c = 0.03(x+y)^3 - 0.6(x+y)^2 + 9.5(x+y) + 7700;$
   $\dfrac{\partial c}{\partial x}, x = 50, y = 80$

For the production functions in Problems 4 and 5, find the marginal productivity functions $\partial P/\partial k$ and $\partial P/\partial l$.

4. $P = 15lk - 3l^2 + 5k^2 + 500$

*5. $P = 2.314l^{0.357}k^{0.643}$

6. **Cobb–Douglas Production Function** In economics, a Cobb–Douglas production function is a production function of the form $P = Al^\alpha k^\beta$, where A, α, and β are constants and $\alpha + \beta = 1$. For such a function, show that

   (a) $\partial P/\partial l = \alpha P/l$
   (b) $\partial P/\partial k = \beta P/k$
   (c) $l\dfrac{\partial P}{\partial l} + k\dfrac{\partial P}{\partial k} = P$. This means that summing the products of the marginal productivity of each factor and the amount of that factor results in the total product P.

In Problems 7–9, q_A and q_B are demand functions for products A and B, respectively. In each case, find $\partial q_A/\partial p_A$, $\partial q_A/\partial p_B$, $\partial q_B/\partial p_A$, $\partial q_B/\partial p_B$ and determine whether A and B are competitive, complementary, or neither.

7. $q_A = 1000 - 50p_A + 2p_B;$ $q_B = 500 + 4p_A - 20p_B$

8. $q_A = 20 - p_A - 2p_B;$ $q_B = 50 - 2p_A - 3p_B$

9. $q_A = \dfrac{100}{p_A\sqrt{p_B}};$ $q_B = \dfrac{500}{p_B\sqrt[3]{p_A}}$

10. **Canadian Manufacturing** The production function for the Canadian manufacturing industries for 1927 is estimated by[8]

$P = 33.0l^{0.46}k^{0.52}$, where P is product, l is labor, and k is capital. Find the marginal productivities for labor and capital, and evaluate when $l = 1$ and $k = 1$.

11. **Dairy Farming** An estimate of the production function for dairy farming in Iowa (1939) is given by[9]

$$P = A^{0.27}B^{0.01}C^{0.01}D^{0.23}E^{0.09}F^{0.27}$$

where P is product, A is land, B is labor, C is improvements, D is liquid assets, E is working assets, and F is cash operating expenses. Find the marginal productivities for labor and improvements.

12. **Production Function** Suppose a production function is given by $P = \dfrac{kl}{2k + 3l}$.

   (a) Determine the marginal productivity functions.
   (b) Show that when $k = l$, the marginal productivities sum to $\dfrac{1}{5}$.

*13. **M.B.A. Compensation** In a study of success among graduates with master's of business administration (M.B.A.) degrees, it was estimated that for staff managers (which include accountants, analysts, etc.), current annual compensation (in dollars) was given by

$$z = 43,960 + 4480x + 3492y$$

where x and y are the number of years of work experience before and after receiving the M.B.A. degree, respectively.[10] Find $\partial z/\partial x$ and interpret your result.

[8]P. Daly and P. Douglas, "The Production Function for Canadian Manufactures," *Journal of the American Statistical Association*, 38 (1943), 178–86.

[9]G. Tintner and O. H. Brownlee, "Production Functions Derived from Farm Records," *American Journal of Agricultural Economics*, 26 (1944), 566–71.

[10]Adapted from A. G. Weinstein and V. Srinivasen, "Predicting Managerial Success of Master of Business Administration (M.B.A.) Graduates," *Journal of Applied Psychology*, 59, no. 2 (1974), 207–12.

14. Status A person's general status S_g is believed to be a function of status attributable to education, S_e, and status attributable to income, S_i, where S_g, S_e, and S_i are represented numerically. If

$$S_g = 7\sqrt[3]{S_e}\sqrt{S_i}$$

determine $\partial S_g/\partial S_e$ and $\partial S_g/\partial S_i$ when $S_e = 125$ and $S_i = 100$, and interpret your results.[11]

15. Reading Ease Sometimes we want to evaluate the degree of readability of a piece of writing. Rudolf Flesch[12] developed a function of two variables that will do this, namely,

$$R = f(w, s) = 206.835 - (1.015w + 0.846s)$$

where R is called the *reading ease score,* w is the average number of words per sentence in 100-word samples, and s is the average number of syllables in such samples. Flesch says that an article for which $R = 0$ is "practically unreadable," but one with $R = 100$ is "easy for any literate person." (a) Find $\partial R/\partial w$ and $\partial R/\partial s$. (b) Which is "easier" to read: an article for which $w = w_0$ and $s = s_0$, or one for which $w = w_0 + 1$ and $s = s_0$?

16. Model for Voice The study of frequency of vibrations of a taut wire is useful in considering such things as an individual's voice. Suppose

$$\omega = \frac{1}{bL}\sqrt{\frac{\tau}{\pi\rho}}$$

where ω (a Greek letter read "omega") is frequency, b is diameter, L is length, ρ (a Greek letter read "rho") is density, and τ (a Greek letter read "tau") is tension.[13] Find $\partial\omega/\partial b$, $\partial\omega/\partial L$, $\partial\omega/\partial\rho$, and $\partial\omega/\partial\tau$.

17. Traffic Flow Consider the following traffic-flow situation. On a highway where two lanes of traffic flow in the same direction, there is a maintenance vehicle blocking the left lane. (See Figure 17.11.) Two vehicles (*lead* and *following*) are in the right lane with a gap between them. The *subject* vehicle can choose either to fill or not to fill the gap. That decision may be based not only on the distance x shown in the diagram, but also on other factors (such as the velocity of the *following* vehicle). A *gap index g* has been used in

analyzing such a decision.[14, 15] The greater the g-value, the greater is the propensity for the *subject* vehicle to fill the gap. Suppose

$$g = \frac{x}{V_F} - \left(0.75 + \frac{V_F - V_S}{19.2}\right)$$

where x (in feet) is as before, V_F is the velocity of the *following* vehicle (in feet per second), and V_S is the velocity of the *subject* vehicle (in feet per second). From the diagram, it seems reasonable that if both V_F and V_S are fixed and x increases, then g should increase. Show that this is true by applying calculus to the function g. Assume that x, V_F, and V_S are positive.

18. Demand Suppose the demand equations for related products A and B are

$$q_A = e^{-(p_A + p_B)} \quad \text{and} \quad q_B = \frac{16}{p_A^2 p_B^2}$$

where q_A and q_B are the number of units of A and B demanded when the unit prices (in thousands of dollars) are p_A and p_B, respectively.

(a) Classify A and B as competitive, complementary, or neither.
(b) If the unit prices of A and B are $1000 and $2000, respectively, estimate the change in the demand for A when the price of B is decreased by $20 and the price of A is held constant.

*19. Demand The demand equations for related products A and B are given by

$$q_A = 10\sqrt{\frac{p_B}{p_A}} \quad \text{and} \quad q_B = 3\sqrt[3]{\frac{p_A}{p_B}}$$

where q_A and q_B are the quantities of A and B demanded and p_A and p_B are the corresponding prices (in dollars) per unit.

(a) Find the values of the two marginal demands for product A when $p_A = 9$ and $p_B = 16$.
(b) If p_B were reduced to 14 from 16, with p_A fixed at 9, use part (a) to estimate the corresponding change in demand for product A.

20. Joint-Cost Function A manufacturer's joint-cost function for producing q_A units of product A and q_B units of product B is given by

$$c = \frac{q_A^2(q_B^3 + q_A)^{1/2}}{17} + q_A q_B^{1/3} + 600$$

where c is in dollars.

(a) Find the marginal-cost functions with respect to q_A and q_B.
(b) Evaluate the marginal-cost function with respect to q_A when $q_A = 17$ and $q_B = 8$. Round your answer to two decimal places.

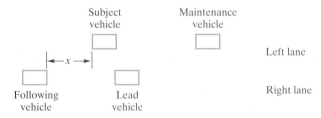

Subject vehicle   Maintenance vehicle

Left lane

Right lane

Following vehicle   Lead vehicle

FIGURE 17.11 Diagram for Problem 17.

[11]Adapted from R. K. Leik and B. F. Meeker, *Mathematical Sociology* (Englewood Cliffs, NJ: Prentice-Hall, Inc., 1975).

[12]R. Flesch, *The Art of Readable Writing* (New York: Harper & Row Publishers, Inc., 1949).

[13]R. M. Thrall, J. A. Mortimer, K. R. Rebman, and R. F. Baum, eds., *Some Mathematical Models in Biology,* rev. ed., Report No. 40241-R-7. Prepared at University of Michigan, 1967.

[14]P. M. Hurst, K. Perchonok, and E. L. Seguin, "Vehicle Kinematics and Gap Acceptance," *Journal of Applied Psychology,* 52, no. 4 (1968), 321–24.

[15]K. Perchonok and P. M. Hurst, "Effect of Lane-Closure Signals upon Driver Decision Making and Traffic Flow," *Journal of Applied Psychology,* 52, no. 5 (1968), 410–13.

(c) Use your answer to part (a) to estimate the change in cost if production of product A is decreased from 17 to 16 units, while production of product B is held constant at 8 units.

21. Elections For the congressional elections of 1974, the Republican percentage, R, of the Republican–Democratic vote in a district is given (approximately) by[16]

$$R = f(E_r, E_d, I_r, I_d, N)$$

$$= 15.4725 + 2.5945 E_r - 0.0804 E_r^2 - 2.3648 E_d$$

$$+ 0.0687 E_d^2 + 2.1914 I_r - 0.0912 I_r^2$$

$$- 0.8096 I_d + 0.0081 I_d^2 - 0.0277 E_r I_r$$

$$+ 0.0493 E_d I_d + 0.8579 N - 0.0061 N^2$$

Here E_r and E_d are the campaign expenditures (in units of \$10,000) by Republicans and Democrats, respectively; I_r and I_d are the number of terms served in Congress, *plus one,* for the Republican and Democratic candidates, respectively, and N is the percentage of the two-party presidential vote that Richard Nixon received in the district for 1968. The variable N gives a measure of Republican strength in the district.

(a) In the Federal Election Campaign Act of 1974, Congress set a limit of \$188,000 on campaign expenditures. By analyzing $\partial R / \partial E_r$, would you have advised a Republican candidate who served nine terms in Congress to spend \$188,000 on his or her campaign?

(b) Find the percentage above which the Nixon vote had a negative effect on R; that is, find N when $\partial R / \partial N < 0$. Give your answer to the nearest percent.

22. Sales After a new product has been launched onto the market, its sales volume (in thousands of units) is given by

$$S = \frac{AT + 450}{\sqrt{A + T^2}}$$

where T is the time (in months) since the product was first introduced and A is the amount (in hundreds of dollars) spent each month on advertising.

(a) Verify that the partial derivative of sales volume with respect to time is given by

$$\frac{\partial S}{\partial T} = \frac{A^2 - 450 T}{(A + T^2)^{3/2}}$$

(b) Use the result in part (a) to predict the number of months that will elapse before the sales volume begins to decrease if the amount allocated to advertising is held fixed at \$9000 per month.

Let f be a demand function for product A and $q_A = f(p_A, p_B)$, where q_A is the quantity of A demanded when the price per unit of A is p_A and the price per unit of product B is p_B. The partial elasticity of demand for A with respect to p_A, denoted η_{p_A}, is defined as $\eta_{p_A} = (p_A/q_A)(\partial q_A/\partial p_A)$. The partial elasticity of demand for A with respect to p_B, denoted η_{p_B}, is defined as $\eta_{p_B} = (p_B/q_A)(\partial q_A/\partial p_B)$. Loosely speaking, η_{p_A} is the ratio of a percentage change in the quantity of A demanded to a percentage change in the price of A when the price of B is fixed. Similarly, η_{p_B} can be loosely interpreted as the ratio of a percentage change in the quantity of A demanded to a percentage change in the price of B when the price of A is fixed. In Problems 23–25, find η_{p_A} and η_{p_B} for the given values of p_A and p_B.

23. $q_A = 1000 - 50 p_A + 2 p_B$; $p_A = 2$, $p_B = 10$

24. $q_A = 60 - 3 p_A - 2 p_B$; $p_A = 5$, $p_B = 3$

25. $q_A = 100/(p_A \sqrt{p_B})$; $p_A = 1$, $p_B = 4$

OBJECTIVE

To find partial derivatives of a function defined implicitly.

17.4 Implicit Partial Differentiation[17]

An equation in x, y, and z does not necessarily define z as a function of x and y. For example, in the equation

$$z^2 - x^2 - y^2 = 0 \tag{1}$$

if $x = 1$ and $y = 1$, then $z^2 - 1 - 1 = 0$, so $z = \pm\sqrt{2}$. Thus, Equation (1) does not define z as a function of x and y. However, solving Equation (1) for z gives

$$z = \sqrt{x^2 + y^2} \quad \text{or} \quad z = -\sqrt{x^2 + y^2}$$

each of which defines z as a function of x and y. Although Equation (1) does not explicitly express z as a function of x and y, it can be thought of as expressing z *implicitly* as one of two different functions of x and y. Note that the equation $z^2 - x^2 - y^2 = 0$ has the form $F(x, y, z) = 0$, where F is a function of three variables. Any equation of the form $F(x, y, z) = 0$ can be thought of as expressing z implicitly as one of a set of possible functions of x and y. Moreover, we can find $\partial z/\partial x$ and $\partial z/\partial y$ directly from the form $F(x, y, z) = 0$.

To find $\partial z/\partial x$ for

$$z^2 - x^2 - y^2 = 0 \tag{2}$$

---

[16] J. Silberman and G. Yochum, "The Role of Money in Determining Election Outcomes," *Social Science Quarterly,* 58, no. 4 (1978), 671–82.

[17] This section can be omitted without loss of continuity.

we first differentiate both sides of Equation (2) with respect to x while treating z as a function of x and y and treating y as a constant:

$$\frac{\partial}{\partial x}(z^2 - x^2 - y^2) = \frac{\partial}{\partial x}(0)$$

$$\frac{\partial}{\partial x}(z^2) - \frac{\partial}{\partial x}(x^2) - \frac{\partial}{\partial x}(y^2) = 0$$

$$2z\frac{\partial z}{\partial x} - 2x - 0 = 0$$

Because y is treated as a constant, $\frac{\partial y}{\partial x} = 0$.

Solving for $\partial z/\partial x$, we obtain

$$2z\frac{\partial z}{\partial x} = 2x$$

$$\frac{\partial z}{\partial x} = \frac{x}{z}$$

To find $\partial z/\partial y$, we differentiate both sides of Equation (2) with respect to y while treating z as a function of x and y and treating x as a constant:

$$\frac{\partial}{\partial y}(z^2 - x^2 - y^2) = \frac{\partial}{\partial y}(0)$$

$$2z\frac{\partial z}{\partial y} - 0 - 2y = 0 \qquad \left(\frac{\partial x}{\partial y} = 0\right)$$

$$2z\frac{\partial z}{\partial y} = 2y$$

Hence,

$$\frac{\partial z}{\partial y} = \frac{y}{z}$$

The method we used to find $\partial z/\partial x$ and $\partial z/\partial y$ is called *implicit partial differentiation*.

● **EXAMPLE 1 Implicit Partial Differentiation**

If $\dfrac{xz^2}{x+y} + y^2 = 0$, *evaluate* $\dfrac{\partial z}{\partial x}$ *when* $x = -1$, $y = 2$, *and* $z = 2$.

Solution: We treat z as a function of x and y and differentiate both sides of the equation with respect to x:

$$\frac{\partial}{\partial x}\left(\frac{xz^2}{x+y}\right) + \frac{\partial}{\partial x}(y^2) = \frac{\partial}{\partial x}(0)$$

Using the quotient rule for the first term on the left, we have

$$\frac{(x+y)\dfrac{\partial}{\partial x}(xz^2) - xz^2\dfrac{\partial}{\partial x}(x+y)}{(x+y)^2} + 0 = 0$$

Using the product rule for $\dfrac{\partial}{\partial x}(xz^2)$ gives

$$\frac{(x+y)\left[x\left(2z\dfrac{\partial z}{\partial x}\right) + z^2(1)\right] - xz^2(1)}{(x+y)^2} = 0$$

Solving for $\partial z/\partial x$, we obtain

$$2xz(x+y)\frac{\partial z}{\partial x} + z^2(x+y) - xz^2 = 0$$

$$\frac{\partial z}{\partial x} = \frac{xz^2 - z^2(x+y)}{2xz(x+y)} = -\frac{yz}{2x(x+y)} \qquad z \neq 0$$

Thus,

$$\left.\frac{\partial z}{\partial x}\right|_{(-1,2,2)} = 2$$

NOW WORK PROBLEM 13

● EXAMPLE 2   **Implicit Partial Differentiation**

If $se^{r^2+u^2} = u \ln(t^2 + 1)$, *determine* $\partial t/\partial u$.

Solution: We consider t as a function of r, s, and u. By differentiating both sides with respect to u while treating r and s as constants, we get

$$\frac{\partial}{\partial u}(se^{r^2+u^2}) = \frac{\partial}{\partial u}(u \ln(t^2 + 1))$$

$$2sue^{r^2+u^2} = u\frac{\partial}{\partial u}(\ln(t^2 + 1)) + \ln(t^2 + 1)\frac{\partial}{\partial u}(u) \qquad \text{(product rule)}$$

$$2sue^{r^2+u^2} = u\frac{2t}{t^2+1}\frac{\partial t}{\partial u} + \ln(t^2 + 1)$$

Therefore,

$$\frac{\partial t}{\partial u} = \frac{(t^2 + 1)(2sue^{r^2+u^2} - \ln(t^2 + 1))}{2ut}$$

NOW WORK PROBLEM 1

Problems 17.4

In Problems 1–11, find the indicated partial derivatives by the method of implicit partial differentiation.

*1. $2x^2 + 3y^2 + 5z^2 = 900$; $\partial z/\partial x$

2. $z^2 - 5x^2 + y^2 = 0$; $\partial z/\partial x$

3. $2z^3 - x^2 - 4y^2 = 0$; $\partial z/\partial y$

4. $3x^2 + y^2 + 2z^3 = 9$; $\partial z/\partial y$

5. $x^2 - 2y - z^2 + x^2yz^2 = 20$; $\partial z/\partial x$

6. $z^3 + 2x^2z^2 - xy = 0$; $\partial z/\partial x$

7. $e^x + e^y + e^z = 10$; $\partial z/\partial y$

8. $xyz + 3y^3x^2 - \ln z^3 = 0$; $\partial z/\partial y$

9. $\ln(z) + 9z - xy = 1$; $\partial z/\partial x$

10. $\ln x + \ln y - \ln z = e^y$; $\partial z/\partial x$

11. $(z^2 + 6xy)\sqrt{x^3 + 5} = 2$; $\partial z/\partial y$

In Problems 12–20, evaluate the indicated partial derivatives for the given values of the variables.

12. $xz + xyz - 5 = 0$; $\partial z/\partial x, x = 1, y = 4, z = 1$

*13. $3xz^2 + 2yz^2 - 7x^4y = 3$; $\partial z/\partial x, x = 1, y = 0, z = 1$

14. $e^{zx} = xyz$; $\partial z/\partial y, x = 1, y = -e^{-1}, z = -1$

15. $e^{yz} = -xyz$; $\partial z/\partial x, x = -e^2/2, y = 1, z = 2$

16. $\sqrt{xz + y^2} - xy = 0$; $\partial z/\partial y, x = 2, y = 2, z = 6$

17. $\ln z = 4x + y$; $\partial z/\partial x, x = 5, y = -20, z = 1$

18. $\dfrac{2r^2s^2}{s^2 + t^2} = t$; $\partial r/\partial t, r = 1, s = 1, t = 1$

19. $\dfrac{s^2 + t^2}{rs} = 10$; $\partial t/\partial r, r = 1, s = 2, t = 4$

20. $\ln(x + y + z) + xyz = ze^{x+y+z}$; $\partial z/\partial x, x = 0, y = 1, z = 0$

21. **Joint-Cost Function**   A joint-cost function is defined implicitly by the equation

$$c + \sqrt{c} = 12 + q_A\sqrt{9 + q_B^2}$$

where c denotes the total cost (in dollars) for producing q_A units of product A and q_B units of product B.

(a) If $q_A = 6$ and $q_B = 4$, find the corresponding value of c.

(b) Determine the marginal costs with respect to q_A and q_B when $q_A = 6$ and $q_B = 4$.

OBJECTIVE

To compute higher-order partial derivatives.

17.5 Higher-Order Partial Derivatives

If $z = f(x, y)$, then not only is z a function of x and y, but also f_x and f_y are each functions of x and y, which may themselves have partial derivatives. If we can differentiate f_x and f_y, we obtain **second-order partial derivatives** of f. Symbolically,

$$f_{xx} \text{ means } (f_x)_x \quad f_{xy} \text{ means } (f_x)_y$$

$$f_{yx} \text{ means } (f_y)_x \quad f_{yy} \text{ means } (f_y)_y$$

In terms of ∂-notation,

$$\frac{\partial^2 z}{\partial x^2} \text{ means } \frac{\partial}{\partial x}\left(\frac{\partial z}{\partial x}\right) \qquad \frac{\partial^2 z}{\partial y\,\partial x} \text{ means } \frac{\partial}{\partial y}\left(\frac{\partial z}{\partial x}\right)$$

$$\frac{\partial^2 z}{\partial x\,\partial y} \text{ means } \frac{\partial}{\partial x}\left(\frac{\partial z}{\partial y}\right) \qquad \frac{\partial^2 z}{\partial y^2} \text{ means } \frac{\partial}{\partial y}\left(\frac{\partial z}{\partial y}\right)$$

CAUTION

For $z = f(x, y)$, $f_{xy} = \partial^2 z/\partial y\,\partial x$.

Note that to find f_{xy}, we first differentiate f with respect to x. For $\partial^2 z/\partial x\,\partial y$, we first differentiate with respect to y.

We can extend our notation beyond second-order partial derivatives. For example, f_{xxy} ($= \partial^3 z/\partial y\,\partial x^2$) is a third-order partial derivative of f, namely, the partial derivative of f_{xx} ($= \partial^2 z/\partial x^2$) with respect to y. The generalization of higher-order partial derivatives to functions of more than two variables should be obvious.

● EXAMPLE 1   **Second-Order Partial Derivatives**

Find the four second-order partial derivatives of $f(x, y) = x^2 y + x^2 y^2$.

Solution: Since

$$f_x(x, y) = 2xy + 2xy^2$$

we have

$$f_{xx}(x, y) = \frac{\partial}{\partial x}(2xy + 2xy^2) = 2y + 2y^2$$

and

$$f_{xy}(x, y) = \frac{\partial}{\partial y}(2xy + 2xy^2) = 2x + 4xy$$

Also, since

$$f_y(x, y) = x^2 + 2x^2 y$$

we have

$$f_{yy}(x, y) = \frac{\partial}{\partial y}(x^2 + 2x^2 y) = 2x^2$$

and

$$f_{yx}(x, y) = \frac{\partial}{\partial x}(x^2 + 2x^2 y) = 2x + 4xy$$

NOW WORK PROBLEM 1 ◖◗◗

The derivatives f_{xy} and f_{yx} are called **mixed partial derivatives.** Observe in Example 1 that $f_{xy}(x, y) = f_{yx}(x, y)$. Under suitable conditions, mixed partial derivatives of a function are equal; that is, the order of differentiation is of no concern. You may assume that this is the case for all the functions that we consider.

● EXAMPLE 2   **Mixed Partial Derivative**

Find the value of $\left.\dfrac{\partial^3 w}{\partial z\,\partial y\,\partial x}\right|_{(1,2,3)}$ if $w = (2x + 3y + 4z)^3$.

Solution:

$$\frac{\partial w}{\partial x} = 3(2x + 3y + 4z)^2 \frac{\partial}{\partial x}(2x + 3y + 4z)$$

$$= 6(2x + 3y + 4z)^2$$

$$\frac{\partial^2 w}{\partial y\,\partial x} = 6 \cdot 2(2x + 3y + 4z)\frac{\partial}{\partial y}(2x + 3y + 4z)$$

$$= 36(2x + 3y + 4z)$$

$$\frac{\partial^3 w}{\partial z\,\partial y\,\partial x} = 36 \cdot 4 = 144$$

Thus,

$$\frac{\partial^3 w}{\partial z\, \partial y\, \partial x}\bigg|_{(1,2,3)} = 144$$

NOW WORK PROBLEM 3

● EXAMPLE 3 **Second-Order Partial Derivative of an Implicit Function**[18]

Determine $\dfrac{\partial^2 z}{\partial x^2}$ *if* $z^2 = xy$.

Solution: By implicit differentiation, we first determine $\partial z/\partial x$:

$$\frac{\partial}{\partial x}(z^2) = \frac{\partial}{\partial x}(xy)$$

$$2z\frac{\partial z}{\partial x} = y$$

$$\frac{\partial z}{\partial x} = \frac{y}{2z} \qquad z \neq 0$$

Differentiating both sides with respect to x, we obtain

$$\frac{\partial}{\partial x}\left(\frac{\partial z}{\partial x}\right) = \frac{\partial}{\partial x}\left(\frac{1}{2}yz^{-1}\right)$$

$$\frac{\partial^2 z}{\partial x^2} = -\frac{1}{2}yz^{-2}\frac{\partial z}{\partial x}$$

Substituting $y/(2z)$ for $\partial z/\partial x$, we have

$$\frac{\partial^2 z}{\partial x^2} = -\frac{1}{2}yz^{-2}\left(\frac{y}{2z}\right) = -\frac{y^2}{4z^3} \qquad z \neq 0$$

NOW WORK PROBLEM 23

Problems 17.5

In Problems 1–10, find the indicated partial derivatives.

*1. $f(x, y) = 6xy^2;$ $f_x(x, y),\ f_{xy}(x, y),\ f_{yx}(x, y)$

2. $f(x, y) = 2x^3y^2 + 6x^2y^3 - 3xy;$ $f_x(x, y),\ f_{xx}(x, y)$

*3. $f(x, y) = 7x^2 + 3y;$ $f_y(x, y),\ f_{yy}(x, y),\ f_{yyx}(x, y)$

4. $f(x, y) = (x^2 + xy + y^2)(x^2 + xy + 1);$ $f_x(x, y),\ f_{xy}(x, y)$

5. $f(x, y) = 9e^{2xy};$ $f_y(x, y),\ f_{yx}(x, y),\ f_{yxy}(x, y)$

6. $f(x, y) = \ln(x^2 + y^2) + 2;$ $f_x(x, y),\ f_{xx}(x, y),\ f_{xy}(x, y)$

7. $f(x, y) = (x + y)^2(xy);$ $f_x(x, y),\ f_y(x, y),\ f_{xx}(x, y),$
   $f_{yy}(x, y)$

8. $f(x, y, z) = x^2y^3z^4;$ $f_x(x, y, z),\ f_{xz}(x, y, z),\ f_{zx}(x, y, z)$

9. $z = e^{\sqrt{x^2+y^2}};$ $\dfrac{\partial z}{\partial y},\ \dfrac{\partial^2 z}{\partial y^2}$

10. $z = \dfrac{\ln(x^2 + 5)}{y};$ $\dfrac{\partial z}{\partial x},\ \dfrac{\partial^2 z}{\partial y\, \partial x}$

In Problems 11–16, find the indicated value.

11. If $f(x, y, z) = 7$, find $f_{yxx}(4, 3, -2)$.

12. If $f(x, y, z) = z^2(3x^2 - 4xy^3)$, find $f_{xyz}(1, 2, 3)$.

13. If $f(l, k) = 3l^3k^6 - 2l^2k^7$, find $f_{klk}(2, 1)$.

14. If $f(x, y) = 3x^3y^2 + xy - x^2y^2$, find $f_{xxy}(5, 1)$ and $f_{xyx}(5, 1)$.

15. If $f(x, y) = y^2e^x + \ln(xy)$, find $f_{xyy}(1, 1)$.

16. If $f(x, y) = x^3 - 6xy^2 + x^2 - y^3$, find $f_{xy}(1, -1)$.

17. **Cost Function** Suppose the cost c of producing q_A units of product A and q_B units of product B is given by

$$c = (3q_A^2 + q_B^3 + 4)^{1/3}$$

and the coupled demand functions for the products are given by

$$q_A = 10 - p_A + p_B^2$$

and

$$q_B = 20 + p_A - 11p_B$$

Find the value of

$$\frac{\partial^2 c}{\partial q_A\, \partial q_B}$$

when $p_A = 25$ and $p_B = 4$.

18. For $f(x, y) = x^4y^4 + 3x^3y^2 - 7x + 4$, show that

$$f_{xyx}(x, y) = f_{xxy}(x, y)$$

19. For $f(x, y) = 8x^3 + 2x^2y^2 + 5y^4$, show that

$$f_{xy}(x, y) = f_{yx}(x, y)$$

---

[18]Omit if Section 17.4 was not covered.

20. For $f(x, y) = e^{xy}$, show that

$$f_{xx}(x, y) + f_{xy}(x, y) + f_{yx}(x, y) + f_{yy}(x, y)$$
$$= f(x, y)((x + y)^2 + 2)$$

21. For $z = \ln(x^2 + y^2)$, show that $\dfrac{\partial^2 z}{\partial x^2} + \dfrac{\partial^2 z}{\partial y^2} = 0$.

[19]**22.** If $3z^2 - 2x^3 - 4y^4 = 0$, find $\dfrac{\partial^2 z}{\partial x^2}$.

*[19]**23.** If $z^2 - 3x^2 + y^2 = 0$, find $\dfrac{\partial^2 z}{\partial y^2}$.

[19]**24.** If $2z^2 = x^2 + 2xy + xz$, find $\dfrac{\partial^2 z}{\partial x \, \partial y}$.

To show how to find the partial derivative of a function of functions by using the chain rule.

17.6 Chain Rule[20]

Suppose a manufacturer of two related products A and B has a joint-cost function given by

$$c = f(q_A, q_B)$$

where c is the total cost of producing quantities q_A and q_B of A and B, respectively. Furthermore, suppose the demand functions for the products are

$$q_A = g(p_A, p_B) \quad \text{and} \quad q_B = h(p_A, p_B)$$

where p_A and p_B are the prices per unit of A and B, respectively. Since c is a function of q_A and q_B, and since both q_A and q_B are themselves functions of p_A and p_B, c can be viewed as a function of p_A and p_B. (Appropriately, the variables q_A and q_B are called *intermediate variables* of c.) Consequently, we should be able to determine $\partial c / \partial p_A$, the rate of change of total cost with respect to the price of A. One way to do this is to substitute the expressions $g(p_A, p_B)$ and $h(p_A, p_B)$ for q_A and q_B, respectively, into $c = f(q_A, q_B)$. Then c is a function of p_A and p_B, and we can differentiate c with respect to p_A directly. This approach has some drawbacks—especially when f, g, or h is given by a complicated expression. Another way to approach the problem would be to use the chain rule (actually *a* chain rule), which we now state without proof.

Chain Rule

Let $z = f(x, y)$, where both x and y are functions of r and s given by $x = x(r, s)$ and $y = y(r, s)$. If f, x, and y have continuous partial derivatives, then z is a function of r and s, and

$$\frac{\partial z}{\partial r} = \frac{\partial z}{\partial x}\frac{\partial x}{\partial r} + \frac{\partial z}{\partial y}\frac{\partial y}{\partial r}$$

and

$$\frac{\partial z}{\partial s} = \frac{\partial z}{\partial x}\frac{\partial x}{\partial s} + \frac{\partial z}{\partial y}\frac{\partial y}{\partial s}$$

Note that in the chain rule, the number of intermediate variables of z (two) is the same as the number of terms that compose each of $\partial z / \partial r$ and $\partial z / \partial s$.

Returning to the original situation concerning the manufacturer, we see that if f, q_A, and q_B have continuous partial derivatives, then, by the chain rule,

$$\frac{\partial c}{\partial p_A} = \frac{\partial c}{\partial q_A}\frac{\partial q_A}{\partial p_A} + \frac{\partial c}{\partial q_B}\frac{\partial q_B}{\partial p_A}$$

● EXAMPLE 1 **Rate of Change of Cost**

For a manufacturer of cameras and film, the total cost c of producing q_C cameras and q_F units of film is given by

$$c = 30q_C + 0.015q_C q_F + q_F + 900$$

---

[19]Omit if Section 17.4 was not covered.

[20]This section can be omitted without loss of continuity.

The demand functions for the cameras and film are given by

$$q_C = \frac{9000}{p_C\sqrt{p_F}} \quad \text{and} \quad q_F = 2000 - p_C - 400p_F$$

where p_C is the price per camera and p_F is the price per unit of film. Find the rate of change of total cost with respect to the price of the camera when $p_C = 50$ and $p_F = 2$.

Solution: We must first determine $\partial c / \partial p_C$. By the chain rule,

$$\frac{\partial c}{\partial p_C} = \frac{\partial c}{\partial q_C}\frac{\partial q_C}{\partial p_C} + \frac{\partial c}{\partial q_F}\frac{\partial q_F}{\partial p_C}$$

$$= (30 + 0.015q_F)\left[\frac{-9000}{p_C^2\sqrt{p_F}}\right] + (0.015q_C + 1)(-1)$$

When $p_C = 50$ and $p_F = 2$, then $q_C = 90\sqrt{2}$ and $q_F = 1150$. Substituting these values into $\partial c / \partial p_C$ and simplifying, we have

$$\left.\frac{\partial c}{\partial p_C}\right|_{\substack{p_C=50 \\ p_F=2}} \approx -123.2$$

NOW WORK PROBLEM 1 ●●●

The chain rule can be extended. For example, suppose $z = f(v, w, x, y)$ and v, w, x, and y are all functions of r, s, and t. Then, if certain conditions of continuity are assumed, z is a function of r, s, and t, and we have

$$\frac{\partial z}{\partial r} = \frac{\partial z}{\partial v}\frac{\partial v}{\partial r} + \frac{\partial z}{\partial w}\frac{\partial w}{\partial r} + \frac{\partial z}{\partial x}\frac{\partial x}{\partial r} + \frac{\partial z}{\partial y}\frac{\partial y}{\partial r}$$

$$\frac{\partial z}{\partial s} = \frac{\partial z}{\partial v}\frac{\partial v}{\partial s} + \frac{\partial z}{\partial w}\frac{\partial w}{\partial s} + \frac{\partial z}{\partial x}\frac{\partial x}{\partial s} + \frac{\partial z}{\partial y}\frac{\partial y}{\partial s}$$

and

$$\frac{\partial z}{\partial t} = \frac{\partial z}{\partial v}\frac{\partial v}{\partial t} + \frac{\partial z}{\partial w}\frac{\partial w}{\partial t} + \frac{\partial z}{\partial x}\frac{\partial x}{\partial t} + \frac{\partial z}{\partial y}\frac{\partial y}{\partial t}$$

Observe that the number of intermediate variables of z (four) is the same as the number of terms that form each of $\partial z / \partial r$, $\partial z / \partial s$, and $\partial z / \partial t$.

Now consider the situation where $z = f(x, y)$ such that $x = x(t)$ and $y = y(t)$. Then

$$\frac{dz}{dt} = \frac{\partial z}{\partial x}\frac{dx}{dt} + \frac{\partial z}{\partial y}\frac{dy}{dt}$$

Use the partial derivative symbols and the ordinary derivative symbols appropriately.

Here we use the symbol dz/dt rather than $\partial z/\partial t$, since z can be considered a function of *one* variable t. Likewise, the symbols dx/dt and dy/dt are used rather than $\partial x/\partial t$ and $\partial y/\partial t$. As is typical, the number of terms that compose dz/dt equals the number of intermediate variables of z. Other situations would be treated in a similar way.

●**EXAMPLE 2   Chain Rule**

a. If $w = f(x, y, z) = 3x^2y + xyz - 4y^2z^3$, *where*

$$x = 2r - 3s \quad y = 6r + s \quad z = r - s$$

determine $\partial w / \partial r$ and $\partial w / \partial s$.

Solution: Since x, y, and z are functions of r and s, then, by the chain rule,

$$\frac{\partial w}{\partial r} = \frac{\partial w}{\partial x}\frac{\partial x}{\partial r} + \frac{\partial w}{\partial y}\frac{\partial y}{\partial r} + \frac{\partial w}{\partial z}\frac{\partial z}{\partial r}$$

$$= (6xy + yz)(2) + (3x^2 + xz - 8yz^3)(6) + (xy - 12y^2z^2)(1)$$

$$= x(18x + 13y + 6z) + 2yz(1 - 24z^2 - 6yz)$$

Also,

$$\frac{\partial w}{\partial s} = \frac{\partial w}{\partial x}\frac{\partial x}{\partial s} + \frac{\partial w}{\partial y}\frac{\partial y}{\partial s} + \frac{\partial w}{\partial z}\frac{\partial z}{\partial s}$$

$$= (6xy + yz)(-3) + (3x^2 + xz - 8yz^3)(1) + (xy - 12y^2z^2)(-1)$$

$$= x(3x - 19y + z) - yz(3 + 8z^2 - 12yz)$$

b. *If* $z = \dfrac{x + e^y}{y}$, *where* $x = rs + se^{rt}$ *and* $y = 9 + rt$, *evaluate* $\partial z/\partial s$ *when* $r = -2, s = 5$, *and* $t = 4$.

Solution: Since x and y are functions of r, s, and t (note that we can write $y = 9 + rt + 0 \cdot s$), by the chain rule,

$$\frac{\partial z}{\partial s} = \frac{\partial z}{\partial x}\frac{\partial x}{\partial s} + \frac{\partial z}{\partial y}\frac{\partial y}{\partial s}$$

$$= \left(\frac{1}{y}\right)(r + e^{rt}) + \frac{\partial z}{\partial y} \cdot (0) = \frac{r + e^{rt}}{y}$$

If $r = -2, s = 5$, and $t = 4$, then $y = 1$. Thus,

$$\left.\frac{\partial z}{\partial s}\right|_{\substack{r=-2 \\ s=5 \\ t=4}} = \frac{-2 + e^{-8}}{1} = -2 + e^{-8}$$

NOW WORK PROBLEM 13 ◖◗◗

● EXAMPLE 3   **Chain Rule**

a. *Determine* $\partial y/\partial r$ *if* $y = x^2 \ln(x^4 + 6)$ *and* $x = (r + 3s)^6$.

Solution: By the chain rule,

$$\frac{\partial y}{\partial r} = \frac{dy}{dx}\frac{\partial x}{\partial r}$$

$$= \left[x^2 \cdot \frac{4x^3}{x^4 + 6} + 2x \cdot \ln(x^4 + 6)\right]\left[6(r + 3s)^5\right]$$

$$= 12x(r + 3s)^5 \left[\frac{2x^4}{x^4 + 6} + \ln(x^4 + 6)\right]$$

b. *Given that* $z = e^{xy}$, $x = r - 4s$, *and* $y = r - s$, *find* $\partial z/\partial r$ *in terms of* r *and* s.

Solution:

$$\frac{\partial z}{\partial r} = \frac{\partial z}{\partial x}\frac{\partial x}{\partial r} + \frac{\partial z}{\partial y}\frac{\partial y}{\partial r}$$

$$= (ye^{xy})(1) + (xe^{xy})(1)$$

$$= (x + y)e^{xy}$$

Since $x = r - 4s$ and $y = r - s$,

$$\frac{\partial z}{\partial r} = [(r - 4s) + (r - s)]e^{(r-4s)(r-s)}$$

$$= (2r - 5s)e^{r^2 - 5rs + 4s^2}$$

NOW WORK PROBLEM 15 ◖◗◗

Problems 17.6

In Problems 1–12, find the indicated derivatives by using the chain rule.

*1. $z = 5x + 3y, x = 2r + 3s, y = r - 2s$; $\partial z/\partial r, \partial z/\partial s$

2. $z = 2x^2 + 3xy + 2y^2, x = r^2 - s^2, y = r^2 + s^2$; $\partial z/\partial r, \partial z/\partial s$

3. $z = e^{x+y}, x = t^2 + 3, y = \sqrt{t^3}$; dz/dt

4. $z = \sqrt{8x + y}, x = t^2 + 3t + 4$,
$y = t^3 + 4$; dz/dt

5. $w = x^2 z^2 + xyz + yz^2$, $x = 5t$,

   $y = 2t + 3$, $z = 6 - t$; dw/dt

6. $w = \ln(x^2 + y^2 + z^2)$,

   $x = 2 - 3t$, $y = t^2 + 3$, $z = 4 - t$; dw/dt

7. $z = (x^2 + xy^2)^3$, $x = r + s + t$,

   $y = 2r - 3s + 8t$; $\partial z/\partial t$

8. $z = \sqrt{x^2 + y^2}$, $x = r^2 + s - t$,

   $y = r - s + t$; $\partial z/\partial r$

9. $w = x^2 + xyz + z^2$, $x = r^2 - s^2$,

   $y = rs$, $z = r^2 + s^2$; $\partial w/\partial s$

10. $w = e^{xyz}$, $x = r^2 s^3$, $y = \ln(r - s)$, $z = \sqrt{rs^2}$; $\partial w/\partial r$

11. $y = x^2 - 7x + 5$, $x = 19rs + 2s^2 t^2$; $\partial y/\partial r$

12. $y = 4 - x^2$, $x = 2r + 3s - 4t$; $\partial y/\partial t$

*13. If $z = (4x + 3y)^3$, where $x = r^2 s$ and $y = r - 2s$, evaluate $\partial z/\partial r$ when $r = 0$ and $s = 1$.

14. If $z = \sqrt{2x + 3y}$, where $x = 3t + 5$ and $y = t^2 + 2t + 1$, evaluate dz/dt when $t = 1$.

*15. If $w = e^{2x+3y}(x^2 + 4z^2)$, where $x = rs$, $y = 2s - 3r$, and $z = r + s$, evaluate $\partial w/\partial s$ when $r = 1$ and $s = 0$.

16. If $y = x/(x - 5)$, where $x = 2t^2 - 3rs - r^2 t$, evaluate $\partial y/\partial t$ when $r = 0$, $s = 2$, and $t = -1$.

17. Cost Function   Suppose the cost c of producing q_A units of product A and q_B units of product B is given by

$$c = (3q_A^2 + q_B^3 + 4)^{1/3}$$

and the coupled demand functions for the products are

given by

$$q_A = 10 - p_A + p_B^2$$

and

$$q_B = 20 + p_A - 11p_B$$

Use a chain rule to evaluate $\dfrac{\partial c}{\partial p_A}$ and $\dfrac{\partial c}{\partial p_B}$ when $p_A = 25$ and $p_B = 4$.

18. Suppose $w = f(x, y)$, where $x = g(t)$ and $y = h(t)$.

   (a) State a chain rule that gives dw/dt.

   (b) Suppose $h(t) = t$, so that $w = f(x, t)$ where $x = g(t)$. Use part (a) to find dw/dt and simplify your answer.

19. (a) Suppose w is a function of x and y, where both x and y are functions of s and t. State a chain rule that expresses $\partial w/\partial t$ in terms of derivatives of these functions.

   (b) Let $w = 2x^2 \ln|3x - 5y|$, where $x = s\sqrt{t^2 + 2}$ and $y = t - 3e^{2-s}$. Use part (a) to evaluate $\partial w/\partial t$ when $s = 1$ and $t = 0$.

20. Production Function   In considering a production function $P = f(l, k)$, where l is labor input and k is capital input, Fon, Boulier, and Goldfarb[21] assume that $l = Lg(h)$, where L is the number of workers, h is the number of hours per day per worker, and $g(h)$ is a labor effectiveness function. In maximizing profit p given by

$$p = aP - whL$$

where a is the price per unit of output and w is the hourly wage per worker, Fon, Boulier, and Goldfarb determine $\partial p/\partial L$ and $\partial p/\partial h$. Assume that k is independent of L and h, and determine these partial derivatives.

17.7 Maxima and Minima for Functions of Two Variables

● ● ●

OBJECTIVE

To discuss relative maxima and relative minima, to find critical points, and to apply the second-derivative test for a function of two variables.

We now extend the notion of relative maxima and minima (or relative extrema) to functions of two variables.

> **DEFINITION**
>
> A function $z = f(x, y)$ is said to have a ***relative maximum*** at the point (a, b)—that is, when $x = a$ and $y = b$—if, for all points (x, y) in the plane that are sufficiently close to (a, b), we have
>
> $$f(a, b) \geq f(x, y) \qquad (1)$$
>
> For a ***relative minimum,*** we replace $\geq$ by $\leq$ in Equation (1).

To say that $z = f(x, y)$ has a relative maximum at (a, b) means, geometrically, that the point $(a, b, f(a, b))$ on the graph of f is higher than (or is as high as) all other points on the surface that are "near" $(a, b, f(a, b))$. In Figure 17.12(a), f has a relative maximum at (a, b). Similarly, the function f in Figure 17.12(b) has a relative minimum when $x = y = 0$, which corresponds to a *low* point on the surface.

Recall that in locating extrema for a function $y = f(x)$ of one variable, we examine those values of x in the domain of f for which $f'(x) = 0$ or $f'(x)$ does not exist. For functions of two (or more) variables, a similar procedure is followed.

---

[21] V. Fon, B. L. Boulier, and R. S. Goldfarb, "The Firm's Demand for Daily Hours of Work: Some Implications," *Atlantic Economic Journal,* XIII, no. 1 (1985), 36–42.

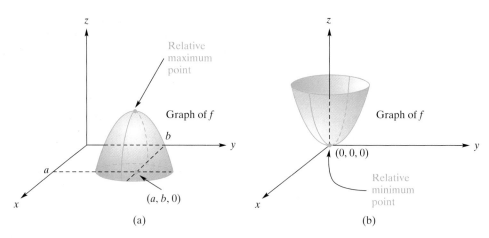

FIGURE 17.12 Relative extrema.

However, for the functions that concern us, extrema will not occur where a derivative does not exist, and such situations will be excluded from consideration.

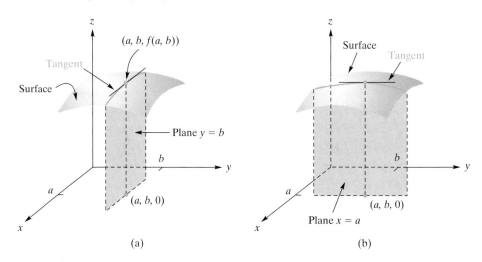

FIGURE 17.13 At relative extremum, $f_x(x, y) = 0$ and $f_y(x, y) = 0$.

Suppose $z = f(x, y)$ has a relative maximum at (a, b), as indicated in Figure 17.13(a). Then the curve where the plane $y = b$ intersects the surface must have a relative maximum when $x = a$. Hence, the slope of the tangent line to the surface in the x-direction must be 0 at (a, b). Equivalently, $f_x(x, y) = 0$ at (a, b). Similarly, on the curve where the plane $x = a$ intersects the surface [Figure 17.13(b)], there must be a relative maximum when $y = b$. Thus, in the y-direction, the slope of the tangent to the surface must be 0 at (a, b). Equivalently, $f_y(x, y) = 0$ at (a, b). Since a similar discussion applies to a relative minimum, we can combine these results as follows:

CAUTION

Rule 1 does not imply that there must be an extremum at a critical point. Just as in the case of functions of one variable, a critical point can give rise to a relative maximum, a relative minimum, or neither. A critical point is only a *candidate* for a relative extremum.

RULE 1

If $z = f(x, y)$ has a relative maximum or minimum at (a, b), and if both f_x and f_y are defined for all points close to (a, b), it is necessary that (a, b) be a solution of the system

$$\begin{cases} f_x(x, y) = 0 \\ f_y(x, y) = 0 \end{cases}$$

A point (a, b) for which $f_x(a, b) = f_y(a, b) = 0$ is called a **critical point** of f. Thus, from Rule 1, we infer that, to locate relative extrema for a function, we should examine its critical points.

Two additional comments are in order: First, Rule 1, as well as the notion of a critical point, can be extended to functions of more than two variables. For example, to locate possible extrema for $w = f(x, y, z)$, we would examine those points for which $w_x = w_y = w_z = 0$. Second, for a function whose domain is restricted, a thorough examination for absolute extrema would include a consideration of boundary points.

● EXAMPLE 1 **Finding Critical Points**

Find the critical points of the following functions.

a. $f(x, y) = 2x^2 + y^2 - 2xy + 5x - 3y + 1$.

Solution: Since $f_x(x, y) = 4x - 2y + 5$ and $f_y(x, y) = 2y - 2x - 3$, we solve the system

$$\begin{cases} 4x - 2y + 5 = 0 \\ -2x + 2y - 3 = 0 \end{cases}$$

This gives $x = -1$ and $y = \frac{1}{2}$. Thus, $\left(-1, \frac{1}{2}\right)$ is the only critical point.

b. $f(l, k) = l^3 + k^3 - lk$.

Solution:

$$\begin{cases} f_l(l, k) = 3l^2 - k = 0 & \text{(2)} \\ f_k(l, k) = 3k^2 - l = 0 & \text{(3)} \end{cases}$$

From Equation (2), $k = 3l^2$. Substituting for k in Equation (3) gives

$$0 = 27l^4 - l = l(27l^3 - 1)$$

Hence, either $l = 0$ or $l = \frac{1}{3}$. If $l = 0$, then $k = 0$; if $l = \frac{1}{3}$, then $k = \frac{1}{3}$. The critical points are therefore $(0, 0)$ and $\left(\frac{1}{3}, \frac{1}{3}\right)$.

c. $f(x, y, z) = 2x^2 + xy + y^2 + 100 - z(x + y - 100)$.

Solution: Solving the system

$$\begin{cases} f_x(x, y, z) = 4x + y - z = 0 \\ f_y(x, y, z) = x + 2y - z = 0 \\ f_z(x, y, z) = -x - y + 100 = 0 \end{cases}$$

gives the critical point $(25, 75, 175)$, as you should verify.

NOW WORK PROBLEM 1 ●●

● EXAMPLE 2 **Finding Critical Points**

Find the critical points of

$$f(x, y) = x^2 - 4x + 2y^2 + 4y + 7$$

Solution: We have $f_x(x, y) = 2x - 4$ and $f_y(x, y) = 4y + 4$. The system

$$\begin{cases} 2x - 4 = 0 \\ 4y + 4 = 0 \end{cases}$$

gives the critical point $(2, -1)$. Observe that we can write the given function as

$$f(x, y) = x^2 - 4x + 4 + 2(y^2 + 2y + 1) + 1$$
$$= (x - 2)^2 + 2(y + 1)^2 + 1$$

and $f(2, -1) = 1$. Clearly, if $(x, y) \neq (2, -1)$, then $f(x, y) > 1$. Hence, a relative minimum occurs at $(2, -1)$. Moreover, there is an *absolute minimum* at $(2, -1)$, since $f(x, y) > f(2, -1)$ for *all* $(x, y) \neq (2, -1)$.

NOW WORK PROBLEM 3 ●●

Although in Example 2 we were able to show that the critical point gave rise to a relative extremum, in many cases this is not so easy to do. There is, however, a second-derivative test that gives conditions under which a critical point will be a relative maximum or minimum. We state it now, omitting the proof.

RULE 2   Second-Derivative Test for Functions of Two Variables

Suppose $z = f(x, y)$ has continuous partial derivatives f_{xx}, f_{yy}, and f_{xy} at all points (x, y) near a critical point (a, b). Let D be the function defined by

$$D(x, y) = f_{xx}(x, y) f_{yy}(x, y) - (f_{xy}(x, y))^2$$

Then

1. if $D(a, b) > 0$ and $f_{xx}(a, b) < 0$, then f has a relative maximum at (a, b);
2. if $D(a, b) > 0$ and $f_{xx}(a, b) > 0$, then f has a relative minimum at (a, b);
3. if $D(a, b) < 0$, then f has a *saddle point* at (a, b) (see Example 4);
4. if $D(a, b) = 0$, then no conclusion about an extremum at (a, b) can be drawn, and further analysis is required.

● EXAMPLE 3   **Applying the Second-Derivative Test**

Examine $f(x, y) = x^3 + y^3 - xy$ for relative maxima or minima by using the second-derivative test.

Solution: First we find critical points:

$$f_x(x, y) = 3x^2 - y \quad f_y(x, y) = 3y^2 - x$$

In the same manner as in Example 1(b), solving $f_x(x, y) = f_y(x, y) = 0$ gives the critical points $(0, 0)$ and $\left(\frac{1}{3}, \frac{1}{3}\right)$.
Now,

$$f_{xx}(x, y) = 6x \qquad f_{yy}(x, y) = 6y \qquad f_{xy}(x, y) = -1$$

Thus,

$$D(x, y) = (6x)(6y) - (-1)^2 = 36xy - 1$$

Since $D(0, 0) = 36(0)(0) - 1 = -1 < 0$, there is no relative extremum at $(0, 0)$. Also, since $D\left(\frac{1}{3}, \frac{1}{3}\right) = 36\left(\frac{1}{3}\right)\left(\frac{1}{3}\right) - 1 = 3 > 0$ and $f_{xx}\left(\frac{1}{3}, \frac{1}{3}\right) = 6\left(\frac{1}{3}\right) = 2 > 0$, there is a relative minimum at $\left(\frac{1}{3}, \frac{1}{3}\right)$. At this point, the value of the function is

$$f\left(\tfrac{1}{3}, \tfrac{1}{3}\right) = \left(\tfrac{1}{3}\right)^3 + \left(\tfrac{1}{3}\right)^3 - \left(\tfrac{1}{3}\right)\left(\tfrac{1}{3}\right) = -\tfrac{1}{27}$$

NOW WORK PROBLEM 7   ●●●

● EXAMPLE 4   **A Saddle Point**

Examine $f(x, y) = y^2 - x^2$ for relative extrema.

Solution: Solving

$$f_x(x, y) = -2x = 0 \quad \text{and} \quad f_y(x, y) = 2y = 0$$

we get the critical point $(0, 0)$. Now we apply the second-derivative test. At $(0, 0)$, and indeed at any point,

$$f_{xx}(x, y) = -2 \quad f_{yy}(x, y) = 2 \quad f_{xy}(x, y) = 0$$

Because $D(0, 0) = (-2)(2) - (0)^2 = -4 < 0$, no relative extremum exists at $(0, 0)$. A sketch of $z = f(x, y) = y^2 - x^2$ appears in Figure 17.14. Note that, for the surface

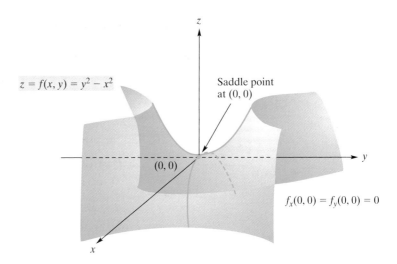

FIGURE 17.14   Saddle point.

The surface in Figure 17.14 is called a hyperbolic paraboloid.

curve cut by the plane $y = 0$, there is a *maximum* at $(0, 0)$; but for the surface curve cut by the plane $x = 0$, there is a *minimum* at $(0, 0)$. Thus, on the *surface*, no relative extremum can exist at the origin, although $(0, 0)$ is a critical point. Around the origin the curve is saddle shaped, and $(0, 0)$ is called a *saddle point* of f.

NOW WORK PROBLEM 11

● EXAMPLE 5   Finding Relative Extrema

Examine $f(x, y) = x^4 + (x - y)^4$ *for relative extrema.*

Solution: If we set

$$f_x(x, y) = 4x^3 + 4(x - y)^3 = 0 \qquad (4)$$

and

$$f_y(x, y) = -4(x - y)^3 = 0 \qquad (5)$$

then, from Equation (5), we have $x - y = 0$, or $x = y$. Substituting into Equation (4) gives $4x^3 = 0$, or $x = 0$. Thus, $x = y = 0$, and $(0, 0)$ is the only critical point. At $(0, 0)$,

$$f_{xx}(x, y) = 12x^2 + 12(x - y)^2 = 0$$
$$f_{yy}(x, y) = 12(x - y)^2 = 0$$

and

$$f_{xy}(x, y) = -12(x - y)^2 = 0$$

Hence, $D(0, 0) = 0$, and the second-derivative test gives no information. However, for all $(x, y) \neq (0, 0)$, we have $f(x, y) > 0$, whereas $f(0, 0) = 0$. Therefore, at $(0, 0)$ the graph of f has a low point, and we conclude that f has a relative (and absolute) minimum at $(0, 0)$.

NOW WORK PROBLEM 13

Applications

In many situations involving functions of two variables, and especially in their applications, the nature of the given problem is an indicator of whether a critical point is in fact a relative (or absolute) maximum or a relative (or absolute) minimum. In such cases, the second-derivative test is not needed. Often, in mathematical studies of applied problems, the appropriate second-order conditions are assumed to hold.

● EXAMPLE 6   **Maximizing Output**

Let P be a production function given by

$$P = f(l, k) = 0.54l^2 - 0.02l^3 + 1.89k^2 - 0.09k^3$$

where l and k are the amounts of labor and capital, respectively, and P is the quantity of output produced. Find the values of l and k that maximize P.

Solution: To find the critical points, we solve the system $P_l = 0$ and $P_k = 0$;

$$P_l = 1.08l - 0.06l^2 \qquad P_k = 3.78k - 0.27k^2$$
$$= 0.06l(18 - l) = 0 \qquad = 0.27k(14 - k) = 0$$
$$l = 0, l = 18 \qquad k = 0, k = 14$$

There are four critical points: $(0, 0)$, $(0, 14)$, $(18, 0)$, and $(18, 14)$.

Now we apply the second-derivative test to each critical point. We have

$$P_{ll} = 1.08 - 0.12l \quad P_{kk} = 3.78 - 0.54k \quad P_{lk} = 0$$

Thus,

$$D(l, k) = P_{ll} P_{kk} - [P_{lk}]^2$$
$$= (1.08 - 0.12l)(3.78 - 0.54k)$$

At $(0, 0)$,

$$D(0, 0) = 1.08(3.78) > 0$$

Since $D(0, 0) > 0$ and $P_{ll} = 1.08 > 0$, there is a relative minimum at $(0, 0)$. At $(0, 14)$,

$$D(0, 14) = 1.08(-3.78) < 0$$

Because $D(0, 14) < 0$, there is no relative extremum at $(0, 14)$. At $(18, 0)$,

$$D(18, 0) = (-1.08)(3.78) < 0$$

Since $D(18, 0) < 0$, there is no relative extremum at $(18, 0)$. At $(18, 14)$,

$$D(18, 14) = (-1.08)(-3.78) > 0$$

Because $D(18, 14) > 0$ and $P_{ll} = -1.08 < 0$, there is a relative maximum at $(18, 14)$. Hence, the maximum output is obtained when $l = 18$ and $k = 14$.

NOW WORK PROBLEM 21   ●●●

● EXAMPLE 7   **Profit Maximization**

A candy company produces two types of candy, A and B, for which the average costs of production are constant at \$2 and \$3 per pound, respectively. The quantities q_A, q_B (in pounds) of A and B that can be sold each week are given by the joint-demand functions

$$q_A = 400(p_B - p_A)$$

and

$$q_B = 400(9 + p_A - 2p_B)$$

where p_A and p_B are the selling prices (in dollars per pound) of A and B, respectively. Determine the selling prices that will maximize the company's profit P.

Solution: The total profit is given by

$$P = \begin{pmatrix} \text{profit} \\ \text{per pound} \\ \text{of A} \end{pmatrix} \begin{pmatrix} \text{pounds} \\ \text{of A} \\ \text{sold} \end{pmatrix} + \begin{pmatrix} \text{profit} \\ \text{per pound} \\ \text{of B} \end{pmatrix} \begin{pmatrix} \text{pounds} \\ \text{of B} \\ \text{sold} \end{pmatrix}$$

For A and B, the profits per pound are $p_A - 2$ and $p_B - 3$, respectively. Thus,

$$P = (p_A - 2)q_A + (p_B - 3)q_B$$
$$= (p_A - 2)[400(p_B - p_A)] + (p_B - 3)[400(9 + p_A - 2p_B)]$$

Notice that P is expressed as a function of two variables, p_A and p_B. To maximize P, we set its partial derivatives equal to 0:

$$\frac{\partial P}{\partial p_A} = (p_A - 2)[400(-1)] + [400(p_B - p_A)](1) + (p_B - 3)[400(1)]$$
$$= 0$$
$$\frac{\partial P}{\partial p_B} = (p_A - 2)[400(1)] + (p_B - 3)[400(-2)] + 400(9 + p_A - 2p_B)](1)$$
$$= 0$$

Simplifying the preceding two equations gives

$$\begin{cases} -2p_A + 2p_B - 1 = 0 \\ 2p_A - 4p_B + 13 = 0 \end{cases}$$

whose solution is $p_A = 5.5$ and $p_B = 6$. Moreover, we find that

$$\frac{\partial^2 P}{\partial p_A^2} = -800 \quad \frac{\partial^2 P}{\partial p_B^2} = -1600 \quad \frac{\partial^2 P}{\partial p_B \partial p_A} = 800$$

Therefore,

$$D(5.5, 6) = (-800)(-1600) - (800)^2 > 0$$

Since $\partial^2 P / \partial p_A^2 < 0$, we indeed have a maximum, and the company should sell candy A at \$5.50 per pound and B at \$6.00 per pound.

NOW WORK PROBLEM 23

● EXAMPLE 8   Profit Maximization for a Monopolist[22]

Suppose a monopolist is practicing price discrimination by selling the same product in two separate markets at different prices. Let q_A be the number of units sold in market A, where the demand function is $p_A = f(q_A)$, and let q_B be the number of units sold in market B, where the demand function is $p_B = g(q_B)$. Then the revenue functions for the two markets are

$$r_A = q_A f(q_A) \quad \text{and} \quad r_B = q_B g(q_B)$$

Assume that all units are produced at one plant, and let the cost function for producing $q = q_A + q_B$ units be $c = c(q)$. Keep in mind that r_A is a function of q_A and r_B is a function of q_B. The monopolist's profit P is

$$P = r_A + r_B - c$$

To maximize P with respect to outputs q_A and q_B, we set its partial derivatives equal to zero. To begin with,

$$\frac{\partial P}{\partial q_A} = \frac{dr_A}{dq_A} + 0 - \frac{\partial c}{\partial q_A}$$
$$= \frac{dr_A}{dq_A} - \frac{dc}{dq} \frac{\partial q}{\partial q_A} = 0 \qquad \text{(chain rule)}$$

[22]Omit if Section 17.6 was not covered.

Because

$$\frac{\partial q}{\partial q_A} = \frac{\partial}{\partial q_A}(q_A + q_B) = 1$$

we have

$$\frac{\partial P}{\partial q_A} = \frac{dr_A}{dq_A} - \frac{dc}{dq} = 0 \qquad (6)$$

Similarly,

$$\frac{\partial P}{\partial q_B} = \frac{dr_B}{dq_B} - \frac{dc}{dq} = 0 \qquad (7)$$

From Equations (6) and (7), we get

$$\frac{dr_A}{dq_A} = \frac{dc}{dq} = \frac{dr_B}{dq_B}$$

But dr_A/dq_A and dr_B/dq_B are marginal revenues, and dc/dq is marginal cost. Hence, to maximize profit, it is necessary to charge prices (and distribute output) so that the marginal revenues in both markets will be the same and, loosely speaking, will also be equal to the cost of the last unit produced in the plant.

NOW WORK PROBLEM 25

Problems 17.7

In Problems 1–6, find the critical points of the functions.

*1. $f(x, y) = x^2 + y^2 - 5x + 4y + xy$

2. $f(x, y) = x^2 + 4y^2 - 6x + 16y$

*3. $f(x, y) = \frac{5}{3}x^3 + \frac{2}{3}y^3 - \frac{15}{2}x^2 + y^2 - 4y + 7$

4. $f(x, y) = xy - x + y$

5. $f(x, y, z) = 2x^2 + xy + y^2 + 100 - z(x + y - 200)$

6. $f(x, y, z, w) = x^2 + y^2 + z^2 - w(x - y + 2z - 6)$

In Problems 7–20, find the critical points of the functions. For each critical point, determine, by the second-derivative test, whether it corresponds to a relative maximum, to a relative minimum, or to neither, or whether the test gives no information.

*7. $f(x, y) = x^2 + 3y^2 + 4x - 9y + 3$

8. $f(x, y) = -2x^2 + 8x - 3y^2 + 24y + 7$

9. $f(x, y) = y - y^2 - 3x - 6x^2$

10. $f(x, y) = 2x^2 + \frac{3}{2}y^2 + 3xy - 10x - 9y + 2$

*11. $f(x, y) = x^2 + 3xy + y^2 + x + 3$

12. $f(x, y) = \frac{x^3}{3} + y^2 - 2x + 2y - 2xy$

*13. $f(x, y) = \frac{1}{3}(x^3 + 8y^3) - 2(x^2 + y^2) + 1$

14. $f(x, y) = x^2 + y^2 - xy + x^3$

15. $f(l, k) = \frac{l^2}{2} + 2lk + 3k^2 - 69l - 164k + 17$

16. $f(l, k) = l^2 + k^2 - 2lk$

17. $f(p, q) = pq - \frac{1}{p} - \frac{1}{q}$

18. $f(x, y) = (x - 3)(y - 3)(x + y - 3).$

19. $f(x, y) = (y^2 - 4)(e^x - 1)$

20. $f(x, y) = \ln(xy) + 2x^2 - xy - 6x$

*21. **Maximizing Output** Suppose

$$P = f(l, k) = 1.08l^2 - 0.03l^3 + 1.68k^2 - 0.08k^3$$

is a production function for a firm. Find the quantities of inputs l and k that maximize output P.

22. **Maximizing Output** In a certain office, computers C and D are utilized for c and d hours, respectively. If daily output Q is a function of c and d, namely,

$$Q = 18c + 20d - 2c^2 - 4d^2 - cd$$

find the values of c and d that maximize Q.

In Problems 23–35, unless otherwise indicated, the variables p_A and p_B denote selling prices of products A and B, respectively. Similarly, q_A and q_B denote quantities of A and B that are produced and sold during some time period. In all cases, the variables employed will be assumed to be units of output, input, money, and so on.

*23. **Profit** A candy company produces two varieties of candy, A and B, for which the constant average costs of production are 60 and 70 (cents per lb), respectively. The demand functions for A and B are given by

$$q_A = 5(p_B - p_A) \quad \text{and} \quad q_B = 500 + 5(p_A - 2p_B)$$

Find the selling prices p_A and p_B that maximize the company's profit.

24. **Profit** Repeat Problem 23 if the constant costs of production of A and B are a and b (cents per lb), respectively.

*25. **Price Discrimination** Suppose a monopolist is practicing price discrimination in the sale of a product by charging different prices in two separate markets. In market A the demand function is

$$p_A = 100 - q_A$$

and in B it is

$$p_B = 84 - q_B$$

where q_A and q_B are the quantities sold per week in A and B, and p_A and p_B are the respective prices per unit. If the monopolist's cost function is

$$c = 600 + 4(q_A + q_B)$$

how much should be sold in each market to maximize profit? What selling prices give this maximum profit? Find the maximum profit.

26. **Profit**   A monopolist sells two competitive products, A and B, for which the demand functions are

$$q_A = 3 - p_A + 2p_B \quad \text{and} \quad q_B = 5 + 5p_A - 2p_B$$

If the constant average cost of producing a unit of A is 3 and a unit of B is 2, how many units of A and B should be sold to maximize the monopolist's profit?

27. **Profit**   For products A and B, the joint-cost function for a manufacturer is

$$c = \frac{3}{2}q_A^2 + 3q_B^2$$

and the demand functions are $p_A = 60 - q_A^2$ and $p_B = 72 - 2q_B^2$. Find the level of production that maximizes profit.

28. **Profit**   For a monopolist's products A and B, the joint-cost function is $c = 2(q_A + q_B + q_A q_B)$, and the demand functions are $q_A = 20 - 2p_A$ and $q_B = 10 - p_B$. Find the values of p_A and p_B that maximize profit. What are the quantities of A and B that correspond to these prices? What is the total profit?

29. **Cost**   An open-top rectangular box is to have a volume of 6 ft³. The cost per square foot of materials is $3 for the bottom, $1 for the front and back, and $0.50 for the other two sides. Find the dimensions of the box so that the cost of materials is minimized. (See Figure 17.15.)

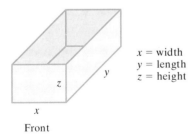

$x = \text{width}$
$y = \text{length}$
$z = \text{height}$

Front

FIGURE 17.15   Diagram for Problem 29.

30. **Collusion**   Suppose A and B are the only two firms in the market selling the same product. (We say that they are *duopolists*.) The industry demand function for the product is

$$p = 92 - q_A - q_B$$

where q_A and q_B denote the output produced and sold by A and B, respectively. For A, the cost function is $c_A = 10q_A$; for B, it is $c_B = 0.5q_B^2$. Suppose the firms decide to enter into an agreement on output and price control by jointly acting as a monopoly. In this case, we say they enter into *collusion*.

Show that the profit function for the monopoly is given by

$$P = pq_A - c_A + pq_B - c_B$$

Express P as a function of q_A and q_B, and determine how output should be allocated so as to maximize the profit of the monopoly.

31. Suppose $f(x, y) = -2x^2 + 5y^2 + 7$, where x and y must satisfy the equation $3x - 2y = 7$. Find the relative extrema of f, subject to the given condition on x and y, by first solving the second equation for y. Substitute the result for y in the given equation. Thus, f is expressed as a function of one variable, for which extrema may be found in the usual way.

32. Repeat Problem 31 if $f(x, y) = x^2 + 4y^2 + 6$, subject to the condition that $2x - 8y = 20$.

33. Suppose the joint-cost function

$$c = q_A^2 + 3q_B^2 + 2q_A q_B + aq_A + bq_B + d$$

has a relative minimum value of 15 when $q_A = 3$ and $q_B = 1$. Determine the values of the constants a, b, and d.

34. Suppose that the function $f(x, y)$ has continuous partial derivatives f_{xx}, f_{yy}, and f_{xy} at all points (x, y) near a critical point (a, b). Let $D(x, y) = f_{xx}(x, y) f_{yy}(x, y) - (f_{xy}(x, y))^2$ and suppose that $D(a, b) > 0$.

(a) Show that $f_{xx}(a, b) < 0$ if and only if $f_{yy}(a, b) < 0$.
(b) Show that $f_{xx}(a, b) > 0$ if and only if $f_{yy}(a, b) > 0$.

35. **Profit from Competitive Products**   A monopolist sells two competitive products, A and B, for which the demand equations are

$$p_A = 35 - 2q_A^2 + q_B$$

and

$$p_B = 20 - q_B + q_A$$

The joint-cost function is

$$c = -8 - 2q_A^3 + 3q_A q_B + 30q_A + 12q_B + \frac{1}{2}q_A^2$$

(a) How many units of A and B should be sold to obtain a relative maximum profit for the monopolist? Use the second-derivative test to justify your answer.
(b) Determine the selling prices required to realize the relative maximum profit. Also, find this relative maximum profit.

36. **Profit and Advertising**   A retailer has determined that the number of TV sets he can sell per week is

$$\frac{5x}{2 + x} + \frac{2y}{5 + y}$$

where x and y represent his weekly expenditures (in dollars) on newspaper and radio advertising, respectively. The profit is $250 per sale, less the cost of advertising, so the weekly profit is given by the formula

$$P = 250 \left[\frac{5x}{2 + x} + \frac{2y}{5 + y} \right] - x - y$$

Find the values of x and y for which the profit is a relative maximum. Use the second-derivative test to verify that your answer corresponds to a relative maximum profit.

37. Profit from Tomato Crop The revenue (in dollars per square meter of ground) obtained from the sale of a crop of tomatoes grown in an artificially heated greenhouse is given by

$$r = 5T(1 - e^{-x})$$

where T is the temperature (in °C) maintained in the greenhouse and x is the amount of fertilizer applied per square meter. The cost of fertilizer is $20x$ dollars per square meter, and the cost of heating is given by $0.1T^2$ dollars per square meter.

(a) Find an expression, in terms of T and x, for the profit per square meter obtained from the sale of the crop of tomatoes.

(b) Verify that the pairs

$$(T, x) = (20, \ln 5) \quad \text{and} \quad (T, x) = (5, \ln \tfrac{5}{4})$$

are critical points of the profit function in part (a). (*Note:* You need not derive the pairs.)

(c) The points in part (b) are the only critical points of the profit function in part (a). Use the second-derivative test to determine whether either of these points corresponds to a relative maximum profit per square meter.

OBJECTIVE

To find critical points for a function, subject to constraints, by applying the method of Lagrange multipliers.

17.8 Lagrange Multipliers

We will now find relative maxima and minima for a function on which certain *constraints* are imposed. Such a situation could arise if a manufacturer wished to minimize a joint-cost function and yet obtain a particular production level.

Suppose we want to find the relative extrema of

$$w = x^2 + y^2 + z^2 \tag{1}$$

subject to the constraint that x, y, and z must satisfy

$$x - y + 2z = 6 \tag{2}$$

We can transform w, which is a function of three variables, into a function of two variables such that the new function reflects constraint (2). Solving Equation (2) for x, we get

$$x = y - 2z + 6 \tag{3}$$

which, when substituted for x in Equation (1), gives

$$w = (y - 2z + 6)^2 + y^2 + z^2 \tag{4}$$

Since w is now expressed as a function of two variables, to find relative extrema we follow the usual procedure of setting the partial derivatives of w equal to 0:

$$\frac{\partial w}{\partial y} = 2(y - 2z + 6) + 2y = 4y - 4z + 12 = 0 \tag{5}$$

$$\frac{\partial w}{\partial z} = -4(y - 2z + 6) + 2z = -4y + 10z - 24 = 0 \tag{6}$$

Solving Equations (5) and (6) simultaneously gives $y = -1$ and $z = 2$. Substituting into Equation (3), we get $x = 1$. Hence, the only critical point of Equation (1) subject to the constraint represented by Equation (2) is $(1, -1, 2)$. By using the second-derivative test on Equation (4) when $y = -1$ and $z = 2$, we have

$$\frac{\partial^2 w}{\partial y^2} = 4 \quad \frac{\partial^2 w}{\partial z^2} = 10 \quad \frac{\partial^2 w}{\partial z \, \partial y} = -4$$

$$D(-1, 2) = 4(10) - (-4)^2 = 24 > 0$$

Thus w, subject to the constraint, has a relative minimum at $(1, -1, 2)$.

This solution was found by using the constraint to express one of the variables in the original function in terms of the other variables. Often this is not practical, but there is another technique, called the method of **Lagrange multipliers,**[23] that avoids this step and yet allows us to obtain critical points.

The method is as follows. Suppose we have a function $f(x, y, z)$ subject to the constraint $g(x, y, z) = 0$. We construct a new function F of *four* variables defined by

---

[23] After the French mathematician Joseph-Louis Lagrange (1736–1813).

the following (where λ is a Greek letter read "lambda"):

$$F(x, y, z, \lambda) = f(x, y, z) - \lambda g(x, y, z)$$

It can be shown that if (a, b, c) is a critical point of f, subject to the constraint $g(x, y, z) = 0$, there exists a value of λ, say, λ_0, such that (a, b, c, λ_0) is a critical point of F. The number λ_0 is called a **Lagrange multiplier.** Also, if (a, b, c, λ_0) is a critical point of F, then (a, b, c) is a critical point of f, subject to the constraint. Thus, to find critical points of f, subject to $g(x, y, z) = 0$, we instead find critical points of F. These are obtained by solving the simultaneous equations

$$\begin{cases} F_x(x, y, z, \lambda) = 0 \\ F_y(x, y, z, \lambda) = 0 \\ F_z(x, y, z, \lambda) = 0 \\ F_\lambda(x, y, z, \lambda) = 0 \end{cases}$$

At times, ingenuity must be used to solve the equations. Once we obtain a critical point (a, b, c, λ_0) of F, we can conclude that (a, b, c) is a critical point of f, subject to the constraint $g(x, y, z) = 0$. Although f and g are functions of three variables, the method of Lagrange multipliers can be extended to n variables.

Let us illustrate the method of Lagrange multipliers for the original situation, namely,

$$f(x, y, z) = x^2 + y^2 + z^2 \quad \text{subject to} \quad x - y + 2z = 6$$

First, we write the constraint as $g(x, y, z) = x - y + 2z - 6 = 0$. Second, we form the function

$$F(x, y, z, \lambda) = f(x, y, z) - \lambda g(x, y, z)$$
$$= x^2 + y^2 + z^2 - \lambda(x - y + 2z - 6)$$

Next, we set each partial derivative of F equal to 0. For convenience, we will write $F_x(x, y, z, \lambda)$ as F_x, and so on:

$$\begin{cases} F_x = 2x - \lambda = 0 & \text{(7)} \\ F_y = 2y + \lambda = 0 & \text{(8)} \\ F_z = 2z - 2\lambda = 0 & \text{(9)} \\ F_\lambda = -x + y - 2z + 6 = 0 & \text{(10)} \end{cases}$$

From Equations (7)–(9), we see immediately that

$$x = \frac{\lambda}{2} \qquad y = -\frac{\lambda}{2} \qquad z = \lambda \qquad \text{(11)}$$

Substituting these values into Equation (10), we obtain

$$-\frac{\lambda}{2} - \frac{\lambda}{2} - 2\lambda + 6 = 0$$
$$-3\lambda + 6 = 0$$
$$\lambda = 2$$

Thus, from Equation (11),

$$x = 1 \quad y = -1 \quad z = 2$$

Hence, the only critical point of f, subject to the constraint, is $(1, -1, 2)$, at which there may exist a relative maximum, a relative minimum, or neither of these. The method of Lagrange multipliers does not directly indicate which of these possibilities occur, although from our previous work, we saw that it is indeed a relative minimum. In applied problems, the nature of the problem itself may give a clue as to how a critical point is to be regarded. Often the existence of either a relative minimum or a relative maximum is assumed, and a critical point is treated accordingly. Actually, sufficient second-order conditions for relative extrema are available, but we will not consider them.

● EXAMPLE 1 **Method of Lagrange Multipliers**

Find the critical points for $z = f(x, y) = 3x - y + 6$, subject to the constraint $x^2 + y^2 = 4$.

Solution: We write the constraint as $g(x, y) = x^2 + y^2 - 4 = 0$ and construct the function

$$F(x, y, \lambda) = f(x, y) - \lambda g(x, y) = 3x - y + 6 - \lambda(x^2 + y^2 - 4)$$

Setting $F_x = F_y = F_\lambda = 0$, we have

$$
\begin{cases}
3 - 2x\lambda = 0 & (12) \\
-1 - 2y\lambda = 0 & (13) \\
-x^2 - y^2 + 4 = 0 & (14)
\end{cases}
$$

From Equations (12) and (13), we can express x and y in terms of λ. Then we will substitute for x and y in Equation (14) and solve for λ. Knowing λ, we can find x and y. To begin, from Equations (12) and (13), we have

$$x = \frac{3}{2\lambda} \quad \text{and} \quad y = -\frac{1}{2\lambda}$$

Substituting into Equation (14), we obtain

$$-\frac{9}{4\lambda^2} - \frac{1}{4\lambda^2} + 4 = 0$$

$$-\frac{10}{4\lambda^2} + 4 = 0$$

$$\lambda = \pm\frac{\sqrt{10}}{4}$$

With these λ-values, we can find x and y. If $\lambda = \sqrt{10}/4$, then

$$x = \frac{3}{2\left(\dfrac{\sqrt{10}}{4}\right)} = \frac{3\sqrt{10}}{5} \quad y = -\frac{1}{2\left(\dfrac{\sqrt{10}}{4}\right)} = -\frac{\sqrt{10}}{5}$$

Similarly, if $\lambda = -\sqrt{10}/4$,

$$x = -\frac{3\sqrt{10}}{5} \quad y = \frac{\sqrt{10}}{5}$$

Thus, the critical points of f, subject to the constraint, are $(3\sqrt{10}/5, -\sqrt{10}/5)$ and $(-3\sqrt{10}/5, \sqrt{10}/5)$. Note that the values of λ do not appear in the answer; they are simply a means to obtain the solution.

NOW WORK PROBLEM 1

● EXAMPLE 2 **Method of Lagrange Multipliers**

Find critical points for $f(x, y, z) = xyz$, where $xyz \neq 0$, subject to the constraint $x + 2y + 3z = 36$.

Solution: We have

$$F(x, y, z, \lambda) = xyz - \lambda(x + 2y + 3z - 36)$$

Setting $F_x = F_y = F_z = F_\lambda = 0$ gives, respectively,

$$\begin{cases} yz - \lambda = 0 \\ xz - 2\lambda = 0 \\ xy - 3\lambda = 0 \\ -x - 2y - 3z + 36 = 0 \end{cases}$$

Because we cannot directly express x, y, and z in terms of λ only, we cannot follow the procedure in Example 1. However, observe that we can express the products yz, xz, and xy as multiples of λ. This suggests that, by looking at quotients of equations, we can obtain a relation between two variables that does not involve λ. (The λ's will cancel.) Proceeding to do this, we write the foregoing system as

$$\begin{cases} yz = \lambda & \text{(15)} \\ \\ xz = 2\lambda & \text{(16)} \\ \\ xy = 3\lambda & \text{(17)} \\ \\ x + 2y + 3z - 36 = 0 & \text{(18)} \end{cases}$$

Dividing each side of Equation (15) by the corresponding side of Equation (16), we get

$$\frac{yz}{xz} = \frac{\lambda}{2\lambda} \quad \text{so} \quad y = \frac{x}{2}$$

This division is valid, since $xyz \neq 0$. Similarly, from Equations (15) and (17), we get

$$\frac{yz}{xy} = \frac{\lambda}{3\lambda} \quad \text{so} \quad z = \frac{x}{3}$$

Now that we have y and z expressed in terms of x only, we can substitute into Equation (18) and solve for x:

$$x + 2\left(\frac{x}{2}\right) + 3\left(\frac{x}{3}\right) - 36 = 0$$

$$x = 12$$

Thus, $y = 6$ and $z = 4$. Hence, $(12, 6, 4)$ is the only critical point satisfying the given conditions. Note that in this situation, we found the critical point without having to find the value for λ.

NOW WORK PROBLEM 7

● EXAMPLE 3 **Minimizing Costs**

Suppose a firm has an order for 200 units of its product and wishes to distribute its manufacture between two of its plants, plant 1 and plant 2. Let q_1 and q_2 denote the outputs of plants 1 and 2, respectively, and suppose the total-cost function is given by $c = f(q_1, q_2) = 2q_1^2 + q_1 q_2 + q_2^2 + 200$. How should the output be distributed in order to minimize costs?

Solution: We minimize $c = f(q_1, q_2)$, given the constraint $q_1 + q_2 = 200$. We have

$$F(q_1, q_2, \lambda) = 2q_1^2 + q_1 q_2 + q_2^2 + 200 - \lambda(q_1 + q_2 - 200)$$

$$\begin{cases} \dfrac{\partial F}{\partial q_1} = 4q_1 + q_2 - \lambda = 0 & \text{(19)} \\ \\ \dfrac{\partial F}{\partial q_2} = q_1 + 2q_2 - \lambda = 0 & \text{(20)} \\ \\ \dfrac{\partial F}{\partial \lambda} = -q_1 - q_2 + 200 = 0 & \text{(21)} \end{cases}$$

We can eliminate λ from Equations (19) and (20) and obtain a relation between q_1 and q_2. Then, solving this equation for q_2 in terms of q_1 and substituting into Equation (21), we can find q_1. We begin by subtracting Equation (20) from Equation (19), which gives

$$3q_1 - q_2 = 0 \quad \text{so} \quad q_2 = 3q_1$$

Substituting into Equation (21), we have

$$-q_1 - 3q_1 + 200 = 0$$
$$-4q_1 = -200$$
$$q_1 = 50$$

Thus, $q_2 = 150$. Accordingly, plant 1 should produce 50 units and plant 2 should produce 150 units in order to minimize costs.

NOW WORK PROBLEM 13

An interesting observation can be made concerning Example 3. From Equation (19), $\lambda = 4q_1 + q_2 = \partial c / \partial q_1$, the marginal cost of plant 1. From Equation (20), $\lambda = q_1 + 2q_2 = \partial c / \partial q_2$, the marginal cost of plant 2. Hence, $\partial c / \partial q_1 = \partial c / \partial q_2$, and we conclude that, to minimize cost, it is necessary that the marginal costs of each plant be equal to each other.

EXAMPLE 4 Least-Cost Input Combination

Suppose a firm must produce a given quantity P_0 of output in the cheapest possible manner. If there are two input factors l and k, and their prices per unit are fixed at p_l and p_k, respectively, discuss the economic significance of combining input to achieve least cost. That is, describe the least-cost input combination.

Solution: Let $P = f(l, k)$ be the production function. Then we must minimize the cost function

$$c = lp_l + kp_k$$

subject to

$$P_0 = f(l, k)$$

We construct

$$F(l, k, \lambda) = lp_l + kp_k - \lambda[f(l, k) - P_0]$$

We have

$$
\begin{cases}
\dfrac{\partial F}{\partial l} = p_l - \lambda \dfrac{\partial}{\partial l}[f(l, k)] = 0 & (22) \\[2ex]
\dfrac{\partial F}{\partial k} = p_k - \lambda \dfrac{\partial}{\partial k}[f(l, k)] = 0 & (23) \\[2ex]
\dfrac{\partial F}{\partial \lambda} = -f(l, k) + P_0 = 0
\end{cases}
$$

From Equations (22) and (23),

$$\lambda = \frac{p_l}{\dfrac{\partial}{\partial l}[f(l, k)]} = \frac{p_k}{\dfrac{\partial}{\partial k}[f(l, k)]} \qquad (24)$$

Hence,

$$\frac{p_l}{p_k} = \frac{\dfrac{\partial}{\partial l}[f(l, k)]}{\dfrac{\partial}{\partial k}[f(l, k)]}$$

We conclude that when the least-cost combination of factors is used, the ratio of the marginal productivities of the input factors must be equal to the ratio of their corresponding unit prices.

NOW WORK PROBLEM 15

Multiple Constraints

The method of Lagrange multipliers is by no means restricted to problems involving a single constraint. For example, suppose $f(x, y, z, w)$ were subject to constraints $g_1(x, y, z, w) = 0$ and $g_2(x, y, z, w) = 0$. Then there would be two lambdas, λ_1 and λ_2 (one corresponding to each constraint), and we would construct the function $F = f - \lambda_1 g_1 - \lambda_2 g_2$. We would then solve the system

$$F_x = F_y = F_z = F_w = F_{\lambda_1} = F_{\lambda_2} = 0$$

EXAMPLE 5 Method of Lagrange Multipliers with Two Constraints

Find critical points for $f(x, y, z) = xy + yz$, *subject to the constraints* $x^2 + y^2 = 8$ *and* $yz = 8$.

Solution: Set

$$F(x, y, z, \lambda_1, \lambda_2) = xy + yz - \lambda_1(x^2 + y^2 - 8) - \lambda_2(yz - 8)$$

Then

$$\begin{cases} F_x &= y - 2x\lambda_1 = 0 \\ F_y &= x + z - 2y\lambda_1 - z\lambda_2 = 0 \\ F_z &= y - y\lambda_2 = 0 \\ F_{\lambda_1} &= -x^2 - y^2 + 8 = 0 \\ F_{\lambda_2} &= -yz + 8 = 0 \end{cases}$$

You would probably agree that this appears to be a challenging system to solve. Thus, ingenuity will come into play. Here is one sequence of operations that will allow us to find the critical points. We can write the system as

$$\begin{cases} \dfrac{y}{2x} = \lambda_1 & (25) \\[2mm] x + z - 2y\lambda_1 - z\lambda_2 = 0 & (26) \\[2mm] \lambda_2 = 1 & (27) \\[2mm] x^2 + y^2 = 8 & (28) \\[2mm] z = \dfrac{8}{y} & (29) \end{cases}$$

Substituting $\lambda_2 = 1$ from Equation (27) into Equation (26) and simplifying gives the equation $x - 2y\lambda_1 = 0$, so

$$\lambda_1 = \frac{x}{2y}$$

Substituting into Equation (25) gives

$$\frac{y}{2x} = \frac{x}{2y}$$

$$y^2 = x^2 \qquad (30)$$

Substituting into Equation (28) gives $x^2 + x^2 = 8$, from which it follows that $x = \pm 2$. If $x = 2$, then, from Equation (30), we have $y = \pm 2$. Similarly, if $x = -2$, then $y = \pm 2$. Thus, if $x = 2$ and $y = 2$, then, from Equation (29), we obtain $z = 4$. Continuing in this manner, we obtain four critical points:

$$(2, 2, 4) \quad (2, -2, -4) \quad (-2, 2, 4) \quad (-2, -2, -4)$$

NOW WORK PROBLEM 9

Problems 17.8

In Problems 1–12, find, by the method of Lagrange multipliers, the critical points of the functions, subject to the given constraints.

*1. $f(x, y) = x^2 + 4y^2 + 6$; $2x - 8y = 20$

2. $f(x, y) = -2x^2 + 5y^2 + 7$; $3x - 2y = 7$

3. $f(x, y, z) = x^2 + y^2 + z^2$; $2x + y - z = 9$

4. $f(x, y, z) = x + y + z$; $xyz = 8$

5. $f(x, y, z) = 2x^2 + xy + y^2 + z$; $x + 2y + 4z = 3$

6. $f(x, y, z) = xyz^2$; $x - y + z = 20 \ (xyz^2 \neq 0)$

*7. $f(x, y, z) = xyz$; $x + 2y + 3z = 18 \ (xyz \neq 0)$

8. $f(x, y, z) = x^2 + y^2 + z^2$; $x + y + z = 3$

*9. $f(x, y, z) = x^2 + 2y - z^2$; $2x - y = 0, \ y + z = 0$

10. $f(x, y, z) = x^2 + y^2 + z^2$; $x + y + z = 4, \ x - y + z = 4$

11. $f(x, y, z) = xy^2z$; $x + y + z = 1, \ x - y + z = 0 \ (xyz \neq 0)$

12. $f(x, y, z, w) = 3x^2 + y^2 + 2z^2 - 5w^2$; $x + 6y + 3z + 2w = 4$

*13. **Production Allocation**   To fill an order for 100 units of its product, a firm wishes to distribute production between its two plants, plant 1 and plant 2. The total-cost function is given by

$$c = f(q_1, q_2) = 0.1q_1^2 + 7q_1 + 15q_2 + 1000$$

where q_1 and q_2 are the numbers of units produced at plants 1 and 2, respectively. How should the output be distributed in order to minimize costs? (You may assume that the critical point obtained does correspond to the minimum cost.)

14. **Production Allocation**   Repeat Problem 13 if the cost function is

$$c = 3q_1^2 + q_1q_2 + 2q_2^2$$

and a total of 200 units are to be produced.

*15. **Maximizing Output**   The production function for a firm is

$$f(l, k) = 12l + 20k - l^2 - 2k^2$$

The cost to the firm of l and k is 4 and 8 per unit, respectively. If the firm wants the total cost of input to be 88, find the greatest output possible, subject to this budget constraint. (You may assume that the critical point obtained does correspond to the maximum output.)

16. **Maximizing Output**   Repeat Problem 15, given that

$$f(l, k) = 20l + 25k - l^2 - 3k^2$$

and the budget constraint is $2l + 4k = 50$.

17. **Advertising Budget**   A computer company has a monthly advertising budget of $60,000. Its marketing department estimates that if x dollars are spent each month on advertising in newspapers and y dollars per month on television advertising, then the monthly sales will be given by $S = 90x^{1/4}y^{3/4}$ dollars. If the profit is 10% of sales, less the advertising cost, determine how to allocate the advertising budget in order to maximize the monthly profit. (You may assume that the critical point obtained does correspond to the maximum profit.)

18. **Maximizing Production**   When l units of labor and k units of capital are invested, a manufacturer's total production q is given by the Cobb–Douglas production function

$q = 6l^{2/5}k^{3/5}$. Each unit of labor costs $25 and each unit of capital costs $69. If exactly $25,875 is to be spent on production, determine the numbers of units of labor and capital that should be invested to maximize production. (You may assume that the maximum occurs at the critical point obtained.)

19. **Political Advertising**   Newspaper advertisements for political parties always have some negative effects. The recently elected party assumed that the three most important election issues X, Y, and Z, had to be mentioned in each ad, with space x, y, and z units, respectively, allotted to each. The combined bad effect of this coverage was estimated by the party's backroom boys as

$$B(x, y, z) = x^2 + y^2 + 2z^2$$

Aesthetics dictated that the total space for X and Y together must be 20, and realism suggested that the total space allotted to Y and Z together must also be 20 units. What values of x, y, and z in each ad would produce the lowest negative effect? (You may assume that any critical point obtained provides the minimum effect.)

20. **Maximizing Profit**   Suppose a manufacturer's production function is given by

$$16q = 65 - 4(l - 4)^2 - 2(k - 5)^2$$

and the cost to the manufacturer is $8 per unit of labor and $16 per unit of capital, so that the total cost (in dollars) is $8l + 16k$. The selling price of the product is $64 per unit.

(a) Express the profit as a function of l and k. Give your answer in expanded form.

(b) Find all critical points of the profit function obtained in part (a). Apply the second-derivative test at each critical point. If the profit is a relative maximum at a critical point, compute the corresponding relative maximum profit.

(c) The profit may be considered a function of l, k, and q (that is, $P = 64q - 8l - 16k$), subject to the constraint

$$16q = 65 - 4(l - 4)^2 - 2(k - 5)^2$$

Use the method of Lagrange multipliers to find all critical points of $P = 64q - 8l - 16k$, subject to the constraint.

Problems 21–24 refer to the following definition. A utility function is a function that attaches a measure to the satisfaction or utility a consumer gets from the consumption of products per unit of time. Suppose $U = f(x, y)$ is such a function, where x and y are the amounts of two products, X and Y. The marginal utility of X is $\partial U / \partial x$ and approximately represents the change in total utility resulting from a one-unit change in consumption of product X per unit of time. We define the marginal utility of Y in similar fashion. If the prices of X and Y are p_x and p_y, respectively, and the consumer has an income or budget of I to spend, then the budget constraint is

$$xp_x + yp_y = I$$

In Problems 21–23, find the quantities of each product that the consumer should buy, subject to the budget, that will allow maximum satisfaction. That is, in Problems 21 and 22, find values of x and y that maximize $U = f(x, y)$, subject to $xp_x + yp_y = I$.

Perform a similar procedure for Problem 23. Assume that such a maximum exists.

21. $U = x^3 y^3$; $p_x = 2, p_y = 3, I = 48$ $(x^3 y^3 \neq 0)$

22. $U = 40x - 5x^2 + 4y - 2y^2$; $p_x = 2, p_y = 3, I = 10$

23. $U = f(x, y, z) = xyz$; $p_x = p_y = p_z = 1$, $I = 100$ $(xyz \neq 0)$

24. Let $U = f(x, y)$ be a utility function subject to the budget constraint $xp_x + yp_y = I$, where $p_x, p_y,$ and I are constants. Show that, to maximize satisfaction, it is necessary that

$$\lambda = \frac{f_x(x, y)}{p_x} = \frac{f_y(x, y)}{p_y}$$

where $f_x(x, y)$ and $f_y(x, y)$ are the marginal utilities of X and Y, respectively. Show that $f_x(x, y)/p_x$ is the marginal utility of one dollar's worth of X. Hence, maximum satisfaction is obtained when the consumer allocates the budget so that the marginal utility of a dollar's worth of X is equal to the marginal utility per dollar's worth of Y. Performing the same procedure as that for $U = f(x, y)$, verify that this is true for $U = f(x, y, z, w)$, subject to the corresponding budget equation. In each case, λ is called the *marginal utility of income*.

OBJECTIVE

To develop the method of least squares and introduce index numbers.

TABLE 17.4

Expenditures x	2	3	4.5	5.5	7
Revenue y		3	6	8	10	11

17.9 Lines of Regression[24]

To study the influence of advertising on sales, a firm compiled the data in Table 17.4. The variable x denotes advertising expenditures in hundreds of dollars, and the variable y denotes the resulting sales revenue in thousands of dollars. If each pair (x, y) of data is plotted, the result is called a **scatter diagram** [Figure 17.16(a)].

From an observation of the distribution of the points, it is reasonable to assume that a relationship exists between x and y and that it is approximately linear. On this basis, we may fit "by eye" a straight line that approximates the given data [Figure 17.16(b)] and, from this line, predict a value of y for a given value of x. The line seems consistent with the trend of the data, although other lines could be drawn as well. Unfortunately, determining a line "by eye" is not very objective. We want to apply criteria in specifying what we will call a line of "best fit." A frequently used technique is called the **method of least squares.**

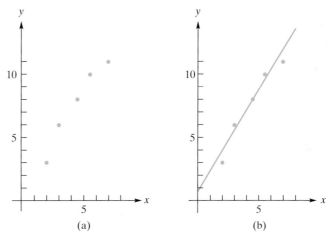

FIGURE 17.16   Scatter diagram and straight-line approximation to data points.

To apply the method of least squares to the data in Table 17.4, we first assume that x and y are approximately linearly related and that we can fit a straight line

$$\widehat{y} = a + bx \tag{1}$$

that approximates the given points by a suitable objective choice of the constants a and b. For a given value of x in Equation (1), $\widehat{y}$ is the corresponding predicted value of y, and $(x, \widehat{y})$ will be on the line. Our aim is that $\widehat{y}$ be near y.

When $x = 2$, the observed value of y is 3. Our predicted value of y is obtained by substituting $x = 2$ in Equation (1), which yields $\widehat{y} = a + 2b$. The error of estimation, or vertical deviation of the point $(2, 3)$ from the line, is $\widehat{y} - y$, which is

$$a + 2b - 3$$

---

[24]This section can be omitted without loss of continuity.

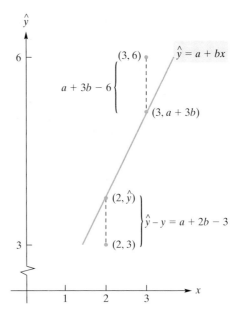

FIGURE 17.17   Vertical deviation of data
points from straight-line approximation.

This vertical deviation is indicated (although exaggerated for clarity) in Figure 17.17. Similarly, the vertical deviation of $(3, 6)$ from the line is $a+3b-6$, as is also illustrated. To avoid possible difficulties associated with positive and negative deviations, we will consider the squares of the deviations and will form the sum S of all such squares for the given data:

$$S = (a + 2b - 3)^2 + (a + 3b - 6)^2 + (a + 4.5b - 8)^2$$
$$+ (a + 5.5b - 10)^2 + (a + 7b - 11)^2$$

The method of least squares requires that we choose as the line of "best fit" the one obtained by selecting a and b so as to minimize S. We can minimize S with respect to a and b by solving the system

$$\begin{cases} \dfrac{\partial S}{\partial a} = 0 \\ \dfrac{\partial S}{\partial b} = 0 \end{cases}$$

We have

$$\frac{\partial S}{\partial a} = 2(a + 2b - 3) + 2(a + 3b - 6) + 2(a + 4.5b - 8)$$
$$+ 2(a + 5.5b - 10) + 2(a + 7b - 11) = 0$$
$$\frac{\partial S}{\partial b} = 4(a + 2b - 3) + 6(a + 3b - 6) + 9(a + 4.5b - 8)$$
$$+ 11(a + 5.5b - 10) + 14(a + 7b - 11) = 0$$

which, when simplified, give

$$\begin{cases} 10a + 44b = 76 \\ 44a + 225b = 384 \end{cases}$$

Solving for a and b, we obtain

$$a = \frac{102}{157} \approx 0.65 \quad b = \frac{248}{157} \approx 1.58$$

From our calculations of $\partial S/\partial a$ and $\partial S/\partial b$, we see that $S_{aa} = 10 > 0$, $S_{bb} = 225$, and $S_{ab} = 44$. Thus $D = S_{aa}S_{bb} - (S_{ab})^2 = 10 \cdot 225 - 44^2 = 314 > 0$. It follows from the second-derivative test of Section 17.7 that S has a minimum value at the critical point. Hence, in the sense of least squares, the line of best fit $\hat{y} = a + bx$ is

$$\hat{y} = 0.65 + 1.58x \tag{2}$$

This is, in fact, the line shown in Figure 17.16(b). It is called the **least squares line of y on** x or the **linear regression line of** y **on** x. The constants a and b are called **linear regression coefficients.** With Equation (2), we would predict that when $x = 5$, the corresponding value of y is $\hat{y} = 0.65 + 1.58(5) = 8.55$.

More generally, suppose we are given the following n pairs of observations:

$$(x_1, y_1), (x_2, y_2), \ldots, (x_n, y_n)$$

If we assume that x and y are approximately linearly related and that we can fit a straight line

$$\hat{y} = a + bx$$

that approximates the data, the sum of the squares of the errors $\hat{y} - y$ is

$$S = (a + bx_1 - y_1)^2 + (a + bx_2 - y_2)^2 + \cdots + (a + bx_n - y_n)^2$$

Since S must be minimized with respect to a and b,

$$\begin{cases} \dfrac{\partial S}{\partial a} = 2(a + bx_1 - y_1) + 2(a + bx_2 - y_2) + \cdots + 2(a + bx_n - y_n) = 0 \\[2mm] \dfrac{\partial S}{\partial b} = 2x_1(a + bx_1 - y_1) + 2x_2(a + bx_2 - y_2) + \cdots + 2x_n(a + bx_n - y_n) = 0 \end{cases}$$

Dividing both equations by 2 and using summation notation, we have

$$\begin{cases} na + \left(\displaystyle\sum_{i=1}^{n} x_i\right) b - \displaystyle\sum_{i=1}^{n} y_i = 0 \\[4mm] \left(\displaystyle\sum_{i=1}^{n} x_i\right) a + \left(\displaystyle\sum_{i=1}^{n} x_i^2\right) b - \displaystyle\sum_{i=1}^{n} x_i y_i = 0 \end{cases}$$

which is a system of two linear equations in a and b, the so-called *normal equations:*

$$\begin{cases} na + \left(\displaystyle\sum_{i=1}^{n} x_i\right) b = \displaystyle\sum_{i=1}^{n} y_i \\[4mm] \left(\displaystyle\sum_{i=1}^{n} x_i\right) a + \left(\displaystyle\sum_{i=1}^{n} x_i^2\right) b = \displaystyle\sum_{i=1}^{n} x_i y_i \end{cases}$$

The coefficients are of course no more than simple sums of values obtained from the observed data. The solution is obtained easily using the techniques of Section 3.4.

$$a = \frac{\left(\displaystyle\sum_{i=1}^{n} x_i^2\right)\left(\displaystyle\sum_{i=1}^{n} y_i\right) - \left(\displaystyle\sum_{i=1}^{n} x_i\right)\left(\displaystyle\sum_{i=1}^{n} x_i y_i\right)}{n \displaystyle\sum_{i=1}^{n} x_i^2 - \left(\displaystyle\sum_{i=1}^{n} x_i\right)^2} \tag{3}$$

$$b = \frac{n \displaystyle\sum_{i=1}^{n} x_i y_i - \left(\displaystyle\sum_{i=1}^{n} x_i\right)\left(\displaystyle\sum_{i=1}^{n} y_i\right)}{n \displaystyle\sum_{i=1}^{n} x_i^2 - \left(\displaystyle\sum_{i=1}^{n} x_i\right)^2} \tag{4}$$

Now we have $S_{aa} = 2n > 0$ and $D = S_{aa}S_{bb} - (S_{ab})^2 = (2n)(2\sum x_i^2) - (2\sum x_i)^2$, independent of (a, b). It can be shown that for distinct x_i and $n \geq 2$ that $D > 0$ so a and b, given by Equations (3) and (4), do indeed minimize S. [For example, when

$n = 2$, $D > 0$ is provably equivalent to $(x_1 - x_2)^2 > 0$, which is true for distinct x_1 and x_2.]

Computing the linear regression coefficients a and b by the formulas of Equations (3) and (4) gives the linear regression line of y on x, namely, $\widehat{y} = a + bx$, which can be used to estimate y for any given value of x.

In the next example, as well as in the exercises, you will encounter **index numbers.** They are used to relate a variable in one period of time to the same variable in another period, called the *base period*. An index number is a *relative* number that describes data that are changing over time. Such data are referred to as *time series*.

For example, consider the time-series data of total production of widgets in the United States for 2002–2006, given in Table 17.5. If we choose 2003 as the base year and assign the index number 100 to it, then the other index numbers are obtained by dividing each year's production by the 2003 production of 900 and multiplying the result by 100. We can, for example, interpret the index 106 for 2006 as meaning that production for that year was 106% of the production in 2003.

TABLE 17.5

Year	Production (in thousands)	Index (based on 2003)
2002	828	92
2003	900	100
2004	936	104
2005	891	99
2006	954	106

In time-series analysis, index numbers are obviously of great advantage if the data involve numbers of great magnitude. But regardless of the magnitude of the data, index numbers simplify the task of comparing changes in data over periods of time.

● EXAMPLE 1 Determining a Linear-Regression Line

By means of the linear-regression line, use the following table to represent the trend for the index of total U.S. government revenue from 1995 to 2000 (1995 = 100).

Year	1995	1996	1997	1998	1999	2000
Index	100	107	117	127	135	150

Source: Economic Report of the President, 2001, U.S. Government Printing Office, Washington, DC, 2001.

Solution: We will let x denote time and y denote the index and treat y as a linear function of x. Also, we will designate 1995 by $x = 1$, 1996 by $x = 2$, and so on. There are $n = 6$ pairs of measurements. To determine the linear-regression coefficients by using Equations (3) and (4), we first perform the arithmetic:

Year	x_i	y_i	$x_i y_i$	x_i^2
1995	1	100	100	1
1996	2	107	214	4
1997	3	117	351	9
1998	4	127	508	16
1999	5	135	675	25
2000	6	150	900	36
Total	21	736	2748	91
	$= \sum\limits_{i=1}^{6} x_i$	$= \sum\limits_{i=1}^{6} y_i$	$= \sum\limits_{i=1}^{6} x_i y_i$	$= \sum\limits_{i=1}^{6} x_i^2$

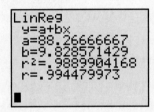

FIGURE 17.18 Linear-regression line
for government revenue.

Hence, by Equation (3),

$$a = \frac{91(736) - 21(2748)}{6(91) - (21)^2} \approx 88.3$$

and by Equation (4),

$$b = \frac{6(2748) - 21(736)}{6(91) - (21)^2} \approx 9.83$$

Thus, the regression line of y on x is

$$\hat{y} = 88.3 + 9.83x$$

whose graph, as well as a scatter diagram, appears in Figure 17.18.

NOW WORK PROBLEM 1

TECHNOLOGY

The TI-83 Plus has a utility that computes the equation of the least squares line for a set of data. We will illustrate by giving the procedure for the six data points (x_i, y_i) of Example 1. After pressing STAT and ENTER, we enter all the x-values and then the y-values. (See Figure 17.19.)

Next, we press STAT and move to CALC. Finally, pressing 8 and ENTER gives the result shown in Figure 17.20. (The number $r \approx 0.99448$ is called the *coefficient of correlation* and is a measure of the degree to which the given data are linearly related.)

FIGURE 17.19 Data of Example 1.

FIGURE 17.20 Equation of least squares line.

Problems 17.9

In this exercise set, use a graphing calculator if your instructor permits you to do so.

In Problems 1–4, find an equation of the least squares linear-regression line of y on x for the given data, and sketch both the line and the data. Predict the value of y corresponding to x = 3.5.

*1.
x	1	2	3	4	5	6
y	1.5	2.3	2.6	3.7	4.0	4.5

2.
x	1	2	3	4	5	6	7
y	1	1.8	2	4	4.5	7	9

3.
x	2	3	4.5	5.5	7
y	3	5	8	10	11

4.
x	2	3	4	5	6	7
y	2.4	2.9	3.3	3.8	4.3	4.9

5. **Demand** A firm finds that when the price of its product is p dollars per unit, the number of units sold is q, as indicated in the following table:

Price, p	10	20	40	50	60	70
Demand, q	75	65	56	50	42	34

Find an equation of the regression line of q on p.

6. **Water and Crop Yield** On a farm, an agronomist finds that the amount of water applied (in inches) and the corresponding yield of a certain crop (in tons per acre) are as given in the following table:

Water, x	8	16	24	32
Yield, y	5.2	5.7	6.3	6.7

Find an equation of the regression line of y on x. Predict y when $x = 20$.

7. **Virus** A rabbit was injected with a virus, and x hours after the injection the temperature y (in degrees Fahrenheit) of the rabbit was measured.[25] The data are given in the following table:

Elapsed Time, x	24	32	48	56
Temperature, y	102.8	104.5	106.5	107.0

Find an equation of the regression line of y on x, and estimate the rabbit's temperature 40 hours after the injection.

[25]R. R. Sokal and F. J. Rohlf, *Introduction to Biostatistics* (San Francisco: W. H. Freeman & Company, Publishers, 1973).

8. Psychology In a psychological experiment, four persons were subjected to a stimulus. Both before and after the stimulus, the systolic blood pressure (in millimeters of mercury) of each subject was measured. The data are given in the following table:

Blood Pressure				
Before Stimulus, x	131	132	135	141
After Stimulus, y	139	139	142	149

Find an equation of the regression line of y on x, where x and y are as defined in the table.

For the time series in Problems 9 and 10, fit a linear-regression line by least squares; that is, find an equation of the linear-regression line of y on x. In each case, let the first year in the table correspond to x = 1.

9.

Production of Product A, 2002–2006 (in thousands of units)

Year	Production
2002	10
2003	15
2004	16
2005	18
2006	21

10. Industrial Production In the following table, let 1975 correspond to $x = 1$, 1977 correspond to $x = 3$, and so on:

Index of Industrial Production—Electrical Machinery (based on 1977)

Year	Index
1975	77
1977	100
1979	126
1981	134

Source: Economic Report of the President, 1988, U.S. Government Printing Office, Washington, DC, 1988.

11. Computer Shipments

(a) Find an equation of the least squares line of y on x for the following data (refer to 2002 as year $x = 1$, and so on):

Overseas Shipments of Computers by Acme Computer Co. (in thousands)

Year	Quantity
2002	35
2003	31
2004	26
2005	24
2006	26

(b) For the data in part (a), refer to 2002 as year $x = -2$, 2003 as year $x = -1$, 2004 as year $x = 0$ and so on. Then $\sum_{i=1}^{5} x_i = 0$. Fit a least squares line and observe how the calculation is simplified.

12. Medical Care For the following time series, find an equation of the linear-regression line that best fits the data (refer to 1983 as year $x = -2$, 1984 as year $x = -1$, and so on):

Consumer Price Index—Medical Care, 1983–1987 (based on 1967)

Year	Index
1983	357
1984	380
1985	403
1986	434
1987	462

Source: Economic Report of the President, 1988, U.S. Government Printing Office, Washington, DC, 1988.

OBJECTIVE

To compute double and triple integrals.

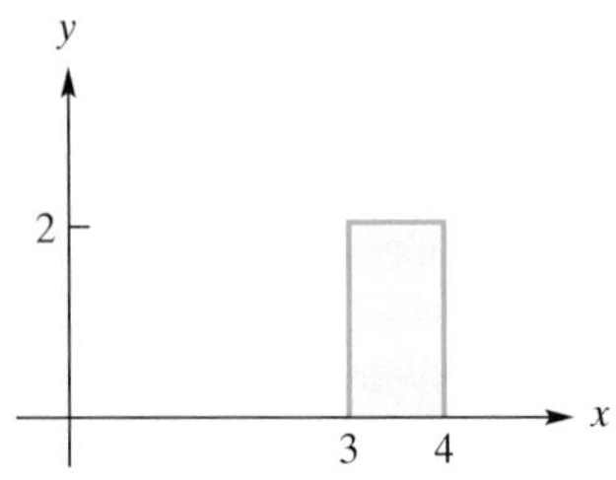

FIGURE 17.21 Region over which $\int_0^2 \int_3^4 xy\,dx\,dy$ is evaluated.

17.10 Multiple Integrals

Recall that the definite integral of a function of one variable is concerned with integration over an *interval*. There are also definite integrals of functions of two variables, called (definite) **double integrals.** These involve integration over a *region* in the plane.

For example, the symbol

$$\int_0^2 \int_3^4 xy\,dx\,dy = \int_0^2 \left(\int_3^4 xy\,dx \right) dy$$

is the double integral of $f(x, y) = xy$ over a region determined by the limits of integration. That region consists of all points (x, y) in the x,y-plane such that $3 \leq x \leq 4$ and $0 \leq y \leq 2$. (See Figure 17.21.)

A double integral is a limit of a sum of the form $\sum f(x, y)\, \Delta x\, \Delta y$, where, in this example, the points (x, y) are in the shaded region. A geometric interpretation of a double integral will be given later.

To evaluate

$$\int_0^2 \int_3^4 xy\,dx\,dy = \int_0^2 \left(\int_3^4 xy\,dx \right) dy$$

we use successive integrations starting with the innermost integral. First, we evaluate

$$\int_3^4 xy\,dx$$

by treating y as a constant and integrating with respect to x between the limits 3 and 4:

$$\int_3^4 xy\,dx = \frac{x^2 y}{2} \Big|_3^4$$

Substituting the limits for the variable x, we have

$$\frac{4^2 \cdot y}{2} - \frac{3^2 \cdot y}{2} = \frac{16y}{2} - \frac{9y}{2} = \frac{7}{2}y$$

Now we integrate this result with respect to y between the limits 0 and 2:

$$\int_0^2 \frac{7}{2} y\,dy = \frac{7y^2}{4} \Big|_0^2 = \frac{7 \cdot 2^2}{4} - 0 = 7$$

Thus,

$$\int_0^2 \int_3^4 xy\,dx\,dy = 7$$

Now consider the double integral

$$\int_0^1 \int_{x^3}^{x^2} (x^3 - xy)\,dy\,dx = \int_0^1 \left(\int_{x^3}^{x^2} (x^3 - xy)\,dy \right) dx$$

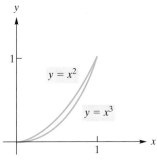

FIGURE 17.22 Region over which $\int_0^1 \int_{x^3}^{x^2} (x^3 - xy)\,dy\,dx$ is evaluated.

Here we integrate first with respect to y and then with respect to x. The region over which the integration takes places is all points (x, y) such that $x^3 \leq y \leq x^2$ and $0 \leq x \leq 1$. (See Figure 17.22.) This double integral is evaluated by first treating x as a constant and integrating $x^3 - xy$ with respect to y between x^3 and x^2, and then integrating the result with respect to x between 0 and 1:

$$\int_0^1 \int_{x^3}^{x^2} (x^3 - xy)\,dy\,dx$$

$$= \int_0^1 \left(\int_{x^3}^{x^2} (x^3 - xy)\,dy \right) dx = \int_0^1 \left(x^3 y - \frac{xy^2}{2} \right) \Big|_{x^3}^{x^2} dx$$

$$= \int_0^1 \left[\left(x^3(x^2) - \frac{x(x^2)^2}{2} \right) - \left(x^3(x^3) - \frac{x(x^3)^2}{2} \right) \right] dx$$

$$= \int_0^1 \left(x^5 - \frac{x^5}{2} - x^6 + \frac{x^7}{2} \right) dx = \int_0^1 \left(\frac{x^5}{2} - x^6 + \frac{x^7}{2} \right) dx$$

$$= \left(\frac{x^6}{12} - \frac{x^7}{7} + \frac{x^8}{16} \right) \Big|_0^1 = \left(\frac{1}{12} - \frac{1}{7} + \frac{1}{16} \right) - 0 = \frac{1}{336}$$

● **EXAMPLE 1 Evaluating a Double Integral**

Find $\displaystyle\int_{-1}^{1}\int_{0}^{1-x}(2x+1)\,dy\,dx$.

Solution: Here we first integrate with respect to y and then integrate the result with respect to x:

$$\int_{-1}^{1}\int_{0}^{1-x}(2x+1)\,dy\,dx = \int_{-1}^{1}\left(\int_{0}^{1-x}(2x+1)\,dy\right)dx$$

$$= \int_{-1}^{1}(2xy+y)\Big|_{0}^{1-x}dx = \int_{-1}^{1}((2x(1-x)+(1-x))-0)\,dx$$

$$= \int_{-1}^{1}(-2x^2+x+1)\,dx = \left(-\frac{2x^3}{3}+\frac{x^2}{2}+x\right)\Big|_{-1}^{1}$$

$$= \left(-\frac{2}{3}+\frac{1}{2}+1\right)-\left(\frac{2}{3}+\frac{1}{2}-1\right)=\frac{2}{3}$$

NOW WORK PROBLEM 9 ●●

● **EXAMPLE 2 Evaluating a Double Integral**

Find $\displaystyle\int_{1}^{\ln 2}\int_{e^y}^{2}dx\,dy$.

Solution: Here we first integrate with respect to x and then integrate the result with respect to y:

$$\int_{1}^{\ln 2}\int_{e^y}^{2}dx\,dy = \int_{1}^{\ln 2}\left(\int_{e^y}^{2}dx\right)dy = \int_{1}^{\ln 2}x\Big|_{e^y}^{2}dy$$

$$= \int_{1}^{\ln 2}(2-e^y)\,dy = (2y-e^y)\Big|_{1}^{\ln 2}$$

$$= (2\ln 2 - 2)-(2-e) = 2\ln 2 - 4 + e$$

$$= \ln 4 - 4 + e$$

NOW WORK PROBLEM 13 ●●

A double integral can be interpreted in terms of the volume of a region between the x,y-plane and a surface $z = f(x, y)$ if $z \geq 0$. In Figure 17.23 is a region whose volume we will consider. The element of volume for this region is a vertical column

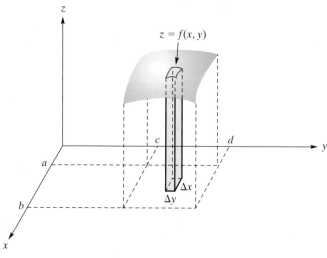

FIGURE 17.23 Interpreting $\int_{a}^{b}\int_{c}^{d} f(x, y)\,dy\,dx$ in terms of volume, where $f(x, y) \geq 0$.

with height approximately $z = f(x, y)$ and base area $\Delta y \, \Delta x$. Thus, its volume is approximately $f(x, y) \, \Delta y \, \Delta x$. The volume of the entire region can be found by summing the volumes of all such elements for $a \leq x \leq b$ and $c \leq y \leq d$ via a double integral:

$$\text{volume} = \int_a^b \int_c^d f(x, y) \, dy \, dx$$

Triple integrals are handled by successively evaluating three integrals, as the next example shows.

● EXAMPLE 3   **Evaluating a Triple Integral**

Find $\displaystyle\int_0^1 \int_0^x \int_0^{x-y} x \, dz \, dy \, dx.$

Solution:

$$\int_0^1 \int_0^x \int_0^{x-y} x \, dz \, dy \, dx = \int_0^1 \int_0^x \left(\int_0^{x-y} x \, dz \right) dy \, dx$$

$$= \int_0^1 \int_0^x (xz) \Big|_0^{x-y} dy \, dx = \int_0^1 \int_0^x (x(x-y) - 0) \, dy \, dx$$

$$= \int_0^1 \int_0^x (x^2 - xy) \, dy \, dx = \int_0^1 \left(\int_0^x (x^2 - xy) \, dy \right) dx$$

$$= \int_0^1 \left(x^2 y - \frac{xy^2}{2} \right) \Big|_0^x dx = \int_0^1 \left[\left(x^3 - \frac{x^3}{2} \right) - 0 \right] dx$$

$$= \int_0^1 \frac{x^3}{2} dx = \frac{x^4}{8} \Big|_0^1 = \frac{1}{8}$$

NOW WORK PROBLEM 21   ●●

Problems 17.10

In Problems 1–22, evaluate the multiple integrals.

1. $\displaystyle\int_0^3 \int_0^4 x \, dy \, dx$

2. $\displaystyle\int_1^4 \int_0^3 y \, dy \, dx$

3. $\displaystyle\int_0^1 \int_0^1 xy \, dx \, dy$

4. $\displaystyle\int_0^2 \int_0^3 x^2 \, dy \, dx$

5. $\displaystyle\int_1^3 \int_1^2 (x^2 - y) \, dx \, dy$

6. $\displaystyle\int_{-2}^3 \int_0^2 (y^2 - 2xy) \, dy \, dx$

7. $\displaystyle\int_0^1 \int_0^2 (x + y) \, dy \, dx$

8. $\displaystyle\int_0^3 \int_0^x (x^2 + y^2) \, dy \, dx$

*9. $\displaystyle\int_1^4 \int_0^{5x} y \, dy \, dx$

10. $\displaystyle\int_1^2 \int_0^{x-1} 2y \, dy \, dx$

11. $\displaystyle\int_0^1 \int_{3x}^{x^2} 14x^2 y \, dy \, dx$

12. $\displaystyle\int_0^2 \int_0^{x^2} xy \, dy \, dx$

*13. $\displaystyle\int_0^3 \int_0^{\sqrt{9-x^2}} y \, dy \, dx$

14. $\displaystyle\int_0^1 \int_{y^2}^y y \, dx \, dy$

15. $\displaystyle\int_{-1}^1 \int_x^{1-x} 3(x + y) \, dy \, dx$

16. $\displaystyle\int_0^3 \int_{y^2}^{3y} 5x \, dx \, dy$

17. $\displaystyle\int_0^1 \int_0^y e^{x+y} \, dx \, dy$

18. $\displaystyle\int_0^1 \int_0^1 e^{y-x} \, dx \, dy$

19. $\displaystyle\int_{-1}^0 \int_{-1}^2 \int_1^2 6xy^2 z^3 \, dx \, dy \, dz$

20. $\displaystyle\int_0^1 \int_0^x \int_0^{x+y} x^2 \, dz \, dy \, dx$

*21. $\displaystyle\int_0^1 \int_{x^2}^x \int_0^{xy} dz \, dy \, dx$

22. $\displaystyle\int_1^e \int_{\ln x}^x \int_0^y dz \, dy \, dx$

23. Statistics   In the study of statistics, a joint density function $z = f(x, y)$ defined on a region in the x,y-plane is represented by a surface in space. The probability that

$$a \leq x \leq b \quad \text{and} \quad c \leq y \leq d$$

is given by

$$P(a \leq x \leq b, c \leq y \leq d) = \int_c^d \int_a^b f(x, y) \, dx \, dy$$

and is represented by the volume between the graph of f and the rectangular region given by

$$a \leq x \leq b \quad \text{and} \quad c \leq y \leq d$$

If $f(x, y) = e^{-(x+y)}$ is a joint density function, where $x \geq 0$ and $y \geq 0$, find

$$P(0 \leq x \leq 2, 1 \leq y \leq 2)$$

and give your answer in terms of e.

24. Statistics In Problem 23, let $f(x, y) = 12e^{-4x-3y}$ for $x, y \geq 0$. Find

$$P(3 \leq x \leq 4, 2 \leq y \leq 6)$$

and give your answer in terms of e.

25. Statistics In Problem 23, let $f(x, y) = 1$, where $0 \leq x \leq 1$ and $0 \leq y \leq 1$. Find $P(x \geq 1/2, y \geq 1/3)$.

26. Statistics In Problem 23, let f be the uniform density function $f(x, y) = 1/8$ defined over the rectangle $0 \leq x \leq 4, 0 \leq y \leq 2$. Determine the probability that $0 \leq x \leq 1$ and $0 \leq y \leq 1$.

17.11 Review

Important Terms and Symbols   Examples

Section 17.1 | **Functions of Several Variables** |
---|---|---
 | $f(x_1, x_2, \ldots, x_n)$ function of n variables | Ex. 1, p. 746
 | three-dimensional coordinates x,y-plane x,z-plane y,z-plane | Ex. 3, p. 749
 | octant   traces | Ex. 5, p. 749
Section 17.2 | **Partial Derivatives** |
 | partial derivative $\dfrac{\partial z}{\partial x} = f_x(x, y)$ $\left.\dfrac{\partial z}{\partial x}\right|_{(a,b)} = f_x(a, b)$ | Ex. 2, p. 752
Section 17.3 | **Applications of Partial Derivatives** |
 | joint-cost function   production function   marginal productivity | Ex. 3, p. 757
 | competitive products   complementary products | Ex. 4, p. 758
Section 17.4 | **Implicit Partial Differentiation** |
 | implicit partial differentiation | Ex. 1, p. 762
Section 17.5 | **Higher-Order Partial Derivatives** |
 | $\dfrac{\partial^2 z}{\partial y \partial x} = f_{xy}$ $\dfrac{\partial^2 z}{\partial x \partial y} = f_{yx}$ $\dfrac{\partial^2 z}{\partial x^2} = f_{xx}$ $\dfrac{\partial^2 z}{\partial y^2} = f_{yy}$ | Ex. 1, p. 764
Section 17.6 | **Chain Rule** |
 | chain rule   intermediate variable | Ex. 1, p. 766
Section 17.7 | **Maxima and Minima for Functions of Two Variables** |
 | relative maxima and minima   critical point | Ex. 1, p. 771
 | second-derivative test for functions of two variables | Ex. 3, p. 772
Section 17.8 | **Lagrange Multipliers** |
 | Lagrange multipliers | Ex. 1, p. 780
Section 17.9 | **Lines of Regression** |
 | method of least squares   linear regression of y on x index numbers | Ex. 1, p. 788
Section 17.10 | **Multiple Integrals** |
 | double integral   triple integral | Ex. 3, p. 793

Summary

We can extend the concept of a function of one variable to functions of several variables. The inputs for functions of n variables are n-tuples. Generally, the graph of a function of two variables is a surface in a three-dimensional coordinate system.

For a function of n variables, we can consider n partial derivatives. For example, if $w = f(x, y, z)$, we have the partial derivatives of f with respect to x, with respect to y, and with respect to z, denoted f_x, f_y, and f_z, or $\partial f/\partial x$, $\partial f/\partial y$, and $\partial f/\partial z$, respectively. To find $f_x(x, y, z)$, we treat y and z as constants and differentiate f with respect to x in the usual way. The other partial derivatives are found in a similar fashion. We can interpret $f_x(x, y, z)$ as the approximate change in w that results from a one-unit change in x when y and z are held fixed. There are similar interpretations for the other partial derivatives. A function of several variables may be defined implicitly. In this case, its partial derivatives are found by implicit partial differentiation.

Functions of several variables occur frequently in business and economic analysis, as well as in other areas of study. If a manufacturer produces x units of product X and y units of product Y, then the total cost c of these units is a function of x and y and is called a joint-cost function. The partial derivatives $\partial c/\partial x$ and $\partial c/\partial y$ are called the marginal costs with respect to x and y, respectively. We can interpret, for example, $\partial c/\partial x$ as the approximate cost of producing an extra unit of X while the level of production of Y is held fixed.

If l units of labor and k units of capital are used to produce P units of a product, then the function $P = f(l, k)$ is called a production function. The partial derivatives of P are called marginal productivity functions.

Suppose two products, A and B, are such that the quantity demanded of each is dependent on the prices of both. If q_A and q_B are the quantities of A and B demanded when the prices of

A and B are p_A and p_B, respectively, then q_A and q_B are each functions of p_A and p_B. When $\partial q_A / \partial p_B > 0$ and $\partial q_B / \partial p_A > 0$, then A and B are called competitive products (or substitutes). When $\partial q_A / \partial p_B < 0$ and $\partial q_B / \partial p_A < 0$, then A and B are called complementary products.

If $z = f(x, y)$, where $x = x(r, s)$ and $y = y(r, s)$, then z can be considered as a function of r and s. To find, for example, $\partial z / \partial r$, a chain rule can be used:

$$\frac{\partial z}{\partial r} = \frac{\partial z}{\partial x} \frac{\partial x}{\partial r} + \frac{\partial z}{\partial y} \frac{\partial y}{\partial r}$$

A partial derivative of a function of n variables is itself a function of n variables. By successively taking partial derivatives of partial derivatives, we obtain higher-order partial derivatives. For example, if f is a function of x and y, then f_{xy} denotes the partial derivative of f_x with respect to y; f_{xy} is called the second-partial derivative of f, first with respect to x and then with respect to y.

If the function $f(x, y)$ has a relative extremum at (a, b), then (a, b) must be a solution of the system

$$\begin{cases} f_x(x, y) = 0 \\ f_y(x, y) = 0 \end{cases}$$

Any solution of this system is called a critical point of f. Thus, critical points are the candidates at which a relative extremum may occur. The second-derivative test for functions of two variables gives conditions under which a critical point corresponds to a relative maximum or a relative minimum. The test states that if (a, b) is a critical point of f and

$$D(x, y) = f_{xx}(x, y) f_{yy}(x, y) - [f_{xy}(x, y)]^2$$

then

1. if $D(a, b) > 0$ and $f_{xx}(a, b) < 0$, then f has a relative maximum at (a, b);
2. if $D(a, b) > 0$ and $f_{xx}(a, b) > 0$, then f has a relative minimum at (a, b);
3. if $D(a, b) < 0$, then f has a saddle point at (a, b);
4. if $D(a, b) = 0$, no conclusion about an extremum at (a, b) can yet be drawn, and further analysis is required.

To find critical points of a function of several variables, subject to a constraint, we can sometimes use the method of Lagrange multipliers. For example, to find the critical points of $f(x, y, z)$, subject to the constraint $g(x, y, z) = 0$, we first form the function

$$F(x, y, z, \lambda) = f(x, y, z) - \lambda g(x, y, z)$$

By solving the system

$$\begin{cases} F_x = 0 \\ F_y = 0 \\ F_z = 0 \\ F_\lambda = 0 \end{cases}$$

we obtain the critical points of F. If (a, b, c, λ_0) is such a critical point, then (a, b, c) is a critical point of f, subject to the constraint. It is important to write the constraint in the form $g(x, y, z) = 0$. For example, if the constraint is $2x + 3y - z = 4$, then $g(x, y, z) = 2x + 3y - z - 4$. If $f(x, y, z)$ is subject to two constraints, $g_1(x, y, z) = 0$ and $g_2(x, y, z) = 0$, then we would form the function $F = f - \lambda_1 g_1 - \lambda_2 g_2$ and solve the system

$$\begin{cases} F_x = 0 \\ F_y = 0 \\ F_z = 0 \\ F_{\lambda_1} = 0 \\ F_{\lambda_2} = 0 \end{cases}$$

Sometimes sample data for two variables, say, x and y, may be related in such a way that the relationship is approximately linear. When such data points (x_i, y_i), where $i = 1, 2, 3, \ldots, n$, are given to us, we can fit a straight line that approximates them. Such a line is the linear-regression line of y on x and is given by

$$\widehat{y} = a + bx$$

where

$$a = \frac{\left(\sum_{i=1}^{n} x_i^2\right)\left(\sum_{i=1}^{n} y_i\right) - \left(\sum_{i=1}^{n} x_i\right)\left(\sum_{i=1}^{n} x_i y_i\right)}{n \sum_{i=1}^{n} x_i^2 - \left(\sum_{i=1}^{n} x_i\right)^2}$$

and

$$b = \frac{n \sum_{i=1}^{n} x_i y_i - \left(\sum_{i=1}^{n} x_i\right)\left(\sum_{i=1}^{n} y_i\right)}{n \sum_{i=1}^{n} x_i^2 - \left(\sum_{i=1}^{n} x_i\right)^2}$$

The $\widehat{y}$-values can be used to predict y-values for given values of x.

When working with functions of several variables, we can consider their multiple integrals. These are determined by successive integration. For example, the double integral

$$\int_1^2 \int_0^y (x + y) \, dx \, dy$$

is determined by first treating y as a constant and integrating $x + y$ with respect to x. After evaluating between the limits 0 and y, we integrate that result with respect to y from $y = 1$ to $y = 2$. Thus,

$$\int_1^2 \int_0^y (x + y) \, dx \, dy = \int_1^2 \left(\int_0^y (x + y) \, dx \right) dy$$

Triple integrals involve functions of three variables and are also evaluated by successive integration.

Review Problems

Problem numbers shown in color indicate problems suggested for use as a practice chapter test.

In Problems 1–4, sketch the given surfaces.

1. $x + y + z = 1$

2. $z = x$

3. $z = y^2$

4. $x^2 + z^2 = 1$

In Problems 5–16, find the indicated partial derivatives.

5. $f(x, y) = 4x^2 + 6xy + y^2 - 1$; $f_x(x, y), f_y(x, y)$

6. $P = l^3 + k^3 - lk$; $\partial P/\partial l, \partial P/\partial k$

7. $z = \dfrac{x}{x + y}$; $\dfrac{\partial z}{\partial x}, \dfrac{\partial z}{\partial y}$

8. $f(p_A, p_B) = 4(p_A - 10) + 5(p_B - 15)$; $f_{p_B}(p_A, p_B)$

9. $f(x, y) = \ln \sqrt{x^2 + y^2}$; $\dfrac{\partial}{\partial y}[f(x, y)]$

10. $w = \dfrac{x}{\sqrt{x^2 + y^2}}$; $\dfrac{\partial w}{\partial y}$

11. $w = e^{x^2 yz}$; $w_{xy}(x, y, z)$

12. $f(x, y) = xy \ln(xy)$; $f_{xy}(x, y)$

13. $f(x, y, z) = (x + y + z)(x^2 + y^2 + z^2)$; $\dfrac{\partial^2}{\partial z^2}(f(x, y, z))$

14. $z = (x^2 - y)(y^2 - 2xy)$; $\partial^2 z/\partial y^2$

15. $w = e^{x+y+z} \ln xyz$; $\partial w/\partial y, \partial^2 w/\partial z \partial x$

16. $P = 100l^{0.11} k^{0.89}$; $\partial^2 P/\partial k \partial l$

17. If $f(x, y, z) = \dfrac{x + y}{xz}$, find $f_{xyz}(2, 7, 4)$

18. If $f(x, y, z) = (6x + 1)e^{y^2 \ln(z+1)}$, find $f_{xyz}(0, 1, 0)$

26 19. If $w = x^2 + 2xy + 3y^2$, $x = e^r$, and $y = \ln(r + s)$, find $\partial w/\partial r$ and $\partial w/\partial s$.

26 20. If $z = \ln(x/y) + e^{xy} - xy$, $x = r^2 + s^2$, and $y = (r + s)^2$, find $\partial z/\partial s$.

26 21. If $x^2 + 2xy - 2z^2 + xz + 2 = 0$, find $\partial z/\partial x$.

26 22. If $z^2 + \ln(yz) + \ln z + x + z = 0$, find $\partial z/\partial y$.

23. Production Function   If a manufacturer's production function is defined by $P = 20l^{0.7} k^{0.3}$, determine the marginal productivity functions.

24. Joint-Cost Function   A manufacturer's cost for producing x units of product X and y units of product Y is given by

$$c = 3x + 0.05xy + 9y + 500$$

Determine the (partial) marginal cost with respect to x when $x = 50$ and $y = 100$.

25. Competitive/Complementary Products
If $q_A = 100 - p_A + 2p_B$ and $q_B = 150 + 3p_A - 2p_B$, where q_A and q_B are the number of units demanded of products A and B, respectively, and p_A and p_B are their respective prices per unit, determine whether A and B are competitive products or complementary products or neither.

26. Innovation   For industry, the following model describes the rate α (a Greek letter read "alpha") at which an innovation substitutes for an established process:[27]

$$\alpha = Z + 0.530P - 0.027S$$

Here, Z is a constant that depends on the particular industry, P is an index of profitability of the innovation, and S is an index of the extent of the investment necessary to make use of the innovation. Find $\partial \alpha/\partial P$ and $\partial \alpha/\partial S$.

27. Examine $f(x, y) = x^2 + 2y^2 - 2xy - 4y + 3$ for relative extrema.

28. Examine $f(w, z) = 2w^3 + 2z^3 - 6wz + 7$ for relative extrema.

29. Minimizing Material   An open-top rectangular cardboard box is to have a volume of 32 cubic feet. Find the dimensions of the box so that the amount of cardboard used is minimized.

30. The function

$$f(x, y) = ax^2 + by^2 + cxy - 10x - 20y$$

has a critical point at $(x, y) = (1, 2)$, and the second-derivative test is inconclusive at this point. Determine the values of the constants a, b, and c.

31. Maximizing Profit   A dairy produces two types of cheese, A and B, at constant average costs of 50 cents and 60 cents per pound, respectively. When the selling price per pound of A is p_A cents and of B is p_B cents, the demands (in pounds) for A and B, are, respectively,

$$q_A = 250(p_B - p_A)$$

and

$$q_B = 32{,}000 + 250(p_A - 2p_B)$$

Find the selling prices that yield a relative maximum profit. Verify that the profit has a relative maximum at these prices.

32. Find all critical points of $f(x, y, z) = xy^2 z$, subject to the condition that

$$x + y + z - 1 = 0 \ (xyz \neq 0)$$

33. Find all critical points of $f(x, y, z) = x^2 + y^2 + z^2$, subject to the constraint $3x + 2y + z = 14$.

34. Surviving Infection   In an experiment,[28] a group of fish was injected with living bacteria. Of those fish maintained at $28°C$, the percentage p that survived the infection t hours after the injection is given in the following table:

t	8	10	18	20	48
p	82	79	78	78	64

Find the linear-regression line of p on t.

[27]A. P. Hurter, Jr., A. H. Rubenstein, et al., "Market Penetration by New Innovations: The Technological Literature," *Technological Forecasting and Social Change,* 11 (1978), 197–221.

[28]J. B. Covert and W. W. Reynolds, "Survival Value of Fever in Fish," *Nature,* 267, no. 5606 (1977), 43–45.

[26]Problem 19–22 refer to Section 17.4 or Section 17.6.

35. Equipment Expenditures   Find the least squares linear-regression line of y on x for the data given in the following table (refer to year 1993 as year $x = 1$, etc.):

Equipment Expenditures of Allied Computer Company, 1993–1998 (in millions of dollars)

Year	Expenditures
1993	15
1994	22
1995	21
1996	26
1997	27
1998	34

In Problems 36–39, evaluate the double integrals.

36. $\displaystyle\int_{1}^{2}\int_{0}^{y} x^2 y^2\, dx\, dy$

37. $\displaystyle\int_{0}^{1}\int_{0}^{y^2} xy\, dx\, dy$

38. $\displaystyle\int_{1}^{4}\int_{x^2}^{2x} y\, dy\, dx$

39. $\displaystyle\int_{0}^{1}\int_{\sqrt{x}}^{x^2} 7(x^2 + 2xy - 3y^2)\, dy\, dx$

Data Analysis to Model Cooling[29]

In Chapter 15 you worked with Newton's law of cooling, which can be used to describe the temperature of a cooling body with respect to time. Here you will determine that relationship in an empirical way via data analysis. This will illustrate how mathematical models are designed in many real-world situations.

Suppose that you want to create a mathematical model of the cooling of hot tea after it is placed in a refrigerator. To do this you place a pitcher containing hot tea and a thermometer in the refrigerator and periodically read and record the temperature of the tea. Table 17.6 gives the collected data, where T is the Fahrenheit temperature t minutes after the tea is placed in the refrigerator. Initially, that is, at $t = 0$, the temperature is $124°F$; when $t = 391$, then $T = 47°F$. After being in the refrigerator overnight, the temperature is $45°F$. Figure 17.24 gives a graph of the data points (t, T) for $t = 0$ to $t = 391$.

TABLE 17.6

Time t	Temperature T	Time t	Temperature T
0 min	124°F	128 min	64°F
5	118	144	62
10	114	178	59
16	109	208	55
20	106	244	51
35	97	299	50
50	89	331	49
65	82	391	47
85	74	Overnight	45

The pattern of these points strongly suggests that they nearly lie on the graph of a decreasing exponential function, such as the one shown in Figure 17.24. In particular, because the overnight temperature is $45°F$, this exponential function should have $T = 45$ as a horizontal asymptote. Such a function has the form

$$\widehat{T} = Ce^{at} + 45 \tag{1}$$

where $\widehat{T}$ gives the predicted temperature at time t, and C and a are constants with $a < 0$. (Note that since $a < 0$, then as $t \to \infty$, you have $Ce^{at} \to 0$, so $Ce^{at} + 45 \to 45$.)

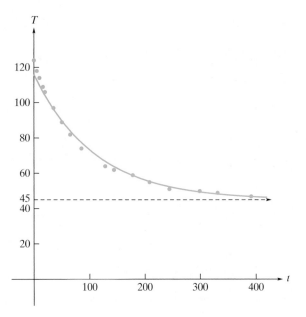

FIGURE 17.24 Data points and exponential approximation.

Now the problem is to find the values of C and a so that the curve given by Equation (1) best fits the data. By writing Equation (1) as

$$\widehat{T} - 45 = Ce^{at}$$

and then taking the natural logarithm of each side, we obtain a linear form:

$$\ln(\widehat{T} - 45) = \ln(Ce^{at})$$

$$\ln(\widehat{T} - 45) = \ln C + \ln e^{at}$$

$$\ln(\widehat{T} - 45) = \ln C + at \tag{2}$$

Letting $\widehat{T}_l = \ln(\widehat{T} - 45)$, Equation (2) becomes

$$\widehat{T}_l = at + \ln C \tag{3}$$

Because a and $\ln C$ are constants, Equation (3) is a linear equation in $\widehat{T}_l$ and t. This means that for the original data, if you plot the points $(t, \ln(T - 45))$, they should nearly lie on a straight line. These points are shown in Figure 17.25, where T_l represents $\ln(T - 45)$. Thus for the line given by Equation (3) that predicts $T_l = \ln(T - 45)$, you can assume that it is the linear regression line of T_l on t. That is, a and $\ln C$ are linear regression coefficients. By using the formulas

[29]Adapted from Gloria Barrett, Dot Doyle, and Dan Teague, "Using Data Analysis in Precalculus to Model Cooling," *The Mathematics Teacher*, 81, no. 8 (November 1988), 680–84. By permission of the National Council of Teachers of Mathematics.

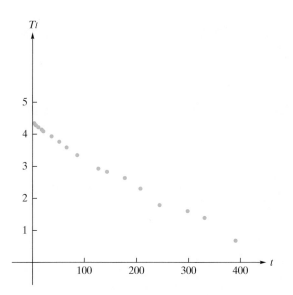

FIGURE 17.25 The points (t, T_l), where $T_l = \ln(T - 45)$, nearly lie on a straight line.

for these coefficients and a calculator, you can determine that

$$a = \dfrac{17\left(\sum\limits_{i=1}^{17} t_i\, T_{l_i}\right) - \left(\sum\limits_{i=1}^{17} t_i\right)\left(\sum\limits_{i=1}^{17} T_{l_i}\right)}{17\left(\sum\limits_{i=1}^{17} t_i^2\right) - \left(\sum\limits_{i=1}^{17} t_i\right)^2} \approx -0.00921$$

and

$$\ln C = \dfrac{\left(\sum\limits_{i=1}^{17} t_i^2\right)\left(\sum\limits_{i=1}^{17} T_{l_i}\right) - \left(\sum\limits_{i=1}^{17} t_i\right)\left(\sum\limits_{i=1}^{17} t_i\, T_{l_i}\right)}{17\left(\sum\limits_{i=1}^{17} t_i^2\right) - \left(\sum\limits_{i=1}^{17} t_i\right)^2}$$

$$\approx 4.260074$$

Since $\ln C \approx 4.260074$, then $C \approx e^{4.260074} \approx 70.82$. Thus from Equation (1),

$$\widehat{T} = 70.82 e^{-0.00921t} + 45$$

which is a model that predicts the temperature of the cooling tea. The graph of this function is the curve shown in Figure 17.24.

Problems

1. Plot the following data points on an x,y-coordinate plane:

x	0	1	4	7	10
y	15	12	9	7	6

Suppose that these points nearly lie on the graph of a decreasing exponential function with horizontal asymptote $y = 5$. Use the technique discussed in this mathematical snapshot to determine the function.

2. Suppose that observed data follow a relation given by $y = C/x^r$, where $x, y, C > 0$. By taking the natural logarithm of each side, show that $\ln x$ and $\ln y$ are linearly related. Thus, the points $(\ln x, \ln y)$ lie on a straight line.

3. Use Newton's law of cooling (see Section 15.6) and the data points $(0, 124)$ and $(128, 64)$ to determine the temperature T of the tea discussed in the snapshot at time t. Assume that the ambient temperature is $45°F$.

4. Try obtaining the final regression equation in the snapshot using the regression capability of a graphing calculator. First use linear regression. How does your result compare with the one in the snapshot? Then try skipping the linear-form transformation and perform an exponential regression. What difficulty do you encounter, if any? How could it be overcome?

APPENDIX A

Compound Interest Tables

$$r = 0.005$$

n	$(1+r)^n$	$(1+r)^{-n}$	$a_{\overline{n}|r}$	$s_{\overline{n}|r}$
1	1.005000	0.995025	0.995025	1.000000
2	1.010025	0.990075	1.985099	2.005000
3	1.015075	0.985149	2.970248	3.015025
4	1.020151	0.980248	3.950496	4.030100
5	1.025251	0.975371	4.925866	5.050251
6	1.030378	0.970518	5.896384	6.075502
7	1.035529	0.965690	6.862074	7.105879
8	1.040707	0.960885	7.822959	8.141409
9	1.045911	0.956105	8.779064	9.182116
10	1.051140	0.951348	9.730412	10.228026
11	1.056396	0.946615	10.677027	11.279167
12	1.061678	0.941905	11.618932	12.335562
13	1.066986	0.937219	12.556151	13.397240
14	1.072321	0.932556	13.488708	14.464226
15	1.077683	0.927917	14.416625	15.536548
16	1.083071	0.923300	15.339925	16.614230
17	1.088487	0.918707	16.258632	17.697301
18	1.093929	0.914136	17.172768	18.785788
19	1.099399	0.909588	18.082356	19.879717
20	1.104896	0.905063	18.987419	20.979115
21	1.110420	0.900560	19.887979	22.084011
22	1.115972	0.896080	20.784059	23.194431
23	1.121552	0.891622	21.675681	24.310403
24	1.127160	0.887186	22.562866	25.431955
25	1.132796	0.882772	23.445638	26.559115
26	1.138460	0.878380	24.324018	27.691911
27	1.144152	0.874010	25.198028	28.830370
28	1.149873	0.869662	26.067689	29.974522
29	1.155622	0.865335	26.933024	31.124395
30	1.161400	0.861030	27.794054	32.280017
31	1.167207	0.856746	28.650800	33.441417
32	1.173043	0.852484	29.503284	34.608624
33	1.178908	0.848242	30.351526	35.781667
34	1.184803	0.844022	31.195548	36.960575
35	1.190727	0.839823	32.035371	38.145378
36	1.196681	0.835645	32.871016	39.336105
37	1.202664	0.831487	33.702504	40.532785
38	1.208677	0.827351	34.529854	41.735449
39	1.214721	0.823235	35.353089	42.944127
40	1.220794	0.819139	36.172228	44.158847
41	1.226898	0.815064	36.987291	45.379642
42	1.233033	0.811009	37.798300	46.606540
43	1.239198	0.806974	38.605274	47.839572
44	1.245394	0.802959	39.408232	49.078770
45	1.251621	0.798964	40.207196	50.324164
46	1.257879	0.794989	41.002185	51.575785
47	1.264168	0.791034	41.793219	52.833664
48	1.270489	0.787098	42.580318	54.097832
49	1.276842	0.783182	43.363500	55.368321
50	1.283226	0.779286	44.142786	56.645163

$$r = 0.0075$$

n	$(1+r)^n$	$(1+r)^{-n}$	$a_{\overline{n}|r}$	$s_{\overline{n}|r}$
1	1.007500	0.992556	0.992556	1.000000
2	1.015056	0.985167	1.977723	2.007500
3	1.022669	0.977833	2.955556	3.022556
4	1.030339	0.970554	3.926110	4.045225
5	1.038067	0.963329	4.889440	5.075565
6	1.045852	0.956158	5.845598	6.113631
7	1.053696	0.949040	6.794638	7.159484
8	1.061599	0.941975	7.736613	8.213180
9	1.069561	0.934963	8.671576	9.274779
10	1.077583	0.928003	9.599580	10.344339
11	1.085664	0.921095	10.520675	11.421922
12	1.093807	0.914238	11.434913	12.507586
13	1.102010	0.907432	12.342345	13.601393
14	1.110276	0.900677	13.243022	14.703404
15	1.118603	0.893973	14.136995	15.813679
16	1.126992	0.887318	15.024313	16.932282
17	1.135445	0.880712	15.905025	18.059274
18	1.143960	0.874156	16.779181	19.194718
19	1.152540	0.867649	17.646830	20.338679
20	1.161184	0.861190	18.508020	21.491219
21	1.169893	0.854779	19.362799	22.652403
22	1.178667	0.848416	20.211215	23.822296
23	1.187507	0.842100	21.053315	25.000963
24	1.196414	0.835831	21.889146	26.188471
25	1.205387	0.829609	22.718755	27.384884
26	1.214427	0.823434	23.542189	28.590271
27	1.223535	0.817304	24.359493	29.804698
28	1.232712	0.811220	25.170713	31.028233
29	1.241957	0.805181	25.975893	32.260945
30	1.251272	0.799187	26.775080	33.502902
31	1.260656	0.793238	27.568318	34.754174
32	1.270111	0.787333	28.355650	36.014830
33	1.279637	0.781472	29.137122	37.284941
34	1.289234	0.775654	29.912776	38.564578
35	1.298904	0.769880	30.682656	39.853813
36	1.308645	0.764149	31.446805	41.152716
37	1.318460	0.758461	32.205266	42.461361
38	1.328349	0.752814	32.958080	43.779822
39	1.338311	0.747210	33.705290	45.108170
40	1.348349	0.741648	34.446938	46.446482
41	1.358461	0.736127	35.183065	47.794830
42	1.368650	0.730647	35.913713	49.153291
43	1.378915	0.725208	36.638921	50.521941
44	1.389256	0.719810	37.358730	51.900856
45	1.399676	0.714451	38.073181	53.290112
46	1.410173	0.709133	38.782314	54.689788
47	1.420750	0.703854	39.486168	56.099961
48	1.431405	0.698614	40.184782	57.520711
49	1.442141	0.693414	40.878195	58.952116
50	1.452957	0.688252	41.566447	60.394257

		$r = 0.01$		
n	$(1+r)^n$	$(1+r)^{-n}$	$a_{\overline{n}|r}$	$s_{\overline{n}|r}$
1	1.010000	0.990099	0.990099	1.000000
2	1.020100	0.980296	1.970395	2.010000
3	1.030301	0.970590	2.940985	3.030100
4	1.040604	0.960980	3.901966	4.060401
5	1.051010	0.951466	4.853431	5.101005
6	1.061520	0.942045	5.795476	6.152015
7	1.072135	0.932718	6.728195	7.213535
8	1.082857	0.923483	7.651678	8.285671
9	1.093685	0.914340	8.566018	9.368527
10	1.104622	0.905287	9.471305	10.462213
11	1.115668	0.896324	10.367628	11.566835
12	1.126825	0.887449	11.255077	12.682503
13	1.138093	0.878663	12.133740	13.809328
14	1.149474	0.869963	13.003703	14.947421
15	1.160969	0.861349	13.865053	16.096896
16	1.172579	0.852821	14.717874	17.257864
17	1.184304	0.844377	15.562251	18.430443
18	1.196147	0.836017	16.398269	19.614748
19	1.208109	0.827740	17.226008	20.810895
20	1.220190	0.819544	18.045553	22.019004
21	1.232392	0.811430	18.856983	23.239194
22	1.244716	0.803396	19.660379	24.471586
23	1.257163	0.795442	20.455821	25.716302
24	1.269735	0.787566	21.243387	26.973465
25	1.282432	0.779768	22.023156	28.243200
26	1.295256	0.772048	22.795204	29.525631
27	1.308209	0.764404	23.559608	30.820888
28	1.321291	0.756836	24.316443	32.129097
29	1.334504	0.749342	25.065785	33.450388
30	1.347849	0.741923	25.807708	34.784892
31	1.361327	0.734577	26.542285	36.132740
32	1.374941	0.727304	27.269589	37.494068
33	1.388690	0.720103	27.989693	38.869009
34	1.402577	0.712973	28.702666	40.257699
35	1.416603	0.705914	29.408580	41.660276
36	1.430769	0.698925	30.107505	43.076878
37	1.445076	0.692005	30.799510	44.507647
38	1.459527	0.685153	31.484663	45.952724
39	1.474123	0.678370	32.163033	47.412251
40	1.488864	0.671653	32.834686	48.886373
41	1.503752	0.665003	33.499689	50.375237
42	1.518790	0.658419	34.158108	51.878989
43	1.533978	0.651900	34.810008	53.397779
44	1.549318	0.645445	35.455454	54.931757
45	1.564811	0.639055	36.094508	56.481075
46	1.580459	0.632728	36.727236	58.045885
47	1.596263	0.626463	37.353699	59.626344
48	1.612226	0.620260	37.973959	61.222608
49	1.628348	0.614119	38.588079	62.834834
50	1.644632	0.608039	39.196118	64.463182

	$r = 0.0125$			
n	$(1+r)^n$	$(1+r)^{-n}$	$a_{\overline{n}|r}$	$s_{\overline{n}|r}$
1	1.012500	0.987654	0.987654	1.000000
2	1.025156	0.975461	1.963115	2.012500
3	1.037971	0.963418	2.926534	3.037656
4	1.050945	0.951524	3.878058	4.075627
5	1.064082	0.939777	4.817835	5.126572
6	1.077383	0.928175	5.746010	6.190654
7	1.090850	0.916716	6.662726	7.268038
8	1.104486	0.905398	7.568124	8.358888
9	1.118292	0.894221	8.462345	9.463374
10	1.132271	0.883181	9.345526	10.581666
11	1.146424	0.872277	10.217803	11.713937
12	1.160755	0.861509	11.079312	12.860361
13	1.175264	0.850873	11.930185	14.021116
14	1.189955	0.840368	12.770553	15.196380
15	1.204829	0.829993	13.600546	16.386335
16	1.219890	0.819746	14.420292	17.591164
17	1.235138	0.809626	15.229918	18.811053
18	1.250577	0.799631	16.029549	20.046192
19	1.266210	0.789759	16.819308	21.296769
20	1.282037	0.780009	17.599316	22.562979
21	1.298063	0.770379	18.369695	23.845016
22	1.314288	0.760868	19.130563	25.143078
23	1.330717	0.751475	19.882037	26.457367
24	1.347351	0.742197	20.624235	27.788084
25	1.364193	0.733034	21.357269	29.135435
26	1.381245	0.723984	22.081253	30.499628
27	1.398511	0.715046	22.796299	31.880873
28	1.415992	0.706219	23.502518	33.279384
29	1.433692	0.697500	24.200018	34.695377
30	1.451613	0.688889	24.888906	36.129069
31	1.469759	0.680384	25.569290	37.580682
32	1.488131	0.671984	26.241274	39.050441
33	1.506732	0.663688	26.904962	40.538571
34	1.525566	0.655494	27.560456	42.045303
35	1.544636	0.647402	28.207858	43.570870
36	1.563944	0.639409	28.847267	45.115505
37	1.583493	0.631515	29.478783	46.679449
38	1.603287	0.623719	30.102501	48.262942
39	1.623328	0.616019	30.718520	49.866229
40	1.643619	0.608413	31.326933	51.489557
41	1.664165	0.600902	31.927835	53.133177
42	1.684967	0.593484	32.521319	54.797341
43	1.706029	0.586157	33.107475	56.482308
44	1.727354	0.578920	33.686395	58.188337
45	1.748946	0.571773	34.258168	59.915691
46	1.770808	0.564714	34.822882	61.664637
47	1.792943	0.557742	35.380624	63.435445
48	1.815355	0.550856	35.931481	65.228388
49	1.838047	0.544056	36.475537	67.043743
50	1.861022	0.537339	37.012876	68.881790

		$r = 0.015$		
n	$(1+r)^n$	$(1+r)^{-n}$	$a_{\overline{n}|r}$	$s_{\overline{n}|r}$
1	1.015000	0.985222	0.985222	1.000000
2	1.030225	0.970662	1.955883	2.015000
3	1.045678	0.956317	2.912200	3.045225
4	1.061364	0.942184	3.854385	4.090903
5	1.077284	0.928260	4.782645	5.152267
6	1.093443	0.914542	5.697187	6.229551
7	1.109845	0.901027	6.598214	7.322994
8	1.126493	0.887711	7.485925	8.432839
9	1.143390	0.874592	8.360517	9.559332
10	1.160541	0.861667	9.222185	10.702722
11	1.177949	0.848933	10.071118	11.863262
12	1.195618	0.836387	10.907505	13.041211
13	1.213552	0.824027	11.731532	14.236830
14	1.231756	0.811849	12.543382	15.450382
15	1.250232	0.799852	13.343233	16.682138
16	1.268986	0.788031	14.131264	17.932370
17	1.288020	0.776385	14.907649	19.201355
18	1.307341	0.764912	15.672561	20.489376
19	1.326951	0.753607	16.426168	21.796716
20	1.346855	0.742470	17.168639	23.123667
21	1.367058	0.731498	17.900137	24.470522
22	1.387564	0.720688	18.620824	25.837580
23	1.408377	0.710037	19.330861	27.225144
24	1.429503	0.699544	20.030405	28.633521
25	1.450945	0.689206	20.719611	30.063024
26	1.472710	0.679021	21.398632	31.513969
27	1.494800	0.668986	22.067617	32.986678
28	1.517222	0.659099	22.726717	34.481479
29	1.539981	0.649359	23.376076	35.998701
30	1.563080	0.639762	24.015838	37.538681
31	1.586526	0.630308	24.646146	39.101762
32	1.610324	0.620993	25.267139	40.688288
33	1.634479	0.611816	25.878954	42.298612
34	1.658996	0.602774	26.481728	43.933092
35	1.683881	0.593866	27.075595	45.592088
36	1.709140	0.585090	27.660684	47.275969
37	1.734777	0.576443	28.237127	48.985109
38	1.760798	0.567924	28.805052	50.719885
39	1.787210	0.559531	29.364583	52.480684
40	1.814018	0.551262	29.915845	54.267894
41	1.841229	0.543116	30.458961	56.081912
42	1.868847	0.535089	30.994050	57.923141
43	1.896880	0.527182	31.521232	59.791988
44	1.925333	0.519391	32.040622	61.688868
45	1.954213	0.511715	32.552337	63.614201
46	1.983526	0.504153	33.056490	65.568414
47	2.013279	0.496702	33.553192	67.551940
48	2.043478	0.489362	34.042554	69.565219
49	2.074130	0.482130	34.524683	71.608698
50	2.105242	0.475005	34.999688	73.682828

		$r = 0.02$		
n	$(1+r)^n$	$(1+r)^{-n}$	$a_{\overline{n}|r}$	$s_{\overline{n}|r}$
1	1.020000	0.980392	0.980392	1.000000
2	1.040400	0.961169	1.941561	2.020000
3	1.061208	0.942322	2.883883	3.060400
4	1.082432	0.923845	3.807729	4.121608
5	1.104081	0.905731	4.713460	5.204040
6	1.126162	0.887971	5.601431	6.308121
7	1.148686	0.870560	6.471991	7.434283
8	1.171659	0.853490	7.325481	8.582969
9	1.195093	0.836755	8.162237	9.754628
10	1.218994	0.820348	8.982585	10.949721
11	1.243374	0.804263	9.786848	12.168715
12	1.268242	0.788493	10.575341	13.412090
13	1.293607	0.773033	11.348374	14.680332
14	1.319479	0.757875	12.106249	15.973938
15	1.345868	0.743015	12.849264	17.293417
16	1.372786	0.728446	13.577709	18.639285
17	1.400241	0.714163	14.291872	20.012071
18	1.428246	0.700159	14.992031	21.412312
19	1.456811	0.686431	15.678462	22.840559
20	1.485947	0.672971	16.351433	24.297370
21	1.515666	0.659776	17.011209	25.783317
22	1.545980	0.646839	17.658048	27.298984
23	1.576899	0.634156	18.292204	28.844963
24	1.608437	0.621721	18.913926	30.421862
25	1.640606	0.609531	19.523456	32.030300
26	1.673418	0.597579	20.121036	33.670906
27	1.706886	0.585862	20.706898	35.344324
28	1.741024	0.574375	21.281272	37.051210
29	1.775845	0.563112	21.844385	38.792235
30	1.811362	0.552071	22.396456	40.568079
31	1.847589	0.541246	22.937702	42.379441
32	1.884541	0.530633	23.468335	44.227030
33	1.922231	0.520229	23.988564	46.111570
34	1.960676	0.510028	24.498592	48.033802
35	1.999890	0.500028	24.998619	49.994478
36	2.039887	0.490223	25.488842	51.994367
37	2.080685	0.480611	25.969453	54.034255
38	2.122299	0.471187	26.440641	56.114940
39	2.164745	0.461948	26.902589	58.237238
40	2.208040	0.452890	27.355479	60.401983
41	2.252200	0.444010	27.799489	62.610023
42	2.297244	0.435304	28.234794	64.862223
43	2.343189	0.426769	28.661562	67.159468
44	2.390053	0.418401	29.079963	69.502657
45	2.437854	0.410197	29.490160	71.892710
46	2.486611	0.402154	29.892314	74.330564
47	2.536344	0.394268	30.286582	76.817176
48	2.587070	0.386538	30.673120	79.353519
49	2.638812	0.378958	31.052078	81.940590
50	2.691588	0.371528	31.423606	84.579401

		$r = 0.025$		
n	$(1+r)^n$	$(1+r)^{-n}$	$a_{\overline{n}|r}$	$s_{\overline{n}|r}$
1	1.025000	0.975610	0.975610	1.000000
2	1.050625	0.951814	1.927424	2.025000
3	1.076891	0.928599	2.856024	3.075625
4	1.103813	0.905951	3.761974	4.152516
5	1.131408	0.883854	4.645828	5.256329
6	1.159693	0.862297	5.508125	6.387737
7	1.188686	0.841265	6.349391	7.547430
8	1.218403	0.820747	7.170137	8.736116
9	1.248863	0.800728	7.970866	9.954519
10	1.280085	0.781198	8.752064	11.203382
11	1.312087	0.762145	9.514209	12.483466
12	1.344889	0.743556	10.257765	13.795553
13	1.378511	0.725420	10.983185	15.140442
14	1.412974	0.707727	11.690912	16.518953
15	1.448298	0.690466	12.381378	17.931927
16	1.484506	0.673625	13.055003	19.380225
17	1.521618	0.657195	13.712198	20.864730
18	1.559659	0.641166	14.353364	22.386349
19	1.598650	0.625528	14.978891	23.946007
20	1.638616	0.610271	15.589162	25.544658
21	1.679582	0.595386	16.184549	27.183274
22	1.721571	0.580865	16.765413	28.862856
23	1.764611	0.566697	17.332110	30.584427
24	1.808726	0.552875	17.884986	32.349038
25	1.853944	0.539391	18.424376	34.157764
26	1.900293	0.526235	18.950611	36.011708
27	1.947800	0.513400	19.464011	37.912001
28	1.996495	0.500878	19.964889	39.859801
29	2.046407	0.488661	20.453550	41.856296
30	2.097568	0.476743	20.930293	43.902703
31	2.150007	0.465115	21.395407	46.000271
32	2.203757	0.453771	21.849178	48.150278
33	2.258851	0.442703	22.291881	50.354034
34	2.315322	0.431905	22.723786	52.612885
35	2.373205	0.421371	23.145157	54.928207
36	2.432535	0.411094	23.556251	57.301413
37	2.493349	0.401067	23.957318	59.733948
38	2.555682	0.391285	24.348603	62.227297
39	2.619574	0.381741	24.730344	64.782979
40	2.685064	0.372431	25.102775	67.402554
41	2.752190	0.363347	25.466122	70.087617
42	2.820995	0.354485	25.820607	72.839808
43	2.891520	0.345839	26.166446	75.660803
44	2.963808	0.337404	26.503849	78.552323
45	3.037903	0.329174	26.833024	81.516131
46	3.113851	0.321146	27.154170	84.554034
47	3.191697	0.313313	27.467483	87.667885
48	3.271490	0.305671	27.773154	90.859582
49	3.353277	0.298216	28.071369	94.131072
50	3.437109	0.290942	28.362312	97.484349

		$r = 0.03$		
n	$(1+r)^n$	$(1+r)^{-n}$	$a_{\overline{n}|r}$	$s_{\overline{n}|r}$
1	1.030000	0.970874	0.970874	1.000000
2	1.060900	0.942596	1.913470	2.030000
3	1.092727	0.915142	2.828611	3.090900
4	1.125509	0.888487	3.717098	4.183627
5	1.159274	0.862609	4.579707	5.309136
6	1.194052	0.837484	5.417191	6.468410
7	1.229874	0.813092	6.230283	7.662462
8	1.266770	0.789409	7.019692	8.892336
9	1.304773	0.766417	7.786109	10.159106
10	1.343916	0.744094	8.530203	11.463879
11	1.384234	0.722421	9.252624	12.807796
12	1.425761	0.701380	9.954004	14.192030
13	1.468534	0.680951	10.634955	15.617790
14	1.512590	0.661118	11.296073	17.086324
15	1.557967	0.641862	11.937935	18.598914
16	1.604706	0.623167	12.561102	20.156881
17	1.652848	0.605016	13.166118	21.761588
18	1.702433	0.587395	13.753513	23.414435
19	1.753506	0.570286	14.323799	25.116868
20	1.806111	0.553676	14.877475	26.870374
21	1.860295	0.537549	15.415024	28.676486
22	1.916103	0.521893	15.936917	30.536780
23	1.973587	0.506692	16.443608	32.452884
24	2.032794	0.491934	16.935542	34.426470
25	2.093778	0.477606	17.413148	36.459264
26	2.156591	0.463695	17.876842	38.553042
27	2.221289	0.450189	18.327031	40.709634
28	2.287928	0.437077	18.764108	42.930923
29	2.356566	0.424346	19.188455	45.218850
30	2.427262	0.411987	19.600441	47.575416
31	2.500080	0.399987	20.000428	50.002678
32	2.575083	0.388337	20.388766	52.502759
33	2.652335	0.377026	20.765792	55.077841
34	2.731905	0.366045	21.131837	57.730177
35	2.813862	0.355383	21.487220	60.462082
36	2.898278	0.345032	21.832252	63.275944
37	2.985227	0.334983	22.167235	66.174223
38	3.074783	0.325226	22.492462	69.159449
39	3.167027	0.315754	22.808215	72.234233
40	3.262038	0.306557	23.114772	75.401260
41	3.359899	0.297628	23.412400	78.663298
42	3.460696	0.288959	23.701359	82.023196
43	3.564517	0.280543	23.981902	85.483892
44	3.671452	0.272372	24.254274	89.048409
45	3.781596	0.264439	24.518713	92.719861
46	3.895044	0.256737	24.775449	96.501457
47	4.011895	0.249259	25.024708	100.396501
48	4.132252	0.241999	25.266707	104.408396
49	4.256219	0.234950	25.501657	108.540648
50	4.383906	0.228107	25.729764	112.796867

		$r = 0.035$		
n	$(1+r)^n$	$(1+r)^{-n}$	$a_{\overline{n}|r}$	$s_{\overline{n}|r}$
1	1.035000	0.966184	0.966184	1.000000
2	1.071225	0.933511	1.899694	2.035000
3	1.108718	0.901943	2.801637	3.106225
4	1.147523	0.871442	3.673079	4.214943
5	1.187686	0.841973	4.515052	5.362466
6	1.229255	0.813501	5.328553	6.550152
7	1.272279	0.785991	6.114544	7.779408
8	1.316809	0.759412	6.873956	9.051687
9	1.362897	0.733731	7.607687	10.368496
10	1.410599	0.708919	8.316605	11.731393
11	1.459970	0.684946	9.001551	13.141992
12	1.511069	0.661783	9.663334	14.601962
13	1.563956	0.639404	10.302738	16.113030
14	1.618695	0.617782	10.920520	17.676986
15	1.675349	0.596891	11.517411	19.295681
16	1.733986	0.576706	12.094117	20.971030
17	1.794676	0.557204	12.651321	22.705016
18	1.857489	0.538361	13.189682	24.499691
19	1.922501	0.520156	13.709837	26.357180
20	1.989789	0.502566	14.212403	28.279682
21	2.059431	0.485571	14.697974	30.269471
22	2.131512	0.469151	15.167125	32.328902
23	2.206114	0.453286	15.620410	34.460414
24	2.283328	0.437957	16.058368	36.666528
25	2.363245	0.423147	16.481515	38.949857
26	2.445959	0.408838	16.890352	41.313102
27	2.531567	0.395012	17.285365	43.759060
28	2.620172	0.381654	17.667019	46.290627
29	2.711878	0.368748	18.035767	48.910799
30	2.806794	0.356278	18.392045	51.622677
31	2.905031	0.344230	18.736276	54.429471
32	3.006708	0.332590	19.068865	57.334502
33	3.111942	0.321343	19.390208	60.341210
34	3.220860	0.310476	19.700684	63.453152
35	3.333590	0.299977	20.000661	66.674013
36	3.450266	0.289833	20.290494	70.007603
37	3.571025	0.280032	20.570525	73.457869
38	3.696011	0.270562	20.841087	77.028895
39	3.825372	0.261413	21.102500	80.724906
40	3.959260	0.252572	21.355072	84.550278
41	4.097834	0.244031	21.599104	88.509537
42	4.241258	0.235779	21.834883	92.607371
43	4.389702	0.227806	22.062689	96.848629
44	4.543342	0.220102	22.282791	101.238331
45	4.702359	0.212659	22.495450	105.781673
46	4.866941	0.205468	22.700918	110.484031
47	5.037284	0.198520	22.899438	115.350973
48	5.213589	0.191806	23.091244	120.388257
49	5.396065	0.185320	23.276564	125.601846
50	5.584927	0.179053	23.455618	130.997910

		$r = 0.04$		
n	$(1+r)^n$	$(1+r)^{-n}$	$a_{\overline{n}|r}$	$s_{\overline{n}|r}$
1	1.040000	0.961538	0.961538	1.000000
2	1.081600	0.924556	1.886095	2.040000
3	1.124864	0.888996	2.775091	3.121600
4	1.169859	0.854804	3.629895	4.246464
5	1.216653	0.821927	4.451822	5.416323
6	1.265319	0.790315	5.242137	6.632975
7	1.315932	0.759918	6.002055	7.898294
8	1.368569	0.730690	6.732745	9.214226
9	1.423312	0.702587	7.435332	10.582795
10	1.480244	0.675564	8.110896	12.006107
11	1.539454	0.649581	8.760477	13.486351
12	1.601032	0.624597	9.385074	15.025805
13	1.665074	0.600574	9.985648	16.626838
14	1.731676	0.577475	10.563123	18.291911
15	1.800944	0.555265	11.118387	20.023588
16	1.872981	0.533908	11.652296	21.824531
17	1.947900	0.513373	12.165669	23.697512
18	2.025817	0.493628	12.659297	25.645413
19	2.106849	0.474642	13.133939	27.671229
20	2.191123	0.456387	13.590326	29.778079
21	2.278768	0.438834	14.029160	31.969202
22	2.369919	0.421955	14.451115	34.247970
23	2.464716	0.405726	14.856842	36.617889
24	2.563304	0.390121	15.246963	39.082604
25	2.665836	0.375117	15.622080	41.645908
26	2.772470	0.360689	15.982769	44.311745
27	2.883369	0.346817	16.329586	47.084214
28	2.998703	0.333477	16.663063	49.967583
29	3.118651	0.320651	16.983715	52.966286
30	3.243398	0.308319	17.292033	56.084938
31	3.373133	0.296460	17.588494	59.328335
32	3.508059	0.285058	17.873551	62.701469
33	3.648381	0.274094	18.147646	66.209527
34	3.794316	0.263552	18.411198	69.857909
35	3.946089	0.253415	18.664613	73.652225
36	4.103933	0.243669	18.908282	77.598314
37	4.268090	0.234297	19.142579	81.702246
38	4.438813	0.225285	19.367864	85.970336
39	4.616366	0.216621	19.584485	90.409150
40	4.801021	0.208289	19.792774	95.025516
41	4.993061	0.200278	19.993052	99.826536
42	5.192784	0.192575	20.185627	104.819598
43	5.400495	0.185168	20.370795	110.012382
44	5.616515	0.178046	20.548841	115.412877
45	5.841176	0.171198	20.720040	121.029392
46	6.074823	0.164614	20.884654	126.870568
47	6.317816	0.158283	21.042936	132.945390
48	6.570528	0.152195	21.195131	139.263206
49	6.833349	0.146341	21.341472	145.833734
50	7.106683	0.140713	21.482185	152.667084

| | | $r = 0.05$ | | |
n	$(1+r)^n$	$(1+r)^{-n}$	$a_{\overline{n}|r}$	$s_{\overline{n}|r}$
1	1.050000	0.952381	0.952381	1.000000
2	1.102500	0.907029	1.859410	2.050000
3	1.157625	0.863838	2.723248	3.152500
4	1.215506	0.822702	3.545951	4.310125
5	1.276282	0.783526	4.329477	5.525631
6	1.340096	0.746215	5.075692	6.801913
7	1.407100	0.710681	5.786373	8.142008
8	1.477455	0.676839	6.463213	9.549109
9	1.551328	0.644609	7.107822	11.026564
10	1.628895	0.613913	7.721735	12.577893
11	1.710339	0.584679	8.306414	14.206787
12	1.795856	0.556837	8.863252	15.917127
13	1.885649	0.530321	9.393573	17.712983
14	1.979932	0.505068	9.898641	19.598632
15	2.078928	0.481017	10.379658	21.578564
16	2.182875	0.458112	10.837770	23.657492
17	2.292018	0.436297	11.274066	25.840366
18	2.406619	0.415521	11.689587	28.132385
19	2.526950	0.395734	12.085321	30.539004
20	2.653298	0.376889	12.462210	33.065954
21	2.785963	0.358942	12.821153	35.719252
22	2.925261	0.341850	13.163003	38.505214
23	3.071524	0.325571	13.488574	41.430475
24	3.225100	0.310068	13.798642	44.501999
25	3.386355	0.295303	14.093945	47.727099
26	3.555673	0.281241	14.375185	51.113454
27	3.733456	0.267848	14.643034	54.669126
28	3.920129	0.255094	14.898127	58.402583
29	4.116136	0.242946	15.141074	62.322712
30	4.321942	0.231377	15.372451	66.438848
31	4.538039	0.220359	15.592811	70.760790
32	4.764941	0.209866	15.802677	75.298829
33	5.003189	0.199873	16.002549	80.063771
34	5.253348	0.190355	16.192904	85.066959
35	5.516015	0.181290	16.374194	90.320307
36	5.791816	0.172657	16.546852	95.836323
37	6.081407	0.164436	16.711287	101.628139
38	6.385477	0.156605	16.867893	107.709546
39	6.704751	0.149148	17.017041	114.095023
40	7.039989	0.142046	17.159086	120.799774
41	7.391988	0.135282	17.294368	127.839763
42	7.761588	0.128840	17.423208	135.231751
43	8.149667	0.122704	17.545912	142.993339
44	8.557150	0.116861	17.662773	151.143006
45	8.985008	0.111297	17.774070	159.700156
46	9.434258	0.105997	17.880066	168.685164
47	9.905971	0.100949	17.981016	178.119422
48	10.401270	0.096142	18.077158	188.025393
49	10.921333	0.091564	18.168722	198.426663
50	11.467400	0.087204	18.255925	209.347996

		$r = 0.06$		
n	$(1+r)^n$	$(1+r)^{-n}$	$a_{\overline{n}|r}$	$s_{\overline{n}|r}$
1	1.060000	0.943396	0.943396	1.000000
2	1.123600	0.889996	1.833393	2.060000
3	1.191016	0.839619	2.673012	3.183600
4	1.262477	0.792094	3.465106	4.374616
5	1.338226	0.747258	4.212364	5.637093
6	1.418519	0.704961	4.917324	6.975319
7	1.503630	0.665057	5.582381	8.393838
8	1.593848	0.627412	6.209794	9.897468
9	1.689479	0.591898	6.801692	11.491316
10	1.790848	0.558395	7.360087	13.180795
11	1.898299	0.526788	7.886875	14.971643
12	2.012196	0.496969	8.383844	16.869941
13	2.132928	0.468839	8.852683	18.882138
14	2.260904	0.442301	9.294984	21.015066
15	2.396558	0.417265	9.712249	23.275970
16	2.540352	0.393646	10.105895	25.672528
17	2.692773	0.371364	10.477260	28.212880
18	2.854339	0.350344	10.827603	30.905653
19	3.025600	0.330513	11.158116	33.759992
20	3.207135	0.311805	11.469921	36.785591
21	3.399564	0.294155	11.764077	39.992727
22	3.603537	0.277505	12.041582	43.392290
23	3.819750	0.261797	12.303379	46.995828
24	4.048935	0.246979	12.550358	50.815577
25	4.291871	0.232999	12.783356	54.864512
26	4.549383	0.219810	13.003166	59.156383
27	4.822346	0.207368	13.210534	63.705766
28	5.111687	0.195630	13.406164	68.528112
29	5.418388	0.184557	13.590721	73.639798
30	5.743491	0.174110	13.764831	79.058186
31	6.088101	0.164255	13.929086	84.801677
32	6.453387	0.154957	14.084043	90.889778
33	6.840590	0.146186	14.230230	97.343165
34	7.251025	0.137912	14.368141	104.183755
35	7.686087	0.130105	14.498246	111.434780
36	8.147252	0.122741	14.620987	119.120867
37	8.636087	0.115793	14.736780	127.268119
38	9.154252	0.109239	14.846019	135.904206
39	9.703507	0.103056	14.949075	145.058458
40	10.285718	0.097222	15.046297	154.761966
41	10.902861	0.091719	15.138016	165.047684
42	11.557033	0.086527	15.224543	175.950545
43	12.250455	0.081630	15.306173	187.507577
44	12.985482	0.077009	15.383182	199.758032
45	13.764611	0.072650	15.455832	212.743514
46	14.590487	0.068538	15.524370	226.508125
47	15.465917	0.064658	15.589028	241.098612
48	16.393872	0.060998	15.650027	256.564529
49	17.377504	0.057546	15.707572	272.958401
50	18.420154	0.054288	15.761861	290.335905

		$r = 0.07$		
n	$(1+r)^n$	$(1+r)^{-n}$	$a_{\overline{n}|r}$	$s_{\overline{n}|r}$
1	1.070000	0.934579	0.934579	1.000000
2	1.144900	0.873439	1.808018	2.070000
3	1.225043	0.816298	2.624316	3.214900
4	1.310796	0.762895	3.387211	4.439943
5	1.402552	0.712986	4.100197	5.750739
6	1.500730	0.666342	4.766540	7.153291
7	1.605781	0.622750	5.389289	8.654021
8	1.718186	0.582009	5.971299	10.259803
9	1.838459	0.543934	6.515232	11.977989
10	1.967151	0.508349	7.023582	13.816448
11	2.104852	0.475093	7.498674	15.783599
12	2.252192	0.444012	7.942686	17.888451
13	2.409845	0.414964	8.357651	20.140643
14	2.578534	0.387817	8.745468	22.550488
15	2.759032	0.362446	9.107914	25.129022
16	2.952164	0.338735	9.446649	27.888054
17	3.158815	0.316574	9.763223	30.840217
18	3.379932	0.295864	10.059087	33.999033
19	3.616528	0.276508	10.335595	37.378965
20	3.869684	0.258419	10.594014	40.995492
21	4.140562	0.241513	10.835527	44.865177
22	4.430402	0.225713	11.061240	49.005739
23	4.740530	0.210947	11.272187	53.436141
24	5.072367	0.197147	11.469334	58.176671
25	5.427433	0.184249	11.653583	63.249038
26	5.807353	0.172195	11.825779	68.676470
27	6.213868	0.160930	11.986709	74.483823
28	6.648838	0.150402	12.137111	80.697691
29	7.114257	0.140563	12.277674	87.346529
30	7.612255	0.131367	12.409041	94.460786
31	8.145113	0.122773	12.531814	102.073041
32	8.715271	0.114741	12.646555	110.218154
33	9.325340	0.107235	12.753790	118.933425
34	9.978114	0.100219	12.854009	128.258765
35	10.676581	0.093663	12.947672	138.236878
36	11.423942	0.087535	13.035208	148.913460
37	12.223618	0.081809	13.117017	160.337402
38	13.079271	0.076457	13.193473	172.561020
39	13.994820	0.071455	13.264928	185.640292
40	14.974458	0.066780	13.331709	199.635112
41	16.022670	0.062412	13.394120	214.609570
42	17.144257	0.058329	13.452449	230.632240
43	18.344355	0.054513	13.506962	247.776496
44	19.628460	0.050946	13.557908	266.120851
45	21.002452	0.047613	13.605522	285.749311
46	22.472623	0.044499	13.650020	306.751763
47	24.045707	0.041587	13.691608	329.224386
48	25.728907	0.038867	13.730474	353.270093
49	27.529930	0.036324	13.766799	378.999000
50	29.457025	0.033948	13.800746	406.528929

		$r = 0.08$		
n	$(1+r)^n$	$(1+r)^{-n}$	$a_{\overline{n}|r}$	$s_{\overline{n}|r}$
1	1.080000	0.925926	0.925926	1.000000
2	1.166400	0.857339	1.783265	2.080000
3	1.259712	0.793832	2.577097	3.246400
4	1.360489	0.735030	3.312127	4.506112
5	1.469328	0.680583	3.992710	5.866601
6	1.586874	0.630170	4.622880	7.335929
7	1.713824	0.583490	5.206370	8.922803
8	1.850930	0.540269	5.746639	10.636628
9	1.999005	0.500249	6.246888	12.487558
10	2.158925	0.463193	6.710081	14.486562
11	2.331639	0.428883	7.138964	16.645487
12	2.518170	0.397114	7.536078	18.977126
13	2.719624	0.367698	7.903776	21.495297
14	2.937194	0.340461	8.244237	24.214920
15	3.172169	0.315242	8.559479	27.152114
16	3.425943	0.291890	8.851369	30.324283
17	3.700018	0.270269	9.121638	33.750226
18	3.996019	0.250249	9.371887	37.450244
19	4.315701	0.231712	9.603599	41.446263
20	4.660957	0.214548	9.818147	45.761964
21	5.033834	0.198656	10.016803	50.422921
22	5.436540	0.183941	10.200744	55.456755
23	5.871464	0.170315	10.371059	60.893296
24	6.341181	0.157699	10.528758	66.764759
25	6.848475	0.146018	10.674776	73.105940
26	7.396353	0.135202	10.809978	79.954415
27	7.988061	0.125187	10.935165	87.350768
28	8.627106	0.115914	11.051078	95.338830
29	9.317275	0.107328	11.158406	103.965936
30	10.062657	0.099377	11.257783	113.283211
31	10.867669	0.092016	11.349799	123.345868
32	11.737083	0.085200	11.434999	134.213537
33	12.676050	0.078889	11.513888	145.950620
34	13.690134	0.073045	11.586934	158.626670
35	14.785344	0.067635	11.654568	172.316804
36	15.968172	0.062625	11.717193	187.102148
37	17.245626	0.057986	11.775179	203.070320
38	18.625276	0.053690	11.828869	220.315945
39	20.115298	0.049713	11.878582	238.941221
40	21.724521	0.046031	11.924613	259.056519
41	23.462483	0.042621	11.967235	280.781040
42	25.339482	0.039464	12.006699	304.243523
43	27.366640	0.036541	12.043240	329.583005
44	29.555972	0.033834	12.077074	356.949646
45	31.920449	0.031328	12.108402	386.505617
46	34.474085	0.029007	12.137409	418.426067
47	37.232012	0.026859	12.164267	452.900152
48	40.210573	0.024869	12.189136	490.132164
49	43.427419	0.023027	12.212163	530.342737
50	46.901613	0.021321	12.233485	573.770156

APPENDIX B

Table of Selected Integrals

Rational Forms Containing $(a + bu)$

1. $\displaystyle \int u^n \, du = \frac{u^{n+1}}{n+1} + C, \; n \neq -1$

2. $\displaystyle \int \frac{du}{a + bu} = \frac{1}{b} \ln |a + bu| + C$

3. $\displaystyle \int \frac{u \, du}{a + bu} = \frac{u}{b} - \frac{a}{b^2} \ln |a + bu| + C$

4. $\displaystyle \int \frac{u^2 \, du}{a + bu} = \frac{u^2}{2b} - \frac{au}{b^2} + \frac{a^2}{b^3} \ln |a + bu| + C$

5. $\displaystyle \int \frac{du}{u(a + bu)} = \frac{1}{a} \ln \left| \frac{u}{a + bu} \right| + C$

6. $\displaystyle \int \frac{du}{u^2(a + bu)} = -\frac{1}{au} + \frac{b}{a^2} \ln \left| \frac{a + bu}{u} \right| + C$

7. $\displaystyle \int \frac{u \, du}{(a + bu)^2} = \frac{1}{b^2} \left(\ln |a + bu| + \frac{a}{a + bu} \right) + C$

8. $\displaystyle \int \frac{u^2 \, du}{(a + bu)^2} = \frac{u}{b^2} - \frac{a^2}{b^3(a + bu)} - \frac{2a}{b^3} \ln |a + bu| + C$

9. $\displaystyle \int \frac{du}{u(a + bu)^2} = \frac{1}{a(a + bu)} + \frac{1}{a^2} \ln \left| \frac{u}{a + bu} \right| + C$

10. $\displaystyle \int \frac{du}{u^2(a + bu)^2} = -\frac{a + 2bu}{a^2u(a + bu)} + \frac{2b}{a^3} \ln \left| \frac{a + bu}{u} \right| + C$

11. $\displaystyle \int \frac{du}{(a + bu)(c + ku)} = \frac{1}{bc - ak} \ln \left| \frac{a + bu}{c + ku} \right| + C$

12. $\displaystyle \int \frac{u \, du}{(a + bu)(c + ku)} = \frac{1}{bc - ak} \left[\frac{c}{k} \ln |c + ku| - \frac{a}{b} \ln |a + bu| \right] + C$

Forms Containing $\sqrt{a + bu}$

13. $\displaystyle \int u\sqrt{a + bu} \, du = \frac{2(3bu - 2a)(a + bu)^{3/2}}{15b^2} + C$

14. $\displaystyle \int u^2 \sqrt{a + bu} \, du = \frac{2(8a^2 - 12abu + 15b^2u^2)(a + bu)^{3/2}}{105b^3} + C$

15. $\displaystyle \int \frac{u \, du}{\sqrt{a + bu}} = \frac{2(bu - 2a)\sqrt{a + bu}}{3b^2} + C$

16. $\displaystyle \int \frac{u^2 \, du}{\sqrt{a + bu}} = \frac{2(3b^2u^2 - 4abu + 8a^2)\sqrt{a + bu}}{15b^3} + C$

17. $\displaystyle\int \frac{du}{u\sqrt{a+bu}} = \frac{1}{\sqrt{a}}\ln\left|\frac{\sqrt{a+bu}-\sqrt{a}}{\sqrt{a+bu}+\sqrt{a}}\right| + C,\ a > 0$

18. $\displaystyle\int \frac{\sqrt{a+bu}\,du}{u} = 2\sqrt{a+bu} + a\int \frac{du}{u\sqrt{a+bu}}$

Forms Containing $\sqrt{a^2 - u^2}$

19. $\displaystyle\int \frac{du}{(a^2-u^2)^{3/2}} = \frac{u}{a^2\sqrt{a^2-u^2}} + C$

20. $\displaystyle\int \frac{du}{u\sqrt{a^2-u^2}} = -\frac{1}{a}\ln\left|\frac{a+\sqrt{a^2-u^2}}{u}\right| + C$

21. $\displaystyle\int \frac{du}{u^2\sqrt{a^2-u^2}} = -\frac{\sqrt{a^2-u^2}}{a^2u} + C$

22. $\displaystyle\int \frac{\sqrt{a^2-u^2}\,du}{u} = \sqrt{a^2-u^2} - a\ln\left|\frac{a+\sqrt{a^2-u^2}}{u}\right| + C,\ a > 0$

Forms Containing $\sqrt{u^2 \pm a^2}$

23. $\displaystyle\int \sqrt{u^2\pm a^2}\,du = \frac{1}{2}\left(u\sqrt{u^2\pm a^2} \pm a^2\ln\left|u+\sqrt{u^2\pm a^2}\right|\right) + C$

24. $\displaystyle\int u^2\sqrt{u^2\pm a^2}\,du = \frac{u}{8}(2u^2\pm a^2)\sqrt{u^2\pm a^2} - \frac{a^4}{8}\ln\left|u+\sqrt{u^2\pm a^2}\right| + C$

25. $\displaystyle\int \frac{\sqrt{u^2+a^2}\,du}{u} = \sqrt{u^2+a^2} - a\ln\left|\frac{a+\sqrt{u^2+a^2}}{u}\right| + C$

26. $\displaystyle\int \frac{\sqrt{u^2\pm a^2}\,du}{u^2} = -\frac{\sqrt{u^2\pm a^2}}{u} + \ln\left|u+\sqrt{u^2\pm a^2}\right| + C$

27. $\displaystyle\int \frac{du}{\sqrt{u^2\pm a^2}} = \ln\left|u+\sqrt{u^2\pm a^2}\right| + C$

28. $\displaystyle\int \frac{du}{u\sqrt{u^2+a^2}} = \frac{1}{a}\ln\left|\frac{\sqrt{u^2+a^2}-a}{u}\right| + C$

29. $\displaystyle\int \frac{u^2\,du}{\sqrt{u^2\pm a^2}} = \frac{1}{2}\left(u\sqrt{u^2\pm a^2} \mp a^2\ln\left|u+\sqrt{u^2\pm a^2}\right|\right) + C$

30. $\displaystyle\int \frac{du}{u^2\sqrt{u^2\pm a^2}} = -\frac{\pm\sqrt{u^2\pm a^2}}{a^2u} + C$

31. $\displaystyle\int (u^2\pm a^2)^{3/2}\,du = \frac{u}{8}(2u^2\pm 5a^2)\sqrt{u^2\pm a^2} + \frac{3a^4}{8}\ln\left|u+\sqrt{u^2\pm a^2}\right| + C$

32. $\displaystyle\int \frac{du}{(u^2\pm a^2)^{3/2}} = \frac{\pm u}{a^2\sqrt{u^2\pm a^2}} + C$

33. $\displaystyle\int \frac{u^2\,du}{(u^2\pm a^2)^{3/2}} = \frac{-u}{\sqrt{u^2\pm a^2}} + \ln\left|u+\sqrt{u^2\pm a^2}\right| + C$

Rational Forms Containing $a^2 - u^2$ and $u^2 - a^2$

34. $\displaystyle\int \frac{du}{a^2-u^2} = \frac{1}{2a}\ln\left|\frac{a+u}{a-u}\right| + C$

35. $\displaystyle\int \frac{du}{u^2-a^2} = \frac{1}{2a}\ln\left|\frac{u-a}{u+a}\right| + C$

Exponential and Logarithmic Forms

36. $\displaystyle\int e^u du = e^u + C$

37. $\displaystyle\int a^u du = \frac{a^u}{\ln a} + C, a > 0, a \neq 1$

38. $\displaystyle\int u e^{au} du = \frac{e^{au}}{a^2}(au - 1) + C$

39. $\displaystyle\int u^n e^{au} du = \frac{u^n e^{au}}{a} - \frac{n}{a}\int u^{n-1} e^{au}\, du$

40. $\displaystyle\int \frac{e^{au}\, du}{u^n} = -\frac{e^{au}}{(n-1)u^{n-1}} + \frac{a}{n-1}\int \frac{e^{au}\, du}{u^{n-1}}, n \neq 1$

41. $\displaystyle\int \ln u\, du = u \ln u - u + C$

42. $\displaystyle\int u^n \ln u\, du = \frac{u^{n+1}\ln u}{n+1} - \frac{u^{n+1}}{(n+1)^2} + C,\ n \neq -1$

43. $\displaystyle\int u^n \ln^m u\, du = \frac{u^{n+1}}{n+1}\ln^m u - \frac{m}{n+1}\int u^n \ln^{m-1} u\, du, m,\ n \neq -1$

44. $\displaystyle\int \frac{du}{u \ln u} = \ln\left|\ln u\right| + C$

45. $\displaystyle\int \frac{du}{a + be^{cu}} = \frac{1}{ac}\left(cu - \ln\left|a + be^{cu}\right|\right) + C$

Miscellaneous Forms

46. $\displaystyle\int \sqrt{\frac{a+u}{b+u}}\, du = \sqrt{(a+u)(b+u)} + (a-b)\ln(\sqrt{a+u} + \sqrt{b+u}) + C$

47. $\displaystyle\int \frac{du}{\sqrt{(a+u)(b+u)}} = \ln\left|\frac{a+b}{2} + u + \sqrt{(a+u)(b+u)}\right| + C$

48. $\displaystyle\int \sqrt{a + bu + cu^2}\, du = \frac{2cu + b}{4c}\sqrt{a + bu + cu^2}$
$$- \frac{b^2 - 4ac}{8c^{3/2}}\ln\left|2cu + b + 2\sqrt{c}\sqrt{a + bu + cu^2}\right| + C, c > 0$$

APPENDIX C

Areas Under the Standard Normal Curve

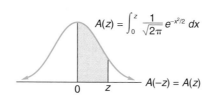

$$A(z) = \int_0^z \frac{1}{\sqrt{2\pi}} e^{-x^2/2} \, dx$$

$$A(-z) = A(z)$$

z	.00	.01	.02	.03	.04	.05	.06	.07	.08	.09
0.0	.0000	.0040	.0080	.0120	.0160	.0199	.0239	.0279	.0319	.0359
0.1	.0398	.0438	.0478	.0517	.0557	.0596	.0636	.0675	.0714	.0753
0.2	.0793	.0832	.0871	.0910	.0948	.0987	.1026	.1064	.1103	.1141
0.3	.1179	.1217	.1255	.1293	.1331	.1368	.1406	.1443	.1480	.1517
0.4	.1554	.1591	.1628	.1664	.1700	.1736	.1772	.1808	.1844	.1879
0.5	.1915	.1950	.1985	.2019	.2054	.2088	.2123	.2157	.2190	.2224
0.6	.2257	.2291	.2324	.2357	.2389	.2422	.2454	.2486	.2517	.2549
0.7	.2580	.2611	.2642	.2673	.2704	.2734	.2764	.2794	.2823	.2852
0.8	.2881	.2910	.2939	.2967	.2995	.3023	.3051	.3078	.3106	.3133
0.9	.3159	.3186	.3212	.3238	.3264	.3289	.3315	.3340	.3365	.3389
1.0	.3413	.3438	.3461	.3485	.3508	.3531	.3554	.3577	.3599	.3621
1.1	.3643	.3665	.3686	.3708	.3729	.3749	.3770	.3790	.3810	.3830
1.2	.3849	.3869	.3888	.3907	.3925	.3944	.3962	.3980	.3997	.4015
1.3	.4032	.4049	.4066	.4082	.4099	.4115	.4131	.4147	.4162	.4177
1.4	.4192	.4207	.4222	.4236	.4251	.4265	.4279	.4292	.4306	.4319
1.5	.4332	.4345	.4357	.4370	.4382	.4394	.4406	.4418	.4429	.4441
1.6	.4452	.4463	.4474	.4484	.4495	.4505	.4515	.4525	.4535	.4545
1.7	.4554	.4564	.4573	.4582	.4591	.4599	.4608	.4616	.4625	.4633
1.8	.4641	.4649	.4656	.4664	.4671	.4678	.4686	.4693	.4699	.4706
1.9	.4713	.4719	.4726	.4732	.4738	.4744	.4750	.4756	.4761	.4767
2.0	.4772	.4778	.4783	.4788	.4793	.4798	.4803	.4808	.4812	.4817
2.1	.4821	.4826	.4830	.4834	.4838	.4842	.4846	.4850	.4854	.4857
2.2	.4861	.4864	.4868	.4871	.4875	.4878	.4881	.4884	.4887	.4890
2.3	.4893	.4896	.4898	.4901	.4904	.4906	.4909	.4911	.4913	.4916
2.4	.4918	.4920	.4922	.4925	.4927	.4929	.4931	.4932	.4934	.4936
2.5	.4938	.4940	.4941	.4943	.4945	.4946	.4948	.4949	.4951	.4952
2.6	.4953	.4955	.4956	.4957	.4959	.4960	.4961	.4962	.4963	.4964
2.7	.4965	.4966	.4967	.4968	.4969	.4970	.4971	.4972	.4973	.4974
2.8	.4974	.4975	.4976	.4977	.4977	.4978	.4979	.4979	.4980	.4981
2.9	.4981	.4982	.4982	.4983	.4984	.4984	.4985	.4985	.4986	.4986
3.0	.4987	.4987	.4987	.4988	.4988	.4989	.4989	.4989	.4990	.4990
3.1	.4990	.4991	.4991	.4991	.4992	.4992	.4992	.4992	.4993	.4993
3.2	.4993	.4993	.4994	.4994	.4994	.4994	.4994	.4995	.4995	.4995
3.3	.4995	.4995	.4995	.4996	.4996	.4996	.4996	.4996	.4996	.4997
3.4	.4997	.4997	.4997	.4997	.4997	.4997	.4997	.4997	.4997	.4998
3.5	.4998	.4998	.4998	.4998	.4998	.4998	.4998	.4998	.4998	.4998

Answers to Odd-Numbered Problems

PROBLEMS 0.1 (page 3)

1. True

3. False; the natural numbers are 1, 2, 3, and so on.

5. True

7. False; $\sqrt{25} = 5$, a positive integer

9. False; we cannot divide by 0.   **11.** True

PROBLEMS 0.2 (page 8)

1. False   **3.** False   **5.** False   **7.** True   **9.** False

11. Distributive   **13.** Associative

15. Commutative and distributive

17. Definition of subtraction   **19.** Distributive

29. -4 **31.** 5   **33.** 8   **35.** -18 **37.** 24

39. 9   **41.** $-7x$ **43.** $6 + y$ **45.** $\dfrac{1}{3}$ **47.** -8

49. -8 **51.** X **53.** $20 + 4x$ **55.** 0   **57.** 5

59. $-\dfrac{3}{2x}$ **61.** $\dfrac{3ab}{c}$ **63.** $\dfrac{by}{x}$ **65.** $\dfrac{10}{xy}$

67. $\dfrac{7}{6}$ **69.** 2   **71.** $\dfrac{17}{12}$ **73.** $\dfrac{6y}{x}$ **75.** $-\dfrac{x^2}{yz}$

77. 0   **79.** 0

PROBLEMS 0.3 (page 14)

1. $2^5 (= 32)$ **3.** w^{12} **5.** $\dfrac{x^8}{y^{14}}$ **7.** $\dfrac{a^{21}}{b^{20}}$

9. $8x^6 y^9$ **11.** x^4 **13.** x^{14} **15.** 5

17. -2 **19.** $\dfrac{1}{2}$ **21.** 7   **23.** 27

25. $\dfrac{1}{4}$ **27.** $\dfrac{1}{16}$ **29.** $5\sqrt{2}$ **31.** $x\sqrt[3]{2}$

33. $4x^2$ **35.** $4\sqrt{2} - 15\sqrt{3} + 4\sqrt[3]{2}$ **37.** $3z^2$

39. $\dfrac{9t^2}{4}$ **41.** $\dfrac{a^5}{b^3 c^2}$ **43.** $\dfrac{5}{m^9}$ **45.** $\dfrac{1}{9t^2}$

47. $5^{1/5} x^{2/5}$ **49.** $x^{1/2} - y^{1/2}$ **51.** $\dfrac{x^{9/4} z^{3/4}}{y^{1/2}}$

53. $\sqrt[3]{(2a - b + c)^2}$ **55.** $\dfrac{1}{\sqrt[5]{x^4}}$ **57.** $\dfrac{3}{\sqrt[5]{w^3}} - \dfrac{1}{\sqrt[5]{27w^3}}$

59. $\dfrac{6\sqrt{5}}{5}$ **61.** $\dfrac{2\sqrt{2x}}{x}$ **63.** $\dfrac{\sqrt[3]{9x^2}}{3x}$ **65.** 2

67. $\dfrac{\sqrt[20]{16a^{10}b^{15}}}{ab}$ **69.** $\dfrac{2x^6}{y^3}$ **71.** 9

73. $\dfrac{64y^6 x^{1/2}}{x^2}$ **75.** xyz **77.** $\dfrac{9}{4}$

79. $\dfrac{4y^4}{x^2}$ **81.** $x^2 y^{5/2}$ **83.** $\dfrac{a^5 c^{14}}{b^{24}}$

85. x^8 **87.** $-\dfrac{4}{s^5}$ **89.** $\dfrac{81x^{12} z^{12}}{16}$

PROBLEMS 0.4 (page 18)

1. $11x - 2y - 3$ **3.** $6t^2 - 2s^2 + 6$

5. $\sqrt{a} + 5\sqrt{3b} - \sqrt{c}$ **7.** $6x^2 - 9xy - 2z + \sqrt{2} - 4$

9. $\sqrt{2y} - \sqrt{3z}$ **11.** $-15x + 15y - 27$

13. $2x^2 - 33y^2 - 7xy$ **15.** $6x^2 + 96$

17. $-40x^3 - 30x^2 + 20x - 50$ **19.** $x^2 + 9x + 20$

21. $w^2 - 3w - 10$ **23.** $10x^2 + 19x + 6$

25. $X^2 + 4XY + 4Y^2$ **27.** $x^2 - 10x + 25$

29. $3x + 10\sqrt{3x} + 25$ **31.** $4s^2 - 1$

33. $x^3 + 4x^2 - 3x - 12$ **35.** $3x^4 + 2x^3 - 13x^2 - 8x + 4$

37. $10x^3 - 52x^2 - 70x$ **39.** $3x^2 + 2y^2 + 5xy + 2x - 8$

41. $8a^3 + 36a^2 + 54a + 27$ **43.** $8x^3 - 36x^2 + 54x - 27$

45. $z - 18$ **47.** $3x^3 + 2x - \dfrac{1}{2x^2}$

49. $x + \dfrac{-3}{x + 5}$ **51.** $3x^2 - 8x + 17 + \dfrac{-37}{x + 2}$

53. $x^2 - 2x + 4 - \dfrac{8}{x + 2}$ **55.** $x - 2 + \dfrac{7}{3x + 2}$

PROBLEMS 0.5 (page 21)

1. $2(ax + b)$ **3.** $5x(2y + z)$

5. $4bc(2a^3 - 3ab^2 d + b^3 cd^2)$ **7.** $(z + 7)(z - 7)$

9. $(p + 3)(p + 1)$ **11.** $(4x + 3)(4x - 3)$

13. $(a + 7)(a + 5)$ **15.** $(x + 3)^2$

17. $5(x + 3)(x + 2)$ **19.** $3(x + 1)(x - 1)$

21. $(6y + 1)(y + 2)$ **23.** $2s(3s + 4)(2s - 1)$

25. $u^{3/5} v(u + 2v)(u - 2v)$ **27.** $2x(x + 3)(x - 2)$

29. $4(2x + 1)^2$ **31.** $x(xy - 7)^2$

33. $(x + 2)(x - 2)^2$ **35.** $(y + 4)^2 (y + 1)(y - 1)$

37. $(b + 4)(b^2 - 4b + 16)$

39. $(x + 1)(x^2 - x + 1)(x - 1)(x^2 + x + 1)$

41. $2(x + 3)^2 (x + 1)(x - 1)$ **43.** $P(1 + r)^2$

45. $(x^2 + 4)(x + 2)(x - 2)$ **47.** $(y^4 + 1)(y^2 + 1)(y + 1)(y - 1)$

49. $(X^2 + 5)(X + 1)(X - 1)$ **51.** $y(x + 1)^2 (x - 1)^2$

PROBLEMS 0.6 (page 26)

1. $\dfrac{a + 3}{a}$ **3.** $\dfrac{x - 5}{x + 5}$ **5.** $\dfrac{3x + 2}{x + 2}$

7. $-\dfrac{y^2}{(y - 3)(y + 2)}$ **9.** $\dfrac{b - ax}{ax + b}$ **11.** $\dfrac{2(x + 4)}{(x - 4)(x + 2)}$

13. $\dfrac{X}{2}$ **15.** $\dfrac{n}{3}$ **17.** $\dfrac{2}{3}$ **19.** $-27x^2$ **21.** 1

23. $\dfrac{2x^2}{x - 1}$ **25.** 1   **27.** $-\dfrac{(2x + 3)(1 + x)}{x + 4}$

29. $x + 2$ **31.** $\dfrac{7}{3t}$ **33.** $\dfrac{1}{1 - x^3}$

35. $\dfrac{2x^2 + 3x + 12}{(2x - 1)(x + 3)}$ **37.** $\dfrac{2(x + 2)}{(x - 3)(x + 1)(x + 3)}$

39. $\dfrac{35 - 8x}{(x - 1)(x + 5)}$ **41.** $\dfrac{x^2 + 2x + 1}{x^2}$

43. $\dfrac{x}{1 - xy}$ **45.** $\dfrac{7x + 1}{5x}$

47. $\dfrac{(x+2)(6x-1)}{2x^2(x+3)}$ **49.** $\dfrac{3(\sqrt[3]{x}-\sqrt[3]{x+h})}{\sqrt[3]{x+h}\,\sqrt[3]{x}}$

51. $2-\sqrt{3}$ **53.** $-\dfrac{\sqrt{6}+2\sqrt{3}}{3}$ **55.** $-4-2\sqrt{6}$

57. $\dfrac{3t-3\sqrt{7}}{t^2-7}$ **59.** $4\sqrt{2}-5\sqrt{3}+14$

PROBLEMS 0.7 (page 34)

1. 0   **3.** $\dfrac{17}{4}$ **5.** -2

7. Adding 5 to both sides; equivalence guaranteed

9. Raising both sides to the third power; equivalence *not* guaranteed

11. Dividing both sides by x; equivalence *not* guaranteed

13. Multiplying both sides by $x-1$; equivalence *not* guaranteed

15. Multiplying both sides by $(2x-3)/2x$; equivalence *not* guaranteed

17. $\dfrac{5}{2}$ **19.** 0   **21.** 1   **23.** $\dfrac{12}{5}$ **25.** -1

27. $-\dfrac{27}{4}$ **29.** $\dfrac{10}{3}$ **31.** 126   **33.** 15   **35.** $-\dfrac{26}{9}$

37. $-\dfrac{37}{18}$ **39.** 192   **41.** $\dfrac{14}{3}$ **43.** 3   **45.** $\dfrac{25}{52}$

47. $\dfrac{1}{5}$ **49.** $\varnothing$ **51.** $\dfrac{29}{14}$ **53.** 2   **55.** 0

57. $\dfrac{7}{2}$ **59.** $\dfrac{1}{8}$ **61.** 3   **63.** $\dfrac{43}{16}$ **65.** $\varnothing$

67. 11   **69.** $\dfrac{68}{3}$ **71.** $-\dfrac{10}{9}$ **73.** 2   **75.** 86

77. $\dfrac{49}{36}$ **79.** $-\dfrac{9}{4}$ **81.** $r=\dfrac{I}{Pt}$ **83.** $q=\dfrac{p+1}{8}$

85. $r=\dfrac{S-P}{Pt}$ **87.** $R=\dfrac{Ai}{1-(1+i)^{-n}}$

89. $t=\dfrac{r-d}{rd}$ **91.** $n=\dfrac{2mI}{rB}-1$ **93.** 170 m

95. $c=x+0.0825x=1.0825x$ **97.** 3 years

99. 44 hours   **101.** 20   **103.** $t=\dfrac{d}{r-c}; c=r-\dfrac{d}{t}$

105. ≈ 84 ft   **107.** 13%   **109.** $-\dfrac{1}{2}$ **111.** 0

PROBLEMS 0.8 (page 42)

1. 2   **3.** 3, 5   **5.** 3, -1 **7.** 4, 9   **9.** ± 2

11. 0, 5   **13.** $\dfrac{1}{2}$ **15.** 1, $\dfrac{2}{3}$ **17.** 5, -2 **19.** 0, $\dfrac{3}{2}$

21. 0, 1, -4 **23.** 0, ± 7 **25.** 0, $\dfrac{1}{2}$, $-\dfrac{4}{3}$

27. 3, ± 2 **29.** 3, 4   **31.** 4, -6 **33.** $\dfrac{3}{2}$

35. $1\pm 2\sqrt{2}$ **37.** No real roots   **39.** $\dfrac{-5\pm\sqrt{57}}{8}$

41. 40, -25 **43.** $\dfrac{-2\pm\sqrt{14}}{2}$ **45.** $\pm\sqrt{3}, \pm\sqrt{2}$

47. 3, $\dfrac{1}{2}$ **49.** $\pm\dfrac{\sqrt{5}}{5}, \pm\dfrac{1}{2}$ **51.** 3, 0

53. $\dfrac{15}{7}, \dfrac{11}{5}$ **55.** $\dfrac{3}{2}, -1$ **57.** 6, -2 **59.** $-\dfrac{1}{2}, 1$

61. 5, -2 **63.** $\dfrac{-9\pm\sqrt{41}}{4}$ **65.** -2 **67.** 6

69. 4, 8   **71.** $\dfrac{5-\sqrt{21}}{2}$ **73.** 0, 4   **75.** 1

77. $\approx 64.15, 3.35$

79. 6 inches by 8 inches

83. 1 year and 10 years; age 23; never

85. (a) 8 s; **(b)** 5.4 s or 2.6 s   **87.** 1.5, 0.75

89. No real root   **91.** 1.999, 0963

MATHEMATICAL SNAPSHOT—CHAPTER 0 (page 44)

1. The results agree.   **3.** The results agree.

PROBLEMS 1.1 (page 51)

1. 120 ft   **3.** $64\dfrac{4}{9}$ oz of A, $80\dfrac{5}{9}$ oz of B

5. $5\dfrac{1}{3}$ oz   **7.** 1 m   **9.** $\approx 13{,}077$ tons

11. \$4000 at 6%, \$16,000 at $7\dfrac{1}{2}$%

13. \$4.25   **15.** 4%   **17.** 80   **19.** \$8000

21. 1209 cartridges must be sold to approximately break even.

23. \$116.25   **25.** 40   **27.** 46,000

29. Either \$440 or \$460   **31.** \$100   **33.** 42

35. 80 ft by 140 ft   **37.** 9 cm long, 4 cm wide

39. \$232,000; $\dfrac{100E}{100-p}$ **41.** 60 acres

43. Either 125 units of A and 100 units of B, or 150 units of A and 125 units of B.

PRINCIPLES IN PRACTICE 1.2

1. 5375

2. $150-x_4\geq 0$; $3x_4-210\geq 0$; $x_4+60\geq 0$; $x_4\geq 0$

PROBLEMS 1.2 (page 58)

1. $(4, \infty)$ **3.** $(-\infty, 4]$ **5.** $\left(-\infty, -\dfrac{1}{2}\right]$

7. $\left(-\infty, \dfrac{2}{7}\right)$ **9.** $(0, \infty)$ **11.** $[2, \infty)$

13. $\left(-\dfrac{2}{7}, \infty\right)$ **15.** $\varnothing$ **17.** $\left(-\infty, \dfrac{\sqrt{3}-2}{2}\right)$

19. $(-\infty, 48)$

21. $(-\infty, -5]$

23. $(-\infty, \infty)$

25. $\left(\dfrac{17}{9}, \infty\right)$

27. $[-12, \infty)$

29. $(0, \infty)$

31. $(-\infty, 0)$

33. $(-\infty, -2]$

35. $600 < S < 1800$ **37.** $x < 70$ degrees

PROBLEMS 1.3 (page 61)

1. 120,001   **3.** 17,000   **5.** 37,500   **7.** $25,714.29

9. 1000   **11.** $t > 36.7$ **13.** At least $67,400

PRINCIPLES IN PRACTICE 1.4

1. $|w - 22\,\text{oz}| \le 0.3\,\text{oz}$

PROBLEMS 1.4 (page 65)

1. 13   **3.** 6   **5.** 7   **7.** $-4 < x < 4$

9. $\sqrt{5} - 2$

11. (a) $|x - 7| < 3$;   (b) $|x - 2| < 3$;   (c) $|x - 7| \le 5$;
     (d) $|x - 7| = 4$;   (e) $|x + 4| < 2$;   (f) $|x| < 3$;   (g) $|x| > 6$;
     (h) $|x - 105| < 3$;   (i) $|x - 850| < 100$

13. $|p_1 - p_2| \le 9$

15. ± 7 **17.** ± 35 **19.** $13, -3$

21. $\dfrac{2}{5}$ **23.** $\dfrac{1}{2}, 3$ **25.** $(-M, M)$

27. $(-\infty, -8) \cup (8, \infty)$ **29.** $(-14, -4)$

31. $(-\infty, 0) \cup (1, \infty)$ **33.** $\left[\dfrac{1}{2}, \dfrac{3}{4}\right]$

35. $(-\infty, 0] \cup \left[\dfrac{16}{3}, \infty\right)$ **37.** $|d - 35.2| \le 0.2\,\text{m}$

39. $(-\infty, \mu - h\sigma) \cup (\mu + h\sigma, \infty)$

PROBLEMS 1.5 (page 69)

1. $12, 17, t$ **3.** 168   **5.** 532   **7.** $\displaystyle\sum_{i=36}^{60} i$

9. $\displaystyle\sum_{j=3}^{8} 5^j$ **11.** $\displaystyle\sum_{i=1}^{8} 2^i$ **13.** 4300   **15.** 5

17. 37,750   **19.** 14,980   **21.** 295,425   **23.** $4\dfrac{23}{25}$

25. $15 - \dfrac{9(n+1)(2n+1)}{2n^2}$

REVIEW PROBLEMS—CHAPTER 1 (page 71)

1. $[-4, \infty)$ **3.** $\left(\dfrac{2}{3}, \infty\right)$ **5.** $\varnothing$

7. $\left(-\infty, \dfrac{5}{2}\right]$ **9.** $(-\infty, \infty)$ **11.** $-2, 5$

13. $(-1, 4)$ **15.** $\left(-\infty, -\dfrac{1}{2}\right] \cup \left[\dfrac{7}{2}, \infty\right)$

17. 775   **19.** 542   **21.** 6000   **23.** $c < \$212,814$

MATHEMATICAL SNAPSHOT—CHAPTER 1 (page 72)

1. 1 hr   **3.** $2M - 240$ min

5. 600; 310   **7.** $t = \dfrac{R(mr - M)}{r - R}$

PRINCIPLES IN PRACTICE 2.1

1. (a) $a(r) = \pi r^2$;   (b) all real numbers;   (c) $r \ge 0$

2. (a) $t(r) = \dfrac{300}{r}$;   (b) all real numbers except 0;   (c) $r > 0$;
     (d) $t(x) = \dfrac{300}{x}$; $t\left(\dfrac{x}{2}\right) = \dfrac{600}{x}$; $t\left(\dfrac{x}{4}\right) = \dfrac{1200}{x}$;   (e) The
     time is scaled by a factor c; $t\left(\dfrac{x}{c}\right) = \dfrac{300c}{x}$

3. (a) 300 pizzas; (b) $21.00 per pizza; (c) $16.00 per pizza

PROBLEMS 2.1 (page 81)

1. $f \ne g$ **3.** $h \ne k$

5. All real numbers except 0   **7.** All real numbers ≥ 3

9. All real numbers   **11.** All real numbers except $-\dfrac{7}{2}$

13. All real numbers except 2

15. All real numbers except 4 and $-\dfrac{1}{2}$

17. $1, 7, -7$ **19.** $-62, 2 - u^2, 2 - u^4$

21. $10, 8v^2 - 2v, 2x^2 + 4ax + 2a^2 - x - a$

23. $4, 0, x^2 + 2xh + h^2 + 2x + 2h + 1$

25. $-\dfrac{2}{27}, \dfrac{3x - 7}{9x^2 + 2}, \dfrac{x + h - 7}{x^2 + 2xh + h^2 + 2}$

27. $0, 256, \dfrac{1}{16}$ **29.** (a) $4x + 4h - 5$;   (b) 4

31. (a) $x^2 + 2hx + h^2 + 2x + 2h$; (b) $2x + h + 2$

33. (a) $3 - 2x - 2h + 4x^2 + 8xh + 4h^2$; (b) $-2 + 8x + 4h$

35. (a) $\dfrac{1}{x + h}$; (b) $-\dfrac{1}{x(x + h)}$ **37.** 5

39. y is a function of x; x is a function of y.

41. y is a function of x; x is not a function of y.

43. Yes   **45.** $V = f(t) = 25,000 + 1700t$ **47.** Yes; P; q

49. 402.72 pounds per week; 935.52 pounds per week; amount supplied increases as the price increases

51. (a) 4;   (b) $8\sqrt[3]{2}$;   (c) $f(2I_0) = 2\sqrt[3]{2}\, f(I_0)$; doubling the intensity increases the response by a factor of $2\sqrt[3]{2}$

53. (a) 3000, 2900, 2300, 2000; 12, 10;
     (b) 10, 12, 17, 20; 3000, 2300

55. (a) -5.13;   (b) 2.64;   (c) -17.43

57. (a) 7.89;   (b) 63.85;   (c) 1.21

PRINCIPLES IN PRACTICE 2.2

1. (a) $p(n) = \$125$;   (b) The premiums do not change;
     (c) constant function

2. (a) quadratic function;   (b) 2;   (c) 3

3. $c(n) = \begin{cases} 3.50n & \text{if } n \leq 5 \\ 3.00n & \text{if } 5 < n \leq 10 \\ 2.75n & \text{if } n > 10 \end{cases}$ **4.** $7! = 5040$

PROBLEMS 2.2 (page 85)

1. Yes   **3.** No   **5.** Yes   **7.** No

9. All real numbers   **11.** All real numbers

13. (a) 3; **(b)** 7   **15. (a)** 7; **(b)** 1

17. 8, 8, 8   **19.** 1, −1, 0, −1   **21.** 7, 2, 2, 2

23. 720   **25.** 2   **27.** n

29. $c(i) = \$4.50$; constant function

31. (a) $C = 850 + 3q$; **(b)** 250

33. $c(n) = \begin{cases} 9.50n & \text{if } n < 12, \\ 8.75n & \text{if } n \geq 12 \end{cases}$ **35.** $\dfrac{9}{64}$

37. (a) All T such that $30 \leq T \leq 39$; **(b)** $4, \dfrac{17}{4}, \dfrac{33}{4}$

39. (a) 1182.74; **(b)** 4985.27; **(c)** 252.15

41. (a) 2.21; **(b)** 9.98; **(c)** −14.52

PRINCIPLES IN PRACTICE 2.3

1. $c(s(x)) = c(x + 3) = 2(x + 3) = 2x + 6$

2. Let the length of a side be represented by the function $l(x) = x + 3$ and the area of a square with sides of length x be represented by $a(x) = x^2$. Then $g(x) = (x + 3)^2 = [l(x)]^2 = a(l(x))$.

PROBLEMS 2.3 (page 90)

1. (a) $2x + 8$; **(b)** 8; **(c)** −2; **(d)** $x^2 + 8x + 15$; **(e)** 3;
  (f) $\dfrac{x + 3}{x + 5}$; **(g)** $x + 8$; **(h)** 11; **(i)** $x + 8$; **(j)** 11

3. (a) $2x^2 - x + 1$; **(b)** $x + 1$; **(c)** $\dfrac{1}{2}$; **(d)** $x^4 - x^3 + x^2 - x$;
  (e) $\dfrac{x^2 + 1}{x^2 - x}$; **(f)** $\dfrac{5}{3}$; **(g)** $x^4 - 2x^3 + x^2 + 1$; **(h)** $x^4 + x^2$;
  (i) 90

5. 6; −32

7. $\dfrac{4}{(t - 1)^2} + \dfrac{14}{t - 1} + 1; \dfrac{2}{t^2 + 7t}$ **9.** $\dfrac{1}{v + 3}; \sqrt{\dfrac{2v^2 + 3}{v^2 + 1}}$

11. $f(x) = x - 7, g(x) = 11x$ **13.** $f(x) = \dfrac{1}{x}, g(x) = x^2 - 2$

15. $f(x) = \sqrt[4]{x}, g(x) = \dfrac{x^2 - 1}{x + 3}$

17. (a) $r(x) = 9.75x$; **(b)** $e(x) = 4.25x + 4500$;
  (c) $(r - e)(x) = 5.5x - 4500$

19. $400m - 10m^2$; the total revenue received when the total output of m employees is sold

21. (a) 14.05; **(b)** 1169.64   **23. (a)** 194.47; **(b)** 0.29

PROBLEMS 2.4 (page 93)

1. $f^{-1}(x) = \dfrac{x}{3} - \dfrac{7}{3}$ **3.** $F^{-1}(x) = 2x + 14$

5. $r(A) = \sqrt{\dfrac{A}{\pi}}$

7. $f(x) = 5x + 12$ is one-to-one

9. $h(x) = (5x + 12)^2$, for $x \geq -\dfrac{5}{12}$, is one-to-one

11. $x = \dfrac{\sqrt{23}}{4} + \dfrac{5}{4}$ **13.** $q = \dfrac{1{,}200{,}000}{p}, \ p > 0$

PRINCIPLES IN PRACTICE 2.5

1. $y = -600x + 7250$; x-intercept $\left(12\dfrac{1}{12}, 0\right)$; y-intercept $(0, 7250)$

2. $y = 24.95$; horizontal line; no x-intercept; y-intercept $(0, 24.95)$

3.

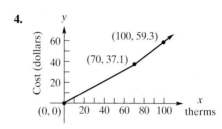

4.

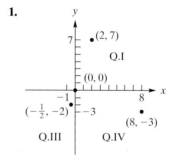

PROBLEMS 2.5 (page 101)

1.

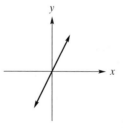

3. (a) 1, 2, 3, 0; **(b)** all real numbers; **(c)** all real numbers;
  (d) −2

5. (a) 0, 1, 1; **(b)** all real numbers; **(c)** all nonnegative real numbers; **(d)** 0

7. $(0, 0)$; function; one-to-one; all real numbers; all real numbers

9. $(0, -5)$, $\left(\dfrac{5}{3}, 0\right)$; function; one-to-one; all real numbers; all real numbers

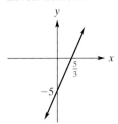

11. $(0, 0)$; function; not one-to-one; all real numbers; all nonnegative real numbers

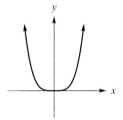

13. Every point on y-axis; not a function of x

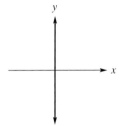

15. $(0, 0)$; function; one-to-one; all real numbers; all real numbers

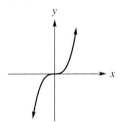

17. $(0, 0)$; not a function of x

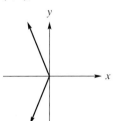

19. $(0, 2)$, $(1, 0)$; function; one-to-one; all real numbers; all real numbers

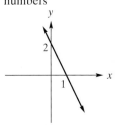

21. All real numbers; all real numbers ≤ 4; $(0, 4)$, $(2, 0)$, $(-2, 0)$

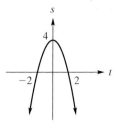

23. All real numbers; 3; $(0, 3)$

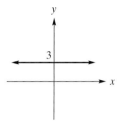

25. All real numbers; all real numbers ≥ -3; $(0, 1)$, $(2 \pm \sqrt{3}, 0)$

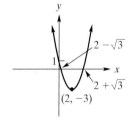

27. All real numbers; all real numbers; $(0, 0)$

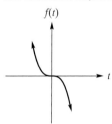

29. All real numbers ≤ -3 and all real numbers ≥ 3; all nonnegative real numbers; $(-3, 0)$, $(3, 0)$

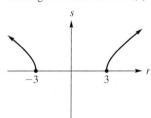

31. All real numbers; all nonnegative real numbers; $(0, 1)$, $\left(\dfrac{1}{2}, 0\right)$

33. All nonzero real numbers; all positive real numbers; no intercepts

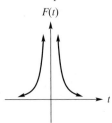

35. All nonnegative real numbers; all real numbers $1 \le c < 8$

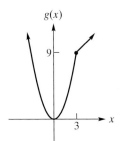

37. All real numbers; all nonnegative real numbers

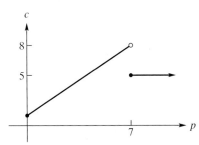

39. (a), (b), (d)

41. $y = 2400 - 275x$, $\left(8\frac{8}{11}, 0\right)$, $(0, 2400)$

43. As price increases, quantity increases; p is a function of q.

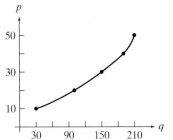

45.

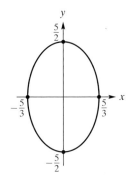

47. 0.39   **49.** $-0.61, -0.04$ **51.** -1.12

53. $-1.70, 0$ **55.** (a) 19.60;   (b) -10.86

57. (a) 5;   (b) 4

59. (a) 28;   (b) $(-\infty, 28]$;   (c) $-4.02, 0.60$

61. (a) 34.21;   (b) 18.68;   (c) $[18.68, 34.21]$;   (d) no intercepts

PROBLEMS 2.6 (page 108)

1. $(0, 0)$; sym. about origin

3. $(\pm 2, 0)$, $(0, 8)$; sym. about y-axis

5. $\left(\pm\frac{5}{4}, 0\right)$; sym. about x-axis, y-axis, origin

7. $(-2, 0)$; sym. about x-axis   **9.** Sym. about x-axis

11. $(-21, 0)$, $(0, -7)$, $(0, 3)$ **13.** $(1, 0)$, $(0, 0)$

15. $\left(0, \frac{3}{8}\right)$

17. $(3, 0)$, $(0, \pm 3)$; sym. about x-axis

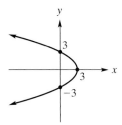

19. $(\pm 2, 0)$, $(0, 0)$; sym. about origin

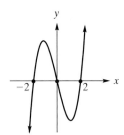

21. $(0, 0)$; sym. about x-axis, y-axis, origin, $y = x$

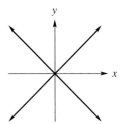

23. $\left(\pm\frac{5}{3}, 0\right)$, $\left(0, \pm\frac{5}{2}\right)$; sym. about x-axis, y-axis, origin

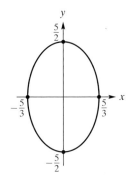

25. **(a)** $(\pm 0.99, 0)$, $(0, 5)$; **(b)** 5; **(c)** $(-\infty, 5]$

27.

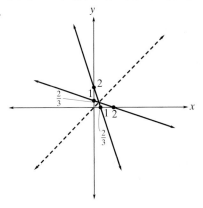

9.

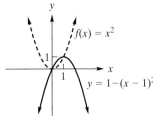

11.

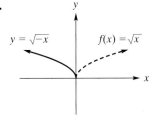

PROBLEMS 2.7 (page 110)

1.

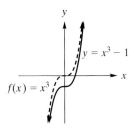

13. Translate 3 units to the left, stretch vertically away from the x-axis by a factor of 2, reflect about the x-axis, and move 2 units upward

15. Reflect about the y-axis and translate 5 units downward

REVIEW PROBLEMS—CHAPTER 2 (page 112)

1. All real numbers except 1 and 5

3. All real numbers

5. All nonnegative real numbers except 1

7. $7, 46, 62, 3t^2 - 4t + 7$ **9.** $0, 2, \sqrt[4]{t-2}, \sqrt[4]{x^3-3}$

11. $\dfrac{3}{5}, 0, \dfrac{\sqrt{x+4}}{x}, \dfrac{\sqrt{u}}{u-4}$

13. $20, -3, -3,$ undefined

15. **(a)** $3 - 7x - 7h$; **(b)** -7

17. **(a)** $4x^2 + 8hx + 4h^2 + 2x + 2h - 5$; **(b)** $8x + 4h + 2$

19. **(a)** $5x + 2$; **(b)** 22; **(c)** $x - 4$; **(d)** $6x^2 + 7x - 3$; **(e)** 10; **(f)** $\dfrac{3x-1}{2x+3}$; **(g)** $6x + 8$; **(h)** 38; **(i)** $6x + 1$

21. $\dfrac{1}{(x+1)^2}, \dfrac{1}{x^2} + 1 = \dfrac{1+x^2}{x^2}$

23. $\sqrt{x^3 + 2}, (x+2)^{3/2}$

25. $(0, 0), (\pm\sqrt{3}, 0)$; sym. about origin

27. $(0, 9), (\pm 3, 0)$; sym. about y-axis

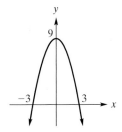

3.

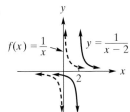

5.

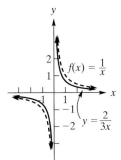

7.

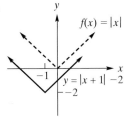

29. $(0, 2)$, $(-4, 0)$; all $u \geq -4$; all real numbers ≥ 0

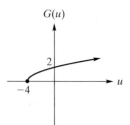

31. $\left(0, \dfrac{1}{2}\right)$; all $t \neq 4$; all positive real numbers

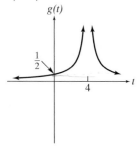

33. All real numbers; all real numbers ≤ 2

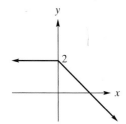

35.

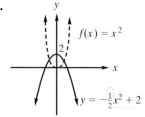

37. (a), (c)   **39.** $-0.67, 0.34, 1.73$

41. $-1.50, -0.88, -0.11, 1.09, 1.40$

43. (a) $(-\infty, \infty)$;   (b) $(1.92, 0)$, $(0, 7)$

45. (a) $0, 2, 4$;   (b) none

MATHEMATICAL SNAPSHOT—CHAPTER 2 (page 114)

1. $2695   **3.** $75,681.50   **5.** Answers may vary

7. $g(x) = \begin{cases} 0.90x & \text{if } 0 \leq x \leq 15{,}100 \\ 0.85x + 755 & \text{if } 15{,}100 < x \leq 61{,}300 \\ 0.75x + 6885 & \text{if } 61{,}300 < x \leq 123{,}700 \\ 0.72x + 10{,}596 & \text{if } 123{,}700 < x \leq 188{,}450 \\ 0.67x + 20{,}018.50 & \text{if } 188{,}450 < x \leq 336{,}550 \\ 0.65x + 26{,}749.50 & \text{if } x > 336{,}550 \end{cases}$

PRINCIPLES IN PRACTICE 3.1

1. -2000; the car depreciated $2000 per year

2. $S = 14T + 8$ **3.** $F = \dfrac{9}{5}C + 32$

4. slope $= \dfrac{125}{3}$; y-intercept $= \dfrac{125}{3}$

5. $9C - 5F + 160 = 0$

6.

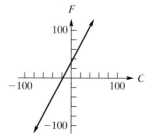

7. The slope of $\overline{AB}$ is 0; the slope of $\overline{BC}$ is 7; the slope of $\overline{CA}$ is 1. None of the slopes are negative reciprocals of each other, so the triangle does not have a right angle. The points do not define a right triangle.

PROBLEMS 3.1 (page 123)

1. 3   **3.** $-\dfrac{1}{2}$ **5.** Undefined   **7.** 0

9. $5x + y - 2 = 0$ **11.** $x + 4y - 18 = 0$

13. $3x - 7y + 25 = 0$ **15.** $4x + y + 16 = 0$

17. $2x - y + 4 = 0$ **19.** $x + 2y + 6 = 0$

21. $y + 3 = 0$ **23.** $x - 2 = 0$ **25.** $4; -6$

27. $-\dfrac{3}{5}; \dfrac{9}{5}$

29. Slope undefined; no y-intercept

31. $3; 0$ **33.** $0; 3$

35. $2x + 3y - 5 = 0$; $y = -\dfrac{2}{3}x + \dfrac{5}{3}$

37. $4x + 9y - 5 = 0$; $y = -\dfrac{4}{9}x + \dfrac{5}{9}$

39. $6x - 8y - 57 = 0$; $y = \dfrac{3}{4}x - \dfrac{57}{8}$

41. Parallel   **43.** Parallel   **45.** Neither

47. Perpendicular   **49.** Perpendicular

51. $y = -\dfrac{1}{4}x + \dfrac{5}{4}$ **53.** $y = 1$ **55.** $y = -\dfrac{1}{3}x + 5$

57. $x = 5$ **59.** $y = -\dfrac{2}{3}x - \dfrac{29}{3}$ **61.** $(5, -4)$

63. -2.9; the stock price dropped an average of $2.90 per year

65. $y = 28{,}000x - 100{,}000$ **67.** $-t + d - 184 = 0$

71. $C = 59.82T + 769.58$ **75.** The slope is 7.1.

PRINCIPLES IN PRACTICE 3.2

1. $x =$ number of skis produced; $y =$ number of boots produced; $8x + 14y = 1000$

2. $p = -\dfrac{3}{8}q + 1025$

3. Answers may vary, but two possible points are $(0, 60)$ and $(2, 140)$.

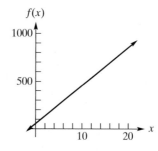

4. $f(t) = 2.3t + 32.2$

5. $f(x) = 70x + 150$

PROBLEMS 3.2 (page 129)

1. $-4; 0$

3. $5; -7$

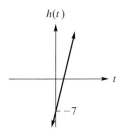

5. $-\dfrac{1}{7}; \dfrac{2}{7}$

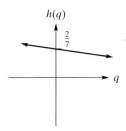

7. $f(x) = 4x$

9. $f(x) = -2x + 4$

11. $f(x) = -\dfrac{2}{3}x - \dfrac{10}{9}$

13. $f(x) = x + 1$

15. $p = -\dfrac{2}{5}q + 28.75;\ \$13.95$

17. $p = \dfrac{1}{4}q + 190$

19. $c = 3q + 10;\ \$115$

21. $f(x) = 0.125x + 4.15$

23. $v = -180t + 1800;$ slope $= -180$

25. $f(x) = 45{,}000x + 735{,}000$

27. $f(x) = 64x + 95$

29. $x + 10y = 100$

31. **(a)** $y = \dfrac{35}{44}x + \dfrac{225}{11}$ **(b)** 52.2

33. **(a)** $p = 0.059t + 0.025;$ **(b)** 0.556

35. **(a)** $t = \dfrac{1}{4}c + 37;$ **(b)** add 37 to the number of chirps in 15 seconds

PRINCIPLES IN PRACTICE 3.3

1. Vertex: $(1, 400)$; y-intercept: $(0, 399)$; x-intercepts: $(-19, 0), (21, 0)$

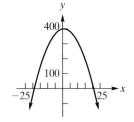

2. Vertex: $(1, 24)$; y-intercept: $(0, 8)$; x-intercepts: $\left(1 + \dfrac{\sqrt{6}}{2}, 0\right), \left(1 - \dfrac{\sqrt{6}}{2}, 0\right)$

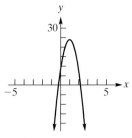

3. 1000 units; \$3000 maximum revenue

PROBLEMS 3.3 (page 136)

1. Quadratic **3.** Not Quadratic **5.** Quadratic

7. Quadratic **9. (a)** $(1, 11)$; **(b)** highest

11. (a) -6; **(b)** $-3, 2$; **(c)** $\left(-\dfrac{1}{2}, -\dfrac{25}{4}\right)$

13. Vertex: $(3, -4)$; intercepts: $(1, 0), (5, 0), (0, 5)$; range: all $y \geq -4$

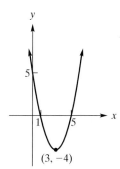

15. Vertex: $\left(-\dfrac{3}{2}, \dfrac{9}{2}\right)$; intercepts: $(0, 0), (-3, 0)$; range: all $y \leq \dfrac{9}{2}$

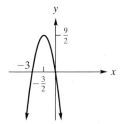

17. Vertex: $(-3, 0)$; intercepts: $(-3, 0), (0, 9)$; range: all $s \geq 0$

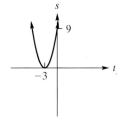

19. Vertex: $(2, -1)$; intercept: $(0, -9)$; range: all $y \leq -1$

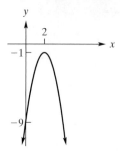

21. Vertex: $(4, -2)$; intercepts: $(4 + \sqrt{2}, 0)$, $(4 - \sqrt{2}, 0)$, $(0, 14)$; range: all $t \geq -2$

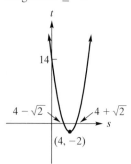

23. Minimum; $\dfrac{808}{49}$ **25.** Maximum; -10

27. $g^{-1}(x) = 1 + \sqrt{x - 3},\ x \geq 3$

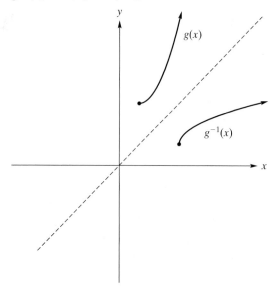

29. 20 units; $2000 maximum revenue

31. 200 units; $240,000 maximum revenue

33. Vertex: $(9, 225)$; y-intercept: $(0, 144)$; x-intercepts: $(-6, 0)$, $(24, 0)$

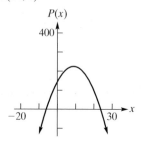

35. 70 grams   **37.** ≈ 134.86 ft; ≈ 2.7 sec

39. Vertex: $\left(\dfrac{5}{2}, 116\right)$; y-intercept: $(0, 16)$,

x-intercepts: $\left(\dfrac{5 + \sqrt{29}}{2}, 0\right)$, $\left(\dfrac{5 - \sqrt{29}}{2}, 0\right)$

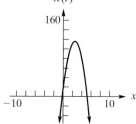

41. 125 ft $\times$ 250 ft   **43.** $(1.11, 2.88)$

45. (a) 0; **(b)** 1; **(c)** 2   **47.** 4.89

PRINCIPLES IN PRACTICE 3.4

1. $120,000 at 9% and $80,000 at 8%

2. 500 of species A and 1000 of species B

3. Infinitely many solutions of the form $A = \dfrac{20,000}{3} - \dfrac{4}{3}r$, $B = r$ where $0 \leq r \leq 5000$

4. $\dfrac{1}{6}$ lb of A; $\dfrac{1}{3}$ lb of B; $\dfrac{1}{2}$ lb of C

PROBLEMS 3.4 (page 146)

1. $x = -1,\ y = 1$ **3.** $x = 3,\ y = -1$

5. $u = 6,\ v = -1$ **7.** $x = -3,\ y = 2$

9. No solution   **11.** $x = 12,\ y = -12$

13. No solution   **15.** $x = \dfrac{1}{2},\ y = \dfrac{1}{2},\ z = \dfrac{1}{4}$

17. $x = 2,\ y = -1,\ z = 4$

19. $x = 1 + 2r,\ y = 3 - r,\ z = r$; r is any real number

21. $x = -\dfrac{1}{3}r,\ y = \dfrac{5}{3}r,\ z = r$; r is any real number

23. $x = \dfrac{3}{2} - r + \dfrac{1}{2}s,\ y = r,\ z = s$; r and s are any real numbers

25. $533\frac{1}{3}$ gal of 20% solution, $266\frac{2}{3}$ gal of 35% solution

27. 0.5 lb of cotton; 0.25 lb of polyester; 0.25 lb of nylon

29. ≈ 285 mi/h (speed of airplane in still air), ≈ 23.2 mi/h (speed of wind)

31. 240 units of early American, 200 units of Contemporary

33. 800 calculators at Exton plant, 700 at Whyton plant

35. 4% on first $100,000, 6% on remainder

37. 190 boxes, 760 clamshells

39. 100 chairs, 100 rockers, 200 chaise lounges

41. 10 semiskilled workers, 5 skilled workers, 55 shipping clerks

45. $x = 3,\ y = 2$ **47.** $x = 8.3,\ y = 14.0$

PROBLEMS 3.5 (page 149)

1. $x = -1 + \sqrt{13},\ y = 5 - 2\sqrt{13};\ x = -1 - \sqrt{13},\ y = 5 + 2\sqrt{13}$

3. $p = -3,\ q = -4;\ p = 2,\ q = 1$

5. $x = 0,\ y = 0;\ x = 1,\ y = 1$

7. $x = 4,\ y = 8;\ x = -1,\ y = 3$

9. $p = 0,\ q = 0;\ p = 1,\ q = 1$

11. $x = \sqrt{17}, y = 2; x = -\sqrt{17}, y = 2; x = \sqrt{14}, y = -1;$
$x = -\sqrt{14}, y = -1$

13. $x = 7, y = 6$

15. At $(10, 8.1)$ and $(-10, 7.9)$ **17.** Three

19. $x = -1.3, y = 5.1$ **21.** $x = 1.76$ **23.** $x = -1.46$

PROBLEMS 3.6 (page 156)

1. The equilibrium point is $(100, 7)$.

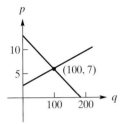

3. $(5, 212.50)$ **5.** $(9, 38)$ **7.** $(15, 5)$

9. The break-even quantity is 2500 units.

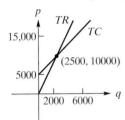

11. Cannot break even at any level of production

13. Cannot break even at any level of production

15. (a) $12; (b) $12.18

17. 5840 units; 840 units; 1840 units

19. $4

21. Total cost always exceeds total revenue—no break-even point.

23. Decreases by $0.70 **25.** $P_A = 8; P_B = 10$

27. 2.4 and 11.3

REVIEW PROBLEMS—CHAPTER 3 (page 158)

1. 9 **3.** $y = -2x - 1; 2x + y + 1 = 0$

5. $y = \dfrac{1}{2}x - 1; x - 2y - 2 = 0$ **7.** $y = 4; y - 4 = 0$

9. $y = \dfrac{2}{5}x - 3; 2x - 5y - 15 = 0$ **11.** Perpendicular

13. Neither **15.** Parallel **17.** $y = \dfrac{3}{2}x - 2; \dfrac{2}{3}$

19. $y = \dfrac{4}{3}; 0$

21. $-5; (0, 17)$

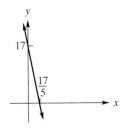

23. $(3, 0), (-3, 0), (0, 9); (0, 9)$

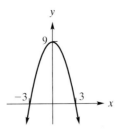

25. $(5, 0), (-1, 0), (0, -5); (2, -9)$

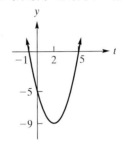

27. $-7; (0, 0)$

29. $(0, -3); (-1, -2)$

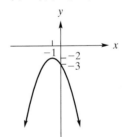

31. $x = \dfrac{17}{7}, y = -\dfrac{8}{7}$ **33.** $x = 2, y = -\dfrac{9}{5}$

35. $x = 8, y = 4$ **37.** $x = 0, y = 1, z = 0$

39. $x = \dfrac{-5 + \sqrt{65}}{4}, y = \dfrac{-21 + 5\sqrt{65}}{8};$
$x = \dfrac{-5 - \sqrt{65}}{4}, y = \dfrac{-21 - 5\sqrt{65}}{8}$

41. $x = -2 - 2r, y = 7 + r, z = r; r$ is any real number

43. $x = r, y = r, z = 0; r$ is any real number

45. $a - 4b = -7; 13$ **47.** $f(x) = -\dfrac{4}{3}x + \dfrac{19}{3}$

49. 50 units; $5000 **51.** ≈ 6.55 **53.** 1250 units; $20,000

55. 2.36 tons per square km **57.** $x = 7.29, y = -0.78$

59. $x = 0.75, y = 1.43$

MATHEMATICAL SNAPSHOT—CHAPTER 3 (page 161)

1. $2337.50

3. Usage between 950 and 1407.14 minutes

5. Usage between 2200 and 4200 minutes

PRINCIPLES IN PRACTICE 4.1

1. The shape of the graphs are the same. The value of A scales the ordinate of any point by A.

2.

Year	Multiplicative Increase	Expression
0	1	1.1^0
1	1.1	1.1^1
2	1.21	1.1^2
3	1.33	1.1^3
4	1.46	1.1^4

1.1; The investment increases by 10% every year.
$(1 + 1(0.1) = 1 + 0.1 = 1.1)$

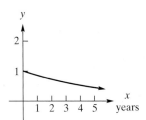

Between 7 and 8 years.

3.

Year	Multiplicative Decrease	Expression
0	1	0.85^0
1	0.85	0.85^1
2	0.72	0.85^2
3	0.61	0.85^3

0.85; The car depreciates by 15% every year.
$(1 - 1(0.15) = 1 - 0.15 = 0.85)$

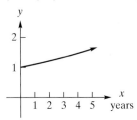

Between 4 and 5 years.

4. $y = 1.08^{t-3}$; Shift the graph 3 units to the right.
5. $3684.87; $1684.87   **6.** 117 employees

7.

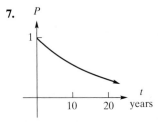

PROBLEMS 4.1 (page 173)

1.

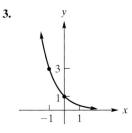

3.

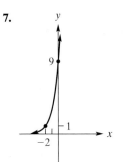

5.

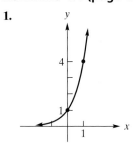

7.

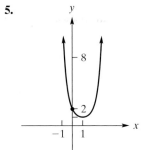

9.

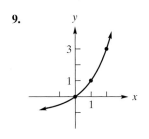

11.

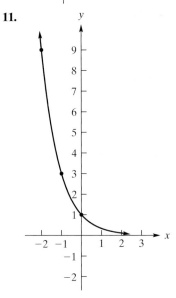

13. B **15.** 138,750

17. $\dfrac{1}{2}, \dfrac{3}{4}, \dfrac{7}{8}$

19. (a) $6014.52;   (b) $2014.52
21. (a) $1964.76;   (b) $1264.76
23. (a) $11,983.37;   (b) $8983.37
25. (a) $6256.36;   (b) $1256.36
27. (a) $9649.69;   (b) $1649.69
29. $8253.28
31. (a) $N = 400(1.05)^t$;   (b) 420;   (c) 486
33.

Year	Multiplicative Increase	Expression
0	1	1.3^0
1	1.3	1.3^1
2	1.69	1.3^2
3	2.20	1.3^3

1.3; The recycling increases by 30% every year.
$(1 + 1(0.3) = 1 + 0.3 = 1.3)$

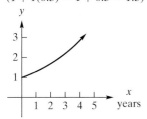

Between 4 and 5 years.

35. 334,485   **37.** 4.4817   **39.** 0.4966
41.

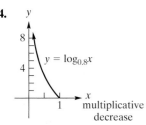

43. 0.2240

45. $(e^k)^t$, where $b = e^k$

47. (a) 12; **(b)** 8.8; **(c)** 3.1; **(d)** 22 hours
49. 32 years   **51.** 0.1465   **55.** 3.17   **57.** 4.2 min
59. 17

PRINCIPLES IN PRACTICE 4.2

1. $t = \log_2 16$; $t = $ the number of times the bacteria have doubled.

2. $\dfrac{I}{I_0} = 10^{8.3}$

3.

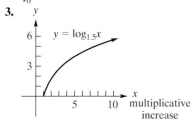

4.

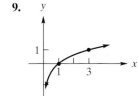

5. Approximately 13.9%   **6.** Approximately 9.2%

PROBLEMS 4.2 (page 180)

1. $\log 10,000 = 4$ **3.** $2^6 = 64$ **5.** $\ln 20.0855 = 3$
7. $e^{1.09861} = 3$
9.

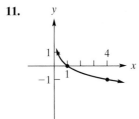

11.

13.

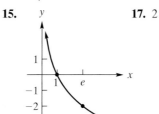

15.   **17.** 2   **19.** 3

21. 1   **23.** -2 **25.** 0   **27.** -3
29. 81   **31.** 125   **33.** $\dfrac{1}{10}$ **35.** e^{-3}
37. 2   **39.** 6   **41.** $\dfrac{1}{27}$ **43.** 2
45. $\dfrac{5}{3}$ **47.** 4   **49.** $\dfrac{\ln 2}{3}$ **51.** $\dfrac{5 + \ln 3}{2}$
53. 1.60944   **55.** 2.00013   **57.** $y = \log_{1.10} x$
59. 3
61. (a) $2N_0$; **(b)** k is the time it takes for the population to double.
63. 36.1 minutes   **65.** $z = y^{3/2}$
67. (a) $(0, 1)$; **(b)** $[-0.37, \infty)$ **69.** 1.10
71. 1.41, 3.06

PRINCIPLES IN PRACTICE 4.3

1. $\log(900,000) - \log(9000) = \log\left(\dfrac{900,000}{9000}\right)$
$$= \log(100) = 2$$
2. $\log(10,000) = \log(10^4) = 4$

PROBLEMS 4.3 (page 185)

1. $a + b + c$ **3.** $a - b$ **5.** $3a - b$ **7.** $2(a + b)$
9. $\dfrac{b}{a}$ **11.** 48   **13.** -7 **15.** 5.01
17. -2 **19.** 2   **21.** $\ln x + 2\ln(x + 1)$
23. $2\ln x - 3\ln(x + 1)$ **25.** $4[\ln(x + 1) + \ln(x + 2)]$
27. $\ln x - \ln(x + 1) - \ln(x + 2)$
29. $\dfrac{1}{2}\ln x - 2\ln(x + 1) - 3\ln(x + 2)$
31. $\dfrac{2}{5}\ln x - \dfrac{1}{5}\ln(x + 1) - \ln(x + 2)$
33. $\log 24$ **35.** $\log_2 \dfrac{2x}{x + 1}$ **37.** $\log_2(10^5 \cdot 13^2)$
39. $\log[100(1.05)^{10}]$ **41.** $\dfrac{81}{64}$ **43.** 1
45. $\dfrac{5}{2}$ **47.** ± 2 **49.** $\dfrac{\ln(2x + 1)}{\ln 2}$
51. $\dfrac{\ln(x^2 + 1)}{\ln 3}$ **53.** $y = \ln \dfrac{z}{7}$ **59.** $\log x = \dfrac{\ln x}{\ln 10}$
61. $\ln 3$

PRINCIPLES IN PRACTICE 4.4

1. 18   **2.** Day 20

3. The other earthquake is 67.5 times as intense as a zero-level earthquake.

PROBLEMS 4.4 (page 190)

1. 3   **3.** 2.75   **5.** −3   **7.** 2

9. 0.125   **11.** 1.099   **13.** 0.028   **15.** 5.140

17. −0.073   **19.** 2.322   **21.** 0.942   **23.** 0.483

25. 2.496   **27.** 1003   **29.** 2.222   **31.** 3.082

33. 1.353   **35.** 0.5   **37.** $S = 12.4A^{0.26}$

39. (a) 100;   **(b)** 46   **41.** 20.5

43. $p = \dfrac{\log(80 - q)}{\log 2}$; 4.32   **49.** 3.33

REVIEW PROBLEMS—CHAPTER 4 (page 192)

1. $\log_3 243 = 5$

3. $81^{1/4} = 3$ **5.** $\ln 54.598 = 4$ **7.** 3

9. −4   **11.** −2   **13.** 4   **15.** $\dfrac{1}{32}$

17. −1   **19.** $3(a + 1)$ **21.** $\log \dfrac{7^3}{5^2}$ **23.** $\ln \dfrac{x^2 y}{z^3}$

25. $\log_2 \dfrac{x^{9/2}}{(x + 1)^3(x + 2)^4}$ **27.** $3 \ln x + 2 \ln y + 5 \ln z$

29. $\dfrac{1}{3}(\ln x + \ln y + \ln z)$ **31.** $\dfrac{1}{2}(\ln y - \ln z) - \ln x$

33. $\dfrac{\ln(x + 5)}{\ln 3}$ **35.** 1.8295   **37.** $2y + \dfrac{1}{2}x$

39. $2x + 1$ **41.** $y = e^{x^2 + 2}$

43.

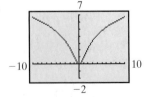

45. 5

47. 1   **49.** 10

51. $\dfrac{1}{3e^2}$

53. 0.880

55. −3.222

57. −1.596

59. (a) $3829.04;   **(b)** $1229.04

61. 14%

63. (a) $P = 6000(0.995)^t$;   **(b)** 5707

65. (a) 10 mg;   **(b)** 4.4;   **(c)** 0.2;   **(d)** 1.7;   **(e)** 5.6

67. (a) 6;   **(b)** 28   **71.** $(-\infty, 0.37]$

73. 2.53

75.

MATHEMATICAL SNAPSHOT—CHAPTER 4 (page 194)

1. (a) $P = \dfrac{T(e^{kl} - 1)}{1 - e^{-dkl}}$;   **(b)** $d = \dfrac{1}{kI} \ln\left[\dfrac{P}{P - T(e^{kl} - 1)}\right]$

3. (a) 156;   **(b)** 65

PRINCIPLES IN PRACTICE 5.1

1. 4.9%   **2.** 7 years, 16 days   **3.** 7.7208%

4. 11.25% compounded quarterly has the better effective rate of interest. The $10,000 investment is slightly better over 20 years.

PROBLEMS 5.1 (page 200)

1. (a) $11,105.58;   **(b)** $5105.58

3. 3.023%   **5.** 4.081%

7. (a) 10%;   **(b)** 10.25%;   **(c)** 10.381%;   **(d)** 10.471%;   **(e)** 10.516%

9. 8.08%   **11.** 8.0 years   **13.** $10,282.95

15. $38,503.23   **17. (a)** 18%;   **(b)** $19.56%

19. $3198.54

21. 8% compounded annually   **23. (a)** 4.93%;   **(b)** 4.86%

25. 11.61%   **27.** 6.29%

PROBLEMS 5.2 (page 204)

1. $2261.34   **3.** $1751.83   **5.** $5821.55

7. $4862.31   **9.** $6838.95   **11.** $11,381.89

13. $14,091.10   **15.** $1238.58   **17.** $3244.63

19. (a) $515.62;   **(b)** profitable   **21.** Savings account

23. $226.25   **25.** 9.55%

PROBLEMS 5.3 (page 207)

1. $5819.97; $1819.97   **3.** $1456.87

5. 4.08%   **7.** 3.05%   **9.** $109.42   **11.** $778,800.78

13. (a) $59,081;   **(b)** $19,181   **15.** 4.88%

17. 16 years

19. (a) $1072.51;   **(b)** $1093.30;   **(c)** $1072.18

21. (a) $9458.51;   **(b)** This strategy is better by $26.90.

PRINCIPLES IN PRACTICE 5.4

1. 48 ft, 36 ft, 27 ft, $20\dfrac{1}{4}$ ft, $15\dfrac{3}{16}$ ft

2. $500(1.5), 500(1.5)^2, 500(1.5)^3, 500(1.5)^4, 500(1.5)^5, 500(1.5)^6$ or 750, 1125, 1688, 2531, 3797, 5695

3. 35.72 m   **4.** $176,994.65   **5.** 6.20%

6. $101,925; $121,925   **7.** $723.03

8. $13,962.01   **9.** $45,502.06   **10.** $48,095.67

PROBLEMS 5.4 (page 216)

1. 64, 32, 16, 8, 4   **3.** 100, 102, 104.04   **5.** $\dfrac{21,044}{16,807}$

7. 1.11111   **9.** 18.664613   **11.** 8.213180

13. $2950.39   **15.** $29,984.06   **17.** $8001.24

19. $90,231.01   **21.** $204,977.46   **23.** $24,594.36

25. $5106.27   **27.** $458.40

29. (a) $3048.85;   **(b)** $648.85   **31.** $3474.12

33. $1725   **35.** 102.91305   **37.** 55,466.57

39. $131.34   **41.** $1,872,984.02

43. $205,073; $142,146   **45.** $181,269.25

PROBLEMS 5.5 (page 221)

1. $273.42   **3.** $502.84

5. **(a)** $221.43; **(b)** $25; **(c)** $196.43

7.

Period	Prin. Outs. at Beginning	Interest for Period	Pmt. at End	Prin. Repaid at End
1	5000.00	350.00	1476.14	1126.14
2	3873.86	271.17	1476.14	1204.97
3	2668.89	186.82	1476.14	1289.32
4	1379.57	96.57	1476.14	1379.57
Total		904.56	5904.56	5000.00

9.

Period	Prin. Outs. at Beginning	Interest for Period	Pmt. at End	Prin. Repaid at End
1	900.00	22.50	193.72	171.22
2	728.78	18.22	193.72	175.50
3	553.28	13.83	193.72	179.89
4	373.39	9.33	193.72	184.39
5	189.00	4.73	193.73	189.00
Total		68.61	968.61	900.00

11. 11   **13.** $1606

15. **(a)** $2089.69; **(b)** $1878.33; **(c)** $211.36; **(d)** $381,907

17. 23   **19.** $113,302   **21.** $38.64

REVIEW PROBLEMS—CHAPTER 5 (page 223)

1. $\dfrac{665}{81}$ **3.** 8.5% compounded annually

5. $586.60   **7.** **(a)** $1997.13; **(b)** $3325.37

9. $2506.59   **11.** $886.98   **13.** $314.00

15.

Period	Prin. Outs. at Beginning	Interest for Period	Pmt. at End	Prin. Repaid at End
1	15,000.00	112.50	3067.84	2955.34
2	12,044.66	90.33	3067.84	2977.51
3	9067.15	68.00	3067.84	2999.84
4	6067.31	45.50	3067.84	3022.34
5	3044.97	22.84	3067.81	3044.97
Total		339.17	15,339.17	15,000.00

17. $1279.36

MATHEMATICAL SNAPSHOT—CHAPTER 5 (page 224)

1. $26,102.13

3. When investors expect a drop in interest rates, long-term investments become more attractive relative to short-term ones.

PRINCIPLES IN PRACTICE 6.1

1. 3×2 or 2×3 **2.** $\begin{bmatrix} 1 & 2 & 4 & 8 & 16 \\ 1 & 2 & 4 & 8 & 16 \\ 1 & 2 & 4 & 8 & 16 \end{bmatrix}$

PROBLEMS 6.1 (page 231)

1. **(a)** 2×3, 3×3, 3×2, 2×2, 4×4, 1×2, 3×1, 3×3, 1×1;
(b) B, D, E, H, J; **(c)** H, J upper triangular; D, J lower triangular; **(d)** F, J; **(e)** G, J

3. 6   **5.** 4   **7.** 0   **9.** 7, 2, 1, 0

11. $\begin{bmatrix} 1 & 4 & 7 & 10 & 13 \\ -1 & 2 & 5 & 8 & 11 \\ -3 & 0 & 3 & 6 & 9 \end{bmatrix}$ **13.** 120 entries, 1, 0, 1, 0

15. **(a)** $\begin{bmatrix} 0 & 0 & 0 & 0 \\ 0 & 0 & 0 & 0 \\ 0 & 0 & 0 & 0 \\ 0 & 0 & 0 & 0 \end{bmatrix}$; **(b)** $\begin{bmatrix} 0 & 0 & 0 & 0 & 0 & 0 \\ 0 & 0 & 0 & 0 & 0 & 0 \\ 0 & 0 & 0 & 0 & 0 & 0 \\ 0 & 0 & 0 & 0 & 0 & 0 \\ 0 & 0 & 0 & 0 & 0 & 0 \\ 0 & 0 & 0 & 0 & 0 & 0 \end{bmatrix}$

17. $\begin{bmatrix} 6 & 2 \\ -3 & 4 \end{bmatrix}$ **19.** $\begin{bmatrix} 1 & 3 & -4 \\ 3 & 2 & 5 \\ 7 & -2 & 0 \\ 3 & 0 & 1 \end{bmatrix}$

21. **(a)** A and C; **(b)** all of them

25. $x = 6$, $y = \dfrac{2}{3}$, $z = \dfrac{7}{2}$ **27.** $x = 0$, $y = 0$

29. **(a)** 7; **(b)** 3; **(c)** February; **(d)** deluxe blue; **(e)** February; **(f)** February; **(g)** 38

31. -2001 **33.** $\begin{bmatrix} 3 & 1 & 1 \\ 1 & 7 & 4 \\ 4 & 3 & 1 \\ 2 & 6 & 2 \end{bmatrix}$

PRINCIPLES IN PRACTICE 6.2

1. $\begin{bmatrix} 230 & 220 \\ 190 & 255 \end{bmatrix}$ **2.** $x_1 = 670$, $x_2 = 835$, $x_3 = 1405$

PROBLEMS 6.2 (page 237)

1. $\begin{bmatrix} 4 & -3 & 1 \\ -2 & 10 & 5 \\ 10 & 5 & 3 \end{bmatrix}$ **3.** $\begin{bmatrix} -5 & 5 \\ -9 & 5 \\ 5 & 9 \end{bmatrix}$ **5.** $[-4 \quad -2 \quad 10]$

7. Not defined   **9.** $\begin{bmatrix} -12 & 36 & -42 & -6 \\ -42 & -6 & -36 & 12 \end{bmatrix}$

11. $\begin{bmatrix} 3 & -5 & 6 \\ -2 & 8 & 0 \\ 5 & 10 & 15 \end{bmatrix}$ **13.** $\begin{bmatrix} 6 & 5 \\ -2 & 3 \end{bmatrix}$ **15.** O

17. $\begin{bmatrix} 66 & 51 \\ 0 & 9 \end{bmatrix}$ **19.** Not defined   **21.** $\begin{bmatrix} -22 & -15 \\ -11 & 9 \end{bmatrix}$

23. $\begin{bmatrix} 21 & \frac{29}{2} \\ \frac{19}{2} & -\frac{15}{2} \end{bmatrix}$ **29.** $\begin{bmatrix} 4 & 7 \\ 2 & -3 \\ 20 & 2 \end{bmatrix}$ **31.** $\begin{bmatrix} -1 & 5 \\ 6 & -8 \end{bmatrix}$

33. Impossible   **35.** $x = \dfrac{90}{29}$, $y = -\dfrac{24}{29}$

37. $x = 6$, $y = \dfrac{4}{3}$ **39.** $x = -6$, $y = -14$, $z = 1$

41. $\begin{bmatrix} 45 & 75 \\ 1760 & 1520 \\ 35 & 35 \end{bmatrix}$ 　　**43.** 1.1

45. $\begin{bmatrix} 15 & -4 & 26 \\ 4 & 7 & 30 \end{bmatrix}$ 　　**47.** $\begin{bmatrix} -10 & 22 & 12 \\ 24 & 36 & -44 \end{bmatrix}$

PRINCIPLES IN PRACTICE 6.3

1. $5780 　　**2.** $22,843.75

3. $\begin{bmatrix} 1 & \frac{8}{5} \\ 1 & \frac{1}{3} \end{bmatrix} \begin{bmatrix} y \\ x \end{bmatrix} = \begin{bmatrix} \frac{8}{5} \\ \frac{5}{3} \end{bmatrix}$

PROBLEMS 6.3 (page 248)

1. -12 　　**3.** 19 　　**5.** 1 　　**7.** 2×2; 4

9. 3×5; 15 　　**11.** 2×1; 2 　　**13.** 3×1; 3 　　**15.** 3×1; 3

17. $\begin{bmatrix} 1 & 0 & 0 & 0 \\ 0 & 1 & 0 & 0 \\ 0 & 0 & 1 & 0 \\ 0 & 0 & 0 & 1 \end{bmatrix}$ 　　**19.** $\begin{bmatrix} 12 & -12 \\ 10 & 6 \end{bmatrix}$ 　　**21.** $\begin{bmatrix} 23 \\ 50 \end{bmatrix}$

23. $\begin{bmatrix} 1 & -4 & 2 \\ 2 & 2 & 4 \\ -3 & -2 & 3 \end{bmatrix}$ 　　**25.** $[-4 \quad 5 \quad -1 \quad -18]$

27. $\begin{bmatrix} 4 & 6 & -4 & 6 \\ 6 & 9 & -6 & 9 \\ -8 & -12 & 8 & -12 \\ 2 & 3 & -2 & 3 \end{bmatrix}$ 　　**29.** $\begin{bmatrix} 78 & 84 \\ -21 & -12 \end{bmatrix}$

31. $\begin{bmatrix} -5 & -8 \\ -5 & -20 \end{bmatrix}$ 　　**33.** $\begin{bmatrix} z \\ y \\ x \end{bmatrix}$ 　　**35.** $\begin{bmatrix} 2x_1 + x_2 + 3x_3 \\ 4x_1 + 9x_2 + 7x_3 \end{bmatrix}$

37. $\begin{bmatrix} 0 & 0 & 0 \\ 0 & -1 & 1 \\ 1 & 2 & 0 \end{bmatrix}$ 　　**39.** $\begin{bmatrix} -1 & -20 \\ -2 & 23 \end{bmatrix}$

41. $\begin{bmatrix} \frac{7}{3} & 0 & 0 \\ 0 & \frac{7}{3} & 0 \\ 0 & 0 & \frac{7}{3} \end{bmatrix}$ 　　**43.** $\begin{bmatrix} -1 & 5 \\ 2 & 17 \\ 1 & 31 \end{bmatrix}$

45. Impossible 　　**47.** $\begin{bmatrix} 0 & 0 & -4 \\ 2 & -1 & -2 \\ 0 & 0 & 8 \end{bmatrix}$ 　　**49.** $\begin{bmatrix} -1 & 2 \\ 1 & 0 \end{bmatrix}$

51. $\begin{bmatrix} 0 & 3 & 0 \\ -1 & -1 & 2 \end{bmatrix}$ 　　**53.** $\begin{bmatrix} 2 & 0 & 0 \\ 0 & 2 & 0 \\ 0 & 0 & 2 \end{bmatrix}$ 　　**55.** $\begin{bmatrix} 1 & -1 & 0 \\ 0 & 1 & 1 \end{bmatrix}$

57. $\begin{bmatrix} 6 & -7 \\ -7 & 9 \end{bmatrix}$ 　　**59.** $\begin{bmatrix} 3 & 1 \\ 2 & -9 \end{bmatrix} \begin{bmatrix} x \\ y \end{bmatrix} = \begin{bmatrix} 6 \\ 5 \end{bmatrix}$

61. $\begin{bmatrix} 2 & -1 & 3 \\ 5 & -1 & 2 \\ 3 & -2 & 2 \end{bmatrix} \begin{bmatrix} r \\ s \\ t \end{bmatrix} = \begin{bmatrix} 9 \\ 5 \\ 11 \end{bmatrix}$ 　　**63.** $2075

65. $828,950

67. **(a)** $180,000, $520,000, $400,000, $270,000, $380,000, $640,000; 　**(b)** $390,000, $100,000, $800,000;

(c) $2,390,000; 　**(d)** $\frac{110}{239}, \frac{129}{239}$

71. $\begin{bmatrix} 72.82 & -9.8 \\ 51.32 & -36.32 \end{bmatrix}$ 　　**73.** $\begin{bmatrix} 15.606 & 64.08 \\ -739.428 & 373.056 \end{bmatrix}$

PRINCIPLES IN PRACTICE 6.4

1. 5 blocks of A, 2 blocks of B, and 1 block of C

2. 3 of X; 4 of Y; 2 of Z

3. $A = 3D$; $B = 1000 - 2D$; $C = 500 - D$; $D =$ any amount between 0 and 500

PROBLEMS 6.4 (page 257)

1. Not reduced 　　**3.** Reduced 　　**5.** Not reduced

7. $\begin{bmatrix} 1 & 0 \\ 0 & 1 \end{bmatrix}$ 　　**9.** $\begin{bmatrix} 1 & 2 & 3 \\ 0 & 0 & 0 \\ 0 & 0 & 0 \end{bmatrix}$ 　　**11.** $\begin{bmatrix} 1 & 0 & 0 & 0 \\ 0 & 1 & 0 & 0 \\ 0 & 0 & 1 & 0 \\ 0 & 0 & 0 & 1 \end{bmatrix}$

13. $x = \frac{220}{13}$, $y = -\frac{30}{13}$ 　　**15.** No solution

17. $x = -\frac{2}{3}r + \frac{5}{3}$, $y = -\frac{1}{6}r + \frac{7}{6}$, $z = r$, where r is any real number

19. No solution 　　**21.** $x = -3$, $y = 2$, $z = 0$

23. $x = 2$, $y = -5$, $z = -1$

25. $x_1 = r$, $x_2 = 0$, $x_3 = 0$, $x_4 = 0$, $x_5 = r$, where r is any real number

27. Federal, $72,000; state, $24,000

29. A, 2000; B, 4000; C, 5000

31. **(a)** 3 of X, 4 of Z; 2 of X, 1 of Y, 5 of Z; 1 of X, 2 of Y, 6 of Z; 3 of Y, 7 of Z; **(b)** 3 of X, 4 of Z; **(c)** 3 of X, 4 of Z; 3 of Y, 7 of Z

33. **(a)** Let s, d, g represent the numbers of units S, D, G respectively. The six combinations are given by:

$$\begin{array}{c|cccccc} s & 5 & 4 & 3 & 2 & 1 & 0 \\ d & 8 & 7 & 6 & 5 & 4 & 3 \\ g & 0 & 1 & 2 & 3 & 4 & 5 \end{array}$$

(b) The combination $s = 0$, $d = 3$, $g = 5$

PRINCIPLES IN PRACTICE 6.5

1. Infinitely many solutions: $x + \frac{1}{2}z = 0$, $y + \frac{1}{2}z = 0$; in parametric form: $x = -\frac{1}{2}r$, $y = -\frac{1}{2}r$, $z = r$, where r is any real number

PROBLEMS 6.5 (page 263)

1. $w = -1 + 7r$, $x = 2 - 5r$, $y = 4 - 7r$, $z = r$, (where r is any real number)

3. $w = -s$, $x = -3r - 4s + 2$, $y = r$, $z = s$ (where r and s are any real numbers)

5. $w = -2r + s - 2$, $x = -r + 4$, $y = r$, $z = s$ (where r and s are any real numbers)

7. $x_1 = -2r + s - 2t + 1$, $x_2 = -r - 2s + t + 4$, $x_3 = r$, $x_4 = s$, $x_5 = t$ (where $r, s,$ and t are any real numbers)

9. Infinitely many 　　**11.** Trivial solution

13. Infinitely many 　　**15.** $x = 0$, $y = 0$

17. $x = -\frac{6}{5}r$, $y = \frac{8}{15}r$, $z = r$

19. $x = 0$, $y = 0$ 　　**21.** $x = r$, $y = -2r$, $z = r$

23. $w = -2r$, $x = -3r$, $y = r$, $z = r$

PRINCIPLES IN PRACTICE 6.6

1. Yes 　　**2.** MEET AT NOON FRIDAY

3. $E^{-1} = \begin{bmatrix} \frac{2}{3} & -\frac{1}{6} & -\frac{1}{3} \\ -\frac{1}{3} & \frac{5}{6} & -\frac{1}{3} \\ -\frac{1}{3} & -\frac{1}{6} & \frac{2}{3} \end{bmatrix}$; F is not invertible.

4. A: 5000 shares; B: 2500 shares; C: 2500 shares

PROBLEMS 6.6 (page 269)

1. $\begin{bmatrix} -1 & 1 \\ 7 & -6 \end{bmatrix}$

3. Not invertible

5. $\begin{bmatrix} 1 & 0 & 0 \\ 0 & -\frac{1}{3} & 0 \\ 0 & 0 & \frac{1}{4} \end{bmatrix}$

7. Not invertible

9. Not invertible (not a square matrix)

11. $\begin{bmatrix} 1 & -2 & 1 \\ 0 & 1 & -2 \\ 0 & 0 & 1 \end{bmatrix}$

13. $\begin{bmatrix} 1 & 0 & 2 \\ 0 & 1 & 0 \\ 3 & 0 & 7 \end{bmatrix}$

15. $\begin{bmatrix} 1 & -\frac{2}{3} & \frac{5}{3} \\ -1 & \frac{4}{3} & -\frac{10}{3} \\ -1 & 1 & -2 \end{bmatrix}$

17. $\begin{bmatrix} \frac{11}{3} & -3 & \frac{1}{3} \\ -\frac{7}{3} & 3 & -\frac{2}{3} \\ \frac{2}{3} & -1 & \frac{1}{3} \end{bmatrix}$

19. $x_1 = 10$, $x_2 = 20$

21. $x = 17$, $y = -20$

23. $x = 2$, $y = 1$

25. $x = -3r + 1$, $y = r$

27. $x = 0$, $y = 1$, $z = 2$

29. $x = 1$, $y = \frac{1}{2}$, $z = \frac{1}{2}$

31. No solution

33. $w = 1$, $x = 3$, $y = -2$, $z = 7$

35. $\begin{bmatrix} -\frac{1}{6} & -\frac{1}{3} \\ \frac{1}{6} & -\frac{2}{3} \end{bmatrix}$

37. (a) 40 of model A, 60 of model B;
(b) 45 of model A, 50 of model B

39. (b) $\begin{bmatrix} 4 & 6 \\ 7 & 10 \end{bmatrix}$

41. Yes

43. D: 5000 shares; E: 1000 shares; F: 4000 shares

45. (a) $\begin{bmatrix} 2.05 & 1.28 \\ 0.73 & 1.71 \end{bmatrix}$; **(b)** $\begin{bmatrix} \frac{84}{41} & \frac{105}{82} \\ \frac{30}{41} & \frac{70}{41} \end{bmatrix}$

47. $\begin{bmatrix} 2.75 & -1.59 & -1.11 \\ -0.48 & 1.43 & 0.00 \\ -1.22 & 0.32 & 2.22 \end{bmatrix}$

49. $w = 14.44$, $x = 0.03$, $y = -0.80$, $z = 10.33$

PROBLEMS 6.7 (page 271)

1. $\begin{bmatrix} 1290 \\ 1425 \end{bmatrix}$; 1405

3. (a) $\begin{bmatrix} 134.29 \\ 162.25 \\ 234.35 \end{bmatrix}$; **(b)** $\begin{bmatrix} 68.59 \\ 84.50 \\ 108.69 \end{bmatrix}$

5. $\begin{bmatrix} 1301 \\ 1215 \\ 1188 \end{bmatrix}$

7. $\begin{bmatrix} 1382 \\ 1344 \\ 1301 \end{bmatrix}$

9. 736.39 units of coal, 536.29 units of steel, 699.96 units of railroad services

REVIEW PROBLEMS—CHAPTER 6 (page 276)

1. $\begin{bmatrix} 3 & 8 \\ -16 & -10 \end{bmatrix}$

3. $\begin{bmatrix} 1 & 42 & 5 \\ 2 & -18 & -7 \\ 1 & 0 & -2 \end{bmatrix}$

5. $\begin{bmatrix} 11 & -4 \\ 8 & 11 \end{bmatrix}$

7. $\begin{bmatrix} 6 \\ 32 \end{bmatrix}$

9. $\begin{bmatrix} -1 & -2 \\ 2 & 1 \end{bmatrix}$

11. $\begin{bmatrix} 2 & 0 \\ 0 & 9 \end{bmatrix}$

13. $x = 3$, $y = 21$

15. $\begin{bmatrix} 1 & 0 \\ 0 & 1 \end{bmatrix}$

17. $\begin{bmatrix} 1 & 0 & 0 \\ 0 & 1 & 0 \\ 0 & 0 & 1 \end{bmatrix}$

19. $x = 0$, $y = 0$

21. No solution

23. $\begin{bmatrix} -\frac{3}{2} & \frac{5}{6} \\ \frac{1}{2} & -\frac{1}{6} \end{bmatrix}$

25. No inverse exists

27. $x = 0$, $y = 1$, $z = 0$

29. $\mathbf{A}^2 = \begin{bmatrix} 0 & 0 & 1 \\ 0 & 0 & 0 \\ 0 & 0 & 0 \end{bmatrix}$, $\mathbf{A}^3 = \mathbf{O}$, $\mathbf{A}^{1000} = \mathbf{O}$, no inverse

31. (a) Let x, y, z represent the weekly doses of capsules of brands, I, II, III, respectively. The combinations are given by:

	x	y	z
combination 1	4	9	0
combination 2	3	6	1
combination 3	2	3	2
combination 4	1	0	3

(b) Combination 4:
$x = 1$, $y = 0$, $z = 3$

33. $\begin{bmatrix} 215 & 87 \\ 89 & 141 \end{bmatrix}$

35. $\begin{bmatrix} 39.7 \\ 35.1 \end{bmatrix}$

MATHEMATICAL SNAPSHOT—CHAPTER 6 (page 278)

1. $151.40

3. It is not possible; different combinations of lengths of stays can cost the same.

PRINCIPLES IN PRACTICE 7.1

1. $2x + 1.5y > 0.9x + 0.7y + 50$, $y > -1.375x + 62.5$; sketch the dashed line $y = -1.375x + 62.5$ and shade the half plane above the line. In order to produce a profit, the number of magnets of types A and B produced and sold must be an ordered pair in the shaded region.

2. $x \geq 0$, $y \geq 0$, $x + y \geq 50$, $x \geq 2y$; The region consists of points on or above the x-axis and on or to the right of the y-axis. In addition, the points must be on or above the line $x + y = 50$ and on or below the line $x = 2y$.

PROBLEMS 7.1 (page 284)

1.

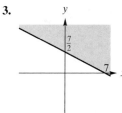

3.

5.

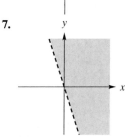

7.

9.

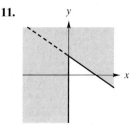

11.

13.

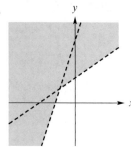

15.

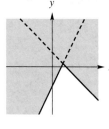

17.

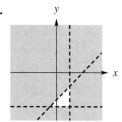

19.

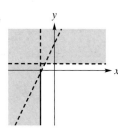

21.

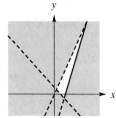

23.

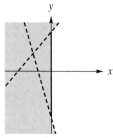

25.

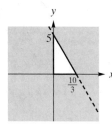

27.

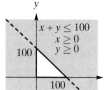

x: number of lb from A
y: number of lb from B

29. $x \geq 0, y \geq 0, 3x + 2y \leq 240, 0.5x + y \leq 80$

PROBLEMS 7.2 (page 291)

1. $P = 112\frac{1}{2}$ when $x = \frac{45}{2}, y = 0$

3. $Z = -10$ when $x = 2, y = 3$

5. No optimum solution (empty feasible region)

7. $Z = 3$ when $x = 0, y = 1$

9. $C = \frac{23}{3}$ when $x = \frac{7}{3}, y = \frac{1}{3}$

11. No optimum solution (unbounded)

13. 10 trucks, 20 spinning tops; $110

15. 4 units of food A, 4 units of food B; $8

17. 10 tons of ore 1, 10 tons of ore II, $1100

19. 6 chambers of type A and 10 chambers of type B

21. **(c)** $x = 0, y = 200$

23. $Z = 15.54$ when $x = 2.56, y = 6.74$

25. $Z = -75.98$ when $x = 9.48, y = 16.67$

PRINCIPLES IN PRACTICE 7.3

1. Ship $10t + 15$ TV sets from C to A, $-10t + 30$ TV sets from C to B, $-10t + 10$ TV sets from D to A, and $10t$ TV sets from D to B, for $0 \leq t \leq 1$; minimum cost $780

PROBLEMS 7.3 (page 296)

1. $Z = 33$ when $x = (1 - t)(2) + 5t = 2 + 3t$,
$y = (1 - t)(3) + 2t = 3 - t$, and $0 \leq t \leq 1$

3. $Z = 84$ when $x = (1 - t)\left(\frac{36}{7}\right) + 6t = \frac{6}{7}t + \frac{36}{7}$,
$y = (1 - t)\left(\frac{4}{7}\right) + 0t = \frac{4}{7} - \frac{4}{7}t$, and $0 \leq t \leq 1$

PRINCIPLES IN PRACTICE 7.4

1. 0 players of Type 1, 72 players of Type 2, 12 players of Type 3; maximum profit of $20,400

PROBLEMS 7.4 (page 307)

1. $Z = 8$ when $x_1 = 0, x_2 = 4$

3. $Z = 2$ when $x_1 = 0, x_2 = 1$

5. $Z = 28$ when $x_1 = 3, x_2 = 2$

7. $Z = 20$ when $x_1 = 0, x_2 = 5, x_3 = 0$

9. $Z = 2$ when $x_1 = 1, x_2 = 0, x_3 = 0$

11. $Z = \frac{72}{13}$ when $x_1 = \frac{22}{13}, x_2 = \frac{50}{13}$

13. $W = 13$ when $x_1 = 1, x_2 = 0, x_3 = 3$

15. $Z = 600$ when $x_1 = 4, x_2 = 1, x_3 = 4, x_4 = 0$

17. 0 from A, 2400 from B; $1200

19. 0 chairs, 300 rockers, 100 chaise lounges; $10,800

PRINCIPLES IN PRACTICE 7.5

1. $35 - 7t$ of device 1, $6t$ of device 2, 0 of device 3, for $0 \leq t \leq 1$

PROBLEMS 7.5 (page 313)

1. Yes; for the table, x_2 is the entering variable and the quotients $\frac{6}{2}$ and $\frac{3}{1}$ tie for being the smallest.

3. No optimum solution (unbounded)

5. $Z = 16$ when $x_1 = \frac{8}{7}t, x_2 = 2 + \frac{4}{7}t$, and $0 \leq t \leq 1$

7. No optimum solution (unbounded)

9. $Z = 13$ when $x_1 = \frac{3}{2} - \frac{3}{2}t, x_2 = 6t, x_3 = 4 - 3t$, and $0 \leq t \leq 1$

11. $15,200. If x_1, x_2, x_3 denote the number of chairs, rockers, and chaise lounges produced, respectively, then
$x_1 = 100 - 100t$,
$x_2 = 100 + 150t$,
$x_3 = 200 - 50t$, and
$0 \leq t \leq 1$

PRINCIPLES IN PRACTICE 7.6

1. Plant I: 500 standard, 700 deluxe; plant II: 500 standard, 100 deluxe; $89,500 maximum profit

PROBLEMS 7.6 (page 323)

1. $Z = 7$ when $x_1 = 1, x_2 = 5$

3. $Z = 4$ when $x_1 = 1, x_2 = 2, x_3 = 0$

5. $Z = 28$ when $x_1 = 8, x_2 = 2, x_3 = 0$

7. $Z = -17$ when $x_1 = 3, x_2 = 2$

9. No optimum solution (empty feasible region)

11. $Z = 2$ when $x_1 = 6$, $x_2 = 10$

13. 166 Standard bookcases, 0 Executive bookcases

15. 30% in A, 0% in AA, 70% in AAA; 6.6%

PROBLEMS 7.7 (page 329)

1. $Z = 14$ when $x_1 = 7$, $x_2 = 0$

3. $Z = 216$ when $x_1 = 18$, $x_2 = 0$, $x_3 = 0$

5. $Z = 4$ when $x_1 = 0$, $x_2 = 0$, $x_3 = 4$

7. $Z = 0$ when $x_1 = 3$, $x_2 = 0$, $x_3 = 1$

9. $Z = 28$ when $x_1 = 3$, $x_2 = 0$, $x_3 = 5$

11. Install device A on kilns producing 700,000 barrels annually, and device B on kilns producing 2,600,000 barrels annually

13. To Columbus, 150 from Akron and 0 from Spring field; to Dayton, 0 from Akron and 150 from Spring field; $1050

15. **(a)** Column 3: 1, 3, 3; column 4: 0, 4, 8; **(b)** $x_1 = 10$, $x_2 = 0$, $x_3 = 20$, $x_4 = 0$; **(c)** 90 in.

PRINCIPLES IN PRACTICE 7.8

1. Minimize $W = 60{,}000y_1 + 2000y_2 + 120y_3$ subject to
$$300y_1 + 20y_2 + 3y_3 \geq 300$$
$$220y_1 + 40y_2 + \quad y_3 \geq 200$$
$$180y_1 + 20y_2 + 2y_3 \geq 200$$
and $y_1, y_2, y_3 \geq 0$

2. Maximize $W = 98y_1 + 80y_2$ subject to
$$20y_1 + \quad 8y_2 \leq 6$$
$$6y_1 + 16y_2 \leq 2$$
and $y_1, y_2 \geq 0$

3. 5 device 1, 0 device 2, 15 device 3

PROBLEMS 7.8 (page 337)

1. Minimize $W = 5y_1 + 3y_2$ subject to
$$y_1 - y_2 \geq 1$$
$$y_1 + y_2 \geq 2$$
$$y_1, y_2 \geq 0$$

3. Maximize $W = 8y_1 + 2y_2$ subject to
$$y_1 - y_2 \leq 1$$
$$y_1 + 2y_2 \leq 8$$
$$y_1 + y_2 \leq 5$$
$$y_1, y_2 \geq 0$$

5. Minimize $W = 13y_1 - 3y_2 - 11y_3$ subject to
$$-y_1 + y_2 - y_3 \geq 1$$
$$2y_1 - y_2 - y_3 \geq -1$$
$$y_1, y_2, y_3 \geq 0$$

7. Maximize $W = -3y_1 + 3y_2$ subject to
$$-y_1 + y_2 \leq 4$$
$$y_1 - y_2 \leq 4$$
$$y_1 + y_2 \leq 6$$
$$y_1, y_2 \geq 0$$

9. $Z = \dfrac{43}{5}$ when $x_1 = 0$, $x_2 = \dfrac{4}{5}$, $x_3 = \dfrac{7}{5}$

11. $Z = 26$ when $x_1 = 6$, $x_2 = 1$

13. $Z = 14$ when $x_1 = 1$, $x_2 = 2$

15. $250 on newspaper advertising, $1400 on radio advertising; $1650.

17. 20 shipping clerk apprentices, 40 shipping clerks, 90 semiskilled workers, 0 skilled workers; $1200

REVIEW PROBLEMS—CHAPTER 7 (page 339)

1.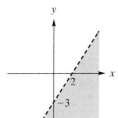

3.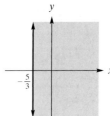

5.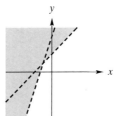

7.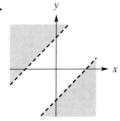

9.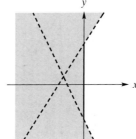

11. $Z = 3$ when $x = 3$, $y = 0$

13. $Z = -2$ when $x = 0$, $y = 2$

15. $Z = \dfrac{70}{9}$ when $x = \dfrac{20}{9}$, $y = \dfrac{10}{9}$

17. $Z = 36$ when $x = 2 + 2t$, $y = 3 - 3t$, and $0 \leq t \leq 1$

19. $Z = 32$ when $x_1 = 8$, $x_2 = 0$

21. $Z = \dfrac{5}{3}$ when $x_1 = 0$, $x_2 = 0$, $x_3 = \dfrac{5}{3}$

23. $Z = 24$ when $x_1 = 0$, $x_2 = 12$

25. $Z = \dfrac{7}{2}$ when $x_1 = \dfrac{5}{4}$, $x_2 = 0$, $x_3 = \dfrac{9}{4}$

27. No optimum solution (unbounded)

29. $Z = 70$ when $x_1 = 35$, $x_2 = 0$, $x_3 = 0$

31. 0 units of X, 6 units of Y, 14 units of Z; $398

33. 500,000 gal from A to D, 100,000 gal from A to C, 400,000 gal from B to C; $19,000

35. 10 kg of food A only

37. $Z = 129.83$ when $x = 9.38$, $y = 1.63$

MATHEMATICAL SNAPSHOT—CHAPTER 7 (page 342)

1. 2 minutes of radiation   **3.** Answers may vary.

PROBLEMS 8.1 (page 350)

1.

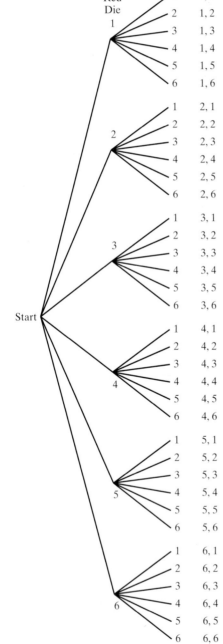

Assembly line	Finishing line	Production route
A	D	AD
	E	AE
B	D	BD
	E	BE
C	D	CD
	E	CE

6 possible production routes

3.

Green Die   Result

Red Die

Start

36 possible results

5. 20   **7.** 96   **9.** 1024   **11.** 120   **13.** 720

15. 720   **17.** 1000; error message is displayed

19. 6   **21.** 336   **23.** 1296   **25.** 1320

27. 360   **29.** 720   **31.** 2520; 5040   **33.** 624

35. 120   **37. (a)** 11,880; **(b)** 19,008   **39.** 48

41. 4320

PROBLEMS 8.2 (page 360)

1. 15   **3.** 1   **5.** 360   **9.** 2380

11. 715   **13.** $\dfrac{74!}{10! \cdot 64!}$ **15.** 56   **17.** 415,800

19. 35   **21.** 720   **23.** 1680   **25.** 252

27. 756,756   **29. (a)** 210; **(b)** 770   **31.** 17,325

33. (a) 1; **(b)** 1; **(c)** 18   **35.** 3744   **37.** 5,250,960

PRINCIPLES IN PRACTICE 8.3

1. 10,586,800

PROBLEMS 8.3 (page 368)

1. {9D, 9H, 9C, 9S}

3. {1H, 1T, 2H, 2T, 3H, 3T, 4H, 4T, 5H, 5T, 6H, 6T}

5. {64, 69, 60, 61, 46, 49, 40, 41, 96, 94, 90, 91, 06, 04, 09, 01, 16, 14, 19, 10}

7. (a) {RR, RW, RB, WR, WW, WB, BR, BW, BB};
(b) {RW, RB, WR, WB, BR, BW}

9. Sample space consists of ordered sets of six elements and each element is H or T; 64.

11. Sample space consists of ordered pairs where first element indicates card drawn and second element indicates number on die; 312.

13. Sample space consists of combinations of 52 cards taken 10 at a time; $_{52}C_{10}$.

15. {1, 3, 5, 7, 9}   **17.** {7, 9}   **19.** {1, 2, 4, 6, 8, 10}

21. S **23.** E_1 and E_4, E_2 and E_3, E_3 and E_4

25. E and G, F and I, G and H, G and I

27. (a) {HHH, HHT, HTH, HTT, THH, THT, TTH, TTT};
(b) {HHH, HHT, HTH, HTT, THH, THT, TTH};
(c) {HHT, HTH, HTT, THH, THT, TTH, TTT}; **(d)** S;
(e) {HHT, HTH, HTT, THH, THT, TTH}; **(f)** $\emptyset$;
(g) {HHH, TTT}

29. (a) {ABC, ACB, BAC, BCA, CAB, CBA};
(b) {ABC, ACB}; **(c)** {BAC, BCA, CAB, CBA}

PROBLEMS 8.4 (page 379)

1. 750   **3. (a)** 0.8; **(b)** 0.4   **5.** No

7. (a) $\dfrac{5}{36}$; **(b)** $\dfrac{1}{12}$; **(c)** $\dfrac{1}{4}$; **(d)** $\dfrac{1}{36}$; **(e)** $\dfrac{1}{2}$; **(f)** $\dfrac{1}{2}$; **(g)** $\dfrac{5}{6}$

9. (a) $\dfrac{1}{52}$; **(b)** $\dfrac{1}{4}$; **(c)** $\dfrac{1}{13}$; **(d)** $\dfrac{1}{2}$; **(e)** $\dfrac{1}{2}$; **(f)** $\dfrac{1}{52}$; **(g)** $\dfrac{4}{13}$;
(h) $\dfrac{1}{26}$; **(i)** 0

11. (a) $\dfrac{1}{624}$; **(b)** $\dfrac{4}{624} = \dfrac{1}{156}$; **(c)** $\dfrac{8}{624} = \dfrac{1}{78}$;
(d) $\dfrac{39}{624} = \dfrac{1}{16}$

13. (a) $\dfrac{4 \cdot 3 \cdot 2}{132,600} = \dfrac{1}{5525}$; **(b)** $\dfrac{13 \cdot 12 \cdot 11}{132,600} = \dfrac{11}{850}$

15. (a) $\dfrac{1}{8}$;   **(b)** $\dfrac{3}{8}$;   **(c)** $\dfrac{1}{8}$;   **(d)** $\dfrac{7}{8}$

17. (a) $\dfrac{4}{5}$;   **(b)** $\dfrac{1}{5}$

19. (a) 0.1;   **(b)** 0.35;   **(c)** 0.7;   **(d)** 0.95;   **(e)** 0.1, 0.35, 0.7, 0.95

21. $\dfrac{1}{10}$ **23. (a)** $\dfrac{1}{2^{10}} = \dfrac{1}{1024}$;   **(b)** $\dfrac{11}{1024}$

25. $\dfrac{13 \cdot {}_4C_4 \cdot 12 \cdot {}_4C_1}{{}_{52}C_5} = \dfrac{13 \cdot 12 \cdot 4}{{}_{52}C_5}$

27. (a) $\dfrac{6545}{161{,}700} \approx 0.040$;   **(b)** $\dfrac{4140}{161{,}700} \approx 0.026$

29. \$19.34   **31.** $\dfrac{1}{9}$

33. (a) 0.51;   **(b)** 0.44;   **(c)** 0.03   **35.** 4:1   **37.** 7:3

39. $\dfrac{7}{12}$ **41.** $\dfrac{2}{7}$ **43.** 3:1

PROBLEMS 8.5 (page 391)

1. (a) $\dfrac{1}{5}$;   **(b)** $\dfrac{4}{5}$;   **(c)** $\dfrac{1}{4}$;   **(d)** $\dfrac{1}{2}$;   **(e)** 0   **3.** 1

5. 0.43   **7. (a)** $\dfrac{1}{2}$;   **(b)** $\dfrac{2}{3}$

9. (a) $\dfrac{2}{3}$;   **(b)** $\dfrac{1}{2}$;   **(c)** $\dfrac{1}{3}$;   **(d)** $\dfrac{1}{6}$

11. (a) $\dfrac{5}{8}$;   **(b)** $\dfrac{35}{58}$;   **(c)** $\dfrac{11}{39}$;   **(d)** $\dfrac{8}{25}$;   **(e)** $\dfrac{10}{47}$;   **(f)** $\dfrac{25}{86}$

13. (a) $\dfrac{1}{2}$;   **(b)** $\dfrac{4}{9}$ **15.** $\dfrac{2}{3}$

17. (a) $\dfrac{1}{2}$;   **(b)** $\dfrac{1}{4}$ **19.** $\dfrac{2}{3}$ **21.** $\dfrac{1}{11}$

23. $\dfrac{1}{6}$ **25.** $\dfrac{2}{3}$ **27.** $\dfrac{1}{13}$ **29.** $\dfrac{40}{51}$

31. $\dfrac{8}{16{,}575}$ **33.** $\dfrac{1}{5525}$ **35.** $\dfrac{2}{17}$

37. (a) $\dfrac{47}{100}$;   **(b)** $\dfrac{27}{47}$ **39. (a)** $\dfrac{3}{4}$;   **(b)** $\dfrac{3}{5}$

41. $\dfrac{9}{20}$ **43.** $\dfrac{1}{4}$ **45.** $\dfrac{7}{300}$ **47.** 0.049

49. (a) 0.06;   **(b)** 0.155   **51.** $\dfrac{4}{31}$

PROBLEMS 8.6 (page 401)

1. (a) $\dfrac{1}{4}$;   **(b)** $\dfrac{5}{6}$;   **(c)** $\dfrac{1}{3}$;   **(d)** $\dfrac{2}{3}$;   **(e)** $\dfrac{1}{12}$;   **(f)** $\dfrac{1}{2}$;   **(g)** $\dfrac{1}{3}$

3. $\dfrac{7}{18}$ **5.** Independent   **7.** Independent

9. Dependent   **11.** Dependent

13. (a) Independent;   **(b)** dependent;   **(c)** dependent;   **(d)** no

15. Dependent   **17.** $\dfrac{1}{18}$ **19.** $\dfrac{1}{25}$ **21.** $\dfrac{3}{676}$

23. (a) $\dfrac{3}{10}$;   **(b)** $\dfrac{1}{40}$;   **(c)** $\dfrac{1}{10}$

25. (a) $\dfrac{2}{5}$;   **(b)** $\dfrac{1}{5}$;   **(c)** $\dfrac{7}{15}$;   **(d)** $\dfrac{13}{15}$;   **(e)** $\dfrac{2}{15}$

27. (a) $\dfrac{7}{54}$;   **(b)** $\dfrac{35}{162}$ **29.** $\dfrac{139}{361}$ **31.** $\dfrac{3}{200}$

33. (a) $\dfrac{1}{1728}$;   **(b)** $\dfrac{3}{8}$

35. (a) $\dfrac{15}{1024}$;   **(b)** $\dfrac{1}{64}$;   **(c)** $\dfrac{53}{512}$ **37.** 0.0106

PROBLEMS 8.7 (page 409)

1. $P(E \mid D) = \dfrac{1}{4}$, $P(F \mid D') = \dfrac{4}{7}$ **3.** $\dfrac{175}{386} \approx 0.453$.

5. (a) $\dfrac{258}{937} \approx 0.275$;   **(b)** $\dfrac{14}{3021} \approx 0.005$

7. $\dfrac{5}{8}$ **9.** $\dfrac{6}{31}$ **11.** $\dfrac{114}{119} \approx 0.958$

13. $\dfrac{27}{49} \approx 55.1\%$ **15.** $\dfrac{3}{4}$ **17.** $\dfrac{24}{29} \approx 0.828$

19. $\dfrac{4}{5}$ **21.** $\dfrac{14}{15} \approx 0.933$

23. (a) $\dfrac{27}{100} = 0.27$;   **(b)** $\dfrac{15}{27} \approx 0.556$;   **(c)** $\dfrac{17}{100} = 0.17$

25. (a) 0.18;   **(b)** 0.23;   **(c)** 0.59;   **(d)** high quality

27. $\dfrac{7}{9} \approx 0.78$

REVIEW PROBLEMS—CHAPTER 8 (page 414)

1. 336   **3.** 36   **5.** 17,576,000   **7.** 32

9. 210   **11.** 462   **13. (a)** 2024;   **(b)** 253

15. 34,650   **17.** 1260

19. (a) $\{1, 2, 3, 4, 5, 6, 7\}$;   **(b)** $\{4, 5, 6\}$;   **(c)** $\{4, 5, 6, 7, 8\}$;   **(d)** $\emptyset$;   **(e)** $\{4, 5, 6, 7, 8\}$;   **(f)** no

21. (a) $\{R_1R_2R_3,\ R_1R_2G_3,\ R_1G_2R_3,\ R_1G_2G_3,\ G_1R_2R_3,$ $G_1R_2G_3,\ G_1G_2R_3,\ G_1G_2G_3\}$;   **(b)** $\{R_1R_2G_3,\ R_1G_2R_3,\ G_1R_2R_3\}$;   **(c)** $\{R_1R_2R_3,\ G_1G_2G_3\}$

23. 0.2   **25.** $\dfrac{45}{512}$ **27. (a)** $\dfrac{4}{25}$;   **(b)** $\dfrac{2}{15}$

29. (a) $\dfrac{1}{8}$;   **(b)** $\dfrac{3}{16}$ **31.** $3:5$ **33.** $\dfrac{6}{7}$

35. $\dfrac{10}{13}$ **37.** 0.42   **39. (a)** $\dfrac{2}{11}$;   **(b)** $\dfrac{1}{18}$

41. $\dfrac{1}{4}$ **43. (a)** $\dfrac{1}{3}$;   **(b)** independent   **45.** Dependent

47. (a) 0.0081;   **(b)** 0.2646;   **(c)** 0.3483

49. $\dfrac{22}{45}$ **51.** $\dfrac{1}{4}$

53. (a) 0.01625;   **(b)** $\dfrac{3}{13} \approx 0.23$

MATHEMATICAL SNAPSHOT—CHAPTER 8 (page 418)

1. ≈ 0.645

PROBLEMS 9.1 (page 426)

1. $\mu = 1.7$; $\text{Var}(X) = 1.01$; $\sigma \approx 1.00$

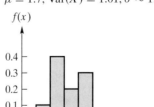

3. $\mu = \dfrac{9}{4} = 2.25$; $\text{Var}(X) = \dfrac{11}{16} = 0.6875$; $\sigma \approx 0.83$

5. (a) 0.1;   **(b)** 5.8;   **(c)** 1.56

7. $E(X) = \dfrac{3}{2} = 1.5$; $\sigma^2 = \dfrac{3}{4} = 0.75$; $\sigma \approx 0.87$

9. $E(X) = \dfrac{6}{5} = 1.2; \sigma^2 = \dfrac{9}{25} = 0.36; \sigma = \dfrac{3}{5} = 0.6$

11. $f(0) = \dfrac{1}{10}, f(1) = \dfrac{3}{5}, f(2) = \dfrac{3}{10}$

13. (a) $-\$1.38$ (a loss); **(b)** $-\$2.75$ (a loss)

15. $\$101.43$ **17.** $\$3.00$ **19.** $\$410$

21. Loss of $\$0.25$; $\$1$

PRINCIPLES IN PRACTICE 9.2

1.

x	$P(x)$
0	$\dfrac{2401}{10,000}$
1	$\dfrac{4116}{10,000}$
2	$\dfrac{2646}{10,000}$
3	$\dfrac{756}{10,000}$
4	$\dfrac{81}{10,000}$

PROBLEMS 9.2 (page 432)

1. $f(0) = \dfrac{16}{25}; f(1) = \dfrac{8}{25}; f(2) = \dfrac{1}{25}; \mu = \dfrac{2}{5}; \sigma = \dfrac{2\sqrt{2}}{5}$

3. $f(0) = \dfrac{1}{27}, f(1) = \dfrac{2}{9}, f(2) = \dfrac{4}{9}, f(3) = \dfrac{8}{27}; \mu = 2; \sigma = \dfrac{\sqrt{6}}{3}$

5. 0.001536 **7.** $\dfrac{96}{625} = 0.1536$ **9.** $\dfrac{3}{16}$

11. $\dfrac{165}{2048} \approx 0.081$ **13.** $\dfrac{1225}{3456} \approx 0.3545$ **15.** 0.002

17. (a) $\dfrac{9}{64}$; **(b)** $\dfrac{5}{32}$ **19.** $\dfrac{2048}{3125} \approx 0.655$ **21.** 0.7599

23. $\dfrac{13}{16}$ **25.** $\dfrac{512}{2187} \approx 0.234$

PROBLEMS 9.3 (page 440)

1. No **3.** No **5.** Yes **7.** $a = \dfrac{1}{3}, b = \dfrac{3}{4}$

9. $a = 0.3, b = 0.6, c = 0.1$ **11.** Yes **13.** No

15. $\mathbf{X}_1 = \begin{bmatrix} \dfrac{11}{12} \\ \dfrac{1}{12} \end{bmatrix}, \mathbf{X}_2 = \begin{bmatrix} \dfrac{25}{36} \\ \dfrac{11}{36} \end{bmatrix}, \mathbf{X}_3 = \begin{bmatrix} \dfrac{83}{108} \\ \dfrac{25}{108} \end{bmatrix}$

17. $\mathbf{X}_1 = \begin{bmatrix} 0.42 \\ 0.58 \end{bmatrix}, \mathbf{X}_2 = \begin{bmatrix} 0.416 \\ 0.584 \end{bmatrix}, \mathbf{X}_3 = \begin{bmatrix} 0.4168 \\ 0.5832 \end{bmatrix}$

19. $\mathbf{X}_1 = \begin{bmatrix} 0.26 \\ 0.28 \\ 0.46 \end{bmatrix}, \mathbf{X}_2 = \begin{bmatrix} 0.164 \\ 0.302 \\ 0.534 \end{bmatrix}, \mathbf{X}_3 = \begin{bmatrix} 0.1766 \\ 0.3138 \\ 0.5096 \end{bmatrix}$

21. (a) $\mathbf{T}^2 = \begin{bmatrix} \dfrac{5}{8} & \dfrac{3}{8} \\ \dfrac{3}{8} & \dfrac{5}{8} \end{bmatrix}, \mathbf{T}^3 = \begin{bmatrix} \dfrac{7}{16} & \dfrac{9}{16} \\ \dfrac{9}{16} & \dfrac{7}{16} \end{bmatrix}$; **(b)** $\dfrac{3}{8}$; **(c)** $\dfrac{9}{16}$

23. (a) $\mathbf{T}^2 = \begin{bmatrix} 0.50 & 0.23 & 0.27 \\ 0.40 & 0.69 & 0.54 \\ 0.10 & 0.08 & 0.19 \end{bmatrix}, \mathbf{T}^3 = \begin{bmatrix} 0.230 & 0.369 & 0.327 \\ 0.690 & 0.530 & 0.543 \\ 0.080 & 0.101 & 0.130 \end{bmatrix}$;

(b) 0.40; **(c)** 0.369

25. $\begin{bmatrix} \dfrac{4}{7} \\ \dfrac{3}{7} \end{bmatrix}$ **27.** $\begin{bmatrix} \dfrac{3}{7} \\ \dfrac{4}{7} \end{bmatrix}$ **29.** $\begin{bmatrix} 0.5 \\ 0.25 \\ 0.25 \end{bmatrix}$

31. (a) $\begin{array}{c} \\ \text{Flu} \\ \text{No Flu} \end{array} \begin{array}{c} \text{Flu} \quad \text{No Flu} \\ \begin{bmatrix} 0.1 & 0.2 \\ 0.9 & 0.8 \end{bmatrix} \end{array}$; **(b)** 37, 36

33. (a) $\begin{array}{c} \\ \text{A} \\ \text{B} \end{array} \begin{array}{c} \text{A} \quad \text{B} \\ \begin{bmatrix} 0.7 & 0.4 \\ 0.3 & 0.6 \end{bmatrix} \end{array}$; **(b)** 0.61

35. (a) $\begin{array}{c} \\ \text{D} \\ \text{R} \\ \text{O} \end{array} \begin{array}{c} \text{D} \quad \text{R} \quad \text{O} \\ \begin{bmatrix} 0.8 & 0.1 & 0.3 \\ 0.1 & 0.8 & 0.2 \\ 0.1 & 0.1 & 0.5 \end{bmatrix} \end{array}$; **(b)** 0.19; **(c)** 40%

37. (a) $\begin{array}{c} \\ \text{A} \\ \text{Compet.} \end{array} \begin{array}{c} \text{A} \quad \text{Compet.} \\ \begin{bmatrix} 0.8 & 0.3 \\ 0.2 & 0.7 \end{bmatrix} \end{array}$; **(b)** 65%; **(c)** 60%

39. (a) $\begin{array}{c} \\ 1 \\ 2 \end{array} \begin{array}{c} 1 \quad 2 \\ \begin{bmatrix} \dfrac{5}{7} & \dfrac{3}{7} \\ \dfrac{2}{7} & \dfrac{4}{7} \end{bmatrix} \end{array}$; **(b)** 59.18% in compartment 1, 40.82% in compartment 2; **(c)** 60% in compartment 1, 40% in compartment 2

41. (a) $\begin{bmatrix} \dfrac{2}{3} \\ \dfrac{1}{3} \end{bmatrix}$; **(b)** $33\dfrac{1}{3}\%$

REVIEW PROBLEMS—CHAPTER 9 (page 444)

1. $\mu = 1.5, \text{Var}(X) = 0.65, \sigma \approx 0.81$

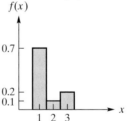

3. (a) $f(0) = \dfrac{1}{12}, f(1) = f(2) = f(3) = f(4) = f(5) = \dfrac{1}{6}, f(6) = \dfrac{1}{12}$; **(b)** 3

5. $-\$0.10$ (a loss) **7. (a)** $\$176$; **(b)** $\$704,000$

9. $f(0) = 0.522, f(1) = 0.368, f(2) = 0.098, f(3) = 0.011, f(4) = 0.0005; \mu = 0.6; \sigma \approx 0.71$

11. $\dfrac{1}{64}$ **13.** $\dfrac{8}{81}$ **15.** $\dfrac{2072}{3125}$

17. $a = 0.3, b = 0.2, c = 0.5$

19. $\mathbf{X}_1 = \begin{bmatrix} 0.10 \\ 0.15 \\ 0.75 \end{bmatrix}, \mathbf{X}_2 = \begin{bmatrix} 0.130 \\ 0.155 \\ 0.715 \end{bmatrix}, \mathbf{X}_3 = \begin{bmatrix} 0.1310 \\ 0.1595 \\ 0.7095 \end{bmatrix}$

21. (a) $\mathbf{T}^2 = \begin{bmatrix} \dfrac{19}{49} & \dfrac{15}{49} \\ \dfrac{30}{49} & \dfrac{34}{49} \end{bmatrix}, \mathbf{T}^3 = \begin{bmatrix} \dfrac{109}{343} & \dfrac{117}{343} \\ \dfrac{234}{343} & \dfrac{226}{343} \end{bmatrix}$; **(b)** $\dfrac{15}{49}$; **(c)** $\dfrac{234}{343}$

23. $\begin{bmatrix} \dfrac{1}{2} \\ \dfrac{1}{2} \end{bmatrix}$

25. (a) 76%; **(b)** 74.4% Japanese, 25.6% non-Japanese; **(c)** 75% Japanese, 25% non-Japanese

MATHEMATICAL SNAPSHOT—CHAPTER 9 (page 446)

1. 7

3. Against Always Defect: $\begin{bmatrix} 0 & 0 & 0 & 0 \\ 1 & 0.1 & 1 & 0.1 \\ 0 & 0 & 0 & 0 \\ 0 & 0.9 & 0 & 0.9 \end{bmatrix}$;

Against Always Cooperate: $\begin{bmatrix} 1 & 0.1 & 1 & 0.1 \\ 0 & 0 & 0 & 0 \\ 0 & 0.9 & 0 & 0.9 \\ 0 & 0 & 0 & 0 \end{bmatrix}$

Against regular Tit-for-tat: $\begin{bmatrix} 1 & 0.1 & 0 & 0 \\ 0 & 0 & 1 & 0.1 \\ 0 & 0.9 & 0 & 0 \\ 0 & 0 & 0 & 0.9 \end{bmatrix}$

PRINCIPLES IN PRACTICE 10.1

1. The limit as $x \to a$ does not exist if a is an integer, but it exists if a is any other value.

2. $\frac{4}{3}\pi$ cc   **3.** 3616   **4.** 20   **5.** 2

PROBLEMS 10.1 (page 457)

1. (a) 1; **(b)** 0; **(c)** 1

3. (a) 1; **(b)** does not exist; **(c)** 3

5. $f(-0.9) = -3.7$, $f(-0.99) = -3.97$, $f(-0.999) = -3.997$, $f(-1.001) = -4.003$, $f(-1.01) = -4.03$, $f(-1.1) = -4.3$; -4

7. $f(-0.1) \approx 0.9516$, $f(-0.01) \approx 0.9950$, $f(-0.001) \approx 0.9995$, $f(0.001) \approx 1.0005$, $f(0.01) \approx 1.0050$, $f(0.1) \approx 1.0517$; 1

9. 16   **11.** 20   **13.** -47 **15.** $-\dfrac{5}{2}$ **17.** 0

19. 5   **21.** -2 **23.** 3   **25.** 5   **27.** $\dfrac{1}{6}$

29. $-\dfrac{1}{5}$ **31.** $\dfrac{11}{9}$ **33.** 4   **35.** $2x$ **37.** -3

39. $2x$ **41.** $3x^2 - 8x$ **43.** $\dfrac{1}{4}$

45. (a) 1; **(b)** 0   **47.** 11.00   **49.** 4.00

51. Does not exist

PRINCIPLES IN PRACTICE 10.2

1. $\lim\limits_{x \to \infty} p(x) = 0$. The graph starts out high and quickly goes down toward zero. Accordingly, consumers are willing to purchase large quantities of the product at prices close to 0.

2. $\lim\limits_{x \to \infty} y(x) = 500$. The greatest yearly sales they can expect with unlimited advertising is $500,000.

3. $\lim\limits_{x \to \infty} C(x) = \infty$. This means that the cost continues to increase without bound as more units are made.

4. The limit does not exist; $250.

PROBLEMS 10.2 (page 465)

1. (a) 2; **(b)** 3; **(c)** does not exist; **(d)** $-\infty$; **(e)** ∞; **(f)** ∞; **(g)** ∞; **(h)** 0; **(i)** 1; **(j)** 1; **(k)** 1

3. 1   **5.** $-\infty$ **7.** $-\infty$ **9.** ∞ **11.** 0

13. ∞ **15.** 0   **17.** ∞ **19.** 0   **21.** 1

23. 0   **25.** ∞ **27.** 0   **29.** $-\dfrac{2}{5}$ **31.** $-\infty$

33. $\dfrac{2}{5}$ **35.** $-\infty$ **37.** $\dfrac{16}{3}$ **39.** $-\dfrac{1}{2}$ **41.** ∞

43. ∞ **45.** ∞ **47.** Does not exist

49. ∞ **51.** 0   **53.** 1

55. (a) 1; **(b)** 2; **(c)** does not exist; **(d)** 1; **(e)** 2

57. (a) 0; **(b)** 0; **(c)** 0; **(d)** $-\infty$; **(e)** $-\infty$

59.

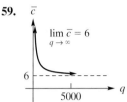

61. 50,000   **63.** 20

65. 1, 0.5, 0.525, 0.631, 0.912, 0.986, 0.998; conclude limit is 1

67. 0   **69. (a)** 11; **(b)** 9; **(c)** does not exist

PROBLEMS 10.3 (page 471)

7. Continuous at -2 and 0   **9.** Discontinuous at ± 3

11. Continuous at 2 and 0   **13.** f is a polynomial function

15. f is a rational function and the denominator is never zero.

17. None   **19.** $x = -4$ **21.** None   **23.** $x = -5, 3$

25. $x = 0, \pm 1$ **27.** None   **29.** $x = 0$ **31.** None

33. $x = 2$

35. Discontinuities at $t = 1, 2, 3, 4$

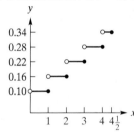

37. Yes, no, no

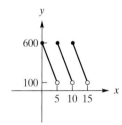

PRINCIPLES IN PRACTICE 10.4

1. $0 < x < 4$

PROBLEMS 10.4 (page 475)

1. $(-\infty, -1), (4, \infty)$ **3.** $[-2, 5]$

5. $\left(-\dfrac{7}{2}, -2\right)$ **7.** No solution

9. $(-\infty, -6], [-2, 3]$ **11.** $(-\infty, -4), (0, 5)$

13. $[0, \infty)$ **15.** $(-\infty, -5], [-3, 0]$

17. $(-\infty, -3), (0, 3)$ **19.** $(1, \infty)$

21. $(-\infty, -5), [-2, 1), [3, \infty)$ **23.** $(-5, -1)$

25. $(-\infty, -1 - \sqrt{3}], [-1 + \sqrt{3}, \infty)$

27. Between 37 and 103, inclusive

29. 17 in. by 17 in.   **31.** $(-\infty, -7.72]$

33. $(-\infty, -0.5), (0.667, \infty)$

REVIEW PROBLEMS—CHAPTER 10 (page 477)

1. -5 **3.** 2   **5.** x **7.** $-\dfrac{8}{3}$ **9.** 0

11. $\dfrac{2}{7}$ **13.** Does not exist   **15.** -1

17. $\dfrac{1}{9}$ **19.** $-\infty$ **21.** ∞

23. $-\infty$ **25.** 1   **27.** $-\infty$ **29.** 8   **31.** 23

35. Continuous everywhere; f is a polynomial function.

37. $x = -3$ **39.** None   **41.** $x = -4, 1$ **43.** $x = -2$

45. $(-\infty, -6), (2, \infty)$ **47.** $(-\infty, 7]$

49. $(-\infty, -5), (-1, 1)$ **51.** $(-\infty, -4), [-3, 0], (2, \infty)$

53. 1.00   **55.** 0   **57.** $[2.00, \infty)$

MATHEMATICAL SNAPSHOT—CHAPTER 10 (page 478)

1. 5.3%

3. An exponential model assumes a fixed repayment rate.

PRINCIPLES IN PRACTICE 11.1

1. $\dfrac{dH}{dt} = 40 - 32t$

PROBLEMS 11.1 (page 488)

1. (a)

x-value of Q	-3	-2.5	-2.2	-2.1	-2.01	-2.001
m_{PQ}	19	15.25	13.24	12.61	12.0601	12.0060

(b) We estimate that $m_{\text{tan}} = 12$.

3. 1   **5.** 3   **7.** -4 **9.** 0   **11.** $2x + 4$

13. $6q + 2$ **15.** $-\dfrac{6}{x^2}$ **17.** $\dfrac{1}{2\sqrt{x+2}}$ **19.** -4

21. 0   **23.** $y = x + 4$ **25.** $y = 4x + 2$

27. $y = -3x + 9$ **29.** $\dfrac{r}{r_L - r - \dfrac{dC}{dD}}$ **31.** $-3.000, 13.445$

33. $-5.120, 0.038$

35. For the x-values of the points where the tangent to the graph of f is horizontal, the corresponding values of $f'(x)$ are 0. This is expected because the slope of a horizontal line is zero and the derivative gives the slope of the tangent line.

37. $20x^4 - 9x^2$

PRINCIPLES IN PRACTICE 11.2

1. $50 - 0.6q$

PROBLEMS 11.2 (page 496)

1. 0   **3.** $6x^5$ **5.** $80x^{79}$ **7.** $18x$ **9.** $56w^6$

11. $\dfrac{8}{3}x^3$ **13.** $\dfrac{7}{25}t^6$ **15.** 1   **17.** $8x - 2$

19. $4p^3 - 9p^2$ **21.** $3x^2 - \dfrac{1}{2\sqrt{x}}$

23. $-39x^2 + 28x - 2$ **25.** $-8x^3$ **27.** $-\dfrac{4}{3}x^3$

29. $16x^3 + 3x^2 - 9x + 8$ **31.** $\dfrac{6}{5}x^3 + 7x^2$ **33.** $\dfrac{3}{5}x^{-2/5}$

35. $\dfrac{3}{4}x^{-1/4} + \dfrac{10}{3}x^{2/3}$ **37.** $\dfrac{11}{2}x^{-1/2}$ or $\dfrac{11}{2\sqrt{x}}$ **39.** $2r^{-2/3}$

41. $-4x^{-5}$ **43.** $-3x^{-4} - 5x^{-6} + 12x^{-7}$

45. $-x^{-2}$ or $-\dfrac{1}{x^2}$ **47.** $-40x^{-6}$ **49.** $-4x^{-4}$

51. $-\dfrac{1}{2}t^{-2}$ **53.** $\dfrac{1}{7} - 7x^{-2}$

55. $-3x^{-2/3} - 2x^{-7/5}$ **57.** $-\dfrac{1}{3}x^{-5/3}$

59. $-x^{-3/2}$ **61.** $\dfrac{5}{2}x^{3/2}$

63. $9x^2 - 20x + 7$ **65.** $45x^4$

67. $\dfrac{1}{3}x^{-2/3} - \dfrac{10}{3}x^{-5/3} = \dfrac{1}{3}x^{-5/3}(x - 10)$

69. $3 + \dfrac{2}{q^2}$ **71.** $2x + 4$ **73.** 1

75. $4, 16, -14$ **77.** $0, 0, 0$ **79.** $y = 13x + 2$

81. $y = -\dfrac{3}{16}x + \dfrac{1}{2}$ **83.** $y = x + 3$

85. $(0, 0), \left(\dfrac{5}{3}, \dfrac{125}{54}\right)$ **87.** $(3, -3)$

89. 0   **91.** The tangent line is $y = 9x - 16$.

PRINCIPLES IN PRACTICE 11.3

1. 2.5 units

2. $\dfrac{dy}{dt} = 16 - 32t$; $\left.\dfrac{dy}{dt}\right|_{t=0.5} = 0$ feet/s

When $t = 0.5$ the object reaches its maximum height.

3. 1.2 and 120%

PROBLEMS 11.3 (page 504)

1.

Δt	1	0.5	0.2	0.1	0.01	0.001
$\Delta s / \Delta t$	9	8	7.4	7.2	7.02	7.002

We estimate the velocity when $t = 1$ to be 7.0000 m/s. Using differentiation the velocity is 7 m/s.

3. (a) 70 m;   **(b)** 25 m/s;   **(c)** 24 m/s

5. (a) 8 m;   **(b)** 6.1208 m/s;   **(c)** 6 m/s

7. (a) 2 m;   **(b)** 10.261 m/s;   **(c)** 9 m/s

9. $\dfrac{dy}{dx} = \dfrac{25}{2}x^{3/2}$; 337.50   **11.** 0.27

13. $dc/dq = 10$; 10

15. $dc/dq = 0.2q + 3$; 4

17. $dc/dq = 2q + 50$; 80, 82, 84

19. $dc/dq = 0.02q + 5$; 6, 7

21. $dc/dq = 0.00006q^2 - 0.02q + 6$; 4.6, 11

23. $dr/dq = 0.8$; 0.8, 0.8, 0.8

25. $dr/dq = 250 + 90q - 3q^2$; 625, 850, 625

27. $dc/dq = 6.750 - 0.000656q$; 5.438;

$\bar{c} = \dfrac{-10,484.69}{q} + 6.750 - 0.000328q$; 0.851655

29. $P = 5,000,000R^{-0.93}$; $dP/dR = -4,650,000R^{-1.93}$

31. (a) -7.5; **(b)** 4.5

33. (a) 1; **(b)** $\dfrac{1}{x+4}$; **(c)** 1; **(d)** $\dfrac{1}{9} \approx 0.111$; **(e)** 11.1%

35. (a) $6x$; **(b)** $\dfrac{6x}{3x^2+7}$; **(c)** 12; **(d)** $\dfrac{12}{19} \approx 0.632$; **(e)** 63.2%

37. (a) $-3x^2$; **(b)** $-\dfrac{3x^2}{8-x^3}$; **(c)** -3; **(d)** $-\dfrac{3}{7} \approx -0.429$;
(e) -42.9%

39. 9.5; 12.8%

41. (a) $dr/dq = 30 - 0.6q$; **(b)** $\dfrac{4}{45} \approx 0.089$; **(c)** 9%

43. $\dfrac{0.432}{t}$ **45.** $\$3125$ **47.** $\$5.07$/unit

PRINCIPLES IN PRACTICE 11.4

1. $\dfrac{dR}{dx} = 6.25 - 6x$

2. $T'(x) = 2x - x^2$; $T'(1) = 1$

PROBLEMS 11.4 (page 513)

1. $(4x+1)(6) + (6x+3)(4) = 48x + 18 = 6(8x+3)$

3. $(5-3t)(3t^2-4t) + (t^3-2t^2)(-3) = -12t^3 + 33t^2 - 20t$

5. $(3r^2-4)(2r-5) + (r^2-5r+1)(6r) = 12r^3 - 45r^2 - 2r + 20$

7. $8x^3 - 10x$

9. $(x^2+3x-2)(4x-1) + (2x^2-x-3)(2x+3)$
$= 8x^3 + 15x^2 - 20x - 7$

11. $(w^2+3w-7)(6w^2) + (2w^3-4)(2w+3)$
$= 10w^4 + 24w^3 - 42w^2 - 8w - 12$

13. $(x^2-1)(9x^2-6) + (3x^3-6x+5)(2x) - 4(8x+2)$
$= 15x^4 - 27x^2 - 22x - 2$

15. $\dfrac{3}{2}\left[\left(5p^{\frac{1}{2}}-2\right)(3) + (3p-1)\left(5\cdot\dfrac{1}{2}p^{-\frac{1}{2}}\right)\right]$
$= \dfrac{3}{4}\left(45p^{\frac{1}{2}} - 12 - 5p^{-\frac{1}{2}}\right)$

17. 0 **19.** $18x^2 + 94x + 31$

21. $\dfrac{(x-1)(5) - (5x)(1)}{(x-1)^2} = -\dfrac{5}{(x-1)^2}$

23. $\dfrac{65}{3x^6}$ **25.** $\dfrac{(x-1)(1) - (x+2)(1)}{(x-1)^2} = -\dfrac{3}{(x-1)^2}$

27. $\dfrac{(z^2-4)(-2) - (6-2z)(2z)}{(z^2-4)^2} = \dfrac{2(z^2-6z+4)}{(z^2-4)^2}$

29. $\dfrac{(x^2-5x)(16x-2) - (8x^2-2x+1)(2x-5)}{(x^2-5x)^2}$
$= \dfrac{-38x^2 - 2x + 5}{(x^2-5x)^2}$

31. $\dfrac{(2x^2-3x+2)(2x-4) - (x^2-4x+3)(4x-3)}{(2x^2-3x+2)^2}$
$= \dfrac{5x^2 - 8x + 1}{(2x^2-3x+2)^2}$

33. $-\dfrac{100x^{99}}{(x^{100}+7)^2}$ **35.** $2v + \dfrac{8}{v^2}$

37. $\dfrac{15x^2 - 2x + 1}{3x^{4/3}}$ **39.** $\dfrac{4}{(x-8)^2} + \dfrac{2}{(3x+1)^2}$

41. $\dfrac{[(x+2)(x-4)](1) - (x-5)(2x-2)}{[(x+2)(x-4)]^2} = \dfrac{-(x^2-10x+18)}{[(x+2)(x-4)]^2}$

43. $\dfrac{[(t^2-1)(t^3+7)](2t+3) - (t^2+3t)(5t^4-3t^2+14t)}{[(t^2-1)(t^3+7)]^2}$
$= \dfrac{-3t^6 - 12t^5 + t^4 + 6t^3 - 21t^2 - 14t - 21}{[(t^2-1)(t^3+7)]^2}$

45. $3 - \dfrac{2x^3 + 3x^2 - 12x + 4}{[x(x-1)(x-2)]^2}$ **47.** $\dfrac{2a}{(a-x)^2}$

49. -6 **51.** $y = -\dfrac{3}{2}x + \dfrac{15}{2}$

53. $y = 16x + 24$ **55.** 1.5

57. 1 m, -1.5 m/s   **59.** $\dfrac{dr}{dq} = 50 - 0.02q$

61. $\dfrac{dr}{dq} = \dfrac{216}{(q+2)^2} - 3$ **63.** $\dfrac{dC}{dI} = 0.672$

65. $\dfrac{7}{6}$; $-\dfrac{1}{6}$ **67.** 0.615; 0.385

69. (a) 0.32; **(b)** 0.026 **71.** $\dfrac{dc}{dq} = \dfrac{6q(q+4)}{(q+2)^2}$

73. $\dfrac{9}{10}$ **75.** $\dfrac{0.7355}{(1+0.02744x)^2}$ **77.** $-\dfrac{1}{120}$

PRINCIPLES IN PRACTICE 11.5

1. $288t$

PROBLEMS 11.5 (page 521)

1. $(2u-2)(2x-1) = 4x^3 - 6x^2 - 2x + 2$

3. $\left(-\dfrac{2}{w^3}\right)(-1) = \dfrac{2}{(2-x)^3}$

5. 0 **7.** 0 **9.** $18(3x+2)^5$

11. $30x^2(3+2x^3)^4$

13. $200(3x^2 - 16x + 1)(x^3 - 8x^2 + x)^{99}$

15. $-6x(x^2-2)^{-4}$

17. $-\dfrac{10}{7}(2x+5)(x^2+5x-2)^{-12/7}$

19. $\dfrac{1}{2}(10x-1)(5x^2-x)^{-1/2}$

21. $\dfrac{1}{2}(2x-1)^{-3/4}$ **23.** $\dfrac{12}{5}x^2(x^3+1)^{-3/5}$

25. $-6(4x-1)(2x^2-x+1)^{-2}$ **27.** $-2(2x-3)(x^2-3x)^{-3}$

29. $-36x(9x^2+1)^{-3/2}$ **31.** $\dfrac{7}{3}(7x)^{-2/3} + \sqrt[3]{7}$

33. $(x^2)[5(x-4)^4(1)] + (x-4)^5(2x) = x(x-4)^4(7x-8)$

35. $4x^2\left[\dfrac{1}{2}(5x+1)^{-\frac{1}{2}}(5)\right] + (\sqrt{5x+1})(8x)$
$= 10x^2(5x+1)^{-\frac{1}{2}} + 8x\sqrt{5x+1}$

37. $(x^2+2x-1)^3(5) + (5x)[3(x^2+2x-1)^2(2x+2)]$
$= 5(x^2+2x-1)^2(7x^2+8x-1)$

39. $(8x-1)^3[4(2x+1)^3(2)] + (2x+1)^4[3(8x-1)^2(8)]$
$= 16(8x-1)^2(2x+1)^3(7x+1)$

41. $12\left(\dfrac{x-3}{x+2}\right)^{11}\left[\dfrac{(x+2)(1) - (x-3)(1)}{(x+2)^2}\right] = \dfrac{60(x-3)^{11}}{(x+2)^{13}}$

43. $\dfrac{1}{2}\left(\dfrac{x-2}{x+3}\right)^{-1/2}\left[\dfrac{(x+3)(1) - (x-2)(1)}{(x+3)^2}\right]$
$= \dfrac{5}{2(x+3)^2}\left(\dfrac{x-2}{x+3}\right)^{-1/2}$

45. $\dfrac{(x^2+4)^3(2)-(2x-5)[3(x^2+4)^2(2x)]}{(x^2+4)^6}$

$=\dfrac{-2(5x^2-15x-4)}{(x^2+4)^4}$

47. $\dfrac{(3x-1)^3[40(8x-1)^4]-(8x-1)^5[9(3x-1)^2]}{(3x-1)^6}$

$=\dfrac{(8x-1)^4(48x-31)}{(3x-1)^4}$

49. $6\{(5x^2+2)[2x^3(x^4+5)^{-1/2}]+(x^4+5)^{1/2}(10x)\}$
$=12x(x^4+5)^{-1/2}(10x^4+2x^2+25)$

51. $8+\dfrac{5}{(t+4)^2}-(8t-7)=15-8t+\dfrac{5}{(t+4)^2}$

53. $\dfrac{(x^3-5)^5[(2x+1)^3(2)(x+3)(1)+(x+3)^2(3)(2x+1)^2(2)]}{-(2x+1)^3(x+3)^2[5(x^3-5)^4(3x^2)]}$ over $(x^3-5)^{10}$

55. 0 **57.** 0 **59.** $y=4x-11$

61. $y=-\dfrac{1}{6}x+\dfrac{5}{3}$ **63.** 96% **65.** 130

67. ≈ 13.99

69. (a) $-\dfrac{q}{\sqrt{q^2+20}}$; (b) $-\dfrac{q}{100\sqrt{q^2+20}-q^2-20}$;

(c) $100-\dfrac{q^2}{\sqrt{q^2+20}}-\sqrt{q^2+20}$

71. -481.5 **73.** $\dfrac{dc}{dq}=\dfrac{5q(q^2+6)}{(q^2+3)^{3/2}}$ **75.** $48\pi(10)^{-19}$

77. (a) $-0.001416x^3+0.01356x^2+1.696x-34.9,\ -256.238$
(b) $-0.016;\ -1.578\%$

79. -4 **81.** 40 **83.** $86,111.37$

REVIEW PROBLEMS—CHAPTER 11 (page 524)

1. $-2x$ **3.** $\dfrac{\sqrt{3}}{2\sqrt{x}}$ **5.** 0

7. $28x^3-18x^2+10x=2x(14x^2-9x+5)$

9. $4s^3+4s=4s(s^2+1)$ **11.** $\dfrac{2x}{5}$

13. $(x^3+7x^2)(3x^2-2x)+(x^3-x^2+5)(3x^2+14x)$
$=6x^5+30x^4-28x^3+15x^2+70x$

15. $100(2x^2+4x)^{99}(4x+4)=400(x+1)(2x^2+4x)^{99}$

17. $-\dfrac{6}{(2x+1)^2}$

19. $(8+2x)(4)(x^2+1)^3(2x)+(x^2+1)^4(2)$
$=2(x^2+1)^3(9x^2+32x+1)$

21. $\dfrac{(z^2+4)(2z)-(z^2-1)(2z)}{(z^2+4)^2}=\dfrac{10z}{(z^2+4)^2}$

23. $\dfrac{4}{3}(4x-1)^{-2/3}$

25. $-\dfrac{1}{2}(1-x^2)^{-3/2}(-2x)=x(1-x^2)^{-3/2}$

27. $(x-6)^4[3(x+5)^2]+(x+5)^3[4(x-6)^3]=$
$(x-6)^3(x+5)^2(7x+2)$

29. $\dfrac{(x+6)(5)-(5x-4)(1)}{(x+6)^2}=\dfrac{34}{(x+6)^2}$

31. $2\left(-\dfrac{3}{8}\right)x^{-11/8}+\left(-\dfrac{3}{8}\right)(2x)^{-11/8}(2)$

$=-\dfrac{3}{4}(1+2^{-11/8})x^{-11/8}$

33. $\dfrac{\sqrt{x^2+5}(2x)-(x^2+6)(1/2)(x^2+5)^{-1/2}(2x)}{x^2+5}=\dfrac{x(x^2+4)}{(x^2+5)^{3/2}}$

35. $\left(\dfrac{3}{5}\right)(x^3+6x^2+9)^{-2/5}(3x^2+12x)$

$=\dfrac{9}{5}x(x+4)(x^3+6x^2+9)^{-2/5}$

37. $-3z^2+4z-1$ **39.** $y=-4x+3$

41. $y=\dfrac{1}{12}x+\dfrac{4}{3}$ **43.** $\dfrac{5}{7}\approx 0.714;\ 71.4\%$

45. $dr/dq=20-0.2q$ **47.** $0.569,\ 0.431$

49. $dr/dq=500-0.2q$

51. $dc/dq=0.125+0.00878q;\ 0.7396$

53. 84 eggs/mm   **55.** (a) $\dfrac{4}{3}$; (b) $\dfrac{1}{24}$

57. 8π ft^3/ft   **59.** $4q-\dfrac{10,000}{q^2}$

61. (a) -315.456; (b) -0.00025; (c) no, since $dr/dm<0$ when $m=240$

63. 0.305 **65.** -0.32

MATHEMATICAL SNAPSHOT—CHAPTER 11 (page 526)

1. The slope is greater—above 0.9. More is spent; less is saved.
3. Spend $705, save $295   **5.** Answers may vary.

PRINCIPLES IN PRACTICE 12.1

1. $\dfrac{dq}{dp}=\dfrac{12p}{3p^2+4}$ **2.** $\dfrac{dR}{dI}=\dfrac{1}{I\ln 10}$

PROBLEMS 12.1 (page 533)

1. $\dfrac{4}{x}$ **3.** $\dfrac{3}{3x-7}$ **5.** $\dfrac{2}{x}$ **7.** $-\dfrac{2x}{1-x^2}$

9. $\dfrac{24X^5+6X^2}{4X^6+2X^3}=\dfrac{3(4X^3+1)}{X(2X^3+1)}$

11. $t\left(\dfrac{1}{t}\right)+(\ln t)(1)=1+\ln t$

13. $\dfrac{2x^3}{2x+5}+3x^2\ln(2x+5)$ **15.** $\dfrac{8}{(\ln 3)(8x-1)}$

17. $2x\left[1+\dfrac{1}{(\ln 2)(x^2+4)}\right]$

19. $\dfrac{z\left(\frac{1}{z}\right)-(\ln z)(1)}{z^2}=\dfrac{1-\ln z}{z^2}$

21. $\dfrac{(\ln x)^2(2x)-(x^2+3)2(\ln x)\frac{1}{x}}{(\ln x)^4}=\dfrac{2x^2\ln x-2(x^2+3)}{x(\ln x)^3}$

23. $\dfrac{3(2x+4)}{x^2+4x+5}=\dfrac{6(x+2)}{x^2+4x+5}$

25. $\dfrac{9x}{1+x^2}$ **27.** $\dfrac{2}{1-l^2}$ **29.** $\dfrac{x}{1-x^4}$

31. $\dfrac{4x}{x^2+2}+\dfrac{3x^2+1}{x^3+x-1}$

33. $\dfrac{26}{x}+\dfrac{65}{3(5x+2)}$ **35.** $\dfrac{2(x^2+1)}{2x+1}+2x\ln(2x+1)$

37. $\dfrac{3(1+\ln^2 x)}{x}$ **39.** $\dfrac{4\ln^3(ax)}{x}$

41. $\dfrac{x}{2(x-1)}+\ln\sqrt{x-1}$ **43.** $\dfrac{3}{2x\sqrt{4+3\ln x}}$

45. $y=5x-20$ **47.** $\dfrac{\ln(3)-1}{\ln^2 3}$ **49.** $\dfrac{25}{7}$

51. $\dfrac{dq}{dp} = \dfrac{20}{2p+1}$ **53.** $\dfrac{6a}{(T-a^2+aT)(a-T)}$

57. $-1.65, 1.65$

PRINCIPLES IN PRACTICE 12.2

1. $\dfrac{dT}{dt} = Cke^{kt}$

PROBLEMS 12.2 (page 537)

1. $5e^x$ **3.** $4xe^{2x^2+3}$ **5.** $-5e^{9-5x}$

7. $(6r+4)e^{3r^2+4r+4} = 2(3r+2)e^{3r^2+4r+4}$

9. $x(e^x) + e^x(1) = e^x(x+1)$ **11.** $2xe^{-x^2}(1-x^2)$

13. $\dfrac{e^x - e^{-x}}{3}$ **15.** $(6x^2)5^{2x^3}\ln 5$ **17.** $\dfrac{2e^{2w}(w-1)}{w^3}$

19. $\dfrac{e^{1+\sqrt{x}}}{2\sqrt{x}}$ **21.** $5x^4 - 5^x \ln 5$ **23.** $\dfrac{2e^x}{(e^x+1)^2}$

25. 1 **27.** $2xe^{x^2\ln x^2}(1 + \ln x^2)$ **29.** $-e$

31. $y - e^{-2} = e^{-2}(x+2)$ or $y = e^{-2}x + 3e^{-2}$

33. $dp/dq = -0.015e^{-0.001q}, -0.015e^{-0.5}$

35. $dc/dq = 10e^{q/700}$; $10e^{0.5}$; $10e$ **37.** -5

39. e **41.** $100e^{-2}$ **47.** $-b(10^{A-bM})\ln 10$

51. 0.0036 **53.** $-0.89, 0.56$

PROBLEMS 12.3 (page 543)

1. -3, elastic   **3.** -1, unit elasticity

5. $-\dfrac{53}{52}$, elastic   **7.** $-\left(\dfrac{150}{e}-1\right)$, elastic

9. -1, unit elasticity   **11.** -2, elastic

13. $-\dfrac{1}{2}$, inelastic

15. $|\eta| = \dfrac{10}{3}$ when $p = 10$, $|\eta| = \dfrac{3}{10}$ when $p = 3$, $|\eta| = 1$ when $p = 6.50$

17. $-1.2, 0.6\%$ decrease   **23.** **(c)** $b = 0$

25. **(a)** $\eta = -\dfrac{207}{15} = -13.8$, elastic; **(b)** 27.6%; **(c)** increase, since demand is elastic

27. $\eta = -1.6$; $\dfrac{dr}{dq} = 30$

29. Maximum at $q = 5$; minimum at $q = 95$

PRINCIPLES IN PRACTICE 12.4

1. $\dfrac{dP}{dt} = 0.5(P - P^2)$

2. $\dfrac{dV}{dt} = 4\pi r^2 \dfrac{dr}{dt}$ and $\dfrac{dV}{dt}\bigg|_{r=12} = 2880\pi$ in^3/minute

3. The top of the ladder is sliding down at a rate of $\dfrac{9}{4}$ feet/second.

PROBLEMS 12.4 (page 548)

1. $-\dfrac{x}{4y}$ **3.** $\dfrac{7}{3y^2}$ **5.** $-\dfrac{\sqrt[3]{y^2}}{\sqrt[3]{x^2}}$ **7.** $-\dfrac{y^{1/4}}{x^{1/4}}$

9. $-\dfrac{y}{x}$ **11.** $\dfrac{11-y}{x-1}$ **13.** $\dfrac{4y-2x^2}{y^2-4x}$

15. $\dfrac{4y^{3/4}}{2y^{1/4}+1}$ **17.** $\dfrac{1-15x^2y^4}{20x^5y^3+2y}$ **19.** $\dfrac{xe^y-y}{x(\ln x - xe^y)}$

21. $-\dfrac{e^y}{xe^y+1}$ **23.** $6e^{3x}(1+e^{3x})(x+y)-1$

25. $-\dfrac{3}{5}$ **27.** $0; -\dfrac{4x_0}{9y_0}$ **29.** $y = -4x - 3$

31. $\dfrac{dq}{dp} = -\dfrac{1}{2q}$ **33.** $\dfrac{dq}{dp} = -\dfrac{(q+5)^3}{40}$

35. $-\lambda I$ **37.** $-\dfrac{f}{\lambda}$ **39.** $\dfrac{3}{8}$

PROBLEMS 12.5 (page 552)

1. $(x+1)^2(x-2)(x^2+3)\left[\dfrac{2}{x+1} + \dfrac{1}{x-2} + \dfrac{2x}{x^2+3}\right]$

3. $(3x^3-1)^2(2x+5)^3\left[\dfrac{18x^2}{3x^3-1} + \dfrac{6}{2x+5}\right]$

5. $\dfrac{\sqrt{x+1}\sqrt{x^2-2}\sqrt{x+4}}{2}\left[\dfrac{1}{x+1} + \dfrac{2x}{x^2-2} + \dfrac{1}{x+4}\right]$

7. $\dfrac{\sqrt{1-x^2}}{1-2x}\left[\dfrac{x}{x^2-1} + \dfrac{2}{1-2x}\right]$

9. $\dfrac{(2x^2+2)^2}{(x+1)^2(3x+2)}\left[\dfrac{4x}{x^2+1} - \dfrac{2}{x+1} - \dfrac{3}{3x+2}\right]$

11. $\dfrac{1}{2}\sqrt{\dfrac{(x+3)(x-2)}{2x-1}}\left[\dfrac{1}{x+3} + \dfrac{1}{x-2} - \dfrac{2}{2x-1}\right]$

13. $x^{x^2+1}\left(\dfrac{x^2+1}{x} + 2x\ln x\right)$ **15.** $\dfrac{x^{1/x}(1-\ln x)}{x^2}$

17. $2(3x+1)^{2x}\left[\dfrac{3x}{3x+1} + \ln(3x+1)\right]$

19. $4e^x x^{3x}(4+3\ln x)$ **21.** 12 **23.** $y = 96x + 36$

25. $y = (4e + 2e\ln 2)x - 2e - 2e\ln 2$

27. $\dfrac{1}{3e^{1.3}}$ **29.** 0.1% decrease

PRINCIPLES IN PRACTICE 12.6

1. 43 and 1958

PROBLEMS 12.6 (page 556)

1. 0.25410 **3.** 1.32472 **5.** -0.68233 **7.** 0.33767

9. 1.90785 **11.** 4.141 **13.** -4.99 and 1.94

15. 13.33 **17.** 2.880 **19.** 3.45

PRINCIPLES IN PRACTICE 12.7

1. $\dfrac{d^2h}{dt^2} = -32$ feet/sec^2 (*Note*: Negative values indicate the downward direction.)

2. $c''(3) = 14$ dollars/unit^2

PROBLEMS 12.7 (page 560)

1. 24 **3.** 0 **5.** e^x **7.** $3 + 2\ln x$

9. $-\dfrac{60}{q^7}$ **11.** $-\dfrac{1}{4(9-r)^{3/2}}$ **13.** $\dfrac{8}{(2x+3)^3}$

15. $\dfrac{4}{(x-1)^3}$ **17.** $-\left[\dfrac{1}{x^2} + \dfrac{1}{(x+6)^2}\right]$

19. $e^z(z^2+4z+2)$ **21.** 275 **23.** $-\dfrac{1}{y^3}$

25. $-\dfrac{4}{y^3}$ **27.** $\dfrac{1}{8x^{3/2}}$ **29.** $\dfrac{2(y-1)}{(1+x)^2}$

31. $\dfrac{y}{(1-y)^3}$ **33.** $\dfrac{25}{32}$ **35.** $300(5x-3)^2$

37. 0.6 **39.** ± 1 **41.** -4.99 and 1.94

REVIEW PROBLEMS—CHAPTER 12 (page 561)

1. $3e^x + 0 + e^{x^2}(2x) + (e^2)x^{e^2-1} = 3e^x + 2xe^{x^2} + e^2x^{e^2-1}$

3. $\dfrac{1}{3r^2 + 7r + 1}(6r + 7) = \dfrac{6r + 7}{3r^2 + 7r + 1}$

5. $e^{x^2+4x+5}(2x + 4) = 2(x + 2)e^{x^2+4x+5}$

7. $e^x(2x) + (x^2 + 2)e^x = e^x(x^2 + 2x + 2)$

9. $\dfrac{\sqrt{(x-6)(x+5)(9-x)}}{2}\left[\dfrac{1}{x-6} + \dfrac{1}{x+5} + \dfrac{1}{x-9}\right]$

11. $\dfrac{e^x\left(\dfrac{1}{x}\right) - (\ln x)(e^x)}{e^{2x}} = \dfrac{1 - x\ln x}{xe^x}$

13. $\dfrac{2}{q+1} + \dfrac{3}{q+2}$

15. $(4x + 2)(\ln 2)2^{2x^2+2x-5}$

17. $\dfrac{4e^{2x+1}(2x - 1)}{x^2}$

19. $\dfrac{16}{(8x + 5)\ln 2}$

21. $\dfrac{1 + 2l + 3l^2}{1 + l + l^2 + l^3}$

23. $(x + 1)^{x+1}[1 + \ln(x + 1)]$

25. $\dfrac{1}{t} + \dfrac{1}{2} \cdot \dfrac{1}{4 - t^2} \cdot (-2t) = \dfrac{1}{t} - \dfrac{t}{4 - t^2}$

27. $y\left[\dfrac{1}{2}\left(\dfrac{1}{x^2 + 1}\right)(2x) + \dfrac{1}{3}\left(\dfrac{1}{x^2 + 2}\right)(2x)\right.$

$\left. - \dfrac{2}{5}\left(\dfrac{1}{2x^3 + 6x}\right)(6x^2 + 6)\right]$

$= y\left[\dfrac{x}{x^2 + 1} + \dfrac{2x}{3(x^2 + 2)} - \dfrac{6(x^2 + 1)}{5(x^3 + 3x)}\right],$

where y is as given in the problem

29. $(x^x)^x(x + 2x\ln x)$ **31.** 4   **33.** -2

35. $y = 6x + 6(1 - \ln 2)$ or $y = 6x + 6 - \ln 64$

37. $(0, 4\ln 2)$ **39.** 2   **41.** 2   **43.** $-\dfrac{y}{x + y}$

45. $\dfrac{xy^2 - y}{2x - x^2y}$ **47.** $\dfrac{4}{9}$ **49.** $\dfrac{dy}{dx} = \dfrac{y + 1}{y}; \dfrac{d^2y}{dx^2} = -\dfrac{y + 1}{y^3}$

51. $f'(t) = 0.008e^{-0.01t} + 0.00004e^{-0.0002t}$

53. 1.13   **55.** $\eta = -1$, unit elasticity

57. $\eta = -0.5$, demand is inelastic

59. $-\dfrac{9}{16}, \approx \dfrac{3}{8}\%$ increase   **61.** 1.7693

MATHEMATICAL SNAPSHOT—CHAPTER 12 (page 566)

1. 305 units

3. Answers may vary.

PRINCIPLES IN PRACTICE 13.1

1. There is a relative maximum when $q = 2$, and a relative minimum when $q = 5$.

2. The drug is at its greatest concentration 2 hours after injection.

PROBLEMS 13.1 (page 576)

1. Dec. on $(-\infty, -1)$ and $(3, \infty)$; inc. on $(-1, 3)$; rel. min. $(-1, -1)$; rel. max. $(3, 4)$

3. Dec. on $(-\infty, -2)$ and $(0, 2)$; inc. on $(-2, 0)$ and $(2, \infty)$; rel. min. $(-2, 1)$ and $(2, 1)$; no rel. max

5. Inc. on $(-3, 1)$ and $(2, \infty)$; dec. on $(-\infty, -3)$ and $(1, 2)$; rel. max. when $x = 1$; rel. min. when $x = -3, 2$

7. Dec. on $(-\infty, -1)$; inc. on $(-1, 3)$ and $(3, \infty)$; rel. min. when $x = -1$

9. Inc. on $(-\infty, 0)$ and $(0, \infty)$; no rel. min. or max

11. Inc. on $\left(-\infty, \dfrac{1}{2}\right)$; dec. on $\left(\dfrac{1}{2}, \infty\right)$; rel. max. when $x = \dfrac{1}{2}$

13. Dec. on $(-\infty, -5)$ and $(1, \infty)$; inc. on $(-5, 1)$; rel. min. when $x = -5$; rel. max. when $x = 1$

15. Dec. on $(-\infty, -1)$ and $(0, 1)$; inc. on $(-1, 0)$ and $(1, \infty)$; rel. max. when $x = 0$; rel. min. when $x = \pm 1$

17. Inc. on $\left(-\infty, \dfrac{1}{3}\right)$ and $(2, \infty)$; dec. on $\left(\dfrac{1}{3}, 2\right)$; rel. max. when $x = \dfrac{1}{3}$; rel. min. when $x = 2$

19. Inc. on $\left(-\infty, -\dfrac{2}{3}\right)$ and $\left(\dfrac{5}{2}, \infty\right)$; dec. on $\left(-\dfrac{2}{3}, \dfrac{5}{2}\right)$; rel. max. when $x = -\dfrac{2}{3}$; rel. min. when $x = \dfrac{5}{2}$

21. Inc. on $(-\infty, 5 - \sqrt{3})$ and $(5 + \sqrt{3}, \infty)$; dec. on $(5 - \sqrt{3}, 5 + \sqrt{3})$; rel. max. when $x = 5 - \sqrt{3}$; rel. min. when $x = 5 + \sqrt{3}$

23. Inc. on $(-\infty, -1)$ and $(1, \infty)$; dec. on $(-1, 0)$ and $(0, 1)$; rel. max. when $x = -1$; rel. min. when $x = 1$

25. Dec. on $(-\infty, -4)$ and $(0, \infty)$; inc. on $(-4, 0)$; rel. min. when $x = -4$; rel. max. when $x = 0$

27. Inc. on $(-\infty, -\sqrt{2})$ and $(0, \sqrt{2})$; dec. on $(-\sqrt{2}, 0)$ and $(\sqrt{2}, \infty)$; rel. max. when $x = \pm\sqrt{2}$; rel. min. when $x = 0$

29. Inc. on $(-1, 0)$ and $(1, \infty)$; dec. on $(-\infty, -1)$ and $(0, 1)$; rel. max. when $x = 0$; rel. min. when $x = \pm 1$

31. Dec. on $(-\infty, 1)$ and $(1, \infty)$; no rel. extremum

33. Dec. on $(0, \infty)$; no rel. extremum

35. Dec. on $(-\infty, 0)$ and $(4, \infty)$; inc. on $(0, 2)$ and $(2, 4)$; rel. min. when $x = 0$; rel. max. when $x = 4$

37. Inc. on $(-\infty, -3)$ and $(-1, \infty)$; dec. on $(-3, -2)$ and $(-2, -1)$; rel. max. when $x = -3$; rel. min. when $x = -1$

39. Dec. on $\left(-\infty, \dfrac{-2 - \sqrt{29}}{5}\right)$ and $\left(\dfrac{-2 + \sqrt{29}}{5}, \infty\right)$; inc. on $\left(\dfrac{-2 - \sqrt{29}}{5}, \dfrac{-2 + \sqrt{29}}{5}\right)$; rel. min. when $x = \dfrac{-2 - \sqrt{29}}{5}$; rel. max. when $x = \dfrac{-2 + \sqrt{29}}{5}$

41. Inc. on $(1, \infty)$; dec. on $(-\infty, 1)$; rel. min. when $x = 1$

43. Inc. on $(-\infty, 0)$, $\left(0, \dfrac{18}{7}\right)$, and $(6, \infty)$; dec. on $\left(\dfrac{18}{7}, 6\right)$; rel. max. when $x = \dfrac{18}{7}$; rel. min. when $x = 6$

45. Dec. on $(-\infty, \infty)$; no rel. extremum.

47. Dec. on $\left(0, \dfrac{3\sqrt{2}}{2}\right)$; inc. on $\left(\dfrac{3\sqrt{2}}{2}, \infty\right)$; rel. min. when $x = \dfrac{3\sqrt{2}}{2}$

49. Dec. on $(-\infty, 0)$; inc. on $(0, \infty)$; rel. min. when $x = 0$

51. Dec. on $(0, 1)$; inc. on $(1, \infty)$; rel. min. when $x = 1$

53. Dec. on $\left(-\infty, \frac{3}{2}\right)$; inc. on $\left(\frac{3}{2}, \infty\right)$; rel. min. when $x = \frac{3}{2}$; intercepts: $(-2, 0), (5, 0), (0, -10)$

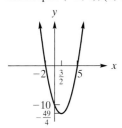

55. Dec. on $(-\infty, -1)$ and $(1, \infty)$; inc. on $(-1, 1)$; rel. min. when $x = -1$; rel. max. when $x = 1$; sym. about origin; intercepts: $(\pm\sqrt{3}, 0)$. $(0, 0)$

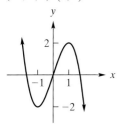

57. Inc. on $(-\infty, 1)$ and $(2, \infty)$; dec. on $(1, 2)$; rel. max. when $x = 1$; rel. min. when $x = 2$; intercept: $(0, 0)$

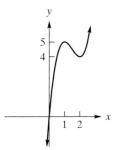

59. Inc. on $(-2, -1)$ and $(0, \infty)$; dec. on $(-\infty, -2)$ and $(-1, 0)$; rel. max. when $x = -1$; rel. min. when $x = -2, 0$; intercepts: $(0, 0), (-2, 0)$

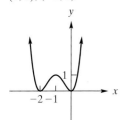

61. Dec. on $(-\infty, -2)$ and $\left(-\frac{1}{2}, 1\right)$; inc. on $\left(-2, -\frac{1}{2}\right)$ and $(1, \infty)$; rel. min. when $x = -2, 1$; rel. max. when $x = -\frac{1}{2}$; intercepts: $(1, 0), (-2, 0), (0, 4)$

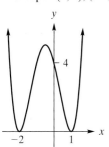

63. Dec. on $(1, \infty)$; inc. on $(0, 1)$; rel. max. when $x = 1$; intercepts: $(0, 0), (4, 0)$

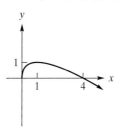

65.

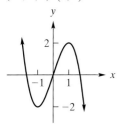

69. Never

71. 40   **75.** **(a)** 25,300; **(b)** 4; **(c)** 17,200

77. Rel. min.: $(-3.83, 0.69)$

79. Rel. max.: $(2.74, 3.74)$; rel. min.: $(-2.74, -3.74)$

81. Rel. min.: 0, 1.50, 2.00; rel. max.: 0.57, 1.77

83. **(a)** $f'(x) = 4 - 6x - 3x^2$; **(c)** Dec.: $(-\infty, -2.53), (0.53, \infty)$; inc.: $(-2.53, 0.53)$

PROBLEMS 13.2 (page 580)

1. Maximum: $f(3) = 6$; minimum: $f(1) = 2$

3. Maximum: $f(-1) = \frac{19}{6}$; minimum: $f(0) = 1$

5. Maximum: $f(3) = 84$; minimum: $f(1) = -8$

7. Maximum: $f(-2) = 56$; minimum: $f(-1) = -2$

9. Maximum: $f(\sqrt{2}) = 4$; minimum: $f(2) = -16$

11. Maximum: $f(0) = f(3) = 2$; minimum: $f\left(\frac{3\sqrt{2}}{2}\right) = -\frac{73}{4}$

13. Maximum: $f(-26) = f(28) = 9$; minimum: $f(1) = 0$

15. **(a)** $-3.22, -0.78$; **(b)** 2.75; **(c)** 9; **(d)** 14,283

PROBLEMS 13.3 (page 586)

1. Conc. up $\left(-\infty, -\frac{3}{4}\right), (0, \infty)$; conc. down $\left(-\frac{3}{4}, 0\right)$; inf. pt. when $x = -\frac{3}{4}, 0$

3. Conc. up $(-\infty, 1), (1, 7)$; conc. down $(7, \infty)$; inf. pt. when $x = 7$

5. Conc. up $(-\infty, -\sqrt{2}), (\sqrt{2}, \infty)$; conc. down $(-\sqrt{2}, \sqrt{2})$; no inf. pt.

7. Conc. down $(-\infty, \infty)$

9. Conc. down $(-\infty, -1)$; conc. up $(-1, \infty)$; inf. pt. when $x = -1$

11. Conc. down $\left(-\infty, \dfrac{5}{6}\right)$; conc. up $\left(\dfrac{5}{6}, \infty\right)$; inf. pt. when $x = \dfrac{5}{6}$

13. Conc. up $(-\infty, -2)$, $(2, \infty)$; conc. down $(-2, 2)$; inf. pt. when $x = \pm 2$

15. Conc. up $(-\infty, 0)$; conc. down $(0, \infty)$; inf. pt. when $x = 0$

17. Conc. up $\left(-\infty, -\dfrac{7}{2}\right)$, $\left(\dfrac{1}{3}, \infty\right)$; conc. down $\left(-\dfrac{7}{2}, \dfrac{1}{3}\right)$; inf. pt. when $x = -\dfrac{7}{2}, \dfrac{1}{3}$

19. Conc. down $(-\infty, 0)$, $\left(\dfrac{3 - \sqrt{5}}{2}, \dfrac{3 + \sqrt{5}}{2}\right)$; conc. up $\left(0, \dfrac{3 - \sqrt{5}}{2}\right)$, $\left(\dfrac{3 + \sqrt{5}}{2}, \infty\right)$; inf. pt. when $x = 0$, $\dfrac{3 \pm \sqrt{5}}{2}$

21. Conc. up $(-\infty, -\sqrt{5})$, $(-\sqrt{2}, \sqrt{2})$, $(\sqrt{5}, \infty)$; conc. down $(-\sqrt{5}, -\sqrt{2})$, $(\sqrt{2}, \sqrt{5})$; inf. pt. when $x = \pm\sqrt{5}, \pm\sqrt{2}$

23. Conc. down $(-\infty, 1)$; conc. up $(1, \infty)$

25. Conc. down $(-\infty, -1/\sqrt{3})$, $(1/\sqrt{3}, \infty)$; conc. up $(-1/\sqrt{3}, 1/\sqrt{3})$; inf. pt. when $x = \pm 1/\sqrt{3}$

27. Conc. down. $(-\infty, -3)$, $\left(-3, \dfrac{2}{7}\right)$; conc. up $\left(\dfrac{2}{7}, \infty\right)$; inf. pt. when $x = \dfrac{2}{7}$

29. Conc. up $(-\infty, \infty)$

31. Conc. down $(-\infty, -2)$; conc. up $(-2, \infty)$; inf. pt. when $x = -2$

33. Conc. down $(0, e^{3/2})$; conc. up $(e^{3/2}, \infty)$; inf. pt. when $x = e^{3/2}$

35. Int. $(-2, 0)$, $(3, 0)$, $(0, -6)$; dec. $\left(-\infty, \dfrac{1}{2}\right)$; inc. $\left(\dfrac{1}{2}, \infty\right)$; rel. min. when $x = \dfrac{1}{2}$; conc. up $(-\infty, \infty)$

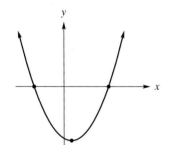

37. Int. $(0, 0)$, $\left(\dfrac{5}{2}, 0\right)$; inc. $\left(-\infty, \dfrac{5}{4}\right)$; dec. $\left(\dfrac{5}{4}, \infty\right)$; rel. max. when $x = \dfrac{5}{4}$; conc. down $(-\infty, \infty)$

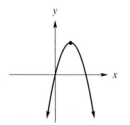

39. Int. $(0, -19)$; inc. $(-\infty, 2)$, $(4, \infty)$; dec. $(2, 4)$; rel. max. when $x = 2$; rel. min. when $x = 4$; conc. down $(-\infty, 3)$; conc. up $(3, \infty)$; inf. pt. when $x = 3$

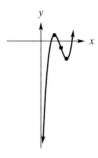

41. Int. $(0, 0)$, $(\pm 2\sqrt{3}, 0)$; inc. $(-\infty, -2)$, $(2, \infty)$; dec. $(-2, 2)$; rel. max. when $x = -2$; rel. min. when $x = 2$; conc. down $(-\infty, 0)$; conc. up $(0, \infty)$; inf. pt. when $x = 0$; sym. about origin

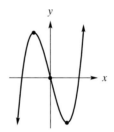

43. Int. $(0, -3)$; inc. $(-\infty, 1)$, $(1, \infty)$; no rel. max. or min.; conc. down $(-\infty, 1)$; conc. up $(1, \infty)$; inf. pt. when $x = 1$

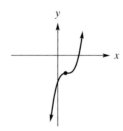

45. Int. $(0, 0)$, $(4/3, 0)$; inc. $(-\infty, 0)$, $(0, 1)$; dec. $(1, \infty)$; rel. max. when $x = 1$; conc. up $(0, 2/3)$; conc. down $(-\infty, 0)$, $(2/3, \infty)$; inf. pt. when $x = 0$, $x = 2/3$

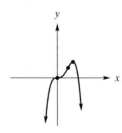

47. Int. $(0, -2)$; dec. $(-\infty, -2)$, $(2, \infty)$; inc. $(-2, 2)$; rel. min. when $x = -2$; rel. max. when $x = 2$; conc. up $(-\infty, 0)$; conc. down $(0, \infty)$; inf. pt. when $x = 0$

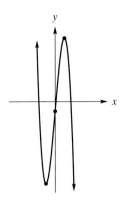

49. Int. $(0, -2)$, $(1, 0)$; inc. $(-\infty, 1)$, $(1, \infty)$; conc. down $(-\infty, 1)$; conc. up $(1, \infty)$; inf. pt. when $x = 1$

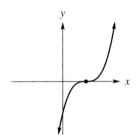

51. Int. $(0, 0)$, $(\pm\sqrt[4]{5}, 0)$; dec. $(-\infty, -1)$, $(1, \infty)$; inc. $(-1, 1)$; rel. min. when $x = -1$; rel. max. when $x = 1$; conc. up $(-\infty, 0)$; conc. down $(0, \infty)$; inf. pt. when $x = 0$; sym. about origin.

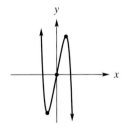

53. Int. $(0, 1)$, $(1, 0)$; dec. $(-\infty, 0)$, $(0, 1)$; inc. $(1, \infty)$; rel. min. when $x = 1$; conc. up $(-\infty, 0)$, $(2/3, \infty)$; conc. down $(0, 2/3)$; inf. pt. when $x = 0$, $x = 2/3$

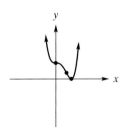

55. Int. $(0, 0)$, $(\pm 2, 0)$; inc. $(-\infty, -\sqrt{2})$, $(0, \sqrt{2})$; dec. $(-\sqrt{2}, 0)$, $(\sqrt{2}, \infty)$; rel. max. when $x = \pm \sqrt{2}$; rel. min. when $x = 0$;

conc. down $(-\infty, -\sqrt{2/3})$, $(\sqrt{2/3}, \infty)$; conc. up $(-\sqrt{2/3}, \sqrt{2/3})$; inf. pt. when $x = \pm\sqrt{2/3}$; sym. about y-axis

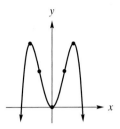

57. Int. $(0, 0)$, $(8, 0)$; dec. $(-\infty, 0)$, $(0, 2)$; inc. $(2, \infty)$; rel. min. when $x = 2$; conc. up $(-\infty, -4)$, $(0, \infty)$; conc. down $(-4, 0)$; inf. pt. when $x = -4$, $x = 0$

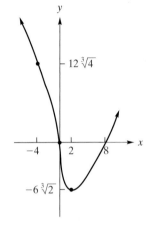

59. Int. $(0, 0)$, $(-4, 0)$; dec. $(-\infty, -1)$; inc. $(-1, 0)$, $(0, \infty)$; rel. min. when $x = -1$; conc. up $(-\infty, 0)$, $(2, \infty)$; conc. down $(0, 2)$; inf. pt. when $x = 0$, $x = 2$

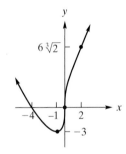

61. Int. $(0, 0)$, $(1728, 0)$; inc. $(0, 512)$; dec. $(-\infty, 0)$, $(512, \infty)$; rel. min. when $x = 0$; rel. max. when $x = 512$; conc. down $(-\infty, 0)$, $(0, \infty)$

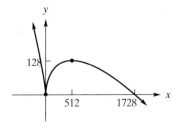

63.

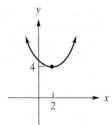

65.

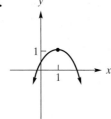

69.

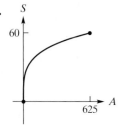

73. (b)

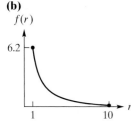

(c) 0.26

75. Two   **77.** Above tangent line; concave up

79. −2.61, −0.26

PROBLEMS 13.4 (page 589)

1. Rel. min. when $x = \dfrac{5}{2}$; abs. min.

3. Rel. max. when $x = \dfrac{1}{4}$; abs. max.

5. Rel. max. when $x = -5$; rel. min. when $x = 1$

7. Rel. min. when $x = 0$; rel. max. when $x = 2$

9. Test fails, when $x = 0$ there is a rel. max. by first-deriv. test

11. Rel. max. when $x = -\dfrac{1}{3}$; rel. min. when $x = \dfrac{1}{3}$

13. Rel. min. when $x = -5, -2$; rel. max. when $x = -\dfrac{7}{2}$

PROBLEMS 13.5 (page 598)

1. $y = 1, x = 1$ **3.** $y = \dfrac{1}{3}, x = \dfrac{5}{3}$

5. $y = 0, x = 0$ **7.** $y = 0, x = 1, x = -1$

9. None   **11.** $y = 2, x = 2, x = -3$

13. $y = 2, x = -\sqrt{5}, x = \sqrt{5}$ **15.** $y = 5, x = 3$

17. $y = -x + 1, x = 0, x = -1$ **19.** $y = \dfrac{1}{4}, x = -\dfrac{1}{2}$

21. $y = \dfrac{1}{2}, x = -\dfrac{4}{3}$ **23.** $y = 4$

25. Dec. $(-\infty, 0), (0, \infty)$; conc. down $(-\infty, 0)$; conc. up $(0, \infty)$; sym. about origin; asymptotes $x = 0, y = 0$

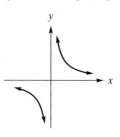

27. Int. $(0, 0)$; dec. $(-\infty, 1), (1, \infty)$; conc. up $(1, \infty)$; conc. down $(-\infty, 1)$; asymptotes $x = 1, y = 1$

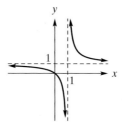

29. Dec. $(-\infty, -1), (0, 1)$; inc. $(-1, 0), (1, \infty)$; rel. min. when $x = \pm 1$; conc. up $(-\infty, 0), (0, \infty)$; sym. about y-axis; asymptote $x = 0$

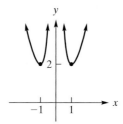

31. Int. $(0, -1)$; inc. $(-\infty, -1), (-1, 0)$; dec. $(0, 1), (1, \infty)$; rel. max. when $x = 0$; conc. up $(-\infty, -1), (1, \infty)$; conc. down $(-1, 1)$; asymptotes $x = 1, x = -1, y = 0$; sym. about y-axis

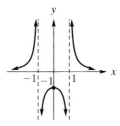

33. Int. $(-1, 0), (0, 1)$; inc. $(-\infty, 1), (1, \infty)$; conc. up $(-\infty, 1)$; conc. down $(1, \infty)$; asymptotes $x = 1, y = -1$

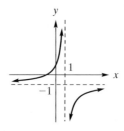

35. Int. $(0, 0)$; inc. $\left(-\infty, -\dfrac{8}{7}\right)$, $(0, \infty)$; dec. $\left(-\dfrac{8}{7}, -\dfrac{4}{7}\right)$, $\left(-\dfrac{4}{7}, 0\right)$; rel. max. when $x = -\dfrac{8}{7}$; rel. min. when $x = 0$; conc. down $\left(-\infty, -\dfrac{4}{7}\right)$; conc. up $\left(-\dfrac{4}{7}, \infty\right)$; asymptotes $x = -\dfrac{4}{7}$; $y = \dfrac{1}{7}x - \dfrac{4}{49}$

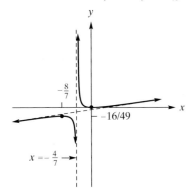

37. Int. $\left(0, -\dfrac{9}{8}\right)$; inc. $\left(-\infty, -\dfrac{2}{3}\right)$, $\left(-\dfrac{2}{3}, \dfrac{1}{3}\right)$; dec. $\left(\dfrac{1}{3}, \dfrac{4}{3}\right)$, $\left(\dfrac{4}{3}, \infty\right)$; rel. max. when $x = \dfrac{1}{3}$; conc. up $\left(-\infty, -\dfrac{2}{3}\right)$, $\left(\dfrac{4}{3}, \infty\right)$; conc, down $\left(-\dfrac{2}{3}, \dfrac{4}{3}\right)$; asymptotes $y = 0$, $x = -\dfrac{2}{3}$, $x = \dfrac{4}{3}$

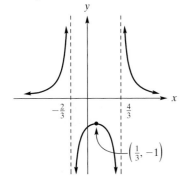

39. Int. $\left(-\dfrac{1}{3}, 0\right)$, $\left(0, \dfrac{1}{4}\right)$; dec. $\left(-\infty, -\dfrac{4}{3}\right)$, $\left(\dfrac{2}{3}, \infty\right)$; inc. $\left(-\dfrac{4}{3}, \dfrac{2}{3}\right)$; rel. min. when $x = -\dfrac{4}{3}$; conc. down $\left(-\infty, -\dfrac{7}{3}\right)$; conc. up $\left(-\dfrac{7}{3}, \dfrac{2}{3}\right)$, $\left(\dfrac{2}{3}, \infty\right)$; inf. pt. when $x = -\dfrac{7}{3}$ asymptotes $x = \dfrac{2}{3}$, $y = 0$

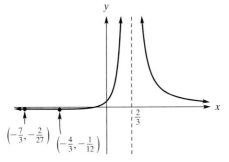

41. Int. $(-1, 0)$, $(1, 0)$; inc. $(-\sqrt{3}, 0)$, $(0, \sqrt{3})$; dec. $(-\infty, -\sqrt{3})$, $(\sqrt{3}, \infty)$; rel. max. when $x = \sqrt{3}$; rel. min. when $x = -\sqrt{3}$; conc. down $(-\infty, -\sqrt{6})$, $(0, \sqrt{6})$; conc. up $(-\sqrt{6}, 0)$, $(\sqrt{6}, \infty)$; inf. pt. when $x = \pm\sqrt{6}$; asymptotes $x = 0$, $y = 0$; sym. about origin

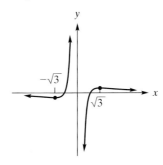

43. Int. $(0, 1)$; inc. $(-\infty, -2)$, $(0, \infty)$; dec. $(-2, -1)$, $(-1, 0)$; rel. max. when $x = -2$; rel. min when $x = 0$; conc. down $(-\infty, -1)$; conc. up $(-1, \infty)$; asymptotes $x = -1$; $y = x$

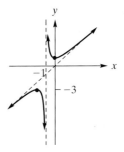

45. Int. $(0, 5)$; dec. $\left(-\infty, -\dfrac{1}{3}\right)$, $\left(-\dfrac{1}{3}, \dfrac{1}{3}\right)$; inc. $\left(\dfrac{1}{3}, 1\right)$, $(1, \infty)$; rel. min. when $x = \dfrac{1}{3}$; conc. down $\left(-\infty, -\dfrac{1}{3}\right)$, $(1, \infty)$; conc. up $\left(-\dfrac{1}{3}, 1\right)$; asymptotes $x = -\dfrac{1}{3}$, $x = 1$, $y = -1$

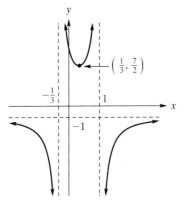

47.

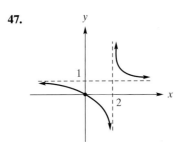

49.

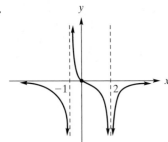

51. $x = -\dfrac{a}{b}$; $y = \dfrac{1}{b}$

55. $x \approx \pm 2.45$, $x \approx 0.67$, $y = 2$ **57.** $y \approx 0.48$

PROBLEMS 13.6 (page 607)

1. 41 and 41   **3.** 300 ft by 250 ft   **5.** 100 units

7. \$15   **9. (a)** 110 grams; **(b)** $51\dfrac{9}{11}$ grams

11. 500 units; price $= \$60$; profit $= \$11,900$

13. \$22   **15.** 120 units; \$86,000   **17.** 625 units; \$4

19. \$17; \$86,700   **21.** 4 ft by 4 ft by 2 ft

23. $\dfrac{L}{6}$ in.; $\dfrac{2L^3}{27}$ in.^{3}

27. 130 units, $p = \$340$, $P = \$36,980$; 125 units, $p = \$350$, $P = \$34,175$

29. 250 per lot (4 lots)   **31.** 35

33. 60 mi/h   **35.** 8; \$3400

37. $5 - \sqrt{3}$ tons; $5 - \sqrt{3}$ tons   **41.** 10 cases; \$50.55

REVIEW PROBLEMS—CHAPTER 13 (page 612)

1. $y = 3$, $x = 4$, $x = -4$

3. $y = \dfrac{5}{9}$, $x = -\dfrac{2}{3}$

5. $x = 0$

7. $x = -\dfrac{15}{8}$, -1

9. Inc. $(-1, 7)$; dec. on $(-\infty, -1)$ and $(7, \infty)$

11. Dec. on $(-\infty, -\sqrt{6})$, $(0, \sqrt{3})$, $(\sqrt{3}, \sqrt{6})$; inc. on $(-\sqrt{6}, -\sqrt{3})$, $(-\sqrt{3}, 0)$, $(\sqrt{6}, \infty)$

13. Conc. up on $(-\infty, 0)$ and $\left(\dfrac{1}{2}, \infty\right)$; conc. down on $\left(0, \dfrac{1}{2}\right)$

15. Conc. down on $\left(-\infty, \dfrac{1}{2}\right)$; conc. up on $\left(\dfrac{1}{2}, \infty\right)$

17. Conc. up on $\left(-\infty, -\dfrac{7}{12}\right)$, $\left(-\dfrac{1}{2}, \infty\right)$; conc. down on $\left(-\dfrac{7}{12}, -\dfrac{1}{2}\right)$

19. Rel. max. at $x = 1$; rel. min. at $x = 2$

21. Rel. min. at $x = -1$

23. Rel. max. at $x = -\dfrac{2}{5}$; rel. min. at $x = 0$

25. At $x = 3$

27. At $x = 1$

29. At $x = 0$, $3 \pm \sqrt{3}$

31. Maximum: $f(2) = 16$; minimum: $f(1) = -1$

33. Maximum: $f(0) = 0$; minimum: $f\left(-\dfrac{6}{5}\right) = -\dfrac{1}{120}$

35. (a) f has no relative extrema; **(b)** f is conc. down on $(1, 3)$; inf. pts.: $(1, 2e^{-1})$, $(3, 10e^{-3})$

37. Int. $(-4, 0)$, $(6, 0)$, $(0, -24)$; inc. $(1, \infty)$; dec. $(-\infty, 1)$; rel. min. when $x = 1$; conc. up $(-\infty, \infty)$

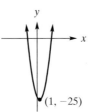

39. Int. $(0, 20)$, inc. $(-\infty, -2)$, $(2, \infty)$; dec. $(-2, 2)$; rel. max. when $x = -2$; rel. min. when $x = 2$; conc. up $(0, \infty)$; conc. down $(-\infty, 0)$; inf. pt. when $x = 0$

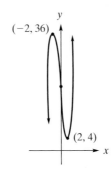

41. Int. $(0, 0)$, $(-1, 0)$, $(1, 0)$; inc. $\left(-\infty, -\dfrac{\sqrt{3}}{3}\right)$, $\left(\dfrac{\sqrt{3}}{3}, \infty\right)$; dec. $\left(-\dfrac{\sqrt{3}}{3}, \dfrac{\sqrt{3}}{3}\right)$; conc. down $(-\infty, 0)$; conc. up $(0, \infty)$; inf. pt. when $x = 0$; sym. about origin

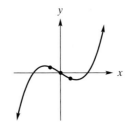

43. Int. $(-5, 0)$; inc. $(-10, 0)$; dec. $(-\infty, -10)$, $(0, \infty)$; rel. min. when $x = -10$; conc. up $(-15, 0)$, $(0, \infty)$; conc. down $(-\infty, -15)$; inf. pt. when $x = -15$; horiz. asym. $y = 0$; vert. asym. $x = 0$

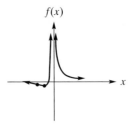

45. Int. $(0, 0)$, inc. $\left(-\infty, -\dfrac{1}{6}\right)$; dec. $\left(-\dfrac{1}{6}, \dfrac{1}{3}\right)$, $\left(\dfrac{1}{3}, \infty\right)$; rel. max. when $x = -\dfrac{1}{6}$; conc. up $\left(-\infty, \dfrac{1}{3}\right)$, $\left(\dfrac{1}{3}, \infty\right)$;

conc. down $\left(-\dfrac{1}{3}, \dfrac{1}{3}\right)$; inf. pt. when $x = -\dfrac{1}{3}$; horiz. asym.
$y = 0$; vert. asym. $x = \dfrac{1}{3}$

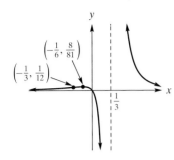

47. Int. $(0, 1)$; inc. $(0, \infty)$; dec. $(-\infty, 0)$; rel. min. when $x = 0$; conc. up $(-\infty, \infty)$; sym. about y-axis

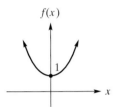

49. (a) False; **(b)** false; **(c)** true; **(d)** false; **(e)** false
51. $q > 2$
57. Rel. max. $(-1.32, 12.28)$; rel. min. $(0.44, 1.29)$
59. $x \approx -0.60$ **61.** 20   **63.** 200
65. \$4624   **67.** 100 ft by 200 ft
69. (a) 200 stands at \$120 per stand; **(b)** 300 stands

MATHEMATICAL SNAPSHOT—CHAPTER 13 (page 616)

1. Figure 13.75 shows that the population reaches its final size in about 45 days.
3. The tangent line will not coincide exactly with the curve in the first place. Smaller time steps could reduce the error.

PROBLEMS 14.1 (page 622)

1. $5\,dx$ **3.** $\dfrac{2x^3}{\sqrt{x^4 - 9}}\,dx$ **5.** $-\dfrac{2}{x^3}\,dx$
7. $\dfrac{2x}{x^2 + 7}\,dx$ **9.** $3e^{2x^2+3}(12x^2 + 4x + 3)\,dx$
11. $\Delta y = -0.14$, $dy = -0.14$
13. $\Delta y = -0.28$, $dy = -0.3$
15. $\Delta y \approx 0.049$, $dy = 0.050$ **17. (a)** -1; **(b)** 2.9
19. $\dfrac{577}{34} \approx 16.97$ **21.** $4\dfrac{1}{32}$ **23.** -0.03 **25.** 1.001
27. $\dfrac{1}{2}$ **29.** $\dfrac{1}{6p(p^2 + 5)^2}$ **31.** $-p^2$
33. $\dfrac{1}{36}$ **35.** $-\dfrac{4}{5}$ **37.** -17; -19.3
39. 2.04   **41.** 0.7
43. $(1.69 \times 10^{-11})\pi$ cm^3 **45. (c)** 42 units

PRINCIPLES IN PRACTICE 14.2

1. $\displaystyle\int 28.3\,dq = 28.3q + C$
2. $\displaystyle\int 0.12t^2\,dt = 0.04t^3 + C$
3. $\displaystyle\int -\dfrac{480}{t^3}\,dt = \dfrac{240}{t^2} + C$
4. $\displaystyle\int (500 + 300\sqrt{t})\,dt = 500t + 200t^{3/2} + C$
5. $S(t) = 0.7t^3 - 32.7t^2 + 491.6t + C$

PROBLEMS 14.2 (page 628)

1. $7x + C$ **3.** $\dfrac{x^9}{9} + C$ **5.** $-\dfrac{5}{6x^6} + C$
7. $-\dfrac{2}{9x^9} + C$ **9.** $-\dfrac{4}{3t^{3/4}} + C$ **11.** $4t + \dfrac{t^2}{2} + C$
13. $\dfrac{y^6}{6} - \dfrac{5y^2}{2} + C$ **15.** $t^3 - 2t^2 + 5t + C$
17. $(7 + e)x + C$ **19.** $\dfrac{x^2}{14} - \dfrac{3x^5}{20} + C$
21. $\pi e^x + C$ **23.** $\dfrac{x^{9.3}}{9.3} - \dfrac{9x^7}{7} - \dfrac{1}{x^3} - \dfrac{1}{2x^2} + C$
25. $-\dfrac{4x^{3/2}}{9} + C$ **27.** $\dfrac{x^{3/4}}{3} + C$
29. $\dfrac{x^4}{12} + \dfrac{3}{2x^2} + C$ **31.** $\dfrac{w^3}{2} + \dfrac{2}{3w} + C$
33. $\dfrac{3}{10}u^2 - \dfrac{4}{5}u + C$ **35.** $\dfrac{u^{e+1}}{e + 1} + e^u + C$
37. $\dfrac{4x^{3/2}}{3} - \dfrac{12x^{5/4}}{5} + C$
39. $-\dfrac{3x^{5/3}}{25} - 7x^{1/2} + 3x^2 + C$
41. $\dfrac{x^4}{4} - x^3 + \dfrac{5x^2}{2} - 15x + C$ **43.** $\dfrac{2x^{5/2}}{5} + 2x^{3/2} + C$
45. $\dfrac{27}{4}u^4 + 18u^3 + 18u^2 + 8u + C$
47. $\dfrac{2v^3}{3} + 3v + \dfrac{1}{2v^4} + C$
49. $\dfrac{z^3}{6} + \dfrac{5z^2}{2} + C$ **51.** $x + e^x + C$
53. No, $F(x) - G(x)$ might be a nonzero constant
55. $\dfrac{1}{\sqrt{x^2 + 1}} + C$

PRINCIPLES IN PRACTICE 14.3

1. $N(t) = 800t + 200e^t + 6317.37$
2. $y(t) = 14t^3 + 12t^2 + 11t + 3$

PROBLEMS 14.3 (page 633)

1. $y = \dfrac{3x^2}{2} - 4x + 1$ **3.** 60
5. $y = -\dfrac{x^4}{4} + \dfrac{2x^3}{3} + x + \dfrac{19}{12}$
7. $y = \dfrac{x^4}{12} + x^2 - 5x + 13$ **9.** $p = 0.7$
11. $p = 275 - 0.5q - 0.1q^2$ **13.** $c = 1.35q + 200$

15. $8079.17

17. $G = -\dfrac{P^2}{50} + 2P + 20$

21. $80 ($dc/dq = 27.50$ when $q = 50$ is not relevant to problem)

PRINCIPLES IN PRACTICE 14.4

1. $T(t) = 10e^{-0.5t} + C$

2. $35 \ln |t + 1| + C$

PROBLEMS 14.4 (page 639)

1. $\dfrac{(x+5)^8}{8} + C$

3. $\dfrac{(x^2+3)^6}{6} + C$

5. $\dfrac{3}{5}(y^3 + 3y^2 + 1)^{5/3} + C$

7. $-\dfrac{5(3x-1)^{-2}}{6} + C$

9. $\dfrac{1}{3}(2x - 1)^{3/2} + C$

11. $\dfrac{(7x-6)^5}{35} + C$

13. $\dfrac{(5u^2 - 9)^{15}}{150} + C$

15. $\dfrac{3}{5}(27 + x^5)^{4/3} + C$

17. $e^{3x} + C$

19. $e^{t^2 + t} + C$

21. $\dfrac{1}{14}e^{7x^2} + C$

23. $-\dfrac{4}{3}e^{-3x} + C$

25. $\ln |x + 5| + C$

27. $\ln |x^3 + x^4| + C$

29. $-\dfrac{3}{4}(z^2 - 6)^{-4} + C$

31. $4 \ln |x| + C$

33. $\dfrac{1}{3} \ln |s^3 + 5| + C$

35. $-\dfrac{5}{2} \ln |4 - 2x| + C$

37. $\dfrac{2}{15}(5x)^{3/2} + C = \dfrac{2\sqrt{5}}{3}x^{3/2} + C$

39. $\sqrt{x^2 - 4} + C$

41. $\dfrac{1}{2}e^{y^4+1} + C$

43. $-\dfrac{1}{6}e^{-2v^3+1} + C$

45. $-\dfrac{1}{5}e^{-5x} + 2e^x + C$

47. $-\dfrac{1}{2}(7 - 2x^2 - 5x)^4 + C$

49. $\dfrac{1}{3} \ln |x^3 + 6x| + C$

51. $2 \ln |3 - 2s + 4s^2| + C$

53. $\dfrac{1}{4} \ln(2x^2 + 1) + C$

55. $\dfrac{1}{27}(x^3 - x^6)^{-9} + C$

57. $\dfrac{1}{4}(x^4 + x^2)^2 + C$

59. $\dfrac{7}{4}(4 - x - x^2)^{-4} + C$

61. $\dfrac{1}{6}e^{4x^3 + 3x^2 - 4} + C$

63. $-\dfrac{1}{25}(8 - 5x^2)^{5/2} + C$

65. $\dfrac{(2x)^{3/2}}{3} - \sqrt{2x} + C = \dfrac{2\sqrt{2}}{3}x^{3/2} - \sqrt{2}x^{1/2} + C$

67. $\dfrac{x^5}{5} + \dfrac{2x^3}{3} + x + C$

69. $\dfrac{1}{2} \ln(x^2 + 1) - \dfrac{1}{6(x^6 + 1)} + C$

71. $\dfrac{1}{2} \ln |4x + 1| + \dfrac{4}{21}(x^3 - x^6)^{-7} + C$

73. $\dfrac{2}{9}(3x + 1)^{3/2} - \dfrac{1}{2} \ln(x^2 + 3) + C$ **75.** $2e^{\sqrt{x}} + C$

77. $-\dfrac{1}{4}e^{-x} + \dfrac{1}{4}e^x + C$

79. $\dfrac{1}{4} \ln^2(x^2 + 2x) + C$

81. $y = -\dfrac{1}{6}(3 - 2x)^3 + \dfrac{11}{2}$

83. $y = -\ln |x| + \dfrac{5}{2}x - \dfrac{1}{2} = \ln |1/x| + \dfrac{5}{2}x - \dfrac{1}{2}$

85. $160e^{0.05t} + 190$

87. $\dfrac{Rr^2}{4K} + B_1 \ln |r| + B_2$

PROBLEMS 14.5 (page 643)

1. $\dfrac{x^5}{5} + \dfrac{4}{3}x^3 - 2 \ln |x| + C$

3. $\dfrac{1}{3}(2x^3 + 4x + 1)^{3/2} + C$

5. $-6\sqrt{2 - 3x} + C$

7. $\dfrac{4^{7x}}{7 \ln 4} + C$

9. $7x^2 - 4e^{(1/4)x^2} + C$

11. $x^2 - 3x + \dfrac{2}{3} \ln |3x - 1| + C$

13. $\dfrac{5}{14} \ln(7e^{2x} + 4) + C$

15. $-\dfrac{1}{7}e^{7/x} + C$

17. $\dfrac{5}{2}x^2 - \dfrac{45}{2} \ln |x^2 + 9| + C$

19. $\dfrac{2}{9}(\sqrt{x} + 2)^3 + C$

21. $3(x^{1/3} + 2)^5 + C$

23. $\dfrac{1}{2}(\ln^2 x) + C$

25. $\dfrac{1}{3} \ln^3(r + 1) + C$

27. $\dfrac{3^{\ln x}}{\ln 3} + C$

29. $\dfrac{2}{3}e^{(x^3 + 1)/2} + C$

31. $8 \ln |\ln(x + 3)| + C$

33. $\dfrac{x^2}{2} + x + \ln |x^2 - 3| + C$

35. $\dfrac{2}{3} \ln^{3/2}[(x^3 + 1)^2] + C$

37. $\dfrac{\sqrt{x^4 - 4x}}{2} - (\ln 7)x + C$

39. $x^2 - 8x - 6 \ln |x| - \dfrac{2}{x^2} + C$

41. $x - \ln |x + 1| + C$

43. $\sqrt{e^{x^2} + 2} + C$

45. $-\dfrac{(e^{-x} + 6)^3}{3} + C$

47. $\dfrac{1}{5}(x^2 + e)^{5/2} + C$

49. $\dfrac{1}{36\sqrt{2}}[(8x)^{3/2} + 3]^{3/2} + C$

51. $-\dfrac{2}{3}e^{-\sqrt{s^3}} + C$

53. $\dfrac{x^3}{3} + x + C$

55. $\dfrac{\ln^2 x}{2} + x + C$

57. $p = -\dfrac{200}{q(q+2)}$

59. $c = 20 \ln |(q + 5)/5| + 2000$

61. $C = 2(\sqrt{I} + 1)$

63. $C = \dfrac{3}{4}I - \dfrac{1}{3}\sqrt{I} + \dfrac{71}{12}$

65. (a) 140 per unit; **(b)** $14,000; **(c)** $14,280

67. $2500 - 800\sqrt{5} \approx \$711$ per acre

69. $I = 3$

PRINCIPLES IN PRACTICE 14.6

1. $5975

PROBLEMS 14.6 (page 651)

1. $\dfrac{2}{3}$ square unit

3. $\dfrac{15}{32}$ square unit

5. $S_n = \dfrac{1}{n}\left[4\left(\dfrac{1}{n}\right) + 4\left(\dfrac{2}{n}\right) + \cdots + 4\left(\dfrac{n}{n}\right)\right] = \dfrac{2(n+1)}{n}$

7. (a) $S_n = \dfrac{n+1}{2n} + 1$; **(b)** $\dfrac{3}{2}$

9. $\dfrac{1}{2}$ square unit

11. $\dfrac{1}{3}$ square unit   **13.** 1 square unit   **15.** 20

17. -18 **19.** $\dfrac{5}{6}$ **21.** 0   **23.** $\dfrac{11}{4}$

25. 14.7 square units   **27.** 2.4   **29.** -25.5

PRINCIPLES IN PRACTICE 14.7

1. $32,830

2. $28,750

PROBLEMS 14.7 (page 657)

1. 15 **3.** $\dfrac{15}{2}$ **5.** -20 **7.** $\dfrac{7}{3}$

9. $\dfrac{15}{2}$ **11.** $\dfrac{4}{3}$ **13.** $\dfrac{768}{7}$ **15.** $\dfrac{5}{3}$

17. $\dfrac{32}{3}$ **19.** $-\dfrac{1}{6}$ **21.** $4\ln 8$ **23.** e^5

25. $\dfrac{5}{3}(e-1)$ **27.** $\dfrac{3}{4}$ **29.** $\dfrac{38}{9}$ **31.** $\dfrac{15}{28}$

33. $\dfrac{1}{2}\ln 3$ **35.** $\dfrac{1}{2}\left(e+\dfrac{1}{e}-2\right)$

37. $\dfrac{1}{e^3}-\dfrac{3}{e}-\dfrac{3}{2e^2}+\dfrac{3}{\pi}+\dfrac{3}{2\pi^2}-\dfrac{1}{\pi^3}$ **39.** $\dfrac{e^3}{2}(e^{12}-1)$

41. $6+\ln 19$ **43.** $\dfrac{47}{12}$ **45.** $6-3e$ **47.** 7

49. 0 **51.** $\alpha^{5/2}T$ **53.** $\displaystyle\int_a^b(-Ax^{-B})\,dx$

55. $8639 **57.** 1,973,333 **59.** $220 **61.** $1367.99

63. 696;492 **65.** $2Ri$ **69.** 0.05 **71.** 3.52 **73.** 14.34

PRINCIPLES IN PRACTICE 14.8

1. 76.90 feet **2.** 5.77 grams

PROBLEMS 14.8 (page 663)

1. 413 **3.** $0.340;\ \dfrac{1}{3}\approx 0.333$ **5.** $\approx 0.767;\ 0.750$

7. $0.883;\ 2-\ln 3\approx 0.901$ **9.** 2,115,215 **11.** 3.0 square units

13. $\dfrac{8}{3}$ **15.** 0.771 **17.** $\dfrac{35}{6}$ km^2

PROBLEMS 14.9 (page 667)

*In Problems **1–33**, answers are assumed to be expressed in square units.*

1. 8 **3.** $\dfrac{87}{2}$ **5.** 8 **7.** $\dfrac{19}{3}$ **9.** 9

11. $\dfrac{50}{3}$ **13.** $\dfrac{9}{2}$ **15.** $\dfrac{123}{4}$ **17.** $\dfrac{32}{3}$ **19.** 1

21. 18 **23.** $\dfrac{26}{3}$ **25.** $\dfrac{3}{2}\sqrt[3]{2}$ **27.** e

29. $\dfrac{3}{2}+2\ln 2=\dfrac{3}{2}+\ln 4$ **31.** 68

33. 2 **35.** 19 square units

37. **(a)** $\dfrac{1}{16}$; **(b)** $\dfrac{3}{4}$; **(c)** $\dfrac{7}{16}$

39. **(a)** $\ln\dfrac{7}{3}$ **(b)** $\ln 5-1$; **(c)** $2-\ln 4$

41. 1.89 square units **43.** 11.41 square units

PROBLEMS 14.10 (page 675)

1. Area $=\displaystyle\int_{-2}^{3}[(x+6)-x^2]\,dx$

3. Area $=\displaystyle\int_0^3[2x-(x^2-x)]\,dx+\int_3^4[(x^2-x)-2x]\,dx$

5. Area $=\displaystyle\int_0^1[(y+1)-\sqrt{1-y}]\,dy$

7. Area $=\displaystyle\int_1^2[(7-2x^2)-(x^2-5)]\,dx$

*In Problems **9–33**, answers are assumed to be expressed in square units.*

9. $\dfrac{4}{3}$ **11.** $\dfrac{4\sqrt{2}}{3}$ **13.** $8\sqrt{6}$ **15.** 40 **17.** $\dfrac{125}{6}$

19. 9 **21.** $\dfrac{125}{12}$ **23.** $\dfrac{256}{9}$ **25.** $\dfrac{44}{3}$

27. $\dfrac{4}{3}(5\sqrt{5}-2\sqrt{2})$ **29.** $\dfrac{1}{2}$ **31.** $\dfrac{255}{32}-4\ln 2$

33. 12 **35.** $\dfrac{14}{45}$ **37.** $\dfrac{3}{2m^3}$ square units

39. $2^{4/3}$ **41.** 4.76 square units **43.** 6.17 square units

PROBLEMS 14.11 (page 677)

1. CS $= 25.6$, PS $= 38.4$

3. CS $= 50\ln 2-25$, PS $= 1.25$

5. CS $= 225$, PS $= 450$ **7.** $426.67

9. $254,000 **11.** CS ≈ 1197, PS ≈ 477

REVIEW PROBLEMS—CHAPTER 14 (page 680)

1. $\dfrac{x^4}{4}+x^2-7x+C$ **3.** $\dfrac{256}{3}$

5. $-3(x+5)^{-2}+C$ **7.** $2\ln|x^3-6x+1|+C$

9. $\dfrac{11\sqrt[3]{11}}{4}-4$ **11.** $\dfrac{y^4}{4}+\dfrac{2y^3}{3}+\dfrac{y^2}{2}+C$

13. $\dfrac{10}{7}t^{7/10}-\dfrac{6}{5}t^{5/6}+C$ **15.** $\dfrac{1}{3}\ln\dfrac{57}{5}$

17. $\dfrac{2}{27}(3x^3+2)^{3/2}+C$ **19.** $\dfrac{1}{2}(e^{2y}+e^{-2y})+C$

21. $\ln|x|-\dfrac{2}{x}+C$ **23.** 111 **25.** $\dfrac{35}{3}$

27. $4-3\sqrt[3]{2}$ **29.** $\dfrac{3}{t}-\dfrac{2}{\sqrt{t}}+C$ **31.** $\dfrac{3}{2}-5\ln 2$

33. $4(x^{3/2}+1)^{3/2}+C$ **35.** 1 **37.** $\dfrac{(1+e^{2x})^4}{8}+C$

39. $\dfrac{2\sqrt{10^{3x}}}{\ln 10}+C$ **41.** $y=\dfrac{1}{2}e^{2x}+3x-1$

*In Problems **43–57**, answers are assumed to be expressed in square units.*

43. $\dfrac{4}{3}$ **45.** $\dfrac{16}{3}$ **47.** $\dfrac{125}{6}$ **49.** $6+\ln 4$

51. $\dfrac{2}{3}$ **53.** 36 **55.** $\dfrac{243}{8}$ **57.** $e-1$

59. $p=100-\sqrt{2q}$ **61.** $1483.33 **63.** 0.5507

65. 15 square units **67.** CS $=166\dfrac{2}{3}$, PS $=53\dfrac{1}{3}$

73. 15.08 square units **75.** CS ≈ 1148, PS ≈ 251

MATHEMATICAL SNAPSHOT—CHAPTER 14 (page 682)

1. **(a)** 475; **(b)** 275

3. **(a)** $2,002,500; **(b)** 18,000; **(c)** $111.25

PRINCIPLES IN PRACTICE 15.1

1. $S(t)=-40te^{0.1t}+400e^{0.1t}+4600$

2. $P(t)=0.025t^2-0.05t^2\ln t+0.05t^2(\ln t)^2+C$

PROBLEMS 15.1 (page 688)

1. $\dfrac{2}{3}x(x+5)^{3/2}-\dfrac{4}{15}(x+5)^{5/2}+C$

3. $-e^{-x}(x+1)+C$ **5.** $\dfrac{y^4}{4}\left[\ln(y)-\dfrac{1}{4}\right]+C$

7. $x[\ln(4x) - 1] + C$

9. $x(2x + 3)^{3/2} - \dfrac{1}{5}(2x + 3)^{5/2} + C = \dfrac{3}{5}(2x + 3)^{3/2}(x - 1) + C$

11. $-\dfrac{x}{10(5x + 2)^2} - \dfrac{1}{50(5x + 2)} + C$

13. $-\dfrac{1}{x}(1 + \ln x) + C$ **15.** $e^2(3e^2 - 1)$

17. $\dfrac{1}{2}(1 - e^{-1})$, parts not needed

19. $2(9\sqrt{3} - 10\sqrt{2})$

21. $3x(x - 2)\ln(x - 2) - \dfrac{3}{2}x^2 + C$

23. $e^x(x^2 - 2x + 2) + C$

25. $\dfrac{x^3}{3} + 2e^{-x}(x + 1) - \dfrac{e^{-2x}}{2} + C$

27. $\dfrac{e^{x^2}}{2}(x^2 - 1) + C$

29. $\dfrac{2^{2x-1}}{\ln 2} + \dfrac{2^{x+1}x}{\ln 2} - \dfrac{2^{x+1}}{\ln^2 2} + \dfrac{x^3}{3} + C$

31. $2e^3 + 1$ square units **33.** $\left[\dfrac{8}{3}\ln(2) - \dfrac{7}{9}\right]$ square units

37. $\displaystyle\int f^{-1}(x)\,dx = xf^{-1}(x) - F(f^{-1}(x)) + C$

PRINCIPLES IN PRACTICE 15.2

1. $r(q) = \dfrac{5}{2}\ln\left|\dfrac{3(q + 1)^3}{q + 3}\right|$

2. $V(t) = 150t^2 - 900\ln(t^2 + 6) + C$

PROBLEMS 15.2 (page 694)

1. $\dfrac{12}{x + 6} - \dfrac{2}{x + 1}$ **3.** $2 + \dfrac{8}{x + 2} - \dfrac{18}{x + 3}$

5. $\dfrac{1}{x + 2} + \dfrac{2}{(x + 2)^2}$ **7.** $\dfrac{3}{x} - \dfrac{2x}{x^2 + 1}$

9. $2\ln|x| + 3\ln|x - 1| + C = \ln|x^2(x - 1)^3| + C$

11. $-3\ln|x + 1| + 4\ln|x - 2| + C = \ln\left|\dfrac{(x - 2)^4}{(x + 1)^3}\right| + C$

13. $\dfrac{1}{4}\left[\dfrac{3x^2}{2} + 2\ln|x - 1| - 2\ln|x + 1|\right] + C$

$\qquad = \dfrac{1}{4}\left(\dfrac{3x^2}{2} + \ln\left[\dfrac{x - 1}{x + 1}\right]^2\right) + C$

15. $2\ln|x| - \dfrac{7}{3}\ln|x + 1| + \dfrac{1}{3}\ln|x - 2| + C = \ln\left|\dfrac{x^2\sqrt[3]{x - 2}}{\sqrt[3]{(x + 1)^7}}\right| + C$

17. $\ln|x^6 + 2x^4 - x^2 - 2| + C$, partial fractions not required.

19. $\dfrac{4}{x - 2} - 5\ln|x - 1| + 7\ln|x - 2| + C = \dfrac{4}{x - 2} + \ln\left|\dfrac{(x - 2)^7}{(x - 1)^5}\right| + C$

21. $4\ln|x| - \ln(x^2 + 4) + C = \ln\left[\dfrac{x^4}{x^2 + 4}\right] + C$

23. $-\dfrac{1}{2}\ln(x^2 + 1) - \dfrac{2}{x - 3} + C$

25. $5\ln(x^2 + 1) + 2\ln(x^2 + 2) + C = \ln[(x^2 + 1)^5(x^2 + 2)^2] + C$

27. $\dfrac{3}{2}\ln(x^2 + 2) - \dfrac{1}{x^2 + 2} + C$

29. $18\ln(4) - 10\ln(5) - 8\ln(3)$

31. $11 + 24\ln\dfrac{2}{3}$ square units

PROBLEMS 15.3 (page 699)

1. $\dfrac{x}{9\sqrt{9 - x^2}} + C$ **3.** $-\dfrac{\sqrt{16x^2 + 3}}{3x} + C$

5. $\dfrac{1}{6}\ln\left|\dfrac{x}{6 + 7x}\right| + C$ **7.** $\dfrac{1}{3}\ln\left|\dfrac{\sqrt{x^2 + 9} - 3}{x}\right| + C$

9. $\dfrac{1}{2}\left[\dfrac{4}{5}\ln|4 + 5x| - \dfrac{2}{3}\ln|2 + 3x|\right] + C$

11. $\dfrac{1}{15}(3x - \ln[5 + 2e^{3x}]) + C$

13. $7\left[\dfrac{1}{5(5 + 2x)} + \dfrac{1}{25}\ln\left|\dfrac{x}{5 + 2x}\right|\right] + C$

15. $1 + \ln\dfrac{4}{9}$

17. $\dfrac{1}{2}(x\sqrt{x^2 - 3} - 3\ln|x + \sqrt{x^2 - 3}|) + C$

19. $\dfrac{1}{144}$ **21.** $e^x(x^2 - 2x + 2) + C$

23. $\dfrac{\sqrt{5}}{2}\left(-\dfrac{\sqrt{5x^2 + 1}}{\sqrt{5}x} + \ln|\sqrt{5}x + \sqrt{5x^2 + 1}|\right) + C$

25. $\dfrac{1}{9}\left(\ln|1 + 3x| + \dfrac{1}{1 + 3x}\right) + C$

27. $\dfrac{1}{\sqrt{5}}\left(\dfrac{1}{2\sqrt{7}}\ln\left|\dfrac{\sqrt{7} + \sqrt{5}x}{\sqrt{7} - \sqrt{5}x}\right|\right) + C$

29. $\dfrac{4}{81}\left[\dfrac{(3x)^6\ln(3x)}{6} - \dfrac{(3x)^6}{36}\right] + C = x^6[6\ln(3x) - 1] + C$

31. $4(9x - 2)(1 + 3x)^{3/2} + C$ **33.** $\dfrac{1}{2}\ln|2x + \sqrt{4x^2 - 13}| + C$

35. $-\dfrac{\sqrt{16 - 9x^2}}{8x} + C$

37. $\dfrac{1}{2\pi}(4\sqrt{x} - \ln|\pi + 7e^{4\sqrt{x}}|) + C$

39. $\dfrac{1}{2}\ln(x^2 + 1) + C$ **41.** $(2x^2 + 1)^{3/2} + C$

43. $\ln\left|\dfrac{x - 3}{x - 2}\right| + C$ **45.** $\dfrac{x^4}{4}\left[\ln(x) - \dfrac{1}{4}\right] + C$

47. $\dfrac{2}{9}e^{3x^2}(3x^2 - 1) + C$ **49.** $x(\ln x)^2 - 2x\ln(x) + 2x + C$

51. $\dfrac{2}{3}(9\sqrt{3} - 10\sqrt{2})$ **53.** $2(2\sqrt{2} - \sqrt{7})$

55. $\dfrac{7}{2}\ln(2) - \dfrac{3}{4}$ **57.** $\ln\left|\dfrac{q_n(1 - q_0)}{q_0(1 - q_n)}\right|$

59. (a) \$7558.09; **(b)** \$16,930.75

61. (a) \$5481; **(b)** \$535

PROBLEMS 15.4 (page 702)

1. $\dfrac{7}{3}$ **3.** -1 **5.** 0

7. 13 **9.** \$11,050 **11.** \$3155.13

PRINCIPLES IN PRACTICE 15.5

1. $I = I_o e^{-0.0085x}$

PROBLEMS 15.5 (page 708)

1. $y = -\dfrac{1}{x^2 + C}$ **3.** $y = (x^2 + 1)^{3/2} + C$

5. $y = Ce^x, C > 0$ **7.** $y = Cx, C > 0$

9. $y = \sqrt[3]{3x - 2}$

11. $y = \ln\dfrac{x^3 + 3}{3}$

13. $y = \dfrac{48(3x^2 + 2)^2}{4 + 31(3x^2 + 2)^2}$

15. $y = \sqrt{\left(\dfrac{3x^2}{2} + \dfrac{3}{2}\right)^2 - 1}$

17. $y = \ln\left(\dfrac{1}{2}\sqrt{x^2 + 3}\right)$

19. $c = (q + 1)e^{1/(q+1)}$

21. 120 weeks

23. $N = 40,000e^{0.018t}$; $N = 40,000(1.2)^{t/10}$; 57,600

25. $2e^{1.14882}$ billion

27. 0.01204; 57.57 sec

29. 2900 years

31. $N = N_0 e^{k(t - t_0)}$, $t \ge t_0$

33. 12.6 units

35. $A = 400(1 - e^{-t/2})$, 157 g/m^2

37. (a) $V = 60,000e^{\frac{t}{9.5}\ln(389/600)}$ (b) June 2028

PROBLEMS 15.6 (page 714)

1. 69,200   **3.** 500   **5.** 1990   **7.** (b) 375

9. 3:21 A.M   **11.** $155,555.56

13. $N = M - (M - N_0)e^{-kt}$

PRINCIPLES IN PRACTICE 15.7

1. 20 ml

PROBLEMS 15.7 (page 718)

1. $\dfrac{1}{18}$ **3.** Div   **5.** $\dfrac{1}{e}$ **7.** Div   **9.** $\dfrac{1}{2}$

11. 0   **13.** (a) 800; (b) $\dfrac{2}{3}$ **15.** 4,000,000

17. $\dfrac{1}{3}$ square unit   **19.** 20,000 increase

REVIEW PROBLEMS—CHAPTER 15 (page 720)

1. $\dfrac{x^2}{4}[2\ln(x) - 1] + C$

3. $2\sqrt{13} + \dfrac{8}{3}\ln\left(\dfrac{3 + \sqrt{13}}{2}\right)$

5. $\ln|3x + 1| + 4\ln|x - 2| + C$

7. $\dfrac{1}{2(x + 2)} + \dfrac{1}{4}\ln\left|\dfrac{x}{x + 2}\right| + C$

9. $-\dfrac{\sqrt{9 - 16x^2}}{9x} + C$

11. $\dfrac{3}{2}\ln\left|\dfrac{x - 3}{x + 3}\right| + C$

13. $e^{7x}(7x - 1) + C$

15. $\dfrac{1}{4}\ln|\ln x^2| + C$

17. $x - \dfrac{3}{2}\ln|3 + 2x| + C$

19. $2\ln|x| + \dfrac{3}{2}\ln(x^2 + 1) + C$

21. $2\sqrt{x + 1}[\ln(x + 1) - 2] + C$ **23.** 34

25. $y = Ce^{x^3 + x^2}$, $C > 0$ **27.** $\dfrac{2}{3}$ **29.** Div

31. 144,000   **33.** 0.0005; 90%

35. $N = \dfrac{450}{1 + 224e^{-1.02t}}$ **37.** 4:16 P.M.

39. 0.95

41. (a) 207, 208; (b) 157, 165; (c) 41, 41

MATHEMATICAL SNAPSHOT—CHAPTER 15 (page 722)

1. 114; 69   **5.** Answers may vary

PRINCIPLES IN PRACTICE 16.1

1. $\dfrac{1}{3}$ **2.** 0.607

3. Mean 5 years, standard deviation 5 years

PROBLEMS 16.1 (page 731)

1. (a) $\dfrac{5}{12}$; (b) $\dfrac{11}{16} = 0.6875$; (c) $\dfrac{13}{16} = 0.8125$; (d) $-1 + \sqrt{10}$

3. (a) $f(x) = \begin{cases} \dfrac{1}{3}, & \text{if } 1 \le x \le 4 \\ 0, & \text{otherwise} \end{cases}$

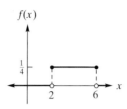

(b) $\dfrac{2}{3}$; (c) 0; (d) $\dfrac{5}{6}$; (e) $\dfrac{1}{3}$; (f) 0; (g) 1; (h) $\dfrac{5}{2}$;

(i) $\dfrac{\sqrt{3}}{2}$; (j) $F(x) = \begin{cases} 0, & \text{if } x < 1 \\ \dfrac{x - 1}{3}, & \text{if } 1 \le x \le 4 \\ 1, & \text{if } x > 4 \end{cases}$

$P(X < 2) = \dfrac{1}{3}$, $P(1 < X < 3) = \dfrac{2}{3}$

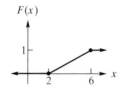

5. (a) $f(x) = \begin{cases} \dfrac{1}{b - a}, & \text{if } a \le x \le b \\ 0, & \text{otherwise} \end{cases}$

(b) $\dfrac{a + b}{2}$; (c) $\sigma^2 = \dfrac{(b - a)^2}{12}$, $\sigma = \dfrac{b - a}{\sqrt{12}}$

7. (a) $-e^{-12} + e^{-3} \approx 0.04978$; (b) $-e^{-12} + 1 \approx 0.99999$;
(c) $e^{-18} \approx 0.00000$; (d) $-e^{-3} + 1 \approx 0.95021$;

(f) $F(x) = \begin{cases} 0, & \text{if } x < 0 \\ 1 - e^{-3x}, & \text{if } x \ge 0 \end{cases}$

9. (a) $\dfrac{1}{8}$; (b) $\dfrac{5}{16}$; (c) $\dfrac{39}{64} \approx 0.609$; (d) 1; (e) $\dfrac{8}{3}$;

(f) $\dfrac{2\sqrt{2}}{3}$; (g) $2\sqrt{2}$; (h) $\dfrac{7}{16}$

11. $\dfrac{7}{10}$; 5 min   **13.** $e^{-3} \approx 0.050$

PROBLEMS 16.2 (page 736)

1. (a) 0.4554; (b) 0.3317; (c) 0.8907; (d) 0.9982;
(e) 0.8972; (f) 0.4880

3. 0.13   **5.** -1.08 **7.** 0.34

9. (a) 0.9970; (b) 0.0668; (c) 0.0873

11. 0.3085   **13.** 0.997   **15.** 8   **17.** 95%   **19.** 90.82%

21. (a) 1.7%; (b) 85.6

PRINCIPLES IN PRACTICE 16.3

1. 0.0396

PROBLEMS 16.3 (page 740)

1. 0.9207; 0.0122   **3.** 0.0430; 0.9232   **5.** 0.7507

7. 0.4129   **9.** 0.5; 0.0287   **11.** 0.0336

REVIEW PROBLEMS—CHAPTER 16 (page 741)

1. (a) 2;   **(b)** $\dfrac{9}{32}$;   **(c)** $\dfrac{3}{4}$,

    (d) $F(x) = \begin{cases} 0, & \text{if } x < 0 \\ \dfrac{x}{3} + \dfrac{2x^3}{3}, & \text{if } 0 \le x \le 1 \\ 1, & \text{if } x > 1 \end{cases}$

3. (a) $\dfrac{10}{3}$;   **(b)** $\sqrt{\dfrac{25}{18}} \approx 1.18$ **5.** 0.3085

7. 0.2417   **9.** 0.7734   **11.** 0.9817   **13.** 0.0228

MATHEMATICAL SNAPSHOT—CHAPTER 16 (page 832)

1. The result should correspond to the known distribution function.

3. Answers may vary

PRINCIPLES IN PRACTICE 17.1

1. (a) \$3260;   **(b)** \$4410

PROBLEMS 17.1 (page 750)

1. 3   **3.** 8   **5.** −1   **7.** 88   **9.** 3

11. $e^{x_0+h+y_0}$ **13.** 2000   **15.** $y = 2$ **17.** $z = 6$

19. 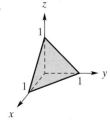   **21.**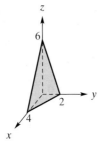

23.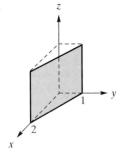

25.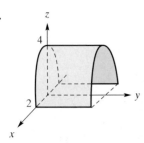

27.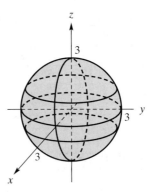

PROBLEMS 17.2 (page 754)

1. $f_x(x, y) = 8x$; $f_y(x, y) = 6y$

3. $f_x(x, y) = 0$; $f_y(x, y) = 2$

5. $g_x(x, y) = 12x^3 y + 2y^2 - 5y + 8$;
$\quad g_y(x, y) = 3x^4 + 4xy - 5x - 9$

7. $g_p(p, q) = \dfrac{q}{2\sqrt{pq}}$; $g_q(p, q) = \dfrac{p}{2\sqrt{pq}}$

9. $h_s(s, t) = \dfrac{2s}{t - 3}$; $h_t(s, t) = -\dfrac{s^2 + 4}{(t - 3)^2}$

11. $u_{q_1}(q_1, q_2) = \dfrac{1}{2(q_1 + 2)}$; $u_{q_2}(q_1, q_2) = \dfrac{1}{3(q_2 + 5)}$

13. $h_x(x, y) = (x^3 + xy^2 + 3y^3)(x^2 + y^2)^{-3/2}$;
$\quad h_y(x, y) = (3x^3 + x^2 y + y^3)(x^2 + y^2)^{-3/2}$

15. $\dfrac{\partial z}{\partial x} = 5ye^{5xy}$; $\dfrac{\partial z}{\partial y} = 5xe^{5xy}$

17. $\dfrac{\partial z}{\partial x} = 5\left[\dfrac{2x^2}{x^2 + y} + \ln(x^2 + y)\right]$; $\dfrac{\partial z}{\partial y} = \dfrac{5x}{x^2 + y}$

19. $f_r(r, s) = \sqrt{r + 2s}\,(3r^2 - 2s) + \dfrac{r^3 - 2rs + s^2}{2\sqrt{r + 2s}}$;

$\quad f_s(r, s) = 2(s - r)\sqrt{r + 2s} + \dfrac{r^3 - 2rs + s^2}{\sqrt{r + 2s}}$

21. $f_r(r, s) = -e^{3-r}\ln(7 - s)$; $f_s(r, s) = \dfrac{e^{3-r}}{s - 7}$

23. $g_x(x, y, z) = 6x^2 y^2 + 2y^3 z$;
$\quad g_y(x, y, z) = 4x^3 y + 6xy^2 z$;
$\quad g_z(x, y, z) = 2xy^3 + 8z$

25. $g_r(r, s, t) = 2re^{s+t}$;
$\quad g_s(r, s, t) = (7s^3 + 21s^2 + r^2)e^{s+t}$;
$\quad g_t(r, s, t) = e^{s+t}(r^2 + 7s^3)$

27. 50   **29.** $\dfrac{1}{\sqrt{14}}$ **31.** 0   **33.** 26

39. $-\dfrac{ra}{2\left[1 + a\dfrac{n-1}{2}\right]^2}$

PROBLEMS 17.3 (page 759)

1. 20   **3.** 1374.5

5. $\dfrac{\partial P}{\partial k} = 1.487902\left(\dfrac{l}{k}\right)^{0.357}$; $\dfrac{\partial P}{\partial l} = 0.826098\left(\dfrac{k}{l}\right)^{0.643}$

7. $\dfrac{\partial q_A}{\partial p_A} = -50; \dfrac{\partial q_A}{\partial p_B} = 2; \dfrac{\partial q_B}{\partial p_A} = 4; \dfrac{\partial q_B}{\partial p_B} = -20;$ competitive

9. $\dfrac{\partial q_A}{\partial p_A} = -\dfrac{100}{p_A^2 p_B^{1/2}}; \dfrac{\partial q_A}{\partial p_B} = -\dfrac{50}{p_A p_B^{3/2}}; \dfrac{\partial q_B}{\partial p_A} = -\dfrac{500}{3 p_B p_A^{4/3}};$

$\dfrac{\partial q_B}{\partial p_B} = -\dfrac{500}{p_B^2 p_A^{1/3}};$ complementary

11. $\dfrac{\partial P}{\partial B} = 0.01 A^{0.27} B^{-0.99} C^{0.01} D^{0.23} E^{0.09} F^{0.27};$

$\dfrac{\partial P}{\partial C} = 0.01 A^{0.27} B^{0.01} C^{-0.99} D^{0.23} E^{0.09} F^{0.27}$

13. 4480; if a staff manager with an M.B.A. degree had an extra year of work experience before the degree, the manager would receive $4480 per year in extra compensation.

15. (a) $-1.015; -0.846;$
   (b) One for which $w = w_0$ and $s = s_0$.

17. $\dfrac{\partial g}{\partial x} = \dfrac{1}{V_F} > 0$ for $V_F > 0$. Thus if x increases and V_F and V_s are fixed, then g increases.

19. (a) When $p_A = 9$ and $p_B = 16$, $\dfrac{\partial q_A}{\partial p_A} = -\dfrac{20}{27}$ and $\dfrac{\partial q_A}{\partial p_B} = \dfrac{5}{12}$
   (b) Demand for A decreases by approximately $\dfrac{5}{6}$ unit.

21. (a) No; **(b)** 70%

23. $\eta_{p_A} = -\dfrac{5}{46}, \eta_{p_B} = \dfrac{1}{46}$ **25.** $\eta_{p_A} = -1, \eta_{p_B} = -\dfrac{1}{2}$

PROBLEMS 17.4 (page 763)

1. $-\dfrac{2x}{5z}$ **3.** $\dfrac{4y}{3z^2}$ **5.** $\dfrac{x(yz^2+1)}{z(1-x^2y)}$

7. $-e^{y-z}$ **9.** $\dfrac{yz}{1+9z}$ **11.** $-\dfrac{3x}{z}$

13. $-\dfrac{1}{2}$ **15.** $-\dfrac{4}{e^2}$ **17.** 4 **19.** $\dfrac{5}{2}$

21. (a) 36; **(b)** With respect to q_A, $\dfrac{60}{13}$; with respect to q_B, $\dfrac{288}{65}$

PROBLEMS 17.5 (page 765)

1. $6y^2; 12y; 12y$ **3.** $3; 0; 0$

5. $18xe^{2xy}; 18e^{2xy}(2xy+1); 72x(1+xy)e^{2xy}$

7. $3x^2y + 4xy^2 + y^3; 3xy^2 + 4x^2y + x^3; 6xy + 4y^2; 6xy + 4x^2$

9. $\dfrac{zy}{\sqrt{x^2+y^2}}; \dfrac{z}{(x^2+y^2)^{\frac{3}{2}}}\left[x^2+y^2\sqrt{x^2+y^2}\right]$

11. 0 **13.** 744 **15.** $2e$ **17.** $-\dfrac{1}{8}$

23. $-\dfrac{y^2+z^2}{z^3} = -\dfrac{3x^2}{z^3}$

PROBLEMS 17.6 (page 768)

1. $\dfrac{\partial z}{\partial r} = 13; \dfrac{\partial z}{\partial s} = 9$ **3.** $\left[2t + \dfrac{3\sqrt{t}}{2}\right]e^{x+y}$

5. $5(2xz^2 + yz) + 2(xz + z^2) - (2x^2z + xy + 2yz)$

7. $3(x^2 + xy^2)^2(2x + y^2 + 16xy)$

9. $-2s(2x + yz) + r(xz) + 2s(xy + 2z)$

11. $19s(2x - 7)$ **13.** 324 **15.** $\dfrac{40}{e^9}$

17. When $p_A = 25$ and $p_B = 4$, $\dfrac{\partial c}{\partial p_A} = -\dfrac{1}{4}$ and $\dfrac{\partial c}{\partial p_B} = \dfrac{5}{4}$

19. (a) $\dfrac{\partial w}{\partial t} = \dfrac{\partial w}{\partial x}\dfrac{\partial x}{\partial t} + \dfrac{\partial w}{\partial y}\dfrac{\partial y}{\partial t};$ **(b)** $-\dfrac{20}{3\sqrt{2}+15e}$

PROBLEMS 17.7 (page 776)

1. $\left(\dfrac{14}{3}, -\dfrac{13}{3}\right)$

3. $(0, -2), (0, 1), (3, -2), (3, 1)$

5. $(50, 150, 350)$

7. $\left(-2, \dfrac{3}{2}\right)$, rel. min. **9.** $\left(-\dfrac{1}{4}, \dfrac{1}{2}\right)$, rel. max.

11. $\left(\dfrac{2}{5}, -\dfrac{3}{5}\right); D = -5 < 0$ no relative extremum

13. $(0, 0)$, rel. max.; $\left(4, \dfrac{1}{2}\right)$, rel. min.; $\left(0, \dfrac{1}{2}\right), (4, 0)$, neither

15. $(43, 13)$, rel. min. **17.** $(-1, -1)$, rel. min.

19. $(0, -2), (0, 2)$, neither **21.** $l = 24, k = 14$

23. $p_A = 80, p_B = 85$

25. $q_A = 48, q_B = 40, p_A = 52, p_B = 44$, profit = 3304

27. $q_A = 4, q_B = 3$ **29.** 1 ft by 2 ft by 3 ft

31. $\left(\dfrac{105}{37}, \dfrac{28}{37}\right)$, rel. min. **33.** $a = -8, b = -12, d = 33$

35. (a) 2 units of A and 3 units B;
   (b) Selling price for A is 30 and selling price for B is 19. Relative maximum profit is 25.

37. (a) $P = 5T(1 - e^{-x}) - 20x - 0.1T^2;$
   (c) Relative maximum at $(20, \ln 5)$; no relative extremum at $\left(5, \ln\dfrac{5}{4}\right)$

PROBLEMS 17.8 (page 784)

1. $(2, -2)$ **3.** $\left(3, \dfrac{3}{2}, -\dfrac{3}{2}\right)$ **5.** $\left(0, \dfrac{1}{4}, \dfrac{5}{8}\right)$

7. $(6, 3, 2)$ **9.** $\left(\dfrac{2}{3}, \dfrac{4}{3}, -\dfrac{4}{3}\right)$ **11.** $\left(\dfrac{1}{4}, \dfrac{1}{2}, \dfrac{1}{4}\right)$

13. Plant 1, 40 units; plant 2, 60 units

15. 74 units (when $l = 8, k = 7$)

17. $15,000 on newspaper advertising and $45,000 on TV advertising

19. $x = 5, y = 15, z = 5$ **21.** $x = 12, y = 8$

23. $x = \dfrac{100}{3}, y = \dfrac{100}{3}, z = \dfrac{100}{3}$

PROBLEMS 17.9 (page 789)

1. $\hat{y} = 0.98 + 0.61x; 3.12$ **3.** $\hat{y} = 0.057 + 1.67x; 5.90$

5. $\hat{q} = 80.5 - 0.643p$ **7.** $\hat{y} = 100 + 0.13x; 105.2$

9. $\hat{y} = 8.5 + 2.5x$

11. (a) $\hat{y} = 35.9 - 2.5x;$ **(b)** $\hat{y} = 28.4 - 2.5x$

PROBLEMS 17.10 (page 793)

1. 18 **3.** $\dfrac{1}{4}$ **5.** $\dfrac{2}{3}$ **7.** 3

9. $\dfrac{525}{2}$ **11.** $-\dfrac{58}{5}$ **13.** 9 **15.** -1

17. $\dfrac{e^2}{2} - e + \dfrac{1}{2}$ **19.** $-\dfrac{27}{4}$ **21.** $\dfrac{1}{24}$

23. $e^{-4} - e^{-2} - e^{-3} + e^{-1}$ **25.** $\dfrac{1}{3}$

REVIEW PROBLEMS—CHAPTER 17 (page 796)

1.

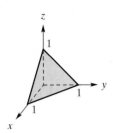

3.

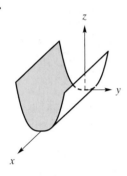

5. $8x + 6y;\ 6x + 2y$

7. $\dfrac{y}{(x+y)^2};\ -\dfrac{x}{(x+y)^2}$

9. $\dfrac{y}{x^2 + y^2}$

11. $2xze^{x^2yz}(1 + x^2yz)$

13. $2x + 2y + 6z$

15. $e^{x+y+z}\left[\ln xyz + \dfrac{1}{y}\right];\ e^{x+y+z}\left[\ln xyz + \dfrac{1}{x} + \dfrac{1}{z}\right]$

17. $\dfrac{1}{64}$

19. $2(x+y)e^r + 2\left(\dfrac{x+3y}{r+s}\right);\ 2\left(\dfrac{x+3y}{r+s}\right)$

21. $\dfrac{2x + 2y + z}{4z - x}$

23. $\dfrac{\partial P}{\partial l} = 14l^{-0.3}k^{0.3};\ \dfrac{\partial P}{\partial k} = 6l^{0.7}k^{-0.7}$

25. Competitive   **27.** $(2, 2)$, rel. min.

29. 4 ft by 4 ft by 2 ft

31. A, 89 cents per pound; B, 94 cents per pound

33. $(3, 2, 1)$ **35.** $\hat{y} = 12.67 + 3.29x$

37. $\dfrac{1}{12}$ **39.** $\dfrac{1}{30}$

MATHEMATICAL SNAPSHOT—CHAPTER 17 (page 798)

1. $y = 9.50e^{-0.22399x} + 5$ **3.** $T = 79e^{-0.01113t} + 45$

Index